T. Shaw

T. Shaw

This volume is dedicated to
Raymond Cecil Moore
February 20, 1892   *April 16, 1974*
Founder of the *Treatise on Invertebrate Paleontology*

TREATISE ON
INVERTEBRATE PALEONTOLOGY

Prepared under Sponsorship of
The Geological Society of America, Inc.

The Paleontological Society   The Society of Economic Paleontologists and Mineralogists
The Palaeontographical Society   The Palaeontological Association

RAYMOND C. MOORE
Founder

RICHARD A. ROBISON and CURT TEICHERT
Editors

JACK D. KEIM, LAVON MCCORMICK, ROGER B. WILLIAMS
Assistant Editors

Part A

INTRODUCTION

FOSSILIZATION (TAPHONOMY) BIOGEOGRAPHY AND BIOSTRATIGRAPHY

By W. A. BERGGREN, A. J. BOUCOT, M. F. GLAESSNER, HELMUT HÖLDER, M. R. HOUSE, VALDAR JAANUSSON, E. G. KAUFFMAN, BERNHARD KUMMEL, A. H. MÜLLER, A. W. NORRIS, A. R. PALMER, ADOLF PAPP, C. A. ROSS, J. R. P. ROSS, and J. A. VAN COUVERING

THE GEOLOGICAL SOCIETY OF AMERICA, INC.
and
THE UNIVERSITY OF KANSAS
BOULDER, COLORADO, and LAWRENCE, KANSAS
1979

© 1979 by The University of Kansas
AND
The Geological Society of America, Inc.

All Rights Reserved

Library of Congress Catalogue Card
Number: 53-12913
I S B N 0-8137-3001-5
TIVPA 1-569 (1979)

Text Composed by
THE UNIVERSITY OF KANSAS PRINTING SERVICE
Lawrence, Kansas

Illustrations and Offset Lithography
THE MERIDEN GRAVURE COMPANY
Meriden, Connecticut

Binding
TAPLEY-RUTTER COMPANY
Moonachie, New Jersey

Published 1979

Distributed by the Geological Society of America, Inc., 3300 Penrose Place, Boulder, Colo. 80301, to which all communications should be addressed.

The *Treatise on Invertebrate Paleontology* has been made possible by (1) grants of funds from The Geological Society of America through the bequest of Richard Alexander Fullerton Penrose, Jr., for initial preparation of illustrations, and partial defrayment of organizational expenses in 1948-1957, and again since 1971, and from the United States National Science Foundation, awarded annually since 1959, for continuation of the *Treatise* project; (2) contribution of the knowledge and labor of specialists throughout the world, working in cooperation under sponsorship of The Geological Society of America, The Paleontological Society, The Society of Economic Paleontologists and Mineralogists, The Palaeontographical Society, and The Palaeontological Association; and (3) acceptance by The University of Kansas of publication without any financial gain to the University.

TREATISE ON INVERTEBRATE PALEONTOLOGY

RAYMOND C. MOORE, Founder

R. A. ROBISON, Editor-in-chief

JACK D. KEIM, LAVON MCCORMICK, ROGER B. WILLIAMS, Assistant Editors

Advisers: J. C. FRYE, J. T. DUTRO, JR., N. J. SILBERLING (The Geological Society of America); T. W. AMSDEN, T. J. M. SCHOPF (The Paleontological Society); R. L. BATTEN, R. J. CUFFEY (The Society of Economic Paleontologists and Mineralogists); R. V. MELVILLE, M. K. HOWARTH (The Palaeontographical Society); F. C. DILLEY, M. R. HOUSE (The Palaeontological Association); E. E. ANGINO (The University of Kansas).

PARTS

Parts of the *Treatise* are distinguished by assigned letters with a view to indicating their systematic sequence while allowing publication of units in whatever order each is made ready for the press. The volumes are cloth-bound with title in gold on the cover. Copies are available on orders sent to the Publication Sales Department, The Geological Society of America, 3300 Penrose Place, Boulder, Colorado 80301. Special discounts are available to members of sponsoring societies under arrangements made by appropriate officers of these societies, to whom inquiries should be addressed.

VOLUMES ALREADY PUBLISHED
(Previous to 1978)

Part C. PROTISTA 2 (Sarcodina, chiefly "Thecamoebians" and Foraminiferida), xxxi + 900 p., 5,311 fig., 1964.

Part D. PROTISTA 3 (chiefly Radiolaria, Tintinnina), xii + 195 p., 1,050 fig., 1954.

Part E. ARCHAEOCYATHA, PORIFERA, xviii + 122 p., 728 fig., 1955.

Part E, Volume 1. ARCHAEOCYATHA, Second Edition (Revised and Enlarged), xxx + 158 p., 871 fig., 1972.

Part F. COELENTERATA, xvii + 498 p., 2,700 fig., 1956.

Part G. BRYOZOA, xii + 253 p., 2,000 fig., 1953.

Part H. BRACHIOPODA, xxxii + 927 p., 5,198 fig., 1965.

Part I. MOLLUSCA 1 (Mollusca General Features, Scaphopoda, Amphineura, Monoplacophora, Gastropoda General Features, Archaeogastropoda, mainly Paleozoic Caenogastropoda and Opisthobranchia), xxiii + 351 p., 1,732 fig., 1960.

Part K. MOLLUSCA 3 (Cephalopoda General Features, Endoceratoidea, Actinoceratoidea, Nautiloidea, Bactritoidea), xxviii + 519 p., 2,382 fig., 1964.

Part L. MOLLUSCA 4 (Ammonoidea), xxii + 490 p., 3,800 fig., 1957.

Part N. MOLLUSCA 6 (Bivalvia), Volumes 1 and 2 (of 3), xxxviii + 952 p., 6,198 fig., 1969; Volume 3, iv + 272 p., 742 fig., 1971.

Part O. ARTHROPODA 1 (Arthropoda General Features, Protarthropoda, Euarthropoda General Features, Trilobitomorpha), xix + 560 p., 2,880 fig., 1959.

Part P. ARTHROPODA 2 (Chelicerata, Pycnogonida, Palaeoisopus), xvii + 181 p., 565 fig., 1955.

Part Q. ARTHROPODA 3 (Crustacea, Ostracoda), xxiii + 442 p., 3,476 fig., 1961.

Part R. ARTHROPODA 4 (Crustacea exclusive of Ostracoda, Myriapoda, Hexapoda), Volumes 1 and 2 (of 3), xxxvi + 651 p., 1,762 fig., 1969.

Part S. ECHINODERMATA 1 (Echinodermata General Features, Homalozoa, Crinozoa, exclusive of Crinoidea), xxx + 650 p., 2,868 fig., 1967 [1968].

Part T. ECHINODERMATA 2 (Crinoidea), Volumes 1-3, xxxviii + 1027 p., 4,833 fig., 1978.

Part U. ECHINODERMATA 3 (Asterozoans, Echinozoans), xxx + 695 p., 3,485 fig., 1966.

Part V. GRAPTOLITHINA, xvii + 101 p., 358 fig., 1955.

Part V. GRAPTOLITHINA, Second Edition (Revised and Enlarged), xxxii + 163 p., 507 fig., 1970.

Part W. MISCELLANEA (Conodonts, Conoidal Shells of Uncertain Affinities, Worms, Trace Fossils, Problematica), xxv + 259 p., 1,058 fig., 1962.
Part W. MISCELLANEA (Supplement 1). Trace Fossils and Problematica, Second Edition (Revised and Enlarged), xxi + 269 p., 912 fig., 1975.

THIS VOLUME

Part A. INTRODUCTION, xxiii + 569 p., 371 fig., 1979.

VOLUMES IN PREPARATION (1978)

Part B. PROTISTA 1 (Chrysomonadida, Coccolithophorida, Charophyta, Diatomacea, etc.).
Part J. MOLLUSCA 2 (Gastropoda, Streptoneura exclusive of Archaeogastropoda, Euthyneura).
Part M. MOLLUSCA 5 (Coleoidea).
Part R. ARTHROPODA 4, Volume 3 (Hexapoda).
Part X. ADDENDA, INDEX.
Part E. PORIFERA, Volume 2 (revised edition).
Part F. COELENTERATA (supplement) (Anthozoa, Rugosa and Tabulata, revised edition).
Part G. BRYOZOA (revised edition).
Part L. MOLLUSCA 4 (supplement) (Ammonoidea, revised edition).
Part Q. ARTHROPODA 3 (Ostracoda, revised edition).
Part W. MISCELLANEA (supplement 2) (Conodonts, revised edition).

CONTRIBUTING AUTHORS

[Arranged by countries and institutions; an alphabetical list follows. An asterisk preceding name indicates author working on revision of or supplement to a published *Treatise* volume.]

AUSTRALIA

South Australia Geological Survey (Adelaide)
N. H. Ludbrook
University of Adelaide
M. F. Glaessner
University of Queensland (Brisbane)
Dorothy Hill

AUSTRIA

Universität Wien (Paläontologisches Institut)
Adolf Papp

BELGIUM

Unattached
*Charles Grégoire (Bruxelles)
Université de Liège
Georges Ubaghs

CANADA

Geological Survey of Canada (Dartmouth, Nova Scotia)
G. L. Williams
Geological Survey of Canada (Ottawa)
J. A. Jeletzky, D. J. McLaren, G. W. Sinclair

Institute of Sedimentary & Petroleum Geology (Geological Survey of Canada, Calgary)
A. W. Norris
University of British Columbia (Vancouver)
V. J. Okulitch
University of Saskatchewan (Saskatoon)
W. A. S. Sarjeant
Royal Ontario Museum (Toronto)
D. H. Collins

DENMARK

Universitet København
*Eckart Håkansson, H. Wienberg Rasmussen
Marine Biologisk Laboratorium (Helsingør)
*Claus Nielsen

FRANCE

Université de Paris
Colette Dechaseaux (Laboratoire de Paléontologie des Vertébrés), Geneviève Lutaud (Laboratoire de Cytologie)
Université Paris-Sud (Orsay)
Michel Roux

GABON
Shell Gabon (Port Gentil)
  *R. A. Pohowsky

GERMAN DEMOCRATIC REPUBLIC
Bergakademie Freiberg (Fachbereich Geowissenschaften)
  A. H. Müller

GERMANY, FEDERAL REPUBLIC OF
Friedrich Wilhelms Universität (Bonn)
  H. K. Erben, *K. J. Müller
Natur-Museum und Forschungs-Institut Senckenberg (Frankfurt)
  Herta Schmidt, Wolfgang Struve
Philipps Universität (Marburg)
  Gerhard Hahn, *Maurits Lindström, *Willi Ziegler
Unattached
  Hertha Sieverts-Doreck (Stuttgart-Möhringen)
Universität Münster
  Helmut Hölder
Universität Tübingen
  *Jürgen Kullmann, *Adolf Seilacher
Universität Würzburg
  Klaus Sdzuy

ITALY
Universitá Modena
  Eugenia Montanaro Gallitelli
Universitá di Roma
  Franco Rasetti

JAPAN
Saito Ho-on Kai Museum of Natural History (Sendai)
  Kotora Hatai
University of Tokyo
  Tetsuro Hanai

NETHERLANDS
Rijksmuseum van Natuurlijke Historie (Leiden)
  L. B. Holthuis
Vrije Universiteit Amsterdam
  A. J. Breimer, M. J. S. Rudwick

NEW ZEALAND
Auckland Institute and Museum
  A. W. B. Powell
Dominion Museum (Wellington)
  R. K. Dell
New Zealand Geological Survey (Lower Hutt)
  C. A. Fleming
Unattached
  John Marwick (Havelock North)

NORWAY
Unattached
  Tron Soot-Ryen (Hosle)
Universitet Oslo
  Gunnar Henningsmoen, Leif Størmer (Institutt for Geologi)

POLAND
Pánstwowe Wydawnictwo Naukowe (Warszawa)
  Gertruda Biernat, Adolf Riedel

SAUDI ARABIA
Arabian American Oil Company (Dhahran)
  A. L. Bowsher

SWEDEN
Naturhistoriska Museet Göteborg
  Bengt Hubendick
Naturhistoriska Riksmuseet Stockholm
  Valdar Jaanusson
Universitet Lund
  Gerhard Regnéll
Universitet Stockholm
  Ivar Hessland
Universitet Uppsala
  R. A. Reyment

SWITZERLAND
Universität Basel
  Manfred Reichel

UNITED KINGDOM
British Museum (Natural History) (London)
  *P. L. Cook, Isabella Gordon, *M. K. Howarth, S. M. Manton, N. J. Morris, C. P. Nuttall
British Petroleum Company (Middlesex)
  F. E. Eames
Institute of Geological Sciences (London)
  Raymond Casey, R. V. Melville
Iraq Petroleum Company (London)
  G. F. Elliott
Queen's University of Belfast
  Margaret Jope, *R. E. H. Reid, A. D. Wright
Unattached
  Dennis Curry (Middlesex), Sir James Stubblefield (London), R. P. Tripp (Sevenoaks, Kent), C. W. Wright (Dorset), Sir Maurice Yonge (Edinburgh)

University of Birmingham
*Anthony Hallam, L. J. Wills

University of Cambridge
H. B. Whittington

University College London
*J. H. Callomon, *D. T. Donovan

University College of Swansea (Swansea, Wales)
D. V. Ager, *J. S. Ryland

University of Durham
*G. P. Larwood

University of Glasgow
W. D. I. Rolfe, John Weir, Alwyn Williams

University of Hull
*M. R. House

University of Leicester
P. C. Sylvester-Bradley

University of Manchester
E. R. Trueman

University of Southampton
*R. L. Austin

UNITED STATES OF AMERICA

Academy of Natural Sciences of Philadelphia (Pennsylvania)
A. A. Olsson, Robert Robertson

American Museum of Natural History (New York)
R. L. Batten, W. K. Emerson, N. D. Newell

Appalachian State University (Boone, North Carolina)
*F. K. McKinney

Brown University (Providence, Rhode Island)
R. D. Staton

California Academy of Sciences (San Francisco)
Eugene Coan, Barry Roth, A. G. Smith

California Institute of Technology (Pasadena)
H. A. Lowenstam

Carnegie Museum (Pittsburgh, Pennsylvania)
Juan Parodiz

Cornell University (Ithaca, New York)
W. S. Cole, J. W. Wells

Exxon Production Research Company (Houston, Texas)
H. H. Beaver, R. M. Jeffords, S. A. Levinson, L. A. Smith, Joan Stough

Field Museum of Natural History (Chicago)
Fritz Haas, G. A. Solem

Florida State University (Tallahassee)
W. H. Heard

Getty Oil Company (Houston, Texas)
Lavon McCormick

Harvard University (Cambridge, Massachusetts)
Kenneth Boss, F. M. Carpenter, W. J. Clench, H. B. Fell, Bernhard Kummel, Ruth Turner

Illinois State Geological Survey (Urbana)
M. L. Thompson

Indiana Geological Survey (Bloomington)
R. H. Shaver

Kansas Geological Survey (Lawrence)
D. E. Nodine Zeller

Kent State University (Kent, Ohio)
A. H. Coogan

Louisiana State University (Baton Rouge)
W. A. van den Bold, H. B. Stenzel

New Mexico Institute Mining & Technology (Socorro)
Christina Lochman-Balk

New York State Museum (Albany)
D. W. Fisher

Ohio State University (Columbus)
*S. M. Bergström, Aurèle La Rocque, W. C. Sweet

Oklahoma Geological Survey (Norman)
T. W. Amsden, R. O. Fay

Oregon State University (Corvallis)
A. J. Boucot, J. G. Johnson

Paleontological Research Institution (Ithaca, New York)
K. V. W. Palmer

Princeton University (Princeton, New Jersey)
A. G. Fischer

Professional Geophysics, Inc. (Oklahoma City, Oklahoma)
J. A. Eyer

Queens College of the City of New York (Flushing)
*R. M. Finks

Radford College (Radford, Virginia)
R. L. Hoffman

St. Mary's College of California (St. Mary's)
A. S. Campbell

San Diego Natural History Museum (San Diego, California)
George Radwin

San Francisco State University (San Francisco, California)
Y. T. Mandra

Smithsonian Institution (Washington, D.C.)
F. M. Bayer, R. H. Benson, *R. S. Boardman, *A. H. Cheetham, A. H. Clarke, Jr., G. A. Cooper, T. G. Gibson, R. E. Grant, E. G. Kauffman, P M. Kier, R. B. Manning, David Pawson, H. A. Rehder

Southern Illinois University (Carbondale)
*John Utgaard

Southern Methodist University (Dallas, Texas)
A. L. McAlester

Southwest Missouri State University (Springfield)
*J. F. Miller

Stanford University (Stanford, California)
A. Myra Keen

State University of New York (Stony Brook)
A. R. Palmer

Syracuse University (Syracuse, New York)
J. C. Brower, *O. B. Nye

Tulane University (New Orleans, Louisiana)
Emily Vokes, H. E. Vokes

Unattached
R. Wright Barker (Bellaire, Texas), J. W. Hedgpeth (Santa Rosa, Calif.), H. S. Puri (Tallahassee, Florida)

United States Geological Survey (Washington, D.C.)
J. M. Berdan, R. C. Douglass, Anita Epstein, Mackenzie Gordon, Jr., *J. E. Hazel, *O. L. Karklins, K. E. Lohman, N. F. Sohl, I. G. Sohn, E. L. Yochelson
(Menlo Park, Calif.)
Dwight Taylor

University of Alaska (Fairbanks)
C. D. Wagner

University of California (Berkeley)
J. W. Durham

University of California (Los Angeles)
A. R. Loeblich, Jr., W. P. Popenoe, Helen Tappan

University of California (San Diego, La Jolla)
R. R. Hessler, W. A. Newman

University of Cincinnati (Ohio)
K. E. Caster, D. L. Meyer

University of Colorado (Boulder)
J. A. Van Couvering

University of Florida (Gainesville)
H. K. Brooks, F. G. Thompson

University of Illinois (Urbana)
*D. B. Blake, *Philip Sandberg, H. W. Scott

University of Indiana (Bloomington)
N. Gary Lane

University of Iowa (Iowa City)
W. M. Furnish, B. F. Glenister, *Gilbert Klapper, H. L. Strimple

University of Kansas (Lawrence)
A. B. Leonard, R. A. Robison, A. J. Rowell, R. H. Thompson

University of Massachusetts (Amherst)
C. W. Pitrat

University of Miami (Florida)
W. W. Hay, Donald Moore

University of Michigan (Ann Arbor)
J. B. Burch, R. V. Kesling, D. B. Macurda, C. P. Morgan, F. H. T. Rhodes

University of Minnesota (Minneapolis)
F. M. Swain

University of Missouri (Columbia)
R. E. Peck

University of North Carolina (Wilmington)
V. A. Zullo

University of Rochester (Rochester, New York)
Curt Teichert

University of Texas (Arlington)
B. F. Perkins

University of Texas (Austin)
J. T. Sprinkle

University of Wisconsin (Madison)
*D. L. Clark

University of Wyoming (Laramie)
D. W. Boyd

Western Reserve University (Cleveland, Ohio)
F. G. Stehli

Western Washington University (Bellingham)
C. A. Ross, J. R. P. Ross

Wichita State University (Kansas)
Paul Tasch

Woods Hole Oceanographic Institute (Massachusetts)
W. A. Berggren

Wright State University (Dayton, Ohio)
*T. S. Wood

DECEASED

W. J. Arkell, R. S. Bassler, H. Boschma, M. N. Bramlette, C. C. Branson, O. M. Bulman, André Chavan, L. R. Cox, L. M. Davies, Harriet Exline, D. L. Frizzell, Julia Gardner, Walter Häntzschel, G. D. Hanna, H. J. Harrington, W. H. Hass, H. L. Hawkins, L. G. Hertlein, H. V. Howe, B. F. Howell, L. H. Hyman, J. B. Knight, M. W. deLaubenfels, Marius Lecompte, A. K. Miller, R. C. Moore, H. M. Muir-Wood, Alexander Petrunkevitch, Chr. Poulsen, Emma Richter, Rudolf Richter, O. H. Schindewolf, W. K. Spencer, M. A. Stainbrook, L. W. Stephenson, E. C. Stumm, O. W. Tiegs, Johannes Wanner, J. M. Weller, T. H. Withers, Arthur Wrigley

Alphabetical List

Ager, D. V., London (Univ. College of Swansea)
Amsden, T. W., Norman, Okla. (Oklahoma Geol. Survey)
Arkell, W. J. (deceased)
*Austin, R. L., Southampton, Eng. (Univ. Southampton)
Barker, R. W., Bellaire, Texas (unattached)
Bassler, R. S. (deceased)
Batten, R. L., New York (American Museum Nat. History)
Bayer, F. M., Washington, D.C. (Smithsonian Inst.)
Beaver, H. H., Houston, Texas (Exxon Production Research Company)
Benson, R. H., Washington, D.C. (Smithsonian Inst.)
Berdan, J. M., Washington, D.C. (U.S. Geol. Survey)
Berggren, W. A., Woods Hole, Mass. (Woods Hole Oceanographic Inst.)
*Bergström, S. M., Ohio (Ohio State Univ.)
Biernat, Gertruda, Warszawa, Poland (Pánstwowe Wydawnictwo Naukowe)
*Blake, D. B., Urbana, Ill. (Univ. Illinois)
*Boardman, R. S., Washington, D.C. (Smithsonian Inst.)
Bold, W. A. van den, Baton Rouge, La. (Louisiana State Univ.)
Boschma, H. (deceased)
Boss, Kenneth, Cambridge, Mass. (Harvard Univ.)
Boucot, A. J., Corvallis, Ore. (Oregon State Univ.)
Bowsher, A. L., Dhahran, Saudi Arabia (Arabian American Oil Co.)
Boyd, D. W., Laramie, Wyo. (Univ. Wyoming)
Bramlette, M. N. (deceased)
Branson, C. C. (deceased)
Breimer, A. J., Amsterdam, Netherlands (Inst. Aardwetensch. Vrije Univ.)
Brooks, H. K., Gainesville, Fla. (Univ. Florida)
Brower, J. C., Syracuse, N.Y. (Syracuse Univ.)
Bulman, O. M. B. (deceased)
Burch, J. B., Ann Arbor, Mich. (Univ. Michigan)
*Callomon, J. H., London (Univ. College)
Campbell, A. S., St. Mary's, Calif. (St. Mary's College)
Carpenter, F. M., Cambridge, Mass. (Harvard Univ.)

Casey, Raymond, London (Inst. Geol. Sciences)
Caster, K. E., Cincinnati, Ohio (Univ. Cincinnati)
Chavan, André (deceased)
*Cheetham, A. H., Washington, D.C. (Smithsonian Inst.)
*Clark, D. L., Madison, Wis. (Univ. Wisconsin)
Clarke, A. H., Jr., Washington, D.C. (Smithsonian Inst.)
Clench, W. J., Cambridge, Mass. (Harvard Univ.)
Coan, Eugene, San Francisco, Calif. (California Acad. Sci.)
Cole, W. S., Ithaca, N.Y. (Cornell Univ.)
Collins, D. H., Toronto, Ontario, Canada (Royal Ontario Museum)
Coogan, A. H., Kent, Ohio (Kent State Univ.)
*Cook, P. L., London (British Museum Nat. History)
Cooper, G. A., Washington, D.C. (Smithsonian Inst.)
Cox, L. R. (deceased)
Curry, Dennis, Middlesex, Eng. (unattached)
Davies, L. M. (deceased)
Dechaseaux, Colette, Paris (Laboratoire de Paléontologie des Vertébrés)
Dell, R. K., Wellington, N.Z. (Dominion Museum)
*Donovan, D. T., London, Eng. (Univ. College)
Douglass, R. C., Washington, D.C. (U.S. Geol. Survey)
Durham, J. W., Berkeley, Calif. (Univ. California)
Eames, F. E., Middlesex, Eng. (British Petroleum Co.)
Elliott, G. F., London (Iraq Petroleum Co.)
Emerson, W. K., New York (American Museum Nat. History)
*Epstein, Anita, Washington, D.C. (U.S. Geol. Survey)
Erben, H. K., Bonn, West Germany (Friedrich Wilhelms Univ.)
Exline, Harriet (deceased)
Eyer, J. A., Oklahoma City, Okla. (Professional Geophysics, Inc.)
Fay, R. O., Norman, Okla. (Oklahoma Geol. Survey)
Fell, H. B., Cambridge, Mass. (Harvard Univ.)
*Finks, R. M., Flushing, N.Y. (Queens College)
Fischer, A. G., Princeton, N.J. (Princeton Univ.)

Fisher, D. W., Albany, N.Y. (New York State Museum)
Fleming, C. A., Lower Hutt, N.Z. (New Zealand Geol. Survey)
Frizzell, D. L. (deceased)
Furnish, W. M., Iowa City, Iowa (Univ. Iowa)
Gardner, Julia (deceased)
Gibson, T. G., Washington, D.C. (Smithsonian Inst.)
Glaessner, M. F., Adelaide, S. Australia (Univ. Adelaide)
Glenister, B. F., Iowa City, Iowa (Univ. Iowa)
Gordon, Isabella, London (British Museum Nat. History)
Gordon, Mackenzie, Jr., Washington, D.C. (U.S. Geol. Survey)
Grant, R. E., Washington, D.C. (Smithsonian Inst.)
*Grégoire, Charles, Bruxelles, Belgium (unattached)
*Håkansson, Eckart, København (Univ. København)
Haas, Fritz, Chicago, Ill. (Field Museum Nat. History)
Häntzschel, Walter (deceased)
Hahn, Gerhard, Marburg (Philipps Univ.)
*Hallam, Anthony, Oxford, Eng. (Univ. Birmingham)
Hanai, Tetsuro, Tokyo (Univ. Tokyo)
Hanna, G. D. (deceased)
Harrington, H. J. (deceased)
Hass, W. H. (deceased)
Hatai, Kotora, Sendai, Japan (Saito Ho-on Kai Museum Nat. History)
Hawkins, H. L. (deceased)
Hay, W. W., Miami, Fla. (Univ. Miami)
*Hazel, J. E., Washington, D.C. (U.S. Geol. Survey)
Heard, W. H., Tallahassee, Fla. (Florida State Univ.)
Hedgpeth, J. W., Santa Rosa, Calif. (unattached)
Henningsmoen, Gunnar, Oslo (Univ. Oslo)
Hertlein, L. G. (deceased)
Hessland, Ivar, Stockholm, Sweden (Univ. Stockholm)
Hessler, R. R., La Jolla, Calif. (Scripps Inst. Oceanography)
Hill, Dorothy, Brisbane, Australia (Univ. Queensland)
Hölder, Helmut, Münster, Germany (Univ. Münster)
Hoffman, R. L., Radford, Va. (Radford College)
Holthuis, L. B., Leiden, Netherlands (Rijksmuseum van Natuurlijke Historie)
House, M. R., Kingston upon Hull, Eng. (Univ. Hull)
*Howarth, M. K., London (British Museum Nat. History)
Howe, H. V. (deceased)
Howell, B. F. (deceased)
Hubendick, Bengt, Göteborg, Sweden (Naturhistoriska Museet)
Hyman, L. H. (deceased)
Jaanusson, Valdar, Stockholm (Naturhistoriska Riksmuseet)
Jeffords, R. M., Houston, Texas (Exxon Production Research Company)
Jeletzky, J. A., Ottawa, Ontario, Canada (Geol. Survey Canada)
Johnson, J. G., Corvallis, Ore. (Oregon State Univ.)
Jope, Margaret, Belfast, N. Ireland (Queen's Univ. of Belfast)
*Karklins, O. L., Washington, D.C. (U.S. Geol. Survey)
Kauffman, E. G., Washington, D.C. (Smithsonian Inst.)
Keen, A. Myra, Stanford, Calif. (Stanford Univ.)
Kesling, R. V., Ann Arbor, Mich. (Univ. Michigan)
Kier, P. M., Washington, D.C. (Smithsonian Inst.)
*Klapper, Gilbert, Iowa City, Iowa (Univ. Iowa)
Knight, J. B. (deceased)
*Kullmann, Jürgen, Tübingen, W. Germany (Univ. Tübingen)
Kummel, Bernhard, Cambridge, Mass. (Harvard Univ.)
Lane, N. Gary, Bloomington, Ind. (Univ. Indiana)
La Rocque, Aurèle, Columbus, Ohio (Ohio State Univ.)
*Larwood, G. P., Durham, Eng. (Univ. Durham)
Laubenfels, M. W. de (deceased)
Lecompte, Marius (deceased)
Leonard, A. B., Lawrence, Kans. (Univ. Kansas)
Levinson, S. A., Houston, Texas (Exxon Production Research Company)
*Lindström, Maurits, Marburg, Germany (Philipps Univ.)
Lochman-Balk, Christina, Socorro, N. Mex. (New Mexico Inst. Mining & Technology)
Loeblich, A. R., Jr., Los Angeles, Calif. (Univ. California)
Lohman, K. E., Washington, D.C. (U.S. Geol. Survey)
Lowenstam, H. A., Pasadena, Calif. (California Inst. Technology)
Ludbrook, N. H., Adelaide, S. Australia (South Australia Geol. Survey)
Lutaud, Geneviève, Paris (Laboratoire Cytologie, Univ. Paris)
McAlester, A. L., Dallas, Texas (Southern Methodist Univ.)
McCormick, Lavon, Houston, Texas (Getty Oil Company)
*McKinney, F. K., Boone, N. Car. (Appalachian State Univ.)
McLaren, D. J., Ottawa, Ontario, Canada (Geol. Survey Canada)
Macurda, D. B., Ann Arbor, Mich. (Univ. Michigan)
Mandra, Y. T., San Francisco, Calif. (San Francisco State Univ.)
Manning, R. B., Washington, D.C. (Smithsonian Inst.)
Manton, S. M., London (British Museum Nat. History)

Marwick, John, Havelock North, N.Z. (unattached)
Melville, R. V., London (Inst. Geol. Sciences)
Meyer, D. L., Cincinnati, O. (Univ. Cincinnati)
Miller, A. K. (deceased)
*Miller, J. F., Springfield, Mo. (Southwest Missouri State Univ.)
Montanaro Gallitelli, Eugenia, Modena, Italy (Univ. Modena)
Moore, Donald, Miami, Fla. (Univ. Miami, Inst. Marine Sci.)
Moore, R. C. (deceased)
Morgan, C. P., Ann Arbor, Mich. (Univ. Michigan)
Morris, N. J., London (British Museum Nat. History)
Müller, A. H., Freiberg, German Democratic Republic (Fachbereich Geowiss.)
*Müller, K. J., Bonn, West Germany (Friedrich Wilhelms Univ.)
Muir-Wood, H. M. (deceased)
Newell, N. D., New York (American Museum Nat. History)
Newman, W. A., La Jolla, Calif. (Scripps Inst. Oceanography)
*Nielsen, Claus, Helsingør, Denmark (Marine Biologisk Lab.)
Norris, A. W., Calgary, Alberta, Canada (Geol. Survey Canada)
Nuttall, C. P., London (British Museum Nat. History)
*Nye, O. B., Syracuse, N.Y. (Syracuse Univ.)
Okulitch, V. J., Vancouver, Canada (Univ. British Columbia)
Olsson, A. A., Coral Gables, Fla. (Acad. Nat. Sci. Philadelphia)
Palmer, A. R., Stony Brook, Long Island, N.Y. (State Univ. New York)
Palmer, K. V. W., Ithaca, N.Y. (Paleont. Research Inst.)
Papp, Adolf, Wien, Austria (Univ. Wien)
Parodiz, Juan, Pittsburgh, Pa. (Carnegie Museum)
Pawson, David, Washington, D.C. (Smithsonian Inst.)
Peck, R. E., Columbia, Mo. (Univ. Missouri)
Perkins, B. F., Arlington, Texas (Univ. Texas at Arlington)
Petrunkevitch, Alexander (deceased)
Pitrat, C. W., Amherst, Mass. (Univ. Massachusetts)
*Pohowsky, R. A., Port Gentil, Gabon (Shell Gabon)
Popenoe, W. P., Los Angeles, Calif. (Univ. Calif.)
Poulsen, Chr. (deceased)
Powell, A. W. B., Auckland, N.Z. (Auckland Inst. & Museum)
Puri, H. S., Tallahassee, Fla. (unattached)
Radwin, George, San Diego, Calif. (San Diego Nat. History Museum)
Rasetti, Franco, Rome, Italy (Univ. Roma)
Rasmussen, H. Wienberg, København, Denmark (Univ. København)
Regnéll, Gerhard, Lund, Sweden (Univ. Lund)

Rehder, H. A., Washington, D.C. (Smithsonian Inst.)
Reichel, Manfred, Basel, Switz. (Univ. Basel)
*Reid, R. E. H., Belfast, N. Ireland (Queen's Univ. Belfast)
Reyment, R. A., Uppsala, Sweden (Univ. Uppsala)
Rhodes, F. H. T., Ann Arbor, Mich. (Univ. Michigan)
Richter, Emma (deceased)
Richter, Rudolf (deceased)
Riedel, Adolf, Warszawa, Poland (Pánstwowe Wydawnictwo Naukowe)
Robertson, Robert, Philadelphia, Pa. (Acad. Nat. Sci.)
Robison, R. A., Lawrence, Kans. (Univ. Kansas)
Rolfe, W. D. I., Glasgow, Scot. (Univ. Glasgow)
Ross, C. A., Bellingham, Wash. (Western Washington Univ.)
Ross, J. R. P., Bellingham, Wash. (Western Washington Univ.)
Roth, Barry, San Francisco, Calif. (California Acad. Sci.)
Roux, Michel, Orsay, France (Univ. Paris-Sud)
Rowell, A. J., Lawrence, Kans. (Univ. Kansas)
Rudwick, M. J. S., Amsterdam (Vrije Univ.)
*Ryland, J. S., Swansea, Wales (Univ. College)
*Sandberg, Philip, Urbana, Ill. (Univ. Illinois)
Sarjeant, W. A. S., Saskatoon, Canada (Univ. Saskatchewan)
Schindewolf, O. H. (deceased)
Schmidt, Herta, Frankfurt, Germany (Natur Museum u. Forsch.-Inst. Senckenberg)
Scott, H. W., Urbana, Ill. (Univ. Illinois)
Sdzuy, Klaus, Würzburg, Germany (Univ. Würzburg)
*Seilacher, Adolf, Tübingen, West Germany (Univ. Tübingen)
Shaver, R. H., Bloomington, Ind. (Indiana Geol. Survey & Univ. Indiana)
Sieverts-Doreck, Hertha, Stuttgart-Möhringen, Germany (unattached)
Sinclair, G. W., Ottawa, Ontario, Canada (Geol. Survey Canada)
Smith, A. G., San Francisco, Calif. (California Acad. Sci.)
Smith, L. A., Houston, Texas (Exxon Production Research Company)
Sohl, N. F., Washington, D.C. (U.S. Geol. Survey)
Sohn, I. G., Washington, D.C. (U.S. Geol. Survey)
Solem, G. A., Chicago, Ill. (Field Museum Nat. History)
Soot-Ryen, Tron, Hosle, Nor. (unattached)
Spencer, W. K. (deceased)
Sprinkle, J. T., Austin, Texas (Univ. Texas)
Stainbrook, M. A. (deceased)
Staton, R. D., Providence, R.I. (Brown Univ.)
Stehli, F. G., Cleveland, Ohio (Western Reserve Univ.)
Stenzel, H. B., Baton Rouge, La. (Louisiana State Univ.)
Stephenson, L. W. (deceased)

Størmer, Leif, Oslo (Univ. Oslo)
Stough, Joan, Houston, Texas (Exxon Production Research Company)
Strimple, H. L., Iowa City, Iowa (Univ. Iowa)
Struve, Wolfgang, Frankfurt, Germany (Natur-Museum u. Forsch.-Inst. Senckenberg)
Stubblefield, Sir James, London (unattached)
Stumm, E. C. (deceased)
Swain, F. M., Minneapolis, Minn. (Univ. Minnesota)
Sweet, W. C., Columbus, Ohio (Ohio State Univ.)
Sylvester-Bradley, P. C., Leicester, Eng. (Univ. Leicester)
Tappan, Helen, Los Angeles, Calif. (Univ. California)
Tasch, Paul, Wichita, Kans. (Wichita State Univ.)
Taylor, Dwight, Menlo Park, Calif. (U.S. Geol. Survey)
Teichert, Curt, Rochester, N.Y. (Univ. Rochester)
Thompson, F. G., Gainesville, Fla. (Univ. Florida)
Thompson, M. L., Urbana, Ill. (Illinois State Geol. Survey)
Thompson, R. H., Lawrence, Kans. (Univ. Kansas)
Tiegs, O. W. (deceased)
Tripp, R. P., Sevenoaks, Kent, Eng. (unattached)
Trueman, E. R., Manchester, Eng. (Univ. Manchester)
Turner, Ruth, Cambridge, Mass. (Harvard Univ.)
Ubaghs, Georges, Liège, Belgium (Univ. Liège)
*Utgaard, John, Carbondale, Ill. (Southern Illinois Univ.)
Van Couvering, John, Boulder, Colo. (Univ. Colorado)
Vokes, Emily, New Orleans, La. (Tulane Univ.)
Vokes, H. E., New Orleans, La. (Tulane Univ.)
Wagner, C. D., Fairbanks, Alaska (Univ. Alaska)
Wanner, Johannes (deceased)
Weir, John, Tayport, Fife, Scotland (Univ. Glasgow)
Weller, J. M. (deceased)
Wells, J. W., Ithaca, N.Y. (Cornell Univ.)
Whittington, H. B., Cambridge, Eng. (Univ. Cambridge)
Williams, Alwyn, Glasgow, Scot. (Univ. Glasgow)
Williams, G. L., Dartmouth, Nova Scotia (Geol. Survey Canada)
Wills, L. J., Birmingham, Eng. (Univ. Birmingham)
Withers, T. H. (deceased)
*Wood, T. S., Dayton, Ohio (Wright State Univ.)
Wright, A. D., Belfast, N. Ireland (Queen's Univ. Belfast)
Wright, C. W., Dorset, Eng. (unattached)
Wrigley, Arthur (deceased)
Yochelson, E. L., Washington, D.C. (U.S. Geol. Survey)
Yonge, Sir Maurice, Edinburgh, Scotland (unattached)
Zeller, D. E. Nodine, Lawrence, Kans. (Kansas Geol. Survey)
*Ziegler, Willi, Marburg, Germany (Philipps Univ.)
Zullo, V. A., Wilmington, N.C. (Univ. North Carolina)

EDITORIAL PREFACE

Preparation of this volume was postponed until most volumes in the series were either published or well on their way toward completion. Initial organization and assignment of authors for Part A began about 1968 under the direction of CURT TEICHERT. Some authors soon submitted manuscripts, but for a variety of reasons, preparation of other manuscripts was delayed. With the death of R. C. MOORE in 1974, other *Treatise* responsibilities required TEICHERT's attention, and major responsibility for the Part A project was transferred to R. A. ROBISON for completion.

As originally conceived, this volume was intended to include three major parts: 1) a general chapter on fossilization, 2) discussions of the evolution and geographic distribution of invertebrate faunas, and 3) stratigraphic distribution charts compiled for all invertebrate groups covered by the *Treatise*. The first part was assigned to Prof. A. H. MÜLLER, the second part was subdivided with each Phanerozoic system and the Precambrian being assigned to either one or two authors, and work on the third part was begun by *Treatise* staff members.

Because some authors, again for a variety of reasons, were unable to complete original assignments, a few replacements were made during 1974 and 1975. At the same time it was decided that publication of the stratigraphic range charts had become impractical, partly because delays had made compilations from the earliest *Treatise* volumes out-of-date, and partly because of lack of information for some invertebrates covered in a few first-edition volumes still in preparation. Therefore, all authors dealing with evolution and biogeography were asked to add a discussion of biostratigraphy to their respective chapters, a request was made for updating of those manuscripts already received, and the stratigraphic range charts were abandoned for Part A. Response was positive, and for this the editors express appreciation, particularly to those authors who completed their assignments promptly and were patient and cooperative when subsequent revisions were solicited.

The chapter on the Precambrian differs from other chapters in this volume by the addition of a section with systematic descriptions of metazoans as well as comprehensive generic lists of other fossils. Some Ediacaran taxa were assigned to the Lower Cambrian in previous *Treatise* volumes, but an assignment to the Upper Precambrian is now generally accepted. Other recently described taxa are unrecorded in the *Treatise*. Therefore, the updating and assembly of these descriptions and lists should be of value to many readers. Features of style in zoological nomenclature that apply to these Precambrian taxa have been discussed in the prefaces of other published *Treatise* parts.

The aim of the *Treatise on Invertebrate Paleontology*, as originally conceived and consistently pursued, is to present the most comprehensive and authoritative, yet compact statement of knowledge concerning invertebrate fossil groups that can be formulated by collaboration of competent specialists in seeking to organize what has been learned of this subject up to the year of publication of each individual part. Such work has value in providing a most useful summary of the collective results of multitudinous investigations and thus constitutes an indispensable text and reference book for all persons who wish to know about remains of invertebrate organisms preserved in rocks of the earth's crust. This applies to neozoologists as well as paleozoologists and to beginners in study of fossils as well as to thoroughly trained, long-experienced professional workers, including teachers, stratigraphical geologists, and individuals engaged in research on fossil invertebrates. The making of a reasonably complete inventory of present knowledge of invertebrate paleontology is yielding needed foundation for future research.

The *Treatise* is divided into parts which bear index letters, each except the initial and concluding ones being defined to include designated groups of invertebrates. The chief purpose of this arrangement is to provide for independence of the several parts as regards date of publication, because it was judged desirable to print and distribute each segment as soon as possible after it is ready for press. Pages in each part bear the assigned index letter joined with numbers beginning with 1 and running consecutively to the end of the part. In numerous cases materials for individual parts were so voluminous that these parts had to be published in two or even three volumes. In such cases, pagination is continuous through successive volumes.

A generous grant of $35,000 was made in 1948 by the Geological Society of America for initial work in preparing *Treatise* illustrations. Additional grants were made by The Geological Society of America in 1971 ($6,200), 1972 ($6,000), $7,000 each year for 1973 and 1974, and $20,000 each for 1975-1978. Administration of expenditures has been in charge of the editors and most of the work by photographers and artists has been done under their direction at the University of Kansas, but sizable parts of this program have also been carried forward in Washington, London, Ottawa, and many other places.

In December, 1959, the National Science Foundation of the United States, through its Division of Biological and Medical Sciences and the Program Director for Systematic Biology, made a grant in the amount of $210,000 for the purpose of aiding the completion of yet-unpublished volumes of the *Treatise*. Payment of this sum was provided to be made in installments distributed over a five-year period, with administration of disbursements handled by the University of Kansas. An additional grant (No. GB 4544) of $102,800 was made by the National Science Foundation in January, 1966, for the two-year period 1966-1967, and this was extended for the calendar year 1968 by payment of $25,700 in October, 1967. This grant was extended further by payments of $57,800 in 1968 for calendar year 1969, and $66,000 each for calendar years 1970-1972. For the years 1973-1977 grants totaled $197,400. These funds were used primarily to maintain editorial operations at the University of Kansas and to provide assistance to authors needed in preparation of manuscripts and illustrations. Grateful acknowledgement to the Foundation is expressed on behalf of the societies sponsoring the *Treatise,* the University of Kansas, and innumerable individuals benefited by the *Treatise* project.

ABBREVIATIONS

Abbreviations used in this division of the *Treatise,* except for those in the references or peculiar to specific figures, are explained in the following alphabetically arranged list.

aff., *affinis* (related to)
Afr., Africa, -an
Am., America, -an
approx., approximately
Arg., Argentina
auctt., *auctorum* (of authors)

B.P., before present
by, billion years

C, centigrade
C., Central
ca., *circa* (about)
Can., Canada
CCD, carbonate compensation depth
cf., *confer* (compare)
Charn., Charnian
cm, centimeter(s); cm^3, cubic centimeter(s)
Co., Company
Coll., Collection
Congr., Congress

Dept., Department
Dev., Devonian
diagramm., diagrammatic
DSDP, Deep Sea Drilling Project

E., East, eastern
ed., editor
e.g., *exempli gratia* (for example)
emend., *emendatus (-a)* (emended, emendation)
Eng., England
et al., *et alii* (and others, persons)
etc., *et cetera* (and others, objects)
et seq., *et sequens (-tes)* (and those that follow)

fig., figure(s)

gen., genus
Geol., Geology, Geological
gm, gram
Gr., Group

i.e., *id est* (that is)

IGC, International Geological Congress
indet., *indeterminata* (indeterminate)
INQUA, International Union for Quaternary Research
int., internal
Internatl., International
I.U.G.S., International Union of Geological Sciences

Jour., Journal

K., kaiserlich, königlich
kg, kilogram(s)
km, kilometer(s)

L., Lower
Loc., locality
low., lower

M, monotypy, monotype
M., Middle
m, meter(s)
Ma, mega-annum, unit of years multiplied by 10^6, measured from 1950 A.D. pastward
Miss., Mississippian
mm, millimeters(s); mm^2, square millimeter(s)
Mt., Mount
Mus., Museum
my, million years
mya, million years ago
MYBP, million years before present

N., North, northern
n(.), new
N.Car., North Carolina
no., number
nom. nud., *nomen nudum* (naked name)
nom. transl., *nomen translatus* (transferred name)
nom. van., *nomen vanum* (vain, void name)
nr., near
NW, northwest

obj. syn., objective synonym

OD, original designation
Ord., Ordovician

p., page(s)
Penn., Pennsylvania
Perm., Permian
pers. commun., personal communication
pl., plate
Precam., Precambrian

R., river
reconstr., reconstruction
reg., region
Repts., Reports

S., South, southern
SE, southeast
ser., series
Sib., Siberia
Sil., Silurian
s.l., *sensu lato* (in the wide sense, broadly defined)
sp., species
Spec., Special
sp. nov., new species
sq km, square kilometer(s)
s.s., *sensu stricto* (in the strict sense, narrowly defined)
SSR, Soviet Socialist Republic
SW, southwest
Swed., Sweden

Trias., Triassic

U., Upper
Univ., University
unpub., unpublished
up., upper
USA, United States of America
U.S.S.R., USSR, Union of Soviet Socialist Republics

v., volume
Vend., Vendian
vs., versus

W., West, western

Yudom., Yudomian

REFERENCES TO LITERATURE

Each part of the *Treatise* is accompanied by a list, or lists, of references to the paleontological literature. In *Treatise* parts published in the 1950's and early 1960's, these lists were highly selective, consisting primarily of recent and comprehensive monographs, but also including some important older works. In time, however, *Treatise* authors and readers pressed for more exhaustive documentation, and for volumes published from about 1965 the aim has been to provide documentation, complete with author, publication year, and page number, for all publications to which reference is made anywhere in the text.

The following is a statement of the full names of serial publications which are cited in abbreviated form in the lists of references in the present volume. The information thus provided should be useful in library research work. The list is alphabetized according to the serial titles which were employed at the time of original publication. Those following in parentheses are those under which the publication may be found currently in the *Union List of Serials,* the United States Library of Congress listing, and most library card catalogues. The names of serials published in Cyrillic are transliterated; in the reference lists these titles, which may be abbreviated, are accompanied by transliterated authors' names and titles, with English translation of the title. The place of publication is added (if not included in the serial title).

The method of transliterating Cyrillic letters that is adopted as "official" in the *Treatise* is that suggested by the Geographical Society of London and the U.S. Board on Geographic Names. It follows that names of some Russian authors in transliterated form derived in this way differ from other forms, possibly including one used by the author himself. In *Treatise* reference lists the alternative (unaccepted) form is given enclosed by square brackets (e.g., Chernyshev [Tschernyschew], T.N.).

List of Serial Publications

Académie Impériale des Sciences, St. Pétersbourg (Akademiya Nauk, SSSR. Leningrad), Mémoires.
Académie Polonaise des Sciences (Polska Akademia Umiejętności), Bulletin. Warsaw.
Académie Royale de Belgique, Classe des Sciences, Bulletins; Mémoires. Brussels.
Académie des Sciences, Agriculture, Commerce, Belles-lettres et Arts du Département de la Somme (Académie des Sciences, des Lettres, et des Arts d'Amiens), Mémoires.
Académie des Sciences de Paris, Comptes-rendus.
Acta Geologica Polonica. Warsaw.
Acta Geologica Sinica (Ti Chih Hsüeh Pao). Peking.
Acta Geologica Taiwanica, Series 1. Taipei, Formosa.
Acta Palaeontologica Polonica. Warsaw.
Acta Palaeontologica Sinica (Ku Sheng Wu Hsüeh Pao). Peking.
Acta Universitatis Stockholmiensis (Stockholm Contributions in Geology).
Akademie der Wissenschaften in Göttingen, mathematisch-physikalische Klass, Nachrichten.
Akademie der Wissenschaften und der Literatur zu Mainz, mathematisch-naturwissenschaftliche Klasse, Abhandlungen. Wiesbaden.
[K.] Akademie der Wissenschaften zu Wien, mathematisch-naturwissenschaftliche Klasse, Anzeiger; Denkschriften; Sitzungsberichte. Vienna.

Akademiya Nauk Gruzinskoi SSR, Soobshcheniya. Tiflis.
Akademiya Nauk Kazakhskoi SSR, Institut Geologicheskikh Nauk, Trudy. Alma-Ata.
Akademiya Nauk SSSR, Dalnevostochnyy Nauchnyy Tsentr. Vladivostok.
Akademiya Nauk, SSSR, Doklady; Trudy; Vestnik. Moscow.
Akademiya Nauk SSSR, Geologicheskiy Institut, Trudy. Moscow.
Akademiya Nauk SSSR, Izvestiya, Seriya Biologicheskaya; Izvestiya, Seriya Geologicheskaya. Moscow.
Akademiya Nauk SSSR, Komi Filial, Trudy. Syktyvkar.
Akademiya Nauk SSSR, Mezhduvedomstvennyi Geofizicheskiy Komitet. Moscow.
Akademiya Nauk SSSR, Paleontologicheskiy Institut, Trudy. Moscow.
Akademiya Nauk SSSR, Paleontologicheskiy Zhurnal. Moscow.
Akademiya Nauk SSSR, Sibirskoe Otdelenie, Institut Geologii i Geofiziki, Trudy. Novosibirsk.
Akademiya Nauk, SSSR, Zoologicheskiy Zhurnal. Moscow.
Akademiya Nauk Ukrainskoy SSR, Dopovidi. Kiev.
Akademiya Nauk Uzbekskoy SSR, Doklady. Tashkent.

Allgemeine Schweizerische Gesellschaft für die gesammten Naturwissenschaften, Neue Denkschriften (Schweizerische Naturforschende Gesellschaft). Zurich.

Ameghiniana (Asociacion Paleontologica Argentina). Buenos Aires.

American Association for the Advancement of Science, Publications. Washington, D.C.

American Association of Petroleum Geologists, Bulletin; Memoirs; Special Publication. Tulsa, Oklahoma.

American Chemical Society, Journal. Baltimore, Maryland.

American Geologist. Minneapolis, Minnesota.

American Journal of Science. New Haven, Connecticut.

American Museum of Natural History, Bulletin. New York, New York.

American Philosophical Society, Proceedings. Philadelphia, Pennsylvania.

American Scientist: The Sigma Xi Quarterly. New Haven, Connecticut.

American Zoologist. Utica, New York.

Annales des Mines. Paris.

Annales de Paléontologie, Invertébrés. Paris.

Annales des Sciences Naturelles. Paris.

Antarctic Research Series, Antarctic Oceanology. American Geophysical Union, Washington, D.C.

Archiv für Hydrobiologie und Planktonkunde. Stuttgart.

Archiv für Mikroskopische Anatomie. Bonn.

Archiv für Molluskenkunde. Frankfurt am Main.

Archives Géologiques du Viet-Nam. Saigon.

Arctic Institute of North America, Technical Paper. Montreal.

Arkticheskiy i Antarkticheskiy Nauchno-issledovatelskiy Institut, Trudy. Leningrad.

Asociacion Geologica Argentina, Revista. Buenos Aires.

Asociacion Venezolana de Geologia, Mineria y Petroleo, Boletin Informativo. Chacao.

Ateneo Parmense. Parma.

Australia Bureau of Mineral Resources, Geology and Geophysics, Bulletin; Reports. Canberra.

Australia, Geological Society of, Journals. Adelaide.

Australian Journal of Science. Sydney.

Bayerisch Akademie der Wissenschaften, Abhandlungen. Munich.

Bayerische Staatssammlung für Paläontologie und Historische Geologie, Mitteilungen. Munich.

Beiträge zur Mineralogie und Petrographie. Berlin.

Beiträge zur Mineralogie und Petrologie. Berlin, Heidelberg.

Beiträge zur Paläontologie und Geologie Österreich-Ungarns und des Orients. Vienna.

Biological Reviews (Cambridge Philosophical Society). Cambridge, England.

Biologisches Zentralblatt. Erlangen, Leipzig.

Boletín de Geologie Publicacion Especial. Caracas.

Boreas. Oslo.

Brigham Young University, Geology Studies. Provo, Utah.

British Museum (Natural History), Geology, Bulletin. London.

Bulletins of American Paleontology. Ithaca, New York.

Bureau de Recherches Géologiques et Minières, Bulletin; Mémoires. Paris.

Canada, Geological Survey of, Department of Mines and Resources, Mines and Geology Branch, Bulletins; Economic Geology Reports; Memoirs; Papers; Report of Activities; Summary Reports. Ottawa.

Canadian Journal of Earth Sciences. National Research Council, Canada. Ottawa.

Canadian Petroleum Geology, Bulletin. Alberta Society of Petroleum Geologists. Calgary.

Cape Town, University of, Department of Geology, Chamber of Mines Precambrian Research Unit, Bulletin. Cape Town.

Centralblatt für Mineralogie, Geologie und Paläontologie. Stuttgart.

Centre National de la Recherche Scientifique, Groupe Français des Argiles, Bulletin. Paris.

Česká Akademie Věd a Umění v Praze, Třiada II. Matematicko-přírodnická, Rozpravy. Prague.

China, Geological Survey of, Palaeontologia Sinica, Memoirs. Peking.

Compagnie Française des Pétroles, Notes et Mémoires. Paris.

Congrès et Colloques de l'Université de Liège.

Congrès pour l'Advancement des Études de Stratigraphie du Carbonifère, 2nd, Heerlen, 1935. Compte Rendu.

Congrès pour l'Advancement des Études de Stratigraphie et de Géologie du Carbonifère, 4th, Heerlen, 1958; 5th, Paris, 1963; 6th, Sheffield, 1967; 7th, Krefeld, 1971. Compte Rendu.

Connecticut Academy of Arts and Sciences, Memoirs. New Haven, Connecticut.

Copeia. New York, New York.

Cushman Foundation for Foraminiferal Research, Contributions. Washington, D.C.

Decheniana. Varhandlungen des Naturhistorischen Vereins der Rheinlande und Westfalens. Bonn.

Deep-Sea Research (Papers in Marine Biology and Oceanography). London.

Deutsche Akademie der Wissenschaften zu Berlin, Abhandlungen; Monatsberichte, Geologie und Mineralogie.

Deutsche Geologische Gesellschaft, Zeitschrift. Berlin.

Deutsche Hydrographische Zeitschrift. Hamburg.

Dijon Université, Publications.

Dresden, Staatliches Museum für Mineralogie und Geologie, Jahrbuch. Dresden-Leipzig.

Earth and Planetary Science, Letters. Amsterdam.

Earth-Science Reviews. Amsterdam.

Eclogae Geologicae Helvetiae (Schweizerische Geologische Gesellschaft). Basel.

Ecology. Brooklyn.

Eesti NSV Teaduste Akadeemia (Loodusuurijate Selts) Toimetised.
Electromedica. Erlangen.
Erdöl und Kohle (Erdgas–Petrochemie Vereinigt mit Brennstoff–Chemie). Hamburg.
Evolution. Lancaster, Pennsylvania.
Experientia. Basel.
Fieldiana, Geology Memoirs. Chicago, Illinois.
Forschungen und Fortschritte: Korrespondenzblatt der Deutschen Wissenschaft und Technik. Berlin.
Forschungsberichte des Landes Nordrhein-Westfalen. Berlin.
Fortschritte der Geologie und Paläontologie. Berlin.
Fortschritte der Geologie von Rheinland und Westfalen. Krefeld.
Fossils and Strata (Universitetsforlaget). Oslo.
Freiberger Forschungshefte. Berlin.
Genetik. Jena.
Géobios. Lyon.
Geochimica et Cosmochimica Acta. Oxford, New York, Braunschweig.
Geologica (Schriftenreihe der Geologischen Institute der Universitäten Berlin). Berlin.
Geologica Hungarica, Series Geologica; Series Palaeontologica. Budapest.
Geologica et Palaeontologica. Marburg.
Geological Association of Canada, Proceedings; Special Paper. Toronto.
Geological Journal. Liverpool.
Geological Journal (Liverpool Geological Society; Manchester Geological Association). Berkenhead, England.
Geological Magazine. London, Hertford.
Geological Newsletter. Antwerp.
Geological Society of America, Abstracts with Programs; Bulletin; Memoir; Programs of Annual Meetings; Special Paper. Boulder, Colorado.
Geological Society of Australia, Journal; Special Publications. Adelaide.
Geological Society of Japan, Journal. Tokyo.
Geological Society of London, Journal; Memoirs; Proceedings; Quarterly Journal; Quarterly Report; Special Report; Transactions.
Geological Society of South Africa. Johannesburg.
Geologicheskiy Komitet, Trudy. Leningrad.
Geologie (Zeitschrift für das Gesamtgebiet der Geologie und Mineralogie sowie der angewandten Geophysik). Berlin.
Geologie en Mijnbouw. Den Haag.
Geologische Blätter für Nordost-Bayern und angrenzende Gebiete. Erlanden.
Geologische Bundesanstalt Wien, Jahrbuch; Verhandlungen.
Geologische Gesellschaft in Wien, Mitteilungen.
Geologische Landesanstalt (Geologischer Dienst), Abhandlungen. Berlin.
Geologische Mitteilungen (Zeitschrift für Allgemeine, Regionale und Angewandte Geologie). Aachen.
[K.K.] Geologische Reichsanstalt Wien (see Geologische Bundesanstalt Wien), Abhandlungen.

Geologische Rundschau. Geologische Vereinigung, Berlin, Leipzig, Stuttgart.
Geologisches Jahrbuch, Beihefte. Hannover.
Geologisches Staatsinstitut in Hamburg, Mitteilungen.
Geologiska Föreningens i Stockholm Förhandlingar.
Geologists' Association, Proceedings. London.
Geologiya i Geofizika. Novosibirsk.
Geology. Geological Society of America, Boulder, Colorado.
Geophysical Journal, Research. Royal Astronomical Society, London.
Geotektonika. Moscow.
Geotimes. American Geological Institute, Washington, D.C.
Gesellschaft zur Beförderung der gesammten Naturwissenschaften zu Marburg, Schriften; Sitzunsberichte.
Gesellschaft Naturforschender Freunde Berlin, Sitzunsberichte.
Gesellschaft für Naturkunde in Württemberg, Jahreshefte (Verein für Vaterländische Naturkunde in Württemberg). Stuttgart.
Gesellschaft der Wissenschaften zu Göttingen, mathematisch-physikalische Klasse, Abhandlungen; Nachrichten.
Giornale di Geologia, Annali del Museo Geologia di Bologna.
Giornale di Scienze Naturali ed Economiche di Palermo.
Great Britain, Geological Survey of, Memoirs. London.
Gulf Coast Association of Geological Societies, Transactions. Houston, Texas.
Harvard University, Museum of Comparative Zoology, Breviora; Bulletin. Cambridge, Massachusetts.
Helgoländer Wissenschaftliche Meeresuntersuchungen, Biologische Anstalt Helgoland; List, Sylt.
Hessisches Landesamt für Bodenforschung, Abhandlungen; Notizblatt. Wiesbaden.
Hiroshima University (Japan) Journal of Science.
Hokkaido University, Journals of the Faculty of Science. Sapporo.
Ideen des Exakten Wissens. Stuttgart.
Illinois State Geological Survey, Bulletin. Urbana, Illinois.
Image. London.
India, Geological Survey of, General Report; Memoirs; Records. Calcutta.
Indian Science Congress Association, Proceedings. Calcutta.
Indiana Academy of Science, Proceedings. Brookville, Indiana.
Indiana Geological Survey, Bulletin. Bloomington, Indiana.
Institut Grand-Ducal de Luxembourg, Section des Sciences Naturelles, Physiques et Mathématiques, Comptes-rendus.
Institut Royal des Sciences Naturelles de Belgique, Bulletin. Brussels.

International Geological Congress; 15th Session, Pretoria; 16th Session, Washington, D.C.; 20th Session, Mexico City; 21st Session, Copenhagen; 22nd Session, New Delhi; 23rd Session, Prague; 24th Session, Montreal; 25th Session, Sydney; Comptes Rendus; Proceedings; Reports.
International Geology Review. Washington, D.C.
International Union for Quaternary Research, 10th Congress, Abstracts. Birmingham, England.
Inter-Nord. Paris.
Iowa, Geological Survey of. Iowa City, Iowa.
Iran, Geological Survey of, Report. Teheran.
Israel Geological Survey, Bulletins. Jerusalem.
Izvestiya Sibirskogo Otdeleniya Akademyi Nauk SSSR.
Japanese Journal of Geology and Geography. Science Council of Japan, Tokyo.
Journal of Foraminiferal Research. Washington, D.C.
Journal of Geography (Chigaku Zasshi). Tokyo.
Journal of Geology. Chicago, Illinois.
Journal of Geophysical Research. Washington, D.C.
Journal of Geosciences. Osaka City University, Osaka.
Journal für Ornithologie. Leipzig.
Journal of Paleontology. Tulsa, Oklahoma.
Journal of Petrology. London.
Journal of Sedimentary Petrology. Tulsa, Oklahoma.
Kansas State Geological Survey, Bulletin. Lawrence, Kansas.
Kyushu University, Memoirs of the Faculty of Science. Fukuoka.
Leicester Literary and Philosophical Society, Transactions.
Leidse Geologische Mededeelingen. Leiden.
Leningradskiy Gosudarstvennyi Universitet, Uchenye Zapiski, Seriya Geologo-Pochvennykh Nauk, Vestnik.
Lethaia. Oslo.
Linnean Society of New South Wales, Proceedings. Sydney.
Louvain Université, Institut Géologique, Mémoires.
Lvovskiy Universitet, Paleontologicheskiy Sbornik. Lvov.
Lyon Université, Annales.
Magyar Állami Földtani Intézet, Evkönyve. Budapest.
Marine Geology. Amsterdam.
Maroc Service Géologique du Division des Mines et de la Géologie, Notes et Mémoires. Rabat.
Matériaux pour la Carte Géologique de l'Algérie. Alger, Macon.
Meddelelser om Grønland. Kommissionen for Videnskabelige Undersøgelser i Grønland, Copenhagen.
The Mercian Geologist. East Midlands Geological Society, Nottingham, England.
Meyniana, Kiel Universität Geologisches Institut.
Mezhvedomstvennyi Stratigraficheskiy **Komitet**, Trudy. Moscow.
Micropaleontologist. New York, New York.
Micropaleontology. American Museum of Natural History, New York, New York.
Mijnbouwkunde Nederlandisch-Oost-Indïe (s'Gravenhage), Jaarboek. Copenhagen.
Mijnwezen in Nederlandisch-Oost-Indïe, Jaarboek. The Hague.
Monitore Zoologico Italiano. Florence.
Moskovskogo Obshchestva Ispytatelei Prirody, Byulletin.
Mountain Geologist. Rocky Mountain Association of Geologists, Denver, Colorado.
Münsterische Forschungen zur Geologie und Paläontologie. Marburg.
Musée Royal d'Histoire Naturelle de Belgique, Bulletin; Mémoires. Brussels.
Museum für Bergbau, Geologie, und Technik, Landesmuseum Joanneum. Graz.
Museum (National) d'Histoire Naturelle, Bulletin. Paris.
Musk-Ox. Saskatoon.
Nanking Institute of Geology and Paleontology (Nan King Ti Chi Ku Sheng Wuh Yan Jiow Shoo), Memoirs.
National Academy of Sciences, Proceedings. Washington, D.C.
Natur und Museum (Natur und Volk), Senckenbergische Naturforschende Gesellschaft. Frankfurt am Main.
Nature. London.
Nature. Paris.
Nature, Physical Science. London.
Naturforschende Gesellschaft in Basel, Berichte; Verhandlungen.
[K.K.] Naturhistorisches Hofmuseum, Annalen. Vienna.
Naturhistorisches Museum in Wien, Annalen.
Die Naturwissenschaften. Berlin.
Nauchno-Issledovatelskiy Institut Geologii Arktiki, Trudy. Leningrad.
Nederlandisch Geologisch-Mijnbouwkundig Genootschap (Geologisch-Mijnbouwkundig Genootschap voor Nederland en Koloniën), Jaarboek. Amsterdam.
Neftyanoy Geologo-Razvedochnyy Institut, **Trudy**. Moscow.
Neues Jahrbuch für Geologie und Paläontologie (Neues Jahrbuch für Mineralogie, Geologie und Paläontologie), Abhandlungen; Beilage-Bände; Monatshefte. Stuttgart.
New York Academy of Sciences, Annals.
New York State Museum and Science Service, Geological Survey Map and Chart Series. Albany, New York.
New Zealand Geological Survey, Paleontological Bulletins. Wellington.
New Zealand Journal of Geology and Geophysics. Wellington.
New Zealand Journal of Science and **Technology**. Wellington.
Newsletters on Stratigraphy. E. J. Brill, London.

Niedersächsischer Geologischer Verein, Jahresbericht. Hanover.
Norsk Geologisk Tidsskrift. Norsk Geologisk Forening, Oslo.
Norsk Polarinstituut, Skrifter. Oslo.
North American Paleontological Convention, Proceedings. Chicago, Illinois.
Notes et Mémoires sur le Moyen-Orient. Paris.
Nova Acta Leopoldina, Neue Folge (Kaiser Leopoldina-Carolina Deutsche Akademie der Naturforscher, Abhandlungen). Halle.
Oberhessische Gesellschaft fur Natur- und Heilkunde, Berichte. Giessen.
Oberrheinische Geologischer Verein, Jahresberichte und Mitteilungen. Stuttgart.
Oklahoma, Geological Survey, Bulletin; Oklahoma Geology Notes. Norman, Oklahoma.
Österreichische Akademie der Wissenschaften, mathematisch-naturwissenschaftliche Klasse, Denkschriften. Vienna.
Pacific Geology. Tokyo.
Pacific Science Congress, 9th, Thailand, Proceedings.
Pakistan Geological Survey, Records. Karachi.
Palaeobiologica. Vienna.
Palaeogeography, Palaeoclimatology, Palaeoecology. Amsterdam.
Palaeontographica. Stuttgart.
Palaeontographical Society, Monograph. London.
Palaeontologia Africana. University of Witwatersrand, Bernard Price Institute for Palaeontological Research, Johannesburg.
Palaeontologia Indica (Memoirs of the Geological Survey of India). Calcutta.
Palaeontologia Polonica. Warsaw.
Palaeontologia Sinica, Geological Survey of China. Peking.
Palaeontological Association, Special Paper. London.
Paläontologie von Timor. Stuttgart.
Paläontologische Zeitschrift. Berlin, Stuttgart.
Palaeontology. Palaeontological Association, London.
Paleobiology. Menlo Park, California.
Paleontological Society of Japan, Special Paper. Tokyo.
Paleontologicheskiy Zhurnal. Akademiya Nauk SSSR, Moscow.
Paleontologiya SSSR. Akademiya Nauk SSSR, Paleontologicheskiy Institut, Moscow.
Pan-American Geologist. Des Moines, Iowa.
Peking, National University, Contributions from the Geological Institute.
Petroleum Geology of Taiwan. Chinese Petroleum Corporation, Chinese Petroleum Exploration Division, Miaoli.
Philosophical Magazine. London.
[K.] Preussische Geologische Landesanstalt und Bergakademie, Abhandlungen; Jahrbuch. Berlin.
Problemy Sovetskoi Geologii. Moscow, Leningrad.
Przeglad Geologiczny. Warsaw.
Quarterly Journal of Microscopical Science. London.
Quaternary Research. New York, London.
Queensland, Geological Survey of, Publication. Brisbane.
Queensland Museum, Memoirs. Brisbane.
Reichsamt für Bodenforschung, Jahrbuch. Berlin.
Review of Palaeobotany and Palynology. Amsterdam.
Reviews of Geophysics and Space Physics. Washington, D.C.
Revista del Museo de la Plata.
Rivista Italiana di Paleontologia e Stratigrafica. Milan.
Rivista Mineraria Siciliana. Palermo.
Royal Irish Academy, Proceedings. Dublin.
Royal Society of Canada, Transactions. Ottawa.
Royal Society of Edinburgh, Transactions.
Royal Society of London, Philosophical Transactions; Proceedings.
Royal Society of New South Wales, Journal; Proceedings. Sydney.
Royal Society of New Zealand, Transactions. Wellington.
Royal Society of South Australia, Transactions. Adelaide.
Royal Society of Tasmania, Papers and Proceedings. Hobart.
Royal Society of Victoria, Proceedings. Melbourne.
Schweizerische Naturforschende Gesellschaft, Abhandlungen. Basel.
Science. American Association for the Advancement of Science, Washington, D.C.
The Science Teacher. Normal, Illinois.
Scientia Sinica, Academia Sinica. Peking.
Search. London.
Sedimentology: Journal of the International Association of Sedimentology. Amsterdam.
Senckenbergiana Lethaea (Senckenbergische Naturforschende Gesellschaft, Wissenschaftliche Mitteilungen). Frankfurt am Main.
Senckenbergische Naturforschende Gesellschaft, Abhandlungen. Frankfurt am Main.
Service de la Carte Géologique de l'Algérie, Publications. Algiers.
Service des Mines et de la Carte Géologique du Maroc, Notes et Mémoires. Rabat.
Service des Mines Madagascar, Annales Géologiques. Tananarive.
Sibirskiy Nauchno-issledovatelskiy Institut Geologii, Geofiziki i Mineralnogo Serya, Trudy. Moscow.
Slovene Academy of Arts and Sciences (Slovenska Akademiya Znanosti in Umetnosti). Ljubljana.
Smithsonian Contributions to Paleobiology. Washington, D.C.
Smithsonian Miscellaneous Collections. Washington, D.C.
Sociedad Geológica del Perú, Boletín. Lima.
Società Geologica Italiana, Bollettino. Rome.
Società dei Naturalisti e Matematici di Modena, Atti.
Società Paleontologica Italiana, Bollettino. Modena.
Société Géologique de Belgique, Annales. Liège.

Société Géologique de France, Bulletin; Compte Rendu des Séances; Mémoires. Paris.
Société Géologique du Nord, Annales. Lille.
Société Impériale des Naturalistes de Moscou, Bulletin.
Société Linnéenne du Nord de la France, Mémoires. Amiens.
Société Paléontologique de la Suisse, Mémoires (Schweizerische Paläontologische Gesellschaft). Zurich.
Society of Economic Paleontologists and Mineralogists, Special Papers; Special Publications. Tulsa, Oklahoma.
South African Museum, Annals. Cape Town.
South Australian Museum, Records. Adelaide.
South Australian Naturalist. Adelaide.
Sovetskaya Geologiya. Moscow.
Special Papers in Palaeontology (Palaeontological Association). London.
Der Steinbruch. Berlin.
Stockholm Contributions in Geology (Acta Universitatis Stockholmiensis).
Systematics Association, Publications. London.
Tectonophysics. Amsterdam.
Tohoku University, Science Reports. Sendai.
Tsentralniy Nauchno-issledovatelskiy Geologo-razvedochnyy Institut (TNIGRI), Trudy. Moscow.
Tuatara, Victoria University College, Biological Society, Journal. Wellington.
United States Geological Survey, Bulletin; Professional Papers. Washington, D.C.
Université de Grenoble, Faculté des Sciences, Travaux du Laboratoire de Géologie.
University of California Publications in Geological Sciences. Berkeley, Los Angeles, California.
University of Chiba, College of Arts and Sciences, Journals.
University of Kansas, Department of Geology, Special Publications. Lawrence, Kansas.
University of Kansas Paleontological Contributions, Article; Paper. Lawrence, Kansas.
University of Kyoto, Faculty of Science, Memoirs.
University of Oregon Publications, Geology Series. Eugene, Oregon.
University of Texas, Bulletin. Austin, Texas.
University of Texas, Bureau of Economic Geology, Circular; Publication. Austin, Texas.
University of Tokyo, Faculty of Science, Journal.
University of Uppsala, Geological Institution, Bulletin.
University of Wyoming Contributions to Geology. Laramie, Wyoming.
Ussher Society, Proceedings. London.
Ústředního Ústavu Geologickeho, Rozpravy; Sborník; Věstník. Prague.
Videnskabelige Meddelelser fra Dansk Naturhistorisk Forening. Copenhagen.
Voprosy Mikropaleontologii. Akademiya Nauk SSSR Otdelenie Nauk o Zemle, Geologicheskii Institut, Moscow.
Vsesoyuznyy Nauchno-issledovatelskiy Geologicheskiy Institut (VSEGEI), Trudy. Moscow.
Vsesoyuznyy Neftyanoy Nauchno-issledovatelskiy Geologo-razvedochnyy Institut (VNIGRI), Trudy. Leningrad.
Washington Academy of Sciences, Journal. Washington, D.C.
Western Australia, Geological Survey of, Bulletin; Report. Perth.
Yorkshire Geological Society, Proceedings. Manchester, Leeds.
Zapiski Mineralogicheskaga Obshchestva. St. Petersburg.
Zavod za Geološka i Geofizička Istraživanja. Belgrade.
Zeitschrift für Geologische Wissenschaften (Gesellschaft für Geologische Wissenschaften der DDR). Berlin.
Zeitschrift für Zoologische Systematik und Evolutionsforschung. Frankfurt am Main.
Zentralblatt für Geologie und Paläontologie (Zentralblatt für Mineralogie, Geologie und Paläontologie). Stuttgart.
Zentralblatt für Mineralogie (Zentralblatt für Mineralogie, Geologie und Paläontologie). Stuttgart.
[K.K.] Zoologisch-botanische Gesellschaft in Wien, Verhandlungen.

SOURCES OF ILLUSTRATIONS

At the end of each figure caption, the name of the author of the illustration and the date of publication are given, reference being made to publications cited in the reference lists. Although original sources do not always produce the best illustrations, they are, historically speaking, definitive and are commonly selected by *Treatise* authors. Previously unpublished illustrations are indicated by the name of the author and the letter n ("new"). Repository and museum catalogue numbers are given where appropriate.

STRATIGRAPHIC DIVISIONS

As commonly cited in the *Treatise,* classification of rocks forming the geologic column is reasonably uniform and firm throughout most of the world as regards major divisions (e.g., series, systems, and rocks representing eras), but it may be variable and unfirm as regards minor divisions (e.g., substages, stages, and subseries), which tend to be provincial in application. Users of the *Treatise* have suggested the desirability of publishing reference lists showing the stratigraphic arrangement of at least the most commonly cited divisions. Accordingly, a tabulation of European and North American units, which generally follows usage by authors of this volume, is given here. Some of the minor divisions are applicable also to other continents.

Generally Recognized Divisions of Geologic Column

EUROPE	NORTH AMERICA
CENOZOIC ERATHEM	**CENOZOIC ERATHEM**
QUATERNARY SYSTEM	**QUATERNARY SYSTEM**
Holocene Series (Recent)	Holocene Series (Recent)
Pleistocene Series	Pleistocene Series
TERTIARY SYSTEM[1]	**TERTIARY SYSTEM**[1,2]
Pliocene Series	Pliocene Series
Astian-Piacenzian Stage	Upper Pliocene Stage
Tabanian-Zanklian Stage	Middle Pliocene Stage
	Lower Pliocene Stage
Miocene Series	Miocene Series
Messinian Stage	
Tortonian Stage	Clovellian Stage
Serravallian Stage	Ducklakean Stage
Langhian Stage	
Burdigalian Stage	Napoleonvillean Stage
Aquitanian Stage	Anahaucan Stage
Oligocene Series	Oligocene Series
Chattian Stage	Chackasawayan Stage
Rupelian Stage	
Lattorfian Stage	Vicksburgian Stage
Eocene Series	Eocene Series
Wemmelian Stage	Jacksonian Stage
Lutetian Stage	Claibornian Stage
Cuisian Stage	Sabinian Stage
Paleocene Series	Paleocene Series
Ilerdian Stage	
Thanetian Stage + Montian	Midwayan Stage
Danian Stage	
MESOZOIC ERATHEM	**MESOZOIC ERATHEM**
CRETACEOUS SYSTEM	**CRETACEOUS SYSTEM**
Upper Cretaceous Series	Upper Cretaceous Series
Maastrichtian Stage	Maastrichtian Stage
Campanian Stage	Campanian Stage
Santonian Stage	Santonian Stage
Coniacian Stage	Coniacian Stage
Turonian Stage	Turonian Stage
Cenomanian Stage	Cenomanian Stage
Lower Cretaceous Series	Lower Cretaceous Series
Albian Stage	Albian Stage
Aptian Stage	Aptian Stage
Barremian Stage	Barremian Stage

    Hauterivian Stage
    Valanginian Stage
    Berriasian Stage

JURASSIC SYSTEM
Upper Jurassic Series
  Tithonian Stage
  Kimmeridgian Stage
  Oxfordian Stage
Middle Jurassic Series
  Callovian Stage[3]
  Bathonian Stage
  Bajocian Stage[4]
Lower Jurassic Series (Liassic)
  Toarcian Stage
  Pliensbachian Stage
  Sinemurian Stage
  Hettangian Stage

TRIASSIC SYSTEM
Upper Triassic Series
  Rhaetian Stage
  Norian Stage
  Carnian Stage
Middle Triassic Series
  Ladinian Stage
  Anisian Stage
Lower Triassic Series

  Scythian Stage

PALEOZOIC ERATHEM
PERMIAN SYSTEM
Upper Permian Series
  Tatarian Stage
  Kazanian Stage[5]
Lower Permian Series
  Artinskian Stage[5]
  Sakmarian Stage
  Asselian Stage

CARBONIFEROUS SYSTEM
Silesian Subsystem

  Stephanian Series[6]

  Westphalian Series[6]

  Namurian Series[6]

Dinantian Subsystem

  Visean Series[6]

  Tournaisian Series[6]

    Hauterivian Stage
    Valanginian Stage
    Berriasian Stage

JURASSIC SYSTEM
Upper Jurassic Series
  Portlandian Stage
  Kimmeridgian Stage
  Oxfordian Stage
Middle Jurassic Series
  Callovian Stage[3]
  Bathonian Stage
  Bajocian Stage[4]
Lower Jurassic Series (Liassic)
  Toarcian Stage
  Pliensbachian Stage
  Sinemurian Stage
  Hettangian Stage

TRIASSIC SYSTEM
Upper Triassic Series
  Rhaetian Stage
  Norian Stage
  Carnian Stage
Middle Triassic Series
  Ladinian Stage
  Anisian Stage
Lower Triassic Series (Scythian)
  Spathian Stage
  Smithian Stage
  Dienerian Stage
  Griesbachian Stage

PALEOZOIC ERATHEM
PERMIAN SYSTEM
Upper Permian Series
  Ochoan Stage
  Guadalupian Stage
Lower Permian Series
  Leonardian Stage
  Wolfcampian Stage

PENNSYLVANIAN SYSTEM
  Virgilian Series[6]
  Missourian Series[6]
  Desmoinesian Series[6]
  Atokan Series[6]
  Morrowan Series[6]
MISSISSIPPIAN SYSTEM
  Chesterian Series[6]
  Meramecian Series[6]
  Osagian Series[6]
  Kinderhookian Series[6]

DEVONIAN SYSTEM 　Upper Devonian Series 　　Famennian Stage 　　Frasnian Stage 　Middle Devonian Series 　　Givetian Stage 　　Eifelian Stage 　Lower Devonian Series 　　Emsian Stage 　　Siegenian Stage 　　Gedinnian Stage	**DEVONIAN SYSTEM** 　Chautauquan Series[7] 　　Bradfordian Stage[7] 　　Cassadagan Stage[7] 　Senecan Series[7] 　　Cohocton Stage[7] 　　Fingerlakesian Stage[7] 　Erian Series[7] 　　Taghanican Stage[7] 　　Tioughniogan Stage[7] 　　Cazenovian Stage[7] 　Ulsterian Series 　　Onesquethawan Stage[7] 　　Deerparkian Stage[7] 　　Helderbergian Stage[7]
SILURIAN SYSTEM 　Pridolian Series 　Ludlovian Series 　Wenlockian Series 　Llandoverian Series	**SILURIAN SYSTEM** 　Pridolian Series 　Ludlovian Series 　Wenlockian Series 　Llandoverian Series
ORDOVICIAN SYSTEM 　Ashgillian Series 　Caradocian Series 　Llandeilian Series 　Llanvirnian Series 　Arenigian Series 　Tremadocian Series[8]	**ORDOVICIAN SYSTEM** 　Cincinnatian Series (Upper Ordovician) 　　Richmondian Stage 　　Maysvillian Stage 　　Edenian Stage 　Champlainian Series 　　(Middle Ordovician) 　　Mohawkian Stage 　　Chazyan Stage 　　Whiterockian Stage 　Canadian Series (Lower Ordovician)
CAMBRIAN SYSTEM 　Upper Cambrian Series (Merioneth) 　Middle Cambrian Series (St. David's) 　Lower Cambrian Series (Comley)	**CAMBRIAN SYSTEM** 　Upper Cambrian Series (Croixian) 　　Trempealeauan Stage 　　Franconian Stage 　　Dresbachian Stage 　Middle Cambrian Series (Albertan) 　Lower Cambrian Series (Waucoban)
ROCKS OF PRECAMBRIAN ERAS 　**PROTEROZOIC ERATHEM** 　　Dalradian, Eocambrian, 　　Vendian, Riphean, 　　and equivalents	**ROCKS OF PRECAMBRIAN ERAS** 　**PROTEROZOIC ERATHEM** 　　Algonkian, Beltian, 　　Hadrynian, Helikian, 　　Aphebian, and equivalents

Richard A. Robison and Curt Teichert

---

[1] For convenience Miocene and Pliocene are often assigned to the Neogene Subsystem, and Paleocene, Eocene, and Oligocene are assigned to the Paleogene Subsystem.
[2] Follows essentially Gulf Coast usage.
[3] Included in Upper Jurassic by some authors.
[4] Lower part includes Aalenian of some authors.
[5] The Kungurian Stage of some authors is a probable facies of the upper Artinskian and lower Kazanian stages.
[6] These units have been designated as stages in previous *Treatise* volumes.
[7] Applies essentially to eastern United States; in western North America, European stage terminology is used.
[8] Tremadocian is placed in Cambrian by some authors.

PART A

INTRODUCTION

FOSSILIZATION (TAPHONOMY) BIOGEOGRAPHY AND BIOSTRATIGRAPHY

By W. A. BERGGREN, A. J. BOUCOT, M. F. GLAESSNER, HELMUT HÖLDER,
M. R. HOUSE, VALDAR JAANUSSON, E. G. KAUFFMAN, BERNHARD KUMMEL,
A. H. MÜLLER, A. W. NORRIS, A. R. PALMER, ADOLF PAPP, C. A. ROSS,
J. R. P. ROSS, and J. A. VAN COUVERING

CONTENTS

	PAGE
FOSSILIZATION (TAPHONOMY) (A. H. Müller)	A2
BIOGEOGRAPHY AND BIOSTRATIGRAPHY	A79
Precambrian (M. F. Glaessner)	A79
Cambrian (A. R. Palmer)	A119
Ordovician (Valdar Jaanusson)	A136
Silurian (A. J. Boucot)	A167
Devonian in the Eastern Hemisphere (M. R. House)	A183
Devonian in the Western Hemisphere (A. W. Norris)	A218
Carboniferous (C. A. Ross)	A254
Permian (C. A. Ross and J. R. P. Ross)	A291
Triassic (Bernhard Kummel)	A351
Jurassic (Helmut Hölder)	A390
Cretaceous (E. G. Kauffman)	A418
Tertiary (Adolf Papp)	A488
Quaternary (W. A. Berggren and J. A. Van Couvering)	A505
INDEX	A544

FOSSILIZATION (TAPHONOMY)[1]

By A. H. Müller

[Sektion Geowissenschaften der Bergakademie, Freiberg, German Democratic Republic]

CONTENTS

	PAGE
INTRODUCTION	A2
BIOSTRATINOMY	A3
General Discussion	A3
Necrotic Processes (Death in general——Causes of death——The death struggle and its traces——Rigor mortis and its traces)	A5
Fate of Organic Material (General discussion——Effects of degassing on organic material covered by sediment——Effects of degassing on exposed organic material——Role of scavengers)	A15
Sequence of Decay of Soft Parts Associated with Preservable Hard Parts (Selective Decomposition)	A20
Fate of Preservable Hard Parts before Final Burial (General discussion——Allochthonous burial)	A22
FOSSIL DIAGENESIS	A48
General Discussion	A48
Formation of Steinkerns and Early Diagenesis	A49
Deformation of Fossils (General discussion——Volume decrease of sediment and its effects——Collapse under own weight——Deformation by tectonic stress)	A53
Selective Dissolution during Diagenesis	A58
Molecular Rearrangements in the Course of Diagenesis (General discussion——Recrystallization in particular grain growth——Formation of concretions and their importance in fossilization——Transformation of polymorphous substances into stable modifications——Metasomatism of fossils)	A61
Preservation of Structural Soft Parts (Preservation of soft parts in Eocene coal——Phosphatization of soft parts)	A69
REFERENCES	A72

INTRODUCTION

The manner in which organisms become fossilized is one of the largest and most diverse areas of research in general paleontology. In essence, the concept of fossilization encompasses what EFREMOV (1940) called "taphonomy" ($\tau\acute{\alpha}\phi$os = burial, $\nu\acute{o}\mu$os = law), which is the study of the transition of all or part of an organism, and its lebensspuren (ichnia), from the biosphere into the lithosphere. Taphonomy studies the introduction of the remains of organisms and their traces into the rock record.

---

[1] Manuscript received October, 1973. This contribution was translated from the original German by W. G. HAKES and CURT TEICHERT.

EFREMOV (1940, p. 85) originally defined the major concern of taphonomy to be "the study of the transition (in all details) of animal remains from the biosphere into the lithosphere, i.e., the study of a process in the upshot of which the organisms pass out of the different parts of the biosphere and, being fossilized, become part of the lithosphere."

Preservation is dependent upon the nature of the organism's body, biotope, rates of sedimentation, as well as the embedding medium; however, the real objects of study in taphonomy are biologic structures that are morphologically recognizable. Such studies are related to the destruction or preservation of an organism, and as a rule, proportionately very little of its entire mass will be preserved. Only the calcareous, siliceous, and chitinous hard parts stand much chance of preservation. The organic material eventually decomposes, becomes structureless, and is chemically altered, and the study of this organic material, which is known to influence the character of sediments, remains primarily the work of chemists, petrographers, and geologists.

Probably the first definition of the concept of fossilization was presented by D'ORBIGNY (1849):

> The term fossilization is understood by us to embrace all changes, more or less, through which the body of a living or extinct organism passes from one epoch, such as the present, into another one,

thereby leaving in the strata permanent traces of its characteristic form. We include here a group of observations which are of great importance, but, nevertheless, have been completely neglected by paleontologists.

Fossilization, or rather the process of fossilization, is the kind of phenomenon through which an organized body more or less loses its original, characteristic composition and is converted into a new substance which displays, in the form of an organized body, characters of chemical composition or texture somewhat different from those of the original body. (Translated by W. G. HAKES.)

In 1869, D'ARCHIAC proposed a short and significant definition for fossilization:

> We shall designate under the name "fossilization" the different modifications that the bodies of organisms undergo during their stay in the rocks of the earth. These modifications are frequent, numerous, and highly variable in nature. This has caused many eminent zoologists to make mistakes. Sometimes modifications are so severe that the body's characteristics are completely obliterated. (Translated by CURT TEICHERT.)

The most important aspects of fossilization are biostratinomy and fossil diagenesis. These subjects have been thoroughly discussed by DEECKE (1923), QUENSTEDT (1927), MÜLLER (1951b, 1963), PAVONI (1959), ROLFE and BRETT (1969), and SEILACHER (1973). Some of the possible fates of skeletal material after the death of the organism are shown in Figure 1.

BIOSTRATINOMY

GENERAL DISCUSSION

The term biostratinomy was proposed as "biostratonomy" by WEIGELT (1919)[1] and was originally defined as the study of the manner in which fossils become oriented and arranged in rocks. Today, it has a broader meaning but still embraces the basic concepts developed by WEIGELT. The study of biostratinomy begins with the death struggle of an organism and ends with the final burial and arrangement of the dead or dying animal or its disarticulated remains.

The fate of materials produced and discarded during the life cycles of organisms must also be included in biostratinomy. Examples are the exuviae of crabs, trilobites, and other arthropods, in addition to sporomorphs and fallen leaves. It is important to determine whether an assemblage is autochthonous (*in situ*) or whether it is allochthonous and has undergone postmortem transport and sorting. In both cases, the relative position of assemblages can be entirely or partially altered by internal and

---

[1] WOLFF (1954) suggested that the spelling be changed from biostratonomy to biostratinomy in order to make the spelling analogous to stratigraphy. Later, W. KNOCKE (Hamburg) (in VOIGT, 1962, p. 30) pointed out that biostratinomy is the only possible spelling for etymological reasons, but many authors have retained the spelling biostratonomy for reasons of euphony.

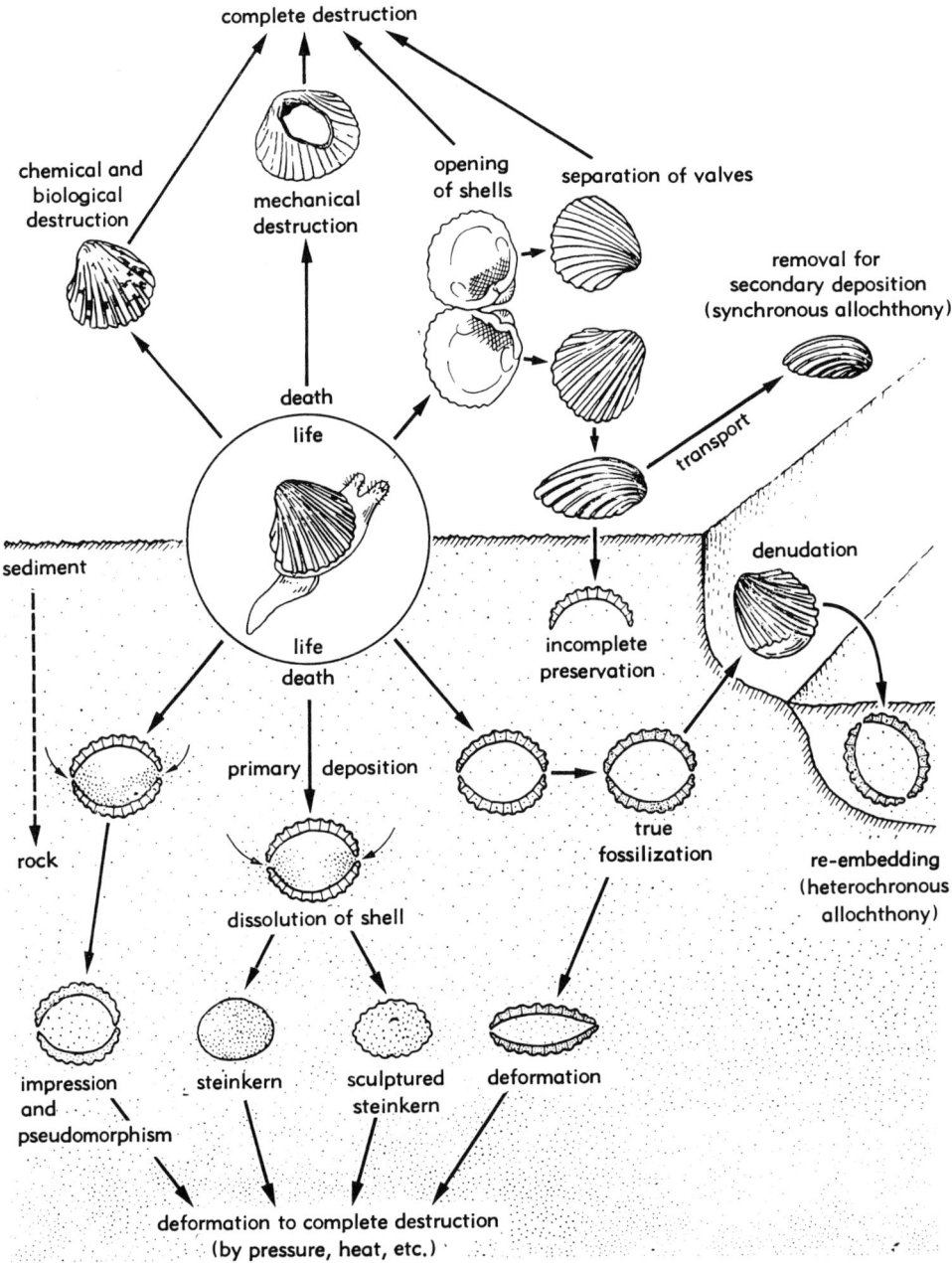

Fig. 1. Diagrammatic representation of fate of skeletal material after the death of the animal (after Thenius, 1963).

external forces such as currents, gravity, buoyancy due to gas entrapment during decay, and predation by scavengers. It follows that the biostratinomer must pay special attention to the original deposit and the types of preservation there. He must also study his material both quantitatively and systematically. Application of the principle of actualism is especially important. Attention must be paid to the recent environment either by direct observation or experimentation, in order to develop a firm base for our knowledge of biostratinomy. In this way, its effectiveness as a tool in the interpretation of ancient environments will be enhanced.

The chance of an organism being preserved as a fossil is dependent upon the conditions of sedimentation, which can, in turn, be related to the paleogeography of an area. Animals living in high altitudes, which are subject to active erosion, are seldom preserved. On the other hand, organisms that live in low-lying areas, or in streams, rivers, swamps, seas, or oceans, stand the best chance of being covered by sediment and eventually being preserved.

The first systematic, biostratinomic observations were made by WALTHER. Later, the trend for this type of investigation was set by WEIGELT, who is considered to be the founder of biostratinomy, and since then, the number of biostratinomic investigations has increased.

At first, incorporation of invertebrates in sediment and the mechanical principles that govern their deposition and arrangement received special attention. WEIGELT (1927) has added much to our knowledge of the manner in which vertebrates become buried in sediment, and since the publication of his work on the Kupferschiefer flora (WEIGELT, 1928b) interest has also grown in the biostratinomic investigations of fossilized plants.

Biostratinomy originally dealt with idiobiology (namely, the study of organisms as individuals). However, after the work of WASMUND (1926), biostratinomy evolved a biosociologic-biocenologic approach that permitted the interpretation of mode and direction of transport and cause of death in allochthonous fossil assemblages.

The cause of death of a fossilized animal, the manner of its decay, decomposition, and burial in the sediment can be accurately interpreted if its morphology, mode of life, and relationship to the surrounding environment are known. For this reason, it is just as important to have an understanding of ecology and behavior of the animals as to understand geological conditions such as sedimentation. In general, conditions of preservation for aquatic animals are better than those for land dwellers. A similar relationship applies to sessile (attached) or infaunal animals when compared to vagile animals.

The practical importance of biostratinomic knowledge is obvious. In questionable cases, it enables the recognition of *in situ* positions and, by interpretations of current conditions, the original orientation of economic deposits. An outstanding example of this type of investigation was the development of the iron ore deposits of Salzgitter (Germany) by biostratinomic methods (WEIGELT, 1923).

In principle, every organic body is preservable, no matter what the proportion of its hard and soft parts. Preservation, in the final analysis, is dependent only upon the dominant physico-chemical and biological conditions to which an organism is subjected. In most cases, the fate of both hard and soft parts is different, and as a rule, the soft parts decompose and decay, leading to a relatively large loss of body material. This process commonly leads to a more or less complete segregation of the more resistant hard parts, which can be broken or macerated depending upon the chemical or physical processes to which they are exposed. It is, therefore, justifiable to discuss the biostratinomy of soft parts and plantlike substances separately from the biostratinomy of hard parts, although transitions between these conditions exist.

Studies of recent processes which especially belong to biostratinomy, are called "actuopaleontology" (RICHTER, 1929; DALQUIST & MAMAY, 1963; SCHÄFER, 1962, 1972).

NECROTIC PROCESSES

DEATH IN GENERAL

In any given case, it is very difficult to determine accurately the exact moment of death, as death is seldom instantaneous. It is almost always linked to a long or short period of progressive deterioration of the metabolic processes, which occurs long before the organism can be considered a cadaver. In higher organisms, cells and cell complexes continue to live for a while after the heart stops beating and, therefore, the end of cardial activity cannot be considered to record the exact moment of death. This is especially true when no external signs of life activity can be observed. In vertebrates, the brain center, which regulates heart activity and breathing, is first affected (Korschelt, 1924). Autolysis leads to an important change in the chemistry of tissues, can influence microbial decay, and is important in the initiation of the partial death and natural rejection of certain body parts. This is especially significant during the molting of arthropods in which up to 90 percent of the covering cuticle is resorbed by enzymatic autolysis (Richards, 1951).

The beginning and end of dying, of course, cannot be determined for fossil material; however, analysis of the manner of burial may lead to useful conclusions about the cause of death, the death struggle, and the death spasms of an animal.

The processes occurring during the death of an organism are called "necrosis." The duration of necrosis varies with the organism and may last for only a few seconds in small organisms, but in large animals may last a considerable time.

CAUSES OF DEATH

GENERAL DISCUSSION

The cause of death is generally quite complex, and in most cases, two or more of the following factors will take part.

External (allogenic) forces. 1) Suffocation—affects all aerobic organisms, if adequate oxygen supply is cut off; 2) starvation—death is caused by prolonged absence of sufficient and properly balanced nutrients; 3) lack of water—death may result from a change in osmotic pressure within an organism; 4) freezing; 5) rise of an animal's temperature above a given level; 6) poisoning through assimilation of toxic, chemically active substances; 7) absence of an adequate light supply, especially important to plants; 8) mechanical vibration, disintegration, and crushing of the entire body or important body parts due to tumbling, lacerating, puncturing, striking, and pinching of important veins; 9) increase or decrease of pressure; 10) infestation by parasites or bacteria.

Internal (autogenic) forces. 1) Old age; 2) organic disease. In this connection it is important to note that for pathological and physiological death, there is no important singular cause.

The number of individuals in a population may be considerably reduced by the regular recurrence of yearly phenomena (dry summer and cold winter periods) or by irregular occurrences (catastrophic climatic fluctuations, disease, volcanic eruptions, floods, severe parasitic activity, and mass appearance of natural enemies). Parasitism and the appearance of natural enemies are commonly the result of a rapid increase in number of individuals within the community that is being attacked. Chetverikov (1905) and Timofeeff-Ressovsky (Timofeeff-Ressovsky, Voroncov, & Jablokov, 1975) referred to such situations as "population waves" *(Populationswellen)*. Mass mortality and the subsequent reduction in population size generally have the effect of increasing evolutionary rates.

DEATH DUE TO OVERGROWTH BY OTHER ORGANISMS

The following are causes of death that are well documented in the fossil record:

Numerous sessile or slightly vagile organisms are able to escape from the danger of overgrowth by organisms of their own or other species by means of various adaptations; however, the animals are often overwhelmed and die. These are mostly animals of flat shape that do not possess the ability to rise coral-like above the sediment. Listed below are examples of free-living, slightly vagile organisms:

1) Massive colonization of oyster banks by *Mytilus edulis*: the meshwork created by the byssal threads of the attached *Mytilus* will entrap enough sediment to suffocate the underlying oyster population and, eventually, the *Mytilus* colonies themselves. These mussel beds then form potential sites for future oyster colonization.

2) *Mytilus edulis* overgrowing *Cardium edule*: living *Cardium* shells are commonly used as sites for colonization by mussels, and it is common to find valves of *Cardium* "spun" together by the byssal threads of *Mytilus edulis*. Death by suffocation therefore results.

3) *Mytilus edulis* overgrowing *Balanus* colonies: *Mytilus* can so completely cover a barnacle colony that the cirri of the barnacles cannot penetrate the byssal threads of

Mytilus. Balanus, therefore, dies of starvation (SEWERTZOFF, 1934).

The mutual overgrowth and encrustation of other animals by sessile organisms has been called immuration by VYALOV (1961). This is a special case in which encrustation occurs only during the life span of the encrusted animal. Encrusting epibionts, in particular, are calcareous algae, poriferans, stromatoporoids, corals, bryozoans, brachiopods, serpulids, bivalves, gastropods, and cirripeds. Following are some examples:

1) Coral covering hippuritids: It is probable that the corals attach themselves to the opercula of the hippuritids at the beginning of their life cycle (DEECKE, 1923).

2) Oysters growing on *Balanus*: Barnacles will become completely embedded in the prismatic layer of the bivalves. When the barnacles die and the scutum and tergum pairs fall off, the oyster shells can then begin to fill the hole left by the barnacles in which frequently organic and inorganic bodies, which were either washed in or were in search of shelter, can also be found.

3) When thick shells of oysters or spondylids are sectioned, it is common to find remains of many organisms that have been killed by overgrowth of the bivalve shell. As a rule, such overgrown organisms are well preserved.

It is especially important that this process permits the preservation of organisms consisting of chitin, spongin, plant substance, and so forth, which are not preserved under normal conditions of sedimentation. In this way, the presence of algae, sea grasses, hydrozoans, and ctenostomate bryozoans can be demonstrated (VOIGT, 1966; 1968a). Some examples are:

1) Leaf of a sea grass (?*Thalassocharis*) (Fig. 2), serving as a substrate for oysters and showing molds of prodissoconchs of several young oysters, from Upper Cretaceous rocks of the Netherlands.

2) Blades of a sea weed with cheilostome and ctenostome Bryozoa (Fig. 3), serving as substrate for an *Exogyra* from Upper Cretaceous rocks of Belgium. The arrow points to *Stolonicella schindewolfi* VOIGT, flanked on both sides by the delicate ribbonlike zoaria of *Taeniocellaria setifera* VOIGT. On the lower border of *S. schindewolfi*,

FIG. 2. Leaf of sea grass, ?*Thalassocharis*, serving as substrate for oysters and exhibiting molds of prodissoconchs of several young oysters, from Upper Cretaceous (upper Maastrichtian, Tuffkreide), St. Pietersberg near Maastricht, Netherlands, ×6.6 (Voigt, 1966).

transverse structures of the setae are clearly recognizable.

3) Immuration and molding of the bryozoan *Onychocella cyclostoma* (GOLDFUSS) by *Reteporidea cancellata* (GOLDFUSS) from Upper Cretaceous rocks of the Netherlands. Figure 4 shows the frontal membranes of the closed opesia with the outline of the semicircular operculum.

DEATH FROM SMOTHERING BY SEDIMENT

The chance of burial by sediment is especially great for organisms firmly attached to substrates. Slightly vagile infauna also stand a risk of burial, and are quite commonly embedded in life position; however, such animals are only seldom preserved as fossils because their remains are repeatedly redeposited in shallow-water sediments. Vagile animals are in less danger. Thus, infaunal gastropods are very adept in burrowing upward through thick accumulations of sediment. On the other hand, it has been observed that the starfish, *Asterias rubens*, was unable to penetrate 60 cm of sediment deposited during a storm in less than half an hour (SCHÄFER, 1962; 1972). This is even more serious when great numbers of these animals are affected and get into each other's way. Their remains are then commonly found with their arms balled up, at an angle to the bedding plane, or overturned. *A. rubens* will often crawl along bedding planes until it becomes stuck and dies. As the sediment impedes forward movement, one or two arms will usually point forward and the rest will drag behind, or else all arms may be directed backward. This type of orientation

Fig. 3. Blade of sea weed with cheilostome and ctenostome Bryozoa serving as substrate for an *Exogyra*, from Upper Cretaceous (upper Maastrichtian, Tuffkreide), Albert-Kanal near Néercanne, Belgium, ×8.2 (Voigt, 1966). Arrow points to *Stolonicella schindewolfi* VOIGT, flanked on both sides by delicate ribbon-like zooaria of *Taeniocellaria setifera* VOIGT. Of particular interest are the transverse structures of the setae, clearly visible near the lower edge.

can occur on the upper surfaces of bedding planes.

Ophiuroids continually work their way upward in the sediment after burial, and in the process become trapped under single bivalve shells, oriented concave side downward. Fossil examples are known from the Upper Muschelkalk of Germany (MÜLLER, 1969a).

Among the Echinoidea, the Irregularia live more or less deeply buried in the sediment, and are in danger of being smothered by it. *Echinocardium* is not able to free itself if it is suddenly buried by 30 or more centimeters of fine sand, which can commonly happen in shallow seas where storms bury numerous individuals at the same time (SCHÄFER, 1962; 1972).

If brachyurans are buried alive and embedded in the sediment, their claws can commonly be found opened slightly and their ambulatory appendages raised above their cephalothorax. This position indicates that the animals raised their legs to lift themselves in the sediment but no longer had the strength to do so (SCHÄFER, 1962; 1972).

DEATH BY OXYGEN DEFICIENCY (ASPHYXIATION)

Aerobic organisms will die if deprived of oxygen. Their needs vary according to degree of activity and environmental conditions. In any given case, oxygen requirements can be extremely different, and such variations are usually determined by the affinity of special hemoglobins for oxygen. The affinity for oxygen in lower animals is many times that in higher ones, whereas conditions are the reverse with regard to carbon dioxide. Humans require an oxygen to carbon dioxide ratio of 250 to 1. In contrast, *Arenicola marina* needs only a ratio of 50 or 70 to 1. Many lower animals, especially polychaetes, can tolerate considerable decreases in the amount of oxygen that they take in without the slightest organic damage. As a rule, such lower animals are endangered only if the oxygen supply is completely cut off.

Many marine polychaetes have the ability to maintain their oxygen requirement by optionally assuming an anoxybiotic existence. The hemoglobin of such animals has an increased affinity for oxygen and the breakdown of stored glycogen enables them to survive in oxygen-deficient environments, in addition to oxygen-rich environments. This is an important fact in the interpretation of ancient environments, as such organisms will succumb only under extremely adverse conditions. This is true for sediments containing FeS_2, if the overlying water is devoid of free oxygen. However, the absence of organic remains does not necessarily mean that a body or layer of water is rich in H_2S (i.e., deficient in oxygen). The absence of the remains of organisms can be linked to other ecological factors such as the lack of suitable nutrients. Environments can be inhospitable only if H_2S is present in the sediment and in the overlying water body. Such types of sediments are most likely to be found in quiet restricted areas with virtually no current activity. In regions where currents do exist, continual poisoning of the water does not occur.

The oxygen content of water is drastically reduced by the decay products of organic material, in particular ammonia and hydrogen sulfide. Reduction in the amount of oxygen is especially severe if there is not a continuous and sufficient supply of fresh water, and if oxygen is not replaced by the activity of plants. If a water body freezes over and the ice is covered with snow, water plants, especially chlorophyll-containing plankton, are able to produce oxygen only if the snow cover is thin. Unless there is influx of oxygen from the outside, the oxygen content is further depleted by decay processes. During winter, many organisms hibernate and thus reduce their oxygen requirements. However, due to the above-mentioned processes, the oxygen content of water is decreased and even can drop below the low level required during hibernation. This will awaken the organisms. Once these animals have awakened, their increased muscular activity increases their oxygen demands in an already oxygen-deficient environment. Because of the ice covering they are unable to reach the surface and the atmosphere, and they perish.

It is well known that mass mortalities of fish, oysters, and crabs occur each year in late summer in Offats Bayou, a narrow marine bay near Galveston Island on the

FIG. 4. Immuration and molding of the bryozoan *Onychocella cyclostoma* (GOLDFUSS) by *Reteporidea cancellata* (GOLDFUSS) *intra vitam*, with the frontal membranes of the closed opesia reflecting the semicircular operculum. From Upper Cretaceous (upper Maastrichtian), Grube Curfs near Berg in the area around Maastricht, Netherlands, ×16 (Voigt, 1968a).

Gulf Coast of North America (GUNTER, 1938). The hydrogen sulfide released during decay of the carcasses causes the water of the bay to "boil" over areas as large as 10 m in diameter and to change in color to a turbid red or black. This process begins at the innermost end of the bay and finally spreads into the entire upper part. After a few days, the water becomes clear again, and may or may not remain in this condition for long. Circulation within the bay and water exchange with West Galveston Bay proceed very slowly and cannot be considered extensive, because the mouth of the narrow bayou is extremely shallow. Tides have little effect on the bayou, and denser water settles into the deepest part at the inner end, where it becomes stagnant. It is in these deep areas during the summer months that accumulated organic material decomposes under anaerobic conditions. The gases produced by this process migrate upward through the sediment and enter the upper water layers, causing the death of organisms living there. Presence of hydrogen sulfide plays an important role, as does the anaerobic condition of the upwelling water. In the summer of 1940, the normal mass mortality did not occur, probably because major storms that struck the Texas coast sufficiently aerated the waters of the bayou.

As death approaches in hermit crabs (pagurids), the oxygen-rich currents that circulate through their borrowed snail shells begin to diminish. This reduces the amount of available oxygen in their domiciles and causes the animals to vacate the premises in search of oxygen. Therefore, it is not surprising that the bodies of fossil and recent hermit crabs are invariably found separate from the shells they once inhabited. A similar phenomenon occurs with crustaceans that live in their own burrows or inhabit burrows of other animals. Approaching death weakens the thalassinid shrimp, *Callianassa*, to such an extent that it can no longer maintain adequate currents for oxygen circulation, and therefore leaves its burrow. This helps to explain why crustacean remains are seldom found in supposed fossil *Callianassa* burrows such as *Ophiomorpha* and *Thalassinoides* (SCHÄFER, 1962; 1972).

DEATH BY SUBMERGENCE OR ENTRAPMENT IN SOFT, INCOHERENT, YIELDING, OR ADHESIVE SUBSTRATES

By observing recent examples, it is known that the death of animals by submergence or entrapment is possible in a variety of natural substances. Although numerous examples have been discovered in the geologic record, only a few are discussed here.

Mud and silt. Muddy substrates can become highly indurated during desiccation; however, their ability to remain firm and to support loads can be quite transitory and can be rapidly lost, as a thorough soaking is usually sufficient to return them to a pliable condition. Animals whose weight exceeds the load capacity of these substrates will sink more or less deeply into the substrate. In many cases, the animals will not remain there and will easily free themselves.

Salt pans and salt-covered muds are widely distributed in the recent environment (i.e., the sabhkas of North Africa, the takyre in Turkestan, and the kewirs of Iran). Within such areas, the exposed parts of partially buried bodies are frequently destroyed by exposure to the atmosphere, removed by scavengers, and eventually transported away by currents. In the Tendaguru beds of East Africa, dinosaur remains are preserved in this manner. Masses of limbs, pelvic bones, and shoulder blades can be found, but only few backbones have been preserved.

Thixotropy plays a large role in these processes; and it is well known that some viscose media can, under mechanical influences, undergo a decrease in viscosity. "Thixotropic consistency" falls somewhere between plasticity and flowability. In such a state, a quasi-solid material that has the potential for flowage exists. The development of thixotropic substrates is dependent upon an increase in the content of clay and water within the sediment.

A potentially thixotropic medium can remain relatively coherent for long periods of time, as it requires a significant change in the physical condition of the sediment to initiate thixotropy. This may be brought about by earth load, and increase in pore water pressure. Especially important is the original water content of the sediment. An animal, beginning to sink in a thixotropic, viscose, muddy medium, will thoroughly agitate that medium in an attempt to free itself. However, this disturbance decreases the viscosity, causing the animal to sink even deeper. Vibration can also play an important part. A stampeding herd of animals can change the physical condition of a thixotropic substrate in which they thereby become entrapped. However, a single animal can easily negotiate such a substrate.

Crude oil, asphalt, and tar. Tar is the hard residue of petroleum with a paraffin base. Asphalt corresponds similarly to the asphaltic oils. Tar and asphalt are both formed when crude oil containing air comes in contact with the atmosphere. Their consistencies can vary from unctuous and soft to hard and weakly brittle. Either in a liquid or later in a more or less soft, solid state, pools of these substances function as traps for unsuspecting animals. Such pools attract animals because of their shiny surfaces, and the smell of hydrogen sulfide attracts scavengers. The struggle of the entrapped animals releases hydrogen sulfide already present in the tar or asphalt, the animals remain stuck, continue to sink, and their bodies attract predators and scavengers. Death by sinking in tar or asphalt is very similar to death caused by sinking into mud, silt, swamps, or quicksand.

Natural upwelling of light paraffin oil is, in general, quite local and inconspicuous, whereas the appearance of the heavy asphaltic oil is more obvious. Light oil is evaporated and easily washed out. The heavier, asphaltic oil appears as dark, tarlike masses, which will undergo a transformation to brittle asphalt. A flowable, dark oil occurs at the center of the asphaltic oil pools, and this is surrounded by a rim of hardening asphalt.

There are numerous examples of pools of asphalt and tar entrapping animals in the recent environment (i.e., Mesopotamia, Syria, Sakhalin, Canada, California, and the Caspian Sea). The best known of these are the asphalt pits of Rancho La Brea in Los Angeles, California. These tar pools have existed since the Pleistocene and are still present, covering an area of 70,000 square meters. The asphalt contained in these pools contains numerous remains of Pleistocene animals (WOODWARD & MARCUS, 1973).

At La Brea, in addition to insects, large numbers of predators (saber-toothed tigers, wolves, leopards, and lynx) and herbivores (deer, horses, camels, elephants, and bison) are found.

The asphalt of McKittrick, California, is another example of an entire fauna such as that preserved at Rancho La Brea.

Occurrences of paraffin seepages are not as well known as those of asphaltic oils. This is primarily because paraffin oils evaporate easily, have a relatively small carbohydrate content, and are easily washed away during precipitation.

Quicksand. Very fine-grained, thixotropic sands are called quicksand. In contrast to other sandy substrates in which animals can become entrapped, quicksand requires a steady upwelling of water. The danger of quicksand and its ability to flow around and engulf an object is directly related to the maintenance of a sufficiently renewable water supply in the deposit. It is the presence of water that decreases grain-to-grain contact. This renders the mass adhesionless and can be easily displaced. The weight of heavy animals will push apart the sand grains, which then flow back over the engulfed animal, and thus trap it.

The most dangerous type of quicksand is formed where water rises from consider-

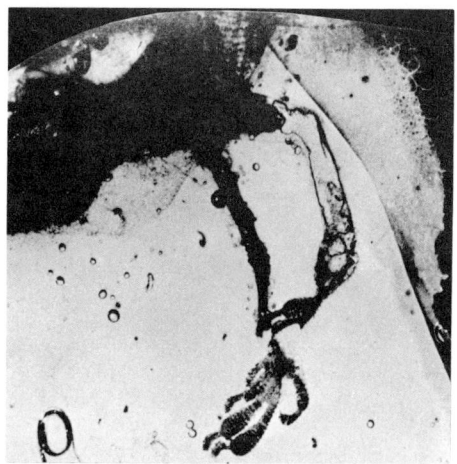

FIG. 5. Part of the head, torso, and forward limbs of *Anolis electrum* LAZELL, preserved in resin (Oligocene or Miocene), Chiapas, Mexico; length of forward limbs about 15 mm (Lazell, 1965).

able depth under hydrostatic pressure. However, this environment is dangerous mainly to larger animals that have fallen in quicksand. They will thrash about trying to free themselves from the mire and therefore quickly weaken and die. The danger of entrapment of smaller invertebrates in quicksand is less.

Death in resin. When the bark of a conifer is cut, resin will flow from the wound. This makes it possible to observe recent resin deposits which serve as traps for small flying insects and to compare them with ancient deposits. The best known of these ancient deposits is the Eocene amber of the Baltic region.

Considerable strength is required for an insect caught in amber to free itself Frequently the animal will damage parts of its body (legs, antennae, wings) during the struggle. There is only one known case of a larger animal preserved in amber and that is a small lizard. However, it is quite possible that this lizard was either dead or dying when it was embedded in the deposit. Figure 5 shows the skeleton of the lizard, *Anolis electrum* LAZELL, which was discovered in either the Oligocene or Miocene amber of Chiapas, Mexico (LAZELL, 1965). The type of preservation and the pronounced broadening at the ends of the toes with a large number of suction cups

Fig. 6. Myriapod, preserved in Eocene amber from the Baltic; length about 5 mm (Müller, 1963).

strongly suggests an arboreal existence for this animal.

If resin has been sufficiently warmed by the sun, it will flow quite easily. Insects embedded in this medium leave little, if any, trace of a death struggle (Fig. 6); however, if the resin is cool (not directly exposed to the sun, possibly in the shadow),

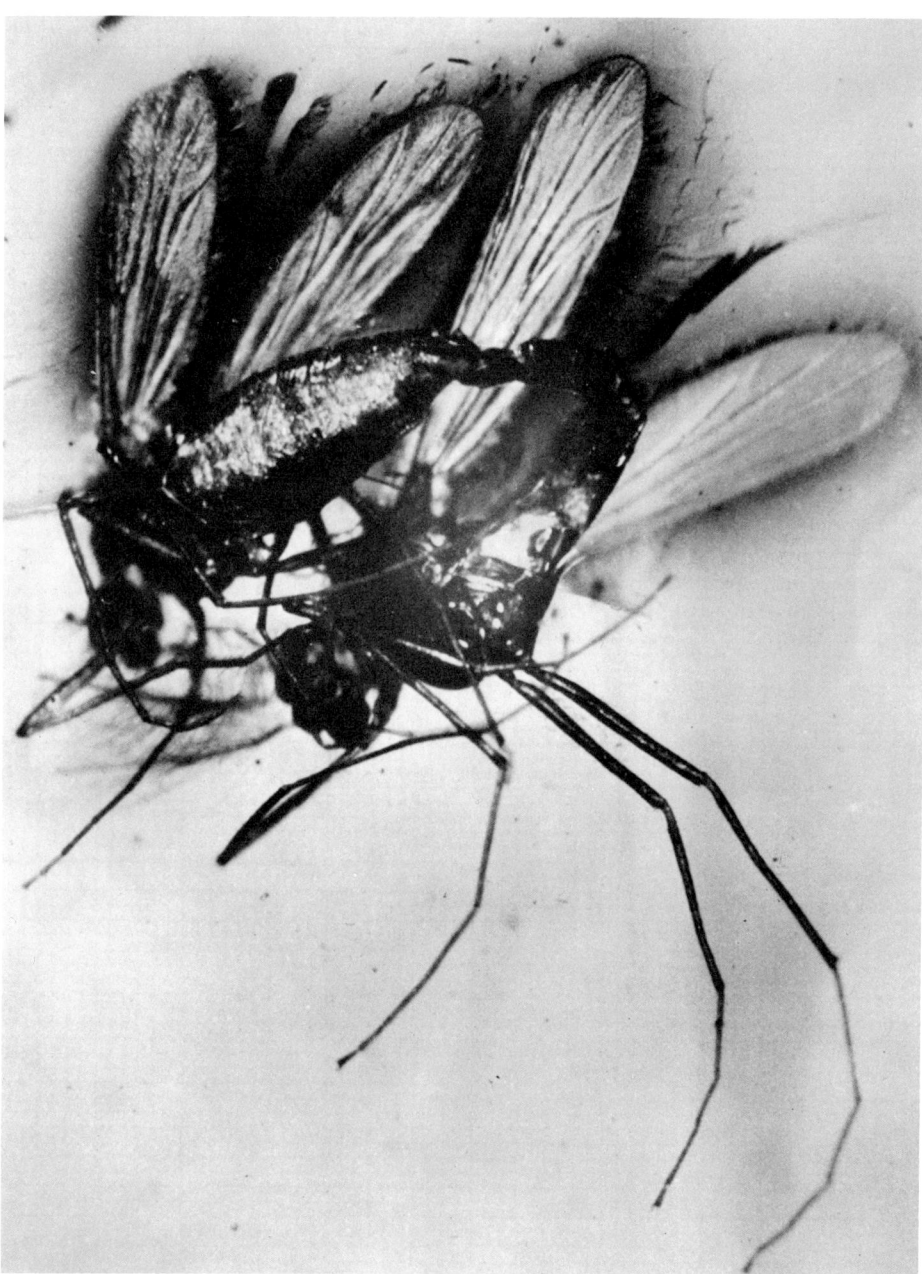

Fig. 7. Flies (Nematocera) in copulation, preserved in Eocene amber from the Baltic (E. Voigt in Müller & Zimmerman, 1962).

it is no longer fluid. Animals entrapped in this type of medium will put up a struggle for varying lengths of time. Examples of rapid embedding have also been observed in amber. Insects engaged in copulation have been embedded so rapidly that they were unable to separate (Fig. 7), and ants have been found with larvae still in their jaws.

When beetles become entrapped in resin,

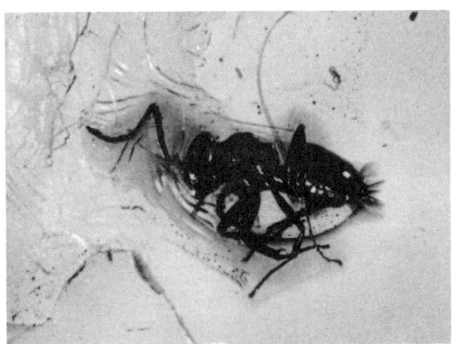

Fig. 8. Ant with its repichnia (locomotion trails), preserved in Eocene amber from the Baltic; length about 2.5 mm (Sektion Geowissenschaften, Bergakademie Freiberg, 249/1; Müller, n).

they violently struggle to free themselves. In many cases, this struggle is so severe that its traces can be found preserved in the amber and the trachea on both sides of the beetle are filled with small air bubbles. These bubbles probably represent the last breath of the animal. Quite frequently ants are embedded during their search for food (Fig. 8). Spiders and flies have been found preserved together with a spider's web. Such finds probably represent the hunter and its prey. In other cases, only the wings and legs of flies are found wrapped in tiny packages, apparently the remains of the spider's dinner.

DEATH BY DESICCATION

The drying up of rivers and lakes is a common phenomenon in arid and semiarid regions. This has also been observed in temperate regions. Wherever desiccation occurs, it often poses mortal danger to many animals, even in the temperate belts.

The concentration of animals in residual puddles of water results in selection according to age, species, and size. Relatively tall animals perish before small ones do, and animals that cannot burrow into the mud die before those that can. Exposed cadavers of vertebrates exhibit positions characteristic of desiccation, with the spinal column flexed dorsally. If firmly held in the sediment, the friable remains of certain organisms (such as the carapace of a crab) will split open upon desiccation.

Trusheim (1938) described a shallow basin, filled with shale and sandstone in the Upper Triassic near Ebrach (Franken). This ancient basin is thought to be analogous to a residual puddle in the recent record. The shale contains many remains of *Triops cancriformis minor* and the undersides of the sandstone beds are covered with a network of delicate ridges, interpreted as the infillings of desiccation cracks that formed in the mud. Therefore, the basin must have been thoroughly desiccated prior to the deposition of each successive bed of sandstone. Each new water current delivered into the basin coarse sandstone, which grades upward into fine sand and, finally, clay. Apparently the death of these organisms was not the result of sudden, catastrophic burial by sand, but was caused by prior desiccation of the mud layers. It is common knowledge that recent *Triops* cannot live out of water.

DEATH BY FLOODING

At the present time, the catastrophic results of sudden floods are well known. Land animals, unless they drown immediately, crowd together on high areas, where they, too, can be killed if the water level continues to rise; however, this manner of death is very difficult to prove in the fossil record and can rarely be interpreted as such.

THE DEATH STRUGGLE AND ITS TRACES

Traces of the death struggle not only tell us something about the last few moments of the life of an animal, but also can yield information about the cause of its death. In many cases, it is possible to judge whether the medium in which animals are preserved was a foreign and hostile environment for the victim. Following are some examples.

Numerous trails of the limulid, *Mesolimulus walchi*, have been discovered on the lower surface of the lithographic limestone of Solnhofen. Usually, the producers of these trails can be found at their terminations. Figure 9 shows such an animal which, during its death struggle, beat its telson several times into the mud.

In the same beds, fossilized cuttlefish (*Trachyteuthis hastiformis* Rüppel), have been found, which apparently moved their tentacles across the sediment before dying, leaving numerous impressions. Close to the mouth of the cuttlefish are pieces of another cuttlefish, and it is thought that these fragments may have been the partially digested portions of a cuttlefish regurgitated during the death struggle of the animal.

Many traces of the death struggle of insects are known from amber deposits. If the insects were caught by only their wings or legs, they struggled violently to free themselves and left very characteristic traces. Click beetles (Elateridae) will attempt to free themselves by excited thrashing while other beetles will either spin or swim in circles trying to escape.

RIGOR MORTIS AND ITS TRACES

Shortly after the death of an animal, rigor mortis normally begins, following a period of the death struggle. Rigor mortis spreads slowly, and is typically accompanied by alteration in the position of single parts of cadavers. Thereafter, the effects will gradually diminish. Animals must be rapidly buried after rigor mortis has set in if the rigor mortis position is to be preserved.

In vertebrates, rigor mortis commonly begins eight to ten hours after death and slowly spreads through the body, commencing in the head region, then moving to the neck, and finally the trunk. After about 10 to 20 hours, its effects will vanish and decomposition becomes noticeable. Occasionally, rigor mortis will set in immediately after death of the animal, especially after great physical exertion.

Toward the end of rigor mortis, the clearly visible process of decomposition begins. If the organism is not well buried, decomposition (i.e., muscles) is extensive, and the typical displacement of body parts associated with rigor mortis will not be preserved. In vertebrates, the most common position is a dorsal bend of the spinal column. This shrinkage can be caused by rigor mortis or by a postmortem shortening of the soft parts, or by desiccation. Therefore, the exact cause for the curvature of the spine in vertebrates cannot always be accurately determined.

It follows that observations on the death position of animal remains do not allow

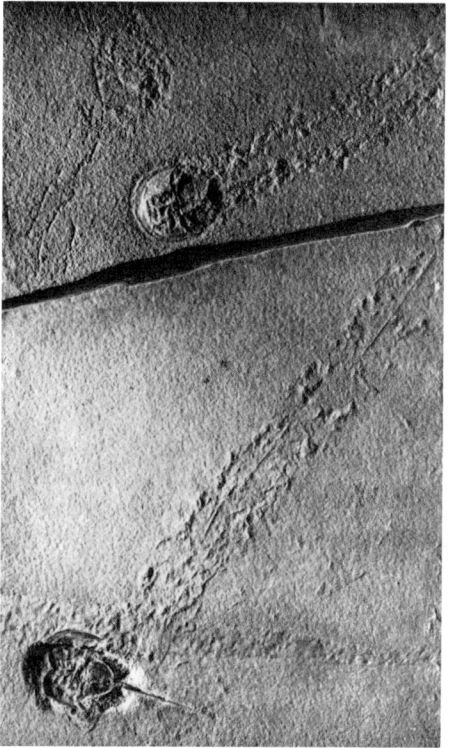

FIG. 9. *Mesolimulus walchi* (DESMAREST) preserved at end of its trail in Solnhofen Limestone (Malm zeta, lower Portlandian), Solnhofen, Bavaria, approx. ×0.1 (Richter, 1954).

one to draw conclusions about the life positions of the animals. The death position gives us information only about the moment of death and subsequent alterations of the cadaver.

Glide marks produced by ophiuroids and other well-articulated, lightweight animals can only be made during rigor mortis on suitable substrates. An example of this is in *Geocoma* sp. in the Solnhofen Limestone (Malm Zeta, lower Tithonian) (BARTHEL, 1972).

FATE OF ORGANIC MATERIAL
GENERAL DISCUSSION

Biostratinomic information is not supplied only by the study of hard parts of animals ("true fossils"). There is much to be learned by studying the soft parts of organisms and the products of their decay. After embedding in sediment, cadavers are exposed to interactions with the medium of deposition (air, water, etc.) and the sediment. Variations in these factors, therefore, change the course of the processes of decay and the nature of the end products. If these reactions are known for given conditions, it is possible to draw conclusions concerning different types of fossil soft parts or traces of them.

The process of decomposition of the organic material of plants and animals begins immediately after death. In vertebrates decomposition becomes noticeable only after rigor mortis has ended. The decomposition of soft parts is a chemical process, carried out primarily by bacterial action, and can be divided into two basic groups: 1) aerobic—decay in the presence of ample oxygen, and 2) anaerobic—decay in the absence of oxygen.

Organic material is rapidly destroyed during aerobic decay. The end products of the process are simple chemical compounds that have been extensively resynthesized by microorganisms, leaving only fossilizable body fossils. This process is, therefore, of special geologic interest.

Anaerobic decay occurs under conditions of incomplete reduction in a sealed environment in which concentration of carbon and nitrogen takes place. Even very resistant materials can be metabolized (STEVENSON, 1961). Aerobic decay represents a slow but complete oxidation of body materials. The products of aeorbic decay are, therefore, of less importance than those of anaerobic decay. They consist mostly of simple gases and fluids composed of hydrogen, carbon, nitrogen, sulfur, and phosphorus.

In nature, transitional situations are observed, and aerobic and anaerobic decay is known to occur simultaneously in the same object. Also, after only partial aerobic decay, a body may be transported into a medium of anaerobic conditions, or vice versa. As a rule, however, the process of decomposition is terminated under anaerobic conditions. After embedding, anaerobic decomposition begins. Presence of bacteria and the results of their activity have been demonstrated in many fossils, especially in bones and coprolites.

EFFECTS OF DEGASSING ON ORGANIC MATERIAL COVERED BY SEDIMENT

Gases are produced during the decomposition of organic materials covered by sediment. Under given conditions, these gases rise through the sediment, leaving traces of their movement behind them.

If decay progresses subaqueously, many of these gases are commonly trapped in pockets, such as the abdominal cavities or shells of animals, producing buoyancy. These gases escape into the sediment, and their pathways may be preserved, if the sediment has adequate viscosity and firmness.

Long vertical or inclined cavities in the sediment are called "degassing canals." Their diameters may be several centimeters, but as a rule, lie between 5 and 10 mm. Their lengths can vary and are dependent upon the surface tension of the gas, the thickness of the sediment, the height of water column, and the viscosity and permeability of the sediment. Degassing canals as long as 20 meters have been observed.

In the recent environment, degassing canals are especially well known from sapropelic environments. In sedimentary rocks, it is frequently difficult to recognize degassing canals because they can easily be mistaken for burrows of different types of animals. The same is true for the canals formed in the intertidal zone where air bubbles rise vertically in the sediment. Here, air bubbles are produced when desiccated sand, filled with air, is submerged. On beaches, degassing canals occasionally terminate in small knolls or funnels created by the escaping gas.

If the viscosity and the water and sediment load are greater than the pressure of the gas produced by organic decay, gases remain in the sediment, where they commonly accumulate in rounded or lens-shaped cavities. These cavities can be preserved if conditions of the sediment are suitable, giving rise to vesicular structure. Very large pockets of gases occasionally lead to formation of menisci. Such examples have been found in Walvis Bay in Southwest Africa and in many lakes in northern Germany where entrapped areas of hydrogen sulfide lead to uparching of small islands.

Uparching and collapse phenomena caused by decomposition gases will occur in discrete cavities of sufficient size. In vertebrates the abdominal cavity is commonly filled by gas; in mollusks, the shell.

If gases escape through rupture, or increase in permeability of the walls of the cavity, and if the sediment covering is not sufficiently lithified, collapse can occur. If, however, the cavity is preserved, then it is gradually filled with minerals and becomes a druse. The precipitation of these void-filling cements is commonly initiated in the bottom of the cavity as the trapped gases rise to the top of the dome-shaped structure. Such voids commonly remain open. Frequently large voids, such as those formed by brachiopod shells, are filled with sediment, and a small druse or void can develop near the tops of such shells (DEECKE, 1923; QUENSTEDT, 1927; TEICHERT, 1930). Thus, geopetal fabrics are produced, which can serve as geologic "bubble levels" (*Wasserwaagen*), making it possible to determine the original horizontal position in disturbed strata and in unoriented hand specimens. Figure 10 shows a cross section through a 3 centimeters-thick terebratulid bed from the Muschelkalk of Gorazde near Gogolin in Silesia. Some of the terebratulids have been preserved in life positions with their beaks pointing downward. The figure also shows that the animals were embedded in the lower part of the bed and then covered with sediment. In each shell, a void filled with drusy calcite is preserved, where decomposition gases accumulated. However, during diagenesis, geopetal fabrics can be displaced to such an extent that it becomes impossible to determine their relationship to the original bedding. Due to dewatering, shrinkage cracks appear in clay and other substances. This is an irreversible reaction in which sediment and parts of shells can be loosened and displaced sideways, even assuming a nested position (HOLLMANN, 1968a). In such a situation, geopetal fabrics that have formed prior to fossilization may also be used as ancient "bubble levels." This is especially important if the life position of the animal does not coincide with the stable embedding position (KRANTZ, 1972).

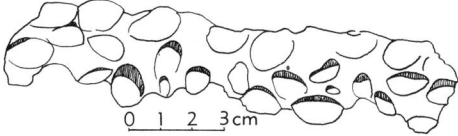

FIG. 10. Cross section through a thin bank of sediment containing *Coenothyris vulgaris* (VON SCHLOTHEIM), Middle Triassic (Muschelkalk), Gorazde, Poland (Müller, 1951b). The upper portions of many shells have been filled by sparry calcite (lined pattern) which serves as a geologic "bubble level."

These fossil "bubble levels" are also found in coprolites (Fig. 11, 12).

EFFECTS OF DEGASSING ON EXPOSED ORGANIC MATERIAL

If organic substance on the sediment surface or just beneath the sediment surface is exposed to sufficient free oxygen to decompose, all traces of such material will be almost completely lost. After decomposition, the only residuals are simple mineral substances and gases. The decomposition of the animal stops as soon as the remains are covered by a sufficient amount of sediment, and the decomposition products will float and be transported away.

Processes that happen as a consequence of the decomposition of exposed organic material, and the effects of which can still be seen in rock, can also be observed. Decaying organisms, which lie either on the sediment or are lightly covered with sediment, although originally denser than water, rise in the water because of gas produced during decomposition, and become transported pseudoplanktonically. It is necessary, however, that the buoyancy results from the entrapment of gases in a sufficiently large body cavity within the organism.

Special conditions exist as long as soft parts of invertebrates remain in their shells. This commonly occurs with cephalopod shells that sink to the bottom before the body is decomposed (STÜRMER, 1967, 1968a,b; ZEISS, 1969; LEHMANN, 1967b; LEHMANN & WEITSCHAT, 1973). Shells of recent nautiloids containing soft parts are only rarely found. This is probably the result of an early postmortem detachment of the animal's body from the living chamber.

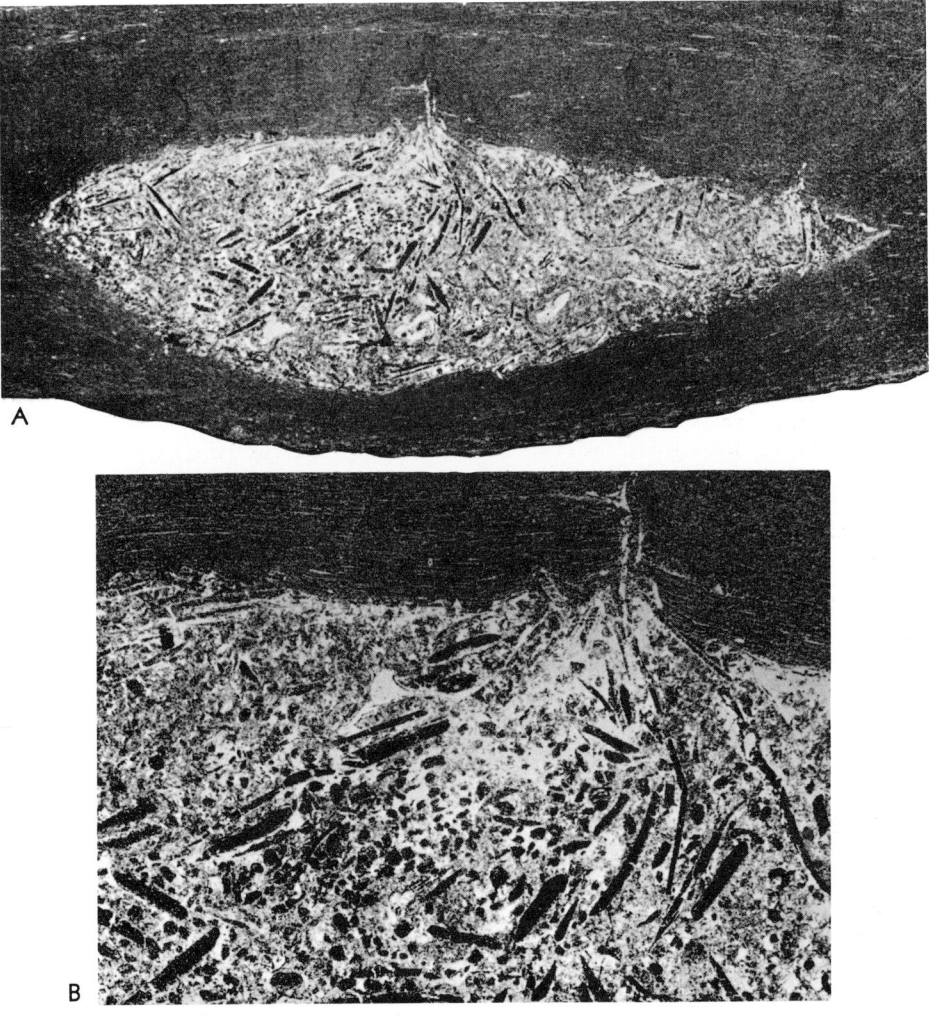

Fig. 11. Cross section through a coprolite (*A*) displaying place of escape for once contained gases produced during decomposition (*B*, enl.), Upper Carboniferous (Staunton, Logan Quarry Shale), Logan Quarry near Bloomington, Illinois. Width of structure is approx. 50 mm (Zangerl, 1971).

The empty shells then drift for long periods of time and are finally cast upon a shore, sometimes in great numbers. The shells of ammonites and nautiloids that leaked and sank to the bottom probably no longer contained soft parts. The shells sank to the bottom in a vertical position and later became inclined as bouyancy diminished. Thus, many perisphinctids in the Solnhofen Limestone are found lying on their sides on bedding planes and have next to them an impression of the venter made when the shells first settled on the sediment (ROTHPLETZ, 1909) (Fig. 13). Ammonites with especially broad venters or long, lateral spines occasionally can be found vertically embedded (Fig. 14). REYMENT (1958) conducted numerous experiments with models of different types of chambered cephalopod shells in order to determine the relationship between the size of the living chamber and the volume of gas contained in the chambers. REYMENT (1970) was able to demonstrate that when

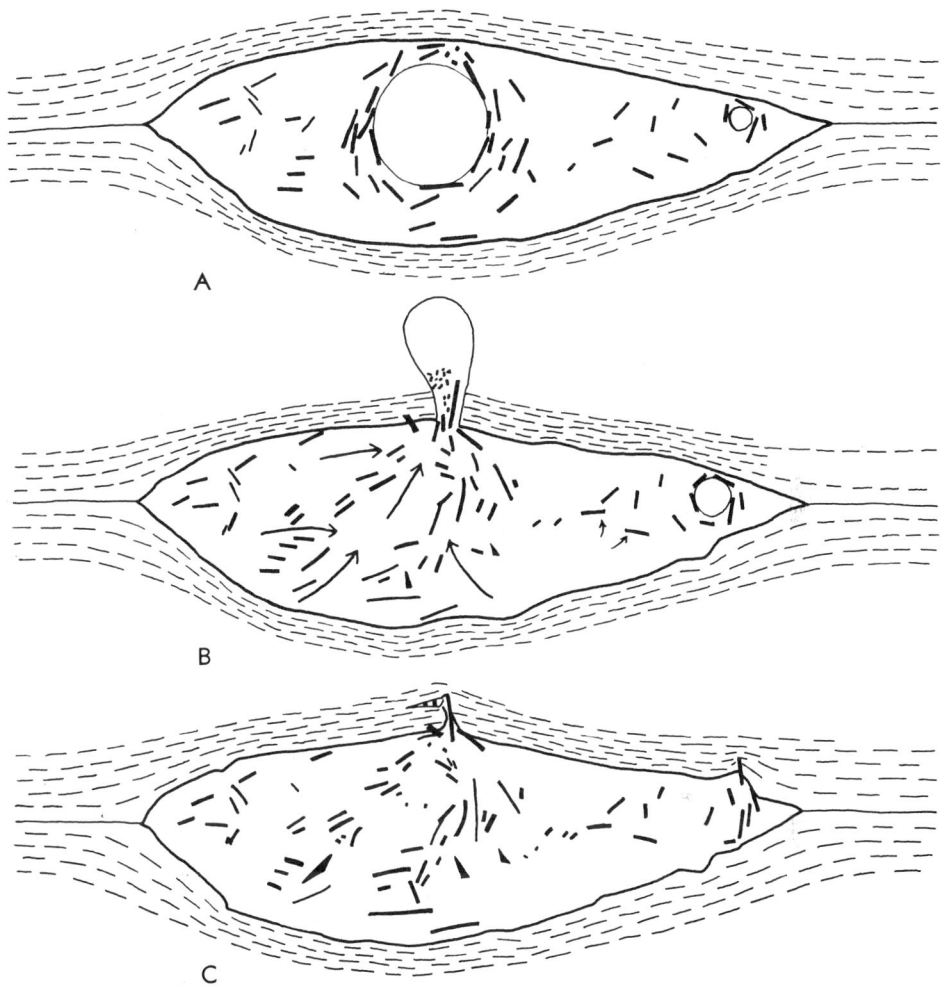

FIG. 12. Schematic explanation for Figure 11.——*A*. During aerobic decomposition, a gas bubble develops. ——*B*. The gas bubble escapes from the coprolite through about 2 mm of overlying mud.——*C*. As a result of the rupturing of the coprolite, the fish scales within it are subsequently aligned and compressed. Width of coprolite approx. 45 mm (Zangerl, 1971).

oxycones or cadicones sank to the bottom, they remained in a vertical position for three to four days until the phragmocones filled with water and tipped over on their sides. The closer the shell form approached that of the cadicone, the greater the possibility that the shell would remain vertically embedded in the sediment. Some examples of vertically embedded Ammonoidea and Nautilida can be found in the Paleozoic and Mesozoic rocks of Europe (MÜLLER, 1951b; REYMENT, 1970; Voss- MERBÄUMER, 1972). When ammonoid shells roll along the sea bed, they make distinct skip marks in the sediment, which have been misinterpreted (KOLB, 1961, 1967; SEILACHER, 1963).

Concretionary consolidation during early diagenesis may simulate the effects of buoyancy, if the shell happens to have been covered by a thin sediment layer. This may be observed in many ammonites in the Solnhofen Limestone in which the buildup is several millimeters higher than under the

Fig. 13. Imprint of an ammonite which at first stood in a vertical position on the sediment (top) and then fell over on its side, from Solnhofen Limestone (Malm zeta, lower Portlandian), Solnhofen, Bavaria. Width approx. 190 mm (Rothpletz, 1909).

chambered portion of the shell in which buoyancy existed while the soft parts of the animal remained in the shell during burial (MAYR, 1966).

ROLE OF SCAVENGERS

As a cause of destruction or alteration of soft parts, the role of scavengers is next in importance after the decay of organic material by bacteria. Such animals are: birds of prey (vultures), foxes, wolves, hyenas, many fish (piranha, eels), polychaete worms, snails, crabs, many insects, and insect larvae (ants, maggots). Direct proof of their existence in the fossil record is found only in exceptional cases (mummies; inclusions in resin). In Siberia, cadavers of mammoths partially devoured by wolves and foxes have been found in permafrost. In the ocean, isopods and decapods rapidly destroy dead animals. On land, maggots are particularly effective.

The activity of scavengers and other animals may be estimated indirectly from the amount of coprogenic materials within the sediment. This material influences the composition of the sediment because the amount of excrement produced during the lifetime of an animal is far greater than its body mass. This is especially true for the often very resistant excrement of worms, snails, bivalves, and other invertebrates, which in both fossil and recent deposits may constitute up to 50 percent of the sediment (HÄNTZSCHEL, EL-BAZ, & AMSTUTZ, 1968; SCHÄFER, 1962, 1972).

Some animals that bore into hard parts of other organisms either chemically or mechanically receive their nourishment from the organic material contained in these hard parts (BOEKSCHOTEN, 1966; BROMLEY, 1970; SOGNNAES, 1963), but the physical strength of the bored shells is weakened and their physical destruction accelerated.

SEQUENCE OF DECAY OF SOFT PARTS ASSOCIATED WITH PRESERVABLE HARD PARTS (SELECTIVE DECOMPOSITION)

The skeletal parts of echinoderms are held together only by skin and connective tissue and are easily disarticulated and scattered. If complete specimens are preserved, it is probably because they were buried rapidly and the remains of the animals were resting in an environment free of currents.

Echinoid spines will drop off, seven to ten days after death, due to the progressive decay of soft parts in an aqueous, aerobic environment. If the spines are still attached to the shell, then burial must have occurred while the echinoid was living or soon after death. After 12 days, the apical region of the animal begins to fall apart, starting from the inside outward. After 17 days, only the jaw apparatus and disarticulated elements remain.

The taphonomy of the starfish *Asterias rubens* has been studied by SCHÄFER (1962; 1972) in marine waters of 18°C. After five days the body of the starfish becomes blown up to the bursting point and is easily moved by bottom currents. The integument on the dorsal side of arms and skeleton will then begin to separate from the body, starting at the arm tips. The dorsal integument may then be carried away by currents or it forms folds on the decomposing body. If

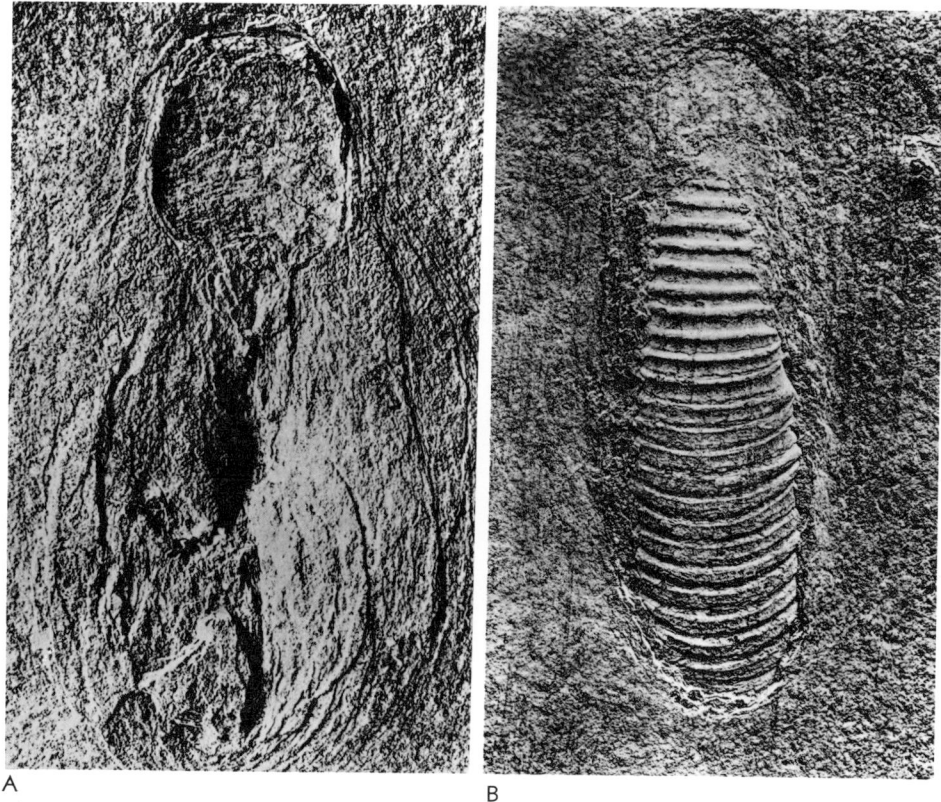

Fig. 14. Vertically embedded ammonite shell; under side *(A)*; upper side of a 12 mm-thick bed *(B)*, Solnhofen Limestone (Malm zeta, lower Portlandian) from Bavaria. Diameter approx. 210 mm (Rothpletz, 1909).

starfish are buried and killed by a thick covering of sediment, the dorsal and ventral skeletons are commonly found wedged together in a tangled heap.

Gas-filled cadavers of holothurians are not buoyant, but are rolled by currents like barrels. Complete decomposition of the thick, hard, resistant dermis may take several weeks. In the absence of currents, the sclerites are deposited in small clusters in the sediment.

SCHÄFER (1962, 1972) has demonstrated that recent coleoids become buoyant immediately after death and will remain afloat at the surface for many days, depending upon the water temperature. The body of the animal is almost totally decomposed before parts of it begin to fall to the bottom. This explains why the soft parts of coleoids are rarely preserved even in fine-grained sediment under anaerobic conditions generally favorable for the preservation of soft parts. Soft parts or the impressions of coleoid soft parts are almost always found in nearshore sediments which were periodically dried out. Only during these periods of exposure are the animals dense enough to sink into the sediment, and then be completely buried by sediment (ABEL, 1916; NAEF, 1922; JELETZKY, 1966).

The drifting coleoids quickly lose their very porous, mostly calcareous cuttlebone, which can float for considerable distances and can serve as a site for attachment of many kinds of organisms. The cuttlebone is commonly destroyed by the action of boring algae or by the pecking of sea birds. Larger fragments rarely stay afloat for more than eight days. In contrast to shells of *Nautilus* and of ammonites, the cuttlebones of sepiids do not settle to the bottom of the sea, but may accumulate in large numbers along beaches. The prominent, mostly chitinous cuttlebones of the Teuthoidea are not buoyant and sink immediately to the bottom where they are embedded together in areas with entire bodies of the organisms, unless strong currents are present. The ink sac is very durable and is occasionally found inside the animal (ABEL, 1916; NAEF, 1922; EHRENBERG, 1942; JELETZKY, 1966; STÜRMER, 1965, fig. 5).

FATE OF PRESERVABLE HARD PARTS BEFORE FINAL BURIAL

GENERAL DISCUSSION

Accumulations of hard parts of animals become sedimentary constituents and therefore obey the laws of sedimentation. Therefore, it is possible to draw many conclusions about the type of sedimentation by studying the interrelationships of the preservable parts of organisms. It is also necessary to study the relationship of hard parts to the surrounding rocks before and during the embedding, as well as the mechanical and chemical changes of the hard parts induced by biologic and inorganic factors.

Accumulations of fossils are burial assemblages or thanatocoenoses (WASMUND, 1926). There are basically three types of thanatocoenoses: 1) autochthonous—life assemblages; 2) allochthonous—assemblages that have been transported over varying distances; 3) assemblages of living organisms that have been washed into hostile environments where they died.

The composition of a thanatocoenosis is strongly determined by the nature of the constituents of the biocoenosis. In **autochthonous thanatocoenoses**, the composition of life and death assemblages are identical, and they reflect the conditions of the biocoenosis. They allow certain conclusions as to the factors that influenced the biocoenosis such as conditions of feeding, respiration, and adaptation. Autochthonous thanatocoenoses are limited in space and dependent on the ecologic conditions of the environment, and must be understood as parts of ancient biocoenoses. Associations of trace fossils (ichnocoenoses, proposed by DAVITASHVILI, 1945) can be considered as biocoenoses as well as thanatocoenoses because, as a rule, they are autochthonous.

Allochthonous thanatocoenoses (taphocoenoses, QUENSTEDT, 1927) are formed by the transport of material and the processes of demixing, sorting, and destruction that occur during transport. These processes also act upon the inorganic, clastic constituents of the surrounding sediment. Thus, the resistivity of skeletal material against attack by chemical and physical forces prior to burial is of utmost importance. The composition of a thanatocoenosis can be drastically changed by the elimination of certain forms, and the individuals found in thanatocoenoses are commonly those elements of the biocoenoses that were most resistant to selective abrasion and diagenesis. In many cases, thanatocoenoses contain

TABLE 1. *Different Types of Fossil Accumulations (Fossil-Lagerstätten) (modified from Seilacher, 1970; Reineck & Singh, 1973).*

1. Enriched deposits (*Konzentrat-Lagerstätten*). Concentrated accumulations of disarticulated skeletal material.
    a. Condensation deposits. Concentration of skeletal material due to slow rates or the absence of sedimentation. Examples: submarine cave deposits, condensation horizons.
    b. Placer deposits. Concentration of hard parts due to transport and sedimentary sorting. Examples: Bone beds, allochthonous amber deposits.
    c. Sedimentation traps. Inorganic filling of cavities. Examples: terrestrial and submarine fissure fillings, burrow fillings.
2. Conservation accumulations (*Konservat-Lagerstätten*). Characterized by the complete or partial preservation of soft parts and the common preservation of complete, articulated skeletons.
    a. Stagnation deposits. Accumulation of organic remains in sapropelic sediments where water layer directly above sediment surface is anoxic. Examples: black shales, lithographic limestones.
    b. Conservation traps. Rapid sinking and embedding of organic remains into preservational medium or cavities. Examples: amber, peat, asphalt.
    c. Burial deposits. Rapid embedment of organic remains in reducing sediments. Examples: Hunsrückschiefer, Sendenhorst fish deposits.

autochthonous as well as allochthonous elements. Accumulations of products of special life cycles such as exuviae, leaves, and sporomorphs have been called **pseudocoenoses** (MARTINSSON, 1955).

Naturally, the degree of transportation can be subject to important change, and many different types of sorting can result. On the other hand, tectonic or astro-climatic cycles and rhythms can create repetitive sequences of conditions of burial. The fact that 44 percent of the animals in the animal kingdom lack hard parts explains incompleteness of the fossil record as well as the fact that fossil plants are much rarer than fossil animals.

SEILACHER (1970) proposed a classification of fossil deposits which he called "*Fossil-Lagerstätten.*" These deposits are arranged according to their mode of formation and the general shape of accumulation (Table 1).

ALLOCHTHONOUS BURIAL

GENERAL DISCUSSION

Transportation causes mechanical destruction of organic material as well as segregation of particles according to shape, hardness, and weight. Small bioclastic particles, just as other sedimentary materials, may be redeposited several times. There is a direct relationship between quality of preservation and time of transport. However, particles may even be further destroyed *in situ* by abrasion in turbulent water. In general, the destiny of hard parts is dependent upon the relationships between the rate of deposition and reworking.

SORTING BY TRANSPORTATION

This process can be defined as the sorting and deposition that results from differences in form, hardness, weight, and durability of shell material and the related chemical and mechanical processes. It can be simulated by the following processes: 1) biological causes, 2) selective removal of certain ontogenetic stages through death, and 3) diagenetic processes after embedding.

Diagenetic processes can wholly or partially (often selectively) destroy, or render unrecognizable, the fossil contents of a rock.

Selection by death occasionally plays a large role. Thus, WASMUND (1926) observed how in early spring, numerous, immature specimens of *Limnaea stagnalis* and *Limnaea auricularis* daily formed allochthonous thanatocoenoses along the shore of Lake Constance. Accumulations of adult shells were rare. In general, only single shells were washed ashore.

The reason for these accumulations is related to the climatic conditions of early spring. During warm days, the animals move upward *en masse* in the sediment and appear at the surface, where they are killed by frost during the following night. WASMUND (1926) also reported observation of a *Spülsaum* of dead larvae of ephemeral flies in a sand bar of the delta of the Rhine on the Austrian shore of Lake Constance. Larvae crawled onto south-facing cliffs, thus escaping from the cool water into the sun and warm breezes. They were then killed by night frosts and added to the enlarging *Spülsaum*. Both of these allochthonous thanatocoenoses were thus caused by local meteorologic catastrophes.

The shells of gastropods are commonly removed by hermit crabs (Paguridae) (SCHÄFER, 1955). Hermit crabs prefer particular shells, and this simulates selective sorting by transportation. Accumulations of only large cutting and crushing teeth of skates and sharks are another example. These animals shed their teeth in the same area for many years. The growth and eventual loss of teeth is related to the reproductive process and change of teeth usually occurs during mating, which commonly takes place in the same area, year after year (SCHÄFER, 1962; 1972). The smaller grasping and crushing teeth fall out all year long and are found widely scattered over the sea floor. Mechanical sorting can also be simulated if organisms of the same age and size settle and colonize a particular substrate.

Marine invertebrates, such as clams, snails, and crabs can be transported by other animals to the land and become embedded in lacustrine or terrestrial sediments far from the sea. This, for instance, is done by the sea gull *Larus argentatus* which feeds mainly on clams and snails, but also on crabs and fishes. During the mating season such prey is offered ceremoniously by males to females. These, and indigestible, regurgitated remains, pile up in the nesting territories and represent a special case of autochthonous sedimentation. Small clams which have been swallowed whole and regurgitated are of an even size and such accumulations can be mistaken for results of current sorting (Teichert & Serventy, 1947; Goethe, 1958).

Wind occasionally plays an important role in the transportation and sorting of organic remains. This is especially true for small bones in which the periosteum and marrow have disappeared. The periosteum and marrow together can account for up to two-thirds of the total weight loss of the bone. The pneumatic bones of birds are especially light. A humerus of the crow, *Corvus frugilegus*, which was 5.4 cm long, weighed only 1.32 gm. On salt-encrusted sand flats of the island Alte Mellum near Wilhelmshaven, vertebrae of seals have been seen to be moved by wind at velocities at which small valves of *Macoma* and *Mactra* as well as the rounded shells of *Littorina* were not moved.

It is necessary to be especially careful in the study of separation of clam shells, the right and left valves of which are often deposited in different places. This is easily explained if form and weight of the two valves differ much, as, e.g., those of *Gervillia (Hoernesia) socialis* of the Upper Muschelkalk. Even in equivalve clams, differences in form and weight of the valves exist. In equivalve heterodonts and schizodonts, the right valve is generally heavier because it carries large hinge teeth. These examples demonstrate that it is necessary to exercise caution in the evaluation of fossil concentrations, especially those consisting of individuals or fragments of equal sizes. In such cases it is necessary to have independent evidence of mechanical sorting by currents. In *Cardium edule*, the weight difference between right and left valves ranges from 0.72 to 4.38 percent (Klähn, 1932). The specific weight even in different valves of the same species varies so much that it must be considered when studying the sorting process affecting bivalve shells.

The significance of these factors related to the sorting of bivalves is enhanced by the hinge teeth and chondrophores acting as anchors. The right valves of desmodonts do not possess a chondrophore and therefore are more easily transported than the left valves which have chondrophores and, therefore, offer more resistance to movement. If a large colony of mussels (*Mya arenaria*) is churned up by waves, the chondrophore-bearing left valves are transported lesser distances than the right valves which lack chondrophores. Large accumulations of these shells can form belts 10 to 50 m wide consisting of up to 90 percent of one valve and 10 percent of the other. An admixture of different types of valves usually results from the addition of new shell material from another nearby colony. In tidal channels, it is quite common to find valves without chondrophores transported over long distances (Richter, 1922).

Durability strongly influences the preservation of shell material in the face of mechanical processes, both within and on the sediment. Some shell parts may be completely destroyed as, for example, in some bivalves of which only one valve remained for a long time. Such valves may even now be rarities (e.g., the right valve of the pelecypod, *Eopecten albertii*, found in the Muschelkalk in the Germanic basin). Small shell components are more easily destroyed than large ones (Hallam, 1967).

The presence of worn-down, delicate, or easily fragmented hard parts, which have most probably been redeposited together with the surrounding sediment, may make it possible to draw conclusions about the length of transport. Only very durable skeletal material can survive long transportation in coarse-grained sediment. Mechanical abrasion and faceting will quickly destroy any fragments having little durability. Thin and friable bones of animals found in Pleistocene caves are examples of skeletal material that has not been transported great distances. Observations of redepositing have shown that such occurrences are geographically local and are accomplished during short periods of inundation. This is especially true for fragile, allochthonous remains of recent mollusks.

As a rule, animals living on the sediment stand a better chance of being redeposited than those living within the sediment; however, destruction is not necessarily the same for all hard parts because infaunal organisms tend to have less durable shells than do members of the epifauna.

TABLE 2. *Floatability Constants for some of the Most Common North Sea Bivalves (after Tauber, 1942).*

Number of Specimens Studied	Species	Average Surface Area (mm²)	Average Weight (gm)	Floatability Constant
12	*Cardium edule*	1,930	2.85	169
8	*Mya arenaria*	7,258	7.10	255
15	*Mytilus edulis*	1,962	1.30	377
20	*Macoma baltica*	330	0.27	305

MECHANICAL DESTRUCTION OF HARD PARTS PRIOR TO BURIAL

Hard parts of organisms may be destroyed by biologic or by entirely inorganic processes. Biologic processes as well as inorganic, chemical processes are discussed at length elsewhere.

The degree of physical destruction of shell material allows one to draw conclusions as to the mechanical processes in operation during the period of shell destruction. This is especially true for conditions of sedimentation. Therefore, it is not only necessary to know what kind of destruction occurred but also the speed at which it took place. These mechanical processes can be divided into two basic types: 1) polishing or abrasion, and 2) breakage or crushing. Because of their abundance and availability, the hard parts of mollusks have received considerable attention in the literature with respect to these destructive processes. Therefore, they are discussed in detail below.

Ambient polishing. If one considers the appearance of the destroyed shell material and the composition of the allochthonous thanatocoenosis, the diversity in preservation is conspicuous (e.g., in shell accumulations along swash marks *(Spülsaum)* or along sandbars). The following are causes responsible for this condition: 1) different resistivity of the shells against chemical, mechanical, and biological destruction (CHAVE, 1964), 2) differences in the ability to float, and 3) different behavior on the sedimentary surface (i.e., rolling, gliding, etc.).

A basis for estimating floatability was developed by TAUBER (1942) who divided the surface area of the shell (in mm²) by the four-fold value of its weight in grams.

The area is determined by rolling the shell along millimeter graph paper and counting the number of covered millimeter squares. Table 2 shows the floatability constants of some common clams found along the North Sea.

Because of their greater floatability, shells of *Mytilus edulis* and *Macoma baltica* are more easily moved by transporting media than those of either *Cardium edule* or *Mya arenaria*. Therefore, shells of the two last-mentioned species, as a rule, move with the sand inside a sand bank whereas *Mytilus edulis* and *Macoma baltica* are commonly transported across the sand in the turbulent zone of shells (KLÄHN, 1932). This explains the observation by KLÄHN (1932) that shells of *Cardium* and *Mya* are commonly strongly polished, whereas those of *Mytilus* and *Macoma* rarely are polished at all.

Differences in floatability of shells is a result of the relative resistivity of different bodies. This is the reason for differences in degree and type of mechanical, and, to some extent, biological destruction of shells.

If biological factors are ignored, the mechanical destruction of shell material in coarse-grained sediment (gravel) results generally in total destruction of the shells, whereas in finer grained, sandy sediment (especially for grain sizes between 0.5 and 0.005 mm), polishing takes place. In addition to sand, shell debris also acts as a polishing agent, whereas the mechanical effect of the silt and clay fraction on shell material is minimal.

Ambient polishing can result from 1) mass transportation in churned-up sand bodies, and 2) rolling of shells over an abrading sediment surface. Mass transport in the marine environment is characteristic for sand banks and giant ripples. The second method of polishing occurs mainly in the surf zone, but also in the terrestrial environment, especially in dune and desert areas.

During polishing of bivalve and gastropod shells, the periostracum and the fine details of the shell surface are destroyed. In bivalves the ornamentation and hinge teeth are also lost, and the edge of the shell is rounded. The valves of bivalved shells exposed to ambient polish soon become separated. In addition to being destroyed by organic decay, the ligaments are gradually loosened, and torn or filed through.

KLÄHN (1932, 1936) has made quantitative investigations on the type, manner, and duration of reworking, and mechanical destruction of mollusk shells. In general, the percentage of reworked shells is less for small forms than for large ones (see also HALLAM, 1967). The relationship is just the opposite of that for the chemical dissolution, which is inversely proportional to size of shell materials. The reason for this phenomenon is clear. Smaller forms have a relatively high floatability and sink slowly. The more bulky the shell, the greater the tendency for it to remain near the sea bottom where polishing occurs. However, correlation between the degree of polish and the initial weight of the shell may also be negative. This is especially true when there is a disproportionate increase in shell thickness with increasing length. Also, angular sand grains are more effective polishers than rounded ones, and fine grains more effective polishers than coarse grains, because they move more swiftly with the water. Single valves are more rapidly polished than closely packed accumulations of shells that protect each other from destruction. Experiments performed on the tumbling of pea-sized shell material show that more abrasion occurs within the first nine hours of tumbling than later (HALLAM, 1967). The sculpture on the convex side of the shell is usually the first to be worn away, and this usually occurs within the first nine hours of rolling. On the other hand, structures of the protected, concave surfaces of shells are commonly preserved for long periods of time. Dense, fine-grained skeletal parts are more resistant than porous ones, and rounded parts more than flat bodies (e.g., snail and bivalve shells).

Faceting. In addition to, or in place of ambient polish, shells may be polished on only one side. This process is called faceting, and the shape and position of the facets is determined by the abrading medium and by the properties of the abraded body. Commonly, faceted objects appear broken. MÜLLER (1951b) distinguished three types of facets: anchor-facets, roll-facets, and glide-facets.

1) Anchor-facets. Anchor-faceting (or sand-polishing in the sense of PAPP, 1941) occurs when hard parts of organisms are firmly anchored in and project slightly above the surface of the substrate. If abrasive materials move across them, they are faceted. This process is similar to glacial erosion. Faceting of this sort differs from roll-faceting and glide-faceting in that the shell material is firmly embedded in the substrate.

A unidirectional current produces a single, oriented facet. If currents are changeable or the object shifts its position, many different facets may be produced on the same object. In addition, anchor-faceting can be produced on the opposite side of the anchored object, if this has been flipped over by currents. The resulting facet will lie more or less parallel to the first facet. If an object has been shifted several times, it may become firmly wedged between pebbles, and different sides may then be faceted, and the facets need not necessarily be parallel. One object may have anchor-, roll-, and glide-facets.

As carrying media for the abrasive material, water, air, and ice are of primary importance; coarse-grained sand is the most important abrasive. Because coarse-grained materials are commonly transported only in shallow water, anchor-faceting is especially characteristic of beach and nearshore deposits. On land, this type of faceting occurs in dune areas, as described by PAPP (1941) for shells of *Pisidium* and *Dreissena* that had been transported by the wind.

Because faceting can take place only in a flowing medium, shells are oriented in their most stable position. Faceting then affects the upper side of the shell. Dish-shaped bodies (single valves of mollusks, patellids, etc.) are turned on their valve margins; high-spired gastropods are mostly turned on their apertures, and cone-shaped bodies (e.g., shells of *Trochus*), whose basal diameter is equal to or greater than their height come to rest on their base. In such spiral forms, the tips of the cones are polished first. The position of facets is, therefore, dependent upon the shape and statics of the body being polished.

With progressive polishing or sandblasting, bowl-shaped objects (such as single valves of bivalves and patellids) assume a ring-shaped appearance. The resulting rings are very fragile and easily break up into shell debris. Fossil examples have been described by PRATJE (1929) from the upper Pliocene (Red Crag) of Foxhall near Ipswich, England, and also from the Vienna Basin. Such occurrences can be documented if shells are preserved, especially in younger strata. Otherwise, they may be mistaken for accumulations of debris caused by accumulation of decomposition gases in the arches of the shells.

2) Roll-facets. Another type of faceting can develop by the continual shaking of bivalve shells with sand, as has been demonstrated experimentally by KLÄHN (1932).

Where roll-faceting occurs in nature, it is found under conditions that correspond to the above-mentioned experiment. These are: 1) relatively coarse-grained sand substrates, 2) relatively strong turbidity, especially in nearshore environments, and 3) especially strong, fluctuating rates of movement that persistently cause rolling and tumbling (i.e., tides, strong longshore currents).

Roll-faceting acts especially on bivalve shells, but it is also common in shells of snails and other hard-shelled animals. Generally, in rolled bivalve shells abrasion begins at the umbo. PRATJE (1929) called this type of abrasion "umbo-faceting." Similar abrasions may be found in the middle of the valves of bivalves such as *Scrobicularia plana* and *Mactra corallina cinerea*. HOLLMANN (1968c) called these "median-facets." Shells abraded in this manner first take on a rounded appearance and then the abraded area progresses from the hinge out following growth lines. Eventually the shell takes on a horseshoe-shaped appearance. On this basis, one may distinguish between roll-faceting, round hole-faceting, and horseshoe-faceting (TAUBER, 1942).

The conditions for the formation of roll-facets are present mostly on sandy beaches in the intertidal zone. In this area, faceting of this type is so common that the term "tidal-faceting" has been proposed and, if observed in ancient sediments, presence of tidal flat conditions may be deduced (KÜPPER, 1933); however, because such forms are also observed elsewhere, under conditions of strong movement and an abrasion substrate, the term "roll-facets" seems more appropriate. Roll-faceting is also known from tideless coasts or coasts where tides are weak, such as the Baltic, Adriatic, and Black Seas. In these regions, abrasion is caused by fluctuating coastal currents. As might be expected, the occurrence of roll-faceting is much less frequent than in areas of stronger tidal currents. PRATJE (1929) cited fossil examples from the English Red Crag and noted that they appeared absent in the Mediterranean Vienna Basin. PAPP (1941) studied abraded parts of *Dreissena* and *Pisidium* and observed that holes were worn in the umbal area of the shell. He deduced that such faceting was the result of movement of the shells over the sands of wandering dunes. Thus, roll-faceting has been shown to occur in marine to continental conditions.

The position of the facet in relation to the shell margin depends on the roundness of the shell. In bivalves, the angle between the facets and the shell margin is generally large. In *Cardium edule* it varies between 30 to 90 degrees. KLÄHN (1932) performed an experiment to study roll-faceting, using the following materials: 10 cm^3 of sand with a grain size between 0.25 and 0.12 mm, 100 cm^3 of water, and *Cardium* shells 1.8 to 2.4 cm in length. This mixture was then tumbled for 92 hours, at 106 revolutions per minute, in a tumbling machine. The first effect on the shells was the development of a small hole in the umbonal region, which gradually increased in size as did the angle of the facet in relation to the shell margin. After 160 hours of tumbling, the angle had increased to 50 degrees. The process took much longer for smaller shells.

3) Glide-facets. Glide-faceting results when hard parts of animals glide over abrasive substrates such as coarse-grained

sand. As in anchor-faceting, a flow medium must be present for glide-faceting to occur; however, glide-faceting results when shells are not anchored in the substrate. Under these conditions, cone-shaped shells, with basal diameter greater than height, glide along on the larger apertural end, which is worn away. In some cases, faceting is so severe than only the points of the shells remain. This is characteristically observed on shells of *Trochus* and similarly shaped shells in which glide-faceting was first reported.

A special case of glide-facets is produced by pagurids. Most hermit crabs inhabit the littoral zone where in places they are so abundant that they occupy practically all empty gastropod shells. The crabs are generally quite active and drag the shells they inhabit along the ground so that they become faceted, especially on sandy substrates (pagurid-faceting). Most severely affected are usually the parts near the aperture, in dextrally coiled shells the left outer side of the last whorl. This characteristic position makes it possible to demonstrate fossil occurrences of pagurids, even if the crabs themselves are not preserved.

Table 3 shows the different types of faceting observed on short-conical gastropod shells.

TABLE 3. *Faceting Present in Short-conical Gastropod Shells.*

a) Glide-faceting
   Parallel to the base on the apertural side
b) Anchor-faceting
   Parallel to the base on the pointed side
c) Pagurid-faceting
   Somewhat to the side of the oldest whorl on the apertural side of the shell and somewhat higher on the shell than the initial glide-facet

FRACTURING

If biologic factors are excluded, mechanical fracturing of shells occurs mostly through the interaction of shells with coarse-grained sedimentary particles over which the shells are shoved, skipped, and rolled. In flowing water, mechanical fracturing occurs in river beds and along pebbly or gravelly shorelines of the sea. The action of the waves alone may have similar effects. Complete destruction of organic debris gen-

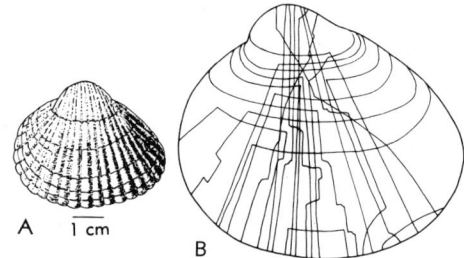

FIG. 15. Fracture deformation of recent *Cardium edule* LINNÉ (*A, B*) collected from broken shell masses along the shore, Scheveningen and Kattwijk, Netherlands (Hollmann, 1968b).

erally does not occur below effective wave base at 50 to 60 m. Tidal and "exchange" currents, which occur at greater depths, do not have sufficient strength to fracture shells, such as those of mollusks.

When the fragments of shells have not been completely separated, their arrangement can sometimes simulate the impression in the sediment of artificially crushed shells.

KLÄHN (1932) demonstrated quantitatively that if bivalve shells are shaken together with pebbles, abrasion plays only a subordinate role in the destruction of the shells. Because of the relatively large size of the pebbles more uniform surfaces are present that can strike against and therefore break the shell material. On the other hand, if bivalve shells are initially tumbled in sand or a similar medium, the dominant process of abrasion will give way to breakage. As the shells become increasingly thinner due to the polishing process, they become more easily fractured. Cracks appear and eventually they fall apart, and the pieces are deposited as accumulations of shell fragments.

Mollusk shells are mostly fractured by hard impacts. The broken edges are, as a rule, very irregular and follow structurally weak parts of the shell. Determining factors are their laminated structure, differences in thickness and convexity, relative size, ornamentation, and microstructure (KESSEL, 1938; SCHÄFER, 1962; 1972; HOLLMANN, 1968b). The nature of primary accumulations of shell debris is determined by the mechancial resistivity of the shells and varies from species to species.

Bivalve valves with well-developed, radial

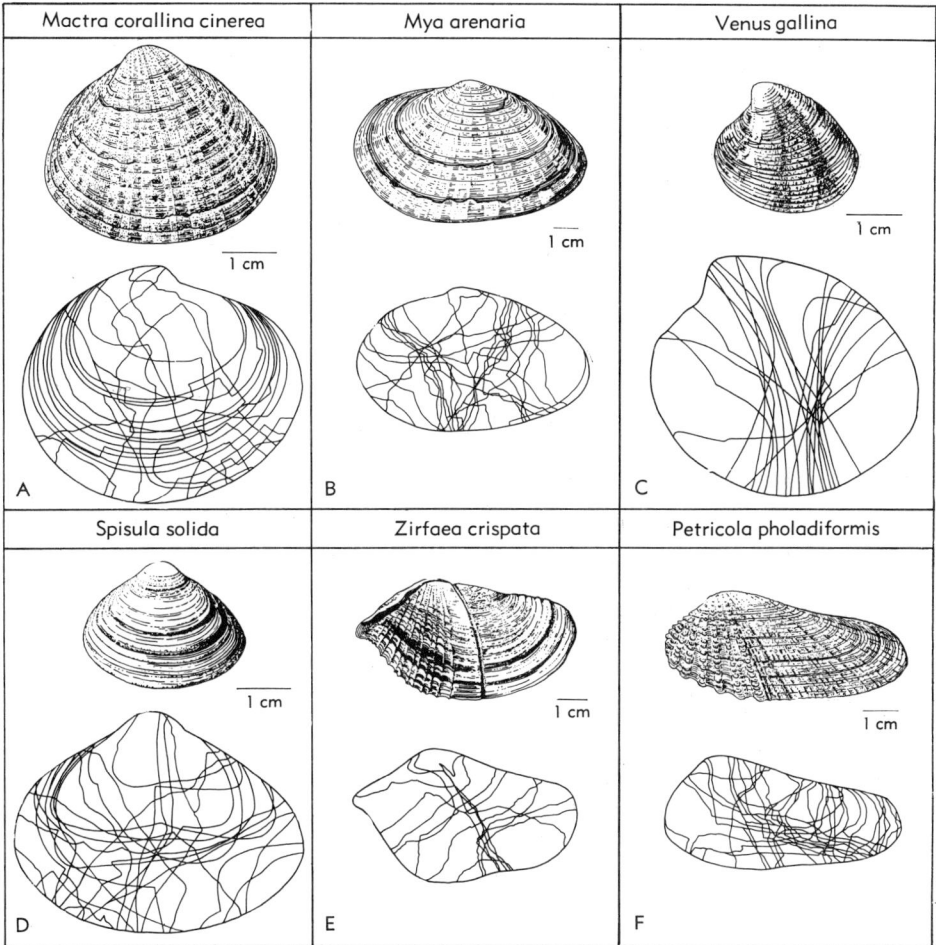

FIG. 16. Fracture patterns developed in different types of recent bivalve shells *(A-F)*, strand line deposits from Scheveningen and Kattwijk, Netherlands (Hollmann, 1968b).

ribs generally break parallel to the ribs. Examples are shells of *Cardium echinatum* and *C. edule* (Fig. 15,*A,B*). Fracture will be perpendicular to weakly developed radial ribs, as in shells of *Venus gallina* (Fig. 16,*C*). Forms with strongly developed growth lines tend to fracture parallel to the growth lines (i.e., *Mactra corallina*, Fig. 16,*A*). Oval-shaped or arched fracture patterns are common in forms such as *Spisula* (Fig. 16,*D*). Transverse fractures and curved fractures with sharply defined boundaries cut across the umbonal regions of shells of *Mactra corallina*, but bypass those of shells of *Spisula*. Thin-walled bivalve shells with distinct, concentric growth lines commonly break along the growth lines. Small rings remain, which, at first glance, resemble anchor-facets. *Ostrea* shells break along the shell margins and are progressively worn away until only the ligament area remains. Cracks in the shells of *Mya arenaria* and *Scrobicularia plana* follow the prismatic layer along the growth lines. The left valve of *Mya* (Fig. 16,*B*) will splinter and produce wavy or serrated lines of fracture, essentially in the ligamental groove. Generally, splintery, ragged, or wavy fractures prevail. In such forms as *Pholas dactylus*, *Cyprina islandica*, and many other bivalves, destruction of the shells is so complete that only the hinge areas remain. The shells of elongate thin-shelled forms such as *Ensis, Barnea,* and *Petricola pholadiformis* (Fig. 16,*E,F*) break along the most strongly arched parts. Shells of *Mactra subtruncata* and *Macoma baltica* seldom break in this manner, and those of *Donax vittatus* almost never do. Species differ in regard to the strength of their hinge and umbonal areas and these parts are often found in rich concentrations.

The destruction of gastropod shells frequently begins with breakage of the aperture and a flattening of the whorls. Details of the process of destruction depend on sculpture and thickness of the shells. As a rule, the especially resistant apertures of shells of *Nassa, Murex, Buccinum,* and similar forms are preserved. Columellas are among the especially resistant parts and may be preserved even when washed around for some time by the surf. In short, conical shells like *Littorina* and *Lunatia,* the apex is commonly the most resistant part. Small breakages can be enlarged to "spiral facts" along the areas of greatest curvature of the whorls and can extend across several whorls (HOLLMANN, 1968b).

Echinoid tests break up along irregular fractures as long as the epidermis and spine muscles are still preserved. When decomposition of the soft tissues is complete, the tests break up along the sutures, especially along the ambulacral and interambulacral areas.

Among crustaceans, especially those having durable carapaces, only the hardest and most compact parts commonly remain under conditions of turbulence. For example, in lobsters, the rostrum with parts of the cephalon and eye sockets, the mandibles, the clawed extremities of the dactylus, the single clawed teeth of the chelae, the ventral, thorned segments of the abdomen, and the basipodites of both uropods are commonly preserved (SCHÄFER, 1962; 1972).

SELECTIVE DISSOLUTION OF HARD PARTS PRIOR TO BURIAL

Skeletal material is commonly destroyed prior to burial by selective dissolution. In sea water, this happens under conditions of supersaturation with respect to calcium carbonate. The normal compensation depth, at which calcareous skeletal material enters solution, lies at about 4,000 meters. At this

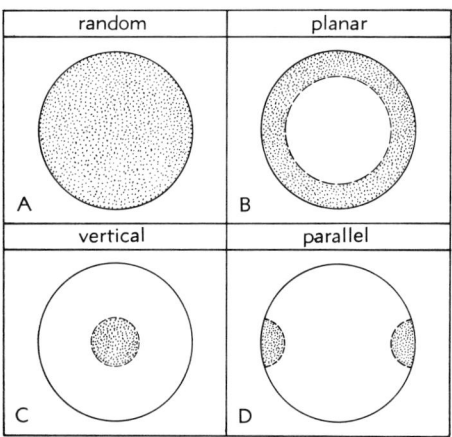

FIG. 17. Diagrammatic representation of orientation patterns (*A-D*) as they appear in stereographic projections (Toots, 1965a).

depth, increased hydrostatic pressure and low temperatures result in a high concentration of dissolved carbon dioxide. The corrosion of calcium carbonates in water where $CaCO_3$ is precipitated fluctuates with pH (7.7 to 8.3) (CLOUD, 1962), and this is thought to be partially related to the liberation of carbon dioxide by plants during the night. The solubility of calcium carbonate increases with salinity and decreases with temperature. The compensation depth in the Antarctic Ocean is only 500 meters (KENNETT, 1966). Calcareous skeletons of microfossils may be partially or completely dissolved in living animals (JARKE, 1961). This process is known as subsolution (HOLLMANN, 1962, 1964) and can result in all stages of preservation from "ghosts" to complete destruction (RICHTER, 1931). No compensation depth is known for siliceous skeletons.

RANDOM EMBEDDING

Random embedding occurs when objects display no preferred orientation (random orientation). If their distribution is plotted as a stereographic projection, no particular orientation is evident (Fig. 17,*A*).

Random embedding has received very little notice in the literature. Generally, it was regarded as a matter of course and was thought to have occurred under conditions of quiet sedimentation. TOOTS (1965b) first proved this assumption to be

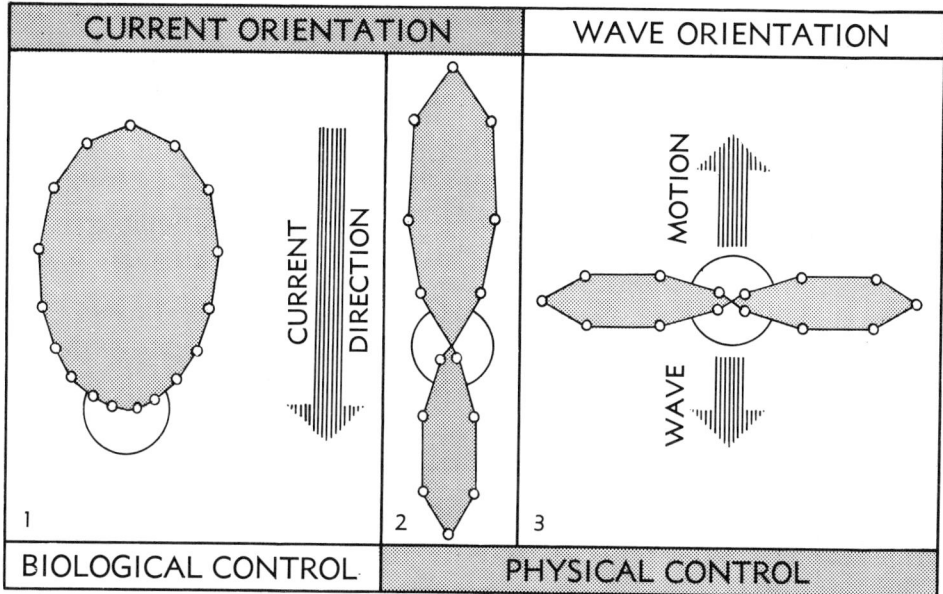

Fig. 18. Orientation of organisms within bedding plane. Arrow designates current direction for *1* and *2*.——*1*. Biologically controlled current orientation: suspension feeders are oriented with only a broad peak into direction of current.——*2*. Physically controlled current orientation: elongate particles arranged so that a sharp major peak points into the current and a smaller counterpeak in the opposite direction.——*3*. Transverse wave orientation: elongate particles aligned with two equal peaks perpendicular to wave movement and occasional smaller peaks parallel to wave movement (after Seilacher, 1973).

incorrect and demonstrated that the process requires special interpretation. Random orientation may result under the following conditions:

1) If organic remains are caught in steep-walled pockets, which act as traps, this results in primary random embedding. But the evidence for this is not unambiguous because gravity plays a part (MÜLLER, 1951b; RICHTER, 1942; TOOTS, 1965a,b).

2) Comparatively rarely, contemporary deformation may be due to a plastic flow of the entire sedimentary mass.

3) Bioturbation. The degree of reworking is dependent upon the size, type, and number of infaunal elements and the time spent in reworking the sediment (MÜLLER, 1951b; QUENSTEDT, 1927; RICHTER, 1936; REINECK, 1967).

ORIENTED EMBEDDING: MOVEMENT AROUND HORIZONTAL (EINKIPPUNG) OR VERTICAL AXES (EINSTEUERUNG)

The types of movement of different media (water, ice, air, mud, or flowing sand) determine the movements of enclosed objects. Such movements can lead to oriented embedding. For example, oblong objects can be oriented either parallel or perpendicular to currents (Fig. 18). The orientation of such objects depends only upon conditions of equilibrium and friction, not upon the original life positions of the organisms. The preservation of fossils and the deposition of them in a fixed position enables conclusions to be drawn about the direction of their movement and final deposition. From this information, it is then possible to determine the direction of the depositing water or wind currents. Such interpretations can be reinforced by the study of independent evidence. This evidence can be sedimentologic (sedimentary structures or the alignment of coarse-grained sedimentary particles). It is also possible to observe the reactions of animals to currents, such as the rheotactic alignments of lebensspuren (SEILACHER, 1953, fig. 5) and the growth orientations of epizoans (SEILACHER, 1960b; MÜLLER, 1963) and certain types of borings (SEILACHER, 1968b; 1969) (Fig. 18).

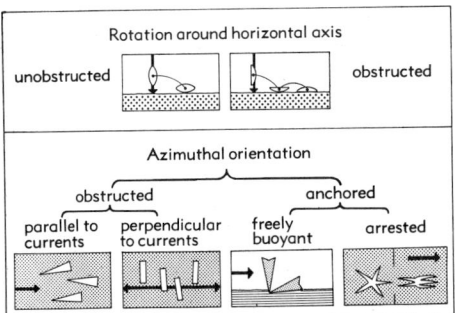

Fig. 19. Types of oriented embedding of skeletal material (after Geyer, 1973). See also Table 4.

On the other hand, it is possible to observe the effect of currents on objects that have sunk into the sediment or are only stuck in the sediment. In either case, objects will acquire the most stable position possible in relation to the forces acting upon them.

As a rule, intraformational folding results in rotation of incorporated fossil remains around a horizontal axis; however, in such cases differential movements often take place. The resulting movements can be very complex and rotation may occur around the vertical as well as inclined axes. Rotation can develop through differential, local carbonate dissolution within the sediment causing settling of the sediment. These and similar processes that result in oriented positions are called oriented embedding (*Einregelung*) (MÜLLER, 1951b; RICHTER, 1942; SHROCK, 1948; TOOTS, 1965a). Current mechanics have been applied to the study of oriented embedding of skeletal hard parts by investigators such as TRUSHEIM (1931), JOHNSON (1957), BRENCHLEY and NEWALL (1970), ABBOTT (1974), and FUTTERER (1974, 1977).

Following RICHTER (1922; 1937; 1942), two major categories of oriented embedding are recognized: *Einkippung* (rotation around a horizontal axis) and *Einsteuerung* (rotation around a vertical axis or azimuthal orientation). Within these categories, a number of subcategories have been defined (Fig. 19, Table 4). Stereographic diagrams are used to record the positions of specific assemblages (Fig. 17,*B-D*). This facilitates the recognition of combinations and transitions of the basic types of embedding described above.

Oriented embedding of bowl-shaped objects. Bowl-shaped objects are common as fossils. Most of them are the shells of brachiopods, bivalves, gastropods, and also arthropods, especially trilobites and crustaceans. Among brachiopods and bivalves, only single valves are, as a rule, truly bowl shaped, but in certain concavo-convex or plano-convex forms (e.g., many strophominids), the entire shell is bowl shaped. We shall first consider the case of *Einkippung*.

If shells sink freely through quiet water, they will be oriented convex side downward (CLIFTON, 1971). A similar orientation will result if shells settle freely through the air, but the aquatic medium is more important geologically. Under conditions of free sinking, the following conditions of free *Einkippung* have been observed:

1) Skeletal parts of dead animals, attached to seaweed or free living, which were pseudoplanktonically transported;

2) Bowl-shaped shells that, once settled, were stirred up by currents and then sank again;

3) Bowl-shaped objects washed up against resistant objects will eventually drift across the obstruction and settle in quiet water on the lee side. In German, this is called *Stillwasserfallen* (still water traps). Two types of orientation can result from *Stillwasserfallen*. Whether the resulting position of the shells is convex side down or arranged in random orientation depends upon the inclination and roughness of the sediment surface, and whether the object is moved by rolling, shoving, or sinking.

Convex-side-down orientations can normally be maintained only in quiet water, in the absence of sufficiently strong, lateral currents. Only shells with a low center of gravity may remain in this position also in the presence of stronger currents. An example from the fossil record is the carapace of *Triarthrus eatoni* described by BEECHER (1894) from the Ordovician of New York, which was found convex downward in an orientation similar to the life position of certain brachiopods and bivalves.

Bowl-shaped shells, whose morphology approaches that of a calot, can be tilted by currents if they are lying convex downward on the sediment. If such tilting occurs

TABLE 4. *Different Types of Shell Orientation (prepared from Richter, 1942).*

A. Orientation with respect to the horizontal axis (*Einkippung*) This type of orientation results when objects are rotated around a more or less straight axis and become inclined. Characteristic of different processes and can be divided into the following two groups:	
1. "Unobstructed" Inclination	2. "Obstructed" Inclination
Results when bodies are to move freely in a medium solely under the influences of hydrodynamic resistance, gravity, and buoyancy. They then can sink unobstructed to the bottom. Example: Bivalve valves sinking to the bottom of a body of water.	Results when shells are under the influence of sufficiently strong lateral currents acting upon the surface of another medium with a higher density, such as water moving over sediment. Example: Plasters of bivalve valves oriented in the convex-up position.

B. Azimuthal Orientation (*Einsteuerung*)			
This type of orientation occurs when shells are rotated around their vertical axis, and on flat or fairly flat surfaces, this can result in a right-left orientation. Azimuthal orientation can be divided into two basic groups:			
1. "Unanchored" Azimuthal Orientation		2. "Anchored" Orientation (Pivotal Orientation)	
Resistance to motion is influenced only by the natural hydrodynamic drag of the bodies, and that determines the orientation.		Results with anchored or attached objects whose point of anchoring is within or outside the body.	
a) freely buoyant	b) hindered	a) freely buoyant	b) hindered
Freely moving within a medium.	Resting on the ground or the sediment, either in one place or sliding along.	Organisms growing together or hanging from one another.	Resting on the ground or the sediment, either in one place or sliding along. Sliding of the object results in the orientation of the shells which acts the same as anchoring.
	Dragging		
	Results when an object is not completely lying on the bottom. A condition results that lies in between buoyancy and sliding, and is therefore between hindered and unhindered.		

when the shells are in contact with a hard substrate, it is said to be obstructed (*gehemmt*). This can happen on land caused by wind, but occurs most frequently in turbulent water. In water-laid deposits, the convex-side-up orientation of shells is, therefore, one of the most important criteria for turbulence and should be considered the rule. RICHTER (1942) called it *Einkippungsregel* (Fig. 20). Often the orientation of shells can be used to clarify difficult stratigraphic and tectonic problems, especially for the recognition of overturned position of strata. REINECK et al. (1967) have demonstrated with the help of box cores that convex-side-up orientation is also predominant in deep parts of the North Sea. The same is true for empty bivalve valves on the continental shelf (EMERY, 1968).

As a rule, weakly curved, bowl-shaped objects are stable convex side up, but stongly curved, almost high-spired, bowl-shaped objects are most stable convex side down. Examples can be found in the fossil record. Callovian specimens of *Gryphaea dilatata* found redeposited in recent strandline deposits near Houlgate are generally oriented convex upward, according to the *Einkippungsregel*. Convex-side-down orientation, as observed for bivalved shells and also those of *Exogyra columba* in the Cenomanian of Saxony, have retained their life positions (PFANNENSTIEL, 1930; HÄNTZSCHEL, 1924).

Azimuthal orientation of bowl-shaped bodies is most often found under water, but it can also be observed subaerially and perhaps even in more viscose substances such as soft mud. The orientation depends upon the degree of streamlining of the body. The shape of the object determines the hydrodynamic drag and the direction in which it will be oriented. Azimuthal orien-

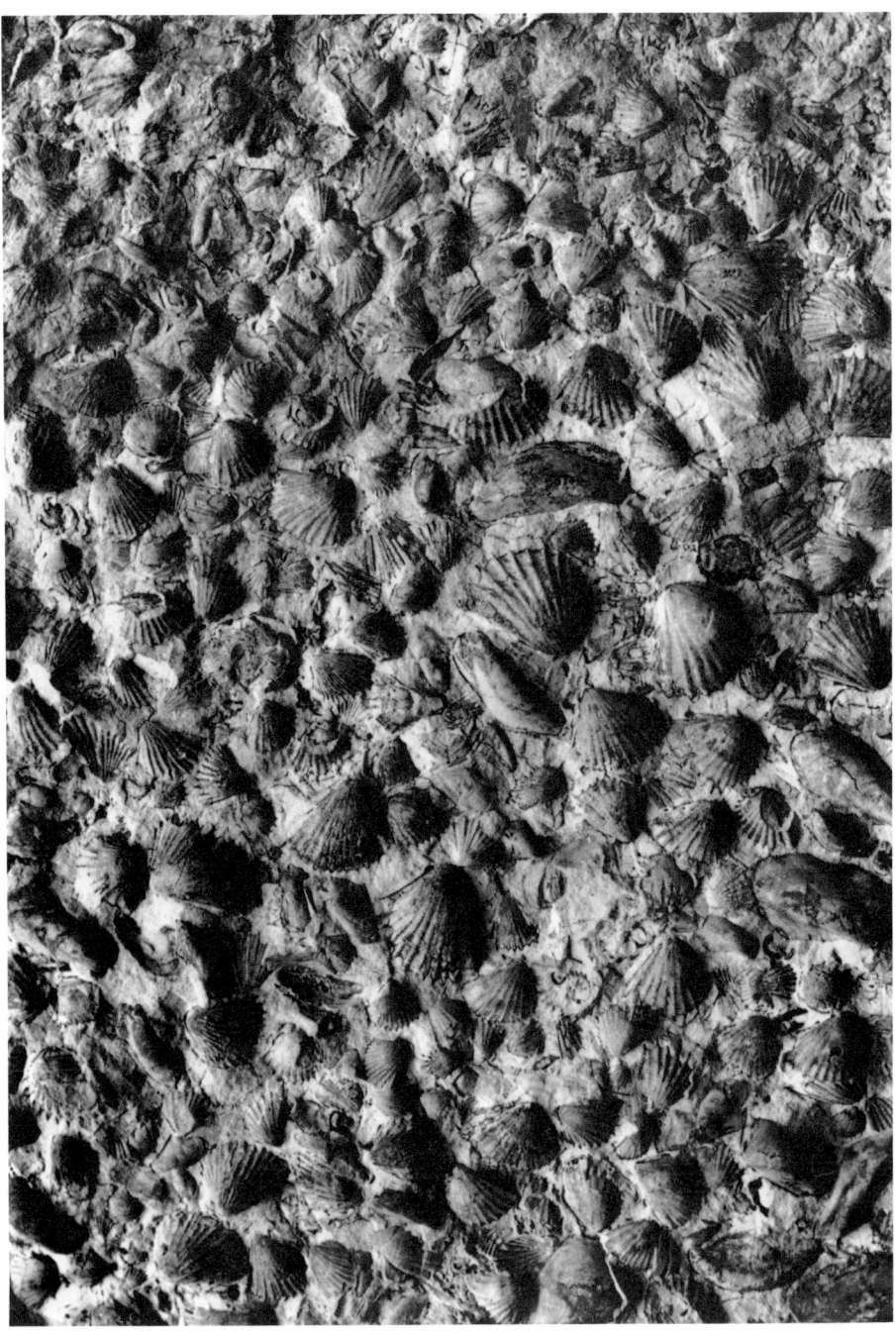

Fig. 20. Plaster of bivalve shells composed essentially of single valves of *Costatoria goldfussi* (von Alberti) from the lower Keuper (Grenzdolomit) in Thuringia, approx. ×1.1 (Sektion Geowissenschaften, Bergakademie Freiberg, 249/2; Müller, n).

tation is best developed when shells have a long and a short axis and are aligned along that axis. Teardrop-shaped objects are best suited to attain azimuthal orientation.

Unobstructed azimuthal orientation has not yet been observed geologically. Possibly it could occur in terrestrial muds and flowing sands or in submarine slides. Obstructed azimuthal orientation occurs when bodies lie upon hard substrates where their transportation is slowed down by friction. In such circumstances, the thickest part of the shell is oriented against the current, although occasionally, and completely unexpectedly, the opposite orientation can occur. TRUSHEIM (1931) performed flume experiments with shells and observed that their orientation may be influenced by minute differences in shape that are often difficult to observe and may even be destroyed during fossilization. Such differences in orientation were observed for single valves of *Cardium echinatum* and *C. edule*, and of *Mactra corallina* and *M. solida*. FUTTERER (1974) studied azimuthal orientation of single valves of recent *Cardium edule* and *C. echinatum* in a flume. These experiments were concerned with the orientation of the shells in relation to changing centers of gravity and it was discovered that a shift of the center of gravity by a few millimeters can lead to very different, opposing orientations of the shells. Apparently, minute differences in shell shapes are important, and it is possible to distinguish different species by the manner in which they are oriented. A considerable amount of caution must therefore be taken when applying results obtained on recent material to the study of fossil conditions. Generally, other criteria than shape should be used. Different modes of orientation of bowl-shaped shells are illustrated in Figure 21, in which rose diagrams and orientation quotients for different biotic constituents of shell plasters in the upper Buntsandstein are compared.

Right and left valves of more or less equivalve bivalves commonly show azimuthal orientations in opposite directions (see also Fig. 22). Even if both valves of a bivalve shell are held together by the ligament, azimuthal orientation can still result. However, such occurrences can only happen if both valves can move with respect

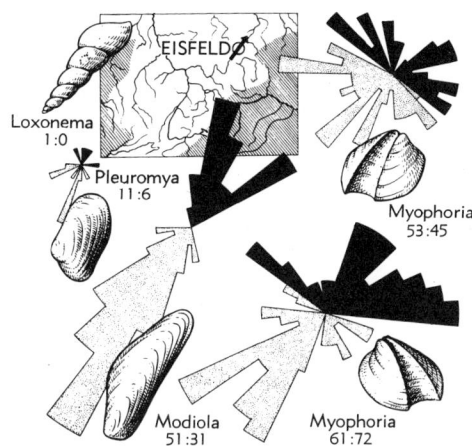

FIG. 21. Rose diagrams and ratios of single valve azimuthal orientations in shell plasters, Lower Triassic (upper Buntsandstein, Röt), Eisfeld, Thuringia, approx. ×0.5 (Seilacher, 1960a).

to one another and if they lie convex side up on the sediment.

1) The shells of *Mytilus, Modiola, Arca, Venerupis, Petricola, Ensis, Phaxas,* and *Donax* are oblong, and their hinge parts are straight. Therefore, these shells behave like sled runners with the shell edges lying flat on the substrate and the umbo pointing toward the current.

2) The shells of *Cardium, Scrobicularia,* and *Mactra* have rounded outlines and even when gaping widely, rest on only one of the valves, while the other valve points upward like the open lid of a can. The umbonal region, being heaviest, always points toward the currents.

It has been observed in nature and with controlled experiments that the thickest end of the shell is positioned against the current. For single bivalve valves, this means that the umbonal region points toward the current. Therefore, this kind of orientation can be used to determine current directions. The situation is different for shells that are embedded convex side up. Bowl-shaped shells with projections or spines that extend over the base of the shell or its margin may be anchored if the substrate is suitable. The anchor always points toward the current. Thus, anchored shells can also be used to determine current directions.

The different conditions of oriented embedding for complete shells and single

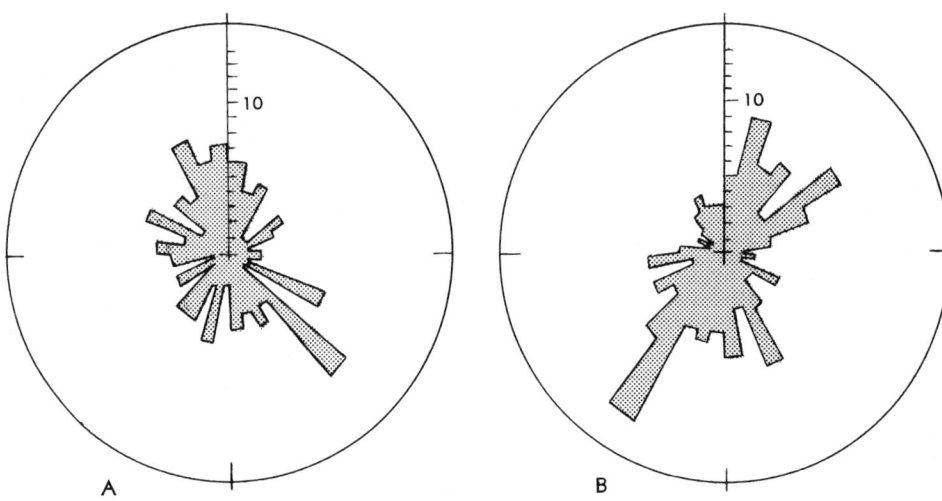

FIG. 22. Rose diagrams for 302 azimuthally oriented, convex-up, embedded single valves of *Costatoria goldfussi* (von ALBERTI), from the lower Keuper (Grenzdolomit, Ku2) at ?Reisdorf, Thuringia (Müller, n).——*A*. Relationships of 149 left valves (=47.6%).——*B*. Relationships of 153 right valves (=52.4%). Compare with Figure 20.

valves of *Schizothaerus nuttali* and *Protothaca staminea* that were exposed to fluctuating current directions and speeds in muddy sands are shown in Figures 23 and 24. The orientation of concavo-convex particles deposited from experimental turbidity currents is shown by MIDDLETON (1967).

Oriented embedding of cone-shaped bodies. Cone-shaped shells of fossil and living organisms are commonly preserved in sediments. Examples are found among oblong foraminifers, tentaculites, volborthellids, belemnites, styliolinids, conularids, high-spired gastropods, and orthoconic ectocochlian cephalopods. Examples are also found among vertebrate remains. Tilting (*Einkippung*) will be considered first. This is called "unobstructed" if shells are transported without contact with the ground or other substrates such as air, water, and very soft, muddy, or sandy sediment, under the influence of only the shape of the shell, gravity, and buoyancy. In general, we are then concerned with orientation perpendicular to the smallest cross section of the object. If conical objects sink freely, they commonly become oriented with the heavy, basal portion of the shell pointing downward; however, if the surrounding medium is viscous and the height of fall too small, this type of orientation will not result.

Examples of unobstructed oriented embedding of conical fossils can only be observed in sediments of sufficient density and strength to preserve the evidence. Such sediments must be so thoroughly saturated with water that shells can sink in them with their long axis vertical and short axis horizontal. If the sediment becomes more cohesive, this position is retained. In some cases, the orientation of the fossil is in the life position. Probably, the shells of cerithiid gastropods, preserved with the apex pointing downward in unbedded, fine-grained sandstones of the Sarmatian of Wiesen (Burgenland), can be considered as an example (KREJCI-GRAF, 1932). It is assumed that these high-spired gastropods were aligned by currents in flowing sands, and were kept in this position when the sand consolidated.

Obstructed, oriented embedding of conical objects results when horizontally acting forces interact with the objects that are in contact with the substrate, when the cone height is less than the diameter of its base.

Examples can be found in the patellids,

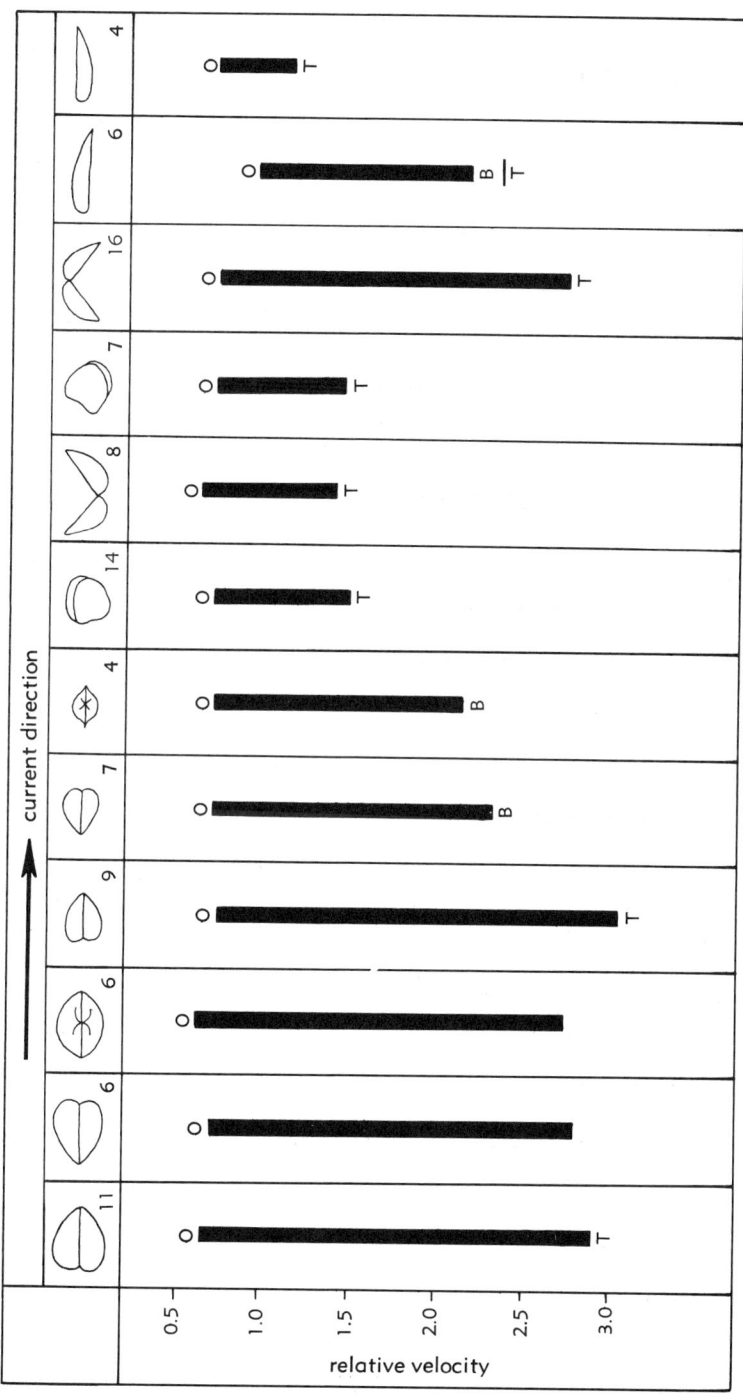

FIG. 23. Reaction of complete shells and single valves of recent *Schizothaerus nuttali* and *Protothaca staminea* exposed to differing conditions of current directions and velocities in muddy sediments (modified from Johnson, 1957). [Explanation: *O*, sand begins to move near the shells; *B*, sand covers the shells; *T*, the shells begin to move. Numbers near the margins of the shells indicate the number of times the experiment was carried out for each shell.]

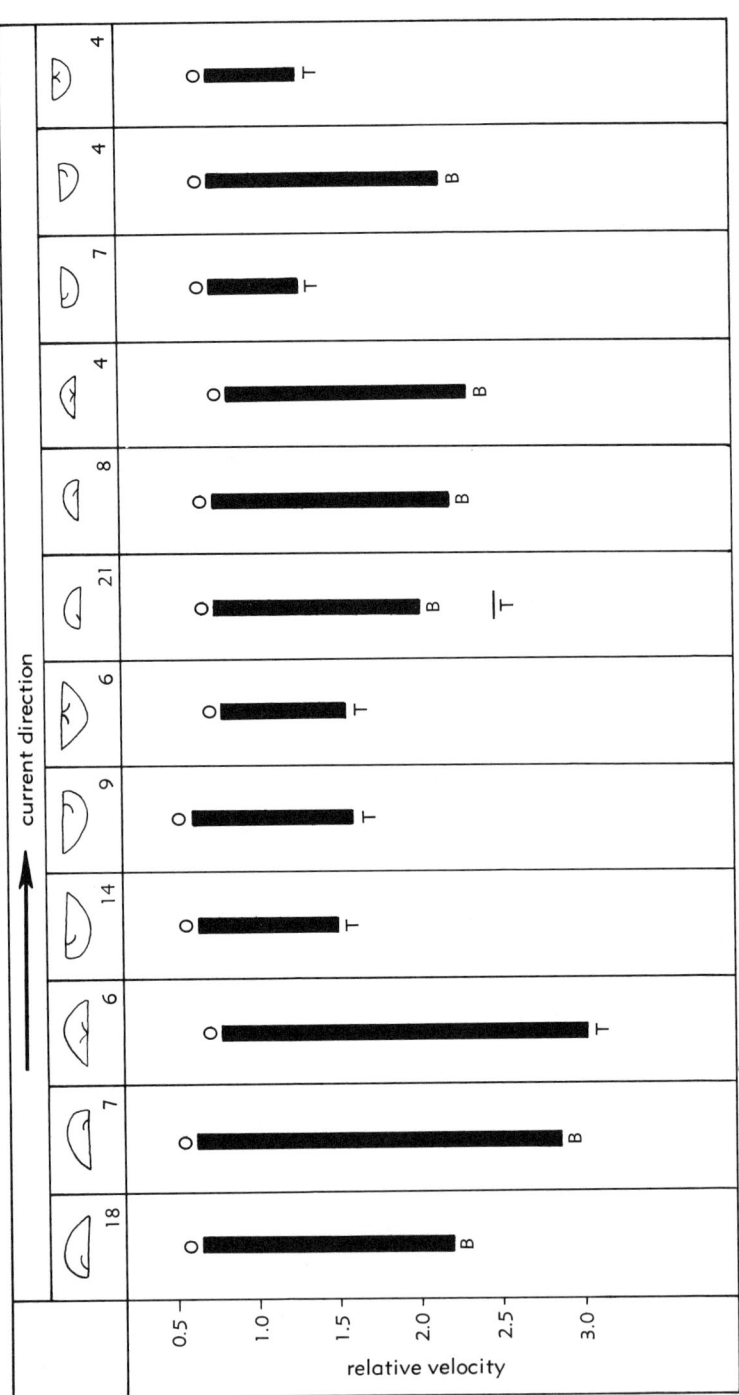

FIG. 24. Reaction of complete shells and single valves of recent *Schizothaerus nuttali* and *Protothaca staminea* exposed to differing conditions of current directions and velocities in muddy sediments (modified from Johnson, 1957). [Explanation: *O*, sand begins to move near the shells; *B*, sand covers the shells; *T*, the shells begin to move. Numbers near the margins of the shells indicate the number of times the experiment was carried out for each shell.]

discinids, and trochids. Shells of *Trochus* can become abraded to such an extent that only the apices remain. Frequently all that remains of gastropod shells with large apertures such as *Murex, Buccinum, Purpura,* and *Aporrhais* are small cap-shaped or "ear"-shaped structures.

The orientation by currents of high-spired, conical objects in which shell height is greater than base diameter is such that their long axes are parallel to the direction of the currents (*Längs-Einsteuerung*). Rose diagrams of such orientations commonly show the presence of two statistical maxima. As a rule, one of the maxima will be much larger and oppose the other. This suggests that even under conditions of unidirectional flow, transported objects are dynamically active and constantly change their orientation from spire upstream to spire downstream and vice versa (Fig. 25). Transverse wave orientation is characterized by two equal peaks, sometimes with an additional smaller peak in the direction of wave progression (SEILACHER, 1960a) (Fig. 18,*1*).

KRINSLEY (1960) studied the azimuthal orientation of 106 orthoconic nautiloids on a bedding plane of the Middle Silurian Waukeshaw Limestone near Lemont, Illinois. The orientations of the nautiloids were plotted in 20 degree classes as shown by rose diagram in Figure 26. Two directions of orientations are obvious, a dominant west-northwest direction and a secondary one at right angles to the first.

Possibly the shells were transported by strong currents and the apices became embedded in the sediment, thus being protected from mechanical erosion and subaqueous solution. Strong currents must have played a large role as shown by the overturning of coral colonies. KRINSLEY concluded, therefore, that predominant winds and currents from west-northwest were responsible for this orientation.

RUEDEMANN (1897) reported parallel, azimuthal orientation of orthoconic nautiloids and monograptids in the Silurian Utica Shale of New York. Drag marks could be seen behind the fossils. The apical ends of the nautiloids and many of the siculae of the graptolites pointed in the direction against the current as was also indicated

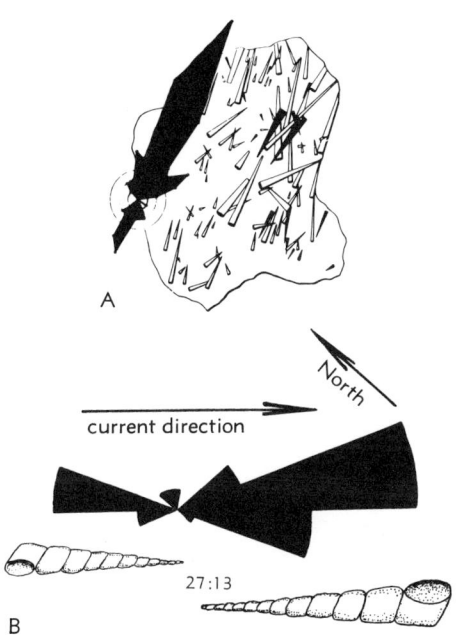

FIG. 25. Orientation of cone-shaped bodies.——*A*. Azimuthal orientation of orthoconic nautiloids from Upper Devonian (Kellwasserkalk), Bicken, Rhineland; approx. ×0.4 (Seilacher, 1960a).——*B*. Rose diagrams for the orientation of 40 gastropod shells (?*Turritella* sp.) in cross-bedded sandstone from Upper Cretaceous (Mesaverde Formation), Carbon County, Wyoming (Toots, 1965a).

by drag marks in the prolongation of the nautiloid shells.

Similar parallel orientation of orthoconic nautiloids is known from the Silurian Budany Limestone of Czechoslovakia. The celphalopods of the Devonian Hlubočepy Limestone, on the other hand, show two dominant directions of orientation at right angles to each other. In the latter case, the orientation was thought to have been caused by cross waves and low velocity currents (PETRÁNEK & KOMÁRKOVÁ, 1953). KAY (1945) observed two directions of orientation for Ordovician orthoconic cephalopods on St. Joseph Island, Ontario, and Cumberland Head, Vermont. The smaller shells were generally aligned at right angles to the larger ones. KAY suggested that the larger shells were oriented perpendicular to the

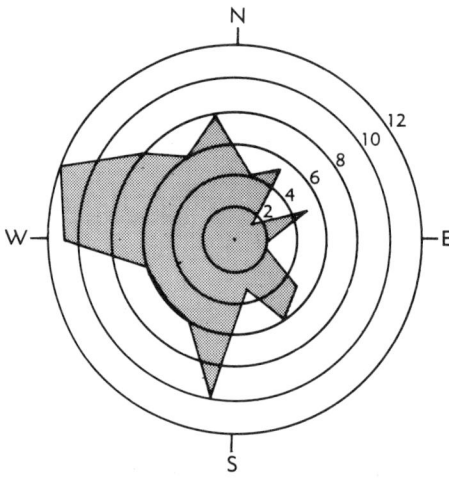

Fig. 26. Rose diagram showing the azimuthal orientation of 106 orthoconic nautiloids along a bedding plane; specimens from Middle Silurian, Lemont, Illinois. The apical ends point in the indicated compass direction (Krinsley, 1960).

(1957, pl. 2, fig. 3) has figured a remarkable example of tentaculites from the Upper Devonian of the USSR that display the same orientation. Possibly the latter orientation was the result of the mechanical interaction of shells as they were deposited along the shoreline.

An example of wind-oriented gastropod shells as indicators to determine paleowind directions has been documented by ERICKSON (1971).

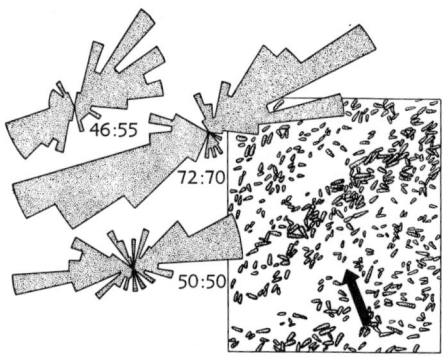

Fig. 27. One of the few known cases of conical bodies (*Haplostiche* sp.) oriented with their long axes perpendicular to the current direction (indicated by arrow). Preserved on the lower surface of a ripple marked bed, from Lower Cretaceous, Texas, ×0.7 (Seilacher, 1960a).

shoreline and the smaller ones parallel to it. Other examples of orientation of conical shells have been discussed by KINDLE (1938), RUTSCH (1937), TRUSHEIM (1931), REYMENT (1968), DIXON (1970), K. BRENNER (1976), and FUTTERER (1976, 1977).

Only few examples of embedding and orientation of high-spired, conical bodies perpendicular to the current direction have been described in the literature. SEILACHER (1960a) studied the distribution of the foraminifer, *Haplostiche*, on ripple-marked bedding planes. The long axes of the foraminifers were observed to lie parallel to the ripples as shown in three separate rose diagrams in Figure 27. The orientations of the shells displayed two large and opposing maxima, which were interpreted to represent the most stable positions. SEILACHER considered the deposit to be an example of wave orientation (*Quer-Einsteuerung*). Another example would be the well-known mass occurrences of belemnites (*Schlachtfelder*) where the rostra are oriented perpendicular to the current (QUENSTEDT, 1927; SEILACHER, 1960a). GEKKER

Oriented embedding of barrel-shaped bodies. Elongate, barrel-shaped bodies in contact with a substrate will roll under the influence of currents and become oriented with their long axes perpendicular to the direction of the current. Of course this can occur only under ideal conditions where both the substrate surface and the barrel-shaped bodies are smooth. If the substrate is uneven, the rolling bodies are obstructed or anchored to the substrate. This is the most common condition, as perfectly smooth substrates rarely exist in nature. Slight differences in relief of the substrate may obstruct or impede the movement of bodies, which may become anchored and then pivot around the anchoring point, thus becoming oriented more or less parallel to the transporting current.

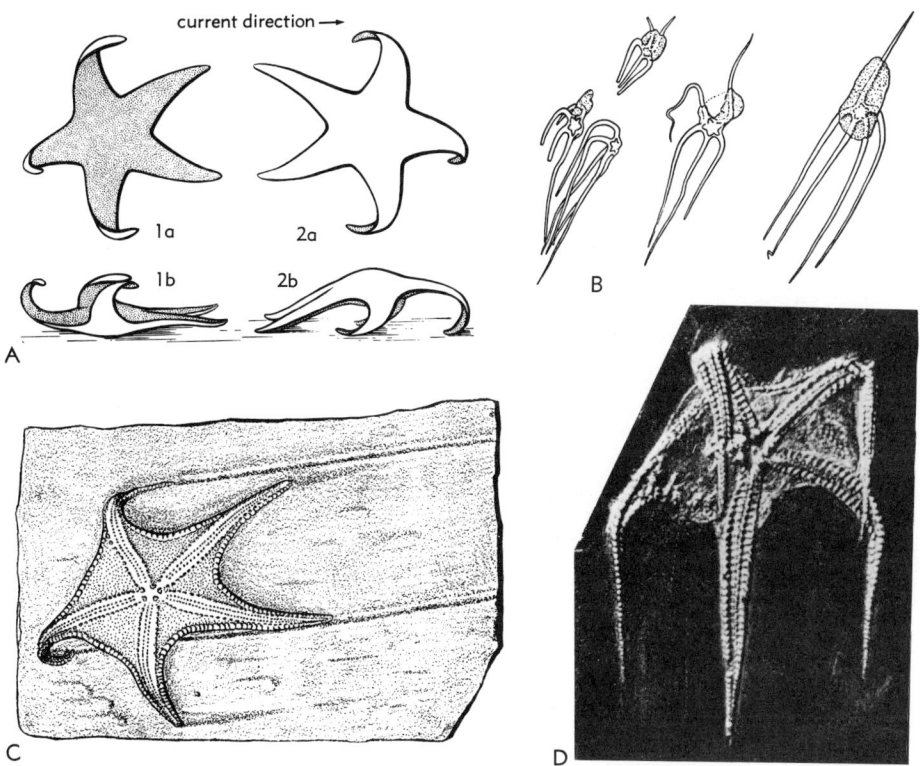

Fig. 28.——*A*. Two possible azimuthal orientations for dead starfish (Seilacher, 1960a).——*B*. Azimuthal orientation of tests of *Furcaster* transported by currents, after partial decomposition, Lower Devonian (Bundenbacher Schiefer), Bundenbach, Rhineland, approx. ×0.23 (Seilacher, 1960a).——*C. Euzonosoma tischbeiniana* (ROEMER) in "umbrella" position. The straight grooves are tool marks produced by the drooping arms. Lower Devonian (Bundenbacher Schiefer), Bundenbach, Rhineland, ×0.36 (Seilacher, 1960a).——*D. Euzonosoma tischbeiniana* (ROEMER) displaying azimuthal orientation and the beginning of rotation around a horizontal axis. Lower Devonian (Bundenbacher Schiefer), Bundenbach, Rhineland, approx. ×0.54 (von Königswald, 1930).

The orientation of these objects in water is always parallel to the shoreline. Most common examples of this type are seen in plant material (branches, pieces of tree trunks, chaff); however, random orientation can also occur.

The orientation of rod-shaped coprolites has been interpreted to demonstrate the presence or absence of waves during deposition of the Upper Pennsylvanian Rock Lake Shale in northeastern Kansas (HAKES, 1976). Wave-oriented coprolites were found to be aligned essentially parallel to ripple crests, whereas coprolites deposited in areas of little wave activity were found to be randomly oriented within bedding planes.

Oriented embedding of bodies with long, flexible, or projecting parts. In this category belong echinoderms (asterozoans, crinoids), vertebrates, and plants. Good examples of *Einkippung* are found among fossil asterozoans. Commonly, obstructed embedding of these objects occur if some of the arms are bent back over the central disc. VON KÖNIGSWALD (1930) described two transitional types of embedding of asterozoans from the Lower Devonian Hunsrückschiefer of the Rhenish Schiefergebirge:

1) Initial inclined position: The body does not move, but some of the arms are bent back across it (Fig. 28,*A,C,D*).

2) Final position: In addition to the

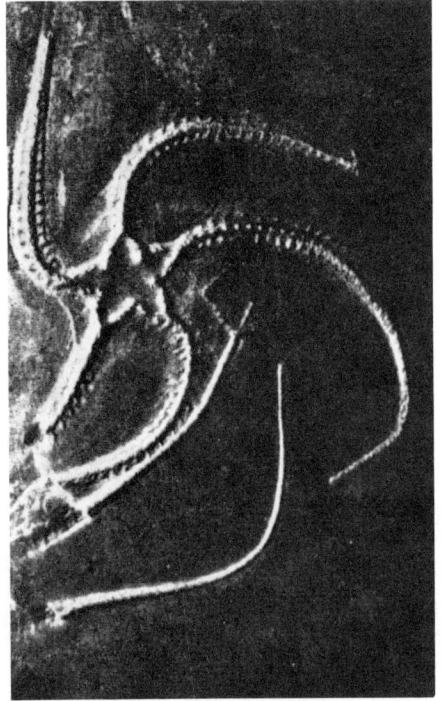

FIG. 29. *Furcaster palaeozoicus* STUERTZ in a "twirl" position. Lower Devonian (Bundenbacher Schiefer), Bundenbach, Rhenish Schiefergebirge, approx. ×1 (von Königswald, 1930).

arms the entire body is moved in the same direction but only for a short distance.

Another type of orientation occurs when the body of a starfish has been carried along by currents, and its arms are dragged over the sediment (SEILACHER, 1960a). Figure 28,*C*, shows drag marks produced by the arms of the animal as straight, parallel furrows. In the specimens of *Furcaster* shown in Figure 28,*B*, the central discs have partly disintegrated and have been moved by the current.

The bodies of starfish are flipped over and tumbled by currents, which accounts for the bending of the arms and their occasional orientation in a direction against the current. Previously this orientation has often been explained as being due to counter (tidal?) currents.

Tentaculites and orthoconic cephalopods do not usually show oriented embedding in the Hunsrückschiefer, which is probably explained by the fact that currents responsible for the transportation and oriented embedding of starfish may be quite weak. As is well known, skeletal elements of echinoderms are composed of a meshwork of calcite crystals having considerable porosity. Therefore, these skeletons have an extremely low specific gravity and can be easily transported by very weak currents that are not sufficiently strong to move other larger and heavier objects. The transportation of echinoderm particles is also facilitated by the entrapment of decomposition gases within their tiny cavities. After embedding, the cavities of the skeletal meshwork are filled in by secondary calcite and each skeletal element becomes one calcite crystal with its own cleavage plane.

Embedding and anchoring of objects are dependent upon the shape of the surface (*Standfläche*) on which they come to rest. This surface is usually small in the case of long, flexible, and projecting body parts such as extremities, tentacles, tails, or necks, which are generally oriented parallel to the current. Commonly they nestle against well-anchored parts, which form a kind of facets ("*Anspülen einer Facette*") (WEIGELT, 1927, p. 119).

Asterozoans commonly display anchored or inhibited azimuthal orientation, if the central disc is firmly anchored in the substrate and only the arms are oriented by currents (Fig. 28,*A,D*). However, if arms are bent over the body of the starfish, this is a form of tilted orientation called "umbrella position" (Fig. 28,*D*). If all arms of starfish are spirally bent in the same direction by currents, this is referred to as *Quirllage* ("twirl position") (Fig. 29). This occurs mainly in forms with long, moveable arms, but is also present in starfish with less well-developed arms. Similarly, the spiral arrangement of distarticulated skeletal elements of *Palaeoniscus* specimens in the Kupferschiefer were apparently embedded by small eddies in rather shallow water.

The orientation of sessile, stalked crinoids can be used to determine current direction. This is especially true if the animals are anchored to the sediment at the time of burial, and if root, stem, and crown are not disarticulated. The animals then become oriented in the direction of the

Fig. 30. *Encrinus carnalli* Beyrich displaying more or less azimuthal orientation by currents. Middle Triassic (lower Muschelkalk, Schaumkalk Zone), Freiberg a. d. Unstrut, East Germany, approx. ×0.38 (Müller & Zimmermann, 1962).

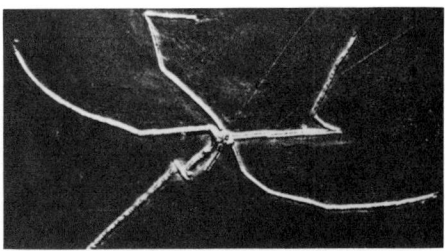

Fig. 31. *Agriocrinus frechi* JAEKEL, stem bent immediately below calyx. Arms diverging 180° caused by reversal of current direction. Lower Devonian (Bundenbacher Schiefer), Bundenbach, Rhenish Schiefergebirge, approx. ×0.4 (von Königswald, 1930).

current. In fairly strong currents, they are quickly disarticulated; however, if the currents are relatively weak and unidirectional, a progressive alternation of orientation can be developed. Initially, the crinoids are deposited in a tangled, disordered mass, but as current intensities increase, a parallel arrangement will eventually develop (Fig. 30). In the latter case, the stalks are commonly embedded in proximity, parallel to each other on a single bedding plane. The appearance of tangled masses of crinoids in the Liassic *Posidonia* Shale of southern Germany indicates low energy conditions, and the more parallel arrangement of them indicates occasional, stronger currents. If the stalk of crinoids is separated from the bottom, it drags along and acts as a drift anchor.

Instructive examples of the orientation of body parts in different directions due to changing current conditions are known in the Hunsrück Shale (VON KÖNIGSWALD, 1930). For example, the crinoid, *Triacrinus*, has long, rigid arms that were aligned by steady currents in close packing. If the current is reversed 180 degrees, the arms are pushed widely apart (Fig. 31). Because of this, current pressure is increased so that the stalk breaks just below the calyx and the same may happen to single arms, because they are relatively rigid.

Changes in current direction are excellently displayed when several layers of fossils are deposited on top of one another on a single bedding plane. The differences in orientation of the fossils can then be as much as 180 degrees. Commonly, in the first layer the fossils are oriented by the dominant current and their orientation is only slightly changed by later changes in current direction (Fig. 32,*A,B*). In graptolite shales, two or, rarely, more layers of oriented graptolites are present, in which the graptolites cross each other at uniform angles.

Rapid changes in current direction are considered to be tidal in origin.

Paleowind directions may be determined by measuring the azimuthal orientation of broken and fallen tree trunks, if they are oriented parallel to each other. This has been found in occurrences within, or stratigraphically above, central European brown coal deposits.

Azimuthal orientation *(Einsteuerung)* of planispiral ammonoids has been studied by BRENNER (1976), and that of *Nautilus* and *Spisula* by FUTTERER (1976, 1977).

Marks in the sediment produced by the remains of organisms. In German, the term "*Marken*" implies inorganic structures produced on substrates by mechanical means. Their development differs from that of lebensspuren (trace fossils) because the forces involved are entirely inorganic. Dead animals or solitary parts of animals can be moved by currents and, when they come in contact with the sediment, produce drag, roll, or prod marks. Their importance is obvious, because, along with other sedimentologic and biostratinomic indicators, they furnish information on nature, direction, and strength of ancient currents. The following are examples.

1) Drag marks (*Schleifmarken*) with chevron-like rills of the "chloephycus" type (Fig. 33) (see HÄNTZSCHEL, 1975, p. *W*171). If objects are dragged across fairly viscose

(See facing page.)

FIG. 32. Examples of orientation by currents.——*A*. Two successive deposits of *Furcaster palaeozoicus* STUERTZ, oriented by two different current directions. Lower Devonian (Bundenbacher Schiefer), Bundenbach, Rhenish Schiefergebirge, approx. ×0.3 (von Königswald, 1930).——*B*. Arms of *Furcaster palaeozoicus* STUERTZ oriented by currents, from Lower Devonian (Bundenbacher Schiefer), Bundenbach, Rhenish Schiefergebirge, approx. ×1.7 (Müller, 1963).

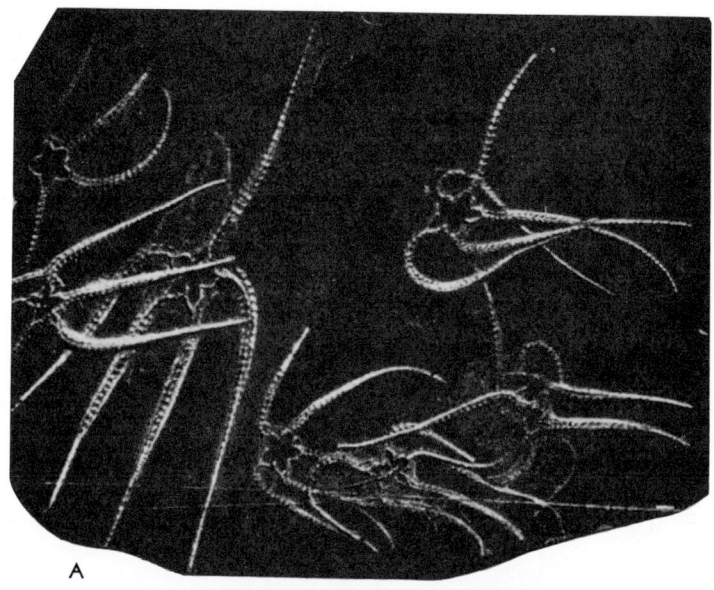

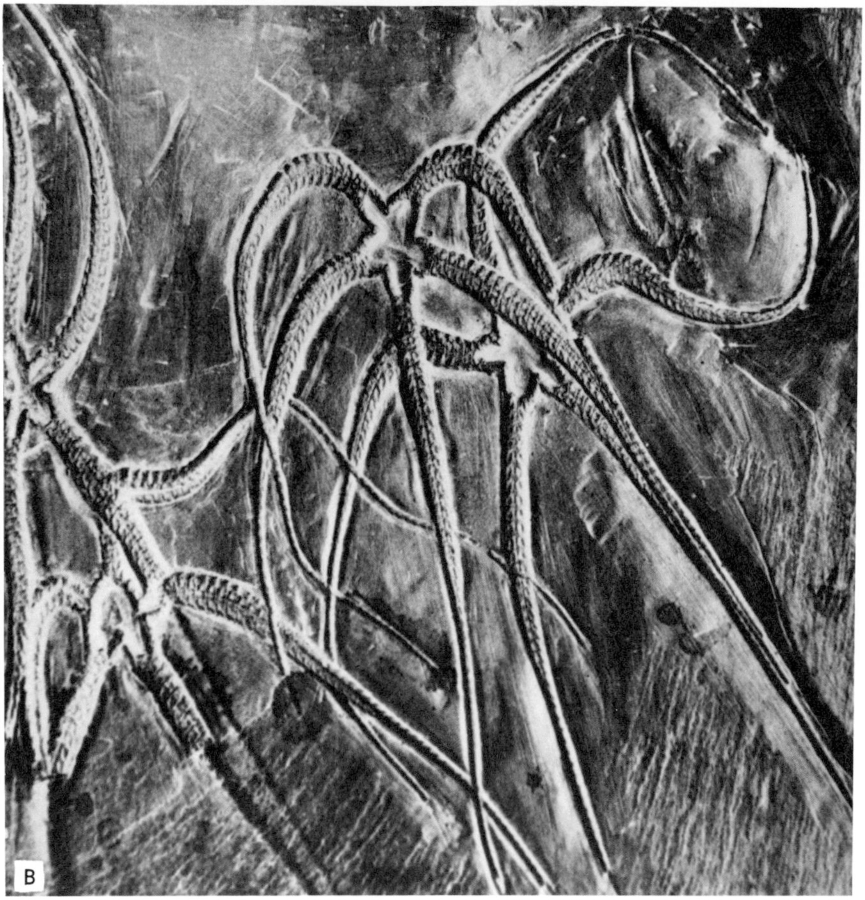

FIG. 33. The drag mark "chloephycus" displayed on an upper bedding plane. The feathery appearance is caused by rhythmic movements of an unknown object along the sediment surface. Lower Permian (lower Rotliegendes, Oberhöfer Schichten), Friedrichroda, Thuringia, ×1.9 (Müller, 1971).

substrates, small chevronlike wrinkles can result. The points of the chevrons point downcurrent, and they can be used as current-direction indicators. These structures are always found on the bottom of beds, and if other indicators of current direction are associated with them, these structures can be used to interpret the direction of flow.

2) Drag marks of medusae, e.g., *Rhizostomites admirandus* (HAECKEL) from the Solnhofen Limestone of southern Germany (JANICKE, 1969; KOLB, 1951). The marks shown in Figure 34 originate in the upper right. From that point, as many as six parallel drag marks are developed that continue for about 12 cm. Toward the center of the picture they change into an entangled mass of fine ridges running to the upper left where the outline of part of the medusa's body is found.

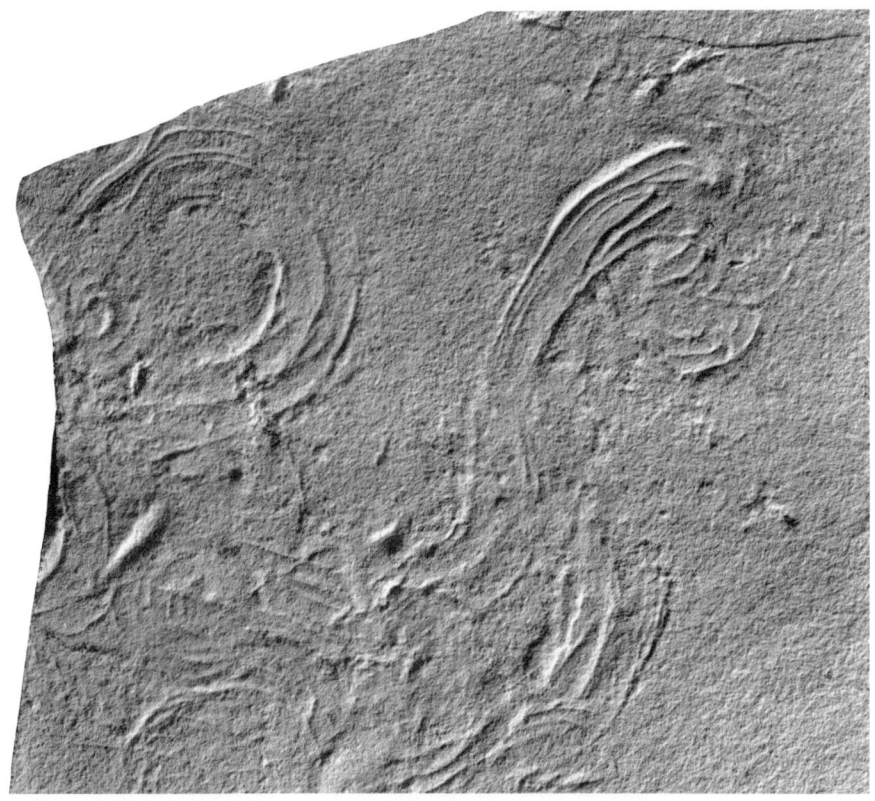

FIG. 34. Drag marks of medusa, *Rhizostomites admirandus* HAECKEL, Solnhofen Limestone (Malm zeta, lower Portlandian), Gungolding, Bavaria; width of section, ×0.45 (Janicke, 1969; photo by courtesy of Bayer. Staatssammlung für Paläontologie u. Historische Geologie, München).

Fig. 35. Roll marks of ?fish vertebrae parallel to the current direction and perpendicular to the trend of the ripples (upper right). Impression of other ?skeletal parts can be seen in the upper part of figure, also aligned with the currents. Oligocene (Flysch), Engi-Matt, Switzerland (Pavoni, 1959).

3) Roll marks of probable fish vertebrae (Fig. 35) in the Oligocene of Switzerland (Pavoni, 1959) and roll marks of ammonite shells that have been moved across the substrate by uniform currents (Barthel, 1964; Janicke, 1969; Seilacher, 1963).

4) Prodmarks of reeds from the Upper Triassic (Keuper) of Germany (Fig. 36). As with prod marks produced by parts of trilobites, ammonite shells, or fragments of those, the objects were carried by currents, oriented in cross or oblique orientation. The resulting marks on the substrate are depressions, which can be used to indicate the general direction of the carrying currents (Janicke, 1969; Seilacher, 1963). In general, the up-current side of these sedimentary structures are flatter than the leeward sides.

5) Changing current conditions can be clearly recognized from marks made by pivoting plant remains (i.e., seaweed, small branches) in anchored azimuthal orientation in both subaqueous (Langerfeldt, 1935)

Fig. 36. Prod marks produced by reeds, from Upper Triassic (Middle Keuper, Schilfsandstein), Sternenfels, Württemberg; approx. ×0.7 (Linck, 1956).

and continental environments (GERHARZ, 1966; MÜLLER, 1967). The tethered objects produce scraping marks in the sediment, which stand a good chance of preservation. They are rarely observed under water, and are much more common on land where they are formed under conditions of rapidly changing wind directions. Because wind direction can change 360 degrees in a short time, the resulting marks left in the substrate can be circular; however, the predominating wind direction is recognizable from the greater intensity of the scraping marks on the lee side. In the fossil record such marks can be used as additional criteria in the interpretation of paleowind directions.

Such marks produced by plants are extremely common in fine, soft sand and are known to have diameters up to 1.5 m, under as much as 2 cm of water cover (MÜLLER, 1967). Their appearance is extremely variable and simple circular impressions may be found next to entire sets of circles, and smooth areas can be located next to wavy or jagged margins. Sometimes, perfect, wheel-shaped structures of concentric rings are produced. Spokelike structures (prod and drag marks) can result if the winds blow steady from one direction for some period of time. Such circular structures are particularly common along beaches where sea and land breezes alternate daily and in slightly indurated sediments, such as boulder clay. They can also be formed on vertical faces, although they are mostly cut in the sediment by pivoting of parts of plants around an anchoring point under the influence of strong, fairly unidirectional winds (Fig. 37). They can also be produced by the bodies of dead animals, such as fish cadavers which are anchored by their tails, swinging from side to side as described from the Solnhofen Limestone in southern Germany (BARTHEL, 1966).

Fig. 37. Circular structures produced on the vertical faces of dune sand by plants blown by winds. Recent example, from Baltic Sea near Wustrow; ×0.25 (Müller, n).

FOSSIL DIAGENESIS

GENERAL DISCUSSION

As sediment accumulates, its increasing weight causes changes in pressure, temperature, and volume which initiate disturbances of the equilibrium that developed between component elements during deposition. This results in rearrangement of constituents until they are again in equilibrium with the surrounding energy conditions. Diagenesis can be defined as the processes acting upon sediment and its constituents from deposition to eventual alteration by metamorphism or weathering. There is no sharp boundary between diagenesis and metamorphism. The same conditions of temperature and pressure that can cause typical metamorphic textures in bedded salt have insignificant effects when they act on limestone. It is just as difficult to draw a line between diagenesis and weathering processes, because the transition is gradual. Once the dead organism and its parts are embedded in the sediment, they become part of it and participate in the diagenesis of the sediment. They undergo alterations, the accurate knowledge of which is important for the proper interpretation of fossils. A study of the older literature reveals that sometimes new "species" were based on products of diagenetic alterations.

According to differences in pressure, temperature, time, and chemical composition of rocks, diagenetic processes progress at very different rates.

For a given sediment, increased deposition tends to continually reduce pore space in the underlying material, and the pore water, except for a very small amount of moisture, is lost through compaction. The composition of this water that migrates upward depends upon that of the surrounding rock and changes generally soon after deposition. Solutions migrate by diffusion because the dissolved substances tend to distribute themselves evenly throughout the available pore space. Although the internal transport of fluids within the pore spaces of the sediment is small, over geologic time it can be quite substantial. The rate of flow is dependent upon both permeability and grain-size of the sediment and reactions will occur through the mixing of introduced fluids with connate water. Changes in pressure, temperature, and composition of the dissolved matter can lead to precipitation of dissolved minerals. It is by this process that voids are filled, for example, inside the camerae of ammonites and nautiloids that have not been infilled by sediment and in which drusy calcite or quartz may develop (Fig. 38).

As grain size changes, solubility increases and large grains continue to grow at the expense of smaller ones that are being dissolved. Void spaces are occluded as minerals are precipitated within them. Organic materials are altered or destroyed. In addition, metasomatic processes commonly occur. Similarly, unstable polymorph substances are changed to their more stable forms. If fossils in sedimentary rocks are exposed to weathering, the equilibrium reached during diagenesis is disturbed. Among processes responsible for such changes are hydrolysis, hydration, oxidation, and reduction, and attacks by acid or alkaline solutions. If temperature and pressure are sufficiently increased, diagenetic processes change to metamorphic processes, leading to extensive changes in mineralogy and texture of the rock (WINKLER, 1964). As stated before, there is no well-defined dividing line between the two processes. Fossils are known to be preserved in regions of contact and regional metamorphism, and can play an important role in the interpretation of such metamorphic strata (BUCHER, 1953; RINEHART et al., 1959).

Of the many diagenetic processes, a few are discussed below because they are important for accurate interpretation and understanding of fossils.

FORMATION OF STEINKERNS AND EARLY DIAGENESIS

The body of an organism can be completely preserved as a body fossil. If it is destroyed within the sediment, and if the sediment is sufficiently lithified, and no secondary minerals are being precipitated, a void is created that may be preserved as such. The inner surface of the void is an impression of the outside of the destroyed object.

Solution of, for example, bivalve shells and subsequent filling of such voids by drusy formations, such as calcite, dolomite, or siderite, is relatively rare (BATHURST, 1967).

If the sediment is not sufficiently competent to maintain these voids, they may collapse; however, if the shells initially possess an internal void or if such voids are created during early diagnesis by removal or decomposition of soft parts, they are commonly filled by sediment. It is by this process that true steinkerns develop that reflect the internal morphology of the shell (Fig. 39).

Steinkerns are produced by many types of sedimentary infilling and have an amazing variety of shapes and textures. Frequently they undergo internal sedimentation and develop geopetal structures. Commonly, smaller fossils and burrows are found in them (Fig. 40). Study of steinkerns allows important conclusions to be drawn with regard to the life habits of an individual organism and the particular conditions of sedimentation. For example, calcilutite and pelletal limestones can occur next to marly limestones and shell debris deposits.

Fig. 38. Medial section through the phragmacone of the ammonite, *Ceratites*, in which the walls of the camera are lined with transparent drusy calcite. Middle Triassic (upper Muschelkalk, middle Ceratites Schichten), Thuringia; ×2.5 (Müller & Zimmermann, 1962).

Fig. 39. Fragments of crinoid columnals (*Ctenocrinus typus* BRONN). The lumen filled with sediment, prior to dissolution of the columnal, producing a true steinkern which reflects the soft part morphology of the animal. Lower Devonian (Spiriferensandstein), Eifel, West Germany, approx. ×2.75 (Müller, 1963).

Multiple phases of fine-grained sedimentation can occur within shells by recurrent water movement. Material in suspension then settles within the shell, especially if the currents flowing out of the shell are weak. In planispiral ammonite shells that are being filled with sediment through their apertures, the sediment in the camerae becomes increasingly finer and more homogeneous toward the apical end. Many camerae receive no sediment and remain empty and most shells are not entirely filled. In order to become completely filled, a hole at the apical end of the shell would have to exist (SEILACHER, 1968a).

When shell substance is gradually destroyed, the surrounding sediment may settle slightly, and the steinkern may be pressed against the impression made by the outer surface of the shell, so that this is then imprinted on the steinkern. Such structures are called *Prägekerne* or *Skulptursteinkerne*. Their analysis helps in the interpretation of early diagenetic processes in the surrounding sediment as well as the manner of dissolution and decomposition of the organism itself. Sculptured steinkerns are most frequently formed as the result of the destruction of organisms with calcareous hardparts, in particular, bivalves.

The dissolution of shell material can begin on either the interior or exterior surface of the shell. Dissolution has been shown to be extremely strong in fractured shell material. The synchronous dissolution of shell material can also lead to the decrease in shell thickness and the loss of sculpture. Shells are most easily reduced in thickness at the umbonal region where they were originally the thickest. This type of shell destruction is mostly the result of processes occurring outside the sediment, prior to burial. In both these cases of shell reduction, it is thought that the conchiolin and the periostracum are destroyed. In oxygen-rich environments, the periostracum is generally

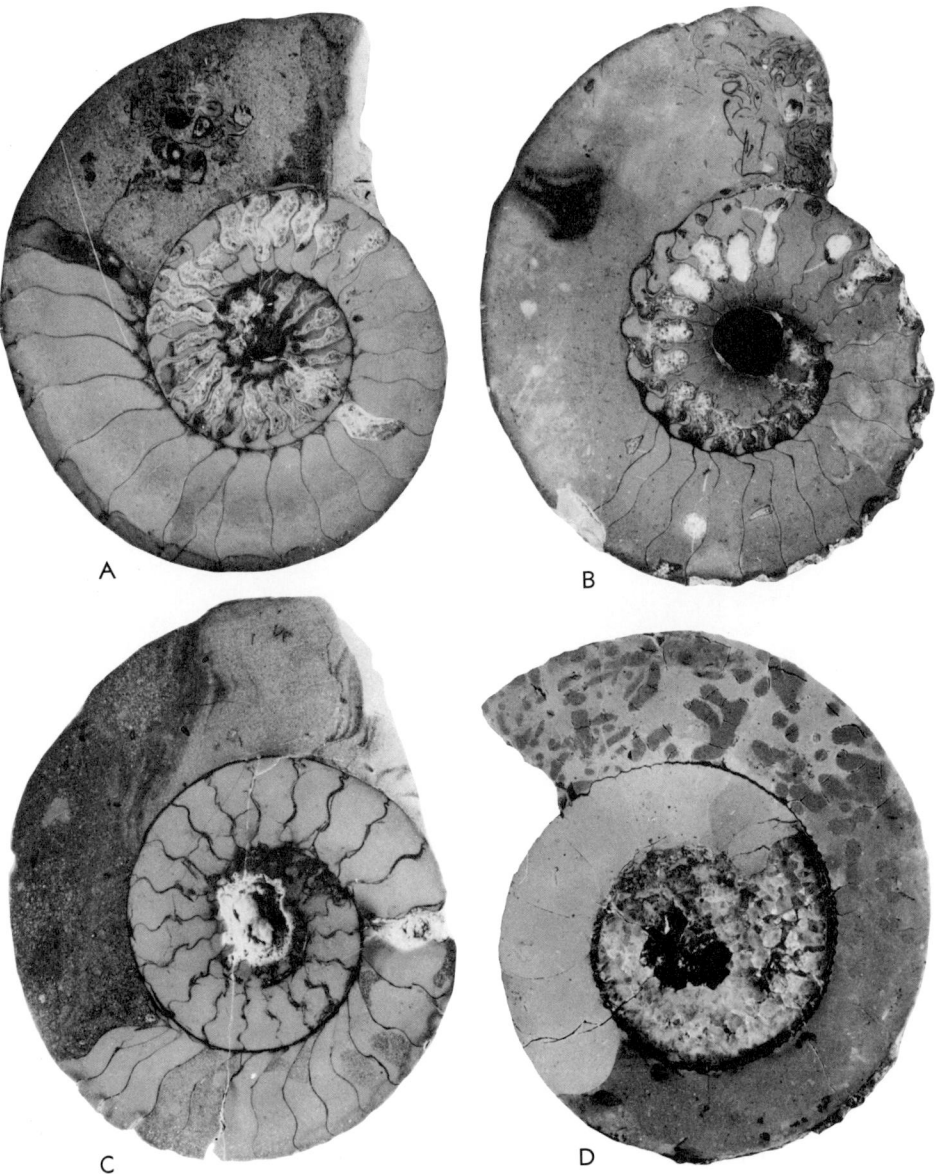

Fig. 40. Differential filling of ammonite shells by sediment (medial sections).——*A. Ceratites evolutus* PHILIPPI, upper Muschelkalk (*C. evolutus* Zone), Ohrdraf, Thuringia; diameter 95 mm.——*B. Ceratites evolutus* PHILIPPI, upper Muschelkalk (*C. evolutus* Zone), Ballenstedt, Harz Mt.; diameter 100 mm.——*C. Ceratites* cf. *C. compressus* PHILIPPI, upper Muschelkalk (*C. compressus* Zone), Remda, Thuringia; diameter 85 mm.——*D. "Perisphinctes"* sp., lower Malm (Oxfordian), Staffelstein, Franconia; diameter 65 mm (Sektion Geowissenschaften, Bergakademie Freiberg, 249/3-6; Müller, n).

destroyed first; the inorganic constituents remain. In oxygen-deficient environments, especially those with high concentrations of hydrogen sulfide, destruction begins with the dissolution of calcareous components. The oxidation of sulfides frees sulfuric acid, which dissolves calcareous material and hinders additional bacterial destruction of organic substances, as decay-producing bacteria are extremely sensitive to the presence of oxygen.

In anaerobic environments, the periostracum protects the underlying shell material, but there is a considerable difference in the solubility of the prismatic and the nacreous layers. The prismatic layer of the shell, secreted by the mantle edge, is generally calcite, whereas the nacreous or porcellaneous layer of the shell consists of more soluble aragonite, which is secreted by the mantle surface. The shells of snails and dimyarian bivalves, whose shells are composed entirely or predominantly of aragonite, are therefore most commonly preserved as steinkerns. The shells of monomyarians are predominantly of calcite and their shells are generally preserved in their entirety.

If the voids created by solution of skeletal material are preserved, solution must have occurred after lithification of the sediment. Occurrences of sculptured steinkerns show that the solution of the shells proceeded outward from the inner side. They can only be formed if the sediment has not become entirely lithified, when the void can be closed under the weight of the settling sediment. During this process, the steinkern moves outward with the progressive solution of the inner side of the shell. Imprinting of the outer ornamentation on the steinkern occurs only after the shell substance has been completely removed.

Sculptured steinkerns of ammonites are known from the *Posidonia Shale* (Lias ε) and the Solnhofen Limestone in southern Germany. They show details of ornamentation without preservation of sutures.

Aragonite can be replaced in skeletal material by the action of marine boring algae, and probably fungi and bacteria. Algal filaments penetrate the shell material centripetally and eventually die and decompose. The resulting cavities can be filled with micritic cements. When this process is repeated, as has been observed in mollusk shells from the Bahamas, a micrite envelope or rind will develop around the shell, which for unknown reasons is not dissolved during diagenesis. Thus, an impression of the original shell is preserved, which later may be filled with calcite crystals (BATHURST, 1966; WOLF, 1965).

FIG. 41. "Metallofact," casting of recent myriapod in molten aluminum, ×1 (Müller, 1963).

Fossils can also be preserved in lava flows. For example, tree trunks which were surrounded by lava can leave many details of the bark clearly imprinted in the surrounding rock, and in some cases, the cell structure of trees has been preserved ("lava trees"). Even a rhinoceros has been preserved in basalt (CHAPPELL et al., 1951). Of particular interest, is the "metallofact" shown in Figure 41, which is a myriapod that was rapidly covered by molten aluminum, thereby preserving its shape.

DEFORMATION OF FOSSILS

GENERAL DISCUSSION

Fossils are commonly found to be deformed by mechanisms that either deform them plastically or fracture them. Of course there are many transitional cases between these two end results. Possible causes for such deformations are: 1) volume decrease of surrounding sediment, 2) collapse due to overburden, and 3) tectonic movements.

VOLUME DECREASE (SETTLING) OF SEDIMENT AND ITS EFFECTS

GENERAL DISCUSSION

Settling of sediments may be caused by the following processes: 1) reduction of interstitial liquid (mostly water), 2) leaching of relatively soluble constituents, and 3) removal of organic matter by degassing. Compacting of sediments can occur during sedimentation under the influence of outside factors such as water movement. As a consequence, pore volume is reduced. Since the wettability of coarse-grained sands

Fig. 42. *Lytoceras* sp.——*A*. Preservation in limestone (Lias α, Braunschweig).——*B*. Specimen deformed by compaction and dewatering of clay-rich sediments (Lias ε, Posidonienschiefer), southern Germany. Shell walls are fractured and laterally displaced during sediment compaction. Scale in cm (Müller, 1963).

is small, maximum compaction may result at this stage. Thus, sands deposited under turbulent conditions, as a rule, will not suffer further compaction, except when they contain a significant amount of organic substances, or easily soluble components, or hollow fossil remains. But even in such conditions, deformation of fossils will occur only occasionally, because hydrostatic pressure is reduced by the friction and mutual support of sand grains.

One can think of porous sand as being composed of numerous, vertical prisms, with their sides in contact, along which downward movement takes place. Since the cross sections of these prisms are relatively large, only fairly large-sized objects can be deformed. Most objects that have relatively small cross sections remain undeformed in such sands. However, deformations due to the weight of the object itself are in a different category. They can occur, if organic tissues are decomposed and yield.

What is true for sand deposits is not necessarily valid for boulder beds. As do single grains in sandstone, pebbles and boulders in conglomerates can support each other to form arches. Little or no deformation should, therefore, be expected, just as in sands; however, the opposite appears to be the case. The reason for this can be found in stress changes that develop, as in sand, when, under load pressure, edges and corners of grains break off. This happens most commonly if the interstices are not entirely filled with sand or some other material.

Porous sediments, which are very fine grained or are composed of colloidal constituents, have considerable water retention because of their large internal surfaces. This is especially true when clay minerals of the montmorillinite group compose a large portion of the sediment. Also, clay commonly contains large amounts of organic material and is, therefore, subject to more settling than other types of sediments (Fig. 42,*B*).

The settling of clayey sediment is caused by increasing sediment accumulation and load pressure, which results in dewatering of the clay particles. During dewatering, the migration of pore fluids out of the sediments not only causes a reduction in volume but also the migration of dissolved substances, which are deposited elsewhere. In this way concretions can be formed, which may be of limonite, phosphorite, marcasite, pyrite, barite, or calcium carbonate (ILLIES, 1950).

During dewatering, lime mud may lose as much as three-fifths of its volume and one might, therefore, expect to find considerable settling effects in such sediments; however, fossils preserved in pure limestone, free from clay particles, silica, or organic material, generally show few signs of early diagenetic deformation (Fig. 42,*A*). This is probably because dewatering and solidification of very stable carbonate mud take place at a very early stage, and the concomitant volume and stress changes do not affect the texture of the rock and its contained fossils. In contrast to clayey and marly sediments, calcareous muds are not subject to later volume reduction by compaction (MÜLLER, 1951a). They pass rapidly through the plastic state, before the deposition of a large amount of overlying sediments. As soon as such sediments have been lithified, no further diagenetic deformation can take place.

The deformation of ammonites with rounded shells and wide living chambers, such as *Ceratites nodosus*, depends upon their mode of embedding. If the shell is oriented parallel to the bedding, it is generally flattened. Shells that are embedded vertical or oblique to the bedding are broadened. Oblique deformation can sometimes be mistaken for tectonic deformation of the shells. Therefore, taxonomic identifications can become more difficult because the cross section of the whorls may be changed during deformation. The diameters of vertically embedded ceratites can be shortened by as much as 30 percent, with little indication of such compaction in the surrounding rock.

If outlines and cross sections of undeformed fossils are well known, the study of deformed specimens may lead to conclusions about the degree and cause of diagenetic deformation. If forms are not known in their undeformed state, then accurate analyses cannot be made and misidentifications may result. This has undoubtedly led to the establishment of many unnecessary species in the literature, especially when many of these fossils have undergone plastic deformation (REGINEK, 1917).

FIG. 43. *Psiloceras (Caloceras) johnstoni* (SOWERBY), showing early diagenetic deformation; from Lower Liassic (α1b, lower Hettangian), Göttingen, West Germany. Largest diameter, 95 mm (Langheinrich, 1966).

PLASTIC DEFORMATION

If an object does not possess some original elasticity, plastic deformation can only occur after diagenetic decomposition and leaching have made it pliable. In general, plastic deformation of steinkerns occurs only when the hard parts have been dissolved prior to lithification of the surrounding sediment. In lower Liassic rocks (α1b, lower Hettangian) of Göttingen, the ammonite shells are "stretched out" along bedding planes, having been deformed during early diagenesis (Fig. 43) (LANGHEINRICH, 1966). In this example, the amount of deformation (as determined by comparing long and short axis of a deformation ellipse) is 84 percent. There is no evidence of tectonism in these deposits, and, therefore, deformation must have occurred during early diagenesis. Thus, all observations concerning decalcification of hard parts due to decomposition of organic substances in or on unconsolidated sediments are of considerable importance in order to understand the process of plastic deformation. This type of deformation occurs primarily in sediments with a high clayey content, and practically not at all in sandstone and pure limestone.

The amount of decaying material must be somewhat substantial. HECHT (1933) studied experimentally the effects of skeletal material during the decay process. He added a number of *Mytilus* valves of known

TABLE 5. *Solution of Bivalve Shells in a Medium of Decaying Soft Parts (after Hecht, 1933).*

Time in Days	Original Weight (grams)	Subsequent Weight (grams)	Weight Lost (grams)	Percent Loss
7	0.8471	0.7205	0.1266	14.9
14	0.8471	0.6204	0.2167	24.5
16	1.5603	1.3870	0.1733	11.1
16	1.8466	1.3870	0.4590	24.86
31	1.6628	1.4650	0.1978	11.86
32	1.6628	1.4824	0.1804	10.85

weight to the flesh of 100 specimens of the same species. At intervals the shells were weighed to determine the amount of calcareous material lost due to the action of acids produced during decay. The results are shown in Table 5. In general, the greater the content of organic material within the sediment, the greater the possibility for the dissolution of calcareous shell material. The most important chemical agent seems to be the carbon dioxide produced during decay in addition to sulfuric acid, which is the oxidation product of hydrogen sulfide normally present in the sediment.

Because the periostracum protects the outer shell layer from solution, the unprotected, internal layers are relatively rapidly dissolved, and thin shells may be reduced to a chalky substance within a few weeks. In thicker shells the periostracum, which consists of conchiolin, begins to peel off after eight weeks without having undergone any changes. It is thrown into folds, parallel to the growth lines, and can be easily detached. In addition to bivalve shells (Fig. 44), arthropod carapaces are sometimes found having such wrinkled surfaces. If the shell or cuticle are not completely dissolved, but have become flexible, impressions of foreign objects or other parts of the same organism are occasionally found on the periostracum.

Hard parts that have become soft and flexible under the influence of organic decomposition products are easily subject to plastic deformation. If the rocks containing them have not been tectonically deformed, their occurrence always indicates that solution has been at work. As an example, fossilized skeletons can be found with individual bones bent and superimposed upon each other on a single bedding plane. In such cases the bones were made soft by diagenetic processes prior to compaction of the sediment. Examples are found among the reptilian remains in the *Posidonia* Shale near Boll and Holzmaden (Württemberg).

DEFORMATION BY FRACTURE

Deformation by fracture can occur if 1) the object has insufficient primary elasticity, and 2) an originally elastic body has become brittle by decomposition and leaching processes.

FIG. 44. Wrinkled but undestroyed periostracum of *Inoceramus dubius* from Upper Liassic (Lias ε, lower Toarcian, Posidonienschiefer), Goslar, Harz Mt., ×0.93 (Müller, 1963).

If during compression of the sediment, a shell cannot expand laterally, then it may break under pressure, and the individual shell fragments are thrust over each other.

The cracks that are formed during fracture deformation of shell material are usually delicate. If the broken fragments suffer no displacement and if they are recemented by mineral precipitation, plastic deformation may be simulated, and imprints of the cracks may be seen on steinkerns. Such specimens show clearly that fracturing preceded solution.

An example of fracturing deformation is shown in Figure 45,*A-C*. If a shell has essentially a circular outline, it is possible to divide the fracture systems into two basic types: 1) concentric fractures, running parallel to the shell outline, and 2) radial fractures that originate from the center of the shell.

Shells with oblong, oval outlines such as the bivalves *Mytilus*, *Modiola*, and *Inoceramus* (Fig. 45,*C*), display characteristic fracture. Commonly a main fracture line runs along the central crest of the shell (Fig. 45,*B*). Objects usually break up along cracks that are oriented at right angles to their sides.

Experimentally, individual shells of bivalve mollusks and brachiopods are capable (according to species) of supporting normal loads of 500-4,000 kg/m^2 before breaking. Natural examples of shell packings could therefore support without collapsing a sedimentary overburden comparable in thickness to one meter. Thus, there may be ample time for cementation to begin, and further strengthen the packings, before critical overburden thicknesses are reached (ALLEN, 1974). Pressure mechanics have recently been applied to the study of skeletal hard parts by BRENNER and EINSELE (1976).

In some cases the result of sedimentary pressure can also be recognized in the absence of shell deformation. This is common in bivalves preserved with articulated valves where one valve is pushed over the other. This is especially common in isodonts and desmodonts such as *Pleuromya*, *Mya*, and *Gresslya*. Without contemporaneous deformation of the valves, this position is observed when compaction of the sediment has led only to a shifting of the

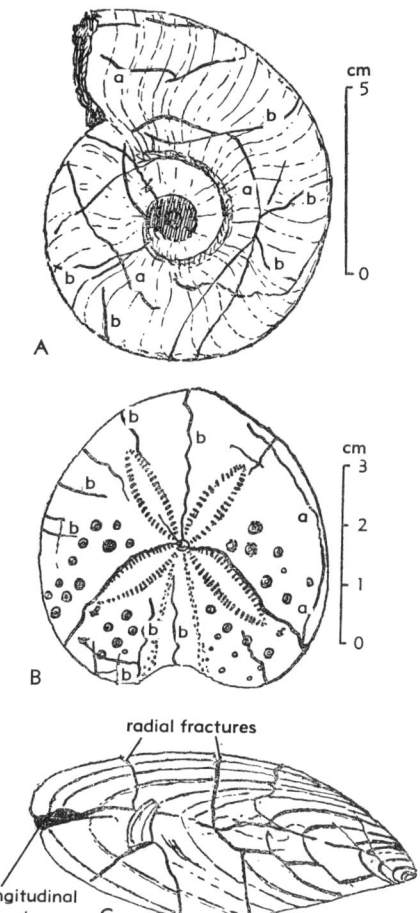

FIG. 45. Types of deformation by fracture of fossils with rounded outlines (*a*, concentric fracture; *b*, radial fracture) (Müller, 1951b).——*A. Leioceras opalinum* REIN., lower Dogger, Goslar.——*B. Echinospatagus hofmanni* GOLDFUSS, upper Oligocene, Bünde near Herford.——*C. Inoceramus labiatus* VON SCHLOTHEIM, lower Turonian, Salzgitter.

shells, not to fracturing or plastic deformation. It is not observed in clayey deposits and is restricted to limestone and calcareous marl, for example, the Ceratite beds of the upper Muschelkalk.

COLLAPSE UNDER OWN WEIGHT

If a shell collapses under its own weight, the degree of deformation is determined by its inherent statics and mechanics. Plastic deformation commonly occurs: 1) in elastic

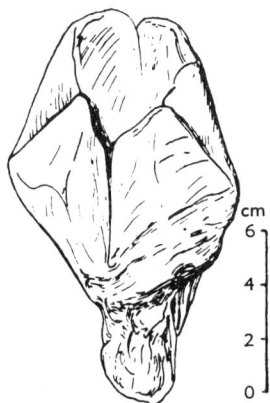

Fig. 46. *Conularia tulipa* RICHTER & RICHTER crumpled like a paper bag by pressure. Lower Devonian (Bundenbacher Schiefer), Bundenbach, Rhineland (Richter, 1931).

forms, especially thin-walled shells (Fig. 46), and 2) in shells that have become flexible as a result of the processes of decomposition and leaching.

On the other hand, deformation by fracture is to be expected if hard parts become brittle during decomposition or leaching.

DEFORMATION BY TECTONIC STRESS

Tectonic movements, especially during periods of intense orogeny, can result in considerable deformation of fossils. However, whereas pressure exerted during settling of unconsolidated sediment acts perpendicular to the bedding planes, causing changes in both the shape and volume of shells, tectonic pressure may act in any direction and generally leads to changes in shape and not volume. It is generally quite easy at the outcrop to tell the two types of deformation apart; however, in unoriented hand specimens this may prove difficult. Just as in pelomorphic deformation, tectonic deformation may result in a variety of forms that can simulate a variety of species depending on the original position of the fossils in the rock. Thus, it could be shown, in some cases, that several hundred "new" species were proposed without taxonomic justification (REGINEK, 1917; FANCK, 1929; BREDDIN, 1964). This is true for mollusks, and for other organisms, such as graptolites. So many types of deformation can result in these deposits, that mistakes are inevitable (Fig. 47,*A-F*).

A distorted fossil can be restored accurately if the degree of deformation, δ, is known. "δ" is the ratio between the long and the short axes of the deformation ellipse within a given bedding plane, and frequently, it is only a minimum number if the fossils are less easily deformed than the surrounding rock. In the case of steinkerns and molds, both are composed of the same material as the surrounding rock, and they therefore become deformed to the same degree as the rock.

Original right angles in fossils (e.g., between hinge and median lines in some ribbed bivalves and brachiopods; between axis and posterior margin of cephalon and thoracic segments in trilobites) may be deformed into oblique angles which then correspond to the conjugate diameters of the deformation ellipse. If two such oblique angles can be measured for two fossils lying in the same bedding plane, the deformation ellipse can be transformed into a circle and the degree of deformation can be determined (Fig. 48,*A,B*). LANGHEINRICH (1968) has published a review of the several methods by which the deformation degree δ in a particular bedding plane may be determined from changes of angles.

If the degree of deformation is known, it is possible to use different methods of restoration of the fossils. Such techniques have been described by GRÄF (1958), BAUMANN (1958), DE VRIES (1959), SDZUY (1962), BREDDIN (1964), STEHN (1968), and others.

SELECTIVE DISSOLUTION DURING DIAGENESIS

Skeletal material exposed at the water-sediment interface is unstable and is selec-

FIG. 47. Tectonic deformation of graptolites, Lower Silurian (Valentian), Weinbergbruch (vineyard quarry) near Hohenleuben (eastern Thuringia); direction of deformation (*b*-axis) for *A-C* marked by double arrow (Schauer, 1971).——*A-D. Monograptus turriculatus* BARRANDE; *A*, tectonic "long form" (deformation parallel to *b*-axis); *B*, tectonic "inclined form"; *C*, tectonic "short form," deformation

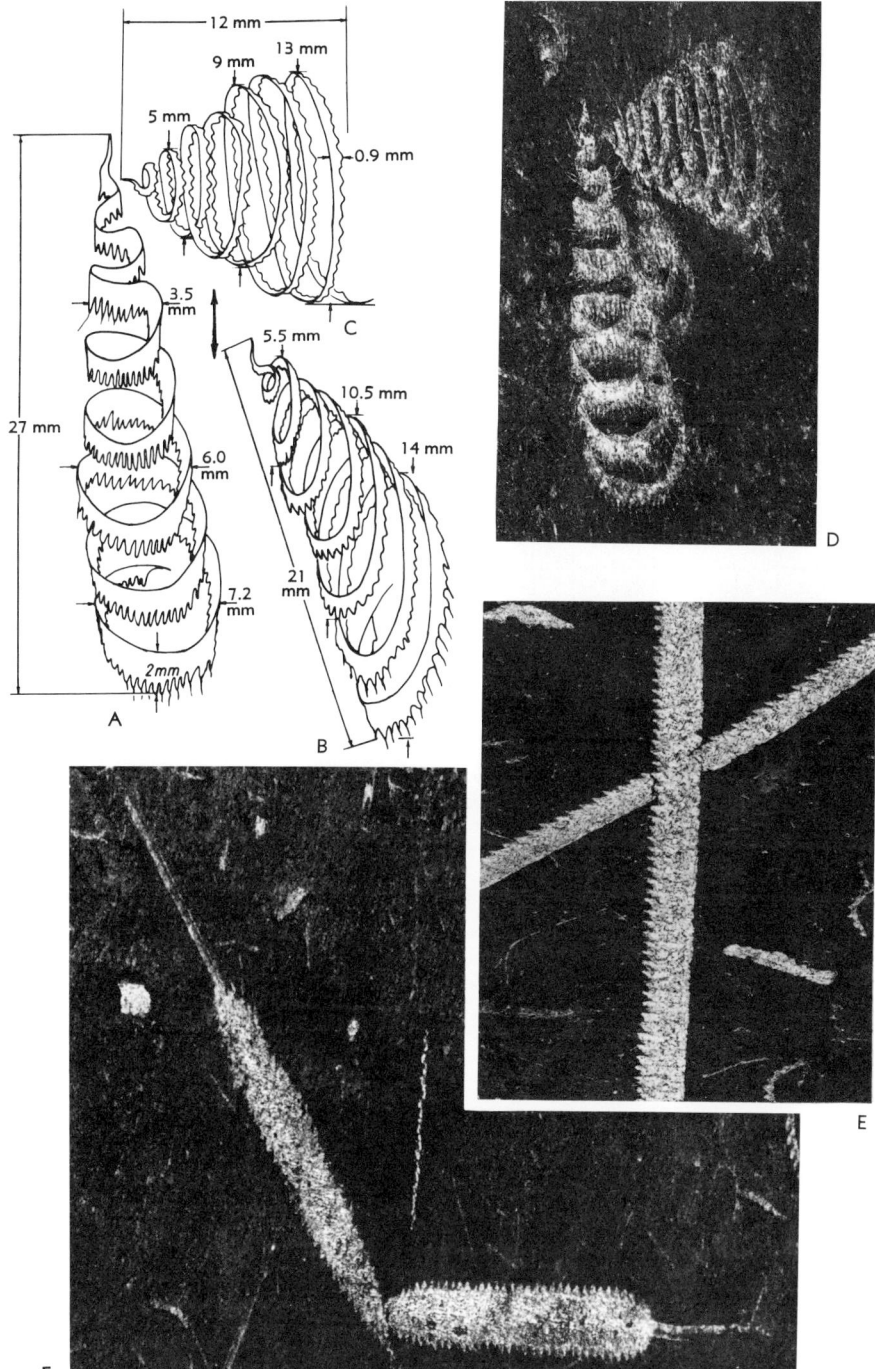

Fig. 47. *(Explanation continued from facing page.)* perpendicular to *b*-axis; *D*, long and short form on one bedding plane, approx. ×1.9.——*E, Pristiograptus nudus nudus* LAPWORTH, approx. ×1.9.——*F. Petalograptus altissimus* ELLES & WOOD, approx. ×1.3.

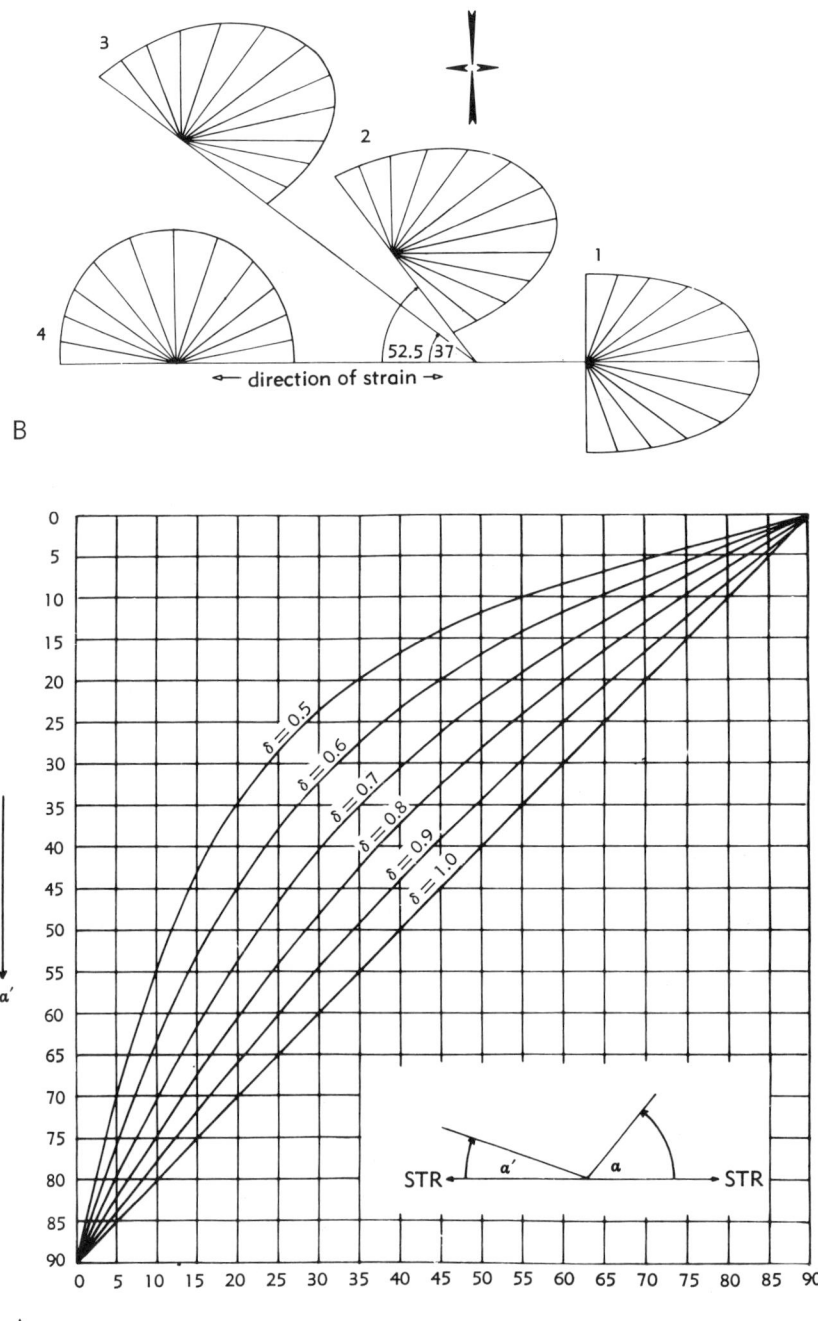

Fig. 48.——*A*. Graphic representation for determination of δ. α and α′ indicate the angle between the direction of elongation *(STR)* and the two lines, respectively, within the fossil, which formed a right angle before deformation (Mädler, 1938).——*B*. Four deformed bivalves (δ = 0.75) having original width-length ratio of 1.36, with their shell margins oriented at different angles (α) to the direction of

tively dissolved according to the solubility of its component substances. This process has been studied in recent and fossil material and much of the work has been substantiated by studies in phase chemistry. LOWENSTAM (1963) summarized the mineralogy of skeletal materials.

It must be remembered that the solubility, and thus preservability, of skeletal elements are influenced by the microstructure of their elements and the content of easily destroyed organic matter such as is found between the prisms in the shells of clams (KESSEL, 1938).

Selective solution of skeletal material also occurs during diagenesis, mainly under load pressure, by an upward migration of pore fluids caused by the pressure of accumulating sediment. By this process a "solubility front" develops, especially in the transition between oxidation and reduction zones, under conditions of low pH (JARKE, 1961). Under anaerobic conditions, hydrogen sulfide can have the same effect on skeletal material as carbon dioxide in sea water. Therefore, hydrogen sulfide need not be oxidized to form sulfuric acid. This explains absence of gypsum, which otherwise would be expected to form during dissolution of calcareous skeletal material (MOSEBACH, 1952).

In the bituminous *Posidonia* Shale (Lias ε) of southern Germany, the aragonitic portions of ammonites have been dissolved, but the calcitic aptychi, the siphuncles, and the periostraca are preserved. Selective dissolution can also occur under aerobic conditions as can be seen in the Solnhofen Limestone (Malm Z, lower Portlandian) in southern Germany, where aragonitic shells of ammonites have been entirely destroyed. As a rule, only faint outlines of the shells remain; however, the aptychi and siphuncles are well preserved.

The speed at which shell material is dissolved depends on the effective surface area. Thin shells may be completely dissolved when remnants of thicker shells are still present. Phosphatic and horny skeletal elements (such as brachiopod shells, gastropod opercula, hooks of coleoids, scolecodonts, arthropods, graptolites, bones, teeth, and scales of vertebrates) may be preserved in acidic environments in which calcareous hard parts are completely destroyed.

MOLECULAR REARRANGEMENTS IN THE COURSE OF DIAGENESIS

GENERAL DISCUSSION

Molecular rearrangements occur during diagenesis when, under the influence of temperature and pressure, existing imbalances are corrected and individual sedimentary particles attain a state of minimal surface energy. Several possibilities exist of which the following are important: 1) recrystallization, especially grain growth, aggrading neomorphism; 2) concentration of thinly disseminated substances with or without chemical reactions (metasomatism); and 3) transformation of substances from unstable into more stable forms.

RECRYSTALLIZATION IN PARTICULAR GRAIN GROWTH

Recrystallization occurs within the sediment, in regions of pore water movement, and is controlled by solubility, grain size, and temperature. During this process, organic hard parts incorporated in the sedimentary matrix are also recrystallized. Such diagenetic changes usually result in changes in grain size, form, and orientation of particular mineral species and their polymorphs (FOLK, 1965). Commonly, the process is characterized by the increase in size of larger grains at the expense of smaller ones. Because the ratio of surface to volume is less in large crystals than in smaller ones, an upper limit of grain growth exists under any given set of circumstances. Organic hard parts can be so intensively recrystallized by grain growth that they are often unrecognizable, except for a thin "dusk line" which represents the shell margin (FOLK, 1965).

(Continued from facing page.)
elongation. The angles (α) of 0, 37, 52.5, and 90 degrees correspond to deformation of 0, 45, 60, and 90 degrees, respectively, corresponding to shells numbered 4, 3, 2, and 1 (Langheinrich, 1968).

Fig. 49. *Aspidura* sp., in space not completely filled with sediment underneath a convex-up bivalve shell; the dorsal plates of the ophiuroid carry large, sparry calcite crystals. Middle Triassic (upper Muschelkalk), Württemberg; width approx. 30 mm (Müller & Zimmermann, 1962).

Small, fragile organisms, consisting of calcareous material, are commonly destroyed during the initial stages of diagenesis. Coccoliths are especially susceptible to complete recrystallization, either by load pressure or shearing stress, and quickly lose their characteristic shape. During the diagenetic alteration of a marl into a marly limestone, coccoliths can become encrusted with slowly growing crystals to such an extent that they are no longer recognizable. As a general rule, the older the sediment, the less likely the chance of finding identifiable coccoliths. Recrystallization of shell material can also result, without appreciable accompanying grain growth, in the solid state through solutions, depending on the porosity and permeability of the rocks (Füchtbauer & Goldschmidt, 1964, p. 195). During such processes primary textures are generally preserved, because the changes generally take place within crystal lattices (Bathurst, 1958).

If recrystallization is very extensive, it leads to the development of a coarsely crystalline mosaic texture and the complete obliteration of original, internal structure. This can commonly occur during the inversion of aragonite to low Mg-calcite or during aggrading neomorphism of small calcite crystals. Such textures are not only found in fossils, but also in the matrix of carbonate rocks where sharp boundaries exist between matrix and shells (Hollmann, 1968a).

Processes of recrystallization and grain growth aid in the natural weathering of fossils when they can be easily separated from the sedimentary matrix by chisel or acid. Bivalve shells and echinoderm plates are naturally quite sturdy, but diagenetic changes may increase their durability. Because the rate of weathering is dependent upon grain size, durable shells weather at a much slower rate than matrix of similar mineral composition. The motion of pore fluids is particularly important in the transporting and removal of constituents. Fluids preferentially move along crystal boundaries, and with special intensity along boundaries between shells and matrix, where small cavities eventually develop around the shells, facilitating their removal from the rock.

It is sometimes possible to determine the length of time during which recrystallization took place if the fossils suffered fracturing during diagenesis. It must be remembered, however, that fissures caused by early diagenetic settling may be closed by later recrystallization.

Echinoderm tests are especially susceptible to recrystallization. Each plate is composed of a meshlike, porous skeleton of a calcite crystal whose crystallographic orientation is determined by the structure of the animal's body (Macurda & Meyer, 1975). During early diagenesis, these pores are filled by optically continuous calcite cement, until each skeletal element forms a single calcite crystal with characteristic cleavage. If space is available, epitaxial overgrowths commonly form around the plates and assume scalenohedral shapes. This process can severely alter the appearance of an echinoderm, and new species have been introduced for diagenetically altered specimens.

Hollow echinoid tests, such as occasionally seen in chalk, are formed through oriented growth of the plates, each of which has a calcite crystal growing on the inside; all the crystals keep growing until the inner lumen is filled. The same explanation can also be applied to the so-called "crystal apples" where calcite crystals have completely filled the inside of the test of *Echinosphaerites*. Echinoderm tests can also change their appearances by outward growth of calcite crystals. Figure 49 shows a specimen of *Aspidura* from the Muschelkalk in which the dorsal plates bear calcite scalenohedra (Müller, 1969a).

An especially interesting phenomenon occurs when repeated precipitation of silica takes place in the interior of an echinoid test. In the specimen shown in Figure 50, the test of an echinoid was first filled to a certain level by chalky sediment, which was changed metasomatically to chert. Later, calcite crystals grew inward from the plates, until the remaining void was filled with light-colored, rather coarse-grained quartz. Finally, the remains of the calcareous skeleton were dissolved (Fig. 51).

Thin sections of limestone samples, for example, red limestones of the Oxfordian, Kimmeridgian, and Portlandian show Fora-

Fig. 50. *Echinocorys* sp.; internal structure is due to repeated precipitation of silicic acid (compare with Fig. 51). Upper Cretaceous (lower Maastrichtian, chalk facies), Rügen, DDR; largest diameter, 80 mm (Müller & Zimmermann, 1962).

minifera and other microfossils whose shells have been altered by calcite overgrowths (HOLLMANN, 1964). Also, the original thickness of shells or carapaces may be

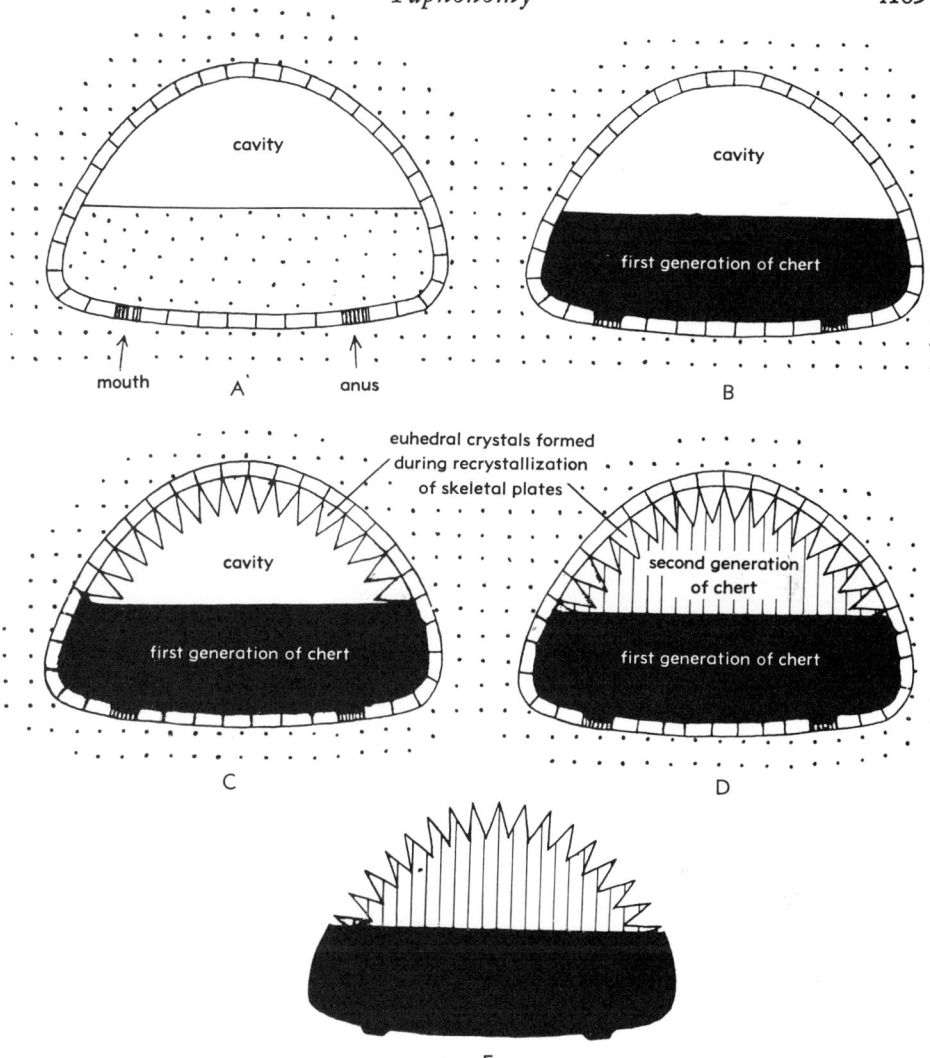

Fig. 51. Schematic diagram demonstrating concretion development *(A-E)* similar to that in Fig. 50 (Müller, n).

altered by recrystallization, and occasionally a considerable increase in thickness can result.

FORMATION OF CONCRETIONS (GEODES) AND THEIR IMPORTANCE IN FOSSILIZATION

Concretions can be divided into three separate groups, according to the time of their formation:

1) *Syngenetic*: formed during sedimentation, frequently characterized by excellent preservation of organic remains within the concretions.

2) *Diagenetic*: formed either shortly after the initial deposition of the sediment or somewhat later.

3) *Epigenetic*: formed after deposition of the surrounding sediment, bedding planes passing from the enclosing rock into the concretion.

Concretions that form around organic remains are of considerable interest to the paleontologist, because they are commonly formed before the initiation of plastic deformation, caused by sediment settling (e.g.,

Fig. 52. *Palaeoniscus macropomus* Agassiz, with head and anterior part of torso preserved, of early diagenetic, calcareous concretion *("Ilmenauer Schwiele")* (Müller, 1962a).——*A.* Right side.——*B.* Ventral side.——*C.* Dorsal side. Upper Permian (lower Zechstein, Kupferschiefer), Sturmheide b. Ilmenau, Thuringia. Length, 85 mm.

Müller, 1962a; Zangerl & Richardson, 1963; Zangerl, 1971). This is also shown by the fact that the interiors of many enclosed fossils contain voids or infillings of unconsolidated sediment. If fossils inside concretions are deformed, it must be concluded that the concretions were formed during or after the settling process.

Voigt (1968b) described concretions that have grown in several phases conditioned by interruptions in sedimentation or by coalescence of concretions of different ages and called these "hiatus concretions." In some cases, concretions may be exposed and corroded or bored on the seafloor. When sedimentation resumes, growth of the concretions continues (Kennedy & Klinger, 1972). This is true for concretions formed syngenetically or by early diagenetic processes in which the organisms around which the concretions grew are well preserved, in many cases with soft parts, or at least their impression.

The material for concretionary growth moves either actively by diffusion or passively with the pore water through the sediment. First of all, the possibility of material migration by diffusion is considered.

Concretions of syngenetic and diagenetic origin are commonly closely connected to the presence of organic remains that form the center of the concretions.

"Thin-walled" concretions tend to outline the enclosed objects. "Thick-walled" concretions tend to be more rounded. Spherical concretions are formed under conditions of hydrostatic pressure. Flattened concretions are commonly formed in connection with inhomogeneities within the sediment, caused by compaction.

It is possible to determine the relative time at which a concretion began to develop by comparing the amount of compaction in the surrounding rock with the degree of deformation of the organism contained within the concretions. Fossils preserved in strongly compacted sediments tend to be completely flattened along bedding planes or when they are attached to the surface of concretions; however, within concretions fossils are usually well preserved, displaying many morphologic details (Fig. 52), because the top of the concretion either protrudes through the sediment surface or merely has less sedimentary load on it because of its geometry. A slow rate of formation of a concretion is indicated, if, for example, the inner volutions of an ammonite shell at its core are better preserved than the outer ones. In this case, the outer whorls were exposed to corrosion by pore water for a longer time than the inner ones.

In some cases, the destruction of fossils by diagenetic processes is obvious. ILLIES (1949) observed such occurrences in Eocene deposits near Havighorst, Schleswig-Holstein, West Germany, where a bivalve plaster could be traced through several concretions in one bed where the fossils in the surrounding rocks had been completely destroyed. Apparently, these concretions developed before the shells buried in the surrounding sediments were destroyed.

Concretions containing fossils may be split by shrinkage cracks and the fragments displaced vertically or horizontally. Internal shrinkage cracks develop if the concretion consists of a considerable amount of clay or other material of little coherence and high water content. The resulting cracks run either radially or concentrically, and commonly follow the inner and outer boundaries of shells. The subsequent dewatering resulting from irreversible chemical processes leads to formation of shrinkage cracks that are then filled by mineral precipitation (Fig. 53).

The inner whorls of planispiral ammonites may become detached during this process and shrinkage cracks can develop covered by several generations of matlike overgrowths of calcite. Such overgrowths of calcite can make the thickness of the inner whorls appear greater than the outer whorls by more than a millimeter (HOLLMANN, 1968a). Because it is difficult to distinguish between the steinkern and the calcite filling within these diagenetically produced voids, the impression may be created that the size of the shell is due to anomalous growth (Fig. 54).

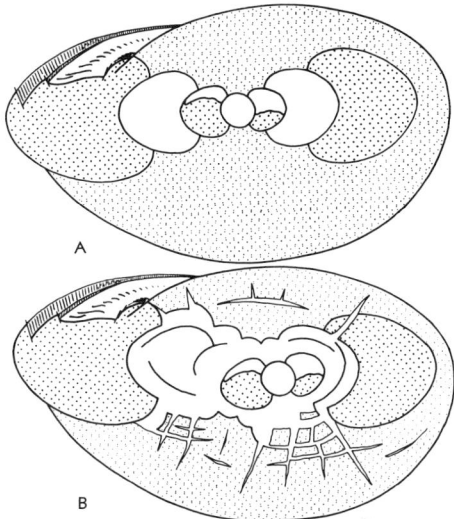

FIG. 53. Cross-sectional, schematic view of septarian concretion with enclosed ammonite (Hollmann, 1968a).——*A*. Prior to the development of shrinkage cracks, camerae serve as "bubble levels."——*B*. Radial and concentric shrinkage cracks have formed from dewatering of the clay minerals which are later filled with mineral precipitates. These cracks commonly conform to inner boundaries of the shell.

TRANSFORMATION OF POLYMORPHOUS SUBSTANCES INTO STABLE MODIFICATIONS

A common process during diagenesis is the transformation of minerals into their more stable polymorphs. Perhaps the most important of these transformations is the inversion of aragonite to calcite. This is important because aragonite occurs in the skeletal material of many organisms, such as corals, scaphopods, gastropods, bivalves, nautiloids, ammonoids, and otoliths (HALL & KENNEDY, 1967; KENNEDY & HALL, 1967; KENNEDY & TAYLOR, 1968). It has been demonstrated that synthetic aragonite, in water at 40° C, can invert to calcite within a few hours (WRAY & DANIELS, 1957); however, organically precipitated aragonite can be preserved for long periods of time.

Aragonite is frequently preserved in fossils of Cenozoic age and more rarely in Mesozoic forms. Commonly, such fossils are preserved in muddy or marly sediments. For example, aragonite is preserved in corals of the marly Triassic Zlambach beds and

Fig. 54. *Microderoceras birchi* (SOWERBY), a so-called double ammonite; shell of the steinkern is 8 mm thicker than that of the shell, void between the two filled with calcite. Middle Liassic (lower Pliensbachian), Lyme Regis, Dorset, England. Diameter of ammonite approx. 11 cm (Sektion Geowissenschaften, Bergakademie Freiberg, 249/7; Müller, n).

in the hinge of hippuritids of the Cretaceous Gosau beds, both of the eastern Alps (ZAPFE, 1936). There is an apparent correlation between the preservation of aragonitic material and the surrounding sediments. Aragonite is almost totally absent from hard parts in Upper Cretaceous limestone that is poor in organic substance, but is quite common in the organic- and pyrite-rich shales of the Gault.

Aragonite is exceedingly rare in Paleozoic sediments, but is found under special types of preservation. Examples exist in the Kendrick Shale (YOCHELSON et al., 1967), the Buckhorn asphalt in the Pennsylvanian of North America (STEHLI, 1956; LOWENSTAM, 1963; GRÉGOIRE & TEICHERT, 1965), and in the Upper Oil Shale Group (Lower Carboniferous) of Scotland (HALLAM & O'HARA, 1962) in which the clayey matrix contains up to 2.9 percent carbon.

It can be concluded, therefore, that fossil aragonite is best preserved in marly, muddy, and bituminous sediments (HALL & KENNEDY, 1967; KENNEDY & HALL, 1967; KENNEDY & TAYLOR, 1968). Water acts as a catalytic agent in the conversion of aragonite to calcite, and it is the organic material, the conchiolin, which surrounds the individual crystals of the shell, that protects the aragonite. Experiments have shown that dry organic aragonite is stable for an almost infinite period of time, whereas in the presence of water at 10° C., it will invert to calcite in a few million years, and in water at 50° C., in 100,000 years (BROWN et al., 1962). GRÉGOIRE (1959a,b) studied the chemical and structural composition of the organic material of recent mollusks and discovered its composition to be water-insoluble keratine, which completely surrounds the aragonite crystals. This material is rapidly broken down into amino acids that are water-soluble. ABELSON (1957) found that the proteins in the shells of Pleistocene specimens of *Mercenaria mercenaria* had degenerated to peptides and amino acids, and that in Miocene forms only residual amino acids were present.

The stabilizing effect of organic substances is also shown by the occurrence of aragonite in the Wealden beds of the Dalum oil field (northwest Germany), where aragonite is found in the shells of cyrenids and gastropods that apparently had been engulfed by oil or were embedded in clay very early in their diagenetic history (FÜCHTBAUER & GOLDSCHMIDT, 1964). The same kinds of shells in clay-free, permeable sediments are completely calcitized. It is probable that very early migration of oil occurred, because the coquinas have high porosity and are underlain by highly bituminous shales that probably represent the source rock.

The conversion of aragonite to calcite leads to the destruction of fine structures within the hard parts of organisms, and all that usually remains are relicts of the original shell structure (BATHURST, 1967; FOLK, 1965). During this process, both color and structure of the shell can change. The original aragonitic shells of ammonites turn glossy white and sparry, or they become micro- to cryptocrystalline with a transparent, light-brown color. The original lamellar structure of the shell is preserved only indistinctly.

METASOMATISM OF FOSSILS

Metasomatism can be defined as the gradual replacement of one mineral by another mineral of a different chemical composition (e.g., the replacement of calcite by quartz or vice versa), which occurs frequently during fossilization. If a molecule-for-molecule replacement takes place, the fine structure of the shell is generally completely preserved. If the morphology of the shell is preserved, the process is called pseudomorphism. Commonly metasomatism results in volume changes, accompanied by expansion, folding, compression, cracking, and porosity changes. Occasionally, such changes may be responsible for misidentifications of fossils.

Metasomatism is often a selective process, which may result in the replacement of unstable minerals by more stable ones (OGOSE, 1956; PAINE, 1937). An example is dolomitization, a process during which aragonitic skeletal parts are affected first, then the rock matrix, and, finally, any parts consisting of original calcite; however, this process is usually accompanied by loss of fine shell structures (for exceptions see PAPP, 1939).

Frequently, calcium carbonate is replaced by pyrite or marcasite (GRIPP & TUFAR, 1965). This process always begins on the surface of the fossils and takes place if acid solutions containing iron ions and hydrogen sulfides react with the carbonate and become neutralized. An example of pseudomorphism of cassiterite (SnO_2) after calcite are crinoid stems in the Permian limestone in the Emmaville district, New South Wales, Australia. The metasomatism was caused by the migration of elements from the Upper Permian granite body, which impregnated the surrounding rocks with cassiterite-containing quartz (LAWRENCE, 1953).

PRESERVATION OF STRUCTURAL SOFT PARTS

Although paleozoology has historically been concerned with the investigation of perservable hard parts or organisms, the preservation of soft parts with fine organic structures is by no means unusual. The following examples are well known: 1) preservation of delicate soft parts and appendages in the Middle Cambrian Burgess Shale, British Columbia (see WHITTINGTON, 1971, for earlier literature); 2) preservation of appendages and soft parts of Cambrian and Ordovician trilobites (RAYMOND, 1920); 3) preservation in limonite-goethite of soft-bodied worms in the Devonian of New York (CAMERON, 1967); 4) preservation in pyrite of soft parts of Devonian cephalopods (STÜRMER, 1969, 1973; RIETSCHEL, 1968; ZEISS, 1969); 5) preservation of various soft parts of Mesozoic ammonoids, e.g., egg cases (?) (LEHMANN, 1966; MÜLLER, 1969b), ink sacs (LEHMANN, 1967b), crop and gills (LEHMANN & WEITSCHAT, 1973); 6) remains of soft parts in the Solnhofen Limestone of Bavaria (REIS, 1893); 7) phosphatized ostracodes from Lower Cretaceous (Aptian, Albian) freshwater beds of Brazil with completely preserved musculature and male sexual organs preserved in the erectile position (BATE, 1971); 8) soft parts and

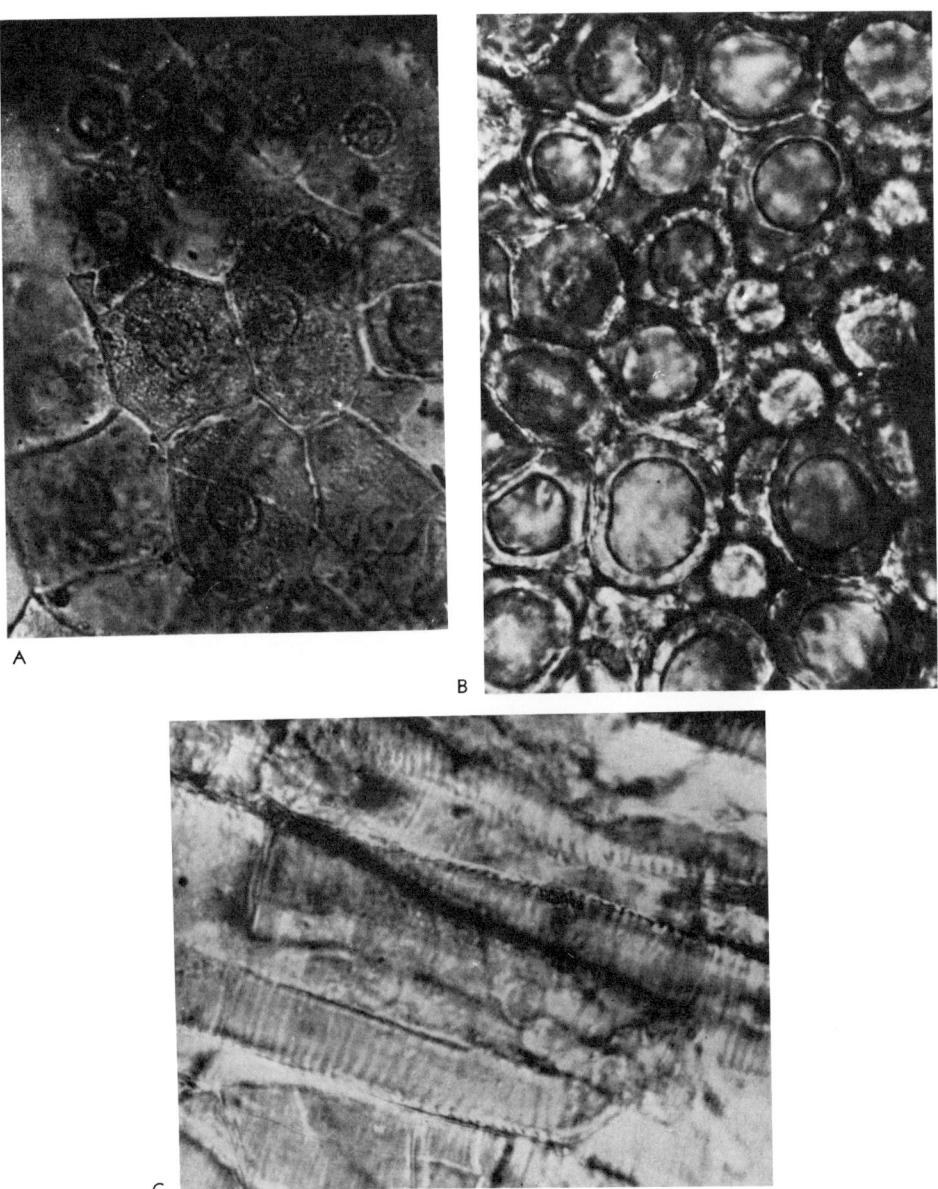

Fig. 55. Preserved soft parts of organisms from Middle Eocene Brown Coal, Geisel Valley near Halle/Saale.——*A.* Large, plate-like epithelial cells, with nuclei, from a frog epidermis, ×747 (Voigt, 1935).——*B. Cecilionycteris prisca* Heller, connective tissue in the ear, ×900 (Voigt, 1936).——*C.* Parallel trachea of *Eopyrophoras* sp., ×900 (Voigt, 1938b).

fine organic structure from the chalk of Europe (W. Wetzel, 1913, 1937); 9) soft parts from the Eocene brown coal of the Geisel Valley near Halle (Voigt, 1935; 1936; 1938a,b; 1950; 1956; 1957; H. Brenner, 1939); 10) soft parts preserved in amber (Voigt, 1937, 1938c) and similar fossil and subfossil remains (Bachofen-Echt in Abel, 1935; p. 601-619; Lazell, 1965; Schlee, 1973); 11) mummified ostracodes in the Pleistocene of Alaska (Schmidt & Sellmann, 1966).

PRESERVATION OF SOFT PARTS IN EOCENE COAL

Specimens preserved in the Eocene brown coal of the Geisel Valley near Halle have been extensively investigated by Voigt (1936; 1938a,b; 1950; 1957) (Fig. 55,*A-C*). With the exception of mumified remains known from Pleistocene and younger deposits, this material shows better preservation than that from any other locality. The Eocene fossils were preserved in bogs, and it is believed that the soft parts were preserved by acidic waters in the absence of bacteria and oxygen. The bones were preserved because of the neutralizing effect of calcium carbonate solutions. In some, soft parts such as muscles and corium, secondary silicification has been observed. As a rule, the tissues of smaller animals were best preserved, because solutions with dissolved minerals were able to penetrate them more rapidly than those of larger organisms. An example is the preservation of the delicate skin of frogs, whereas physically more durable structures, such as hooves, beaks, and horns were not preserved. The following soft parts have been found and studied microscopically: fat cells of reptiles and mammals; epidermis of frogs, bats, and artiodactyles; different types of mammal hair; the connective tissue of the corium of fish, frogs, reptiles, and mammals (Fig. 55,*B*); epithelial cells with nuclei of frogs (Fig. 55,*A*); melanophores of frogs and fish; blood vessels with erythrozytes of lizards; hyaline cartilage with cartilage cells of hyracotheriids; musculature of roaches and crabs, fat cells of roaches and larvae of Diptera.

Such observations are especially valuable if they yield diagnostic features that cannot be seen in hard parts alone. One of the few fossil representatives of the Nematophora, *Gordius tenuifibrosus* Voigt, was accidentally discovered on the basis of the structure of a single, 15 mm-long fragment of subcuticular tissue (Voigt, 1938a).

PHOSPHATIZATION OF SOFT PARTS

The phosphatization of soft parts is rather sporadic in occurrence, and found in sedimentary deposits since the Early Carboniferous. Typically, fish are preserved in this way, especially the musculature, stomach and intestinal contents, cutis, testicles, and spinal cord.

Reis (1893) studied in great detail the phosphatized musculature of worms (annelids), coleoid cephalopods, insects, fish, and reptiles from the Solnhofen Limestone.

Reis described the fossilization of organic material by phosphate which he called *Myo-* and *Zoophosphorit*. The composition of the material was about 70 percent $Ca_3P_2O_8$ and 6 to 6.5 percent CaF_2.

Specimens in the Solnhofen Limestone, on the other hand, are composed of 97 to 98 percent $Ca(MgK_2Na_2)CO_3$, small amounts of SiO_2, Al_2O_3, and so on, traces of P_2O_5 and no fluorite. It is quite obvious that the limestone itself does not contain sufficient $Ca_3P_2O_8$ and CaF_2 for phosphorization of fossils and, therefore, these minerals must have been concentrated by diagenetic processes. These have been found together with the abdomens of embedded cadavers of flesh- and bone-eating animals such as sharks, bony fish, and dibranchiate cephalopods. In these, the phosphorization of muscle tissue is most common; however, the carnivores of the genus *Lepidotus* and the pycnodontids are very seldom phosphorized. Phosphorization has never been observed in cephalopods with external shells.

Pyritized soft parts of ectocochleate cephalopods (orthocerids, bactritids, goniatites) have been discovered by X-radiography in Middle Devonian Wissenbach Shale of Germany, especially the intestines, funnel, and arms (Stürmer, 1967, 1968a, 1968b; Zeiss, 1969). The intestines, especially the terminal part, are longer than those of living cephalopods, though generally similar in morphology. The arms of coiled forms are more strongly differentiated than those of orthoconic forms possessing an umbrella-like sail of which the arms are extensions. In one case more than 10 arms have been observed, supporting the conclusion that, judging from the number of their arms, the ammonoids, bactritids, and probably also the orthocerids have to be included in the dibranchiate coleoids. This conclusion is supported by finds of heterodont radulae in Paleozoic and Mesozoic ammonoids (Closs, 1967; Lehmann, 1967a) as well as by arguments presented by Teichert et al. (1964).

REFERENCES

Abbott, B. M., 1974, *Flume studies on the stability of model corals as an aid to quantitative palaeoecology:* Palaeogeography, Palaeoclimatology, Palaeoecology, v. 15, p. 1-27.

Abel, Othenio, 1916, *Paläobiologie der Cephalopoden aus der Gruppe der Dibranchiaten:* vii + 281 p., 101 text-fig., Fischer (Jena).——1935, *Vorzeitliche Lebensspuren:* xv + 644 p., 530 text-fig., Gustav Fischer (Jena).

Abelson, P. H., 1957, *Some aspects of paleobiochemistry:* New York Acad. Sci., Ann., v. 69, p. 276-285.

Allen, I. R. L., 1974, *Packing and resistance to compaction of shells:* Sedimentology, v. 21, p. 71-86, text-fig. 1-9.

Archiac, Adolphe d', 1869, *Introduction à l'étude de la paléontologie stratigraphique:* v. 2, 616 p., 3 pl., F. Savy (Paris).

Barthel, K. W., 1964, *Die Entstehung der Solnhofener Plattenkalke (unteres Untertithon):* Bayer. Staatssamml. Paläontologie, Historische, Geologie, Mitteil., v. 4, p. 37-69, 4 pl.——1966, *Concentric marks: current indicators:* Jour. Sed. Petrology, v. 36, p. 1156-1162, text-fig. 1-7.——1972, *The genesis of the Solnhofen lithographic limestone (low. Tithonian): further data and comments:* Neues Jahrb. Geologie, Paläontologie, Monatsh., 1972, p. 133-145, text-fig. 1-4.

Bate, R. H., 1971, *Phosphatized ostracods from the Cretaceous of Brazil:* Nature, v. 230, p. 397-398, text-fig. 1.

Bathurst, R. G. C., 1958, *Diagenetic fabrics in some British Dinantian limestones:* Liverpool & Manchester Geol. Jour., v. 2, p. 11-36, text-fig. 1, 2, 1 pl.——1966, *Boring algae, micrite envelopes and lithification of molluscan biosparites:* Jour. Geology, v. 5, p. 15-32.——1967, *The replacement of aragonite by calcite in the molluscan shell wall:* in Approaches to paleoecology, J. Imbrie, & N. D. Newell (eds.), p. 357-376, 1 text-fig., 4 pl., 2 tables, J. Wiley & Sons (New York).

Baumann, E., 1958, *Affine Entzerrung mit einfachen optischen Mitteln:* Schweiz. Paläont. Abhandl., v. 73, p. 19-21, text-fig. 1-4.

Beecher, C. E., 1894, *On the mode of occurrence and the structure and development of Triarthrus becki:* Am. Geologist, v. 13, p. 38-43.

Boekschoten, G. J., 1966, *Shell borings of sessile epibiontic organisms as palaeogeological guides (with examples from the Dutch coast):* Palaeogeography, Palaeoclimatology, Palaeoecology, v. 2, p. 333-379, text-fig. 1-16, pl. 1-3.

Breddin, Hans, 1956, *Die tektonische Deformation der Fossilien im Rheinischen Schiefergebirge:* Deutsche Geol. Gesell., Zeitschr., v. 106, p. 227-305, text-fig. 1-40, 5 pl.——1964, *Die tektonische Deformation der Fossilien und Gesteine in der Molasse von St. Gallen (Schweiz):* Geol. Mitteil., v. 4, p. 1-68, text-fig. 1-32.

Brenchley, P. J., & Newall, G., 1970, *Flume experiments on the orientation and transport of models and shell valves:* Palaeogeography, Palaeoclimatology, Palaeoecology, v. 7, p. 185-220.

Brenner, H., 1939, *Paläophysiologische Untersuchungen an der fossilen Muskulatur aus der eozänen Braunkohle des Geiseltales bei Halle (Saale):* Nova Acta Leopoldina, n. ser. 7, p. 95-118, 4 pl.

Brenner, K., 1976, *Ammoniten-Gehäuse als Anzeiger von Palaeoströmungen:* Neues Jahrb. Geol. Paläont. Abh., Bd. 151, p. 101-118, text-fig. 1-15.

——, & Einsele, G., 1976, *Schalenbruch im Experiment:* Zentralbl. Geol. Paläont., pt. 2, Paläont., v. 1976, p. 349-354, text-fig. 1-6.

Bromley, R. G., 1970, *Borings as trace fossils and Entobia cretacea Portlock, as an example:* in Trace fossils, T. P. Crimes, & J. C. Harper (eds.), p. 49-90, text-fig. 1-4, pl. 1-5, Seel House Press (Liverpool).

Brown, W. H., Fyfe, W. S., & Turner, F. J., 1962, *Aragonite in California glaucophane schists, and the kinetics of the aragonite-calcite transformation:* Jour. Petrology, v. 3, p. 566-582.

Bucher, W. H., 1953, *Fossils in metamorphic rocks:* Geol. Soc. America, Bull., v. 64, p. 275-300, 997-999.

Cameron, Barry, 1967, *Fossilization of an ancient (Devonian) soft-bodied worm:* Science, v. 155, p. 1246-1248, text-fig. 1, 2.

Chamberlain, J. A., 1976, *Flow patterns and drag coefficients of cephalopod shells:* Palaeontology, v. 19, p. 539-563.

Chappell, W. M., Durham, J. W., & Savage, D. E., 1951, *Mold of a rhinoceros in basalt, Lower Grand Coulee, Washington:* Geol. Soc. America, Bull., v. 62, p. 907-918, text-fig. 1-4, pl. 1, 2.

Chave, K. E., 1964, *Skeletal durability and preservation:* in Approaches to paleoecology, John Imbrie, and N. D. Newell (eds.), p. 377-387, Wiley & Sons (New York).

Clifton, H. E., 1971, *Orientation of empty pelecypod shells and shell fragments in quiet water:* Jour. Sed. Petrology, v. 41, p. 671-682, text-fig. 1-15.

Closs, Darcy, 1967, *Goniatiten mit Radula und Kieferapparat in der Itararé-Formation von Uruguay:* Paläont. Zeitschr., v. 41, p. 19-37, pl. 1-3.

Cloud, P. E., 1962, *Environment of calcium carbonate deposition west of Andros Island, Bahamas:* U.S. Geol. Survey, Prof. Paper 350, 138 p.

Dalquist, W. W., & Mamay, S. H., 1963, *A remarkable concentration of Permian amphibian remains in Haskell County, Texas:* Jour. Geology, v. 71, p. 641-644.

Davitashvili, L. Sh., 1945, *Tsenozy zhivykh organizmov i organicheskikh ostatkov:* Akad. Nauk Gruzin. SSSR, Soobshch., v. 6, no. 7, p. 527-534. [*Biocoenoses of living organisms and of organic remains.*] (Traduction no. 2077, Bureau de Recherches géologique, géophysique et minières, Paris.)

Deecke, Wilhelm, 1923, *Die Fossilisation:* vii + 216 p., Borntraeger (Berlin).

Dixon, O. A., 1970, *Nautiloids and current ripples as paleo-current indicators in Upper Ordovician limestones, Anticosti Island, Canada:* Jour. Sed. Petrology, v. 40, p. 682-687, text-fig. 1-6.

Efremov, J. A., 1940, *Taphonomy; a new branch of geology:* Pan-Am. Geologist, v. 74, p. 81-93.

Ehrenberg, Kurt, 1942, *Über einen möglicherweise von einem Tintenbeutel herrührenden Abdruck bei einem Orthoceras sp. aus dem böhmischen Obersilur:* Palaeobiologica, v. 7, p. 404-427.

Emery, K. O., 1968, *Positions of empty pelecypod valves on the continental shelf:* Jour. Sed. Petrology, v. 38, no. 4, p. 1264-1269, text-fig. 1, 2.

Erickson, J. M., 1971, *Wind-oriented gastropod shells as indicators of paleowind direction:* Jour. Sed. Petrology, v. 41, p. 589-593, text-fig. 1-5.

Fanck, Arnold, 1929, *Die bruchlose Deformation von Fossilien durch tektonischen Druck und ihr Einfluss auf die Bestimmung der Arten. Beobachtet und bearbeitet an den Pelecypoden der St. Galler Meeresmolasse:* 59 p., 9 text-fig., Diss. Univ. Zürich, Gebr. Fretz (Zürich).

Folk, R. L., 1965, *Some aspects of recrystallization in ancient limestones:* in Dolomitization and limestone diagenesis, a symposium, L. C. Pray, & R. C. Murray (eds.), Soc. Econ. Paleontologists, Mineralogists, Spec. Publ. no. 13, p. 16-48, text-fig. 1-14, tables 1-7.

Füchtbauer, Hans, & Goldschmidt, Herta, 1964, *Aragonitische Lumachellen im bituminösen Wealden des Emslandes:* Beiträge Mineralogie, Petrographie, v. 10, p. 184-197, text-fig. 1-8.

Futterer, Elke, 1974, *Untersuchungen zum Einsteuerungverhalten der Enzelklappen von Cardium echinatum L. und Cardium edule L. im Strömungskanal:* Neues Jahrb. Geologie, Paläontologie, Monatsh. 1974, p. 449-455, text-fig. 1-5.
———1976, *Rezente Schille: Transport und Einregelung tierischer Hartteile im Strömungskanal:* Zentralbl. Geol. Paläont., pt. 2, Paläont., v. 1976, p. 267-271, text-fig. 1-4.———1977, *Einregelung, Transport und Ablagerung biogener Hartteile im Strömungskanal:* 133 p., 26 text-fig., Univ. Tübingen Diss.

Gekker [Hecker], R. F., 1957, *Vvedenie v paleockologiyu:* 126 p., 27 text-fig., 20 pl., Gosudar. Nauch.-Tekh. Izd. Lit. Geol. i Okhrane Nedr (Moskva). [*Introduction to paleoecology.*]

Gerharz, R., 1966, *Sandwheels as a wind effect during transitory seasons:* Neues Jahrb. Geologie, Paläontologic, Monatsh., v. 1966, p. 550-553, text-fig. 1-3.

Geyer, O. F., 1973, *Grundzuege der Stratigraphie und Fazieskunde; Erster Band, Palaeontologische Grundlagen I; Das geologische Profil; Stratigraphie und Geochronologie:* 279 p., E. Schweizerbartsche Verlagsbuchhandlung (Naegele u. Obermiller) (Stuttgart).

Goethe, Friedrich, 1958, *Anhäufungen unversehrter Muscheln durch Silbermöven:* Natur u. Volk, v. 88, p. 181-187, 5 text-fig.

Gräf, I. E., 1958, *Tektonisch deformierte Goniatiten aus dem Westfal der Bohrung Rosenthal (Schacht Sophia-Jacoba V) im Erkelenzer Steinkohlenrevier:* Neues Jahrb. Geologie, Paläontologie, Monatsh. 1958, p. 68-95, text-fig. 1-8, 6 tables.

Grégoire, Charles, 1959a, *A study on the remains of organic components in fossil mother-of-pearl:* Inst. Royal Sci. Nat. Belgique, Bull., v. 35, no. 13, 14 p., 8 pl.———1959b, *Conchiolin remnants in mother-of-pearl from fossil Cephalopoda:* Nature, v. 184, Suppl., p. 1157-1158, text-fig. 1, 2.

———, & Teichert, Curt, 1965, *Conchiolin membranes in shell and cameral deposits of Pennsylvanian cephalopods, Oklahoma:* Oklahoma Geol. Notes, v. 25, p. 175-201, text-fig. 1, 11 pl.

Gripp, Karl, & Tufar, Werner, 1965, *Pyrit-Fossilien aus dem Unter-Eozän von Johannistal bei Heiligenhafen (Ost-Holstein):* Meyniana, v. 15, p. 29-40, pl. 1-4.

Gunter, Gordon, 1938, *Notes on invasion of fresh water by fishes of the Gulf of Mexico with special reference to the Mississippi-Atchafalaya River system:* Copeia, 1938, no. 2, p. 69-72.

Häntzschel, Walter, 1924, *Die Einbettungslage von Exogyra columba im sächsischen Cenoman-Quader:* Senckenbergiana, v. 6, p. 223-225.——— 1975, *Trace fossils and Problematica:* in Treatise on invertebrate paleontology, Curt Teichert (ed.), Part W (Miscellanea, Suppl. 1), 2nd edit. (rev. & enl.), 269 p., 110 text-fig., Geol. Soc. America & Univ. Kansas (Boulder, Colo. & Lawrence, Kans.).

———, El-Baz, Farouk, & Amstutz, G. C., 1968, *Coprolites: An annotated bibliography:* Geol. Soc. America, Mem. 108, 132 p., 6 text-fig., 11 pl.

Hakes, W. G., 1976, *Trace fossils and depositional environment of four clastic units, Upper Pennsylvanian megacyclothems, northeast Kansas:* Univ. Kansas Paleont. Contrib., Art. 63, 46 p., 11 text-fig., 13 pl.

Hall, A., & Kennedy, W. J., 1967, *Aragonite in fossils:* Royal Soc. London, Proc., v. 168B, p. 377-412.

Hallam, Anthony, 1967, *The interpretation of size-frequency distributions in molluscan death assemblages:* Palaeontology, v. 10, p. 25-42, text-fig. 1-11.

———, & O'Hara, M. J., 1962, *Aragonitic fossils in the Lower Carboniferous of Scotland:* Nature, v. 195, p. 273-274.

Hecht, Franz, 1933, *Der Verbleib der organischen Substanz der Tiere bei meerischer Einbettung:* Senckenbergiana, v. 15, p. 165-249, text-fig. 1-19.

Hollmann, Rudolf, 1962, *Über Subsolution und die "Knollenkalke" des Calcare Ammonitico Rosso im Monte Baldo (Malm, Norditalien):* Neues Jahrb. Geologie, Paläontologie, Monatsh. 1962, p. 163-179, text-fig. 1-8.——1964, *Subsolutions-Fragmente (Zur Biostratinomie der Ammonoidea in Malm des Monte Baldo/Norditalien):* Neues Jahrb. Geologie, Paläontologie, Abhandl., v. 119, no. 1, p. 22-82, text-fig. 1-7, 4 pl., 1 table.——1968a, *Diagenetische Gehäuse-Hypertrophie an Ammoniten aus dem Oberjura Ostafrikas:* Neues Jahrb. Geologie, Paläontologie, Abhandl., v. 130, no. 3, p. 305-334, text-fig. 1-4, tables 26-30. ——1968b, *Zur Morphologie rezenter Mollusken-Bruchschille:* Paläont. Zeitschr., v. 42, p. 217-235, 14 text-fig., 2 pl.——1968c, *Über Schalenabschliffe bei Cardium edule aus der Königsbucht bei List auf Sylt:* Helgol. Wiss. Meeresunters., v. 18, p. 169-193, 21 text-fig.

Illies, Hennig, 1949, *Die Lithogenese des Untereozäns in Nordwest-Deutschland:* Geol. Staatsinstitut Hamburg, Mitteil., v. 18, p. 7-45, 9 text-fig., 2 pl.——1950, *Über die erdgeschichtliche Bedeutung der Konkretionen:* Deutsche Geol. Gesell., Zeitschr., v. 101, p. 95-98.

Janicke, Volkmar, 1969, *Untersuchungen über den Biotop der Solnhofener Plattenkalke:* Bayer. Staatssamml. Paläontologie, Historische, Geologie, Mitteil., v. 9, p. 117-181, text-fig. 1-21, 5 pl.

Jarke, J., 1961, *Beobachtungen über Kalkauflösung an Schalen von Mikrofossilien in Sedimenten der westlichen Ostsee:* Deutsche Hydrogr. Zeitschr., v. 14, p. 6-11, 1 pl.

Jeletzky, J. A., 1966, *Comparative morphology, phylogeny, and classification of fossil Coleoidea:* Univ. Kansas Paleont. Contrib., Mollusca, Art. 7, 162 p., 15 text-fig., 25 pl.

Johnson, R. G., 1957, *Experiments on the burial of shells:* Jour. Geology, v. 65, p. 527-535, text-fig. 1,2.

Kay, Marshall, 1945, *Paleogeographic and palinspastic maps:* Am. Assoc. Petroleum Geologists, Bull., v. 29, p. 426-450, text-fig. 1-12.

Kennedy, W. A., & Hall, A., 1967, *The influence of organic matter on the preservation of aragonite in fossils:* Geol. Soc. London, Proc., vol. 1967, no. 1643, p. 253-255.

——, & Taylor, J. D., 1968, *Aragonite in rudistes:* Geol. Soc. London, Proc., vol. 1968, no. 1645, p. 325-331.

Kennedy, W. J., & Klinger, H. C., 1972, *Hiatus concretions and hardground horizons in the Cretaceous of Zululand (South Africa):* Palaeontology, v. 15, pt. 4, p. 539-549, text-fig. 1-3, pl. 106-108.

Kennett, J. P., 1966, *Foraminiferal evidence of a shallow calcium carbonate solution boundary, Ross Sea, Antarctica:* Science, v. 153, p. 191-193.

Kessel, E., 1938, *Über Erhaltungsfähigkeit mariner Molluskenschalen in Abhängigkeit von der Struktur:* Arch. Molluskenkunde, v. 70, p. 248-254, text-fig. 1, 2.

Kindle, E. M., 1938, *A pteropod record of current direction:* Jour. Paleontology, v. 12, p. 515-516, text-fig. 1.

Klähn, Hans, 1932, *Der quantitative Verlauf der Aufarbeitung von Sanden, Geröllen und Schalen in wässerigem Medium:* Neues Jahrb. Mineralogie, Geologie, Paläontologie, Beil.-Bd. 67 B I, p. 313-412, text-fig. 1-5, tables 1-42.——1936, *Die Anlösungsgeschwindigkeit kalkiger anorganischer und organischer Körper innerhalb eines wässerigen Mediums:* Centralbl. Mineralogie, Geologie, Paläontologie, v. 1936 A, p. 328-348.

Königswald, G. H. R. von, 1930, *Die Arten der Einregelung ins Sediment bei den Seesternen und Seelilien des unterdevonischen Bundenbacher Schiefers:* Senckenbergiana, v. 12, p. 338-360, text-fig. 1-19, 7 pl.

Kolb, Anton, 1951, *Die erste Meduse mit Schleifspur aus den Solnhofener Schiefern:* Geol. Blätter. Nordost-Bayern, v. 1, p. 63-69, text-fig. 1, pl. 1.——1961, *Die Ammoniten als Dibranchiata:* Geol. Blätter. Nordost-Bayern, v. 11, p. 1-26, text-fig. 1-5, pl. 1, 2.——1967, *Ammoniten-Marken aus den Solnhofener Schiefern. (Ein Beweis für die Octopoden-Organisation der Ammoniten):* Geol. Blätter. Nordost-Bayern, v. 17, p. 21-37.

Korschelt, E., 1924, *Lebensdauer, Altern und Tod.* 3. Aufl.: viii + 451 p., 221 text-fig., Fischer (Jena).

Krantz, Renate, 1972, *Die sponge-gravels von Faringdon (England):* Neues Jahrb. Geologie, Paläontologie, Abhandl., v. 140, p. 207-231, text-fig. 1-11.

Krejci-Graf, Karl, 1932, *Senkrechte Regelung von Schneckengehäusen:* Senckenbergiana, v. 14, p. 295-299, text-fig. 1.

Krinsley, David, 1960, *Orientation of orthoceracone cephalopods at Lemont, Illinois:* Jour. Sed. Petrology, v. 30, p. 321-323, text-fig. 1.

Küpper, Heinrich, 1933, *Schleifspuren an Gastropodenschalen, ein Hinweis auf den Ebbe-Flut-Bereich:* Nat. Museums Wien, Ann., v. 1932/1933, p. 271-274, 1 pl.

Langerfeldt, J., 1935, *Untermeerische Scharrkreise im Schlick:* Natur u. Volk, v. 65, p. 458-461, text-fig. 1-4.

Langheinrich, Gunter, 1966, *Syndiagenetische Fossildeformationen im untersten Lias (Hettangium) von Göttingen:* Neues Jahrb. Geologie, Paläontologie, Monatsh. 1966, p. 666-680, text-fig. 1-5. ——1968, *Die Bestimmung des Deformationsgrades von tektonisch deformierten Fossilien mit primär rechten Winkeln:* Geologie, v. 17, p. 1086-1095, text-fig. 1-5.

Lawrence, L. J., 1953, *The replacement of crinoid stems and gastropods by cassiterite at Emmaville, New South Wales:* Royal Soc. New South Wales, Jour., Proc., v. 86, p. 119-122, 1 pl.

Lazell, J. D., 1965, *An Anolis (Sauria, Iguanidae) in amber:* Jour. Paleontology, v. 39, p. 379-382, pl. 54.

Lehmann, Ulrich, 1966, *Dimorphismus bei Ammoniten der Ahrensburger Lias-Geschiebe:* Paläont. Zeitschr., v. 40, p. 26-55, text-fig. 1-11, pl. 1,2.——1967a, *Ammoniten mit Kieferapparat und Radula aus Lias-Geschieben:* Paläont. Zeitschr., v. 41, p. 38-45, pl. 4.——1967b, *Ammoniten mit Tintenbeutel:* Paläont. Zeitschr., v. 41, p. 132-136, text-fig. 1-4.

——, & Weitschat, Wolfgang, 1973, *Zur Anatomie und Ökologie von Ammoniten: Funde von Kropf und Kiemen:* Paläont. Zeitschr., v. 47, p. 69-76, text-fig. 1, pl. 11.

Linck, Otto, 1956, *Drift-Marken von Schachtelhalm-Gewächsen aus dem mittleren Keuper (Trias):* Senckenbergiana Lethaea, v. 37, p. 39-51, text-fig. 1-3, pl. 1,2.

Lowenstam, H. A., 1963, *Biologic problems relating to the composition and diagenesis of sediments:* in The earth sciences, problems and progress in current research, T. W. Donnelly (ed.), Rice Univ. Semicentenn. Publ., p. 137-195, Chicago Univ. Press (Chicago).

Macurda, D. B., Jr., & Meyer, D. L., 1975, *The microstructure of the crinoid endoskeleton:* Univ. Kansas, Paleont. Contrib., Paper 74, p. 1-22, pl. 1-30.

Mädler, K., 1938, *Eine natürliche Katzenmumie:* Natur u. Volk, v. 68, p. 574-576, text-fig. 1,2.

Martinsson, Anders, 1955, *Studies on the ostracode family Primitiopsidae:* Univ. Uppsala, Geol. Inst., Bull., v. 36, p. 1-33.

Mayr, F. X., 1966, *Zur Frage des "Auftriebes" und der Einbettung bei Fossilien der Solnhofener Schichten:* Geol. Blätter Nordost-Bayern, v. 16, p. 102-107, text-fig. 1, 2.

Middleton, G. V., 1967, *The orientation of concavo-convex particles deposited from experimental turbidity currents:* Jour. Sed. Petrology, v. 37, p. 229-232, text-fig. 1, 2.

Mosebach, Rudolf, 1952, *Wässerige H_2S-Lösungen und das Verschwinden kalkiger tierischer Hartteile aus werdenden Sedimenten:* Senckenbergiana, v. 33, p. 13-22, text-fig. 1, tables 1, 2.

Müller, A. H., 1950, *Stratonomische Untersuchungen im oberen Muschelkalk des Thüringer Beckens:* Geologica, v. 4, 74 p., 10 text-fig., 11 pl.——1951a, *Diagenetische Untersuchungen in der obersenonen Schreibkreide von Rügen:* Geol. Dienst Berlin, Abhandl., v. 228, 29 p., 9 text-fig., 4 pl.——1951b, *Grundlagen der Biostratonomie:* Deutsche Akad. Wiss. Berlin, Abhandl., Kl. f. Math. u. allg. Nat., v. 1950, no. 3, 147 p., text-fig. 1-79.——1962a, *Körperlich erhaltene Fische (Palaeoniscoidea) aus dem Zechstein (Kupferschiefer) von Ilmenau (Thüringen):* Geologie, v. 11, p. 845-856, text-fig. 1-11.——1962b, *Fossil oder Pseudofossil?:* Geologie, v. 10, p. 1204-1213, text-fig. 1-3, pl. 1,2.——1963, *Lehrbuch der Paläozoologie. I. Allgemeine Grundlagen:* 1st edit. (1957), 322 p., 177 text-fig.; 2nd edit. (1963), 405 p., 228 text-fig., G. Fischer (Jena).——1967, *Besondere Lebensspuren und Sedimentmarken auf rezenten Flugsanddünen:* Natur u. Museum, v. 97, p. 354-366, text-fig. 1-14.——1969a, *Zur Ökologie und Fossilisation triadischer Ophiuroidea (Echinodermata):* Deutsche Akad. Wiss. Berlin, Monatsber., v. 11, p. 386-398, pl. 1-4.——1969b, *Ammoniten mit "Eierbeutel" und die Frage nach dem Sexualdimorphismus der Ceratiten (Cephalopoda):* Deutsche Akad. Wiss. Berlin, Monatsber., v. 11, p. 411-420, text-fig. 1-3.——1971, *Miscellanea aus dem limnisch-terrestrischen Unterperm (Rotliegenden) von Mitteleuropa: Teil 1:* Deutsche Akad. Wiss. Berlin, Monatsber., v. 13, p. 937-948, text-fig. 1-4, pl. 1-3.——1976, *Zur Taphonomie, Ichnologie und Ökologie triadischer Ophiuroidea (Echinodermata):* Zeitschr. Geol. Wiss., v. 4, p. 1399-1411, 4 pl.

——, & Zimmermann, Helmut, 1962, *Aus Jahrmillionen. Tiere der Vorzeit:* 409 p., 290 text-fig., G. Fischer (Jena).

Naef, Adolf, 1922, *Die fossilen Tintenfische. Eine paläozoologische Monographie:* vi + 322 p., 102 text-fig., Fischer (Jena).

Ogose, Sunao, 1956, *On some problems concerning the so-called "fossil-enclosure":* Geol. Soc. Japan, Jour., v. 67, p. 585-600, text-fig. 1-8.

Orbigny, Alcide d', 1849, *Cours élémentaire de Paléontologie et de Géologie stratigraphique. Vol. 1. Chapitre de la Fossilization:* p. 34-69, V. Masson (Paris).

Paine, Gaylord, 1937, *Fossilization of bone:* Am. Jour. Sci., v. (5) 34, p. 148-157.

Papp, Adolph, 1939, *Beobachtungen über Sedimentsonderungen und Spülsäume an Binnenmeeren:* Senckenbergiana, v. 21, p. 113-118, text-fig. 1-3.——1941, *Beobachtungen über Aufarbeitung von Molluskenschalen in Gegenwart und Vergangenheit:* Zool.-bot. Gesell. Wien, Verhandl., v. 88/89, p. 231-236, text-fig. 1-7.

Pavoni, Nazario, 1959, *Rollmarken von Fischwirbeln aus den oligozänen Flyschschiefern von Angi-Matt (Kt. Glarus):* Eclogae Geol. Helvetiae, v. 52, p. 941-949, text-fig. 1-3.

Petránek, Jan, & Komárková, Eva, 1953, *Orientace schránek hlavanožců ve vápecích Barrandienu a její paleogeografický význam:* Ústred. Ústavu Geol., Sborník, v. 20, p. 129-148. [*The orientation of the shells of cephalopods in the limestones of the Barrandian and its paleogeographical significance.*]

Pfannenstiel, Max, 1930, *Über die Einbettungslage der Gryphaea dilatata im Callov der Normandie und im heutigen Strandsediment:* Senckenbergiana, v. 12, p. 126-139, text-fig. 1-8.

Pratje, Otto, 1929, *Fazettieren von Molluskenschalen:* Paläont. Zeitschr., v. 11, p. 151-169, text-fig. 1-73.

Quenstedt, Werner, 1927, *Beiträge zum Kapitel Fossil und Sediment vor und bei der Einbettung:* Neues Jahrb. Mineralogie, Geologie, Paläontologie

(Pompeckj-Festband), v. 58B, p. 353-432.
Raymond, P. E., 1920, *The appendages, anatomy, and relationships of trilobites:* Connecticut Acad. Arts Sci., Mem., v. 7, 169 p., 46 text-fig., 11 pl.
Reginek, Hans, 1917, *Die pelomorphe Deformation bei den jurassischen Pholodomyen und ihr Einfluss auf die bisherige Unterscheidung der Arten:* Soc. Paléont. Suisse, Mém. 42, no. 3, p. 1-67, pl. 1-4.
Reineck, H.-E., 1967, *Parameter von Schichtung und Bioturbation:* Geol. Rundschau, v. 56, p. 420-438, text-fig. 1-11.
———, Gutmann, W. F., & Hertweck, Günther, 1967, *Das Schlickgebiet südlich Helgoland als Beispiel rezenter Schelfablagerungen:* Senckenbergiana Lethaea, v. 48, p. 219-275, text-fig. 1-12, pl. 1-7.
———, & Singh, I. B., 1973, *Depositional sedimentary environments with reference to terrigenous clastics:* 439 p., 579 text-fig., Springer (New York).
Reis, O. M., 1893, *Untersuchungen über die Petrificirung der Muskulatur:* Archiv f. Mikrosk. Anatomie, v. 41, p. 492-584, 3 pl.
Reyment, R. A., 1958, *Some factors in the distribution of fossil cephalopods:* Acta Univ. Stockholm, Contrib. Geology, v. 1, p. 97-184, text-fig. 1-24, 7 pl.——1968, *Orthoconic nautiloids as indicators of shoreline surface currents:* Jour. Sed. Petrology, v. 38, p. 1387-1389, text-fig. 1-3.—— 1970, *Vertically inbedded cephalopod shells. Some factors in the distribution of fossil cephalopods, 2:* Palaeogeography, Palaeoclimatology, Palaeoecology, v. 7, p. 103-111, pl. 1-3.
Richards, A. G., 1951, *The integument of arthropods:* 411 p., Minnesota Univ. Press (Minneapolis).
Richter, Rudolf, 1922, *Flachseebeobachtungen zur Paläontologie und Geologie III-VI:* Senckenbergiana, v. 4, p. 103-141, pl. 3.——1929, *Gründung und Aufgaben der Forschungsstelle für Meeresgeologie "Senckenberg" in Wilhelmshaven:* Natur u. Museum, v. 59, p. 1-30.—— 1931, *Tierwelt und Umwelt im Hunsrückschiefer; zur Entstehung eines schwarzen Schlammgesteins:* Senckenbergiana, v. 13, p. 299-342, text-fig. 1-16.
———1936, *Marken und Spuren im Hunsrückschiefer. II Schichtung und Grundleben:* Senckenbergiana, v. 18, p. 215-244, text-fig. 1-4.
———1937, *Die "Salter'sche Einbettung" als Folge und Kennzeichen des Häutungsvorganges:* Senckenbergiana, v. 19, p. 413-431, text-fig. 1-3.
———1942, *Die Einkippungsregel:* Senckenbergiana, v. 25, p. 181-206, text-fig. 1.——1954, *Fährte eines "Riesenkrebses" im Rheinischen Schiefergebirge:* Natur u. Volk, v. 84, p. 261-269, text-fig. 1-5.
Rietschel, Siegfried, 1968, *Bedeutung, Muttergestein und Fundumstände der Weichkörper Wissenbacher Cephalopoden:* Natur u. Museum, v. 98, p. 409-412, text-fig. 1-4.

Rinehart, C. D., Ross, D. C., & Hubek, N. K., 1959, *Paleozoic and Mesozoic fossils in a thick stratigraphic section in the eastern Sierra Nevada, California:* Geol. Soc. America, Bull., v. 70, p. 941-946, text-fig. 1, 2.
Rolfe, W. D., & Brett, D. W., 1969, *Fossilization processes:* in Organic geochemistry, methods and results, G. Eglinton & M. T. J. Murphy (ed.), p. 213-244, 9 text-fig., Springer-Verlag (Berlin).
Rothpletz, August, 1909, *Über die Einbettung der Ammoniten in den Solnhofener Schichten:* K. Bayer. Akad. Wiss., Abhandl., v. 24, 2. Kl, 2. Abt., p. 311-337, text-fig. 1-3, 3 pl.
Ruedemann, Rudolf, 1897, *Evidence of current action in the Ordovician of New York:* Am. Geologist, v. 19, p. 367-391.
Rutsch, Rolf, 1937, *Ein Fall von Einregelung bei Mollusken aus dem Vindobonien des bernischen Seelandes:* Eclogae Geol. Helvetiae, v. 29, p. 600-607, 1 pl.
Schäfer, Wilhelm, 1955, *Fossilisationsbedingungen der Meeressäuger und Vögel:* Senckenbergiana Lethaea, v. 36, p. 1-25, text-fig. 1-3, 2 pl.—— 1962, *Aktuo-Paläontologie nach Studien in der Nordsee:* 668 p., 277 text-fig., 36 pl., Kramer (Frankfurt a. M.).——1972, *Ecology and palaeoecology of marine environments:* G. Y. Craig (ed.), 568 p., 277 text-fig., 39 pl., Oliver & Boyd (Edinburgh).
Schauer, M., 1971, *Biostratigraphie und Taxionomie der Graptolithen des tieferen Silurs unter besonderer Berücksichtigung der tektonischen Deformation:* Freiberger Forsch.-Hefte, v. C273, 185 p., 35 text-fig., 45 pl.
Schlee, Dieter, 1973, *Harzkonservierte fossile Vogelfedern aus der untersten Kreide:* Jour. Ornithologie, v. 114, p. 207-219, text-fig. 1,2.
Schmidt, R. A. M., & Sellmann, P. V., 1966, *Mummified Pleistocene ostracods in Alaska:* Science, v. 153, p. 167-168, text-fig. 1.
Sdzuy, Klaus, 1962, *Über das Entzerren von Fossilen (mit Beispielen aus der unterkambrischen Saukianda-Fauna):* Paläont. Zeitschr., v. 36, p. 275-284, text-fig. 1-8, pl. 1.
Seilacher, Adolf, 1953, *Studien zur Palichnologie. II. Die fossilien Ruhespuren (Cubichnia):* Neues Jahrb. Geologie, Paläontologie, Abhandl., v. 98, p. 87-124, text-fig. 1-5, pl. 7-13.——1960a, *Strömungsanzeichen im Hunsrückschiefer:* Hess. Landesamt Bodenforsch., Notizbl., v. 88, p. 88-106, text-fig. 1-13, pl. 12, 13.——1960b, *Epizoans as a key to ammonoid ecology:* Jour. Paleontology, v. 34, p. 189-193, text-fig. 1-3.
———1963, *Umlagerung und Rolltransport von Cephalopoden-Gehäusen:* Neues Jahrb. Geologie, Paläontologie, Monatsh., 1963, p. 593-615, text-fig. 1-9.——1968a, *Sedimentationsprozesse im Ammonitengehäusen:* Akad. Wiss. Lit., Abhandl., math.-nat., Kl., v. 1967, no. 9, p. 191-203, text-fig. 1-5, pl. 1.——1968b, *Swimming habits of belemnites—recorded by boring barnacles:* Palae-

ogeography, Palaeoclimatology, Palaeoecology, v. 4, p. 279-285, text-fig. 1-5.——1969, *Paleoecology of boring barnacles:* Am. Zoologist, v. 9, p. 705-719, text-fig. 1-8, pl. 1-5.——1970, *Begriff und Bedeutung der Fossil-Lagerstäten:* Neues Jahrb. Geologie, Paläontologie, Monatsh., 1970, p. 34-39.——1973, *Biostratinomy: The sedimentology of biologically standardized particles:* in Evolving concepts in sedimentology, R. N. Ginsburg (ed.), p. 159-177, text-fig. 1-16, Johns Hopkins Univ. Press (Baltimore).

Sewertzoff, S. A., 1934, *Vom Massenwechel bei den Wildtieren:* Biol. Zentralbl., v. 54, p. 337-364, text-fig. 1.

Shrock, R. R., 1948, *Sequence in layered rocks:* 507 p., illus., McGraw-Hill (New York).

Sognnaes, R. F., 1963, *Mechanisms of hard tissue destruction:* Am. Assoc. Advanc. Sci., Publ., no. 75, xiv + 714 p., illus.

Stehli, F. G., 1956, *Shell mineralogy in Paleozoic invertebrates:* Science, v. 123, p. 1031-1032.

Stehn, O. E., 1968, *Zur Fossildeformation im Rheinischen Schiefergebirge:* Forsch.-Ber. d. Landes Nordrhein-Westfalen, v. 1968, 31 p., text-fig. 1-19.

Stevenson, F. J., 1961, *Some aspects of the distribution of biochemicals in geologic environments:* Geochimica et Cosmochimica Acta, v. 19, p. 261-271.

Stürmer, Wilhelm, 1965, *Röntgenaufnahmen von einigen Fossilien aus dem Geologischen Institut der Universität Erlangen-Nürnberg:* Geol. Blätter Nordost-Bayern, v. 15, p. 217-223, text-fig. 1-7, 1 pl.——1967, *Röntgenaufnahmen von Fossilien:* Image, v. 1967 (2), p. 25-32.——1968a, *Einige Beobachtungen an devonischen Fossilien mit Röntgenstrahlen:* Natur u. Museum, v. 98, p. 413-417, text-fig. 1-5.——1968b, *Blick ins Unsichtbare:* Festschrift des Kronberg-Gymnasiums Aschaffenburg, p. 195-206, text-fig. 1-9.——1969, *Pyrit-Erhaltung von Weichteilen bei devonischen Cephalopoden:* Paläont. Zeitschr., v. 43, p. 10-12, pl. 1-3.——1973, *Neue Ergebnisse der Paläontologie durch Röntgenuntersuchungen:* Die Naturwissenschaften, v. 60, p. 407-411, text-fig. 1-6.

Tauber, A. F., 1942, *Postmortale Veränderungen an Molluskenschalen und ihre Auswertbarkeit für die Erforschung vorzeitlicher Lebensräume:* Palaeobiologica, v. 7, p. 448-495, text-fig. 1-15, pl. 1-4.

Teichert, Curt, 1930, *Über die Möglichkeit der syngenetischen Entstehung einiger Metallsulfide in Kalken durch die konzentrierende Tätigkeit der Organismen:* Centralbl., Mineralogie, Geologie, Paläontologie, v. 1930 B, p. 49-70.

——, & Serventy, D. L., 1947, *Deposits of shells transported by birds:* Am. Jour. Sci., v. 245, p. 322-328.

——, et al., 1964, *Endoceratoidea—Actinoceratoidea—Nautiloidea:* in Treatise on invertebrate paleontology, R. C. Moore (ed.), Part K, p. K13-K466, text-fig. 1-337, Geol. Soc. America & Univ. Kansas Press (New York; Lawrence, Kansas).

Thenius, Erich, 1963, *Versteinerte Urkenden:* 174 p., 77 text-fig., Springer (Berlin, Göttingen, Heidelberg). [2nd ed., 1971, 184 p., 89 text-fig.]

Timofeeff-Ressovsky, N. V., Voroncov, N. N., & Jablokov, A. N., 1975, *Kurzer Grundriss der Evolutionstheorie:* in Genetik, v. 7, 360 p., 132 text-fig.

Toots, Heinrich, 1965a, *Orientation and distribution of fossils as environmental indicators:* Wyoming Geol. Assoc. Guidebook, 19th Ann. Field Conf., p. 219-229, text-fig. 1-6, 2 tables.——1965b, *Sequence of disarticulation in mammalian skeletons:* Wyoming Univ., Contrib. Geology, v. 4, p. 37-39, text-fig. 1-3.

Trusheim, Ferdinand, 1931, *Versuche über Transport und Ablagerung von Mollusken:* Senckenbergiana, v. 13, p. 124-139, text-fig. 1-3.——1938, *Triopsiden (Crust. Phyll.) aus dem Keuper Frankens:* Paläont. Zeitschr., v. 19 (1937), p. 198-216, text-fig. 1-10, pl. 13, 14.

Voigt, Ehrhard, 1935, *Die Erhaltung von Epithelzellen mit Zellkernen, von Chromatophoren und Corium in fossiler Froschhaut aus der mitteleozänen Braunkohle des Geiseltales:* Nova Acta Leopoldina, n. ser., v. 3, p. 339-360, 5 pl.——1936, *Weichteile an Säugetieren aus der eozänen Braunkohle des Geiseltales:* Nova Acta Leopoldina, n. ser., v. 4, p. 301-310, 2 pl.——1937, *Paläohistologische Untersuchungen an Bernsteineinschlüssen:* Paläont. Zeitschr., v. 19, p. 35-46.——1938a, *Ein fossiler Saitenwurm (Gordius tenuifibrosus n. sp. aus der Eozänen Braunkohle des Geiseltales:* Nova Acta Leopoldina, n. ser., v. 5, p. 351-360, 2 pl.——1938b, *Weichteile an fossilen Insekten aus der eozänen Braunkohle des Geiseltales bei Halle (Saale):* Nova Acta Leopoldina, n. ser., v. 6, p. 1-38, 3 text-fig., 7 pl.——1938c, *Eine neue Methode zur mikroskopischen Untersuchung von Bernsteineinschlüssen:* Forsch. u. Fortschritte, v. 14, p. 55-56.——1950, *Mikroskopische Untersuchungen an fossilen tierischen Weichteilen und ihre Bedeutung für Systematik und Biologie:* Deutsche Geol. Gesell., Zeitschr., v. 101, p. 99-104.——1956, *Der Nachweis des Phytals durch Epizoen als Kriterien der Tiefe vorzeitlicher Meere:* Geol. Rundschau, v. 45, p. 97-119, text-fig. 1-5, pl. 1-4.——1957, *Ein parasitischer Nematode in fossiler Coleopteren-Muskulatur as der eozänen Braunkohle des Geiseltales bei Halle (Saale):* Paläont. Zeitschr., v. 31, p. 35-39, pl. 1.——1962, *Johannes Weigelt als Paläontologe:* Hamburg. Geol. Staatsinst., Mitteil., v. 31, p. 25-50, text-fig. 1.——1966, *Die Erhaltung vergänglicher Organismen durch Abformung infolge Inkrustation durch sessile Tiere, unter besonderer Berücksichtigung einiger Bryozoen und Hydrozoen aus der oberen Kreide:* Neues Jahrb. Geologie, Paläontologie, Abhandl.,

v. 125, p. 401-422, text-fig. 1-6, pl. 1-5.——1968a, *Über Immuration bei fossilen Bryozoen dargestellt an Funden aus der oberen Kreide:* Akad. Wiss. Göttingen, Math.-phys. Kl., Nachricht., v. 1968, no. 4, p. 47-63, 4 pl.——1968b, *Über Hiatus-Konkretionen (dargestellt an Beispielen aus dem Lias):* Geol. Rundschau, v. 58, p. 281-296, text-fig. 1-8.

Vossmerbäumer, Herbert, 1972, *Cephalopoden im Muschelkalk Mainfrankens. Ein biostratinomischer Beitrag.:* Geol. Blätter Nordost-Bayern, v. 22, p. 8-25, text-fig. 1-6, pl. 1-3.

Vries, Peter de, 1959, *Optisches Entzerrungsgerät für tektonisch verformte Fossilien:* Geologie, v. 8, p. 448-449, text-fig. 1-3.

Vyalov [Vialov], O. S., 1961, *Javlenic žibbyx immurazii v priode:* Akad. Nauk Ukrains. RSR, Dopovidi, v. 11, p. 1510-1512 (Engl. summary). [*Phenomena of vital immuration in nature.*]

Wasmund, Erich, 1926, *Biocoenose und Thanatocoenose. Biosoziologische Studie über Totengesellschaften:* Archiv Hydrobiologie, v. 17, 116 p., 4 pl.

Weigelt, Johannes, 1919, *Geologie und Nordseefauna:* Der Steinbruch, v. 14, p. 228-231, 244-246, text-fig. 1-10.——1923, *Angewandte Geologie und Paläontologie der Flachseegesteine und das Erzlager von Salzgitter:* Fortschr. Geologie Palaeontologie, no. 4, p. 1-128, pl. 1-14. 1927, *Rezente Wirbeltierleichen und ihre paläobiologische Bedeutung:* xvi + 227 p., 28 text-fig., 37 pl., Max Weg (Leipzig).——1928a, *Ganoidfischleichen im Kupferschiefer und in der Gegenwart:* Palaeobiologica, v. 1, p. 323-356, text-fig. 1-16, pl. 1-4.——1928b, *Die Pflanzenreste des mitteldeutschen Kupferschiefers und ihre Einschaltung ins Sediment:* Fortschr. Geologie Palaeontologie, v. 6, pt. 19, p. I-IV, 395-592, text-fig. 1-14, pl. 1-35.

Wetzel, Otto, 1933, *Die in organischer Substanz erhaltenen Mikrofossilien des baltischen Kreidefeuersteins:* Palaeontographica, v. 77, p. 147-186; v. 78, p. 1-110, text-fig. 1-15, pl. 1-7.

Wetzel, Walter, 1913, *Über ein Kieselholzgeschiebe mit Teredonen aus den Holtenauer Kanalaufschlüssen:* 6. Niedersächs. Geol. Vereins, Jahresber., p. 20-59.——1937, *Die koprogenen Beimengungen mariner Sedimente und ihre diagnostische und lithogenetische Bedeutung:* Neues Jahrb. Mineralogie, Geologie, Paläontologie, Beil-Bd. 78 B, p. 109-122.

Whittington, H. B., 1971, *The Burgess Shale: History of research and preservation of fossils:* North American Paleont. Convention, Proc., Symposium 1969, Chicago, v. 2, Part I, p. 1170-1201, text-fig. 1-24, Allen Press (Lawrence, Kans.).

Winkler, H. G. F., 1964, *Das T-P-Feld der Diagenese und niedrigtemperierten Metamorphose auf Grund von Mineralreaktionen:* Beitr. Mineralogie, Petrologie, v. 10, p. 70-93.

Wolf, Karl, 1965, *"Grain-diminution" of algal colonies to micrite:* Jour. Sed. Petrology, v. 35, p. 420-427, text-fig. 1-7.

Wolff, Erwin, 1954, *Taxionomie, Stratigraphie und Stratinomie (nicht Taxonomie, Stratographie und Stratonomie) und Verkürzungen wie Palichnologie, Palökologie:* Senckenbergiana Lethaea, v. 35, p. 115-117.

Woodward, G. D., & Marcus, L. F., 1973, *Rancho La Brea fossil deposits: A re-evaluation from stratigraphic and geological evidence:* Jour. Paleontology, v. 40, p. 54-69, text-fig. 1-6.

Wray, J. L., & Daniels, F., 1957, *Precipitation of calcite and aragonite:* Am. Chem. Soc., Jour., v. 79, p. 2031-2034.

Yochelson, E. L., White, J. S., Jr., & Gordon, M., Jr., 1967, *Aragonite and calcite in mollusks from the Pennsylvanian Kendrick Shale (of Jillson) in Kentucky:* U.S. Geol. Survey, Prof. Paper 575D, p. 76-78, 1 pl.

Zangerl, Rainer, 1971, *On the geologic significance of perfectly preserved fossils:* North American Paleont. Convention, Proc., v. 2, Part I, p. 1207-1222, text-fig. 1-8, Allen Press (Lawrence, Kans.).——, & Richardson, E. S., 1963, *The paleoecological history of two Pennsylvanian black shales:* Fieldiana, Geol. Mem., v. 4, p. 1-352, text-fig. 1-51, pl. 1-55.

Zapfe, Helmuth, 1936, *Die Erhaltungsmöglichkeit von Aragonit im Fossilisationsprozess, untersucht mit Hilfe des Reagens von Feigel und Leitmeier:* Akad. Wiss. Wien, Anzeiger, v. 11, 2 p.

Zeiss, Arnold, 1969, *Weichteile ectocochleater paläozoischer Cephalopoden in Röntgenaufnahmen und ihre paläontologische Bedeutung:* Paläont. Zeitschr., v. 43, p. 13-27, text-fig. 1-6, pl. 1-3.

Precambrian

BIOGEOGRAPHY AND BIOSTRATIGRAPHY

PRECAMBRIAN[1]

By M. F. Glaessner
[University of Adelaide]

CONTENTS

	PAGE
Introduction	A80
Stratigraphic Distribution	A82
Geographic Distribution	A83
Chronostratigraphic Data	A85
Paleoecology and Taphonomy	A86
Evolutionary Significance	A87
Suprageneric Taxa of Precambrian Metazoa	A90
Systematic Descriptions	A91
?Phylum Porifera Grant, 1872	A91
Class, Order, and Family Uncertain	A91
Phylum Coelenterata Frey & Leuckart, 1847	A91
Class Hydrozoa Owen, 1843	A91
Class Scyphozoa Götte, 1887	A92
Class Conulata Moore & Harrington, 1956	A94
Medusae of Uncertain Affinities	A94
Problematical Coelenterata: "Petalonamae" Pflug, 1970	A96
Phylum Annelida Lamarck, 1809	A102
Class Polychaeta Grube, 1850	A102
Phylum Arthropoda Siebold & Stannius, 1845	A104
Superclass Trilobitamorpha Størmer, 1944 (or Chelicerata Heymons, 1901)	A104
Class and Order Uncertain	A104
Superclass Crustacea Pennant, 1777	A105
Class Branchiopoda Latreille, 1817	A105
?Phylum Pogonophora Johannson, 1938	A107
Phylum Uncertain	A107
Class, Order, and Family Uncertain	A107
Taxa with Doubtful Invertebrate Affinities	A108
Names Given to Precambrian Nonmetazoan and Trace Fossils	
Filamentous, Coccoid, and Other Microscopic Algae	A108
Sporomorphs and Acritarchs	A109
Stromatolites	A109
Doubtful Taxa	A110

---

[1] Manuscript received February, 1976.

Problematical Fossils (Including Microphytoliths) A111
Megascopic Algae A111
Trace Fossils A112

REJECTED AND UNRECOGNIZABLE TAXA A112

REFERENCES A113

INTRODUCTION

Precambrian time comprises about seven-eighths of the time since the planet Earth came into being, or about five-sixths of the time represented by actually or potentially fossiliferous rocks. The ratio of fossil taxa of Precambrian age to younger ones can only be a minute fraction. The main reasons for this disproportionality have now become clearer. Although it is true that most Precambrian rocks have been so much altered that organic remains are not preserved in them, intensive exploration of less altered Precambrian sediments has indicated that during the greater part of Precambrian time organic evolution advanced only slowly. One-half of the entire time span of evolution for which there is fossil evidence produced only Procaryota, non-nucleated cells without organelles. The oldest "fossils" are microscopic bodies of simple structure and mostly spheroidal shape from rocks 3.1 to 3.4 billion years old. Their biogenic origin is at present controversial. Permineralized, cellular remains of Procaryota, the blue-green algae (Cyanophyta) and bacteria, have been found in many cherts of Proterozoic age. Stromatolites, large, layered, biogenic sedimentary structures in carbonates and siliceous sediments, are known from rocks about 3 billion years old. They are abundantly represented in Proterozoic sediments, commencing with some that are about 2.3 billion years old. "Stromatolitic, microbial biocoenoses of this time and the later Precambrian were based on filamentous photoautotrophs" (SCHOPF, 1975). They became markedly less abundant during the Paleozoic, but are still being formed today in restricted environmental conditions that are unfavorable for Metazoa. In some stromatolitic rocks the fossilized cells of blue-green algae and bacteria that caused their deposition have been observed.

Some 20 different Proterozoic fossil microbiota are known, mostly from cherts associated with algal-laminated (stromatolitic) carbonate rocks. The best known are those described by BARGHOORN and TYLER (1965) and J. W. SCHOPF (1968). Eucaryotic cells, with traces of nuclei and evidence of cell division, appeared in middle to late Proterozoic time. SCHOPF (1972) claimed that spore-like cells about one billion years old may represent an early stage in the establishment of meiotic cell division, documenting the origin of sexuality that tended to spread variability, enhance selection and accelerate evolution, but this interpretation has been challenged (KNOLL & BARGHOORN, 1975). His later assessment (SCHOPF, 1975) "that the development of the megascopic, multicellular, eukaryotic level of organization was a relatively recent innovation, possibly occurring about 800-700 m.y. ago" is more likely. Sporomorphs or acritarchs of simple morphology are widespread and common in some Upper Precambrian shale and limestone, particularly those known to be less than 800 my[1] old.

Many biologists consider it as likely that the first "animal" Protista developed aerobic respiration and "cell-eating," herbivorous, and "carnivorous" habits almost as soon as eukaryotic cells evolved. STANLEY (1973) has drawn attention to the increase in organic diversity, which according to ecological theory would have followed the advent of heterotrophy and the development of new trophic levels after the long, purely resource-limited reign of relatively undiffer-

[1] In conformity with current *Treatise* style, abbreviations of units of measure are written without periods, including "my" and "by" for "million years" and "billion years"; however, the author of this chapter disapproves of this style and prefers "m.y." and "b.y."

entiated autotrophic prokaryotes; however, the chances of preservation of amoeboid or ciliate cells that lack resistant cell walls are infinitely less than those of fossilization of plant Protista. The long evolutionary pathways to the first animal Protista and from them to the first fossilizable Metazoa are unlikely to be documented in the rocks. At the present state of our knowledge all known Precambrian microfossils must be considered as plants in the widest sense of the term and therefore as outside the scope of this *Treatise*. No skeleton-forming Protista (Radiolaria, Foraminiferida) of Precambrian age are known. There are reports (DUNN, 1964; VOLOGDIN & DROZDOVA, 1970) of Proterozoic sponge spicules, which, if confirmed, would make the Parazoa the earliest animals with fossil representatives. There are also reports of bioturbation traces and burrows filled with fecal matter in Precambrian rocks from southern Norway and from the Soviet Union (SINGH, 1969; SABRODIN, 1971), which are more than 900 my old, and of "feeding burrows" in rocks about 750 my old (SQUIRE, 1973). Their interpretation as traces of ancient Metazoa requires confirmation.

From a practical, pragmatic viewpoint the known diversity of Precambrian fossils can be divided into the following categories: 1) filamentous and coccoid algal (and possibly fungal) cells, 2) sporomorphs and acritarchs, 3) megascopic algae, 4) stromatolites (and oncolites), 5) trace fossils, 6) problematic fossils (including microphytoliths), 7) metazoan body fossils.

Fossils in these categories are, in general, prepared and investigated according to different technical methods. Those in the first category are studied in thin sections by microscopy. Those in the second category have organic walls and are revealed and studied by palynological methods involving acid treatment of the rock matrix and concentration. Megascopic algae are investigated by other paleobotanic methods. Stromatolites are studied by and reconstructed from serial sections, in addition to studies of the microstructure of their layers. These categories of fossils are outside the scope of invertebrate paleontology. Trace fossils have been reviewed in a recently published volume of this *Treatise* (HÄNTZSCHEL, 1975) and the few names proposed for Precambrian genera are listed below (p. *A*112). A number of names, also listed below (p. *A*111), have been applied to fossils whose biogenic origin is here considered as problematic. Their substance may be partly organic but they do not necessarily represent formerly living organisms. Also included in this category are "microphytoliths" that were formed probably under the influence of some ill-defined organic activity on sedimentary processes, grading into the formation of oolites and spherulites. A few of these structures resemble fecal pellets; if they are that, their originators must be Metazoa.

The following discussion concerns mainly the last category, the body fossils of metazoan origin and their taxonomy. It is followed by lists of generic names that have been proposed for objects in the other categories and finally by a list of rejected and unrecognizable taxa. Many names have been given to configurations in Precambrian rocks that resemble metazoan or other organic remains, but which are now known or at least generally believed to be of chemical (concretionary) or mechanical origin. It will be noted that few of these names are of recent date. This hopefully indicates a significant advance in the approach to Precambrian paleontology since it was reviewed a decade ago (GLAESSNER, 1966) and a welcome clarification of the fossil record of Precambrian time.

ACKNOWLEDGMENTS

The author expresses his sincere thanks to those who have supplied him with valuable information: Dr. M. A. FEDONKIN, Moscow; Dr. T. D. FORD, Leicester; Dr. A. S. HOROWITZ, Indiana University; Dr. R. F. JENKINS, Adelaide; Professor B. M. KELLER, Moscow; Dr. V. M. PALIY, Kiev; Dr. J. W. SCHOPF, Los Angeles; Professor B. S. SOKOLOV, Moscow and Novosibirsk; Dr. MARY WADE, Brisbane; Dr. M. R. WALTER, Canberra.

STRATIGRAPHIC DISTRIBUTION

The first definite metazoan body fossils appear in uppermost Precambrian strata (Upper Proterozoic; Vendian, upper Vendomian, or Terminal Riphean in the Russian literature; see Fig. 1). None of them is demonstrably older than the youngest Precambrian glacigene rocks ("tillites"). These fossils constitute a distinctive assemblage of marine animals that are characterized by the absence of mineralized skeletons or shells and by the prevalence of coelenterates, the presence of diversified polychaete annelids, and the rare occurrence of arthropods that are markedly more primitive than those occurring in Cambrian faunas. These assemblages document a definite pre-Cambrian level of metazoan evolution. They are collectively referred to as the Ediacaran faunas, after the locality where the first abundant finds were made by R. C. SPRIGG in 1947. At the present time it is not possible to divide the occurrences of Precambrian bodily preserved Metazoa into a number of zonal assemblages defining a sequence of stratigraphic intervals preceding the Cambrian, but with the rapid increase in discoveries of Late Precambrian Metazoa and the advance in dating of sedimentary rocks containing them, this may become possible. The first discoverer of the fossils at Ediacara in South Australia considered their age to be Early Cambrian because this was the age conventionally and conveniently assigned to the rocks containing them, the Pound Quartzite, which had been placed at the base of the Cambrian in the regional stratigraphic scheme. There are now three basic reasons for placing these fossiliferous rocks below the Cambrian. Firstly, they do not contain any fossils that are found together with Cambrian faunas. Secondly, the Pound Quartzite is separated from rocks containing the first Lower Cambrian fossils by profound regional unconformities. Thirdly, similar fossils are found elsewhere, also not in synchronous association with, but below strata containing Lower Cambrian fossils. Most of these fossiliferous rocks cannot be dated precisely by geochronological methods, but all available evidence places them in the latest Precambrian (Vendian).

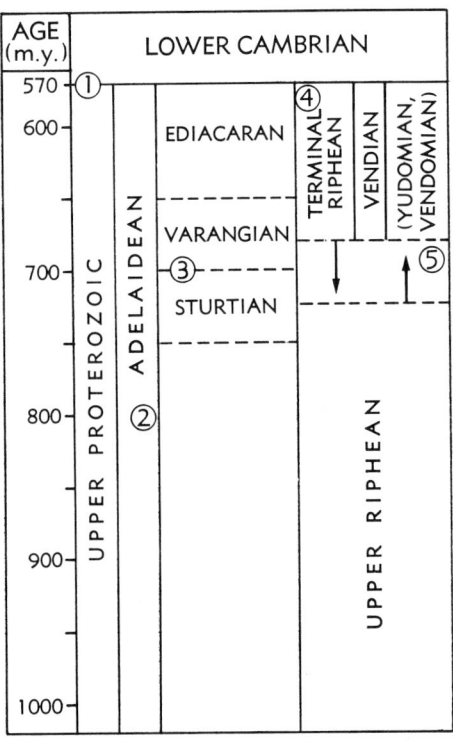

Fig. 1. Proposed subdivisions of the upper Proterozoic (Glaessner, n). (Numbers correspond to numbered statements below.)

1. Agreement on the definition of the Precambrian-Cambrian boundary is being sought by the International Commission on Stratigraphy through the efforts of a Working Group. Pending completion and adoption of its final report, position and dating of this boundary remain uncertain. There is substantial agreement that the Ediacaran faunas pre-date it and that the first appearance of trilobite body fossils is in the Lower Cambrian. A boundary stratotype is being sought between these two biohorizons.
2. The dating of the lower boundary of the Adelaidean is being investigated.
3. The general use of the three divisions in this column (as Periods) was proposed by W. B. HARLAND and K. N. HEROD (Geol. Jour., Spec. Issue 6, Liverpool 1975, p.205). Their acceptance is conditional on agreement on boundary stratotypes and their dating relies on long-range correlations. Use of the terms Eocambrian or Infracambrian for their combined time span is not recommended. Both carry the undesirable implication of being subdivisions of the Cambrian. The term Infracambrian is said to have been applied originally to significantly older rocks.
4. The stratotype of the Riphean is in Bashkiria, on the western slope of the Ural Mountains. Paleontological studies have led to a four-fold division and to adoption by a number of Russian authors of "Terminal Riphean" for its uppermost part. This is correlated with the Vendian of the Russian platform and the Yudomian of Siberia. Some Russian stratigraphers (KELLER, SEMIKHATOV, and others) believe that adoption of the combined term Vendomian based on a combination of stratotypes will avoid existing difficulties

GEOGRAPHIC DISTRIBUTION

Australia. The fossiliferous layers in the Pound Quartzite at Ediacara in South Australia (Fig. 2, Loc. 1) have yielded over 1,500 specimens. Two-thirds of them are coelenterates, most of them medusoids; not less than one-quarter are annelid worms and five percent are arthropods. Similar but poorer faunas are now known throughout the Flinders Ranges of South Australia, an area of approximately 100×200 km (Fig. 2, Loc. 2). All fossils come from one stratigraphic horizon of varying thickness, up to a maximum thickness of 112 m. An isolated locality in northwest South Australia, Punkerri Hills (130°25′ E, 27°40′ S, Fig. 2 Loc. 3), has yielded a large external mold of a remotely *Pteridinium*-like petaloid, probably representing the genus *Charniodiscus*; a single, smaller, similar specimen was found at the base of the Arumbera Sandstone south of Alice Springs in central Australia (Fig. 2, Loc. 4). At other localities in the same formation or its equivalents, the medusoids *Hallidaya* and *Skinnera* and abundant molds of *Arumberia* occur.

Southwest Africa. A rich fauna is preserved locally in the Upper Clastic Member of the Kuibis Formation at the base of the Nama Group (Fig. 2, Loc. 5) where *Rangea*, *Pteridinium*, *Namalia*, possibly a sprigginid worm and, in the uppermost part, Erniettidae occur in considerable numbers. The first two of these fossils are also recorded from the basal clastic member of the next higher Schwarzrand Formation. Above them, but in the same formation, a medusoid was found that was first recorded as *Cyclomedusa* but later as *Eoporpita*. Limestone members intercalated in both formations contain abundant worm tubes of the genus *Cloudina*. In the Nasep Quartzite Member of the Schwarzrand Formation a medusoid and the genus *Nasepia* were found. The upper part of the Nama Group contains only trace fossils. The fossiliferous outcrops of the Lower Nama Group extend over an area that is approximately equal to that of the fossiliferous Pound Quartzite in South Australia.

England. In the Precambrian of Charnwood Forest near Leicester (Fig. 2, Loc. 6), about 20 fossils were found on bedding planes of slaty tuffaceous siltstones (Woodhouse Beds) and described as *Charnia* and *Charniodiscus*. Both resemble fossils from Ediacara. Concentrically ribbed medusoid casts also occur.

Scandinavia. In northern Sweden, north and south of Lake Torneträsk (STRAND & KULLING, 1972), two medusoid specimens with strong and regular concentric ribs were found in sandy shale, in what appear to be the youngest Precambrian strata, below Lower Cambrian with *Platysolenites, Volborthella*, and hyolithids (Fig. 2, Loc. 7).

Northern Russia. In a core taken at 1,552 m from a bore at Yarensk, some 750 km northeast of Moscow (Fig. 2, Loc. 9), a specimen of *Vendia* was found in siltstone of the Valdai "Series" (upper Vendian). A rich fauna of about the same age was found on the coast of the White Sea near Arkhangelsk (Fig. 2, Loc. 13). It includes *Pteridinium, Dickinsonia*, and several new genera, one of which resembles *Vendia*. These beds, which contain also acritarchs of the genus *Leiospheridia*, are considered as upper Vendian (KELLER et al., 1974). Another locality is farther east, on the river Pesha (Fig. 2, Loc. 12), where *Glaessnerina* was found.

Southwestern U.S.S.R. In outcrops along the river Dniestr, on the southwestern border of the Ukrainian massif (Fig. 2, Loc. 8), *Cyclomedusa* occurs together with other medusoid remains and trace fossils in siltstones and sandstones that are equivalents of the Valdai "Series" of the Vendian (ZAIKA-NOVATSKIY et al., 1968; PALIY, 1969).

Northern Siberia. A specimen of *Glaessnerina* was found in the sandy dolomites

---

(Continued from facing page.)

about defining the base of the Vendian.
5. The boundary between Upper and Terminal Riphean is stratigraphically and geochronologically uncertain. Absence of glacigene deposits in some and presence of one or of two glaciations in other areas cause problems as they have been used in boundary definitions. Long-range correlations used in assigning dates to boundaries and intervals lead to other problems.

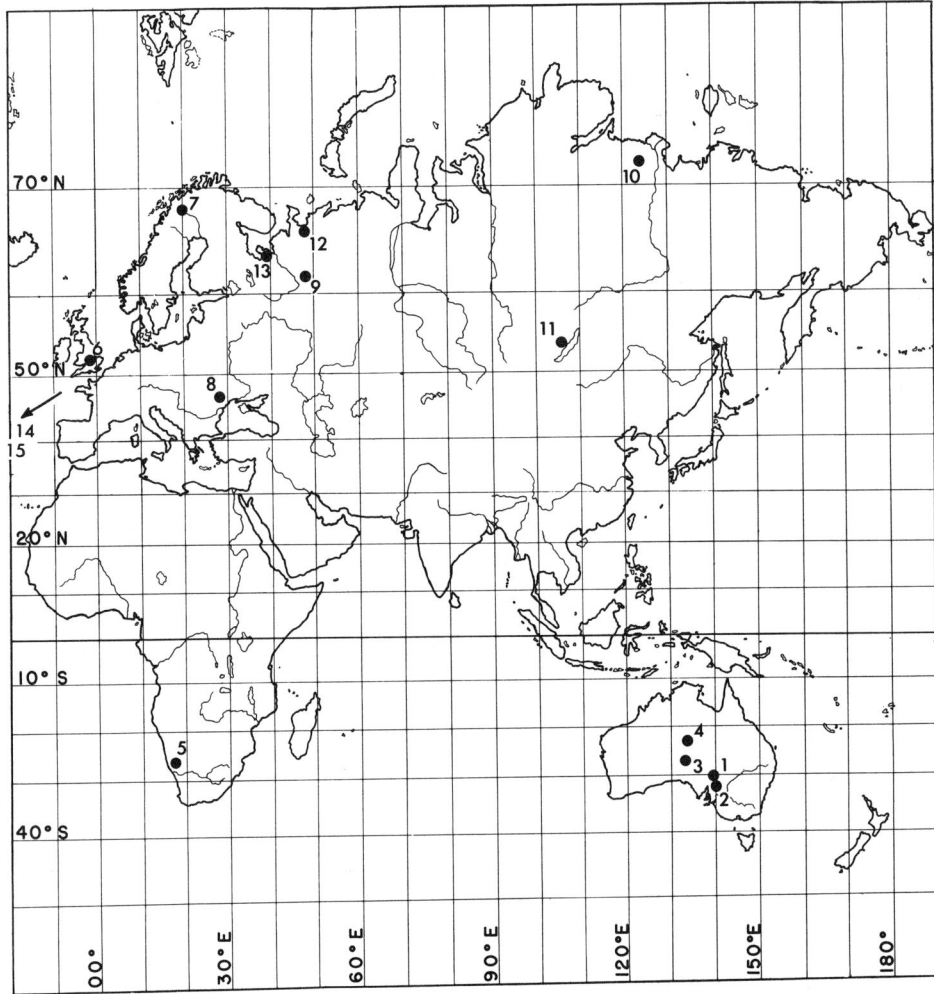

Fig. 2. Geographic distribution of Upper Precambrian fossil localities. The position of the North American localities (14, Newfoundland; 15, North Carolina) relative to Europe in Late Precambrian time is not precisely known but believed to have been at no great distance in the direction of the arrow (Glaessner, n). [For explanation of numbers see text, p. A83.]

of the Khatyspit Formation, in the uppermost Precambrian of the Olenek uplift (Fig. 2, Loc. 10) (Sokolov, 1973).

Lake Baikal region. Several fragmentary fossils were found west of Lake Baikal (Fig. 2, Loc. 11) in the Irkutsk "Series" at the top of the Precambrian sequence (Sokolov, 1973).

North America. Fossils occur abundantly in tuffaceous shales of the Conception Group near Mistaken Point, Avalon Peninsula, southeastern Newfoundland (Fig. 3). This fauna has yet to be described (Misra, 1969). Large fossil worms were recently found in tuffaceous sediments of Late Precambrian age near Durham, North Carolina, and have been described by Cloud et al. (1976).

Fig. 3. Cast of impression of undescribed fossil (?hydrozoan colony) from the Upper Precambrian of Newfoundland (Misra, 1969), ×0.85.

CHRONOSTRATIGRAPHIC DATA

The ages of few of the known assemblages of Late Precambrian fossils have been fixed directly by radiometric methods. By the use of combined data from known ages of tillitic rocks occurring in some sequences below the fossiliferous sediments, and from their relations to overlying Lower Cambrian strata, a probable age range of the known occurrences of Precambrian metazoan body fossils can be deduced. These approximate datings confirm a rough correlation of these fossiliferous rocks but they do not support any biostratigraphic subdivision at the present state of our knowledge.

The oldest occurrence of metazoans in Australia is significantly younger than the youngest tillitic rocks. Such rocks have been dated directly only in northwestern Australia where no Precambrian metazoan fossils are known. The minimum age of the youngest glaciation in this area is 660 to 670 my. This is compatible with the dating of other upper Proterozoic rocks in central and South Australia. A correlation of the glacigene strata in these areas is widely accepted. A claim for alternation of tillitic and fossiliferous rocks in the Nama Group in southwest Africa was made by GERMS (1972a). The Nama Group has not yet been dated directly but it is known to be older than 510 and significantly younger than 720 my. The fossiliferous lower part may correspond, in part, to the interval from 600 to 650 my. This is in agreement with the polar wandering curve based on the latest paleomagnetic data (McELHINNY et al., 1974) and the pre-Nama age of the Numees Tillite. The widespread glacigene rocks ("Varangian") found around the Baltic and Ukrainian shields in northeastern Europe occur below the fossiliferous strata in the Vendian sequence wherever the stratigraphic relations between them can be observed. The K-Ar dating of glauconites from the fossiliferous Upper Precambrian strata of the Soviet Union (northern Russia and Baikal area) gave ages near or slightly younger than 600 my. If corrected for the use of the generally accepted decay constant ($\lambda_e = 0.585 \times 10^{-10}$), these ages would be approximately 575 to 580 my. The fossiliferous strata in the Ukraine are considered to be of about the same, middle to late Vendian, age. The northern Siberian occurrence is in a rock series that was dated on glauconite at 670 my. This should be similarly corrected to about 640 my. The age of the fossiliferous strata of the Conception Group of southeastern Newfoundland has not been determined directly. It is believed by ANDERSON (1972) to be 610 to 630 my. The use of a slightly different decay constant for Rb^{87} (1.42×10^{-11}), which is now recommended, may reduce these numbers to about 590 to 610 my. The latest find of fossils in North Carolina (CLOUD et al., 1976) was made in rocks that are known to be 620 my old. There is reason to doubt the previously accepted date of 680 my for the age of the fossiliferous Charnian rocks in England (DUNNING, 1975). Except for this doubtful instance, the stratigraphic positions of rocks with Ediacaran assemblages of fossil Metazoa tend to fall in the age range from 575 to 640 my. This equals the length of the Tertiary Period. Only a few trace fossils may be older, but none is likely to be older than 1,000 my, except the as yet unconfirmed finds of sponge spicules in the Middle Proterozoic rocks of eastern Siberia and Australia, which are about 1,500 my old.

PALEOECOLOGY AND TAPHONOMY

With a few localized exceptions, metazoan fossils of Precambrian age are rare. This is only partly due to peculiarities of the Precambrian environment. Prior to the acquisition of mineralized skeletons, relatively few fossils would have been preserved in the sediments. Shells and skeletons were an evolutionary novelty essentially of Cambrian and later age. The total number of Metazoa able to leave traces of their locomotion or feeding on the surface of sediments or within them (bioturbation) must have been smaller in Late Precambrian time than in the Early Cambrian. In many sequences of unaltered clastic rocks extending across the Precambrian-Cambrian boundary, for example in the United States, in Scandinavia, and in Australia, there are few and relatively undifferentiated trace fossils in the Precambrian, contrasting with their abundance and diversity in the Lower Cambrian. A lower level of evolutionary differentiation compared with that in the early Paleozoic undoubtedly contributed to the poverty of the Precambrian fossil record. Nonenvironmental factors make it difficult to deduce the nature of environmental changes at the beginning of the Cambrian directly from the results of comparative paleoecological studies. Almost any conceivable change in the environment claimed to be unique in character or magnitude has been held responsible for the poverty of the Precambrian fossil record: absence of sedimentation during the supposed "Lipalian interval"; dominance of shallow or of deep water; change in salinity or in temperature, or in the distribution of continents and oceans; and changes in the composition of the atmosphere. With the exception of the now disproved "Lipalian interval," all other factors could have contributed but they were not uniquely active during the Precambrian-Cambrian transition.

The assumption that the amount of oxygen in the atmosphere reached a significant threshold value (one percent of the present level) at the beginning of the Cambrian was based originally on the then current belief that prior to that point in time there was no animal life and no oxydative respiration, and that the absence of the ozone shield permitted ultraviolet radiation to penetrate most of the photic zone, excluding from it phytoplankton and thus reducing primary nutrient supply from the ocean and its floor. This view is no longer tenable.

A distinctive phase in the history of the Metazoa, older than the earliest Cambrian, is now known. It is characterized by the following ecologically significant features: 1) exclusively microphagous feeding and absence of large predators; 2) presence of numerous surface- and bottom-feeding coelenterates; of benthonic, probably detritus-feeding worms; of the first few arthropods; 3) occupation of a variety of marine habitats such as near-shore areas with sandy and silty bottom (Australia, southwest Africa), shallow-water muddy areas (eastern Europe), deeper water with turbidity current activity (Newfoundland), and occasionally areas of carbonate deposition (southwest Africa, possibly northern Siberia). The biogenic reef environment is notably absent. Algal mats and columnar stromatolitic structures could flourish in normal marine littoral to sublittoral environments without being affected by grazing. No clear environmental grounds for the absence of tissue mineralization can be deduced from present observations and the lack of basic differences between Precambrian and younger sediments (apart from the absence of skeletal carbonate rocks) makes their existence unlikely. It is probable that the complex biochemical basis for the formation of mineralized shells and skeletons had not evolved by the time of the first appearance in the record of megascopic Metazoa. This evolutionary novelty was heralded by the appearance of the first agglutinated, calcareous worm tubes (*Cloudina*).

It must be remembered that large numbers of marine invertebrates without mineralized tissues still flourish in the present oceans (Siphonophora, Actiniaria, Ctenophora, Chaetognatha, Aschelminthes, Platyhelminthes, Sipunculida, Nudibranchia, Aplacophora, Annelida, Euphausiacea, Tunicata, etc.). Most of them, although almost certainly quite ancient, have left no fossil record or only an insignificant one, except the annelids that acquired in Late Precambrian time the ability to build calcareous tubes. Collagen fibers and probably chitinous deposits accounted for the stiffness of many of the "soft-bodied" organisms in the Ediacara fauna, such as medusae, chondrophoran hydrozoans, and a conulariid. In "Petalonamae," silicified fibrous tissues have been observed but other microscopic structures described in this group could well be the result of alteration by silicification and weathering. Where the fossils are preserved as casts and molds, the mechanical deformability of soft bodies in more or less compressible sediments makes interpretation difficult. Partly decomposed bodies are preserved, and in others shrinkage or contraction of muscles occurred after the first contact between body and sediment. Even the escape of organic decomposition products to the sediment surface can produce trace fossils (in the sense of "*Spurenfossilien*" or signs of former presence of life, not of "*Lebensspuren*" or traces of life activities), for example, *Pseudorhizostomites* (Wade, 1968). The amazing variability of many of the Ediacara and most of the Nama fossils may be largely due to taphonomic factors; however, it is possible that the characters of medusae from Ediacara, and of "Petalonamae" from both regions, may not have been as stable as had been expected by the first observers who named them. It is not yet possible to assess the relative significance of phenotypic and genetic varibility in these ancient fossil organisms or to state unequivocally that variability was uncommonly great compared with that of the present fauna.

Much work remains to be done before the composition of these assemblages is sufficiently well known to be analyzed in terms of community structure and ecological factors. Though they are thanatocoenoses, the fact that they consist of more or less easily decomposable soft-bodied animals means that at each locality they are strictly contemporaneous and that most of them must have lived within short distances of their place of burial. The "Petalonamae" are almost certainly sessile coelenterates and many of them were embedded where they lived. Medusae and chondrophores drifted inshore from the sea in swarms, as they still do today. Worms and small arthropods fed on the organic detritus and may have provided food for the large medusae. Meander traces of feeding on sedimentary surfaces prove that the detritus was abundant and that it was exploited, as was the organic content of the sediments, though apparently to a limited extent. This limited picture will be augmented by an analysis of other biofacies when the faunas from Newfoundland and other newer localities are described. It is not likely to be changed fundamentally. (For imaginative, comprehensive views on Precambrian evolution and environments see Cloud, 1968, 1974, 1976a,b; Fischer, 1972).

EVOLUTIONARY SIGNIFICANCE

The Ediacaran faunas are the earliest known assemblages of Metazoa. They differ fundamentally in their composition from Cambrian and younger assemblages; they appear to have been preceded by a long, undocumented phase of metazoan evolution. The most advanced members are relatively undifferentiated arthropods. They include *Parvancorina*, possibly ancestors of notostracan Crustacea, and *Praecambridium* and *Vendia*, which resemble either primitive Trilobitomorpha or Merostomata. The level of differentiation indicated by the first Cambrian Crustacea, trilobites, and Merostomata had obviously not yet been reached but the distinction of the first from the other two lines is already indicated. In this context it is significant to recall MANTON's (1969, p. R53) remark: "The head shields and limbs of the Merostomata and Trilobita have more in common than either had with the heads and limbs of Crustacea." Pending the discovery of the limbs of *Praecambridium* it can be said that there is now some paleontological evidence for MANTON's view of the early relations and differences between arthropod lineages (MANTON, 1969, p. R8, Fig. C).

In contrast with the arthropods, the annelids are represented by four very different orders. Two of them are still extant and may be considered as relatively primitive polychaetes. The other two are represented only by simple or complexly structured tubes that are so different from younger ones that these orders must be considered as extinct after the Early Cambrian.

It has long been known that the Ediacaran faunas are dominated by coelenterates, in number of taxa as well as of individuals. At first it seemed easy to distinguish between hydrozoan and scyphozoan medusae and to place most of the fossils in extant orders (see this *Treatise*, Part F, 1956). This is no longer considered as justifiable, for two reasons. Firstly, the preservation of these soft-bodied organisms is such that diagnostic characters are not usually ascertainable, and secondly, the taxonomic characters by which orders are defined in the living fauna may not have existed in Precambrian time. It is now obvious that as much evolutionary change has occurred at lower grades of metazoan organization, such as the coelenterate grade, as at some higher ones. Notwithstanding these difficulties, some answers to phylogenetic problems posed by living coelenterates can now be based on the results of studies of Ediacaran fossils. Chondrophoran hydrozoans have been recognized because of the preservability of their distinctive chitinous pneumatophores. The genera *Ovatoscutum* and *Chondroplon* had many Paleozoic successors. Among them was probably the late Early or early Middle Cambrian *Velumbrella* with its stiffly radiating ribs indicating floating rather than medusoid contractile swimming locomotion. Other chondrophorans are *Archaeonectris* (Ord.-L.Sil.), *Discophyllum* (Ord.), *Palaeonectris* (Dev.), *Plectodiscus* and *Paropsonema* (Sil.-U.Dev.). Compared with this relative abundance, the sole living genera *Velella* and *Porpita* are seen as survivors of an ancient and formerly diverse group of hydrozoans that early discovered the surface of the sea as a feeding area and adapted to it. Diversity was strikingly reduced in late Paleozoic time by competition or predation. The resemblance of the floats of *Ovatoscutum* and *Chondroplon* to those of the surviving Porpitidae is so specific that we can be sure of the hydrozoan affinities of these Precambrian fossils, without confirmation from polyp morphology. *Conomedusites* is now considered as related to the Conulariida, a subclass of the Scyphozoa. This placement was confirmed by comparison of Paleozoic conulariids with living chitinous *Stephanoscyphus*, a coronatan polyp form (WERNER, 1966). Though this comparison was disputed by KOZŁOWSKI (1968), early growth stages of conulariids found by him are identical with hydrozoan hydrorhizae (GLAESSNER, 1971). This establishes an early link between the Hydrozoa and the Scyphozoa to which several Ediacaran medusae are assigned. Whether other medusae, including the great majority of specimens in the Ediacaran assemblages, have hydrozoan or scyphozoan affinities cannot be established.

In the absence of polyps the precise zoological affinities of a large number of Precambrian, sessile benthonic, colonial forms with strong resemblances to some living Pennatulacea cannot be established. There are no reasons why they should be placed in a higher than coelenterate structural grade and there is no justification for speculation about the possibility of evolutionary links between these Metazoa and unspecified Metaphyta as suggested repeatedly by PFLUG (1974a) following the now generally discounted views of HARDY (1953). PFLUG proposed the name "Petalonamae" for this group of intergrading but highly diversified fossil organisms. Some of them have striking and specific structural resemblances to living Pennatulacea, but in the absence of traces of fossil polyps, the possibility of convergence between Precambrian Hydrozoa and later octocorallian Anthozoa cannot be excluded. The silicified and calcified cellular tissues described by PFLUG in some "Petalonamae" are here considered as the effects of fossilization. No spicules are preserved and fractureless folding-over or bending of many of these leaf-shaped fossils proves the absence of biogenic mineralization of their tissues. If the "Petalonamae" include ancestral Pennatulacea, the Anthozoa must have separated early from the Hydrozoa, before these evolved a medusoid stage as a locomotive and reproductive stage in their life cycle. The Scyphozoa appear to have evolved from Hydrozoa not long before the time of the Ediacaran faunas, and to have elaborated the medusoid at the expense of the polyp stage.

The early Metazoa were almost certainly planula-like and very small. Their chances of preservation were minimal, also because of the absence of resistant integuments or cell walls, or of readily mineralized mucous sheaths that enabled Precambrian plant microfossils to be preserved in large numbers. The undetected and probably undetectable evolution from amoeboid or ciliate cells to cell colonies and to tissue grade could well have taken as much time as the entire Phanerozoic phase of metazoan evolution. Between the attainment of tissue grade and the level of the Ediacaran faunas lies the evolution of the mesoderm and the coelom, which must have preceded the observed diversification of the annelids and arthropods. These processes could have occurred in late Proterozoic (late Riphean) time, within a time span of some 300 my, preceding the oldest Ediacaran fauna. It is possible that further discoveries or reinterpretations will extend through this time interval the range of either such body fossils as have been found at Ediacara, or definite trace fossils, or bioturbation, or microcoprolites. Such finds have been announced, but not yet convincingly described or stratigraphically documented.

The scarcity of Precambrian traces of animal life compared with Cambrian and later ones may be indicative of an abrupt increase in total metazoan abundance (biomass) about the beginning of Cambrian time. It was probably related to changing oceanographic conditions. About that time, but not simultaneously, many (but by no means all) invertebrates acquired mineralized tissues (shells or skeletons). Some polychaete worms (*Cloudina, Anabarites*) appear to have been the first to take this step in evolution by building calcareous tubes. At the beginning of Cambrian time they were followed by archaeocyathans, brachiopods, gastropods, Hyolithelminthes, Tommotiidae, and Hyolithida (of uncertain affinities), later by trilobites, crustaceans, and echinoderms. It should be noted that the unique *Xenusion* (see this *Treatise,* p. O19-O20), formerly considered as a Precambrian fossil, has now been convincingly traced to a basal Cambrian source bed (JAEGER & MARTINSSON, 1967), and that the hyolithid *Wyattia* (TAYLOR, 1966) is probably also of earliest Cambrian age. Some groups of invertebrates are so highly differentiated at the time of their first appearance in the Early Cambrian that the existence of unknown ancestors in Late Precambrian time has been assumed. Mollusca, small brachiopods, and various small arthropods could have been present in Ediacaran time without possessing mineralized shells and without leaving fossil evidence. Some Precambrian tracks show a degree of complexity that suggests mollusks as their originators. The earliest trilobites were markedly thin-shelled. The fact that numerous marine arthropods with unmineralized, thin, chitinous exoskeletons still exist in large numbers, whereas their fossil representatives are unknown, suggests why the Precambrian ancestors of the trilobites have not been fossilized. In contrast to this group, the earliest Cambrian Archaeocyatha (ROZANOV, 1973) are rare and structurally primitive and may not have had a long history. The early stages in the evolution of the echinoderms (UBAGHS, 1971) are problematic. They are highly differentiated in early Cambrian time and yet it is difficult to postulate their existence in the Precambrian as unmineralized, minute, and unfossilizable creatures, without their distinctive calcite skeleton. One genus in the Ediacara fauna, *Tribrachidium*, has no trace of calcareous plates, yet resembles nothing as much as some Edrioasteroidea; however, they are not considered as primitive echinoderms. The problem remains unsolved. So does the question of the origin of the deuterostomatous phyla Hemichordata and Chordata. It has been claimed that the organic-walled tubes of the Sabelliditida that are found in the Riphean and Vendian represent the Pogonophora (SOKOLOV, 1967, 1972). They have been considered as aberrant "Deuterostomata" but zoologists are no longer unanimous about the placing of the Pogonophora in the system of the Metazoa (NØRREVANG, 1975).

Although Precambrian fossils are rarely mentioned in current discussions on problems of metazoan evolution and relationships, the Late Precambrian faunas can contribute important data. Their effective use will depend on speedy and comprehensive description of the collected specimens and also on a deeper understanding of the chemical and mechanical alteration of their soft, organic material in the course of fossilization.

SUPRAGENERIC TAXA OF PRECAMBRIAN METAZOA

Figures in parentheses indicate numbers of included genera.

?Phylum Porifera
  Class, order, family uncertain (1)
Phylum Coelenterata
  Subphylum Cnidaria
    Class Hydrozoa
      Order Hydroida
        Suborder Chondrophorina
          Family Chondroplidae (2)
          Family Porpitidae (1)
    Class Scyphozoa
      Family uncertain (5)
    Class Conulata
      Order Conulariida
        Suborder Conchopeltina
          Family Conchopeltidae (1)
  Medusae of uncertain affinities (8)
  Problematical Coelenterata
    Family Pteridiniidae (1)
    Family Rangeidae (1)
    Family Charniidae (3)
    Family Erniettidae (5)
    Family uncertain (4)
Phylum Annelida
  Class Polychaeta
    Order Cribricyathea
      ?Family Vologdinophyllidae (1)
    Order uncertain
      Family Dickinsoniidae (1)
      Family Sprigginidae (2)
      Family Anabaritidae (1)
Phylum Arthropoda
  Superclass Trilobitomorpha
    (or Chelicerata)
    Class and order uncertain
      Family Vendomiidae (4)
  Superclass Crustacea
    Class Branchiopoda
      Order unknown
        Family Parvancorinidae (1)
  Doubtful Arthropoda (1)
?Phylum Pogonophora
  Order Sabelliditida
    Family Saarinidae (1)
    Family Sabelliditidae (1)
Phylum uncertain (3)
Doubtful invertebrates
    Family Suvorovellidae (2)
    Family uncertain (1)

Precambrian   A91

SYSTEMATIC DESCRIPTIONS[1]

?Phylum PORIFERA Grant, 1872

Class, Order, and Family UNCERTAIN

Tyrkanispongia VOLOGDIN & DROZDOVA, 1970, p. 197 [*T. tenua; OD]. Siliceous, straight or curved, fragmentary spicules, hollow, with pointed, rounded or narrowed ends, diameter 40-110 μm, occurring together with hook-shaped and globular siliceous bodies of similar size. [The fragmentary preservation and the unusual shape of spicules with narrowed or hooked tips does not support unequivocal assignment of the *"Tyrkanispongia"* assemblage to the Porifera. The rocks containing it are 1,500-1,550 my old. The only other documented Precambrian occurrence of apparent sponge spicules are triact-like shapes observed in thin sections of cherts from the Carpentarian of northern Australia, about 1,500 my old (DUNN, 1964). In the absence of clear axial canals it is difficult to distinguish supposed sponge spicules from glass shards of volcanic origin where only thin sections are available.] *U.Precam.(Gonam "Ser."),* E.Sib. (Uchur R.).

Phylum COELENTERATA Frey & Leuckart, 1847

[See this *Treatise*, 1956, p. F9]

Class HYDROZOA Owen, 1843

[See this *Treatise*, 1956, p. F67]

Order HYDROIDA Johnston, 1836

[See this *Treatise*, 1956, p. F83]

Suborder CHONDROPHORINA Chamisso & Eysenhardt, 1821

[See this *Treatise*, 1956, p. F148]

The status of this suborder is uncertain. It is widely accepted that it does not belong to the order Siphonophorida to which it

FIG. 4. Chondroplidae (p. *A91-A92*).

had been subordinated in earlier zoological classifications, but to the order Hydroida. Within it, relations to the suborder Athecata are so close that some authors included the chondrophorans in the athecate hydrozoans. One or two living monotypic families are recognized and several Paleozoic genera are included in them; for *Palaeoscia* CASTER, 1942, see also HÄNTZSCHEL, 1975, p. W147.

Family CHONDROPLIDAE Wade, 1971

[Chondroplidae WADE, 1971, p. 188]

Float chambered, bilaterally symmetrical, axis narrow, chambers narrowing from one end of the axis to the other, leaving a notch between them either at the narrow or at both ends. *U.Precam.(Vend.).*

Chondroplon WADE, 1971, p. 184 [*C. bilobatum; OD]. Float large, bilobed, with rounded outline, axis strongly marked as ventral groove and probably as blunt dorsal keel, initial chamber large and elongate, early chambers annular, later chambers leaving peripheral notch where they are wide, last-formed chambers leaving notch also at opposite end where they are narrow and disposed transversely; their outer ends form a scalloped margin; chambers higher than wide, separated by depressed sutures. [Conspicuous, large, radial folds on the surface of the only known specimen were considered as fortuitous by WADE.] *U.Precam.*, S.Australia.——FIG. 4,*1*. *C. bilobatum,* Ediacara; holotype, ×0.32 (Wade, 1971).

Ovatoscutum GLAESSNER & WADE, 1966, p. 612 [*O. concentricum; OD]. Float shieldlike, with weakly marked axis, initial chamber small, oval,

---

[1] After completion of the typescript for this chapter, the discovery of a Precambrian microscopic flatworm (*Brabbinithes churkini* ALLISON, Class Turbellaria) was announced (ALLISON, 1975). It is 0.45 mm long, apparently known only from one thin section, and comes from the Tindir Group of eastern Alaska, which is correlated with the Rapitan Formation in northwestern Canada; its age is less than 850 my. Regrettably, the available morphological data are not sufficiently well defined to support the author's far-reaching evolutionary conclusions. CLOUD, WRIGHT, & GLOVER (1976) now prefer an interpretation of this fossil as an hexactinellid sponge spicule.

surrounded concentrically by narrow, elongate, sinuous chambers, leaving a somewhat ill-defined triangular notch in peripheral margin where they are narrowest; notch is situated axially but does not reach center; chambers separated by deep, narrow grooves. U.Precam., S.Australia.——FIG. 5,1. *O. concentricum, Ediacara; holotype, ×0.8 (Glaessner & Wade, 1966).

Family PORPITIDAE Brandt, 1835

[See this *Treatise*, 1956, p. F150]

Eoporpita WADE, 1972, p. 198 [*E. medusa; OD]. Circular or elliptical in outline, two groups of club-shaped polypides, outer ones of nearly constant lengths and in several series, inner series with inwardly reducing size, grouped around single, large, central cone; above them, the remains of a delicate float with numerous narrow, concentric, annular, chambers surrounding small, circular, central chamber; aboral side of disc showing faint radial striae. [It seems reasonable to consider the outer series as dactylozooids, the inner series as gonozooids, and the larger central cone as the gastrozooid, when comparing these fossils with the living *Porpita* LINNÉ.] U.Precam., S.Australia. ——FIG. 6,1. *E. medusa, Ediacara; paratype, ×0.6 (Wade, 1972a).

1   Eoporpita

FIG. 6. Porpitidae (p. *A92*).

Class SCYPHOZOA Götte, 1887

[For diagnosis see this *Treatise*, 1956, p. F27]

The following genera are assigned to the Scyphozoa with varying degrees of uncertainty because of incomplete preservation and morphological and evolutionary remoteness from other fossil and living genera. They are not at present placed in orders or families. Among the Precambrian "Medusae of uncertain affinities" (p. *A94*), *Rugoconites* has been provisionally restored as a scyphozoan (WADE, 1972).

Family UNCERTAIN

Albumares FEDONKIN in KELLER & FEDONKIN, 1976, p. 38 [*A. brunsae; OD]. Disc shield-like, lobate, with 3 circumoral ridges narrowing radially; delicate gastrovascular canals, 3 on each lobe, dividing dichotomously 4 times each toward the periphery; over 100 short, very thin, marginal tentacles. Resembling *Skinnera* which differs in absence of fine, dichotomously branching canals, lobes and tentacles. U.Precam.(U.Vend.), N.Russia.

Brachina WADE, 1972, p. 207 [*B. delicata; OD]. Discoid medusa with numerous small marginal lappets attached to an outer ring; on exumbrellar side two concentric outer grooves, an inner groove and a central peak; on subumbrellar side probably with a small, conical manubrium, outwardly branching and inwardly anastomosing gastrovascular canals, and apparently an annular gonad. U.Precam., S.Australia.——FIG. 7,3. *B. delicata, Brachina Gorge; 3a, reconstr., ×0.5; 3b, holotype, int. mold, ×1.5 (Wade, 1972a).

Hallidaya WADE, 1969, p. 356 [*H. brueri; OD]. Discoid medusa with truncated margin, low-domed or flat, with scattered "nuclei" (3-13) near center;

1   Ovatoscutum

FIG. 5. Chondroplidae (p. *A91-A92*).

Precambrian

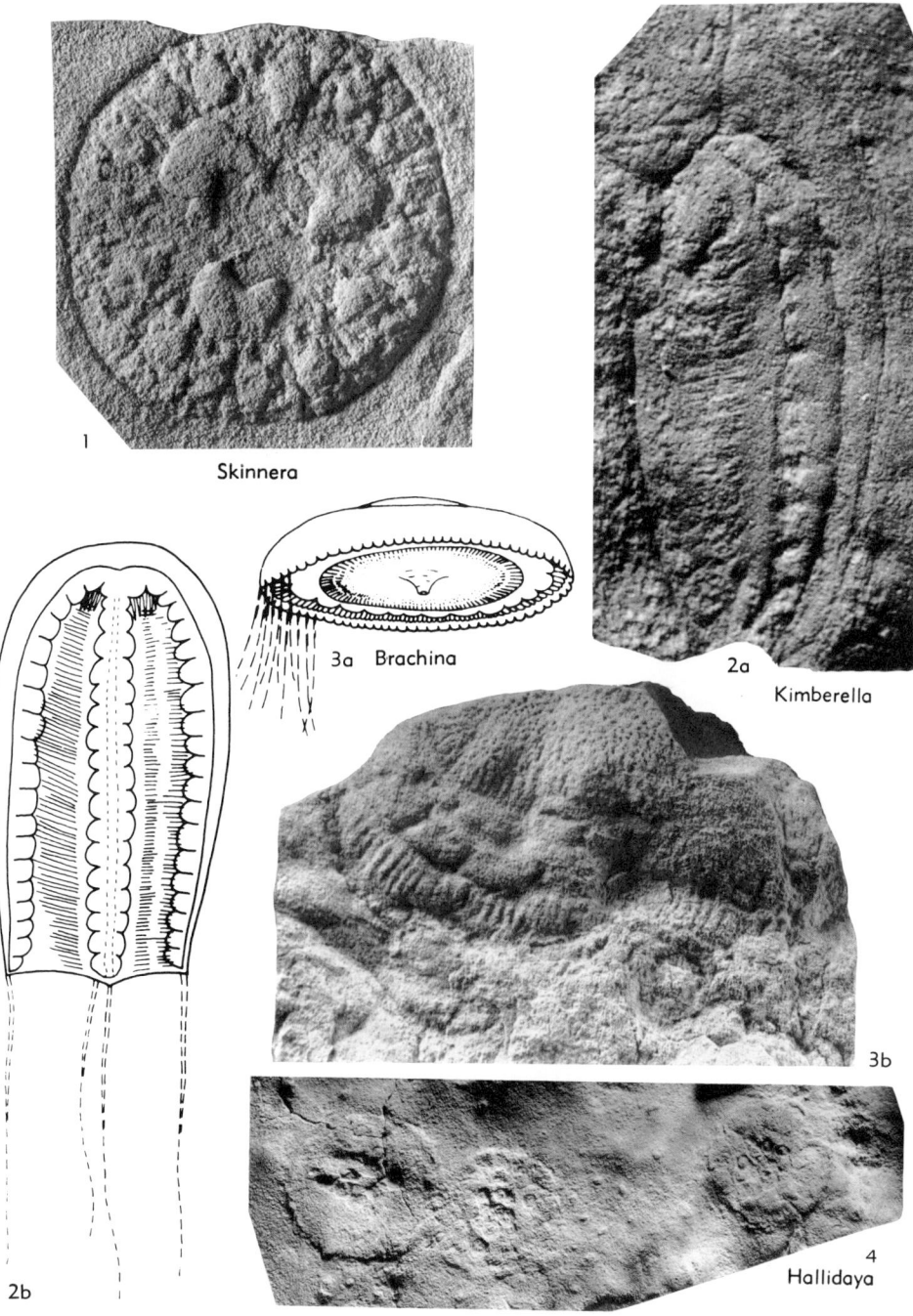

Fig. 7. Family uncertain (p. *A*92-*A*94).

some specimens with a rayed subcentral impression that could indicate a mouth, and dichotomous, radial furrows near periphery. *Uppermost Precam.*, C.Australia.——FIG. 7,4. **H. brueri*, Mt. Skinner; paratype, ×1 (Wade, 1969).

Kimberella WADE, 1972, p. 215 [*pro Kimberia* GLAESSNER & WADE, 1966 (*non* COTTON & WOODS, 1935)] [**Kimberia quadrata* GLAESSNER & WADE, 1966; OD]. Elongate, slender bell, probably squarish in transverse section, with 4 pouched gonads attached to radial canals projecting into cavity of bell; gastric filaments present adapically; conspicuous transversely striated zones are explained as representing contracted subumbrellar muscle bands. *U.Precam.*, S.Australia.——FIG. 7,2. **K. quadrata* (GLAESSNER & WADE), Ediacara; *2a*, holotype, ×1 (Glaessner & Wade, 1966); *2b*, reconstr. (Wade, 1972a).

Skinnera WADE, 1969, p. 361 [**S. brooksi*; OD]. Disc-shaped, probably originally plano-convex, with 3 large inner "pouches" and, connected with them by paired canals, 15 outer (secondary) "pouches" symmetrically placed near periphery. [According to WADE, "the shape of the internal system of spaces and canals closely parallels the gastrovascular system of a medusa of the scyphozoan grade of complexity."] *Uppermost Precam.*, C.Australia.——FIG. 7,1. **S. brooksi*, Mt. Skinner; holotype, ×2 (Wade, 1969).

Class CONULATA
Moore & Harrington, 1956

[*nom. transl.* GLAESSNER, 1971, p. 15 (*ex* Subclass Conulata MOORE & HARRINGTON, 1956, p. F28)] [For diagnosis see Subclass Conulata, this *Treatise*, 1956, p. F28]

Order CONULARIIDA
Miller & Gurley, 1896

[For diagnosis see Suborder Conulariina, this *Treatise*, 1956, p. F58]

Suborder CONCHOPELTINA
Moore & Harrington, 1956

[For diagnosis see this *Treatise*, 1956, p. F57]

Family CONCHOPELTIDAE
Moore & Harrington, 1956

[For diagnosis see this *Treatise*, 1956, p. F57]

Conomedusites GLAESSNER & WADE, 1966, p. 608 [**C. lobatus*; OD]. Theca forming low cone with concentric rugosities, divided by 4 deep radial grooves; peripheral margin correspondingly with 4 lobes, each of which may be further subdivided by a shallow indentation; fringe of rather thick tentacles may be preserved around peripheral margin. *U.Precam.*, S.Australia.——FIG. 8,1. **C. lobatus*, Ediacara; ×0.7 (Glaessner, 1971).

1   Conomedusites

FIG. 8. Conchopeltidae (p. *A94*).

MEDUSAE OF UNCERTAIN AFFINITIES

A large number of Precambrian medusa-like fossils cannot be recognized with certainty or even high probability as either hydrozoan or scyphozoan in their structure and affinities. Some are rather featureless and difficult to distinguish from discoidal basal attachment structures known in some Precambrian benthonic cnidarians (see *Charniodiscus*). Others are distinctive, but only their exumbrellar side is known (*Mawsonites, Planomedusites, Rugoconites*). Others again are so variable, either inherently (*Ediacaria*) or because of varying degrees of decomposition (*Pseudorhizostomites*), that their true morphology cannot be discerned and diagnosed.

Cyclomedusa SPRIGG, 1947, p. 220 [**C. davidi*; OD] [=*Madigania* SPRIGG, 1949, p. 93 (*non* WHITLEY, 1945); *Tateana* SPRIGG, 1949, p. 86; *Spriggia* SOUTHCOTT, 1958, p. 59]. Outline subcircular, surface of disc with several to many concentric grooves separating slightly elevated areas (rugae); their arrangement indicates an originally conical shape of the center or, in some species, of most of the body. Many specimens show fine, straight, radial grooves interpreted as gastrodermal canals showing on surface because of partial composite molding. [The oral surfaces of these commonly occurring medusae are unknown. They may have been attached by their aboral cones. WADE (1972a, p. 204) placed *Cyclomedusa* under "Class Hydrozoa?" A very similar genus, *Tirasiana*, was named by PALIY in a dissertation abstract dated 1975 but it remained unpublished until BEKKER (1977) published *T. concentralis* n. sp. from the Vendian of the W. Ural.]

Fig. 9. Medusae of uncertain affinities (p. *A94-A96*).

U.Precam., S.Australia-SW.Afr.-S.USSR-N.Swed.-E.Sib.——Fig. 9,*2a*. *C. plana* GLAESSNER & WADE, U.Precam., SW.USSR (Dnjestr R., Ukraine); ×0.67 (Zaika-Novatskii *et al.,* 1968).——Fig. 9,*2b*. *C. gigantea* SPRIGG, U.Precam., Australia (Ediacara); holotype, ×1 (Sprigg, 1949).——Fig.

9,2c,d. *C. davidi, U.Precam., S.Australia (Ediacara); ×1.4 (Sprigg, 1949).

Ediacaria SPRIGG, 1947, p. 215 [*E. flindersi; OD] [=Protodipleurosoma SPRIGG, 1949, p. 79; ?Beltanella SPRIGG, 1947, p. 218]. Exumbrellar surface showing central disc and an outer ring; sharp circular furrow may mark edge of gastric cavity; radial furrows mostly confined to outer ring; on subumbrellar surface a rounded central mouth without appendages; one specimen appears to show numerous, long, fine, peripheral tentacles. U.Precam., S.Australia.——FIG. 9,1. E. sp., Ediacara; 3 specimens on slab, ×0.3 (fig. by Wade, Glaessner, n).

Lorenzinites GLAESSNER & WADE, 1966, p. 608 [*L. rarus; OD]. Small central disc from which lobes radiate to length equal to or exceeding that of radius of disc, broadening and flattening at outer ends. [As only one specimen is known, the status of this taxon is doubtful. For a discussion of similar Phanerozoic fossils, which may be trace fossils, see HÄNTZSCHEL, 1975, p. W144.] U.Precam., S.Australia.

Mawsonites GLAESSNER & WADE, 1966, p. 607 [*M. spriggi; OD]. Large, with smooth conical center; greater part of surface strongly sculptured with large, irregular bosses that increase in size outward to form radially elongate lobes separated by furrows leading to the lobate peripheral margin. Subumbrellar surface unknown: U.Precam., S.Australia.——FIG. 9,3. *M. spriggi, Ediacara; paratype, ×0.67 (Glaessner & Wade, 1966).

Medusinites GLAESSNER & WADE, 1966, p. 605 [*Medusina asteroides SPRIGG, 1949; OD]. Small, subcircular, discoidal bodies with central disc separated by deep circular groove from large outer ring with radius greater than that of central disc and with radial grooves irregularly preserved on it; there is a very narrow marginal flange; faint concentric markings occur more commonly on central disc; subumbrellar surface unknown. U.Precam., S.Australia-S.USSR.——FIG. 10,1. *M. asteroides, S.Australia(Ediacara); 1a,b, ×1 (Glaessner & Wade, 1966).

Planomedusites SOKOLOV, 1972 [*P. grandis; M]. Large, saucer shaped, surface smooth, edge raised as a narrow rim, apparently surrounded by narrow, thin flange. U.Precam.(Vend.), S.USSR.——FIG. 10,3. *P. grandis; Dnjestr R., Ukraine; holotype, ×0.77 (Sokolov, 1972).

Pseudorhizostomites SPRIGG, 1949, p. 87 [*P. howchini; OD] [=Pseudorhopilema SPRIGG, 1949, p. 88]. Furrows radiating outward from variably shaped center; bifurcating or anastomosing and becoming finer outward; no distinct peripheral margin; fluted passage extending vertically upward from center through overlying sediment suggests that shape of fossils was influenced by decay and escape of decay products of a medusoid which is probably not yet known in better preservation. U.Precam., S.Australia.——FIG. 10,2. *P. howchini, Ediacara; 2a,b, ×1.3 (Sprigg, 1949).

Rugoconites GLAESSNER & WADE, 1966, p. 610 [*R. enigmaticus; OD]. Low conical body with circular to oval peripheral margin; a few furrows diverge from small polygonal center, repeatedly branching dichotomously and anastomosing; furrows either few and coarse or numerous and fine; peripheral margin with narrow flange and numerous fine tentacles. U.Precam., S.Australia.——FIG. 10,4. *R. enigmaticus, Ediacara; holotype, ×1 (Glaessner & Wade, 1966).

PROBLEMATICAL COELENTERATA

"PETALONAMAE" Pflug, 1970

[Petalonamidae PFLUG 1970a, p. 258; Phylum Petalonamae PFLUG, 1972b, p. 158]

Leaflike structures ("petaloids"), often with a median line or zone and lateral grooves and ribs on each leaf, disposed as primary, secondary and occasionally tertiary branches; petaloids may occur in clusters or be joined to form fanlike composite structures ("flabella" and "petalodia"), or be bent into bun or bag shapes; their external layer may show fibrous microstructure; they appear to represent colonial organisms. [According to PFLUG (1970a,b), the petaloids are typically linked in groups but the evidence for dendroid branching, and for the assumed extensive structural homologies between the groupings of petaloids in flabella composing the petalodia in different genera is inconclusive. PFLUG has built up an elaborate and complex terminology that is not intended to be purely descriptive, but which tends to explain conceptual "structural plans." A discussion of these theoretical views is considered to be outside the scope of this *Treatise* (see PFLUG, 1970a-1974b, for details); however, it is convenient to group the interconnected taxa listed below provisionally under the name proposed by PFLUG (which was accepted by GERMS, 1972a, and GLAESSNER & WALTER, 1975), without defining its status in classification and nomenclature, until clear distinctions between observable and hypothetically postulated characters can be drawn.] U.Precam.

Family PTERIDINIIDAE Richter, 1955

[=Class Pteridiniomorpha PFLUG, 1972b, p. 158]

Leaflike, elongate structures ("petal-

FIG. 10. Medusae of uncertain affinities (p. *A*94-*A*96).

Fig. 11. Pteridiniidae *(1)*; Rangeidae *(2)* (p. *A96-A99*).

oids"), roughly bilaterally symmetrical, probably composite, elastically deformed during embedding, with distinct median groove often showing a series of small rhombical elements ("commissurae"); more conspicuous are transverse, sharply incised, primary furrows and convex branches; a marginal zone, which is almost smooth, may be present. [The evidence for strictly defined dendroid branching modes proposed by PFLUG is unconvincing. These fossils, together with *Rangea*, were considered by RICHTER (1955) as Gorgonaria and by GLAESSNER (in GLAESSNER & DAILY, 1959) as Pennatulacea.] *U.Precam.(Vend.)*.

Pteridinium GÜRICH, 1933, p. 144 [*P. simplex*; OD] [=*Pteridium* GÜRICH, 1930, p. 637 *(nom. nud.)* (non SCOPOLI, 1777); ?*Onegia* SOKOLOV, 1976]. Characters of family. *U.Precam.(Nama Gr.)*, SW.Afr.; *U.Precam.(up. Vend.)*, N.USSR-S.Australia.——FIG. 11,*1*. **P. simplex*, Nama Gr., SW.Afr.; *1a*, specimen closely resembling neotype, ×0.8 (Glaessner, 1963); *1b*, group of specimens on slab in State Museum, Windhook) (Glaessner, n).

Family RANGEIDAE Glaessner, new

[=Class Rangeomorpha PFLUG, 1972b, p. 158]

Leaflike rounded structures ("petaloids"), roughly bilaterally symmetrical, probably composite; with median groove or track and lateral primary branches that are divided into small, chevron-shaped secondary branches. *U.Precam.*

Rangea GÜRICH, 1930, p. 680 [*R. schneiderhöhni*; OD]. Characters of family. *U.Precam.(Nama Gr.)*, SW.Afr.——FIG. 11,*2*. **R. schneiderhöhni*, ×1.6 (Pflug, 1970b).

Family CHARNIIDAE Glaessner, new

Leaflike bodies ("petaloids") single, elongate, often with narrow stem and expanded discoidal base, with median groove or track; secondary branches disposed as convex, parallel structures between primary branches. *U.Precam.*

Charnia FORD, 1958, p. 212 [*C. masoni*; OD]. Narrow petaloids with sinuous median line and sharply defined primary grooves forming acute angles with corresponding secondary grooves and branches; these are therefore in almost transverse position on the petaloids. *U.Precam.(Charn.)*, Eng.——FIG. 12,*3*. **C. masoni*; plaster cast of holotype, ×0.45 (Glaessner, n; Leicester City Museum no. 279,1958).

Charniodiscus FORD, 1958, p. 213 [**C. concentricus*; OD] [=*Charnia* FORD, 1958, p. 212 *(partim); Arborea* GLAESSNER & WADE, 1966, p. 618]. Elongate petaloids with broad "dorsal" and slightly narrower "ventral" median track, extending downward into a stalk ending in an expanded discoidal base; secondary branches on flangelike expansions on the "ventral" faces of primary branches. [FORD (1958) distinguished *Charniodiscus concentricus* from *Charnia masoni* because only in one specimen (the holotype of the former) were "frond and disc apparently associated" and there "it could be interpreted as a distinct type of frond. . . ." Hence he diagnosed *Charniodiscus* as a "disc-like organism." In 1963 he figured this entire specimen as *C. concentricus* "with *Charnia masoni* frond attached." Dr. R. J. F. JENKINS (pers. commun.) agrees with the alternative view that this is a distinct type of frond. He considers that therefore the entire specimen becomes the holotype of *Charniodiscus concentricus* and concludes also that this frond shows the characters of *Arborea*, which consequently becomes a synonym of *Charniodiscus*.] *U.Precam.*, Eng.-S.Australia.——FIG. 12,*2a*. **C. concentricus*, Charn., Eng.; plaster cast of holotype with frond attached, approx. ×0.3 (Glaessner, n; Leicester Univ. no. 2383/1-2).——FIG. 12,*2b,c*. *C. arboreus* (GLAESSNER); *2b*, S.Australia (Ediacara), discoidal base and stalk, ×0.3 (Glaessner, n; Adelaide Univ., unnumbered specimen); *2c*, S.Australia (Bunyeroo Gorge), plaster cast of impression on lower bedding plane showing discoidal base inferred to have been attached to adjacent complete petaloid, about ×0.2 (fig. by R. J. F. Jenkins, Glaessner, n; Adelaide Univ.).

Glaessnerina GERMS, 1973, p. 5 [**Rangea grandis* GLAESSNER & WADE, 1966; OD]. Resembling *Charnia* but secondary grooves and ribs form large angles with primary grooves and hence in markedly oblique to almost longitudinal position on the petaloids. *U.Precam.*, S.Australia-N.USSR-N.Sib.——FIG. 12,*1*. *G. sibirica* (SOKOLOV), N.Sib.; ×0.67 (Sokolov, 1972).

Family ERNIETTIDAE Pflug, 1972

[Erniettidae PFLUG, 1972b, p. 163] [=Class Erniettomorpha PFLUG, 1972b, p. 158; Family Ernionormidae PFLUG, 1972b, p. 158]

Rounded, bun- or bag-shaped, ellipsoidal or cylindrical bodies with external ribs generally divided by a median groove disposed so as to suggest derivation from or interpretation as folded petaloids similar to those of *Pteridinium*. *U. Precam.* [PFLUG (1972b) has studied these fossils in great detail. He considered not only major but also minor differences in shape as taxonomic characters, distinguishing 13 genera (with 28 species)

FIG. 12. Charniidae (p. A99).

in five subfamilies, four families and two orders of a class Erniettomorpha; however, many apparent differences could be the results of postmortal deformation whereas others appear to be due to variability of growth and preservation. Some genera have type specimens that are so poorly preserved as to be unrecognizable (see list on p. A112). It seems appropriate to separate the tall, complexly sutured forms at subfamily level from those more or less close to *Ernietta*.]

Precambrian

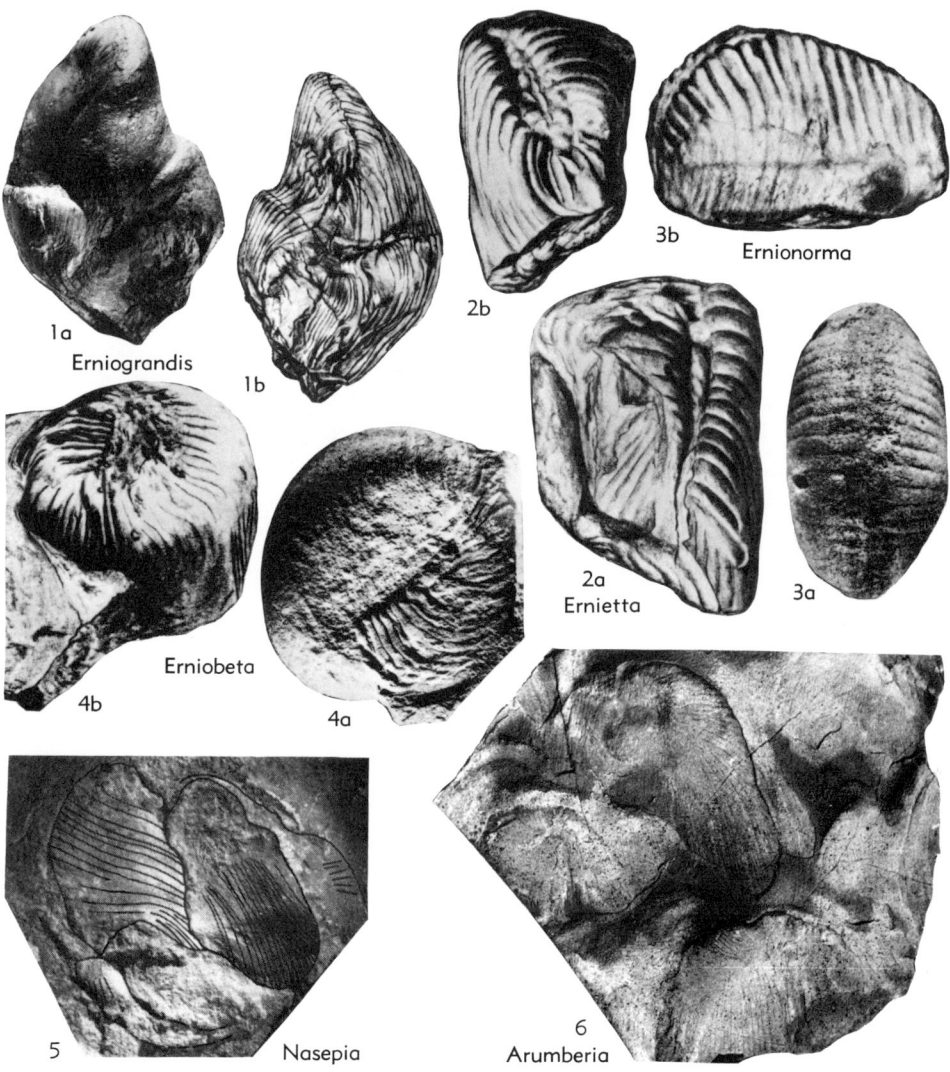

Fig. 13. Erniettidae *(1-4)*; Family uncertain *(5,6)* (p. *A*99-*A*102).

Subfamily ERNIETTINAE Pflug, 1972

[Erniettinae Pflug, 1972b, p. 163] [=Ernionorminae Pflug, 1972b, p. 160; Erniodiscinae Pflug, 1972b, p. 158]

Body flattened, round to subcylindrical, with ribs more or less clearly divided by median zigzag line. *U.Precam.*

Ernietta Pflug, 1966, p. 19 [**E. plateauensis*; OD]. Body compressed or bent into U-shape; ribs strongly developed, separated by zigzag median line; resembling a folded petaloid of *Pteridinium*. *U.Precam.*, SW.Afr.——FIG. 13,*2*. **E. plateauensis*; *2a,b*, ×1 (Pflug, 1972b).

Erniofossa Pflug, 1972, p. 159 [**E. prognatha*; OD] [=?*Erniodiscus* Pflug, 1972b, p. 158; ?*Erniaster* Pflug, 1972b, p. 159]. Body with rounded to elliptic basal outline, upper surface convex, with central depression or flattened; general shape varying from discoidal to cylindrical, median groove short or indistinct. *U.Precam.* *(Nama Gr.)*, SW.Afr.

Ernionorma Pflug, 1972, p. 160 [**E. abyssoides*; OD] [=*Erniobaris* Pflug, 1972b, p. 161]. Body with elliptic basal outline, highly convex, with median groove between distinct lateral ribs. *U. Precam.(Nama Gr.)*, SW.Afr.——FIG. 13,*3*. *E.*

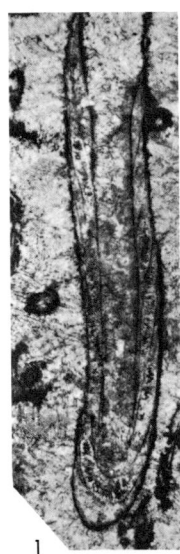

1
Cloudina

Fig. 14. Vologdinophyllidae (p. *A*102).

corrector PFLUG; *3a,* ×1; *3b,* ×1 (Pflug, 1972b).

Subfamily ERNIOBETINAE Pflug, 1972

[Erniobetinae PFLUG, 1972b, p. 165]

Body columnar, tall, ribbed, with median line near apex and with transverse, incised sutures. *U.Precam.*

Erniobeta PFLUG, 1972, p. 166 [**E. scapulosa*; OD]. Body columnar, ribbed, top surface convex, with median groove, sutures more or less distinct; occurring in "colonies." *U.Precam.(Nama Gr.),* SW. Afr.——FIG. 13,*4. E. forensis* PFLUG; *4a,b,* ×1 (Pflug, 1972b).
Erniograndis PFLUG, 1972, p. 165 [**E. sandalix*; OD]. Body tall, ribbed, bulbous, transversely sutured, open end narrowed, median line subapical. *U.Precam.(Nama Gr.),* SW.Afr.——FIG. 13,*1. *E. sandalix; 1a,b,* ×0.3 (Pflug, 1972b).

Family UNCERTAIN

Arumberia GLAESSNER & WALTER, 1975, p. 61 [**A. banksi*; OD]. Hollow, compressible, ribbed, conical to cylindrical bodies, attached by blunt apex; ribs may bifurcate. *Uppermost Precam.,* C.Australia.——FIG. 13,*6. *A. banksi,* Laura Creek (nr. Alice Springs); casts of several specimens on lower bedding surface; ×0.2 (Glaessner & Walter, 1975).
Baikalina SOKOLOV, 1972 [**B. sessilis*; M]. Bag-shaped body, narrow at base, with mm-sized, longitudinal, flat ribs. *U.Precam.(U.Vend.),* S.Sib.
Namalia GERMS, 1968, p. 53 [**N. villiersiensis*; OD]. Conical to cylindrical, with rounded cross section, longitudinal ribs, blunt apex; occurring in "colonies." *U.Precam.(Nama Gr.),* SW.Afr.
Nasepia GERMS, 1972, p. 7 [**N. altae*; OD]. Leaflike bodies with fine ribs subparallel to long axis and with clearly marked margins. *U.Precam. (Nama Gr.),* SW.Afr.——FIG. 13,*5. *N. altae;* ×0.6 (Germs, 1973a).

UNRECOGNIZABLE AND REJECTED GENERA ASSIGNED BY PFLUG (1972b) TO "ERNIETTOMORPHA"

Erniocarpus PFLUG, 1972, p. 164 [**E. carpoides*; OD]. Single, weathered, discoidal specimen.
Erniocentris PFLUG, 1972, p. 159 [**E. centriformis*; OD]. Single specimen with concentric ribbing. Possibly a concretion.
Erniocoris PFLUG, 1972, p. 164 [**E. orbiformis*; OD]. Single weathered specimen superficially resembling mold of bivalve shell.
Erniopelta PFLUG, 1972, p. 162 [**E. scrupula*; OD]. Convex bodies with obscure and irregular sculpture.
Erniotaxis PFLUG, 1972, p. 165 [**E. segmentrix*; OD]. Fragmentary molds consisting of few ribs only, placed by PFLUG (1972b) in a monotypic family Erniotaxidae and subfamily Erniotaxinae.

Phylum ANNELIDA Lamarck, 1809

Class POLYCHAETA Grube, 1850

[see this *Treatise,* 1962, p. *W*148]

Order CRIBRICYATHEA Vologdin, 1961

[*nom. transl.* GLAESSNER, 1976a (*ex* Class Cribricyathea VOLOGDIN, 1961, p. 177); see HILL, 1972, p. *E*134]

?Family VOLOGDINOPHYLLIDAE Radugin, 1964

[see HILL, 1972, p. *E*138]

Cloudina GERMS, 1972b, p. 752 [**C. hartmannae* (=*C. hartmanae* GLAESSNER, 1976a, *emend.*); OD]. Tubes sinuous, conical to almost cylindrical, walls with outer layer covered with close-set transverse annular ridges and grooves; the main layer consisting of stacked, inverted cones sloping inward toward the apex, incomplete in transverse section (forming half-rings); inner surface of tube elliptical, smooth. [For a discussion of the relations of *Cloudina* to the Cribricyathea and of this order to the Polychaeta, see GLAESSNER, 1976a.] *U.Precam.,* SW.Afr.; ?*L.Cam.,* S.Am.(Arg.).——FIG. 14,*1. *C. hartmanae;* U.Precam., SW.Afr.; ×0.3 (Germs, 1972b).

Order UNCERTAIN

Family DICKINSONIIDAE Harrington & Moore, 1955

Body elliptical to elongate, segments numerous, generally widening outwardly, anterior segments fused medially. *U. Precam.*

[The monotypic family Dickinsoniidae was placed in an order Dickinsoniida and a class Dipleurozoa of the Coelenterata by HARRINGTON & MOORE (1955) (see also this *Treatise*, 1956, p. F24-F25). This was based on a series of misunderstandings starting with SPRIGG's (1947) faulty reconstruction of *Dickinsonia* as "symmetrical across both longitudinal and transverse planes." Since the close relationship between *Dickinsonia* and the living polychaete *Spinther* has been established by the study of much additional material, there is no need for higher taxa above family rank for the former genus. The supposed Silurian "Dipleurozoa" named *Rutgersella* by JOHNSON and FOX (1968) are unrelated to *Dickinsonia* and are now considered by CLOUD (1973) as based on pyrite rosettes. The family Dickinsoniidae can be accommodated in the order Amphinomorpha in CLARK's (1969) system of the Polychaeta, in view of its proximity to the Spintheridae. WADE (1972b) suggested the use of the order Dickinsoniida for both Dickinsoniidae and Spintheridae. The classification of the Polychaeta above family level is in a state of flux and (with one exception) higher taxa will not be used here for the known Precambrian polychaetes.]

Dickinsonia SPRIGG, 1947, p. 221 [**D. costata*; OD] [=*Papilionata* SPRIGG, 1947, p. 223]. Broad, flat, with numerous, short segments; anterior body segments fused pre-orally along median line; segmental furrows depressed dorsally and ventrally; neuropodia reduced, notopodial-elytral ridges well developed; filled intestine with intestinal caeca may be preserved. *U.Precam.*, S.Australia-N.USSR.
——FIG. 15,2. **D. costata*, S.Australia(Ediacara); 2a, ×0.64; 2b,c, ×0.8 (Sprigg, 1949).

Family SPRIGGINIDAE Glaessner, 1958

[See HOWELL, 1962, *Treatise*, p. W154]

Spriggina GLAESSNER, 1958, p. 158 [**S. floundersi*; OD]. Prostomium horseshoe-shaped, strongly sclerotized, with a sharp medio-posterior semi-circular impression on its margin, without external segmentation. Body flexible, rather flat, consisting of up to 40 segments, tapering gently to a rounded, minutely segmented posterior end; neuropodia with acicular setae, a double series of medio-dorsal paired convexities separated by a sagittal groove probably represent dorsal longitudinal mus-

1 Marywadea

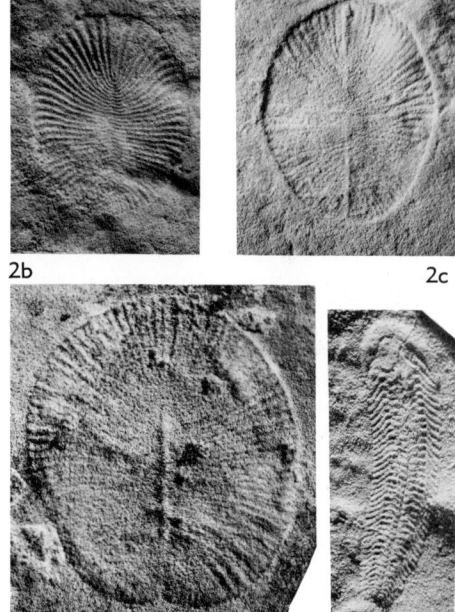

2a Dickinsonia 2b 2c 3 Spriggina

FIG. 15. Dickinsoniidae *(2)*; Sprigginidae *(1,3)* (p. *A103-A104*).

cles; convexity variably placed near the first trunk segments suggests a well-developed pharynx. *U. Precam.*, S.Australia.——FIG. 15,*3*. **S. floundersi*, Ediacara; ×1 (Glaessner, n; Univ. Adelaide coll. no. F17354).

Marywadea GLAESSNER, 1976, p. 169 [**Spriggina ? ovata* GLAESSNER & WADE, 1966, p. 622; OD]. Prostomium half-moon shaped, not wider than the body with appendages, integument thin; up to 50 short, broad segments bearing long, curved setae; behind the prostomium a pair of elongate impressions suggesting teeth; posterior end of body broadly rounded. *U.Precam.*, S.Australia-?SW.Afr. ——FIG. 15,*1*. **M. ovata* (GLAESSNER & WADE), Australia (Ediacara); holotype, ×3.2 (Glaessner & Wade, 1966).

Family ANABARITIDAE Glaessner, new

[=Angustiochreidae VALKOV & SYSOIEV, 1970, p. 96 (*invalid name*)]

Small, straight or curved, conical tubes of calcareous composition, with rounded to triangular or stellate cross section; three or more evenly spaced straight or curved longitudinal grooves and corresponding internal ribs or rows of spines. [VALKOV & SYSOIEV (1970, p. 97) erected a new genus *Angustiochrea* for Lower Cambrian tubular fossils and placed it together with (among others) the closely similar *Anabarites* MISSARZHEVSKY in a new family Augustiochreidae (p. 96), expressly based on *Anabarites* as "type genus," in violation of the International Code of Zoological Nomenclature. They also erected a new order Angustiochreida.] *U.Precam.*, *?L.Cam.*

Anabarites MISSARZHEVSKY in VORONOVA & MISSARZHEVSKY, 1969, p. 209 [**A. trisulcatus*; OD] [=*Angustiochrea* VALKOV & SYSOIEV, 1970, p. 97

FIG. 16. Anabaritidae (p. *A*104).

(type, *A. lata*; OD)]. Small, thin-walled, elongate-conical tubes, trilobed to triangular or triradiate in transverse section. [The synonymy of *Angustiochrea* and *Anabarites* was suggested to me by V. V. MISSARZHEVSKY (pers. commun., May, 1975) and appears justified.] *Uppermost Precam. and basal Cam.*, N.&E.Sib.——FIG. 16,*1*. **A. trisulcatus*, U.Precam.(Vend.), Anabar reg.; ×25 (Matthews & Missharzhevsky, 1975).

Phylum ARTHROPODA
Siebold & Stannius, 1845

[see this *Treatise*, 1959, p. *O*4]

Superclass TRILOBITOMORPHA Størmer, 1944 (or CHELICERATA Heymons, 1901)

[See this *Treatise*, 1959, p. *O*22; 1955, p. *P*1, and 1969, p. *R*13]

Class and Order UNCERTAIN

Family VENDOMIIDAE
Keller in Keller & Fedonkin, 1976

Small, elongate, discoidal body with broadly arcuate anterior margin; head shield large, followed by up to five chevron-shaped segments and a small telson. [In the absence of any traces of appendages it cannot be decided whether this family should be placed in the Trilobitomorpha or Chelicerata; the general morphology of the body resembles both.] *U.Precam.*

Vendomia KELLER in KELLER & FEDONKIN, 1976, p. 43 [**V. menneri*; OD]. Horseshoe-shaped cephalic area occupying ⅔ of length, separated from trunk area consisting of 5 somites, decreasing in size posteriorly, with median groove; telson not clearly observable. *U.Precam.(U.Vend.)*, N.USSR.

Onega FEDONKIN in KELLER & FEDONKIN, 1976, p. 42 [**O. stepanovi*; OD]. Elliptic flat body with sharp outline; wide, smooth, marginal area, wider at anterior (?) end which is crescent-shaped; behind it an axially segmented series of 5 paired lobes, transversely elongate, slightly curved towards anterior (?) end, with deep, wide axial groove, length of segments decreases posteriorly. *U.Precam.(U.Vend.)*, N.USSR.

Praecambridium GLAESSNER & WADE, 1966, p. 623 [**P. sigillum*; OD]. Body small, dorsal side with horseshoe-shaped head region bearing a heart-shaped glabellar area surrounded by surface crenulations (?digestive caeca) and followed by 3 to 5 chevron-shaped segmental ridges and a triangular terminal somite. *U.Precam.*, S.Australia.——FIG.

17,1. *P. sigillum;* diagramm. reconstr., ×7.5 (GLAESSNER & WADE, 1971).

Vendia KELLER in ROZANOV et al., 1969, p. 175 [**V. sokolovi;* OD]. Body elongate, disc shaped; horseshoe-shaped cephalic area followed by 5 narrow, inverted V-shaped segments diverging outward and backward, weak and uneven median ridge. *U.Precam.(Vend.)*, N.USSR.——FIG. 17,2. **V. sokolovi,* Yarensk; ×5 (SOKOLOV, 1972).

Superclass CRUSTACEA
Pennant, 1777

Class BRANCHIOPODA
Latreille, 1817

(see this *Treatise,* 1969, p. R131)

Order UNCERTAIN

Family PARVANCORINIDAE
Glaessner, new

Shieldlike carapace elongate, with faint marginal raised rim and distinctly elevated, anterolateral and median, smooth, dorsal ridges; about five pairs of stout anterior appendages are followed by up to 20 pairs of posterior, undifferentiated, filiform appendages. [Some resemblance with *Vachonisia* LEHMANN, but also with *Marrella* WALCOTT and *Mimetaster* GÜRICH, should be noted. The observation of trilobitomorph legs in the Devonian genus *Vachonisia* appears to require its transfer to the Marellomorpha from the Crustacea Branchiopoda (see STÜRMER & BERGSTRÖM, 1976). Similarity of *Parvancorina* with Marellomorpha may suggest that it is close to the ancestors of the Crustacea, the derivation of which from Trilobitomorpha was suggested by HESSLER and NEUMAN (1975). DELLE CAVE and SIMONETTA (1975) indicated similarities with *Skania fragilis* WALCOTT, 1931. If confirmed, this could extend the range of the family to *M.Cam.*] *U.Precam.*

Parvancorina GLAESSNER in GLAESSNER & DAILY, 1959, p. 187 [**P. minchami;* OD]. Characters of family (see Fig. 18). *U.Precam.,* S.Australia.

DOUBTFUL ARTHROPODA

Velancorina PFLUG, 1966, p. 17 [**V. martina;* OD]. Outline arcuate anteriorly, converging posteriorly, several marginal ridges and furrows, two stronger

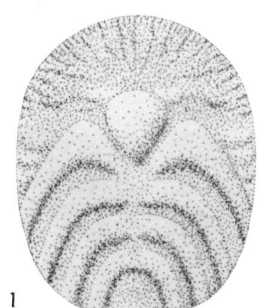

1 Praecambridium

2 Vendia

FIG. 17. Vendomiidae (p. *A104-A105*).

central ridges separated by a pronounced median furrow, not extending to margins; some fine longitudinal striae, no pronounced transverse sculpture except short grooves near axis. *U.Precam.,* SW.Afr.; *L.Cam.,* W.Can. [YOUNG (1972) described a "*Rusophycus* sp." from the Lower Cambrian. It lacks the characteristic transverse sculpture. It does not show the transverse grooves described by PFLUG that are shown in his drawing of *Velancorina* but not discernible in his photographs; the Canadian fossil is otherwise indistinguishable from *Velancorina.* Although not a trilobite "resting trail," this fossil with its bilaterally symmetrical shieldlike outline and median groove may represent the ventral impression of a primitive arthropod.]

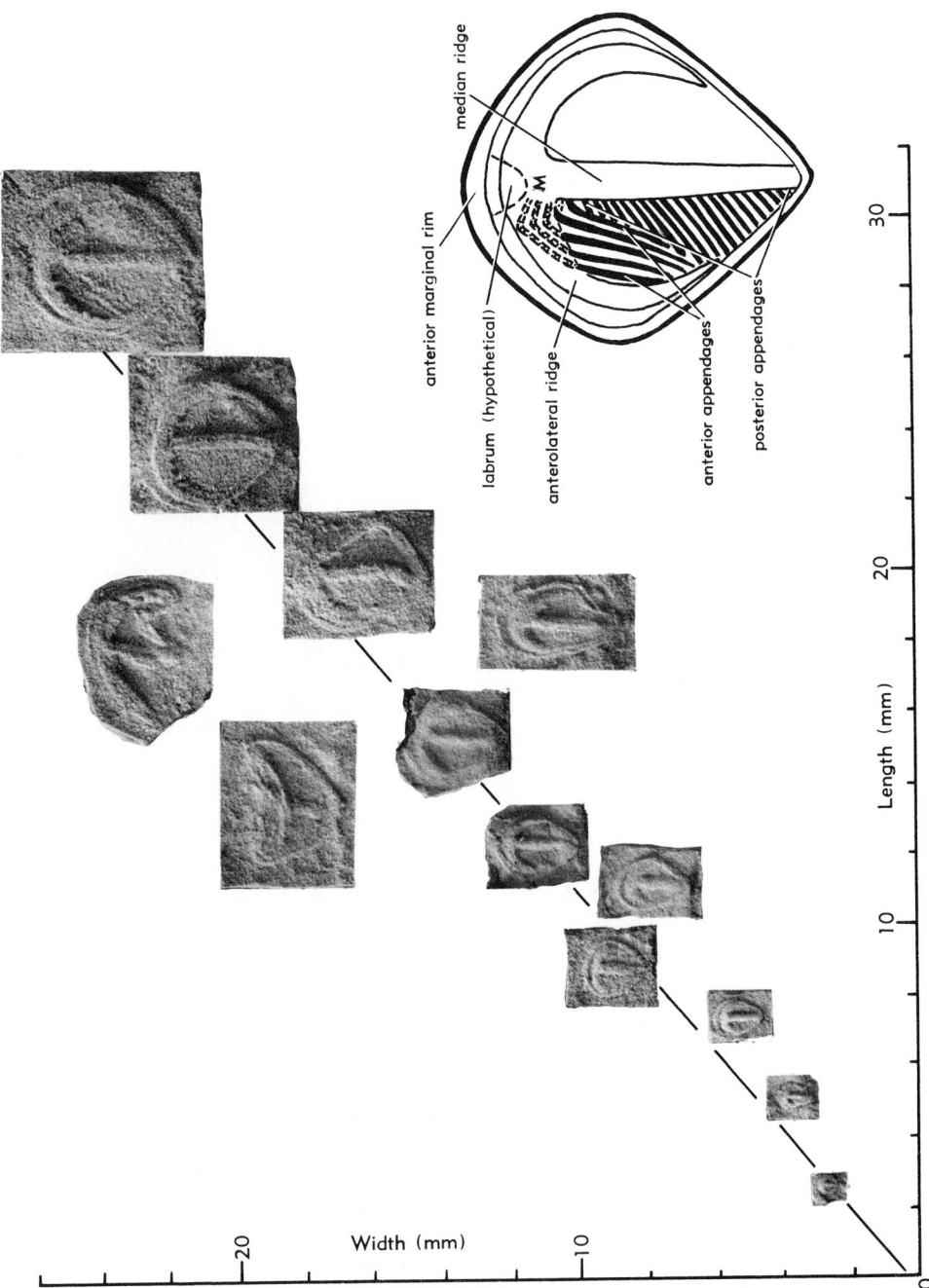

FIG. 18. Growth series of *Parvancorina minchami* GLAESSNER from S. Australia (Ediacara) plotted on expanded scale (×5) of length and width, with illustrations of specimens (×1) placed in approximate coordinate positions. Strictly nonallometric growth is not implied; the carapace seems to be more elongate in early growth stages and as wide as long in late growth stages when the entire material is considered. Biometric tests tend to be invalidated by distortion of the flexible carapaces; the specimens

?Phylum POGONOPHORA Johannson, 1938

Order SABELLIDITIDA Sokolov, 1965

Elastic, thin, very long, slender, cylindrical tubes; walls often fibrous with smooth terminal portions, or irregularly wrinkled transversely; not branching; black, brown, or translucent yellow, organic (combustible). *U.Precam.-L.Cam.*

[This order is considered by SOKOLOV as representing Pogonophora. It should be noted that zoologists tend to disagree as to whether the Pogonophora is a separate phylum or where it should be placed in the classification of the Metazoa, since significant similarities between them and polychaete annelids have been discovered (see NØRREVANG, 1975).]

Family SAARINIDAE Sokolov, 1965

Tubes very thin, translucent, consisting of funnel-shaped narrow rings. *U.Precam. (Vend.)-L.Cam.*

Calyptrina SOKOLOV, 1965, p. 91 [*C. partita*; OD]. Tubes light yellow or colorless, consisting of narrow, transverse rings with rounded, projecting edges. *U.Precam.(Vend.)*, Sib.; *L.Cam.*, N.USSR.

Family SABELLIDITIDAE Sokolov, 1965

Tubes long and thin, black, brown or light yellow, elastic and collapsible, transversely wrinkled, or smooth at one end. *U.Precam.(up.Vend.)-basal Cam.*

Paleolina SOKOLOV, 1965, p. 90 [*P. evenkiana*; OD]. Tubes very thin, narrow, semi-transparent to transparent, yellow; walls smooth or sharply and irregularly transversely wrinkled; diameter 1.0-1.2 mm, length to 120 mm. *U.Precam.(up. Vend.)-basal Cam.*, Sib.-N.USSR.——FIG. 19,*1*. **P. cf. P. evenkiana*, int. molds, up.Vend., S.Sib.; ×5 (Sokolov, 1975).

1
Paleolina

FIG. 19. Sabelliditidae (p. *A*107).

Phylum UNCERTAIN

Class, Order, and Family UNCERTAIN

Redkinia SOKOLOV, 1976, p. 141 [*R. spinosa*; OD]. Brown to black, chitinoid, blade-like fossils with 13-15 curved spines about 0.5 mm long along one edge, which is up to 3.5 mm long. Possibly annelid jaws. Considered by SOKOLOV as possibly legs of Protonychophora. *U.Precam.(Vend.,* Redkino Ser.), Nepeitsino bore, central Russia.

Tribrachidium GLAESSNER in GLAESSNER & DAILY, 1959, p. 389 [*T. heraldicum*; OD]. Disc shaped, slightly convex, with steeply sloping peripheral margin; one side (oral?) has three raised arms (brachia) radiating at equal angles, curving clockwise (in artificial casts representing bodies as deposited) to join periphery of disc where they taper.

(Continued from facing page.)

above the line are shortened by overfolding, that below the line is lengthened by lateral compression. Reconstruction (below, right) shows appendages on left only (*M*, hypothetical position of mouth, flanked by proximal parts of anterior appendages). The adaxial and abaxial ends of appendages are not preserved, probably mainly because of the greater rigidity of the ridges (specimens selected, cast in latex and photographed by Dr. M. WADE) (Glaessner, n; specimen numbers, from left to right, 806/2, P 14245/2, P 12774, P 13815, P 14252/1, P 14245/1, P 14251, P 14206, P 14204, 943, 543, P 14190, P 12901/1; specimens prefixed "P" are in collections of S. Austral. Mus., others are in collections of Dept. Geol., Univ. Adelaide).

TAXA WITH DOUBTFUL INVERTEBRATE AFFINITIES

Family SUVOROVELLIDAE Vologdin & Maslov, 1960

Calcareous, non-porous, double-walled discoidal or flatly conical skeletons; no structural elements between the walls; external wall may be sculptured with raised rhomboidal areas. *Uppermost Precam.(low. Yudom.).*

Tribrachidium

FIG. 20. Family uncertain (p. *A*107-*A*108).

Small, Y-shaped groove (?mouth) is rarely seen in center between arms; attached to each arm on its convexly curved side is a small, raised area ("bulla"). Distal 0.7 of each arm bears short, stout tentacles on outer side and tip. Fine, long, straight or gently curved bristlelike structures may extend from crest and concave side of each arm across tips of adjoining arms. Opposite (?aboral) side of disc shows only a few concentric grooves. *U.Precam.,* S.Australia.——FIG. 20,*1*. **T. heraldicum,* Ediacara; ×1 (Glaessner & Daily, 1959).

Vermiforma CLOUD, in CLOUD et al., 1976, p. 405 [**V. antiqua*; OD]. Wormlike, coiled and looped impressions, 1.5-2 cm wide, up to 1-1.1 m long, with scalloped or scaly surface texture; on bedding surfaces of meta-tuffaceous sediments. *U.Precam.,* USA(N.Car.). [This genus was classified by its authors as Annelida, class, order, and family unknown.]

Suvorovella VOLOGDIN & MASLOV, 1960, p. 691 [**S. aldanica*; OD]. Skeleton saucer shaped, diameter to 30 mm, surface with small raised rhombs in intersecting curved rows. *Uppermost Precam. (low.Yudom.),* E.Sib.

Majella VOLOGDIN & MASLOV, 1960, p. 692 [**M. verkhojanica*; OD]. Skeleton irregularly discoidal or saucer shaped, consisting of two flat walls, often with irregular concentric wrinkles. *Uppermost Precam.(low.Yudom.),* E.Sib.

Family UNCERTAIN

Petalostroma PFLUG, 1973, p. 192 [**P. kuibis*; OD]. Saucer-shaped bodies up to tens of cm in size, without internal cavities; surface irregularly wrinkled radially and concentrically, with fibrous and cellular external tissues; skeleton said to consist of organic and carbonate material. [PFLUG (1973) claims to have observed petaloids consisting of microscopic tubular structures. His proposed homologies and transitions between Petalonamae and Petalostromae are unconvincing.] *U.Precam.* (Nama Gr.), SW.Afr.

NAMES GIVEN TO PRECAMBRIAN NONMETAZOAN AND TRACE FOSSILS

Only taxa of generic rank are included, generally without reference to synonymy. Genera based on Cambrian or younger species are generally excluded.

FILAMENTOUS, COCCOID, AND OTHER MICROSCOPIC ALGAE

(This list possibly also includes fungal or bacterial remains.)

Anabaenidium SCHOPF, 1968 [**A. johnsonii*]
Animikiea BARGHOORN, 1965 [**A. septata*]
Antigus BUTIN, 1959 [**A. cusarandicus*]
Archaeogloeocapsa REITLINGER, 1956 [**A. povarovkensis*]
Archaeonema SCHOPF, 1968 [**A. longicellularis*]
Archaeorestis BARGHOORN, 1965 [**A. schreiberensis*]
?**Archaeosphaeroides** SCHOPF & BARGHOORN, 1967 [**A. barbertonensis*]
Archaeotrichion SCHOPF, 1968 [**A. contortum*]
Bigeminococcus SCHOPF & BLACIC, 1971 [**B. lamellosus*]
Biocatenoides SCHOPF, 1968 [**B. sphaerula*]
Calyptothrix SCHOPF, 1968 [**C. annulata*]
Caryosphaeroides SCHOPF, 1968 [**C. pristina*]
?**Catinella** PFLUG, 1966a [**C. polymorpha*]
Caudiculophycus SCHOPF, 1968 [**C. rivularioides*]
Cephalophytarion SCHOPF, 1968 [**C. grande*]

Chlamydomonopsis EDHORN, 1973 [*C. primordialis]
Contortothrix SCHOPF, 1968 [*C. vermiformis]
Cumulosphaera EDHORN, 1973 [*C. lamellosa]
Cyanonema SCHOPF, 1968 [*C. attenuata]
Entosphaeroides BARGHOORN, 1965 [*E. amplus]
Eoastrion BARGHOORN, 1965 [*E. simplex]
Eobacterium BARGHOORN & SCHOPF, 1966 [*E. isolatum]
Eoepiphyton BUTIN, 1959 [*E. jalgamicum]
Eomycetopsis SCHOPF, 1968 [*E. robusta]
Eosphaera BARGHOORN, 1965 [*E. tyleri]
Eotetrahedrion SCHOPF & BLACIC, 1971 [*E. princeps]
Eozygion SCHOPF & BLACIC, 1971 [*E. grande]
?Fibularix PFLUG, 1965 [*F. funicula]
Filamentella PFLUG, 1965 [*F. plurima]
Filiconstrictosus SCHOPF & BLACIC, 1971 [*F. majusculus]
Glenobotrydion SCHOPF, 1968 [*G. aenigmatis]
Globophycus SCHOPF, 1968 [*G. rugosus]
Gloeodiniopsis SCHOPF, 1968 [*G. lamellosa]
Gunflintia BARGHOORN, 1965 [*G. minuta]
Halythrix SCHOPF, 1968 [*H. nodosa]
Heliconema SCHOPF, 1968 [*H. australiensis]
Huroniospora BARGHOORN, 1965 [*H. microreticulata]
Kakabekia BARGHOORN, 1965 [*K. umbellata]
Millaria PFLUG, 1966a [*M. implexa]
Montanella PFLUG, 1965 [*M. beltensis]
Myxococcoides SCHOPF, 1968 [*M. minor]
Obconiphycus SCHOPF & BLACIC, 1971 [*O. amadeus]
Oscillatoriopsis SCHOPF, 1968 [*O. obtusa]
Palaeoanacystis SCHOPF, 1968 [*P. vulgaris]
Palaeolyngbya SCHOPF, 1968 [*P. barghoorniana]
Palaeomicrocoleus KORDE in VOLOGDIN & KORDE, 1965 [*P. gruneri]
Palaeopleurocapsa KNOLL, BARGHOORN, & GOLUBIĆ, 1975 [*P. wopfnerii]
Palaeorivularia KORDE, 1965 [*P. ontarica]
Palaeoscytonema EDHORN, 1973 [*P. moorhousei]
Palaeospiralis EDHORN, 1973 [*P. canadensis]
Palaeospirulina EDHORN, 1973 [*P. arcuata]
Partitiofilum SCHOPF & BLACIC, 1971 [*P. gongyloides]
Petraphera L. NAGY, 1974 [*P. vivescenticula]
Phanerosphaerops SCHOPF & BLACIC, 1971 [*P. capitaneus]
Polycellaria PFLUG, 1965 [*P. bonnerensis]
Primorivularia EDHORN, 1973 [*P. thunderbayensis]
?Protorivularia BUTIN, 1959 [*P. onega]
Ramsaysphaera PFLUG, 1976 [*R. ramses] [The systematic placement of this 3.4 by old fossil is uncertain]
Scintilla PFLUG, 1966 [*S. perforata]
Siphonophycus SCHOPF, 1968 [*S. kestron]
Sphaerocongregus MOORMAN, 1974 [*S. variabilis] [According to VIDAL (1976), an acritarch, and a synonym of Bavlinella SHEPELEVA, 1962]
Sphaerophycus SCHOPF, 1968 [*S. parvum]
Tenuofilum SCHOPF, 1968 [*T. septatum]
Tormentella PFLUG, 1966a [*T. tubiformis]
Tricellaria PFLUG, 1965 [*T. deylensis]
Veteronostocale SCHOPF & BLACIC, 1971 [*V. amoenum]
Zosterosphaera SCHOPF, 1968 [*Z. tripunctata]

SPOROMORPHS AND ACRITARCHS

It is not possible to give a complete list of taxa established for Precambrian sporomorphs and acritarchs because their nomenclature is exceedingly confused. The most common acritarchs from the Upper Precambrian are Sphaeromorphitae. Many of them have been placed in the Paleozoic genus *Leiosphaeridia* EISENACK, 1958, with the following tentative synonymy (after VOLKOVA et al., 1968, with additions): [=*Botholigotriletum* TIMOFEEV, 1958, *Ocridoligotriletum* TIMOFEEV, 1958; *Stenozonoligotriletum* TIMOFEEV, 1958; *Trachyoligotriletum* TIMOFEEV, 1958; *Protoleiosphaeridium* TIMOFEEV, 1959; *Leiosphaeridium* TIMOFEEV, 1959; *Lopholigotriletum* TIMOFEEV, 1959; *Leiopsophosphaera* NAUMOVA, 1960, *Wendiella* TIMOFEEV, 1960 (nom. nud.); *Kildinella* SHEPELEVA & TIMOFEEV, 1963; *Turuchania* RUDAVSKAYA, 1964; *Protosphaeridium* TIMOFEEV, 1966; *?Menneria* LOPUKHIN, 1971]. Others, mostly larger, are placed in the genus *Chuaria* WALCOTT, 1899 [*C. cicularis*] [=*Fermoria* CHAPMAN, 1935; *Protobolella* CHAPMAN, 1935; *Vindhyanella* SAHNI, 1936; *Krishnania* SAHNI & SRIVASTAVA, 1954; *Kildinella* SHEPELEVA & TIMOFEEV, 1963 (partim)]. (Synonymy from FORD & BREED, 1973.) A similar form, ranging up to 44 mm in diameter, is *Beltanelloides* SOKOLOV, 1965 [*B. sorichevae*] [=*Beltanelliformis* MENNER, 1963 (nom. nud., figured but not described until 1968) (type, *B. brunsae*)]. Some of these Upper Precambrian fossils have been mistakenly considered as Metazoa or Protozoa.

STROMATOLITES

A comprehensive bibliography, together with other information on stromatolites, is included in a monographic work by WALTER (1976).

Acaciella WALTER, 1972 [*Cryptozoon australicum HOWCHIN, 1914]

Alcheringa WALTER, 1972 [*A. narrina]
Aldania KRYLOV, 1969 [*Gymnosolen sibericus YAKOVLEV, 1934]
Alternella RAABEN, 1972 [*A. hyperboreica]
Anabaria KOMAR, 1964 [*A. radialis]
Archaeozoon MATHEWS, 1890 [*A. acadiense]
Baicalia KRYLOV, 1963 [*Collenia baicalica MASLOV, 1937]
Basisphaera WALTER, 1972 [*B. irregularis]
Boxonia KOROLJUK, 1960 [*B. gracilis]
Calevia BUTIN, 1959 [*C. olenica]
Carelozoon METZGER, 1924 [*C. jatulicum] [=Carelosoon BUTIN, 1966 (nom. van.)] [See HÄNTZSCHEL, 1975, p. W182. Recent work has made it clear that material from Karelia, identical with the original material from Finland, represents stromatolites]
Collenella KOMAR, 1964 [*C. cormosa]
Collenia WALCOTT, 1914 [*C. undosa]
Colleniella KOROLJUK, 1960 [*C. idensis]
Columnacollenia KOROLJUK, 1960 [No type species designated]
Columnaefacta KOROLJUK, 1960 [*C. elongata]
Columnaria VOLOGDIN, 1962 [No type species designated]
Compactocollenia KOROLJUK, 1960 [No type species designated]
Conophyton MASLOV, 1937 [*C. lituum]
Conusella GOLOVANOV, 1970 [*C. regularis]
Dabania SHENFIL, 1972 [*D. chopichica]
Dgerbia DOLNIK, 1974 [*D. grumulosa] [=Djerbia AUCTT. (nom. van.)]
Eucapsiphora CLOUD & SEMIKHATOV, 1969 [*E. paradisa]
Gaia KRYLOV, 1975 [*G. irkuskanica]
Georginia WALTER, 1972 [*G. howchini]
Gongylina KOMAR, 1964 [*G. differenciata]
Gruneria CLOUD & SEMIKHATOV, 1969 [*G. biwabikia]
Gymnosolen STEINMANN, 1911 [*G. ramsayi]
Iliella KRYLOV, 1975 [*I. kotuikanica]
Inzeria KRYLOV, 1963 [*I. tjomusi]
Irregularia KOROLJUK, 1960 [No species mentioned]
Jacutophyton SHAPOVALOVA, 1968 [*J. ramosum]
Jurusania KRYLOV, 1963 [*J. cylindrica]
Kasaia BERTRAND-SAFATI, 1972 [*K. convexa]
Katavia KRYLOV, 1963 [*K. karatavica]
Katernia CLOUD & SEMIKHATOV, 1969 [*K. africana]
Kotuikania KOMAR, 1964 [*K. torulosa]
Kulparia PREISS & WALTER, in WALTER, 1972 [*K. kulparensis PREISS, 1973]
Kurtunia SHENFIL, 1972 [*K. uluntuica]
Kussiella KRYLOV, 1963 [*Collenia kussiensis; (=C. kussiensis MASLOV MS., nom. nud.)]
Lenia DOLNIK, 1971 [*L. jacutica]
Linella KRYLOV, 1967 [*L. ukka]
Malginella KOMAR & SEMIKHATOV, 1970 [?]
Microstylus KOMAR, 1966 [*M. perplexus]
Minjaria KRYLOV, 1963 [*M. uralica]
Nouatila BERTRAND-SARFATI, 1972 [*N. frutectosa]
Nucleella KOMAR, 1966 [*N. figurata]
Olenia BUTIN, 1960 [*O. rasus]
Omachtenia NUZHNOV, 1967 [*O. omachtensis]
?Palia BUTIN, 1966 [*P. septentrionalis]
Paniscollenia KOROLJUK, 1960 [*P. vulgaris]
Parmites RAABEN, 1964 [*P. concrescens]
Patomia KRYLOV, 1967 [*P. ossica]
Pilbaria WALTER, 1972 [*P. perplexa]
Pitella SEMIKHATOV, 1962 [*P. lanceolata]
Planocollina KOROLJUK, 1960 [*P. serrata]
Platella KOROLJUK, 1963 [No species mentioned]
Poludia RAABEN, 1964 [*P. polymorpha]
Pseudokussiella KRYLOV, 1963 [*P. aii]
Ramulus RAABEN, 1972 [*R. sociabilis]
Sacculia KOROLJUK, 1960 [*S. ovata]
Segosia BUTIN, 1966 [*S. columnaris]
Serizia BERTRAND-SARFATI, 1972 [*S. radians]
Stratifera KOROLJUK, 1963 [*S. rara]
Sundia BUTIN, 1966 [*S. ramosa]
Svetliella SHAPOVALOVA, 1968 [*S. svetlica]
Tarioufetia BERTRAND-SARFATI, 1972 [*T. hemispherica]
Tenupalusella GOLOVANOV, 1970 [*T. bracteata]
Tifounkeia BERTRAND-SARFATI, 1972 [*T. ramificata]
Tilemsina BERTRAND-SARFATI, 1970 [*T. divergens]
Tinnia DOLNIK, 1971 [*T. patomica]
Tungussia SEMIKHATOV, 1962 [*T. nodosa]

DOUBTFUL TAXA

Most investigators consider that genera established on the basis of microstructures from stromatolites in carbonate preservation are of questionable value because these structures are affected by diagenetic processes. A reexamination of this material may lead to a reassessment as there is now growing interest in algal assemblages that form stromatolites. VOLOGDIN (1962) considered these taxa as Cyanophyta except *Pustularia, Rubellophyton* (possibly Rhodophyta), and *Tubulistroma* (possibly Chlorophyta).

Abruptophycus VOLOGDIN, 1962 [*A. compositus]
Amplectostroma VOLOGDIN, 1966b [*A. ramificata]
Angarophycus VOLOGDIN, 1962 [*A. depictus]
Antiquophytolithus VOLOGDIN, 1962 [*A. filamentaris]
Azyrtalia VOLOGDIN, 1969b [*A. zonulata]
Borlogella VOLOGDIN, 1962 [*B. multifaria]
Bulbistroma VOLOGDIN, 1962 [*B. curtothallum]
Bursiphycus VOLOGDIN, 1962 [*B. bullatus]
Cirriphycus VOLOGDIN, 1962 [*C. ordinatus]
Columnaria VOLOGDIN, 1962 [*C. turuchanica]
Crispophycus VOLOGDIN, 1962 [*C. sibiricus]
Crustophycus VOLOGDIN, 1962 [*C. angaricus]
Cyanostroma VOLOGDIN, 1962 [*C. turuchanicum]
Cystostroma VOLOGDIN, 1962 [*C. varians]

Echaninia VOLOGDIN, 1969a [*E. mucosa]
Fibrostroma VOLOGDIN, 1962 [*F. fibrillatum]
Fillostroma VOLOGDIN, 1962 [*F. moticum]
Grabauella VOLOGDIN, 1964 [*G. dependitis]
Granifer VOLOGDIN, 1955 [*G. conicus]
Jatuliana KORDE in VOLOGDIN & KORDE, 1965 [*J. furcata]
Lamellophycus VOLOGDIN, 1962 [*L. aculeatus]
Lamellostroma VOLOGDIN, 1962 [*L. vesiculare]
Leiostroma VOLOGDIN, 1966b [*L. elegantα]
Leptotrichomaria VOLOGDIN, 1962 [*L. intermissa]
Lermontovaephycus VOLOGDIN, 1962 [*L. lamellosus]
Lopatinella VOLOGDIN, 1962 [*L. bipartita]
Mucostroma VOLOGDIN, 1966b [*M. carelica]
Murandavia VOLOGDIN, 1965b [*M. amurica] [=?Kareliana KORDE in VOLOGDIN & KORDE, 1965]
Nerusiandella VOLOGDIN, 1962 [*N. faveolata]
Papulophycus VOLOGDIN, 1962 [*P. pennatus]
Perennaria VOLOGDIN, 1962 [*P. ambigua]
Pilostroma VOLOGDIN, 1962 [*P. grumosum]
Plexostroma VOLOGDIN, 1962 [*P. pleurotropum]
Praechroococcus VOLOGDIN, 1962 [*P. catervatus]
Protoepiphyton VOLOGDIN, 1962 [*P. curtofiligerum]
Pustularia VOLOGDIN, 1955 [*P. taeniata]
Ramulostroma VOLOGDIN, 1962 [*R. ramulosum]
Rubellophyton VOLOGDIN, 1966b [*R. rameus]
Sarmaella VOLOGDIN, 1962 [*S. vesiculosa]
Scandophycus VOLOGDIN, 1962 [*S. crispobilis]
Sphaerothallus VOLOGDIN, 1962 [*S. spissus]
Telastroma VOLOGDIN, 1962 [*T. tenuirimulatum]
Thysanoplanta VOLOGDIN & TITORENKO, 1966 [*T. filamentosa]
Trichostroma VOLOGDIN, 1962 [*T. capilliforme]
Tschichatschevia VOLOGDIN, 1955 [*Conophyton lituus MASLOV] [obj. syn. of Conophyton]
Tubulistroma VOLOGDIN, 1962 [*T. scrofulosum]
Vesicularia VOLOGDIN, 1962 [*V. nidifica]
Vittophyton VOLOGDIN, 1962 [*V. parvum]

PROBLEMATIC FOSSILS (INCLUDING MICROPHYTOLITHS)

Mostly microscopic structures, often associated with stromatolites, occasionally grading into sedimentary structures resembling oolites or coprolites, of more or less questionable organic origin. Some are associated with volcanic rocks and may be droplets of abiogenic organic compounds.

Agamus VOLOGDIN, 1970 [*A. shungiticus]
Ambigolamellatus ZHURAVLEVA, 1968 [*A. horridus]
Antholithina CHOUBERT & H. & G. TERMIER, 1951 [*A. rosacea] [see HÄNTZSCHEL, 1975, p. W169]
Aseptalia VOLOGDIN in VOLOGDIN & STRYGIN, 1969 [*A. ukrainika] [2 billion years old; certainly not metazoan as claimed]

Asterosphaeroides REITLINGER, 1959 [No type species designated]
Birrimarnoldia HOVASSE & COUTURE, 1961 [*Arnoldia antiqua HOVASSE] [see HÄNTZSCHEL, 1975, p. W155]
Cayeuxipora GRAINDOR, 1957 [No type species designated] [see HÄNTZSCHEL, 1975, p. W155]
Cayeuxistylus GRAINDOR, 1957 [No species designated] [see HÄNTZSCHEL, 1975, p. W155]
Conferta KLINGER, 1968 [*C. rara]
Crenulata BERTRAND-SARFATI, 1972 [*C. gigantea]
Foninia KORDE, 1973 [*F. fasciculata]
Globoidella MILSTEIN, 1970 [*G. jusmastachica]
Gonamophyton VOLOGDIN & DROZDOVA, 1964b [*G. ovale]
Gorlovella VOLOGDIN, 1970 [*G. obvoluta]
?Ladogaella VOLOGDIN, 1967 [*L. variabilis]
Marenita KORDE, 1973 [*M. kundatica]
Medullarites NAROZHNYKH in NAROZHNYKH & RABOTNOV, 1965 [No type species designated]
Nelcanella VOLOGDIN & DROZDOVA, 1964a [*N. stellata]
Protospira VOLOGDIN in VOLOGDIN & STRYGIN, 1969 [*P. strygini]
Ptilophyton VOLOGDIN, 1967 [*P. makarovae]
Radiosus ZHURAVLEVA, 1964 [*R. limpidus]
Tazenakhtia CHOUBERT & H. & G. TERMIER, 1951 [*T. aenigmatica] [see HÄNTZSCHEL, 1975, p. W179]
Tubiphyton CHOUBERT & H. & G. TERMIER, 1951 [*T. taghdoutensis] [see HÄNTZSCHEL, 1975, p. W179]
Vallenia RAUNSGAARD PEDERSEN, 1967 [*V. erlingi] [see HÄNTZSCHEL, 1975, p. W167]
Vermiculites REITLINGER, 1959 [non Vermiculites ROUAULT, 1850, nec BRONN, 1848]
Vermiculus BERTRAND-SARFATI, 1972 [*V. contortus]
Vesicophyton VOLOGDIN in VOLOGDIN & DROZDOVA, 1969 [*V. punctatum]
Vesicularites REITLINGER, 1959 [*V. flexuosus]
Volvatella NAROZHNYKH, 1967 [*V. obsoleta]

MEGASCOPIC ALGAE

Aataenia GNILOVSKAYA, 1976 [*A. reticularis]
Eoholynia GNILOVSKAYA, 1975 [*E. mosquensis]
Grypania WALTER, OEHLER, & OEHLER, 1976 [*Helminthoidichnites? spiralis WALCOTT, 1899 (non H. FITCH, 1850)]
Laminarites BRONGNIART, 1828 [see HÄNTZSCHEL, 1975, p. W186] [The name L. antiquissimus EICHWALD, 1856, is often used for megascopic plant remains, probably algae, from the uppermost Precambrian of eastern Europe; they are unlikely to represent this genus and are specifically unidentifiable].
Lanceoforma WALTER, OEHLER, & OEHLER, 1976 [*L. striata]

Papillomembrana SPJELDNAES, 1963 [*P. compta*] [This fossil, though not strictly megascopic, with a diameter of 0.5 mm, is included here because it is thought to represent a dasycladacean alga from the Upper Precambrian, Norway].

Proterotainia WALTER, OEHLER, & OEHLER, 1976 [*P. montana*]

Timanella VOLOGDIN in VOLOGDIN & KOCHETKOV, 1966 [*T. gigas*] [Described as Chlorophyta Dasycladacea, 10-20 cm long, from Lower Cambrian-Proterozoic of N.USSR].

Tyrasotaenia GNILOVSKAYA, 1971 [*T. podolica*] [? Phaeophyta]

Vendotaenia GNILOVSKAYA, 1971 [*V. antiqua*] [? Phaeophyta]

TRACE FOSSILS

Archaeichnium GLAESSNER, 1963 [*A. haughtoni*] [see HÄNTZSCHEL, 1975, p. W37]

Buchholzbrunnichnus GERMS, 1973 [*B. kroeneri*]

Bunyerichnus GLAESSNER, 1969 [*B. dalgarnoi*] [see HÄNTZSCHEL, 1975, p. W49]

Harlaniella SOKOLOV, 1973 [*H. podolica*]

Nenoxites FEDONKIN, 1976 [*N. curvus*]

Suzmites FEDONKIN, 1976 [*S. volutatus*]

Torrowangea WEBBY, 1970 [*T. rosei*] [see HÄNTZSCHEL, 1975, p. W117] [The Precambrian age of the Lintiss Vale Formation (W. New South Wales, Australia) was disputed by DAILY (1973) but insisted on by WEBBY (1973), who accepted correlation of the Lintiss Vale with the Uratanna Formation, generally considered to be post-Ediacaran].

REJECTED AND UNRECOGNIZABLE TAXA

[With few exceptions, these taxa are included in *Treatise* Part W, Supplement 1 (HÄNTZSCHEL, 1975), to which page references are given and where relevant bibliographic references can be found.]

Amanlisia LEBESCONTE, 1891 [*A. simplex*] [p. W180]

Archaeophyton BRITTON, 1888 [*A. newberryanum*] [p. W169]

Archaeosphaerina DAWSON, 1875 [No species named] [p. W169]

Aristophycus MILLER & DYER, 1878 [*A. ramosum*] [p. W169]

Armelia LEBESCONTE, 1891 [*A. barrandei*] [p. W180]

Aspidella BILLINGS, 1872 [*A. terranovica*] [p. W171]

Atikokania WALCOTT, 1912 [*A. lawsoni*] [p. W171]

Beaumontia DAVID, 1928 [*B. eckersleyi*] [p. W180] [=*Beaumontella* DAVID, 1928 (nom. null.)]

Beltina WALCOTT, 1899 [*B. danai*] [p.W182]

Botswanella PFLUG & STRÜBEL, 1969 [Considered by the authors as postsedimentary products of iron bacteria; the name is not meant as biological nomenclature]

Camasia WALCOTT, 1914 [*C. spongiosa*] [p. W171]

Caragassia VOLOGDIN, 1965a [*C. krassevi*] [Casts of mud flakes]

Collinsia BAIN, 1927 [*C. mississagiense*] [p.W173]

Copperia WALCOTT, 1914 [*C. tubiformis*] [p. W173] [=*Cooperia* CHOUBERT & H. & G. TERMIER (nom. null.)]

Corycium SEDERHOLM, 1911 [*C. enigmaticum*] [p. W184] [Accumulation of organic carbon in a distinctive form; not a biosystematic taxon]

Ctenichnites MATTHEW in SELWYN, 1890 [No species mentioned] [p. W173]

Eospicula DE LAUBENFELS, 1955 [*E. cayeuxi*] [p. E33]

Eozoon DAWSON, 1865 [*E. canadense*] [p. W173]

Gakarusia HAUGHTON, 1964 [*G. addisoni*] [p. W147]

Gallatinia WALCOTT, 1914 [*G. pertexta*] [p. W175]

Greysonia WALCOTT, 1914 [*G. basaltica*] [p. W175]

Ikeya VOLOGDIN, 1965a [*I. tumida*] [Casts of mud flakes]

Iyaia VOLOGDIN, 1965a [*I. sayanica*] [Casts of mud flakes]

Kempia BAIN, 1927 [*K. huronense*] [p. W175]

Kinneya WALCOTT, 1914 [*K. simulans*] [p. W176]

Manchuriophycus ENDO, 1933 [*M. yamamotoi*] [p. W176]

Mawsonella CHAPMAN, 1927 [*M. wooltanensis*] [Described as green alga, now considered as intraformational carbonate breccia]

Medusichnites MATTHEW, 1891 [No species] [p. W175]

Montfortia LEBESCONTE, 1887 [p. W190]

Neantia LEBESCONTE, 1887 [p. W176]

Newlandia WALCOTT, 1914 [*N. frondosa*] [p. W176]

Orthogonium GÜRICH, 1933 [*O. parallelum*] [p. W186]

Palaeotrochis EMMONS, 1856 [No type species designated] [p. W177]

Protadelaidea TILLYARD, 1936 [*P. howchini*] [p. W177]

Protoniobia SPRIGG, 1949 [*P. wadea*] [see HAR-

Rington & Moore, 1956, p. F159]
Reynella David, 1928 [*R. howchini] [p. W178]
Rhysonetron Hofmann, 1967 [*R. lahtii] [p. W178]
Sayanella Vologdin, 1966a [*S. akshanica] [Bedding plane features of mechanical origin]
Telemarkites Dons, 1959 [*T. enigmaticus] [p. W179] [Concretions which according to some authors may have been formed under the influence of syngenetic organic activity]

REFERENCES

Allison, C. W., 1975, *Primitive fossil flatworm from Alaska: New evidence bearing on ancestry of the Metazoa:* Geology, v. 3, p. 649-652, text-fig. 1.

Anderson, M. M., 1972, *A possible time span for the Late Precambrian of the Avalon Peninsula, southeastern Newfoundland in the light of worldwide correlation of fossils, tillites, and rock units within the succession:* Canadian Jour. Earth Sci., v. 9, p. 1710-1726.

Barghoorn, E. S., & Schopf, J. W., 1966, *Microorganisms three billion years old from the Precambrian of South Africa:* Science, v. 152, p. 758-763.

―――, & Tyler, S. A., 1965, *Microorganisms from the Gunflint Chert:* Science, v. 147, p. 563-577, text-fig. 1-10.

Bekker, Yu. K., 1977, *Pervye paleontologicheskie nakhodki v rifee Urala:* Akad. Nauk SSSR, Izvestiya, Ser. Geol., No. 3, p. 90-100. [The first paleontological finds in the Riphean of the Ural.]

Bertrand-Sarfati, J., 1972, *Stromatolites columnaires du Précambrien supérieur:* Centre Rech. Zones Arides, sér. géol. no. 14, CNRS Paris, 235 p., 30 pl.

Bondesen, E., Raunsgaard Pedersen, K. T., & Jørgensen, O., 1967, *Precambrian organisms and the isotopic composition of organic remains in the Ketilidian of south-west Greenland:* Medd. Grønland, v. 164, p. 5-41, pl. 1-13.

Butin, R. V., 1966, *Iskopaemye vodorosli proterozoya karelii:* in Ostatki organismov i problematika proterozoiskykh obrazovanii Karelii: Geol. Inst. Petrozavodsk, p. 34-63. [Fossil algae from the Proterozoic of Karelia, in Remains of organisms and problematica from Proterozoic formations of Karelia.]

Choubert, Georges, & Termier, Henri, & Geneviéve, 1951, *Les calcaires précambriens de Taghdout et leurs organismes problématiques:* Serv. Géol. Maroc, Notes Mém., v. 85, p. 9-34.

Clark, R. B., 1969, *Systematics and phylogeny: Annelida, Echiura, Sipuncula:* in Chemical zoology, M. Florkin, & B. T. Scheer (eds.), v. 4, p. 1-68, Academic Press (New York & London).

Cloud, P. E., Jr., 1968, *Pre-Metazoan evolution and the origins of the Metazoa:* in Evolution and environment, T. Drake (ed.), 72 p., 11 text-fig., Yale Univ. Press (New Haven and London).

―――1973, *Pseudofossils: a plea for caution:* Geology, v. 1, p. 123-127, text-fig. 1-7.―――1974, *Evolution of ecosystems:* Am. Scientist, v. 62, p. 54-66.―――1976a, *Beginnings of biospheric evolution and their biogeochemical consequences:* Paleobiology, v. 2, p. 351-387, 5 pl., 2 fig.―――1976b, *Major features of crustal evolution:* Geol. Soc. S. Afr., Annex to v. 79 (Alex L. du Toit Memorial Lecture No. 14), 33 p.

―――, Wright, James, & Glover, Lynn, 1976, *Traces of animal life from 620-million-year-old rocks in North Carolina:* Am. Scientist, v. 64, p. 396-406, 11 text-fig.

Daily, Brian, 1973, *Discovery and significance of basal Cambrian Uratanna Formation, Mt. Scott Range, Flinders Ranges, South Australia:* Search, v. 4, p. 202-205.

Delle Cave, L., & Simonetta, A. M., 1975, *Notes on the morphology and taxonomic position of Aysheaia (Onycophora?) and of Skania (undetermined phylum):* Monitore Zool. Ital. (n. ser.) v. 9, p. 67-81.

Downie, Charles, 1974, *Acritarchs near the Precambrian-Cambrian boundary. A preliminary report:* Rev. Palaeobotany, Palynology, v. 18, p. 57-60.

Dunn, P. R., 1964, *Triact spicules in Proterozoic rocks of the Northern Territory of Australia:* Geol. Soc. Australia, Jour., v. 11, pt. 2, p. 195-197, pl. 1.

Dunning, F. W., 1975, *Precambrian craton of central England and Welsh Borders; in Precambrian. A correlation of Precambrian rocks in the British Isles:* Geol. Soc. London, Spec. Rept. no. 6, p. 83-95.

Edhorn, A. -S., 1973, *Further investigations of fossils from the Animikie, Thunder Bay, Ontario:* Geol. Assoc. Canada, Proc., v. 25, p. 37-66, pl. 1-10.

Eichwald, Edouard d', 1856, *Beitrag zur geographischen Verbreitung der fossilen Thiere Russlands. Alte Periode:* Soc. Impér. Naturalistes Moscou, Bull., v. 29, p. 406-453.

Fedonkin, M. A., 1976, *Sledy mnogokletochnykh iz valdayskoy serii:* Akad. Nauk SSSR, Izvestiya, Ser. Geol. 4, p. 129-132, text-fig. 1-5. [Tracks of multicellular organisms from the Valday Series.]

Fischer, A. G., 1972, *Atmosphere and the evolution of life:* Main currents in modern thought, v. 28, no. 5, 9 p., 1 text-fig.

Ford, T. D., 1958, *Pre-Cambrian fossils from*

Charnwood Forest: Yorkshire Geol. Soc., Proc., v. 31, p. 211-217, text-fig. 1-3.──1963, *The Pre-Cambrian fossils of Charnwood Forest:* Leicester Lit. Philos. Soc., Trans., v. 57, p. 57-62, pl. 1.

──, & **Breed, W. J.,** 1973, *The problematical fossil Chuaria:* Palaeontology, v. 16, no. 3, p. 535-550, pl. 61-63.

Germs, G. J. B., 1968, *Discovery of a new fossil in the Nama System, South West Africa:* Nature, v. 219, p. 53-54, fig. 1, 2.──1972a, *The stratigraphy and paleontology of the Lower Nama Group, South West Africa:* Univ. Cape Town Dept. Geology, Chamber of Mines Precambrian Research Unit, Bull. 12, 250 p.──1972b, *New shelly fossils from the Nama Group, South West Africa:* Am. Jour. Sci., v. 272, p. 752-761, text-fig. 1-4.──1973a, *A reinterpretation of Rangea schneiderhoehni and the discovery of a related new fossil from the Nama Group, South West Africa:* Lethaia, v. 6, p. 1-10, text-fig. 1, 2.── 1973b, *Possible sprigginid worms and a new trace fossil from the Nama Group, South West Africa:* Geology, v. 1, p. 69-70.

Glaessner, M. F., 1958, *New fossils from the base of the Cambrian in South Australia:* Royal Soc. S. Australia, Trans., v. 81, p. 185-188, 1 pl.── 1963, *Zur Kenntnis der Nama-Fossilien Südwest-Afrikas:* Naturhist. Mus. Wien, Ann., v. 66, p. 113-120, 3 pl. (pl. 1, fig. 1).──1966, *Precambrian palaeontology:* Earth Sci. Rev., v. 1, p. 29-50, text-fig. 1, 2.──1969, *Trace fossils from the Precambrian and basal Cambrian:* Lethaia, v. 2, p. 369-393.──1971, *The genus Conomedusites Glaessner & Wade and the diversification of the Cnidaria:* Paläont. Zeitschr., v. 45, no. 1/2, p. 7-17, 1 pl.──1976a, *Early Phanerozoic annelid worms and their geological and biological significance:* Geol. Soc. London, Jour., v. 132, no. 3, p. 259-275, 2 pl.──1976b, *A new genus of Late Precambrian polychaete worms from South Australia:* Royal Soc. S. Australia, Trans., v. 100, p. 169-170.

──, & **Daily, Brian,** 1959, *The geology and Late Precambrian fauna of the Ediacara Fossil Reserve:* S. Australian Museum, Rec., v. 13, no. 3, p. 369-401, pl. 42-47.

──, & **Wade, Mary,** 1966, *The late Precambrian fossils from Ediacara, South Australia:* Palaeontology, v. 9, pt. 4, p. 599-628, pl. 97-103. ──1971, *Praecambridium—a primitive arthropod:* Lethaia, v. 4, p. 71-77, text-fig. 1-4.

──, & **Walter, M. R.,** 1975, *New Precambrian fossils from the Arumbera Sandstone, Northern Territory, Australia:* Alcheringa, v. 1, p. 11-28, text-fig. 1-4.

Gnilovskaya, M. B., 1971, *Drevneyshie vodnye rasteniya venda Russkoy Platformy (Pozdniy dokembriy):* Paleont. Zhurnal, 1971, no. 3, p. 101-107, pl. 11. [*The most ancient Vendian water plants of the Russian platform (Late Precambrian).*]──1975, *Novye dannye o prirode vendotenid:* Akad. Nauk SSSR, Doklady, Ser. Geol., v. 221, no. 4, p. 953-955, pl. 1. [*New data on the nature of vendotaenids.*]──1976, *Drevneyshie Metaphyta:* Internatl. Geol. Congr., 25th Sess., Rept. Soviet Geologists, Paleont. Marine Geol. "Nauka," Moscow, p. 10-14, pl. 1. [*The oldest Metaphyta.*]

Graindor, M. J., 1957, *Cayeuxidae nov. fam., organismes à squelette du briovérien:* Acad. Sci. Paris, Comptes Rendus, v. 244, p. 2075-2077.

Gürich, Georg, 1930, *Die bislang ältesten Spuren von Organismen in Südafrika:* 15th Internatl. Geol. Congr. South Africa 1929, Comptes Rendus, p. 670-680, text-fig. 1-5.

──, 1933, *Die Kuibis-Fossilien der Nama-Formation von Südwestafrika:* Paläont. Zeitschr., v. 15, p. 137-154.

Häntzschel, Walter, 1975, *Trace fossils and problematica:* in Treatise on invertebrate paleontology, Part W, suppl. 1, Curt Teichert (ed.), 269 p., 912 fig., Geol. Soc. America, Univ. Kansas (Boulder, Colo.; Lawrence, Kans.).

Hardy, A. C., 1953, *On the origin of the Metazoa:* Quart. Jour. Micros. Sci., v. 94, p. 441-443.

Harrington, H. J., & **Moore, R. C.,** 1955, *Fossil jellyfishes from Kansas Pennsylvanian rocks and elsewhere:* Kansas Geol. Survey, Bull. 114, pt. 5, p. 153-163, pl. 1, 2.

──, & ──, 1956, *Medusae incertae sedis and unrecognizable forms:* in Treatise on invertebrate paleontology, R. C. Moore (ed.), Part F, p. F153-F161, text-fig. 122-131, Geological Soc. America & Univ. Kansas Press (New York; Lawrence, Kans.).

Hessler, R. R., & **Newman, W. A.,** 1975, *A trilobitomorph origin for the Crustacea:* Fossils & Strata, No. 4, p. 437-459, text-fig. 1-12.

Hill, Dorothy, 1972, *Archaeocyatha:* in Treatise on invertebrate paleontology, Part E (revised), Curt Teichert (ed.), 158 p., 871 text-fig., Geol. Soc. America, Univ. Kansas (Boulder, Colo.; Lawrence, Kans.).

Hovasse, R., & **Couture, R.,** 1961, *Nouvelle découverte dans l'Antécambrien de la Côte d'Ivoire, de Birrimarnoldia antiqua (gen. nov.) = Arnoldia antiqua:* Acad. Sci. Paris, Comptes Rendus, v. 252, p. 1054-1056.

Howell, B. F., 1962, *Worms:* in Treatise on invertebrate paleontology, R. C. Moore (ed.), Part W, p. W144-W177, text-fig. 85-108, Geol. Soc. America, & Univ. Kansas Press (New York; Lawrence, Kans.).

Jaeger, H., & **Martinsson, Anders,** 1967, *Remarks on the problematic fossil Xenusion auerswaldae:* Geol. Foren. Förhandl., v. 88, p. 435-452, text-fig. 1-5.

Johnson, H., & **Fox, S. K., Jr.,** 1968, *Diplurozoa from Lower Silurian of North America:* Science, v. 162, p. 119-120, fig. 1-3.

Keller, B. M., Menner, V. V., Stepanov, V. A., &

Chumakov, N. M., 1974, *Novye nakhodki Metazoa v Vendomii Russkoy platformy:* Akad. Nauk SSSR, Izvestiya, Ser. Geol., no. 12, p. 130-134, 1 pl. [*New finds of Metazoa in the Vendomian of the Russian platform.*]

———, & Fedonkin, M. A., 1976, *Novye nakhodki okamenelostey v valdayskoy serii dokembriya po r. Syuzme:* Akad. Nauk SSSR, Izvestiya, Ser. Geol. 3, p. 38-44, 2 pl. [*New finds of fossils in the Precambrian Valday Series on the R. Syuzma.*]

Knoll, A. H., & Barghoorn, E. S., 1975, *Precambrian eucaryote organisms: a reassessment of the evidence:* Science, v. 190, no. 4209, p. 52-54.

———, ———, & Golubic, S., 1975, *Palaeopleurocapsa wopfnerii gen. et sp. nov.: a Late Precambrian alga and its modern counterpart:* Natl. Acad. Sci. USA, Proc., v. 72, p. 2488-2492.

Kozłowski, Roman, 1968, *Nouvelles observations sur les Conulaires:* Acta Palaeont. Polonica, v. 8, p. 497-529.

McElhinny, M. W., Giddings, J. W., & Embleton, B. J. J., 1974, *Palaeomagnetic results and late Precambrian glaciations:* Nature, v. 248, p. 557-561.

Manton, S. M., 1969, *(a) Classification of Arthropoda, (b) Evolution and affinities of Oncychophora, Myriapoda, Hexapoda, and Crustacea:* in Treatise on invertebrate paleontology, Part R, Arthropoda 4, R. C. Moore (ed.), (a) p. R3-R15, (b) p. R15-R57, Geol. Soc. America, Univ. Kansas (Boulder, Colo.; Lawrence, Kans.).

Matthews, S. C., & Missarzhevsky, V. V., 1975, *Small shelly fossils of Late Precambrian and Early Cambrian age: a review of recent work:* Geol. Soc. London, Quart. Jour., v. 131, p. 289-304.

Misra, S. B., 1969, *Late Precambrian (?) fossils from southeastern Newfoundland:* Geol. Soc. America, Bull., v. 80, p. 2133-2140, text-fig. 1, 2, pl. 1-8.

Moorman, Mary, 1974, *Microbiota of the late Proterozoic Hector Formation, southeastern Alberta, Canada:* Jour. Paleontology, v. 48, no. 3, p. 524-539, text-fig. 1-3, pl. 1-3.

Nagy, L. A., 1974, *Transvaal stromatolite: first evidence for the diversification of cells about 2.2×10^9 years ago:* Science, v. 183, p. 514-516.

Narozhnykh, L. I., 1967, *Onkolity i katagrafii Yudomskoy svity Uchuro-Mayskogo rayona:* Akad. Nauk SSSR, Doklady, v. 173, p. 887-890. [*Oncolites and catagraphites of the Yudoma Suite in the Uchur-Maya area.*]

———, & Rabotnov, V. T., 1965, *Stratigrafiya i novyye formy organicheskykh ostatkov Rifeya (Siniya) i Yudomskogo kompleksa severnogo sklona Aldanskogo shchita:* Akad. Nauk SSSR, Doklady, v. 160, p. 910-913. [*Stratigraphy and new fossil remains from the Riphean (Sinian) and the Yudoma Complex on the northern slope of the Aldan Shield.*]

Nørrevang, Axel, 1975, *The phylogeny and systematic position of the Pogonophora:* Zeitschr. Zool. Syst. Evolutionsforsch, Sonderheft, 143 p., 104 text-fig.

Oehler, J. H., 1976, *Experimental studies in Precambrian paleontology: Structural and chemical changes in blue-green algae during simulated fossilization in synthetic chert:* Geol. Soc. America, Bull., v. 87, p. 117-129, text-fig. 1-14.

Paliy, V. M., 1969, *O novom vide tsiklomedus iz venda Podolii:* Lvov Univ., Paleont. Sbornik, v. 1, no. 6, p. 110-113, 1 fig. [*On a new species of Cyclomedusa from the Vendian of Podolia.*]

Pflug, H. D., 1964, *Niedere Algen und ähnliche Kleinformen aus dem Algonkium der Belt-Serie:* Oberhess. Ges. Natur- und Heilkunde, Bericht, n. ser., Naturwiss., Abt., v. 33, no. 4, p. 403-411, 1 pl.———1965, *Organische Reste aus der Belt Serie (Algonkium) von Nordamerika:* Paläont. Zeitschr., v. 39, no. 1-2, p. 10-25.——— 1966a, *Einige Reste niederer Pflanzen aus dem Algonkium:* Palaeontographica, v. 117, Abt. B, p. 59-74.———1966b, *Neue Fossilreste aus den Nama-Schichten in Südwest-Afrika:* Paläont. Zeitschr., v. 40, no. 1-2, p. 14-25, pl. 1, 2.——— 1968, *Gesteinsbildende Organismen aus dem Molar Tooth Limestone der Beltserie (Präkambrium):* Palaeontographica, v. 121, Abt. B, Lief 4-6, p. 134-141, pl. 1, 2.———1970a, *Zur Fauna der Nama-Schichten in Südwest-Afrika. I. Pteridinia, Bau und systematische Zugehörigkeit:* Palaeontographica, v. 134, Abt. A, no. 4-6, p. 153-262, text-fig. 1-14, pl. 21-23.———1970b, *Zur fauna der Nama-Schichten in Südwest-Afrika. II. Rangeidae, Bau und systematische Zugehörigkeit:* Palaeontographica, v. 135, Abt. A, no. 3-6, p. 198-231, text-fig. 1-12, pl. 33-35.———1972a, *Systematik der jung-präkambrischen Petalonamae Pflug 1970:* Paläont. Zeitschr., v. 46, no. 1-2, p. 56-67.———1972b, *Zur Fauna der Nama-Schichten in Südwest-Afrika. III. Erniettomorpha, Bau und Systematik:* Palaeontographica, v. 139, Abt. A, p. 134-170, text-fig. 1-9, pl. 27-39.———1973, *Zur Fauna der Nama-Schichten in Südwest-Afrika. IV. Mikroskopische Anatomie der Petalo-Organismen:* Palaeontographica, v. 144, Abt. A, p. 166-202, text-fig. 1-10, pl. 35-43.——— 1974a, *Vor- und Frühgeschichte der Metazoen:* Neues Jahrb. Geologie, Paläontologie, Abt., v. 145, no. 3, p. 328-347.———1974b, *Feinstruktur und Ontogenie der jung-präkambrischen Petalo-Organismen:* Paläont. Zeitschr., v. 48, no. 1/2, p. 77-109.———1976, *Ramsaysphaera ramses n. gen. n. sp. aus den Onverwacht-Schichten (Archaikum) von Süd-Afrika:* Palaeontographica, v. 158, Abt. B (Lfg. 5,6), p. 130-168.

———, & Strübel, G., 1969, *Algen und Bakterien in Präkambrischen Konkretionen:* Palaeontographica, v. 127, Abt. B, Lief. 1-6, p. 143-158.

Reitlinger, E. A., 1956, *Mikroskopicheskie organicheskiye (?) ostatki serdobskoy serii:* Akad. Nauk SSSR, Doklady, v. 111, p. 1098-1100. [*Microscopic organic (?) remains from the Serdobsk Series.*]——1959, *Atlas mikroskopicheskikh organicheskikh ostatkov i problematiki drevnikh tolshch Sibiri:* Akad. Nauk SSSR, Trudy, Geol. Inst., v. 25, 59 p., pl. 1-22. [*Atlas of microscopic organic remains and problematica from ancient rocks of Siberia.*]

Richter, Rudolf, 1955, *Die ältesten Fossilien Süd-Afrikas:* Senckenberg. Lethaea, v. 36, p. 243-289, pl.

Rozanov, A. Yu., 1973, *Zakonomernosti morfologicheskoy evolyutsii arkheotsiat i voprosy yarusnogo raschleneniya nizhnego kembriya:* Akad. Nauk USSR, Trudy, Geol. Inst., v. 241, 164 p. [*Regularities in the morphological evolution of regular Archaeocyatha and the problems of the Lower Cambrian Stage division.*]

——, Missarzhevsky, V. V., Volkova, N. A., Voronova, L. G., Krylov, I. N., Keller, B. M., Korolyuk, I. K., Lendzion, K., Michniak, R., Pychova, N. G., & Sidorov, A. D., 1969, *Tommotskii yarus i problema nizhnei granitsy kembriya:* Akad. Nauk USSR, Trudy, Geol. Inst., v. 206, 380 p., 79 text-fig., 15 tables. [*Tommotian stage and the Cambrian lower boundary problem.*]

Sabrodin [Zabrodin], Wladimir, 1971, *Leben im Präkambrium:* Ideen d. exakten Wissens 12/71, p. 835-842.

Schopf, J. M., 1970, *Precambrian microfossils:* in Aspects of palynology, R. H. Tschudy, & R. A. Scott (eds.), p. 145-161, 8 text-fig., Wiley Interscience (London, New York).

Schopf, J. W., 1968, *Microflora of the Bitter Springs Formation, Late Precambrian, Central Australia:* Jour. Paleontology, v. 42, no. 3, p. 651-688, pl. 77-86.——1972, *Evolutionary significance of the Bitter Springs (Late Precambrian) microflora:* Internatl. Geol. Congress 24th Sess., sec. 1, p. 68-77, 53 text-fig.——1975, *Precambrian paleobiology: problems and perspectives:* Annual Rev. Earth & Planetary Sci., v. 3, p. 213-249.

——, & Barghoorn, E. S., 1967, *Alga-like fossils from the Early Precambrian of South Africa:* Science, v. 156, p. 508-512.

——, & Blacic, J. M., 1971, *New microorganisms from the Bitter Springs Formation (Late Precambrian) of the north-central Amadeus Basin, Australia:* Jour. Paleontology, v. 45, no. 6, p. 925-960, pl. 105-113.

——, Haugh, B. N., Molnar, R. E., & Satterthwait, D. F., 1973, *On the development of metaphytes and metazoans:* Jour. Paleontology, v. 47, no. 1, p. 1-9.

Singh, I. B., 1969, *Primary sedimentary structures in Precambrian quartzites of Telemark, southern Norway, and their environmental significance:* Norsk Geol. Tidsskr., v. 49, no. 1, p. 1-31.

Sokolov, B. S., 1965, *Drevneyshie otlozheniya rannego kembriya i Sabelliditidy:* in Vses. simpozium po paleontologii dokembriya (Tesizy dokladov), p. 78-91, Akad. Nauk SSSR, Sibir. Otdel., Inst. Geol. Geofiz. (Novosibirsk). [*The oldest deposits of the early Cambrian and the sabelliditids,* in All-Union symposium on the paleontology of the Precambrian. Abstracts.]——1967, *Drevneyshiye pogonofory:* Akad. Nauk SSSR, Doklady, v. 177, p. 201-204. [*The oldest Pogonophora.*]——1972, *Vendskii etap v istorii zemli:* Akad. Nauk SSSR, M.G.K. 24 Sess., Doklady, Sov. Geol., Probl. 7, p. 114-123, 5 pl. [*Vendian Stage in the earth history.*]——1973, *Vendian of northern Eurasia:* Am. Assoc. Petrol. Geologists, Mem. 19, p. 204-218, 6 text-fig.——1975, *O paleontologicheskikh nakhodkakh v dousolskikh otlozheniyakh Irkutskogo Amfiteatra:* in Analogi Vendskogo kompleksa v Sibiri, Akad. Nauk SSSR, Sibir. Otdel., Inst. Geol. Geofiz., Trudy, v. 232, p. 112-117, pl. 1-3. [*On paleontological discoveries in pre-Usolje deposits of Irkutsk Amphitheatre,* in Analogues of Vendian complex in Siberia.]——1976, *Organicheskiy mir zenli na puti k fanerozoiskoy differentsiatsii:* Akad. Nauk SSSR, Vestnik, no. 1, p. 126-143. [*The organic world on the path of Phanerozoic differentiation.*]

Southcott, R. V., 1958, *South Australian jellyfish:* South Australian Naturalist, v. 32, p. 53-81.

Spjeldnaes, Nils., 1963, *A new fossil (Papillomembrana sp.) from the Upper Precambrian of Norway:* Nature, v. 200, p. 63-64.

Sprigg, R. C., 1947, *Early Cambrian (?) jellyfishes from the Flinders Ranges, S. Australia:* Royal Soc. S. Australia, Trans., v. 71, p. 212-224, pl. 5-8.——1949, *Early Cambrian "jellyfishes" of Ediacara, South Australia, and Mount John, Kimberley District, Western Australia:* Royal Soc. S. Australia, Trans., v. 73, p. 72-99, pl. 9-21.

Squire, A. D., 1973, *Discovery of Late Precambrian trace fossils in Jersey, Channel Islands:* Geol. Mag., v. 110, no. 3, p. 223-226.

Stanley, S. M., 1973, *An ecological theory for the sudden origin of multicellular life in the Late Precambrian:* Natl. Acad. Sci. USA, Proc., v. 70, no. 5, p. 1486-1489.

Strand, Trygve, & Kulling, Oscar, 1972, *The Scandinavian Caledonides:* 302 p., Wiley-Interscience Publishers (London-New York).

Stürmer, Wilhelm, & Bergström, Jan, 1976, *The arthropods Mimetaster and Vachonisia from the Devonian Hunsruck Shale:* Paläont. Zeitschr., v. 50, p. 78-111, pl. 9-18.

Taylor, M. E., 1966, *Precambrian mollusc-like fossils from Inyo County, California:* Science, v. 153, p. 198-201.

Ubaghs, Georges, 1971, *Diversité et spécialisation des plus anciens échinodermes que l'on connaisse:* Biol. Rev., v. 46, no. 1, p. 157-200.

Valkov, A. K., & Sysoiev, B. A., 1970, *Angustio-*

kreidy kembriya sibiri: in Stratigrafiya Proterozoya i kembriya vostoka Sibirskoy Plaformy, Akad. Nauk SSSR, Yakut. Branch Sib. Div. Int. Geol., p. 94-100. [*The Angustiochreida of the Siberian Cambrian,* in Stratigraphy and paleontology of the Proterozoic and Cambrian of the eastern part of the Siberian platform.]

Vidal, Gonzalo, 1976, *Late Precambrian microfossils from the Visingsö beds in southern Sweden:* Fossils & Strata No. 9, p. 1-57, text-fig. 1-23.

Volkova, N. A., Zhuravleva, Z. A., Zabrodin, V. E., & Klinger, B. S., 1968, *Problematika pogranichnykh sloyev rifeya i kembriya Russkoy platformy, Urala i Kazakhstana:* Akad. Nauk SSSR, Trudy, v. 188, p. 5-106. [*Problematica of the boundary strata of the Riphean and Cambrian of the Russian platform, the Urals and Kazakhstan.*]

Vologdin, A. G., 1961, *Arkheotsiaty i ikh stratigraficheskoe znachenie:* Mezhdunar. geol. Kongr. 20th sess., Simpoziuma po kembriyu, v. 3, p. 173-177. [*Archaeocyatha and their stratigraphical significance.*]———1962, *Drevneyshiye vodorosli SSSR:* 656 p., 80 pl., Akad. Nauk (Moskva), [*The oldest algae of the USSR.*]———1965a, *Otkrytiye ostatkov ogromnykh pantsirnykh zhivotnykh v karagasskoy svite Vostochnogo Sayana:* Akad. Nauk SSSR, Doklady, v. 161, p. 216-220. [*Discovery of the remains of large carapacial animals in the Karagas Suite of the Eastern Sayan.*]———1965b, *K otkrytiyu ostatkov vodorosley v murandavskoy svite proterozoya Malogo Khingana (DVK):* Akad. Nauk SSSR, Doklady, v. 164, p. 677-680. [*Toward discovery of algal remains in the Proterozoic Murandav Suite of the Lesser Khingan (Far Eastern Region).*]———1966a, *Ostatki protomedus iz nizov karagaskoy svity Vostochnogo Sayana:* Akad. Nauk SSSR, Doklady, v. 167, p. 434-436. [*Remains of protomedusae from the base of the East Sayan Karagas Suite.*]———1966b, *Ostatki mikrovodorosley is proterozoya Karelii:* in Ostatki organizmov i problematika proterozoiskykh obrazovanii Karelii: Geol. Inst. Petrozavodsk, p. 65-95. [*Remains of microscopic algae from the Proterozoic of Karelia,* in Remains of organisms and problematica from Proterozoic formations of Karelia.]———1967, *Ostatki organizmov iz ladozhskoy serii proterozoya Karelii:* Akad. Nauk SSSR, Doklady, v. 175, p. 217-220. [*Remains of organisms from the Proterozoic Ladoga Series of Karelia.*]———1969a, *Novye sinezelenyye vodorosli dokembryskogo vozrasta iz Batenevskogo kryazha:* Akad. Nauk SSSR, Doklady, v. 187, p. 440-442. [*New Precambrian cyanophyceans from Batenev Ridge.*]———1969b, *K otkrytiyu vodorosley semeystva Rivulariaceae v pozdnem dokembrii:* Akad. Nauk SSSR, Doklady, v. 187, p. 1162-1163. [*A discovery of algae of the Family Rivulariaceae in the Upper Precambrian.*]———1970, *Ostatki organizmov iz shungitov dokembriya Karelii:* Akad. Nauk SSSR, Doklady, v. 193, p. 258-261. [*Organic remains from shungites of the Precambrian of Karelia.*]———, & Drozdov, N. A., 1964a, *Neskolko vidov vodorosley iz gonamskoy svity uchurskoy serii proterozoya Ayano-Mayskogo rayona Dalnego Vostoka:* Akad. Nauk SSSR, Doklady, v. 159, p. 114-116. [*Several species of algae from the Gonam Suite of the Proterozoic Uchur Series, Ayan-Maya District, Far East.*]———1964b, *Iskopayemaya sinezelenaya vodorosl v pozdnedokembriyskykh otlozheniyakh Dalnego Vostoka:* Akad. Nauk SSSR, Doklady, v. 159, p. 576-578. [*Cyanophycean algae in Upper Precambrian sediments of the Far East.*]———1969, *Vodorosli sem. Gloeocapsaceae v osadkakh dokembriya Batenevskogo Kryazha:* Akad. Nauk SSSR, Doklady, v. 186, p. 1419-1421. [*Algae of the Family Gloeocapsaceae in the Precambrian of the Batenev Ridge.*]———1970, *Novaya nakhodka drevneyshey fauny:* Akad. Nauk SSSR, Doklady, v. 190, p. 195-197. [*A new find of the oldest fauna.*]

———, & Kochetkov, O. S., 1966, *Ob otkrytii ostatkov giganticheskikh sifoney v drevnikh sloyakh Timanskogo kryazha:* Akad. Nauk SSSR, Doklady, v. 169, p. 672-675. [*Discovery of the remains of gigantic siphonean algae in the ancient strata of Timan Ridge.*]

———, & Korde, K. B., 1965, *Neskolko vidov drevnikh Cyanophyta i ikh tsenosy:* Akad. Nauk SSSR, Doklady, v. 164, p. 207-210. [*Several species of ancient Cyanophyta and their coenoses.*]

———, & Maslov, A. B., 1960, *O novoy gruppe iskopayemykh organizmov iz nizov yudomskoy svity sibirskoy platformy:* Akad. Nauk SSSR, Doklady, v. 134, p. 691-693. [*A new group of fossil organisms from the base of the Yudoma series of the Siberian platform.*]

———, & Strygin, A. I., 1969, *Otkrytie ostatkov organismov v verkhney svite krivorozhskoy serii dokembriya Ukrainy:* Akad. Nauk SSSR, Doklady, v. 188, p. 446-449. [*A discovery of organic remains in the Upper Suite of the Krivoy Rog Series in the Precambrian of the Ukraine.*]

———, & Titorenko, T. N., 1966, *Proterozoyskiye vodorosli s reki Kurtun, yugo-zapadnoye Pribaykalye:* Akad. Nauk SSSR, Doklady, v. 166, p. 193-196. [*Proterozoic algae from the Kurtun River, southwest Baikal region.*]

Voronova, L. G., & Missarzhevsky, V. V., 1969, *Nakhodki vodoroslei i trubok chervei v pogranichnykh sloyakh kembriya i dokembriya na severe sibirskoy platformy:* Akad. Nauk SSSR, Doklady, v. 184, p. 207-210. [*Finds of algae and worm tubes in boundary layers of Cambrian and Precambrian in the north of the Siberian platform.*]

Wade, Mary, 1968, *Preservation of soft-bodied animals in Precambrian sandstones at Ediacara,*

South Australia: Lethaia, v. 1, 1968, p. 238-267, 29 text-fig.—— 1969, *Medusae from uppermost Precambrian or Cambrian sandstones, central Australia:* Palaeontology, v. 12, no. 3, p. 351-365, pl. 68,69.——1971, *Bilateral Precambrian chondrophores from the Ediacara fauna, South Australia:* Royal Soc. Victoria, Proc., v. 84, pt. 1, p. 183-188, pl. 6.——1972a, *Hydrozoa and Scyphozoa and other medusoids from the Precambrian Ediacara fauna, South Australia:* Palaeontology, v. 15, no. 2, p. 197-225, pl. 40-43.—— 1972b, *Dickinsonia: polychaete worms from the Late Precambrian Ediacara fauna, South Australia:* Queensland Museum, Mem., v. 16, no. 2, p. 171-190.

Walter, M. R. (ed.), 1976, *Stromatolites:* in Developments in sedimentology, v. 20, xii + 790 p., Elsevier (Amsterdam, New York).

——, **Oehler, J. H., & Oehler, D. Z.,** 1976, *Megascopic algae 1300 million years old from the Belt Supergroup, Montana: A reinterpretation of Walcott's Helminthoidichnites:* Jour. Paleontology, v. 50, p. 872-881, pl. 1, 2.

Webby, B. D., 1970, *Late Precambrian trace fossils from New South Wales:* Lethaia, v. 3, p. 79-109.

——1973, *Trace fossils from the Lintiss Vale Formation of New South Wales: a late Precambrian fauna:* Search, v. 4, p. 494-496.

Werner, Bernhard, 1966, *Stephanoscyphus (Scyphozoa, Coronatae) and seine direkte Abstammung von den fossilen Conulata:* Helgoländer Wiss. Meeresunters., v. 13, p. 317-347.

Young, F. G., 1972, *Early Cambrian and older trace fossils from the southern Cordillera of Canada:* Canadian Jour. Earth Sci., v. 9, p. 1-17, 10 text-fig.

Zaika-Novatskiy, V. S., Velikanov, V. A., & Koval, A. P., 1968, *Pervyy predstavitel ediakarskoy fauny v vende Russkoy platformy (Verkhniy Dokembriy):* Paleont. Zhurnal, 1968, no. 2, p. 132-134, text-fig. 1. [*The first representative of the Ediacara fauna in the Vendian of the Russian platform (Upper Precambrian).*]

Zhuravleva, Z. A., 1964, *Onkolity i katagrafii rifeya i nizhnego kembriya Sibiri i ikh stratigraficheskoye znacheniye:* Akad. Nauk SSSR, Trudy, Geol. Inst., v. 114, p. 1-73, pl. 1-24. [*Riphean and Lower Cambrian oncolites and catagraphites of Siberia and their stratigraphic importance.*]

CAMBRIAN[1]

By A. R. Palmer

[State University of New York, Stony Brook]

CONTENTS

	PAGE
INTRODUCTION	A119
CAMBRIAN BIOTA	A120
Cambrian of North America	A120
Cambrian of Eurasia and Africa	A124
Cambrian of Australia	A127
Miscellaneous Cambrian Areas	A128
BIOGEOGRAPHIC "PROVINCES"	A128
Early Cambrian Biogeography	A129
Middle and Late Cambrian Biogeography	A130
CAMBRIAN BIOSTRATIGRAPHY	A130
North America	A131
Northern Europe	A132
Central and Southern Europe	A132
Soviet Union	A133
China	A133
Australia	A134
REFERENCES	A134

INTRODUCTION

Cambrian rocks of marine origin are widely distributed in Europe, Asia, Australia, and North America, and they are found in parts of North Africa, the Cordilleran region of South America, New Zealand, and Antarctica, indicating that much of the present land area of the world was inundated during at least part of Cambrian time. Because strong evidence now indicates that the present distribution and composition of the continents is related to Mesozoic and younger movements of continental blocks, the present distribution of Cambrian rocks and faunas does not accurately reflect global Cambrian biogeography and paleogeography. Nevertheless, observations that can be made about Cambrian faunal and stratigraphic relations on the present continental blocks must be considered in any attempt to construct early Paleozoic global geography and environmental distributions. In the following pages the general characteristics of Cambrian biota are outlined and the present distribution of Cambrian outcrops and major faunal elements are presented. Finally, a global synthesis is suggested and the current state of Cambrian biostratigraphy is reviewed.

---

[1] Manuscript received August, 1969; revised manuscript received September, 1975.

CAMBRIAN BIOTA

At the beginning of Cambrian time, the seas of the world were already populated by a diverse biota that included representatives of one or more classes of most of the major invertebrate phyla. With the exception of the Archaeocyatha, which flourished in the Early Cambrian but almost completely disappeared from younger rocks, all of the phyla present at the beginning of the Cambrian have survived to the present time. Among these, the principal phyla, in order of stratigraphic importance, are the Arthropoda, Brachiopoda, Mollusca, Echinodermata, and Porifera. Lesser phyla are the Coelenterata, Annelida, Hemichordata, and Protista. The Bryozoa have no unequivocal Cambrian record.

The Arthropoda of the Cambrian include, first and foremost, the Trilobita. This was the largest and most diverse group of Cambrian organisms and apparently occupied most normal marine environments. It included open-ocean planktonic representatives, as well as probably restricted vagile benthos. Trilobites are the most commonly encountered Cambrian fossils and are the group from which almost all biogeographic data have been derived (SDZUY, 1958; LOCHMAN-BALK & WILSON, 1958; KOBAYASHI, 1967; REPINA, 1968; PALMER, 1969, 1972, 1973; COWIE, 1971). All other Arthropoda are rare and insignificant elements of the Cambrian faunas.

The Brachiopoda are represented by both inarticulate and articulate forms. Inarticulate brachiopods are numerous and the phosphatic shells of lingulides, paterinides, and acrotretides are found in many Cambrian rocks. The Acrotretida, particularly those obtained from insoluble residues of limestones, have considerable potential for future biostratigraphic and biogeographic studies. Articulate brachiopods, although locally abundant, are not common.

Among the Mollusca, the subphylum Cyrtosoma is well represented from the earliest Cambrian onward; however, several distinctive subgroups (e.g., Stenothecoididae, Helcionellidae, Pelagiellida) became extinct before the Ordovician. Cephalopoda appeared only as rare forms in the latest Cambrian. The subphylum Diasoma is comparatively poorly represented in Cambrian rocks by rare ribeirioids and by the Early Cambrian supposed bivalve, *Fordilla*. The next younger record of the Bivalvia is in rocks of Ordovician age.

Hyolitha, regarded by some as a small phylum separate from the Mollusca and related to the Sipunculoidea, are common in many parts of the Cambrian where they often are found in high concentrations.

Echinodermata were abundant in the Cambrian seas and contributed significant quantities of skeletal debris to Cambrian sediments, but only rarely are articulated skeletons preserved. Echinozoans, crinozoans (?), blastozoans, and homalozoans all have records from the Early or Middle Cambrian.

Porifera, represented by rare individuals, and moderately abundant siliceous or calcareous spicules, are present in most Cambrian areas.

Coelenterata are represented by rare Scyphozoa and Hydrozoa. No Cambrian record for the Anthozoa is given.

Many Cambrian sediments show signs of active burrowing inhabitants. Some of the burrows have been attributed to the Annelida, and a few tube-forming annelids have been recognized in some Early Cambrian deposits.

Conodonts, the biologic affinities of which are uncertain, are known from rocks as old as Early Cambrian and Hemichordata represented by the dendroid graptolites have been found in rocks as old as Middle Cambrian.

Typical Cambrian assemblages, particularly in carbonate rocks, yield several species of trilobites, lingulide, paterinide and acrotretide brachiopods, hexactinellid or chancelloriid sponge spicules, or both kinds, hyolithids or small coiled mollusks, and disarticulated echinoderms.

CAMBRIAN OF NORTH AMERICA

Cambrian rocks are sufficiently widespread in North America to permit a reasonably accurate evaluation of the broader aspects of Cambrian biogeography for this

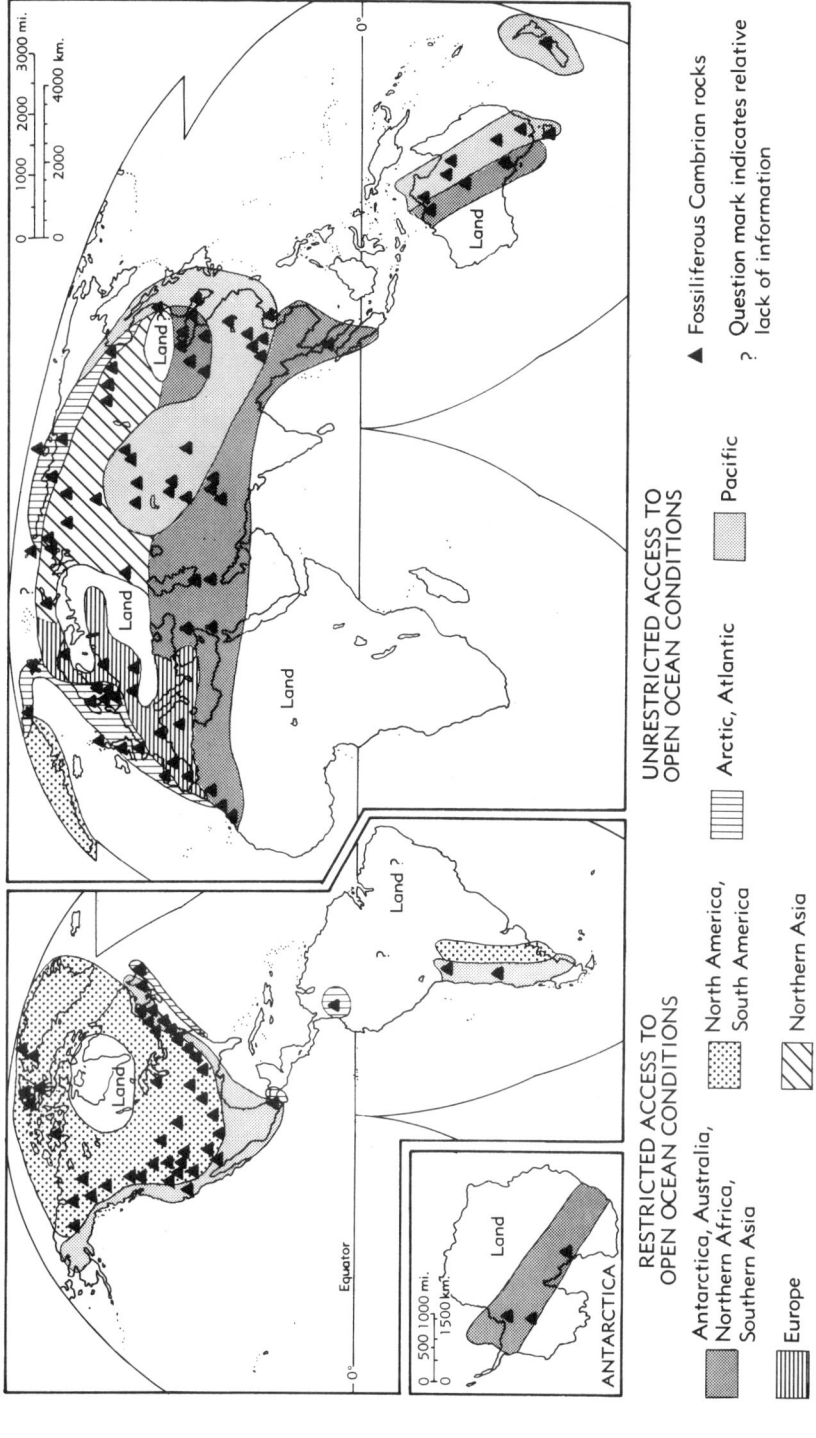

FIG. 1. Distribution of fossiliferous Cambrian outcrop areas and general distribution of the principal paleogeographic and biogeographic regions during the Cambrian Period ("Cambrian Period," Encyclopaedia Britannica, published with permission; modified figure).

continent. Their outcrops reflect a generally concentric distribution of Cambrian biotas about the continental center in north-central Canada.

Outcrops are found in the Cordilleran region from Alaska and northwestern Canada to northern Mexico, Appalachian region of eastern United States, maritime provinces of Canada, eastern and northern coasts of Greenland, several Arctic islands, several isolated regions in central United States, and one small area of southern Mexico (see Fig. 1). Some subsurface information is available for many parts of the United States and the plains of western Canada. A major land area extended southwestward from the Hudson Bay region of Canada to south-central United States throughout the Early and Middle Cambrian. Its extension in the United States was gradually submerged during the Late Cambrian and by the beginning of the Ordovician most of the United States was submerged beneath a shallow sea.

LOCHMAN-BALK and WILSON (1958) in their pioneering synthesis of North American Cambrian biogeography recognized three apparently concentric biofacies realms characterized by both tectonic and environmental criteria: a cratonic realm characteristic of the shallow shelves; an extracratonic-intermediate realm characteristic of the miogeosynclines; and an extracratonic-euxinic realm characteristic of the eugeosynclines. The faunas of the first two realms have been traditionally representative of the Pacific province of North America. The faunas of the extracratonic-euxinic realm have been traditionally representative of the Atlantic province. Subsequent work in western United States and Alaska, however, has led to an alternative interpretation of concentric faunal relationships around North America.

In western North America, a broad belt of carbonate sediments, largely reflecting extremely shallow-water conditions across a broad carbonate bank, paralleled the western cratonic margin during most of Middle and Late Cambrian time. The carbonate belt separated an inner region of light-colored terrigenous sediments ("inner detrital belt") generally reflecting shallow-water conditions, from an outer region of dark-gray or black silty and shaly sediments ("outer detrital belt"), commonly associated with dark-colored thin-bedded limestone, that reflects deeper water conditions. A similar tripartite facies pattern is present in eastern United States, and this pattern may have existed around much of North America. The shallow carbonate banks served as an effective barrier separating two major faunal regions: an inner region, including the inner bank margins and the inner detrital belt; and an outer region, including the outer bank margin and the outer detrital belt. The inner region corresponds approximately to the cratonic realm of LOCHMAN-BALK and WILSON, and the outer region corresponds to their extracratonic-intermediate realm. Their extracratonic-euxinic realm, which has been documented only in extreme eastern North America, seems to be unrelated to the remainder of the continent (see p. *A*124).

Throughout Cambrian time, wherever faunal documentation is adequate, the trilobite faunas became increasingly varied and cosmopolitan toward the most peripheral regions. The faunas toward the continental interior consist largely of endemic species and genera of polymerid trilobites. The faunas in the peripheral regions include, in addition to endemic American families, significant numbers of Eodiscidae, Oryctocephalidae, and Pagetiidae in parts of the Early Cambrian and early Middle Cambrian, and a variety of common agnostids from the middle Middle Cambrian through the Late Cambrian. Many of these, both genera and species, are found on other continents. Where carbonate banks existed as significant barriers to easy migration in the seas bordering the continent, the differences between the peripheral and inner faunas were accentuated. During the late Middle Cambrian and through much of the Late Cambrian, the contrast is so striking that precise correlation of faunal sequences between the areas is difficult.

In the Early Cambrian, the oldest beds are fossiliferous shale and sandstone units and no areal differentiation in the faunas has been noted. The most fossiliferous beds are in western North America and contain species of *Fallotaspis, Daguinaspis,* and *Holmia,* typical of similar facies in North Africa and northwestern Europe, as well as species of *Nevadia* that are restricted to North America. Slightly later, areas of carbonate sedimentation developed in the peripheral regions, in some cases forming substantial carbonate banks locally rich in archaeocyathids. With the appearance of these areas of carbonate sedimentation, a subtle gradation in character of the Early Cambrian faunas from the shaly and sandy inner regions extended onto and across the carbonate areas, and the Early Cambrian biota increased seaward in richness and variety. The inner regions had small trilobite faunas composed almost wholly of olenellids, including species of *Olenellus, Nevadella,* and *Bristolia*. In the carbonate areas and at their outer margins, several groups of simple ptychopariids, oryctocephalids, corynexochids, and eodiscids, as well as many nontrilobite organisms, and the olenellid *Wanneria,* appeared with *Olenellus* or *Nevadella*. Among the more peripheral trilobite faunas, eodiscid genera (e.g., *Calodiscus, Serrodiscus*) were also widely distributed in Eurasia. Other indications of intercontinental exchange of faunas in the peripheral regions are shown by the occurrence of the same or related genera of unusual spinose eodiscids in both the Taconic region of eastern United States and the Nuneaton area of central England, and the strong affinity of some of the Early Cambrian trilobites of Alaska and northern Canada with the faunas of Siberia.

In the early Middle Cambrian, the pattern of increasingly varied and cosmopolitan faunas in the peripheral regions was well established. The shales of the inner regions typically contained a small selection of corynexochid genera, including *Albertella* and the slightly younger *Glossopleura*. In the areas of carbonate sedimentation, these genera were associated with a variety of other corynexochid genera and a rich assortment of ptychopariid trilobites, all of which were endemic to North America. Toward the outer margins of the carbonate areas and in the deeper water areas beyond, Pagetiidae, Oryctocephalidae, Agnostidae, and Dorypygidae appeared and genera of these families were also widespread on other continents. The deeper water sediments beyond the outer edge of the banks at this time were the repository for the rich and varied Burgess Shale fauna. In eastern North America early Middle Cambrian faunas are lacking in all areas of deeper water sediments. Throughout this region, very little stratigraphic distance separates beds bearing late Middle Cambrian faunas from underlying Lower Cambrian beds.

During late Middle Cambrian and early Late Cambrian time the faunas of the inner region, including the inner margins of the banks, were strikingly different from those of the outer margin of the banks and the deeper waters beyond. The carbonate banks had by then become extensive around much of North America and seem to have been very effective barriers to free exchange of trilobites. The inner trilobite faunas during the late Middle Cambrian are dominated by simple ptychopariid trilobites, together with a few species of *Bathyuriscus* and *Kootenia,* all of which are endemic forms. The outer trilobite faunas include largely different genera and species of ptychopariid trilobites, together with some Dorypygidae and abundant agnostids. Most of the agnostid genera and species and the ubiquitous paradoxidid, *Centropleura,* are found on other continents and provide important means for intercontinental correlation of the upper Middle Cambrian (ROBISON, 1964).

An extensive transgression into the continental interior near the end of the Middle Cambrian established the general facies patterns that persisted to the end of the period. The trilobite genera that are dominant in the faunas of the broad inner detrital belt during the early Late Cambrian, such as *Crepicephalus, Lonchocephalus,* and *Menomonia,* became minor elements in the faunas of the carbonate banks and almost completely disappeared in the faunas of the outer detrital belt. In contrast, *Tricrepicephalus* and species of the Kingstoniidae, Llanoaspidae, and Blountiidae, which are rare in the sandy facies of the inner detrital belt, are common in the carbonate belt, and true *Cedaria,* which is probably not congeneric with the "*Cedaria*" species of the inner detrital belt, is found only in the faunas of the outer detrital belt associated with early species of *Glyptagnostus* and other widespread agnostids.

During late Dresbachian and early Franconian time, the only extensive Cambrian marine regression in North America took place. This did not significantly influence the faunal patterns but did affect the record of this time in the inner regions. A hiatus in the continental interior separates the early representatives of the principal families of this time, the Elviniidae and Pterocephaliidae, from their descendants and the resultant "faunal break" has been the traditional boundary between the Dresbachian and Franconian stages. In the peripheral regions the Elviniidae and Pterocephaliidae are well represented by complete evolutionary sequences and they are associated with species of *Glyptagnostus, Pseudagnostus, Aspidagnostus,* and the ubiquitous genus *Irvingella,* which are found on most other continents.

During the later part of the Cambrian, facies conditions similar to those preceding the late Dresbachian regression were restored. With them came a contrast in trilobite faunas between the major facies belts that is comparable to the earlier Late Cambrian contrasts. In the inner detrital belt, species of *Conaspis, Ptychaspis, Dikelocephalus,* and Saukiidae are common. In the carbonate belt, these are not significant faunal elements and the Parabolinoididae, Idahoiidae, and Eurekiidae are characteristic. The outer detrital belt faunas of both the east and the west during this time contain a strikingly different suite of trilobites, many of which are representative of the *Hungaia magnifica* fauna and such cosmopolitan agnostids as *Lotagnostus, Pseudagnostus,* and *Geragnostus.* In Alaska and western Nevada these are associated with species of the Ceratopygidae and *Hedinaspis,* which are widespread in Asia. In western Newfoundland, they are associated with rare species of olenids that are abundant in eastern Newfoundland and western Europe. This indicates some degree of access between adjacent major faunal regions, but the foreign trilobites seem to have been "tourists" and they never established a serious base in North America. With only a few exceptions, the principal cosmopolitan trilobite genera represent the Agnostida and it is largely through this group that intercontinental correlation and evaluation of Cambrian events is possible.

CAMBRIAN OF EURASIA AND AFRICA

Unlike North America, Eurasia is a complex and perhaps composite continent that includes three major regions distinct both faunally and in the character of their stratigraphy. Europe, together with the Mediterranean region and North Africa, but exclusive of northwestern Scotland, constitutes one region; the Soviet Union east of the Urals constitutes a second region; and southeastern Asia, including China and Korea, constitutes a third region.

EUROPE-MEDITERRANEAN-NORTH AFRICA

In this region (see Fig. 1), outcrops of Cambrian rocks are known along the eastern margin of the Caledonian mountains in Norway and Sweden, the lowlands of southern Sweden and the islands of Bornholm and Öland in the Baltic Sea, the southern coast of the Gulf of Finland, southern Poland, western Czechoslovakia and adjacent parts of East and West Germany to the north, central England and the coastal regions of north and south Wales, northwestern Scotland, the Normandy region of western France, the Montagne Noire in southern France, northern and southern Spain and a small area in southern Portugal, southern Sardinia, northwestern Africa, eastern Turkey, and the mountains east of the Dead Sea. Subsurface information is available for southern Sweden, Poland, and western Soviet Union. Except for southern Sweden and North Africa, outcrops showing continuous successions through significant parts of the Cambrian System are rare. In most regions the exposures are poor, or strongly disturbed by later tectonic movements, or both.

As a result of these deficiencies, as well as significant local variations in stratigraphic detail and considerable distances between major outcrop areas, only the most general aspects of European Cambrian paleogeography can be defined. To the northeast, a large positive area variously called the Baltic or Balto-Samartian shield persisted throughout the Cambrian Period. Similarly, most of North Africa and southern Arabia constituted positive areas flanking the Cambrian marine regions on the south. Smaller positive areas have been postulated south of the Cambrian outcrop areas of Czechoslovakia and Poland, west of Normandy, and in northern and southern Spain.

From late Early Cambrian through the Late Cambrian, the region shares many common polymerid trilobite genera that are rare or absent in other parts of Eurasia. Some differentiation into northern and southern regions has been suggested (SDZUY, 1958), but the similarities of the faunas of this region far outweigh the differences, and the scattered nature of the fossiliferous areas makes differentiation into meaningful subregions difficult.

During the Early Cambrian, an extensive development of limestone took place in southern Europe and North Africa, clearly distinguishing this region from northern Europe, which lacked carbonate sedimentation. Most of the Early Cambrian limestone sequences have varying quantities of associated archaeocyathids. After the Early Cambrian, all of the European-Mediterranean-North African region is characterized by the absence or poor development of carbonate sediment. The Early Cambrian trilobite faunas are characterized by species of the Protolenidae and Ellipsocephalidae, and by olenellids that are largely different from those of North America. However, some of the olenellids (e.g., *Fallotaspis, Daguinaspis, Holmia*) have been reported from western North America, and *Fallotaspis* is also found in Siberia. *Redlichia*, which is common in the Lower Cambrian of China, is a rare element of the southern European faunas. The Early Cambrian Eodiscidae, which are particularly well developed in central England and in Spain, are also found in parts of North America and Siberia.

The Middle Cambrian faunas are particularly characterized by the Paradoxididae and Conocoryphidae. In southern and central Europe, various ptychopariid genera representing the Saoinae are also characteristic. In Sweden, which has the only rich development of Middle Cambrian faunas in northern Europe, black shales and thin associated limestones have abundant agnostids that are found on most other continents, and associated Anomocaridae, Solenopleuridae, and Agraulidae, some of which are also found outside of the European region.

Late Cambrian faunas are known only in northern and central Europe, where they are dominated by the Olenidae. In Scandinavia and Great Britain, where the Late Cambrian faunas are found in black shale or black limestone, the olenids are associated with agnostid genera and species found on most other continents, and a few other forms, such as *Irvingella*, that have wide geographic distribution. In contrast, trilobites other than olenids are almost completely absent from the sandy facies of Poland.

The Cambrian faunas of Europe provide important data that cannot be ignored in discussions of plate tectonics. Almost all of the common genera of northwestern Europe are found in eastern Newfoundland, Nova Scotia, and New Brunswick in eastern Canada, and the Early Cambrian genera are also found in eastern Massachusetts in northeastern United States. These genera occur in the same faunal assemblages as in Europe, and have no admixture of North American forms. The North American rocks in which they are found include most of the typical lithologies described from the Cambrian of Great Britain in approximately the same stratigraphic succession. With little doubt these rocks and their faunas are "un-American." Despite the occurrence of the ubiquitous paradoxidid *Centropleura* in western North America, the easternmost North American faunas as a whole are not found in the west and do not constitute a peripheral North American facies as suggested by LOCHMAN-BALK and WILSON (1958). The seaway that separated these "Atlantic Province" areas from the rest of North America was probably the same seaway that separated the Cambrian of England and Wales from that of northwest Scotland where the Lower Cambrian terrigenous rocks and overlying thick succession of carbonates are entirely of North American aspect (PALMER, 1969).

In contrast, the close faunal and stratigraphic affinities of Spain to North Africa, as well as to central Europe and the eastern Mediterranean, raise some real problems concerning significant separation of Gondwanaland from Laurasia through the present Mediterranean region.

EASTERN SOVIET UNION

In the Soviet Union east of the Urals (Fig. 1), outcrops are known in the Arctic islands of Novaya Zemlya, Severnaya Zemlya, and Bennett Island of the New Siberian Islands; the Taymyr Peninsula; the northeastern part of the central Siberian Plateau between the Kotuy and Lena rivers; the western part of the plateau region near the lower reaches of the Yenesei River; a broad belt extending from the Aldan River basin on the east to the vicinity of the junction of the Angara and Yenesei rivers; the complexly folded areas of southern Siberia; and Kazakhstan, Kirgizstan, and Tadzhikstan.

In contrast to Europe, carbonate sediments and volcanic activity played important roles in the development of Cambrian facies and their related faunas. These may be responsible for many of the faunal contrasts between the two regions.

In a very general way, the Soviet Union east of the Urals can be divided into three large facies regions with contrasting marine sedimentary sequences. Details of tectonically positive areas related to these sedimentary sequences are not yet clear.

A southern region includes the complexly folded and faulted areas of southern Siberia, Kazakhstan, Kirgizstan, and Tadzhikstan. This was a region of active volcanism throughout most of the Cambrian Period. It projects westward to the southern Urals where Cambrian volcanics are also known. Toward the southern part of this region, in Kirgizstan and Tadzhikstan (referred to by Russians as Middle Asia), volcanism seems to have been less intense than in Kazakhstan.

The central region includes most of the rest of the mainland outcrops of the Soviet Union except for some eastern tributaries of the Aldan River, the Olenek uplift, and the Kharaulak Mountains near the mouth of the Lena River, and the eastern part of Taymyr Peninsula. All Cambrian sections of this region are dominated by limestone and dolomite or evaporites.

The northern and northeastern regions include the areas excluded from the central region described in the preceding paragraph, and the Arctic islands. This region is characterized by predominantly shaly or sandy sedimentary sequences, with or without associated thin-bedded cherty or pyritic limestones.

In the Early Cambrian, the southern part of the central region was an area of restricted environments characterized by limestone-dolomite-evaporite sequences and an endemic trilobite complex. This area was flanked on the east and south by an area of archaeocyathid bioherms and a different, also largely endemic, trilobite complex. Still farther east, in sequences of limestone and terrigenous rocks, and southward in the volcanic regions, a third trilobite complex has been recognized (REPINA, 1968). Eodiscina, including both endemic genera and the widespread *Serrodiscus, Calodiscus, Triangulaspis,* and *Hebediscus* are characteristic of the third complex. Relatively rare olenellids and *Redlichia* are found also in this complex, and in association with the archaeocyathid bioherms, in areas interpreted as normal marine environments.

During the Middle and Late Cambrian, the central carbonate region continued to support a varied trilobite fauna composed largely of endemic genera and species of ptychopariid trilobites. In the volcanic region to the south, and particularly in black-shale and thin-bedded limestone areas to the east and north, ubiquitous agnostid genera and species became increasingly abundant, associated in the Middle Cambrian with Paradoxididae and Oryctocephalidae, and, in the Late Cambrian, with Olenidae and Pterocephaliidae. These faunas have many almost cosmopolitan elements and can be related easily to the faunas in similar facies elsewhere in the world. Intercalations of the sediments and faunas of this facies into the principal carbonate sequences along the margins of the central carbonate region provide some slight help in relating the largely endemic carbonate facies faunas to those in other parts of the world.

SOUTHEASTERN ASIA

In southeastern Asia (Fig. 1), Cambrian outcrops are known in South and North Korea and adjacent parts of southeastern Siberia, many parts of eastern China, the mountains of northern and northwestern China, South China and the boundary region between China and North Viet Nam, and the small island of Turatao off the west coast of Thailand. Early Cambrian seas transgressed northward toward a major land area located in eastern Mongolia and perhaps also northeast of Korea. Other major land areas are not well documented.

Two principal facies regions have been described by KOBAYASHI (1967): the Hwang-ho facies, principally distributed in north China and also recognized in the China-North Viet Nam border region; and the Yangtze or Machari facies of east-central and western China and South Korea.

During the Early Cambrian, the region of the Hwang-ho facies was the site of shale deposition and is characterized by the presence of species of *Redlichia*. The region of the Yangtze facies included significant areas of carbonate sedimentation with locally flourishing archaeocyathids in east-central China, and the associated trilobites included a few Eodiscidae, in addition to *Redlichia* and some Protolenidae. No Olenellidae have been reported from the Early Cambrian of China.

During the Middle Cambrian, the Hwang-ho facies reflects shallow-water carbonate sedimentation, grading northward in north China and Korea into increasingly terrigenous sediments. The trilobite faunas are largely endemic and include such typical genera as *Amphoton*, *Solenoparia*, and *Anomocarella*. The contrasting Yangtze and Machari facies are characterized by shaly and silty sequences with associated thin-bedded pyritic limestone indicative of deeper water conditions. In northwestern China, volcanic rocks are associated with this facies, and the trilobite faunas are characterized by cosmopolitan agnostid genera.

Toward the end of the Middle Cambrian and in the early Late Cambrian, the Yangtze facies spread into parts of the northern region where it is represented by a variety of genera of the Damesellidae.

During the remainder of the Late Cambrian, the regions of the Hwang-ho facies were again dominated by endemic trilobites, including Chuangiidae and, later, genera such as *Asioptychaspis* and *Quadraticephalus*, which are related to North American Ptychaspididae and Saukiidae. The area of the Yangtze facies continued to have a cosmopolitan agnostid fauna, including such genera as *Glyptagnostus* and *Lotagnostus*, associated with Ceratopygidae.

CAMBRIAN OF AUSTRALIA

The only other major area of the world for which regional data are available is Australia. Cambrian outcrops are known (Fig. 1) in northwestern Queensland, broad areas of the Northern Territory, the northeastern part of Western Australia, south-central South Australia, small outcrops in western New South Wales, Victoria, and Tasmania. ÖPIK (1957) postulated a narrow north-south seaway through the middle of Australia in Early Cambrian time, and a complex marine region covering most of eastern Australia in Middle and Late Cambrian times. Western Australia was a land area throughout most of the Cambrian.

During Early Cambrian time, South Australia was a region of carbonate sedimentation with locally rich archaeocyathid faunas. The few trilobites in the sequence include species of *Redlichia* and Protolenidae.

During Middle Cambrian time, Queensland and the northern part of the Northern Territory was a region of limestone and shale sedimentation that supported a rich and varied fauna of trilobites including many agnostids and oryctocephalids and the distinctive paradoxidid *Xystridura*. To the south, in Victoria and Tasmania, thick sequences of shales and interbedded volcanics are known. These contain a few fossiliferous intervals characterized by cosmopolitan agnostids in Tasmania and by agnostids and polymerid genera (*e.g., Fouchouia, Amphoton, Dinesus*) typical of eastern Asia, in Victoria.

In the early part of the Late Cambrian, the faunas of western Queensland contained a rich association of endemic genera together with Damesellidae and other trilobites typical of eastern Asia, many cosmopolitan agnostid genera, and a few widespread polymerids such as *Irvingella* and *Erixanium*. Younger Late Cambrian faunas recorded from sandstones in the Northern Territory include Ptychaspididae and Saukiidae similar to forms from China. To the east, contemporaneous faunas include abundant agnostids and genera with both Chinese and western American affinites.

MISCELLANEOUS CAMBRIAN AREAS

Cambrian rocks are known without much regional context (Fig. 1) in western Argentina, southern Bolivia, and eastern Colombia, New Zealand, Antarctica, the Himalayan region of India and Pakistan, northern and southern Iran, and the island of Spitsbergen in the North Atlantic.

In New Zealand, a Middle Cambrian trilobite fauna has its affinities entirely with faunas of Queensland in Australia. In the Antarctic, Early, Middle, and Late Cambrian trilobites and Early Cambrian archaeocyathids have their affinities with the faunas of Australia and Asia. In contrast, the Early, Middle, and Late Cambrian faunas of northwestern Argentina are essentially the same as those of western United States. The latest Cambrian trilobite faunas of Bolivia have their closest affinities with the faunas of southern Mexico which, in turn, have strong affinities to both North America and northwestern Europe. In Colombia, a small collection with *Paradoxides* suggests affinities to northwestern Europe. The faunas of the Himalayan region and Iran include Early and Late Cambrian trilobites closely related to those of eastern Asia and, in the Middle Cambrian, ubiquitous representatives of the Oryctocephalidae. The Early Cambrian fauna of Spitzbergen includes Olenellidae related to those of North America.

BIOGEOGRAPHIC "PROVINCES"

The basic clue to understanding the biogeographic framework of the Cambrian on a global scale lies in an appreciation of the distribution of trilobites at various taxonomic levels. At the specific level, the only group with wide geographic distribution is the Agnostida. At the generic level, the most widely distributed trilobites are the Agnostida and polymerid forms commonly associated with them. Regions poor in Agnostida commonly have geographically restricted genera. Many trilobite families have worldwide distribution but are restricted to particular environmental areas.

Further important data are provided by the faunal distribution patterns around North America. These show that the regions rich in Agnostida and their associates, particularly in the Middle and Late Cambrian, are in the peripheral marine areas on the outer side of the carbonate banks—the regions with unrestricted access to the open ocean. In the protected marine areas on or behind the carbonate banks, the Agnostida are not abundant and most trilobite genera are typically North American.

The open ocean also served as a genetic reservoir. Several times during the Cambrian, the polymerid trilobite faunas of the carbonate banks and the protected areas behind the banks in North America were virtually annihilated by abrupt changes in environmental conditions—perhaps temperature—that left no record in the sediments beyond an abrupt nonevolutionary change in the trilobite faunas. The changes took place first in the peripheral regions beyond the carbonate banks, thus indicating that the source for the new faunas was the oceanic region. Furthermore, the incoming elements of each new fauna had their greatest affinities with the incoming elements of the fauna that followed the previous annihilation. This similarity was not superficial, and supports the idea that the source of genetic continuity was in the oceanic region. Additional support comes from the fact that long-ranging genera, such as *Ogygopsis* and *Zacanthoides*, and long-ranging families, such as the Oryctocephalidae and Pagetiidae, are typical of the unrestricted environments beyond the carbonate banks.

Neither the geotectonic criteria of geosyncline versus craton (LOCHMAN-BALK & WILSON, 1958) nor the lithofacies pattern of carbonate banks and inner and outer detrital belts (PALMER, 1969) described above for North America, are applicable on a worldwide basis to explain the general trilobite distribution. The major faunal contrasts on the largest scale are between those areas that had unrestricted access to the open ocean, and those areas where such access was restricted either by a carbonate barrier, or by undefined modifications of environmental parameters such as temperature and salinity that were related to broad expanses of shallow sea over either carbonate or terrigenous substrates. Areas of the first type are the agnostid-rich areas that share many common faunal characteristics on a global scale. Areas of the second type are those where endemic polymerid genera dominate. If trilobite distributions are viewed in this context of contrasting marine environments—restricted versus unrestricted access to open ocean conditions—then a reasonable explanation for both the intracontinental diversity and intercontinental similarity of the trilobite faunas can be found. TAYLOR (1973) has suggested that an additional strong factor in the geographic control of the trilobite faunas here referred to as open-ocean may be temperature. This suggestion should be given serious consideration in future compilations of Cambrian biogeography.

Figure 1 shows the general distribution of persistent areas of open-ocean and restricted conditions during Cambrian time. The margins between these areas fluctuated throughout the Cambrian and in addition were not sharply defined. Thus, boundaries on the maps indicate only an approximate average position on a shifting spectrum of conditions.

Within this broad framework, both the open-ocean regions and the restricted regions supported biotas of limited extent that define "provinces." Because very little is known about the precise habitat requirements of almost all trilobites, the "provinces" are, again, only crude generalizations that outline regions sharing certain distinctive taxa.

EARLY CAMBRIAN BIOGEOGRAPHY

The described trilobite faunas of the Early Cambrian show many contrasts that could be attributed to "provincial" differences (COWIE, 1971); however, many of these reflect differences in the environments available for sampling and the inadequacy of the Early Cambrian record on a global scale. The rich and varied invertebrate faunas of the Asiatic Soviet Union are associated with broad areas of carbonate banks where margins were exposed to open ocean conditions. Most western North American and Arctic Early Cambrian faunas are in restricted regions associated with terrigenous sequences of the inner detrital belt or the inner margins of the carbonate belt; however, sequences representing the outer detrital belt and outer margin of the carbonate belt have recently been described from northwestern Canada. The faunas of southwestern Europe and North Africa are associated with terrigenous sequences but they seem to have had better access to open-ocean conditions than most of the North American faunas.

Two "provinces"—an "olenellid province" and a "redlichiid province"—characterized by trilobite families typical for the restricted regions can be recognized. In regions with better access to the open ocean, representatives of both families are known. The "olenellid province" includes North America, South America, and northwestern Europe. The "redlichiid province" includes China, southern Asia eastward from the Mediterranean region, Australia, and Antarctica. In southwestern Europe and adjacent parts of Africa, and in the Asiatic Soviet Union, elements of both "provinces" are found.

The only other Early Cambrian group for which biogeography has been evaluated on a global scale is the Archaeocyatha (ZHURAVLEVA, 1968; HILL, 1972). These are almost completely confined to regions interpreted here as open ocean. During much of this time, many families and genera were common to all areas. During the middle Early Cambrian, when the Archaeocyatha reached their evolutionary peak, the complexes of Australia and Antarctica included a significant number of forms not known in the Northern Hemisphere. Within the eastern Soviet Union consistent differences existed during the early and middle Early Cambrian between the archaeocyathid complexes of the volcanic regions in the south and the nonvolcanic limestone regions to the north.

MIDDLE AND LATE CAMBRIAN BIOGEOGRAPHY

During the Middle and Late Cambrian, "provincial" differences are shown in both the restricted and the open-ocean regions. In the restricted regions, four provincial areas typified by many endemic genera and species can be recognized: 1) the inner detrital belt and adjacent margins of the carbonate belt of North America; 2) the sandy facies of central Europe; 3) the carbonate banks of the Siberian platform; and 4) the Hwang-ho facies of China. The Late Cambrian sandy facies of central Australia seems to have a close relationship to the Hwang-ho facies.

In the regions with unrestricted access to the open ocean, the number of provincial areas is less and they are much less well defined. Three provincial regions focused on western Europe, North America, and southeast Asia-Australia can be recognized. The western European "province" is characterized by the Olenidae, Conocoryphidae, and Paradoxididae. Significant elements of the faunas of this province are found in the open-ocean regions of the Asiatic Soviet Union. The North American "province" is characterized by Oryctocephalidae, certain Corynexochida (*Bathyuriscus, Ogygopsis, Zacanthoides*), Marjumiidae, Pterocephalidae, Richardsonellidae, and Catillicephalidae. However, some of the typical elements of this province are found in the open-ocean regions of the Asiatic Soviet Union and Australia. The southeast Asia-Australia "province" is characterized by Dameselllidae, certain Corynexochida (*Amphoton*), Anomocarellidae, Ceratopygidae, and Xystridurinae. Some elements of these faunas are found in the open-ocean regions of Asiatic Soviet Union and northwestern North America.

CAMBRIAN BIOSTRATIGRAPHY

For almost a century, attempts have been made to divide rocks of Cambrian age into ever smaller chronostratigraphic units using primarily stratigraphic ranges of trilobites. In some areas, and for different times within the Cambrian Period, fossils other than trilobites have also been used. Lower Cambrian sequences on the Baltic shield include zones based on characteristic and common occurrences of *Mobergella* and *Volborthella* —fossils whose biologic affinities are still being debated. The lowermost Cambrian beds of the Siberian platform, which include faunas assigned to several zones of the Tommotian Stage, are characterized by abundant nontrilobite fossils. Some of these fossils are now being found in Lower Cambrian beds of the Baltic region, England, and Australia. In the eastern part of the Soviet Union, archaeocyathids have been effectively used to characterize divisions of the Lower Cambrian. Recently, conodonts have been found in sufficient abundance in Upper Cambrian beds to characterize zones that will be important in discussions of intercontinental correlation of the Cambrian-Ordovician boundary. Acrotretide brachiopods have the potential to be useful biostratigraphic tools in Middle and Upper Cambrian carbonate sequences in many areas; however, at the present time, effective interregional biostratigraphic synthesis for the Cambrian System must be based on trilobites.

In each region where reasonably detailed work has been done, successions of assemblage zones have been established. In some regions, these have been grouped into stages. Neither zones nor stages have consistent interregional utility now. Even the Lower, Middle, and Upper Cambrian series, which have received different local names in some areas, can only be approximately correlated on a global scale. The basal boundary of the Cambrian is a special philosophical problem and is hotly debated. The Lower-Middle Cambrian boundary and the Cambrian-Ordovician boundary are both subject to international disagreement about correlation involving possible discrepancies of the order of a stage. The Middle-Upper Cambrian boundary seems to be the least disputed, but even there possible intercontinental discrepancies exist.

The reason for the difficulty in correlation of Cambrian faunal sequences is the fact that trilobite distribution is strongly facies controlled. This is well demonstrated by stratigraphic analyses of Cambrian faunal differences in the Cordilleran region of western North America, in eastern China, and along the southern boundary of the Siberian platform. Polymerid trilobite faunas in the Cordilleran region of North America, for example, represent three or four distinct depositional environments: the restricted shelf, with subtle contrasts between regions of clastic and carbonate sedimentation; the ocean-facing shelf margin; the shelf slope; and the deep shelf or open ocean. Due to shifting sites of these environments the often distinctly different faunas representing them may be stacked in different orders, in different stratigraphic sections. Alternatively, different biofacies dominated the same time intervals in different geographic areas. Thus, in order to work out a meaningful biostratigraphy, the biofacies and lithofacies relationships of the Cambrian faunas must be more clearly established. This work is still in progress.

The "classical" North American trilobite zonation reviewed by LOCHMAN-BALK and WILSON (1958) is a typical example of the problems introduced by nonrecognition of trilobite biofacies. The Lower Cambrian zonation and the Middle Cambrian zonation up through the *Glossopleura* Zone is representative only of the restricted-shelf biofacies; the remainder of the Middle Cambrian zones represent the shelf-margin and shelf-slope biofacies. There is no quarrel with the faunal sequence, which has been adequately tested and is generally reliable; however, the biostratigraphy of the restricted shelf region must be separated from the biostratigraphies of the ocean-facing and oceanic regions. Subsequent integration of the different biostratigraphies will then permit maximum utilization of faunal information for correlation purposes.

Empirical observations indicate that a useful global biostratigraphy will probably have to be based on the faunas of the oceanic regions, which are dominated by Agnostida, and that precise interregional correlation between faunas of the restricted-shelf regions will have to be based on fortuitous interlayering of these faunas with distinctive elements of the oceanic faunas.

One additional factor that may be important for problems of intercontinental correlation is that of extinction events referred to earlier (p. *A*128). In the North American Cambrian sequences at least three, perhaps as many as six, major continent-wide extinctions affected the trilobite faunas of the shelf regions. The extinction events mark the boundaries of biostratigraphic units called biomeres, which may include several trilobite zones. The causes of the extinctions are postulated to be abrupt cooling events, but conclusive evidence of this hypothesis must still be obtained. Similar and apparently synchronous events have been reported at several levels in Australia, but are seemingly absent from the Siberian and Baltic regions. Further work on the significance of extinctions at biomere boundaries may provide a basis for global biostratigraphic units of stage magnitude. For the moment, local zonal and stage schemes will have to remain as the only available, although inadequate, descriptions of Cambrian biostratigraphy. These schemes, for each of the principal Cambrian regions of the world, are discussed below.

NORTH AMERICA

The sequence of zones given in Table 1 has wide applicability in North America, but it cannot be applied uniformly to all Cambrian regions.

The Lower Cambrian subdivisions are recognizable only in the Cordilleran region. Early Cambrian localities in the Appalachian region, the Canadian Arctic islands, and Greenland have faunas that can be related only in a general way to the Cordilleran zonal scheme. Most of the faunas from those regions can be included in the *Bonnia-Olenellus* Zone but exact positioning within this rather broad zone is not yet possible.

The Middle Cambrian subdivisions are also primarily recognizable in the Cordilleran region. Early Middle Cambrian faunas assignable to the *Plagiura-Poliella* and *Albertella* Zones have not been recognized in eastern North America. In the deeper water facies, there seems to be an anomalous hiatus at this time. The *Glossopleura* Zone is known in the restricted-shelf facies in the Appalachian region and in Greenland, but is also absent from the deeper water facies adjacent to the shelf. Faunas repre-

TABLE 1. *Cambrian Biostratigraphy of North America.*

Upper Cambrian Series
Trempealeauan Stage
Saukia Zone
Franconian Stage
Saratogia Zone
Taenicephalus Zone
Elvinia Zone
Dresbachian Stage
Dunderbergia Zone
Prehousia Zone
Dicanthopyge Zone
Aphelaspis Zone
Crepicephalus Zone
Cedaria Zone
Middle Cambrian Series
Bolaspidella Zone
Bathyuriscus-Elrathina Zone
Glossopleura Zone
Albertella Zone
Plagiura-Poliella Zone
Lower Cambrian Series
Bonnia-Olenellus Zone
Nevadella Zone
Fallotaspis Zone

senting most of the younger zones are known from the restricted-shelf and deeper water facies of the Appalachian region and have recently been discovered in the Canadian Arctic islands and Greenland.

Precise intercontinental correlation is only possible for parts of the upper Middle Cambrian and Upper Cambrian, where agnostid trilobites are particularly abundant and varied and for the uppermost Cambrian where conodont biostratigraphy is beginning to produce useful results.

NORTHERN EUROPE

The zonal succession shown in Table 2 can be used effectively from Poland across northern Europe to Great Britain. In addition, many of the elements of this succession are found in the coastal Cambrian exposures of eastern North America from Newfoundland to New England, in the Middle and Upper Cambrian of northeastern Siberia, and in the Middle Cambrian of southern Europe and North Africa so that reasonably precise correlations can be effected among parts of these regions. The nearly cosmopolitan agnostid trilobites permit correlation with several parts of the North American, Australian, and Chinese sequences. Throughout this region, Lower and Middle Cambrian beds are separated by a hiatus.

TABLE 2. *Cambrian Biostratigraphy of Northern Europe (Martinsson, 1974).*

Upper Cambrian Series
Acerocare Zone
Peltura scarabaeoides Zone
Peltura minor Zone
Protopeltura praecursor Zone
Leptoplastus Zone
Parabolina spinulosa Zone
Homagnostus obesus Zone
Agnostus pisiformis Zone
Middle Cambrian Series
Paradoxides forchhammeri Stage
Lejopyge laevigata Zone
Jincella brachymetopa Zone
Paradoxides paradoxissimus Stage
Ptychagnostus lundgreni-Goniagnostus nathorsti Zone
Ptychagnostus punctuosus Zone
Hypagnostus parvifrons Zone
Tomagnostus fissus-Ptychagnostus atavus Zone
Ptychagnostus gibbus Zone
Eccaparadoxides oelandicus Stage
Eccaparadoxides pinus Zone
Eccaparadoxides insularis Zone
Hiatus
Lower Cambrian Series
Strenuaeva linnarssoni Zone
Holmia kjerulfi Zone
Volborthella-Schmidtiellus mickwitzi Zone
Mobergella holsti Zone

CENTRAL AND SOUTHERN EUROPE

Within this region, the only biostratigraphic zonation of more than local value is that of the Middle Cambrian (Table 3). The sequence established by SDZUY (1972) for Spain is applicable in the Montagne Noire of southern France, Sardinia, and North Africa. Upper Cambrian beds are poorly fossiliferous or absent. Lower Cambrian beds are variably fossiliferous but only in North Africa has a zonal sequence been established (HUPÉ, 1952) and its regional applicability has not yet been demonstrated.

TABLE 3. *Cambrian Biostratigraphy of Spain, France, Sardinia, and North Africa (Sdzuy, 1972).*

Upper Cambrian Series (not described, poorly represented)
Middle Cambrian Series
Solenopleuropsis Zone
Pardailhania Zone
Badulesia Zone
Acadolenus Zone
Conocoryphe ovata Zone
Paradoxides (Acadoparadoxides) mureroensis Zone
Lower Cambrian Series (various local sequences)

SOVIET UNION

Because of the vast area of eastern Soviet Union that contains Cambrian rocks, no single scheme has yet been established at any level in the intrasystemic biostratigraphic hierachy. Each major outcrop area has its own local zonation. Table 4 shows the stages that are most commonly used, but critiques of the Lower and Middle Cambrian stage structure by ROZANOV (1973) and SAVITSKIY (1969) have pointed out many difficulties with this scheme.

TABLE 4. *Commonly Used Cambrian Stages of the Soviet Union.*

Upper Cambrian Series
Shidertinian Stage
Tuorian Stage
Middle Cambrian Series
Amgan Stage
Mayan Stage
Lower Cambrian Series
Lenian Stage
Botomian Stage
Atdabanian Stage
Tommotian Stage

On the northeastern and southeastern margins of the Siberian platform, and in the orogenic belts of Kazakhstan, the Middle and Upper Cambrian faunas contain agnostids and other trilobites that permit some precise correlation with other major world Cambrian areas. The area of the Kharaulakh Mountains near the mouth of the Lena River has a particularly significant interrelationship between western American and western Europe Cambrian faunal elements.

CHINA

The biostratigraphy of China has been summarized by Lu and others (1974). Three trilobite biofacies, designated the Northern Type, Transitional Type, and Southeastern Type have been recognized. The formal stage and zone nomenclature encountered in many publications about the Cambrian of China (Table 5) applies pri-

TABLE 5. *Cambrian Stages and Zones of the Northern Type Biofacies Region of China.*

Upper Cambrian Series
Fengshan Stage
Tellerina-Calvinella Zone
Ellesmeroceras-Dictyella Zone
Quadraticephalus Zone
Ptychaspis-Tsinania Zone
Changshan Stage
Kaolishania Zone
Changshania Zone
Chuangia Zone
Kushan Stage
Drepanura Zone
Blackwelderia Zone
Middle Cambrian Series
Changhia Stage
Damesella Zone
Taitzuia Zone
Amphoton Zone
Crepicephalina Zone
Liaoyangaspis Zone
Hsuchuan Stage
Bailiella Zone
Poriagraulos Zone
Sunaspis Zone
Kochaspis hsuchuangensis Zone
Lower Cambrian Series
Maochuan Stage
Shantungaspis Zone
Manto Stage
Redlichia chinensis Zone
Tsangpin Stage
*Megapaleolenus Zone
*Paleolenus Zone
*Drepanuroides Zone
*Malungia Zone
*Unnamed zones—including interval with pre-trilobite shelly fossils.

* Recognized only in the Yangtze area.

marily to the shallow-water, normal marine faunas of Northern Type. Faunas of Southeastern Type, which include beds rich in agnostids and eodiscids, have not been formally subdivided biostratigraphically. These faunas contain many elements permitting precise correlation of parts of the enclosing formations with those of other parts of the world that have comparable facies. The faunas of Northern Type have many elements also found in the Australian Cambrian and some precise correlations are possible between similar facies of these two areas. Localities with Transitional Type faunas contain interbedding or admixtures of the Northern and Southeastern type faunas that greatly facilitate intercontinental correlation of many parts of the Chinese Cambrian.

AUSTRALIA

Most of the published detailed biostratigraphy of the Australian Cambrian is concentrated in the late Middle Cambrian and lower Upper Cambrian where an elaborate sequence of zones has been proposed by ÖPIK (1957, 1961, 1967). Lower and Upper Cambrian beds have been assigned to Australian stages (ÖPIK, 1968; JONES et al., 1971), but details of the zonations have not yet been published. The present biostratigraphic breakdown of the Australian Cambrian is shown in Table 6. The rich Middle and Upper Cambrian agnostid sequences of Queensland provide many opportunities for intercontinental correlation. Associated polymerid trilobites, many of Chinese aspect, provide a key for correlation of some of the Chinese faunas with the restricted-shelf faunas of Siberia and North America.

Pending resolution of problems of detailed intercontinental correlation in the vicinity of the upper boundary of the Middle Cambrian, some parts of the Mindyallan Stage may be reassignable to the Middle Cambrian (JAGO & DAILY, 1975).

TABLE 6. *Cambrian Biostratigraphy of Australia.*

Upper Cambrian Series
  Payntonian Stage
    Zones not established
  Unnamed pre-Payntonian and post-Idamean stages
  Idamean Stage
    Irvingella tropica-Agnostotes inconstans Zone
    Erixanium sentum Zone
    Corynexochus plumula Zone
    Glyptagnostus reticulatus Zone
  Mindyallan Stage
    Glyptagnostus stolidotus Zone
    Cyclagnostus quasivespa Zone
    Erediaspis eretes Zone
    Damesella torosa-Ascionepea janitrix Zone
Middle Cambrian Series
  Stage(s) undesignated
    Holteria arepo Zone
    Proampyx agra Zone
    Ptychagnostus cassis Zone
    Ptychagnostus nathorsti Zone
    Ptychagnostus punctuosus Zone
    Euagnostus opimus Zone
    Ptychagnostus atavus Zone
  Templetonian Stage
    Ptychagnostus gibbus Zone
    "*Dinesus-Xystridura*" Zone
  Ordian Stage
    Zones not designated
Lower Cambrian Series
  Stages not designated
    Zones not designated (Faunal assemblages 1 to 9 of DAILY, in ÖPIK et al., 1957).

REFERENCES

Cowie, J. W., 1971, *Lower Cambrian faunal provinces:* in Faunal provinces in space and time, F. A. Middlemiss et al., p. 31-46, Seel House Press (Liverpool).

Daily, Brian, 1957, *The Cambrian in South Australia:* in The Cambrian geology of Australia, A. A. Öpik et al., Austral. Bur. Min. Res. Bull. 49, p. 91-147.

Hill, Dorothy, 1972, *Archaeocyatha:* in Treatise on invertebrate paleontology, Part E, vol. 1, 2nd edit. (revised), Curt Teichert (ed.), xxx + 158 p., 871 fig., Geol. Soc. America & Univ. Kansas Press (Boulder, Colo.; Lawrence, Kans.).

Hupé, Pierre, 1952, *Contribution a l'étude du Cambrien inférieur et du Précambrien III de l'Anti-Atlas Marocain:* Serv. Mines Carte Géol. Maroc, Notes Mém., v. 103, p. 1-402.

Jago, J. A., & Daily, Brian, 1975, *The trilobite*

Lejopyge Hawle and Corda *and the Middle-Upper Cambrian boundary:* Palaeontology, v. 18, p. 527-550, pl. 62-63.

Jones, P. J., et al., 1971, *Late Cambrian and Early Ordovician stages in western Queensland:* Geol. Soc. Australia, Jour., v. 18, p. 1-32.

Kobayashi, Teiichi, 1967, *The Cambrian of eastern Asia and other parts of the continent:* Univ. Tokyo, Jour. Fac. Sci., sec. II, v. 16, pt. 3, p. 381-534.

Lochman-Balk, Christina, & Wilson, J. L., 1958, *Cambrian biostratigraphy in North America:* Jour. Paleontology, v. 32, no. 2, p. 321-350.

Lu Yen-hao, et al., 1974, *Sheng wuh hwan jing kong jyh lun jyi chyi tzai harn wu jih sheng wuh ti jseng shyue shang her ku donq wuh ti li shang de ing yonq:* Nan king ti chi ku sheng wuh yan jiou shoo jyi kan (Nanking Inst. Geology and Paleontology), Mem., no. 5, p. 27-116. [*Bio-environmental control hypothesis and its application to Cambrian biostratigraphy and paleozoogeography.*]

Martinsson, Anders, 1974, *The Cambrian of Norden:* in Cambrian of the British Isles, Norden and Spitzbergen, p. 185-284, John Wiley & Sons (New York).

Öpik, A. A., 1957, *Cambrian geology of Queensland:* in The Cambrian geology of Australia, A. A. Öpik et al., Austral. Bur. Min. Res. Bull. 49, p. 1-24.———1961, *The geology and paleontology of the headwaters of the Burke River, Queensland:* Austral. Bur. Min. Res. Bull. 53, 249 p., 24 pl.———1967, *The Mindyallan fauna of northwestern Queensland:* Austral. Bur. Min. Res. Bull. 74, v. 1, 404 p.; v. 2, 166 p., 67 pl.——— 1968, *The Ordian Stage of the Cambrian and its Australian Metadoxidae:* Austral. Bur. Min. Res. Bull. 92, p. 133-169.

Palmer, A. R., 1969, *Cambrian trilobite distributions in North America and their bearing on the Cambrian paleogeography of Newfoundland:* Am. Assoc. Petroleum Geologists, Mem. 12, p. 139-144.———1972, *Problems in Cambrian biogeography:* 24th Internatl. Geol. Congress, Proc., sec. 7, p. 310-315 (Montreal).———1973, *Cambrian trilobites:* in Atlas of paleobiogeography, A. Hallam (ed.), p. 3-11, Elsevier Sci. Publ. Co. (London).

Repina, L. N., 1968, *Biogeografiya rannego kembriya Sibiri po trilobitam:* 23rd Internatl. Geol. Congress (Prague), Doklady Sovetskikh Geologov, p. 46-56, Izdatelstvo "Nauka" (Moskva) (English abstract). [*Biogeography of Early Cambrian Siberian trilobites.*]

Robison, R. A., 1964, *Middle-Upper Cambrian boundary in North America:* Geol. Soc. America, Bull., v. 75, p. 987-994, text-fig. 1, 2.

Rozanov, A. Yu., 1973, *Zakonomernosti morfologicheskoy evolyutsii Arkheotsiat i voprosy yarusnogo raschleneniya Nizhnego Kembriya:* Akad. Nauk SSSR, Geol. Inst., Trudy, v. 241, 164 p. [*Principles of morphological evolution of the Archaeocyatha and questions of the stage subdivision of the Lower Cambrian.*]

Savitskiy, V. E., 1969, *O yarusnom raschlenenii srednego kembriya Sibiri i nekotorykh obshchikh voprosakh razrabotki etalonnoy shkaly yarusnykh podrazdeleniy:* Sibir. Nauchno-Issledov. Inst. Geologii Geofiziki Mineral. Syrya (SNIIGGIMS) Trudy, v. 94, p. 140-149. [*Subdivision of the Middle Cambrian of Siberia into stages and problems of establishing a standard scale for stage divisions.*]

Sdzuy, Klaus, 1958, *Tiergeographie und Paläogeographie im europäischen Mittelkambrium:* Geol. Rundschau, v. 47, p. 450-462.———1972, *Das Kambrium der acadobaltischen Faunenprovinz:* Zentralbl. Geologie Paläontologie, Teil II, Jahrg. 1972, no. 1/2, p. 1-91 (July).

Taylor, M. E., 1973, *Late Cambrian biofacies in the western United States:* Geol. Soc. America, Abstr. with Programs, v. 5, no. 7, p. 836-837.

Zhuravleva, I. T., 1968, *Biogeografiya i geokhronologiya rannego Kembriya po Arkheotsiatam:* 23rd Internatl. Geol. Congress, Prague, 1968, Doklady Sovetskikh Geologov, p. 33-45, Izdatelstvo "Nauka" (Moskva) (English abstract). [*Biogeography and geochronology of the Early Cambrian based on Archaeocyatha.*]

*A*136 *Introduction—Biogeography and Biostratigraphy*

ORDOVICIAN[1]

By VALDAR JAANUSSON

[Naturhistoriska Riksmuseet, Stockholm]

CONTENTS

	PAGE
INTRODUCTION	*A*136
LOWER BOUNDARY OF THE ORDOVICIAN	*A*137
SHELLY FAUNAS	*A*139
Tremadocian	*A*139
Arenigian (Post-Tremadocian Canadian)	*A*140
Llanvirnian (Including Whiterock and Kunda Stages)	*A*142
Middle Ordovician	*A*143
Upper Ordovician (Excluding the Hirnantian)	*A*149
Uppermost Upper Ordovician (Hirnantian)	*A*153
PLANKTONIC GRAPTOLITES	*A*154
CONODONT FAUNAS	*A*156
BIOGEOGRAPHIC EVALUATION OF FAUNAL DIFFERENTIATION	*A*157
BIOGEOGRAPHIC CHANGES LEADING TO THE COSMOPOLITAN SILURIAN FAUNA	*A*160
REFERENCES	*A*162

INTRODUCTION

Ordovician time, with a duration of approximately 60 million years, had a considerable biogeographic differentiation. This can be traced in the distribution of most groups of organisms, although many are comparatively rare or have not yet been intensely studied. In this contribution, distributional data of shelly groups (mainly trilobites and articulate brachiopods), planktonic graptolites, and conodonts are treated separately. Owing to limited space available, documentation is restricted to a minimum.

Distribution of various faunas is illustrated by the known occurrence of selected taxa plotted on maps rather than by showing the distribution of inferred biogeographic units or faunas. After considerable experimenting, the map of the modern world was chosen for illustrating Ordovician distribution patterns, although during Ordovician time the location of oceans and continental lithospheric plates obviously was completely different. Not only is the former geographic position of many individual Ordovician lithospheric plates uncertain, but in several cases it is not known what constituted an individual plate. For this reason the use of any of the proposed geographical models for the Ordovician would introduce considerable speculation. The situation will not improve until more paleomagnetic data are available from Ordovician rocks in various parts of the world. The maps (Fig. 2, 4, 6-13) show the distribution

---

[1] Manuscript received November, 1969; revised manuscript received December, 1975.

of land and sea, roughly reconstructed, based on available paleogeographic maps (KELLER & PREDTECHENSKY in VINOGRADOV, 1968; LIU, 1958, complemented with JEN, 1964, and HU et al., 1965; WOLFART, 1967; CUERDA, 1973; LEGRAND, 1974; and others).

According to all recent reconstructions of Ordovician geography, Africa, Arabian Peninsula, peninsular India, South America, Antarctica, and Australia formed a single huge continental plate, Gondwanaland. The Ordovician South Pole was located somewhere in northwestern Africa (SMITH et al., 1973) and thus much of the plate was situated in a cold climatic zone. Much of Ordovician Gondwanaland was dry land and it may have been the main source of terrigenous sediments to southern Europe, where the Ordovician sequence consists of clastic rocks; however, in some reconstructions (WHITTINGTON & HUGHES, 1972, 1973) southern Europe is considered to have constituted a separate lithospheric plate, separated from Gondwanaland by a wide proto-Tethyan ocean and from northeastern Europe by a mid-European ocean.

In the rest of the Ordovician world, land areas were scattered and formed archipelagos rather than continents. At least three individual continental lithospheric plates are distinguished: 1) North America and Arctic islands, 2) Russian platform and adjoining areas to the west, and 3) Asia. On these plates the Ordovician epicontinental deposits consist mostly of carbonates. On most reconstructions of Ordovician geography much of all three plates is in tropical and subtropical latitudes. The boundary between the north European and south European plates is variously drawn along the Alpine chain (SMITH et al., 1973; WILLIAMS, 1973), the southwestern margin of the Russian platform (WHITTINGTON & HUGHES, 1972, 1973), or roughly along the northern limit of the Variscan Mountains (BURRETT, 1973). This part of Europe has widely distributed fossiliferous Ordovician rocks, and differences in position of the plate boundary have a profound effect on biogeographic interpretations. North America and northern Europe were separated by the Iapetus Ocean (HARLAND & GAYER, 1972; Proto-Atlantic Ocean, WILSON, 1966; WHITTINGTON & HUGHES, 1972) that is generally considered to have decreased in width during the Ordovician. Asia, with the western plate margin along the present Ural Mountains, is tentatively treated as a single lithospheric plate, or up to five separate plates, variously situated relative to the other plates. Paleomagnetic data are available only for Siberia.

ACKNOWLEDGMENTS

During the completion of various versions of this contribution since 1968, many colleagues generously helped by providing unpublished information as well as inaccessible publications. For this I am particularly indebted to M. K. APOLLONOV (Alma-Ata), S. M. BERGSTRÖM (Columbus), W. T. DEAN (Ottawa), R. MÄNNIL (Tallinn), R. B. NEUMAN (Washington, D.C.), ZOYA E. PETRUNINA (Novokuznetsk), R. J. ROSS (Denver), and A. D. WRIGHT (Belfast).

LOWER BOUNDARY OF THE ORDOVICIAN

In accordance with general practice (except among British geologists), the Tremadoc Series is here treated as the lowermost Ordovician. The intercontinental correlation of the lower boundary of the Tremadocian presents problems and must be briefly discussed because of its biogeographic implications. More exactly, on account of differences in correlation, the biogeographic conclusions presented in this chapter differ in several respects from those given for the Tremadocian by WHITTINGTON and HUGHES (1974).

The lower boundary of the Tremadocian is traditionally drawn at the level of appearance in northern Europe of the first dendroid graptolites with a "free" sicula indicating a change from sedentary to planktonic mode of life (Fig. 1). The correlation of this horizon in regions without graptolites is difficult. The particular problem is whether and how much of the

British series		Atlantic graptolite zones, Oslo region, Norway		North Atlantic conodont zones	North American midcontinent conodont zones	North American series	Trilobite zones, western North America	
TREMADOCIAN	UPPER	Ceratopyge beds	Poorly graptolitiferous, zonal classification not yet well established	Paltodus deltifer	Chosonodina herfurthi	CANADIAN	D	Leiostegium - Kainella
							C	Paraplethopeltis
	LOWER	Dictyonema Shale	Anisograptus	Cordylodus angulatus	Cordylodus angulatus		B	Symphysurina
			Dictyonema norvegicum	-----?-----	-----------			
			Dictyonema flabelliforme					
			Dictyonema sociale	?	Cordylodus proavus		A	Missisquoia
			Dictyonema parabola			U.€.	Corbinia apopsis	

Fig. 1. Correlation of the Tremadocian Series (Jaanusson, n).

North American Trempealeauan Stage and comparable beds elsewhere containing saukiid trilobites, generally referred to the Upper Cambrian, are equivalent to the Lower Tremadocian. Records of saukiids and *Richardsonella* associated with early Tremadocian trilobites in Oaxaca, Mexico (ROBISON & PANTOJA-ALOR, 1968), initially seemed to indicate that a substantial part of the Trempealeauan is equivalent to the early Tremadocian; however, recent studies on conodonts do not confirm such a correlation. In the conodont fauna, the entry of *Cordylodus* forms a clearly recognizable level and has been suggested for use as a tentative additional criterion of the base of the Ordovician (MILLER, 1969). The earliest Ordovician conodont faunas are largely cosmopolitan and closely similar successions of species have been reported from widely separated regions (MILLER, 1969, 1970; DRUCE & JONES, 1971; JONES, 1971, etc.). It is now known (MILLER, 1970) that *Cordylodus* enters at the base of the *Corbinia apopsis* Subzone which, by tradition, is regarded as the top of the Upper Cambrian Trempealeauan Stage but could as well be included in the Ordovician.

The exact level of the entry of *Cordylodus* in the North European sequence is not known because *Dictyonema* shales and associated rocks are mostly devoid of identifiable conodonts. There the earliest known Ordovician conodont fauna is from the upper Maardu (*"Obolus"*) Sandstone of northwestern Estonia (VIIRA, 1966), below the local lithostratigraphic *Dictyonema* Shale, and associated with *Dictyonema flabelliforme flabelliforme*. The level is probably fairly high within a complete *Dictyonema* Shale sequence. The assemblage is comparable to that from the upper part of the North American *Symphysurina* Zone that is thus obviously of Lower and not Upper Tremadocian age. The overlap between the Tremadocian and the Trempealeauan, if present at all, seems to be inconsiderable (cf. also MILLER et al., 1974). For these reasons the Tsinaniid and *Rasettia* provinces distinguished by WHITTINGTON & HUGHES (1974) are here regarded as Upper Cambrian and not Tremadocian (see also SHERGOLD, 1975).

The Ordovician Period began with a transgression that was one of the most extensive in the Paleozoic. Over almost the whole of Europe and northern Africa there is a conspicuous break at the base of the

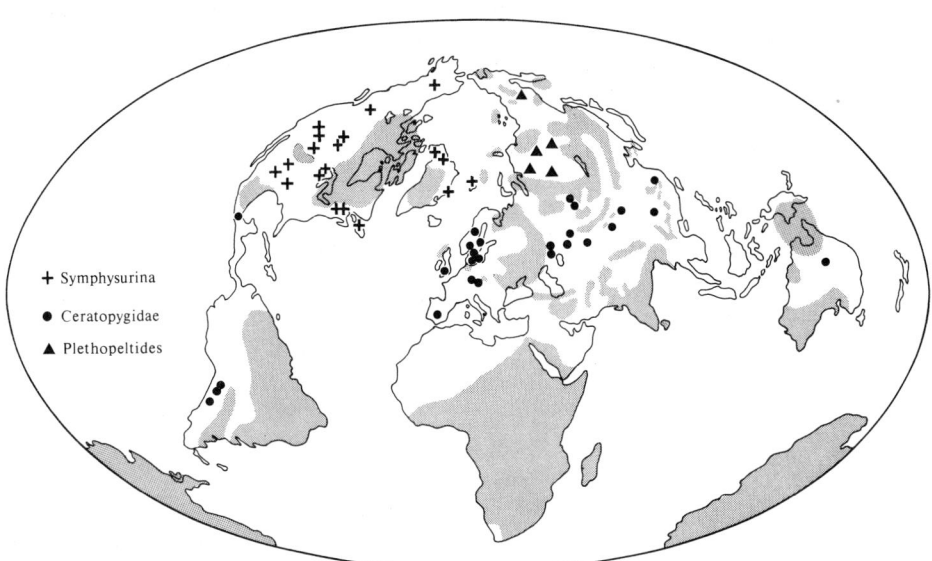

Fig. 2. Distribution of selected Tremadocian trilobite taxa. The asaphid *Symphysurina* characterizes the North American fauna, the plethopeltid *Plethopeltides* the possibly separate Tungusian fauna, and ceratopygids the Southern fauna (Jaanusson, n). [Shaded area in this as well as in other maps indicates probable land areas. For probable boundaries between the main lithospheric plates, see Figure 13.]

Lower Tremadocian, and what appear to be continuous sequences from the Cambrian to Ordovician are known only in some areas of southern Scandinavia and along a belt to central Poland. In most of continental Europe and northern Africa the break comprises the entire Upper Cambrian and on the Russian platform east of the Baltic also the Middle Cambrian. In extensive Arctic areas, such as the Canadian Arctic Archipelago, Greenland, with the exception of Washington Land in western North Greenland, and Spitzbergen, Lower Tremadocian rests on Middle Cambrian, in southern Mexico on Precambrian, and in the allochthonous Taconic sequence of New York on Lower Cambrian. The widespread occurrence of the break suggests an eustatic control of the transgression. The early Tremadocian transgression opened new communications between shelf areas. The most important of these was along the "Paleotethys" from northwestern Africa over southern Europe, which may have been a prerequisite for the development of Whittington's (1966) "Southern Fauna."

SHELLY FAUNAS

TREMADOCIAN

Among Tremadocian shelly fossils, only trilobites are widely distributed and reasonably well studied. Among trilobites two main faunas can be distinguished, here provisionally termed the North American (*Hystricurus* fauna, Whittington, 1966; *Highgatella* Province, Whittington & Hughes, 1974) and Southern Faunas (Ceratopygid Province, Whittington & Hughes, 1974). The distribution of the faunas largely follows that of *Symphysurina* and Ceratopygidae (Fig. 2). Compared with the distribution of Upper Cambrian trilobites the degree of biogeographic differentiation seems to have decreased. Moreover, biogeographic affinities have changed in several areas. Most notably, the fauna in western Argentina is now related to European-southwestern Siberian faunas whereas in the Upper Cambrian it has close North American affinities.

The North American Fauna occupied North America, Greenland, and Spitzbergen and is characterized by *Symphysurina* (Fig. 2) and its allies, *Clelandia* and its allies, *Missisquoia, Highgatella,* and other genera. In comparison with many southern faunas the taxonomic diversity is small. The earliest undoubted leperditiacean ostracodes have been described from the post-*Symphysurina* Tremadocian beds of Vermont (CREATH & SHAW, 1968), and throughout the Ordovician Period this group remained one of the distinctive elements of North American and related faunas.

The Southern Fauna extended over a vast area (see the distribution of the Ceratopygidae, Fig. 2), from the present South America and southern Mexico in the west over Morocco, Wales, and Scandinavia to southeastern China in the east. The fauna is characterized by ceratopygids, dikelokephalinids, nileids, orometopids, *Macropyge,* and other trilobites. Some areas have a high taxonomic diversity (some 75 trilobite genera in the Sayan-Altai mountain region; PETRUNINA, 1966).

The known Tremadocian shelly fauna from the Siberian platform and northeastern Siberia indicates that a separate Lower Tremadocian Tungusian fauna may be distinguishable (cf. the distribution of *Plethopeltides,* Fig. 2; other distinctive genera include *Diceratocephalina, Pseudoacrocephalites,* and *Dolgeuloma*). The known Tremadocian trilobite faunas from these regions and from northern Korea, northern China, and Australia, however, are small and are at present difficult to evaluate biogeographically.

In eastern North America a belt in eastern New Brunswick, Nova Scotia, and Newfoundland has yielded a Cambrian fauna with close affinities to that in northern Europe. On eastern Nova Scotia the sequence includes Tremadocian strata and the small shelly fauna encountered there is astonishingly similar to that of Wales. The current explanation of the faunal and lithological similarity is that the areas in eastern North America mentioned above belonged to the North European plate. During closing of the Iapetus Ocean, the North American and North European plates collided, and the areas became welded to the North American plate. When the present Atlantic Ocean was initiated, this part of the original North European plate followed North America.

ARENIGIAN (POST-TREMADOCIAN CANADIAN)

At the base of the Arenigian the biogeographic diversity increased. The region with the North American Fauna retained its biogeographic identity and boundaries. In the post-Tremadocian Canadian carbonate sequence of North America, Greenland, and Spitzbergen, the trilobite fauna is characterized by various bathyurids and hystricurids (Bathyurid Fanua; WHITTINGTON, 1963) (Fig. 3). Further distinctive forms include the gastropod *Ceratopea* as well as leperditiacean ostracodes. The Durness Limestone in northern Scotland has yielded distinctive trilobites and cephalopods of the North American Fauna (very little faunal information is available on the probable Tremadocian portion of the limestone) suggesting that this part of Scotland, and probably also northern Ireland, may originally have been a part of the North American lithospheric plate.

The known Upper Canadian shelly fauna of the Siberian platform, Taymyr Peninsula, and northeastern Siberia is in several respects close to that of North America (WHITTINGTON & HUGHES, 1972; included in American-Siberian Biogeographic Region, CHUGAEVA, 1968, 1973) but with endemic elements such as the bathyurid trilobite *Biolgina* (Fig. 4), *Prodalmanitina,* and the widespread brachiopod *Angarella.* Many characteristic North American forms, including leperditiacean ostracodes, have not been found.

In the Balto-Scandian region a fauna developed that includes numerous endemic elements (Asaphid Fauna, WHITTINGTON, 1963, 1973, and WHITTINGTON & HUGHES, 1972; Baltic Province, WILLIAMS, 1973; Baltic Biogeographic Region, CHUGAEVA, 1968, 1973; Balto-Scandian fauna, JAANUSSON, 1973a). During the Arenigian the fauna extended from the Vaygach Island and the Pay Khoy Peninsula in the north to central Poland in the south and from the Oslo

British series	Atlantic graptolite zones, Oslo region, Norway	Balto-Scandian series and stages	North Atlantic conodont zones	Pacific graptolite zones, North America		Trilobite zones, western North America
ARENIGIAN	Didymograptus hirundo	VOLK-HOVIAN	Microzarkodina parva — Prioniodus triangularis	Isograptus caduceus	CHAMP.	WHITEROCKIAN
	Didymograptus extensus — Phyllograptus angustifolius elongatus	ONTIKAN	Prioniodus evae	Didymograptus bifidus	CANADIAN SERIES	J Pseudocybele nasuta
	Phyllograptus densus			Didymograptus protobifidus		I Presbynileus ibexensis
	Didymograptus balticus	BILLINGENIAN	Prioniodus elegans	Tetragraptus fruticosus		H Trigonocerca typica
				Tetragraptus approximatus		G2 Protopliomerella contracta
	Tetragraptus approximatus	HUNNE-BERGIAN	Paroistodus proteus			G1 Hintzeia celsaora
						F Rossaspis superciliosa
	Poorly graptolitiferous, zonal classification not yet well established			Clonograptus		E Tesselecauda

FIG. 3. Correlation of the Arenigian Series and equivalent strata (Jaanusson, n).

region in the west to the Moscow basin in the east. Information on the Arenigian fauna from the Ural Mountains is scanty but the brachiopods from the southern Ural Mountains include Balto-Scandian elements together with endemic forms (ANDREEVA, 1972). Characteristic elements of the Balto-Scandian fauna include a number of asaphid genera, and the illaenid *Panderia*. *Agerina* is a possible bathyurid, but hystricurids, calymenaceans, and trinucleids are lacking. The porambonitacean brachiopods that frequently dominate Arenigian faunas elsewhere are represented only by porambonitids and angusticardiniids; characteristic North American-Siberian taxa, such as polytoechiids and finkelnburgiids, are absent. The brachiopod fauna is in places dominated by a variety of gonambonitids, *Productorthis*, and *Paurorthis*.

The Arenigian "southern" fauna (Calymenid-Trinucleid Province, WHITTINGTON, 1963; Southern Region, WHITTINGTON, 1966; Sino-European Biogeographic Region, CHUGAEVA, 1968; Paleotethyan Region, CHUGAEVA, 1973) is less homogeneous than the Tremadocian "southern" fauna. What is here provisionally termed as the Mediterranean fauna (Tethyan fauna, DEAN, 1967; *Selenopeltis* fauna, WHITTINGTON, 1966; WHITTINGTON & HUGHES, 1972; Mediterranean Province, HAVLÍČEK, 1974) extended from Wales in the north to Morocco in the south and from the Iberian Peninsula in the west over Bohemia possibly as far as the Pamir in the east. The fauna is characterized by a variety of calymenaceans, trinucleids, taihungshaniids, and cyclopygids. Even within the region, the fauna is not particularly homogeneous (DEAN, 1967; HAVLÍČEK, 1976). The known Arenigian fauna from central and southern China is small and difficult to evaluate biogeographically but occurrences of *Taihungshania* (Fig. 4), *Hanchunglithus*, and *Neseuretus* links it with the Mediterranean fauna.

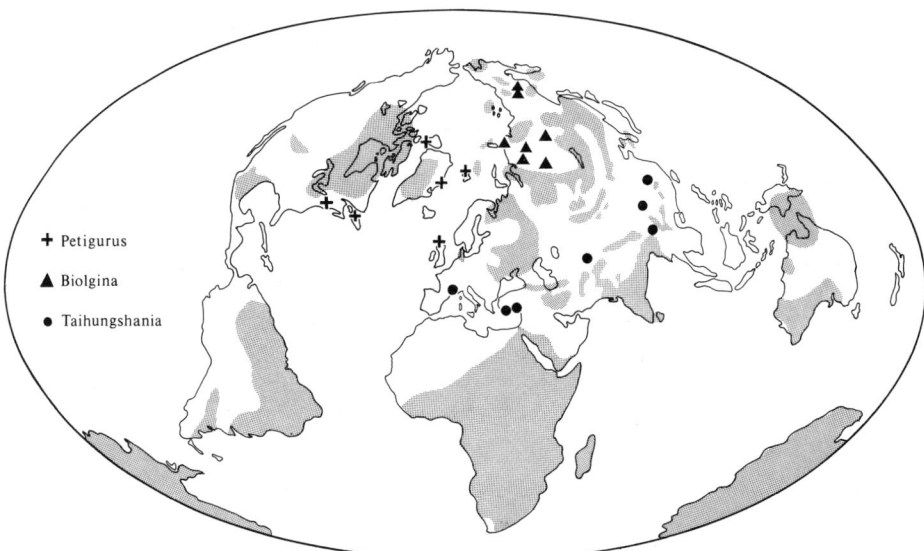

FIG. 4. Distribution of selected Arenigian (post-Tremadocian Canadian) trilobite genera. Occurrence of the bathyurid *Petigurus* is restricted to the North American fauna, the bathyurid *Biolgina* characterizes the Tungusian fauna, and the taihungshaniid *Taihungshania* is one of the southern elements with a trans-Eurasian distribution (Jaanusson, n).

These are examples of distribution of various taxa from Europe to China or, in terms of present-day geography, of the Ordovician "transeurasiatic migration route" for which there are numerous examples from different Ordovician epochs (KOBAYASHI, 1971; BURRETT, 1973).

Within the Mediterranean fauna WILLIAMS (1973) distinguished Celtic (Anglesey and southeastern Ireland) and Anglo-Welsh (Shropshire and Montagne Noire) Arenigian provinces, based on cluster analysis of brachiopod genera. The evaluation of this classification is difficult because the taxonomic diversity is mostly low and only four assemblages are available for analysis.

The Arenigian trilobite fauna of South America includes endemic elements (the asaphid subfamily Thysanopyginae). WHITTINGTON and HUGHES (1972) included it together with Australian faunas in a separate *Asaphopsis* Province. The Australian faunas of this age have not been described in detail and may turn out to include a stronger endemic component than known at present. BURRETT (1973) doubted that close affinities exist between the Australian and South American Arenigian and Llanvirnian faunas.

Information on Arenigian shelly faunas in central Asia, Kazakhstan, and southwestern Siberia is at present very limited.

LLANVIRNIAN (INCLUDING WHITEROCK AND KUNDA STAGES)

At the end of the Canadian Epoch much of the North American craton and Appalachian miogeosyncline emerged, causing a break in deposition. Carbonate sedimentation was more continuous along the western margin of the craton from southern Nevada to Yukon, in the south (Oklahoma), and in the northeast (western Newfoundland). There the gap is filled by beds with the so-called Whiterockian fauna (COOPER, 1956) (Fig. 5). During the Whiterockian Age a number of trilobites of possible "southern" and Balto-Scandian origin entered North America (*Cybelurus, Raymondaspis, Nileus*). Other new forms include *Illaenus, Endymionia*, and *Ectenonotus* among trilobites, and *Orthidiella, Rhysostrophia*, and earliest triplesiaceans among articulate brachiopods. Except for the possible upper

Canadian *Polydesmia* from Manchuria, the Whiterockian beds contain the earliest actinoceroid cephalopods and the latest cyrtoconic ellesmerocerids. Close equivalents to the Whiterockian faunas are known in Spitzbergen (FORTEY & BRUTON, 1973), western Ireland (WILLIAMS, 1972), and northeastern Siberia (CHUGAEVA, 1973; ORADOVSKAYA, 1973). On the Siberian platform this fauna has not been recognized and SIDYARENKO and KANYGIN (1965) have suggested that there is a break in the sequence, roughly corresponding to the beds with the Whiterockian fauna in northeastern Siberia. If this is true, then the Siberian platform behaved during this time much the same as did the North American craton.

Kazakhstan has a southern Tremadocian fauna, but in beds roughly comparable to the Llanvirnian the general affinities of the fauna are with North America (NIKITIN, 1972; WHITTINGTON & HUGHES, 1972). The fauna has a strong endemic component however, that may increase in importance when more material has been studied. The same applies also to Gornyi Altay in southwestern Siberia.

In the Balto-Scandian region the Kunda Stage is the equivalent of the lower, main part of the Whiterockian. There the fauna retained its provincial character with numerous endemic taxa (*Asaphus, Megistaspidella, Cyrtometopus,* and others among trilobites, gonambonitids and *Lycophoria* [Fig. 6] among brachiopods). In the west, the trilobite fauna of the Otta Conglomerate in the Caledonidian eugeosynclinal belt of southcentral Norway is decidely of the Balto-Scandian type (e.g., *Neoasaphus, Megistaspidella*) and the postulated strong North American affinities of its gastropod fauna (YOCHELSON, 1963) are questionable. On the other hand, the Trondheim region of the Scandinavian Caledonides has a fauna related to the North American Whiterockian fauna (NEUMAN & BRUTON, 1974). WILSON (1966) suggested that this part of Scandinavia was originally part of the North American lithospheric plate (see also WHITTINGTON & HUGHES, 1972).

Based on cluster analysis of brachiopods, WILLIAMS (1973) concluded that the Llanvirnian Baltic Province extended westward to Anglesey and Maine. Brachiopod assemblages on Anglesey (BATES, 1968) and in the Magog belt of the northern Appalachians from Maine to Newfoundland (NEUMAN, 1970) are unusual in that taxa otherwise characteristic for different faunas and ages are associated in the same beds. In beds of roughly Whiterockian age Balto-Scandian taxa (e.g., gonambonitids, *Ahtiella*) occur together with polytoechiids (elsewhere in North America not known above Canadian), taxa that characterize the middle Ordovician Scoto-Appalachian fauna (e.g., *Christiania, Eoplectodonta*), and genera that have not yet been found elsewhere (NEUMAN, 1972). Biogeographic classification of these exotic assemblages is difficult at present. NEUMAN (1972) suggested that these faunas originally inhabited the ocean floor around dominantly volcanic islands and that this environment was the site of evolution of many stocks that later spread to continental platforms. On Anglesey the Arenigian and Llanvirnian brachiopod assemblages are associated with a trilobite fauna of Mediterranean type (WHITTINGTON & HUGHES, 1972), which further complicates the matter.

The distribution of the Llanvirnian Mediterranean (cf. that of *Placoparia,* Fig. 6) and other southern faunas agrees in the main with those of the Arenigian.

MIDDLE ORDOVICIAN

For the purpose of this contribution the middle part of the Ordovician, comprising the interval from about the upper *Didymograptus murchisoni* Zone to the base of the *Pleurograptus linearis* Zone, is informally termed the middle Ordovician. It should be emphasized that this term is here not used as designating a formal series or epoch and that at present there does not exist any international agreement as to the definition of a middle Ordovician series.

During the middle Ordovician, sea invaded successively the southern and central Appalachian miogeosyncline and extensive cratonic areas of North America. This was associated with a differentiation of the fauna in which the element in the main miogeosynclinal belt came to differ in several respects from that of the cratonic region. Roughly comparable differentiation existed

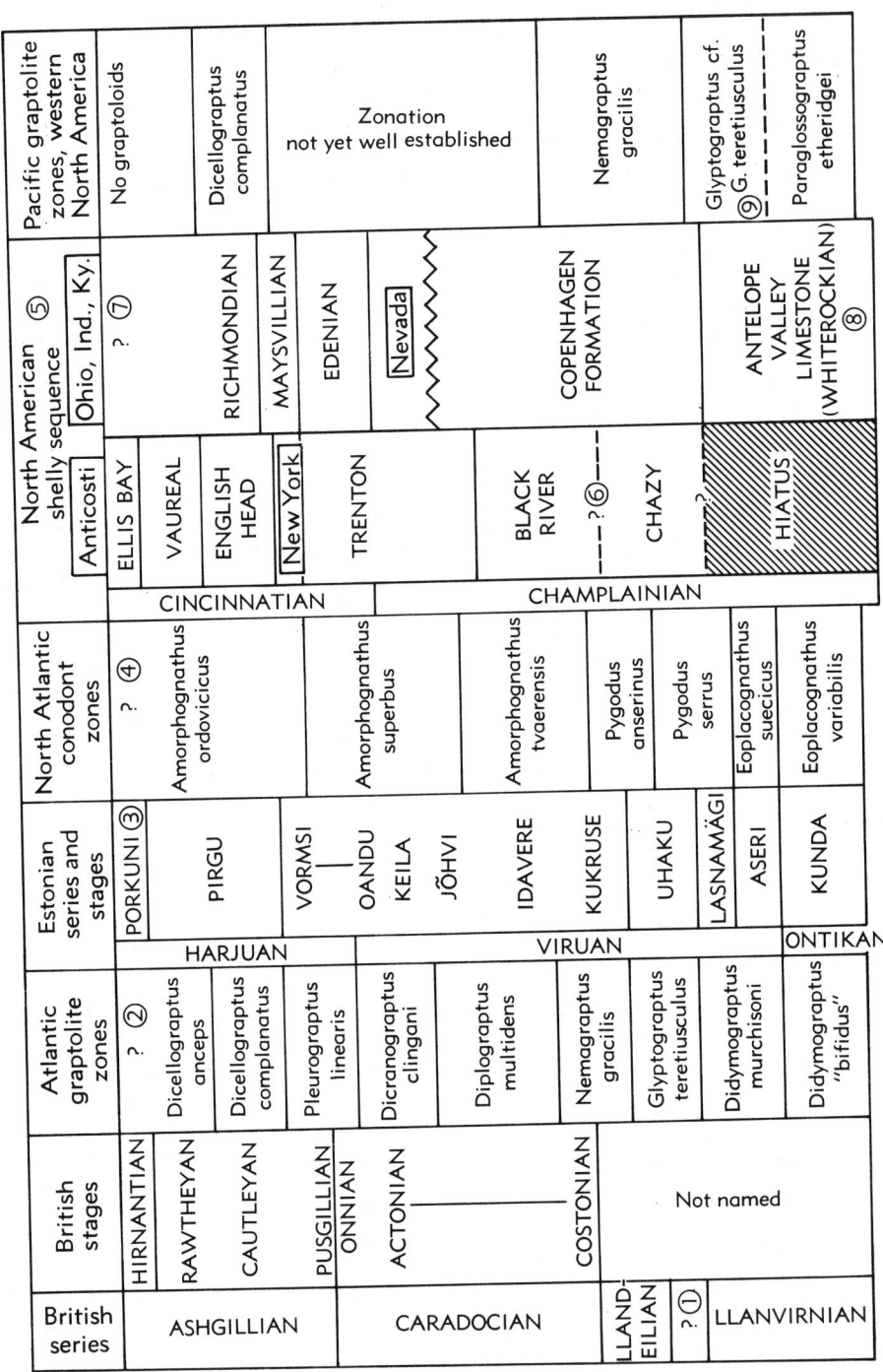

Fig. 5. Correlation of the middle and Upper Ordovician (Jaanusson, n). [Numbers in circles refer to notes at foot of facing page.]

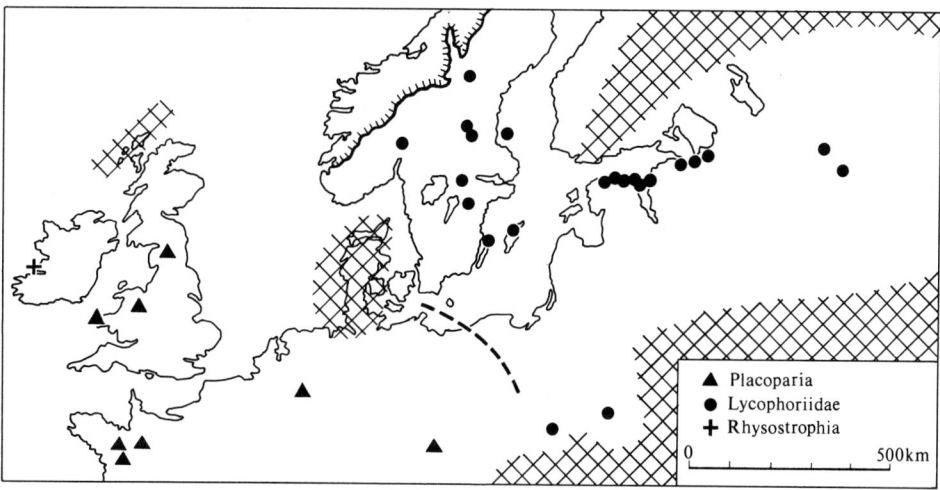

FIG. 6. Distribution of selected Llanvirnian taxa in northern Europe. Occurrence of the brachiopod family Lycophoriidae is restricted to the Balto-Scandian fauna and the trilobite subfamily Placopariinae is a member of the Mediterranean fauna (Llanvirnian representatives of *Placoparia* are known also from Spain, Portugal, Morocco, and Kazakhstan; finds in southwestern Turkey came from a somewhat uncertain horizon). The porambonitacean brachiopod *Rhysostrophia* in Ireland indicates Whiterockian affinities (JAANUSSON, n). [Dashed line indicates the western boundary of the Russian platform (so-called Tornquist's Line); cross-hatched pattern represents probable land areas.]

also in earlier Ordovician deposits along the Cordilleran geosyncline (Ross, 1975) and in parts of the northern Appalachians.

In the southern Appalachians many genera do not reach westward beyond the Helena-Saltville thrust (McLAUGHLIN, 1973), which also is an important biogeographic boundary in the conodont faunas (BERGSTRÖM, 1971). Such genera are *Christiania, Bimuria, Bilobia, Ptychoglyptus, Glyptambonites, Titanambonites, Cyphomena, Productorthis, Taphrorthis, Laticrura,* and *Kul-*

FIG. 5. *(Continued from facing page.)*

1. The lower boundary of the type Llandeilian is within the *Glyptograptus teretiusculus* Zone as distinguished in Sweden (BERGSTRÖM, 1971). The upper boundary of Llanvirnian is defined as that of the Zone of *Didymograptus murchisoni*. Thus, a portion of the British sequence, corresponding to the basal *Glyptograptus teretiusculus* Zone, is at present not included in the British serial classification.
2. In Europe no distinctive graptolite fauna has been found in beds of Hirnantian age. In Kazakhstan, beds with a comparable macrofauna have yielded *Glyptograptus persculptus*, the index fossil of the lowermost Silurian graptolite zone; however, whether this species has a longer range than previously believed or at least the upper part of the Hirnantian belongs to the *Glyptograptus persculptus* Assemblage-zone is at present not known.
3. Equivalent beds in the central Balto-Scandian confacies belt contain a *Dalmanitina-Hirnantia* fauna and are known as *Dalmanitina* beds or Tommarpian Stage (a junior synonym of Hirnantian Stage).
4. In beds of undoubted Hirnantian age no representative conodont fauna has yet been described in detail.
5. Increasing bulk of evidence indicates that COOPER's (1956) stages Ashbyan, Porterfieldian, and Wildernessian, each defined in a separate belt in the southern Appalachians, are largely contemporaneous (BERGSTRÖM, 1971). They reflect spatial faunal differentiation rather than faunal changes in time.
6. Exact position of the boundary between Chazyan and Blackriveran stages with respect to North Atlantic conodont zones is not known at present.
7. No undoubted Hirnantian equivalents at present can be distinguished in Ohio, Indiana, and Kentucky. Whether this depends upon a break in the sequence, poorly fossiliferous condition of the uppermost Richmondian beds, or biogeographic differentiation is not known. Beds of Hirnantian age are developed in western Illinois and eastern Missouri (Edgewood Limestone and its equivalents).
8. According to the evidence from conodonts, the top of the Antelope Valley Formation is comparable to the lower Lasnamägian of Balto-Scandia (BERGSTRÖM, ETHINGTON, & JAANUSSON, 1973); however, it is not clear whether the Whiterockian Stage of COOPER (1956) should be considered to coincide with the extent of the Antelope Valley Limestone or defined in the Ikes Canyon section of the Toquima Range based on the succession of faunal zones listed by COOPER. In the latter case, the upper boundary of the Whiterockian is probably below the base of the Lasnamägian Stage.
9. Graptolite faunas strongly indicate that the uppermost Darriwilian Zone of *Glyptograptus teretiusculus* of Australia is equivalent to the Zone of *Didymograptus murchisoni* (JAANUSSON, 1960). The correlation of the North American Zone of *Glyptograptus* cf. *G. teretiusculus* is more difficult to determine because the known fauna comprises few species; however, it may be largely of a comparable age. The Scandinavian equivalent to the Zone of *Glyptograptus teretiusculus* is characterized by the appearance of *Dicellograptus, Dicranograptus,* and *Nemagraptus.* The equivalent beds of North America probably have been included in the Zone of *Nemagraptus gracilis.*

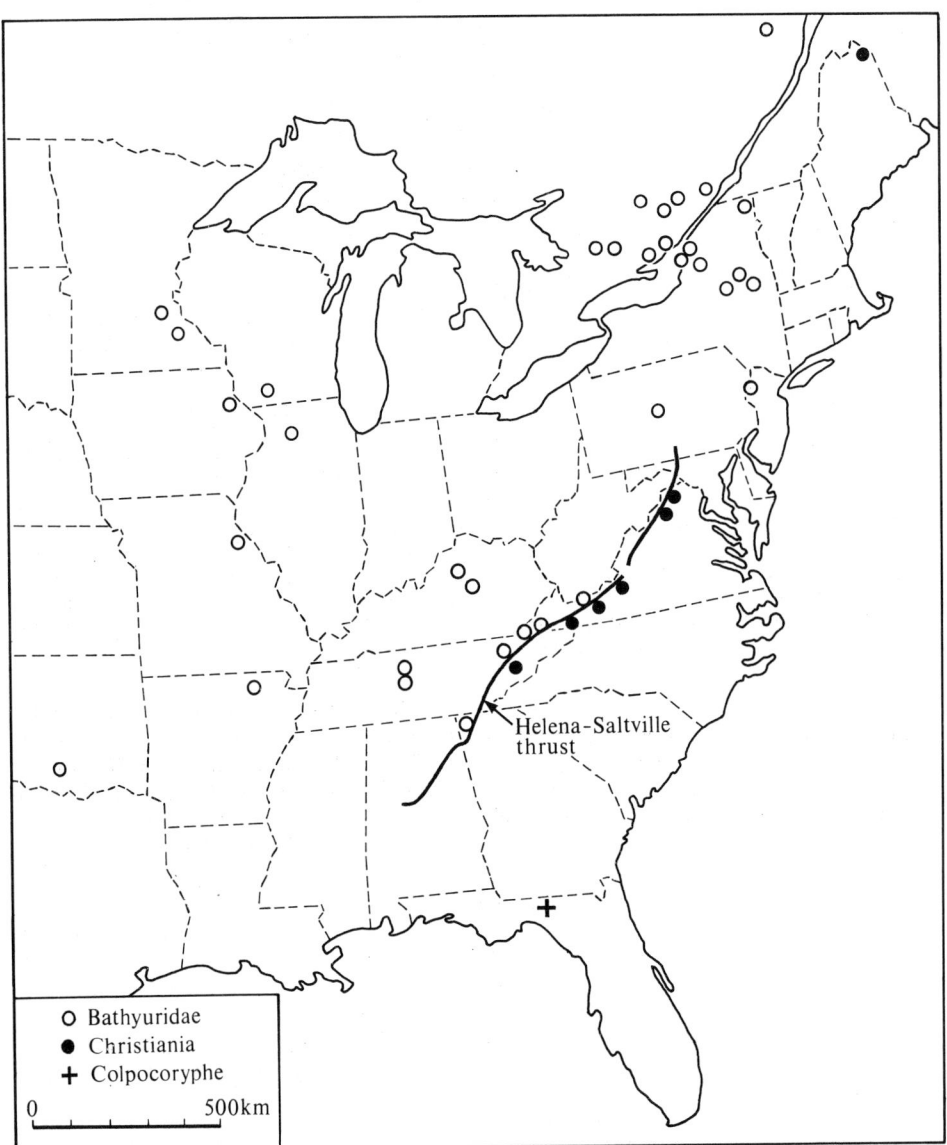

Fig. 7. Distribution of bathyurid trilobites and the strophomenacean brachiopod family Christianiidae in eastern North America in beds equivalent to the Zone of *Nemagraptus gracilis* and the lower part of the Zone of *Diplograptus multidens*. The bathyurids characterize the North American Midcontinent fauna and the christianiids the Scoto-Appalachian fauna. The subsurface occurrence of the calymenacean *Colpocoryphe* in northern Florida is also indicated although it probably comes from somewhat lower beds (Jaanusson, n).

Iervo. Other taxa reach farther to the west, but not beyond the St. Paul-St. Clair thrusts (e.g., *Palaeostrophomena, Isophragma, Leptellina, Cyrtonotella,* styginid trilobites). On the other hand, several taxa that are widely distributed on the midcontinent have not been found east of the Helena-Saltville thrust. Such taxa are bathyurid trilobites (Fig. 7), the articulate brachiopods *Strophomena* and *Ancistrorhynchia,* and *Gonioceras*

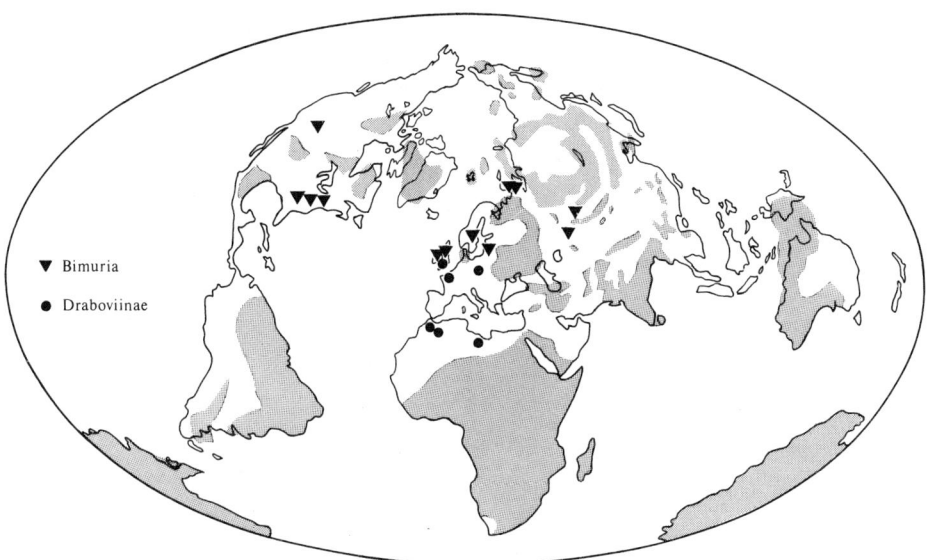

Fig. 8. Distribution of the plectambonitacean brachiopod family Bimuriidae (indicative of the Scoto-Appalachian and related faunas) and the enteletacean brachiopod subfamily Draboviinae (a member of the Mediterranean fauna) in beds equivalent to the Zone of *Nemagraptus gracilis* and the lower part of the Zone of *Diplograptus multidens*. In the Balto-Scandian region *Bimuria* is known only from the central confacies belt (Jaanusson, n).

among cephalopods.

A fauna very similar to the Appalachian fauna east of the Helena-Saltville thrust is known from the Girvan district of southern Scotland (WILLIAMS, 1962, 1969; TRIPP, 1962, 1965, 1967), and the term Scoto-Appalachian fauna (WHITTINGTON & WILLIAMS, 1955) can be used as a designation of middle Ordovician faunas of similar type elsewhere (JAANUSSON, 1973a; not in the wide sense applied to this term by WILLIAMS, 1973). The middle Ordovician fauna of Scoto-Appalachian type, better defined in brachiopod than trilobite assemblages, has a wide distribution. In North America a fauna of this type has an amphicratonic distribution in that several distinctive genera are known also from the western side of the craton (for example, in the Copenhagen Formation in Nevada). A related fauna occurs also in the Novaya Zemlya-Pay Khoy region (BONDAREV, 1968) and it can be followed as far as to Gornyi Altay in southwestern Siberia (LEVITSKIY, 1963; cf. distribution of *Bimuria*, Fig. 8). Also in parts of Kazakhstan the brachiopod fauna has Scoto-Appalachian affinities.

Cratonic North America, the Canadian Arctic Archipelago, and Greenland were inhabited by a fauna (North American Midcontinent fauna) with mostly smaller taxonomic diversity than in the Scoto-Appalachian area. Middle Ordovician bathyurid trilobites (Fig. 9) are not known elsewhere in the world.

The middle Ordovician fauna of the Siberian platform and the southern "structural-facial zone" of the Taymyr Peninsula is similar to the North American Midcontinent fauna, but has monorakid trilobites as a distinctive element (Fig. 9) and lacks enteletacean brachiopods. For convenience this fauna can be termed the Tungusian fauna or, combined with the fauna in the North American continental interior, the North American Midcontinent-Tungusian fauna (JAANUSSON, 1973a). Monorakids are known also from northeastern Siberia, Chukot Peninsula (ORADOVSKAYA, 1970), and Novosibirskoe Ostrova. All these regions have yielded middle Ordovician faunas similar to that of the Siberian platform. Monorakids have also been reported from the Sayan Mountains and Tuva, but the known

FIG. 9. Distribution of selected trilobite taxa in beds equivalent to the Zone of *Nemagraptus gracilis* and the lower part of the Zone of *Diplograptus multidens*. Bathyurids characterize the North American Midcontinent fauna and monorakids the Tungusian fauna. In this interval *Chasmops* is restricted to the Balto-Scandian region, but attains a somewhat wider distribution in the upper part of the middle Ordovician. Cyclopygids are a southern element with a trans-Eurasian distribution and *Pliomerina* characterizes the southeastern Asian-Australian fauna (Jaanusson, n).

middle Ordovician faunas from these areas are far too small for biogeographic conclusions. A possibly related genus *(Isalaux)* occurs on the Siberian platform and in Colorado.

Based mainly on cephalopods, the North American affinities of the fauna in northern China and northern Korea have repeatedly been pointed out; however, very little information on trilobites and brachiopods is available from these regions.

During the middle Ordovician the provincial character of the Balto-Scandian fauna became progressively less pronounced. The region still possessed a number of taxa that are not known elsewhere (*Asaphus* and six additional asaphid genera; some other genera, such as *Chasmops,* Fig. 9, and *Estoniops,* were endemic for Balto-Scandia during the early part of the middle Ordovician, but then spread to the southern part of British Isles). Within the region, faunal differentiation increased considerably (JAANUSSON, 1976). In the central Balto-Scandian confacies belt[1] a successively increasing Scoto-Appalachian influence is apparent in the brachiopod faunas, whereas the North Estonian belt retained some of the provincial character (gonambonitid brachiopods such as *Estlandia, Clitambonites,* apatorthids). The Balto-Scandian fauna can be followed to the central Ural Mountains (ANZYGIN in VARGANOV, 1973) in the east and to Moldavia in the south.

During late middle Ordovician time a notable exchange took place between some faunas: several Mediterranean trilobite genera (WHITTINGTON, 1966) and *Platystrophia* entered the North American midcontinent; Balto-Scandian and Scoto-Appalachian genera appeared in England and Wales; and close to the end of the epoch, the Balto-Scandian *Asaphus (Neoasaphus)* reached as far as southwestern China. A remarkable invasion of new faunal elements took place at the end of the epoch in northwestern Estonia, as well as in the Mjøsa and

---

[1] A *confacies belt* differs from adjacent contemporaneous belts in lithology and fauna, and although lithology, as well as fauna, changes with time within the belt, the geographic position and individuality of the belt remains roughly the same during appreciable time. For further discussion and examples, see JAANUSSON, 1976.

Langesund-Skien districts of the Oslo region, Norway. The invading taxa include leperditiacean ostracodes (Mjøsa district), telotremate brachiopods *(Rhynchotrema, Rostricellula, Zygospira)*, *Bumastoides* and *Encrinuroides* among trilobites, and the earliest stromatoporoids for the region. These elements are mainly North American and the associated conodont fauna is of North American Midcontinent type (SWEET & BERGSTRÖM, 1974). The new fauna is mostly associated with bahamitic carbonate sediments, previously unknown from the Balto-Scandian region (JAANUSSON, 1973b). This fauna is poorly represented in contemporaneous beds elsewhere in the Balto-Scandian region and most of it soon disappeared. The invasion and associated sediments suggest a shift to subtropical or tropical temperatures in shallow-water areas of northern Balto-Scandia during a relatively short time.

The middle Ordovician trilobite fauna of Australia, Southeast Asia, and Kazakhstan has general northern ("remopleuridid," WHITTINGTON & HUGHES, 1972) affinities, but it includes endemic elements (see distribution of *Pliomerina*, Fig. 9) and has been distinguished as the *Heptabronteus-Pliomerina* Province (WEBBY, 1974; *Encrinurella* fauna, WHITTINGTON, 1966). Further endemic genera are the blind cheirurid *Prosopiscus*, the raphiophorid *Ampyxinella*, and others (see also LU, 1975). Brachiopods from these regions are still very poorly described.

Lower Ordovician trilobite faunas, up to the base of the *Nemagraptus gracilis* Zone of western England, Wales, and southern Ireland, have strong Mediterranean affinities (WHITTINGTON & HUGHES, 1972). The middle and upper middle Ordovician faunas, on the other hand, are linked to the Scoto-Appalachian fauna (WHITTINGTON & HUGHES, 1972), although there still is a considerable Mediterranean component. Otherwise, the Mediterranean fauna *(Selenopeltis* fauna, WHITTINGTON, 1966, WHITTINGTON & HUGHES, 1972; Anglo-French and Bohemian Provinces, WILLIAMS, 1973; Mediterranean Province, HAVLÍČEK, 1976) occupies the same area as earlier, from Morocco in the west over Bohemia to Turkey in the east. The fauna is characterized by cyclopygid (Fig. 9) and homalonotid trilobites, several endemic dalmanitids, and a variety of enteletacean brachiopods such as *Svobodaina* (SPJELDNAES, 1967) and Draboviinae (Fig. 8). The distribution of several taxa follows the "trans-Eurasian migration route" from northwestern Africa and southern Europe to central and southern China. Examples are cyclopygids (Fig. 9) among trilobites, porambonitids among brachiopods, and *Aristocystites* and *Sinocystis* among cystoids.

The few middle Ordovician trilobites known from South America have uncertain biogeographic affinity.

UPPER ORDOVICIAN (EXCLUDING THE HIRNANTIAN)

Over wide areas, Upper Ordovician (from the zone of *Pleurograptus linearis,* inclusively, to the top of the system) deposits are either missing (northern China, northern Korea, the Tarim platform) or with a break at the top (e.g., the Siberian platform, western continental Europe, Greenland, and parts of North America). In South America undoubted Upper Ordovician is known only from the Precordillera of western Argentina.

Most authors have concluded that faunal provinciality decreased progressively during the middle and Late Ordovician (WHITTINGTON & HUGHES, 1972; WILLIAMS, 1973) so that during the Ashgillian a cosmopolitan fauna began to emerge (WILLIAMS, 1973), culminating with the latest Ordovician *Dalmanitina (Mucronaspis)-Hirnantia* fauna that is regarded by some as worldwide. The general trend during this time toward reduced provinciality is fairly evident but there still is a considerable biogeographic differentiation up to the end of the period.

Based on statistical analysis of 14 lists of genera, WHITTINGTON and HUGHES (1972) suggested that in Ashgillian trilobite faunas two provinces can be distinguished, one (*Selenopeltis* Province) restricted to Morocco and Bohemia and the other (Remopleuridid Province) comprising the rest of the samples analyzed (North America, northern Europe, Kazakhstan, and China). Their interpretation of the statistical results may be questioned. The only sample from the

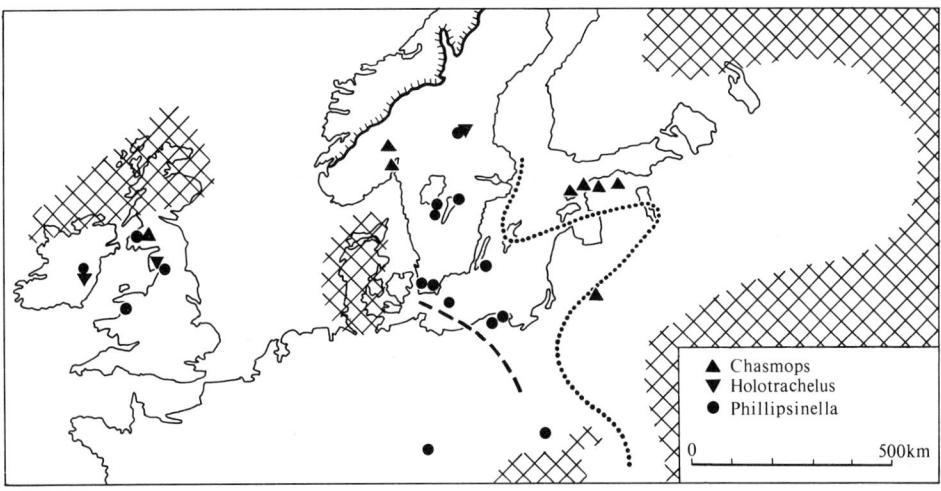

FIG. 10. Distribution of selected lower and middle Ashgillian trilobites in northern Europe. *Phillipsinella* characterizes faunas with Mediterranean affinities, whereas *Holotrachelus* is mostly confined to limestones with a Hiberno-Salairian fauna. The main occurrence of *Chasmops* in northern Europe is in limestones of the North Estonian and Lithuanian belts and the Oslo region. Ashgillian *Phillipsinella* is known also from Carnic Alps, Kazakhstan, and Uzbekistan. [Western boundary of the Russian platform indicated by dashed line and eastern boundary of the central Balto-Scandian confacies indicated by dotted line (Jaanusson, n).]

North American Midcontinent (Iowa) was found to be linked at low dissimilarity indices over Anticosti, and several other samples, to Poland and several other North European samples; however, none of the 16 genera in the Maquoketa Shale of Iowa (WALTER, 1926; generic names updated herein) is known from the lower and middle Ashgillian of Poland (about 37 genera; KIELAN, 1959), close to the other end of the chain of samples linked at low dissimilarity indices. In this case two completely different faunas are included in the same province because they have come to be linked over a chain of samples that happen to include transitional ("mingled") faunas.

The trilobite fauna of the Maquoketa Shale contains a distinctive assemblage of genera *(Anataphrus, Ectenaspis, Thaleops, Bumastoides, Ceraurus, Remipyga,* etc.*)* that, in the Upper Ordovician, is not known outside North America except for occasional occurrences in the Taymyr Peninsula and northeastern Siberia. Most of the genera continue from the middle Ordovician within the same regions. Indeed, in the carbonate deposits of North America and the Arctic Islands, the North American Midcontinent fauna continues into the Upper Ordovician with few modifications. Distinctive taxa include, in addition to several trilobite genera (for *Isotelus,* see Fig. 11), *Zygospira* among brachiopods (JAANUSSON, 1973a, fig. 3; known also from the Altay region and other places), leperditiacean ostracodes (known also from northeastern Siberia), and aulacerid stromatoporoids (also on Siberian platform). Rhynchotrematid brachiopods are common in many areas.

On the Siberian platform and Taymyr Peninsula a related fauna ranges from middle to Upper Ordovician without great change. Monorakid trilobites continue to be characteristic. They occur also in northeastern Siberia (see Fig. 11) but the fauna there includes also *Eospirigerina, Ptychoglyptus,* and large pentameraceans *(Tcherskidium;* NIKOLAEV et al., 1974).

In northern Europe, where Upper Ordovician rocks are widespread, three spatially different lower and middle Ashgillian faunas can be distinguished (Fig. 10):

1) Sequences that consist predominantly of mudstone, have a high taxonomic diversity of trilobites, a limited brachiopod fauna, and few other fossils. The important, in

Fig. 11. Distribution of selected trilobite taxa in beds equivalent to the lower and middle Ashgillian. The asaphid *Isotelus* is widely distributed in the North American Midcontinent fauna and monorakids are confined to the Tungusian fauna. The encrinurid subfamily Dindymeninae is restricted to the Mediterranean fauna and hammatocnemids extend from central Poland to central China. Ashgillian *Isotelus* has been recorded also from some additional areas but figured specimens indicate that other genera are probably involved. Middle Ordovician dindymenines are known from Bohemia, Kazakhstan, and western Pamir (Jaanusson, n).

places main, component of the fauna consists of taxa that in the middle Ordovician occurred in the Mediterranean (Bohemian) region (KIELAN, 1959), such as Dindymeninae, Cyclopygidae, Ectillaeninae, Dionidae, and *Dalmanitina (Mucronaspis)*, and continued to be characteristic for the Mediterranean region also during the Late Ordovician. Trinucleids were common. That during this time Bohemia and central Poland should belong to different provinces (WHITTINGTON & HUGHES, 1972) is questionable, and may depend upon which index values are selected in the statistical analysis. The contemporaneous brachiopod fauna in Scania, Sweden, is of Bohemian type (SHEEHAN, 1973). Thus, in Europe a fauna of Mediterranean type spread northward to the British Isles (as far as western Ireland and Scotland) and to the central Balto-Scandian confacies belt (cf. the distribution of *Phillipsinella*, Fig. 10, and of *Dindymene*, Fig. 11). Similar faunas are known in the Percé district, Quebec (*Stenopareia* fauna; LESPÉRANCE, 1968), and particularly in the Magog belt of Maine (NEUMAN, 1970).

2) Limestones that are associated with mudstone have mostly a patchy distribution and a varied fauna. Both trilobites and brachiopods have a high taxonomic diversity. The trilobite fauna has little in common with that of contemporaneous mudstone facies close-by. In fact, in statistical analysis the magnitude of the difference may be that between different faunal provinces (WHITTINGTON & HUGHES, 1972, with regard to the Boda Limestone in the Siljan district, Sweden). The Boda Limestone (carbonate mounds with stromatactis) lacks all the taxa mentioned as characteristic for the Mediterranean mudstone fauna as well as *Chasmops*. The brachiopod fauna, too, is very different from that of Mediterranean mudstones. It has much in common with the middle Ordovician Scoto-Appalachian fauna (*Christiania, Bimuria, Ptychoglyptus, Dolerorthis, Kullervo*), but includes new elements such as dicoelosiids, *Eospirigerina*, and meristellids. Limestone facies with a similar fauna (see also the distribution

of *Holotrachelus,* Fig. 10) are known in Ireland (Portane and Chair of Kildare limestones), northern England (Keisley Limestone), and Salair Mountains in southwestern Siberia (so-called Weberian Limestone). JAANUSSON (1973a) termed this type of Ordovician fauna the Hiberno-Salairian fauna. In several limestone facies the fauna of Hiberno-Salairian type is mingled with Mediterranean elements and, conversely, Hiberno-Salairian elements are in places found associated with a preponderantly Mediterranean fauna. This may, in part at least, be a consequence of the occurrence of transitional lithologies, reflecting transitional environments; however, a biogeographic gradient may also be involved because limestone with the Hiberno-Salairian fauna, as well as mingled assemblages, occurs only in a belt along the northern boundary of the distribution of the Mediterranean fauna (Fig. 10). Geographic proximity of the different faunas may be a factor contributing to mingling.

3) The North Estonian carbonate confacies belt still retains its provincial character. No Mediterranean element had entered the area. The fauna is rich, particularly in corals, but the trilobite fauna has a low diversity with *Chasmops* as one of the commonest elements (Fig. 10). The brachiopod fauna is varied, in several respects close to the Hiberno-Salairian fauna *(Dicoelosia, Eospirigerina)*, but it lacks the Scoto-Appalachian imprint. Endemic forms include *Equirostra, Ilmarinia,* and *Apatorthis,* all descendents of the middle Ordovician fauna of the belt. The trilobite fauna in the 5a-limestone belt in the Oslo region, Norway, is in many respects similar to that of the North Estonian belt, as is much of the rest of the fauna, but it lacks the North Estonian endemics.

The Balto-Scandian coral fauna in the Upper Ordovician limestone facies has been considered to belong to a separate Baltic province (LELESHUS, 1970; European Province, KALJO et al., 1970) or form together with the central Asiatic and Chinese faunas the Eurasiatic province (KALJO & KLAAMANN, 1973). It is interesting to note that although the Ural Mountains show Balto-Scandian affinities in the trilobite and brachiopod faunas (VARGANOV et al., 1973), its coral faunas are regarded as intermediate between the Arctic (North America, northeastern Siberia, Soviet Arctic) and the Siberian (Siberian platform, southwestern Siberia) provinces (LELESHUS, 1970) or they are included in the Americo-Siberian province (KALJO et al., 1970). Also, the trilobite and brachiopod faunas of the Ural Mountains include, in each Ordovician epoch, some genera that are not known in the East Baltic or Scandinavia, but occur either in North America *(Hypodicranotus)* or the Siberian platform *(Dolgeuloma, Angarella, Xenelasmella, Cyrtophyllum),* or both.

In the southern and central Appalachians the Scoto-Appalachian fauna disappeared close to the end of the middle Ordovician and all Upper Ordovician faunas from the region are of the Midcontinent type; however, an Upper Ordovician brachiopod fauna resembling the Scoto-Appalachian fauna occurs in some areas in the periphery of the continent: east-central Alaska, Klamath Mountains in northern California, and Percé district in Quebec. ROZMAN (1968, 1970) suggested that the Alaskan and northeast Siberian Upper Ordovician brachiopod faunas belong to a separate Kolyma-Alaskan biogeographic belt but the evidence is inconclusive. All these peripheral North American Upper Ordovician brachiopod faunas have close affinities with the Hiberno-Salairian fauna.

The Late Ordovician trilobite fauna of Bukantau (ABDULLAEV, 1972), central Tien Shan (ABDULLAEV in ABDULLAEV & KHALETSKAYA, 1970), parts of Kazakhstan (APOLLONOV, 1974), and southwestern China (Szechuan-Kueichou border; SHENG, 1964) show affinities with the Mediterranean fauna. Several taxa have a "trans-Eurasian" distribution (cyclopygids, *Nankinolithus*). A characteristic early and middle Ashgillian element in Kazakhstan, Uzbekistan, and southwestern China is the family Hammatocnemidae (Fig. 11), endemic in some of the regions also in the late middle Ordovician. It reached central Poland in the middle Ashgillian (KIELAN, 1959), but has not been found in the rest of Europe. From Australia and South America no contemporaneous shelly fauna is known.

In parts of Kazakhstan and Uzbekistan

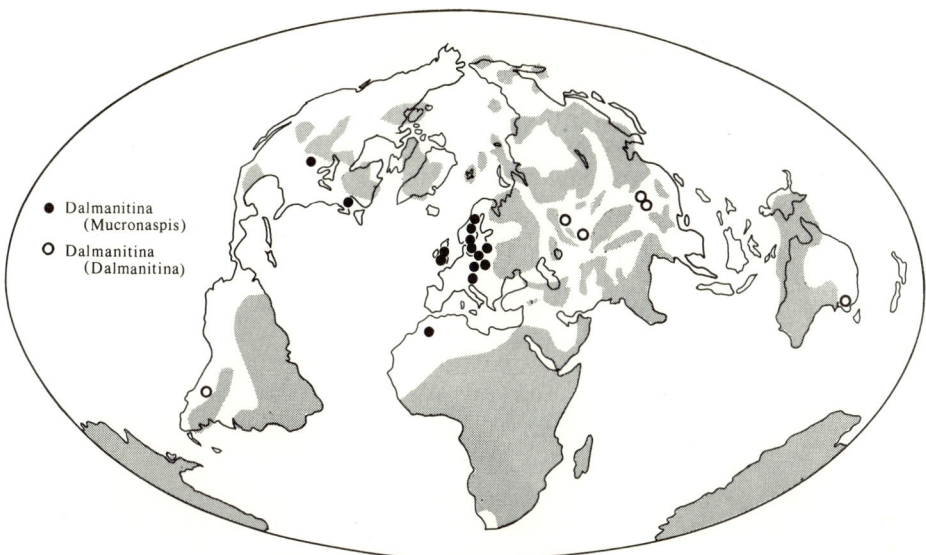

FIG. 12. Distribution of the trilobites *Dalmanitina (Dalmanitina)* and *Dalmanitina (Mucronaspis)* in the Hirnantian (Jaanusson, n).

the fauna in middle Ashgillian limestones shows some Hiberno-Salairian affinities (e.g., *Holotrachelus, Eospirigerina*), but the brachiopod fauna has not yet been described in detail. The coral faunas seem to indicate that Kazakhstan, Soviet Central Asia, and parts of China formed during the Late Ordovician either a separate province (the central Asian Province; KALJO et al., 1970; LELESHUS, 1970) or a part of the Euroasiatic province (KALJO & KLAAMANN, 1973).

UPPERMOST UPPER ORDOVICIAN (HIRNANTIAN)

The distribution of the so-called *Dalmanitina-Hirnantia* fauna in the uppermost Ordovician has recently been the subject of much discussion in connection with the Ordovician glaciation. Distribution of the fauna coincides largely with that of the earlier mudstone fauna of Mediterranean type, from Percé in Quebec over British Isles to the Central Balto-Scandian confacies belt in the north and over central Poland and Bohemia to Morocco in the south (see *Dalmanitina (Mucronaspis)*; Fig. 12). The trilobite assemblage has been claimed to extend eastward to Kazakhstan (APOLLONOV, 1974), southern China (SHENG, 1964), and Australia (WHITTINGTON & HUGHES, 1972, fig. 12). The dominating elements in the European assemblage, however, are *Dalmanitina (Mucronaspis) mucronata* and *Brongniartella* whereas in Kazakhstan and China they are replaced by *Dalmanitina (Dalmanitina)* and *Platycoryphe*, and it is *Dalmanitina (Dalmanitina)* that reached Australia (CAMPBELL, 1973) and the Precordillera of western Argentina (BALDIS & BLASCO, 1975). The *Hirnantia* assemblage of brachiopods is characterized by *Hirnantia sagittifera, Kinnella, Eostropheodonta, Plectothyrella,* and some other genera. Evidence of this fauna east of Europe, or possibly Kazakhstan, is at present tenuous. The appearance of the *Dalmanitina-Hirnantia* assemblage does not denote any major change in the fauna because most genera and several species are known in earlier beds (LESPÉRANCE, 1974) with the Mediterranean fauna. In many areas the taxonomic diversity of trilobites decreased considerably (LESPÉRANCE, 1974).

In limestone facies with a Hiberno-Salairian and related fauna, beds equivalent to the Hirnantian, such as the upper part of the Boda carbonate mounds in Sweden and the 5b-calcareous sandstone of the Oslo region, Norway, do not differ faunally very much from the underlying beds and have almost nothing in common with the *Dalmanitina-Hirnantia* assemblage. New elements include large pentameracean brachiopods *(Holorhynchus, Proconchidium)*. Uppermost Ordovician beds characterized by these pentamerids have a wide distribution, from the eugeosynclinal Caledonian deposits in Västerbotten, northern Sweden, and western and parts of the central Balto-Scandian belts over Ural Mountains to southern Tien Shan (NIKIFOROVA & SAPELNIKOV, 1973) and Kazakhstan. In Kazakhstan the uppermost Upper Ordovician limestone facies have yielded at least seven different pentameracean genera (SAPELNIKOV & RUKAVISHNIKOVA, 1975), but it is not always clear whether they all have come from beds equivalent to the Hirnantian. A comparable uppermost Ordovician limestone with large pentameraceans *(Eoconchidium)* is known also from northeastern Siberia. Many of the uppermost Ordovician limestone beds with pentameraceans are poor in trilobites but have in places a rich coral fauna.

Differences between the fauna of the North Estonian belt, equivalent to the Hirnantian (Porkuni Stage), and the other faunas are the same as in underlying beds. Endemic relicts still occur *(Chasmops, Conolichas, Ilmarinia, Vellamo)*, and the fauna has no Mediterranean elements.

The Edgewood Limestone and its equivalents in southwestern Illinois, eastern Missouri, and southern Oklahoma contain a varied brachiopod fauna showing some affinities to the *Hirnantia* fauna (AMSDEN, 1974). The Ordovician age of this fauna has been in doubt, but in Illinois these beds have yielded also an asaphid trilobite (SAVAGE, 1917; most probably *Anataphrus*), and other indications suggesting a Hirnantian age. The fauna includes *Eospirigerina, Cryptothyrella, Dicoelosia,* and a pentameracean, all belonging to subfamilies or families not known in the Upper Ordovician Midcontinent fauna of North America, but in part widely distributed in earlier beds with the Hiberno-Salairian and related faunas. Associated forms include *Hirnantia* and *Dalmanitina (Mucronaspis)* known in earlier beds of the Mediterranean region. The Ellis Bay Formation of Anticosti (roughly of Hirnantian age) still has a strong Midcontinent component (e.g., "*Brachyaspis,*" *Remipyga, Vellamo, Dinorthis*), but mingled with genera having a wide distribution in Hiberno-Salairian and related faunas *(Eospirigerina,* meristellids), or of unknown origin *(Protatrypa)*.

No contemporaneous fauna is known from the Siberian platform where uppermost Ordovician deposits may be missing. Thus, it seems that the North American Midcontinent and Tungusian faunas virtually ceased to exist as biogeographic units before the end of the Ordovician, although some of their elements (such as *Anataphrus* and *Strophomena* in the Edgewood fauna) still lingered as relicts into the Hirnantian. The biogeographic situation of Anticosti during the Hirnantian may be comparable to that of the North Estonian belt in Balto-Scandia: a marginal platform area with numerous relicts.

PLANKTONIC GRAPTOLITES

Rich and varied planktonic graptolite faunas are confined to dark shale, and these sediments have a sporadic distribution. Thus, our main knowledge of graptolite faunas is restricted to limited geographical regions, mostly outside cratonic platforms.

As in the shelly faunas, the degree of biogeographic differentiation varies with time. In the Tremadocian and lowermost Arenigian, provincialism is relatively weak although not yet well understood. The differentiation is greatest in the middle and upper Arenigian and Llanvirnian. From the *Nemagraptus gracilis* Zone (as defined in Scania) to the top of the Ordovician the graptolite faunas are almost cosmopolitan and, although provincial trends do occur (RIVA, 1969), they are almost exclusively at the species level and at present difficult to define.

Biogeography of the Ordovician graptolites has recently been treated in several papers (BULMAN, 1964, 1971; SKEVINGTON, 1969, 1973, 1974; BOUČEK, 1972; MU, 1974). Generally two Arenigian and Llanvirnian provinces are distinguished.

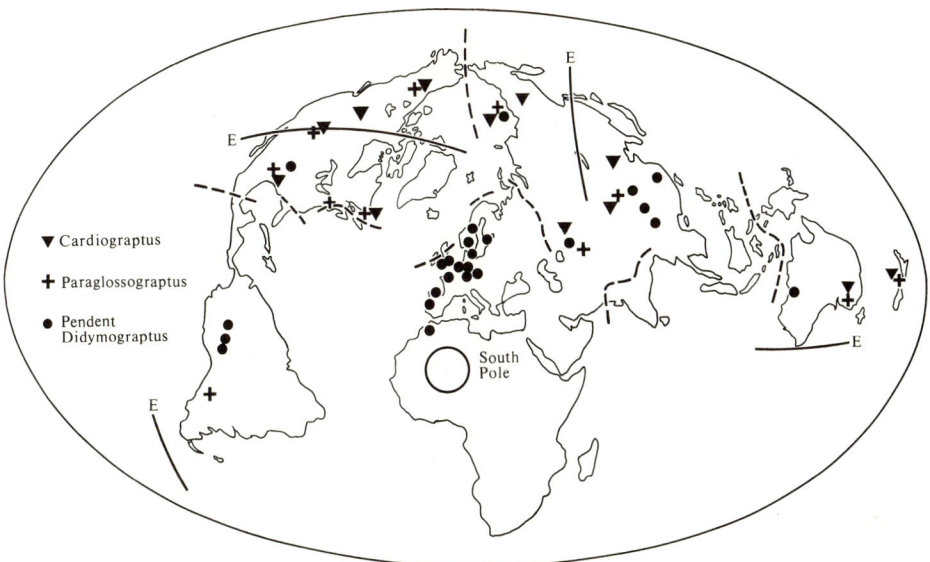

Fig. 13. Distribution of some graptoloids in the Llanvirnian and its equivalents. *Paraglossograptus* and *Cardiograptus* characterize the Pacific fauna and the main distribution of pendent species of *Didymograptus* is in the Atlantic fauna. [Dashed lines indicate probable boundaries between main lithospheric plates (according to several writers the number of separate Ordovician cratonic lithospheric plates is far greater but in these cases the position of plate boundaries is uncertain). Lines marked with E indicate the approximate positions of the Ordovician equators for North America, South America, Siberia, and Australia, based on paleomagnetic data (SMITH et al., 1973). Probable Ordovician position of the South ("Gondwana") Pole indicated by a circle (Jaanusson, n).]

The fauna of the Pacific province is developed in southeastern Australia (Victoria and New South Wales), New Zealand, Texas (Marathon region), Cordilleran North America (from Nevada to Yukon), the Canadian Arctic Archipelago, Appalachians (Georgia, New York, Quebec, Newfoundland), western Ireland (Galway and Mayo), northern Taymyr Peninsula, Bennett Island, northeastern Siberia, northwestern (Chilianshan, Ordos) and north-central China (Honan), Kirgizistan, and Kazakhstan (Chu-Ili Mountains).

The graptolite fauna of the Atlantic (or European; BULMAN, 1971) province occurs in Scandinavia (e.g., Scania, Oslo region in Norway, Västergötland and Jämtland in Sweden), subsurface eastern Latvia and eastern Moscow basin, Wales, England (Lake district and subsurface London platform), southern and eastern Ireland, in limited areas on continental Europe (e.g., Belgium, central Poland, Bohemia), and northern Africa.

The distribution of graptolite faunas in South America is particularly interesting. The Llanvirnian fauna in Peru and Bolivia is decidedly Atlantic whereas farther southward, in the Precordillera of western Argentina, the affinities of the fauna (CUERDA, 1973) are Pacific. The situation is somewhat similar also in China with the Pacific fauna in northern China but an Atlantic fauna in southwestern China (MU, 1974; Szechuan, northern Kueichou, Yunnan). In southeastern China (Anhui, Chekian) the fauna is preponderantly Pacific (MU, 1974) but mingled with Atlantic elements.

The middle and late Arenigian Pacific fauna is characterized by *Goniograptus, Sigmagraptus, Skiagraptus, Apiograptus, Oncograptus,* and *Cardiograptus*. A profusion of various forms of *Isograptus* is characteristic for the upper part of the sequence. The Atlantic fauna differs by lack of Pacific elements rather than by endemics, and by the order of appearance and disappearance of various taxa in relation to other taxa.

Azygograptus and some multiramous dichograptids may be endemic.

The Llanvirnian Pacific fauna contains a number of genera restricted to the province, such as *Paraglossograptus* (Fig. 13), *Pseudobryograptus, Cardiograptus* (Fig. 13), and *Brachiograptus*. The Llanvirnian Atlantic fauna is distinguished by the abundance of pendent *Didymograptus* (Fig. 13) that is rare or missing in the contemporaneous Pacific fauna.

The Balto-Scandian Arenigian and Llanvirnian graptolite assemblages are not quite like those of Wales and northern England. BERRY (1960) suggested that Scandinavia belonged to a separate faunal region, but differences are mostly in quantitative composition of the assemblages and may characterize one of several subprovinces that at present are difficult to define; however, during the time corresponding to the zone of *Glyptograptus teretiusculus* in Scania, provincial features increased in importance in Balto-Scandia and a separate province may have come into existence. The fauna is characterized by *Gymnograptus* associated with a complex of species not known outside Balto-Scandia. *Gymnograptus* has been found also in central China (Szechuan-Kueichou border) in a contemporaneous assemblage poor in species. The correlation of this Balto-Scandian fauna with those of the other areas is otherwise notoriously difficult. It is probable that contemporaneous beds elsewhere form the lower part of the undifferentiated zone of *Nemagraptus gracilis* (JAANUSSON, 1960).

CONODONT FAUNAS

Our knowledge of conodont faunas comes almost exclusively from carbonate rocks. As a consequence, little information on conodonts is available from wide areas without carbonate deposits (in the Ordovician, for example, from much of the region with a Mediterranean shelly fauna). Information on the distribution of Ordovician conodonts is also incomplete because in many areas with suitable rocks systematic work on conodonts has barely started (e.g., southwestern Siberia, Kazakhstan, eastern Asia, Ural Mountains).

Available evidence on the biogeography of Ordovician condonts has been summarized by BERGSTRÖM (1971, 1973), SWEET et al. (1971), BARNES et al. (1973), SWEET & BERGSTRÖM (1974), and BARNES & FÅHRAEUS (1975). The lower Tremadocian conodont fauna has a low taxonomic diversity and seems to have an almost cosmopolitan distribution. A well-defined biogeographic differentiation begins with the Arenigian and can be followed throughout the middle and Upper Ordovician. In most papers two provinces or faunas have been distinguished. The North American Midcontinent province (SWEET et al., 1959) is developed in North American cratonic and inner miogeosynclinal areas from Chihuahua in Mexico in the south to Ellesmereland on the north and from inner miogeosynclinal belts of the Appalachians in the east to eastern Nevada in the west. Faunas of the same type are known from northern Scotland (Durness Limestone), eugeosynclinal sequence of Norway (Hølonda Limestone in the Trondheim region, BERGSTRÖM, 1971), Korea, Siberian platform, and northeastern Siberia. In the latest middle Ordovician the fauna temporarily invaded parts of Balto-Scandia (BERGSTRÖM, 1971; SWEET & BERGSTRÖM, 1974).

Conodont faunas of the North Atlantic province (BERGSTRÖM, 1971; Anglo-Scandinavian-Appalachian Province, SWEET et al., 1959; European Province, BERGSTRÖM & SWEET, 1966) have been reported from Balto-Scandia, British Isles, and scattered areas in continental Europe. In North America this type of fauna has an amphicratonic distribution, occurring along the Appalachians as well as the Cordillera. In the Lower Ordovician of the Appalachians the North Atlantic conodonts are known only in limited eastern areas with "exotic" rocks (Newfoundland, Pennsylvania; BERGSTRÖM et al., 1973). In the middle Ordovician the North Atlantic fauna spread westward to the Helena-Saltville thrust of the southern Appalachians (BERGSTRÖM, 1971) and into corresponding belts farther to the north as far as Newfoundland. The fauna is also known from Texas (Marathon region) and in the Cordillera from Nevada to Yukon and Alaska. A Lower Ordovician conodont fauna with North Atlantic affinities has been described from Precordilleran Argentina (SERPAGLI, 1974).

In Australia, collections from New South Wales and Queensland show North American Midcontinent affinities (SWEET & BERGSTRÖM, 1974), whereas Lower Ordovician conodonts from the Canning basin (McTAVISH, 1973) are mostly related to those from the North Atlantic province. Undescribed collections from the middle Ordovician of central Australia and Tasmania are characterized by species largely unknown in either the Midcontinent or North Atlantic provinces, which led BERGSTRÖM (1971) to postulate that a separate Australian province may be recognizable.

In the Early Ordovician simple-cone species apparently dominate Midcontinent faunas to the virtual exclusion of other types, whereas in the North Atlantic province taxa with ramiform-element apparatuses form an important component in the conodont faunas. Middle and Upper Ordovician faunas of Midcontinent type are composed largely of "fibrous" conodonts and "nonfibrous" ramiform-element genera (e.g., *Phragmodus, Plectodina*). Genera with platform-type skeletal elements are rare. North Atlantic fauna mostly lacks the ramiform-element "fibrous" conodonts and includes a far greater variety of taxa with platform-type skeletal elements (e.g., *Eoplacognathus, Pygodus*).

The well-defined provinciality of the conodont faunas ended with the Ordovician, and the succeeding Silurian fauna is described as cosmopolitan.

BIOGEOGRAPHIC EVALUATION OF FAUNAL DIFFERENTIATION

North America has one of the simplest geological structures of all continental lithospheric plates and provides valuable clues for interpretation of faunal differentiation. The amphicratonic distribution of the shelly faunas of the Scoto-Appalachian and Hiberno-Salairian types (peripheral faunas) relative to the Midcontinent fauna is analogous to the apparently concentric arrangement of Middle and Late Cambrian biofacies realms (for summaries, see PALMER, 1969, 1972, 1973, 1974; COOK & TAYLOR, 1975) and is probably controlled by the same main factors. The same spatial distribution shows also the North Atlantic conodont fauna relative to the North American Midcontinent fauna. The Midcontinent faunas inhabited a wide carbonate platform in the continental interior, whereas the peripheral faunas occupied the outer, oceanward margin of the carbonate platform and extended in places in what has been termed the eugeosynclinal zone (Magog belt, etc.). An outer Ordovician biofacies realm, corresponding to that occupying much of the Cambrian "outer detrital belt," is characterized by planktonic graptolite fauna that has no exact counterpart in the Cambrian.

The similarity to the concentric arrangement of Cambrian biofacies realms was greatest during the Early Ordovician. The extensive Ordovician carbonate deposits of the continental interior and much of the Appalachians reflect shallow-water conditions over wide areas and the fauna that inhabited the region includes numerous endemic supraspecific taxa. The peripheral shelly fauna is best preserved in the west (Ross, 1975). In the middle and Late Ordovician the conditions were somewhat different. The depositional environment of the Midcontinent region was more varied and not so markedly associated with shallow water as during the Late Cambrian and Early Ordovician. The Midcontinent shelly fauna is far less specialized than in earlier epochs and differs from the faunas of the Scoto-Appalachian and Hiberno-Salairian type by lack of taxa widely distributed in the peripheral faunas rather than by endemic elements.

The bulk of evidence indicates that the difference between the peripheral and Midcontinent shelly faunas was primarily due to ecological factors; however, those environmental factors that caused the differentiation are not always clear. With respect to the Cambrian peripheral faunas it has been stressed that they are found in former shelf-margin to open-sea areas with "unrestricted access to open ocean conditions" (PALMER, 1973) where widespread to cosmopolitan forms either lived or were transported. According to WHITTINGTON and HUGHES (1972), comparable Ordovician faunas have occupied shelf slopes, which probably implies that they lived in deeper water than contemporaneous faunas on the continental platform; however, water depth was not always the main ecological factor that controlled the distribution of Ordovician peripheral shelly faunas because in places these faunas are associated with sediment indicating shallow-water environment or at least deposition within the photic zone. Also, in parts of the Girvan district, Scotland, the Scoto-Appalachian fauna is associated with sediment of very shallow water (WILLIAMS, 1962). The difference in water temperature between open ocean and relatively shallow water upon a platform may constitute a possibly important ecologic factor. On several platform areas, such as the North American continental interior, Canadian Arctic Archipelago, northwestern Greenland, Severnaya Zemlya, Siberian platform, and northern China (Shensi), presence of Ordovician evaporites indicates that in some areas evaporation was temporarily greater than precipitation. This, in turn, indicates that water salinity on the platforms was higher than in the oceans, at least temporarily (Ross, 1976). This may have influenced faunal differentiation.

In North America, distribution of the North Atlantic conodont fauna follows closely that of the peripheral shelly faunas, indicating that the same factors may have controlled both faunal differentiations. SWEET and BERGSTRÖM (1974) and SERPAGLI (1974) suggested that distribution of the conodont faunas was controlled by water temperature, the North Atlantic fauna forming a warm-temperate fauna on both sides of a continental platform inhabited by a tropical-subtropical fauna. According to these authors, most, if not all, conodonts were planktonic or nektonic. BARNES et al., (1973) and BARNES and FÅHRAEUS (1975) suggested that the majority of Ordovician conodonts were benthonic or nektobenthonic. The Midcontinent province was largely restricted to equatorial regions characterized by raised salinity and temperature, whereas the North Atlantic faunas represented a normal marine, virtually cosmopolitan province (BARNES & FÅHRAEUS, 1975).

If the peripheral Ordovician faunas of North America were ecologically controlled, it follows that a similar ecological control has affected faunas of the same type elsewhere in the world, complicating the biogeographical evaluation of the distributional data in the same way as some Cambrian faunas do. A further complication arises from ecologic zonation (in paleontological literature often inappropriately termed "communities") within a region. Examples are described from shelly faunas (for trilobites, see WEBBY, 1974; FORTEY, 1975; APOLLONOV, 1975) and conodont faunas (SEDDON & SWEET, 1971; BARNES et al., 1973; BARNES & FÅHRAEUS, 1975), and there exists clear evidence for a roughly similar zonation in graptolite faunas. Much of the ecological differentiation in shelly faunas is usually attributed to depth zonation, but it is often not clear whether the main ecological factor responsible for differentiation was depth or physical properties of the substrate, or an intricate combination of both these and possibly some additional factors. In the Upper Ordovician of northern Europe the distribution of various shelly faunas tends to be patchy (Fig. 10). In the Siljan district, Sweden, for example, carbonate mounds with a rich Hiberno-Salairian fauna form patches surrounded by contemporaneous mudstones and calcareous mudstones with a completely different fauna of the same Mediterranean type as further southward (cf. distribution of *Phillipsinella*, Fig. 10). A similar patchy distribution of Upper Ordovician limestone facies and associated fauna seems to prevail in Ireland and parts of northern England, although there the limestones have not yet proved to represent stromatactis carbonate mounds and some mingling of the faunas took place. In these cases the main ecological factor controlling faunal distribution may be physical properties of the substrate (whereby major ecologically important differences in the substrate do not necessarily follow the petrographic classification of rocks). Faunas of the Mediterranean type are commonly associated with terrigenous sediments, mostly former mud bottoms, and the type of substrate may have controlled the distribution of many elements in these faunas.

Much further work on Ordovician faunas is needed in order to understand what is ecologically and what is distributionally controlled. For this reason the neutral term "fauna" or "type of fauna" is used in this paper rather than formal categories, such as "faunal region" or "province," applied in biogeographic classification.

Evidence accumulated during the last 10 years demonstrates that the present position of cratonic lithospheric plates has very little relationship to geographies in the past. Although reconstructions of the geography back to the Permian have been shown to be possible by reversing the data of sea-floor spreading, reconstructions of the conditions before Pangaea are difficult and no satisfactory model has been presented. For the Ordovician reliable paleomagnetic data are still too few for presenting a coherent picture of latitudinal positions of various lithospheric plates (for a recent discussion, see BRIDEN et al., 1973). During the last few years much attention has been focused on biogeographic data as a tool for determining the longitudinal geographic position of the cratonic lithospheric plates during Ordovician time (WHITTINGTON & HUGHES, 1972, 1973, 1975; BURRETT, 1973; WILLIAMS, 1973; Ross, 1975). The reconstructions are based on the assumption that oceans were the major barriers to migrations of shallow-water faunas and that the degree of faunal resemblance is proportional to the width of the ocean. The classical examples from the Ordovician faunas pertaining to continental drift in the northern Atlantic region have been referred to in appropriate places of the text. In other parts of the world interpretation of available data on faunal similarities or dissimilarities is difficult at present. Oceanic barriers are not the only cause of dissimilarity between faunas. For example, in shelf areas where major cold and warm oceanic currents meet, the effect on faunas may be of comparable magnitude.

In modern seas the primary factor regulating the distribution of faunal provinces is temperature. Within a temperature zone, further biogeographic differentiation is due to the lack or restriction of communications between seas, in modern time as well as in the immediate past. According to paleomagnetic data, the Ordovician north ("Pacific") pole was situated somewhere in the present southwestern Pacific ocean, far away from any continental plates. This implies that no northern ice cap existed because water in the polar region had free exchange with the water of a vast ocean. This, in turn, suggests that climatic zones of the Ordovician northern hemisphere were probably poorly defined. The Ordovician south ("Gondwana") pole on the other hand, was very likely situated on a continent, more exactly somewhere in northwestern Africa (SMITH et al., 1973). An ice cap was probably present, and the reported widespread occurrence of the middle and Late Ordovician glaciation phenomena in northwestern Africa (BEUF et al., 1971) may have been in part associated with this ice cap. Glacial deposits (Pakhuis Tillite) that may be roughly contemporaneous have been reported from western South Africa. The Ordovician southern hemisphere presumably had well-defined climate zones. BURRETT (1973) suggested that climate was not the major control of faunal distribution in the Ordovician, chiefly because most plates do not appear to show any obvious latitudinal zonation. Several regions in the Ordovician southern hemisphere (Balto-Scandia, England, China) do show some spatial faunal zonation, although it is not easy to prove that the cause was climatic.

Paleomagnetic data indicate that the Ordovician equator passed across North America and the Siberian platform (Fig. 13). Thus, the shelly faunas of North American Midcontinent-Tungusian type inhabited warm to tropical seas. According to paleomagnetic evidence, Australia was situated just north of the Ordovician equator in the range of warm to tropical temperatures. No paleomagnetic data are available from east Asia.

In northern Africa fauna of the Mediterranean type occurs close to the probable Ordovician south ("Gondwana") pole and in part within the region of Ordovician glaciation. A brachiopod assemblage, probably of latest Ordovician age and comparable to the *Hirnantia* fauna, has been recorded also from western Cape Province in South Africa (COCKS et al., 1970) above the Pakhuis Tillite. Upper Ordovician glacigene deposits have been recorded as far to the north as Normandy, and even from Scotland (for a general review, see HARLAND, 1972b). This indicates that the Mediterranean fauna lived mainly in cold water (SPJELDNAES, 1961; HAVLIČEK, 1974; ROSS, 1975).

Distribution of the Pacific graptolite fauna is associated with lithospheric plates, which, according to paleomagnetic or other evidence, were situated in the region of warm to tropical climate (e.g., North America, Siberia, Australia-New Zealand). Thus, this fauna probably represents a warm-water planktonic fauna (SKEVINGTON, 1974). Graptolites in areas with the Mediterranean shelly fauna belong to the Atlantic province that obviously extended into cold water.

SKEVINGTON (1974) suggested that the cosmopolitan distribution of Ordovician graptoloids from the *Nemagraptus gracilis* Zone onward is due to disappearance of graptolites from regions with cold water (northern Africa, southern Europe) so that from then on all Ordovician graptoloid faunas, with rare exceptions, were confined to the tropical zone. Based on current models of Ordovician geography, this conclusion is possible only if the boundary between the South and North European lithospheric plates is drawn along the Alpine chain (as done by SKEVINGTON, following SMITH et al., 1973), and Bohemia is included in the North European plate. Evidence from shelly faunas does not support this plate boundary. The rarity of middle and Late Ordovician graptolites in northern Africa and southern Europe is more likely due to the lack of suitable sediments for preservation.

Distribution of graptolite faunas in South America suggests the presence of latitudinal faunal zonation. In all recent reconstructions of Ordovician geography, this continent forms part of Gondwanaland and is oriented with Patagonia toward the equator so that the Precordillera of Argentina with a Pacific fauna reaches low latitudes. Peru and Bolivia with an Atlantic fauna are situated at temperate latitudes. Thus, the faunas with Pacific affinities may have inhabited warm water and the Atlantic fauna a temperate water. A similar suggestion with respect to climate was put forth by SERPAGLI (1973, 1974), who showed that the Lower Ordovician limestones of the Precordillera are in part bahamitic, indicating deposition in warm water. The middle Ordovician marine glacigene deposits in northernmost Argentina and Bolivia may be an indicator of temporary cool water in areas with an Atlantic graptolite fauna.

The position of the Russian platform and Scandinavia close to the equator in almost all recent reconstructions of Ordovician geography does not fit into this model at all. The Balto-Scandian region has an Atlantic graptolite fauna and its shelly fauna differs from that of other presumed warm-water faunas. Based on faunal (TROEDSSON, 1928; SPJELDNAES, 1961; SKEVINGTON, 1974) and lithological (LINDSTRÖM, 1972; JAANUSSON, 1973b) evidence, it has been suggested that the Balto-Scandian region was during at least most of Ordovician time within the temperate or unspecified cold-climate zone. If this was the case, the zone possibly embraced also southwestern China (and the Tarim platform?). Northern China, with a Pacific graptolite fauna, evaporites, extensive carbonate sequence, and shelly faunas showing some North American affinities, probably was within the zone of warm to tropical water.

BIOGEOGRAPHIC CHANGES LEADING TO THE COSMOPOLITAN SILURIAN FAUNA

The Ordovician biogeographic differentiation was greatest during the Arenigian-Llanvirnian and their equivalents. During the middle and Late Ordovician the provinciality decreased successively in the shelly faunas (WHITTINGTON, 1966; WHITTINGTON & HUGHES, 1972) ultimately followed by a cosmopolitan fauna in the Early Silurian. In this process of successively decreasing provinciality the greatest single step was at the boundary between the Ordovician and Silurian (at the top of the Hirnantian).

Diversity of the Ordovician trilobite fauna of Mediterranean type decreased at about the base of the Hirnantian by extinction of many taxa (Cyclopygidae, Dionidae, Remopleurididae, Hammatocnemidae, Dindymeninae, Ectillaeninae, and others). A further wave of extinction at the top of the Hirnantian (Trinucleidae, Agnostida, Philipsinellidae) virtually eliminated the Mediterranean fauna. In the region with the

Ordovician Mediterranean fauna the Lower Silurian deposits consist almost exclusively of graptolitic shale that is practically devoid of shelly fossils. This makes one wonder whether or not elimination of habitats was a contributing factor in the extinction of the Mediterranean shelly fauna. In areas with Ordovician cold-water fauna no Llandoverian shelly fauna is known (except the *Clarkeia* fauna, the appearance of which is difficult to date, COCKS, 1972; COCKS & MCKERROW, 1973). The rarity of preserved remains of a cold-water benthic fauna undoubtedly exaggerates the cosmopolitan nature of the known Early Silurian shelly fauna.

The North American Midcontinent-Tungusian faunas began to lose their biogeographic identity before the Hirnantian. In the equivalents to the Hirnantian the importance of some of its distinctive elements (e.g., *Anataphrus*, "*Brachyaspis*") seems to have been reduced to the status of relicts. Large areas of epicontinental seas previously occupied by these faunas emerged and this may have contributed to extinction by elimination of habitats.

Extinction during the middle and late Ashgillian and their equivalents affected all groups of organisms and all faunas. Of some 38 trilobite families known in the Ashgillian, only 14 continued into the Silurian (and only one added). Of about 70 Upper Ordovician genera of tabulate and heliolitid corals about 50 became extinct before the Silurian (KALJO & KLAAMANN, 1973). Extinction affected also brachiopods (Clitambonitacea and Porambonitacea, as well as many families), cephalopods (Endoceratoidea, with the possible exception of the enigmatic *Humeoceras*), stromatoporoids (Aulaceridae) and other groups. Thus, the physical event or combination of events that triggered the extinction had a profound effect. The change to the cosmopolitan Silurian shelly fauna was associated with a considerable loss of overall taxonomic diversity.

The Hirnantian and Llandoverian trilobite faunas have a low taxonomic diversity and the center of origin of the Silurian fauna is not obvious. The Silurian brachiopod fauna is largely based on the peripheral Ordovician faunas of Hiberno-Salairian type (see p. *A*157) where many of the taxa that became worldwide in the Silurian have a wide distribution (e.g., Atrypidae, Dicoelosiidae, Pentameracea, Meristellidae, *Dolerorthis*). During the Hirnantian elements of this fauna invaded parts of the North American Midcontinent and Anticosti. The relation of these peripheral faunas to temperature or ecologic zonation is not clear. Balto-Scandia and possibly also the Kazakhstan-Tien Shan region may have occupied the temperate climatic zone. If this was the case, the cosmopolitan spread of many of its elements during the Silurian might indicate that the Silurian climate became more uniformly temperate. On the other hand, the Silurian conodont fauna developed mainly from the presumably tropical-subtropical North American Midcontinent fauna (SWEET & BERGSTRÖM, 1974).

It has been suggested that the major event causing the faunal change from Ordovician to Silurian was the Late Ordovician glaciation (SHEEHAN, 1973; BERRY & BOUCOT, 1973). It may have affected the faunas in two ways. Firstly, accumulation of precipitation in glaciers caused eustatic lowering of sea level and widespread regression of shelf seas resulting in elimination of habitats of shallow marine faunas. Secondly, the glaciation caused cooling of the oceans and extinction of stenothermal organisms; however, Pleistocene glaciations have not produced effects of a comparable magnitude on marine faunas, indicating that at the transition from Ordovician to Silurian other factors were involved. Another explanation of the change to a cosmopolitan fauna is that because of continental drift oceans between cratonic lithospheric plates decreased in width so much that they did not act any more as distributional barriers (WHITTINGTON & HUGHES, 1972); however, in this case biogeographic changes are expected to have been more gradual than they appear to have been and also not so contemporaneously worldwide. It is probable that the faunal changes from Ordovician to Silurian were caused by a combined effect of several factors whereby the relative importance of individual factors is at present difficult to determine.

REFERENCES

Abdullaev, R. N., 1972, *Trilobity verkhnego ordovika Bukantau:* in Novye dannye po faune paleozoya i mezozoya Uzbekistana, p. 103-126, pl. 44-49, Akad. Nauk Uzbek. SSR, Inst. Geol. Geofiz. (Tashkent). [*Upper Ordovician trilobites of Bukantau.*]

——, & Khaletskaya, O. N., 1970, *Nizhniy paleozoy Chatkalskogo khrebta, trilobity i graptolity ordovika Pskemskogo khrebta:* Akad. Nauk Uzbek. SSR, Inst. Geol. Geofiz., 104 p., 8 pl. [*Lower Paleozoic of the Chatkal Range, Ordovician trilobites and graptolites of the Pskem Range.*]

Amsden, T. W., 1974, *Late Ordovician and Early Silurian articulate brachiopods from Oklahoma, southwestern Illinois, and eastern Missouri:* Oklahoma Geol. Survey, Bull. 119, 154 p., pl. 1-28.

Andreeva, O. N., 1972, *Brakhiopody kuraganskoy svity ordovika Yuzhnogo Urala:* Paleont. Zhurnal 1972, p. 45-56, pl. 7, 8. [*Brachiopods of the Ordovician Kuragan Formation in southern Urals.*]

Apollonov, M. K., 1974, *Ashgillskie trilobity Kazakhstana:* Akad. Nauk Kazakh. SSR, Inst. Geol. Nauk, 136 p., 21 pl. [*Ashgillian trilobites of Kazakhstan.*]——1975, *Ordovician trilobite assemblages of Kazakhstan:* Fossils and Strata, no. 4, p. 375-380.

Baldis, B. A., & Blasco, Graciela, 1975, *Primeros trilobites ashgillianos del Ordovicico Sudamericano:* Actas I Congr. Argent. Paleont. Bioestr., v. 1, p. 33-48.

Barnes, C. R., & Fåhraeus, L. E., 1975, *Provinces, communities, and the proposed nektobenthic habit of Ordovician conodontophorids:* Lethaia, v. 8, p. 133-149.

——, Rexroad, C. B., & Miller, J. F., 1973, *Lower Paleozoic provincialism:* in Conodont paleozoology, F. H. T. Rhodes (ed.), Geol. Soc. America, Spec. Paper 141, p. 157-190.

Bates, D. E. B., 1968, *The Lower Palaeozoic brachiopod and trilobite faunas of Anglesey:* Brit. Museum (Nat. History), Bull., Geol., v. 16, p. 127-199, pl. 1-14.

Bergström, S. M., 1971, *Conodont biostratigraphy of the Middle and Upper Ordovician of Europe and eastern North America:* in Symposium on conodont biostratigraphy, W. C. Sweet & S. M. Bergström (eds.), Geol. Soc. America, Mem. 127, p. 83-157.——1973, *Ordovician conodonts:* in Atlas of palaeobiogeography, Anthony Hallam (ed.), p. 47-58, Elsevier (Amsterdam).

——, Epstein, Anita G., & Epstein, J. B., 1973, *Early Ordovician North Atlantic Province conodonts in eastern Pennsylvania:* U.S. Geol. Survey, Prof. Paper 800-D, p. D37-D44.

——, Ethington, R. L., & Jaanusson, Valdar, 1973, *On the stage subdivision of the North American lower Middle Ordovician: Age of strata at the top of Whiterock reference sequences in Nevada:* Geol. Soc. America, Abstracts with Programs, North-Central Section Ann. Mtg. 1973, p. 299.

——, & Sweet, W. C., 1966, *Conodonts from the Lexington Limestone (Middle Ordovician) of Kentucky and its lateral equivalents in Ohio and Indiana:* Bull. Am. Paleontology, v. 50, p. 271-441.

Berry, W. B. N., 1960, *Correlation of Ordovician graptolite-bearing sequences:* Internatl. Geol. Congress, 21st Sess., Proc. Sec. 7, p. 97-108 (Copenhagen).

——, & Boucot, A. J., 1973, *Glacio-eustatic control of Late Ordovician-Early Silurian platform sedimentation and faunal changes:* Geol. Soc. America, Bull., v. 84, p. 275-284.

Beuf, Serge, Biju-Duval, Bernard, de Charpal, Olivier, Rognon, Pierre, Gariel, Olivier, & Bennacef, Abdelkrim, 1971, *Les grès du Paléozoique inférieur au Sahara (Sedimentation et discontinuités; évolution structurale d'un craton):* Inst. Français Pétrole, Publ. Coll. Sci. et Techn. Pétrole no. 18, iv + 464 p.

Bondarev, V. I., 1968, *Stratigrafiya i kharakternye brakhiopody ordovikskikh otlozheniy yuga Novoy Zemli, ostrova Vaygasch i severnogo Pay-Khoya:* Nauchno-Issledov. Inst. Geol. Arktiki, Trudy, v. 157, p. 3-144, pl. 1-13. [*Stratigraphy and characteristic brachiopods of the Ordovician deposits of southern Novaya Zemlya, the Island of Vaygach, and northern Pay-Khoy.*]

Bouček, Bedřich, 1972, *The palaeogeography of Lower Ordovician graptolite faunas: a possible evidence of continental drift:* Internatl. Geol. Congress, 24th Sess., Proc. Sec. 7, p. 266-272 (Montreal).

Briden, J. C., Morris, W. A., & Piper, J. D. A., 1973, *Palaeomagnetic studies in the British Caledonides—VI. Regional and global implications:* Geophys. Jour. Res., abstr. Soc. 34, p. 107-134.

Bulman, O. M. B., 1964, *Lower Palaeozoic plankton:* Geol. Soc. London, Quart. Jour., v. 120, p. 455-476.——1971, *Graptolite faunal distribution:* in Faunal provinces in space and time, F. A. Rawson, P. F. Middlemiss, & G. Newall (eds.), Geol. Jour., Spec. Issue 4, p. 47-60.

Burrett, Clive, 1973, *Ordovician biogeography and continental drift:* Palaeogeography, Palaeoclimatology, Palaeoecology, v. 13, p. 161-201.

Campbell, K. S. W., 1973, *A species of the trilobite Dalmanitina (Dalmanitina) from Australia:* Geol. Fören. Stockholm, Förhandl., v. 95, p. 69-77.

Chugaeva, M. N., 1968, *Biogeograficheskie oblasti kontsa rannego ordovika (Areniga) po trilobitam:* Internatl. Geol. Congress, 23rd Sess., Rept. Soviet Geologists, Probl. 9, p. 63-68 (Moscow). [*Biogeographical areas at the end of Early Ordovician (Arenig) on the basis of trilobites.*] (In Russian with English summary.)——1973, *Biogeografiya kontsa rannego ordovika:* in Biostratigrafiya nizhney chasti ordovika severovostoka SSSR i biogeografiya kontsa rannego ordovika, Akad. Nauk Geol. Inst., Trudy, v. 213, p. 238-280. [*Biogeography of the uppermost Lower Ordovician,* in Biostratigraphy of the lower part of the Ordovician in the northeast of the USSR and biogeography of the uppermost Lower Ordovician.]

Cocks, L. R. M., 1972, *The origin of the Silurian Clarkeia shelly fauna of South America, and its extension to West Africa:* Palaeontology, v. 15, p. 623-630.

——, Brunton, C. H. C., Rowell, A. J., & Rust, I. C., 1970, *The first Lower Palaeozoic fauna proved from South Africa:* Geol. Soc. London, Quart. Jour., v. 125, p. 583-603.

——, & McKerrow, W. S., 1973, *Brachiopod distributions and faunal provinces in the Silurian and Lower Devonian:* Spec. Papers in Palaeontology, no. 12, p. 291-304.

Cook, H. E., & Taylor, M. E., 1975, *Early Paleozoic continental margin sedimentation, trilobite biofacies, and the thermocline, western United States:* Geology, v. 3, p. 559-562.

Cooper, G. A., 1956, *Chazyan and related brachiopods:* Smithson. Misc. Coll., v. 127, pt. 1, p. 1-1024; pt. 2, p. 1025-1245, pl. 1-209.

Creath, W. G., & Shaw, A. B., 1966, *Paleontology of northwestern Vermont. XIII. Isochilina from the Ordovician Highgate Formation:* Jour. Paleontology, v. 40, p. 1312-1330, pl. 161-163.

Cuerda, A. J., 1973, *Resena del ordovicio Argentina:* Ameghiniana, v. 10, 272-312.

Dean, W. T., 1967, *The distribution of Ordovician shelly faunas in the Tethyan region:* in Aspects of Tethyan biogeography, C. G. Adams, & D. V. Ager (eds.), Systematics Assoc. Publ. 7, p. 11-44.

Druce, E. C., & Jones, P. J., 1971, *Cambro-Ordovician conodonts from the Burke River structural belt:* Australia Bur. Min. Resources, Bull. 110, 167 p., pl. 1-20.

Fortey, R. A., 1975, *Early Ordovician trilobite communities:* Fossils and Strata, no. 4, p. 339-360.

——, & Bruton, D. L., 1973, *Cambrian-Ordovician rocks adjacent to Hinlopenstretet, North Ny Friesland, Spitsbergen:* Geol. Soc. America, Bull., v. 84, p. 2227-2242.

Harland, W. B., 1972, *The Ordovician ice age:* Geol. Mag., v. 109, p. 451-456.

——, & Gayer, R. A., 1972, *The Arctic Caledonides and earlier oceans:* Geol. Mag., v. 109, p. 289-314.

Havlíček, Vladimír, 1974, *Some problems of the Ordovician in the Mediterranean region:* Ustřed. Ustav. Geol., Věstnik, no. 49, p. 343-348.—— 1976, *Evolution of Ordovician brachiopod communities in the Mediterranean Province:* in The Ordovician System: Proceedings of a Palaeontological Association symposium, M. G. Bassett (ed.), p. 349-358, Palaeont. Assoc., Univ. Wales Press (Cardiff).

Hu Bing, Wang Jing-bin, Gao Zhen-jia, & Fang Xiao-di, 1965, [*Problems of the Paleozoics of Tarim platform*]: Ti Chih Hsüeh Pao (Acta Geol. Sinica), v. 45, p. 131-142. [In Chinese.]

Jaanusson, Valdar, 1960, *Graptoloids from the Ontikan and Viruan (Ordov.) limestones of Estonia and Sweden:* Geol. Inst. Univ. Uppsala, Bull., v. 37, p. 289-366.——1973a, *Ordovician articulate brachiopods:* in Atlas of Palaeobiogeography, A. Hallam (ed.), p. 19-25, Elsevier (Amsterdam).——1973b, *Aspects of carbonate sedimentation in the Ordovician of Baltoscandia:* Lethaia, v. 6, p. 11-34.——1976, *Faunal dynamics in the Middle Ordovician (Viruan) of Balto-Scandia:* in Ordovician System: Proceedings of a Palaeontological Association symposium, M. G. Bassett (ed.), p. 301-326, Palaeont. Assoc. (Cardiff).

Jen Chi-shun, 1964, [*Certain geotectonic formations occurring before the Devonian in southeastern China*]: Ti Chih Hsüeh Pao (Acta Geol. Sinica), v. 44, p. 418-431. [In Chinese.]

Jones, P. J., 1971, *Lower Ordovician conodonts from the Bonaparte Gulf Basin and the Daly River Basin, northwestern Australia:* Bur. Min. Resources Australia, Bull. 117, 80 p.

Kaljo, Dimitri, & Klaamann, Einar, 1973, *Ordovician and Silurian corals:* in Atlas of Palaeobiogeography, A. Hallam (ed.), p. 37-45, Elsevier (Amsterdam).

——, ——, & Nestor, Heldur, 1970, *Paleobiogeograficheskiy obzor ordovikskikh i siluriyskikh korallov i stromatoporoidey:* in Distribution and sequence of Paleozoic corals of the USSR, Papers of II All-union Symposium on fossil corals of the USSR, v. 3, p. 6-15. [*Paleobiogeographic survey of Ordovician and Silurian corals and stromatoporoids.*]

Kielan, Zofia, 1959, *Upper Ordovician trilobites from Poland and some related forms from Bohemia and Scandinavia:* Palaeont. Polonica, v. 11, p. 1-198, pl. 1-35.

Kobayashi, Teiichi, 1971, *The Eurasiatic faunal connection in the Ordovician Period:* Bur. Rech. Geol. Min., Mém., no. 73, p. 281-290.

Legrand, Philippe, 1974, *Essai sur la paléogéographie de l'Ordovicien au Sahara Algérien:* Compagnie Francaise des Pétroles, Notes & Mém. 11, p. 121-138, 8 pl.

Leleshus, V. L., 1970, *Paleozoogeografiya ordovika, silura i rannego devona po tabulyatomorfnym korallam i granitsy siluriyskoy sistemy:* Akad. Nauk SSSR, Izvestiya, ser. geol., 1970, p. 184-192. [*Ordovician, Silurian, and Early Devonian paleozoogeography based on tabulatomorph cor-*

als, and boundaries of the Silurian System.]

Lespérance, P. J., 1968, *Faunal affinities of the trilobite faunas, White Head Formation, Percé region, Quebec, Canada:* Internatl. Geol. Congress, 23rd Sess., Proc. Sec. 9, p. 145-159 (Prague).———1974, *The Hirnantian fauna of the Percé area (Quebec) and the Ordovician-Silurian boundary:* Am. Jour. Sci., v. 274, p. 10-30.

Levitskiy, E. S., 1963, *Trilobity srednego ordovika severo-zapada Gornogo Altaya i ikh stratigraficheskoe znachenie:* Avtoreferat dissertatsii, p. 1-23 (Moskva). [*Middle Ordovician trilobites of northwestern Gornyi Altay, and their stratigraphic significance.*]

Lindström, Maurits, 1972, *Ice-marked sand grains in the Lower Ordovician of Sweden:* Geologica et Palaeontologica, v. 6, p. 25-32.

Lu Yen-hao, 1975, *Ordovician trilobite faunas of central and southwestern China:* Chung-kuo ku shêng wu chih (Palaeontologia Sinica), no. 152, p. 1-463, 50 pl. [In Chinese and English.]

Lui Hung-yun, 1958, [*Paleogeographic map of China*]: Science Press (?Peking). [In Chinese.]

McLaughlin, R. E., 1973, *Observations on the biostratigraphy and stratigraphy of Knox County, Tennessee, and vicinity:* Tennessee Div. Geol., Bull. 70, p. 25-62.

McTavish, R. A., 1973, *Prioniodontacean conodonts from the Emanuel Formation (Lower Ordovician) of western Australia:* Geologica et Palaeontologica, v. 7, p. 27-58, 3 pl.

Miller, J. F., 1969, *Conodont fauna of the Notch Peak Limestone (Cambro-Ordovician), House Range, Utah:* Jour. Paleontology, v. 43, p. 413-439, pl. 63-66.———1970, *Conodont zonation of the uppermost Cambrian and lowest Ordovician:* Geol. Soc. America, Abstracts with Programs, Ann. Mtg. 1970, p. 624.

———, Robison, R. A., & Clark, D. L., 1974, *Correlation of Tremadocian conodont and trilobite faunas, Europe and North America:* Geol. Soc. America, Abstracts with Programs, Ann. Mtg. 1974, p. 1048-1049.

Mu, A. T., 1974, *Evolution, classification and distribution of Graptoloidea and graptodendroids:* Scientia Sinica, v. 17, p. 227-238.

Neuman, R. B., 1970, *Paleogeographic implications of Ordovician shelly fossils in the Magog belt of the northern Appalachian region:* in Studies of Appalachian geology; northern and maritime, E. A. Zen et al. (eds.), p. 35-48, Intersci. Publ. (New York).———1972, *Brachiopods of Early Ordovician volcanic islands:* Internatl. Geol. Congress, 24th Sess., Sec. 7, p. 297-302 (Montreal).

———, & Bruton, D. L., 1974, *Early Middle Ordovician fossils from the Hølanda area, Trondheim region, Norway:* Norsk Geol. Tidsskr., v. 54, p. 69-115.

Nikiforova, O. I., & Sapelnikov, V. P., 1973, *Nekotorye drevnie pentameridy Zeravshanskogo khrebta:* Akad. Nauk SSSR, Uralskiy Nauchnyy Tsentr, Trudy, Inst. Geol. Geokhim., v. 99, p. 64-82, pl. 1-6. [*Some ancient pentamerids from the Zeravshan Range.*]

Nikitin, I. F., 1972, *Ordovik Kazakhstana, I, Stratigrafiya:* Akad. Nauk Kazakh. SSR, Inst. Geol. Nauk, 242 p. (with English summary). [*The Ordovician of Kazakhstan, I. Stratigraphy.*]

Nikolaev, A. A., Oradovskaya, M. M., Preobrazhenskiy, B. V., Obut, A. M., Sobolevskaya, R. F., & Kabankov, V. Ya., 1974, *Opornye razrezy paleozoya severo-vostoka SSSR:* Akad. Nauk SSSR, Dalnevostochnyy Nauchnyy Tsentr, 161 p., pl. 1-43 (Magadan). [*Key sections of the Upper Ordovician in the north-east of the USSR.*]

Oradovskaya, M. M., 1970, *Stratigrafiya ordovika i silura Chukotskogo poluostrova:* Akad. Nauk SSSR, Doklady, v. 191, p. 190-193. [*Ordovician and Silurian stratigraphy of the Chukot Peninsula.*]———1973, *Brakhiopody:* in Biostratigrafiya nizhney ordovika severo-vostoka SSSR, i biogeografiya kontsa rannego ordovika, Akad. Nauk SSSR, Trudy, Geol. Inst., v. 213, p. 141-209. [*Brachiopods, in Biostratigraphy of the lower part of the Ordovician in the northeast of the USSR and biogeography of the uppermost Early Ordovician.*]

Palmer, A. R., 1969, *Cambrian trilobite distributions in North America and their bearing on the Cambrian paleogeography of Newfoundland:* in North Atlantic geology and continental drift, G. M. Kay (ed.), Am. Assoc. Petrol. Geologists, Mem. 12, p. 139-144.———1972, *Problems of Cambrian biogeography:* Internatl. Geol. Congress, 24th Sess., Proc. Sec. 7, p. 310-315 (Montreal).———1973, *Cambrian trilobites:* in Atlas of palaeobiogeography, A. Hallam (ed.), p. 3-11, Elsevier (Amsterdam).———1974, *Search for the Cambrian world:* Am. Scientist, v. 62, p. 216-224.

Petrunina, Z. E., 1966, *Trilobity i biostratigrafiya tremadoka zapadnoy chasti Sayano-Altayskoy gornoy oblasti:* Avtoreferat dissertatsii, p. 1-30 (Alma-Ata). [*Tremadocian trilobites and biostratigraphy in the western part of the Sayan-Altai Mountain Region.*]

Riva, John, 1969, *Middle and Upper Ordovician graptolite faunas of St. Lawrence Lowlands of Quebec, and of Anticosti Island:* Am. Assoc. Petrol. Geologists, Mem. 12, p. 513-556.

Robison, R. A., & Pantoja-Alor, Jeres, 1968, *Tremadocian trilobites from the Nochixtlán region, Oaxaca, Mexico:* Jour. Paleontology, v. 42, p. 767-800.

Ross, R. J., Jr., 1975, *Early Paleozoic trilobites, sedimentary facies, lithospheric plates, and ocean currents:* Fossils and Strata, no. 4, p. 307-329.———1976, *Ordovician sedimentation in the western United States:* in The Ordovician System: Proceedings of a Palaeontological Association symposium, M. G. Bassett (ed.), p. 73-105, Palaeont. Assoc. (Cardiff).

Rozman, K. S., 1968, *Yarusnoe raschlenenie verkhnego ordovika i biogeograficheskie osobennosti razvitia pozdneordovikskoy fauny:* Internatl. Geol. Congress, 23rd Sess., Repts. Soviet Geologists, Prob. 9, p. 95-103 (Moskva). (In Russian with English summary.) [*Stage subdivision of the Upper Ordovician and biogeographical peculiarities in the development of Late Ordovician fossils.*]——1970, *Biostratigrafiya i paleobiogeografiya verkhnego ordovika Severo-Vostoka SSSR:* in Biostratigrafiya verkhnego ordovika Severo-Vostoka SSSR, Kh. S. Rozman, V. A. Ivanova, I. N. Krasilova, & E. A. Modzalevskaya (eds.), Akad. Nauk SSSR, Trudy, Geol. Inst., v. 205, p. 212-270. [*Upper Ordovician biostratigraphy and paleobiogeography of the northeast USSR.*]

Sapelnikov, V. P., & Rukavishnikova, T. B., 1975, *Verkhneordovikskie, siluriyskie i nizhnedevonskie pentameridy Kazakhstana:* Akad. Nauk SSSR, Uralskiy Nauchnyy Tsentr, 227 p., pl. 1-43. [*Upper Ordovician, Silurian and Lower Devonian pentamerids of Kazakhstan.*]

Savage, T. E., 1917, *Stratigraphy and paleontology of the Alexandrian Series in Illinois and Missouri:* Illinois Geol. Survey, Bull., v. 23, p. 67-160, pl. 3-9.

Seddon, George, & Sweet, W. C., 1971, *An ecological model for conodonts:* Jour. Paleontology, v. 45, p. 869-880.

Serpagli, Enrico, 1973, *Carbonati di tipo bahamitico nell'Ordoviciano inferiore della Precordillera Argentina e relative osservazioni paleoclimatologiche:* Soc. Nat. Mat. Modena, Atti, v. 104, p. 239-245.——1974, *Lower Ordovician conodonts from Precordilleran Argentina (Province of San Juan):* Soc. Paleont. Italiana, Boll., v. 13, p. 17-98, pl. 7-31.

Sheehan, P. M., 1973, *Brachiopods from the Jerrestad Mudstone (Early Ashgillian, Ordovician) from a boring in southern Sweden:* Geologica et Palaeontologica, v. 7, p. 59-76.——1973, *The relation of Late Ordovician glaciation to the Ordovician-Silurian changeover in North American brachiopod faunas:* Lethaia, v. 6, p. 147-154.

Sheng, Shin-fu, 1964, *Chuan chyan woan au taur shyh san yeh chorng de yan jiow ping tao lun shang aw taur toong de shang shiah jieh shiann wenn ti:* Ku Sheng Wu Hsüeh Pao (Acta Palaeont. Sinica), v. 12, p. 537-552, pl. 1-4. [*Upper Ordovician trilobites of Szechuan-Kweichow with special discussion on the classification and boundaries of the Upper Ordovician.*] (In Chinese with English summary.)

Shergold, J. H., 1975, *Late Cambrian and Early Ordovician trilobites from the Burke River structural belt, western Queensland, Australia:* Bur. Min. Resources, Australia, Bull. 153, v. 1 (text), 251 p., v. 2 (plates), pl. 1-58.

Sidyarenko, A. I., & Kanygin, A. V., 1965, *O stratigraficheskom polozhenii krivolutskogo yarusa Sibirskoy platformy:* Akad. Nauk SSSR, Doklady, v. 161, p. 187-189. [*On the stratigraphical position of the Krivoi Luk Stage of the Siberian platform.*]

Skevington, David, 1969, *Graptolite faunal provinces in the Ordovician of north-west Europe:* in North Atlantic geology and continental drift, G. M. Kay (ed.), Am. Assoc. Petrol. Geologists, Mem. 12, p. 557-562.——1973, *Ordovician graptolites:* in Atlas of palaeobiogeography, A. Hallam (ed.), p. 27-35, Elsevier (Amsterdam).——1974, *Controls influencing the composition and distribution of Ordovician graptolite faunal provinces:* Spec. Papers in Palaeontology, no. 13, p. 59-73.

Smith, A. G., Briden, J. C., & Drewry, G. E., 1973, *Phanerozoic world maps:* Spec. Papers in Palaeontology, no. 12, p. 1-42.

Spjeldnaes, Nils, 1961, *Ordovician climatic zones:* Norsk Geol. Tidsskr. v. 41, p. 45-77.——1967, *The palaeogeography of the Tethyan region during the Ordovician:* in Aspects of Tethyan biogeography, C. G. Adams & D. V. Ager (eds.), Systematics Assoc. Publ. 7, p. 45-57.

Sweet, W. C., & Bergström, S. M., 1974, *Provincialism exhibited by Ordovician conodont faunas:* in Paleogeographic provinces and provinciality, C. A. Ross (ed.), Soc. Econ. Paleontologists & Mineralogists, Spec. Publ. 21, p. 189-202.

——, Ethington, R. L., & Barnes, C. R., 1971, *North American Middle and Upper Ordovician conodont faunas:* in Symposium on conodont biostratigraphy, W. C. Sweet & S. M. Bergström (eds.), Geol. Soc. America, Mem. 127, p. 163-193.

——, Turco, C. A., Warner, Earl, & Wilkie, L. C., 1959, *The American Upper Ordovician standard. I. Eden conodonts from the Cincinnati region of Ohio and Kentucky:* Jour. Paleontology, v. 33, p. 1029-1068.

Tripp, R. P., 1962, *Trilobites from the "Confinis" Flags (Ordovician) of the Girvan district, Ayrshire:* Royal Soc. Edinburgh, Trans., v. 65, p. 1-40, pl. 1-4.——1965, *Trilobites from the Albany division (Ordovician) of the Girvan district, Ayrshire:* Palaeontology, v. 8, p. 577-603, pl. 80-83.——1967, *Trilobites from the Upper Stinchar Limestone (Ordovician) of the Girvan district, Ayrshire:* Royal Soc. Edinburgh, Trans., v. 67, p. 43-93, pl. 1-6.

Troedsson, G. T., 1928, *On the Middle and Upper Ordovician faunas of northern Greenland. II:* Meddel. Grønland, v. 72, p. 1-197, pl. 1-56.

Varganov, V. G., Antsygin, N. Ya., Nasedkina, V. A., Shurygina, M. V., & Militsina, V. S., 1973, *Stratigrafiya i fauna ordovika srednego Urala:* Ural. Terr. Geol. Uprav., 228 p., pl. 1-30, Nedra (Moskva). [*Ordovician stratigraphy and fauna in central Urals.*]

Viira, Viive, 1966, *Rasprostranenie konodontov v*

nizhneordovikskikh otlozheniyakh razreza Suhkrumägi (g. Tallinn): Eesti Teaduste Akad. Toimetised, v. 15, Füüs.-Mat. Tehn. Seeria, no. 1, p. 150-155 (with English summary). [*Distribution of conodonts in the Lower Ordovician sequence of Suhkrumägi* (Tallinn district).]

Vinogradov, A. P. (ed.), 1968, *Atlas litologo-paleogeograficheskikh kart SSSR:* v. 1, 52 maps, Vses. Aerogeol. Trest Minist. Geologii SSSR (Moskva). [*Atlas of the lithological-paleogeographical maps of the USSR* (Precambrian, Cambrian, Ordovician and Silurian).]

Walter, O. T., 1926, *Trilobites of Iowa and some related Paleozoic forms:* Iowa Geol. Survey, v. 31, p. 169-388, pl. 10-27.

Webby, B. D., 1974, *Upper Ordovician trilobites from central New South Wales:* Palaeontology, v. 17, p. 203-252.

Whittington, H. B., 1963, *Middle Ordovician trilobites from Lower Head, western Newfoundland:* Harvard Univ., Museum Comp. Zoology, Bull., v. 129, p. 1-119.——1966, *Phylogeny and distribution of Ordovician trilobites:* Jour. Paleontology, v. 40, p. 696-737.——1973, *Ordovician trilobites:* in Atlas of palaeobiogeography, A. Hallam (ed.), p. 13-18, Elsevier (Amsterdam).

——, & Hughes, C. P., 1972, *Ordovician geography and faunal provinces deduced from trilobite distribution:* Royal Soc. London, Philos. Trans., ser. B, v. 263, p. 235-278.——1973, *Ordovician trilobite distribution and geography:* Spec. Papers Palaeontology, no. 12, p. 235-240.

——1974, *Geography and faunal provinces in the Tremadoc Epoch:* in Paleogeographic provinces and provinciality, C. A. Ross (ed.), Soc. Econ. Paleontologists & Mineralogists, Spec. Publ. 21, p. 203-218.

——, & Williams, Alwyn, 1955, *The fauna of the Derfel Limestone of the Arenig district, North Wales:* Royal Soc. London, Philos. Trans., ser. B, v. 238, p. 397-430.

Williams, Alwyn, 1962, *The Barr and Lower Ardmillan Series (Caradoc) of the Girvan district, southwest Ayrshire, with descriptions of the Brachiopoda:* Geol. Soc. London, Mem. 3, p. 1-267, pl. 1-25.——1969, *Ordovician faunal provinces with reference to brachiopod distribution:* in The Pre-Cambrian and Lower Palaeozoic rocks of Wales, A. Wood (ed.), p. 117-150, Univ. Wales Press (Cardiff).——1972, *An Ordovician Whiterock fauna in western Ireland:* Royal Irish Acad., Proc., v. 72, sec. B, p. 209-219, pl. 10,11.——1973, *Distribution of brachiopod assemblages in relation to Ordovician palaeogeography:* Spec. Papers in Palaeontology, no. 12, p. 241-269.

Wilson, J. T., 1966, *Did the Atlantic close and then re-open?:* Nature, London, v. 211, p. 676-681.

Wolfart, Reinhard, 1967, *Zur Entwicklung der paläozoischen Tethys in Vorderasien:* Erdöl u. Kohle, v. 20, p. 168-180.

Yochelson, E. L., 1963, *Gastropods from the Otta Conglomerate:* Norsk Geol. Tidsskr., v. 43, p. 75-81.

SILURIAN

By A. J. Boucot[1]

[Oregon State University]

CONTENTS

	PAGE
GENERAL TRENDS	A168
CORRELATION AND BIOSTRATIGRAPHY	A170
DISTRIBUTION OF SILURIAN STRATA	A172
North America	A173
Europe	A174
Africa	A174
Asia	A174
Australia, New Zealand, Oceania, Indonesia, and the Philippines	A175
Central and South America	A175
SILURIAN PALEOECOLOGY	A176
SILURIAN FLORA AND FAUNA	A176
Chitinozoans and Acritarchs	A177
Arenaceous Foraminifers	A177
Porifera	A177
Coelenterates	A177
Bryozoans	A178
Brachiopods	A178
Gastropods	A179
Cephalopods	A179
Bivalves	A179
Trilobites	A179
Eurypterids	A179
Ostracodes	A180
Other Arthropods	A180
Echinoderms	A180
Graptolites	A180
Conodonts	A181
Vertebrates	A181
Calcareous Algae	A181
Remains of Land Plant Type	A181
REFERENCES	A181

---

[1] Manuscript received January, 1969; revised manuscript received May, 1975.

GENERAL TRENDS

During Silurian time, a period with a duration of about 30 million years (BOUCOT, 1975), the stratigraphic, paleontologic, and paleogeographic entities developed during the Cambrian-Ordovician were further extended without major changes. The relative time duration of the following major subdivisions of the Silurian have been estimated from considerations of "average" evolutionary rates: early and middle Llandoverian 0.9, late Llandoverian 1.0, Wenlockian 0.9, Ludlovian 1.0, and Pridolian 0.5 (BOUCOT, 1975).

The Silurian was a time of little or no orogeny, except for the later phases of Taconic age orogenies that may have persisted in some areas from the Late Ordovician into the Early Silurian; however, on a worldwide basis the areas affected by Taconic orogeny appear to be very restricted.

Major regression accompanying the continental glaciation affecting much of Africa and South America during the Ashgillian and possibly parts of the earlier Llandoverian occurred in the earlier part of the Silurian. Major transgression on a worldwide scale occurred later in the period (see BOUCOT, 1975, and references therein for details); however, regression occurred in areas subject to isostatic rebound, particularly in Africa.

Volcanism of Silurian age was very limited in distribution, only portions of a few geosynclines exhibiting any significant developments.

The climate of the Silurian is poorly known; however, available evidence (see BOUCOT, 1975, for summary) indicates that the North Silurian realm (North America, Europe except for part of the Mediterranean region, Asia, Australia, and northern and westermost South America) was the site of a "warm" climate contrasting markedly with a "cold" climate present in the Malvinokaffric realm (Africa, Mediterranean Europe, southern two-thirds of South America).

Reefs composed of calcareous algae, stromatoporoids, tabulate corals, and other taxa were prominent in the later Silurian, but virtually absent during the Llandoverian and early Wenlockian (the bulk of the Ordovician reef biota disappeared during the Ashgillian extinction event that coincided with major continental regression and glaciation in parts of Africa and South America). The Late Silurian reefs reached a maximum during the late Wenlockian-Ludlovian and then appear to have declined in importance during the Pridolian. Limestone and secondary dolomite are abundant in the North Silurian realm, but are essentially absent in the Malvinokaffric realm.

Marine evaporites are unknown in the Lower Silurian, possibly as a result of a pluvial regime in the North Silurian realm that corresponded with a glacial or very cold regime in the Malvinokaffric realm, followed by the deposition of widespread evaporites on the North American, Siberian, Australian platforms, and possibly on the Russian platform during the Late Silurian.

Red beds, including material probably weathered out of warm, humid land areas, are characteristic of the North Silurian realm but absent in the Malvinokaffric realm.

Taxonomic diversity in level-bottom communities is very low in the Malvinokaffric realm as contrasted with the North Silurian realm. Overall taxonomic diversity is far higher at species through superfamily levels in the North Silurian realm as compared with the Malvinokaffric realm, although the Late Silurian presence of reef communities in the North Silurian realm serves to exaggerate this effect (BOUCOT, 1975).

Biogeography of the marine shelf and platform biota of the Silurian is characterized by a southern Malvinokaffric realm and a northern North Silurian realm (Fig. 1). The North Silurian realm during the Late Silurian may be divided into the North Atlantic, Uralian-Cordilleran, and Mongolo-Okhotsk regions, followed by the appearance during the Pridolian of faunas presaging those of the Devonian in the Eastern Americas realm and Rhenish-Bohemian region of the Old World realm (see BOUCOT, 1975, for extensive treatment).

The initial Silurian faunas of the lower Llandoverian are essentially relict continu-

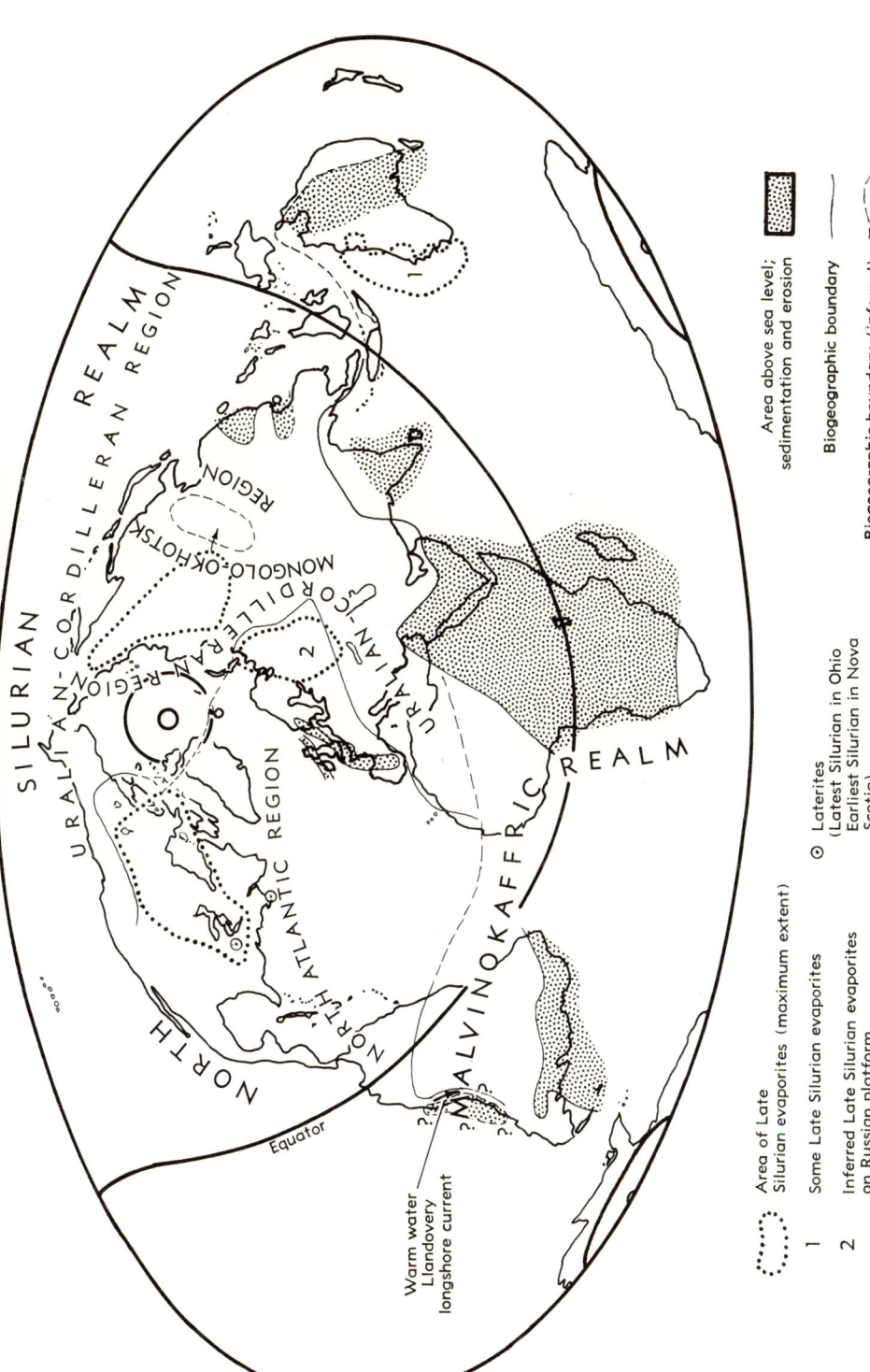

FIG. 1. Silurian biogeographic units and shorelines (modified from Boucot, 1975).

ations of the Ashgillian Old World realm faunas (those of the North American realm became extinct by the end of the Ashgillian). Near the beginning of the late Llandoverian new taxa, some at the superfamily level, migrated from southeastern Kazakhstan and replaced many preexisting Ordovician types (BOUCOT, 1975). A marked enrichment of the Silurian marine fauna during the later Wenlockian was coincident with rapid spread and diversification of reef biotas. At the end of the Ludlovian, marked extinction coincided with the diminution of the reef environment.

The fossil record provides no positive evidence for the existence of either plants or animals on the land or in freshwater during the Silurian, despite an abundance of such evidence for the earliest Devonian. Cutinized trilete spores, spore tetrads, cuticle-like and tracheid-like microfossils are abundant in nearshore facies of the entire Silurian (see work of GRAY cited in BOUCOT, 1975), but it is uncertain whether or not these materials of land-plant types represent plants deposited in the nearshore region after transportation from the land, or whether they represent materials of land-plant type that first developed in nearshore marine and brackish regions.

Silurian graptolites, acritarchs, chitinozoans, and possibly conodonts were depth stratified (see BOUCOT, 1975, for discussion). The benthos of the shelves and platforms was organized into a large number of communities belonging to level-bottom and reef, as well as rocky bottom associations (see BOUCOT, 1975, for an analysis of the changing communities in the level-bottom environment). In general, the intertidal benthic communities have lower diversity than do subtidal communities. Also, a general trend toward smaller shell size from the intertidal to the shelf-margin region is evident in the Silurian (see BOUCOT, 1975, for discussion). The Silurian marine benthos is highly correlated with bottom-sediment type; taxa of rough-water type commonly occur with sand- and granule-size materials, whereas taxa of quiet-water type occur with clay- and silt-size sediment. The photic zone biota differs significantly from that of the subphotic (BOUCOT, 1975).

Much is known about marine filter feeders, suspension feeders, and deposit feeders of the Silurian, but little is known about the carnivores of higher trophic levels and parasites.

We have no solid information about the nature, or even existence, of oceanic benthos, plankton, or nekton during the Silurian, although good evidence is known for a rich neritic planktonic fauna as well as rich shelf-depth benthos. It is logical to deduce that an oceanic fauna was present, however, as there is no rational mechanism for preventing nutrients from having been distributed into the oceanic environment nor for the development of a fauna capable of utilizing those nutrients.

The Silurian benthic faunas are widespread over the vast continental platforms, as are similar Cambrian and Ordovician faunas, in contrast to those of the later Phanerozoic during breakup of the vast pre-Devonian platforms, which were inherited from the worldwide peneplanation of the late Precambrian.

Rates of evolution during the Silurian are easily interpreted in terms of worldwide size of interbreeding populations (see BOUCOT, 1975, for discussion). Extraterrestrial or cataclysmic events are unnecessary to explain the known facts of the fossil record. Major extinction events were absent during the Silurian, although the rate of terminal extinction did fluctuate for one reason or another, as did the rates of phyletic and cladogenetic evolution.

CORRELATION AND BIOSTRATIGRAPHY

Worldwide correlation of Silurian marine beds is presently based almost entirely on brachiopods, graptolites, and conodonts (Fig. 2). These are supplemented locally by chitinozoans, tetracorals, tabulate corals, stromatoporoids, and ostracodes. Little use is currently made of acritarchs, foraminifers, radiolarians, sponges, bryozoans, bivalves, gastropods, nautiloids, trilobites, echinoderms, vertebrates, cutinized spores of land-plant type, as well as many additional groups, although many of them have the

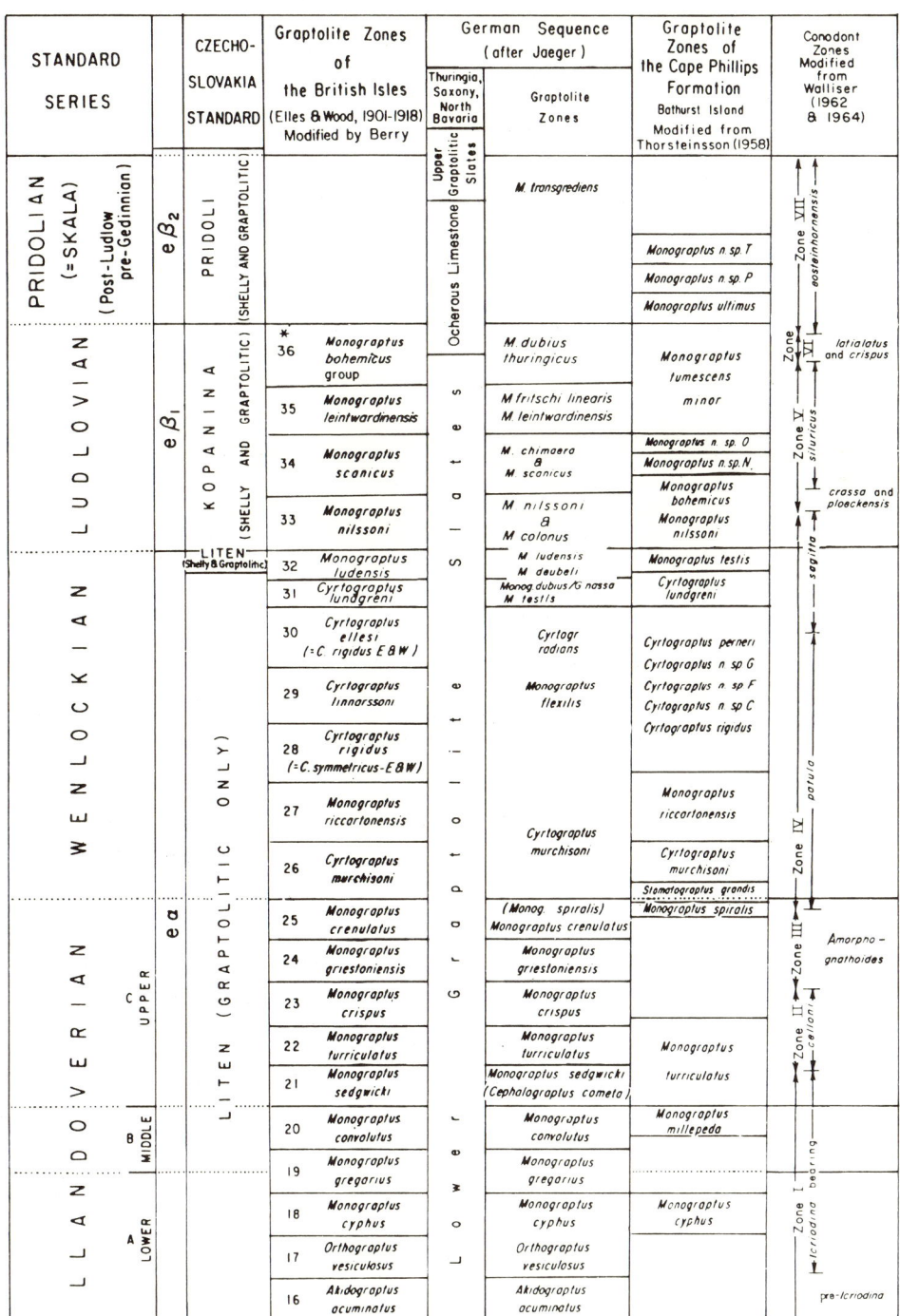

FIG. 2. Zonal schemes for Silurian graptolites and conodonts (modified from Berry & Boucot, 1970, pl. 2).

potential for making important contributions.

Enough work has been done in the Silurian to make it clear that the precision of correlation is, of course, controlled by rates of evolution. *But,* rates of evolution vary from one biologic group to another. Also, biologic groups are not evenly distributed, either locally or worldwide. Thus, we find that the correlation of certain evolving community groups containing taxa characterized by rapid phyletic evolution or rapid cladogenetic evolution (controlled in both cases by sizes of interbreeding populations as the first order control) is excellent, whereas that in slowly evolving groups is correspondingly poor (BOUCOT, 1975). For example, worldwide correlation in the confines of the evolving *Eocoelia* community (*sensu* BOUCOT, 1975, not ZIEGLER, 1965) is very precise as compared with the *Pentamerus* community (*sensu* BOUCOT, 1975, not ZIEGLER, 1965). In the same way correlation employing planktonic communities is most precise for deeper water aggregations, as they sum up the evolution of everything from the surface down, as opposed to shallow-water communities. In other words, a separate correlation scheme must be devised for each evolving community and also for each evolving biogeographic unit that includes its own distinctive communities, with a firm understanding that each of these entities will have its own level of correlation precision; this is the kernel of ecostratigraphic correlation precision.

At the present time, biostratigraphic subdivision of the Silurian is based upon the modified ELLES and WOOD graptolite zones as used by BERRY for those regions yielding graptolites in reasonable abundance (in BERRY & BOUCOT, 1970; BERRY & BOUCOT, eds., 1974; see also BULMAN, 1970). The widespread Silurian shelly faunas commonly lack graptolites in sufficient numbers to make them useful for zonal purposes. In shelly beds recourse at the present time is had to brachiopods as discussed in BERRY and BOUCOT (1970). For the Llandoverian and Wenlockian, the latter only in North America above the lower Wenlockian, the various stricklandiid lineages are employed (see BERRY & BOUCOT, 1970). The stricklandiids are supplemented in the upper Llandoverian-lower Wenlockian by species of *Eocoelia*. Brachiopods useful for the bulk of the Upper Silurian are at present used on a first or last appearance basis as we currently do not have enough information about their evolution. Corals and stromatoporoids are of some use for zonal purposes, although the extensive potential of tabulate corals, stromatoporoids and pelmatozoan columnals has not as yet been taken advantage of outside the Soviet Union. Conodonts have recently become of great use in zoning the Silurian in the shelly carbonate lithofacies. Other groups of marine organisms, including the various arthropods (except for the ostracodes that are of considerable local value in a few regions), and the molluscan classes, are presently of little value for biostratigraphic work within the Silurian, a situation that undoubtedly reflects lack of work upon them rather than any intrinsic biologic characteristic. Chitinozoa are presently beginning to be of value for interregional correlation. The bulk of bivalve taxa appears to have been so environmentally restricted to very shallow waters that it is doubtful if they will ever loom very large for purposes of zonation and correlation within the Silurian.

DISTRIBUTION OF SILURIAN STRATA

Beginning in 1960, W. B. N. BERRY and I made an intensive effort to compile the available data concerning Silurian correlation, paleogeography, and lithofacies on maps at the scale of 1:5,000,000. In this work we have received the aid and encouragement of some hundreds of interested geologists and paleontologists from all over the world. The work is far enough advanced now to enable certain generalizations to be made (see Fig. 1).

1) During most of Silurian time the present-day continental areas of the Northern Hemisphere were subject to relatively shallow-water marine sedimentation. Known Silurian land areas were relatively

minor as contrasted with the vast reaches subjected to marine sedimentation. These land areas included the "Appalachia" of eastern North America; a relatively small island or islands involving portions of southeastern Britain, western Scandinavia, Brittany, and adjacent portions of northwestern Spain; plus a portion of the Siberian platform during post-Wenlockian time; much of the Angaran shield; Africa southeast of the central Sahara, as well as much of Arabia and adjacent Egypt.

2) On the contrary, much of the present-day Southern Hemisphere continental area may have been above sea level including the bulk of Africa, Australia west of the Tasman geosynclinal region (except for the western and northwestern fringe), Antarctica, and eastern South America including the Falkland Islands.

3) The area occupied by Silurian geosynclines is relatively minor (see BERRY & BOUCOT, 1967) as contrasted with the broad platform expanses. The stratified geosynclinal rocks of the Silurian are predominantly of terrigenous, nonvolcanic nature except for the Ural-Kazakhstan geosyncline, which includes a very large proportion of dark-colored volcanic and volcanogenic rocks.

4) Silurian platform rocks may be divided into Platform Carbonates (including a dolomitic suite and a limestone suite) and Platform Mudstones (BERRY & BOUCOT, 1967, 1970).

5) The present-day continental distribution of Silurian strata indicates that during the Silurian significant areas of land and platform, which were subject to shallow-water marine sedimentation, extended out beyond the present-day continental strandlines. Our knowledge of the distribution of Silurian shallow-water animal communities upon the present-day continents helps to reinforce the conclusions based upon lithofacies distributions.

6) All of this information indicates that the present-day Northern Hemisphere continental areas were regions upon which marine faunas faced few obstacles to distribution insofar as land is concerned in breaking up current or temperature patterns. BOUCOT (1974) discussed most of the possible Silurian continental relations.

NORTH AMERICA

North America during the Silurian was characterized by the presence of a vast continental platform, the continuation of that present during the Cambrian-Ordovician, extending from central Nevada and eastern British Columbia to Anticosti Island and Davis Strait in the east, and from El Paso, Texas, north to Cornwallis Island. This vast platform is covered with a veneer of marine strata consisting chiefly of limestone and dolomite, including two restricted areas (Michigan basin and Hudson Bay basin) containing Upper Silurian evaporitic suite rocks. It is doubtful if any extensive land areas were situated upon this platform.

Geosynclinal belts are located around the periphery of the North American platform. The Appalachian geosyncline on the southeast was characterized by a linear land mass twice the length of Cuba separating it from the platform. This land mass, the Appalachia of the literature, steadily diminished in size throughout the period until it was completely, or almost totally, submerged by the end of the period. Within the Appalachian geosyncline a complex Silurian paleogeography changed continually throughout the period (see BERRY & BOUCOT, 1970). A narrow coastal volcanic belt was present from southeastern New Brunswick southwest to the Boston region.

The Ouachita geosyncline to the south is very poorly known, and extends from the subsurface of the Mississippi embayment west-southwest into at least central Chihuahua. Insofar as we have knowledge of the sedimentary rocks, it contains pelitic and arenaceous Silurian strata.

The Cordilleran geosyncline on the west is poorly known as contrasted with the Appalachian, but several linear lithofacies belts can be recognized (BERRY & BOUCOT, 1970). The presence in western North America of extensive transverse faulting of San Andreas type makes the restoration of the original paleogeography of the Cordilleran geosyncline away from the platform margin very difficult. In any case, a variety of rock types is present.

The Franklinian geosyncline to the north can be subdivided into several belts (BERRY & BOUCOT, 1970). The Cape Phillips belt next to the platform margin is relatively similar to the Western Assemblage belt in the same position in the Cordilleran geosyncline.

Off most of East Greenland we lack evidence about Silurian rocks. It is remarkable that during the Silurian the margins of the North American platform remained relatively fixed.

EUROPE

The Silurian of Europe consists of the western, Eurafrican portion of the vast Old World platform, which covers most of Eurafrica and Asia, including the Caledonian geosyncline to the northwest, and the Novaya Zemlya-Ural-southeast Kazakhstan geosyncline to the east, which terminated against the land mass of the Angaran shield in the south.

The Eurafrican portion of the Old World platform may be divided into an eastern Dolomite and Limestone region (the so-called Russian platform) that continues to the southeast into southwestern Asia; a northern and central Platform Mudstone region that extends from Sweden in the north, southward into the central Sahara, as well as from Morocco in the west to Jordan in the east; and a central and southern nonmarine region that extends from the central Sahara to Capetown, and includes much of Arabia (BERRY & BOUCOT, 1972, 1973). The presence of a single shallow-water *Heterorthella* community occurrence in the Capetown region serves to substantiate the African picture. The Eurafrican portion of the Old World platform has commonly in the past had those portions occurring on the sites of younger geosynclinal areas (e.g., the Pyrenees, Alps, Hercynian chains) assigned a geosynclinal character themselves, but no significant difference in lithologic character, thickness, or fauna between these rocks and the Platform Mudstone appears to exist on either side in tectonically nongeosynclinal areas (for example, the Barrandian basin, the Carnic Alps, the Polish subsurface, and Scania as contrasted to the previously cited locales).

The extant portions of the Caledonian geosyncline contain a variety of clastic strata, together with rare, localized volcanics. This geosyncline is relatively narrow.

It is bounded to the northwest in Britain by an extensive land area (BERRY & BOUCOT, 1974), and to the southeast by another land area. This southeastern British land area separated the Platform Mudstone to the east from the geosyncline proper. This southeastern land area can be linked with similar areas in western France, northern Spain, and southwestern Norway to make a relatively linear island or belt of islands.

The Uralian geosyncline (BOUCOT, 1969) is characterized by a tremendous accumulation of greenstone derived from basic volcanics, and bounded to the west and east by banded argillites that further grade laterally into platform carbonate rocks. The width of the Uralian geosyncline diminished by about 50 percent near the end of the Wenlockian.

During the early and middle Llandoverian much of the Russian platform was probably above sea level; late Llandoverian time saw extensive submergence; and Pridolian time, at least in the north and northwest, saw the initiation again of nonmarine conditions.

AFRICA

As noted earlier, the African portion of the Old World platform can be subdivided into a southern nonmarine region extending north into the central Sahara, and a northern half of platform mudstone extending from the central Sahara into the Mediterranean region. African Silurian stratigraphy appears to be very monotonous (BERRY & BOUCOT, 1973).

ASIA

The Silurian of Asia is less well known than that of either North America or Eurafrica. The Asian Silurian can be divided, for convenience, into a northern, Siberian platform separated by the Angaran shield land area (also of platform nature) from the southern Chinese-Southeast Asian platform, which grades farther west through the Himalayan region to join the Eurafrican portion of the Old World platform.

The Siberian platform is characterized by a thin veneer of platform carbonates of Llandoverian-Wenlockian age, overlain by a red-bed and evaporite sequence of later

Silurian age. The southern margin of the Siberian platform (interpreted by BERRY & BOUCOT, 1967, as extending from the Yenissei River east into the Chegitun region of Chukotka, i.e., including the Verkhoyansk Range, Kolyma, and the Chegitun region) onlapped the northern margins of the Angaran shield during the early half of the Silurian and offlapped it during the latter half.

The Angaran shield area is characterized by fringing nonmarine deposits extending from southeastern Kazakhstan easterly through the southern margins of the Siberian platform into the Amur River region. To the south, the Angaran shield is poorly known from a Silurian point of view, but can be inferred to have shown an offlap relationship relative to China from the beginning to the end of the period. The western margins of the land area are situated in the southeastern Kazakhstan region.

The Chinese-Southeast Asian platform was characterized by extensive platform mudstone development during the early half of the period, and was succeeded by clastic shallow-water and limestone strata during the late half. Small land areas were present in southeastern China and Yunnan during the period. From the Shan States south through the Isthmus of Kra as far as Kuala Lumpur there is a western gradation in platform carbonate rocks of Silurian age (BERRY & BOUCOT, 1972).

Geosynclinal rocks may be present in the eastern portions of the Japanese Islands, but are absent elsewhere except along the geosynclinal boundary with the Russian platform from southeast Kazakhstan to Novaya Zemlya.

AUSTRALIA, NEW ZEALAND, OCEANIA, INDONESIA, AND THE PHILIPPINES

The Silurian of the Tasman geosyncline extends north from Tasmania through Victoria, New South Wales, and Queensland, and may be inferred to swing westerly through the mountainous spine of New Guinea into the Vogelkop region. A variety of geosynclinal rocks are present, but are not too well studied on a regional scale. The mainland of Australia is devoid of marine Silurian rocks west of the Tasman geosyncline, although relatively unfossiliferous nonmarine strata may be present. A borehole on Australia's westernmost island, Dirk Hartog, penetrated Silurian fossiliferous carbonate rocks unrelated to those elsewhere (TALENT, BERRY, & BOUCOT, 1975).

Northwestern South Island, New Zealand, contains Silurian clastic rocks in sequence with other geosynclinal early Paleozoic strata; whether or not these rocks are part of the Tasman geosyncline remains to be determined.

Silurian rocks have not been recognized in Oceania, Indonesia, or the Philippines, although their absence may be ascribed to lack of adequate field work rather than other causes.

CENTRAL AND SOUTH AMERICA

The Silurian of South America consists of platform mudstone to the west and north, and is inferred to grade easterly into nonmarine platform beds. Fossiliferous Silurian rocks are presently known from the Merida Andes of Venezuela, the Lake Titicaca region of Peru, much of Bolivia, from northern Argentina south to about Latitude 40, in Paraguay, and in the Amazon and Parnaiba-Maranhao basins of central and eastern Brazil. It is of interest that the Parana basin of southern Brazil and adjacent Uruguay has not yielded Silurian fossils nor have the Paleozoic strata of the Falkland Islands; these two regions may have been nonmarine during the Silurian. Marine Silurian rocks have not been found along the little known coast of southern Chile, but are predicted to occur there. BERRY and BOUCOT (1972), have further summarized the Silurian of South America.

Silurian strata have not been recognized in Central America.

SILURIAN PALEOECOLOGY

Silurian marine paleoecology, both benthic and planktonic, has been recently compiled and reviewed (BOUCOT, 1975). A variety of shelly benthic communities may be encountered on various transects from the shoreline to the continental margin. The precise communities encountered depend on local conditions determined by such factors as turbulent- or quiet-water conditions, hypersaline or normal salinity, or brackish water, light penetration (as deduced from the presence of organisms dependent on light such as calcareous algae), oxygen tenor as it affects either infaunal or epifaunal organisms, and others. The number of such communities encountered on a particular transect depends in part on the uniformity or nonuniformity of the environmental conditions encountered. This refers to level-bottom communities dominated in large part by brachiopods. Additional are level-bottom communities dominated by organisms such as stromatoporoids and tabulate corals. The complex of reef communities is important in the Upper Silurian. Silurian reefs are dominated by calcareous algae, stromatoporoids, and tabulate corals, but include a variety of communities dominated by both attached and vagrant organisms occupying a great variety of niches. Planktonic communities, including such organisms as graptolites and conodonts, and possibly acritarchs, are in large part depth stratified. Global temperature and other biogeographic barriers resulted in a variety of communities developing in isolation from each other although reflecting similar physical conditions in part. Communities of the Malvinokaffric realm probably existed in a relatively low-temperature regime characterized by low taxonomic diversity for the level-bottom benthic organisms. The communities concluded to represent the intertidal region (Benthic Assemblages 1 and 2 of BOUCOT, 1975) are characteristically of low diversity, whereas the photic-zone communities (Benthic Assemblage 3) range from high to low taxonomic diversity, depending on the restrictiveness of the environment, and the subphotic to continental-margin communities may also be either low or high in taxonomic diversity. Beyond the position of the continental margin (Benthic Assemblages 5 and 6) there is little evidence for the presence of shelly organisms, although trace fossils are present. Trace fossils occur from the intertidal Benthic Assemblage 1 position seaward to well beyond Benthic Assemblage 6 in a taxonomically systematic manner. Among closely related taxa, shells in the intertidal zone tend to be significantly larger than those of the subtidal, and in general, shells decrease in size from the shallow-intertidal to the shelf-margin region.

Little is known of trophic relations in the Silurian owing, naturally, to our ignorance of much of the fauna and flora. It is uncertain as to what organisms played the role of phytoplankton, although the acritarchs may possibly have fulfilled this role. The bulk of the benthic marine invertebrates were low-level filter and suspension feeders as well as deposit feeders. We know little about the various higher trophic levels, although it is reasonable to infer that there were a variety of carnivores including such animals as eurypterids and nautiloid cephalopods.

The shelly biomass decreased significantly from the intertidal and shallow-subtidal region through the shelf-margin region where it finally failed.

SILURIAN FLORA AND FAUNA

It is presently uncertain whether there was a nonmarine flora and fauna during the Silurian. By contrast, the marine fauna of the Silurian is rich, and has been subjected to intensive study since interest in fossils first arose over two hundred years ago. In general, Silurian invertebrates form a relatively perfect continuum with those of the Ordovician and the Devonian with no major breaks of the type encountered adjacent to the Permian-Triassic boundary. This situation manifests itself by the difficulty that specialists have in assigning many boundary faunas to the Silurian or adjacent systems. The only break of any consequence, and this is relatively minor when compared with some other parts of the Phanerozoic, is that between the middle and upper Llandoverian. A number of new families first appear in the interval C_1-C_3 of JONES (1925) in the upper Llandoverian. The geologically almost instantaneous appearance of these families is perplexing, but at least in the case of the brachiopods (which see), those of southeastern Kazakhstan provide a satisfactory pre-late Llandoverian source. The Devonian boundary is recognized by the appearance of taxa belonging to several families previously unrecognized in the record insofar as brachiopods are concerned, but most groups, including the brachiopods, manifest a relatively continuous record across this boundary. The early and middle Llandoverian faunas have, as would be expected, a very late Ordovician aspect (they are essentially relict holdovers). To a lesser extent, those of the Pridolian begin to show a Devonian aspect.

Biogeographically the marine fauna of the Silurian is relatively cosmopolitan, especially as contrasted with the immediately preceding and succeeding faunas, which are highly provincial. The bulk of the widespread Silurian invertebrates are concluded to have had their origins in the northwestern part of the Old World during the Late Ordovician (see section on brachiopods), although some taxa belong to widespread Late Ordovician groups. As detailed in the section on the brachiopods, the only regions of strong Silurian provincialism are a restricted area in central Asia and another which includes southern South America and central and southern Africa. During the late half of the Silurian (Wenlockian through Pridolian) a degree of provincialism developed which heralded the high endemism of the succeeding Early Devonian.

When studied on a worldwide basis, the strongly developed faunal differentiation characterizing the Silurian can best be ascribed to the effects of environment. Many characteristic associations of Silurian invertebrate taxa have such an irregular, disjunct worldwide distribution, related in part, however, to geographic factors such as proximity to ancient shorelines or to reef masses, that an environmental control appears to be the only logical conclusion at the present time.

CHITINOZOANS AND ACRITARCHS

Chitinozoa have been little studied except in Europe and North Africa, and minor work has been done in the New World, Asia, and Australia. It is obvious that their distribution will be found to be worldwide, but at present too little information is available with which to make biogeographic conclusions although they are beginning to be of great value. Silurian zonation based on information from acritarchs is at present very crude as contrasted with that derived from brachiopods, ostracodes, graptolites, and conodonts.

ARENACEOUS FORAMINIFERS

Arenaceous foraminifers have been recovered from acid residues in many regions of the world, but knowledge concerning their detailed stratigraphic and geographic distribution is still too fragmentary to form the basis for conclusions of worldwide utility.

PORIFERA

Porifera are widely distributed in marine strata of Silurian age, but are too seldom recognized, collected, and studied to be of much stratigraphic, ecologic, or biogeographic value at the present time.

COELENTERATES

Tetracorals, tabulate corals, and stromatoporoids are among the most abundant fossils in the Silurian fauna (jellyfish are so rarely preserved as to be little more than curiosities). The stromatoporoids appear to be one of the major, if not the major, agents in the formation of Silurian bioherms and biostromes. Algae have not been considered extensively in this facies. Isolated occurrences of stromatoporoids in shallow-water environments are not uncommon.

Stromatoporoidal masses are abundant in platform carbonate environments, and also much of the limestone (admittedly small in volume) found in the geosynclinal terrigenous and volcanic terrains is rich in stromatoporoids. The platform mudstones do not yield a stromatoporoid fauna, except in those rare instances (Prague region, Carnic Alps) where the bottom was shallowed enough by volcanism or other agencies for a shelly fauna to flourish locally. The worldwide distribution of Silurian stromatoporoid taxa is too little known to provide significant biogeographic data. The situation regarding tabulate corals and tetracorals differs little from that for Silurian stromatoporoids except that the former are also very widely distributed in nonbioherm, nonbiostrome platform carbonate and geosynclinal shallow-water beds, and are not major contributors in making up the volume of bioherms and biostromes. It should be noted, however, that within the Soviet Union, where tabulate corals and stromatoporoids have been intensively studied, they have proved to be of great stratigraphic value. Studies elsewhere in Europe are in agreement with the Soviet experience. These groups will be of great worldwide value when studied more fully.

BRYOZOANS

Stony and fenestellid bryozoans are widely distributed within the platform carbonates, but their stratigraphic, ecologic, and biogeographic value is at present poorly known on a worldwide basis. Their abundance, however, augurs well for the future.

BRACHIOPODS

Brachiopods are widely distributed in relatively shallow-water platform and geosynclinal marine Silurian rocks of the world. Origins of the Silurian brachiopod fauna are twofold, including an early Llandoverian complement (BOUCOT, 1968) transitional from the Late Ordovician of northern Europe as far east as the Urals and Kazakhstan, and a late Llandoverian complement transitional from Late Ordovician to middle Llandoverian taxa restricted during this earlier time interval to southeastern Kazakhstan. Ecologically the brachiopods were strongly controlled in their distribution by a variety of factors. For the late Llandoverian in Britain, ZIEGLER (1965) has demonstrated a correlation of brachiopod distribution with depth; ZIEGLER and BOUCOT (in BERRY & BOUCOT, 1970) have shown the same for North America; and BERRY and BOUCOT (1967) have suggested that in Britain and North America, though not always elsewhere, temperature varying with depth is the primary factor in their distribution. Certain taxa appear to have been ecologically restricted to reef environments during the Silurian. BOUCOT (1975) has provided more detail on the Silurian brachiopod-dominated communities.

At the generic level, the Silurian brachiopod fauna is relatively cosmopolitan, contrasting strongly in this regard with the preceding Late Ordovician and succeeding Early Devonian faunas. The only areas of strong provincialism during the Silurian are in South America south of Lake Titicaca and in South Africa where the *Clarkeia* community (a shallow-water entity) is found, and in the narrow belt extending from the Amur River in the east, westward through the southeastern Sayan and Altay, Tuva, and Mongolia to a few parts of southeastern Kazakhstan where the *Tuvaella* community (another shallow-water entity) is found. At the regional level, brachiopods of Wenlockian through Pridolian age in the vast region from the east slope of the Urals and Kazakhstan, Australia to southeastern Alaska, the Yukon, Cornwallis Island, North Greenland, and Nevada display increasing endemism when compared with faunas known from the west slope of the Urals, Europe, the bulk of North America, northern and westernmost South America, and southern Asia. It must be emphasized, however, that this growing endemism is of much smaller magnitude than the succeeding provincialism of the Early Devonian (BOUCOT, 1975).

A simple observation whose significance is hard to assess is the fact that most Silurian brachiopod taxa are represented by smaller specimens than are found in the Lower Devonian, in accord with COPE's Rule.

Brachiopods are now of great service in zoning and correlating the Silurian of the world as well as in making ecologic analyses of community distributions on a worldwide scale.

GASTROPODS

Gastropods are widespread, but low in frequency, in marine Silurian rocks of the world. In an average collection of Silurian invertebrates, it is seldom that more than one or two gastropods occur with a thousand other shells. Poleumitid gastropods and euomphalopterids are the most characteristic Silurian gastropods, persisting as Silurian relicts in the Early Devonian Old World realm. Poleumitids are widespread in platform carbonates, with relatively high abundance in reef and calcareous shale associations, and also occur in many shallow-water terrigenous and carbonate geosynclinal areas. Platyceratids are relatively ubiquitous. Bellerophontids are widespread, but most abundant in Benthic Assemblages 1 and 2 (BOUCOT, 1975). The platform mudstones are very poor in gastropods. Plectonotid gastropods are largely restricted to Benthic Assemblages 1 and 2. Presently known anomalies in the distribution patterns of Silurian gastropods may be most easily ascribed to either environmental factors or lack of information. General lack of interest in Silurian gastropods is not conducive to correcting this situation.

CEPHALOPODS

Nautiloid cephalopods are among the most widely distributed Silurian invertebrates in all marine rock types. They are particularly abundant in the *"Orthoceras"* limestones forming a characteristic, although volumetrically small, part of the Eurafrican platform mudstones, and in the biohermal and biostromal structures of the platform carbonates. An average collection of Silurian invertebrates almost invariably contains a few percent of nautiloid fragments. Unfortunately, the nautiloids have not received the intensive worldwide treatment that they need in order to be of prime importance for ecologic, biostratigraphic, and biogeographic purposes.

BIVALVES

Bivalves are widespread in the marine Silurian of the world. Their most common occurrences are in the bivalve-graptolite-rich platform mudstones of the Eurafrican platform (see BERRY & BOUCOT, 1967), where the so-called "Bohemian forms" occur in relative abundance, in reef environments where the genus *Megalomus* is commonly cited, and in the Benthic Assemblage 1 and 2. Elsewhere, bivalves are relatively rare in Silurian rocks, seldom making up more than one percent of the number of specimens in a collection. In the bivalve-graptolite faunas, as well as in Benthic Assemblages 1 and 2, bivalves may commonly form the most abundant element in the fauna, both in number of species and specimens. Worldwide distribution patterns of Silurian bivalves have been too little investigated to provide any current information of biogeographic importance. Presently known anomalies in distribution pattern can rationally be ascribed to environmental factors rather than to isolating mechanisms.

TRILOBITES

Trilobites are widely distributed in Silurian rocks in almost every marine environment. At present, their stratigraphic and geographic distribution is poorly known in comparison with other major invertebrate groups. Trilobites form on the average about five percent or less of the invertebrate fauna in terms of specimens. It is notable that Silurian, as well as Early and Middle Devonian, homolanotids are almost completely restricted to the relatively shallow-water regions of Benthic Assemblages 1 and 2. The ecologic distribution and zonation of trilobites will undoubtedly prove very valuable, although up to the present no systematic efforts have been made in this direction. Biogeographically, no systematic attempt has been made to synthesize available information regarding trilobites, although scattered comments are consistent with the more extensive data for brachiopods.

EURYPTERIDS

Eurypterids are relatively uncommon in Silurian strata, except in a few specialized facies. Ecologically they are of great potential use in discriminating between certain marine, brackish, and hypersaline environments. Particular assemblages are characteristic of brackish and hypersaline environments (see information in Boucot, 1975). Their occurrences are so few as to make them of little value for biogeographic considerations, but they are of some biostratigraphic value.

OSTRACODES

Ostracodes are widely distributed in marine Silurian rocks of the world. In those regions and parts of the column subjected to intensive study they have proved to have great stratigraphic and biogeographic value. Upper Llandoverian through Pridolian ostracodes of the Appalachians appear to represent a series of relatively endemic faunas known from Anticosti Island southwest through the central Appalachians, including occurrences in the northern Appalachians of Quebec and northern Maine. Ludlovian and Pridolian ostracodes of the Baltic region, Great Britain, Nova Scotia, southeastern New Brunswick, coastal Maine, and eastern Massachusetts, as well as Podolia, form another apparently endemic fauna, which has been zoned stratigraphically in some detail. The lowest upper Llandoverian of the North American platform contains some forms that may be endemic and stratigraphically restricted. Isolated Silurian faunas elsewhere are too little known and unique to afford information for stratigraphic or biogeographic generalizations. In summary, it can be stated that the ostracodes may possibly provide important biogeographic insights within the Silurian (see BERDAN & MARTINSSON, in BERRY & BOUCOT, 1970).

OTHER ARTHROPODS

At present, other groups of Silurian arthropods are too little known to be of stratigraphic or biogeographic value.

ECHINODERMS

Pelmatozoan debris is widespread and volumetrically important in marine Silurian rocks of the world, particularly platform carbonates; however, such material is present in very low abundance in platform mudstones. Silurian crinoids are widespread, particularly in platform carbonates, but are too uncommon on a worldwide basis to contribute much to our knowledge of Silurian biogeography. The same situation obtains with the cystoids and blastoids. Other groups of echinoderms are relatively rare in the Silurian.

Except in certain quiet-water, marly facies (both reef and level-bottom), articulated echinoderms are relatively rare. Twenty years of collecting have provided the writer with less than a dozen specimens from outside these facies, despite the abundance of pelmatozoan debris in many places, which suggests that our knowledge of Silurian echinoderms may be very distorted.

Of biogeographic interest are the European, African, eastern North American, and Oklahoma occurrences of *Camarocrinus* in strata of Pridolian age (except for Tennessee and Missouri where an early Gedinnian age is indicated as well), and its occurrence in southern China and Burma in strata of early Llandoverian age.

Disarticulated pelmatozoan debris has been extensively employed within the Soviet Union for purposes of zonation and correlation, but specialists in other countries have not yet employed or tested this work.

GRAPTOLITES

Together with the conodonts, the graptolites are the most ubiquitous of the well-studied Silurian invertebrates. They are known from every marine environment; presumably owing to their depth-stratified, planktonic mode of life. Graptolites are abundant in platform mudstones (BERRY & BOUCOT, 1967), and every other marine facies, except for the hypersaline, has yielded at least a few graptolites. The popular notion ascribing the graptolites only to the "black shale" and "basin" facies is grossly in error, although statistically it is true that these rocks have yielded the great majority of specimens. They display a certain level of biogeographic differentiation in the Silurian (see BOUCOT, 1975, for summary).

CONODONTS

Silurian conodonts are worldwide in their distribution, having been recovered in abundance by acid treatment from carbonate rocks of the platform carbonate, platform mudstone, and geosynclinal facies (carbonate interbeds with terrigenous and volcanic strata). They have proved of prime importance in the zonation of upper Llandoverian through Pridolian beds (see KLAPPER, BERRY, & BOUCOT, in BERRY & BOUCOT, 1970). Conodonts have been recovered in abundance from strata reflecting so many environments in the Silurian, from the intertidal to shelf margin and regions beyond, that it is tempting to conclude that like the graptolites they probably are the remains of primarily pelagic organisms. Their ubiquitous worldwide occurrence and zonation indicates that like brachiopods and graptolites they were a relatively cosmopolitan group during the Silurian.

VERTEBRATES

Vertebrates obtained from brackish beds are too rare and poorly known at present to be of widespread stratigraphic or biogeographic value; however, where studied, particularly in northern Europe, isolated microscopic denticles, plates, and spines obtained from acid residues are beginning to have stratigraphic value in beds of Ludlovian and Pridolian age. Vertebrates are widespread in North Silurian realm marine beds, although normally they are rare as individuals.

CALCAREOUS ALGAE

Although calcareous algae are widespread and abundant in marine Silurian beds deposited in the photic zone, they have been little studied. Therefore, they are not used for either biostratigraphic or biogeographic purposes, although they have been used to to a limited extent in paleoecologic considerations, chiefly in determining the lower limits of the photic zone, and relative degrees of turbulence as indicated by different types of oncolites.

REMAINS OF LAND PLANT TYPE

Acid-resistant spores, tracheid-like and cuticle-like microfossils, are abundant in shallow-water and nearshore Silurian facies (see GRAY, LAUFELD, & BOUCOT, 1974, for a typical example), but it is highly uncertain as to whether or not the parent organisms lived on land, in the intertidal environment, or in the shallow subtidal region, although the desiccation-resistant structure of the fossils indicates an affinity for at least in part a life style involving the nonaqueous. Macrofossils of land plant type are known from the Wenlockian onward, but their growth site is fully as uncertain as is that of plant microfossils of land type.

REFERENCES

Berry, W. B. N., & Boucot, A. J., 1967, *Pelecypod-graptolite association in the Old World Silurian:* Geol. Soc. America, Bull., v. 78, p. 1515-1522.——1970, *Correlation of the North American Silurian rocks:* Geol. Soc. America, Spec. Paper 102, 189 p.——(eds.), 1972, *Correlation of the South American Silurian rocks:* Geol. Soc. America, Spec. Paper 133, 59 p.——1972, *Correlation of the southeast Asian and near eastern Silurian rocks:* Geol. Soc. America, Spec. Paper 137, 65 p.——1973, *Correlation of the African Silurian rocks:* Geol. Soc. America, Spec. Paper 147, 83 p.——(eds.), 1974, *Correlation of the Silurian rocks of the British Isles:* Geol. Soc. America, Spec. Paper 154, 154 p.

Boucot, A. J., 1968, *Origins of the Silurian fauna:* Geol. Soc. America, Spec. Pap. 121, p. 33-34.

——1969, *The Soviet Silurian: recent impressions:* Geol. Soc. America, Bull., v. 80, p. 1155-1162.——1974, *Early Paleozoic evidence of continental drift: pro and con:* in Plate tectonics—assessments and reassessments, C. F. Kahle (ed.), Am. Assoc. Petrol. Geologists, Mem. 23, p. 273-294.——1975, *Evolution and extinction rate controls:* 427 p., Elsevier (Amsterdam).

Bulman, O. M. B., 1970, *Graptolithina:* in Treatise on invertebrate paleontology, Part V (Revised), Curt Teichert (ed.), 163 p., Geol. Soc. America and Univ. Kansas (Boulder, Colo.; Lawrence, Kans.).

Gray, Jane, Laufeld, Sven, & Boucot, A. J., 1974, *Silurian trilete spores and spore tetrads from Gotland: their implications for land plant evolution:* Science, v. 185, p. 260-263.

Jones, O. T., 1925, *On the geology of the Llandovery district, pt. 1:* Geol. Soc. London, Quart. Jour., v. 81, p. 344-388.

Talent, J. A., Berry, W. B. N., & Boucot, A. J., 1975, *Correlation of the Silurian rocks of Australia, New Zealand, and New Guinea:* Geol. Soc. America, Spec. Paper 150, 108 p.

Ziegler, A. M., 1965, *Silurian marine communities and their environmental significance:* Nature, v. 207, no. 4994, p. 270-272.

DEVONIAN IN THE EASTERN HEMISPHERE[1]

By M. R. House

[University of Hull, England]

CONTENTS

	PAGE
INTRODUCTION	A184
FACIES PATTERNS IN THE EUROPEAN DEVONIAN	A184
STAGES AND ZONES OF THE EUROPEAN DEVONIAN	A187
Stages	A187
Zones	A189
MAIN AREAS OF DEVONIAN ROCKS	A192
Western and Southern Europe	A192
Soviet Union	A192
China, South and Southeast Asia	A193
Australasia	A193
Africa	A193
LOWER DEVONIAN FAUNAS	A194
Europe	A195
Asia	A196
Australasia	A196
Africa	A197
MIDDLE DEVONIAN FAUNAS	A198
Europe	A199
Asia	A201
Australasia	A202
Africa	A203
UPPER DEVONIAN FAUNAS	A203
European Frasnian	A204
European Famennian	A205
Asia	A206
Australasia	A207
Africa	A208
DEVONIAN GLOBAL RECONSTRUCTION	A208
REFERENCES	A210

---

[1] Manuscript received September, 1975.

INTRODUCTION

It was the prescient recognition by WILLIAM LONSDALE that corals from the limestones of South Devon were intermediate in type between those then known from the Silurian rocks of the Welsh Borders and those from the Mountain Limestone [Mississippian] that led, in April, 1839, to the founding of the Devonian System by ADAM SEDGWICK and RODERICK MURCHISON (1839a, p. 259, 354; 1839b, p. 121; LONSDALE, 1840, p. 721). This correlation between the marine rocks of Devon and the Old Red Sandstone of Wales, the Welsh Borders, and Scotland WILLIAM BUCKLAND hailed as "undoubtedly the greatest change which had ever been attempted at one time in the classification of British rocks" (GEIKIE, 1875, p. 269). After spending four months of the summer and autumn of 1839 on the Continent, SEDGWICK and MURCHISON were able to recognize the excellent development of the new system along the Rhine valley and Schiefergebirge. In late 1839, MURCHISON proceeded to Russia where, in the Leningrad region, he observed the intercalation of Devonian marine levels within typical Old Red Sandstone sequences. Thus, in an incredibly short time, the new system was recognized and broadly correlated over wide areas of Europe (Fig. 1). Subsequently, knowledge has grown considerably, owing its growth to no small extent to the German school of geologists and paleontologists centered in the several universities that surround the varied Devonian rocks of the Rhenish Schiefergebirge.

This contribution is concerned solely with invertebrates of the Devonian, especially those that have a bearing on broad biogeographical problems. A recent review of Devonian floras was given by EDWARDS (in HALLAM, 1973) and one on Devonian fish by HALSTEAD and TURNER (in HALLAM, 1973). Invertebrates of the Old Red Sandstone facies are similar to those reviewed for the Western Hemisphere by NORRIS (this volume, p. A218), and especially various groups of arachnids, scorpions, and early insects add interest to the fauna. But for details on these, and the many groups not discussed here, reference should be made to the appropriate volume of the *Treatise*.

FACIES PATTERNS IN THE EUROPEAN DEVONIAN

The existence of distinct facies regimes in the Devonian, as in other systems, heightens the problems of zonal correlation, paleoecological interpretation, and the recognition of faunal provinces. Because a clarification of these is an essential background to biogeographical analysis, and because in the European Devonian literature several peculiar facies terms are in current usage, a general discussion of facies patterns is an important preliminary to the biogeographical review.

The major facies regimes of continental fluvial facies, nearshore clastic facies, and offshore clear-water facies are known in the European Devonian as the Old Red Sandstone, Rhenish and Hercynian or Bohemian magnafacies (ERBEN, 1964). Some difficulties result from different uses of the last two terms (BROUWER in OSWALD, 1968b, p. 1149), and concern has been expressed that the Rhenish-Hercynian distinction, which is clearly seen in the Lower Devonian of Germany, becomes less clear upward in the succession. The truth is, however, that reef and associated carbonates are almost unknown in the Devonian from the late Frasnian onward; excepting this the distinction remains valid. The accompanying facies transect illustrates the model that has been proposed by SCHMIDT, RABIEN, KREBS, and others to explain the relationships of the facies types (Fig. 2). In some usages the Rhenish facies would include all the neritic environments there marked, in others the reef and associated facies would be included with the Hercynian or Bohemian facies.

It is only the carbonate and basinal facies having a Devonian terminology that need be commented upon here.

For the carbonate complexes KREBS (1974) distinguished the *Schwelm* facies as a bank or biostrome regime in which

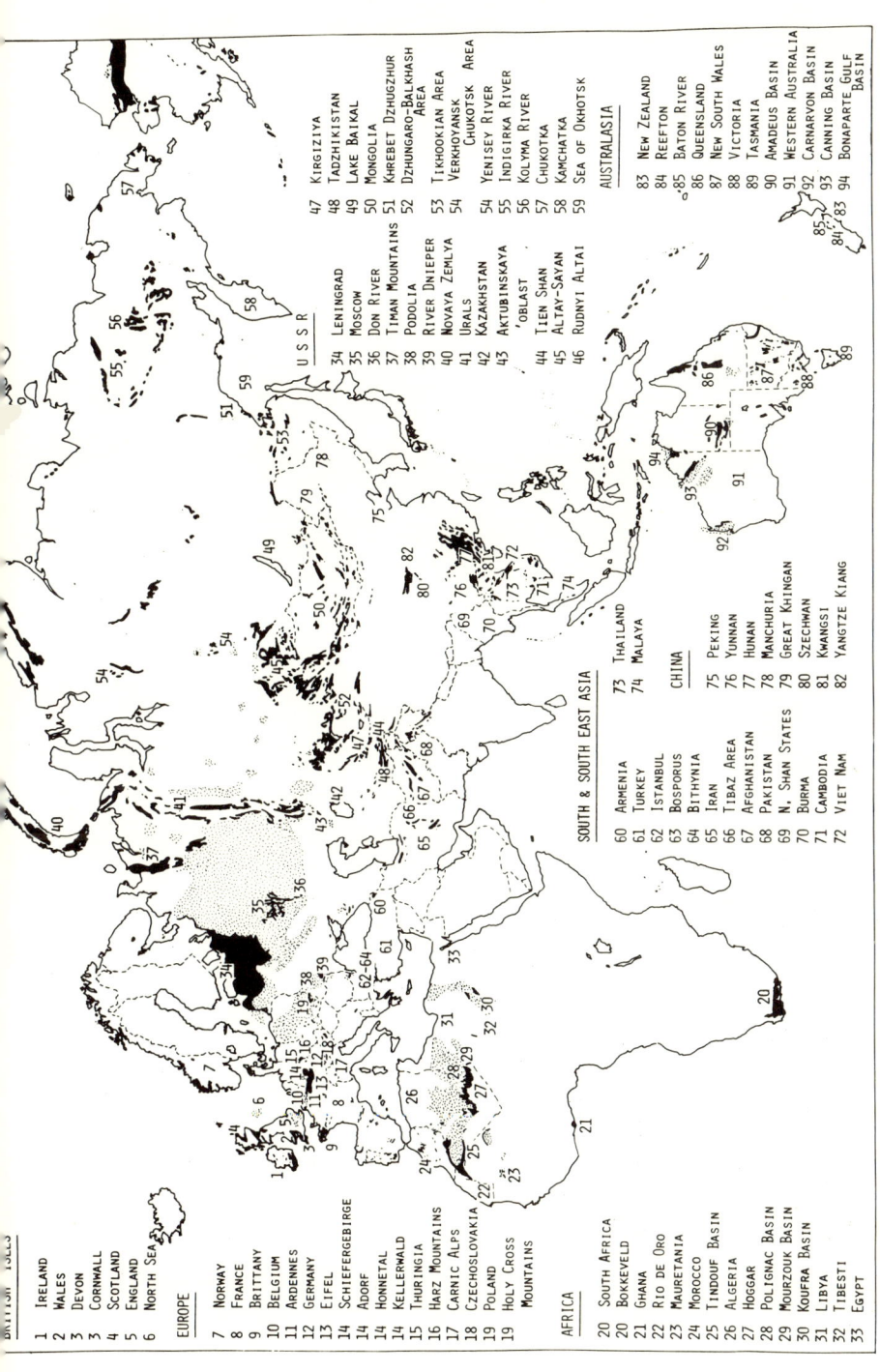

Fig. 1. Main areas of Devonian rocks in the Eastern Hemisphere and localities referred to in the text (House, n). [Outcrops are in black and known subsurface occurrences are stippled.]

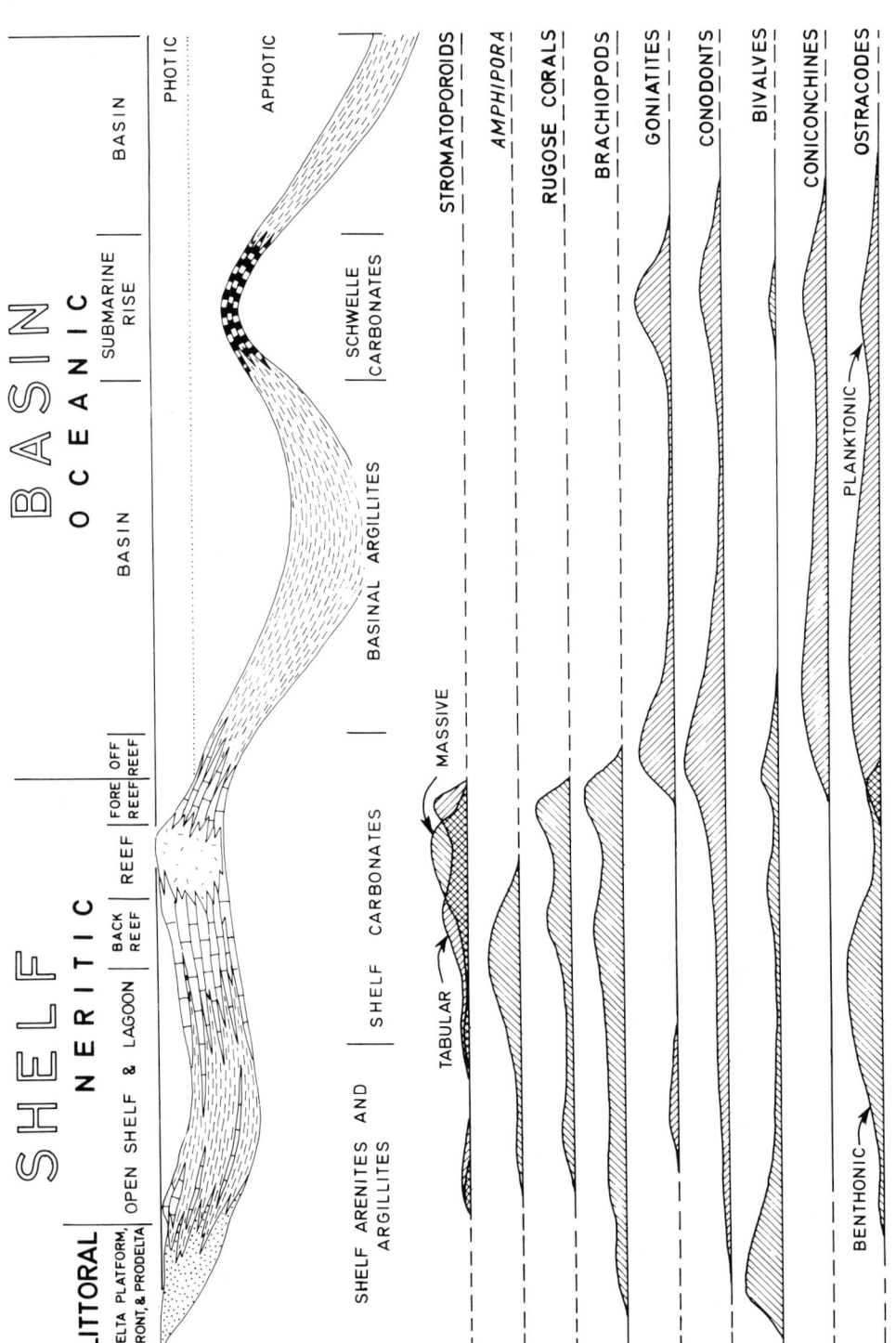

Fig. 2. Distribution of facies types and faunal groups from the Devonian of Europe (after House, 1975b).

the common stromatoporoids and corals lived essentially below the level of strong wave action. This often formed the base on which the true reef, or *Dorp* facies, was developed, usually showing rapid growth of reef organisms, notably stromatoporoids, associated with a subsidence of the platform as a whole. The *Dorp* facies is further subdivided into backreef, reef, and forereef facies. The *Iberg* facies was the final capping phase at the apex of the subsiding reef masses before their final termination (usually in the early and middle Frasnian): by the stage at which this facies was formed, no backreef lagoon existed, and detritus from the capping *Iberg* facies interfingers laterally with basinal argillite within a very short distance.

The basinal argillite areas comprise the *Becken* facies. In the lowest Devonian this is the graptolite shale facies, but with reduction of the graptolites, the nowakiids, and styliolinids, of supposed planktonic habit, characterize the *Styliolinenschiefer* and *Tentaculitenschiefer* of the middle and early Late Devonian. Gradually during the Frasnian these Cricoconarida became extinct or reduced and their place was taken in the planktonic environment by the fingerprint ostracodes of the Entomozoidae characterizing the Late Devonian *Cypridinenschiefer* facies. Tongues of allodapic turbidite limestone from adjacent reefs or crinoid groves project into the lateral *Becken* areas to produce the distinctive *Flinz* facies of alternating limestone and shale units.

The submarine rise or seamount limestones have very much reduced but nearly complete sequences of nautiloid and ammonoid limestones. This is the *Schwelle* facies of pelagic limestones with the distinctive associated lithologies of *Knollenkalk (griotte), Kramenzelstein,* and *Kalkknollenschiefer.*

The European Devonian, has a special place in the international development of Devonian studies, and many of the biofacial, biostratigraphic, and biogeographic concepts owe their origin here. It is appropriate that a discussion of the European schemes of zonation and some of the facies concepts should be described as a preliminary to a broader discussion of the biogeography of Devonian rocks of the Eastern Hemisphere.

STAGES AND ZONES OF THE EUROPEAN DEVONIAN

The diversity of facies represented by the European Devonian has resulted in much effort being expended in the establishment of schemes of biostratigraphic zonation as a basis for interpreting these changes. Despite this there remain problems of correlation between the various facies. There is even no agreed consensus for the stage nomenclature and the definition of stages for the European Devonian, although this is currently the concern of the Devonian Subcommittee of the I.U.G.S. Stage or series names in common use are shown in Figure 3. Some notes on usage are required.

STAGES

The Silurian-Devonian boundary stratotype, accepted at the 1972 meeting of the International Geological Congress at Montreal, is within Bed No. 20 of a section at Klonk, near Suchomasty, Czechslovakia, as described by CHLUPÁČ (1972), at the entry of abundant *Monograptus uniformis* (McLAREN, 1972); this is also taken as the Budnanian-Lochkovian boundary in the Czech terminology. This decision brings international stability to the long-continued boundary problem, but pre-1972 literature may well use different boundary definitions. This is particularly so in the British type area for the Silurian and Devonian where the Ludlow Bone Bed, a marker horizon near the original level proposed by MURCHISON (WHITE, 1950; EARP, 1973) may correlate with the base of the Pridolian and of the *M. ultimus* Zone (WESTOLL in HOUSE, ed., 1977), and the new boundary may lie well up in the Downtonian or near the base of the Dittonian. In the classic sections in Podolia in the Ukraine (NIKIFOROVA & PREDTECHENSKIY, 1968; SOKOLOV, 1972) the new boundary falls within the Borschov, and in other regions similar corrections will be needed. The final report of the

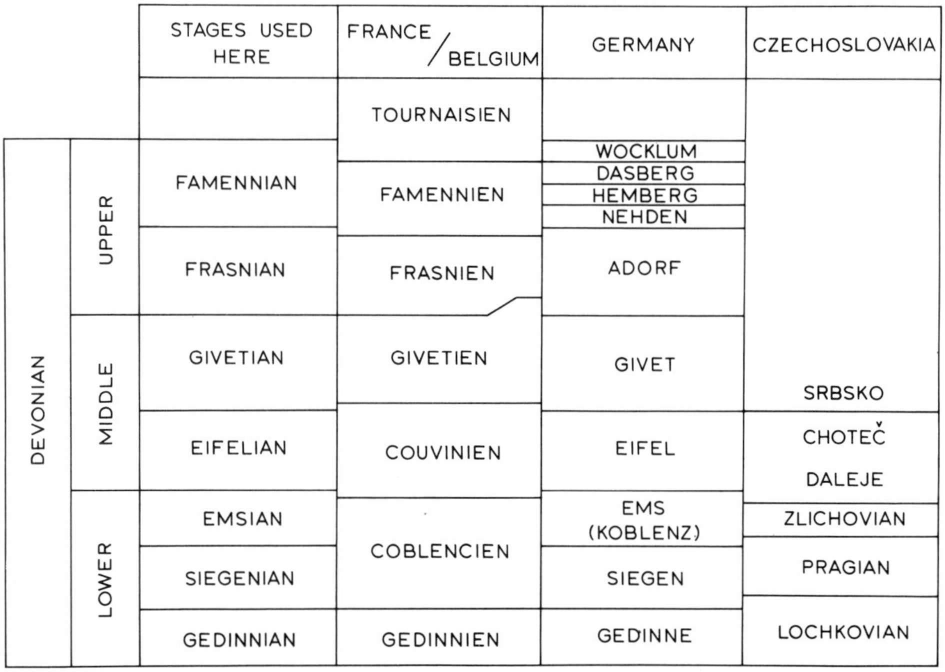

		STAGES USED HERE	FRANCE / BELGIUM	GERMANY	CZECHOSLOVAKIA
			TOURNAISIEN		
DEVONIAN	UPPER	FAMENNIAN	FAMENNIEN	WOCKLUM / DASBERG / HEMBERG / NEHDEN	
		FRASNIAN	FRASNIEN	ADORF	
	MIDDLE	GIVETIAN	GIVETIEN	GIVET	SRBSKO
		EIFELIAN	COUVINIEN	EIFEL	CHOTEČ / DALEJE
	LOWER	EMSIAN	COBLENCIEN	EMS (KOBLENZ)	ZLICHOVIAN
		SIEGENIAN		SIEGEN	PRAGIAN
		GEDINNIAN	GEDINNIEN	GEDINNE	LOCHKOVIAN

FIG. 3. Stage level terminology used in Europe and that used herein (House, n). (For definitions see text.)

Committee on the Silurian-Devonian boundary (MARTINSSON, ed., 1977) gives an international review of the placing of the boundary.

The type of the Gedinnian Stage is in Belgium and it carries a nearshore clastic-facies fauna that appears to correlate with the Lochkovian; it is to be hoped that the base of the Gedinnian will be so defined as to agree with the system boundary. The base of the Siegenian is taken at the base of the Siegener Schichten of the Rhineland and to correspond approximately to the base of the Schistes de Saint Hubert of the Ardennes; regions again of restricted facies faunas. The major problem here is the placing of the Gedinnian-Siegenian boundary in relation to the Lochkovian-Pragian boundary of Czechoslovakia (PŘIBYL & VANEK, 1968); at present the latter boundary is thought to lie perhaps as high as the middle-upper Siegenian boundary of Germany, but the matter is not resolved. The base of the Emsian in Germany is taken at the base of the Singhofen Beds near the type area (KUTSCHER & SCHMIDT, 1958, p. 57 et seq.; ERBEN & ZAGORA in OSWALD, 1968a) and the base of the upper Emsian at the Ems Quartzite. Again there are problems in the correlation of this essentially clastic sequence with the more carbonate-rich rocks of Czechoslovakia. For example, CARLS and others (1972) considered that the Zlichov-Daleje boundary correlates approximately with the lower-upper Emsian boundary, yet others have taken the top of the Zlichov as the top of the Emsian. With such discrepancies, faunal ranges using a "common" stage nomenclature must be treated with the greatest circumspection.

Similar problems surround the Lower-Middle Devonian boundary. The Belgian Couvinian Stage includes the Assise de Bure and equivalents of the Cultrijugatus Schichten; these in Germany are referred to the Emsian, the base of the Eifelian and Middle Devonian being taken in the Eifel at the junction of the Heisdorf and Laucher Schichten. But recently it has become apparent that the latter level may not be the same as that taken in the Schiefergebirge where the *Gracilis*-Grenze (ERBEN, 1962, p. 95) has been used and thought to correspond to the Czech Zlichov-Daleje boundary, a correlation disputed by CARLS and others (1972). The type Givetian (ERRERA, MAMET, & SARTENAER, 1972) in northern France raises less problems since the conodont relations have been established by BULTYNCK (1970). The difficulty here is that the current Couvinian-Givetian boundary seems likely to be younger than the base of the *Cabrieroceras crispiforme* Zone, which has been used to define the junction in the Schiefergebirge, Czechoslovakia (CHLUPÁČ in OSWALD, 1968a, p. 117), North Africa (HOLLARD, 1974), and which has been recognized in North America also (HOUSE, 1962). Only an internationally agreed definition can resolve this problem.

Until very recently, there was considerable agreement on the definition of the Middle-Upper Devonian boundary and on the base of the Belgian type Frasnian, the boundary being drawn at the base of the Assise de Fromelennes (WATERLOT, 1957, p. 199-205) in Belgium and at the base of the *Pharciceras lunulicosta* Zone of Germany (KUTSCHER & SCHMIDT, 1958, p. 309), a level which seems to fall within the conodont *Polygnathus varcus* Zone (HOUSE, 1973a; BENSAID, 1974; BOUCKAERT & STREEL, 1974). Unfortunately, when the primary work on Frasnian conodonts was done on the Adorf type sections (ZIEGLER, 1958, 1962, 1966) the correlation with the goniatite zonation established by WEDEKIND (1913) at that locality was misaligned and it still seems that the 1α limestone of WEDEKIND is placed too high; it is more likely to be represented by the new *Pharciceras*-bearing level described by KULLMAN & ZIEGLER (1970). The result of this error has been that conodont workers have used a boundary higher than the basal *P. lunulicosta* Zone that German geologists have taken as the Middle-Upper Devonian boundary for a considerable time. It seems also to be the reason behind moves in Belgium to redefine the base of the Frasnian at some level near the base of the Assise de Frasnes (HOUSE, 1973a). The lithostratigraphic definition of the base of the Famennian Stage is in southern Belgium at the boundary of the Assise de Matagne and the Assise de Senzeille in the Senzeille railway cutting (SARTENAER, 1970). This seems to be an approximate correlative of the Adorf-Nehden or *Manticoceras-Cheiloceras* Stufen boundary used in Germany. The Belgian boundary seems to lie near the base of the conodont middle *Palmatolepis triangularis* Zone (BOUCKAERT et al., 1972); the German boundary may be near the base of the upper *P. triangularis* Zone (BUGGISCH & CLAUSEN, 1972).

The Devonian-Carboniferous boundary has been defined at the *Wocklumeria-Gattendorfia* Stufen boundary following the Heerlen Conference decision (JONGMANS & GOTHAN, 1937, p. 6) and this was also the recommendation of the Sheffield Conference (GEORGE & WAGNER, 1969, p. xlv). The sections considered particularly important in these discussions are those in the Oberrödinghausen railway cutting in the Hönnetal (SCHINDEWOLF, 1937; VÖHRINGER, 1940; AUSTIN et al., 1970). While there are detailed problems at this level (ALBERTI et al., 1974), the boundary has had the recommendation of the Carboniferous conferences for 40 years and therefore seems the most acceptable, but no recommendation has been made by the Carboniferous Subcommission. Unfortunately, the Belgian and French geologists have not redefined the base of the Tournaisian to correspond with the Heerlen and Sheffield decisions, and their usage has been followed by most Russian authors. It would appear that the basal *Gattendorfia* Stufe agrees approximately with the Tn_{1a}-Tn_{1b} boundary or lies within Tn_{1a} (CONIL & PIRLET, 1970).

These difficulties of definition and usage, and the problems of correlation between the zonal schemes using different groups need to be borne in mind when Devonian literature is read, since quite misleading information on fossil ranges and distribution at particular times can be inferred if the local terminology is misunderstood.

ZONES

Biostratigraphic schemes of zonation have been proposed using a range of invertebrate groups. Again, because these groups tend to be restricted to certain environments, there are substantial problems in the correlation between such schemes. The

STAGES	STUFEN	AMMONOID ZONES	CONODONT ZONES	OSTRACODE ZONES
UPPER DEVONIAN — FAMENNIAN	WOCKLUMERIA	Prionoceras sp.	Protognathodus	
		Cymaclymenia euryomphala		
		Wocklumeria sphaeroides	Spathognathodus costatus	
		Kalloclymenia subarmata		Maternella hemisphaerica and Maternella dichotoma
	CLYMENIA	Gonioclymenia speciosa	Polygnathus styriacus	
		Gonioclymenia hoevelensis		
	PLATYCLYMENIA	Platyclymenia annulata	Scaphignathus velifer	Franklinella intercostata
		Prolobites delphinus		Entomozoe (R) serratostriata
		Pseudoclymenia sandbergeri	Palmatolepis marginifera	Franklinella intercostata Interregnum
	CHEILOCERAS	Sporadoceras pompeckji	Palmatolepis rhomboidea	Entomozoe (R) serratostriata and Entomozoe (N) nehdensis
		Cheiloceras curvispina	Palmatolepis crepida	Ungarella sigmoidale
UPPER DEVONIAN — FRASNIAN	MANTICOCERAS	Crickites holzapfeli	Palmatolepis triangularis	Entomozoe (E) variostriata
			Palmatolepis gigas	Wcicatricosa E barrandi Interregnum
		Manticoceras cordatum	Ancyrognathus triangularis	Waldeckella cicatricosa
			Polygnathus asymmetricus	
		Pharciceras lunulicosta	S. hermanni - P. cristatus	Wcicatricosa E torleyi Interregnum Franklinella torleyi
MIDDLE DEVONIAN — GIVETIAN	MAENIOCERAS	Maenioceras terebratum	Polygnathus varcus	
		Maenioceras molarium		
		Cabrieroceras crispiforme	Icriodus obliquimarginatus	
MIDDLE DEVONIAN — EIFELIAN	ANARCESTES	Pinacites jugleri	Polygnathus kokeliana	
			Spathognathodus bidentatus	
		Anarcestes lateseptatus	Icriodus corniger	
LOWER DEVONIAN — EMSIAN	ANETOCERAS	Sellanarcestes wenkenbachi	non - latericrescid Icriodus - Polygnathus	GRAPTOLITE ZONES
		Mimagoniatites zorgensis	Icriodus bilatericrescens s.s.- Sp. steinhornensis - Polygnathus	
		Anetoceras hunsrueckianum	Icriodus h. curvicauda - Icriodus huddlei s.s.	Monograptus yukonensis
				Monograptus falcarius
LOWER DEVONIAN — SIEGENIAN			Icriodus h. curvicauda - rectangularis s.l. angustoides	Monograptus hercynicus
LOWER DEVONIAN — GEDINNIAN			Ancyrodelloides - Ic. pesavis	Monograptus praehercynicus
			Icriodus w. postwoschmidti	
			Icriodus woschmidti s.s.	Monograptus uniformis

FIG. 4. Zonal schemes for the Devonian using various invertebrate groups. Precise correlation between these should not be inferred (House, n). (For sources see text.)

accompanying table (Fig. 4) is intended to illustrate several of these schemes. Correlation between columns should be regarded as an approximate guide only. In this section reference will be made to some of the more substantial biostratigraphic works that have formed the basis for these zonations.

As for the period up to the close of the Cretaceous, the ammonoid succession has been regarded as giving the Devonian orthochronology, that of the preceding graptolites giving out in the Siegenian where the ammonoids appear. The goniatite and clymenoid zonations result largely from the works of WEDEKIND (1908, 1913, 1914, 1917) followed by that of SCHINDE-WOLF (1923, 1937), SCHMIDT (1921, 1924, 1950), MATERN (1931a,b), CLAUSEN (1969, 1971), BENSAID (1974) and others. These mostly concern the German succession. Reviews have been given by HOUSE (1962, 1973b) and in the revised *Treatise* Part L, Mollusca 4 (in preparation).

The conodont scheme has developed over the last fifteen or so years, and it now provides the most detailed subdivision available for the Upper Devonian. The major works are by BISCHOFF & ZIEGLER (1957), ZIEGLER (1958, 1962, 1966, 1971), WITTEKINDT (1965), KREBS & ZIEGLER (1965), KLAPPER & ZIEGLER (1967), BULTYNCK (1970), and CARLS (1969). A general review of the European conodont sequence

STAGES	STUFEN	TENTACULITED ZONES	W. EUROPE BRACHIOPOD ZONES	USSR BRACHIOPOD ZONES
UPPER DEVONIAN — FAMENNIAN	WOCKLUMERIA			(INCLUDED WITH THE CARBONIFEROUS IN THE USSR)
UPPER DEVONIAN — FAMENNIAN	CLYMENIA		Ptychomaletaechia letiensis	Adolfia talasica
UPPER DEVONIAN — FAMENNIAN	PLATYCLYMENIA		Ptychomaletaechia letiensis	Leiorhynchus (Zigania) ursus and Ptychomaletoechia turanica
UPPER DEVONIAN — FAMENNIAN	CHEILOCERAS		Pt dumonti, Pt gonthieri, Pt omaliusi, Eoparaphorhynchus lentiformis etc	Zilimia polonica and Cyrtospirifer spp.
UPPER DEVONIAN — FRASNIAN	MANTICOCERAS	(FOR DETAILED ZONATION SEE LYASHENKO 1965 ETC)	Cariorhynchus tumida, Hypothyridina cuboides, Cyrtospirifer malaisei, Cyrtospirifer orbelianus, Cyrtospirifer tenticulum	Hypothyridina cuboides and Theodossia anossofi, Cyrtospirifer spp., Mucrospirifer novosibiricus, Uchtospirifer murchisonianus
MIDDLE DEVONIAN — GIVETIAN	MAENIOCERAS	Nowakia otomaria	Mediospirifer mediotextus, Stringocephalus burtini, Undispirifer undiferus	Euryspirifer cheehiei, Stringocephalus burtini, Bornhardtina langurica
MIDDLE DEVONIAN — EIFELIAN	ANARCESTES	Nowakia sulcata, Nowakia holynensis, Nowakia richteri, Nowakia cancellata	Spinatrypa kelusiana, Spinacytia ostiolatus, Euryspirifer intermedius	Lazutkinia mamontovensis, Megastrophia uralensis
LOWER DEVONIAN — EMSIAN	ANETOCERAS	Nowakia barrandei, Nowakia praecursor, Nowakia zlichovensis	Paraspirifer cultrijugatus, Euryspirifer paradoxus, Acrospirifer pellico	Eospirifer superbus, Paraspirifer ? gurjevkensis, Latonotoechia latoma
LOWER DEVONIAN — SIEGENIAN		Nowakia arcuata, Paranowakia intermedia	Acrospirifer primaevus	Spirigerina supramarginalis etc
LOWER DEVONIAN — GEDINNIAN		Paranowakia bohemica, Tentaculites ornatus	Delthyris dumontianus, Howellella group	Howellella group

FIG. 4. *(Continued from facing page.)*

is given by ZIEGLER (1971).

It was not recognized until 1960 that graptoloid graptolites extend up into the Lower Devonian when Lochkovian graptolites were found associated with Rhenish type brachiopods in the Kellerwald (JAEGER, 1965). Notable information has come also from boreholes in Poland (KOREJWO & TELLER, 1965) and from Thuringia, the Carnic Alps, and other regions. In a review of the Lower Devonian graptolite sequence, JAEGER (1973) recognized the zones indicated on the accompanying table.

For the Cricoconarida (dacryoconarids, "tentaculitids," coniconchines) a detailed zonation using several of the various groups has stemmed from the initial studies by G. P. LYASHENKO (1953, 1959) in European Russia. BOUČEK (1964) followed this scheme for the Prague area and Harz, and LARDEUX (1969) for southern Europe and North Africa.

The Late Devonian planktonic ostracodes provide the most useful sequence in the Late Devonian basinal shale facies, and the sequence has been worked out by RABIEN (1954, 1956, 1960, 1970), BLUMENSTENGEL (1959), and GROOS-UFFENORDE & UFFENORDE (1974).

For the brachiopods the situation is more complex. The zonal schemes given here are taken from NALIVKIN, RZHONSNITSKAYA, & MARKOVSKIY (1973). But in reality different brachiopod groups have been used for discrimination without always the formal distinction of zones. The most thorough treatment of Devonian brachiopod ranges has been given by BOUCOT (1975), but since he does not define his stage units there are problems in his collation. Nevertheless, the work of BOUCOT and his colleagues will be heavily drawn on in the later sections where specific reference to monographic works will be made.

For trilobites, the long continued studies by RUDOLF and EMMA RICHTER (especially 1926, 1950, 1952) provide a biostratigraphic foundation, and this work has been continued by ALBERTI (1968, 1973). WEDEKIND, using corals, established a series of zones for the German Middle Devonian (WEDEKIND, 1924-25; WEDEKIND & VOLLBRECHT, 1931-32), and BIRENHEIDE (1968, 1972) has done much to revise this, but schemes using corals have not found favor internationally.

MAIN AREAS OF DEVONIAN ROCKS

Devonian rocks are widely scattered on all of the continents of the Eastern Hemisphere (Fig. 1), but recorded information varies considerably. The most comprehensive collation for most areas is the *International Symposium on the Devonian System*, edited by D. H. OSWALD, published in 1968. For the vast area of Russia there are two volumes of a Silurian-Devonian Boundary Conference (published under the general editorship of D. V. NALIVKIN in 1971 and 1973) and the two volumes of the *Devonskaya Sistema* edited by NALIVKIN, RZHONSNITSKAYA, and MARKOVSKIY (1973). For China the most accessible source is the *Paleogeograficheskiy Atlas Kitaya* (SOKOLOVA, ed., 1962) and a regional stratigraphy (SOKOLOV, 1960). Australia has been reviewed in a series of Devonian correlation charts (STRUSZ, 1972; PICKETT, 1972; ROBERTS et al., 1972).

WESTERN AND SOUTHERN EUROPE

The areas of Old Red Sandstone sedimentation comprise Ireland, Wales, Scotland, and Norway. In southeastern England, the North Sea, and probably a subsurface belt passing east to the Baltic States platform, sediments of continental facies are occasionally intercalated by marine tongues. From southwest England and northwestern France, eastward through southern Belgium, the Eifel, Rhenish Schiefergebirge, and Harz Mountains to the Holy Cross Mountains of southern Poland lay the Hercynian (Armorican or Variscan) geosyncline in Devonian times. Farther south the scattered outcrops mostly occur in the Paleozoic massifs strongly affected by late Paleozoic and subsequent tectonism; these mostly indicate marine deposits of geosynclinal type which, with the Hercynian belt, continue eastward into the complex series of outcrops of central Asia and the Himalaya.

SOVIET UNION

From the extensive outcrops of the Main Devonian Field near Leningrad Devonian rocks extend in the subsurface eastward to rise to the surface in the Timan and Ural mountains. Southward they crop out along the tributaries of the Don River south of Moscow and along the Dnieper valley in Podolia. Across the bulk of the Russian platform the deposits are of Middle and Late Devonian age and are represented by continental and evaporite deposits with marine intercalations. In Podolia a marine Lower Devonian sequence follows conformably upward from the Silurian (KOZLOWSKI, 1929; NIKIFOROVA & PREDTECHENSKIY, 1968). A fuller marine sequence is typical for the Devonian in its development from Novaya Zemlya, the Ural and the Timan mountains. The Ural sequences are broadly eugeosynclinal to the east and miogeosynclinal to the west. This belt formed the Uralian geosyncline in Devonian times, and was folded in the late Paleozoic.

Scattered Devonian outcrops stretch from the Tien Shan, through Kazakhstan, the Altay to Baikal and Mongolia to the Khrebet Dzhugzhur, and along the coast of the Sea of Okhotsk. This belt broadly formed the Angaran geosyncline, bounded to the north by the western Siberian plate and by the Siberian platform in the east. This belt, too, was folded by Hercynian movements.

In most of the Devonian the Angaran geosyncline was linked with the southernmost part of the Ural geosyncline, and there appears to have been an almost continuous link with its northern parts via the Altay-Sayan. The central and southern parts of the Central Asian fold belts are affected by the Cimmerian (Carpathian) folding, and the southern parts also by Alpine folding. These areas are often considered extensions of the Tethyan or Mediterranean geosyncline.

Mostly continental Devonian is known in the subsurface between the Urals and the Yenisey River, which forms the western margin of the Siberian platform. The occurrences in the western and southern parts of the platform are also of continental facies. Those of the northern margins are mostly marine and have facies affinities with the Urals. In the Verkhoyansk-Chukotsk region of northeastern Siberia, especially along the Indigirka and Kolyma rivers, marine deposits are widespread, again with Uralian affinities.

CHINA, SOUTH AND SOUTHEAST ASIA

The Devonian outcrops of Russian Kirgiziya and Tadzhikistan pass eastward in tracts that are interpreted as a northerly Mongolian-Manchurian geosyncline (part of the Angaran geosyncline) crossing Mongolia to the Sea of Okhotsk, and passing north of the broad supposed land area of the Peking platform. Another tract passed south of this platform forming the wide outcrops of continental and marine Devonian in eastern China, especially in the ancient provinces of Yunnan and Hunan. Within this tract several circumscribed positive areas seem to have been active throughout the Devonian, such as the Tarim massif with the Kun Lun Shan geosyncline of northern Tibet south of it, and the Tsaidam massif with the Nan Shan geosyncline between it and the Peking platform (which embraced all but the Manchurian part of Inner Mongolia). A complex archipelago is envisaged for southwestern China along and south of the Yangtze Kiang River. Yet farther south is the broad area with isolated outcrops that stretch from the Bosporus coast of Turkey, through Iran, Afghanistan, Pakistan, the Northern Shan States, Viet Nam, Thailand, and Malaya. This last tract is distinctive for the European nature of many of its faunas.

AUSTRALASIA

In New Zealand the Lower Devonian is known in the Reefton and Baton River areas. Devonian rocks are widespread in eastern Australia in a belt from Queensland to Tasmania that formed part of the complex Tasman geosyncline composed of marine, volcanic, and continental deposits. In Central Australia the Amadeus basin has continental deposits. Bordering the margins of the Precambrian platform of Western Australia are the marine deposits forming the three discrete areas of the Carnarvon, Canning (and Lennard Shelf basin and Fitzroy trough), and Bonaparte Gulf basins. In these it is the Upper Devonian that is best developed.

AFRICA

The main development of Devonian rocks is in North Africa in a broad belt from the Tindouf basin and isolated outcrops in Mauritania, Rio de Oro, and northern Morocco eastward to the northern Hoggar, Polignac basin, Mourzouk basin, and Tibesti and Koufra basins at the Libya-Egypt frontier. Broadly, this North African development is of shelf-facies marine Devonian lapping southward onto the African Precambrian shield. The faunal affinities are almost wholly European, but with Appalachian elements in the west.

By contrast, the Bokkeveld development of Lower Devonian marine faunas in South Africa shows typical elements of the Malvinokaffric province discussed with Antarctica and South America in the Western Hemisphere section. The Devonian of Ghana seems to show similar affinities, but with European elements also (ANDERSON et al., 1966).

Fig. 5. Paleogeographic map for the Early Devonian (Emsian) of the Eastern Hemisphere (based on Boucot & Johnson, 1973; House, n). [Inferred land areas are stippled and areas of marine rocks are horizontally lined.]

LOWER DEVONIAN FAUNAS

In the Lower Devonian (Fig. 5), faunal provincialism has long been recognized. For the Eastern Hemisphere, and excluding, therefore, the Appalachian province, the division is essentially into an Old World province embracing Europe and North Africa, Asia, Australasia, and western North America, and a Malvinokaffric province embracing South Africa, the Falkland (Malvinas) Islands, South America, and Antarctica (the term Austral province was introduced for the latter by J. M. CLARKE in 1913, but since this could be confused with Australasian faunas, RICHTER (1941) proposed the name *"Malvinocaffrisch"* and it is this in anglicized form that is normally used). Paleomagnetic evidence suggests that the Old World province lay in equatorial and temperate latitudes and that the Malvinokaffric province lay nearer the "southern" pole.

The Old World province is commonly divided into a Rhenish-Bohemian subprovince, comprising Europe excluding the Urals, North Africa, parts of the Near East and, for the late Early Devonian, parts of eastern North America; a Uralian subprovince embracing the Urals and Asiatic Russia and the Angaran geosyncline with elements of the Appalachian province in the latter; a Tasman subprovince recognized in eastern Australia; and a New Zealand subprovince including South Island but with some links (*Reeftonia*) with Southeast Asia (BOUCOT, JOHNSON, & TALENT, 1969; BOUCOT, 1974).

This provincialism is best shown by benthonic organisms, notably brachiopods and corals, and probably reached its acme in the Emsian (BOUCOT, 1975).

EUROPE

In this area there are great faunal contrasts between the Rhenish facies of nearshore clastics with abundant brachiopods and bivalves, the Bohemian facies of calcareous rocks, which carry the richest faunas, and the basinal shales, which preserve graptolites and Cricoconarida. Major problems remain on the detailed correlation between these facies. Space permits only reference to a selection of the many major works on these faunas.

The brachiopods have generally been used for the subdivision of the nearshore clastic facies. The Gedinnian carries *Quadrifarius, Proschizophoria, Podolella, Mutationella,* and *Cyrtina* as newly appearing forms (BOUCOT, 1960; JOHNSON & BOUCOT in HALLAM, 1973). Near the beginning of the Siegenian appear genera such as *Anoplotheca, Bifida, Meganteris, Multispirifer, Pradoia, Rhenorensselaeria,* and others. The stratigraphically useful spiriferids of the Siegenian and Emsian have been monographed, especially by SOLLE (1953, 1971). Notable studies on Lower Devonian bivalves are by MAILLIEUX (1937) for the Ardennes and BEUSHAUSEN (1895) for the Schiefergebirge. Trilobite faunas include *Acastella* and *Warburgella* in the Gedinnian, and common phacopids, *Treveropyge* and *Burmeisteria* in the Siegenian and Emsian (ALBERTI, 1968; RICHTER & RICHTER, 1954).

The Bohemian facies of Czechoslovakia is best known through the detailed studies of CHLUPÁČ (1957, 1959, 1960, and in OSWALD, 1968a), who gave stratigraphical precision to many of the faunas monographed by BARRANDE. CHLUPÁČ (1972) has also described the faunas around the stratotype for the Silurian-Devonian boundary where the base of the Devonian and of the Lochkovian is characterized by the appearance of *Monograptus uniformis* and *Warburgella rugulosa rugosa. Scyphocrinites* is abundant just above and below the boundary. The Lochkovian carries a rich fauna of brachiopods including *Howellella, Plectodonta,* stropheodontids, and spiriferids (HAVLÍČEK, 1959), and trilobites such as proetids and scutellids. Merostomes, phyllocarids, bivalves and coral faunas are also richly represented. The overlying Pragian has distinctive brachiopod faunas, including species of *Sieberella, Glossinotoechia, Stegerhynchus, Hysterolites,* and others. The Lower Devonian here has been subdivided by using Cricoconarida (BOUČEK, 1964), and this work has been extended by LARDEUX (1969) to southern Europe and North Africa. In Czechoslovakia (CHLUPÁČ in OSWALD, 1968a, p. 115) the Zlichov-Daleje boundary, marked by the entry of the goniatite *Gyroceratites* and the loss of the bizarre *Caeleceras* and of trilobites, notably *Odontochile,* has been used as the Lower-Middle Devonian boundary, but CARLS and others (1972) have argued that it corresponds rather with the Lower-Upper Emsian boundary of Germany. Representatives of the Bohemian facies also recur in the Harz Mountains (ERBEN, 1953b, 1960).

The basinal shale facies bearing graptolites extends well up into the Lower Devonian (JAEGER, 1965) and *Monograptus hercynicus* also occurs in the late Lochkovian, enabling correlation between the facies as well as internationally where the Lochkovian-Pragian boundary can be identified better than the Gedinnian-Siegenian boundary. In this facies, also, cricoconarids (LARDEUX, 1969) are increasingly important upward in the succession. It is in the Siegenian Hunsrück-Schiefer of this facies that goniatites appear (ERBEN, 1966), particularly *Anetoceras,* and in this unit also is a remarkable pyritic fauna of crinoids, trilobites and other groups revealed in minute detail by X rays (STÜRMER, 1969).

In the Early Devonian of Europe, therefore, three facies regimes carried their own distinctive faunal elements (ERBEN, 1964) and doubtless had independent community evolution. It is this that has led to difficulties in the use internationally of the classical stage divisions adopted here.

Analysis of the faunal communities or assemblages represented in the various facies regimes of the Lower Devonian of Europe has scarcely begun. FUCHS (1971) has made such a study of the upper Siegenian and lower Emsian facies of the Eifel and recognized a seaward sequence of communities dominated by *Mutationella, Trigonorhyn-*

chia-Subcuspidella, Rhenorensselaria, and *Acrospirifer,* followed by a deeper water community of rarer brachiopods, bivalves and orthocones. BOUCOT (1975) has attempted a preliminary review of communities generally in the Lower Devonian and has drawn attention to the rarity in most of these communities of species or genera of value for time discrimination.

ASIA

Here it is convenient to follow the broad groupings adopted by NALIVKIN, RZHONSNITSKAYA, & MARKOVSKIY (1973) and BOUCOT (1975), recognizing the following divisions for convenience: the Uralian region and Arctic Siberia; the Dzhungaro-Balkhash region; the Altay-Sayan area; Mongolia and the Tikhookian area (the Mongolo-Okhotsk region of BOUCOT); China south of the Peking platform; and the tract of outcrops from Turkey to Southeast Asia. These are in part faunal provinces, in part regionally defined areas.

The Uralian area shows close similarity to Europe and belongs to the Old World province as normally defined. Differences in southern areas, especially southeast Kazakhstan, reflects the terrigenous and volcanic facies there. In the more calcareous facies *Karpinskia* is a characteristic brachiopod, but it also occurs in the Carnic Alps and elsewhere. The Arctic area, including Novaya Zemlya, has widely distributed genera such as *Eoglossinotoechia, Howellella,* and *Hebertoechia,* but the combination of *Phragmostrophia* and *Cortezorthis* also occurs and gives a clear link with Cordilleran faunas (JOHNSON & BOUCOT in HALLAM, 1973). The goniatite faunas of the Urals and Novaya Zemlya are very close to European groups (BOGOSLOVSKIY, 1969, 1972). Rich coral faunas are present, especially in the upper Lower Devonian reef complexes, but these are mostly of European types with some endemics.

The Dzhungaro-Balkhash area of Kazakhstan is distinctive for the presence in it of brachiopod genera of Appalachian type, such as *Leptocoelia, Rhytistrophia,* and *Meristella* at levels probably correlating with the Siegenian. These occur with a mixture of European and Tasman subprovince types. Links both west and east, via the Angaran geosyncline, are inferred. The Australasian *Maoristrophia* also occurs here.

The Altay-Sayan area is thought to have been in marine contact with the northern Urals and with the Dzhungaro-Balkhash area during much of the Early Devonian. Here, rich faunal sequences in carbonate facies continue upward from the Silurian (RZHONSNITSKAYA, 1968). The faunas give good links with the Urals, Podolia, and Czechoslovakia.

In the Mongolia and Tikhookian area Lower Devonian faunas are not well described, but they include *Leptocoelia, Costispirifer,* and *Rhytistrophia* emphasizing links with the Appalachian region. Links to the west are indicated by corals such as *Barrandeophyllum* and *Lindstroemia,* and the crinoid *Kuzbassocrinus.*

Very little is known about the Lower Devonian of China (HAMADA in OSWALD, 1968a). Early Devonian fish are recorded in Yunnan and other areas, but records of marine fossils are mostly unspecified (SOKOLOVA, 1962). *Hysterolites* and *"Anastrophia"* suggest that the upper Lower Devonian is present in Old World facies. A varied Lower Devonian fauna has recently been described by MU and others (1974), which includes *Anetoceras, Erbenoceras,* and *Mimagoniatites* mostly of European type. Other faunal elements are also described.

For convenience, the scattered outcrops along the Tethys belt are considered together. The well-known Lower Devonian faunas of Turkish Bithynia (PAECKELMANN, 1925; KULLMANN, 1973) contain many German genera and species, especially in the Emsian carbonates. In the many Devonian outcrops of Iran, Afghanistan, and Pakistan, Lower Devonian marine faunas are only poorly documented, and the same is true in Burma, Thailand, and Malaya. Noteworthy records are of *Icriodus woschmidti* in the Nowshera Formation of Nowshera, Pakistan (STAUFFER in OSWALD, 1968a), and Lower Devonian graptolites from Maymyo, Burma, and Thailand (JAEGER, STEIN, & WOLFART, 1969). These show that at least locally marine sedimentation continued upward from the Silurian.

AUSTRALASIA

The Tasman geosyncline belt of eastern

Australia has given its name to the Tasman subprovince, a subgroup of the Old World province. Links with Europe are substantial, mainly in Rhenish facies, but also in the carbonate Bohemian type (SAVAGE, 1971). The Early Devonian graptolite faunas of southeastern Australia (JAEGER, 1966, 1970) represent the only graptolite faunas of this age in the southern hemisphere, and they allow close correlation with those of Europe. STRUSZ (1972) has recognized four successive brachiopod faunas for correlation purposes in this area. Fauna I (lower Yeringian) commonly contains *Eospirifer parahentius, Isorthis allani, Lissatrypa lenticulata, Maoristrophia banksi,* and *Boucotia*; Fauna II (upper Yeringian) has the association *Acrospirifer lilydalensis, Cymostrophia stephani,* and *Boucotia*; Fauna III (Tabberabberan) contains *Adolfia, Gypidula vultur,* and *Nadiastrophia*; and Fauna IV (Buchanian), contains common *Buchanathyris, Howittia, Malurostrophia,* and *Spinella*. Fauna I is thought to be Lochkovian, Faunas II and III Pragian, and Fauna IV is thought to range from the latest Pragian to the early Eifelian. Part of Faunas III and IV are referred to the Emsian, and it is then that the Tasman province is particularly well delimited with genera such as *Reeftonia, Maoristrophia, Nadiastrophia,* and *Notanoplia* (the first two also occur in New Zealand and in Asiatic Russia and Manchuria; BOUCOT, 1975, p. 315). *Australocoelia* gives a single link with the Malvinokaffric province.

By contrast, the coral faunas of this belt, dealt with in many monographs by DOROTHY HILL, include large numbers of endemic elements. The corals and conodonts (PHILIP & PEDDER, 1967) have formed the basis for the recognition of 11 faunas (A-K) of which A-F are referred to the Lower Devonian. The conodont sequence, too, shows differences from that of Europe, a matter emphasized again by TELFORD (1975), who considered the sequence of faunas better related to those of western North America.

The New Zealand subprovince is based on Lower Devonian faunas from Baton River and Reefton, South Island, New Zealand. The Baton River fauna was dated as Upper Siegenian or Lower Coblencian by SHIRLEY (1938), as Gedinnian by BOUCOT, CASTER, IVES, & TALENT (1963), and WRIGHT (in OSWALD, 1968a) has suggested levels from Gedinnian to Emsian may be represented. Forms such as *Fascicostella, Acrospirifer, Cyrtina, Howellella, Mutationella,* and *Hipparionyx* all have a European aspect. The Reefton Beds, on the other hand, seem to be of Emsian age, and *Acrospirifer coxi* in the fauna is sufficiently close to *A. hercyniae* of Germany to suggest early Emsian. Distinctive genera of the fauna are *Reeftonia, Maoristrophia, Tanerhynchia, Pleurothyrella,* and the endemic *Allenetes*. Because it is only this fauna for which any discreteness can be claimed, elevation of the local early Emsian as a subprovince seems excessive, particularly as elements of the fauna are increasingly being recognized in the Orient.

AFRICA

The South African faunas of the Lower Devonian are of Malvinokaffric province type and hence have their affinity with regions discussed in the Devonian of the Western Hemisphere (NORRIS, herein p. A218). The North African Devonian, on the other hand, is clearly of Old World province type.

The Bokkeveld fauna of South Africa (REED, 1925; BOUCOT, CASTER, IVES, & TALENT, 1963) appears to be of early Emsian age. Characteristic brachiopods are *Australospirifer, Australocoelia, Scaphiocoelia,* and *Pleurothyrella*. This fauna has clear links with Antarctica, the Falkland Islands and South America, and slight links *(Australocoelia)* with Australia. Paleomagnetic evidence supports the view that this is a high latitude fauna.

The Accraian Series of Ghana carries a fauna of bivalves, gastropods, and homalonotids (SAUL, BOUCOT, & FINKS, 1963) and rare mutationellid terebratulids (ANDERSON, BOUCOT, & JOHNSON, 1966). Because *Mutationella* has not been recorded from the North African Lower Devonian, but occurs in the Falkland Islands, this has been taken to indicate affinity with the Malvinokaffric province.

The North African Lower Devonian of Morocco and Algeria (HOLLARD & LEGRAND

Fig. 6. Paleogeographic map for the Middle Devonian of the Eastern Hemisphere (based on Boucot & Johnson, 1973; House, n). [Inferred land areas are stippled and areas of marine rocks are horizontally lined.]

in OSWALD, 1968a) shows intimate affinity with the classical successions of Germany and Czechoslovakia. But whereas in those areas tectonic complications and poor exposure raise many problems, those of North Africa are superbly exposed along enormous spreads of outcrop. The potential for refined biostratigraphy is immense. Also, facies of both the European Rhenish and Bohemian types occur so that evidence from here has contributed to the correlation between the Czech and German stages.

Monograptus uniformis and the associated Silurian-Devonian boundary faunas and the faunal sequences have been well documented by HOLLARD (1968, 1974), and work by specialists on various groups is in progress, and some on brachiopods (DROT, 1971), cricoconarids (LARDEUX, 1969), and trilobites (ALBERTI, 1969) have appeared. During the Early Devonian, evidence of faunal affinity with North America has not been as well documented as in younger divisions, but OLIVER (1976) has drawn attention to the occurrence of eastern North American coral genera in North Africa.

MIDDLE DEVONIAN FAUNAS

By the Middle Devonian little evidence remains for a Malvinokaffric province in the area considered here (Fig. 6). In many areas, extensive transgressions near the Lower-Middle Devonian boundary, such as the movements across the Russian platform and the Onondaga onlap in North America, led to a general decrease in faunal provincialism, and to a mixing where provinces were adjacent. The distribution of the Old World province is as indicated for the Lower Devonian, but the New Zealand subprovince cannot be recognized. Particularly remarkable, and an indication of the reduction of provincialism by the Givetian, is the almost international distribution, in their appropriate different facies, of the giant *Stringocephalus* and its relatives (BOUCOT, JOHNSON, & STRUVE, 1966) and of *Amphipora* (DUNCAN in CLOUD, 1959), witnessing to the cosmopolitan nature of some elements of the carbonate environment. But isolated areas of reef carbonates, in Europe, the Urals, and other areas are associated with some endemism. The mixing at province boundaries is illustrated by the number of Appalachian province forms occurring in North Africa and Spain, and European elements that are found in eastern North America, and the Appalachian elements that, as in the Lower Devonian, extend westward along the Angaran belt to Kazakhstan.

The Middle Devonian represents one of the chief periods of generic and familial diversification known in the Paleozoic. There appears to be some link between this and the widespread distribution of essentially shallow-water carbonates in areas thought to have been near the tropics, and the early Middle Devonian transgressive movements made these more extensive. But the close of the Middle Devonian saw the loss of many groups, and the brachiopod groups Rhynchotrematidae, Stringocephalidae, Dayiacea, Lyssatripidae, and Palaferellidae appear to have become extinct before the earliest Frasnian (HARLAND et al., 1967); of trilobite groups, the Calymenina, homalonotids, and cyphaspids are last recorded in the Middle Devonian. The goniatite break is very important, with the loss of all but two or three genera (HOUSE, 1973b). The pattern of extinction for several groups is shown in Figure 7.

EUROPE

In the European area the distinction between Rhenish and Hercynian facies is still clear, although the former is much less well developed than in the Lower Devonian. Here for the carbonate facies fundamental work has been done in the establishment of ecologically based faunal assemblages, particularly by LECOMPTE (1970) and TSIEN (1971) for the Ardennes, and by STRUVE (1963) for the Eifel. These may be briefly summarized, in shallow- to deep-water sequence, as follows (LECOMPTE's more interpretative terms are given in parentheses): Stromatoporoid *Bankriff* (Turbulent Zone); *Knollen-Blockriff*, and *Rasenriff* with crinoid meadows (Subturbulent Zone); *Rübenriff* (below Turbulent Zone); *Brachiopod-Siedlungen* (Quiescent Zone). The *Knollen-Blockriff* comprises loosely associated cerioid rugose corals and tabulates, the *Rasenriff* fasciculate rugosans and tabulates, and the *Rübenriff* is the zone of horn corals.

In the deeper water Hercynian facies, the basinal argillites are now characterized by styliolinid and nowakiid cricoconarids of supposed planktonic habit and these are abundant in the Tentaculitenschiefer and Styliolinenschiefer units of the Rhenish Schiefergebirge and in equivalent facies in Cornwall and other areas. The pelagic carbonate sequences of the seamounts are rich in goniatites and orthocones, but of types different from those of the Lower Devonian.

In the Middle Devonian brachiopod faunas increased in diversity. The productids entered first apparently in the Eifelian (rather than Emsian; BOUCOT, 1975), and derivatives are distinctive. Other stocks show distinctive genera; however, definition of a Lower-Middle Devonian boundary using brachiopods is difficult. The *Cultrijugatus Schichten*, here referred to the Lower Devonian, are characterized by species of *Paraspirifer* (SOLLE, 1971), but the name-giving species occurs in the *Laucher Schichten*, and it is not a diagnostic form. Locally, especially in the Eifel, various brachiopod groups have proved useful in time discrimination, particularly atrypids and reticulariids (STRUVE, 1966, 1970). One of the most useful genera for the Givetian is *Stringocephalus* and its allies. *Stringocephalus* survived into the earliest Late Devonian as used here, and the extinction corresponded approximately to the Middle-Late Devonian boundary as used here.

Middle Devonian goniatites are characterized by the proliferation of anarcestid and agoniatitid stocks. If CARLS and others (1972) are correct in correlating the *Gyroceratites gracilis* boundary (of ERBEN in ERBEN, 1962, p. 65) with a level in the Emsian, then it is the occurrence of *Sel-*

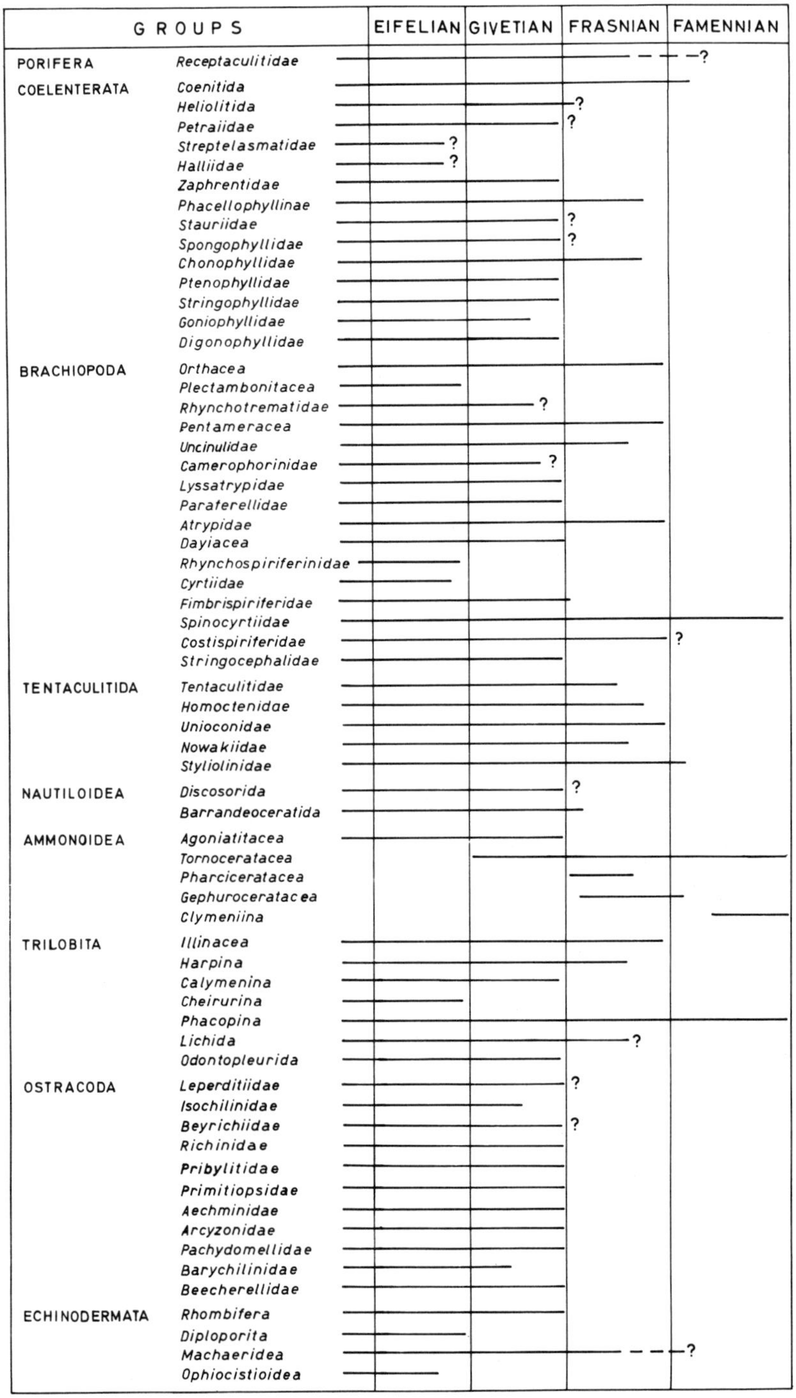

GROUPS		EIFELIAN	GIVETIAN	FRASNIAN	FAMENNIAN
PORIFERA	Receptaculitidae			— —	— —?
COELENTERATA	Coenitida				
	Heliolitida			—?	
	Petraiidae			—?	
	Streptelasmatidae	—?			
	Halliidae	—?			
	Zaphrentidae				
	Phacellophyllinae				
	Stauriidae			—?	
	Spongophyllidae			—?	
	Chonophyllidae				
	Ptenophyllidae				
	Stringophyllidae				
	Goniophyllidae				
	Digonophyllidae				
BRACHIOPODA	Orthacea				
	Plectambonitacea				
	Rhynchotrematidae		—?		
	Pentameracea				
	Uncinulidae				
	Camerophorinidae		—?		
	Lyssatrypidae				
	Paraferellidae				
	Atrypidae				
	Dayiacea				
	Rhynchospiriferinidae				
	Cyrtiidae				
	Fimbrispiriferidae				
	Spinocyrtiidae				
	Costispiriferidae				—?
	Stringocephalidae				
TENTACULITIDA	Tentaculitidae				
	Homoctenidae				
	Unioconidae				
	Nowakiidae				
	Styliolinidae				
NAUTILOIDEA	Discosorida			?	
	Barrandeoceratida				
AMMONOIDEA	Agoniatitacea				
	Tornorataceae				
	Phariceratacea				
	Gephuroceratacea				
	Clymeniina				
TRILOBITA	Illinacea				
	Harpina				
	Calymenina				
	Cheirurina				
	Phacopina				
	Lichida			—?	
	Odontopleurida				
OSTRACODA	Leperditiidae			?	
	Isochilinidae				
	Beyrichiidae			?	
	Richinidae				
	Pribylitidae				
	Primitiopsidae				
	Aechminidae				
	Arcyzonidae				
	Pachydomellidae				
	Barychilinidae				
	Beecherellidae				
ECHINODERMATA	Rhombifera				
	Diploporita				
	Machaeridea			— —	—?
	Ophiocistioidea				

FIG. 7. Pattern of extinctions for selected invertebrate taxa during the Middle and Late Devonian (after House, 1975b).

lanarcestes wenkenbachi (e.g., the *Obere Kondel-Gruppe*) that represents the best fauna taken as terminal Lower Devonian as recommended by SCHMIDT (1926) and widely followed. The upper Eifelian fauna with *Pinacites jugleri* is distinct and quite widely spread in Europe, occurring in the Choteč Limestone in Czechoslovakia. The *Cabrieroceras crispiforme* fauna makes the best marker for the basal Givetian; this occurs in the Kačak Member of the Srbsko Formation of Czechoslovakia, but the relations in Belgium are uncertain because it is not recorded there. The late Givetian is characterized by *Maenioceras, Sobolewia, Agoniatites, Foordites, Holzapfeloceras,* late *Cabrieroceras,* and others. Of these, only *Maenioceras* may just range into the *Phariceras lunulicosta* Zone, taken here as basal Upper Devonian. Useful as these goniatite faunas are for correlation in deeper water facies, their rarity in near-shore and reef facies raises many problems.

The coral *Calceola* has been used as a guide to the Eifelian, and it is widespread in shale facies of that age but is now known to range both lower and higher. The importance of corals generally in calcareous facies in the Middle Devonian of Europe has been indicated in relation to the recognition of assemblages. It was WEDEKIND who did the primary work in Germany (1924-25) and proved the value of this biostratigraphic tool. His scheme recognized five successive faunas throughout the Middle Devonian. From below upward, these include the *Keriophyllum, Astrophyllum, Digonophyllum, Dohmophyllum, Leptoinophyllum, Stenophyllum, Sparganophyllum,* and *Dialytophyllum* zones. The truth is, however, that subsequent work has shown that the ranges of several of the groups used to define these divisions are not as WEDEKIND supposed, and taxonomic revisions have eliminated many of his names (MIDDLETON, 1959; BIRENHEIDE, 1972).

For trilobites there is a break at the Heisdorf-Lauch boundary marked by the disappearance of forms such as *Basidechenella kayseri, Comura defensor,* and *Acastoides henni posthumus* and the appearance of *Pholonyx philonyx, Schizoproetus onyx,* and *Longiproetus cultrijugati.* In higher Eifelian strata scutellids, harpids, phacopids, and thysanopeltids characterize shallower water facies and proetids and subgenera of *Phacops* characterize the deeper water facies. The Givetian limestone facies is characterized by bizarre spinose trilobites such as *Acanthopyge, Radiolichas,* and *Cheirurus,* in addition to traditional groups.

In the basinal argillite sequences, it is the cricoconarids that are important, and detailed zonations have been established, especially for the Eifelian. The earlier faunas have been studied by ALBERTI (1971), and a general review is given by LARDEUX (1969).

As has already been remarked, there have been attempts to collate faunal information in terms of facies faunas, assemblages, or communities for the carbonate facies of the Middle Devonian, but there has been little systematic collation for other facies (HOUSE, 1975b).

ASIA

Treatment here will follow the pattern adopted earlier for the Lower Devonian. Again, the Uralian faunas show close affinity with Europe, but as before those of the Arctic area differ. The Uralian area, embracing now the Altay-Sayan, has the Eifelian brachiopods *Zdmir* [*Conchidiella*] *pseudobaschkirica, Megastrophia uralensis, Carinatina, Janius,* and other European genera. The Altay-Sayan fauna has a number of endemics. In the Givetian, a similar broad pattern is seen, with common European genera such as *Stringocephalus, Bornhardtina, Uncinulus, Schnurella,* and *Emmanuella,* but in the Altay-Sayan the occasional exotic form *Indospirifer* from the east, and rare endemics such as *Urella* occur (NALIVKIN, RZHONSNITSKAYA, and MARKOVSKIY, 1973). In the Givetian, when substantial general transgression had taken place in the Russian area, there seems to be a decrease in provincialism.

The goniatite sequences of the Urals, and indeed of Russia generally, are not well known (BOGOSLOVSKIY in NALIVKIN, ed., 1973, p. 51), although general correlation with Germany is indicated. Little has been achieved in conodont work or in using cricoconarids (LYASHENKO in OSWALD, 1968b).

Paleogeographic distribution of corals, on the other hand, has been reviewed by DUBATOLOV & SPASSKIY (in OSWALD, 1968b) and by SPASSKIY and others (in NALIVKIN, 1973, p. 220, p. 229). The latter authors recognized three broad time divisions of coral faunas in the Russian Middle Devonian, and listed endemics of particular areas. The wide distribution of European forms is striking, but nevertheless they referred Asiatic faunas to a distinct Uralo-Siberian-Asiatic province with only a separate Mongolian fauna (Mongolo-Okhotsk province).

The Dzhungaro-Balkhash region, with *Euryspirifer*, *Acrospirifer*, and *Fimbrispirifer* in the Eifelian and *Mucrospirifer* higher continues its Appalachian province aspect. Records of corals such as *Heliophyllum* (SPASSKIY and others in NALIVKIN, 1973) seem to confirm this, but OLIVER (1976) casts doubt on the *Heliophyllum* determination. Several of the brachiopod genera occur again in the Tikhookian and Mongolian area, however, confirming the easterly link in general terms. Corals thought to be distinctive of this area include *Stellatophyllum*, *Pseudotryplasma*, *Gurjevskiella*, and *Amurolites*, but with *Iowaphyllum* suggesting an easterly link.

The Chinese Middle Devonian is poorly known. Common European coral genera, including *Calceola*, and brachiopods, including *Stringocephalus*, indicate general links with Europe. Apart from the early work on brachiopods by GRABAU (1931-33) and TIEN (1938), the work on Yunnan corals by WANG (1948) and the revisions by HAMADA (in OSWALD, 1968a) seem to be the main sources confirming the general picture. But SPASSKIY and others (in NALIVKIN, 1973, p. 234) united the corals of this area with those of Australia in a common Sino-Australian realm. Rich faunas of corals, tentaculites, and other fossils have been described by MU and others (1974).

The Tethyan belt has scattered Middle Devonian faunas that generally are most comparable with those of Europe. From near Istanbul KULLMANN (1973) recorded *Latanarcestes noeggerati*, *Anarcestes lateseptatus*, and *Pinacites jugleri* in successively higher beds following Emsian-Eifelian transitional faunas. The Middle Devonian is poorly known in Iran, but BRICE (1970) and DURKOOP (1970) have reported a varied fauna of brachiopods and corals from Afghanistan, which have a close link with European and Uralian faunas. REED (1908) recorded both *Calceola* and *Anarcestes* from northern Burma (ANDERSON, BOUCOT, & JOHNSON, 1969), but good Middle Devonian brachiopod faunas are recorded from Viet Nam and Cambodia, assigned mostly to European species, and including *Calceola* and *Stringocephalus*. There are also records of *Stringocephalus* in Malaya (GOBBETT, 1966), and BOUCOT, JOHNSON, & STRUVE (1966) have reviewed the distribution of this genus and its relatives.

AUSTRALASIA

In New Zealand there is no evidence that the Lower Devonian faunal sequence continues into the Middle Devonian. In eastern Australia, however, along the area referred to as the Tasman geosyncline, rocks of this age are well developed. In western Australia, although a Middle Devonian transgressive phase has been reported (TEICHERT, 1974), little has been described; however, *Stringocephalus* and *Stringophyllum* have been reported from the Canning basin (PICKETT, 1972). This section is therefore concerned only with eastern Australia.

The formal biostratigraphic subdivision of PHILIP & PEDDER (1967) has already been noted. In this scheme the Middle Devonian embraces divisions G to K. Most localities are in New South Wales and Queensland and are from discrete outcrops. Fauna G, of *Macgeea touti*, marks the incoming of *Dohmophyllum* and *Stringophyllum*, and Fauna H, of *Taimyrophyllum callosum*, marks the first incoming of *Endophyllum*. The *Xystriphyllum giganteum* fauna (Fauna I) contains the early stringocephalid genus *Bornhardtina*, and the top of this fauna contains conodonts that correlate with the *Polygnathus kockelianus* Zone of Europe. Fauna J (of *Grypophyllum* cf. *G. denckmanni*) and Fauna K (of "*Endophyllum*" *schlueteri*) have been correlated with the European *Polygnathus varcus* Zone (KLAPPER, PHILIP, & JACKSON, 1970). Much work remains to be done on the brachiopod faunas, but the general European affinity of the coral faunas is clear. But as HILL (1957) and SPASSKIY and others (in NALIVKIN, 1973) pointed out, and has been emphasized by PEDDER, there is much endemism in the faunas at generic and especially specific level. Thus, the Tasman subprovince continued into the Middle Devonian.

AFRICA

There is no evidence that the faunas of the South African Bokkeveld beds, or that of the overlying Witteberg beds, extend upward into the Middle Devonian. The same is true of the Lower Devonian of Ghana. In North Africa, on the other hand, the Middle Devonian is superbly developed (HOLLARD & LEGRAND in OSWALD, 1968a). The fauna is extremely close to that of Europe, but there is evidence of faunal links with the Appalachian area.

A sequence of goniatite faunas from near the Lower-Middle Devonian boundary has been established by HOLLARD (1974). The levels are characterized, in sequence, by *Sellanarcestes wenkenbachi*, *Anarcestes lateseptatus lateseptatus*, then *A. lateseptatus plebeius* with *Pinacites jugleri*. The *P. jugleri* faunas HOLLARD referred to the upper Eifelian, as in Germany, and he has disproved the earlier date assigned by SOUGY (1969). A widespread fauna containing *Cabrieroceras crispiforme* is taken to mark the basal Givetian (records of Eifelian *Cabrieroceras* from this area refer to forms best assigned to *Werneroceras*). The detailed affinity of this goniatite sequence with that established in Germany is most striking. The *C. crispiforme* levels seem to correlate exactly with the New York *Werneroceras* Bed.

Among brachiopods, too, typical European forms are well represented (DROT, 1964, 1971), but the coral records of *Heliophyllum* and *Phillipsastrea* are of particular interest (LE MAITRE, 1947, 1952). The former is taken, with otherwise endemic eastern North American forms (OLIVER, 1976), to indicate close links with the Appalachian area. The occurrence of *Phillipsastrea* is interesting, as here and in Spain and England it occurs in the Middle Devonian, but generally in Europe it is an Upper Devonian form (SCRUTTON, 1975).

UPPER DEVONIAN FAUNAS

Approximately at the Middle-Upper Devonian boundary there is widespread evidence of renewed transgression (Fig. 8). In the European area this is represented by the extension of marine sediments well on to cratonic areas (HOUSE, 1975a) and there seems to be a good correlation with the Taghanic onlap of North America (JOHNSON, 1970). Thus, by the middle Frasnian, faunas of the Devonian were probably at their most cosmopolitan. Remarkable changes took place during the late Frasnian. Apparently as a result of continued eustatic rise of sea level, or widespread sea-floor subsidence in shelf areas, there was first a restriction of reef carbonates, then the attenuation of those that remained, and finally this facies almost disappeared. Accompanying this was an associated diminution and extinction of the various specialized reef and associated carbonate organisms. The end of the Frasnian apparently saw the extinction of several groups of brachiopods (Pentameroidea, Atrypoidea, possibly Costispiriferidae, and Orthacea), trilobites (harpids, thysanopeltids, Dalmanitacea, Odontopleuridae, Lichidae), coral and stromatoporoid genera, and almost all groups of Devonian cricoconarids (Uniconidae, Tentaculitidae, Homoctenidae, Nowakiidae) (Fig. 7).

This faunal break between the Frasnian and Famennian is a major Phanerozoic event, and apart from the Permian-Triassic break, is probably the most marked in the Paleozoic as a whole. Documentation is not precise enough to state that all these groups became extinct at the same time, but the main extinctions are clearly late in the Frasnian. Some global cause is required, and MCLAREN (1970) argued, doubtless with tongue in cheek, that a meteorite landing in the Devonian "Pacific" could have set up a tidal wave giving near instantaneous extinction. HOUSE (1975a,b), on the other hand, has pointed out the critical relation between the disappearance of reef carbonates and these extinctions.

Fig. 8. Paleogeographic map for the Late Devonian (Frasnian) of the Eastern Hemisphere (based on House, 1973; House, n). [Inferred land areas are stippled; areas of marine rocks are horizontally lined.]

Once reefs had gone, not only could their specialized inhabitants not survive, but the removal of any protection that reef carbonates may have provided at platform margins would have rendered vulnerable many other shelf environments.

Faunally, as a result of these events, the Famennian represents a discrete interval terminated by further extinctions (Clymenoidea, Phacopidae) at the close of the stage. The almost complete absence of organodetrital carbonates and associated faunas, the replacement of well-known brachiopod groups by a development of rhynchonellids and spiriferids, the entry and diversity of the clymenoid ammonoids, and in the basinal environment, the replacement of coniconchines by the planktonic ostracodes give a quite different stamp to the Famennian. Some element of provincialism remains, for example, the restriction of well-developed ostracode slate facies to Europe and the Urals, but it is quite diminished in comparison with the Lower and Middle Devonian.

EUROPEAN FRASNIAN

The progressive series of transgressions throughout the Frasnian of Europe resulted in limited preservation of nearshore clastic facies. Better represented are carbonate and reef complexes, basinal argillites, and the reduced successions of seamount *Schwellen*. But these, as has already been indicated, show faunal distinction from those of the Middle Devonian following the extinctions of particularly coral, brachiopod, and trilobite groups near the end of the Givetian. As used here the Frasnian is treated as synonymous, or nearly so, with the Adorfian, named from the reduced goniatite-rich sequence of the northern Schiefergebirge.

The type Frasnian area is in southern Belgium where spectacular knoll-like reef masses are developed in the middle Frasnian, although they appear to have been overwhelmed in the *Palmatolepis gigas* Zone with the onset of the terminal Frasnian transgressions. *Cyrtospirifer*, long considered the main guide to the Upper Devonian, enters in the basal Assise de Fromelennes (LECOMPTE in OSWALD, 1968a) and this explains why the base of the Frasnian was defined at that level in 1952. Other so-called guide fossils include *Hypothyridina cuboides* and *Phillipsastrea hennahi*, but both perhaps have their original types from the Givetian (ORCHARD, 1974; SCRUTTON, 1968). BOUCOT (1975) has indicated a range of brachiopod genera restricted to the Frasnian. The type Frasnian reefs were studied by LECOMPTE (1970 and earlier works) and formed part of the basis for the recognition of depth zonation. This approach has been followed by TSIEN (1971).

Reduced successions of the pelagic carbonates have provided the type sequences for both the goniatite and conodont zonations, which were established at Adorf (now Diemelsee). The distinction of the basal Adorfian goniatite faunas is the occurrence of multilobed pharciceratids such as *Pharciceras* and *Synpharciceras*, also *Epitornoceras*, *Ponticeras*, *Probeloceras*, and others unknown in the Middle Devonian. In the middle Fasnian are the typical *Manticoceras* and *Beloceras*. *Crickites holzapfel* and others characterize the late Frasnian. The pyritic or hematitic fauna of contemporary shales is referred to as the Büdesheimer Schiefer. The goniatite sequence was established mainly by WEDEKIND (1913). The conodont sequence was established by ZIEGLER (1958, 1962), but, as remarked before, certain correlation between the two schemes has not been achieved.

The gradual dominance of fingerprint ostracodes over cricoconarids in the photic zones of the basinal regions took place through the Frasnian, a typical succession being described by KREBS and RABIEN (1964) and this is reproducible in other European areas.

One particularly unusual facies, the two Kellwasserkalk bituminous limestones, were recognized by BUGGISCH (1972) over wide areas of Europe and in North Africa. These carry a remarkable fish fauna and goniatites, orthocones, and even coprolites with conodont assemblages (LANGE, 1968). BUGGISCH considered these units were formed at times of transgression during which large quantities of organic matter were deposited. The deepening in pulses that terminated the various periods of reef development in Belgium were of similar type.

A striking coherence exists in the pattern of Frasnian facies and faunas found in western Europe and North Africa. In the Main Devonian Field near Leningrad, and in the Timan Mountains, widespread shallow-water calcareous environments have rather distinctive brachiopod faunas (LYASHENKO, 1959; NALIVKIN, RZHONSNITSKAYA, & MARKOVSKIY, 1973) and goniatite faunas (BOGOSLOVSKIY, 1969, 1971) that have similarities with Cordilleran types. Also, a cricoconarid sequence, more detailed than elsewhere, has been described (LYASHENKO, 1959). The abundance of fish in interleaving horizons makes this a critical area for correlation into the Old Red Sandstone facies (WESTOLL in HOUSE, 1977).

EUROPEAN FAMENNIAN

The almost complete absence of biohermal carbonates is the most distinctive feature of the Famennian biofacies of Europe. From the Holy Cross Mountains of Poland ROZKOWSKA (1969) has described a sparse fauna of four phillipsastreaid individuals, some endophyllids, one genus of cystiphylloids and the earliest known heterophylloids from the early Famennian, and this is the best European Famennian coral fauna known. The limited stromatoporoid fauna of the Etroeungt (LE MAITRE, 1933) represents the best occurrence of that group. The result is a marked contrast with the earlier Devonian.

In the shallower water, clastic-rich facies brachiopod and bivalve faunas are well developed, the former dominated by rhynchonellids (SARTENAER in OSWALD, 1968b) and spiriferines. SARTENAER recognized in France and Belgium his international *Eoparaphorhynchus* and *Basilicorhynchus* zones of the lower Famennian, but higher faunas are not well discriminated. In northern Europe, this facies intertongues with Old Red Sandstone type terrestrial facies in which fish, plant, and spore remains are common.

In the deeper water argillite facies only proetids and phacopids remain of trilobite groups. The most remarkable feature is the dominance of planktonic Ostracoda, which often cover bedding planes in the Cypridinenschiefer facies; with them are bivalves, such as *Karadjalia venusta,* which may have been epiplanktonic on floating weed. In shallower silty facies the infaunal *Sanguinolites* is characteristic.

Occurring in the argillites, especially in hematitic or pyritic facies, but more particularly in the seamount or pelagic carbonates, are the rich and varied ammonoid faunas. The goniatite *Cheiloceras* characterizes the earliest Famennian, and then the clymeniids enter, characterizing the later Famennian and showing remarkable diversity until their sudden extinction in the late *Wocklumeria* Stufe. Goniatites such as *Sporadoceras* and *Imitoceras,* and also nautiloids, accompany them; however, for biostratigraphical purposes, it is the conodont sequence established especially by ZIEGLER (1962, 1971) that has international importance and, for the Famennian, this seems well linked to the ammonoid scale.

Much detailed work has been done in Europe on the correlation of beds near the Devonian-Carboniferous boundary (AUSTIN, et al., 1970; VÖHRINGER, 1960; ZIEGLER, SANDBERG, & AUSTIN, 1974). The boundary, in the German (and Heerlen and Sheffield congresses) definition, is marked by the extinction of both clymeniids and phacopids, whereas in the Belgian and Russian definitions the boundary is difficult to diagnose except at low taxonomic levels.

ASIA

For this enormous area the Frasnian and Famennian are here treated together. In the Upper Devonian of Asia the broad geosynclinal tracts recognized earlier in the Devonian remain broadly discrete. These are 1) the Ural belt, including Novaya Zemlya in the north and the Tien Shan and Kazakhstan in the south, 2) the northern Siberian areas of shallow-water sedimentation conveniently termed the Arctic area, 3) the Altay-Sayan area, 4) the Dzhungaro-Balkhash area, linking through the poorly known regions of China and Mongolia to the east, 5) the Tikhookian region of southeast Russia, 6) the Chinese area south of the Peking platform, and 7) the tract of separated and disjunct outcrops from Turkey to Southeast Asia. This grouping (based in part on NALIVKIN, RZHONSNITSKAYA, & MARKOWSKIY, 1973) is here used for discussion.

In the Frasnian and Famennian, the Uralian and Arctic area, comprising the Atlantic region of Russian authors, is essentially of European type. Even the characteristic basinal and seamount facies are known in the Urals. When faunas were most cosmopolitan, in the Frasnian, typical European genera such as *Manticoceras, Hypothyridina, Ladogia, Theodossia,* and *Mucrospirifer* were widely distributed across this broad area. *Ladogioides,* a Cordilleran type, occurs in northeast Siberia. Endemics were rare; among goniatites there is *Tamarites,* probably earliest Frasnian in age. *Manticoceras* of the middle Frasnian is widely distributed (HOUSE, 1973b). The rich coral faunas (DUBATOLOV & SPASSKIY in NALIVKIN, 1973) show less provincialism than earlier in the Devonian and the typical Frasnian genera of Europe are abundant. The apparent uniqueness of the cricoconarid faunas of the Russian platform and Urals in the Frasnian results mainly from the lack of detailed studies elsewhere.

The Uralian and Arctic areas show faunal differences in the Famennian. The former resembles European biofacies and contains a very similar ammonoid sequence (BOGOSLOVSKIY, 1971) and, although only briefly described, the conodont sequence has broad similarities (KHALYMBADZHA & CHERNYSHEVA, 1970), but the brachiopods *Zilimia* and *Dzieduszichia* occur in addition to common European genera. In the Arctic area Famennian ammonoids are very rare, and distinction is given by the occurrence of *Gastrodetoechia* (according to NALIVKIN, RZHONSNITSKAYA, & MARKOWSKIY, 1973, but not recorded by SARTENAER, 1969), a brachiopod common in western North America.

In the Altay-Sayan area, both in the

Frasnian and Famennian, there are close links faunally to the Urals, but faunas of the Kazakhstan area and the Dzhungaro-Balkhash region contain the eastern brachiopod *Yunnanella,* according to NALIVKIN, RZHONSNITSKAYA, & MARKOVSKIY (1973), but SARTENAER (1971) has reassigned most of the specimens on which the genus was recognized. Other evidence of eastern faunal affinity is the record by BOGOSLOVSKIY (1971) of *Sinotites* in the Aktyubinsk district, a genus otherwise known only in the Great Khingan.

The Tikhookian and Mongolian late Devonian faunas include cyrtospiriferids and other brachiopods, many of which are assigned in the literature to New York species. To the south is the remarkable sequence of the Great Khingan (CHANG, 1958). Here the Lower Suhuho Formation, with *Sinospirifer,* atrypids, *Yunnanella,* and *Nayunella* (SARTENAER, 1971), has been referred to the Frasnian. The overlying Upper Suhuho Formation contains a rich goniatite fauna of early Famennian age, including *Cheiloceras, Sporadoceras,* and *Pseudoclymenia,* whereas *Platyclymenia* indicates a middle Famennian age. CHANG (1960) has described the peculiar goniatites *Sinotites* and *Sunites* from here, which are endemic apart from the Aktubinskayan record mentioned above.

The main Upper Devonian records in China are from Yunnan, Kwangsi, Hunan, and Szechwan. The typical Frasnian goniatites *Manticoceras, Beloceras, Eobeloceras,* and *Ponticeras* occur in this area (CHAO, 1956), but there has been considerable doubt as to whether the Famennian is represented at all, a substantial post-Devonian break being thought to be present above the Frasnian. Whether the *Yunnanella*-bearing rocks, which are very widespread (SARTENAER, 1971), are wholly Frasnian is not clear. The earliest Famennian may be present, but atrypids recorded in association with *Yunnanella* suggest that only Frasnian is involved. This is of some importance because the type species of *Cyrtiopsis* is from this region, and in North America this genus has been taken as a Famennian guide. According to the review by SPASSKIY and others (in NALIVKIN, 1973), the Frasnian coral fauna is broadly similar to that of Asia and Australia. The plant *Leptophloeum* is widespread in more terrestrial facies in China, Japan, and also eastern Australia (HAMADA in OSWALD, 1968a). Evidence of late Famennian in southwest China comes from the record of *Cymaclymenia* and *Parawocklumeria* by Mu and others (1974).

Upper Devonian rocks are widely exposed in scattered outcrops along the Tethys belt. In Iran, brachiopods of Frasnian age, such as *Ripidiorhynchus* and *Cyphoterorhynchus* are widespread (SARTENAER, 1966), *Cyphoterorhynchus* being found also from Armenia to Pakistan. Early Famennian *Gastrodetoechia* faunas occur, and in the Tabaz area there is a fauna of *Platyclymenia, Sporadoceras,* and *Prionoceras* (WALLISER, 1966). Much richer faunas have been described from Afghanistan by BRICE (1970), including a variety of corals from carbonate horizons. A rich fauna of brachiopods and corals is known from the Northern Shan States (REED, 1908), which includes a record of *Phillipsastrea* and brachiopods seemingly of Frasnian age. Scattered records of Late Devonian in Cambodia, Viet Nam, and Thailand exist. An interesting tie with Europe is the occurrence of the trilobite *Cyrtosymbole (Waribole)* from Malaya (KOBAYASHI & HAMADA, 1966). This is an exotic occurrence of a subgenus abundant in western Europe and eastward to Kazakhstan. So much has still to be learned about these faunas that generalizations are premature. Nevertheless, the affinities with faunas of Europe and the southern Uralian belt seem sufficiently strong to confirm the same provincial assignment given to earlier Devonian faunas.

AUSTRALASIA

The Upper Devonian of Australia (ROBERTS et al., 1972) shows a continued development along the Tasman geosyncline, but additionally spectacular reef development occurs marginal to the Precambrian shield in Western Australia.

In eastern Australia terrestrial facies occur in many areas with European fish genera and *Leptophloeum* and other plant remains of Late Devonian type. Good marine faunas in Queensland include Frasnian corals, cyrtospiriferids and other Frasnian brachiopods and typical European conodonts. McKELLAR (1970) has established a series of Famennian zones using productoids, some of which have affinities with the western United States as well as with Europe. Other faunas, including "*Cyrtiopsis,*" *Cyrtospirifer,* and *Tenticospirifer,* may be compared with Chinese and European forms. In New South Wales conodonts have enabled correlation with the European zonation in several places (PHILIP & JACKSON, 1971), the faunas ranging through the Frasnian and Famennian. Also known are the Famennian ammonoid genera *Cheiloceras, Genuclymenia, Platyclymenia,* and *Cymaclymenia* (JENKINS, 1968; PICKETT, 1960).

The remarkable Frasnian reef sequences of the Canning basin (PLAYFORD & LOWRY, 1966) and other developments in the Bonaparte Gulf basin and Carnarvon basin are significant for their incredibly close affinity with European faunas. For many goniatites (GLENISTER, 1958; PETERSEN, 1975) this is at specific level, as it is for the conodonts (DRUCE, 1969; GLENISTER & KLAPPER, 1966), brachiopods (VEEVERS, 1959a,b), and phyllocarid crustaceans and other groups (BRUNTON et al., 1969). *Fitzroyella,* one supposed endemic brachiopod, has since been recorded in Poland (BIERNAT, 1969). Although indicated in the literature, it would appear that the evidence for the stromatoporoid reef facies extending up into the Famennian in this area is very slight; the Famennian reefs are stromatolitic.

AFRICA

It is only in North Africa that Upper Devonian rocks are preserved and there the faunal agreement with Germany is particularly close. This is well illustrated by the ammonoid sequences (PETTER, 1959, 1960), which replicate the standard successions. Goniatite faunas near the Middle-Upper Devonian boundary have been described by BENSAID (1974), who showed that the base of the *Pharciceras lunulicosta* Zone (and *Manticoceras* Stufe) falls within the *Polygnathus varcus* Zone of the conodont zonation, thus solving anomalies in other areas (HOUSE, 1973a). The rare endemic form *Petteroceras* is in a fauna mostly conspecific with German forms. This area has also contributed to the correlation of conodont and ammonoid zonations near the Frasnian-Famennian boundary, and BUGGISCH & CLAUSEN (1972) have demonstrated that the boundary between the *Manticoceras* and *Cheiloceras* Stufen falls at the base of the upper *Palmatolepis triangularis* Zone. Their recognition of the Kellwasser Kalk facies of Germany at the expected levels is another tie.

For other groups also, the congruence with central Europe is apparent. ALBERTI (1973) recognized only a few new forms of trilobites from the Famennian. For corals, provincialism is mostly lost by the Frasnian, so similarities with eastern North America (OLIVER, 1976) are to be expected. North Africa is an area in which much has still to be contributed in detailed biostratigraphy.

DEVONIAN GLOBAL RECONSTRUCTION

Paleomagnetic data allow some approximation to be made on the position of the continents in relation to the magnetic poles and equator in the Devonian. It is assumed that then, as now, the dipole field axis was approximately coincident with the earth's rotational axis. The paleomagnetic evidence cannot give longitude position, and hence positioning of continental masses along lines of latitude is an arbitrary matter and can be changed at will, guided only by the constraints of any geological demands. In the reconstruction given here (Fig. 9), the positioning of the southern continents is that of BRIDEN, DREWRY, & SMITH (1974), and the same authors in HUGHES (1973); the situation is given by them as the same throughout the Devonian. For the northern continents, following the arguments of ROY (1971), the Late Devonian reconstruction brings North America and Europe close together. Russian evidence (KHRAMOV in OSWALD, 1968b) would bring Asia perhaps five degrees or so farther south than shown.

It is clear from this reconstruction that

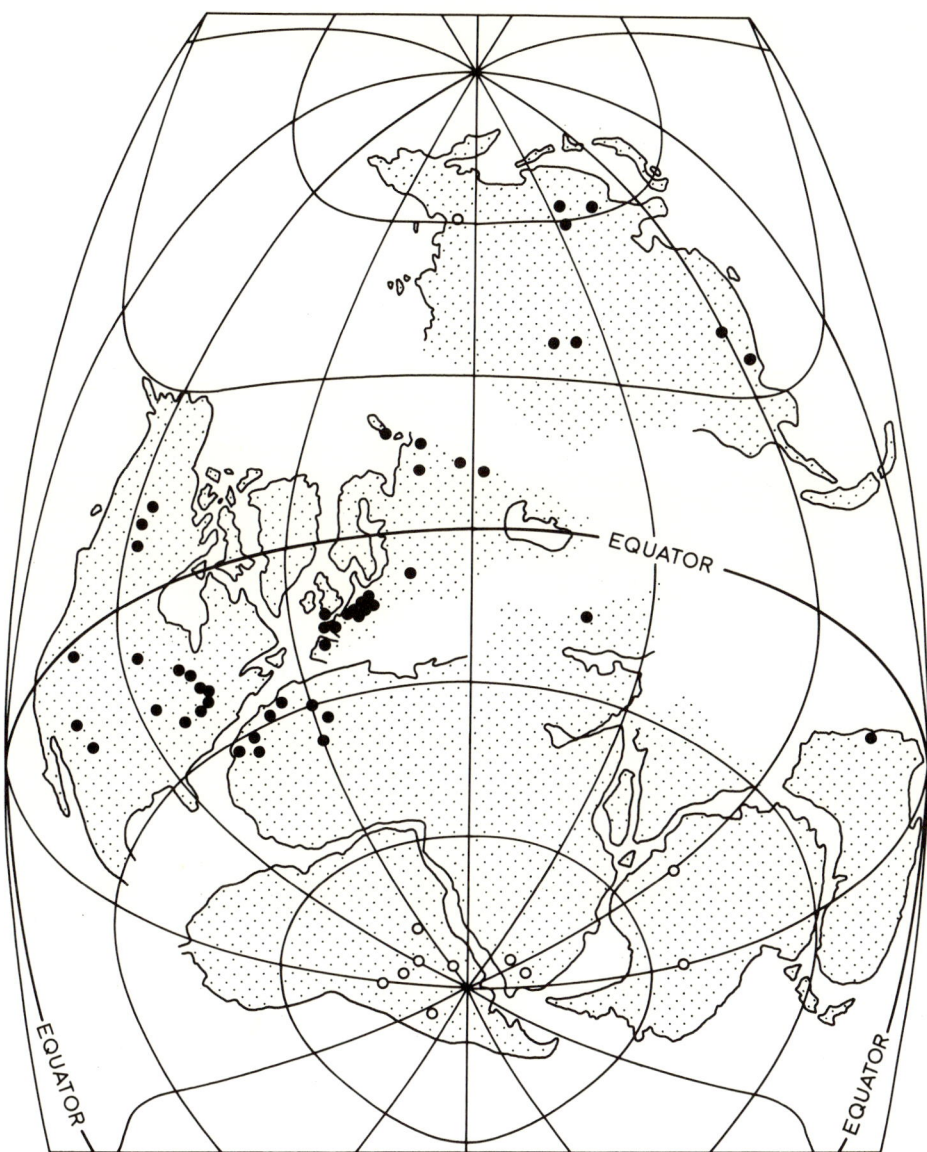

Fig. 9. Global reconstruction for the Devonian using paleomagnetic evidence. The Emsian distribution of Malvinokaffric province brachiopods is shown by open circles. The Frasnian distribution of the goniatite *Manticoceras* is shown by black circles. Note that southeast Asia was probably discrete microplates, the positions of which are uncertain (House, n; for sources see text).

the Malvinokaffric province of the Emsian has essentially a high latitude fauna, and this fact may be the main cause of its uniqueness. Of the divisions of the Old World province, the Rhenish-Bohemian subprovince is clearly tropical (House, 1975a,b), as is the Uralian subprovince, but the Siberian Arctic faunas, and those of the Angaran and Tethyan belts, are progressively of higher latitude. The break between the Appalachian province and that of the Old World may be explained in two ways.

Firstly, some paleomagnetic reconstructions place North America in a much more southerly position than that given here for the Lower Devonian, when the provincialism was best developed. Secondly, the Caledonian orogeny and mid-Acadian orogeny appear to have interposed an oblique mountain barrier stretching from eastern Greenland and Norway southwestward, between eastern North America and northwest Africa. The latter would have been an effective barrier to migration, although in the Frasnian its effects were mostly lost.

The Appalachian province links with the Mongolian, Tikhookian, and Dzhungaro-Balkhash areas are less readily explained, and the problem looks deceptively easy when only present-day projections are used (as in Boucot, 1975, and elsewhere). Cordilleran links with Europe north of the Old Red Sandstone continental area, and with Arctic Siberia, pose fewer problems.

Attempts at global reconstruction using plate tectonic principles have not progressed very far. After the establishment of the Caledonian Mountain belt, the European Hercynian (Armorican, Variscan) geosynclinal belt stretched from Cornwall and Britanny eastward to beyond southern Poland. This has been interpreted as a subduction zone, some authors inferring subduction to the north (Burrett, 1972; Riding, 1974), others to the south (Anderson, 1975). But the review by Krebs & Wachendorf (1973) and the various opinions expressed by others, make the placing of any Benioff Zone uncertain, and the possibility of microplate accretion in the southerly belts adds to this uncertainty. Riding (1974), following many earlier authors, linked the European Hercynian belt with that of the Piedmont and Ouachita belt of eastern North America.

The Ural belt shows some of the characteristics of another subduction belt, but the interpretation of the southern and central Asiatic fold belts in the Devonian is exceedingly difficult, particularly because the northward movement of India through the later Phanerozoic, which is so well documented, may have been associated with microplate accretion. All this will not be resolved until much more detailed paleomagnetic evidence is available.

Using modern-day maps, even more problems relate to the inferred link between Siberia and western North America. Churkin (1973) has argued that the Franklinian geosyncline of the Canadian Arctic Islands may have linked with the ancestral Brooks Range belt of Alaska to join the Chukotka geosyncline of northeast Siberia, and he separated this from the Cordilleran geosyncline, which he linked with the Koryak geosyncline of northern Kamchatka. As has been indicated in the preceding pages, there is faunal evidence for some link of this kind, and the proposed paleomagnetic position for Asia seems sufficiently at variance to make it suspect. For the southern continents, the link of the Tasman geosyncline with the Transantarctic Mountain belt in Antarctica has been long attested.

It is clear that in the Devonian shallow seas were widespread over present continental areas (Fig. 5,6,8). The plate tectonic model requires that ocean areas would also have been oceanic in the Devonian, so enormous spreads of seas covered the Devonian globe. It is also clear that ice caps, if they existed, were unlikely to have been large. It would seem, therefore, that a generally much warmer world climate than at present was partly the cause of the enormous diversity of Devonian marine faunas, and was the background for the remarkable development of vascular plants and early vertebrates in the period.

REFERENCES

Alberti, G. K. B., 1969, *Trilobiten des jüngeren Siluriums sowie des Unter-und Mitteldevons: I, Mit Beiträgen zur Silur-Devon-Stratigraphie einiger Gebiete Marokkos und Oberfranken:* Senckenberg. Naturforsch. Gesell., Abhandl., no. 520, 692 p.

Alberti, Helmut, 1968, *Trilobiten (Proetidae, Otari-onidae, Phacopidae) aus dem Devon des Harzes und des Rheinischen Schiefergebirges:* Beihefte Geol. Jahrb., v. 73, p. 1-147, pl. 1-8.——1973, *Neue Trilobiten (Cyrtosymbolen) aus dem Ober-Devon IV bis VI (Nord Afrika und Mittel-Europa)—Beitrag 1:* Neues Jahrb. Geologie Paläontologie, Abhandl., v. 144, p. 143-180.

——, Groos-Uffenorde, Helga, Streel, Maurice, Uffendorde, Henning, & Walliser, O. H., 1974, *The stratigraphical significance of the Protognathodus fauna from Stockum (Devonian/Carboniferous boundary, Rhenish Schiefergebirge)*: Newsl. Stratigraphy, v. 3, no. 4, p. 263-276.

Anderson, M. M., Boucot, A. J., & Johnson, J. G., 1966, *Devonian terebratulid brachiopods from the Accraian Series of Ghana:* Jour. Paleontology, v. 40, p. 1365-1367, pl. 172.——1969, *Eifelian brachiopods from Padaukpin, northern Shan States, Burma:* Brit. Museum (Nat. History), Bull., Geol., v. 18, p. 105-163.

Anderson, T. A., 1975, *Carboniferous subduction complex in the Harz Mountains, Germany:* Geol. Soc. America, Bull., v. 86, p. 77-82.

Austin, R. L., Conil, R. C. L., Dolby, Grace, Lys, Maurice, Paproth, Eva, Rhodes, F. H. T., Streel, Maurice, Utting, J., & Weyer, Dieter, 1970, *Les couches de passage du Dévonien au Carbonifère de Hook Head (Irlande) au Bohlen (D.D.R.):* Congrès Colloq. Univ. Liège, v. 55, p. 167-177.

Bensaid, Mohamed, 1974, *Etude sur des goniatites à la limite du Dévonien moyen et supérieur du Sud marocain:* Serv. Géol. Maroc, Notes, v. 36, no. 264, p. 81-140, pl. 1-6.

Beushausen, H. E. L., 1895, *Die Lamellibranchiaten des Rheinischen Devon mit Ausschluss der Aviculiden:* Preuss. Geol. Landesanst., Abhandl., n. ser., v. 17, 514 p., 38 pl. (in atlas).

Biernat, Gertruda, 1969, *On the Frasnian brachiopod genus Fitzroyella Veevers from Poland:* Acta Palaeont. Polonica, v. 14, p. 373-394, pl. 1-3 (Pol. and Russ. summaries).

Birenheide, Rudolf, 1968, *Die Typen der Sammlung Wedekind aus der Gattung Plasmophyllum (Rugosa; Mitteldevon):* Senckenberg. Lethaea, v. 49, p. 1-37, pl. 1-3.——1972, *Ptenophyllidae (Rugosa) aus dem W-deutschen Mitteldevon:* Senckenberg. Lethaea, v. 53, p. 405-437, pl. 1-5.

Bischoff, Günther, & Ziegler, Willi, 1957, *Die Conodontenchronologie des Mitteldevons und des tiefsten Oberdevons:* Hess. Landesamt Bodenforsch., Abhandl., v. 22, p. 1-136, pl. 1-21.

Blumenstengel, Horst, 1959, *Über oberdevonische Ostracoden und ihre stratigraphische Verbreitung im Gebiet zwischen Saalfeld und dem Kamm des Thüringer Waldes:* Freiburg. Forsch., v. C 72, p. 53-107, pl. 1,2.

Bogoslovskiy, B. I., 1969, *Devonskie ammonoidei, I, Agoniatity:* Akad. Nauk SSSR, Paleont. Inst., Trudy, v. 124, p. 1-341, pl. 1-29. [*Devonian ammonoids, I, Agoniatitida.*]——1971, *Devonskie ammonoidei, II, Goniatity:* Akad. Nauk SSSR, Paleont. Inst., Trudy, v. 127, p. 1-228, pl. 1-19. [*Devonian ammonoids, II, Goniatitida.*]——1972, *Novye Rannedevonskie golovonogie Novoy Zemly:* Paleont. Zhurnal, 1972, no. 4, p. 44-45, pl. 5. [*New Early Devonian cephalopods from Novaya Zemlya.*]

Bouček, Bedrich, 1964, *The tentaculites of Bohemia; their morphology, taxonomy, ecology, phylogeny, and biostratigraphy:* Cesk. Akad. Věd, 215 p., 40 pl.

Bouckaert, Joseph, & Streel, Maurice (eds.), 1974, *International Symposium on Belgian micropaleontological limits, Namur 1971, General information:* Guidebook, Geol. Survey Belgium, 12 p., 5 charts.

——, Mouravieff, A. N., Streel, Maurice, Thorez, Jacques, & Ziegler, Willi, 1972, *The Frasnian-Famennian boundary in Belgium:* Geologica et Palaeontologia, v. 6, p. 87-92.

Boucot, A. J., 1960, *Lower Gedinnian brachiopods of Belgium:* Inst. Géol. Univ. Louvain, Mém., v. 21, p. 281-325, pl. 9-18.——1974, *Silurian and Devonian biogeography:* in Paleogeographic provinces and provinciality, C. A. Ross (ed.), Soc. Econ. Paleontologists & Mineralogists, Spec. Paper, no. 21, p. 165-176.——1975, *Evolution and extinction rate controls:* x + 427 p., Elsevier (Amsterdam).

——, Caster, K. E., Ives, David, & Talent, J. A., 1963, *Relationships of a new Lower Devonian terebratuloid (Brachiopoda) from Antarctica:* Bull. Am. Paleontology, v. 46, p. 81-151, pl. 16-41.

——, Johnson, J. G., & Struve, Wolfgang, 1966, *Stringocephalus, ontogeny and distribution:* Jour. Paleontology, v. 40, p. 1349-1364, pl. 169-171.

——, ——, & Talent, J. A., 1969, *Early Devonian brachiopod zoogeography:* Geol. Soc. America, Spec. Paper, no. 119, 106 p.

Brice, Denise, 1970, *Étude paléontologique et stratigraphique du Dévonien de l'Afghanistan:* Moyen-Orient, Notes Mém., v. 11, p. 1-364, pl. 1-20.

Briden, J. C., Drewry, G. E., & Smith, A. G., 1974, *Phanerozoic equal-area world maps:* Jour. Geology, v. 82, p. 555-574.

Brunton, C. H. C., Miles, R. S., & Rolfe, W. D. I., 1969, *Gogo Expedition 1967:* Geol. Soc. London, Proc., no. 1655, p. 79-83.

Buggisch, Werner, 1972, *Zur Geologie und Geochemie der Kellwasserkalke und ihrer begleitenden Sedimente (Unteres Oberdevon):* Hess. Landesamt Bodenforsch., Abhandl., v. 62, p. 1-68.

——, & Clausen, C.-D., 1972, *Conodonten- und Goniatiten-Faunen aus dem oberen Frasnium und unteren Famennium Marokkos (Tafilalet, Antiatlas):* Neues Jahrb. Geologie Paläontologie, Abhandl., v. 141, p. 137-167.

Bultynck, P. L., 1970, *Révision stratigraphique et paléontologique de la coupe type du Couvinien à Couvin:* Univ. Louvain, Inst. Geol., Mém., v. 26, p. 1-152, pl. 1-39.

Burrett, C. F., 1972, *Plate tectonics and the Hercynian orogeny:* Nature, v. 239, p. 155-156 (London).

Carls, Peter, 1969, *Stratigraphie und Conodonten des Unter-Devons der östlichen Iberischen Ketten (NE-Spanien):* Neues Jahrb. Geologie Paläon-

tologie, Abhandl., v. 132, p. 155-218.
——, Gandl, Josef, Groos-Uffendorde, Helga, Jahnke, Hans, & Walliser, O. H., 1972, *Neue Daten zur Grenze Unter-Mittel-Devon:* Newsl. Stratigraphy, v. 2, p. 115-147.
Ch'ang An-chin [Chang An-chi], 1958, *Ta Hsing an ling hai shen shih hiu yan ti tseng Ku sheng Wu chün ho Ku ti li ti yuan chiu ping tao lun chung kuo nan Pu Shang ni pen chi hou ch'i ti chien tuan:* Ku Sheng Wu Hsüeh Pao (Acta Palaeont. Sinica), v. 6, p. 71-89. [*Stratigraphy, palaeontology and palaeogeography of the ammonite fauna of the Clymenienkalk from Great Khingan with special reference to the post-Devonian break (hiatus) of South China.*] (In Chinese and English.)——1960, *Ta hsing an ling woan ni pen shyh shin jyu shyr chyuan jyi chiji tzai sheng wuh fen ley shang de yih yih:* Ku Sheng Wu Hsüeh Pao (Acta Palaeont. Sinica), v. 8, p. 180-192; pl. 1. [*New late Upper Devonian ammonite faunas of the Great Khingan and its biological classification.*] (In Chinese and English.)
Chao King-koo, 1956, *Notes on some Devonian ammonoids from southern Kwangsi:* Ku Sheng Wu Hsüeh Pao (Acta Palaeont. Sinica), v. 4, p. 101-116.
Chlupáč, Ivo, 1957, *Faciální vývoj a biostratigrafie středočeského spodního devonu:* Ústred. Ústavu Geol., Sborník, v. 23, Geol., p. 369-485, pl. 1-7.——1959, *Faciální vývoj a biostratigrafie břidlic dalejských a vápenců hlubočepských (eifel) ve středočeském devonu:* Ústred. Ústavu Geol., Sborník, v. 25, Geol., p. 445-511, pl. 1-5.—— 1960, *Stratigrafická studie o vrstvách srbských (givet) ve středočeském devonu:* Ústred. Ústavu Geol., Sborník, p. 143-185, pl. 1,2.——1972, *The Silurian-Devonian boundary in the Barrandian:* Canadian Petrol. Geology, Bull., v. 20, p. 104-174.
Churkin, Michael, 1973, *Paleozoic and precambrian rocks of Alaska and their role in its structural evolution:* U.S. Geol. Survey, Prof. Paper, no. 740, p. 1-65.
Clarke, J. M., 1913, *Fosseis Devonianos do Paraná:* Serv. Geol. Mineral. do Brasil, Mon., v. 1, 353 p., 27 pl.
Clausen, C. D., 1969, *Oberdevonische Cephalopoden aus dem Rheinischen Schiefergebirge. II. Gephuroceratidae, Beloceratidae:* Palaeontographica, Abt. A, v. 132, p. 95-178, pl. 22-26.——1971, *Geschichte, Umfang und Evolution der Gephuroceratidae (Ceph.; Oberdevon.) in heutiger Sicht:* Neues Jahrb. Geologie Paläontologie, Abhandl., v. 137, p. 175-208.
Cloud, P. E., 1959, *Paleoecology—retrospect and prospect:* Jour. Paleontology, v. 33, p. 926-962.
Conil, R. C. L., & Pirlet, H., 1970, *Le Calcaire Carbonifère du Synclinorium de Dinant et le sommet du Famennien:* Congrès Colloq. Univ. Liège, v. 55, p. 47-63.

Drot, Jeannine, 1964, *Rhynchonelloidea et Spiriferoidea siluro-dévonien du Maroc pré-saharien:* Serv. Géol. Maroc, Notes Mém., no. 178, p. 1-238.——1971, *Rhynchonellida siluriens et dévoniens du Maroc Pré-saharien, nouvelles observations:* Serv. Géol. Maroc, Notes Mém., no. 237, p. 65-108.
Druce, E. C., 1969, *Devonian and Carboniferous conodonts from the Bonaparte Gulf Basin, Northern Australia and their use in international correlation:* Bur. Min. Res. Geol. Geophys., Bull. 98, 242 p., 43 pl.
Dürkoop, Arnfrid, 1970, *Brachiopoden aus dem Silur, Devon und Karbon in Afghanistan:* Palaeontographica, Abt. A, v. 134, p. 153-225, pl. 14-19.
Earp, J. R., 1973, *Silurian, Devonian and Old Red Sandstone on Geological Survey maps:* Geol. Mag., v. 110, p. 301-302.
Erben, H. K., 1953a, *Goniatitacea (Ceph.) aus dem Unterdevon und dem unteren Mitteldevon:* Neues Jahrb. Geologie Paläontologie, v. 98, p. 175-225, pl. 17-19.——1953b, *Stratigraphie, Tektonik und Faziesverhältnisse des böhmisch entwickelten Unterdevons im Harz:* Beihefte Geol. Jahrb., v. 9, vii + 98 p.——1960, *Die Grenze Unterdevon/Mitteldeon im Hercyn Deutschlands und des Massif Armoricain—ihre Korrelation mit dem Barrandium:* in J. Svoboda (ed.), Prager Arbeitstagung über die Stratigraphie des Silurs und des Devons (1958), p. 187-207, Nakladatelství Ceskoslovenska Akad. Věd (Prague).——1962, *Unterlagen zur Diskussion der Unter/Mitteldevon-Grenze:* in Internationale Arbeitstagung über die Silur/Devon-Grenze und die Stratigraphie von Silur und Devon, Bonn-Bruxelles 1960, Symposiums-Band 2, H. K. Erben (ed.), p. 62-70, E. Schweizerbart'sche Verlagsbuchhandlung (Stuttgart).——1964, *Facies developments in the marine Devonian of the Old World:* Ussher Soc., Proc., v. 1, p. 92-118.——1966, *Über den Ursprung der Ammonoidea:* Biol. Rev., 41, p. 641-658.
Errera, Michel, Mamet, Bernard, & Sartenaer, P. J. M. J., 1972, *Le Calcaire de Givet et le Givetien à Givet:* Inst. Roy. Sci. Nat. Belgique, Bull., v. 49, p. 1-59.
Fuchs, Günter, 1971, *Faunengemeinschaften und Fazieszonen im Unterdevon der Osteifel als Schlüssel zur Paläogeographie:* Hess. Landesamt Bodenforsch., Notizbl., v. 99, p. 78-105.
Geikie, Archibald, 1875, *Life of Sir Roderick I. Murchison:* v. 1, 387 p.; v. 2, 375 p., John Murray (London).
George, T. N., & Wagner, R. H., 1969, *Report of the International Union of Geological Sciences Subcommission on Carboniferous Stratigraphy:* 6ème Congrès Internatl. Strat. Géol. Carbonif. Compte Rendu, v. 1, p. xlii-xlv (Maastricht).
Glenister, B. F., 1958, *Upper Devonian ammonoids from the Manticoceras Zone, Fitzroy Basin,*

Western Australia: Jour. Paleontology, v. 32, p. 58-96, pl. 5-15.
——, & Klapper, Gilbert, 1966, Upper Devonian conodonts from the Canning Basin, Western Australia: Jour. Paleontology, v. 40, p. 776-842, pl. 85-96.
Gobbett, D. J., 1966, The brachiopod genus Stringocephalus from Malaya: Jour. Paleontology, v. 40, p. 1345-1348, pl. 168.
Grabau, A. W., 1931-33, Devonian Brachiopoda of China. I. Devonian Brachiopoda from Yunnan and other districts in South China: Palaeont. Sinica, ser. B, v. 3, no. 3, p. 1-454 (1931), pl. 1-54 (1933).
Gross-Uffenorde, Helga, & Uffenorde, Henning, 1974, Zur Mikrofauna im höchsten Oberdevon und tiefen Unterkarbon im nördlichen Sauerland: Hess. Landesamt Bodenforsch., Notizbl., v. 102, p. 58-87, pl. 1-6.
Hallam, Anthony (ed.), 1973, Atlas of palaeobiogeography: 531 p., Elsevier (Amsterdam).
Harland, W. B., Holland, C. H., House, M. R., Hughes, N. F., Reynolds, A. B., Rudwick, M. J. S., Satterthwaite, G. E., Tarlo, L. B. H., & Willey, E. C., 1967, The fossil record. A symposium with documentation, jointly sponsored by the Geological Society of London and the Palaeontological Association: vii + 828 p., Geol. Soc. London (London).
Havlíček, Vladimir, 1959, Rhynchonelloidea des böhmischen und mährischen Mitteldevon (Brachiopoda): Ústřed. Ústavu Geol., Rozpravy, v. 25, p. 1-275, pl. 1-28.
Hill, Dorothy, 1957, The sequence and distribution of Upper Palaeozoic coral faunas: Austral. Jour. Sci., v. 19, p. 42-61.
Hollard, Henri, 1974, Recherches sur la stratigraphie des formations du Dévonien moyen, de l'Emsien supérieur au Frasnien, dans le Sud du Tafilalet et dans Ma'der (Anti-Atlas oriental): Serv. Géol. Maroc, Notes, v. 36, no. 264, p. 7-68.
House, M. R., 1962, Observations on the ammonoid succession of the North American Devonian: Jour. Paleontology, v. 36, p. 247-284, pl. 43-48.——1973a, Delimitation of the Frasnian: Acta Geol. Polonica, v. 23, p. 1-14.——1973b, Devonian goniatites: in Atlas of palaeobiogeography, A. Hallam (ed.), p. 97-104, Elsevier (Amsterdam).——1975a, Facies and time in Devonian tropical areas: Yorkshire Geol. Soc., Proc., v. 40, p. 233-288.——1975b, Faunas and time in the marine Devonian: Yorkshire Geol. Soc., Proc., v. 40, p. 490-495.——1977 (ed.), A correlation of the Devonian rocks of the British Isles: Geol. Soc. London (in press).
Hughes, N. F. (ed.), 1973, Organisms and continents through time: a symposium: Spec. Paper in Palaeontology, no. 12, vi + 334 p.
Jaeger, Hermann, 1965, Referate—Symposiums-Band der 2. Internationalen Arbeitstagung über die Silur/Devon-Grenze und die Stratigraphie von Silur und Devon, Bonn-Bruxelles 1960: Geologie, v. 14, p. 348-364.——1966, Two late Monograptus species from Victoria, Australia, and their significance for dating the Baragwanathia flora: Royal Soc. Victoria, Proc., v. 79, p. 393-413, pl. 1-3.——1970, Remarks on the stratigraphy and morphology of Praguian and probably younger monograptids: Lethaia, v. 3, p. 173-182.——1973, O Rannedevonskiy graptolitakh: in Stratigrafiya nizhnego i srednego Devona, D. Nalivkin (general ed.), p. 99-109, Akad. Nauk SSSR (Moskva). [On early Devonian graptolites.]
——, Stein, Volker, & Wolfart, Reinhard, 1969, Fauna (Graptolithen, Brachiopoden) der unterdevonischen Schwarzschiefer Nord-Thailands: Neues Jahrb. Geologie Paläontologie, Abhandl., v. 133, p. 171-190, pl. 14-17.
Jenkins, T. B. H., 1968, Famennian ammonoids from New South Wales: Palaeontology, v. 11, p. 535-548, pl. 104, 105.
Johnson, J. G., 1970, Taghanic onlap and the end of North American Devonian provincialism: Geol. Soc. America, Bull., v. 81, p. 2077-2106.
Jongmans, W. J., & Gothan, Walter, 1937, Betrachtungen über die Ergebnisse des zweiten Kongresses für Karbon-Stratigraphie: 2ème Congr. Étud. Strat. Carb., Compte Rendu, v. 1, p. 1-40.
Khalymbadzha, V. G., & Chernysheva, N. G., 1970, Konodonty Ancyrodella iz devonskikh otlozheniy Volg-Kamskogo kraya i ikh stratigraficheskoye znacheniye: in Biostratifrafiya i paleontologiya otlozheniy vostoka Russkoy platformy i zapadnogo Priuralya, no. 1, p. 81-103, Kazan. Gos. Univ. (Kazan). [Conodonts of the genus Ancyrodella from Devonian deposits of the Volga-Kama region and their stratigraphic significance.]
Klapper, Gilbert, Philip, G. M., & Jackson, J. H., 1970, Revision of the Polygnathus varcus group (Conodonta, Middle Devonian): Neues Jahrb. Mineralogie Paläontologie, Monatsh. 1970/11, p. 650-667.
——, & Ziegler, Willi, 1967, Evolutionary development of the Icriodus latericrescens group (Conodonta) in the Devonian of Europe and North America: Palaeontographica, Abt. A, v. 127, p. 68-83, pl. 8-11.
Kobayashi, Teiichi, & Hamada, Takashi, 1966, A new proetoid trilobite from Perlis, Malaysia (Malaya): Japan. Jour. Geology Geography, v. 37, p. 87-92, pl. 2.
Korejwo, Krystyna, & Teller, Lech, 1965, Upper Silurian non-graptolite fauna from the Chełm borehole (eastern Poland): Acta Geol. Polonica, v. 14, p. 233-302, pl. 1-26 (with Pol. summary).
Kozłowski, Roman, 1929, Brachiopodes Gothlandiens de la Podolie Polonaise: Palaeont. Polonica, v. 1, 254 p., 26 pl.
Krebs, Wolfgang, 1974, Devonian carbonate complexes of central Europe: Soc. Econ. Paleontolo-

gists & Mineralogists, Spec. Publ., no. 18, p. 155-208.

———, & Rabien, Arnold, 1964, *Zur Biostratigraphie und Fazies der Adorf-Stufe bei Donsbach:* Hess. Landesamt Bodenforsch. Wiesbaden, Notizbl., v. 92, p. 75-119, pl. 6,7.

———, & Wachendorf, Horst, 1973, *Proterozoic-Palaeozoic geosynclinal and orogenic evolution of central Europe:* Geol. Soc. America, Bull., v. 84, p. 2611-2630.

———, & Ziegler, Willi, 1965, *Über die Mitteldevon/Oberdevon-Grenze in der Riffazies bei Aachen:* Fortschr. Geol. Rheinland Westfalen, v. 9, p. 731-754, pl. 1,2.

Kullmann, Jürgen, 1973, *Goniatite-coral associations from the Devonian of Istanbul, Turkey:* in Paleozoic of Istanbul, O. Kaya (ed.), Ege Univ., Fen Fak., Kitaplar Ser., Jeol., no. 40, p. 97-116 (Izmir).

———, & Ziegler, Willi, 1970, *Conodonten und Goniatiten von der Grenze Mittel-/Oberdevon aus dem Profil am Martenberg (Ostrand des Rheinischen Schiefergebirges):* Geologica et Palaeontologica, v. 4, p. 73-85, pl. 1.

Kutscher, Fritz, & Schmidt, Hermann, 1958, *Lexique stratigraphique International, Europe, Fasc. 5, Allemagne, 5b, Dévonien:* 386 p., Louis Jean-Gap (Paris).

Lange, F. G., 1968, *Conodonten-Gruppenfunde aus Kalken des tieferen Oberdevon:* Geologica et Palaeontologica, v. 2, p. 37-57.

Lardeux, Hubert, 1969, *Les tentaculites d'Europe occidentale et d'Afrique du nord:* Cent. Natl. Rech. Sci., 238 p. (Paris).

Lecompte, Marius, 1970, *Die Riffe im Devon der Ardennen und ihre Bildungsbedingungen:* Geologica et Palaeontologica, v. 4, p. 25-71.

Le Maitre, Dorothée, 1933, *Description des stromatoporoides de l'Assise Etroeungt:* Soc. Géol. France, Mém., n. ser., v. 9, p. 1-32.———1947, *Contribution à l'étude du Dévonien: 2, Le récif corraligène de Ouihalane:* Serv. Géol. Maroc, Notes Mém., no. 67, p. 1-115.———1952, *La faune du Dévonien inferieur et moyen de la Sahara et des abords de l'Erg Djemel (Sud Oranais):* Mat. Cart. Géol. Algérie, sér. 1, Paléont., no. 12, 170 p., 22 pl.

Lonsdale, William, 1840, *Notes on the age of the limestones of South Devonshire:* Geol. Soc. London, Trans., ser. 2, v. 5, p. 721-738.

Lyashenko, A. I., 1959, *Atlas brakhiopod i stratigrafiya Devonskikh otlozheniy tsentralnykh oblastey Russkoy platformy:* Min. Geol. Okhran. Nedr. SSSR, VNIGNI, 451 p., 87 pl. [*Atlas of the brachiopods and stratigraphy of the Devonian deposits of the central part of the Russian platform.*]

Lyashenko, G. P., 1953, *O stratigraficheskom znachenii tentakulitov:* Akad. Nauk SSSR, Doklady, v. 91, p. 371-374. [*On the stratigraphical significance of tentaculites.*]———1959, *Konikonkhii Devona tsentralnykh i vostochnykh oblastey Russkoy platformy:* 153 p., 31 pl., Lengostoptekhizdat, VNEGNI (Leningrad). [*Devonian coniconchs of the central and eastern parts of the Russian platform.*]

McKellar, R. G., 1970, *The Devonian productoid brachiopod faunas of Queensland:* Geol. Survey Queensland, Publ., no. 342, Paleont. Paper 18, p. 1-40.

McLaren, D. J., 1970, *Time, life and boundaries:* Jour. Paleontology, v. 44, p. 801-815.———1972, *Report from the Committee on the Silurian-Devonian Boundary and Stratigraphy to the President of the Commission on Stratigraphy:* Geol. Newsletter, v. 1972, p. 268-288.

Maillieux, Eugène, 1937, *Les Lamellibranchs du Dévonien Inférieur de l'Ardenne:* Musée Royal Histoire Nat. Belgique, Mém., no. 81, 273 p., 14 pl.

Martinsson, Anders (ed.), 1977, *The Silurian-Devonian boundary. Final report on the Silurian-Devonian boundary within IUGS Commission on Stratigraphy and a state of the art report for Project Ecostratigraphy:* Int. Union Geol. Sci., Ser. A, 349 p.

Matern, Hans, 1931a, *Die Goniatiten-Fauna der Schistes de Matagne in Belgien:* Musée Royal Histoire Nat. Belgique, Bull., v. 7, p. 1-15.——— 1931b, *Das Oberdevon der Dill-Mulde:* Preuss. Geol. Landesanst., Abhandl., n. ser., v. 134, p. 1-139, pl. 1-4.

Middleton, G. V., 1959, *Devonian tetracorals from South Devonshire, England:* Jour. Paleontology, v. 33, p. 138-160, pl. 27.

Mu En-chih, Li Chi-chin, Ko Mei-yu, Chen Hsü, Ni Yü-nan, Lin Yao-k'un, & Mu Hsi-nan, 1974, *Hsi Nan ti chü ti tseng Ku sheng wu shoǔ ts'e: Chung kuo Ke Hsüeh yuan nan Ching ti chih ku sheng wù yan chiu suo pien chu:* 454 p., 202 pl., Ke Hsüeh ch'u Păn she (Nanking). [*A handbook of the stratigraphy and paleontology in southwest China.*]

Nalivkin, D. V. (general ed.), 1971, *Granitsa Silura i Devona i biostratigrafiya Silura, Trudy III Mezh. Simp. I:* 283 p., Akad. Nauk SSSR (VSEGEI), (Leningrad). [*The Silurian and Devonian boundary and Silurian biostratigraphy, Transactions 3rd International Symposium.*]——— 1973, (general ed.), *Stratigrafiya nizhnego i srednego Devona. Trudy III Mezh. Simp. po Granitse Siluri i Devona, 2:* 292 p., Akad. Nauk SSSR (Leningrad). [*Stratigraphy of the Lower and Middle Devonian. Transactions 3rd International Symposium on the Silurian-Devonian boundary and the stratigraphy of the Lower and Middle Devonian.*]

———, Rzhonsnitskaya, M. A., & Markovskiy, B. P. (ed.), 1973, *Stratigrafiya SSSR, Devonskaya Sistema:* v. 1, 519 p., v. 2, 376 p., Izd. Nedra (Moskva). [*Stratigraphy of USSR, Devonian System.*]

Nikiforova, O. I., & Predtechenskiy, N. N., 1968, *A guide to the geological excursion on Silurian and Lower Devonian deposits of Podolia (Middle Dnestr River)*: 58 p., Akad. Nauk SSSR (VSEGEI) (Leningrad).

Oliver, A. W., 1976, *Biogeography of Devonian rugose corals*: Jour. Paleontology, v. 50, p. 365-373.

Orchard, M. J., 1974, *Famennian conodonts and cavity infills in the Plymouth Limestone (S. Devon)*: Ussher Soc., Proc., v. 3, p. 49-54.

Oswald, D. H. (ed.), 1968a,b, *International Symposium on the Devonian System, Calgary, Alberta, 1967*: Alberta Soc. Petrol. Geologists, v. 1, 1055 p. [1968a]; v. 2, 1377 p. [1968b].

Paeckelmann, Werner, 1925, *Beiträge zur Kenntnis des Devons am Bosporus, insbesondere in Bithynien*: Preuss. Geol. Landesanst., Abhandl., n. ser., v. 98, p. 1-152, pl. 1-6.

Petersen, M. S., 1975, *Upper Devonian (Famennian) ammonoids from the Canning Basin, Western Australia*: Jour. Paleontology, v. 49, suppl. to no. 5, Paleont. Soc. Mem. 8, 55 p., pl. 1-7.

Petter, Germaine, 1959, *Goniatites Dévoniennes du Sahara*: Serv. Carte Géol. l'Algérie Publ., n. sér., Paléont. Mém. no. 2, p. 1-313, pl. 1-26.—— 1960, *Clymènes du Sahara*: Serv. Carte Géol. l'Algérie, Publ., n. sér., Paléont. Mém. no. 6, p. 1-76, pl. 1-8.

Philip, G. M., & Jackson, J. H., 1971, *Late Devonian conodonts from the Luton Formation, northern New South Wales*: Linnean Soc. New S. Wales, Proc., v. 96, p. 66-76.

——, & Pedder, A. E. H., 1967, *A correlation of some Devonian limestones of New South Wales and Victoria*: Geol. Mag., v. 104, p. 232-239.

Pickett, J. W., 1960, *A cylmeniid from the Wocklumeria zone of New South Wales*: Palaeontology, v. 3, p. 237-241, pl. 41.—— 1972, *Correlation of the Middle Devonian formations of Australia*: Geol. Soc. Australia, Jour., v. 18, p. 457-466.

Playford, P. E., & Lowry, D. C., 1966, *Devonian reef complexes of the Canning Basin, Western Australia*: Geol. Survey W. Australia, Bull., no. 118, 150 p. (with separate atlas of plates).

Přibyl, Alois, & Vaněk, Jiři, 1968, *Biostratigraphische Studie über die Fauna des Budňaniums bis Pragiums im Hinblick auf die Grenze zwischen Silur und Devon im Barrandium und in den übrigen europäischen Gebieten*: Neues Jahrb. Geologie Paläontologie, n. ser., v. 7, p. 413-440.

Rabien, Arnold, 1954, *Zur Taxionomie und Chronologie der oberdevonischen Ostracoden*: Hess. Landesamt Bodenforsch., Abhandl., v. 9, p. 1-268, pl. 1-5.—— 1956, *Die Stratigraphie und Fazies des Ober-Devons in der Waldecker Hauptmulde*: Hess. Landesamt Bodenforsch., Abhandl., v. 16, p. 1-83, pl. 1-3.—— 1960, *Zur Ostracoden-Stratigraphie an der Devon/Karbon-Grenze im Rheinischen Schiefergebirge*: Fortschr. Geol. Rheinland Westfalen, v. 3, p. 61-106, pl. 103.—— 1970, in *Erläuterungen zur geologischen Karte von Hessen 1:25,000*; Blatt Nr. 5215 Dillenburg, H. J. Lippert, Hans Hentschel, Arnold Rabien, et al., Hess. Landesamt Bodenforsch., Abhandl., p. 78-83, 103-235.

Reed, F. R. C., 1908, *The Devonian faunas of the northern Shan States*: Palaeont. Indica, n. ser., v. 2, Mem. 5, 183 p., 20 pl.—— 1925, *Revision of the fauna of the Bokkeveld beds*: South African Museum, Ann., v. 22, p. 27-225.

Richter, Rudolf, & Richter, Emma, 1926, *Die Trilobiten des Oberdevons. Beiträge zur Kenntnis devonischer Trilobiten. IV*: Preuss. Geol. Landesanst., Abhandl., n. ser., v. 99, p. 1-314, pl. 1-12.—— 1941, *Devon*: Geol. Jahresb., v. 3A, p. 31-43.—— 1950, *Arten der Dechenellinae (Tril.)*: Senckenberg. Lethaea, v. 31, p. 151-184, pl. 1-4.—— 1952, *Phacopacea von der Grenze Emsium/Eiflium (Tril.)*: Senckenberg. Lethaea, v. 33, p. 79-108, pl. 1-4.—— 1954, *Die Trilobiten des Ebbe-Sattels und zu vergleichende Arten*: Senckenberg. Naturforsch. Gesell., Abhandl., v. 488, p. 1-76, pl. 1-6.

Riding, Robert, 1974, *Model of the Hercynian foldbelt*: Earth Planet. Sci. Letters, v. 24, p. 125-135.

Roberts, John, Jones, P. J., Jell, J. S., Jenkins, T. B. H., Marsden, M. A. H., McKellar, R. G., McKelvey, B. C., & Seddon, G., 1972, *Correlation of the Upper Devonian rocks of Australia*: Geol. Soc. Australia, Jour., v. 18, p. 467-490.

Roy, J. L., 1971, *A pattern of rupture of the eastern North American-western European block*: Earth Planet. Sci. Letters, v. 14, p. 104-114.

Różkowska, Maria, 1969, *Famennian tetracorralloid and heterocoralloid fauna from the Holy Cross Mountains (Poland)*: Acta Palaeont. Polonica, v. 14, p. 5-198, pl. 1-8 (Pol. and Russ. summaries).

Rzhonsnitskaya, M. A., 1968, *Biostratigrafiya Devona Okrani Kuznetskogo Basseyna*: Vses. Nauchno-Issledov. Geol. Inst. (VSEGEI), Trudy, 288 p. [*Devonian biostratigraphy of the outlying areas of the Kuznetsk basin.*]

Sartenaer, P. J. M. J., 1966, *Frasnian Rhynchonellida from the Ozbak-Kuh and Tabas regions (East Iran)*: Geol. Survey Iran, Rept. no. 6, p. 25-53, pl. 1,2.—— 1969, *Late Upper Devonian (Famennian) rhynchonellid brachiopods from Western Canada*: Geol. Survey Canada, Bull., v. 169, p. 1-269, pl. 1-19.—— 1970, *Le contact Frasnien-Famennien dans la région de Houyet-Han-sur-Lesse*: Soc. Géol. Belgique, Ann., v. 92, p. 345-357.—— 1971, *Redescription of the brachiopod genus Yunnanella Grabau, 1923 (Rhynchonellidae)*: Smithson. Contrib. Paleobiology, no. 3, p. 203-218.

Saul, J. M., Boucot, A. J., & Finks, R. M., 1963, *Fauna of the Accraian Series (Devonian of*

Ghana) *including a revision of the gastropod* Plectonotus: Jour. Paleontology, v. 37, p. 1042-1053, pl. 135-138.

Savage, N. M., 1971, *Brachiopoda from the Lower Devonian Mandagery Park Formation, New South Wales:* Palaeontology, v. 14, p. 387-422, pl. 69-71.

Schindewolf, O. H., 1923, *Beiträge zur Kenntnis der Paläozoicums in Oberfranken, Ostthüringen und dem Sächsischen Vogtlande. I. Stratigraphie und Ammoneenfauna des Oberdevons von Hof a. S.:* Neues Jahrb. Mineralogie Geologie Paläontologie, v. 49, p. 250-358, 393-509, pl. 14-18. ——1937, *Zur Stratigraphie und Paläontologie der Wocklumer Schichten (Oberdevon):* Preuss. Geol. Landesanst., Abhandl., n. ser., v. 178, p. 1-132, pl. 1-4.

Schmidt, Hermann, 1921, *Das Oberdevon-Culm-Gebiet von Warstein i. W. und Belecke:* Preuss. Geol. Landesanst., Jahrb., v. 41, p. 254-339, pl. 6,7. ——1924, *Zwei Cephalopodenfaunen aus der Devon-Carbon-Grenze im Sauerland:* Preuss. Geol. Landesanst, Jahrb., v. 44, p. 98-171, pl. 6-8. ——1926, *Beobachtungen über mitteldevonische Zonen-Goniatiten:* Senckenberg. Lethaea, v. 1, p. 291-295. ——1950, *Werneroceras crispiforme Kayser und andere Goniatiten des Eifeler Mitteldevon:* Senckenberg. Lethaea, v. 31, p. 89-94.

Scrutton, C. T., 1968, *Colonial Phillipsastraeidae from the Devonian of south-east Devon, England:* Brit. Museum (Nat. History), Bull., Geol., v. 15, p. 181-281, pl. 1-18. ——1975, *Preliminary observations on the distribution of Devonian rugose coral faunas in south-west England:* Akad. Nauk SSSR, Trudy, Inst. Geol. Geofiz., v. 202, no. 2, p. 131-140.

Sedgwick, Adam, & Murchison, R. I., 1839a, London, Edinburgh, Dublin Philos. Mag., 1839 (April), p. 259. ——1839b, *On the classification of the older rocks of Devon and Cornwall:* Geol. Soc., London, Proc., v. 3, p. 121-123. ——1840, *On the physical structure of Devonshire, and on the subdivisions and geological relations of its older stratified deposits, & c.:* Geol. Soc., London, Trans., ser. 2, v. 5, p. 633-704, pl. 50-58.

Shirley, Jack, 1938, *The fauna of the Baton River beds (Devonian) New Zealand:* Geol. Soc. London, Quart. Jour., v. 94, p. 459-506.

Sokolov, D. S. (ed.), 1960, *Regionalnaya stratigrafiya Kitaya :* [transl. into Russian from the Chinese edition of 1956, Li Ssu-kuang (dir.)], 659 p., Izd. Inostran. Lit. (Moskva). [*Regional stratigraphy of China.*] ——1962, *Paleogeograficheskiy Atlas Kitaya:* [transl. from Chinese atlas by Liu Huang-yung], 119 p., Izd. Inostran. Lit. (Moskva). [*Atlas of paleogeography of China.*]

Sokolov, V. S. (ed.), 1972, *Opornyy razrez Silura i Nizhnego Devona Podolii:* Mezh. Strat. Kom. SSSR, Akad. Nauk SSSR, Trudy, v. 5, p. 1-262. [*Silurian and Lower Devonian key sections of Podolia.*]

Solle, Gerhard, 1953, *Die Spiriferen der Gruppe arduennensis-intermedius im Rheinischen Devon:* Hess. Landesamt Bodenforsch., Abhandl., v. 5, p. 1-156, pl. 1-18. ——1971, *Brachyspirifer und Paraspirifer im Rheinischen Devon:* Hess. Landesamt Bodenforsch., Abhandl., v. 59, p. 1-163, pl. 1-20.

Sougy, J. A. M., 1969, *Présence inattendue de Pinacites jugleri (Roemer) dans un calcaire situé à la base des siltstones de Tighirt (Couvinien inférieur du Zemmour noir, Mauritanie septentrionale):* Soc. Géol. France, Bull., sér. 7, v. 11, p. 268-272.

Strusz, D. L., 1972, *Correlation of the Lower Devonian rocks of Australasia:* Geol. Soc. Australia, Jour., v. a8, p. 427-455.

Struve, Wolfgang, 1963, *Das Korallen-Meer der Eifel vor 300 Millionen Jahren-Funde, Deutungen, Probleme:* Natur und Museum, v. 93, p. 237-276. ——1966, *Einige Atrypinae aus dem Silurium und Devon:* Senckenberg. Lethaea, v. 47, p. 123-163, pl. 15,16. ——1970, *"Curvate Spiriferen" der Gattung Rhenothyris und einige andere Reticulariidae aus dem Rheinischen Devon:* Senckenberg. Lethaea, v. 51, p. 449-577, pl. 1-15.

Stürmer, Wilhelm, 1969, *Röntgenuntersuchung an paläontologischen Präparaten:* Electromedica, v. 1969(2), p. 48-51.

Teichert, Curt, 1974, *Marine sedimentary environments and their faunas in the Gondwana area:* Am. Assoc. Petrol. Geologists, Mem., no. 23, p. 361-394.

Telford, P. G., 1975, *Lower and Middle Devonian conodonts from the Broken River Embayment, North Queensland, Australia:* Spec. Paper Palaeontology, no. 15, 96 p., 16 pl.

Tien, C. C., 1938, *Devonian Brachiopoda of Hunan:* Palaeont. Sinica, ser. B, v. 4, p. 1-192, pl. 1-22.

Tsien, H. H., 1971, *The Middle and Upper Devonian reef-complexes of Belgium:* Petrol. Geology Taiwan, no. 8, p. 119-173.

Veevers, J. J., 1959a, *Devonian brachiopods from the Fitzroy Basin, Western Australia:* Bur. Min. Res. Geol. Geophys. Australia, Bull., no. 45, p. 1-220. ——1959b, *Devonian and Carboniferous brachiopods from north-western Australia:* Bur. Min. Res. Geol. Geophys. Australia, Bull., no. 55, p. 1-42.

Vöhringer, E., 1960, *Die Goniatiten der unterkarbonischen Gattendorfia-Stufe im Hönnetal (Sauerland):* Fortschr. Geol. Rheinland Westfalen, v. 3, p. 107-196, pl. 1-7.

Walliser, O. H., 1966, *Preliminary notes on Devonian, Lower and Upper Carboniferous goniatites in Iran:* Geol. Survey Iran, Rept. no. 6, p. 7-24, pl. 1-4.

Wang, H. C., 1948, *The Middle Devonian rugose corals of eastern Yunnan:* Natl. Univ. Peking, Contrib. Geol. Inst., no. 55, p. 1-46, pl. 1-3.

Waterlot, Gerard (ed.), 1957, *Lexique stratigraphique International, Europe, Fasc. 4, France, Belgique, Pays Bas, Luxembourg, 4a, Antecambrien, Paléozoique Inférieur:* 432 p., Louis-Jean.-Gap (Paris).

Wedekind, Rudolf, 1908, *Die Cephalopodenfauna des höheren Oberdevon am Enkeberg:* Neues Jahrb. Mineralogie Geologie Paläontologie, v. 26, p. 565-635, pl. 35-45.——1913, *Die Goniatitenkalke des unteren Oberdevon von Martenberg bei Adorf:* Gesell. Natur. Freunde, Berlin, Sitzungsber., 1913, no. 1, p. 23-77, pl. 4-7.——1914, *Monographie der Clymenien des Rheinischen Gebirges:* K. Gesell. Wiss. Göttingen, Abhandl., Math.-Phys., n. ser., v. 10, no. 1, p. 1-73, pl. 1-7.——1917, *Die Genera der Palaeoammonoidea (Goniatiten):* Palaeontographica, v. 62, p. 85-184, pl. 14-22.——1924-25, *Das Mitteldevon der Eifel:* Gesell. Beförderung Gesamt. Naturwiss., Schriften, v. 14, no. 3 (1924), p. 1-93, pl. 1,2; v. 14, no. 4 (1925), p. 1-85, pl. 1-17.

——, & **Vollbrecht, Emmi,** 1931-32, *Die Lytophyllidae des mittleren Mitteldevon der Eifel:* Palaeontographica, v. 75 (1931), p. 81-110, pl. 15-46, v. 76 (1932), p. 96-120, pl. 9-14.

White, E. I., 1950, *The vertebrate faunas of the Lower Old Red Sandstone of the Welsh Borders:* Brit. Museum (Nat. History), Bull., v. A1, p. 51-67.

Wittekindt, Hanspeter, 1965, *Zur Conodontenchronologie des Mitteldevons:* Fortschr. Geologie Rheinland Westfalen, v. 9, p. 621-643, pl. 1-3.

Ziegler, Willi, 1958, *Conodontenfeinstratigraphische Untersuchungen an der Grenze Mitteldevon/Oberdevon und in der Adorfstufe:* Hess. Landesamt Bodenforsch., Notizbl., v. 87, p. 7-77, pl. 1-12.——1962, *Taxionomie und Phylogenie oberdevonischer Conodonten und ihre stratigraphische Bedeutung:* Hess. Landesamt Bodenforsch., Abhandl., v. 38, 166 p., 14 pl.——1966, *Eine Verfeinerung der Conodonten-Gliederung an der Grenze Mittel-/Oberdevon:* Fortschr. Geologie Rheinland Westfalen, v. 9, p. 647-676, pl. 1-6.——1971, *Conodont stratigraphy of the European Devonian:* in Symposium on conodont biostratigraphy, W. C. Sweet & S. M. Bergström (eds.), Geol. Soc. America, Mem. 127, p. 227-284.

——, **Sandberg, C. A.,** & **Austin, R. L.,** 1974, *Revision of the Bispathodus group (Conodonta) in the Upper Devonian and Lower Carboniferous:* Geologica et Palaeontologica, v. 8, p. 97-112, pl. 1-3.

DEVONIAN IN THE WESTERN HEMISPHERE[1]

By A. W. Norris

[Geological Survey of Canada, Calgary, Alberta]

CONTENTS

	PAGE
Introduction	A219
Main Areas of Devonian Rocks	A219
North America	A219
Greenland	A222
South America and Falkland Islands	A222
Antarctica	A222
Silurian-Devonian Boundary	A223
Lower Devonian Faunas (Gedinnian, Siegenian, and Emsian)	A223
Eastern North America	A223
Western North America	A226
Northwestern North America	A226
Northern North America (Arctic Archipelago)	A227
South America and Falkland Islands	A230
Antarctica	A231
Provincialism and Ammonoid Distribution	A231
Middle Devonian Faunas (Eifelian and Givetian)	A234
Eastern North America	A234
Western North America	A235
Northwestern North America	A238
Northern North America	A240
Provincialism and Ammonoid Distribution	A242
Upper Devonian Faunas (Frasnian and Famennian)	A243
Eastern North America	A243
Western North America	A244
Northern North America	A245
South America	A246
Provincialism and Ammonoid Distribution	A246
Devonian Continental Deposits and Associated Invertebrate Faunas	A247
References	A248

---

[1] Manuscript received January, 1969; revised manuscript received September, 1975.

INTRODUCTION

The Devonian System was proposed by SEDGWICK and MURCHISON in 1839 for marine rocks in Devon, England, that were the lateral equivalent of the Old Red Sandstone of Scotland and the Welsh Borders (HOUSE, 1964, p. 262). In North America the Devonian System was recognized in 1847 as a result of a visit and the subsequent writings of DE VERNEUIL, whose views were made known by HALL (COOPER et al., 1942, p. 1729). Since 1847, marine and continental Devonian deposits have been recognized in many widespread areas of the Western Hemisphere.

Radiometric evidence on the limits of the Devonian Period on the geological time scale, summarized by FRIEND and HOUSE (1964, p. 233-236), suggests that the base and top of the Devonian should be dated at about 395 and 345 million years, respectively. Using cumulative thicknesses, they estimated the ages of the base of the Middle Devonian at about 370 million years and that of the base of the Upper Devonian at about 358 million years. Accordingly, the approximate durations for the Early, Middle, and Late Devonian are 25, 12 and 13 million years, respectively.

Recent precise radiometric age estimates of the base of the Devonian in eastern North America (BOTTINO & FULLAGAR, 1966) and the top of the Devonian in Victoria, Australia (MCDOUGALL et al., 1966), were 413 ±5 my and 363 ±6 my, respectively.

More recently the International Geochronological Commission (IUGS, 1967) proposed recommendations for a standard global chronostratigraphic (geochronologic) scale in which the averaged datings for the base and top of the Devonian were given as 405 and 350 million years, respectively, with a duration of 55 million years.

Hypothetical Devonian paleolatitudes have been illustrated by JOHNSON (1970c, p. 2088, fig. 6), SMITH et al. (1973, fig. 12), and others. JOHNSON showed the equator extending diagonally across North America from near the north end of the Gulf of California in the west to the southern part of James Bay in the east.

Paleogeographic and biogeographic data have been used by numerous workers in reconstructions involving the plate tectonic theory. Various data suggest that the Old and New Worlds were juxtaposed during the Devonian, or nearly so at the beginning of the Devonian (JOHNSON & DASCH, 1972). Regarding the southern continents, the existence of Gondwanaland accounts for the Malvinokaffric distributions in South America, and the Falkland Islands and Antarctica in the Western Hemisphere (JOHNSON & BOUCOT, 1973, p. 95).

The main areas of Devonian rocks in the Western Hemisphere, both in outcrop and the subsurface, are shown in a general way in Figure 1. The distribution of Lower, Middle, and Upper Devonian rocks is shown in Figures 2 to 5. These illustrate the known present distribution for North America and Greenland as generalized facies maps from which more interpretive paleogeographic maps may be drawn if so desired. For South America, where the Devonian sediments consist mainly of clastic rocks and where the geology is known in considerably less detail, the distribution of Lower, Middle, and Upper Devonian is shown on paleogeographic maps. The standard Devonian ammonoid and conodont zones of Europe and North America, along with some other zones and ranges of selected species in Devonian rocks of North America, are shown in Figures 6 and 7.

MAIN AREAS OF DEVONIAN ROCKS

NORTH AMERICA

In North America two main areas of Devonian rocks incompletely surround the Canadian shield on its south, southwest, and north sides (Fig. 1). One is centered in northeastern United States and adjoining eastern Canada, and the other extends southwestward through the Arctic Islands into northern continental Canada and then south-southeastward to north-central United States. These two main areas are separated from one another by a northeast-trending transcontinental arch which extends from New Mexico to Hudson Bay. Thick, but areally restricted remnants of the Cordil-

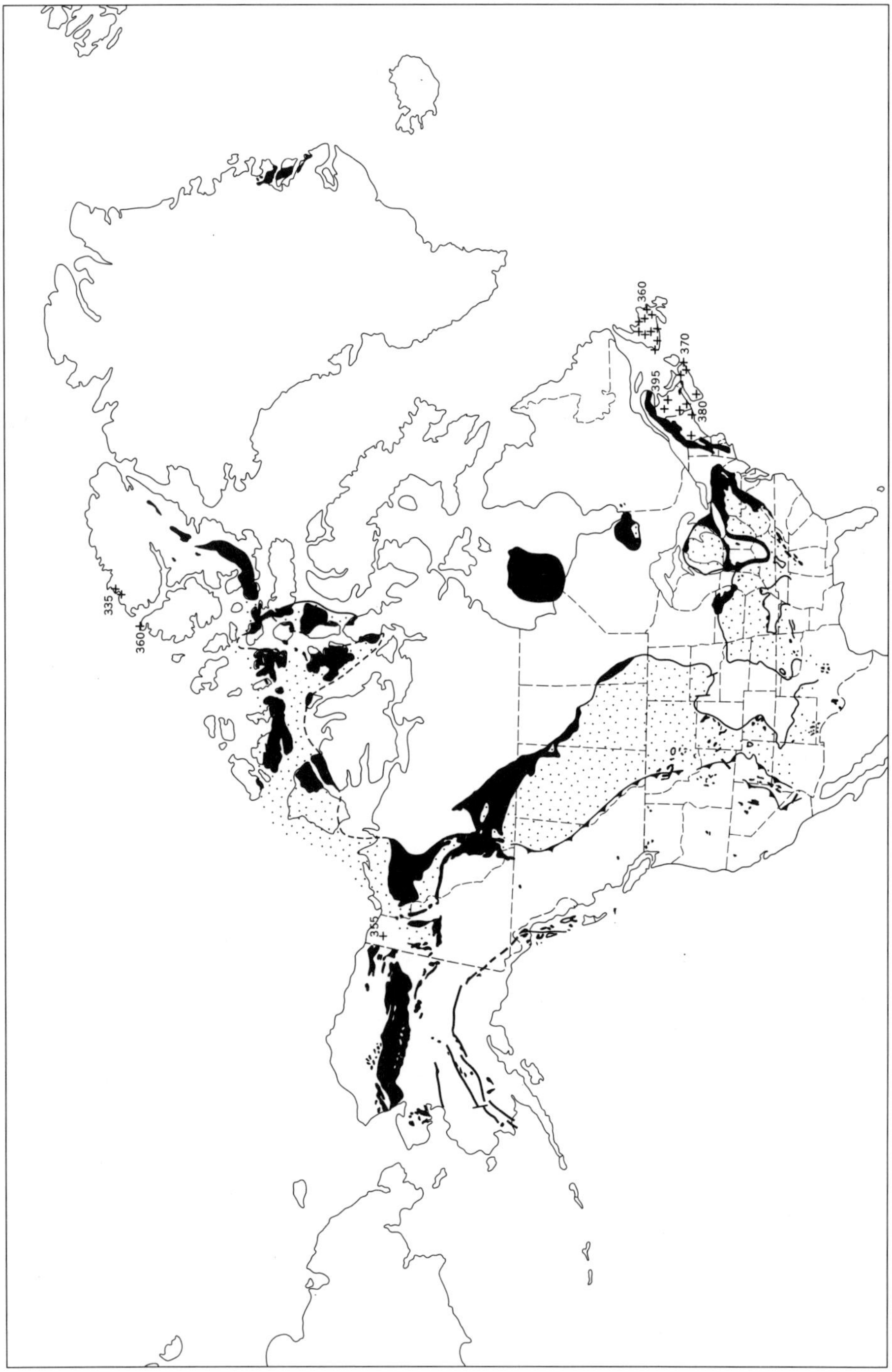

A220 *Introduction—Biogeography and Biostratigraphy*

Fig. 1. *(See facing page.)*

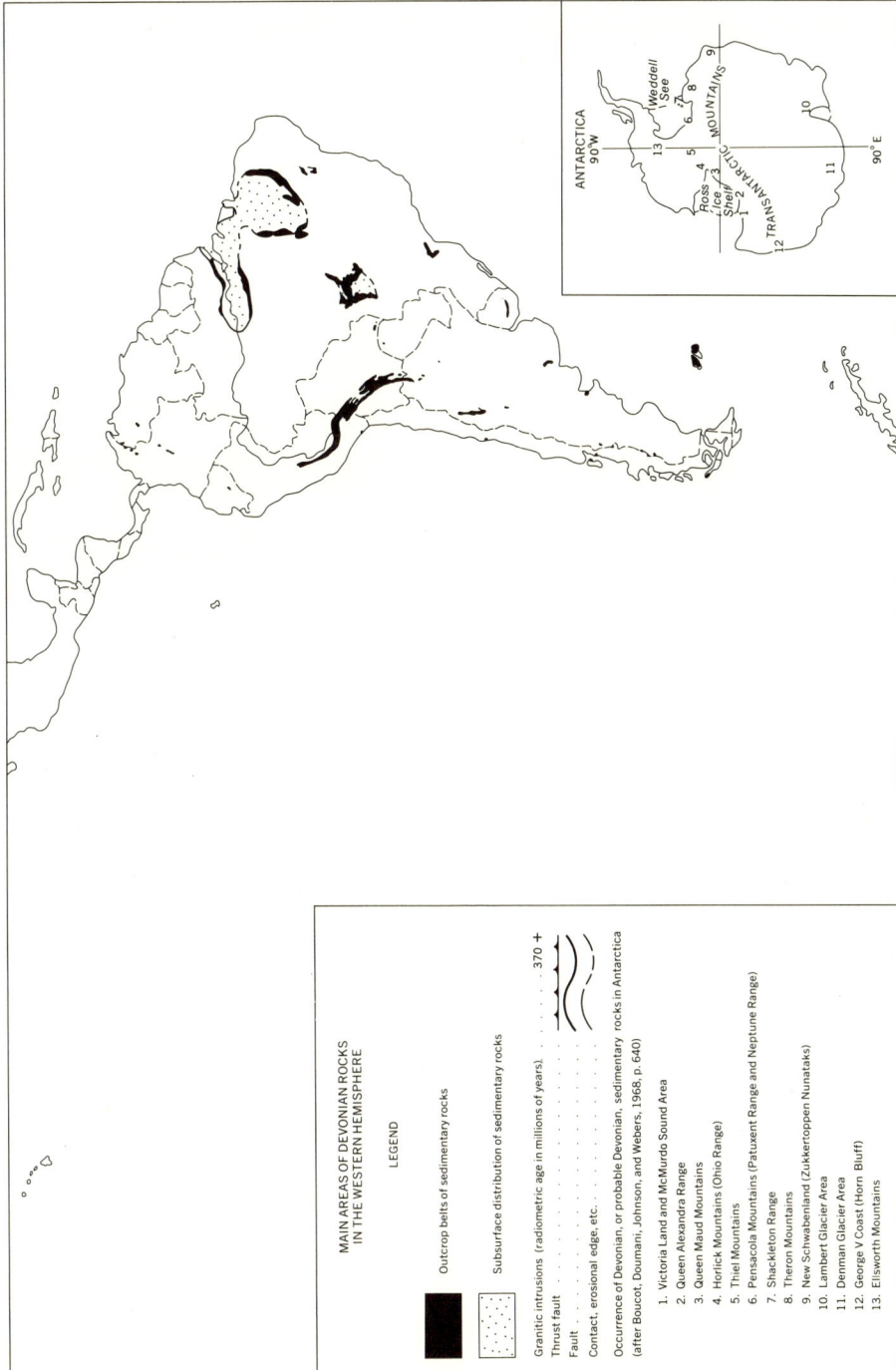

FIG. 1. Main areas of Devonian rocks in the Western Hemisphere (Norris, n).

leran geosyncline are present in the Nevada-Idaho basin of west-central United States.

Devonian rocks are widespread also in Alaska but are incompletely known. The rocks of southeastern Alaska are of eugeosynclinal origin, those in northern Alaska are in a miogeosynclinal belt, and a possible intervening shelf-like environment is suggested for the rocks in the vicinity of the Porcupine and Kuskokwim rivers (GRYC et al., 1968).

Devonian rocks are present also in two intracratonic basins; the Moose River and Hudson Bay in the Hudson platform, which are lithologically and faunally related to the Appalachian sequence of eastern North America (SANFORD & NORRIS, 1975).

GREENLAND

In the folded belt of central East Greenland, bordering the Greenland shield, continental rocks interbedded with volcanic rocks are present in an elongated basin (Fig. 1) where they overlie Caledonian folded formations (BUTLER, 1961). The sequence is 7,000 to 8,000 meters thick and contains rocks of late Middle (Givetian) and Late Devonian (Frasnian and Famennian) ages dated on rich vertebrate faunas (ALLEN et al., 1968).

Besides the vertebrates, the arthropod *Estheria* is the only invertebrate found in the East Greenland Devonian succession, and fossil plants, although locally abundant, are generally poorly preserved (HALLER, 1971, p. 246). The exceedingly rich vertebrate faunas of East Greenland, summarized by JARVIK (1961, 1963), have attracted considerable interest. Among the large collections of vertebrate remains are forms that are unique to East Greenland and forms showing the transition from fish to tetrapod (e.g., *Ichthyostegalia*).

Devonian continental beds are present also on Spitsbergen (FRIEND, 1961) and on Bear Island (HOLTEDAHL, 1919) which, along with the Greenland deposits, are interpreted as remnants of the Old Red Sandstone continent (HOUSE, 1968b).

SOUTH AMERICA AND FALKLAND ISLANDS

The distribution (Fig. 1) and paleogeography of Devonian rocks of South America and the Falkland Islands (Fig. 2 to 5) have been described by WEEKS (1947), HARRINGTON (1962, 1968), and others.

Numerous scattered outcrop belts of Devonian rocks of geosynclinal origin are present in the Andean orogenic belt that borders the entire western margin of the continent, and these occur in Venezuela, Colombia, Ecuador, Peru, Bolivia, Argentina, and Chile. Devonian deposits in the pericratonic basins that lie between the Andean folded belt and stable cratonic areas occur in eastern Bolivia, Paraguay, and the Falkland Islands. Large outcrop belts of Devonian rocks are present also in the intercratonic basins, which include the Amazonas basin of northern Brazil, the Parnaíba and São Francisco basins of eastern Brazil, and the Paraná basin of southern Brazil and Uruguay.

A peculiarity of the Devonian of South America is that carbonate rocks are confined to the Andes of the northern part of the continent in the Colombian-Venezuelan frontier area; southward this limy facies gives way to clastic rocks. CASTER (1952) has postulated that temperature was probably the critical lime-controlling factor, and that the Devonian sediments of a large part of South America were deposited under a cool regime that inhibited the precipitation of calcium carbonate.

The Devonian sequence of the Falkland Islands has been described by BAKER (1923), and summarized by HARRINGTON (1968). Lithologically, this sequence is strikingly similar to that of South Africa, but the Lower Devonian marine faunas are related more closely to those of Brazil. The lower two-thirds of the succession represent shallow-marine deposits, whereas the upper third is of continental, fluviatile origin.

ANTARCTICA

Devonian rocks of Antarctica have been described by BOUCOT et al. (1968). They indicated that Devonian, or probable Devonian, rocks had been recognized from only thirteen scattered localities in ice-free areas around the edge of the continent (Fig. 1). Five of the localities are in or near the Ross Ice Shelf, four are in or near the Filchner Ice Shelf, and the remaining four localities are widely separated in East Antarctica.

The Devonian of Antarctica is part of a succession of sedimentary strata belonging to the Beacon Group. Fish remains, assigned to the Upper or Middle Devonian (WOODWARD, 1921) have been collected from central Victoria Land. A Lower Devonian marine fauna has been described by BOUCOT et al. (1963) and by DOUMANI et al. (1965) from the Horlick Formation in the Ohio Mountains.

SILURIAN-DEVONIAN BOUNDARY

Most workers throughout the world now agree in placing the Silurian-Devonian boundary at the base of the Gedinnian. This boundary coincides with the base of the *Monograptus uniformis* Zone and is close to the base of the *Icriodus woschmidti* Zone. In shelly successions with corals, brachiopods, and trilobites the boundary is recognizable by the disappearance of pentamerids, *Atrypella, Gracianella*, halysitids, and *Encrinurus*, and by the appearance of terebratulids, *Cyrtina*, and common *Schizophoria* (BERDAN et al., 1969).

Recently this boundary was fixed by international agreement on a stratotype at Klonk near Suchomasty, Bohemia, Czechoslovakia. The horizon chosen is immediately below the first occurrence of *Monograptus uniformis*, within bed 20 at Klonk (CHLUPÁČ, 1972, p. 111, 113; McLAREN, 1972). The base of the Gedinnian coincides with the base of the Lochkovian (in NALIVKIN, 1973, p. 12).

In the standard section of eastern North America the base of the Devonian, defined as the base of the Gedinnian, is located at or near the base of the Helderbergian, which traditionally has been regarded as the lowermost Lower Devonian stage (BERDAN et al., 1969).

LOWER DEVONIAN FAUNAS
(GEDINNIAN, SIEGENIAN, AND EMSIAN)

EASTERN NORTH AMERICA

The Lower Devonian marine sedimentary sequence of the Appalachian area contains coral, brachiopod, and trilobite faunas of pronounced provincial aspect. Four brachiopod zones, based on rensselariid evolution, were recognized by BOUCOT and JOHNSON (1967, 1968) as follows: 1) *Nanothyris* Zone, occurring in the Manlius-Coeymans and Kalkberg-New Scotland intervals (Gedinnian age); 2) *Rensselaeria* Zone, in the Becraft-Port Ewen of the upper Helderbergian and in the Oriskany (Siegenian age); 3) *Etymothyris* Zone; and 4) *Amphigenia* (small form) Zone. The latter two zones occur in the Esopus and Schoharie-Bois Blanc intervals (Emsian age).

Brachiopod assemblage zones were recognized by BOUCOT and JOHNSON (1967, 1968) also in those areas where the rensselaeriid zones are poorly represented or absent.

Notable rugose coral assemblages were reported by OLIVER (1968) in the Helderberg (Gedinnian and lower Siegenian) and middle Onesquethaw (Emsian). Rugose coral genera occurring in stromatoporoid biostrome facies of the Helderberg include: *Spongyphylloides, Chlamydophyllum*, and *Lyrielasma*; in argillaceous facies: *Cyathophyllum, Enterolasma, Heterophrentis, Siphonophrentis, Lindostroemia*, and *Syringaxon*; and in calcarenite facies, characteristic genera are *Briantelasma, Fletcherina, Nalivkinella, Pseudoblothrophyllum*, and *Aknisophyllum* (OLIVER, 1968).

Few corals are known from the lower part of the Onesquethaw (Esopus), but the small assemblage contains the oldest known "*Billingsastraea*" (=*Asterobillingsa*) (OLIVER, 1968, p. 740).

The Schoharie (Emsian) fauna includes the largest and most widespread of the rugose coral assemblages. Some of the more important forms include *Acinophyllum davisi, Acrophyllum oneidaense, Edaphophyllum sulcatum*, and *Kionelasma* (OLIVER, 1968, p. 740).

Lower Devonian trilobites in eastern North America also display marked provinciality and their provincial distribution is similar to that of the brachiopods (ORMISTON, 1972). Gedinnian endemic genera include *Cordania, Roncellia, Neoprobolium*,

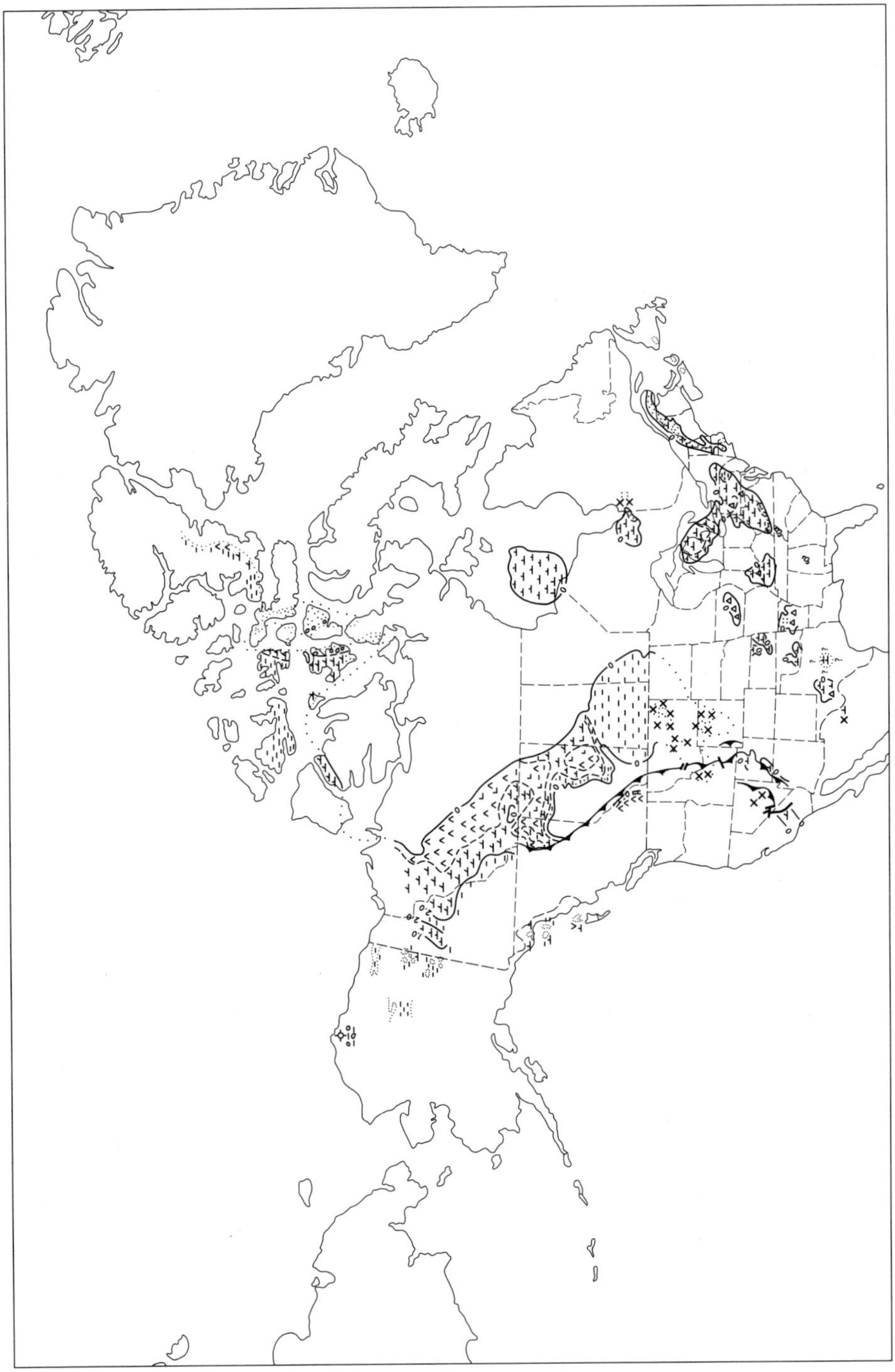

A224 Introduction—Biogeography and Biostratigraphy

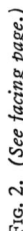

Fig. 2. *(See facing page.)*

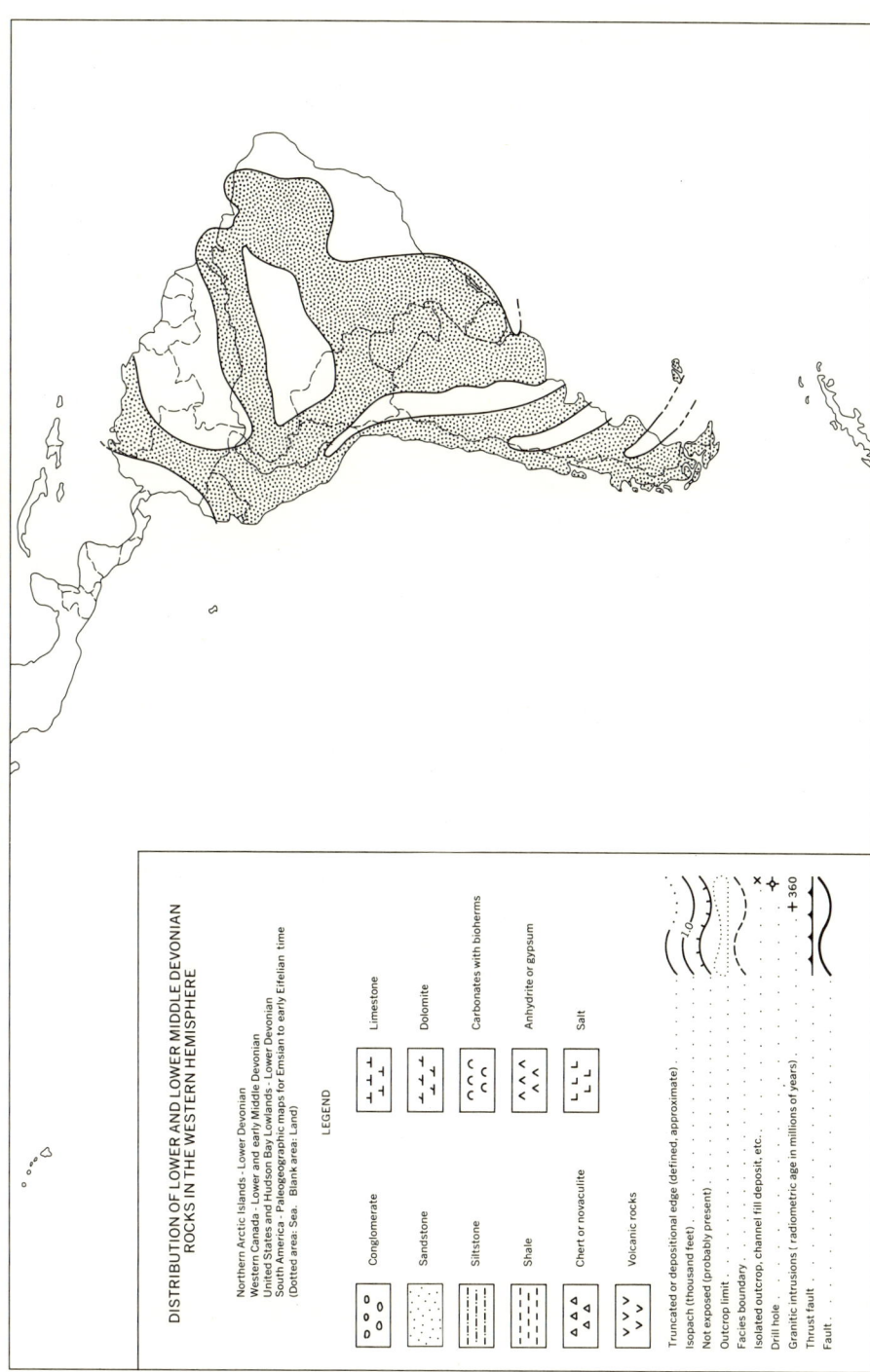

Fig. 2. Distribution of Lower (including some lower Middle) Devonian rocks in the Western Hemisphere (Norris, n).

Phacopina, Kosovopeltis, Homalonotus, Dicranurus, Ceratonurus, Odontochile, Dalmanites, and *Echinolichas.* Siegenian trilobites are less well known. During the Emsian, and continuing into the Eifelian, trilobite provinciality in eastern North America became even more pronounced and is marked by a host of endemic genera including *Odontocephalus, Anchiopsis, Synphoria, Synphoroides, Trypaulites, Phacops, Odontochile, Dalmanites, Corycephalus, Greenops?, Dechenellurus, Terataspis, Echinolichas, Mystrocephala, Crassiproetus,* and *Isoprusia* (ORMISTON, 1972, p. 599).

Gedinnian conodonts representative of the *Icriodus (=Pedavis) pesavis* faunas of Nevada are recorded from New York (KLAPPER et al., 1971, p. 291). Emsian conodont faunas in eastern United States are characterized by the highest occurrence of *Icriodus latericrescens huddlei* and the lowest occurrence of *I. latericrescens robustus* (KLAPPER et al., 1971, p. 292).

Monograptids are rare in the Lower Devonian of the Appalachian succession (JOHNSON & MURPHY, 1968).

WESTERN NORTH AMERICA

Throughout the Great Basin of Nevada, Utah, and Idaho the Lower Devonian is represented almost wholly by marine carbonate rocks that were deposited on a broad shelf. In the abundantly fossiliferous Lower Devonian strata of Nevada eight distinctive faunal assemblages based on brachiopods have been recognized. These are the *Gypidula pelagica* beds (lower lower Gedinnian), *Quadrithyris* Zone (upper lower Gedinnian), *Spinoplasia* Zone (upper Gedinnian), *Oriskania* beds (lower Siegenian), *Trematospira* Zone (upper Siegenian), *Acrospirifer kobehana* Zone (upper upper Siegenian), *Eurekaspirifer pinyonensis* Zone (lower Emsian), and *Elythyna* beds (upper Emsian) (JOHNSON, BOUCOT, & MURPHY, 1968; JOHNSON & BOUCOT, 1968; JOHNSON, 1975).

Important forms from the lower *Gypidula pelagica* beds include the lowest *Cyrtina* and *Schizophoria, Icriodus woschmidti,* and monograptids representing the *Monograptus uniformis* Zone (JOHNSON & MURPHY, 1969). The widely distributed index trilobite, *Warburgella rugulosa,* occurs in upper *Gypidula pelagica* beds, above beds with *M. uniformis* (ALBERTI et al., 1971) and above beds containing *I. woschmidti* (JOHNSON, BOUCOT, & MURPHY, 1973, p. 11).

Succeeding beds contain two graptolite zones, *Monograptus praehercynicus* and *M. hercynicus* (JOHNSON & MURPHY, 1969); the latter zone is recognized by the presence of *M. hercynicus nevadensis* (BERRY, 1967, 1968). *Quadrithyris* Zone brachiopods, *M. hercynicus nevadensis,* and the conodonts *Icriodus pesavis pesavis* and *Spathognathodus johnsoni* are closely associated and treated as one fauna (JOHNSON & MURPHY, 1969, p. 1279).

Monograptus thomasi and *M. yukonensis* occur together and are associated with the conodont *Eognathodus sulcatus* immediately underlying beds of the *Spinoplasia* Zone (BERRY & MURPHY, 1972; JOHNSON, 1975). The *Spinoplasia* Zone contains a shelly fauna of Appalachian affinity (JOHNSON, 1965, p. 374) dated as late Helderbergian (Port Ewen).

An interval above the *Spinoplasia* Zone has been referred to by JOHNSON (1975) as "beds with *Oriskania*," but its fauna has not been described.

The succeeding *Trematospira* Zone includes the highest beds in Nevada assignable to the Siegenian. The conodont *Eognathodus sulcatus,* a form that overlaps the lower range of *Monograptus yukonensis* in the Yukon, occurs in the zone.

The *Acrospirifer kobehana* Zone is characterized by many forms that range from underlying to overlying zones and newly introduced elements.

The *Eurekaspirifer pinyonensis* Zone contains a rich megafauna of corals, brachiopods, trilobites, and mollusks, and has yielded the only Lower Devonian goniatites in the western United States (JOHNSON, 1970b, p. 58).

The *Elythyna* beds contain abundant *E. "undifera"* and are correlated with the upper part of the Sawkill Stage of New York (JOHNSON, 1970b).

NORTHWESTERN NORTH AMERICA

Lower Devonian strata are distributed widely in the Cordilleran folded belt of northeastern British Columbia, Northwest

Territories, and Alaska. In the Royal Creek section of northern Yukon Territory, the Silurian-Devonian boundary is drawn between beds with *Atrypella* cf. *A. tenuis* and *Gypidula pelagica*. In the Lower Devonian of that area, LENZ (1966, 1968) recognized four brachiopod faunal units and one graptolite zone, and KLAPPER (1969) has outlined the associated conodonts. The four brachiopod faunal units in ascending sequence comprise: *Gypidula* cf. *G. pelagica, Spirigerina, Gypidula* sp. 1-*Davidsoniatrypa* sp., and *Sieberella* cf. *S. weberi-Nymphorhynchia pseudolivonica*.

In the *Gypidula* cf. *G. pelagica* unit the brachiopod genera *Cyrtina* and *Schizophoria* appear at or near the base, and the conodonts *Icriodus woschmidti* and *Ozarkodina remscheidensis* occur together in the unit, an association similar to that known from Nevada.

The *Spirigerina* unit is characterized by *Spirigerina* cf. *S. supramarginalis, Toquimaella kayi*, and *Ogilviella rotunda*. It is correlated with the *Quadrithyris* Zone of Nevada, which JOHNSON (1975) dated as late Lochkovian and correlated with a level near the middle of the type Gedinnian.

An unnamed interval above the *Spirigerina* unit at Royal Creek is characterized by the first appearance of *Eognathodus sulcatus* and the presence of *Icriodus latericrescens* subspecies B, forms which are associated with the *Spinoplazia* Zone of Nevada (KLAPPER, 1969).

The *Gypidula* sp. 1-*Davidsoniatrypa* unit, as emended by LENZ (1968), contains a rich and varied fauna. The principal conodont species in the unit is *Eognathodus sulcatus*, which also occurs in the *Trematospira* Zone of Nevada (KLAPPER, 1969, p. 7). The upper part of the unit is overlapped by the *Monograptus yukonensis* Zone (JACKSON & LENZ, 1963), which also occurs in Alaska (CHURKIN & BRABB, 1965, 1968), in the Canadian Arctic Islands (THORSTEINSSON in BERDAN et al., 1969), and elsewhere.

The *Sieberella* cf. *S. weberi-Nymphorhynchia pseudolivonica* unit succeeds the *Monograptus yukonensis* Zone. The unit contains forms ranging from below, as well as newly appearing species which include the two name bearers as well as *Janius sergaensis, Strophonella?, Cortezorthis* cf. *C. bathurstensis*, and a distinctive echinoderm ossicle with a double axial canal (LENZ, 1968). Associated conodonts include *Polygnathus dehiscens* and *Pandorinellina exigua* dated as early Emsian.

Regionally the *Sieberella* cf. *S. weberi-Nymphorhynchia pseudolivonica* unit correlates with the lower part of the richly fossiliferous carbonate and shale Michelle Formation (NORRIS, 1968a, 1968b; LUDVIGSEN, 1970). Brachiopods in the Michelle include *Cortezorthis* cf. *C. cortezensis, Carinapyga loweryi* and *Schizophoria* cf. *S. nevadensis*, which suggest a correlation with the *Eurekaspirifer pinyonensis* Zone of Nevada of mid-Emsian age (LUDVIGSEN, 1970). Important trilobites in the formation include *Lacunoporaspis norrisi* (most abundant) and *Ricticuloharpes* cf. *R. reticulatus* (ORMISTON, 1971). Conodonts in the Michelle studied by FÅHRAEUS (1971) were assigned to the *Polygnathus dehiscens* fauna. The dacryoconarid fauna of the Michelle was correlated by LUDVIGSEN (1970, 1972) with the *Guerichina strangulata* Zone of late Pragian (early Emsian) age. The ammonoid *Teicherticeras lenzi* described by HOUSE (in HOUSE & PEDDER, 1963) occurs in the lower part of the formation.

Succeeding Lower Devonian faunas in the lower part of the carbonate Ogilvie Formation are correlated also with the *Eurekaspirifer pinyonensis* Zone of Nevada. Two Lower Devonian conodont faunal units were recognized by KLAPPER (in PERRY et al., 1974) in the Ogilvie Formation comprising *Polygnathus perbonus perbonus* and *P. perbonus*, n. subspecies.

Three Lower Devonian graptolite zones and one informal unit were recognized by LENZ and JACKSON (1971) in northwestern Canada. These are, in ascending sequence: *Monograptus uniformis* and *M. hercynicus* zones of Lochkovian age, and beds with *M. thomasi* and the *M. yukonensis* Zone of Pragian age.

NORTHERN NORTH AMERICA (ARCTIC ARCHIPELAGO)

Lower Devonian carbonates and shales are distributed widely in the Franklinian miogeosyncline of the Canadian Arctic Is-

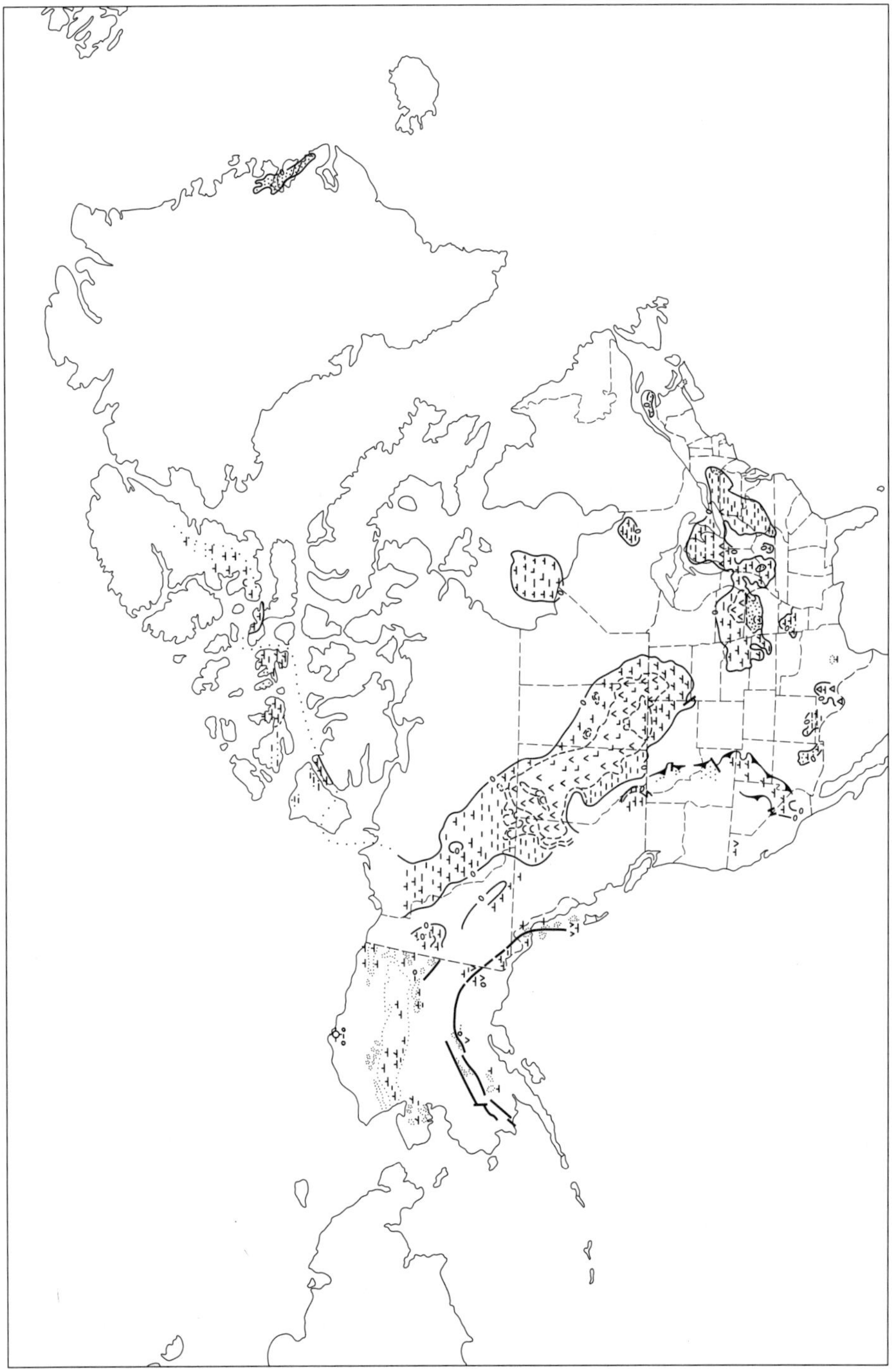

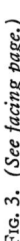

Fig. 3. (See facing page.)

FIG. 3. Distribution of Middle Devonian rocks in the Western Hemisphere (Norris, n).

lands. The graptolite sequence of the Cape Phillips Formation, and partly equivalent Bathurst Island Formation, contain a faunal succession that is continuous across the Silurian-Devonian boundary. Beds containing Pridolian (Late Silurian) monograptids are succeeded by beds containing a succession of monograptids similar to that found in northwestern Canada which include *Monograptus uniformis, Monograptus* of the *M. hercynicus* type, *M.* cf. *M. thomasi,* and *M. yukonensis* (THORSTEINSSON in BERDAN et al., 1969; THORSTEINSSON in McGREGOR & UYENO, 1972).

The lower Gedinnian trilobite, *Warburgella rugulosa canadensis,* has been described by ORMISTON (1967) from limestones on Baillie Hamilton Island. From Devon Island an equivocal Gedinnian shelly fauna consisting of *Cyrtina, Schizophoria,* and other fossils has been described from the Sutherland River Formation (BOUCOT et al., 1960; BERDAN et al., 1969). From the same island, the brachiopod *Toquimaella kayi* associated with the conodont *Icriodus pesavis pesavis* was reported by JOHNSON (1967) from the lower part of the Stuart Bay Formation, indicating correlation with the *Quadrithyris* Zone (mid-Gedinnian) of Nevada. From the Stuart Bay Formation on Bathurst Island, LENZ (1973) has described a brachiopod fauna containing species in common with the *Spirigerina* fauna of the Yukon and the coeval *Quadrithyris* Zone of Nevada. Brachiopods of the *Quadrithyris* Zone were recognized by JOHNSON (1975) also on Prince of Wales and Cornwallis Islands.

From the lower part of the Bathurst Island Formation on Bathurst Island, UYENO (in McGREGOR & UYENO, 1972) reported *Eognathodus sulcatus* and *Ozarkodina remscheidensis* suggesting an early Siegenian age.

The Stuart Bay Formation of Bathurst Island is characterized by *Pandorinellina expansa* and by "two-hole" echinoderm ossicles (McGREGOR & UYENO, 1972; UYENO & MASON, 1975). *Polygnathus dehiscens,* suggesting an early Emsian age, occurs in the lower part of the formation; and *P. perbonus* and other conodonts indicating a middle to late Emsian age occur in the upper part of the formation and in the overlying Eids Formation. A sparse trilobite fauna has been identified by ORMISTON (1967) from the Stuart Bay Formation.

A rich brachiopod fauna associated with the conodont *Polygnathus perbonus perbonus* of late Emsian age has been reported by ORMISTON (in KLAPPER, 1969) from the lower part of the Blue Fiord Formation on Devon Island. Brachiopods of probable Emsian age occur also in the Disappointment Bay Formation on Cornwallis and Bathurst Islands (JOHNSON, 1971b, p. 3268).

Brachiopods described by BRICE and MEATS (1977) from the lower part of the Blue Fiord Formation on Ellesmere, Devon, and Bathurst Islands are related closely to forms in the *Eurekaspirifer pinyonensis* Zone of Nevada of middle or late Emsian age.

Trilobites of Emsian age described by ORMISTON (1967) from the Eids Formation include *Platyscutellum brevicaulis, Cornuproetus tozeri, Harpes* cf. *H. macrocephalus,* and other forms. Elements in this fauna closely resemble forms from Europe and the Urals.

SOUTH AMERICA AND FALKLAND ISLANDS

Two distinct provincial faunal assemblages are known in the Devonian of South America. One, characterizing the Lower Devonian, belongs to the Malvinokaffric realm, a name proposed by RUDOLF RICHTER (1941) and RUDOLF and EMMA RICHTER (1942), and derived from Malvinas (Falkland) Islands and the "Kaffric" (South African) Bokkeveld beds. The other, developed in the Lower and Middle Devonian of Venezuela and Colombia, has strong relationships with the Appohimchi subprovince of eastern North America.

The Early Devonian (?early Emsian) Malvinokaffric fauna is characterized by distinctive genera of brachiopods and trilobites unknown in Northern Hemisphere assemblages. Among the brachiopods, *Australospirifer, Australocoelia, Notiochonetes,* and *Scaphiocoelia* are some of the typical representatives. Among the trilobites, *Calmonia, Paracalmonia, Metacryphaeus, Pennaia, Bainella, Tibagya,* and *Probolops* are the most conspicuous. The fauna is characterized also by an abundance of large palaeoneilid and nuculitid bivalves, and by a scarcity of corals, bryozoans, aviculids, pterinids, platyceratids, cephalopods, cystoids, and crinoids (HARRINGTON, 1968, p. 663). In South America, elements of this fauna have been found in the southern two-thirds of the continent; in Peru, Paraguay, Argentina, Paraná basin of Brazil (CLARKE, 1913), Bolivia (WOLFART & VOGES, 1968), and the Falkland Islands (CLARKE, 1913; SHIRLEY, 1964). All of the Lower Devonian (Gedinnian to Emsian) is represented in Bolivia according to WOLFART and VOGES (1968), based on a detailed study of trilobites.

The contemporaneous (?early Emsian) Amazon-Colombian subprovince of the Eastern Americas realm is found in northern South America, mainly in Colombia and Venezuela. The Eastern Americas realm elements have been noted by CASTER (1939), AMOS and BOUCOT (1963), and others. Characteristic brachiopod genera include *Leptocoelia, Megakozlowskiella, Eodevonaria (arcuata)* type, *Prionothyris, Amphigenia, Pentagonia,* and others (BOUCOT, JOHNSON, & TALENT, 1968). The associated corals were studied by SCRUTTON (1973), who concluded that they have strong affinities with the upper Onesquethaw (early Middle Devonian) of eastern North America. The brachiopods, dated by BOWEN (1972) as late Early Devonian, were considered by JOHNSON (in SCRUTTON, 1973) to represent an overlap of late Early and early Middle Devonian forms.

ANTARCTICA

A Malvinokaffric realm faunal assemblage of probable early Emsian age has been described by BOUCOT et al. (1963) and DOUMANI et al. (1965) from the Horlick Formation in the Ohio Range of Antarctica. Brachiopods are the most abundant fossils and include species of *Pleurothyrella, Australospirifer,* and *Tanerhynchia.* Other forms include the trilobite *Burmeisteria,* and a profusion of bivalves and gastropods that are indistinguishable from South American and South African species.

PROVINCIALISM AND AMMONOID DISTRIBUTION

Silurian shelly faunas, particularly brachiopods, were relatively cosmopolitan, but gave way to moderately provincial faunas during early Gedinnian time. Provinciality became more pronounced in the late Gedinnian when two distinct faunal realms developed, the Eastern Americas and the Old World, each characterized by distinctive brachiopod assemblages (BOUCOT, JOHNSON, & TALENT, 1968). The Appohimchi subprovince of the Eastern Americas realm extended in North America from Gaspé Peninsula to New Mexico. The Cordilleran region of the Old World realm extended from Nevada through the Yukon to the Arctic Archipelago.

During Siegenian time brachiopod provinciality increased and part of Nova Scotia became joined to the Rhenish-Bohemian region of the Old World realm. Western North America remained a part of the Old World realm, except for Nevada, which became an Appalachian enclave (Nevadian subprovince of Eastern Americas realm) in the late Siegenian (BOUCOT, JOHNSON, & TALENT, 1968).

During the early Emsian, marine deposition was more widespread and brachiopod provincialism increased by the addition of the Malvinokaffric realm which, in the Western Hemisphere, covered parts of the southern two-thirds of South America and part of Antarctica. Appohimchi subprovince influence in Nevada ceased as endemic new forms appeared and were joined by brachiopod genera of the Old World realm, the mixture characterizing a Cordilleran region. During the early Emsian, Uralian brachiopod elements from the region bordering the Siberian platform mingled with Cordilleran forms in the northern Yukon and Canadian Arctic, and this mixture is referred to as the Cordilleran-Uralian region of the Old World realm (BOUCOT, JOHNSON, & TALENT, 1968; BOUCOT, 1975).

Beginning approximately in late Emsian or early Eifelian time the Malvinokaffric realm disappeared in Antarctica and remained only in the deeper parts of the basins of southern South America.

Trilobite provincialism in North America

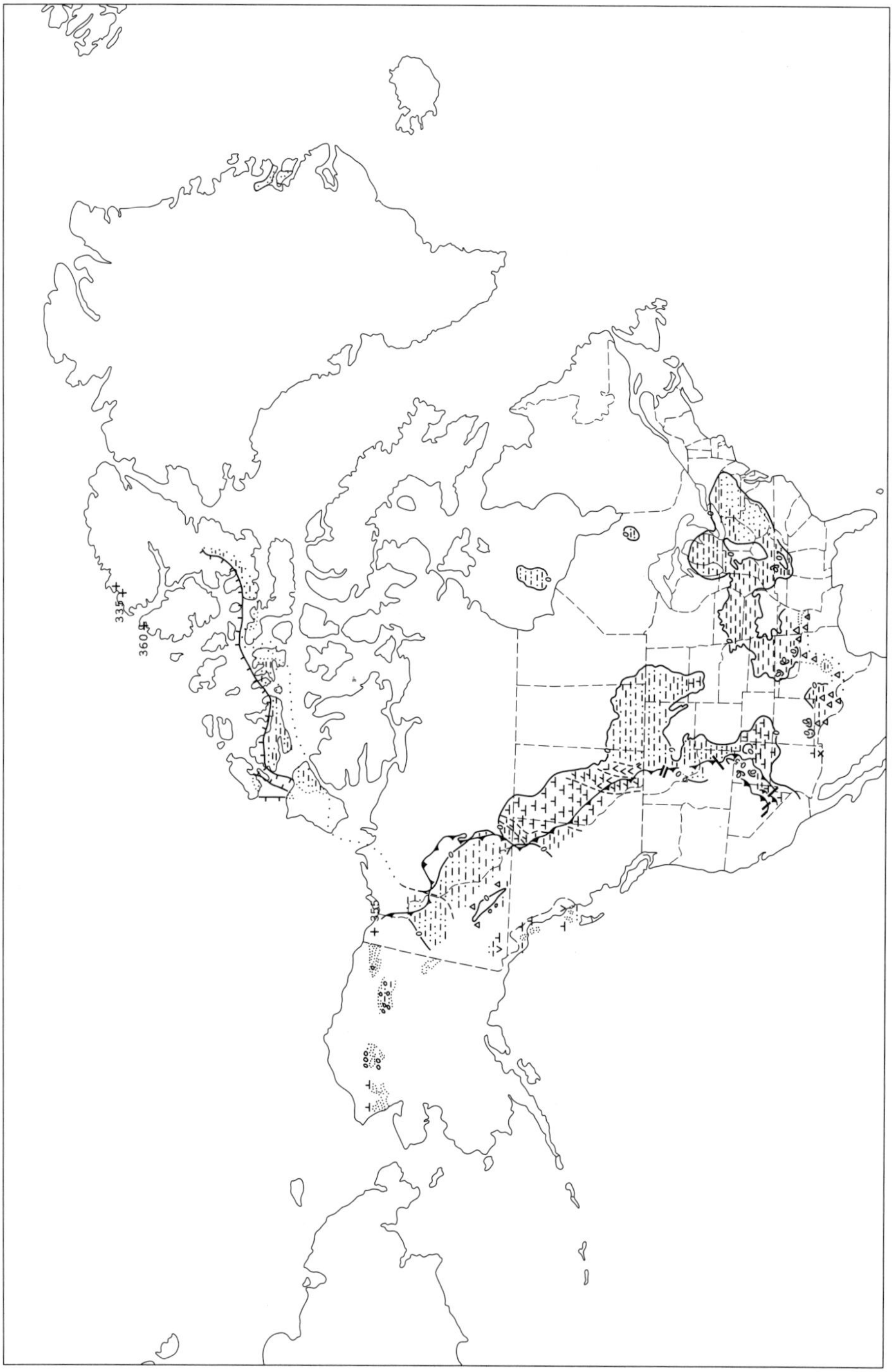

Fig. 4. (See facing page.)

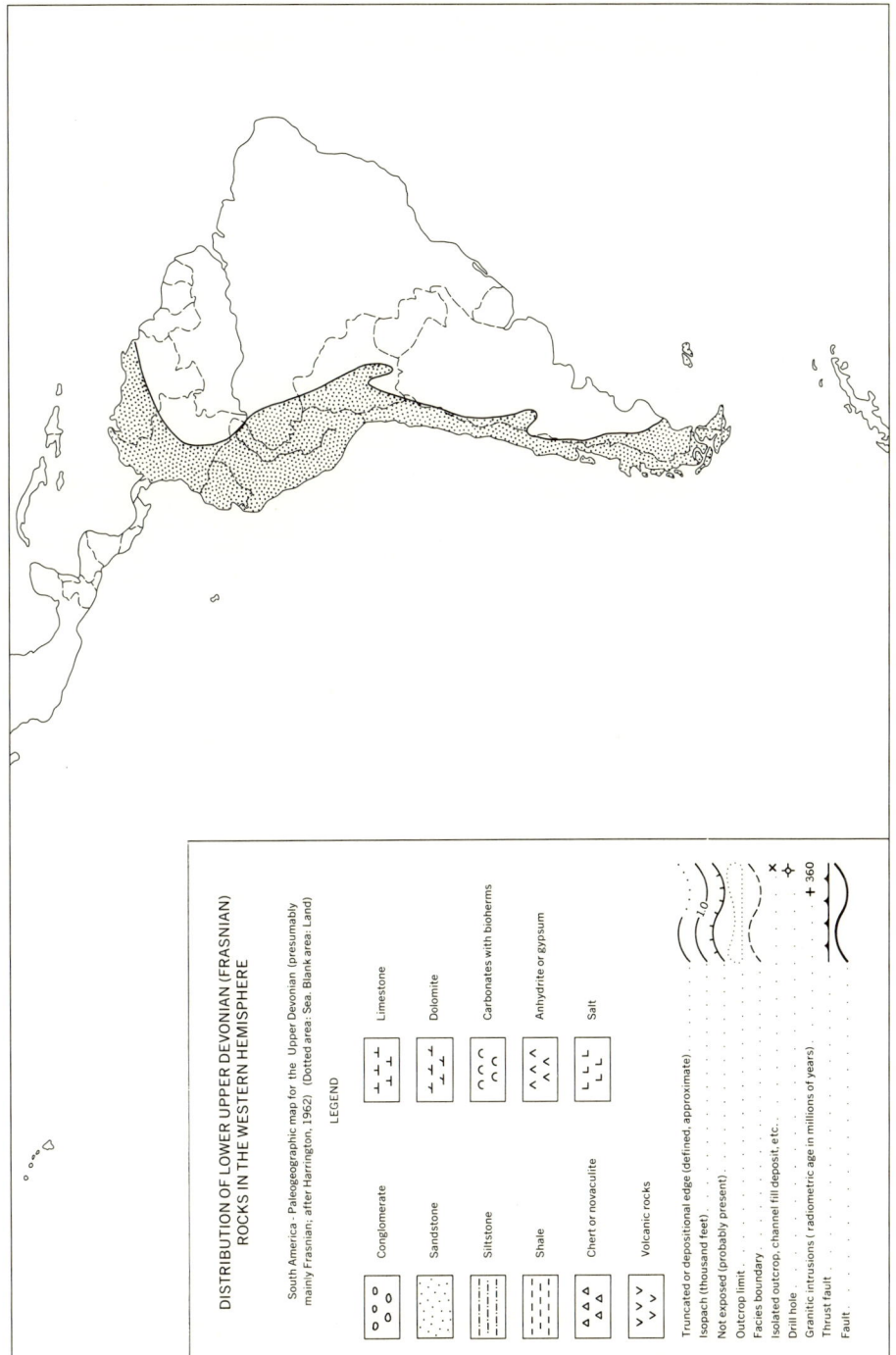

Fig. 4. Distribution of lower Upper Devonian (Frasnian) rocks in the Western Hemisphere (Norris, n).

during the Gedinnian and Siegenian is somewhat similar to that of brachiopods (ORMISTON, 1972). During Emsian to Eifelian time provincialism among North American trilobite faunas was most pronounced and at least five biogeographic subdivisions were recognized by ORMISTON (1972), comprising the "Appalachian Province," "Old World Province," "Cordilleran Subprovince," "Siberian-Canadian Subprovince," and the "Uralian Subprovince."

In South America, WOLFART and VOGES (1968) recognized two major subdivisions of trilobite provincialism in the Malvinokaffric realm, a South American region in the north and a South African-Malvinian-?West Antarctic region in the south. They further recognized two minor subdivisions of the Malvinokaffric realm occurring in the Andean geosyncline and shelf areas, each characterized by distinctive trilobite assemblages.

Goniatites appear in the mid-Siegenian as simple primitive types which diversified rapidly (HOUSE, 1967). The richest Lower Devonian ammonoid faunas of the Siegenian and especially the Emsian occur in northern Europe, characterizing the *Mimosphinctes*-Stufe (HOUSE, 1964, p. 263). Only a few occurrences are recorded from North America; *Teicherticeras* and *Anetoceras* occur together in Nevada (HOUSE, 1962), and *Teicherticeras* (HOUSE & PEDDER, 1963) and *Anetoceras* (NORRIS, 1968a) occur separately in the northern Yukon.

MIDDLE DEVONIAN FAUNAS (EIFELIAN AND GIVETIAN)

EASTERN NORTH AMERICA

The Middle Devonian of eastern North America is represented by the Onondaga Limestone, Hamilton Group, Tully Limestone, and their equivalents. Characteristic brachiopods of the Onondaga Limestone and its equivalents include *Coelospira, Levenea, Protoleptostrophia, Amphigenia,* "*Leptocoelia*" of the *acutiplicata* type, *Longispina, Megakozlowskiella, Elytha, Pentagonia,* and *Centronella* (BOUCOT, JOHNSON, & TALENT, 1969, p. 25). Within the Onondaga Limestone, the Edgecliff, Moorehouse, and Seneca members have distinct assemblages of rugose corals, and many species range throughout all three members and into the overlying Hamilton Group (OLIVER, 1968).

Conodonts in the basal Edgecliff Member of the Onondaga Formation are peculiar to North America, and cannot be dated precisely. The Edgecliff is characterized by a lack of *Polygnathus* and presence of *Icriodus latericrescens robustus* as its only index species (KLAPPER, 1971, p. 60; ORR, 1971, p. 10). A *Polygnathus costatus patulus-P. linguiformis cooperi* fauna occurs in the Nedrow and lower Moorehouse members of the Onondaga Limestone, associated with the Eifelian ammonoid *Foordites* in the upper part of the Nedrow (KLAPPER, 1971, p. 60). The succeeding *Polygnathus robusticostatus* fauna is characterized by the name-giver and *P. angusticostatus,* as well as rare occurrences of *P. linguiformis linguiformis* α morphotype and *P. costatus patulus,* which occur in the upper part of the Moorehouse Member. The *P. costatus costatus-P.* aff. *P. trigonicus* fauna occurs in the uppermost bed of the Moorehouse and throughout the Seneca Member of the Onondaga Limestone. The succeeding *P. pseudofoliatus-P.* aff. *P. eiflius* fauna occurs in the *Werneroceras* bed of the Union Springs Member and the Cherry Valley Member of the Marcellus Formation. According to KLAPPER (1971, p. 60), the conodonts of this fauna indicate correlation with the upper Eifelian and upper Couvinian, rather than with the Givetian as suggested by ammonoid evidence (HOUSE, 1962, p. 253-254).

The Givetian in New York embraces roughly the Hamilton Group and Tully Limestone. These sediments form a great clastic wedge filling a geosyncline in eastern New York that thins westward onto the shelf areas of the Midwest where the sequence changes mainly to carbonate rocks. In New York the Hamilton Group has been subdivided into four formations in ascending sequence as follows: Marcellus, Skaneateles, Ludlowville, and Moscow (COOPER, 1933-34). These units are traced westward by means of fossiliferous limestone lenses, the most important of which is the Centerfield Member at the base of the Ludlowville Formation (COOPER & WARTHIN, 1942; COOPER, 1957).

Important brachiopods that range throughout the Hamilton include *Spinocyrtia granulosa, Mucrospirifer mucronatus, Mediospirifer audaculis,* and *Tropidoleptus carinatus.* Brachiopods confined to the Centerfield Limestone at the base of the Ludlowville include *Fimbrispirifer venustus* and *Pentagonia bisulcata. Pustulatia pustulosa* first appears in the Centerfield and ranges upward into rocks of Moscow age. *Spinocyrtia marcyi* appears at the base of the Moscow and ranges upward into beds of Finger Lakes age.

Stringocephalus was reported by COOPER and PHELAN (1966) for the first time in the Midwest from the Miami Bend Formation of Indiana associated with *Subrensselandia* and other megafossils. Conodonts in the Miami Bend are within the zone of *Icriodus latericrescens latericrescens* below the lowest position of *Polygnathus varcus* (ORR, 1969, p. 337), suggesting an early Givetian age.

Much systematic work has been done on the coral assemblages of the Hamilton and Transverse groups but their sequences have not been worked out (OLIVER, 1968, p. 743).

Conodonts of probable early Givetian age, and characterized by the lowest occurrences of *Icriodus latericrescens latericrescens* (below the first appearance in sequence of *Polygnathus varcus*) are known in the Skaneateles Formation of New York and in the lower part of the Traverse Group of Michigan, and elsewhere (KLAPPER et al., 1971, p. 296).

The succeeding *Polygnathus varcus* Zone is characterized in North America by the association of *P. varcus, Icriodus latericrescens latericrescens,* and *Polygnathus linguiformis linguiformis,* which occur in New York in the interval from the Centerfield Limestone (earliest occurrence of *P. varcus*) to the top of the Tully Formation (KLAPPER et al., 1971, p. 297).

In central New York, COOPER and WILLIAMS (1935) divided the Tully into three members: the Tinkers Falls at the base containing abundant *Rhyssochonetes aurora* and "*Stropheodonta*" *tulliensis*; the Apulia abounding in *Hypothyridina venustula, Schizophoria tulliensis,* and *R. aurora*; and the West Brook, containing a great variety of Hamilton species including crinoids, brachiopods, corals, and numerous mollusks.

Conodonts of the uppermost Middle Devonian *Schmidtognathus hermanni-Polygnathus cristatus* Zone are well developed in a limestone at the base of the New Albany Shale in southern Indiana (KLAPPER et al., 1971, p. 297). Conodonts of this zone have been recognized also in beds in eastern Wisconsin, southern Illinois, Iowa, and elsewhere.

The problem of the boundary between the Middle and Upper Devonian is as yet unresolved. COOPER and OLIVER considered the Tully brachiopods and corals to be Middle Devonian types (COOPER, 1968; OLIVER et al., 1968). HOUSE (1962, p. 256) assigned the Tully to the Upper Devonian (Frasnian), because of the presence, near the top of the formation, of *Pharciceras* and tornoceratids with lingulate lateral lobes. KLAPPER and ZIEGLER (1967) reported that the conodonts in the Tully are confined to the *Polygnathus varcus* Zone of Europe, and dated them as late Givetian.

Many workers in North America place the Middle-Upper Devonian boundary at the base of or within beds carrying the *Pandorinellina insita* fauna, which succeeds the *Schmidtognathus hermanni-Polygnathus cristatus* Zone (KLAPPER et al., 1971, p. 286).

WESTERN NORTH AMERICA

A number of zonal schemes have been proposed by various authors including JOHNSON (1966), JOHNSON and BOUCOT (1968), and others for the Middle Devonian succession of western United States.

A zonal sequence based on brachiopods proposed by JOHNSON and BOUCOT (1968, p. 69) for the Middle Devonian of central Nevada includes the following in ascending sequence: the *Leptathyris circula* Zone with cricoconarids identified by BOUČEK including *Nowakia otomari, Variatellina pseudogeinitziana,* and *Striatostyliolina striata*; the *Warrenella kirki* Zone with *Cabrieroceras* cf. *C. crispiforme* in the lower faunule; beds with *Warrenella franklini* associated with *Leiorhynchus castanea, Parastringocephalus, Subrensselandia, Mimatrypa* cf. *M. insquamosa,* and *Schizophoria mc-*

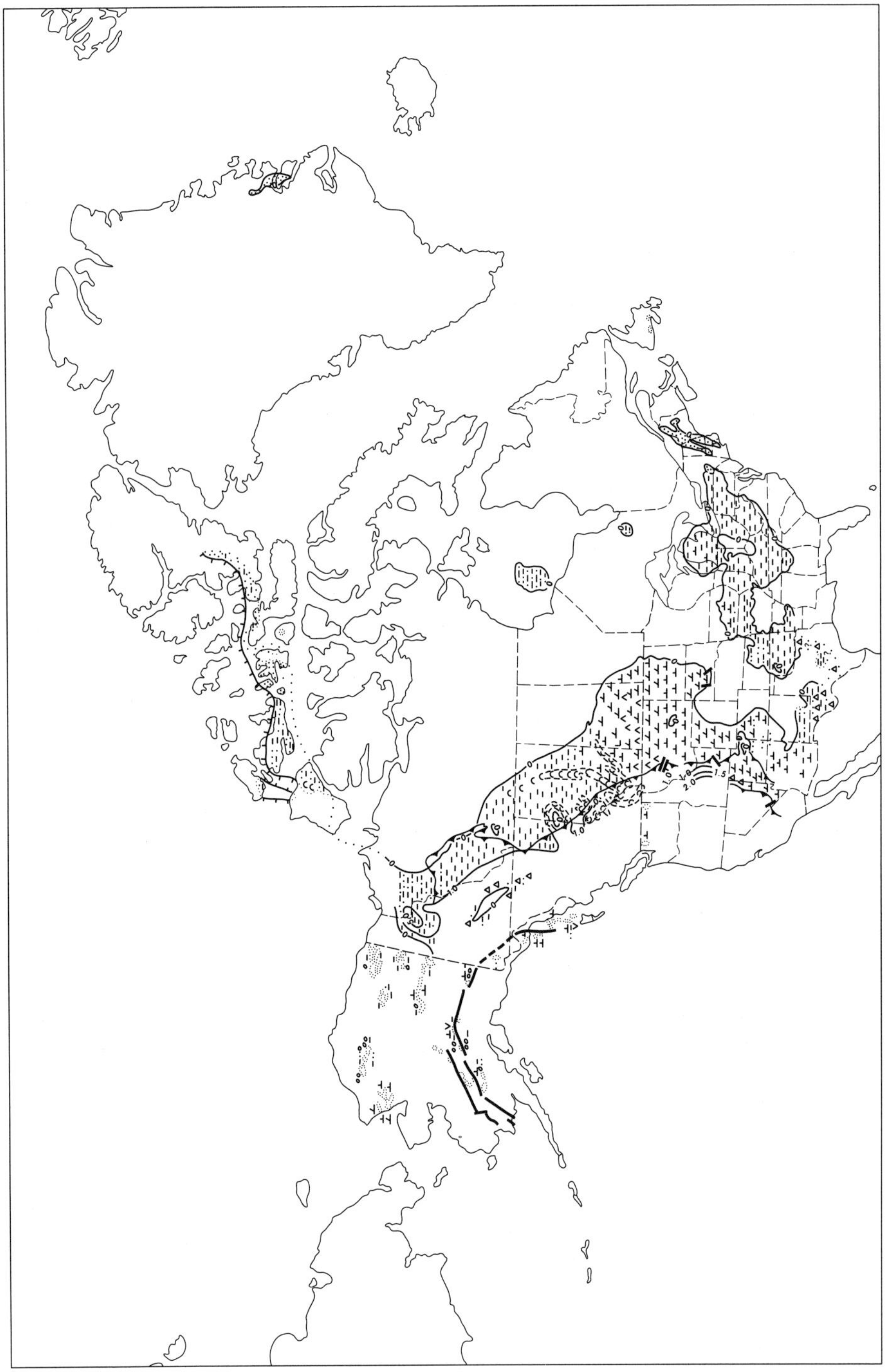

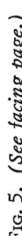

Fig. 5. *(See facing page.)*

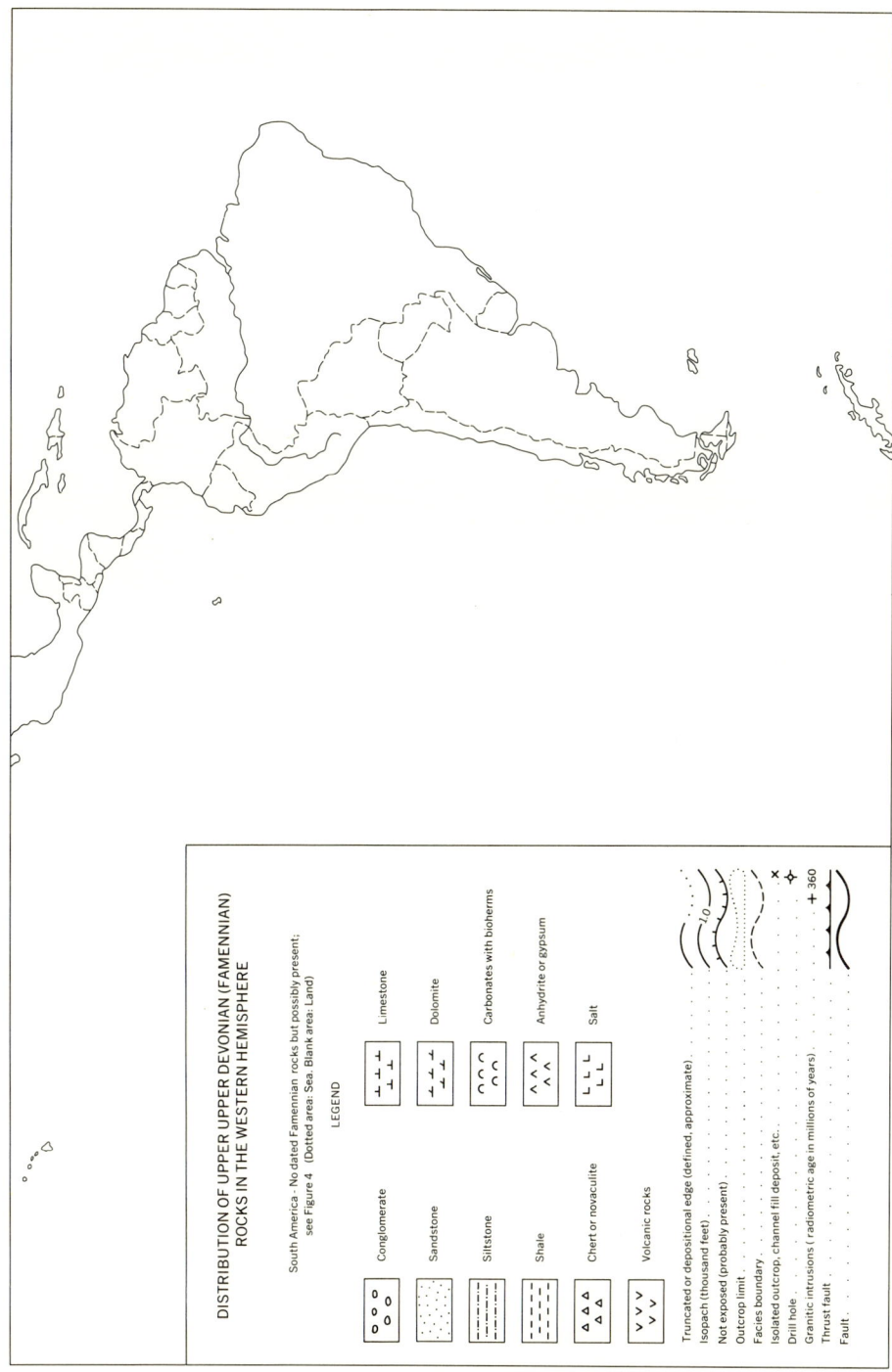

Fig. 5. Distribution of upper Upper Devonian (Famennian) rocks in the Western Hemisphere (Norris, n).

farlani; and *Warrenella occidentalis* Zone containing *Schmidtognathus hermanni-Polygnathus cristatus* Zone conodonts identified by KLAPPER, indicating a late Givetian age.

The two brachiopod faunal units above the *Warrenella kirki* Zone were later subdivided by JOHNSON (1970c, p. 2089) into three units comprising, in ascending sequence, the *Leiorhynchus castanea* Zone, the *Rhyssochonetes aurora* fauna, and the *Leiorhynchus hippocastanea* Zone.

The *Leptathyris circula* Zone with its basal *Pentamerella* Subzone contains a large brachiopod fauna described by JOHNSON (1966, 1970a). An interval containing abundant two-hole echinoderm ossicles occurs at the base of the zone in some sections of Nevada (JOHNSON, 1971a, p. 304). Conodonts in the zone are assignable to the upper Eifelian *Polygnathus kockelianus* Zone.

The succeeding lower *Warrenella kirki* Subzone, as restricted by JOHNSON (1971a), is characterized by *Spinulicosta muirwoodi, Leptathyris index,* and the goniatite mentioned above.

The upper *Warrenella kirki* Subzone contains a less diversified fauna and is characterized by a great abundance of *W. kirki kirki* and *Leiorhynchus miram alpha* (JOHNSON, 1971a, p. 304-305).

The *Leiorhynchus castanea* Zone of Nevada is characterized by *Parastringocephalus* cf. *P. dorsalis, Subrensselandia nolani, Warrenella* cf. *W. franklini, Polygnathus varcus,* and other fossils (JOHNSON, 1969; 1970c, p. 2087). It was dated by JOHNSON as pre-Taghanic Givetian.

The succeeding *Rhyssochonetes aurora* fauna includes *R. aurora solex, Leiorhynchus* sp. aff. *L. mesacostale,* and *Polygnathus varcus.* It is correlated with the Lower Taghanic of New York (JOHNSON, 1970c, p. 2087).

The *Leiorhynchus hippocastanea* Zone contains *Warrenella occidentalis, Hadrorhynchia sandersoni,* abundant *Emanuella* cf. *E. meristoides,* and a conodont fauna of the *Schmidtognathus hermanni-Polygnathus cristatus* Zone. This is the highest brachiopod fauna of Middle Devonian aspect in Nevada and it is correlated with the Middle Taghanic of New York (JOHNSON, 1970c).

NORTHWESTERN NORTH AMERICA

Faunal assemblages of late Early Devonian (Emsian) to late Middle Devonian (Eifelian) age that are widespread in the northern Yukon comprise, in ascending sequence: *Moelleritia canadensis, Gasterocoma? bicaula, "Schuchertella" adoceta,* and *Radiastraea verrilli.*

The very large and distinctive ostracode, *Moelleritia canadensis* (COPELAND, 1962), is generally the only megafossil found in dolomitic rocks in the upper half of the Gossage Formation and upper part of the Bear Rock Formation in northern Yukon Territory and adjacent District of Mackenzie. Although formerly dated as Eifelian, conodonts recovered from the Bear Rock Formation (UYENO & MASON, 1975, p. 720) suggest that *M. canadensis* probably is confined to beds of late Emsian to early Eifelian age.

Echinoderm ossicles with double and crosslike axial canals, named *Gasterocoma? bicaula* by JOHNSON & LANE (1969), occur in a variety of facies in beds immediately above *M. canadensis,* but also overlap this form. In Nevada, *G.? bicaula* occurs at the base of the *Leptathyris circula* Zone of Eifelian age. In northern Yukon the acme zone of these ossicles is above *M. canadensis,* but they range down to the *Sieberella-Nymphorhynchia pseudolivonica* unit of mid- or early Emsian age, and range upward into beds of early Eifelian age.

In succeeding beds two widely distributed, richly fossiliferous associations that occur typically in the Hume Formation of the central Mackenzie River region are present. These are characterized by *"Schuchertella" adoceta* and *Radiastraea verrilli* (approximately equivalent to *"Carinatina" dysmorphostrota* and *"Spinulicosta" stainbrooki* zones, respectively). These faunal assemblages have been listed and commented upon by CRICKMAY (1966), CALDWELL (1971), LENZ and PEDDER (1972), PEDDER (1975), and others.

The *"Schuchertella" adoceta* Zone contains many brachiopods that extend into overlying zones, but the corals *"Microcyclus" multiradiatus, Radiastraea trochomisca, R. verrilli,* and *Taimyrophyllum*

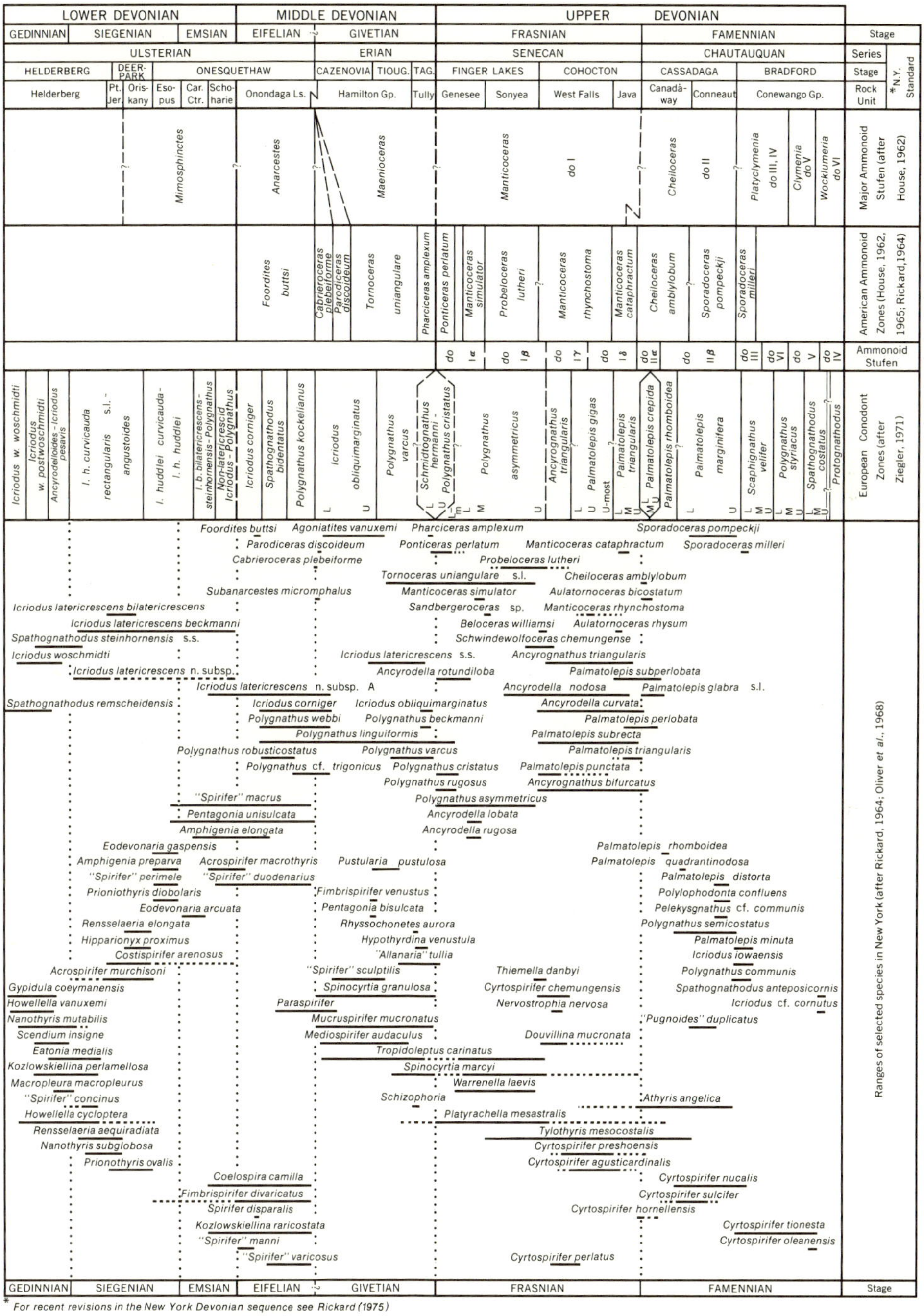

* For recent revisions in the New York Devonian sequence see Rickard (1975)

FIG. 6. Main Devonian faunal zones, and ranges of selected species in New York (Norris, n).

triadorum are diagnostic of the zone (PEDDER, 1975, p. 572). Conodonts in the zone suggest an Eifelian, probably early Eifelian, age (UYENO in LENZ & PEDDER, 1972), and Eifelian goniatites have been recorded by HOUSE and PEDDER (1963) in the stratigraphically lower Funeral Formation.

The *"Carinatina" dysmorphostrota* Zone contains a rich fauna including *Spinatrypa borealis, S. andersonensis, S. coriacea*; and the corals *Radiastraea tapetiformis, Taimyrophyllum stirps, Aphroidophyllum howelli, A. meeki,* and *Mackenziephyllum insolitum* (PEDDER, 1975, p. 572). Conodonts identified by UYENO (in LENZ & PEDDER, 1972) from the zone indicate an Eifelian age.

Upper Middle Devonian (Givetian) zones, or "hemerae," recognized by CRICKMAY (1966) in western Canada comprise: *Desquamatia (Variatrypa) arctica, Stringocephalus glaphyrus, S. chasmognathus, S. aleskanus, S. axius, Leiorhynchus hippocastanea, Emanuella vernilis,* and *Desquamatia (Independatrypa) independensis.*

An alternative zonal scheme proposed by PEDDER (1975) for Givetian strata of the central Mackenzie valley area is as follows, in ascending sequence: *Leiorhynchus castanea, Ectorensselandia laevis, Stringocephalus aleskanus, Leiorhynchus hippocastanea* and *Grypophyllum mackenziense*.

Desquamatia (V.) arctica occurs in the Elm Point of Manitoba, lower Methy of Alberta and Saskatchewan, lower Pine Point of the Great Slave Lake area, upper Hume and lower Hare Indian of the Mackenzie area, and is present also in the Hudson platform and Michigan. Conodonts associated with *D. (V.) arctica* in Manitoba (UYENO in NORRIS & UYENO, 1972) and in the Hudson platform (UYENO in SANFORD & NORRIS, 1975) are dated as late Eifelian and probably younger.

Stringocephalus, a guide fossil for the Givetian, is widely distributed in western North America (BOUCOT, JOHNSON & STRUVE, 1966, p. 1358, fig. 2) and was discovered recently in Indiana of the midcontinent area (COOPER & PHELAN, 1966). *Stringocephalus* and the closely related genus *Geranocephalus* are extremely variable forms and some of the named species are difficult to differentiate and correlate precisely. The earliest forms in western Canada appear in the Winnipegosis, Methy, and Pine Point formations and include *Stringocephalus glaphyrus* and *S. sapiens*. The *Stringocephalus* and associated mollusks in the Miami Bend Formation of Indiana are closely related to forms in the Winnipegosis Formation of Manitoba. Some of the later stringocephalids include *Stringocephalus asteius* and *S. alaskanus*, which occur in the Ramparts Limestone of the Mackenzie region.

The *Leiorhynchus hippocastanea* Zone occurs in post-*Stringocephalus* beds below the reefs of the Ramparts Formation of the Mackenzie valley. Diagnostic brachiopods include *Schizophoria mcfarlani, Stelckia galearius, Hadrorhynchia sandersoni, "Atrypa" percrassa, Warrenella occidentalis timetea,* and other forms (PEDDER, 1975, p. 574).

The *Grypophyllum mackenziense* Zone occurs in the upper reefal part of the Ramparts Limestone and is approximately equivalent to the *Emanuella vernilis* Zone of the Slave Point Formation of the Great Slave Lake area and the *Desquamatia (I.) independensis* Zone of the Swan Hills Formation of northern Alberta. Conodonts of the *Schmidtognathus hermanni-Polygnathus cristatus* Zone of late Givetian age occur with both the *Leiorhynchus hippocastanea* and *G. mackenziense* faunas.

NORTHERN NORTH AMERICA

The Middle Devonian Series in the Canadian Arctic is represented by marine clastic carbonate rocks, including some remarkably persistent formations.

The upper two-thirds of the poorly fossiliferous, clastic Eids Formation of Bathurst Island contains the *Icriodus corniger-I. curvirostratus-I. introlevatus* assemblage and *Polygnathus linguiformis linguiformis,* which UYENO (in McGREGOR & UYENO, 1972) considered to be of mid-Couvinian age.

The Blue Fiord Formation contains abundant megafossils of which stromatoporoids, corals, brachiopods and trilobites are important elements. The upper range of the "two-hole" echinoderm ossicle, *Gasterocoma? bicaula,* extends into the lower part of the formation on Bathurst Island (McGREGOR & UYENO, 1972, table 1). Brachio-

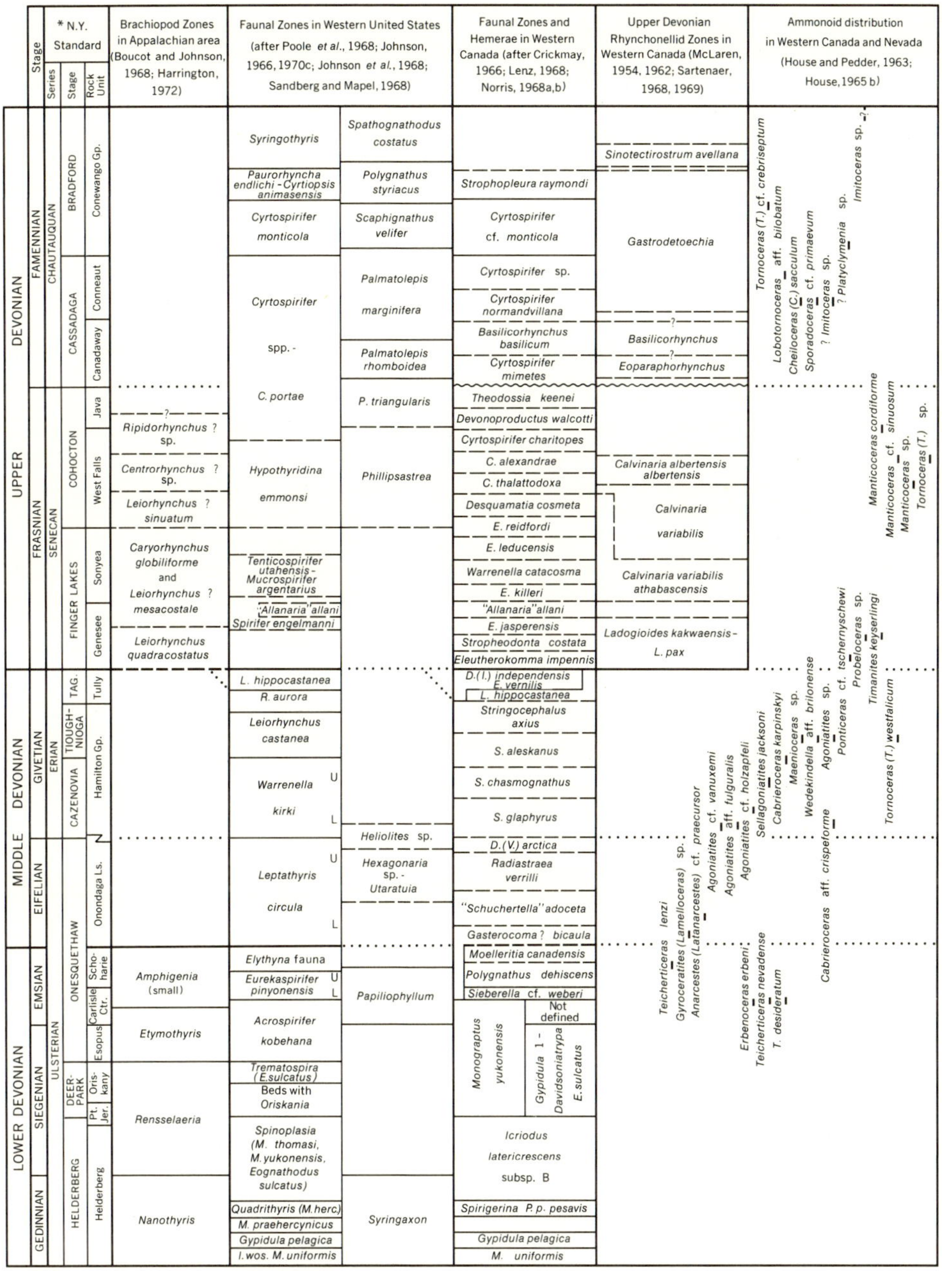

* For recent revisions in the New York Devonian sequence see Rickard (1975)

FIG. 7. Some Devonian faunal zones of Appalachian area, western United States, and western Canada, and ammonoid distribution in Devonian rocks of western Canada and Nevada (Norris, n).

pods from the middle and upper parts of the formation were correlated by BRICE and MEATS (1977) with the Nevada zones of *Leptathyris circula* (Eifelian) and lower part of *Warrenella kirki* (late Eifelian or Givetian), respectively. ORMISTON (1967) has indicated that the Blue Fiord trilobite fauna is highly diverse. Species that are identical with Eifelian forms from Europe include *Harpes macrocephalus, Astycorphe cimelia, Leonaspis elliptica,* and *Otarion belanops.* About half of the Blue Fiord trilobites are dechenellids, including *Dechenella (D.) mclareni, D. (D.) tesca, Deltadechenella bathurstensis,* and many others. Associated corals, according to ORMISTON (1967), consist entirely of genera present in beds of Eifelian age in Germany. Conodonts recorded by UYENO (in MCGREGOR & UYENO, 1972) include a form close to *Icriodus corniger* and *Eognathodus bipennatus,* indicating an Eifelian age.

The succeeding Bird Fiord Formation has a relatively restricted fauna. Corals are generally rare; brachiopods include *"Camarotoechia princeps," Emanuella,* and abundant *Atrypa*; trilobites are very restricted, comprising only five genera of which four are rare. The presence of *Ancyropyge manitobensis* is the best evidence furnished by trilobites for the Givetian age of the greater part of the Bird Fiord Formation. In Manitoba this form occurs in the Winnipegosis Formation with *Stringocephalus.*

Brachiopods of the Bird Fiord Formation recently studied by BRICE and MEATS (1977) were correlated with the upper *Warrenella kirki* Subzone (early Givetian) of Nevada.

Conodonts recovered by UYENO (in MCGREGOR & UYENO, 1972) from the lower part of the Bird Fiord Formation of Bathurst Island are nondiagnostic. However, *Icriodus* cf. *I. obliquimarginatus* occurs in the upper part of the formation, a form commonly associated with *Polygnathus varcus,* and indicating a Givetian age.

On Melville Island, in a shale-sandstone facies, the Givetian brachiopod *Leiorhynchus castanea* was found, indicating marine connection with the mainland of western Canada. *Stringocephalus* has not been found in the Canadian Arctic, presumably because of provincialism and unfavorable facies during this interval.

PROVINCIALISM AND AMMONOID DISTRIBUTION

During the Eifelian and most of the Givetian, excluding the Taghanic Stage, the contrast between the brachiopod faunas of the Eastern Americas and Old World realms continued to be pronounced (JOHNSON & BOUCOT, 1973). The Malvinokaffric realm seas had retreated from Antarctica and remained only in the deeper parts of the troughs of southern South America. The Cordilleran region of the Old World realm in western North America extended from southeastern California to the Northwest Territories (JOHNSON & BOUCOT, 1973). Throughout most of Middle Devonian time the Eastern Americas realm in eastern North America and northern South America continued to bear a distinctly endemic fauna.

Pronounced trilobite provinciality continued from the Emsian into the Eifelian and, in western North America, exhibited even greater provinciality than the contemporaneous brachiopods (ORMISTON, 1972). A decrease in trilobite provinciality occurred in the Givetian, which is marked by the southward migration of *Ancyropyge* from the Arctic Islands into the Northwest Territories, Manitoba, Nevada, and Michigan. This increasing trend towards cosmopolitanism is marked also by the appearance of *Scutellum* in the Appalachian area, and the wide distribution of *Dechenella* in western Canada and the Great Basin (ORMISTON, 1972).

Ammonoid dispersion increased in the middle Devonian, especially during the Givetian. Eifelian ammonoids recorded from North America include *Gyroceratites (Lamelloceras)* and *Anarcestes* from the Northwest Territories (HOUSE & PEDDER, 1963). *Foordites* from the New York Onondaga Limestone and from Virginia (HOUSE, 1962), and *Foordites* and *Cabrieroceras* from the Columbus Limestone of Ohio (SWEET & MILLER, 1956). HOUSE (1964, p. 265) pointed out that a distinctive feature of North American Eifelian ammonoid faunas is the apparent absence of *Anarcestes* of the *A. lateseptatus* group and *Pinacites,* genera which in Europe characterize the lower and upper Eifelian, respectively.

Givetian ammonoid faunas in North America are much more widespread and begin with the entry of *Cabrieroceras plebeiforme* (HOUSE, 1964). From the Northwest Territories *Wedekindella, Maenioceras,*

Agoniatites, Tornoceras, and *Cabrieroceras* are recorded (HOUSE & PEDDER, 1963; HOUSE, 1964), all of which are closely related to European forms. Of interest is the apparent absence of Middle Devonian goniatites from the more southerly Rocky Mountains and the platform areas of the Plains.

The Givetian goniatites which occur in the New York-Pennsylvania embayment do not show typical European features (HOUSE, 1964, p. 265). From this area *Tornoceras (Tornoceras), Agoniatites,* and other forms are known from the Cherry Valley Limestone, but typical genera (e.g., *Maenioceras, Wedekindella, Sobolewia*) are lacking. From farther south in Virginia, however, *Sobolewia* and *Maenioceras,* which resemble a mid-Givetian fauna from Devon are recorded (HOUSE, 1962). From this distribution HOUSE (1964) suggested that there was trans-Arctic contact between Europe, Russia, and western Canada, and perhaps trans-Atlantic contact between Europe and Virginia. During the same time little connection was made across the shelf areas over much of northern United States and southern Canada.

The first goniatite from the Devonian of South America was described recently by LEANZA (1968). It was named *Tornoceras baldisi* and was collected from the Chavela Formation outcropping in San Juan Province of Argentina. LEANZA (1968) dated the new species as Late Devonian but, according to W. W. NASSICHUK (pers. commun., October, 1973), the suture line of the new form closely resembles that of *Tornoceras uniangulare aldense* (HOUSE, 1965a), which occurs in the Middle Devonian (Givetian) Alden Marcasite of New York State.

UPPER DEVONIAN FAUNAS (FRASNIAN AND FAMENNIAN)

EASTERN NORTH AMERICA

The Upper Devonian of the New York standard section is subdivided into the Senecan Series (Frasnian) with the Finger Lakes and Cohocton stages, and the Chautauquan Series (Famennian) embracing the Cassadaga and Bradford stages. The rocks of this interval are almost entirely clastic and make up more than two-thirds of the volume of sedimentary rocks in the Appalachian basin (RICKARD, 1964). The position of the base of the Upper Devonian is not agreed upon but generally it is taken at the base of the Geneseo Shale.

Conodonts and ammonoids are the principal fossils used for correlating the New York Upper Devonian with Europe. Conodonts of the upper part of the *Schmidtognathus hermanni-Polygnathus cristatus* Zone occur at the base of the Genesee Group. The presence of the *Pandorinellina insita* has not been demonstrated in New York (RICKARD, 1975). Conodonts of the lowermost *Mesotaxis asymmetrica* Zone occur in the Lodi Limestone. The lower *M. asymmetrica* Zone begins in the lower Penn Yan Shale where *Ancyrodella rotundiloba* first appears. The base of the middle *M. asymmetrica* Zone appears in the upper West River Shale where it is marked by *Palmatolepis punctata*. Conodonts of the upper *M. asymmetrica* Zone occur in the lower Rhinestreet Shale; and the "*Ancyrognathus" triangularis* Zone occurs in the middle and upper parts of the Rhinestreet. Upper *Palmatolepis triangularis* Zone conodonts occur in the upper Hanover and Dunkirk shales (RICKARD, 1975, p. 9).

Ammonoid zones in rocks of the lower Upper Devonian (Senecan Series) comprise in ascending sequences: *Epitornoceras peracutum, Ponticeras perlatum, Manticoceras styliophillum, Sandbergeroceras syngonum, Probeloceras lutheri, P. strix, Manticoceras rhynchostoma,* and *Crickites* cf. *C. holzapfeli* (HOUSE & KIRCHGASSER in RICKARD, 1975). *Probeloceras strix* is a new zone described by KIRCHGASSER (1975) that occurs in the upper Cashaqua and lower Rhinestreet shales of New York.

Well-known brachiopods in rocks of the Finger Lakes Stage include: *Mucrospirifer posterus,* limited to the Genesee Group; *Warrenella laevis* and *Spinocyrtia marcyi,* ranging throughout the stage; and *Tylothyris mesocostalis,* appearing in the Sonyea Group and ranging into higher beds.

The brachiopod genus *Cyrtospirifer* first appears in the lower part of the West Falls

Group with *C. chemungensis* and *C. perlatus* as important species; and *C. angusticardinalis* and *C. preshoensis* make their appearance in the upper part of the group.

The Frasnian-Famennian boundary in Belgium traditionally has been placed at the base of the Assise de Senseille, which appears to correspond to a level slightly below the base of the *Cheiloceras*-Stufe (HOUSE in RICKARD, 1975). Opinions differ on the position of the *Manticoceras-Cheiloceras* boundary relative to the conodont zonation. In New York this boundary is placed at or near the base of the Upper *triangularis* Zone (RICKARD, 1975). HOUSE (1967, 1973) recorded that *Crickites*, indicative of the upper *Manticoceras*-Stufe, occurs 30 feet below the top of the Hanover Shale, and KLAPPER et al. (1971) referred the upper Hanover to the upper *triangularis* Zone.

Ammonoids and conodonts are rarer in the upper Upper Devonian (Chautauquan Series) and correlation with the Famennian of Europe becomes increasingly more difficult.

Conodont zones represented in rocks of the Cassadaga Stage include the upper *Palmatolepis triangularis, P. crepida, P. rhomboidea,* and *P. marginifera,* but the precise limits of the zones are unknown (RICKARD, 1975).

The ammonoid *Cheiloceras amblylobum,* associated with conodonts of the *Palmatolepis crepida* Zone, occurs in the Gowanda Shale. The youngest ammonoid zone, *Sporadoceras milleri,* present in Pennsylvania, occurs in the lower part of the Bradford Stage which correlates with part of Zone doIII of Europe (OLIVER et al., 1968, p. 1033).

Scattered conodonts indicate that succeeding rocks of the Bradford Stage may include equivalents of the European ammonoid Zones doIV to doVI (OLIVER et al., 1968, p. 1035).

Brachiopods in rocks of the Bradford Stage include *Cyrtospirifer tionesta* that ranges throughout the stage; and *C. oleanensis* and *"Camarotoechia" allegania* that appear in the upper part of the stage.

WESTERN NORTH AMERICA

Various megafaunal zonal schemes have been proposed for the abundantly fossiliferous Frasnian rocks of western Canada by a number of workers including CRICKMAY (1966), MAURIN and RAASCH (1972), McLAREN (1954, 1962), RAASCH in DOOGE (1966), WARREN and STELCK (1950), and others. Frasnian rhynchonellid brachiopods studied by McLAREN (1954, 1962) are useful zonal fossils, especially for the lower three-quarters of the Frasnian succession. McLAREN's zones are listed on Figure 7, as well as an alternative zonal scheme proposed by CRICKMAY (1966).

The *Eleutherokomma impennis* to *E. killeri* zones occur typically in the Waterways (Beaverhill Lake) Formation of northern Alberta. Elements of these zones are recognized in Manitoba, Saskatchewan, the Alberta Rocky Mountains, and elsewhere.

Conodonts in the Waterways Formation studied by UYENO (1974) comprise the *Pandorinellina insita* fauna, and Lower and Middle *Mesotaxis asymmetrica* zones which correlate with the lower *Manticoceras*-Stufe of Europe.

The genus *Ladogioides,* commonly associated with *Eleutherokomma impennis* and *Platyterorhynchus russelli,* is also a valuable zone marker for the base of the Frasnian succession in northern Alberta, northeastern British Columbia, and the southern Northwest Territories.

McLAREN's (1962) *Calvinaria variabilis athabascensis* Zone is not contiguous with the *Ladogioides* Zone, but is restricted to the Maligne and lower Perdrix formations in the Alberta Rockies and the Cooking Lake Formation of central Alberta.

The *Calvinaria variabilis insculpta* Zone of McLAREN (1962) occurs in the upper Perdrix Formation and its equivalents in the Alberta Rocky Mountains, and in the Duvernay Formation of the central Alberta subsurface. *Leiorhynchus* (=*Caryorhynchus* of some authors) *carya* commonly is associated with the zone fossil and represents the last occurrence of the genus in the area. *Eleutherokomma* disappears late in the *C. variabilis insculpta* Zone, within the *"Desquamatia" cosmeta* Zone of CRICKMAY (1966). *Cyrtospirifer* first appears in the lower part of the *C. variabilis insculpta* Zone, within the *E. reidfordi* Zone of CRICKMAY (1966), suggesting a correlation with the lower part of the Cohocton Stage of

New York of mid-Frasnian age.

The *Calvinaria albertensis albertensis* Zone of McLaren (1962) is represented in the lower and middle Mount Hawk Formation and its equivalents in the Alberta Rocky Mountains and in the Ireton Formation of the Alberta subsurface.

In the upper Frasnian succession, Crickmay's (1966) zones of *Cyrtospirifer charitopes, Devonoproductus walcotti,* and *Theodossia keenei* are represented. The lower two zones occur typically in the Redknife Formation and the upper zone occurs in the Kakisa Formation of the Trout River outcrop area described by Belyea and McLaren (1962). The fauna of the *T. keenei* Zone is separated from overlying faunas by a profound break that is remarkably widespread throughout western and northern Canada. This break at the end of the Frasnian is marked by the disappearance of numerous corals, stromatoporoids, stropheodontids, pentamerids, and atrypids.

Fossiliferous Frasnian rocks of western United States are less abundant and less continuous than those in western Canada. Various tentative megafaunal zones have been proposed by Poole et al. (1968), Johnson (1970c), Baars (1972), and others. Brachiopod zones used by Poole et al. (1968) in southwestern United States are listed on Figure 7.

Six brachiopod zones were recognized by Crickmay (1966) within Famennian rocks of western Canada comprising in ascending sequence: *Cyrtospirifer mimetes, Basilicorhynchus basilicum, Cyrtospirifer normandvillana, C.* sp., *C.* cf. *C. monticola,* and *Strophopleura raymondi.* Another zonal scheme based on rhynchonellids and proposed by Sartenaer (1968, 1969) for the same interval is as follows: zones of *Eoparaphorhynchus, Basilicorhynchus, Gastrodetoechia,* and *Sinotectirostrum avellana.*

The *Eoparaphorhynchus* Zone is restricted to the lower part of the lower Famennian and is represented in the Trout River and Sassenach formations, and in the upper part of the Winterburn Group. The zone is represented by *E. maclareni* in the Mackenzie River area and Alaska; by *E. lentiformis* in the area east and north of Jasper in the Rocky Mountains; and by *E. walcotti* in Nevada and Utah.

The succeeding *Basilicorhynchus* Zone is restricted to the upper part of the lower Famennian and occurs in the Tetcho Formation, in the lower part of the Wabamun Group, and in the upper Sassenach and lower Palliser formations.

The *Gastrodetoechia* Zone extends from the middle Famennian to the lower part of the upper Famennian. The zone is present in the Rocky Mountains and Northwest Territories of Canada, and in Idaho, Montana and Utah of the United States. The zone contains a large brachiopod fauna, as well as ammonoids described by House (in House & Pedder, 1963), which include species of *Cheiloceras, Tornoceras* and *Platyclymenia.*

The *Sinotectirostrum avellana* Zone occurs in the upper part of the upper Famennian and is known only in the North Nahanni and Root Rivers areas of the Northwest Territories and from wells in northeastern British Columbia. The zone corresponds to Hume's (1922) *Athyris angelica* Zone.

Famennian brachiopod assemblages recognized by Poole et al. (1968) in southwestern United States comprise in ascending sequence: *Cyrtospirifer* spp.-*C. portae, C. monticola, Paurorhyncha endlichi-Cyrtiopsis animasensis,* and *Syringothyris* spp.

NORTHERN NORTH AMERICA

Upper Devonian rocks in the eastern Arctic Islands are predominantly clastic strata of nonmarine origin. Marine Frasnian rocks, however, have been reported from scattered localities in the western Arctic Islands, notably Banks, Melville, and Prince Patrick islands. On Banks Island the early Frasnian is indicated by the presence of *Ladogioides pax, Eleutherokomma* cf. *E. impennis,* and *"Allanaria" allani*; and the mid-Frasnian by *Cyrtospirifer thalattodoxa* and *Calvinaria albertensis.* A rich fauna of middle to late Frasnian age occurs in reefs on northeastern Banks Island and includes *Hexagonaria, Phillipsastrea, Theodossia,* and others (Klovan & Embry, 1971).

Marine Famennian megafossils of the Canadian Arctic Islands have been summarized by McLaren (in Kerr, McGregor & McLaren, 1965). They occur in the upper part of the Griper Bay Formation on Bathurst, Cameron, Byam Martin, and Melville islands. Brachiopods of this assemblage represent a single fauna and include *Acanthatia* sp., *Basilicorhynchus* sp., *Ptychomaletoechia?* sp., and possibly *Cyrtospirifer* sp. Conodonts associated with the brachiopods on Byam Martin Island include *Palmatolepis perlobata* and *Icriodus costatus*. The former ranges from the upper *Palmatolepis triangularis* Zone through the lower *"Spathognathodus" costatus* Zone (MCGREGOR & UYENO, 1972).

SOUTH AMERICA

In Late Devonian time the sea withdrew from most of the extra-Andean regions of South America, but seems to have persisted in the Andean belt. The upper barren parts of the Devonian clastic successions in western Venezuela, southern Peru, central Bolivia, and western Argentina may belong in the Upper Devonian (HARRINGTON, 1962, p. 1786). Recently, COUSMINER (1964, p. 33-35) reported lower Upper Devonian palynomorphs from the upper part of the Devonian succession of Bolivia. Sparsely fossiliferous rocks present also in the Parnaíba basin of Brazil are probably of early Frasnian age (HARRINGTON, 1968, p. 662). A few authors, notably MAACK (1964), have argued that South America was glaciated during the Late Devonian. MAACK (1964, p. 291, 292) cited as evidence widespread Upper Devonian tillites present in northeastern Brazil and northwestern Argentina.

PROVINCIALISM AND AMMONOID DISTRIBUTION

Beginning in late Eifelian time, routes for faunal migration were reestablished between the midcontinent area and western North America via the Williston basin, as indicated by an intermingling of Eastern Americas and Old World shelly elements in Indiana, Michigan, northern Ontario, and northern Manitoba. The separation between the Old World and Eastern Americas realms apparently was breached completely by onlap of epicontinental seas during the Taghanic (JOHNSON, 1970c), resulting in a single cosmopolitan brachiopod fauna in the marine areas of North America and elsewhere during the Frasnian (JOHNSON & BOUCOT, 1973). Important widespread brachiopod genera of this fauna include *Calvinaria, Cariniferella, Devonoproductus, Douvillina, Eleutherokomma, Hypothyridina, Nervostrophia, "Spirifer"* of the *orestes*-type, *Tenticospirifer,* and *Theodossia* (JOHNSON & BOUCOT, 1973, p. 94).

The tubular stromatoporoid *Amphipora* abundant in western North America and the Old World reached the midcontinent and Appalachian areas during this interval (DUNCAN in CLOUD, 1959, p. 948, fig. 12); and similar evidence showing intermingling of coral genera was cited by OLIVER (1975).

In contrast to the cosmopolitanism shown by Frasnian brachiopods, trilobites are very scarce at this time; *Scutellum* and a few phacopids only are known in western North America (ORMISTON, 1972, p. 602). The radiation in the cyrtosymbolinids that took place in Europe did not reach North America because of the shutting-off of the migration route from Europe to North America by a continental clastic wedge which was deposited across northern Alaska and the Arctic Islands during the Frasnian and Famennian (ORMISTON, 1972, p. 602).

At the end of Frasnian time a number of important invertebrate groups became extinct, including the brachiopod taxa Atrypoidea, Pentameroidea, Stropheodontidae, Delthyridae, and Meristellinae. Among the trilobites, the harpids and thysanopeltids disappeared in the Late Devonian, probably in late Frasnian time (HOUSE, 1967).

The principal brachiopod genera that survived the extinction at the end of the Frasnian and continued into the Famennian include: *Attribonium, Aulacella* or *Rhipidomella,* "*Chonetes*" or *Retichonetes, Crurithyris, Cupularostrum* or *Ptychomaletoechia, Cyrtina, Cyrtospirifer, Productella, Schizophoria,* and *Steinhagella* (JOHNSON & BOUCOT, 1973). Along with these survivors the Famennian brachiopod fauna comprises principally new rhynchonellid, productid, athyridid and spiriferid genera, some of which are widely distributed (SARTENAER, 1969), indicating continued cosmopolitanism (JOHNSON & BOUCOT, 1973).

The ammonoid genus *Manticoceras,*

guide fossil for the Frasnian, is more widely distributed than any other, but is absent from South America (House, 1964). Ties between North America and Europe are indicated by occurrences of early Frasnian forms in several places. *Timanites* is recorded from Alberta, and *Koenenites* is known from Michigan (Miller, 1938). *Pharciceras* is the European guide genus of the lower Frasnian, and it, or its close allies, occur in the upper Tully Limestone of New York (House, 1962, 1968a). The most widespread distribution of *Manticoceras* apparently occurred in the middle Frasnian, where the genus is known from the Northwest Territories, Alberta, Ontario, including the Hudson Bay Lowlands (Miller, 1938, p. 115, 116), Michigan, Iowa, Utah, New Mexico, Missouri, Ohio, Virginia, and especially New York. These occurrences form a link with those along the northern Siberian coast (House, 1964, p. 266). The richest North American Frasnian goniatite faunas occur in New York where a succession of species closely allied to those of Europe may be recognized.

In North America, *Cheiloceras* is known from New York (House, 1962) and western Canada (House & Pedder, 1963), and is apparently almost as widespread as *Manticoceras* geographically, but is sparser numerically (House, 1964). This sparseness marks the beginning of geographical restriction of later Devonian ammonoid faunas.

Clymeniids entered with the faunas of the *Platyclymenia*-Stufe, a fauna well known from Europe and North Africa. In North America representatives of this fauna occur in the Three Forks Shale of Montana and include *Tornoceras, Platyclymenia, Rectoclymenia,* and *Raymondiceras* (House, 1964). The latter genus is recorded also from New Mexico (Miller & Collinson, 1951). Of interest is the apparent absence of this fauna from eastern North America, except for *Sporadoceras milleri* from Pennsylvania, which may occur at this level (House, 1962, p. 263).

The goniatites and clymeniids of the uppermost Famennian, including the *Clymenia* and *Wocklumeria* zones known in England, Germany, France, North Africa, and elsewhere, are sparingly represented in North America and the traces found are equivocal. *Cyrtoclymenia* and *Cymaclymenia* are known from Ohio, and *Cyrtoclymenia* from New Mexico (House, 1962), and also from Alaska (Sable & Dutro, 1961). None of the North American occurrences can be dated certainly as latest Famennian, and all may well be mid-Famennian in age (House, 1954, p. 268).

DEVONIAN CONTINENTAL DEPOSITS AND ASSOCIATED INVERTEBRATE FAUNAS

Nonmarine Devonian rocks occur in several well-known areas in the Western Hemisphere, the main areas being in the Appalachians of the eastern United States and Canada, in Arctic Canada, and in eastern Greenland; these are commonly referred to as "Old Red Sandstone" basins of deposition. Devonian continental beds also are represented possibly around the edges of the cratonic basins of South America and Antarctica where clastic sequences are largely barren or sparsely fossiliferous and difficult to date. Smaller, thinner, and more isolated occurrences are known also in the southern part of the Hudson Bay Lowlands (Sanford & Norris, 1975), in Montana and Wyoming (Sandberg, 1961), in Arizona (Teichert & Schopf, 1958), in the Rocky Mountains of Alberta (Aitken, 1966), and elsewhere. Many of these deposits contain abundant plant and fish remains but lack identifiable invertebrate fossils.

Invertebrate fossils in Devonian continental beds are generally scanty in variety and numbers. A few of the more important forms include the brachiopod genus *Lingula*, several mollusks (e.g., *Amnigenia*), leperditiid ostracodes, the euryptids *Pterygotus* and *Eurypterus*, ceratiocarid crustaceans, and the arthropod *Estheria*.

REFERENCES

Aitken, J. D., 1966, *Sub-Fairholme Devonian rocks of the eastern front ranges, southern Rocky Mountains, Alberta:* Geol. Survey Canada, Paper 64-33, 88 p.

Alberti, G. K. B., Hass, W., & Ormiston, A. R., 1971, *Discovery of the trilobite Warburgella rugulosa (Alth, 1874) in Gedinnian strata of central Nevada:* Neues Jahrb. Geologie, Paläontologie, Monatsh. 4, p. 193-194.

Allen, J. R. L., Dineley, D. L., & Friend, P. F., 1968, *Old Red Sandstone basins of North America and northwest Europe:* in International symposium on the Devonian System, Calgary, Alberta, Sept. 1967, D. H. Oswald (ed.), v. 1, p. 69-98, Alberta Soc. Petrol. Geologists (Calgary).

Amos, Arturo, & Boucot, A. J., 1963, *A revision of the brachiopod family Leptocoeliidae:* Palaeontology, v. 6, pt. 3, p. 440-457, pl. 62-65.

Baars, D. A., 1972, *Devonian System:* in Geologic atlas of the Rocky Mountain Region, United States of America, W. W. Mallory (ed.-in-chief), p. 90-99, Rocky Mountain Assoc. Geologists (Denver, Colo.).

Baker, H. A., 1923, *Final report on the geological investigations in the Falkland Islands (1920-1922):* p. 1-38, fig., geol. map, Fol. (London) (not seen by author).

Belyea, H. R., & McLaren, D. J., 1962, *Upper Devonian formations, southern part of Northwest Territories, northeastern British Columbia, and northwestern Alberta:* Geol. Survey Canada, Paper 61-29, 74 p.

Berdan, J. M., Berry, W. B. N., Cooper, G. A., Jackson, D. E., Klapper, Gilbert, Lenz, A. C., Martinsson, Anders, Rickard, L. V., & Thorsteinsson, R., 1969, *Siluro-Devonian boundary in North America:* Geol. Soc. America, Bull., v. 80, no. 11, p. 2165-2174, text-fig. 1, pl. 1.

Berry, W. B. N., 1967, *Monograptus hercynicus nevadensis n. subsp., from the Devonian in Nevada:* U.S. Geol. Survey, Prof. Paper 575-B, p. 26-31.——1968, *Siluro-Devonian graptolite sequence in the Great Basin (abstr.):* Geol. Soc. America, Cordilleran Section, 64th Ann. Meeting, 1968, p. 35-36. [Also in Geol. Soc. America, Abstr. for 1968, Spec. Paper no. 121, p. 483-484 (1969).]

——— & Murphy, M. A., 1972, *Early Devonian graptolites from the Rabbit Hill Limestone in Nevada:* Jour. Paleontology, v. 46, no. 2, p. 261-265, text-fig. 1,2.

Bottino, M. L., & Fullager, P. D., 1966, *Whole rock rubidium-strontium age of the Silurian-Devonian boundary in northeastern North America:* Geol. Soc. America, Bull., v. 77, no. 10, p. 1167-1176.

Boucot, A. J., 1975, *Evolution and extinction rate controls; developments in paleontology and stratigraphy, 1:* 427 p., Elsevier Scient. Publ. Co. (Amsterdam, Oxford, New York).

———, Caster, K. E., Ives, David, & Talent, J. A., 1963, *Relationships of a new Lower Devonian terebratuloid (Brachiopoda) from Antarctica:* Bull. Am. Paleontology, v. 46, no. 207, p. 77-151, pl. 16-41.

———, Doumani, G. A., Johnson, J. G., & Webers, G. F., 1968, *Devonian of Antarctica:* in International symposium on the Devonian System, Calgary, Alberta, Sept. 1967, D. H. Oswald (ed.), v. 1, p. 639-648, Alberta Soc. Petrol. Geologists (Calgary).

———, & Johnson, J. G., 1967, *Paleogeography and correlation of Appalachian province Lower Devonian sedimentary rocks:* in Symposium—Silurian-Devonian rocks of Oklahoma and environs, Tulsa Geol. Soc. Digest, v. 35, p. 35-87, 2 pl.——1968, *Appalachian province Early Devonian palaeogeography and brachiopod zonation:* International symposium on the Devonian System, Calgary, Alberta, Sept. 1967, D. H. Oswald (ed.), v. 2, p. 1255-1267, Alberta Soc. Petrol. Geologists (Calgary).

———, ———, & Struve, Wolfgang, 1966, *Stringocephalus, ontogeny and distribution:* Jour. Paleontology, v. 40, no. 6, p. 1349-1364, pl. 169-171.

———, ———, & Talent, J. A., 1968, *Lower and Middle Devonian faunal provinces based on Brachiopoda:* in International symposium on the Devonian System, Calgary, Alberta, Sept. 1967, D. H. Oswald (ed.), v. 2, p. 1239-1254, Alberta Soc. Petrol. Geologists (Calgary).——1969, *Early Devonian brachiopod zoogeography:* Geol. Soc. America, Spec. Paper 119, 113 p., 20 pl., 6 text-fig., 1 table.

———, Martinsson, Anders, Thorsteinsson, R., Walliser, O. H., Whittington, H. B., & Yochelson, E. L., 1960, *A Late Silurian fauna from the Sutherland River Formation, Devon Island, Canadian Archipelago:* Geol. Survey Canada, Bull. 65, 51 p.

Bowen, J. M., 1972, *Estratigrafía del precretáceo en la parte norte de la Sierra de Perijá:* Bol. Geol., Publ. Esp., Caracas, no. 5, p. 729-760.

Brice, D. & Meats, P., 1977, *Brachiopodes du Devonien Inférieur et Moyen des Formations Blue Fiord et Bird Fiord des Iles Artiques Canadiennes:* Geol. Survey Canada, Bull. (in press).

Bütler, H., 1961, *Devonian deposits of central East Greenland:* in Geology of the Arctic, v. 1, p. 188-196, G. O. Raasch (ed.), Alberta Soc. Petrol. Geologists and Univ. of Toronto Press (Toronto).

Caldwell, W. G. E., 1971, *The biostratigraphy of some Middle and Upper Devonian rocks in the Northwest Territories: an historical review:* The Musk-Ox, Publ. no. 9, 1971, p. 1-20.

Caster, K. E., 1939, *A Devonian fauna from Co-

lombia: Bull. Am. Paleontology, v. 24, no. 83, 218 p., 14 pl.———1952, *Stratigraphic and paleontologic data relevant to the problem of Afro-American ligation during the Paleozoic and Mesozoic:* in Symposium: The problem of land connections across the South Atlantic, ... ; Am. Museum Nat. History, Bull., v. 99, art. 3, p. 105-152.

Chulpáč, Ivo, 1972, *The Silurian-Devonian boundary in the Barrandian; with contributions by H. Jaeger and J. Zikmundova:* Canad. Petrol. Geol. Bull., v. 20, no. 1, p. 104-174, 33 text-fig.

Churkin, Michael, Jr., & Brabb, E. E., 1965, *Ordovician, Silurian, and Devonian biostratigraphy of east-central Alaska:* Am. Assoc. Petrol. Geol., Bull., v. 49, no. 2, p. 172-185, 4 fig., 3 tables.———1968, *Devonian rocks of the Yukon-Porcupine Rivers area and their tectonic relation to other Devonian sequences in Alaska:* in International symposium on the Devonian System, Calgary, Alberta, Sept., 1967, D. H. Oswald (ed.), v. 2, p. 227-258, pl. 1-5, Alberta Soc. Petrol. Geologists (Calgary).

Clarke, J. M., 1913, *Fosseis Devonianos do Paraná:* Serv. Geol. Mineral. do Brasil, Mon., v. 1, 353 p., 27 pl.

Cloud, P. E., 1959, *Paleoecology—retrospect and prospect:* Jour. Paleontology, v. 33, no. 5, p. 926-962, text-fig. 1-16.

Cooper, G. A., 1933-34, *Stratigraphy of the Hamilton Group in eastern New York. Pt. 1:* Am. Jour. Sci., 5th ser., v. 26, p. 537-551 (1933); v. 27, p. 1-12 (1934).———1957, *Paleoecology of Middle Devonian of eastern and central United States:* in Paleoecology, H. S. Ladd (ed.), Geol. Soc. America, Mem. 67, v. 2, p. 249-278, text-fig. 1,2, 1 pl.———1968, *Age and correlation of the Tully and Cedar Valley Formations in the United States:* in International symposium on the Devonian System, Calgary, Alberta, Sept. 1967, D. H. Oswald (ed.), v. 2, p. 701-709, Alberta Soc. Petrol. Geologists (Calgary).

———, Butts, Charles, Caster, K. E., Chadwick, G. H., Goldring, Winifred, Kindle, E. M., Kirk, Edwin, Merriam, C. W., Swartz, F. M., Warren, P. S., Warthin, A. S., & Willard, Bradford, 1942, *Correlation of the Devonian sedimentary formations of North America:* Geol. Soc. America, Bull., v. 53, p. 1729-1794, text-fig. 1, pl. 1.

———, Phelan, Thomas, 1966, *Stringocephalus in the Devonian of Indiana:* Smithson. Misc. Coll., v. 151, no. 1 (pub. 4664), p. 1-20, pl. 1-5.

———, & Warthin, A. S., 1942, *New Devonian (Hamilton) correlations:* Geol. Soc. America, Bull., v. 53, p. 873-888.

———, & Williams, J. S., 1935, *Tully Formation of New York:* Geol. Soc. America, Bull., v. 46, p. 781-868, pl. 54-60.

Copeland, M. J., 1962, *Canadian fossil ostracods, Conchostraca, Eurypterida, and Phyllocarida:* Geol. Survey Canada, Bull. 91, 57 p., 12 pl.

Cousminer, H. L., 1964, *Biostratigraphic value of Devonian palynomorphs in medial South America* (abstr.): Geol. Soc. America, Program, 1964 Ann. Meetings, p. 33-34.

Crickmay, C. H., 1966, *Devonian time in western Canada (Article 10):* 38 p., publ. by author, E. de Mille Books (Calgary, Alberta).

Dooge, Jasper, 1966, *The stratigraphy of an Upper Devonian carbonate-shale transition between the north and south Ram Rivers of the Canadian Rocky Mountains:* Leidse Geol. Meded., v. 39, 53 p.

Doumani, G. A., Boardman, R. S., Rowell, A. J., Boucot, A. J., Johnson, J. G., McAlester, A. L., Saul, John, Fisher, D. W., & Miles, R. S., 1965, *Lower Devonian fauna of the Horlick Formation, Ohio Range, Antarctica:* Antarctic Res. Ser., v. 6, p. 241-281, pl. 1-18.

Fåhraeus, L. E., 1971, *Lower Devonian conodonts from the Michelle and Prongs Creek Formations, Yukon Territory:* Jour. Paleontology, v. 45, no. 4, p. 665-683, pl. 77-78.

Friend, P. F., 1961, *The Devonian stratigraphy of north and central Westspitsbergen:* Yorkshire Geol. Soc., Proc., v. 33, p. 77-118.

———, & House, M. R., 1964, *The Devonian Period:* in The Phanerozoic time-scale, W. B. Harland, A. G. Smith, and B. Wilcock (eds.), Geol. Soc. London, suppl. to Quart. Jour. Geol. Soc. London, v. 120S, p. 233-236, 1 fig.

Gryc, George, Dutro, J. T., Jr., Brosgé, W. G., Tailleur, I. L., & Churkin, Michael, Jr., 1968, *Devonian of Alaska:* in International symposium on the Devonian System, Calgary, Alberta, Sept. 1967, D. H. Oswald (ed.), v. 1, p. 703-716, Alberta Soc. Petrol. Geologists (Calgary).

Haller, John, 1971, *Geology of the East Greenland Caledonides:* 413 p., Interscience Publishers, John Wiley & Sons (London, New York, Sydney, Toronto).

Harrington, H. J., 1962, *Paleogeographic development of South America:* Am. Assoc. Petrol. Geol., Bull., v. 46, no. 10, p. 1773-1814.———1968, *Devonian of South America:* in International symposium on the Devonian System, Calgary, Alberta, Sept. 1967, D. H. Oswald (ed.), v. 1, p. 651-671, Alberta Soc. Petrol. Geologists (Calgary).

Harrington, J. W., 1972, *Rhynchonelloid brachiopod zonation of the New York Senecan (Early Upper Devonian):* 24th Internatl. Geol. Congr., sec. 6, Stratigraphy and Sedimentology, Montreal, 1972, p. 278-284, text-fig. 1, pl. 1,2.

Holtedahl, Olaf, 1919, *On the Paleozoic Series of Bear Island, especially on the Heclahook System:* Norsk Geol. Tiddskr., v. 5, p. 121-248.

House, M. R., 1962, *Observations on the ammonoid succession of the North American Devonian:* Jour. Paleontology, v. 36, no. 2, p. 247-284, text-fig. 1-15, pl. 43-48.———1964, *Devonian Northern Hemisphere ammonoid distribution and marine links:* in Problems of palaeoclimatology, A. E. M. Nairn (ed.), p. 262-269, 299-301, Interscience Publishers (New York, London).—— 1965a, *A study in the Tornoceratidae: the succession of Tornoceras and related genera in the North American Devonian:* Royal Soc. London, Philos. Trans., ser. B, v. 250, p. 79-130, pl. 5-11. ——1965b, *Devonian goniatites from Nevada:* Neues Jahrb. Geologie, Paläontologie, Abh., v. 122, 3, p. 337-342, pl. 32.——1967, *Fluctuations in the evolution of Paleozoic invertebrates:* in The fossil record, A symposium with documentation, W. B. Harland, et al. (ed.), Geol. Soc. London, p. 41-54.——1968a, *Devonian ammonoid zonation and correlations between North America and Europe:* in International symposium on the Devonian System, Calgary, Alberta, Sept. 1967, D. H. Oswald (ed.), v. 2, p. 1061-1068, Alberta Soc. Petrol. Geologists (Calgary).—— 1968b, *Continental drift and the Devonian System:* Inaugural lecture, Univ. Hull, 24 p., 4 text-fig.——1973, *Delimitation of the Frasnian:* Acta Geol. Polonica, v. 23, p. 1-14.

——, & **Pedder, A. E. H.**, 1963, *Devonian goniatites and stratigraphical correlations in western Canada:* Palaeontology, v. 6, pt. 3, p. 491-539, pl. 70-77.

Hume, G. S., 1922, *North Nahanni and Root Rivers Area and Caribou Island, Mackenzie River District:* Geol. Survey Canada, Summ. Rept. 1921, pt. B, p. 67-78.

International Union of Geological Sciences, International Geochronological Commission, 1967, *Proposed recommendations for the world-wide geochronological scale: A comparative survey of geochronological scales recently published for the Phanerozoic (Pz-Mz-Cz):* Internatl. Geol. Rev., v. 9, no. 3, p. 323-326.

Jackson, D. E., & **Lenz, A. C.**, 1963, *A new species of Monograptus from the Road River Formation, Yukon:* Palaeontology, v. 6, pt. 4, p. 751-754.

Jarvik, Erik, 1961, *Devonian vertebrates:* in Geology of the Arctic, G. O. Raasch (ed.), v. 1, p. 197-204, Univ. Toronto Press (Toronto).—— 1963, *The fossil vertebrates from East Greenland and their zoological importance:* Experientia, v. 19, p. 284-289.

Johnson, J. G., 1965, *Lower Devonian stratigraphy and correlation, northern Simpson Park Range, Nevada:* Canad. Petrol. Geol., Bull., v. 13, no. 3, p. 365-381, text-fig. 1-4.——1966, *Middle Devonian brachiopods from the Roberts Mountains, central Nevada:* Palaeontology, v. 9, pt. 1, p. 152-181, pl. 23-28.——1967, *Toquimaella, a new genus of karpinskiinid brachiopod:* Jour. Paleontology, v. 41, no. 4, p. 874-880, text-fig. 1,2, pl. 111.——1969, *Some North American rensselandiid brachiopods:* Jour. Paleontology, v. 43, no. 3, p. 829-837, pl. 105-106.——1970a, *Early Middle Devonian brachiopods from central Nevada:* Jour. Paleontology, v. 44, no. 2, p. 252-264, pl. 51-53.——1970b, *Great Basin Lower Devonian Brachiopoda:* Geol. Soc. America, Mem. 121, 421 p., 74 pl.——1970c, *Taghanic onlap and the end of North American Devonian provinciality:* Geol. Soc. America, Bull., v. 81, no. 7, p. 2077-2105, text-fig. 1-6, pl. 1-4.—— 1971a, *Lower Givetian brachiopods from central Nevada:* Jour. Paleontology, v. 45, no. 2, p. 301-326, text-fig. 1-3, pl. 38-46.——1971b, *Timing and coordination of orogenic, epeirogenic, and eustatic events:* Geol. Soc. America, Bull., v. 82, no. 12, p. 3263-3298, text-fig. 1-17.—— 1975, *Devonian brachiopods from the Quadrithyris Zone (Upper Lochkovian), Canadian Arctic Archipelago:* Geol. Survey Canada, Contrib. Canad. Palaeontology, Bull. 235, p. 5-56, text-fig. 2-11, pl. 2-11.

——, & **Boucot, A. J.**, 1968, *Llandovery to Givetian brachiopod zonal sequence in the Silurian and Devonian of central Nevada (abstr.):* Geol. Soc. America, Cordilleran Sec., 64th Ann. Mtg., 1968, p. 69. [Also in Geol. Soc. America, Spec. Paper 121, Abstr. for 1968, p. 518, 1969.] ——1973, *Devonian brachiopods:* in Atlas of palaeobiogeography, A. Hallam (ed.), p. 89-96, text-fig. 1-6, Elsevier Scient. Publ. Co. (Amsterdam, London, New York).

——, ——, & **Murphy, M. A.**, 1968, *Lower Devonian faunal succession in central Nevada:* in International Symposium on the Devonian System, Calgary, Alberta, Sept. 1967, D. H. Oswald (ed.), v. 2, p. 679-691, Alberta Soc. Petrol. Geologists (Calgary).——1973, *Pridolian and early Gedinnian age brachiopods from the Roberts Mountains Formation of central Nevada:* Univ. California, Publ. Geol. Sci., v. 100, 75 p., 30 pl.

——, & **Dasch, E. J.**, 1972, *Origin of the Appalachian faunal province of the Devonian:* Nature Phys. Sci., v. 236, no. 69, p. 125-126.

——, & **Lane, N. G.**, 1969, *Two new Devonian crinoids from central Nevada:* Jour. Paleontology, v. 43, no. 1, p. 69-73, text-fig. 1, pl. 14.

——, & **Murphy M. A.**, 1968, *Age and position of Lower Devonian graptolite zones in the Appalachian standard succession* (abstr.): Geol. Soc. America, Program Abstracts, Ann. Meeting, Nov. 11-13, 1968, Mexico City, Mexico, p. 151. ——1969, *Age and position of Lower Devonian graptolite zones relative to the Appalachian standard succession:* Geol. Soc. America, Bull., v. 80, no. 7, p. 1275-1282, text-fig. 1,2.

Kerr, J. W., McGregor, D. C., & **McLaren, D. J.**, 1965, *An unconformity between Middle and Upper Devonian rocks of Bathurst Island, with comments on Upper Devonian faunas and microfloras of the Parry Islands:* Canad. Petrol. Geol., Bull., v. 13, no. 3, p. 409-431, text-fig. 1,2, pl. 1-4.

Kirchgasser, W. T., 1975, *Revision of Probeloceras Clarke, 1898 and related ammonoids from Up-*

per Devonian of western New York: Jour. Paleontology, v. 49, no. 1, p. 58-90.

Klapper, Gilbert, 1969, *Lower Devonian conodont sequence, Royal Creek, Yukon Territory, and Devon Island, Canada:* Jour Paleontology, v. 43, no. 1, p. 1-27, text-fig. 1-4, pl. 1-6, 1 table.—— 1971, *Sequence within the conodont genus Polygnathus in the New York lower Middle Devonian:* Geologica et Paleontologica, v. 5, p. 59-79, text-fig. 1, pl. 1-3, 5 tables.

——, Sandberg, C. A., Collinson, Charles, Huddle, J. W., Orr, R. W., Rickard, L. V., Schumacher, Dietmar, Seddon, George, & Uyeno, T. T., 1971, *North American Devonian conodont biostratigraphy:* in Symposium on conodont biostratigraphy, W. C. Sweet, & S. M. Bergström (eds.), Geol. Soc. America, Mem. 127, p. 285-316, text-fig. 1-6.

——, & Ziegler, Willi, 1967, *Evolutionary development of Icriodus latericrescens group (Conodonta) in the Devonian of Europe and North America:* Palaeontographica, Abt. A, v. 127, p. 68-83, pl. 8-11.

Klovan, J. E., & Embry, A. F., III, 1971, *Upper Devonian stratigraphy, northeastern Banks Island, N. W. T.:* Canad. Petrol. Geol., Bull., v. 19, no. 4, p. 705-729, text-fig. 1-10, pl. 1-4.

Leanza, A. F., 1968, *Acerca del descubrimiento de Ammonodeos Devonicos en la Republica Argentina:* Asoc. Geol. Argentina, Revista, v. 23, no. 4, p. 326-330.

Lenz, A. C., 1966, *Upper Silurian and Lower Devonian paleontology and correlations, Royal Creek, Yukon Territory: a preliminary report:* Canad. Petrol. Geol., Bull., v. 14, no. 4, p. 604-612.——1968, *Upper Silurian and Lower Devonian biostratigraphy, Royal Creek, Yukon Territory, Canada:* in International Symposium on the Devonian System, Calgary, Alberta, Sept. 1967, D. H. Oswald (ed.), v. 2, p. 587-599, Alberta Soc. Petrol. Geologists (Calgary).——1973, *Quadrithyris Zone (Lower Devonian) near-reef brachiopods from Bathurst Island, Arctic Canada; with a description of a new rhynchonellid brachiopod Franklinella:* Canad. Jour. Earth Sci., v. 10, no. 9, p. 1403-1409, text-fig. 1, pl. 1,2, 1 table.

——, & Jackson, D. E., 1971, *Latest Silurian (Pridolian) and Early Devonian Monograptus of northwestern Canada:* Geol. Survey Canada, Bull. 192, p. 1-21, text-fig. 1-5, pl. 1,2.

——, & Pedder, A. E. H., 1972, *Lower and Middle Paleozoic sediments and paleontology of Royal Creek and Peel River, Yukon, and Powell Creek, N.W.T.:* 24th Internatl. Geol. Congr., Montreal, Quebec, 1972, Guidebook, Excursion A-14, 43 p., 9 text-fig.

Ludvigsen, Rolf, 1970, *Age and fauna of the Michelle Formation, northern Yukon Territory:* Canad. Petrol. Geol., Bull., v. 18, no. 3, p. 407-429, text-fig. 1-5, pl. 1-4.——1972, *Late Early Devonian dacryoconarid tentaculites, northern Yukon Territory:* Canad. Jour. Earth Sci., v. 9, no. 3, p. 297-318, text-fig. 1-6, pl. 1-3.

Maack, Reinhard, 1964, *Characteristic features of the palaeogeography and stratigraphy of the Devonian of Brazil and South Africa:* in Problems of palaeoclimatology, A. E. M. Nairn (ed.), p. 285-293, 301-302, Interscience Publishers (New York, London).

McDougall, I., Compston, W., & Bofinger, V. M., 1966, *Isotopic age determinations of Upper Devonian rocks from Victoria, Australia: a revised estimate for the age of the Devonian-Carboniferous boundary:* Geol. Soc. America, Bull., v. 77, no. 10, p. 1075-1088.

McGregor, D. C., & Uyeno T. T., 1972, *Devonian spores and conodonts of Melville and Bathurst Islands, District of Franklin:* Geol. Survey Canada, Paper 71-13, 36 p., 8 text-fig., 4 pl.

McLaren, D. J., 1954, *Upper Devonian rhynchonellid zones in the Canadian Rocky Mountains:* in Western Canada sedimentary basin—a symposium, L. M. Clark (ed.), Am. Assoc. Petrol. Geologists, Ralph Leslie Rutherford Memorial Volume, p. 159-181, 1 pl. (Tulsa, Okla.).——1962, *Middle and early Upper Devonian rhynchonelloid brachiopods from western Canada:* Geol. Survey Canada, Bull. 86, 122 p., 29 text-fig., pl. 1-18.——1972 (chairman), *Report from the Committee on the Silurian-Devonian Boundary and Stratigraphy to the President of the Commission on Stratigraphy:* Geol. Newsletter, v. 1972, no. 4, p. 268-288.

Maurin, A. F., & Raasch, G. O., 1972, *Early Frasnian stratigraphy, Kakwa-Cecilia Lakes, British Columbia, Canada:* Compagnie Française des Pétroles, Notes et Mémoires, no. 10, p. 1-80, text-fig. 1-22, pl. 1-12.

Miller, A. K., 1938, *Devonian ammonoids of America:* Geol. Soc. America, Spec. Paper 14, 262 p., 39 pl., 41 text-fig.

——, & Collinson, Chas., 1951, *Lower Mississippian ammonoids of Missouri:* Jour. Paleontology, v. 25, no. 4, p. 454-487, pl. 68-71.

Nalivkin, D. V. (general ed.), 1973, *Stratigrafiya nizhnego i srednego Devona. Trudy III Mezh. Simp. po Granitse Siluri i Devona,* 2: 292 p., Akad. Nauk SSSR (Leningrad). [*Stratigraphy of the Lower and Middle Devonian. Transactions 3rd International Symposium of the Silurian-Devonian Boundary, 2.*]

Norris, A. W., 1968a, *Devonian of northern Yukon Territory and adjacent District of Mackenzie:* in International symposium on the Devonian System, Calgary, Alberta, Sept. 1967, D. H. Oswald (ed.), v. 1, p. 753-780, Alberta Soc. Petrol. Geologists (Calgary).——1968b, *Reconnaissance Devonian stratigraphy of northern Yukon Territory and northwestern District of Mackenzie:* Geol. Survey Canada, Paper 67-53, 287 p.

——, & Uyeno, T. T., 1972, *Stratigraphy and*

conodont faunas of Devonian outcrop belts, Manitoba: Geol. Assoc. Canada, Spec. Paper no. 9 (1971), p. 209-223, text-fig. 1-4, pl. 1-3.

Oliver, W. A., Jr., 1968, *Succession of rugose coral faunas in the Lower and Middle Devonian of eastern North America:* in International symposium on the Devonian System, Calgary, Alberta, Sept. 1967, D. H. Oswald (ed.), v. 2, p. 733-744, Alberta Soc. Petrol. Geologists (Calgary).———1975, *Endemism and evolution of Late Silurian to Middle Devonian rugose corals in eastern North America:* in Ancient Cnidaria, B. S. Sokolov (ed.), v. 2, p. 148-159, "Nauka," Siberian Br. (Novosibirsk) [in English only].

———, deWitt, Wallace, Jr., Dennison, J. M., Hoskins, D. M., & Huddle, J. W., 1968, *Devonian of the Appalachian Basin, United States:* in International symposium on the Devonian System, Calgary, Alberta, Sept. 1967, D. H. Oswald (ed.), v. 1, p. 1001-1040, Alberta Soc. Petrol. Geologists (Calgary).

Ormiston, A. R., 1967, *Lower and Middle Devonian trilobites of the Canadian Arctic Islands:* Geol. Survey Canada, Bull. 153, 148 p., 17 pl.———1971, *Lower Devonian trilobites from the Michelle Formation, Yukon Territory:* Geol. Survey Canada, Bull. 192, p. 27-44, text-fig. 1-6, pl. 3,4.———1972, *Lower and Middle Devonian trilobite zoogeography in northern North America:* 24th Internatl. Geol. Congr., Montreal, Canada, 1972, sec. 7, Paleont., p. 594-604, text-fig. 1,2, pl. 1.

Orr, R. W., 1969, *Stratigraphy and correlation of Middle Devonian strata in the Longansport Sag, north-central Indiana:* Indiana Acad. Sci., Proc., 1968, v. 78, p. 333-341.———1971, *Conodonts from Middle Devonian strata of the Michigan Basin:* Indiana Geol. Survey, Bull. 45, p. 1-110, text-fig. 1-4, pl. 1-6.

Pedder, A. E. H., 1975, *Revised megafossil zonation of Middle and lowest Upper Devonian strata, central Mackenzie Valley:* Geol. Survey Canada, Rept. of Activities, April to Oct. 1974, Paper 75-1, Pt. A, p. 571-576.

Perry, D. G., Klapper, Gilbert, & Lenz, A. C., 1974, *Age of the Ogilvie Formation (Devonian), northern Yukon: based primarily on the occurrence of brachiopods and conodonts:* Canad. Jour. Earth Sci., v. 11, no. 8, p. 1055-1097.

Poole, F. G., Baars, D. L., Drewes, Harald, Hayes, P. T., Ketner, K. B., McKee, E. G., Teichert, Curt, & Williams, J. S., 1968, *Devonian of the southwestern United States:* in International symposium on the Devonian System, Calgary, Alberta, Sept. 1967, D. H. Oswald (ed.), v. 1, p. 879-912, Alberta Soc. Petrol. Geologists (Calgary).

Richter, Rudolf, 1941, *Devon, Die Abgrenzung der Formation:* Geol. Jahresb., pt. 3A, p. 31-42.

———, & Richter, Emma, 1942, *Die Trilobiten der Weismes-Schichten am Hohen Venn, mit Bemerkungen über die Malvinocaffrische Provinz:* Senckenbergiana, v. 25, no. 1/3, p. 156-179.

Rickard, L. V., 1964, *Correlation of the Devonian rocks in New York State:* New York State Museum, Sci. Serv. Geol. Survey, Map and Chart Ser. no. 4.———1975, *Correlation of the Silurian and Devonian rocks in New York State:* New York State Museum Sci. Serv., Map and Chart Ser. no. 24.

Sable, E. G., & Dutro, J. T., 1961, *New Devonian and Mississippian Formations in DeLong Mountains, northern Alaska:* Am. Assoc. Petrol. Geol., Bull., v. 45, no. 5, p. 585-593.

Sandberg, C. A., 1961, *Widespread Beartooth Butte Formation of Early Devonian age in Montana and Wyoming and its paleogeographic significance:* Am. Assoc. Petrol. Geol., Bull., v. 45, no. 8, p. 1301-1309.

———, & Mapel, W. J., 1968, *Devonian of northern Rocky Mountains and Plains:* in International symposium on the Devonian System, Calgary, Alberta, Sept. 1967, D. H. Oswald (ed.), v. 1, p. 843-877, Alberta Soc. Petrol. Geologists (Calgary).

Sanford, B. V., & Norris, A. W., 1975, *Devonian stratigraphy of the Hudson platform:* Geol. Survey Canada, Mem. 379, pt. 1, 124 p., 22 text-fig., 19 pl., 21 tab., pt. 2, 248 p.

Sartenaer, P. J. M., 1968, *Famennian rhynchonellid brachiopod genera as a tool for correlation:* in International symposium on the Devonian System, Calgary, Alberta, Sept. 1967, D. H. Oswald (ed.), v. 2, p. 1043-1060, Alberta Soc. Petrol. Geologists (Calgary).———1969, *Late Upper Devonian (Famennian) rhynchonelloid brachiopods from western Canada:* Geol. Survey Canada, Bull. 169, 269 p., 6 fig., 41 text-fig., 19 pl.

Scrutton, C. T., 1973, *Palaeozoic coral faunas from Venezuela, II. Devonian and Carboniferous corals from the Sierra de Perijá:* Brit. Museum (Nat. History) Bull., Geol., v. 23, no. 4, p. 223-281, pl. 1-10.

Shirley, J., 1964, *The distribution of Lower Devonian faunas:* in Problems of palaeoclimatology, A. E. M. Nairn (ed.), p. 255-261, 297-299, Interscience Publishers (New York, London).

Smith, A. G., Briden, J. C., & Drewey, G. E., 1973, *Phanerozoic world maps:* in Organisms and continents through time: a symposium, 1972, N. E. Hughes (ed.), Spec. Paper Palaeontology, no. 12, p. 1-42.

Sweet, W. C., & Miller, A. K., 1956, *Goniatites from the Middle Devonian Columbus Limestone of Ohio:* Jour. Paleontology, v. 30, no. 4, p. 811-817, text-fig. 1, pl. 94.

Teichert, Curt, & Schopf, J. M., 1958, *A Middle or Lower Devonian psilophyte flora from central Arizona and its paleogeographic significance:* Jour. Geology, v. 66, no. 2, p. 208-217.

Uyeno, T. T., 1974, *Conodonts of the Waterways Formation (Upper Devonian) of northeastern and central Alberta:* Geol. Survey Canada, Bull. 232, 93 p., 8 pl.

———, & **Mason, David,** 1975, *New Lower and Middle Devonian conodonts from northern Canada:* Jour. Paleontology, v. 49, no. 4, p. 710-723, text-fig. 1,2, 1 pl.

Warren, P. S., & **Stelck, C. R.,** 1950, *Succession of Devonian faunas in Western Canada:* Royal Soc. Canada, Trans., 3rd ser., v. 44, sec. 4, p. 61-78.

———1956, *Devonian faunas of Western Canada:* Geol. Assoc. Canada, Spec. Paper no. 1, 15 p., 29 pl.

Weeks, L. G., 1947, *Paleogeography of South America:* Am. Assoc. Petrol. Geol., Bull., v. 31, no. 7, p. 1194-1241, text-fig. 1-17.

Wolfart, Reinhard, & **Voges, A.,** 1968, *Beitrage zur Kenntnis des Devons von Bolivien:* Geol. Jahrb., Beihefte 74, 241 p., 10 text-fig., 30 pl., 6 tables.

Woodward, A. S., 1921, *Fish-remains from the Upper Old Red Sandstone of Granite Harbour, Antarctica:* British Antarctic ("Terra Nova") Exped., 1910, Nat. History Rept. Geology, 1 (2), p. 51-62.

Ziegler, Willi, 1971, *Conodont stratigraphy of the European Devonian:* in Symposium on conodont biostratigraphy, W. C. Sweet, & S. M. Bergström (eds.), Geol. Soc. America, Mem. 127, p. 227-284 (1970).

CARBONIFEROUS[1]

By CHARLES A. ROSS
[Western Washington University, Bellingham]

CONTENTS

	PAGE
INTRODUCTION	A254
BIOSTRATIGRAPHIC AND BIOGEOGRAPHIC ANALYSIS	A260
Foraminifers	A261
Corals	A268
Brachiopods	A273
Cephalopods	A275
Conodonts	A279
Insects	A281
BIOGEOGRAPHIC INTERPRETATION	A283
REFERENCES	A286

INTRODUCTION

The name Carboniferous was introduced by CONYBEARE and PHILLIPS (1822) for the English succession that included the Old Red Sandstone, Mountain Limestone, Millstone Grit, and Coal Measures. Later SEDGWICK and MURCHISON (1839) removed the Old Red Sandstone from the Carboniferous to establish the Devonian System. In 1841 MURCHISON established the Permian System in the Province of Perm, Russia, as overlying the Carboniferous. Since then the lower and upper boundaries of the Carboniferous with adjacent systems have presented problems. The Old Red Sandstone (Devonian) of Devonshire is mainly nonmarine, is structurally deformed, and lacks diagnostic fossils of worldwide biostratigraphic value. The overlying Permian (Magnesian Limestone) in England also has few abundant fossils and the Russian Carboniferous and Permian sections are far away and have different faunal and floral assemblages than the Carboniferous and Permian of England. Furthermore, MURCHISON's original definition of the Permian System left some doubt concerning the placement of beds above the Moscovian Stage and below the Artinskian Stage, beds having marine faunas that are not present in the predominantly nonmarine sections in northwestern Europe.

The recognition of lateral changes in facies in the upper part of the Devonian from continental beds in Devonshire into marine strata bearing marine faunas in the lower Rhine valley led to the acceptance of locating a succession in marine strata that could serve as a type section to define the Devonian-Carboniferous boundary. Although the Rhine valley sections contain a variety of cephalopods and conodonts, faunas of the carbonate facies are poorly represented. The Belgium succession includes bioherms and reefs in this interval, and after considerable study the lower boundary was agreed upon by the first *Congrès de stratigraphie et de géologie du Carbonifère,* Heerlen, 1927, as the base of the Tournaisian Stage in Belgium. At the second Heerlen congress (1935) the base was reassigned as the first appearance of a zonal guide fossil, the ammonoid *Gattendorfia subinvoluta,* a useful index fossil in the German strati-

---

[1] Manuscript received October, 1975.

graphic section, which was thought to be equivalent to the Etroeungt beds at the base of the Tournaisian Stage. Later study has demonstrated that the lower beds in the type sections of the Tournaisian include strata older than the zone of *Gattendorfia subinvoluta* and span nearly the entire latest Devonian zone of *Wocklumeria*. The Subcommission on Carboniferous Stratigraphy at the Seventh *Congrès International de stratigraphie et de géologie du Carbonifère* (1971) adopted several working agreements that attempted to resolve this boundary problem. These in effect are as follows:

1. The base of the *Gattendorfia subinvoluta* Zone will continue to serve as the accepted base of the Carboniferous.
2. Because the geographic distribution of the zone name-bearer is restricted, the evolution (first appearance) of the conodont *Siphonodella sulcata* from its Devonian ancestor (which occurs at the same stratigraphic position as the appearance of *G. subinvoluta* in the German succession) will be used on a worldwide basis as the base of the Carboniferous.
3. The designation of a stratotype will be postponed until a search has been made for a suitable, continuous, marine stratigraphic succession, preferably in the region of the original chronostratigraphic units (stages).
4. The boundary of the lowest Carboniferous stage is to be modified to correspond with the lower boundary of the system.

The upper boundary of the Carboniferous has been placed at several positions although there are at least six major recognizable, worldwide, marine, faunal biostratigraphic zones between the top of the Moscovian Stage and the base of the Artinskian Stage. Presently most European and North American biostratigraphers follow the recommendations of the All Soviet Committee on the Carboniferous-Permian boundary (1952) in placing the top of the Carboniferous at the top of the Gshelian Stage. In the past the Carboniferous-Permian boundary has been placed as high as the top of the Sakmarian Stage (or base of the Artinskian Stage) or lower at the top of the Asselian Stage or even within the Asselian. Some biostratigraphers have continued to use the term Permo-Carboniferous for the interval between the Gshelian and Artinskian. Thus, the Autunian Series of Europe and the lower part of the Dunkardian Series and the Wolfcampian Series and even the lowest parts of the Leonardian Series in North America at times have been correlated with the Carboniferous or "Permo-Carboniferous" based on these various boundary positions. The top of the Carboniferous has only infrequently been placed below the top of the Gshelian Stage in the Soviet Union.

Subdivisions.—Three different schemes of subdivision of Carboniferous rocks are extensively used (Fig. 1): the western European classic sections with subdivision into Lower and Upper parts; the southern Urals, Russian platform, and greater Donetz basin sections with subdivision into Lower, Middle, and Upper parts; and the central North American sections recognizing the Mississippian and Pennsylvanian as independent systems. Other sets of subdivisions proposed for various eastern Asia and Gondwana regions probably are equally useful. Correlations between different Carboniferous biogeographic realms and regions are still incomplete and Figure 1 indicates a generally accepted correlation of the three sets of series and stages.

In northwestern Europe the Carboniferous System is divided into two subsystems, the Dinantian below and Silesian above. The Dinantian takes its name from the Dinant area of Belgium where type sections for the mainly carbonate series, Tournaisian (MORTELMANS, 1969) and Visean, are located (GEORGE, 1969). The Silesian takes its name from the coal basins of the Silesian region of Poland. Three series are recognized: the Namurian (at the base), Westphalian, and Stephanian, and the successions are largely nonmarine or paralic with a few marine bands (transgressions) and a number of deeply weathered, flinty underclays (seatearths). In practice the Namurian (HUDSON, 1945; RAMSBOTTOM, 1969) and most of the Westphalian marine bands are correlated with the British succession which has a slightly more complete fauna than the type succession on the European continent (Fig. 1). The Westphalian (CALVER, 1969) is subdivided into four stages (A, B, C, and

Western Europe Type Sections				USSR Reference Sections		North American Reference Sections		North American Type Sections	
Permian	Autunian	B		Sakmarian		Wolfcampian		Dunkardian	
		A		Asselian					
	—280mya—							—--Monongahelan--—	
Carboniferous	Silesian	Stephanian	C	Upper	Orenburgian	Pennsylvanian	Upper	Virgilian	
			B		Gshelian				Conemaughan
			A		Kasimovian			Missourian	
		Cantabrian							
		Westphalian	D	Middle Moscovian	Upper	Myachkovian	Middle	Desmoinesian	Alleghenyan
			C			Podolian			
			—300mya—						
			B		Lower	Kashirian			Pottsvillean
			A			Verean		Atokan	Atoka Fm.
	Namurian		C	Bashkirian		Melekesian	Lower Morrowan	Bloyd Shale	
						Cheremshanian			
						Prikamian			
		B	Marsdenian			Severokeltmenian		Hale Fm.	
			Kinderscoutian			Krasnopolyanian			
			—320mya—						
			Alportian	Serpukhovian		Protvinian			hiatus
		A	Chokierian			Stesheyian		Imo Sh.	Chesterian
			Arnsbergian			Tarusian		Pitkin Fm.	Elvira Gr. Grove Church Kinkaid
			Pendleian					Fayetteville Sh.	Homberg Gr. Glen Dean Ls.
	Dinantian	Visean	C superior	Lower Visean		Okian	Mississippian	Batesville Ss.	Gasper Gr.
			V₃ inferior					Ruddell Sh.	Meramecian Ste. Genevieve
			B						St. Louis Ls.
			A			Yasnopolyanian		Moorefield Sh.	Salem Ls.
			—330mya—						Warsaw Ls.
			V₂ B						Keokuk Ls.
			A			Malinovian			Osagean Burlington Ls.
			V₁ B					Boone Fm.	Fern Glen Fm.
			A						"Sedalia" Ls.
		Tournaisian	Tn₃ C	Tournaisian		Chernyshinian			Chouteau Ls.
			B					Walls Ferry Ls.	
			A						Kinderhookian Hannibal Fm.
			Tn₂ C						
			B			Likhvinian			
			A						"Glen Park" Ls.
			—345mya—						
			Tn₁₍b₎						
Devonian	"Etroeungt"							Louisiana Ls.	
	Famennian							Devonian	

Fig. 1. Comparison of Carboniferous stratigraphic terminology in western Europe, European USSR and North America, and correlation of boundaries. Biostratigraphic zone fossils for ammonoids, foraminiferids, brachiopods, corals, and megaflora indicate approximate relationship between zonations based on different

Carboniferous

Ammonoid Zones		Foraminiferid Zones	Brachiopod Zones	Coral Zones	Megafloral Zones	
Arkansas	Europe					
Properrinites				Proto-wentzelella & Diphystrotion	Callipteris spp.	
	Propopanoceras	Pseudoschwagerina			Danaeites spp.	
Vidrioceras Schistoceras Agathiceras Uddenites Emilites	Shumardites Artinskia Daixites	Daixina	Lissochonetes & Chonetes transversalis	Tschus-sovskenia, Eolitho-strotionella, Gshelia, Timania, Bothro-phyllum	Odontopteris spp.	Lescuropteris spp.
	Neo-dimorphoceras	Triticites				
Eothalasoceras & Prouddenites	Paraschistoceras Eoschistoceras	Monti-parus	Kansan-ella	Chonetina	Lophophyllidium	
Wellerites		Fusulinella & Fusulina	Mesolobus mesolobus & Marginifera muricatina	Cystophora & Arachnastraea	Chaetetes & Provincial species of caniniids and amygdalophylloids	Neuropteris flexuosa
Owenoceras	Politoceras	Beedeina				Neuropteris rarinervis
Paralegoceras texanum	Anthracoceras	Fusulinella	Mesolobus striatus			Neuropteris tenuifolia
Winslowoceras henbesti	G₂	Profusulinella		Marginifera haydenensis & M. nana		Megalopteris spp.
Diaboloceras varicostatum	Gastrioceras					
D. neumeieri	G₁	Eoschu-bertella	Hustedia miseri			N. tennesseeana Mariopteris pygmea
Axinolobus modulus	R₂					M. pottsvillea & Aneimites
Branner. branneri Arkanites relictus	Reticuloceras	Pseudo-staff-ella	Miller-ella			N. pocahontas & M. eremopteroides
Reticul. tiro	R₁					
	H₂ Homoceras H₁					Fryopsis spp. & Sphenopteridium spp.
Cravenoceras miseri C. involutum C. richardsonianum	E₂ Eumorphoceras	Eosigmoilina Zellerina Asteroarchaediscus	Composita subquadrata Spirifer increbes-cens	Clisiophyllum		
Tumulites varians	E₁					A break in zonation because of a general lack of plant fossils in this interval
Goniatites granosus	P₂	Neoarchaediscus Bradyina	Productus cestriensis	Palaeosmilia, Lonsdaleia, Lithostrotion, Lithostrotionella, Arachnolasma		
Goniatites multiliratus	Goniatites					
Goniatites americanus	P₁	Archaediscus Endothyranopsis	Othotetes kaskaskiensis			
	B₂					
	Pe δ					
Beyrichoceras hornerae	Beyrichoceras		Brachythyris subcardiformis			
	Pe γ Pericyclus B₁	Dainella & Eoparastaffella	Syringothyris textus			
Ammo. ballardensis	Pe β	Spinoendothyra & Latiendothyra	Orthotetes keokuk Brachythyris suborbicularis	Enygmophyllum, Keyserlingophyllum Uralinia, Amygdalophyllum, Lithostrotion, Caninia, Amplexus, Cyathoxonia, Lithostrotionella Zaphrentites	"Lepidodendropsis"	Triphyl-lopteris
cf. Mero. drostei	Pe α					
Muenst. pfefferae Muensteroceras arkansanum						
		Septabrunsiina & Chernyshinella	Leptaena analoga Brachythyris peculiaris			Adiantites spp.
	Gattendorfia subinvoluta	Quasiendothyra				
	Wocklumeria		Productella pyxidata Schuchertella lens Spirifer marionensis			
	Clymenia					

FIG. 1. *(Continued from facing page.)* groups. See Figure 7 for ranges of stratigraphically important genera of conodonts. The megafloral zones are for the Euramerian floral region (modified from Read & Mamay, 1964).

D) by three marine bands and some differences in floral content. The Stephanian has no marine bands in its type area where it is subdivided into three stages (A, B, and C) based on floral differences. The Upper Westphalian and Lower Stephanian were separated by the International Subcommission on Carboniferous Stratigraphy (1967) by using Tonstein 60 as the boundary in the Saar-Lorraine area. A floral succession that either spans Late Westphalian-Early Stephanian time or fills a gap between them has been found in northern Spain and named the Cantabrian Stage (WAGNER, 1969). At present the type section of the Stephanian lies in the coal basin of St. Étienne in Loire and passes transitionally upward into beds having fossil floras of Autunian (Early Permian) appearance (LIABEUF & ALPERN, 1969). Correlations of the Westphalian and Stephanian are based mainly on the distribution of plant leaves, stems, and spores and become complicated by ecological differences in floras between basins of approximately the same age, but in different environmental settings (i.e., elevation, rainfall) (BOUROZ, 1969; DE MAISTRE, 1969).

The North American succession of strata equivalent to the Carboniferous is separated into two local systems, the Mississippian below and the Pennsylvanian above. The Mississippian System, named for a sequence of predominantly limestone, sandstone and shale exposed along the Mississippi valley in Missouri, Illinois, and Iowa, is based on the "Mississippi Group" (WINCHELL, 1869), between the Devonian and the "Coal Measures." The "Coal Measures" were renamed the Pennsylvania Series (WILLIAMS, 1891) for exposures in Pennsylvania. Both system names gained wide usage in North American textbooks early in this century. The type Mississippian strata (Fig. 1) have a clastic and limestone portion at their base (WELLER et al., 1948), the Kinderhookian Series (the original definition included some latest Devonian beds), which represents a shallow-water facies with reworked faunas in part. The mainly carbonate succession is divided into the Osagian Series, which is mainly middle and late Tournaisian, and the Meramecian Series, which includes all except latest Visean equivalents. These two carbonate series are locally dominated by echinoderm faunas that, in general, are not a typical lithofacies and this has made correlation outside the type region difficult. The upper part of the carbonate sequence grades laterally into the basal part of the Chesterian Series. The Chesterian has fifteen or so sequences of transgressive and regressive sediments that are terminated by a major regional unconformity.

The type Pennsylvanian (BRANSON, 1962) is in southwestern Pennsylvania in strata that are predominantly nonmarine with a few marine beds. Because of uncertainly in correlating the strata of the type area with marine and paralic strata in much of the rest of North America, a set of standard reference sections was established in the southern Midcontinent region (in Arkansas, Missouri, Oklahoma, Kansas, and Iowa) (MOORE et al., 1944). These reference sections form the basis for five standard series, in ascending order: Morrowan, Atokan, Desmoinesian, Missourian, and Virgilian. The lower two, Morrowan and Atokan, are located near the southern edge of the Early Pennsylvanian North American craton and are geosynclinal sequences deposited at a time of active tectonic deformation of the main part of the geosyncline to the south (GORDON, 1974). The type section of the Desmoinesian was deposited to the north in Iowa, well on the North American craton, as were the Missourian and Virgilian series in Missouri and Kansas. These higher three series include evidence of numerous transgressions and regressions and complications of repeated cyclothems (WANLESS, 1969). On the cratonic shelf, low tectonic arches and shallow basins (Illinois, Michigan, and Forest City basins) caused sedimentary units to thin and disappear or thicken and change facies (WANLESS, 1969). These standard sections are marine and paralic and are separated from areas to the southwest by block-faulted uplifts and basins and from areas west of the Southern Rockies by the transcontinental arch.

The third set of sections that are used as references for the Carboniferous are those on the Russian platform (DALMATSKAYA et al., 1961), southern Urals, and greater Donetz basin (AYZENVERG, 1969). In these sections the Tournaisian and Visean strata and faunas are similar to those in the Belgian type area and no major correlation problems are apparent. The "Namurian" on the Russian platform, which equates to only the lower part of the Namurian of the Belgium section, was renamed the Serpukhovian Stage by EYNOR in 1970. The Tournaisian, Visean, and Serpukhovian stages are included by Soviet biostratigraphers in their Lower Carboniferous.

A major unconformity separates these Lower Carboniferous units from higher Carboniferous units on the Russian platform (BARKHATOVA et al., 1970); however, the sections near the margins of the geosynclines (Southern Urals) and sedimentary troughs (Donetz basin) have a succession, called the Bashkirian Stage, which marks the base of the Middle Carboniferous. The Bashkirian is overlain by the Moscovian Stage, which covers most of the Russian platform and forms the upper part of the Middle Carboniferous. The Upper Carboniferous, Kasmovian and Gshelian stages, is present on much of the Russian platform where it is overlain by the Asselian Stage (Lower Permian).

As revised in 1974, the Lower Carboniferous of the Russian platform is correlated with the North American Mississippian; the Middle Carboniferous is correlated with the Morrowan, Atokan, and Desmoinesian; and the Upper Carboniferous with the Missourian and Virgilian. Most marine Carboniferous sequences on cratonic shelves and in geosynclines are adaptable to this three-fold subdivision. In the succeeding discussions in this paper, Lower, Middle, and Upper Carboniferous will be used as outlined above.

The Carboniferous histories of Australia, peninsular India, South America east of the Andes, and Antarctica are generally similar, although not as closely similar as their Permian and Triassic histories.

Two regions of deposition were present in Australia during the Carboniferous: an eastern region along the Tasman geosyncline, which was divided into a number of structurally delimited basins; and a western region which included the Carnarvon, Canning and Bonaparte Gulf basins. In the eastern region seven faunal zones are recognizable in the Lower Carboniferous succession (HILL, 1957; CAMPBELL & McKELLAR, 1969; CAMPBELL et al., 1969; ROBERTS, 1975), based primarily on coral and brachiopod ranges. Above these seven faunal zones, one brachiopod zone is recognized in Middle Carboniferous strata and another zone is identified in what are probably Upper Carboniferous strata. These two highest zones are interbedded with rocks having the *Rhacopteris* flora.

The zonation in the eastern region is not applicable to the western basins where different brachiopod zones and conodont zones are used for the Lower Carboniferous succession. Middle and Upper Carboniferous strata in the western region are poorly fossiliferous and pass upward without much depositional change into beds of probably Early Permian age bearing *Eurydesma*.

The eastern Australian Lower Carboniferous includes conglomerate and sandstone beds at its base and a succession of shale with a few limestone beds. The Middle and Upper Carboniferous are mainly clastic rocks and volcanics and include conglomerates, tuffs, flows, tillites, diamictite with a few limestones and marine and nonmarine shales. The Tournaisian and early Visean corals show similarity to those of southeastern Asia, but later Visean and early Namurian corals show considerable endemism (HILL, 1948, 1957, 1973; CAMPBELL & McKELLAR, 1969). This has been interpreted as a gradual change from warm or temperate marine conditions to cooler conditions before the end of the Early Carboniferous. The two marine zones within the predominantly clastic and volcanic beds of Middle and Late Carboniferous age having interbedded tillites and glacial marine beds represent possible times of climatic warming in an otherwise cold interval. The Early Permian *Eurydesma* fauna is closely associated with glacial deposits and is considered to be a cold-water assemblage.

In eastern South America, in Brazil, Argentina, Paraguay, and Uruguay, Lower Carboniferous, and possibly Middle Carboniferous, strata are separated from Upper Carboniferous and Permian strata by a major regional unconformity that forms the sub-Gondwana surface (CASTER, 1952). The relationship between the Lower and Middle Carboniferous of Argentina suggests that a hiatus of some magnitude exists between them. The Lower Carboniferous marine faunas are mainly conulariids, brachiopods, and bivalves (AMOS & SABATTINI, 1969) and are not completely studied. The Middle Carboniferous faunas contain brachiopods, bryozoans, and trilobites, which are similar to Australian faunas of this age. These faunas also include the cephalopods *Anthracoceras* and *Eoasianites*, which are similar to those of the Desmoinesian of North America. Strata of probable Late Carboniferous age in Argentina have plants that may be Stephanian and locally marine faunas of brachiopods, such as *Cancrinella, Crurithyris, Tornquistia, Spirifer,* and *Alispirifer*. The lateral relationships of these Argentinian faunas to nearby glacial deposits remain unanswered; however, *Eoasianites* and other faunas and floras of probable late Middle Carboniferous age are known from beds within the Tubarão glacial beds in Brazil and Uruguay. The coals of the Paraná basin in Brazil are in interglacial deposits that lie between glacial beds and are probably Carboniferous in age rather than Permian; however, the upper part of the tillite succession may be Early Permian in age if the occurrence of *Eurydesma* is a reliable guide.

In the Cape region of South Africa, the Lower Dwyka Shale contains a sparse flora similar to the *Rhacopteris* flora in Middle or Upper Carboniferous strata of eastern Australia. Most of the overlying tillite of the Dwyka is probably Late Carboniferous in age. The *Eurydesma* fauna occurs in mudstone and shale above the tillites of the Dwyka and is considered to be of Early Permian age.

The Carboniferous in other Gondwanan continents or continental fragments is poorly understood. In Madagascar the Carboniferous, if present, seems to be represented by a thin tillite, and in the Falkland Islands, the succession is remarkably similar to that in South America and South Africa. The Lower? Carboniferous contains plant fossils and is overlain by tillites and clastic beds interbedded with marine units of probable Middle and Late Carboniferous age. In Antarctica, Devonian strata having fish remains are unconformably overlain by tillite and coal-bearing sandstone that are considered to be Permian based on the presence of *Glossopteris* without non-Gondwanan plants.

BIOSTRATIGRAPHIC AND BIOGEOGRAPHIC ANALYSIS

The Carboniferous Period, estimated to extend from 345 mya to 280 mya (FRANCIS & WOODLAND, 1964), had a duration of 65 my, which approximates the length of each of the four other long geologic periods, the Cambrian, Ordovician, Cretaceous, and Tertiary. During this long interval of time, distributions of marine invertebrate faunas were strongly influenced by changes in climate; in seaways, their interconnections, and sea levels; in the relative position of land areas; and in tectonic, orogenic, and epeirogenic changes. The details of how these physical changes modified the Carboniferous physical environment are much less thoroughly known that is the actual distribution of most of the fossil groups representing part of the biological environment. At present the mechanisms of the dispersals of faunas between different realms and regions during the Carboniferous are not well understood and require much more study. These dispersals appear to reflect the different threshhold levels of each faunal group to dispersal filters that accompanied changes in paleogeographic climate and environments during the period. This was a period of widespread crustal deformation as well as sea-level and climatic fluctuations, even unconformities on the more stable parts of continental masses are commonly irregular in duration and extent. Later crustal movements have partially obscured Carboniferous tectonic settings. In addition, Carboniferous biostratigraphers started only in the late 1960's to reconsider distributions of pre-Mesozoic faunas and floras in the context of plate tectonic models.

The biogeographic distribution of many invertebrates during the Carboniferous Period (EYNOR et al., 1973; Ross, 1973) is marked by times of widespread dispersals of nearly cosmopolitan faunas but this is contrasted at other times by well-developed provincialism with many endemic genera and families. These distributional patterns have been studied for Protozoa (Foraminiferida), Coelenterata (Rugosa), Brachiopoda (Articulata), Mollusca (Cephalopoda), and Arthropoda (Insecta). Although many questions remain unanswered, a general pattern for marine invertebrates shows increasing provincialism after the Visean. Provincialism was particularly strong by the beginning of the Middle Carboniferous followed by some decrease in the levels of provincialism near the end of Middle Carboniferous (Ross, 1970).

During parts of the Carboniferous, strongly differentiated provincial faunas result in a number of biostratigraphic problems. Within individual biogeographic realms (or regions) correlation is based largely on successful and abundant endemic faunas. Therefore, correlation from one biogeographic realm (or region) to the next is generally more difficult and less certain than within a particular realm because distinctive genera and families may have considerably different stratigraphic and geographic ranges in different realms. At several times during the Carboniferous, incomplete faunal interchange between provinces took place and resulted in the introduction of new lineages into another province and often the extinction, at least within several provinces, of another lineage, presumably by ecologic replacement. The biogeographic distribution of other invertebrate groups is less thoroughly studied, but available data suggest that they also may have a similar pattern. Terrestrial insect distributions are closely comparable to those recorded for plant fossils.

The biostratigraphic distribution of Carboniferous fossils is well known for a number of groups that are extensively used as zonal index fossils. In marine carbonate strata protozoans (Foraminiferida), corals (Rugosa), and brachiopods are commonly abundant and are widely used for correlation because of their rapid evolution or distinctive morphological features or both. Shaly marine sequences may have large numbers of mollusks (Cephalopoda) which, because of their rapid evolution, are valuable guide fossils. Conodonts occur in most Carboniferous marine strata and have been particularly thoroughly studied from Lower Carboniferous strata.

FORAMINIFERS

This group of microfossils is widely used in correlating marine strata, particularly the calcareous foraminifers that are commonly abundant in limestone. A large proportion of the foraminifers used in interregional correlation are nearly cosmopolitan and are the basis for a worldwide correlation scheme (Fig. 2, 3), although numerous species and genera have endemic distributions.

In summarizing the Lower Carboniferous foraminiferal zonation, data were compiled from the publications of RAUZER-CHERNOUSOVA (1948), REYTLINGER (1960, 1962, 1963, 1969), LIPINA (1963, 1964, 1965, 1970, 1973) MAMET (1962, 1968), MAMET and SKIPP (1971), SANDO, MAMET, and DUTRO (1969), CONIL (1964), CONIL and LYS (1968), and CONIL, PAPROTH, and LYS (1968). In the lower Tournaisian (Zones 6 and 7) the earliest Carboniferous foraminiferal zone is identified as Zone 6, which is characterized in Eurasia by an abundance of Quasiendothyridae, which are absent in most of North America. Also present in both regions are species of *Septaglomospiranella*, *Latiendothyra*, and *Earlandia*. This fauna is not diverse and a number of the species are endemic. The next younger Tournaisian foraminiferal zone (Zone 7) has nearly uniform faunas in Eurasia and North America. It is characterized by many genera of the Tournayellidae. Especially abundant are individuals of *Septaglomospiranella* and *Palaeospiroplectammina*, and the first appearance of *Tuberendothyra* of the Endothyridae.

The upper part of the Tournaisian has two zones, the lower, Zone 8, contains the first appearance of *Spinoseptatournayella* and *Spinoendothyra* and an abundance of *Tuberendothyra*. The upper zone, Zone 9, is characterized by an abundance of *Tuberendothyra*, as well as *Spinoendothyra*, and shows a continued increase in faunal diversity. Some genera, which are common to abundant in Eurasia, such as *Brunsia*, *Urbanella*, *Glomospiranella*, and others, are rare or lacking in North America.

Zone 10 is of earliest Visean age. The important faunal elements in North America are *Tournayella*, *Eoforschia* of the group

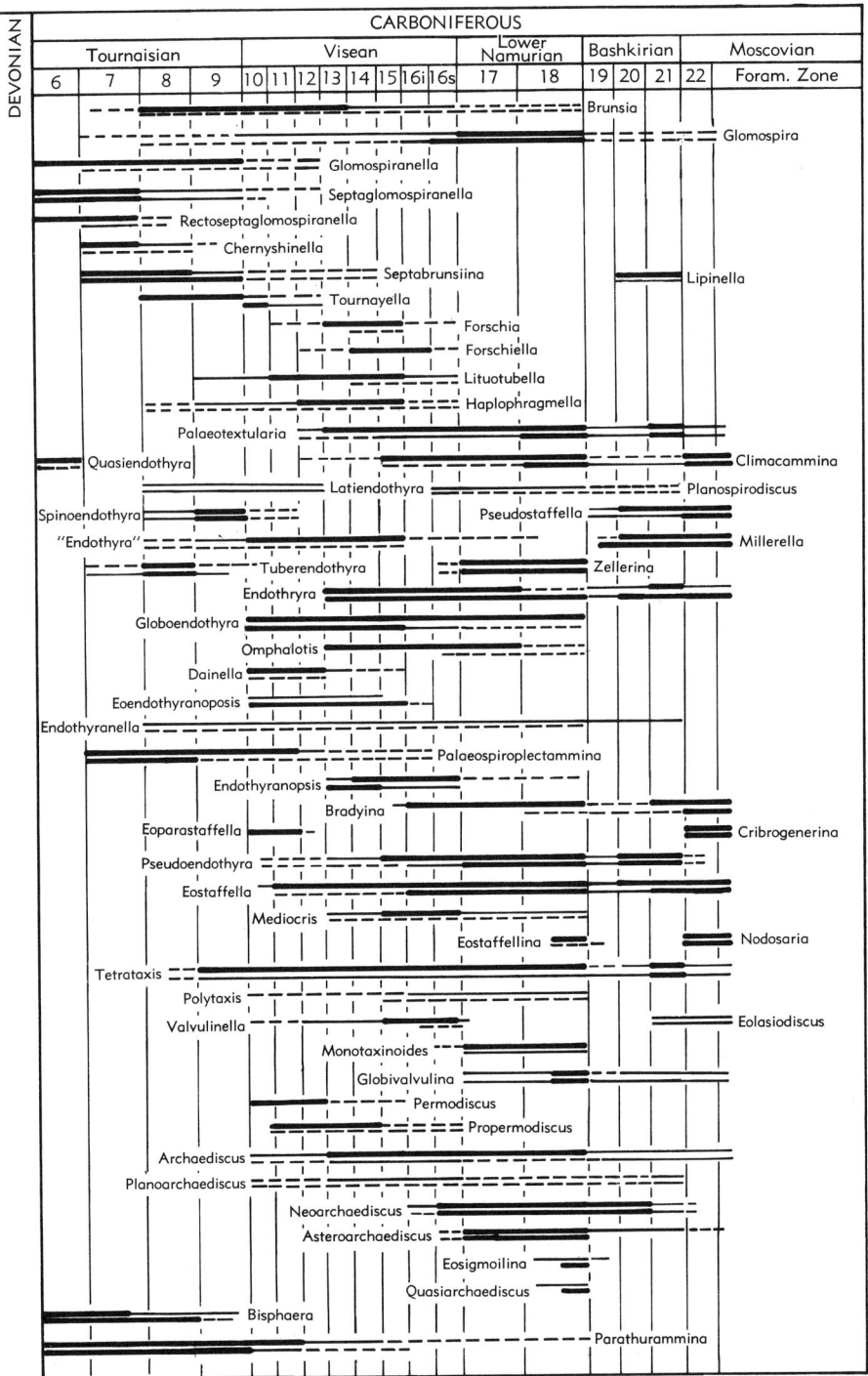

Fig. 2. Stratigraphic range of important genera of Foraminiferida in the Lower and part of Middle Carboniferous. The upper line indicates the range in Europe and the lower line the range in North

E. moelleri, Spinoendothyra (sparse), *Calcisphaera, Pachysphaerica, Tetrataxis, Globoendothyra baileyi, Eoendothyranopsis, Propermodiscus, Archaediscus* of the group *A. kerstovnikovi,* and *Planoarchaediscus.* In Eurasia *Brunsia, Endothyra* of the group *E. prisca, Latiendothyra, Dainella, Eoparastaffella,* primitive *Eostaffella, Permodiscus, Urbanella,* and *Lituotubella* are common but are unreported or sparse in North America.

Zone 11 is late early Visean in age. The important genera in North America are *Tournayella, Eoforschia, Eoforschia* of the group *E. moelleri, Endothyra* of the group *E. prisca, Globoendothyra,* and *Eoendothyranopsis* of the group *E. spiroides, Tetrataxis, Archaediscus* of the group *A. krestovnikovi* and *Planoarchaediscus.* In Eurasia *Forschia* (first appearance), *Lituotubella, Eotextularia, Latiendothyra, Dainella, Eoparastaffella, Polytaxis, Valvulinella* (rare), *Permodiscus,* and *Archaediscus* of the group *A. moelleri* are important parts of this fauna but are sparse or unreported in North America.

Zone 12 is of early middle Visean age. Characteristic genera of Zone 12 in North America are *Tournayella, Eoforschia* of the group *E. moelleri, Palaeotextularia* of the group *P. consobrina, Globoendothyra, Eoendothyranopsis* of the group *E. spiroides, Tetrataxis, Propermodiscus, Archaediscus* of the group *A. krestovnikovi,* and *Planoarchaediscus.* In Eurasia, diagnostic faunal constituents are: *Forschia, Forschiella* (first appearance), *Lituotubella, Haplophragmella, Eotextularia, Cribrostomum, Climacammina* of the group *C. prisca, Endothyra* of the group *E. prisca, Dainella, Eostaffella, Vissariotaxis* (first appearance), *Permodiscus* and *Archaediscus* of the group *A. chernoussovensis* (first appearance).

Late middle Visean (Zone 13) foraminifers include: *Eoforschia, Paleotextularia* of the group *P. consobrina, Climacammina* of the group *C. prisca, Globoendothyra, Eoendothyranopsis, Tetrataxis, Propermodiscus, Archaediscus* of the group *A. krestovnikovi, Archaediscus* of the group *A. moelleri, Archaediscus* of the group *A.* *chernoussovensis* and *Planoarchaediscus,* and the following first appear: *Endothyra* of the group *E. bowmani, Globoendothyra* of the group *G. tomiliensis, Eoendothyranopsis* of the group *E. rarus,* and *Endothyranopsis* of the group *Endothyranopsis compressus.* In Eurasia, important genera include *Forschia, Lituotubella, Haplophragmella, Cribrostomum, Climacammina* of the group *C. prisca, Endothyra* of the group *E. prisca, Dainella, Eostaffella, Valvulinella, Vissariotaxis, Omphalotis, Endostaffella, Endothyra* of the group *E. pauciseptata, Janischewskina,* and *Mediocris* make their first appearance.

Early late Visean foraminifers (Zone 14) in North America are found in the uppermost part of the St. Louis Limestone and the lowermost part of the Ste. Genevieve Limestone and include most of the faunal elements listed under Zone 13 with the addition of *Cribrostomum, Eoendothyranopsis utahensis, Eoendothyranopsis macrus, Mikhailovella,* and *Cribrospira.* Rare specimens of *Forschia* and *Lituotubella* are recognized in the northernmost Cordillera. In Eurasia also the faunas of Zones 13 and 14 are very similar.

Middle upper Visean (Ste. Genevieve in part) (Zone 15) in North America is characterized by abundant *Palaeotextularia* of the group *P. consobrina, Cribrostomum, Endothyra* of the group *E. bowmani, Globoendothyra, Eoendothyranopsis utahensis, Tetrataxis* and the first appearance of *Climacammina* of the group *C. prisca, Palaeotextularia* of the group *P. longiseptata, Climacammina* of the group *C. patula,* and *Endothyranopsis crassus.* In Eurasia, *Forschia, Forshiella, Haplophragmella, Endothyra* of the group *E. prisca, Endothyranopsis compressus, Pseudoendothyra, Eostaffella, Mediocris, Valvulinella, Vissariotaxis, Archaediscus* of several groups, and *Planoarchaediscus karreri* are important.

Late Visean Zone 16i (early Chesterian) is characterized by common to abundant *Cornuspira, Palaeotextularia* of the group *P. consobrina, Climacammina* of the group *C.*

FIG. 2. *(Continued from facing page.)* America (mainly western part). [Thick lines indicate abundant, thin lines common, and dashed lines rare or questionable occurrences. Numbered foraminiferal zones and data from Mamet & Skipp (1971); Ross, n.]

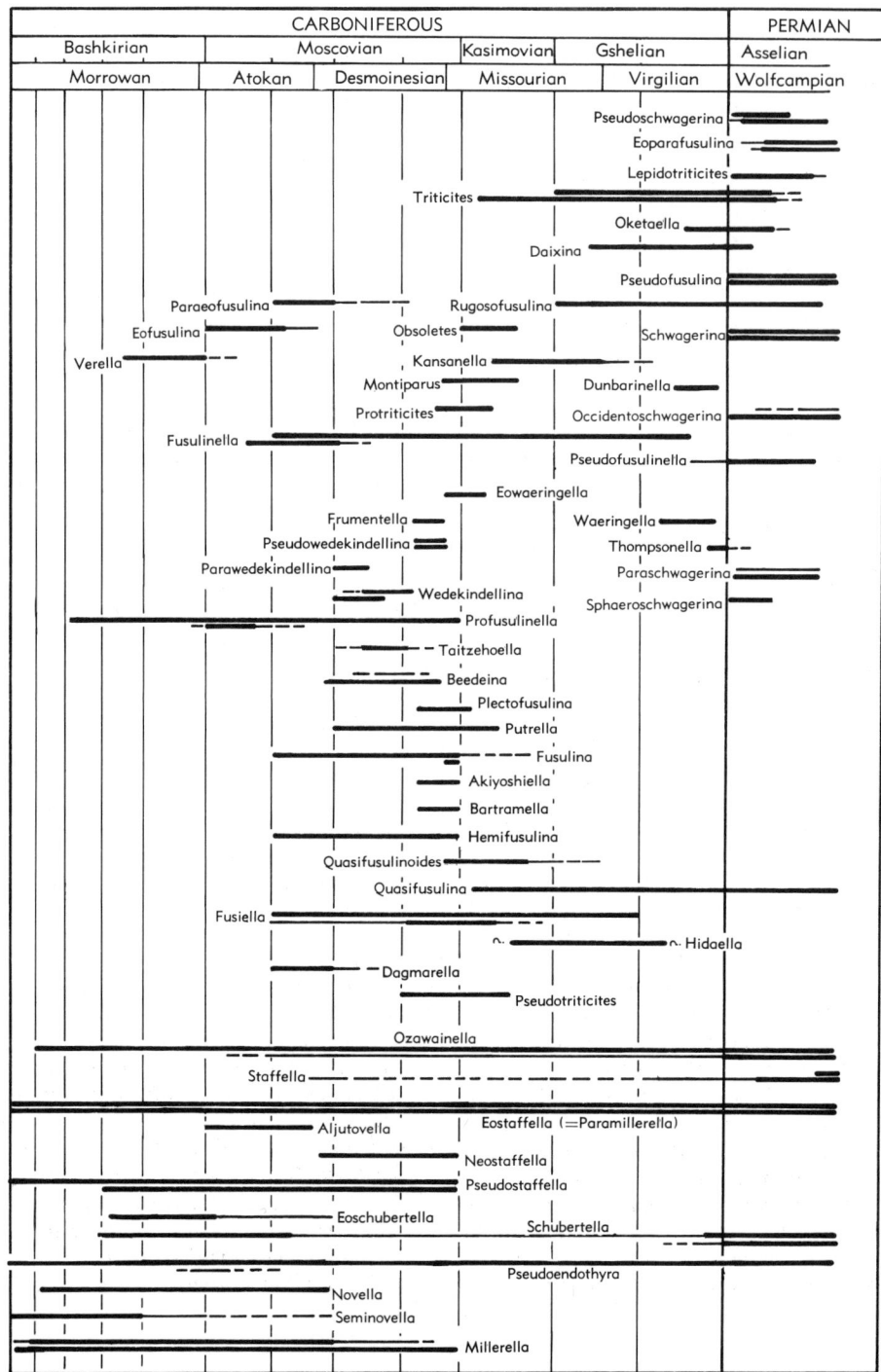

Fig. 3. Range of Middle and Late Carboniferous fusulinacean genera in Arctic-Eurasian province (upper line) and in North American Midcontinent province (lower line) (data from Ayzenverg and others, 1969;

prisca, *Palaeotextularia* of the group *P. longiseptata*, *Climacammina* of the group *C. patula*, *Endothyra bowmani*, *Endothyranopsis crassus*, *Globoendothyra globulus*, *Pseudoendothyra*, *Eostaffella* (= *Paramillerella*), *Tetrataxis*, *Archaediscus* of the group *A. krestovnikovi*, *Archaediscus* of the group *A. moelleri*, *Archaediscus* of the group *A. chernoussovensis*, and *Stacheia*. *Archaediscus* of the group *A. latispiralis* and *Neoarchaediscus* appear for the first time. The Eurasian fauna is similar but differs in retaining *Forschia* (sparse,), *Forschiella* (sparse), *Lituotubella* (sparse), *Haplophragmella*, *Cribrostomum*, *Endothyra* of the group *E. prisca*, *Globoendothyra*, *Omphalotis*, *Cribrospira*, *Bradyina*, and *Mediocris* as important elements.

Latest Visean foraminifers (Zone 16s) in North America are similar to those of the preceding zone (Zone 16i) but include, in addition, the first appearance of *Zellerina* and *Asteroarchaediscus* (very scarce). *Glomospira*, *Planospirodiscus*, and *Neoarchaediscus* become abundant faunal elements. In Eurasia, the fauna is much the same, but primitive ancestral *Loeblichia*? appears. This is the highest zone having *Lituotubella* (very sparse), *Haplophragmella* (sparse), *Cribrostomum*, *Globoendothyra* of the group *G. tomiliensis*, *Endothyranopsis compressus* (sparse), *Janischewskina*, *Valvulinella*, *Howchinia*, *Propermodiscus*, and *Planoarchaediscus* and marks the top of the Visean.

The lowest Namurian foraminifers (Zone 17) occur in the Glen Dean Limestone (Chesterian Series) in North American and include abundant coiled unichambered calcareous Foraminiferida, such as *Pseudoglomospira*, *Trepeilopsis*, and *Palaeotextularia* of the group *P. consobrina*. *Climacammina* of the group *C. prisca*, *Palaeotextularia* of the group *P. longiseptata*, *Climacammina* of the group *C. patula*, *Endothyra* of the group *E. bowmani*, *Pseudoendothyra*, *Eostaffella*, *Tetrataxis*, *Archaediscus* of the group *A. krestovnikovi*, *Archaediscus* of the group *A. chernoussovensis*, *Archaediscus* of the group *A. latispiralis*, *Neoarchaediscus*, *Earlandia*, *Calcisphaera* of the group *C. laevis*, *Stacheia*, and *Tuberitina* remain important faunal elements. To these are added *Monotaxinoides*, *Vissariotaxis*, primitive "*Globivalvulina*," *Mediocris* (very rare), and abundant specimens of *Asteroarchaediscus*. The Eurasian fauna is similar, differing largely in being more abundant and in possessing a few elements such as *Bradyina* and *Loeblichia*, which have not been observed in faunas of this age in North America.

Foraminifers in upper Lower Namurian strata (upper part of the Chesterian) (Zones 18-19) in North America contain most of the faunal elements of Zone 17 and also include the first appearances of highly evolved *Bradyina*, *Pseudoendothyra* of the group *P. kremenskensis*, *Eostaffellina*, "*Globivalvulina*" of the group *G. parva*, *Eosigmoilina*, and *Quasiarchaediscus*. In these zones Eurasian faunas have a few elements, such as *Loeblichia*, which have not been reported from North America. In the Donetz basin and in many other parts of Eurasia, Namurian foraminifers of the *Homoceras* Zone (Zone 19) are characterized by the transitional archaediscid-miliolid *Quasiarchaediscus-Eosigmoilina*? fauna. Although beds of this age are missing in the type Mississippian section, this foraminiferal fauna is reported from the North American Cordillera (MAMET, SKIPP, SANDO, & MAPEL, 1971).

Foraminiferal zones in the Morrowan of the North America Midcontinent region and the Bashkirian of the Donetz basin (Fig. 3) are particularly diverse and differ in faunal composition. The Morrowan is dominated by *Millerella*, *Eoschubertella*, and trilayered *Globivalvulina*, and the Bashkirian has abundant *Bradyina*, *Archaediscus*, *Asteroarchaediscus*, *Neoarchaediscus*, *Pseudoendothyra*, *Eostaffella*, *Novella* (primitive), *Pseudostaffella*, *Schubertella*, *Seminovella*, *Verella*, *Eofusulina*, *Ozawainella*, and *Profusulinella*. The early Bashkirian in the Donetz basin and southern Ural Mountains can be divided on the basis of fusulinaceans

FIG. 3. *(Continued from facing page.)*
Rauzer-Chernousova and others, 1951; Ross, 1967, 1970, 1973; Rozovskaya, 1950, 1969; Thompson, 1948, 1966, 1967; Ross, n). [Thick lines indicate abundant, thin lines common, and dashed lines rare or questionable occurrences.]

into a lower zone having the first appearance of *Pseudostaffella, Seminovella,* and *Millerella,* and a higher zone having the first appearance of *Novella* and *Ozawainella.* The late Bashkirian is characterized by a lowest zone having the first appearance of *Profusulinella,* a middle zone having the first appearance of *Schubertella* and *Verella,* and a highest zone having the first appearance of *Parastaffella.* In addition, species groups of *Palaeotextularia, Climacammina, Endothyra, Bradyina, Pseudoendothyra, Eostaffella (=Paramillerella), Tetrataxis, Eolasiodiscus,* and *Neoarchaediscus* and the first appearance of *Lipinella* are important zonal fossils among the other foraminiferids in the Bashkirian. In the Midcontinent region of North America the Morrowan Series is the time equivalent of most or all of the Bashkirian. The middle part of the type Morrowan has *Millerella* as the dominant foraminifer and in other areas higher parts of this stratigraphic interval have in addition *Eoschubertella* and *Pseudostaffella* (THOMPSON, 1948).

Above the Morrowan and its equivalents, foraminiferal correlations are based largely on members of the superfamily Fusulinacea (THOMPSON, 1948, 1966, 1967; RAUZER-CHERNOUSOVA et al., 1951, 1954, ROZOVSKAYA, 1950, 1969; Ross, 1967, 1970, 1973). A few long-ranging genera of smaller foraminifers appear near the base of the Moscovian and Atokan-Desmoinesian sequences, such as *Nodosaria* and *Cribrogenerina,* but these and their late Carboniferous descendants have not been studied in as much detail as the earlier Carboniferous genera. Although the Midcontinent North American and Arctic-Eurasian fusulinid faunas show general similarities during the remainder of the Carboniferous, the range of many genera differs between the two regions, and a number of endemic genera are reported (Fig. 3) in both regions.

Atokan fusulinids are characterized in North America by *Profusulinella* in the lower part, which evolved with little overlapping range into *Fusulinella* in the upper part to provide a twofold division into a zone of *Profusulinella* and a zone of *Fusulinella.* Several long-ranging genera such as *Millerella, Eostaffella (=Paramillerella), Parastaffella, Pseudostaffella,* and *Eoschubertella* persist from the Morrowan in the Midcontinent North American sections into the Atokan. One lineage of *Fusulinella* evolved through several transitional species into *Beedeina* near the base of the Desmoinesian Series and formed the beginning of the zone of *Beedeina,* which characterizes most of the Desmoinesian strata. *Fusulinella, Eoschubertella,* and *Pseudostaffella* continue into the zone of *Beedeina* but are usually rare above the lower third of the zone. *Wedekindellina* first appears slightly above the base of the zone. In the northern Cordilleran region *Akiyoshiella, Pseudowedekindellina, Eowaeringella,* and *Bartramella* appear in the assemblage. The highest species of *Beedeina* occur near the top of the Desmoinesian Series.

The Atokan-Desmoinesian interval is correlated with the Moscovian Stage in Eurasia and the Arctic. The lower part of the Moscovian contains a fusulinacean assemblage that includes most of the Bashkirian genera, except *Verella,* and the first appearance of *Aljutovella, Hemifusulina,* and *Eofusulina.* The upper part of the Moscovian has the additional genera *Fusulinella, Fusulina, Ozawainella, Fusiella, Wedekindellina, Putrella,* and *Protriticites.*

Except for the lowest part of the Missourian Series, the later part of the Pennsylvanian in North America is characterized by the zone of *Triticites.* The oldest Missourian assemblage includes *Fusulina* (of the Russian-platform type; it is not a *Beedeina*), and associated *Oketaella, Schubertella,* and *Eowaeringella.*

The succeeding zone of *Triticites* includes most of the Missourian and Virgilian series and has the first appearance of *Triticites, Iowanella,* and *Kansanella* in the Midcontinent region. Of these, *Triticites* is long ranging and continues into overlying zones in the Permian. Within the zone *Waeringella* and *Dunbarinella* appear briefly. The top of the zone of *Triticites* is usually placed below the first occurrence of *Schwagerina* or *Pseudofusulina.*

In Eurasia and the Arctic, the zone of *Triticites* in its lower part includes forms ancestral to *Triticites* that have subsequently been given generic status, *Protriticites, Obsoletes, Montiparus,* as well as *Fusulinella, Quasifusulina,* and the last of *Quasifu-*

sulinoides. *Triticites* is advanced but relatively unspecialized when it first appears in the middle part of the zone and early in its history it gives rise to *Daixina* and *Jigulites* and finally to primitive *Pseudofusulina* near the top of the zone (REYTLINGER, 1969). In the Soviet Union, a number of fusulinacean paleontologists include the overlying zone of *"Schwagerina"* (i.e., Asselian Stage) in the Carboniferous, whereas most other paleontologists and biostratigraphers place the Asselian in the basal part of the Permian. The Asselian, as a biostratigraphic unit, is correlated with the lower part of the North American Wolfcampian Series (Neal Ranch Formation), which is considered lowest Permian by most North American paleontologists.

The biogeographic distribution of Carboniferous Foraminiferida has been discussed by EYNOR and others (1973), LIPINA (1973), LIPINA and REYTLINGER (1970), MAMET and SKIPP (1971), Ross (1967, 1973), and THOMPSON (1967).

The epeirogenic sea of the North American craton had an impoverished foraminiferal fauna compared to the seas in the miogeosynclines and on the edges of the craton. The fauna on the craton is less diverse and is characterized by accumulations of monogeneric and even monospecific assemblages and there are fewer individuals. The least diverse and most endemic foraminiferal faunas in North America are in the upper Mississippi valley and Midcontinent regions and this leads to difficulties in extending biozones of the type sections of the Mississippian to the Cordilleran region. The distribution of different families of Foraminiferida is affected by the environments of this epeirogenic sea. Cornuspiridae and Fischerinidae? are evenly distributed in North America, Tournayellidae are mainly restricted to the Cordillera, Forschiidae are rare in the Midcontinent, and Palaeotextulariidae are common in the Visean of the Cordillera but scarce in rocks older than Namurian in the Midcontinent. Endothyridae are widespread in North America and the variety of species makes them stratigraphically useful in the Midcontinent.

Early Carboniferous foraminiferal assemblages of Eurasia and North America show some latitudinal differences but there is no evidence of exclusive provincialism. The most complete assemblages are in miogeosynclines and on shelf margins in the Urals, Donetz, Turkey, North Africa, western Europe, Iran, Vietnam, and Australia, and form a Tethyan realm. The foraminiferal faunas in northern Siberia, Kuznetzk, Japan, and the Cordilleran and Appalachian geosynclines differ from the Tethyan populations in having a much greater percentage of Endothyridae, a sparsity of Forschiidae, Bradyinidae, and Palaeotextulariidae, and a lower species diversity.

Inverse relations exist between the importance of planispirally coiled *Eoendothyranopsis* and of *Eostaffella* and *Pseudoendothyra;* in the European Visean *Eostaffella* and *Pseudoendothyra* are abundant and *Eoendothyranopsis* is scarce whereas in North America the Visean *Eoendothyranopsis* is abundant and *Eostaffella* is sparse. Exchanges of fauna between the two realms were widespread during most of the Early Carboniferous and the North American fauna, despite impoverishment in many elements, always has counterparts in Eurasian faunas.

Middle and Late Carboniferous members of the superfamily Fusulinacea are widely distributed, commonly abundant, and evolved rapidly. Figure 3 shows the stratigraphic and geographic distribution of genera belonging to the five Carboniferous families of fusulinaceans. In the Eurasian-Arctic province many more genera and a more complex zonal sequence of genera in these families arose than in the Midcontinent-Andean province (Ross, 1967). Infrequent and incomplete faunal exchange through a filter region between faunal provinces, such as occurred early in Morrowan time and again early in Atokan time, established the initial fusulinacean stock that produced successive provincial lineages characterizing the zones of *Millerella, Profusulinella, Fusulinella,* and *Beedeina* in the Midcontinent-Andean province.

A nearly independent evolutionary history of the Eurasian-Arctic and Midcontinent-Andean faunal provinces is indicated by the different stratigraphic ranges of *Profusulinella, Fusulinella, Beedeina,* and *Fusulina* in the two provinces. In the Eurasian-Arctic province *Fusulinella* and *Fusulina* appear at about the same time and persist, along with their ancestor, *Profusulinella*, until near the end of the Moscovian. In the Midcontinent province a phylogenetic sequence from *Profusulinella* to *Fusulinella* to *Beedeina* shows little stratigraphic overlap between these three genera. *Fusulina* appears briefly in the basal part of the Missourian (Midcontinent-Andean province) only after the extinction of *Beedeina*. The few genera of Fusulinidae that are nearly cosmopolitan, such as *Pseudostaffella*, *Fusulinella*, *Wedekindellina*, and *Beedeina*, have different species complexes in the different provinces during the Middle Carboniferous. Near the end of the Middle Carboniferous, a few genera, such as *Fusiella*, which is typical of the late Moscovian of the Eurasian-Arctic province, and *Bartramella* appear abruptly in the Desmoinesian in the Midcontinent-Andean province.

In both provinces many lineages of Middle Carboniferous foraminiferids became extinct at about the end of Desmoinesian or Moscovian time and relatively few genera survived into the Late Carboniferous. Most surviving lineages were modified sufficiently to be given new generic names. In the Eurasian-Arctic province rapid evolution produced a sequence of early schwagerinid genera: *Protriticites, Montiparus, Obsoletes,* and finally *Triticites*. *Triticites* is the only genus of this sequence that successfully invaded the Midcontinent province. However, the Midcontinent genera, *Iowanella* and *Kansanella*, are probably derived from the Eurasian-Arctic genus *Montiparus* and not directly from *Triticites*. In Eurasian-Arctic faunas *Fusulinella* persisted in a distinctive facies and gave rise to *Pseudofusulinella* near the end of Late Carboniferous time. *Eowaeringella* appears to be restricted to North America and ranges from the Middle Carboniferous into the Upper Carboniferous; it gave rise to the Virgilian *Waeringella* of the Midcontinent province and possibly gave rise to *Thompsonella*. During the later part of the Late Carboniferous in the European part of the Eurasian-Arctic province *Triticites* evolved into several diverse lineages with several subgenera, such as *Jigulites* and *Rauserites*, and the genera *Pseudofusulina, Rugosofusulina,* and *Daixina*. Some of these are restricted to the Eurasian-Arctic province and, at times, to only certain of its subprovinces. Within the Midcontinent-Andean province species of *Triticites* form several distinctive lineages.

In comparison with the Early and Middle Carboniferous, during Late Carboniferous both the Eurasian-Arctic and Midcontinent-Andean provinces had less diversity. Major morphological changes occurred in only the earliest part and again in the later part of the Late Carboniferous and seem to coincide with increased dispersal.

The patterns of distribution and dispersal and intervals of endemic evolution of Foraminiferida during the Carboniferous (Ross, 1970) are similar to those that have been demonstrated for rugose corals (HILL, 1957, 1973) and ammonoids (RUZHENTSEV, 1966; GORDON, 1970). As with many questions of biogeography, it is difficult to establish causal factors for these similar histories; however, some speculations are possible. LIPINA and REYTLINGER (1970) suggested that the faunal differences between the European and Siberian subprovinces of their Eurasian foraminiferal province coincide with the change from tropical to extra-tropical floral zones and their interpretation is supported by the decrease in foraminiferal diversity and lack of forms having massive shells. Early and Middle Carboniferous paleofloral regions (MEYEN, 1970) of the Euramerican region extending from south-central Asia across Europe into North America have woods lacking annual rings (indicative of tropical climate) and those in the Angaran region have woods with annual rings (indicative of extra-tropical climates).

CORALS

Although Carboniferous coral occurrences (Fig. 1, 4) are considered to be closely related to biofacies, they have been widely used for establishing stratigraphic correlations. Where corals are abundant and several biofacies are represented, correlations within a province are usually consistent with those based on other fossil groups. Corals apparently did disperse slowly and relatively infrequently between some provinces after the Visean and this limits their effective use in younger Carboniferous strata.

Lower Carboniferous coral faunas in North America, Eurasia, and Australia are sufficiently distinct from one another that HILL (1948, 1957, 1973) considered each of these areas a separate zoogeographic region. The Eurasian region includes four major subdivisions: 1) a western European province including part of the maritime provinces of eastern Canada and northwestern Africa, 2) an eastern European province, 3) a Chinese-Japanese province, and 4) a southcentral Asian province, which with additional study may be further subdivided (HILL, 1973).

During earliest Carboniferous time (Tournaisian) the North American zoogeographic region was subdivided by the transcontinental arch that partially separated coral faunas of the stable broad Mississippi valley epicontinental shelf from the faunas of the Cordilleran geosyncline and its marginal shelf. According to HILL (1973), a number of solitary coral genera of the Mississippi valley shelf appear to be related to ceratoid genera of the western European province. In addition, several North American endemic lineages developed (e.g., *Clinophyllum* and *Neozaphrentis*) and some genera continued from Late Devonian time (e.g., *Microcyclus* and *Hadrophyllum*). The North American Cordilleran coral faunas include a number of genera in common with those of the Mississippi valley shelf, such as *Cyathaxonia* and *Homalophyllites*, but also include several eastern European genera such as *Keyserlingophyllum*, *Enygmophyllum*, and *Uralinia?* which suggest a shallow marine connection with the Uralian geosyncline. *Zaphrentites*, *Amplexus*, and *Vesiculophyllum* are also present in this fauna.

Within the Eurasian region, HILL (1973) recognized considerable provincialism. The western European province (including a portion of northwestern Africa) has great diversity and includes an impressive assemblage of corals characterized by *Cravenia*, *Allotropiophyllum*, *Amplexocarinia*, *Amplexus*, *Caninia*, *Caninophyllum*, *Cyathoclisia*, *Koninckophyllum*, *Cryptophyllum*, *Cyathaxonia*, *Lonsdaleia*, *Menophyllum*, *Sychnoelasma*, *Palaeosmilia*, *Rotiphyllum*, *Siphonophyllia*, *Thysanophyllum*, *Zaphrentis*, and a number of tabulate corals (HILL, 1948). The eastern European corals of the

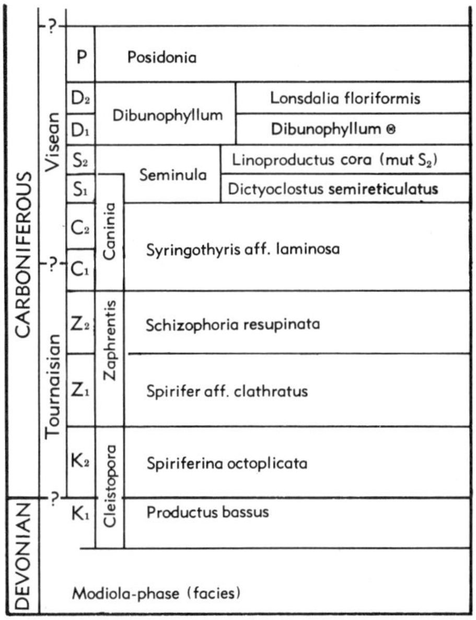

CARBONIFEROUS	Visean	P	Posidonia	
		D₂	Dibunophyllum	Lonsdalia floriformis
		D₁		Dibunophyllum ⊖
		S₂	Seminula	Linoproductus cora (mut S₂)
		S₁		Dictyoclostus semireticulatus
		C₂	Caninia	Syringothyris aff. laminosa
		C₁		
	Tournaisian	Z₂	Zaphrentis	Schizophoria resupinata
		Z₁		Spirifer aff. clathratus
DEVONIAN		K₂	Cleistopora	Spiriferina octoplicata
		K₁		Productus bassus
		Modiola-phase (facies)		

FIG. 4. Coral and brachiopod zonation of the Lower Carboniferous of the Avon section of southern England (Vaughan, 1905, 1915) and approximate correlation with the boundaries of the Tournaisian and Visean stages of Belgium (modified from Vaughan, 1915; George, 1969).

Russian platform (and the margins of that platform) contain different species assemblages that were used by VASILYUK et al. (1970) to identify three subprovinces: one in the north in the region of Novaya Zemlya, another in the Moscow basin, and a third in southern Ural-Donbas (or southshelf edge). In these subprovinces the tabulate corals *Syringopora*, *Tetraporinus*, *Roemeripora*, *Michelinia*, *Emmonsia*, and *Gorskyites* are important and *Caninia*, *Caninophyllum*, *Uralinia*, *Siphonophyllia* and other caninioid genera are present. Farther east the Kazakhstan Tournaisian coral faunas are small in number of species and genera and include a Chinese element, *Cystophrentis*, suggesting connections to the east. The Kuznetzk basin faunas include *Cyathoclisia* and possible ancestors of the Visean *Arachnolasma* and *Yuanophyllum* of China. Late Tournaisian coral faunas became more widely dispersed in the east European-Asian provinces, and of these *Uralinia*, *Keyserlingophyllum*, and *Enygmophyllum* are particularly widespread and characteristic and were dispersed as far as the western Cordilleran province of North America. The southern Ural, Kazakhstan, and Kuznetzk areas have a number of distinctive genera, such as *Bifossularia*, indicating dispersal of some genera was still partially restricted. Kazakhstan had several endemic species or subspecies and Kuznetzk, in addition to endemic species, had several endemic genera, such as *Kuzbasophyllum*, *Adamanophyllum*, and *Tachyphyllum* (DOBROLYUBOVA, et al., 1966). The appearance of *Chia* and *Tetraporinus* in the Donetz basin at this time suggests faunal dispersals along an east-west seaway (or shelf margin) with parts of China. Along the northwestern edge of the Angaran craton IVANOVSKIY (1967) reported a small coral fauna including *Amplexus*, *Campophyllum*, *Tachyphyllum?*, and *Trochophyllum* and on the northeastern edge of the Angaran craton the Tournaisian coral fauna includes *Uralinia*, *Keyserlingophyllum*, *Amplexus*, *Caninia*, *Caninophyllum*, *Rotiphyllum*, *Sychnoelasma*, and *Trochophyllum*.

HILL (1973) recognized a Chinese province in which *Cystophrentis* and *Pseudouralinia* are common together with the widely distributed genera *Caninia* and *Zaphrentites*. In Japan a Tournaisian sequence includes three assemblage zones: *Amygdalophyllum* and *Lithostrotionella* at the base; next, *Amplexus* and *Syringopora*; and *Sugiyamaella* near the Tournaisian-Visean boundary.

Thus far, relatively few coral localities are reported from the Tournaisian of the Australian biogeographic region. The earliest occurrences (which may prove to be latest Devonian) include *Lithostrotion*, *Amygdalophyllum*, *Naoides* (endemic), *Michelinia*, *Yavorskia*, and *Syringopora*. Another locality has *Cladochonus* and *Bibucia* (endemic) and a higher locality has *Amygdalophyllum* and *Merlewoodia* (endemic).

Visean corals were also divided into three zoogeographical regions by HILL (1973): North American region, eastern Australian region, and Eurasian region. The Eurasian region includes the maritime provinces of Canada and Bonaparte basin (northwestern Australia) in addition to much of northwestern Africa. Visean coral faunas are more diverse than those of the Tournaisian and include a number of new introductions and newly evolved genera.

In the Cordilleran province of North America a large representation of species of *Lithostrotionella* is associated with *Lithostrotion*, *Thysanophyllum*, *Sciophyllum*, *Diphyphyllum*, and *?Dorlodotia*. In addition, *Ekvasophyllum*, *Faberophyllum*, *Liardiphyllum*, and *Zaphriphyllum*, along with a number of nondissepimented genera, occur in the northern part of the North American Cordilleran and also in the Taymyr geosyncline north of the Angaran shield. The epicontinental shelf of central North America contained a less diverse fauna than the Cordilleran region and includes at least one genus, *Palaeosmilia*, of Eurasian affinities.

On the basis of Visean corals, the Eurasian faunal region is divisible into a number of provinces and subprovinces. The western European province (including Nova Scotia and northwestern Africa) has a diverse fauna with the characteristic genera *Allotropiophyllum*, *Amplexus*, *Caninia*, *Carruthersella*, *Cravenia*, *Cyathaxonia*, *Dibunophyllum*, *Koninckophyllum*, *Lithostrotion*, *Lonsdaleia*, *Orionastrea*, *Zaphrentites*, and *Palaeosmilia*, but lacks *Lithostrotionella*, *Ekvasophyllum*, *Faberophyllum*, *Liardiphyllum*, *Zaphriphyllum*, and *Gangamophyllum*. The Nova Scotian Visean coral fauna includes the distinctly Eurasian genera *Lonsdaleia* and *Dibunophyllum*. The eastern and southeastern European Visean province includes most of the genera of the western province and, in addition, has some distinctive genera, such as *Gangamophyllum*, *Turbinatocaninia*, and *Paralithostrotion*, as well as a few genera, such as *Eolithostrotionella* and *Melanophyllum*, that suggest connections with parts of central and eastern Asia. The Donetz basin fauna is largely western European in its affinities but does carry representatives from the proto-Tethys geosynclinal margin, such as *Neoclisiophyllum* and *Arachnolasma* of Chinese affinities, *Amygdalophyllum* with Australian affinities, and a number of endemic genera.

Farther south in Turkey, Iran, and southern Soviet Union, Visean corals include a mixture of eastern European and Chinese genera, including *Kueichouphyllum* and

Kueichowpora, and in Ferghana the fauna appears to be a combination of various genera from western European, eastern European, central and eastern Asian, and Australian faunas. In the Visean, other areas in central Asia also include mixtures of genera from western or eastern European or Chinese faunas but with endemic species, as in Pamir and Kazakhstan. In the Kuznetzk basin, endemism is more marked and is indicated by the genera *Kuzbasophyllum*, *Adamanophyllum*, and *Bifossularia*. Also present is the North American *Faberophyllum*. In the eastern part of the Taymyr area (VASILYUK et al., 1970) the mainly Eurasian genus *Lithostrotion* occurs with Chinese tabulate and North American rugose genera.

The Chinese coral province of the Visean has a distinctive mixture of western European, North American, and more or less endemic southeast Asian genera. Subprovinces are apparent. In Hunan, YU (1934, 1937) recognized two assemblages, the *Thysanophyllum* and *Yuanophyllum* assemblage zones, based on endemic species. In the lower zone *Kueichouphyllum*, *Bothrophyllum*, *Caninia*, *Dibunophyllum*, *Kwangsiphyllum*, and *Lithostrotion* are common and in the higher zone *Cyathaxonia*, *Rotiphyllum*, *Zaphrentites*, *Zaphrentoides*, *Heterocaninia*, *Dibunophyllum*, *Clisiophyllum*, *Koninckophyllum*, *Arachnolasma*, *Ekvasophyllum*, *Caninostrotion*, *Lithostrotionella*, *Lithostrotion*, and *Aulina* form the dominant part of the assemblage. The upper part of the Hunan Visean succession has reduced diversity and *Lonsdaleia* and *Arachnastrea* appear in this province for the first time.

In Sinkiang and Kansu, Visean coral faunas include *Siphonophyllia*, *Palaeosmilia*, *Kueichouphyllum*, *Yuanophyllum*, *Arachnolasma*, *Gangamophyllum*, and others with important endemic species and many endemic species of more widespread genera. Elements of this fauna have been traced into Yunnan, Tibet (YANG & WU, 1964), Nepal (FLÜGEL, 1966), and farther along the proto-Tethys. Much of this coral fauna is traceable along the western margin of the South China-Indochina craton (FONTAINE, 1961; SMITH, 1948) and into the Bonaparte Gulf basin of northwestern Australia.

In Japan, Visean corals form a separate subprovince of the Chinese province and are further divisible into an inner and outer zone based on sedimentary environmental features as well as faunal composition. The outer zone of dark limestone facies includes many genera typical of the main part of the Chinese province but also includes *Pseudodorlodotia*, *Sciophyllum*, *Setamainella*, and *Tschussovskenia*; this assemblage may extend into the lower part of the Namurian. The inner zone is largely a reef and slope facies and contains *Nagatophyllum*, *Echigophyllum*, *Taisyakuphyllum*, *Pseudopavona*, and *Akiyosiphyllum*. In this inner zone the fauna ranges from upper Visean into Middle Carboniferous and becomes increasingly endemic.

Eastern Australia forms a separate Visean coral region and has *Lithostrotion*, *Michelinia*, *Caninophyllum*, *Amygdalophyllum*, *Nothaphrophyllum* (endemic), *Symplectophyllum* (endemic), *Merlewoodia* (endemic), *Orionastraea*, *Palaeacis*, *Aphrophylloides*, *Heterophyllia*, *Syringopora*, and *Aphrophyllum*. Although many endemic species and several endemic genera are important constituents of this fauna, the lower endemic Tournaisian genera, such as *Bibucia* and *Naoides*, are not found in Visean faunas, and only the endemic *Merlewoodia* appears as a holdover among the Visean assemblages.

Lower Namurian coral faunas are much less widely distributed than either Tournaisian or Visean faunas and usually are a continuation of upper Visean assemblages that are reduced in diversity and numbers. The central North American region contains coral faunas having *Palaeosmilia*, *Kinkaidia*, *Caninostrotion*, *Koninckophyllum*, and a few other genera, all with distinctive endemic species (EASTON, 1945). The Namurian fauna of the western Cordillera of North America has *Caninia*, *Zaphrentites*, *Lithostrotionella*, *Syringoporella?*, *Hayasakaia?*, and *Siphonophyllia?* (SANDO, 1965; ROWETT, 1969) most of which have endemic species but are predominantly descended from faunas of the Chinese Visean province.

The Eurasian region is still clearly distinguished during the Namurian and is composed of faunal provinces that are related to their Visean predecessors. About a dozen genera range from the Visean into the Namurian of western Europe (HILL, 1973) although species (and individuals) are much less numerous. These genera are *Aulina, Aulophyllum, Carcinophyllum, Clisiophyllum, Dibunophyllum, Koninckophyllum, Lithostrotion, Lonsdaleia, Zaphrentites, Carruthersella, Palaeacis,* and *Palaeosmilia?*. The Donetz basin continues to have a dominantly western European fauna with a few Chinese genera. The eastern European province has *Arachnolasma, Gangamophyllum, Kazachiphyllum, Melanophyllum, Nervophyllum,* and *Paralithostrotion,* in addition to most of the western European genera.

In central Asia, Kazakhstan, and Pamir, the Namurian corals are closely similar to those reported from the eastern European province. The Chinese province carries a continuation of its Visean coral faunas and Namurian coral assemblages are difficult to distinguish from Visean ones. Corals have not been reported from the Namurian strata of Kuznetzk and Australia.

Late Namurian (or Bashkirian) coral faunas are relatively poorly known probably because rocks of this age with a coral-bearing carbonate facies are not extensively exposed. In the Donetz basin and southern Ural regions a continuation of earlier genera, such as *Lithostrotion, Orionastraea,* and *Corwenia,* is accompanied by the appearance of *Eolithostrotionella, Cystophora,* and *Donophyllum*. Endemic complexes of species of caniniid-clisiophyllid corals, in which the axial structures are poorly developed or lack a clearly defined pattern, are common and include species of *Lophophyllum, Neokoninckophyllum, Yuanophylloides, Caninia, Campophyllum, Yakovleviella,* and others, including the endemic genus *Sestrophyllum*. The Chinese genus *Arachnolasma* is present, and in the Donetz basin *Cyathaxonia, Lophophyllidium, Zaphrentites, Allotropiophyllum,* and the endemic *Parastereophrentis, Kumpanophyllum,* and *Cystilophophyllum* have been recorded. In the southern Urals, the endemic *Lytvophyllum* is present at this time. Bashkirian coral faunas from Japan are rare, not widely distributed, and include *Diphyphyllum, Thysanophyllum, Lithostrotionella, Nagatophyllum,* and *Chaetetes*. A *Cyathaxonia*-bearing fauna has been reported from the Atokan of the United States and may be of late Bashkirian age.

Near the western end of the proto-Tethys in northwestern Spain, DE GROOT (1963) reported endemic species of *Lithostrotionella, Dibunophyllum, Clisiophyllum, Koninckophyllum, Pseudozaphrentoides, Koninckocarinia,* and *Carcinophyllum*. Other Bashkirian (or Morrowan) coral faunas are poorly known, but apparently are present in the marine part of the proto-Tethyan geosyncline in Tien Shan along the southern edge of the Angaran craton. In North America, Morrowan corals are widely distributed but are not well known.

Moscovian (Atokan-Desmoinesian) corals are more widely distributed than Bashkirian corals, but have not been thoroughly studied. Most of western Europe had nonmarine deposition during most of this time (Westphalian) but in the eastern European province along the edges of the Russian platform *Cystophora* and *Arachnastraea* are common and *Koninckocarinia, Carniaphyllum, Caninophyllum, Bothrophyllum,* and *Timania* may be present in various localities in the Ural, Donetz basin, Tien Shan, and Kazakhstan regions. The first species of *Durhamina* has been reported from the northern Urals (MINATO & KATO, 1965a). Farther east, the Middle Carboniferous of China contains *Kionophyllum* and *Gshelia*. However, these faunas are poorly known and are apparently not abundant or common. The earliest waagenophyllid, *Huangia*, occurred in China during the Moscovian (MINATO & KATO, 1965b). In Japan the inner zone assemblages with *Nagatophyllum* apparently range from strata of late Visean age upward into strata of Namurian age and probably into strata as young as Moscovian age. In northwestern Spain, DE GROOT (1963) recorded endemic species of *Koninckophyllum, Carcinophyllum, Lithostrotion, Corwenia, Arachnastraea,* and *Lithostrotionella* from Moscovian strata. In North America a number of genera and species were endemic to central North America during the Pennsylvanian including the middle Pennsylvanian (Atokan-Desmoinesian) genera *Empodesma, Stereocorpha, Lophotichium, Barytichisma, Acaciapora,* and *Cumminsia;* endemic species of *Dibunophyllum* and *Neokoninckophyllum;* and also the highest occurrence of

Chaetetes in the North American region.

During Moscovian time (HILL, 1957), a distinctive coral province extended at least from Spitsbergen along the Russian platform and western side of the Ural geosyncline across southern Europe to northwestern Spain. In southeastern Asia several genera, such as *Gshelia, Cyctophora,* and *Arachnastraea,* suggest a separate province at this time.

Post-Moscovian—pre-Permian (i.e., Late Pennsylvanian) coral faunas have not been widely studied and apparently are not common. The eastern European province has a continuation of dominant Moscovian genera, such as *Timania, Gshelia, Bothrophyllum, Caninophyllum, Eolithostrotionella, Tschussovskenia,* and *Protolonsdaleiastraea. Kionophyllum* and *Amandophyllum* spread to the Carnic Alps during latest Carboniferous time. Also, in very Late Pennsylvanian time in southwestern North America *Durhamina, Stereostylus,* and *Leonardophyllum* formed the beginnings of several endemic lineages that extended into Early Permian time. *Lophophyllidium, Syringopora, Aulopora,* and "*Caninia*" also are recorded from the Upper Pennsylvanian at several localities in central North America. Limited knowledge of post-Moscovian (post-Desmoinesian)—pre-Permian corals of the Carboniferous does suggest well-developed provincial and regional faunas. These faunas included new families and genera that led to an increasing diversity marking the beginning of the Permian; however, the details are not presently known.

BRACHIOPODS

Near the end of the Devonian many of the major brachiopod superfamilies and families either became extinct or were greatly reduced in importance. These include the strophomenaceans, punctate and impunctate orthids, pentamerids, atrypaceans, dayiaceans, and early terebratulids (RUDWICK, 1970). The surviving rhynchonellids, spiriferids, and strophomenids (particularly those with tubular spines or shell cementation) evolved rapidly to give the Lower Carboniferous brachiopods high taxonomic diversity (Fig. 5), and they were adapted to a wide range of environmental habitats. Other strophomenids, such as chonetaceans, davidsoniaceans, and strophalosiaceans, spiriferids, atrypids, rhynchonellids, and the few surviving terebratulids also underwent significant taxonomic expansions. In the Early Carboniferous several major families of spiriferids evolved (IVANOVA, 1972): Spiriferidae and Syringothyrididae (both of which may have first evolved in the Late Devonian), and Brachythyrididae, Choristitidae, Paeckelmanellidae, Davidsoninidae, and Spiriferellinidae. Several Carboniferous subfamilies have short stratigraphic ranges. The subfamily Spiriferinae appears to range no higher than the upper part of the Visean. Although present earlier, Neospiriferinae were common in the Middle and Late Carboniferous and gave rise to the Trigonotretinae during the Late Carboniferous. The family Davidsoninidae appeared in the Visean and became extinct by the end of the Namurian (IVANOVA, 1972). Rapid evolution and short stratigraphic ranges of species of the family Choristitidae are useful in subdividing the Middle Carboniferous in Eurasia (KHALYMBADZHA & TIKHVINSKIY, 1967).

During the Tournaisian and Visean, spiriferids were widespread and most groups were cosmopolitan. Beginning in the Namurian their distributions became increasingly more restricted as endemic species and then endemic genera replaced the older more cosmopolitan species and genera. Middle Carboniferous spiriferids are less well known but are commonly abundant and diverse in regions of marine epicontinental deposition (KHALYMBADZHA & TIKHVINSKIY, 1967). However, the number of cosmopolitan species and genera was considerably less than in the Early Carboniferous. The data available for spiriferid distributions in the Late Carboniferous are less complete than for the other parts of the Carboniferous, but a renewed diversification in the Early Permian suggests increasing geographic dispersal during the Late Carboniferous.

The general history of the evolution and geographic distribution of productoid brachiopods (MUIR-WOOD & COOPER, 1960) is similar to that of the spiriferids. Most Early Carboniferous diversity in the productoids was in lineages that first appeared in the Late Devonian. Tournaisian and Visean

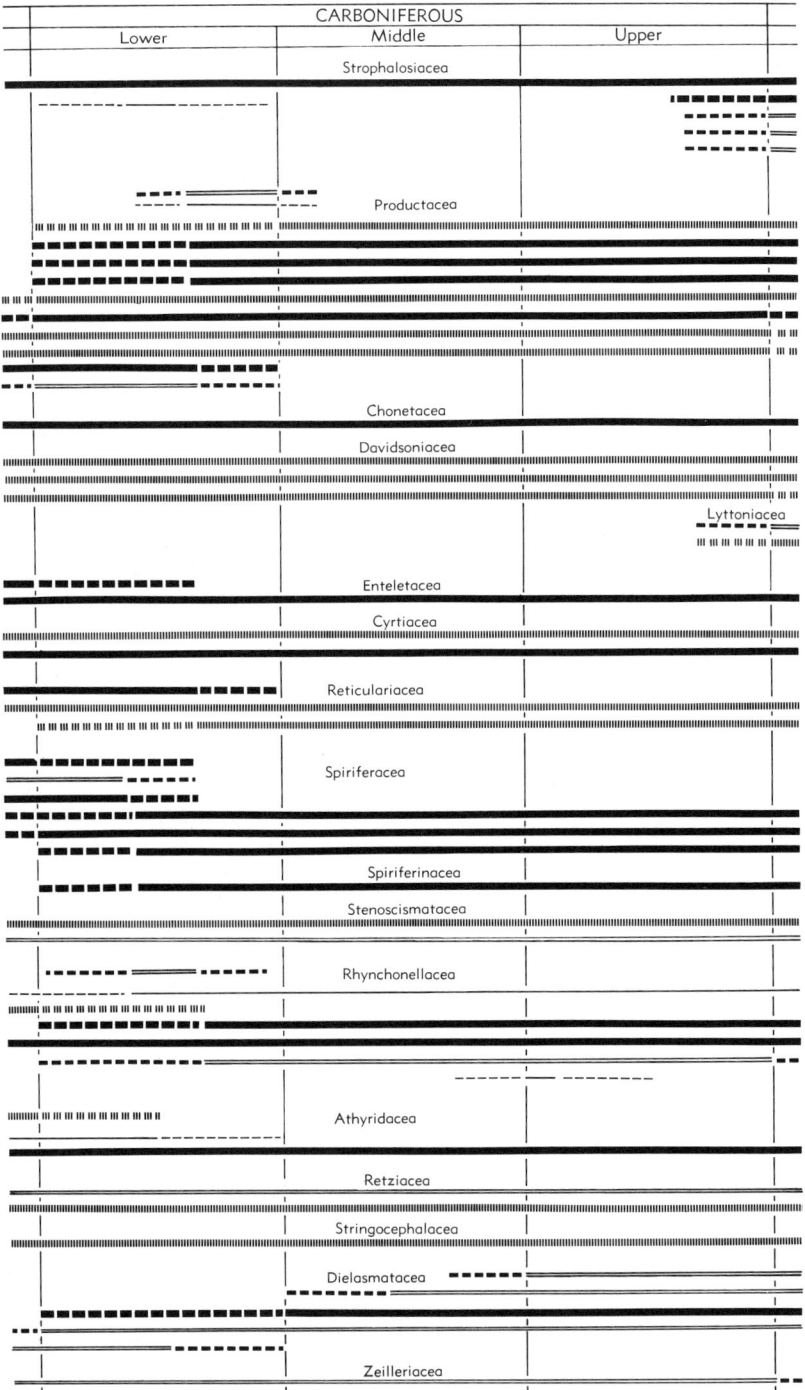

Fig. 5. Stratigraphic ranges of superfamilies and families of brachiopods in the Carboniferous. Dashed lines indicate the limits of uncertainty of the stratigraphic range of each family. The number of genera

productoids evolved into diverse and widespread groups that contained many cosmopolitan elements as well as a number of endemic and locally important genera (e.g., the four genera of the Gigantoproductidae). During the later part of the Early Carboniferous a number of genera and subfamilies became extinct (Sinuatellidae, Productellidae).

Middle Carboniferous productoids had less cosmopolitan distributions. Most Carboniferous genera of the Linoproductidae, except *Striatifera,* had endemic distributions and only near the end of the period or in the Early Permian did this family become widely dispersed. This distributional pattern is also apparent in the Buxtoniidae, Dictyoclostidae, and Echinoconchidae. Several families had subfamilies or genera with disjunct distributions during the Carboniferous: during the Visean the family Strophalosiidae had only one genus in Europe and two in North America; and during the Late Carboniferous the subfamily Echinosteginae had two endemic stocks, one in North America, and other in the Ural region, that gave rise to major lineages that subsequently dispersed in the Early Permian. Several of the surviving Devonian lineages, such as the Leioproductidae, Productininae, Overtonidae, and Productidae, evolved a number of endemic genera in addition to a few cosmopolitan genera in the Lower Carboniferous. The reduction in productoid generic diversity near the beginning of the Middle Carboniferous was followed by a gradual increase in diversity again near the end of Middle Carboniferous time, and it continued through the late Carboniferous into the Early Permian.

CEPHALOPODS

Carboniferous cephalopods were a diverse, rapidly evolving, free swimming, benthonic group that forms one of the important series of stratigraphic index fossils for the period. As with several other groups of Carboniferous fossils, the cephalopods of Europe and central and southwestern United States are known in considerably more detail than in many other parts of the world because of the availability of collections of specimens and more detailed study. Many species and genera are believed to have lived within particular depth intervals and, as a consequence, if a cephalopod assemblage was derived from the overlying water column, deeper water sediments commonly contain a greater diversity than shallower water sediments. On the other hand, many empty shells floated and were carried by currents hundreds, perhaps thousands, of kilometers beyond the biogeographic areas occupied by living specimens, as presently happens to shells of recent *Nautilus.* Thus, a number of cephalopod fossil assemblages, particularly those having an unusually large number of species and genera, is believed not to be living assemblages, but postmortem collections of drifted shells concentrated at shorelines.

Carboniferous cephalopods include the remnants of the subclass Actinoceratoidea, which had a long history in the early and middle Paleozoic, the long-ranging subclass Nautiloidea, and the rapidly diversifying subclass Ammonoidea. Only a few genera of the subclass Actinoceratoidea have been reported from Carboniferous strata (TEICHERT et al., 1964): *Mstikhinoceras* from the Visean of the Russian platform, *Carbactinoceras* and *Aploceras* from Visean strata in western Europe, and *Rayonnoceras,* which is longer ranging and more widespread in both North America and Europe.

In the subclass Nautiloidea three orders occur in the Carboniferous: Orthocerida, Oncocerida, and Nautilida. The Orthocerida includes a smaller superfamily, Orthocerataceae, having long-ranging genera that are mainly European and North American in distribution and a larger superfamily, Pseudorthocerataceae. This latter superfamily includes *Pseudocyrtoceras, Cyrtothoracoceras,* five closely related genera, *Pseudactinoceras, Bergoceras, Campyloceras, Eusthenoceras,* and *Paraloxoceras,* which are known only

---

FIG. 5. *(Continued from facing page.)* in each family is shown by the type of line; less than three (single fine line), three to six (double fine lines), seven to twelve (broken thick line), and more than twelve (solid thick line). (Data from Rudwick, 1970; Ross, n.)

from the Tournaisian and Visean of western Europe, and *Euloxoceras* from the upper Mississippian and Pennsylvanian of central North America. The remaining seven pseudorthoceratid genera are widespread and apparently cosmopolitan. The order Oncocerida has three Lower Carboniferous genera of which *Argocheilus* is known only from China and the other two from North America and Europe.

The order Nautilida has nine families in the Carboniferous, six of which contain a number of genera that appear to be widespread in Europe and North America in Lower Carboniferous strata; for example, in the family Koninckioceratidae, *Millkoninckioceras, Endololobus, Knightoceras, Planetoceras,* and *Temnocheilus;* in the family Trigonocerataceae, *Aphelaeceras, Maccoyoceras, Rineceras, Stroboceras, Thrincoceras,* and *Vestinautilus;* in the family Solenochilidae, *Acanthonautilus;* in the family Liroceratidae, *Liroceras;* and the family Ephippioceratidae, *Ephippioceras.* A few of these genera are recorded elsewhere also, such as *Liroceras* and *Ephippioceras.* Among the nine nautilid families, many other genera are known from Lower Carboniferous strata from limited geographical areas, for example, *Tylonautilus, Duerleyoceras, Lophoceras, Subvestinautilus, Trigonoceras, Discitoceras, Epistroboceras, Leuroceras, Lispoceras, Mesochasmoceras, Pararineceras, Subclymenia, Diorugoceras, Phacoceras, Epidomatoceras,* and *Bistrialites* are known from Europe (mostly western Europe), and *Edaphoceras, Tylodiscoceras, Chouteauoceras,* and *Diodoceras* are known only from central and eastern North America.

Middle and Upper Carboniferous genera of the family Tainoceratidae are widely distributed in North America and Europe, except for *Gzheloceras* from the Urals. In the family Grypoceratidae most genera are nearly cosmopolitan, such as *Domatoceras, Stenopoceras,* and *Titanoceras,* and in the family Solenochilidae *Solenochilus* is also cosmopolitan. A few Pennsylvanian genera seem restricted to North America, including *Coelogasteroceras, Condraoceras* (which appears also in the Permian of Europe), and *Megaglossoceras.* A few additional distributions are of interest largely because they indicate the incompleteness of these records.

Valhallites has been reported only from Siberia and Arkansas, and *Phacoceras* is known not only from the Lower Carboniferous of Europe but also from the Lower Permian of Australia.

The subclass Ammonoidea is widely, and commonly abundantly, distributed in Carboniferous strata and appears to be generally more cosmopolitan than the Actinoceratoidea and Nautiloidea (MILLER & FURNISH, 1957). The evolution and biostratigraphic zonation of Carboniferous ammonoids have been extensively studied in Europe and North America because of the usefulness of this group as stratigraphic index fossils. Based on these studies, ammonoids from the rest of the world have been compared and analyzed and assigned to biostratigraphic units. RUZHENTSEV (1966) recognized nine major biostratigraphic assemblages (Fig. 6) using the stratigraphic ranges of 158 ammonoid genera.

Few recent studies have considered the geographical distributions of Carboniferous Ammonoidea in detail. Data available for Tournaisian and Visean goniatites (BISAT, 1924; DRAHOVZAL, 1972; HODSON & RAMSBOTTOM, 1973) indicate that seven assemblage zones were widely distributed. The oldest of these Carboniferous goniatite faunas appears at the edges of the Hercynian belt in Europe, Kazakhstan, and North Africa, and commonly occurs in strata that rest without hiatus on highest Devonian strata. The succeeding five zones are identified mostly on the basis of ranges of genera of Goniatitidae that are widely distributed in the northern hemisphere, but they are poorly known from the southern hemisphere except Australia where two of the assemblage zones are recorded. Based on wide geographic distribution, these five sets of late Tournaisian and Visean assemblage zones are considered to be cosmopolitan. The highest Visean assemblage zone is likewise widely distributed but the species show provincial distributions.

Provincialism becomes characteristic of ammonoids for the remainder of the Lower and Middle Carboniferous. For example, about half of the Ural faunas are made up of members of the Prolecanitidae, Agathiceratidae, Pronoritidae, Delepinoceratidae, and Ferganoceratidae; forms that are absent in

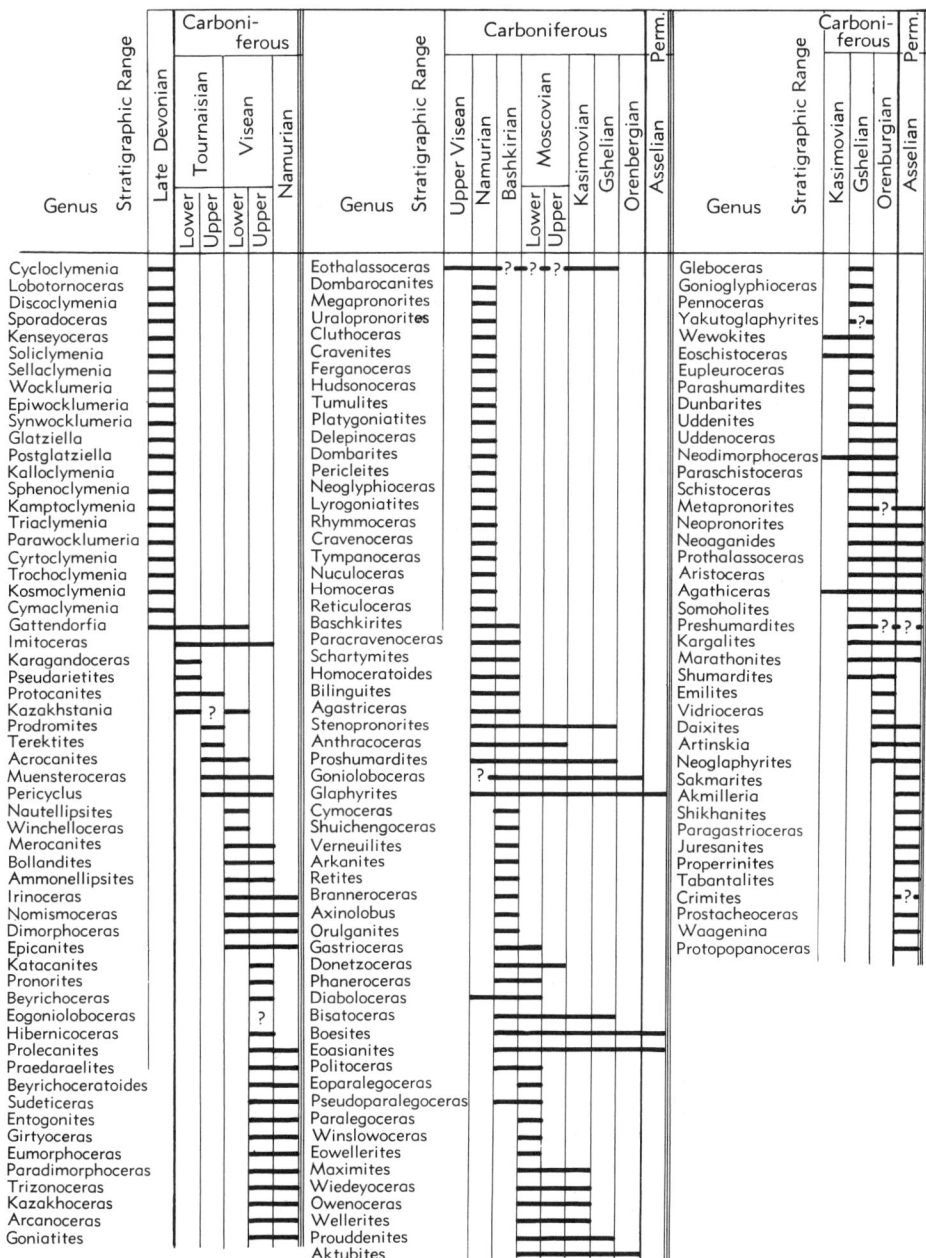

Fig. 6. Ranges of ammonoid genera in the Upper Devonian, Carboniferous, and Lower Permian. (Data from Ruzhentsev, 1960, 1966; Ross, n.)

northwestern Europe. Namurian and early Westphalian ammonoid distributions are characterized by a number of provincial faunas in which generic ranges extend irregularly upward in the succession (RAMSBOTTOM, 1971); some lineages became extinct or evolved into new but endemic genera. Goniatitids persisted in North Africa

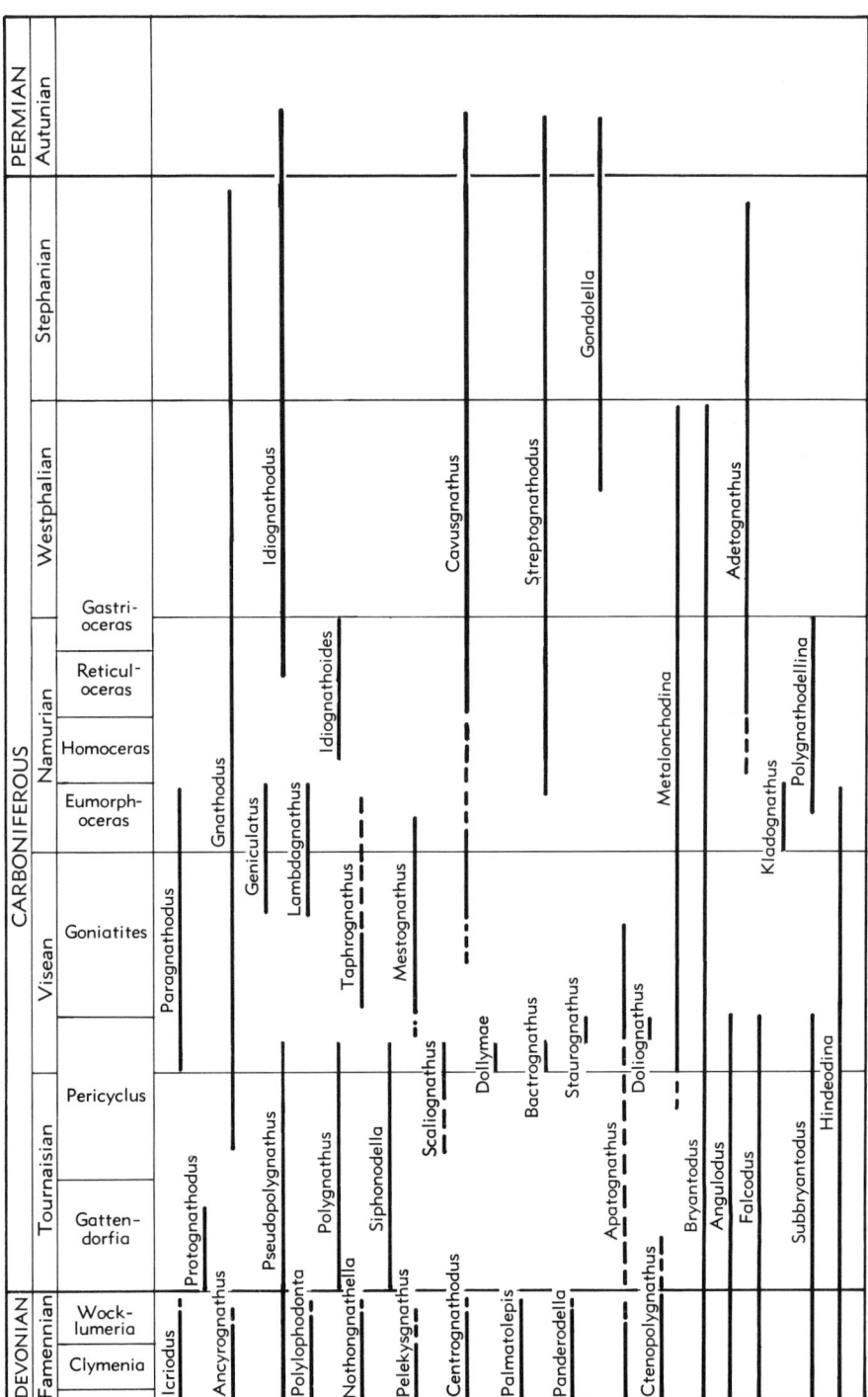

Fig. 7. Stratigraphic range of important genera of conodonts in the Carboniferous. Several genera that

and Arkansas well into the Namurian to give rise to *Proshumardites,* and in the Ural region *Dunbarites* and *Platygoniatites* persisted into the Namurian, whereas in northwestern Europe the only goniatitids to survive are in the base of the Namurian. Although the genus *Cravenoceras* (Zone E_1) is widespread and cosmopolitan, its species tend to be endemic to several provinces. *Eumorphoceras* and *Delepinoceras* are widespread but their contribution to various faunas varies considerably and *Cravenoceratoides* (Zones E_{2b} and E_{2c}) is a common northwestern European genus that has also been recorded from Spain, the Sahara, and central Asia.

At the top of Zone E_2 and in Zone H, goniatite zonal index faunas are lacking in North America, but they are common in northwestern Europe, North Africa, and possibly in the eastern Urals; the zonal genus *Homoceras* has been reported from as far east as the Tien Shan Mountains. Pronoritids are commonly associated with these occurrences except in northwestern Europe.

Reticuloceras is common in the Ural Mountains, Donetz basin, and to the south along the proto-Tethyan geosyncline where it is associated with *Proshumardites* and pronoritids, but *Reticuloceras* is rare in North America where *Syngastrioceras* and *Cymoceras* dominate. In northwestern Europe the goniatite fauna is less diverse and restricted. Later *Reticuloceras* stocks in northwest Europe and North America appear to have had independent histories.

Gastrioceras, Branneroceras, and *Syngastrioceras* form the typical assemblage of Zone G_1 and are nearly cosmopolitan except for northwestern Europe where *Gastrioceras* occurs without the other two typical genera. In succeeding strata with Zone G_2, northwestern European goniatites represent a relict fauna with the exception of *Politoceras,* which was introduced from North America. In North America the Middle Carboniferous is an interval of increasingly diverse goniatite faunas, and a number of genera, such as *Bisatoceras, Goniolaboceras, Gonioglyphioceras, Wiedeyoceras, Dunbarites, Vidrioceras, Paralegoceras, Diabloceras, Winslowoceras,* and *Owenoceras,* are endemic.

Late Carboniferous ammonoids appear to be more cosmopolitan in distribution than those of the Middle Carboniferous. *Dunbarites* was still endemic to south-central North America as were *Eupleuroceras* and *Vidrioceras,* and in the Ural Mountains *Gleboceras* was endemic.

Tournaisian and Visean seas covered a larger proportion of continental platforms than did Namurian and later seas (RAMSBOTTOM, 1971). The most varied cephalopod faunas are in the proto-Tethyan region which was connected at various times to these epicontinental seas. When ammonoid distributions are plotted on reconstructed Carboniferous paleogeographic maps, they occur between Carboniferous latitudes 60°N and 50°S.

CONODONTS

The use of discrete conodont elements in establishing a detailed biostratigraphic zonation for the Devonian and Lower Carboniferous has provided a tool for accurately dating many marine strata that have relatively few other fossils (Fig. 7). Little biological information is available about the organism which bore the conodont elements. Assemblages containing different types of conodont elements are not commonly preserved, so the custom of referring to each type of element as a different form taxon is a useful biostratigraphic expedient. Although many of the conodont guide fossils have an uneven geographic and stratigraphic distribution (RHODES & AUSTIN, 1971), they form a widely used biostratigraphic scheme. The British Tournaisian and Visean are divided into 14 zones based on conodont succession and the comparable sections in North America are divided into 13 zones and in Germany into eight zones.

In North America and in many other

FIG. 7. *(Continued from facing page.)*
range through the Carboniferous, such as *Spathognathodus,* and have species with distinctive stratigraphic ranges, are not shown. (Data from Rhodes and Austin, 1971; Collinson, Rexroad, & Thompson, 1971; Lane, Merrill, Straka, & Webster, 1971; Meischner, 1970; Ross, n.).

parts of the world (SANDBERG et al., 1972), *Siphonodella* has an important succession of species that permits subdivision of the lower part of the Lower Carboniferous. These are succeeded in the upper part of the Lower Carboniferous by a number of species of *Spathognathodus, Polygnathus, Bactrognathus, Taphrognathus, Apatognathus,* and *Cavusgnathus,* which have distinctive stratigraphic ranges.

In England the Lower Carboniferous conodont succession is characterized by a large number of short-ranging species of spathognathodids and pseudopolygnathids. *Siphonodella,* a distinctive genus for the Tournaisian elsewhere, is very rare in England. *Gnathodus* has many short-ranging species in Visean and lower Namurian strata. *Idiognathodus* first appears at about the same horizon as the cephalopod *Reticuloceras* and different species of *Idiognathodus* are distinctive in the upper part of the Namurian. *Gnathodus* becomes increasingly rare in this interval and finally disappears. Westphalian conodont faunas lack *Gnathodus* and are dominated by species of *Idiognathodus.*

In North America, Mississippian conodont zonation has been well studied in the upper Mississippi valley and adjacent areas (COLLINSON, REXROAD, & THOMPSON, 1971) and 16 named conodont zones and several subzones are recognized (Fig. 7). The Kinderhookian Series has six zones based on stratigraphic ranges of species of *Siphonodella, Gnathodus,* and *Protognathodus.* The Osagean Series is subdivided using ranges of species of *Bactrognathus, Doliognathus, Gnathodus, Pseudopolygnathus, Polygnathus,* and *Taphognathus.* The base of the Chesterian Series falls within one of these zones. Species of *Kladognathus* appear in the upper part of the Chesterian and the highest zone in the Mississippi valley region has the first appearance of *Streptognathodus.* The top of the type Mississippian System is marked by a major hiatus; however, at least part of the missing interval is represented by strata in Nevada where a higher conodont zone, *Gnathodus girtyi simplex* Zone, is assigned to the Mississippian.

Studies in North American Pennsylvanian conodont zonation have concentrated on the lower part of the succession, particularly the Morrowan Series (LANE et al., 1971). Four Morrowan zones are recognized based on ranges of species of *Spathognathodus, Idiognathoides,* and *Gnathodus.* In the Midcontinent area, *Idiognathoides* dominates the Morrowan conodont faunas. Several species of *Gondolella* are restricted to the Desmoinesian, Missourian or Virgilian series and a species of *Gnathodus* occurs in the Desmoinesian. *Idiognathodus* and *Streptognathodus* are important in late Carboniferous conodont faunas of the Midcontinent region.

In the central Appalachian Mountains, Pennsylvanian lineages of *Gnathodus* start in the lower part of the Pottsvillean Series and have not been found above the middle part of the Conemaughan Series. Specimens of *Idiognathodus* are the most abundant conodont in the Pottsvillean and Alleghenyan Series but *Streptognathodus* dominates the Conemaughan Series.

In the southwestern part of the Cordilleran miogeosyncline, the first appearance of species of *Streptognathodus* and *Idiognathoides* is in strata of early Morrowan age. A number of species appear in both the Cordillera and the Midcontinent areas but they seem to have different regional stratigraphic ranges and apparently reflect intermittent connection between the two regions.

The Carboniferous succession in Germany is divisible into 14 conodont zones and subzones (MEISCHNER, 1970). Tournaisian strata are divided into five conodont zones based on ranges of species of *Protognathodus, Pseudopolygnathus, Polygnathus,* and *Siphonodella.* Five zones are recognized in Visean strata based on ranges of species of *Scaliognathus, Gnathodus,* and *Paragnathodus,* some of which range upward into the lower part of the Namurian. From strata of the *Homoceras* Zone and higher Namurian and Lower Westphalian, species of *Gnathodus* dominate the conodont faunas. The first appearance of *Idiognathodus* is in the middle part of the *Gastrioceras* Zone at the base of the Westphalian, a stratigraphic position that suggests *Idiognathodus* occurs higher in Germany than England or North America.

Studies of the biogeographic distribution of conodonts (DRUCE, 1973) have come about only since the completion of detailed stratigraphic analyses from a number of different regions. Several Carboniferous genera are apparently geographically restricted to certain regions. The late Tournaisian genera *Bactrognathus* and *Staurognathus* are found only in the Midcontinent and southwestern regions of the United States and other reported occurrences are questionable. *Cavusgnathus* occurs in the Visean of North America, England, Scotland, and Germany, but is rare or lacking in other parts of Europe (DRUCE, 1973); however, this genus may be closely associated with nearshore deposits (LANE et al., 1971) and these distributions may reflect inadequate sampling. Other Tournaisian or Visean genera with restricted distributions include *Clydagnathus* (Western Australia and southern England), *Doliognathus* and *Dollymae* (Europe and central North America), *Mestognathus* (Nova Scotia, Australia, and Europe), and *Taphrognathus* (North America, British Isles, and Australia). Among the Middle and Late Carboniferous conodonts, *Gondolella* was restricted to central North America but became widespread during the Permian Period.

Most other important zonal index genera are cosmopolitan in distribution including *Gnathodus, Idiognathodus, Spathognathus,* and *Siphonodella*. Several genera have patchy distributions but are reported from many different parts of the world; these include *Idiognathoides, Scaphignathus,* and *Streptognathodus*. Conodont distributions in the Carboniferous are not particularly provincial and many of the patterns of distribution may be the result of inadequate sampling of all of the biofacies for each zone (DRUCE, 1973). Most conodont studies have concentrated on Lower Carboniferous (Tournaisian, Visean, and lower Namurian) strata where other faunas generally also suggest cosmopolitan distributions. At present there is insufficient knowledge of Middle and Upper Carboniferous conodonts to evaluate their geographic distributions.

INSECTS

Insect fossils from the base of the Westphalian through the remainder of the Carboniferous have the potential of being extremely useful as index fossils (Fig. 8, 9). They show considerably greater taxonomic diversity and more rapid evolution than do the fossil flora with which they are commonly associated. Many insects have stratigraphic ranges that are closely similar in North America and Europe and relatively few have strongly endemic distributions.

Although relatively few of the 17 orders of Insecta of the Carboniferous have been studied in terms of their world paleogeography, several analyses have examined the evidence from one or two orders in a more limited geographic framework. The early Insecta of the eastern Canadian (Maritime provinces) Carboniferous have a pronounced number of European elements (DURDEN, 1967, 1969) as do the plant fossils (BELL, 1938). Some European elements appear farther to the southwest and some American elements appear to the east in Wales and France, suggesting a transitional belt of provincial differentiation and a lack of major physical barriers. Blattoid (cockroach-like) insects appear in lower or middle Namurian strata of central Europe and upper Namurian of western Europe and eastern Canada. By early Westphalian time, the insect assemblage, characterized by blattoids, dispersed into central eastern North America (Pennsylvania and Maryland) and two parallel bands of geographically distinct assemblages (but with overlapping boundaries) are recognized. By middle Westphalian time, additional assemblages are recognized and have distributions that also form subparallel geographic bands from central North America into western Europe. During the later part of Westphalian time several earlier assemblages were reduced in size but the assemblage from the southern part of western Europe expanded markedly over a much larger area. Several assemblages were missing during this time, but because they (or their descendants) reappear in younger strata, DURDEN (1974) assumed that they were displaced geographically outside present collecting localities.

The early Stephanian of Europe was complicated by the Asturian orogenic phase, which disrupted many of the Westphalian depositional patterns and caused major changes in insect distributions and also in

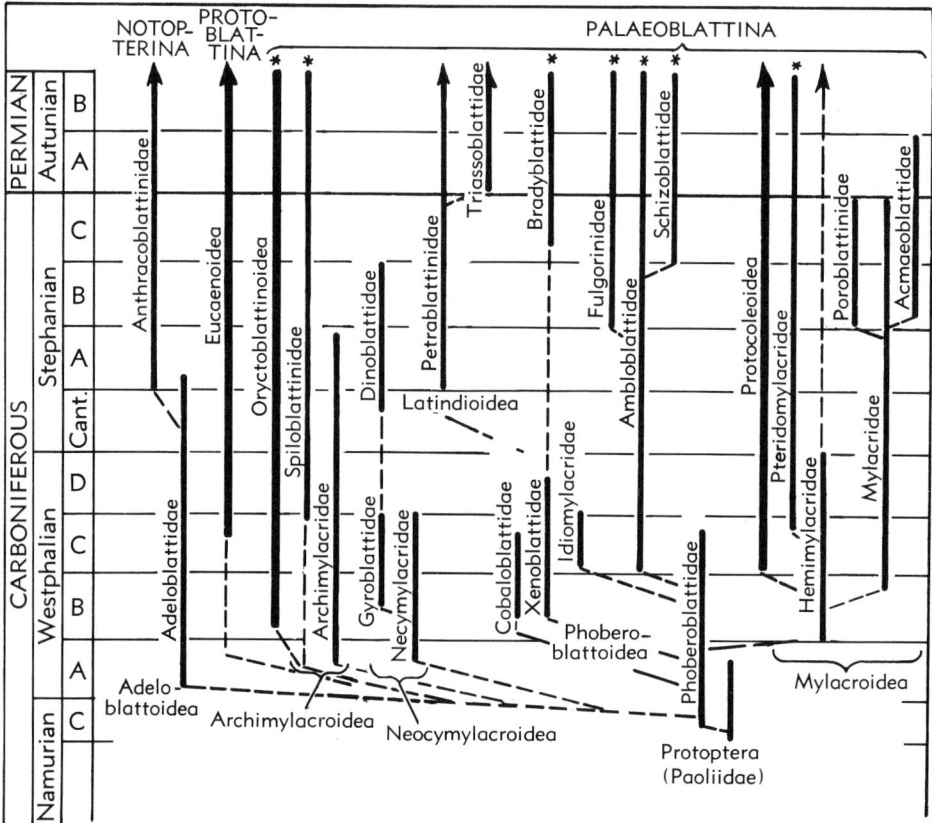

Fig. 8. Ranges of families, superfamilies, and suborders of early blattoid-like insects and phylogenetic relations. Lineages marked with an asterisk became extinct during the Permian. Lineages with arrows have given rise to recent descendants. (Data from Durden, 1969; Ross, n.)

plant distributions (DE MAISTRE, 1970). Southwest European insect assemblages of this age are found as far west as Pennsylvania associated with the remnants of an earlier north-central (Mazonian) North American assemblage. Many of the non-blattoid insect assemblages of the Stephanian include a significant percentage of new groups that have no known immediate ancestors suggesting dispersal from earlier isolated provinces that are not known at present. During the Stephanian an eastern North American (Ottweilerian) assemblage developed and expanded westward into Missouri and eastward into the Saar basin. During middle Stephanian time (Missourian-early Virgilian) the southwest European (Iberian) assemblage occurred in Portugal, southern Pennsylvania, Missouri, and possibly in Brazil. Just to the north of the Iberian assemblage the Ottweilerian assemblage continued to be widespread, reaching from eastern Germany to Pennsylvania and, as mixed faunas, as far west as central Colorado. In Kansas, the beginnings of two assemblages are seen, the Zavjalovian and the Elmoan, which are mixed with Ottweilerian genera. The Elmoan assemblage has many new genera and these are presumed to have dispersed into Kansas at this time from an unknown area to the northwest.

During late Stephanian to early Autunian time the Ottweileran insect assemblage extended from Wettin, Germany, through Pennsylvania into Colorado. At Wettin this assemblage also included forms similar to those reported from strata of about the same age in the Kuznetzk basin, Siberia. The

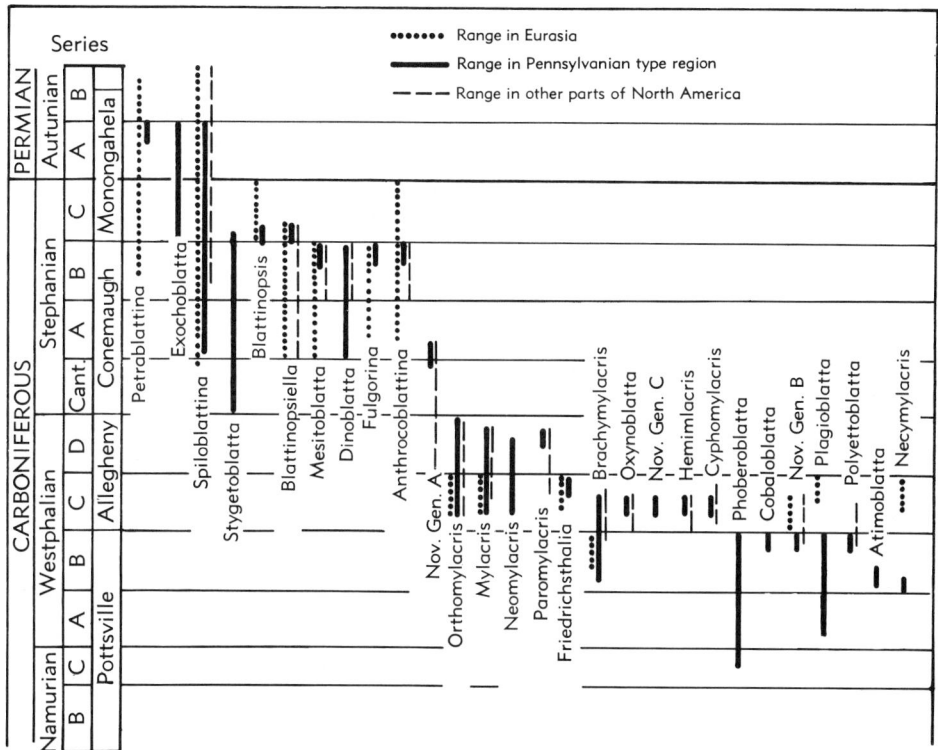

Fig. 9. Ranges of blattoid-like insect genera having multiple occurrences in Eurasia, Pennsylvania (type region), and other parts of North America. (Data from Durden, 1969; Ross, n.)

Iberian assemblage occurred to the south in Texas at this time.

The late Virgilian and Wolfcampian distribution of blattoid-like insects showed a continuation of the Iberian assemblage in Texas. The Lebachian assemblage, a new, rich and varied assemblage, is known mainly from this interval in Germany and West Virginia, with a few representatives in Texas and Kansas. In Colorado parts of this assemblage are mixed with genera of Ottweilerian and Zavjalovian affinities in a predominantly Elmoan fauna. DURDEN (1974) was able to trace many of these groups through the Early Permian and showed that most of the three Early Permian assemblages had their origins during the Late Carboniferous.

BIOGEOGRAPHIC INTERPRETATION

Three general trends appear from the foregoing analyses of the distributions of Carboniferous invertebrate faunas; one has evolutionary implications and the other two have climatic fluctuation and climatic zonation implications. Faunas of Tournaisian and Visean times were relatively widespread; many were nearly cosmopolitan and the remainder had low levels of provincialism. These were also times of widespread, shallow, warm-water carbonate deposition, particularly on continental platforms. As a great percentage of the area of these platforms was submerged, world sea level was apparently relatively high and the marine connections between adjacent platforms were well established. After the Visean, Carboniferous faunas became increasingly provincial with several times of infrequent to frequent dispersal between marine prov-

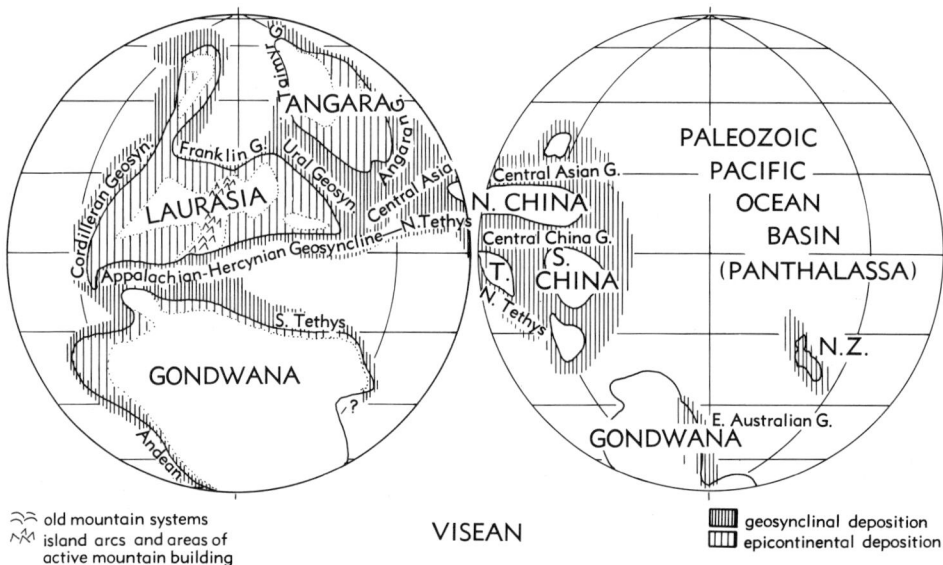

Fig. 10. Reconstruction of Carboniferous paleogeography during an interval of relatively high sea level: middle Early Carboniferous (Visean) (Ross, n). [Explanation: N.Z., New Zealand; T., Tarim stable block.]

inces until the end of the Carboniferous.

The later part of the Early Carboniferous was the beginning of Hercynian mountain building and saw an increase in volcanic activity in parts of Europe and western Siberia. These events are also associated with the beginning of cyclic sedimentation, the reduction in warm-water carbonate deposition, and probably the initiation of glaciation in parts of the Gondwana continent in conjunction with general lowering of world temperatures. During the later part of the Early Carboniferous the average sea level declined and reached a minimum level before the beginning of the Middle Carboniferous. Most of the continental shelves were extensively exposed to erosion at this time (EYNOR et al., 1965). Dark shale, sandstone and limestone are typical deposits and marine connections between regions were mainly by seaways through those geosynclines that had not been disrupted by orogenic activities. Much of the Hercynian orogeny was well advanced by this time and, during the later part of the Early Carboniferous and Middle and Late Carboniferous, this fold belt prevented direct marine connection between western North America and southeastern Europe.

During Middle Carboniferous time, sea level gradually rose accompanied by a series of transgressions and regressions, reaching a comparatively high level during the Desmoinesian when seas covered much of the continental platforms. As world sea level started to fall again near the end of the Middle Carboniferous a number of changes started to take place in marine invertebrate faunas. Many of those faunal groups that had held dominant positions in Middle Carboniferous communities became gradually less important and a number of transitional forms appeared which evolved into new genera, and also new families, in the early part of the Late Carboniferous. The end of the Middle Carboniferous is marked by a reduction in world sea level of almost the same magnitude, but of shorter duration, as that at the end of the Early Carboniferous. A number of previously important genera and families of the Early and Middle Carboniferous did not survive into the Late Carboniferous.

During the Late Carboniferous the new lineages of marine invertebrate faunas evolved into numerous genera and families and gradually formed a large variety of communities. The faunas and communities

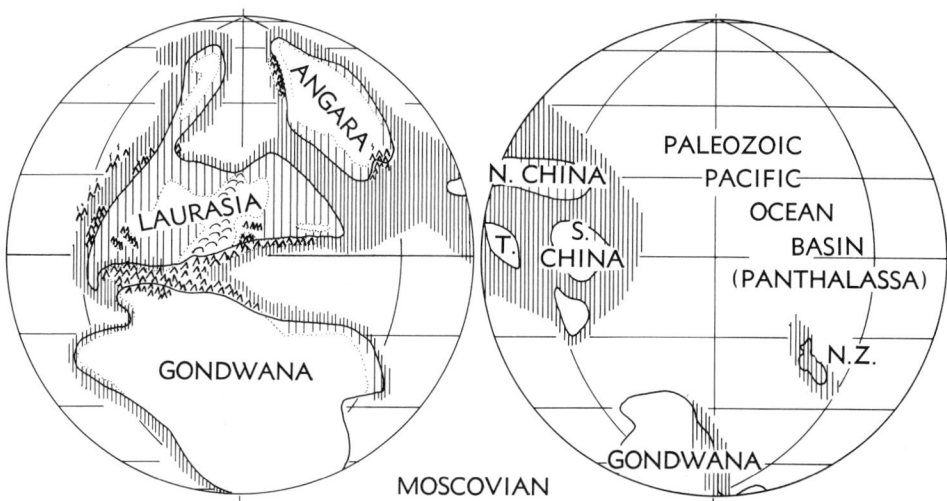

Fig. 11. Reconstruction of Carboniferous paleogeography during an interval of relatively high sea level: late Middle Carboniferous (Moscovian) (Ross, n). [See Fig. 10 for explanation of abbreviations and symbols.]

were not highly specialized, suggesting that they were adapted to fluctuating climatic conditions. Although cyclic sedimentation continued through the Late Carboniferous, local variations in the sedimentary cycles became more pronounced and, in general, the volume of clastic material became greater. This suggests orogenic activity had reached a level that significantly altered local climatic patterns and related erosional rates and that interglacial intervals were warmer, dryer, and probably of longer duration. The faunal changes at the end of the Carboniferous and the beginning of the Permian, while significant, were concurrent with specialization of existing Late Carboniferous families into an increasingly complex set of Permian communities and increased dispersals between faunal provinces.

Climatic zonation in the Carboniferous was first recognized in the distribution of Carboniferous plants, which may exhibit seasonal growth features such as growth rings. Carboniferous floras (MEYEN, 1972) became differentiated into at least two regions near the beginning of the Namurian, i.e., the Euramerian and Angaran regions, characterized largely by different ecological associations of lepidophytes. The Euramerian flora was tropical but it has not been established yet that the Angaran woods were extra-tropical at this time. Middle Carboniferous floras of the Euramerian and Angaran regions show substantial differences in composition and woods from the Angaran region consistently have annual rings. Early and Middle Carboniferous floras from Gondwana are poorly known and fossil wood data are not available for comparison.

Late Carboniferous plants form four paleofloral regions, Euramerian, Cathaysian, Angaran, and Gondwanan (JONGMANS, 1936; HALLE, 1937). The Euramerian and Cathaysian floral regions differ mainly in the presence of large numbers of endemic genera. These two floras are diverse, generally lack annual rings and are considered tropical. The Angaran flora had a few Euramerian elements but endemic genera were dominant by the end of the Late Carboniferous. This flora had annual growth rings and is considered extratropical. The Gondwanan flora is poorly known from strata older than Late Carboniferous; however, by Late Carboniferous time it was widespread and remarkably uniform in species composition. This early Gondwanan flora is associated with tillites and possibly was contemporaneous with glaciation. The flora was deciduous and had annual rings, a relatively low number of genera and species and is considered to be extra-tropical. The Gondwanan flora was isolated from other floras at this time, having few, if any, genera and species in common with the Euramerian, Cathaysian, or Angaran floras.

Some marine invertebrate evidence also suggests well-developed climatic zonation during the Carboniferous. A tropical foraminiferid faunal belt is identified for the Early and Middle Carboniferous based on greater species diversity and dominance of calcareous taxa (MAMET & SKIPP, 1970). The position of this tropical belt corresponds closely to the position of tropical climates as suggested from plant data. Away from this tropical belt foraminiferids gradually become less diverse and arenaceous taxa become dominant. The margins of the Angaran continent and the Gondwanan continent are identifiable by these two changes (YUFEREV, 1969). Also, ammonoid distributions appear to be restricted to a band about 50° to 60° on either side of a Carboniferous paleoequator (RAMSBOTTOM, 1971).

Two paleogeographic maps, one of Visean time (Fig. 10) and the other of Moscovian times of major marine transgressions onto the cratonic platforms. During Visean time, marine connections existed in an east-west direction near the paleoequator from western North America to eastern Asia and this apparently accounts for the nearly cosmopolitan nature of the equatorial marine faunas. By Moscovian time, the Hercynian orogeny had closed off these east-west marine connections along the paleoequator and dispersal of tropical marine faunas between different provinces became more difficult because these organisms had to pass through extratropical areas in order to disperse into other provinces. These generalized maps leave many details to be explained; for example, the Middle Carboniferous marine faunas of one tectonic belt in the western Cordillera of North America are probably tropical whereas faunas in the adjacent tectonic belt are probably nontropical, suggesting a more complicated history of geographic displacement than indicated by these maps. Of particular concern is the need for additional criteria and evidence that would establish the relative positions during the Carboniferous of the cratonic blocks that presently form southern and southeastern Asia.

REFERENCES

Amos, A. J., & Sabattini, N., 1969, *Palaeozoic faunal similitude between Argentina and Australia:* Gondwana Stratigraphy, I.U.G.S. Symposium, Buenos Aires, 1967, UNESCO, Earth Sci. ser. 2, p. 235-248.

Ayzenverg, D. Ye. (ed.), 1969, *Stratigrafiya URSR; v. 5, Karbon:* Akad. Nauk URSR, Inst. Geol. Nauk, Kiev, 412 p. [*Stratigraphy of the Ukraine; v. 5, Carboniferous.*]

Barkhatova, V. P., Kruchinina, O. N., Lyuber, A. A., & Simakova, M. M., 1970, *On the Middle and Upper Carboniferous of the European part of the USSR:* 6ᵉ Congr. Internatl. Strat. Géol. Carbonifère, Sheffield, 1967, Compte Rendu 2, p. 453-458.

Bell, W. A., 1938, *Fossil flora of the Sydney Coal Field, Nova Scotia:* Geol. Survey Canada, Mem., no. 215, 334 p., 109 pl.

Bisat, W. S., 1924, *The Carboniferous goniatites of the north of England and their zones:* Yorkshire Geol. Soc., Proc., v. 20, p. 40-124.

Bouroz, Alexis, 1969, *Les difficultés du choix d'un stratotype dans les séries purement continentales houillères:* 6ᵉ Congr. Internatl. Strat. Géol. Carbonifère, Sheffield, 1967, Compte Rendu 1, p. 183-184.

Branson, C. C. (ed.), 1962, *Pennsylvanian System in the United States:* 508 p., Am. Assoc. Petrol. Geologists (Tulsa, Okla.).

Calver, M. A., 1969, *Westphalian of Britain:* 6ᵉ Congr. Internatl. Strat. Géol. Carbonifère, Sheffield, 1967, Compte Rendu 1, p. 233-254.

Campbell, K. S. W., Dear, J. F., Rattigan, J. H., & Roberts, John, 1969, *Correlation chart for the Carboniferous System in Australia:* Gondwana Stratigraphy, I.U.G.S. Symposium, Buenos Aires, 1967, Earth Sci. ser. 2, p. 471-474.

———, & McKellar, R. G., 1969, *Eastern Australian Carboniferous invertebrates: Sequence and affinities:* in Stratigraphy and paleontology: essays in honour of Dorothy Hill, K. S. W. Campbell (ed.), p. 77-119, National University Press (Canberra, Australia).

Caster, K. E., 1952, *Stratigraphic and paleontologic data relevant to the problem of Afro-American ligation during the Paleozoic and Mesozoic:* in The problem of land connections across the South Atlantic, with special reference to the Mesozoic, E. Mayr (ed.), Am. Museum Nat. History, Bull. 99, Art. 3, p. 105-152.

Collinson, Chas., Rexroad, C. B., & Thompson, T. L., 1971, *Conodont zonation of the North American Mississippian:* Geol. Soc. America, Mem., v. 127, p. 353-394, text-fig. 1-8.

Conil, Raphaël, 1964, *Localités et coupes types pour l'étude du Tournaisien inférieur (révision des limites sous l'aspect micropaléontologique)*: Acad. Royale Belgique (Cl. Sci.), Mém., sér. 2, v. 20, p. 616-655.

———, & Lys, Maurice, 1968, *Utilisation stratigraphique des foraminifères du Dinantien*: Soc. Géol. Belgique, Ann., v. 91, p. 491-536.

———, Paproth, Eva, & Lys, Maurice, 1968, *Mit Foraminiferen gegliederte Profile aus dem nordwest-deutschen Kohlenkalk und Kulm. Mit einem paläontologischen Anhang*: Decheniana, v. 119, p. 51-94.

Conybeare, W. D., & Phillips, Wm., 1822, *Outlines of the geology of England and Wales*: 542 p., George Yard (London).

Dalmatskaya, I. I., Latskova, V. E., Orlova, I. N., Rauzer-Chernousova, D. M., Reytlinger, Ye. A., Safonova, T. P., Semikhatova, E. N., & Chernova, E. I., 1961, *Stratigrafiya srednekamennougolnykh otlozheniy tsentralnoy i vostochnoy chasti Russkoy platformy (na osnove izucheniya foraminifer); 2, Povolzhe i prikame*: Akad. Nauk SSSR, Geol. Inst., Regionalnaya stratigrafiya SSSR, v. 5, 356 p. [*Stratigraphy of the Middle Carboniferous deposits of central and eastern part of the Russian platform (based on foraminiferal studies)*.]

Dobrolyubova, T. A., Kabakovich, N. V., & Sayutina, T. A., 1966, *Korally nizhnego carbona Kuznetskoy kotloviny*: Akad. Nauk SSSR, Paleont. Inst., Trudy, v. 111, p. 1-276. [*Lower Carboniferous corals of Kuznetsk basin.*]

Drahovzal, J. A., 1972, *The Lower Carboniferous ammonoid genus Goniatites*: Internatl. Paleont. Union, Proc., 23rd Sess. Internatl. Geol. Congr. Czechoslovakia 1968, p. 15-52, Pol. Inst. Geol. (Warsaw).

Druce, E. C., 1973, *Upper Paleozoic and Triassic conodont distribution and the recognition of biofacies*: Geol. Soc. America, Spec. Paper, no. 141, p. 191-237, text-fig. 1.

Durden, C. J., 1967, *Faunal affinities of the Acadian Upper Carboniferous insects*: Geol. Assoc. Canada-Mineralog. Assoc. Canada, Internatl. Mtg., Kingston, Ontario, Abstr. Paper, p. 25-26.——— 1969, *Pennsylvanian correlation using blattoid insects*: Canadian Jour. Earth Sci., v. 6, p. 1159-1177.———1974, *Biomerization: an ecologic theory of provincial differentiation*: Soc. Econ. Paleontologists & Mineralogists, Spec. Publ. no. 21, p. 18-53.

Easton, W. H., 1945, *Kinkaid corals from Illinois*: Jour. Paleontology, v. 19, p. 383-389, text-fig. 1-8.

Eynor, O. L., 1970, *Serpukhovskiy yarus i ego polozhenie v kamennougolnoy sisteme*: Problemy stratigrafii Karbona, Akad. Nauk SSSR, Minist. Geol. SSSR, Mezhvedomstvennyy Stratigraficheskiy Komitet SSSR, Trudy, v. 4, p. 107-122. [*Serpukhovian stage and its place in the Carboniferous system.*]

———, Alexandri-Sadova, T. A., Betechtina, O. A., Vasilyuk, N. P., Vdovenko, M. V., Gorak, S. V., Dunaeva, N. N., Radtchenko, G. P., Sergeeva, M. T., Solovjeva, M. N., & Stshegolev, A. K., 1973, *Correlation and evolution of the paleobiogeographic units of the Carboniferous sea and land in the Northern Hemisphere*: 7e Congr. Internatl. Strat. Géol. Carbonifère, Krefeld, 1971, Compte Rendu 2, p. 441-447.

———, Belgovskiy, G. L., & Smirnov, G. A., 1965, *Osnovnyye cherty geologicheskogo razvitiya i paleogeografiya territorii SSSR v kamennougolnom periode*: Sovet. Geol., 1965, no. 8, p. 32-44, text-fig. 1,2. [*Fundamental features of geological development and paleogeography of the territories of the USSR in the Carboniferous Period.*] (Transl. Internatl. Geol. Rev., 1970, v. 12, no. 2, p. 105-113.)

Flügel, Helmut, 1966, *Paläozoische Korallen aus der Tibetischen Zone von Dolpo (Nepal)*: Geol. Bundesanst. Jahrb., v. 12, p. 101-120.

Fontaine, Henri, 1961, *Les madréporaires paléozoïques du Viêt-nam, du Laos et du Cambodge*: Archiv. Géol. Viêt-Nam, v. 5, p. 1-276.

Francis, E. H., & Woodland, A. W., 1964, *The Carboniferous Period*: in The Phanerozoic timescale; a symposium, Geol. Soc. London, Quart. Jour., v. 120, Suppl., p. 221-232.

George, T. N., 1969, *British Dinantian stratigraphy*: 6e Congr. Internatl. Strat. Géol. Carbonifère, Sheffield, 1967, Compte Rendu 1, p. 193-218.

Gordon, Mackenzie, Jr., 1970, *Carboniferous ammonoid zones of the south-central and western United States*: 6e Congr. Internatl. Strat. Géol. Carbonifère, Sheffield, 1967, Compte Rendu 2, p. 817-826.———1974, *The Mississippian-Pennsylvanian boundary in the United States*: 7e Congr. Internatl. Strat. Géol. Carbonifère, Krefeld, 1971, Compte Rendu 3, p. 129-141.

Groot, G. E. de, 1963, *Rugose corals from the Carboniferous of northern Palencia (Spain)*: Leidse Geol. Meded., v. 29, 124 p., 26 pl.

Halle, T. G., 1937, *The relations between the late Palaeozoic floras of eastern and northern Asia*: Deuxième Congr. pour l'Avancement des Études de Stratigraphie et de Géologie du Carbonifère, Heerlen, 1935, Compte Rendu 1, p. 237-245.

Hill, Dorothy, 1948, *The distribution and sequence of Carboniferous coral faunas*: Geol. Mag., v. 85, p. 121-148, text-fig. 1-5.———1957, *The sequence and distribution of upper Palaeozoic coral faunas*: Austral. Jour. Sci., v. 19, no. 3a, p. 42-61, text-fig. 1.———1973, *Lower Carboniferous corals*: in Atlas of palaeobiogeography, A. Hallam (ed.), p. 133-145, text-fig. 1, Elsevier (Amsterdam, London, & New York).

Hodson, F., & Ramsbottom, W. H. C., 1973, *The distribution of Lower Carboniferous goniatite faunas in relation to suggested continental reconstruction for the period*: Palaeontology, Special Paper, no. 12, p. 321-329.

Hudson, R. G. S., 1945, *The goniatite zones of the Namurian:* Geol. Mag., v. 82, p. 1-9.

Ivanova, Ye. A., 1972, *Osnovnye zakonomernosti evolyutsii spiriferid (Brachiopoda):* Paleont. Zhurnal 1972, p. 28-42, text-fig. 1-5. [*The main features of spiriferid evolution (Brachiopoda).*] (Transl. Paleont. Jour. 1973, v. 6, no. 3, p. 309-329.)

Ivanovskiy, A. B., 1967, *Etyudy o rannekamennougolnykh rugozakh:* Akad. Nauk SSSR, Sibir. Otdel., Inst. Geol. i Geofiz., 86 p. [*Study of Early Carboniferous Rugosa.*]

Jongmans, W. J., 1936, *Floral correlations and geobotanic provinces within the Carboniferous:* 16th Internatl. Geol. Congr., Proc., Washington, D.C., 1933, v. 1, p. 519-527.

Khalymbadzha, V. G., & Tikhvinskiy, I. N., 1967, *Smena kompleksov brakhiopod v pozdnem Paleozoye Volgo-Kamskogo kraya:* Paleont. Zhurnal, 1967, no. 2, p. 28-36. [*Succession of brachiopod complexes in the Late Paleozoic of the Volga-Kama Territory.*] (Transl. Paleont. Jour., 1967, no. 2, p. 19-26.)

Lane, H. R., Merrill, G. K., Straka, J. J., II, & Webster, G. D., 1971, *North American Pennsylvanian conodont biostratigraphy:* Geol. Soc. America, Mem., v. 127, p. 395-414.

Liabeuf, J. J., & Alpern, B., 1969, *Étude palynologique du bassin houiller de St. Étienne, stratotype du Stephanien:* 6ᵉ Congr. Internatl. Strat. Géol. Carbonifère, Sheffield, 1967, Compte Rendu 1, p. 155-169.

Lipina, O. A., 1963, *Ob etapnosti razvitiya turneishkikh foraminifer:* Voprosy Mikropaleont., v. 7, p. 13-21. [*On stages in the evolution of the Tournaisian Foraminifera.*]——1964, *Stratigraphie et limites du Tournaisien en U.R.S.S. d'après les Foraminifères:* 5ᵉ Congr. Internatl. Strat. Géol. Carbonifère, Paris, 1963, Compte Rendu 2, p. 539-551.——1965, *Sistematika Turneyellid:* Akad. Nauk SSSR, Geol. Inst., Trudy, v. 130, 114 p., 24 pl. [*Systematics of the Tournayellidae.*]——1970, *Evolyutsiya dvuryadnukh pryamolineynykh rannekamennougolnykh foraminifer:* Voprosy Mikropaleont., v. 13, p. 3-29. [*Evolution of biserial Early Carboniferous Foraminifera.*]——1973, *Zonalnaya stratigrafiya i paleobiogeografiya turne po foraminiferam:* Voprosy Mikropaleont., v. 16, p. 3-35. [*Stratigraphic zonation and paleogeographic distribution of Tournaisian Foraminifera.*]

——, & Reytlinger, Ye. A., 1970, *Stratigraphie zonal et paléozoogéographie du Carbonifère inférieur d'après les foraminifères:* 6ᵉ Congr. Internatl. Strat. Géol. Carbonifère, Sheffield, 1967, Compte Rendu 3, p. 1101-1112.

Maistre, Jacques de, 1970, *Variations latérales de la flore du Carbonifère et les problèmes qu'elles posent pour le choix des stratotypes et la définition de leurs limites:* 6ᵉ Congr. Internatl. Strat.

Géol. Carbonifère, Sheffield, 1967, Compte Rendu 3, p. 1113-1120.

Mamet, B. L., 1962, *Remarques sur la microfaune de Foraminifères du Dinantien:* Soc. Géol. Belgique, Ann., v. 70, no. 2, p. 166-173.——1968, *The Devonian-Carboniferous boundary in Eurasia:* Internatl. Symposium on the Devonian System, v. 2, p. 995-1007, Alberta Soc. Petrol. Geology (Calgary).

——, & Skipp, Betty, 1971, *Lower Carboniferous calcareous Foraminifera: Preliminary zonation and stratigraphic implications for the Mississippian of North America:* 6ᵉ Congr. Internatl. Strat. Géol. Carbonifère, Sheffield, 1967, Compte Rendu 3, p. 1129-1146.

——, ——, Sando, W. J., & Mapel, W. J., 1971, *Biostratigraphy of upper Mississippian and associated Carboniferous rocks in south-central Idaho:* Am. Assoc. Petrol. Geologists, Bull., v. 55, no. 1, p. 20-33.

Meischner, Dieter, 1970, *Conodonten-Chronologie des deutschen Karbons:* 6ᵉ Congr. Internatl. Strat. Géol. Carbonifère, Sheffield, 1967, Compte Rendu 3, p. 1169-1180.

Meyen, S. V., 1972, *On the origin and relationship of the main Carboniferous and Permian floras and their bearing on general paleogeography of this time:* Symposium on Gondwana stratigraphy and palaeontology, South Africa, 2nd mtg., 1970, 9 p.

Miller, A. K., Furnish, W. M., & Schindewolf, O. H., 1957, *Paleozoic Ammonoidea:* in Treatise on invertebrate paleontology, Part L, R. C. Moore (ed.), p. L11-L79, text-fig. 1-123, Geol. Soc. America & Univ. Kansas Press (New York; Lawrence, Kans.).

Minato, Masao, & Kato, Makoto, 1965a, *Durhaminidae (tetracoral):* Hokkaido Univ., Jour. Fac. Sci., ser. 4, Geol. Miner., v. 13, p. 11-86.——1965b, *Waagenophyllidae:* Hokkaido Univ., Jour. Fac. Sci., v. 12, no. 3, 241 p.

Moore, R. C., et al., 1944, *Correlation of Pennsylvanian formations of North America:* Geol. Soc. America, Bull., v. 55, p. 657-706.

Mortelmans, G., 1969, *L'étage Tournaisien dans sa localité-type:* 6ᵉ Congr. Internatl. Strat. Géol. Carbonifère, Sheffield, 1967, Compte Rendu 1, p. 19-43.

Muir-Wood, Helen, & Cooper, G. A., 1960, *Morphology, classification and life habits of the Productoidea (Brachiopoda):* Geol. Soc. America, Mem., v. 81, 447 p., 135 pl.

Murchison, R. I., 1841, *First sketch of some of the principal results of a second geological survey of Russia, in a letter to M. Fischer:* Philos. Mag., v. 19, p. 419-422.

Ramsbottom, W. H. C., 1969, *The Namurian of Britain:* 6ᵉ Congr. Internatl. Strat. Géol. Carbonifère, Sheffield, 1967, Compte Rendu 1, p. 219-232.——1971, *Palaeogeography and gonia-*

tite distribution in the Namurian and early Westphalian: 6° Congr. Internatl. Strat. Géol. Carbonifère, Sheffield, 1967, Compte Rendu 4, p. 1395-1400.

Rauzer-Chernousova, D. M., 1948, *Nekotorye novye vidy foraminifer iz nizhnekamennougolnye otlozheniy Podmoskovnogo basseyna:* Akad. Nauk SSSR, Inst. Geol. Nauk, Trudy, v. 62, ser. geol., no. 19, p. 239-243, pl. 17. [*Some new Foraminifera from the Lower Carboniferous deposits in the vicinity of the Moscow basin.*]

——, Gryzlova, N. D., Kireeva, G. D., Leontovich, G. E., Safonova, T. P., & Chernova, E. I., 1951, *Srednekamennougolnye fuzulinidy Russkoy platformy i sopredelnykh oblastey:* 380 p., 30 text-fig., 58 pl., Akad. Nauk SSSR, Inst. Geol. Nauk (Moskva). [*Middle Carboniferous fusulinids of the Russian platform and nearby districts.*]

——, Reytlinger, Ye. A., Balashova, N. N., Dalmatskaya, I. I., & Chernova, E. I., 1954, *Stratigrafiya srednekamennougolnykh otlozheniy tsentraloy i vostochnoy chasti Russkoy platformy (na osnove izucheniya foraminifer); 1, Moskovskaya sinekliza:* Akad. Nauk SSSR, Inst. Geol. Nauk, Regionalnaya stratigrafiya SSSR, v. 2, 270 p. [Moskva]. [*Stratigraphy of the Middle Carboniferous deposits of central and eastern parts of the Russian platform (based on study of Foraminifera).*]

Read, C. B., & Mamay, S. H., 1964, *Upper Paleozoic floral zones and floral provinces of the United States:* U.S. Geol. Survey, Prof. Paper, no. 454-K, p. K1-K35, 19 pl.

Reytlinger, Ye. A. 1960, *Znachenie foraminifer dlya stratigrafii nizhnego karbona:* Mezhdunar. Geol. Congress, 21st Sess., Probl. 6, p. 56-64. [*Significance of Foraminifera to the stratigraphy of the Lower Carboniferous.*]——1962, *Limits of Lower Carboniferous in the stratigraphic diagram of the USSR based on foraminiferal fauna:* 4° Congr. Internatl. Strat. Géol. Carbonifère, Heerlen, 1958, Compte Rendu 3, p. 591-598. ——1963, *Ob odnom paleontologicheskom kriterii ustanovleniya granits nizhnekamennougolnogo otdela po faune foraminifer:* Voprosy Mikropaleont., v. 7, p. 22-56. [*Concerning a single paleontological criterion establishing the boundary of the Lower Carboniferous subdivisions based on foraminiferal faunas.*]——1969, *Etapnost razvitiya foraminifer i ee znachenie dlya stratigrafii kamennougolnykh otlozheniy:* Voprosy Mikropaleont., v. 12, p. 3-33. [*Stages in the evolutionary development of Foraminifera and their significance for the stratigraphy of Carboniferous deposits.*]

Rhodes, F. H. T., & Austin, R. L., 1971, *Carboniferous conodont faunas of Europe:* Geol. Soc. America, Mem., v. 127, p. 317-352, pl. 1,2.

Roberts, John, 1975, *Early Carboniferous brachiopod zones of eastern Australia:* Geol. Soc. Australia, Jour., v. 22, pt. 1, p. 1-31.

Ross, C. A., 1967, *Development of fusulinid (Foraminiferida) faunal realms:* Jour. Paleontology, v. 41, p. 1341-1354, text-fig. 1-9.——1970, *Concepts in Late Paleozoic correlations:* Geol. Soc. America, Spec. Paper, v. 124, p. 7-36, text-fig. 1-10.——1973, *Carboniferous Foraminiferida:* in Atlas of palaeogeography, A. Hallam (ed.), p. 127-132, Elsevier (Amsterdam, London, & New York).

Rowett, C. L., 1969, *Upper Palaeozoic stratigraphy and corals from the east-central Alaska Range, Alaska:* Arctic Inst. North America, Tech. Paper, no. 23, 120 p.

Rozovskaya, S. Ye., 1950, *Rod Triticites, ego razvitie i stratigraficheskoe znachenie:* Akad. Nauk SSSR, Paleont. Inst., Trudy, v. 26, p. 1-78, pl. 1-10. [*Genus Triticites, its development and stratigraphic significance.*]——1969, *K revisii otryada Fusulinida:* Paleont. Zhurnal, no. 3, p. 34-44. [*Revision of the Order Fusulinida.*] (Transl. Paleont. Jour. 1969, no. 3, p. 317-325.)

Rudwick, M. J. S., 1970, *Living and fossil brachiopods:* 199 p., Hutchinson Univ. Library (London).

Ruzhentsev, V. E., 1960, *Printsipy sistematiki sistema i filogeniya Paleozoyskikh ammonoidey:* Akad. Nauk SSSR, Paleont. Inst., Trudy, v. 83, 331 p., 128 text-fig. [*Systematic principles for the taxonomy and phylogeny of Paleozoic ammonoids.*]——1965, *Osnovnye kompleksy ammonoidey kamennougolnogo perioda:* Paleont. Zhurnal, 1966, no. 2, p. 3-17. [*Principal ammonoid assemblages of the Carboniferous Period.*] (Trans. Internatl. Geol. Review, v. 8, no. 1, p. 48-59.)

Sandberg, C. A., Streel, M., & Scott, R. A., 1972, *Comparison between conodont zonation and spore assemblages at the Devonian-Carboniferous boundary in the western and central United States and in Europe:* 7° Congr. Internatl. Strat. Géol. Carbonifère, Krefeld, 1971, Compte Rendu 1, p. 179-203.

Sando, W. J., 1965, *Revision of some Paleozoic coral species from the Western United States:* U.S. Geol. Survey, Prof. Paper, no. 503-E, p. E1-E38.

——, Mamet, B. L., & Dutro, J. T., Jr., 1969, *Carboniferous megafaunal and microfaunal zonation in the northern Cordillera of the United States:* U.S. Geol. Survey, Prof. Paper, no. 613-E, p. E1-E29.

Sedgwick, Adam, & Murchison, R. I., 1839, *On the older rocks of Devonshire and Cornwall:* Geol. Soc. London, Proc., v. 3, no. 63, p. 121-123.

Smith, Stanley, 1948, *Carboniferous corals from Malaya:* in Malayan Lower Carboniferous fossils and their bearing on the Viséan palaeogeography of Asia, H. M. Muir-Wood (ed.), p. 93-96, Brit. Museum Nat. History (London).

Teichert, Curt, Kummel, Bernhard, Sweet, W. C., Stenzel, H. B., Furnish, W. M., Glenister, B. F., Erben, H. K., Moore, R. C., & Zeller, D. E. N., 1964, *Cephalopoda, general features, Endoceratoidea, Actinoceratoidea, Nautiloidea, and Bactritoidea:* in Treatise on invertebrate paleontology, Part K, R. C. Moore (ed.), 519 p., 369 text-fig., Geol. Soc. America & Univ. Kansas Press (New York & Lawrence, Kans.).

Thompson, M. L., 1948, *Studies of American fusulinids:* Univ. Kansas Paleont. Contrib. 4, Protozoa, Art. 1, 184 p., 38 pl.——1966, *Rasprostranenie profusulinell v Severnoy Amerike:* Akad. Nauk SSSR, Geol. Inst., Voprosy Mikropaleont. v. 10, p. 126-134. [*Distribution of Profusulinella in North America.*]——1967, *American fusulinacean faunas containing elements from other continents:* Univ. Kansas, Dept. Geol. Spec. Publ., no. 2, p. 102-112.

Vasilyuk, N. P., Kachnov, Ye. I., & Pyzhyanov, I. V., 1970, *Paleobiogeograficheskiy ocherk kamennougolnykh i permskikh tselenterat:* in Zakonomernosti rasprostraneniya paleozoyskikh korallov SSSR, D. L. Kalo (ed.), Vses. Simpos. Izuch. Iskop. Korallov SSSR, Trudy II, no. 3, p. 45-60, text-fig. 1-7. [*Paleobiogeographic sketch of Carboniferous and Permian coelenterates.*]

Vaughan, Arthur, 1905, *The palaeontological sequence in the Carboniferous Limestone of the Bristol area:* Geol. Soc. London, Quart. Jour., v. 61, p. 181-307.——1915, *Correlation of Dinantian and Avonian:* Geol. Soc. London, Quart. Jour., v. 71, p. 1-52.

Wagner, R. H., 1969, *Proposal for the recognition of a new 'Cantabrian' Stage at the base of the Stephanian Series:* 6ᵉ Congr. Internatl. Strat. Géol. Carbonifère, Sheffield, 1967, Compte Rendu 1, p. 139-154.

Wanless, H. R., 1969, *Marine and non-marine facies of the Upper Carboniferous of North America:* 6ᵉ Congr. Internatl. Strat. Géol. Carbonifère, Sheffield, 1967, Compte Rendu 1, p. 293-336.

Weller, J. M., et al., 1948, *Correlation of the Mississippian formations of North America:* Geol. Soc. America, Bull., v. 59, p. 91-196.

Williams, H. S., 1891, *Correlation papers: Devonian and Carboniferous:* U.S. Geol. Survey, Bull., v. 80, 279 p.

Winchell, Alexander, 1869, *On the geological age and equivalents of the Marshall Group:* Am. Philos. Soc., Proc., v. 11, p. 57-82.

Yang, K. C., & Wu, W. S., 1964, *The classification and correlation of the Carboniferous System of China:* 5ᵉ Congr. Internatl. Strat. Géol. Carbonifère, Paris, 1963, Compte Rendu 2, p. 853-865.

Yu, C. C., 1934, *Lower Carboniferous corals of China:* Geol. Survey China, Palaeont. Sinica, ser. B, v. 12(1933), no. 3, 221 p., 24 pl.——1937, *The Fenginian (lower Carboniferous) corals of south China:* Acad. Sinica, Inst. Geol., Mem., v. 16, 111 p., 12 pl.

Yuferev, O. V., 1969, *Printsipy paleobiogeograficheskogo rayonirovaniya i podrazdeleniya stratigraficheskoy shkaly:* Geologii i Geofiziki (1969), no. 9, p. 19-28. [*Principles of paleobiogeographic regional classification and subdivision of the stratigraphic scale.*] (Transl. Internatl. Geol. Rev., v. 12, no. 8, 1970, p. 942-948.)

PERMIAN[1]

By CHARLES A. ROSS and JUNE R. P. ROSS
[Western Washington University, Bellingham]

CONTENTS

	PAGE
INTRODUCTION	A291
REGIONAL BIOSTRATIGRAPHY	A293
Permian System in Type Area	A293
Northwestern Europe	A296
Central East Greenland	A297
Other Northern Areas	A298
Tethyan Area	A298
North America	A303
Gondwana Continents	A304
FAUNAL ZONATION AND DISTRIBUTION	A307
Fusulinaceans	A307
Other Foraminiferida	A316
Corals	A316
Bryozoans	A321
Brachiopods	A326
Ammonoids	A330
Conodonts	A337
PALEOGEOGRAPHIC RECONSTRUCTION	A339
SUMMARY AND CONCLUSIONS	A341
REFERENCES	A343

INTRODUCTION

The Permian System was named by MURCHISON in 1841 for strata exposed in a large structural basin on the western flank of the Ural Mountains and takes its name from the Province of Perm about 1,200 km east of Moscow. Along the western flanks of the Urals, Carboniferous strata are overlain by sandstone, conglomerate, evaporite, shale, and marly limestone strata from which MURCHISON identified brachiopods that, in general, are similar to those of Carboniferous faunas elsewhere. They also contain fishes and amphibians similar to those in the Zechstein of the German Dyas, and correlate with the Magnesian Limestone, which in Great Britain overlies the Carboniferous coal measures. MURCHISON realized that the stratigraphic interval above the Carboniferous and below the Triassic must be more significant than suggested by the poorly fossiliferous reddish sandstone, shale, and dolostone of western Europe and that the Russian faunas and stratigraphic succession of the Perm region fitted into

[1] Manuscript received August, 1976.

this part of the geologic column; therefore, the Permian became the only system (except for the North American usage of Mississippian and Pennsylvanian for Carboniferous) to be defined outside of northwestern Europe.

Neither the lower nor upper boundaries of the Permian System were well defined by MURCHISON and this has resulted in some confusion that has been compounded by later, more detailed biostratigraphic studies that have attempted to place the lower boundary at so-called "natural faunal breaks" in the succession. One consequence of this type of faunal approach to studies of Permian strata in the type area is an overemphasis on biostratigraphic units and a neglect of detailed lithologic studies. In the type area the upper boundary of the Permian with the Triassic System is in a clastic redbed continental succession and is defined on vertebrate fossils where these are available. Considering the work of Russian geologists available at the time and the fact that MURCHISON spent only the summers of 1840 and 1841 examining the whole of the geology of Russia in order to prepare his monograph on *The Geology of Russia in Europe and the Ural Mountains*, it is remarkable that there are not more problems with the definition of the Permian System.

As described in more detail by MURCHISON, DE VERNEUIL, and VON KEYSERLING (1845), the strata originally included in the Permian System clearly encompass beds of the Kungurian facies and the Ufimian, Kazanian, and Tatarian stages (DUNBAR et al., 1960). The thick section of Artinskian shale, sandstone, and conglomerate exposed in the southern flanks of the Ural Mountains contains ammonoids (KARPINSKY, 1889) that are considerably more advanced phylogenetically than those known from Upper Carboniferous beds and can be correlated with faunas in other parts of the world that generally have been accepted as of Permian age. KARPINSKY's Artinskian Stage, which he included in the Permian System, has the Sakmarian Limestone at its base. RUZHENTSEV (1936), after a preliminary study of the ammonoids, proposed recognition of the Sakmarian Stage for this limestone and restricted the Artinskian Stage to the beds above the Sterlitamak beds.

MURCHISON considered the lower limestone units exposed on the Ufa plateau and Timan arch to be Upper Carboniferous, on limited fossil evidence. CHERNYSHEV (1902), FREDERIKS (1928, 1932), and TOLSTIKHINA (1935), while studying fossils from these beds, described them as Upper Carboniferous and considered them to constitute the "Uralian Stage" of DE LAPPARENT (1902). Several distinctive fossil zones were eventually recognized and the next to lowest zone was designated as the Zone of "*Schwagerina*" (=*Sphaeroschwagerina*), which was also known from the lower part of the Sterlitamakian Stage in the southern flanks of the Urals. The "Uralian Stage" thus overlapped with the Sterlitamak and Artinskian beds but was not an exact equivalent. The lowest zone of the "Uralian" containing fossils of the Zone of *Triticites* is considered to be Upper Carboniferous, and the name Gzhelian Stage has been adopted for the major part of this Upper Carboniferous sequence.

The question of the placement of the lower boundary of the Permian was discussed by leading Soviet specialists in 1936 prior to the seventeenth International Geological Congress in 1937 and no consensus was reached on whether the base should be at the base of the "*Schwagerina*"-bearing beds, which form the lower part (Asselian) of the Sakmarian Stage, as indicated by RUZHENTSEV's studies, or near the top of the "Uralian" equivalent at a stratigraphic position within the Sterlitamak beds as suggested by RAUZER-CHERNOUSOVA (1940, 1949) based on fusulinacean studies. Earlier, in North America, BEEDE and KNIKER (1924) had recognized the stratigraphic significance of the Zone of *Pseudoschwagerina* and had used that unit to recognize the base of the Permian in North America and elsewhere. It may be argued (DUNBAR, 1940) that at least some Soviet paleontologists had correlated fossils, such as ammonoids, from the Asselian, Sakmarian, and Artinskian with Wolfcampian and Leonardian faunas described from North America. Because these North American faunas had been assigned to the Permian on the strength of their stratigraphic position with respect to the Zone of *Pseudoschwagerina*, many Soviet biostratigraphers decided to assign the Asselian, Sakmarian, and Artinskian stages to the Permian. In so doing a major marine faunal succession was included in the type sections of the Permian System. Chinese biostratigraphers (CHAO, 1965) have apparently maintained that the base of the Permian should not be placed this low and they choose a boundary that lies approximately at the base of the Ufimian Stage or near the base of the "Kungurian" facies.

In 1950, and again in 1960, the Soviet All-Union Institute of Petroleum Geology (VNIGRI) held conferences concerning the boundary between the Carboniferous and Permian and the subdivisions of the Permian System. The major conclusions of these conferences are summarized in Figure 1 (see also MIKLUKHO-MAKLAY, 1963; LIKHAREV, 1966; NALIVKIN, 1973). The base of the Permian System is placed at the base of the fusulinid Zone of *"Schwagerina."* One change in nomenclature that the second of these conferences adopted was the recognition of the Zone of *"Schwagerina"* as the Asselian Stage and the restriction of the Sakmarian Stage to the stratigraphic equivalents of the Tastuba and Sterlitamak beds.

REGIONAL BIOSTRATIGRAPHY

PERMIAN SYSTEM IN TYPE AREA

On the Russian platform and along the western margin of the Ural Mountains (Fig. 1) the Asselian, Sakmarian, and Artinskian stages show numerous facies changes, which also led to early stratigraphic confusion. Large, massive reef mounds are common along the southeastern edge of the Russian platform and these pass eastward into shale and sandstone along the downwarp at the western edge of the Uralian geosyncline. To the west on the Russian platform these stages include limestone, commonly dolomitized or silicified, dolostone, and anhydrite. In the northern part of the Russian platform Asselian and Sakmarian gypsum, anhydrite, and dolostone are widespread and overlain by Artinskian terrestrial red beds. Several structural arches on the Russian platform were exposed during these ages to form low islands. The Uralian geosyncline was in the initial process of being structurally compressed and uplifted, and formed an eastern source of clastic sediments.

Most of the Kungurian facies contains poorly fossiliferous gypsum and red shale (GLUSKO & FEDOROV, 1974). These rock types, although locally missing, are widely distributed on the Russian platform above Artinskian strata. Kungurian marly limestone and dolomitized limestone beds bearing restricted marine faunas are not widely distributed, and contain species having Artinskian affinities as well as other species having Kazanian affinities. NALIVKIN (1973) suggested that many of the evaporite beds presently considered to be Kungurian in age may eventually prove to be in part Artinskian, Ufimian, or Kazanian.

The Ufimian Stage is less continuously distributed and between the Volga River and the Ural Mountains these strata locally appear as red clastic, variegated clay, sandstone, and marly shale beds reaching as much as 150 m in thickness. The fauna is meager and includes fresh-water bivalves and ostracodes in the lacustrine and marly clays. These nonmarine strata of the Ufimian increase to 1,500 m in thickness near the Urals and indicate the presence of the Ural Mountain Range as a source of clastic sediment by Ufimian time. The lateral facies relationships between the Ufimian nonmarine strata, the evaporites of the Kungurian, and the brackish-water deposits of the Kazanian Stage are not well known.

The Kazanian Stage overlies the Kungurian and Ufimian stages, consists of 100 m or less of well-cemented, greenish-gray, impure limestone, clay and marly clay, and has a brackish-water or restricted marine fauna. Lithologically the Kazanian strata resemble the Zechstein beds of western Europe and are traditionally correlated with them. Kazanian strata and faunas are distributed in an elongate basin that extends north-south along the central part of the Russian platform. NALIVKIN (1973) compared these distributions to those of the modern Caspian

PAMIRS	CHINA		JAPAN	AUSTRALIA SYDNEY BASIN		N. AMERICA WEST TEXAS REF. SECS.	
Lower Triassic	Lower Triassic		Lower Triassic	Lower Triassic		?Middle Triassic	
Pamirian	Lopingian	Changhsing Fm.	Mitaian	Newcastle Coal Measures		hiatus	
			— ? —	— ? —		— ? —	
		Wuchiaping Ls.	Kuman	Tomago Coal Measures		Ochoan	
	— ? —					— ? —	
Murgabian	Maokou Ls. (Maokouan)		Akasakan	Mulbring Fm.	Maitland Group / Guadalupian	Capitan Ls. (Capitanian)	
				— ? —			
				Muree Fm.		Word Fm. (Wordian)	
Kubergandinian			Nabeyaman	"Fenestella Zone"	Branxton Fm. / Leonardian	Road Canyon Fm. (Roadian)	
— ? —	— ? —					Cathedral Mt. Fm.	
"Artinskian"	Chihsia Fm. (Chihsian)			Greta Coal Measures		Skinner Ranch Fm.	
— ? —				Farley Fm.			
			Sakamotozawan	Rutherford Fm.	Wolfcampian	Lenox Hills Fm.	
"Sakmarian"	Maping Ls. (Mapingian)			— ? —			
				Allandale Fm.		Neal Ranch Fm.	
				— ? —			
				Lochinvar Fm.			
				— ? —			
			Hikawan				

FIG. 1. Type Permian section, other sections commonly used for regional standard sections, and their approximate correlation. In Turkmen SSR, the Asselian and Sakmarian equivalents commonly are referred

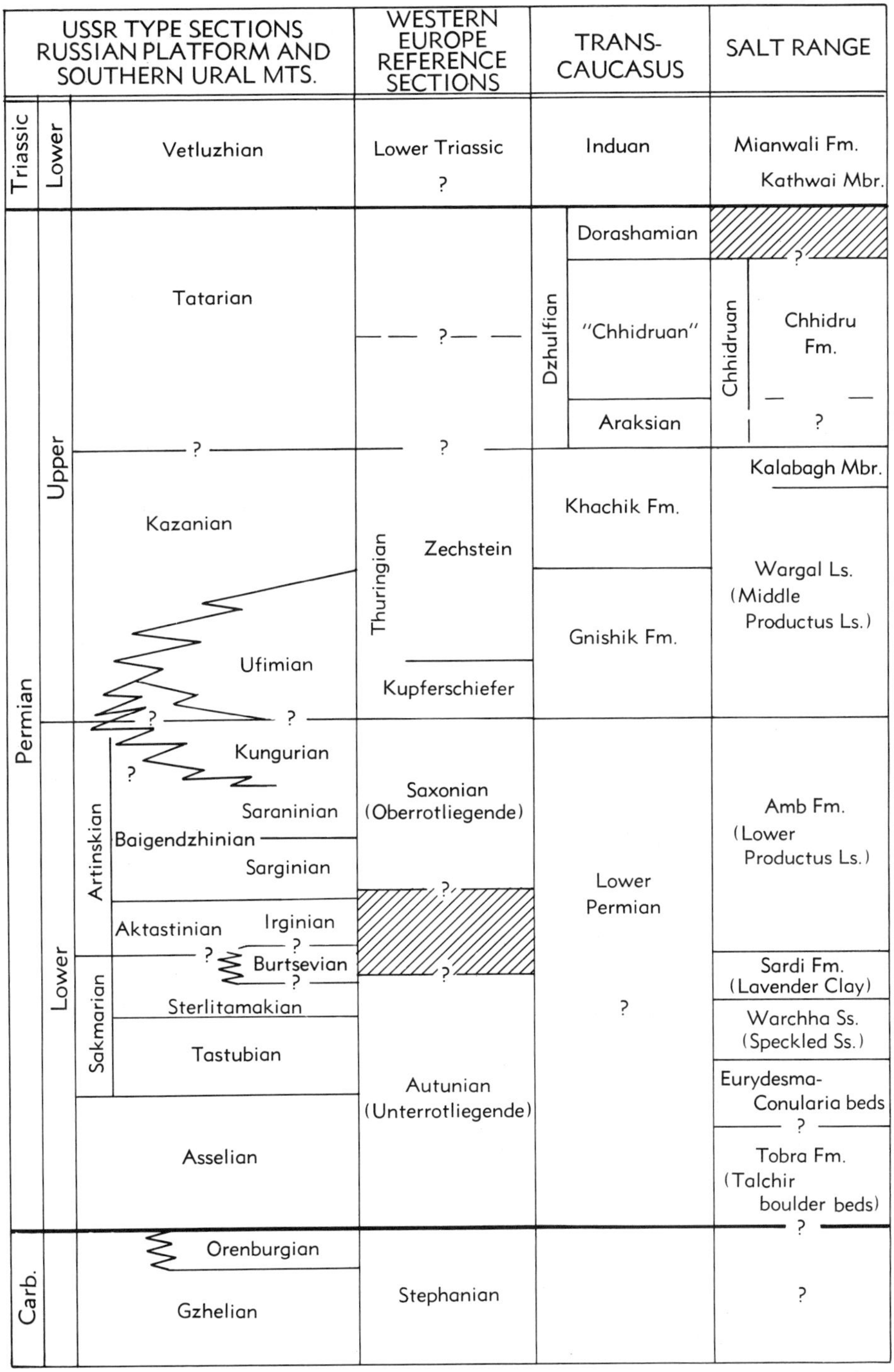

USSR TYPE SECTIONS RUSSIAN PLATFORM AND SOUTHERN URAL MTS.			WESTERN EUROPE REFERENCE SECTIONS	TRANS-CAUCASUS		SALT RANGE	
Triassic	Lower	Vetluzhian	Lower Triassic ?	Induan		Mianwali Fm. Kathwai Mbr.	
Permian	Upper	Tatarian	?	Dzhulfian	Dorashamian		Chhiduan
					"Chhidruan"	Chhidru Fm.	
		?	?		Araksian	?	
		Kazanian	Thuringian Zechstein	Khachik Fm.		Kalabagh Mbr.	
						Wargal Ls. (Middle Productus Ls.)	
		Ufimian		Gnishik Fm.			
		? ?	Kupferschiefer				
	Lower	Artinskian Kungurian Saraninian Baigendzhinian Sarginian	Saxonian (Oberrotliegende)	Lower Permian		Amb Fm. (Lower Productus Ls.)	
		Aktastinian Irginian ? ? Burtsevian ?	? ?			Sardi Fm. (Lavender Clay)	
		Sakmarian Sterlitamakian Tastubian		?		Warchha Ss. (Speckled Ss.)	
			Autunian (Unterrotliegende)			Eurydesma-Conularia beds ?	
		Asselian				Tobra Fm. (Talchir boulder beds)	
Carb.		Orenburgian Gzhelian	Stephanian			?	

FIG. 1. *(Continued from facing page.)*
to the Karachatyrian Stage and the Artinskian equivalents to the Darvazian Stage (Ross & Ross, n).

Sea. The Kazanian fauna is lacking in colonial corals, cephalopods, trilobites, and echinoderms, but has two faunal facies; one in which bivalves predominate, the other in which brachiopods and bryozoans predominate. The main elements of the restricted Kazanian fauna are relicts of the earlier restricted Kungurian fauna and their local geographic and stratigraphic positions in Kazanian strata relate to facies variation in water of low salinity.

In the southeastern part of the Kazanian depositional basin near Orenburg, the strata include more sandstone and, locally, thick evaporite deposits. The eastern edge of the typical Kazanian impure limestone facies intertongues with red clastic beds that represent a coastal plain extending eastward toward the Urals. In the northern part of the Russian platform in Arkhangel Province, faunas from the lower part of the Kazanian include rare goniatites of the genus *Pseudogastrioceras,* or a closely related genus having a ventral sinus, indicative of an age equivalent to the Wordian Stage in North America (KULIKOV, PAVLOV & ROSTEVTSEV, 1973).

The Tatarian Stage is the highest and youngest of the Permian deposits on the Russian platform and is formed of brightly colored, variegated sandstone, conglomerate, and clastic strata that represent fluviatile, eolian, and lacustrine deposits with local evaporite beds. Although generally not richly fossiliferous, the fauna includes fish, bivalves, and ostracodes in the lacustrine deposits, and reptiles and amphibians in other deposits. Locally, plant fossils are common and thin seams of coal are present. Tatarian deposits are irregular in thickness and reach a maximum thickness of only a few hundred meters. These deposits are confined to the depositional basin of the earlier Kazanian sea and were attributed by NALIVKIN (1973) to deposition in arid conditions. These conditions of deposition continued without change into Triassic times over the same part of the Russian platform, and the Lower Triassic succession is called the Vetluzhian Stage. The Triassic fauna is not abundant but includes distinctive ostracodes, fresh-water phyllopods, reptiles, and a new flora. The Tatarian and Vetluzhian are similar in lithology and distribution so that their identification is based on the recognition of nonmarine faunas of different ages.

In summary, the Permian System in its type region on the southern and western margins of the Ural Mountains and on the eastern part of the Russian platform is a complicated set of intertonguing facies. These facies and their faunas reflect the progressive changes of depositional environments from a marine carbonate shelf, shelf edge, and elongate basin setting during the Late Carboniferous and earliest Permian (Asselian to early Artinskian) to a gradual restriction of marine circulation late in Artinskian and Kungurian times with the deposition of evaporites and presence of specialized faunas. This was accompanied by the continued closing and uplift of the Uralian geosyncline, which began at about this time to shed terrestrial debris westward in the form of the Ufimian deposits. By Kazanian time deposition was mainly confined to a shallow interior basin having water of low salinity and a specialized relict fauna derived from Kungurian time. By Tatarian time this interior sea dried up and gave way to fluviatile, eolian, and lacustrine deposits having terrestrial and fresh-water biotas. This set of conditions persisted into Early Triassic time.

NORTHWESTERN EUROPE

Beds above the Carboniferous and below the Triassic are widespread in western Europe (Fig. 1) and are mainly red sandstone (Rotliegende beds) overlain unconformably by beds of conglomerate, chalcopyritic shale, dolomitic limestone, evaporites, and shale (Zechstein beds). Because of this consistent twofold division these beds were once known as the Dyas. The Rotliegende beds are subdivided into a lower part, the Autunian Series (Unterrotliegende), and an upper part, the Saxonian Series (or Oberrotliegende). The Zechstein, which forms the Thuringian Series, is subdivided into a basal conglomerate, a copper-bearing shale (Kupferschiefer), a dolomitic limestone (Zechstein), and an evaporite unit consisting of red clay, thin-bedded dolomitic limestone, anhydrite, halite, sylvite, and magnesium salts.

Autunian rocks are less widely distributed in western Europe than later Permian strata and are closely associated with the Upper Carboniferous Stephanian coal basins. Generally Stephanian deposits are followed with little interruption by Autunian deposits, which are commonly dark-gray to red shale, locally containing volcanics and some thin coal beds. Plant fossils, including *Walchia* and *Callipteris,* and the vertebrates, *Palaeohatteria* and *Archegosaurus,* are common. Most of these basins were tectonically active during the Autunian and in the Saar-Nahe basin as much as 3,000 m of sediment was deposited during this epoch.

Saxonian deposits are more widely distributed and usually unconformably overlie Autunian and older rocks. They are younger than the Saale tectonic phase, last major movement of the Hercynian orogeny, and show a major change in geographic distribution and climates. The Saxonian includes conglomerate and sandstone deposits that reach a maximum thickness of about 500 m.

Thuringian deposits also are widely distributed in northwestern Europe where they unconformably overlie Saxonian beds. The base is marked by a thin conglomerate that contains rare *Cancrinella cancrini,* which is also found in the Kazanian of the Russian platform. Above this conglomerate lies the Kupferschiefer, a thin 0.6-m bed that contains well-preserved fish fossils, such as *Palaeoniscus,* and plants, such as *Voltzia.* The Kupferschiefer is a possible facies equivalent of the Kungurian on the Russian platform. The Zechstein dolomitic limestone, which is 5 to 10 m thick and contains an impoverished fauna having low species diversity but considerable numbers of individuals, is the prominent marker bed in the region. Coelenterates, cephalopods, and echinoderms are rare or lacking and bryozoans, brachiopods, bivalves, and gastropods are abundant. The brachiopods *Horridonia horrida, Strophalosia goldfussi, Dielasma elongata, Pterospirifer alatus, Spiriferina multiplicata, Punctospirifer cristata,* and *Cleiothyridina pectinifera* are characteristic of the Zechstein or its stratigraphic equivalent, the Magnesian Limestone of Great Britain. Small reefs form a distinctive facies and include some brachiopods and numerous fenestrate bryozoans. Conodonts and smaller foraminifers also occur in the Zechstein but their study is still in progress. The upper part of the Thuringian is a complex evaporite facies that includes shale, limestone, and various evaporitic salts, including bitter salts. Although the thickness of these evaporites varies from place to place for several reasons, locally in northern Germany they reach several hundred meters in thickness. In the uppermost part of this evaporitic sequence, sandy shales contain a few marine fossils such as the bivalves *Schizodus* and *Gervilleia* and the brachiopod *Dielasma.*

The extent of the Thuringian sea is well known for northwestern Europe where it covered most of Germany, parts of Poland and Great Britain, and the North Sea. It is difficult to establish whether or not this sea also connected to the east with the Kazanian sea on the Russian platform. Low species diversity and general similarity of faunas in the two regions suggest similar ecological conditions if not some interchange of faunas.

CENTRAL EAST GREENLAND

The Permian of the central East Greenland succession, the Foldvik Creek Formation, has a basal conglomerate lying on a regional unconformity with topographic relief. Above this basal conglomerate, a 30- to 40-m interval has a number of intertonguing lithologies including light-gray dolomitic limestone, gypsum, shale, and near the top, a fish-bearing black *"Posidonia* Shale." Overlying this is the brachiopod-rich *"Martinia* Shale" that grades laterally from gray limestone to calcareous shale and siltstone. In the upper part of this unit is a prominent *"Productus"* bed. The *"Martinia* Shale" is unconformably overlain by the clastic Lower Triassic Kap Stosch Formation, which contains blocks of Foldvik Creek Limestone (TEICHERT & KUMMEL, 1972). On the basis of brachiopods, DUNBAR (1955, 1961) correlated the Foldvik Creek Formation with the Zechstein and Magnesian Limestone, and considered the brachiopods to be at least as young as the Capitanian of the west Texas standard sections. The ammonoids from *Martinia*-bearing limestone include *Medlicottia malm-*

quisti and *Cyclolobus kullingi* (MILLER & FURNISH, 1940b), which were believed to represent a latest Permian fauna; however, subsequent studies have shown that *Cyclolobus* ranges through much of the upper Permian, and primitive species occur as low as the *Timorites* Zone (FURNISH, 1966). *C. kullingi* is a primitive species and therefore may be early Capitanian in age. Corals from the Upper Permian of the Kap Stosch area (FLÜGEL, 1973) include 11 genera and subgenera that are older than Dzhulfian corals and possibly equivalent in age to the *Yabeina* Zone.

Reconstructions of Pangaea for late Paleozoic time place Greenland and parts of North America against northern Europe and the Barents shelf (DIETZ & HOLDEN, 1970). In such a geography the Upper Permian deposits of central East Greenland are in juxtaposition with the northwestern end of the Late Permian Thuringian basin. The faunas from the Permian of central East Greenland are of particular interest because they provide clues for the correlation of units in the Thuringian of northwestern Europe (and possibly the Kazanian of the Russian platform) with other Late Permian successions in Spitsbergen and the Canadian Arctic islands.

OTHER NORTHERN AREAS

The Permian of Pechora and Pay-Khoy of northeastern Europe, north East Greenland, Spitsbergen, the Canadian Arctic islands, Yukon, and east-central Alaska share many features with each other. The Lower Permian parts of these successions commonly have well-developed fusulinid zones, particularly for the Asselian to lower Artinskian equivalents, with one or more evaporite units and clastic marginal facies. These Lower Permian beds also show abrupt changes in thickness and lithologies and include shelf and platform clastics, a few shelf evaporites, shelf-edge carbonate bank and basinal evaporites, limestone and shale. The marked changes in thickness apparently relate to late phases of Hercynian structural adjustments equivalent to those of northwestern Europe during the Autunian, but in a marine shelf environment.

Unconformably above these Lower Permian beds is a succession of limestone, cherty limestone, chert, and sandstone that forms the Brachiopod Cherts of Spitsbergen (see recent summary by GOBBETT, 1963), the Upper Marine Group of north East Greenland (DUNBAR et al., 1962), and the Assistance Formation of the Grinnell Peninsula (HARKER & THORSTEINSSON, 1960). *Streptorhynchus, Derbyia, Dictyoclostus, Muirwoodia, Kochiproductus, Waagenoconcha, Stenoscisma, Rhynchopora, Pterospirifer,* and *Spirifella* are common in these faunas (HARKER & THORSTEINSSON, 1960), and their abundance and widespread distribution led STEPANOV (1957) to propose the name Svalbardian Stage as a replacement for the Kungurian and Ufimian stages. In present usage, the Svalbardian is important as a northern facies (or regional stage) and, although it is difficult to correlate with the Russian platform sections, STEPANOV (1936, 1937) suggested that the faunas had most similarity to those in the upper part of the Kungurian and lower part of the Kazanian. Later, STEPANOV (1957) included the brachiopod fauna from central East Greenland in his Svalbardian although DUNBAR (1955), FREBOLD (1950), and HARKER and THORSTEINSSON (1960) had pointed out some differences in generic and specific composition. Also, FREBOLD, HARKER, and THORSTEINSSON considered that many similarities are present and the faunas are either of somewhat different facies of the same age or the central East Greenland faunas are only slightly younger. In the Canadian Arctic islands, THORSTEINSSON (1974) showed that the Trold Fiord Formation and its apparent lateral equivalent, the Degerböls Formation, unconformably overlies the Assistance Formation and its lateral facies, the Van Haven Formation. The Trold Fiord and Degerböls formations contain a sparse fauna of corals, brachiopods, and ammonoids of probable Guadalupian age. Future studies may demonstrate that part of the central East Greenland faunas are of the same age as those of the Trold Fiord and Degerböls formations.

TETHYAN AREA

The Hercynian orogenic belt extends generally east-west from central western Europe with few interruptions into central Asia and eastern Asia. Although the Hercynian orogeny started in the later part of the Early Carboniferous, a number of strong orogenic pulses occurred during the Middle and Late Carboniferous, and the final major interval of deformation in Europe, the Saale phase, took place in the middle or later part of Artinskian time. This last phase of Hercynian deformation terminated formation of the Autunian coal-basin structures of northwestern Europe and many similar structures in other parts of Europe and southwestern Siberia (LAPKIN & SOLOVYEV, 1969). One result of this final phase was to separate the shallow seas on the northern side of the Hercynian Mountain belt (in which accumulated Kazanian and Zechstein deposits and strata to the north in Greenland, Spitsbergen, the Barents shelf, and northern Canada) from a large significant shelf and Tethyan geosyncline to the south. Although the exact details of the geographic features involved in the separation of the southern geosynclinal seas and the northern epicontinental seas need further study, the faunas on either side attest to a nearly complete isolation after the early part of the Artinskian.

In seas south of the Hercynian orogenic belt a distinctive fusulinacean, coral, and brachiopod fauna rapidly evolved. Later orogenic movements, particularly during Cenozoic times, have greatly complicated the relationships of Tethyan strata, and their interpretation is a major problem in Permian biostratigraphic and biogeographic studies. Several linear belts of Permian rock seem to have been present, each of which may have included shallow reef deposits, shallow clastic deposits, and deeper water clastic and carbonate deposits. These depositional belts fringed a number of small cratonic areas that have since been displaced and moved much closer together (YANSHIN, 1965).

The extent of the Tethyan belt during the Permian (Fig. 1) can be judged by the distribution of its distinctive fauna that extends from as far west as Tunisia, through Sicily, the Carnic Alps, Yugoslavia, Greece, Crimea, Turkey, Iran, Iraq, Soviet Middle Asia, northern Pakistan, Mongolia, China, Indochina, Indonesia, New Zealand, Japan, Soviet Maritime Province, Kamchatka, southern Alaska, British Columbia, Washington, Oregon, and into California. This Tethyan fauna is now fragmented and has been displaced in pieces to many latitudes and different climatic belts, but overall similarity of the fauna suggests free communication and great diversity within warm to tropical waters.

In the southern part of the Soviet Union where a large portion of the Tethyan belt is exposed, considerable effort has been directed toward establishing correlations with the type Permian on the Russian platform and the Ural region with inconclusive results (ARAKELIAN et al., 1964; STEPANOV, 1970). Relatively few species are common to the two areas and regional stages are used widely for the Tethyan realm. The Lower Permian is divided into the Karachatyrian and Darvazian stages (VLASOV, LIKHAREV, & MIKLUKHO-MAKLAY, 1962) and the Upper Permian into the Murgabian and Pamirian stages. The following is a summary of the diagnostic faunas (LIKHAREV & MIKLUKHO-MAKLAY, 1972).

The Karachatyrian Stage contains a much greater diversity of fusulinaceans than the Asselian or Sakmarian, including species of *Ozawainella, Boultonia, Schubertella, Quasifusulina, Occidentoschwagerina, Zellia, Pseudoschwagerina, Schwagerina, Sphaeroschwagerina, Robustoschwagerina, Pseudofusulina, Paraschwagerina, Rugososchwagerina,* and primitive *Parafusulina*. Brachiopods are common and include species of *Isogramma, Enteletes, Dictyoclostus, Terebratuloidea, Stenoscisma, Spirifer, Martinia, Dielasma,* and *Notothyris,* and corals include species of *Tachylasma, Caninia, Timorphyllum, Caninophyllum, Cyathaxonia, Amplexocarinia, Allotropiophyllum,* and *Lophophyllidium*. In addition, gastropods, bivalves, and algae (*Tubiphites, Eugenophyllum,* and *Epimastopora*) are abundant and ammonoids are lacking or not reported.

The Darvazian Stage also includes a diverse fusulinacean fauna including numerous species of *Nankinella, Sphaerulina, Yangchienia, Minojapanella, Kahlerina, Darvasites, Robustoschwagerina, Nagatoella, Pseudofusulina, Chusenella, Parafusulina, Brevaxina, Misellina, Armenina,* and primitive *Cancellina*. Brachiopods include species of *Enteletes, Echinoconchus, Productus (Striatifera?), Wellerella, Martinia, Heterolasmina,* and *Notothyris;* corals include, in the lower part of the stage, species of *Verbeekiella, Cyathocarinia, Sinophyllum, Amplexocarinia,* and in the upper part *Yatsengia* and *Granophyllum.* Ammonoids include *Propinacoceras, Prosicanites, Perrinites, Marathonites,* and *Agathiceras.* Among the algae are *Mizzia, Tubiphites, Epimastopora,* and *Girvanella.*

The Murgabian Stage contains a diverse fusulinacean fauna that is particularly widely dispersed, including numerous species of *Leella, Neofusulinella, Pseudofusulina, Paraverbeekina, Verbeekina, Armenina, Neoschwagerina, Praesumatrina, Sumatrina,* and *Polydiexodina.* Brachiopods include species of *Derbyia, Chonetella, Linoproductus, Dictyoclostus, Productus, Martinifera,* and *Lyttonia.* Corals include species of *Yatsengia, Granophyllum, Waagenophyllum,* and *Polythecalis;* the ammonoids include species of *Neostacheoceras, Tauroceras, Paraceltites,* and *Adrianites;* and algae include species of *Gymnocodium, Permocalculus, Indopolia, Cyrocopora,* and *Vermiporella.*

Although the Pamirian Stage generally is not as complete in the Soviet Union as equivalent beds in China, it does contain a typical high Upper Permian fauna and locally the section is nearly complete. Fusulinaceans and other foraminiferids include species of *Reichelina, Palaeofusulina, Codonofusiella, Lasiodiscus, Colaniella,* and *Pachyphloia.* Brachiopods are represented by species of *Enteletella, Striatifera, Marginifera, Urushtenia, Tschernyschewia, Wellerella, Pugnax, Ambocoelia, Athyris,* and *Hemiptychina.* Ammonoids include species of *Prototoceras, Rotaraxoceras, Araxoceras,* and *Urartoceras;* algae include species of *Gymnocodium, Permocalculus,* and *Vermiporella.*

In the Pamirs, the Pamirian Stage commonly rests unconformably on the Murgabian Stage (LEVEN, 1967) and is unconformably overlain by strata assigned to the Triassic System. In the southeastern Pamirs, a number of lithologic facies are present, which include the zones of *Yabeina, Codonofusiella-Reichelina,* and *Palaeofusulina* of very late Permian age. Elsewhere in the Pamirs the Permian section is not as complete and it is not clear whether this is because of nondeposition or pre-Triassic erosion. Ammonoids are apparently not recorded from these Pamirian beds.

Another classic Upper Permian section, which was believed for a long time to have a transitional boundary with the Triassic, is located in the Trans-Caucasus (Fig. 1) along the Aras (=Araks, Araxes) River near Dzhulfa, Soviet Azerbaijan, and Julfa, Iran. Ammonoids and brachiopods are common in several of these beds (ABICH, 1878) and the section has been restudied in detail by several groups (RUZHENTSEV & SARYCHEVA, 1965; STEPANOV, GOLSHANI, & STÖCKLIN, 1969; KUMMEL & TEICHERT, 1973). The combined list of genera of ammonoids from the Dzhulfian Stage is impressive and appears to represent a nearly complete succession of post-Guadalupian, pre-Triassic ammonoids. Fusulinaceans are common in the lower 2 to 5.5 m of the Dzhulfian Stage and include *Codonofusiella* and *Reichelina* in a fine-grained, bituminous limestone that lithologically contrasts with the underlying Khachik Limestone, which carries a late Guadalupian fauna. Higher in the Dzhulfian *Araxoceras* and *Oldhamina* are common and a few beds containing *Codonofusiella* and *Reichelina* form the succeeding 8 to 20 m. Above these are up to 19 m of coral-, brachiopod-, and ammonoid-bearing beds with the *Vedioceras* and *Haydenella* fauna. Above these are 4.5 m of beds with the ammonoids *Phisonites, Xenaspis,* and *Xenodiscus,* and the brachiopod *Comelicania* and others. Soviet scientists choose to place the top of the Permian at the top of this *Phisonites* unit, but most other scientists also include the succeeding Ali Bashi Formation, about 18 m thick, in the Permian, which includes *Paratirolites* and several other ammonoids (KUMMEL & TEICHERT, 1973). Above this are 10 to 20 m of fine-grained limestone containing the Triassic bivalve *Claraia.*

Two other sections or areas of considerable historical interest to the problems of the uppermost Permian biostratigraphy are in and near the Salt Range of Pakistan (Fig. 1) and in Kashmir (KUMMEL & TEICHERT, 1966, 1970, 1973). The Chhidru Formation ("Upper *Productus* Limestone") of the Salt Range contains a fauna that is similar to that from the lower part of the Dzhulfian Stage in having mainly corals, brachiopods, and ammonoids but no fusulinaceans. The overlying Kathwai Member of the Mianwali Formation (uppermost part of "Upper *Productus* Limestone") contains the ammonoids *Ophiceras* and *Glyptophiceras* which KUMMEL and TEICHERT (1973) considered to be indicative of Early Triassic age, and the brachiopods *Enteletes, Orthothetina, Spinomarginifera*, and *Martinia*, which GRANT (1970) considered to be typical Permian genera and equivalent in age to the Dzhulfian brachiopods listed and described by RUZHENTSEV and SARYCHEVA (1965).

The section at Guryul Ravine, Kashmir (HAYDEN, 1907; MIDDLEMISS, 1910), also has been of considerable interest as a sequence that passes conformably from uppermost Permian into Lower Triassic strata (KUMMEL & TEICHERT, 1973; NAKAZAWA et al., 1975). Here, as in the Salt Range, Triassic ammonoids and bivalves are associated with brachiopods that belong to genera usually considered to be Late Permian. Also, conodonts of the *Anchignathodus typicalis* Assemblage-Zone cross the Permian-Triassic boundary. The Late Permian Zeewan Formation (120 m thick), which rests on the Panjal volcanics, is sandy in its upper part and detailed faunal studies are needed. Fusulinaceans have not been reported.

The Pamirian, Dzhulfian, and Chhidruan stages appear to represent approximately the same time interval. The Pamirian may include somewhat older beds in its lower part than the other two and the Chhidruan may not include beds as young as latest Dzhulfian. Faunal zones that can be laterally traced and that include fossils common in Upper Permian strata in other geographical areas are much needed. Some of the faunas studied are closely associated with particular depositional environments and the lateral equivalency of different facies is commonly difficult to determine. The Lopingian Series in South China offers possibilities for establishing a reliable biostratigraphic framework for this latest Late Permian interval (Fig. 1) (SHENG, 1963; CHAO, 1965).

The Lopingian Series unconformably overlies the Maokou Limestone, which has the Zone of *Yabeina* in its upper part. The series was subdivided by CHAO (1965), and a lower part is made up of the Wuchiaping Limestone and its lateral equivalents, such as the Hoshan Limestone, the Lungtan Coal Series in the lower Yangtze valley, and the Loping Coal Series in north-central Kiangsi. These coal series, reaching 600 to 700 m in thickness, have continental sandstone and shale interbedded with marine shale and thin limestone beds that yield brachiopods and ammonoids. The continental beds include floras of *Taeniopteris, Pecopteris, Sphenopteris, Cladophlebis,* and *Gigantopteris*. Common brachiopods are *Chonetes, Dictyoclostus,* and *Squamularia*. In the lower third of this succession the ammonoids *Anderssonoceras, Prototoceras, Araxoceras, Kiangsiceras, Vescotoceras,* and *Pseudogastrioceras* are common and are similar to those in the lower part of the Dzhulfian Stage. Near the top of this succession, continental beds have *Neuropteris, Lepidodendron,* and *Lobatannularia*.

The upper part of the Lopingian Series, the Changhsing Limestone, is 120 to 150 m thick and has a *Palaeofusulina-Reichelina* fusulinacean fauna in north-central Kiangsi. This limestone is typically developed in the Changhsing coal field of northern Chekiang where it is formed of 25 to 34 m of dark-gray to black limestone and siliceous interbeds. The lower part contains *Pseudotirolites* and *Pseudogastrioceras*. The upper part contains *Stacheoceras, Pachydiscoceras, Rotodiscoceras, Trigonogastrites,* and *Changhsingoceras* in association with *Palaeofusulina, Reichelina,* and the brachiopods *Oldhamina, Dictyoclostus,* and *Hustedia*. In many parts of South China the predominantly limestone facies of the Changhsing changes laterally into the Talung Formation, which consists of siliceous shale and limestone and contains *Palaeofusulina, Pseudogastrioceras, Pseudotirolites,* and other genera of ammonoids. CHAO (1965) estimated that the Changhsing and Talung strata contain more than 100 species and 30 genera of ammonoids, most of which are not described. In South China the Lopingian Series has at least two ammonoid zones, the *Prototoceras-Araxoceras* Zone in the lower part and the *Pseudotirolites-Pleuronodoceras* Zone in the upper part.

Fusulinaceans are common in much of the Lopingian Series of the Permian of South China (SHENG, 1963) and, although not present in some of the Iranian, Soviet, Pakistan, and Kashmir sections, they are more widely distributed (TORIYAMA, 1973) than most of the ammonoid genera listed by CHAO (1965). TORIYAMA (1973) reviewed these high Upper Permian occurrences of fusulinaceans and concluded that three zones could be recognized. A late Guadalupian fusulinacean assemblage-zone includes *Yabeina, Lepidolina, Codonofusiella, Reichelina,* and a large variety of Verbeekininae; a lower Dzhulfian (lower Lopingian) assemblage-zone has *Codonofusiella* and *Reichelina* but without other fusulinaceans (except possibly in Japan where the *Lepidolina kumaensis* Zone may extend into this zone); and an upper Dzhulfian (upper Lopingian) assemblage-zone includes *Palaeofusulina* and *Reichelina*. These fusulinacean zones, as well as those based on ammonoids and brachiopods, are in need of further investigations.

Other well-studied Permian Tethyan sections (Fig. 1) include Japan and the Carnic Alps, which are discussed in more detail under fusulinacean zonation. The Japanese sections were deposited in eight different tectonic settings (TORIYAMA, 1973), each with its own lithofacies and structural history, occurring in an outer tectonic zone, an inner tectonic zone, and a number of massifs and basins. The base of the Permian, drawn at the base of the *Pseudoschwagerina* Zone, is usually unconformable on Carboniferous and older rocks. Four Permian series are commonly recognized:

SERIES	ZONE
Kuman	*Yabeina yasubaensis-Lepidolina toriyamai* (lacks *Neoschwagerina*)
Akasakan	*Neoschwagerina craticulifera-Verbeekina verbeeki*
Nabeyaman	*Parafusulina kaerimizensis-Neoschwagerina simplex*
Sakamotozawan	*Pseudoschwagerina morikawai-Pseudofusulina vulgaris*

In only two or three sections are Triassic strata preserved above these Permian beds. The Kuman Series is usually not present or, if present, may lie either unconformably or in fault contact with the underlying Akasakan Series, suggesting a period of mountain-building activity that may correspond to the Tungwa movements of China. An earlier orogenic pulse also seems to have been widespread during the latter part of the Nabeyaman Epoch. Both orogenic intervals resulted in considerable changes in paleogeography.

Tethyan deposits of southern Europe are shared by Austria, Italy, Yugoslavia, Greece, and Sicily and include a complex sequence of Permian rocks. One of the better known successions is the Lower Permian of the Carnic Alps, which consists of 200 to 270 m of limestone and interbedded shale and 200 to 400 m of massive limestone (FLÜGEL & SCHÖNLAUB, 1972). These form the Rattendorfer Limestone and Shale and the Trogkofel Limestone. The Trogkofel is locally reefoid and also includes a clastic facies and breccia. The overlying beds rest disconformably on the Trogkofel Limestone and include 30 to 40 m of red clastic Grödener Shale. This clastic depositional phase probably is a result of the Saale tectonic pulse that terminated Autunian deposition in northwestern Europe. Thus, the Grödener is approximately equivalent to the lower clastic portion of the Zechstein. Above this are up to 200 m of *Bellerophon*-bearing shale and limestone that are equivalent to the middle and upper part of the Zechstein.

The Rattendorfer and Trogkofel beds contain a diverse, apparently normal marine Tethyan fauna including many fusulinaceans, corals, brachiopods, and cephalopods. This depositional pattern started in the Late Carboniferous with the deposition of the Auernig clastic beds and a few marine limestone beds bearing plants, fusulinaceans, corals, and brachiopods of either Stephanian or Gzhelian affinities. The Rattendorfer beds have abundant and diverse fusulinacean faunas that include typical Asselian and Sakmarian species. The Trogkofel Limestone is particularly rich in Artinskian cephalopods, including *Medlicottia* near the base and *Agathiceras, Popanoceras,* and *Thalassoceras* near the top. The Grödener beds are plant-bearing and are similar in lithology to the Saxonian facies of northwestern Europe. The *Bellerophon*-bearing beds represent a complex of lagoonal facies having a number of different lithologies in which brachiopods, gastropods, bivalves, and rare cephalopods show affinities to both the Zechstein and the Upper Permian of the Salt Range. Northwestward from the Carnic Alps, the Permian strata thin and become nonmarine, and to the southeast, in Yugoslavia, Greece, Crimea, and Turkey, the Upper Permian carries a varied Tethyan fauna that includes fusulinaceans, brachiopods, and other invertebrates.

NORTH AMERICA

Permian strata of North America include shallow marine shelf deposits, evaporites and nonmarine beds in much of the midcontinent and southwestern regions, nonmarine plant-bearing beds in parts of West Virginia and adjacent areas (Dunkard Group), and a complex set of eugeosynclinal and miogeosynclinal sediments in the western Cordillera as far east as Utah, Idaho, and western Alberta (MCKEE et al., 1967).

The Dunkard Group was deposited in a semi-isolated nonmarine basin along the western flank of the Appalachian orogenic belt, and on the basis of plant fossils appears to be no younger than latest Autunian and no older than earliest Permian.

The midcontinent and much of the southwestern region is characterized by a complicated set of lithofacies that were gradually modified by changes in sea level, movements related to tectonics of the craton, and gradual filling-in of basins and troughs. The Ouachita sector of the Appalachian orogenic belt lay as a large S-shaped feature to the south and actively moved against the southern margin of the North American craton, closing a deep marine sea that reached from west Texas into southern Oklahoma and Arkansas. Block faulting and resulting deposition progressively closed off the sea in northern Oklahoma and Kansas by the end of Wolfcampian time and the normal marine faunas, which were rich in fusulinaceans, bryozoans, bivalves, and crinoids gave way to specialized brackish and saline faunas in the later part of the Permian. By the early part of the Leonardian, the east Texas shelf between the Midland basin and the Ouachita belt also progressively developed into a broad, low, deltaic plain and the deposits are rich in nonmarine vertebrate fossils. Shelf-edge carbonate banks of algae, fusulinaceans, bryozoans, brachiopods, and crinoid debris fringed most of the Delaware and Midland basins during the Early Permian (Fig. 1) and these basins had considerable topographic relief; the Midland basin being in part below a sill restricting water circulation and bottom faunas.

During the Leonardian the reefs around the Delaware basin continued to deposit thick sedimentary units. By Guadalupian time the Midland basin was restricted as the Ouachita orogenic belt and debris from it closed the passage at the southern end of the Central Basin platform. A marine channel to the south and southwest supplied the Delaware basin with normal marine water and large marginal reefs continued to form around that basin (KING, 1949; NEWELL et al., 1953). To the south in northern Mexico volcanics finally closed this channel and marked the end of Guadalupian deposition. The faunal zonation of these west Texas reference sections is discussed in the Fusulinacean and Ammonoid sections of this review.

Westward in New Mexico and southeastern Arizona, other fault-bounded blocks show generally similar histories of basin infilling during the Early Permian (Ross, 1973), and the gradual restriction of faunas to very specialized biofacies by Leonardian time. Although carbonates of Guadalupian age are extensive over the southwestern end of the transcontinental arch in the Grand Canyon region, the biofacies were not normal marine, probably somewhat brackish, having a predominantly molluscan fauna. North of the western end of the transcontinental arch, block-faulted marine basins and clastic source areas extended well into Utah during Early Permian time and appear to be the continuation of the structural patterns seen in Arizona, New Mexico, and Texas. These basins and adjacent clastic source areas have many thousands of meters of displacement and the basins are filled mainly with conglomerate, sandstone, and shale, with locally well-developed limestone having a normal marine fauna including fusulinaceans. Farther north in Montana,

Alberta, and Yukon, the western edge of the Permian cratonic shelf was more stable and extensive thin sheets of sandstone are common in the Lower Permian and a few are apparently as young as early Late Permian. Marine faunas are meager in these shelf and shelf-edge clastic belts but a few limestones and silicified limestones contain Early Permian brachiopods, corals, and fusulinaceans.

In central eastern Alaska, Canadian Arctic islands, and northeastern Greenland this shelf clastic facies is well developed and intertongues near the shelf edge with thicker carbonate beds that have well-developed Early Permian fusulinacean, coral, bryozoan, brachiopod, and ammonoid faunas. These faunas are similar to those in Spitsbergen, the northern Urals, and the Russian platforms, but have not been completely studied.

West of the Permian edge of the North American craton several different, highly deformed sedimentary belts have contrasting faunas of Early and early Late Permian age and are presently aligned more or less parallel to the former cratonic edge (MONGER & ROSS, 1971). Some of these, particularly the eastern belt, have Lower Permian shale and limestone deposited on structurally disturbed lower and middle Paleozoic strata. Some of the diverse Early Permian faunas of these strata are similar to those of the reefs and carbonate banks along the Ural geosyncline. A central belt includes thick, extensively developed carbonate banks and reefs that rest on ribbon chert, basaltic tuffs, and greenstone, and are overlain by basalts. These limestones have an abundant Tethyan fauna of fusulinaceans, bryozoans, brachiopods, ammonoids, crinoids, and a rich algal flora, and they were probably deposited in shallow, tropical or subtropical water. Further west, other parts of the Cordillera have a pre-Devonian granitic basement overlain by a thick Lower Permian carbonate sequence with algae, fusulinaceans, corals, bryozoans, brachiopods, and crinoids, which also may represent a part of the Tethyan flora and fauna. Other fusulinaceans having affinities with the west Texas faunas are found in places in British Columbia and Yukon in another one of these Cordilleran tectonic belts. Interpretations of these distributions generally imply a significant geographical rearrangement of late Paleozoic strata during the Mesozoic and early Cenozoic. The youngest well-dated Permian fusulinaceans include primitive *Yabeina* with Wordian ammonoids, suggesting that in much of this region deposition was minor after middle Guadalupian time.

The relationships between these structurally bounded belts that have different sedimentary origins and contrasting faunas are only partially worked out; however, parts or pieces of many of these belts extend from southwestern Alaska through western Canada and western United States into Baja California and Sonora. These belts are principally identified by their faunal affinities and by their stratigraphic and structural relationships to underlying and overlying rock units.

In northern Mexico, east of Sonora, the Permian was deposited in basins much like those in Arizona, New Mexico, and west Texas, and on the flanks of the Ouachita-Marathon orogenic belt that turns south into Mexico. In Chiapas, in southern Mexico, light-gray thick-bedded limestone has yielded an abundant and diverse Lower Permian fauna in an algal-rich limestone, and in Belize dark-gray limestone in dark shale also has Lower Permian fusulinaceans. The remainder of Central America and the Caribbean region is composed of rocks that are mostly Jurassic or younger in age, and Permian strata are not reported.

GONDWANA CONTINENTS

AUSTRALIA

Major sequences of Permian strata are located in four basins in the western part of Australia (Perth, Carnarvon, Canning, and Bonaparte Gulf basins) and in five basins along the Tasman geosyncline in the eastern part of Australia (Tasmanian, Sydney, Maryborough, Bowen, and Yarrol basins). Based on faunal and floral associations (DICKINS, 1970; THOMAS, 1971), six subdivisions are correlated between the western basins and the Bowen basin in the east, and from there to other eastern basins (RUNNEGAR, 1969; DEAR, 1971). The earliest Permian fauna (Stage A, DICKINS, 1970) of

Asselian and early Sakmarian age is characterized by the bivalve *Eurydesma*, which was apparently adapted to cold water. In the west it is known only from the Carnarvon basin, but it is more widely distributed in the eastern basins. Glacial marine tillites, clastic strata, and volcanics are present in this interval in most basins. The second faunal assemblage (Stage B, DICKINS, 1970) of late Sakmarian or early Artinskian age is more widely distributed, suggesting more widespread marine conditions, and in the Carnarvon basin this assemblage contains a few genera of Tethyan affinities. The bivalve *Edmondia* is characteristic of this fauna and is associated with *Oriocrassatella*, *Astartella*, *Pseudomyalina*, and *Atomodesma*. In the Bowen and Sydney basins the bivalves *Astartila?*, *Megadesmus*, and *Eurydesma*, the brachiopods *Ingelarella*, *Strophalosia*, and *Notospirifer*, and the ammonoid *Uraloceras* also occur in this fauna (ARMSTRONG et al., 1967). The third subdivision (Stage C, DICKINS, 1970) is largely nonmarine and dominated by clastic sediments. In many of the basins coal beds formed in this interval and contain the plant fossils *Gangamopteris*, *Glossopteris*, and *Dadoxylon* (DAVID & BROWNE, 1950). The fourth subdivision (Stage D, DICKINS, 1970) includes another widespread marine fauna, which in the western basins has additional genera of Tethyan affinities (TEICHERT, 1974) and locally reaches considerable thickness (up to 2,000 m) in the Carnarvon basin. An older (Stage D_1) and a younger (Stage D_2) part of this faunal assemblage have been recognized. A late Artinskian age was suggested by DICKINS (1970) for the older part and a Kazanian age was indicated for the younger.

The fifth subdivision (Stage E, DICKINS, 1970) is a succession of coal and plant-bearing clastic strata that generally lack marine fossils. The sixth subdivision (Stage F) includes the youngest Permian marine faunas of western Australia, which occur in the upper part of the Liveringa Formation (Hardman Member). This fauna includes many genera of bivalves (such as *Phestia*, *Megadesmus*, *Astartila?*, "*Allorisma*," "*Modiolus*," *Atomodesma*, *Aviculopecten*, *Girtypecten?*, *Acanthopecten*, *Streblopteria*, *Pseudomonotis*, *Schizodus*, *Oriocrassatella*, and *Astartella*) and several genera of gastropods. It is probably Tatarian (Dzhulfian) in age.

Permian ammonoids are rare in most Australian sequences and include 17 species from the western basins and two from the Sydney basin in the east (GLENISTER & FURNISH, 1961). From the western basins several species of *Uraloceras* and *Juresanites* occur in the Holmwood Shale (upper part of Stage A) and suggest a Sakmarian age. *Thalassoceras*, *Metalegoceras*, and *Propopanoceras* in the Nura Nura Member of the Poole Sandstone (Stage B) are of late Sakmarian age. Several ammonoid assemblages of Artinskian (Baigendzhinian) age have been reported and include species of *Neocrimites*, *Propinacoceras*, and *Pseudoschistoceras*. The lower beds of the Liveringa Formation (Lightjack Member) and beds of approximately the same age (Stage D) contain *Pseudoschistoceras*, *Agathiceras*, and possibly *Propinacoceras*, indicating an age near the Early-Late Permian boundary or slightly above. The ammonoid fauna from the Kockatea Shale of the Perth basin is now known to be early Triassic in age. In the Sydney basin in eastern Australia (Fig. 1), the Farley Formation contains *Uraloceras*, brachiopods, gastropods and bivalves, of probable late Sakmarian age; the Branxton Subgroup has *Neocrimites*, corals, and bryozoans of Artinskian (Baigendzhinian) age.

An extensive Late Permian insect fauna is known from the Newcastle Coal Measures near Newcastle and Lake Macquarie in the northern part of the Sydney basin in eastern Australia (TILLYARD, 1917-36). This insect fauna lacks many of the forms usually found in Permian tropical or subtropical moist-forest assemblages and is dominated by the orders Hemiptera and Mecoptera. These insects are distinctively more advanced than those of the Early Permian of Kansas and include many genera that are closely related or ancestral to orders that are typically Mesozoic and younger. This fauna is generally considered to be Tatarian in age based on a few closely similar species that are found in the Russian section. In Tasmania, the Permian-Triassic boundary is drawn at the top of the Cygnet Coal Group (BANKS & NAQVI, 1967).

INDIA AND PAKISTAN

The Permian deposits of India and Pakistan are distributed in a number of fault-bounded basins located along three structural trends in peninsular India, and to the north in elongated fold belts on the flanks of the Himalayan orogenic belt. The successions in peninsular India lie on a Precambrian basement, are called the Gondwana Group, and generally have a tillite or a diamictite at their base. Clastic nonmarine strata dominate the overlying parts of the Gondwana Group (Permian to Early Cretaceous) and have interbedded coal and only a few thin marine beds. Two marine assemblages are recognized, suggesting two different intervals of flooding. The oldest (Asselian or early Sakmarian) has an *Eurydesma* fauna, was deposited within the Talchir Boulder beds, and was probably a cold-water assemblage. Three other localities have more diverse faunas, which are generally considered to be of late Sakmarian age. *Eurydesma* may be present in these three faunal assemblages, along with bryozoans, brachiopods, gastropods, bivalves, ostracodes, and echinoderms (TEICHERT, 1974). Some of these fossil groups include genera known also from Australia and South America and others are nearly cosmopolitan. The faunas suggest that these beds were the result of brief marine flooding of the Gondwana depositional basins by water from the northern edge of the Indian craton.

The Permian strata exposed in the flanks of the Himalayan Mountains appear to represent deposits along the edge of the Indian craton that bordered the southern Tethys. Diamictite in the Salt Range is overlain by marine clastic and carbonate units that carry a fossil succession having mostly Tethyan forms (Zaluch Group = Lower, Middle, and Upper "Productus" Limestone of early reports) (see discussion above on Tethyan Belt).

AFRICA

Central, southern, and parts of northwestern Africa were consolidated by the end of Middle Carboniferous time into a relatively stable platform. Block faulting developed basins for clastic deposition during the Permian and the stratigraphic succession (Karroo Series) that is preserved is similar to that of peninsular India. At the base, the Dwyka Tillite (or Diamictite) Group is widespread and in southwestern Africa marine tongues are found with *Eurydesma, Conularia, Archaeocidaris,* and the fish *Palaeoniscus.* Nonmarine shaly tongues near the base have *Glossopteris* fragments, and shales near the top have *Glossopteris* and the Gondwana aquatic reptile *Mesosaurus,* which is also found in South America. These fossils suggest a Late Carboniferous to Asselian (or early Sakmarian of some authors) age and presumably are equivalent to the earliest Permian assemblages of India and Australia.

Above the Dwyka Group, the Ecca Group is widely distributed and is mainly fine- to coarse-grained sandstone with some shaly beds. In the Karroo trough these strata are mostly gray shale, siltstone, and subgraywacke sandstone having *Gangamopteris, Glossopteris,* and *Cyclondendron.* To the east in Natal, Transvaal, and Orange Free State, coals are well developed and to the west in southwestern Africa red sandstone and shale are typical of the Ecca. Much of the middle part of the Ecca Group represents cyclic deposition associated with several widespread coal beds. Most of these depositional cycles have only the nonmarine portion preserved, but locally some marine shale is preserved above the coal. In the Dundee district in northern Natal, one such occurrence yielded *Paraceltites?,* which suggests a latest Early Permian or Late Permian age for that part of the Ecca (TEICHERT & RILETT, 1974). Northward the Ecca Group thins and lies directly on the pre-Permian platform where the Dwyka Tillite is missing.

The youngest Permian beds are nonmarine and they form the lower part of the Beaufort Group, which passes upward into strata of Triassic age. Six reptilian zones are recognized in the Beaufort but only the lower three are Permian. These are the Zone of *Tapinocephalus,* overlain by the Zone of *Endothiodon,* and finally by the Zone of *Cisticephalus* at the top of the Permian. This lower part of the Beaufort Group seems to be restricted to the Karroo trough.

MADAGASCAR

The western half of Madagascar is underlain by a succession of Gondwana strata that have many features of the southern African, peninsular Indian, and Salt Range successions. The Permian part of this succession is the Sakoa Group and the lower part of the overlying Sakamena Group. A tillite, the lowest formation, is overlain by mainly nonmarine sandstone and shale beds with several marine limestone and shale units. The *Glossopteris-Gangamopteris* flora, associated reptiles, and the occurrence of several coal beds are similar to the Karroo Series of southern Africa. Productid and spiriferid brachiopods are common in the marine shale and limestone units. The lower part of the Sakamena Group in northern Madagascar also has several marine limestone and shale units and these include productid, spiriferid, and atrypid brachiopods, bivalves, and cephalopods, such as *Cyclolobus, Episageceras, Xenodiscus, Propinacoceras,* and *Popanoceras,* that are similar to those of the Chhidru Formation of the Salt Range, Pakistan (TEICHERT, 1974).

SOUTH AMERICA

East of the Andean belt most of South America had a Permian history that is closely similar to that of Africa. By Middle or Late Carboniferous time the shield areas of South America formed a large platform, which, based on geological similarities, appears to have been continuous with that of western Africa. The Gondwana deposits in the Parana basin that lie unconformably on this platform are of glacial origin and are similar to those in southwestern Africa (CASTER, 1952). The Itararé Group at the base has shaly interbeds with paleoniscid fish, ammonoids, and *Glossopteris* and *Rhacopteris* floras that are probably of Middle and Late Carboniferous age. The overlying Guatá Group has local glacial sediment, coal, and marginal marine deposits. *Eurydesma* and *Glossopteris* occur in the upper part of the Guatá and suggest an Early Permian age. Higher, the Iratí Formation has *Mesosaurus,* and the Estrada Nova Group and Rio Do Rasto Group are mainly clastics that contain an extensive plant succession and a few marine tongues having sparse invertebrates. Rynchosaurid reptiles *Cephalonia* and *Scaphonyx* are found in the overlying Santa Maria Redbeds and are of Middle Triassic age.

FAUNAL ZONATION AND DISTRIBUTION

In reviewing the zonation and geographical distribution of Permian faunas, groups have been selected that have been widely used for correlation or that have shown a strong potential use, as in the case of conodonts. Such groups are likely to have received more thorough study and more detailed phylogenetic analysis than other groups. Because of limitation of space and time, the following groups are examined in hierarchical order: fusulinaceans (foraminiferid protozoans), other foraminifers, corals, bryozoans, brachiopods, ammonoids, and conodonts. Other groups that are widely distributed, but which have been less thoroughly studied or less widely used for correlation, such as sponges, bivalves, gastropods, arthropods, and echinoderms, are not discussed. It is recognized, based on a number of local studies, that many of these groups also may have a potential importance in evaluating worldwide biostratigraphic and paleogeographic problems.

FUSULINACEANS

Fusulinaceans are a group of large extinct protozoans that evolved rapidly during the late Paleozoic. Of the more than 80 genera recognized in Permian strata only about a dozen are Carboniferous holdovers (Fig. 2). Two major intervals of unusually rapid evolution are known; the first in Early Permian during the Asselian in all parts of the world, and the second during the later part of the Darvasian and early part of the Murgabian in the Tethyan faunal realm.

The phylogeny and geologic history of fusulinaceans has been summarized by Ross (1967; 1978) and ROZOVSKAYA (1975) and Figure 2 is based on these reports. All of the six Permian families are assigned to

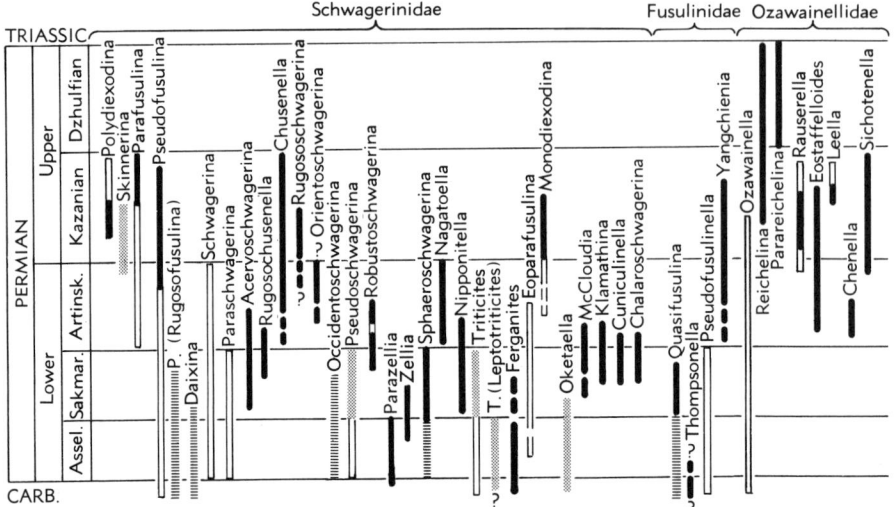

Fig. 2. Ranges of fusulinacean genera arranged by families and their geographic distribution. [Explanation: unshaded, cosmopolitan distribution; black, Tethyan distribution; ruled pattern, Uralian, Franklinian, and northeastern Cordilleran distribution; stippled pattern, midcontinent, southwestern

the one superfamily, Fusulinacea, and are recognized on the basis of differences in wall structure and other internal features. Although the Ozawainellidae evolved into at least nine genera during the Middle Carboniferous and again into three or four genera in the Late Permian, the family is more important because it gave rise to the families Fusulinidae, Schubertellidae and Staffellidae. In their evolutionary development genera of the Fusulinidae become markedly elongate along their axis of coiling. Only one of the three subfamilies, the most primitive group, the Fusulinellinae, ranged into the Permian. It survived until the Late Permian, and retained the basically simple unfolded septa and simple layered-wall structure. By Late Carboniferous time, the Fusulinidae gave rise to the Schwagerinidae, which evolved only slowly during the Late Carboniferous. In Asselian, Sakmarian, and lower Artinskian rocks, genera in the family Schwagerinidae are important zonal fossils, particularly genera with inflated chambers such as *Sphaeroschwagerina, Pseudoschwagerina, Paraschwagerina, Robustoschwagerina, Zellia, Parazellia, Occidentoschwagerina, Acervoschwagerina,* and *Rugososchwagerina,* which have distinctive ranges. Also important are the elongate subcylindrical genera derived from *Pseudofusulina,* such as *Eoparafusulina,* primitive *Parafusulina,* and primitive *Monodiexodina.* Several inflated genera extended into the later part of the Early Permian and *Rugososchwagerina* extended into the early part of the Late Permian. Several of these inflated genera may have been pelagic; however, most other fusulinaceans were benthonic. In southwestern North America, schwagerinids extend through the Guadalupian Series before becoming extinct and include many of the descendants of the Early Permian schwagerinids together with a few Tethyan migrant genera. Advanced species of *Parafusulina* and the first appearance of *Chusenella, Skinnerina, Nipponitella,* and *Polydiexodina* occur in the Guadalupian.

The Schubertellidae, a second family that evolved from the Ozawainellidae, had a conservative Carboniferous history; however, during the Permian it started to expand and one of its lineages, *Boultonia,* eventually gave rise to a burst of Late Permian genera that extended to the end of the Permian. The schubertellids are small, and many became uncoiled and are closely associated with reefs and shallow lagoonal deposits in the later part of the Permian.

Permian
A309

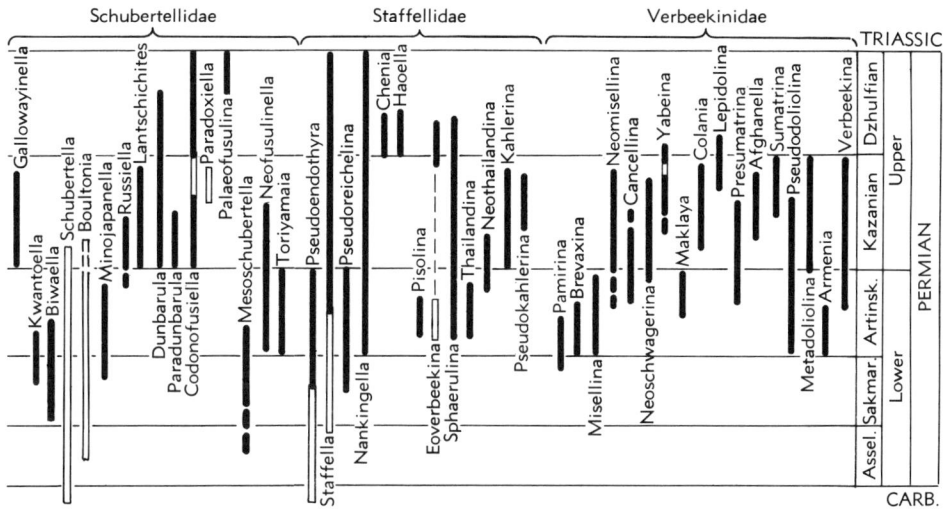

Fig. 2. *(Continued from facing page.)* North American, and northern Andean distribution.] (Data principally from Ross, 1967, and Rozovskaya, 1975; Ross & Ross, n.)

The Staffellidae, which also arose from the Ozawainellidae, had a few, small conservative Carboniferous genera; however, in the later part of the Early Permian this family rapidly evolved. It reached greatest diversity in the early part of the Late Permian, but only one of two genera occur in youngest Permian strata and are associated with back-reef lagoonal sediments.

The Verbeekinidae arose from the Staffellidae and rapidly evolved in the later part of the Early Permian into five or six lineages before becoming extinct about the middle of the Late Permian. Some of its genera are large, subspherical forms with complex internal features.

In addition to their common ancestry, the Staffellidae and Verbeekinidae are similar in that they evolved rapidly in the later part of the Early Permian and early part of the Late Permian, increased in size, and had strongly modified wall structures. The schubertellid subfamily Boultoninae increased in diversity in the early part of the Late Permian but showed an even greater diversity of genera in the latest part of the Permian before becoming extinct. Most of the generic diversity in the Schwagerinidae occurred in the early part of the Early Permian and only a few new genera appeared in the Late Permian. The family Fusulinidae includes one long-ranging lineage that extended into the Late Permian. The family Ozawainellidae had a diversity peak in the early part of the Late Permian. These phylogenetic patterns suggest that the early part of the Early Permian and the early part of the Late Permian were times of major fusulinacean diversification, and later parts of the Early Permian and the latest Permian were times of restriction of previously successful lineages. These evolutionary patterns appear to be associated with changes in geographical distributions and the development of regional and endemic lineages.

FUSULINACEAN ZONES OF THE RUSSIAN PLATFORM, URAL REGION, AND ADJACENT REGIONS

The Late Carboniferous and Early Permian fusulinacean zonation of the Russian platform and Ural region is well known from extensive studies by RAUZER-CHERNOUSOVA (1937, 1940, 1949, 1965), RAUZER-CHERNOUSOVA and others (1958), SYEMINA (1961), SHCHERBOVICH (1969), MIKHAYLOVA (1974), and many others. Although there is a lack of agreement among some Soviet paleontologists as to which of the zones should be used for defining the base of the Permian and subdivisions within the Permian, the actual sequence of zones and the fusulinacean assemblages that form these zones have been extensively studied.

The zones characterized by *Montiparus, Triticites,* and *Pseudofusulina* (*i.e.,* the Kasimovian, Gshelian, and Orenburgian stages) are generally included in the Upper Carboniferous. The overlying zones are included by many in the Lower Permian and that procedure is followed in this review.

Eight fusulinacean assemblage-zones are recognized from the type area of the Lower Permian; however, no fusulinaceans have been reported from Upper Permian strata in this area or in areas that were laterally connected with it to the north (VISSARIONOVA et al., 1949).

ASSELIAN

The Asselian Stage contains three fusulinacean zones. These are separated on the basis of different species of *Sphaeroschwagerina* and combine to make up the "Zone of *Schwagerina*" as commonly used in Soviet literature. The lowest is the Zone of *Sphaeroschwagerina fusiformis* and *Sphaeroschwagerina vulgaris,* which includes about 33 characteristic species of *Pseudofusulina, Pseudofusulinella, Schubertella, Triticites, Jigulites, Daixina, Schwagerina,* and *Pseudoendothyra.* In the middle is the Zone of *Sphaeroschwagerina moelleri* and *Pseudofusulina fecunda,* which includes 57 characteristic species of *Pseudoendothyra, Pseudofusulinella, Schubertella, Fusiella, Triticites, Jigulites, Daixina, Pseudofusulina, Schwagerina, Pseudoschwagerina,* and *Sphaeroschwagerina.* At the top is the Zone of *Sphaeroschwagerina sphaerica* and *Pseudofusulina firma,* which includes 38 characteristic species of *Pseudoendothyra, Pseudofusulinella, Schubertella, Triticites, Pseudofusulina,* and *Sphaeroschwagerina.*

Within the Asselian, species of *Sphaeroschwagerina* show an evolutionary trend toward becoming more globose in outline. *Pseudofusulina* has many lineages of species including some that have strongly rugose outer walls and others that develop heavy secondary deposits.

Unfortunately, many Soviet micropaleontologists prefer to continue to use the generic concept of VON MOELLER (1877) for the genus *Schwagerina* rather than the type specimens illustrated by EHRENBERG (1854, plate xxxvi, x, c, figs. 1-4) as *"Borelis princeps,"* to which VON MOELLER (1877) referred as follows: *"Als eine typische Form derselben sehe ich die Schwagerina princeps Ehrenb. an.*[6]*."* Although VON MOELLER illustrated thin sections of highly inflated forms that he believed were the same as EHRENBERG's species, polished and thin-section study of EHRENBERG's material showed it to be quite different in internal features and probably closely related, if not identical, to *"Fusulina" krotowi* (DUNBAR & SKINNER, 1936; DUNBAR, 1958). Earlier, DUNBAR and SKINNER (1931) had erected *Pseudofusulina* for elongate forms that, as it turned out, had many of the features of EHRENBERG's *Borelis princeps* and, for several years after 1936, *Pseudofusulina* was considered by most non-Soviet micropaleontologists to be a synonym of *Schwagerina.* The North American forms, which previously had been assigned to *Schwagerina* were placed in *Pseudoschwagerina* and it was assumed that the Soviet species also belonged there. Between 1936 and 1976, 10 new genera have been proposed for various inflated schwagerinids, including *Sphaeroschwagerina,* using VON MOELLER's specimens of *"princeps"* as its type. In contrast to *Pseudoschwagerina,* which occurs in many of the same stratigraphic beds with it, *Sphaeroschwagerina* has low, thin-walled juvenile whorls that lack folded septa and massive chomata and the two genera are easily distinguishable.

The result of this nomenclatural confusion over use of the name *Schwagerina* is even more awkward because it is associated with the naming of several distinctive, widely distributed, and easily recognized biostratigraphic zones within the Asselian Stage. Thus the *"Schwagerina"* zones of the Russian platform and Urals are based on species that are neither *Schwagerina* nor *Pseudoschwagerina,* but species of *Sphaeroschwagerina.*

SAKMARIAN

This stage has been divided into two major subdivisions (RAUZER-CHERNOUSOVA

1965), the Tastubian "superzone" and the Sterlitamakian "superzone," which are identified on the basis of different species of *Pseudofusulina*. The lower part of the Tastubian is identified as the Zone of *Pseudofusulina moelleri* and includes about 10 additional species of *Pseudofusulina, Fusiella, Triticites,* and *Pseudoendothyra*. The upper part of the Tastubian is the zone of *Pseudofusulina verneuili* and includes five other species of *Pseudofusulina, Pseudofusulinella, Fusiella, Pseudostaffella,* and *Schubertella*.

The Sterlitamakian is essentially the Zone of *Pseudofusulina plicatissima, P. urdalensis, P. schellwieni, P. intermedia,* and *P. callosa*. It also includes several species of *Rugosofusulina, Daixina,* and *Pseudofusulinella*. The Burtsevian "superzone" is included by many Soviet biostratigraphers as a facies of the upper part of the Sterlitamakian. It includes nearly a dozen closely related species of *Pseudofusulinella,* including *P. concavutas, P. vissarionovae, P. paraconcavutas, P. pseudoconcavutas, P. delicata, P. schellwieni, P. kutkanensis,* and *P. juresanensis*. Although the trend appeared in the Tastubian, the fusulinacean faunas at this time began to show a marked decrease in generic and species diversity; and limestone facies are less widely developed, particularly in many parts of the Russian platform.

ARTINSKIAN

The Irginian Substage includes *Pseudofusulinella concessa, P. paraconcessa, P. solida, P. schellwieni, P. verneuili,* and *Parafusulina lutugini*. *Parafusulina lutugini* is a long-ranging species that has many primitive features of the genus. Its age relationship to other primitive lineages of *Parafusulina,* such as those in the Leonardian Series of North America, is poorly known.

FUSULINACEAN ZONES OF THE TETHYAN REGIONS

The most complete stratigraphic record of Permian fusulinaceans occurs in the Tethyan regions (Fig. 2). These regions were sites of major structural deformation during Permian and post-Permian time so that the structure is usually complex and strata have complicated facies relations because of large and changing topographic relief. Fusulinaceans are well studied from several of these Tethyan regions and, in spite of the structural and stratigraphic difficulties, a fusulinacean zonation has been established that greatly aids the correlation of these rocks with the type Permian (ARAKELIAN et al., 1964). Although not all of the biostratigraphic problems are resolved, particularly near the top of the Permian (NAKAZAWA, ISHII, et al., 1975; NAKAZAWA, KAPOOR, et al., 1975), the Tethyan fusulinacean succession for Japan is the most thoroughly documented (TORIYAMA, 1958, 1963, 1967) and is used in the following discussion of zonation as the reference standard for the regions.

SAKAMOTOZAWAN SERIES

The Sakamotozawan includes the zones of *Pseudoschwagerina morikawai* and *Pseudofusulina vulgaris,* and in its type section is defined as the stratigraphic range of *Pseudoschwagerina*. Three subzones based on species ranges are recognized. The lowest, the Subzone of *Pseudoschwagerina morikawai,* has associated species of *Triticites* and is correlated with the Asselian Stage. It commonly is not present above a widespread Carboniferous-Permian unconformity. The middle, the Subzone of *Pseudofusulina vulgaris,* is distributed more widely in Japan and other Tethyan regions and is associated with more advanced inflated genera such as *Robustoschwagerina,* advanced species of *Paraschwagerina,* species of *Nagatoella,* and holdover species of *Triticites*. This subzone appears to correlate with some part of the Sakmarian Stage. The highest unit is the Subzone of *Pseudofusulina ambigua,* which also has a wide distribution in Tethyan successions. This subzone is marked by a profusion of related species of *Pseudofusulina,* such as *P. krafti, P. globosa, P. fusiformis, P. japonica,* and others. *Nagatoella* and *Schwagerina* are locally common, and *Triticites* is rare. Primitive *Misellina,* one of the early verbeekinids, first appears in the subzone of *Pseudofusulina vulgaris* and is well developed in the Subzone of *Pseudofusulina ambigua*. The fauna with *Misellina* is sufficiently different in its species assemblage to suggest a different depositional facies.

NABEYAMAN SERIES

The Nabeyaman is identified as the Zone of *Parafusulina* and is characterized by species of the *Parafusulina kaerimizensis*, *Parafusulina matsubaishi*, or *Pseudofusulina japonica* stage of development; however, these species and species with these stages of development are known to range also into overlying zones. Much of this zone is equivalent to the Zone of *Cancellina*, which is abundant in some facies. The Zone of *Parafusulina* also includes primitive species of *Neoschwagerina*, marking the Subzone of *Neoschwagerina simplex*, a species that forms an important facies in which *Cancellina*, *Verbeekina*, and *Yangchienia* also are commonly well represented. Because of facies differences, this zone is difficult to interpret at present. In correlation, the Nabeyaman Series usually is placed as equivalent to the middle, or middle and upper parts of the Artinskian Stage and to the middle, or middle and upper parts of the Leonardian.

AKASAKAN SERIES

This series is the Zone of *Neoschwagerina* as characterized by the Subzone of *Neoschwagerina craticulifera* below and the Subzone of *N. margaritae* above. The *Neoschwagerina craticulifera* fauna is widely distributed in Tethyan regions and includes additional species of *Neoschwagerina*, *Afghanella*, *Pseudodoliolina*, *Verbeekina*, *Parafusulina*, and *Pseudofusulina*. The higher *Neoschwagerina margaritae* assemblage also is widely distributed and includes additional species of most of the genera of the lower subzone although species of *Parafusulina* and *Pseudofusulina* become increasingly rare. Elsewhere in the Tethyan regions, *Skinnerina* (primitive *Polydiexodina* of authors) occurs low in this Zone of *Neoschwagerina* and suggests correlation with the lower part of the Guadalupian Series.

KUMAN SERIES

This is the youngest series that is usually recognized in Japan and in most part corresponds to the Assemblage-Zone of *Yabeina* and *Lepidolina* in which *Neoschwagerina* is not recorded. Although the stratigraphic evidence is not entirely clear, it appears that the lower part of this zone is dominated by *Yabeina globosa* and *Lepidolina elongata* and the middle part is dominated by *Lepidolina kumaensis* and *L. shiraiwensis*.

YOUNGEST TETHYAN FUSULINACEAN ZONES

The upper part of the Kuma Formation on which the Kuman Series is defined includes species of *Codonofusiella*, *Rauserella*, *Parareichelina*, *Dunbarula*, *Sichotenella*, *Staffella*, and *Nankinella* along with smaller foraminiferids. Elsewhere in Japan a fauna dominated by *Codonofusiella*, *Reichelina*, and *Palaeofusulina* are known from a few high Permian strata and these are probably equivalent in age to a part of the Loping Series of south China (TORIYAMA, 1973). The most complete development of the youngest Permian fusulinacean zonation is in southern and central China (Nanking, Inst. Geol. and Paleont., 1974; SHENG, 1963, 1965), in southeast Asia (TORIYAMA, 1975; TORIYAMA et al., 1974; LYEM, 1971, 1974), and in southeast Pamir (LEVEN, 1967, 1975). Two assemblage-zones apparently are present, a lower Zone of *Codonofusiella* and *Reichelina* and an upper Zone of *Palaeofusulina* and *Reichelina* (SHENG, 1963; TORIYAMA, 1973). Both *Codonofusiella* and *Reichelina* appear in the underlying Zone of *Yabeina* and *Lepidolina* and extend into younger strata that contain *Palaeofusulina*. As far as can be determined, the extinction of *Palaeofusulina*, *Codonofusiella*, *Reichelina*, and associated genera in the highest Permian fusulinacean zone took place prior to the first appearance of the Triassic fauna.

Permian fusulinaceans are widely distributed in southern and southwestern China (CHEN, 1934, 1956; SHENG, 1963, 1965) and have a zonation that closely parallels that of Japan. The lower part of the Maping Limestone includes one or more Late Carboniferous (as used in this review) species zones of *Triticites*. As these rocks include mainly latest Carboniferous zones (characterized by *Quasifusulina*, *Rugosofusulina*, and *Pseudofusulina*), the unconformity with the underlying Huanglung Series (characterized by *Fusulina* and *Fusulinella*) may represent a long hiatus. The higher parts of the Maping Limestone contain *Pseudoschwagerina*, *Zellia*, *Quasifusulina*, *Rugosofusulina*, *Paraschwagerina*, *Pseudofusulina*, and *Triticites*, which correlate with the Asselian and, possibly, parts of the Sakmarian Stage of the Russian platform. Post-Maping diastrophism resulted in considerable erosion prior to the deposition of the Chihsia Formation. This widespread unconformity is one of the reasons that Chinese geologists consider the Chihsia Formation to be the lowest Permian unit (SHENG & LEE, 1964). The Chihsia Formation contains part of the Zone of *Parafusulina* and includes the Subzone of *Misellina*. The overlying Maokou Formation includes the upper part of the Zone of *Parafusulina*, with the Subzone of *Cancellina*, and the Zones of *Neoschwagerina* and *Yabeina* (CHEN, 1956). The Chihsia and Maokou Formations comprise the Yangsinian Series. Disconformably overlying the Maokou Formation is a thick sequence of limestone of the Lopingian Series that has *Codonofusiella* in the lower 400 m (Wuchiaping Limestone) and *Palaeofusulina* in the upper 100 m. The Lopingian Series lacks *Yabeina* and *Lepidolina* and, therefore, is probably younger than all but the upper part of the Kuman Series of Japan. Shale and limestone with the Triassic *Claraia* bivalve fauna lie above the Lopingian Series.

PAMIRS

The Permian of Soviet Middle Asia, particularly in the Pamir region, has a nearly complete succession that LEVEN (1967, 1975) divided into three series. The lowest is the Yaikian Series having Asselian, Sakmarian, and probably early Artinskian equivalents, that is, the zones of *Sphaeroschwagerina* through to the earliest of the primitive *Parafusulina*, and is equivalent to the upper part of the Maping Limestone of southwestern China. The middle series, the Kushanian, is subdivided into three regional stages, the Chisyanaian Stage (encompassing the Zone of *Misellina*), the Kubergandinian Stage (encompassing the Zone of *Cancellina*), and the Murgabian Stage (including the Zone of *Neoschwagerina*). The highest series, the Pamirian (LEVEN, 1967), was renamed the Arianian (LEVEN, 1975) and three stages were recognized, the Keptenian (Capitanian) (zones of *Yabeina* and *Lepidolina*), the Dzhulfian (Zone of *Paradunbarula*), and the Chansinian (Changhsingian) (Zone of *Palaeofusulina*) at the top. As in the southwestern China section, regional unconformities lie below the Zone of *Misellina* and above the zones of *Yabeina* and *Lepidolina*.

FUSULINACEAN ZONES OF NORTH AMERICA

The succession of fusulinaceans is known for a number of stratigraphic sequences in western and southwestern North America. The standard reference sections in west Texas include the Wolfcampian and Leonardian series of early Permian age in the Glass Mountains and the Guadalupian and Ochoan series of late Permian age in the Guadalupe Mountains.

The usually accepted placement of the lower boundary of the Permian in North America has been at the base of the Zone of *Pseudoschwagerina* (BEEDE & KNIKER, 1924; THOMPSON, 1954) and this corresponds closely to the base of the Asselian Stage on the Russian platform, which also contains primitive *Pseudoschwagerina*, as well as primitive *Sphaeroschwagerina*. In the type section of the Wolfcampian Series, the first *Pseudoschwagerina* appears in the Neal Ranch Formation (Ross, 1959, 1963a) and, using this criterion, the upper beds of the underlying Gaptank Formation (Bed 2 of the Gray Limestone Member and the *Uddenites*-bearing shale of R. E. KING, 1931, 1938) are considered to be Carboniferous. The fusulinacean fauna in Bed 2 is meager and includes advanced species of *Triticites*, including some that are similar to *Daixina* and, possibly *Schwagerina*.

The lower part of the Wolfcampian, the Neal Ranch Formation, contains the Zone of *Pseudoschwagerina uddeni* and associated species of *Paraschwagerina, Pseudofusulina, Schwagerina, Stewartina, Schubertella, Eoparafusulina*, and advanced species of *Triticites* (DUNBAR & SKINNER, 1937; Ross, 1963a). The upper part of the Wolfcampian Series, the Lenox Hills Formation, contains the Zone of *Pseudoschwagerina robusta* and advanced species of *Paraschwagerina, Eoparafusulina, Schwagerina*, and small *Staffella?* (Ross, 1963a).

The type Leonardian Series has been subdivided into three formations, the Skinner Ranch Formation at the base, the Cathedral Mountain Formation, and the Road Canyon

Formation at the top (COOPER & GRANT, 1964, 1966, 1973), which have a thin-bedded lateral facies, the Hess Limestone (Ross, 1960, 1962a). The Skinner Ranch Formation and the lower part of the Hess Limestone are characterized by *Schwagerina hawkinsi, S. crassitectoria, S. guembeli, S. dugoutensis, Parafusulina spissisepta, P. allisonensis,* and *Eoparafusulina linearis*. The Cathedral Mountain Formation contains *Parafusulina durhami* and laterally equivalent strata contain *Schwagerina setum* and *Skinnerella* sp. All the species of *Parafusulina* from the Skinner Ranch and Cathedral Mountain formations have low cuniculi and are considered primitive because these structures appear irregularly in the early volutions. Road Canyon fusulinaceans include *Parafusulina* having well-developed cuniculi, such as *P.* cf. *P. lineata* and *P. sullivanensis*.

The lower part of the Guadalupian Series in the Glass Mountains, the Word Formation, contains *Rauserella, Skinnerina,* and large species of *Parafusulina* with well-developed cuniculi and large proloculi. In the Guadalupe Mountains, the lower part of the Guadalupian also is characterized by large advanced species of *Parafusulina,* such as *P. rothi, P. maleyi,* and rare specimens of *Leella*. The upper part of the Guadalupian, the Capitan Limestone, is characterized by several species of *Polydiexodina* that have a central as well as accessory tunnels. The upper member of this limestone contains a few genera and species with Tethyan affinities, such as *Yabeina texana* and *Codonofusiella paradoxica*. The Ochoan Series is an evaporite succession that lacks fusulinaceans and other normal marine fossils, and our understanding of the correlation of this series is based on its stratigraphic position in one depositional basin where it lies above strata of Capitanian age and below Middle Triassic nonmarine strata.

Elsewhere in North America depositional facies became gradually unsuited for fusulinaceans during the Early Permian as parts of the epicontinental seas became filled in by sediments or were uplifted. This is well shown in the midcontinent region, where fusulinaceans did not survive beyond the Wolfcampian, and in the Midland basin and the east Texas shelf, where they did not survive beyond the Leonardian. The block-faulted basins to the west in New Mexico and southeastern Arizona generally lack fusulinaceans in strata younger than Wolfcampian or early Leonardian, except for one thin limestone unit of local extent that has early Guadalupian species of *Parafusulina* (Ross, 1973).

To the north and west of the transcontinental arch the general stratigraphic range of fusulinaceans is similar to that of Early Permian fusulinaceans, being widespread and becoming progressively restricted before the middle of the period. Only in the strongly deformed structural belts of the western Cordillera do fusulinaceans range well up into the Late Permian. Along this western cratonic shelf Early Permian fusulinaceans from east-central Alaska, northern Canada (THORSTEINSSON, 1974), Greenland (DUNBAR et al., 1962), and the northern part of the Russian platform spread southward and reached into southern Alberta, and a few reached the Basin and Range region in Nevada and Utah (SLADE, 1961).

South of the transcontinental arch, as far as Venezuela, Colombia, Bolivia, and Peru, the Early Permian fusulinid faunas are similar in general composition as far as the limited data permit comparison.

The early Permian of northern California (SKINNER & WILDE, 1965) has a particularly diverse fauna of fusulinaceans and parts of this fauna and later Permian fusulinaceans of both Tethyan and non-Tethyan affinities occur to the north in Oregon, Washington, British Columbia, Yukon, and Alaska in structurally complex rocks (MONGER & Ross, 1971). Primitive species of *Yabeina* and *Waagenoceras* (Ross & NASSICHUK, 1970) suggest that the upper part of the Zone of *Neoschwagerina* is equivalent in age to the early part of the Guadalupian (Word Formation). It is not clear how high in the Permian the succession in these western belts of strata extend, but locally they seem to be at least as young as the Zone of *Yabeina* and *Lepidolina*.

GEOGRAPHIC DISTRIBUTION

Fusulinaceans are recorded from North America, South America, Europe, Asia, and northern Africa, but are unknown from

Australia, Antarctica, and central and southern Africa (Ross, 1967). They are associated with normal marine limestones and are common in bioherms and banks having coral, algal, and echinodermal fragments in three types of depositional settings. The first type is thin-bedded, algal-rich limestone on cratonic shelves usually associated with widespread, rapid transgressions and regressions of the shoreline across areas of low relief during the Early Permian (Ross, 1972). Ecological niches on these shelves were numerous and, as each shelf was more or less isolated from the others, endemic species abound and a few endemic genera also are present. The second type includes reefs and lagoons as well as some deeper water carbonate environments between the reefs on cratonic shelf edges. Different species and even different genera inhabited each of the various shelf and shelf-edge environments. Cratonic shelf and shelf-edge environments are particularly important because their stratigraphic successions are generally uncomplicated by major contemporaneous or later structural deformation so that facies relations can be traced. The third depositional setting is less thoroughly understood but includes thick carbonate deposits with abrupt lateral changes in depositional facies into dark shale, graywacke, and ribbon chert (Monger & Ross, 1971). The carbonate facies are associated with basaltic, noncratonic igneous rocks and are probably closely related to reef environments on former island arcs or oceanic ridges. The abundantly fossiliferous limestone normally includes a wide variety of corals, bryozoans, brachiopods, crinoids, and algae in addition to many species and genera of fusulinaceans, and indicates tropical or subtropical shallow-water deposition. Deposits of this type were usually strongly deformed by contemporaneous or later structural events and their internal stratigraphy is complicated and generally difficult to work out in detail.

Near the end of Carboniferous and the beginning of Permian time, a few generic and species complexes became widely dispersed, perhaps several times within relatively brief intervals (Ross, 1962b). These species groups established lineages in different regions that can be traced through several geologic stages. Early Permian fusulinacean faunas are dominated by genera and species of Schwagerinidae that evolved into many well-defined lineages (Ross, 1967; 1977). Several genera, such as *Sphaeroschwagerina, Pseudoschwagerina, Paraschwagerina, Zellia,* and *Parazellia,* are short ranging and are found in the earlier part of the epoch. *Sphaeroschwagerina, Zellia, Parazellia, Robustoschwagerina, Acervoschwagerina,* and *Biwaella* are commonly found only in the Tethyan-Uralian-Franklinian region and most occurrences outside of that region represent sporadic, brief migrations that did not survive for any appreciable length of time. In contrast, one major lineage, *Pseudoschwagerina,* was present for only the early part of its history in that region and then became confined to the midcontinent-southwestern North American region and to the Andean belt of South America. Although not as completely known as those from North America and Eurasia, these Andean fusulinaceans are most closely related to those from southwestern North America and should be included with that region (Ross, 1963b, 1967).

The later part of the Early Permian Epoch was marked by increasing endemism in species complexes in most genera of Schwagerinidae and also in several other families. Dispersals between different regions became less frequent and before the end of Early Permian time three regions developed strongly differentiated faunas. A Tethyan region, a separate Uralian-Franklinian region, and a southwestern North American-Andean region are recognizable. Separation of the Tethyan region from the Uralian-Franklinian region after the early part of Artinskian time appears sharp because fusulinaceans in these regions became markedly different and faunal correlations between all three regions becomes increasingly difficult as species and generic endemism increases. The Schubertellidae, Ozawainellidae, Staffellidae, and Verbeekinidae evolved rapidly into nearly 40 genera that became dominant in the Tethyan region (Gobbett, 1967). By the beginning of Late Permian time the Tethyan fusulinacean faunas became distinctively endemic and are difficult to correlate with those of other areas. Each of the three regions should be considered a separate faunal realm beginning at this time.

In the Tethyan region *Pseudoendothyra*, a long-ranging conservative genus from the Middle Carboniferous or earlier, began a remarkable evolutionary diversification in the later part of the Early Permian Epoch, giving rise to a dozen genera of advanced Staffellidae. One of these gave rise to *Misellina*, the earliest verbeekinid. The Schubertellidae show a significant but smaller evolutionary burst and a few generic and several species complexes of the Schwagerinidae show increasingly endemic distribution at this time. Although the Verbeekinidae and Schwagerinidae died out and the latest Permian Lopingian faunas were composed of only six Tethyan genera and one new genus, *Palaeofusulina*, all seven of these genera became extinct before the beginning of the Triassic.

Outside the Tethyan realm, fusulinaceans have a much different history in middle and late Artinskian and Late Permian times. In the Uralian-Franklinian realm the number and diversity of fusulinaceans decreased rapidly. The few surviving lineages of Fusulinidae and Staffellidae died out in the region before the Artinskian and *Pseudofusulina, Schwagerina*, and primitive *Parafusulina* continued until about the end of Early Permian time when they also died out. The Late Permian in the Uralian-Franklinian realm lacks a fusulinacean fauna.

In the midcontinent-southwestern North American-Andean region progressive reduction in the extent of epicontinental seas resulted in the fusulinacean distribution being reduced to the southwestern part of North America where, in the later part of Early Permian time, *Schwagerina*, primitive *Parafusulina*, and *Staffella* were dominant. Fusulinidae and early lineages of Verbeekinidae are absent. Representatives of *Schubertella* and *Boultonia* appeared at different times in the region and indicate separate temporary dispersals during the Early Permian. Several other genera, such as *Robustoschwagerina*, temporarily dispersed into the region during the later part of the Early Permian. By the beginning of Late Permian time the fusulinacean genera were greatly reduced to abundant advanced *Parafusulina* and rare *Skinnerina* and *Rauserella*. By the beginning of Capitanian time, these genera were replaced by *Polydiexodina*, a probable emigrant from the Tethyan fauna.

A few other genera, such as *Paradoxiella, Codonofusiella, Yabeina*, and *Leella*, appeared for short intervals in the southwestern North American realm during Late Permian time, but they established no long lineages of species. In general, the southwestern North American realm is identified by species complexes of *Parafusulina* and *Polydiexodina*, which dominated the fusulinacean fauna from the Leonardian to the end of the Guadalupian, and by the general lack of persistent lineages of Verbeekinidae. By latest Permian time, fusulinaceans became extinct in this realm.

OTHER FORAMINIFERIDA

Few studies on nonfusulinacean foraminiferids are available for Permian strata, although arenaceous and smaller calcareous foraminiferids commonly are present in nonfusulinacean-bearing shale and limestone. Lituolidae, Textulariidae, Ammodiscidae, Lagenidae, and Nodosinellidae of Asselian, Sakmarian, and Artinskian age are recorded from the Bashkirian region (LIPINA, 1949; MOROZOVA, 1949). From the Kazanian of the Russian platform 27 genera of arenaceous foraminifers were reported from acid residues (UCHARSKAJA, 1970). In the Maritime Territory of the Soviet Union, NIKITINA (1969) located widely distributed *Hemigordiopsis* in the zones of *Neoschwagerina* and *Yabeina*. The genus is also recorded from northern Caucasus and Cyprus. PANTIC (1970) described *Hemigordiopsis*, other foraminiferids, and algae from the middle and upper parts of the Permian of western Serbia, which includes the Zone of *Reichelina-Codonofusiella* at the top. Although additional studies of smaller foraminiferids are available for various parts of the Permian succession (see SOSNINA, 1965; ISHII et al., 1975; OKIMURA et al., 1975), our understanding of their phylogeny and distributional patterns remains poorly known.

CORALS

Many late Paleozoic coral genera have

long geological ranges and wide geographic distributions, whereas other groups of genera have restricted geological ranges and also restricted geographic distributions (Fig. 3). This pattern of distributions presents difficulties in analyzing the biostratigraphy, interregional correlations, and paleobiogeography of Permian Anthozoa. The comprehensive summations by HILL (1948, 1957, 1958), the most recent compilation reviewing the worldwide distribution of Permian Rugosa, noted the distinct distribution of some waagenophyllid genera and reaffirmed her earlier interpretations that the distribution of rugose corals was strongly influenced by environmental conditions. In relatively shallow, nearshore, clastic marine environments the coral assemblages consisted principally of small, solitary, morphologically simple, nondissepimented corals. In contrast, in relatively deeper water offshore marine environments the coral assemblages were dominated commonly by large solitary or compound dissepimented corals.

MINATO and KATO (1965a, 1965b, 1970) extensively studied the phylogeny and distribution of two distinctive families of Rugosa, the Durhaminidae and Waagenophyllidae (Fig. 3), and they demonstrated the distinct biogeographic distribution of genera in these two families. The durhaminid distribution plotted on a map of the world with present-day geographic locations was in the more northerly subarctic and arctic regions of the northern hemisphere with a southerly extension along the western part of North America, whereas the waagenophyllids were distributed in lower latitudes of Europe, Eurasia, and southern Asia. Earlier detailed investigations on the Permian Anthozoa of the Soviet Union by SOSHKINA et al. (1941), SOKOLOV (1955), and others provided considerable information on faunal assemblages in the Ural Mountain region and the Russian platform. VASILYUK et al. (1970) and SHCHUKINA (1973) have more recently provided additional information on the distribution of faunas, and ROWETT (1972, 1975a) and STEVENS (1975a) have presented further as-

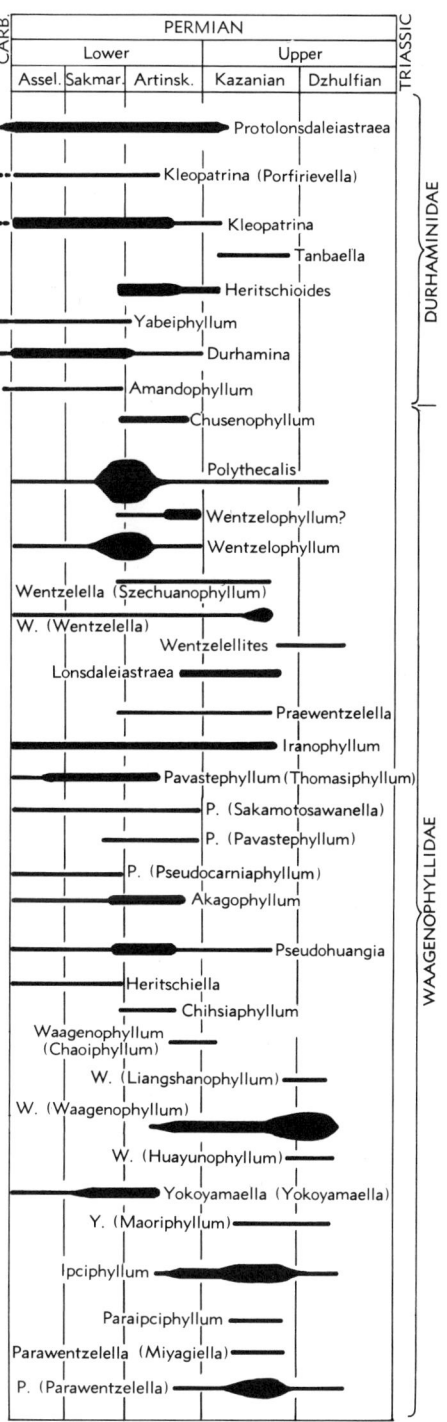

FIG. 3. Ranges of genera of the rugose coral families Durhaminidae and Waagenophyllidae. (Data from Minato & Kato, 1965a and 1965b; Ross & Ross, n.)

pects of the paleobiogeography of Early Permian Rugosa.

Presently, Permian corals are considered to form three faunal provinces, two of which, the Ural-Artinsk and Tethys, are partly identified on the basis of other faunal distributions, particularly fusulinid provinces (see Ross, 1967). The third province includes the more poorly delimited fauna of the midcontinent and southwestern parts of the United States and possibly also Central America and the northern Andes of South America. The Ural-Artinsk province (so named by VASILYUK, 1970) is dominated by durhaminid rugose corals and extends from the southern Ural Mountain region and Russian platform north and westward to the region of Vest Spitsbergen and the Canadian Arctic Islands into western North America and possibly Central America. The coral assemblages of this province were originally identified in the Ural Mountain region and the geographic boundaries have been more clearly defined as more studies have been undertaken. The other distinct province, the Tethys, is dominated by waagenophyllid corals and extends from Tunisia across the Carnic Alps, Donbas, Crimea, Caucasus, Transcaucasus, Iran, Pamirs, Pakistan, Nepal, parts of China, Japan, Maritime Territory of the Soviet Union, Timor, New Zealand, and Australia. The Early Permian faunal assemblages of the midcontinent of the south-central and southwestern parts of the United States appear to form another province that has some endemic forms and a mixing of genera at different times from both the Ural-Artinsk and Tethyan provinces. Mixing of generic assemblages from the Ural-Artinsk and Tethys provinces also occurred in the Early Permian (Sakmarian) in the western (Carnic Alps) and eastern (parts of China and Japan) regions of the Tethys province.

URAL-ARTINSK PROVINCE

In the Ural Mountain region and Russian platform the exceedingly rich, Early Permian (Asselian and Sakmarian) coral faunas have many genera, most of which range upward from the Carboniferous. They include massive colonial forms such as the lonsdaleiid *Thysanophyllum*, the lithstrotionids *Orionastraea* and *Stylastraea*, the durhaminid *Protolonsdaleistraea*, as well as solitary fasciculate caniniids. The durhaminid *Kleopatrina* and the aulophyllid *Protowentzelella*, which appear for the first time in the Urals in the Early Permian, may occur in older Carboniferous strata in other parts (Novaya Zemlya and Alaska) of the Ural-Artinsk province (STEVENS, 1975a). Except for *Protowentzelella* and *Thysanophyllum*, all these genera continue into the Artinskian. In the upper part of Artinskian and in Kungurian strata colonial Rugosa are absent and only small long-ranging solitary forms are present along with tabulates such as *Cladochonus*.

In the late Asselian or early Sakmarian of Vest Spitsbergen, the massive colonial corals include *Thysanophyllum*, *Stylastraea*, *Kleopatrina*, and *Protolonsdaleiastraea*. The Canadian Arctic islands have a fauna that is principally Artinskian in age and have the massive corals *Stylastraea*, *Protolonsdaleiastraea*, and *Kleopatrina*, as well as *Clisiophyllum?* and *Caninia*.

Faunal assemblages of Artinskian age from northern Alaska (Lisburne Peninsula and northeast Brooks Range) are strongly dominated by small, simple nondissepimented corals (ROWETT, 1975b). The fauna includes the polycoeliids *Tachylasma*, *Ufimia*, and *Sochkineophyllum*; the hapsiphyllids *Amplexizaphrentis*, *Allotropiophyllum*, and *Euryphyllum*; the metriophyllid *Stereocorypha*; the laccophyllid *Amplexocarinia*, and cyathopsid *Hornsundia*. These assemblages occur in nearshore clastic marine facies. In east-central Alaska the Artinskian coral fauna is also in the same facies and has small nondissepimented corals such as the lophophyllidiid *Lophophyllidium* and the hapsiphyllids *Euryphyllum*, *Hapsiphyllum*, *Neozaphrentis*, and *Amplexizaphrentis*. In southern and southeastern Alaska, in the Sakmarian and Artinskian, the coral faunas are markedly different from those of northern Alaska. Here they are dominated by large dissepimented solitary and colonial corals that inhabited a deeper water eugeosynclinal environment. The solitary forms include the lophophyllidiids *Lophophyllidium* and *Stereostylus*; the aulophyllids *Clisiophyllum* and *Auloclisia*; and the cyathopsids *Bothro-*

phyllum, Timania, Caninophyllum, and *Caninia.* The compound fasciculate dissepimented corals are the durhaminids *Durhamina* and *Heritschioides.* Massive cerioid corals are rare and are represented by the durhaminid *Protolonsdaleiastraea* and the waagenophyllid *Wentzelella.* The occurrence of a waagenophyllid in this Ural-Artinsk assemblage may represent mixing of the faunas of the Ural-Artinsk and Tethyan assemblages or, as suggested by MONGER and Ross (1971), the Tethyan faunas are part of a tectonic belt that has been structurally moved against the Ural-Artinsk province. Tabulates are also common in this southern region and include the auloporids *Syringopora* and *Cladochonus,* and the favositid *Michelinia,* and the sinoporid *Sinopora.*

In Nevada in the western United States, a coral fauna of probably early Sakmarian age contains the massive colonial lonsdaleiid forms *Thysanophyllum, Sciophyllum, Eastonoides,* and *Lithostrotionella;* the massive colonial durhaminid *Kleopatrina;* the fasiculate durhaminid *Durhamina;* a fasciculate lithostrotionid, and solitary forms such as the cyathopsid *Caninia* and the lophophyllidiid *Stereostylus.*

In the Upper Permian Kazanian deposits of the Russian platform coral faunas are widely distributed and consist of long-ranging solitary forms such as the polycoeliids *Plerophyllum* and *Calophyllum.* In Alaska, in the northeast Brooks Range, strata of probable Kazanian age contain the metriophyllid *Duplophyllum* and the polycoeliid *Pseudobradyphyllum.* In the Kap Stosch area of central East Greenland in beds of late Guadalupian age, 11 rugose genera from six families are all solitary forms and include *Calophyllum, Pentamplexus, Amplexocarinia, Cryptophyllum, Lytvolasma?, Amplexizaphrentis, Bradyphyllum, Sinophyllum, Hapsiphyllum?,* and *Leonardophyllum?* (FLÜGEL, 1973).

Corals have not been reported from the Tatarian of the Russian section or from other latest Permian strata of the Ural-Artinsk province.

TETHYAN PROVINCE

In the Tethyan province colonial and solitary waagenophyllids and solitary zaphrentid and caniniid forms commonly characterize the coral assemblages. Extensive documentation of the waagenophyllids by MINATO and KATO (1965b) provided much of the data in the following summary. Some of waagenophyllid genera such as *Waagenophyllum (Waagenophyllum), Ipciphyllum, Akagophyllum,* and *Polythecalis* have a wide geographic distribution, and, in addition, many waagenophyllid genera have long geological ranges. Some with short geologic ranges and limited geographic distribution may be useful in regional and, possibly interregional correlation. MINATO and KATO (1965b) considered the following coral genera or subgenera to have such short ranges: *Pavastehphyllum (Pseudocarniaphyllum)* and *Heritschiella,* which characterize the Zone of *Pseudoschwagerina* in strata of Asselian or, possibly, Sakmarian age; *Chihsiaphyllum* and *Polythecalis (Chusenophyllum),* which characterize the Zone of *Pseudofusulina* and occur in rocks of usually late Sakmarian, but possibly also early Sakmarian age; *Waagenophyllum (Chaoiphyllum),* which occurs in the Zone of *Parafusulina* of early to middle Artinskian age; *Parawentzelella (Miyagiella), Paraipciphyllum,* and *Wentzellites,* which characterize the Zone of *Neoschwagerina* of early Wordian (early Guadalupian) age; and *Waagenophyllum (Liangschanophyllum), Waagenophyllum (Huayunophyllum),* and *Wentzelloides,* which identify the Zone of *Yabeina* of Capitanian age.

In the Early Permian (Zone of *Pseudoschwagerina*), waagenophyllid corals are widely distributed and a number of genera, such as the solitary forms *Iranophyllum, Iranophyllum (Laophyllum), Pavastehphyllum (Sakamotosawanella), Pavastehphyllum (Pseudocarniaphyllum), Akagophyllum, Pseudohuangia, Yokoyamaella, Polythecalis, Wentzelophyllum, Wentzelella,* and *P. (Pavastehphyllum)* range upward from the Carboniferous. *Pavastehphyllum (sensu lato)* appears to characterize the earliest Permian coral assemblages. In the Donbas and Pamirs, the coral fauna of the Zone of *Pseudoschwagerina* includes the tabulate *Michelinia* and the lophophyllidiid *Lophocarinophyllum,* as well as abundant caniniids and zaphrentids, *Pavastehphyllum,* and the colonial forms *Pseudohuangia* and *Heritschiella* (VASILYUK et al., 1970). In China, the Zone of *Pseudoschwagerina* has *Caninia* and *Pseudocarniaphyllum* (SHENG & LEE, 1964). In the Early Permian of Australia and Timor only solitary nondissepimented corals, such as *Euryphyllum, Tachylasma, Verbeekiella* and *Plerophyllum* and tabulates such as *Cladochonus* and *Thamnopora* occur.

From the Zone of *Pseudofusulina*, usually correlated as late Sakmarian but possibly in part also early Sakmarian, the waagenophyllids *Pseudohuangia* and *Lonsdaleiastraea* identify this interval. During the Artinskian some of the earlier waagenophyllid genera such as *Akagophyllum, Wentzelophyllum, P. (Polythecalis), Yokoyamaella,* and *Pseudohuangia* became rarer and genera such as *Waagenophyllum (Waagenophyllum), Ipciphyllum, Parawentzelella* and *Lonsdaleiastraea* became increasingly significant in the coral assemblages. In the Transcaucasus, Pamirs, and Caucasus, a rugose coral fauna of marked diversity contains *Wentzelophyllum, Yatsengia, Ipciphyllum,* and *Parawentzelella*. In the Pamirs the coral assemblage includes *Carnithiaphyllum, Iranophyllum, Pavastehphyllum (Pavastehphyllum), Yatsengia,* and *Heritschiella:* In China this zone has the tabulate *Hayasakia* and the rugose forms *Stylidophyllum* and *Polythecalis*. In Timor and Australia only small, solitary rugose corals, such as *Euryphyllum, Plerophyllum, Allotropiophyllum,* and *Verbeekiella,* and tabulates are present.

The Zone of *Neoschwagerina*, correlated with the early Wordian (early Guadalupian), has an abundant waagenophyllid, *Ipciphyllum. Waagenophyllum (Waagenophyllum)* continued to diversify with wide distribution of its species. At this time the genus extended into the Caucasus, Crimea, and Pamirs, and in the Pamirs the coral assemblage also included *Yatsengia* and *Iranophyllum*. In the Maritime Territory of the Soviet Union and in China *Wentzelella* is distinctive, but in the Transcaucasus waagenophyllids are poorly developed and the coral assemblage consists mostly of plerophyllids. In the southern part of the Tethys, waagenophyllids reached Timor and New Zealand but did not reach Australia where only small, solitary rugose corals are found.

The Zone of *Yabeina*, correlated with the Capitanian (late Guadalupian), is the time of greatest diversity and abundance of *Waagenophyllum*. At this time *P. (Pavastehphyllum)* reached Tunisia (STEVENS, 1975b). In the latest Permian (Dzhulfian), the coral fauna in most parts of the Tethys consisted of solitary corals and is known as the *Plerophyllum* fauna. Environmental conditions in the Tethys were apparently suitable for only a few corals and as conditions continued to change these corals gradually became extinct. In China, in the region between Shensi and Szechuan provinces, a rich coral assemblage with *Waagenophyllum, Lophophyllidium,* and *Liangshanophyllum* occurs in a limestone with the fusulinids *Codonofusiella* and *Reichelina,* and represents the highest occurrence of an extensive Permian coral assemblage (CHAO, 1965). A primitive species of *Waagenophyllum* has been reported also from the Pamirian of the Pamirs. In many other areas, such as Transcaucasus, Iran (FLÜGEL, 1968), Nepal, Timor, and China, micheliniid tabulates and small plerophyllid corals, such as *Plerophyllum, Pleramplexus,* and *Wannerophyllum,* and polycoeliids *Polycoelia* and *Ufimia* are present. FLÜGEL (1970), in discussing the gradual decline and extinction of the rugose corals in the Upper Permian in the Tethys, considered the genera with more complex morphology to become extinct first. These would be the cerioid forms with dissepiments or with dissepiments and presepiments, for example, *Wentzelella* and *Yokoyamaella (Maoriphyllum)* with dissepiments and *Polythecalis* with dissepiments and presepiments. These genera occur only as high as the top of the Zone of *Yabeina*. This was followed by extinctions in fasciculate forms with dissepiments (for example, *Waagenophyllum,* which extends as high as the Zone of *Codonofusiella*), and finally, extinctions of forms with septal columella (for example, *Lophophyllidium,* which extends to the top of the Zone of *Codonofusiella*).

MIDCONTINENT NORTH AMERICAN PROVINCE

Lower Permian rocks of the midcontinent region of the United States extend from Nebraska into Kansas and Oklahoma. To the southwest in Texas, Lower Permian rocks occur in the north-central and western regions of the state. The coral faunas of the Wolfcampian in shallow, nearshore, cyclothem deposits of Kansas (MOORE & JEFFORDS, 1941) and biohermal limestones in west Texas (Ross & Ross, 1962, 1963) are sparse and are dominated by solitary corals, especially *Lophophyllidium*. In the Lower Permian, however, this genus is not as abundant as it was in the Upper Carboniferous. Other solitary corals are *Neokoninckophyllum, Lophamplexus,* and *Stereostylus*. In Kansas, the waagenophyllid *Heritschiella* indicates a dispersal of a genus in a family that has a distinctively Tethyan distribution. In the Leonardian in the Glass Mountains, Texas, the fauna has *Lophophyllidium, Stereostylus, Amplexocarinia,* and *Durhamina*. The occurrence of *Durhamina* represents mixing of a genus that is widespread in the Ural-Artinsk province. Other Early Permian genera in the midcontinent are the aulophyllids *Palaeosmilia* and *Dibunophyllum*, the timorphyllid *Leonardophyllum*, and the auloporid tabulate *Aulopora*. ROWETT (1975a) noted that these faunas have similarities with those in northwestern South America. Such genera as *Lophophyllidium* and *Stereostylus* are widely distributed in western North America including Alaska, suggesting periodic mixing with faunas from the small *cul-de-sac* Midcontinent North American province bounded by the transcontinental arch on the north and west and by the Marathon-Ouachita orogenic belt to the east and south. The sparse Upper Permian coral faunas of the Guadalupian of west Texas include the aulophyllid *Palaeosmilia* and the tabulate *Cladopora*.

BRYOZOANS

Bryozoans (ectoprocts) are a significant part of Permian faunas in certain environmental settings. The Cystoporata and Cryptostomata are dominant and the Trepostomata are far less abundant than at other times in the Paleozoic (Fig. 4). The Cystoporata with 24 genera in six families show a gradual increase in abundance and diversity through the Permian, becoming restricted in distribution in the later part of the Permian. The Cryptostomata with 57

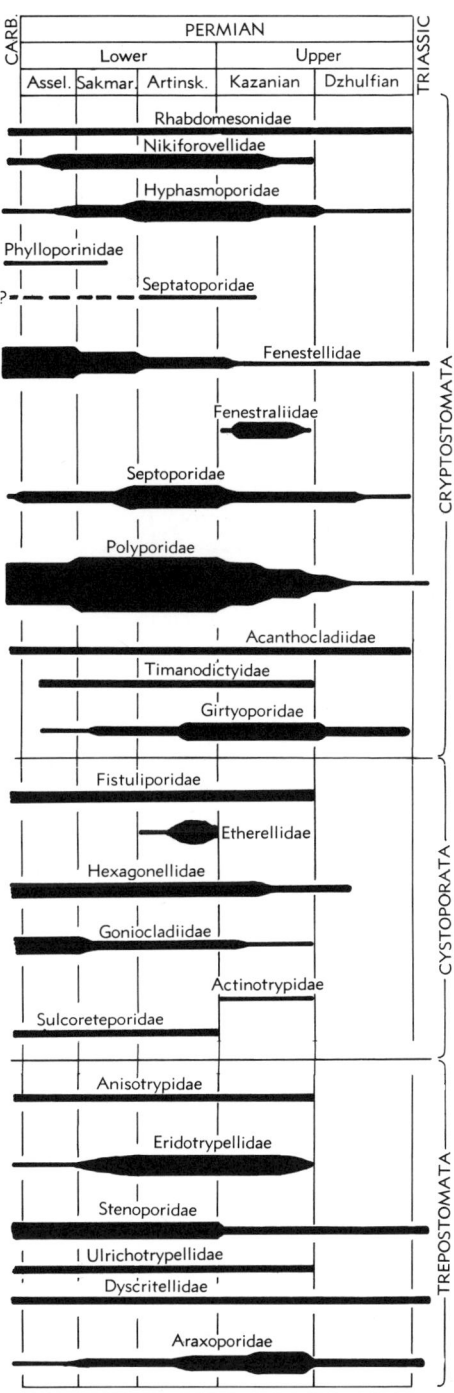

FIG. 4. Range and relative abundance of families of Bryozoa (after J. R. P. Ross, 1978).

genera in 12 families have a pattern of development in the Permian similar to the Cystoporata. The Trepostomata are represented by 19 genera in six families and show a pattern of increased, albeit limited, diversification through the Permian until late Kazanian time. Data for this review came from many reports, but principal references were J. R. P. Ross (1978), Gorjunova (1975), and Morozova (1970). The biogeographic distribution of the bryozoans is summarized for nine different regions in the Permian.

RUSSIAN PLATFORM AND URALIAN SEA

During the Permian, on the Russian platform and adjacent shelf areas a rich fauna of bryozoans continued from the Carboniferous into the later part of the Kazanian. These faunas have been extensively documented by Russian scientists including Nikiforova (1939), Trizna (1950), Shulga-Nesterenko (1952), Trizna and Klautsan (1961), and Morozova (1970). In the Early Permian (Asselian and Sakmarian), fenestrate cryptostomes of the families Fenestellidae and Polyporidae were dominant members of the hydractinoid reefs and adjacent shelf biota. Other groups, represented by cryptostomes such as *Ascopora, Nicklesopora, Streblotrypa, Acanthocladia*, and *Timanodictya*, the stenoporid *Rhombotrypella*, and the cystoporates *Hexagonella, Fistulipora*, and *Ramipora*, were also part of this rich fauna. In the Artinskian, generic diversity was still high with an abundance of fenestellids and polyporids, the trepostome *Rhombotrypella*, the cystoporates *Hexagonella* and *Goniocladia*, and the cryptostomes *Ptylopora* and *Ptyloporella*. During Artinskian time the trepostomes *Pseudobatostomella* and *Stenopora* and the fistuliporid cystoporate *Eridopora* became proportionally more significant in the faunas.

Toward the end of Artinskian time the bryozoans show a marked reduction in generic diversity and this appears to relate to changes resulting from tectonic movements and subsidence in the Uralian Sea and on the Russian platform. *Fenestella, Polypora, Rhombotrypella, Pseudobatostomella*, and *Streblotrypa* are still present but *Clausotrypa* and *Hexagonella* have disappeared. In Kazanian time the distribution of bryozoans shows an increasing generic diversity gradient from the southern part of the Uralian Sea north and westward to the edge of this sea (Morozova, 1970). Eleven genera are present in the southern part of the Uralian Sea in the region of Tatarian SSR, whereas in the north in the Arkhangel region, 26 genera are present. Genera that are well represented in all parts of this Uralian Sea include the trepostomes *Pseudobatostomella, Dyscritella, Rhombotrypella*, and *Tabulipora*, the fenestellids *Wjatkella* and *Fenestella*, and the cryptostomes *Streblascopora, Pinegopora, Parafenestralia*, and *Triznella*. Bryozoans have not been reported in higher units of the Upper Permian on the Russian platform and adjacent regions.

FRANKLINIAN SEA AND ADJACENT SHELF AREAS

Bryozoans from this region that lay north and west of the Uralian Sea stretching from Novaya Zemlya to Spitsbergen, Greenland, and Canadian Arctic islands are poorly known, but they appear to have affinities with the faunas of the Russian platform. In the Sakmarian of northeastern Greenland (Ross & Ross, 1962) the trepostomes *Rhombotrypella, Stenopora*, and *Tabulipora*, and the cryptostomes *Polypora, Fenestella*, and *Timanodictya* occur. A sparse Lower Permian fauna from Novaya Zemlya contains *Pseudobatostomella?, Fenestella, Ramipora*, and *Hyphasmopora*. From Spitsbergen an upper Permian fauna of probable Kazanian age contains *Fenestella, Polypora, Ptylopora, Septopora*, and *Ramipora*.

ZECHSTEIN SEA

In western Europe (England, Germany, Poland, and southern Baltic region), bryozoans, mainly fenestrate cryptostomes, are present in the Upper Permian. The faunas in the Zechstein or equivalents, such as those in the Magnesian Limestone of England, have low generic diversity. The polyporid *Kingopora* and *Fenestella* have a wide distribution. In small reefs in the Zechstein Limestone of Germany the fauna also includes the cryptostomes *Thamniscus, Acan-*

thocladia, *Penniretepora*, and *Synocladia*, the trepostome *Stenopora*, and the hexagonellid cystoporate *Coscinotrypa*.

SOUTHERN REGION OF NORTH AMERICA

Faunas from this region, which are presently poorly known, have been described from the midcontinent region of the United States (MOORE & DUDLEY, 1944) and Texas (GIRTY, 1908). In the Lower Permian of the midcontinent region (Nebraska, Kansas and Oklahoma), cryptostomes are abundant in strata of Wolfcampian age and are represented by *Fenestella, Polypora, Septopora, Thamniscus, Streblotrypa, Rhombopora*, and *Syringoclemis*. The fistuliporid cystoporate *Cyclotrypa* occurs in the Wolfcampian of Nebraska and the hexagonellid cystoporate *Meekopora* in the Wolfcampian of Kansas. These cystoporates are present in the Leonardian together with *Meekoporella* and *Fistulipora*, and all four genera extend up into the Wordian (lower Guadalupian). In the Guadalupian two other cystoporates, *Goniocladia* and *Epiactinotrypa*, also appear. The Guadalupian cryptostomes have marked diversity with *Fenestella, Polypora, Thamniscus, Acanthocladia, Girtyopora*, and *Girtyoporina*. The trepostomes are *Pseudobatostomella?, Paraleioclema?* and *Stenopora*. Higher in the succession in the upper Guadalupian (Capitanian) the fauna has the cystoporates *Fistulipora* and *Goniocladia*, the cryptostomes *Fenestella, Acanthocladia*, and *Girtyoporina*, and the trepostome *Paraleioclema*.

ANDEAN SEA

Two faunas in the Lower Permian of southern Peru occur in strata of Wolfcampian age. They have the cystoporates *Meekopora* and *Goniocladia* and the cystoporates *Fenestella, Polypora, Septopora, Rhombopora*, and *Acanthocladia*.

NORTHERN TETHYAN SEA

The abundant bryozoan faunas of Japan occur in rocks in various structural blocks and basins, and they range in age from Asselian (Zone of *Pseudoschwagerina*) to Pamirian (Zone of *Yabeina-Lepidolina*) (SAKAGAMI, 1970). Bryozoans from the Zone of *Pseudoschwagerina* range in age from Asselian into the Sakmarian and include the cystoporates *Fistulipora, Coscinotrypa*, and *Sulcoretepora*, the trepostomes *Pseudobatostomella, Stenopora*, and *Tabulipora*, and a great many different cryptostomes (e.g., *Fenestella, Penniretepora, Polypora, Anastomopora, Thamniscus, Hayasakapora*, and *Streblascopora*).

In Japan, bryozoans of the Zone of *Parafusulina*, correlated with the Artinskian, include a number of genera that range up from the Zone of *Pseudoschwagerina*, such as *Fistulipora, Pseudobatostomella, Stenopora, Fenestella, Penniretepora, Hayasakapora*, and *Streblascopora*. Hexagonellid cystoporates show considerable diversity with *Meekopora, Meekoporella*, and *Prismopora*. In the Zone of *Neoschwagerina*, correlated with the Wordian of the Guadalupian (Ross & NASSICHUK, 1970), many genera range up from the Zone of *Parafusulina*, and additional genera include the hexagonellid *Fistulamina*, the cryptostomes *Septopora* and *Saffordotaxis*, and the trepostome *Ulrichotrypella*. The youngest bryozoans in the succession occur in the Zone of *Yabeina-Lepidolina* and range in age from upper Guadalupian to possibly Dzhulfian. The cystoporates have high diversity and include *Fistulipora, Meekopora, Prismopora, Sulcoretepora, Coscinotrypa, Goniocladia*, and *Ramipora*. The trepostomes are still sparse and the cryptostomes include *Septopora, Synocladia, Rhabdomeson*, and *Clausotrypa*.

In the Maritime Territory and Kabarovsk region of the Soviet Union the structurally complex suite of rocks has a rich bryozoan fauna (NIKITINA *et al.*, 1970). Little is known of the Lower Permian faunas. In Upper Permian strata of Kazanian age the bryozoan fauna contains *Fenestella, Polypora, Coscinotrypa, Dyscritella, Paraleioclema*, and *Permoleioclema*. From the upper part of the Zone of *Yabeina*, MOROZOVA (1970) listed 30 genera, 18 of which occur in both the Maritime Territory and Kabarovsk region. The fauna common to both areas contains the cystoporates *Fistulamina* and *Fistulipora*, the trepostomes *Dyscritella, Ulrichotrypella, Hinganella, Stenodiscus?, Tabulipora, Paraleioclema*, and *Permoleioclema*, and abundant cryptostomes, such as *Rhabdomeson, Streblascopora, Maychella, Clausotrypa, Fenestella, Polypora, Septopora, Girtyoporina,* and *Girtyopora.*

From the Kolyma and Omolon massifs and adjacent regions, an Upper Permian fauna has a great abundance of cryptostomes, such as *Fenestella, Wjatkella, Maychella, Polypora, Synocladia,* and *Timanodictya,* the cystoporate *Fistulipora,* and the trepostomes *Dyscritella* and *Primorella.*

CENTRAL TETHYAN SEA

The well-developed bryozoan faunas of the Darvas and Pamir regions range in age from Asselian into the Pamirian (GORJUNOVA, 1975). The faunas are abundant in the Lower Permian and sparse in the Upper Permian. In southwestern Darvas, the bryozoans from strata of Asselian and Sakmarian age include the cystoporates *Fistulipora, Actinotrypella, Goniocladia, Ramiporidra,* and *Sulcoretepora,* the trepostomes *Rhombotrypella* and *Primorella,* and the cryptostomes *Rhabdomeson* and *Streblascopora.* Strata of Artinskian age contain a small cystoporate fauna of *Fistulipora, Eridopora, Cyclotrypa,* and *Hexagonella.* Only the two cystoporates *Fistulipora* and *Hexagonella* occur in the Artinskian in central Pamir. In southeastern Pamir, the lower Permian (Sakmarian) has beds filled with the cryptostome *Nikiforovella,* and in the Artinskian, a more diverse fauna includes the cystoporates *Fistulamina* and *Ramiporidra,* the trepostomes *Rhombotrypella* and *Dyscritella,* and abundant representatives of the cryptostomes *Pamirella* and *Streblascopora.*

High in the Permian (Pamirian) of southwestern Darvas the cystoporate *Fistulipora* is the only bryozoan reported. No bryozoans have been recorded from the intervening stages. In central Pamir, where stages are not presently differentiated, the fauna has the cystoporates *Fistulipora* and *Eridopora* and the cryptostome *Ogbinopora.* In southeast Pamir the Murgabian has the cystoporate *Hexagonella* and the trepostome *Araxopora.*

In the Transcaucasus region of the south-central Soviet Union, the Upper Permian fauna is dominated by cryptostomes (MOROZOVA, 1970). The bryozoan assemblages in the lower part of the Guadalupian (Gnishikian) consist of the cryptostomes *Fenestella, Septopora, Polypora, Rhabdomeson, Ogbinopora,* and *Streblascopora,* the cystoporates *Fistulipora, Cyclotrypa, Hexagonella,* and *Sulcoretepora,* and the trepostomes *Paraleioclema* and *Araxopora.* In the upper part of the Guadalupian (Kachikian) the trepostome *Araxopora* is the only genus reported. The Dzhulfian includes the cryptostomes *Synocladia, Polypora, Septopora, Streblotrypa,* and *Girtyoporina,* and the cystoporate *Fistulipora.*

In central-eastern China, the Lower Permian Chihsia Limestone has the cosmopolitan genera *Fistulipora, Fenestella, Polypora,* and *Septopora* (Loo, 1958). The first three of these genera as well as *Stenopora* and *Dyscritella* occur in the Upper Permian in the Maokou Formation of China and the Jisu Honguer Limestone of Inner Mongolia. In addition, in the Upper Permian of China, the cryptostome *Acanthocladia* has been reported, and in Inner Mongolia other genera are the cryptostomes *Rhabdomeson, Streblascopora, Maychella, Girtyopora,* and *Girtyoporina,* the cystoporates *Fistulamina* and *Hexagonella,* and the trepostomes *Tabulipora* and *Paraleioclema* (GRABAU, 1931; MOROZOVA, 1970). In China, in higher units of the Permian, the Loping Series contains *Fistulipora, Polypora, Septopora, Penniretepora, Synocladia, Pseudobatostomella,* and *Paraleioclema.*

SOUTHERN TETHYAN SEA

In central-eastern Afghanistan abundant bryozoans occur in strata ranging in age from Sakmarian into Kazanian. The Lower Permian faunas of Sakmarian and Artinskian age have numerous polyporids and fenestellids such as *Polypora (Pustulopora)* and *Fenestella,* which range through the entire sequence, and *Polypora (Paucipora)* and *Minilya.* Other cryptostomes are *Rhombopora, Saffordotaxis, Septopora,* and *Rhabdomeson.* The cystoporates include *Cyclotrypa, Meekopora, Goniocladia,* and *Sulcoretepora.* The trepostomes have low representation with *Tabulipora, Dyscritella,* and *Rhombotrypella.* The Upper Permian (Kubergandinian) faunas have a number of genera that extend up from the Artinskian, including *Rhombotrypella* and the four cystoporates noted above. The cryptostomes *Ascopora, Streblascopora, Septopora,* and *Acanthocladia* first appear in the Permian. Higher in the Upper Permian (Murgabian), *Tabulipora* reappears and the cystoporates are represented by *Goniocladia, Hexagonella,* and *Coscinotrypa,* and the cryptostomes by *Streblascopora, Rhabdomeson, Thamniscus,* and *Reteporidra.*

In the Salt Range of Pakistan, the recorded Lower Permian (Artinskian) faunal assemblages are not as diverse. Lower Permian (Artinskian) strata contain the cystoporates *Fistulipora* and *Hexagonella,* the trepostomes *Stenopora* and *Stenodiscus,* and the cryptostomes *Fenestella, Polypora, Rhombopora, Thamniscus, Acanthocladia,* and *Girtyopora.* In the Upper Permian, the two cystoporates noted above are associated with *Goniocladia,* and the cryptostomes are limited to *Polypora, Rhombopora,* and *Synocladia.*

In Thailand the Lower Permian bryozoan faunas of late Sakmarian to late Artinskian age have abundant cystoporates and cryptostomes. The cystoporates include *Fistulipora, Coscinotrypa, Hexagonella, Goniocladia, Ramipora, Liguloclema,* and *Sulcoretepora,* and cryptostome assemblages consist of *Fenestella, Polypora, Thamniscus, Pennireteporta, Acanthocladia, Rhabdomeson, Ascopora, Streblascopora, Streblotrypa?, Rhombopora, Timanodictya?,* and *Ogbinopora.* Other members of the faunas are the trepostomes *Dyscritella* and *Leioclema?.* In Malaya a distinctive Upper Permian fauna of Guadalupian age includes *Fenestella, Pseudobatostomella, Araxopora, Paraleioclema,* and *Clausotrypa.*

In western Australia, in the Fitzroy trough of the Canning basin and in the Carnarvon basin (CROCKFORD, 1951; Ross, 1963), Lower Permian (Sakmarian) bryozoan assemblages contain the cystoporates *Evactinostella, Hexagonella,* and *Fistulipora,* the cryptostomes *Streblascopora, Fenestella, Polypora,* and *Lyropora,* and the trepostomes *Dyscritella, Stenopora,* and *Paraleioclema.* In rocks of Artinskian age the fauna shows a marked increase in diversity with the addition of the cystoporates *Prismopora, Fistulamina, Evactinopora, Eridopora, Liguloclema, Etherella, Goniocladia,* and *Ramipora,* and the cryptostomes *Synocladia, Minilya, Streblotrypa, Acanthocladia, Septopora, Saffordotaxis, Rhabdomeson, Megacanthopora?, Callocladia?, Rhombocladia, Streblotrypa,* and *Streblocladia.* In the Bonaparte Gulf basin in strata considered to be in the lower part of the Upper Permian, the sparse fauna consists of *Fistulipora, Rhombopora, Streblotrypa,* and *Ramipora.* In Upper Permian strata of Tatarian age in the Fitzroy trough, the assemblage has *Stenodiscus* and *Dyscritella.*

In Timor, bryozoan faunas that are assignable to stratigraphically identified units range in age from Early Permian into the Late Permian. The sparse fauna from the Bitauni Beds of Artinskian age comprise the cystoporate *Fistulipora,* the cryptostomes *Rhombopora, Streblascopora,* and *Fenestella,* and the trepostomes *Ulrichotrypa* and *Hinganella.* Upper Permian bryozoan faunas occur in the Basleo and Amarassi beds of Kazanian age. In the Basleo Beds, cystoporates are more abundant and include *Fistulipora, Eridopora, Fistulotrypa, Goniocladia,* and *Hexagonella,* and, in addition, the trepostome *Hinganella* and the cryptostomes *Streblascopora* and *Fenestella.* In the overlying Amarassi Beds, the sparse fauna consists of *Fistulipora, Fenestella, Clausotrypa,* and *Stenopora.* Higher in the sequence in strata of probable Kazanian age the fauna has the two cryptostomes *Rhabdomeson* and *Streblotrypella.*

TASMAN GEOSYNCLINE (EASTERN AUSTRALIA)

In northeastern Australia the Permian bryozoan faunas show a pattern of generic diversity comparable to that of Western Australia. In the Lower Permian (Sakmarian) of the Bowen basin and Springsure shelf, Queensland, the fauna is sparse and contains *Fenestella* and *Polypora* (WASS, 1969). In strata of Artinskian age cystoporates and cryptostomes are abundant and include *Fistulipora, Ramipora, Goniocladia, Liguloclema, Saffordotaxis, Rhombopora, Streblascopora, Diploporaria, Pennireteporta, Polypora,* and *Minilya.* The trepostomes are *Stenopora, Dyscritella,* and *Stenodiscus.* In the Upper Permian the fauna of Kazanian age comprises the cryptostomes *Saffordotaxis, Pennireteporta, Ptylopora, Levifenestella, Septatopora, Fenestella,* and *Poly-*

pora, and the trepostomes *Paraleioclema?*, *Stenopora*, and *Stenodiscus*.

In the Sydney basin of southeastern Australia, formations of Sakmarian age have a restricted fauna, including *Stenopora* and *Dyscritella*. In deposits of Artinskian age, species diversity increases, particularly in the genus *Stenopora*; other genera are *Pseudobatostomella?*, *Rhombopora*, *Fenestella*, *Minilya*, and *Polypora*. In Tasmania, rocks of the same age also show great species diversity in *Stenopora* and also contain *Hemitrypa?* and *Stenodiscus*.

BIOGEOGRAPHIC DISTRIBUTION

During a large part of the Early Permian (Asselian to Artinskian) a number of cystoporate genera had cosmopolitan distribution, including the fistuliporids *Fistulipora* and *Cyclotrypa*, the hexagonellids *Hexagonella*, *Coscinotrypa*, and *Meekopora*, and the goniocladiids *Ramipora* and *Goniocladia*. *Cyclotrypa*, *Meekopora*, *Ramipora*, and *Goniocladia* continued to have cosmopolitan distribution in the Kazanian. Other cystoporates form a distinctive part of many Tethyan faunas and in the Early Permian are represented by the hexagonellids *Evactinopora*, *Evactinostella*, *Fistulamina*, *Prismopora*, and the sulcoreteporid *Sulcoretepora*. During the Late Permian, Tethyan cystoporate assemblages included the fistuliporid *Fistulotrypa*, the hexagonellids *Fistulamina*, *Prismopora*, and the sulcoreteporid *Sulcoretepora*.

Trepostomes generally paralleled the cystoporates in being mainly either cosmopolitan or Tethyan. Cosmopolitan genera ranging through the Early and into the Late Permian were *Stenopora*, *Tabulipora*, *Rhombotrypella*, and, possibly, *Pseudobatostomella*. Genera restricted to the Tethyan included *Paraleioclema* in the Early and Late Permian and *Permoleioclema*, *Araxopora*, *Primorella*, *Dyscritellina*, *Arcticopora*, and *Permopora* in the Late Permian. Some genera, such as *Dyscritella*, were widely distributed in the Uralian and Tethyan seas early in the Permian, dispersed into southern Europe in the Artinskian and became restricted to the Tethys in the Late Permian.

The cryptostome genera provide more detailed distributional data for delineating faunas of the Uralian, Franklinian, Tasman, Zechstein, and Tethyan seas, and they exhibit dispersal patterns different from those of the cystoporates and trepostomes. From the Asselian through the Kazanian the fenestellids *Fenestella* and *Penniretepora* and the polyporid *Polypora* were cosmopolitan. The septaporid *Septapora* was cosmopolitan during the Sakmarian, then apparently was restricted to the Tethyan Sea during the Artinskian and had a worldwide distribution again in the Kazanian. The septaporid *Synocladia* first appeared in the Franklinian Sea and gradually dispersed so that it was cosmopolitan by the early part of the Kazanian. In the Uralian Sea, the fenestellids *Ptylopora* and *Ptiloporella* were distinctive genera during the Early Permian and the fenestraliids *Parafenestralia* and *Triznella* were distinctive in the Late Permian. Certain genera apparently were restricted to the Tethyan Sea in the Permian and include the fenestellid *Minilya*, the polyporid *Lyropora*, the nikiforovellid *Nikiforovella*, and the rhabdomesid *Pamirella*. In the Tethyan Sea, the girtyoporid *Hayasakapora* ranged through the Permian, the hyphasmoporid *Ogbinopora* occurred in the Artinskian and the Kazanian, and the girtyoporid *Tavayzopora* was present during the Kazanian. Some cryptostomes displayed a bipolar distribution during the Early Permian, such as the septaporid *Synocladia* and the fenestellid *Diploporaria*, occurring in the Uralian Sea and the Tasman geosyncline, and in the Late Permian the fenestellid *Ptylopora* was present in the Franklinian Sea and Tasman geosyncline. Dispersal of some genera, such as the timanodictyid *Timanodictya* and the rhabdomesid *Ascopora*, from the Uralian Sea into the Tethyan Sea occurred during the Early Permian. Other genera, such as the nikiforovellid *Clausotrypa*, were present in the Uralian Sea in the Early Permian and the Tethyan Sea in the Late Permian.

BRACHIOPODS

The evolution and dispersal of Permian brachiopods were extremely complex and their taxonomy, phylogeny, and stratigraphic and geographic distributions are known in only broad outline on a worldwide basis. Thus, considerable uncertainty exists concerning the placement of many genera and some families into a generally accepted scheme of classification (WILLIAMS et al., 1965; RUDWICK, 1970; WATERHOUSE & BONHAM-CARTER, 1975). These problems in classification introduce many additional uncertainties for establishing stratigraphic and geographic distributions of families and superfamilies (Fig. 5) in addition to those uncertainties associated directly with the stratigraphic record.

The most diverse order during the Permian was the Strophomenida followed by the Spiriferida and Rhynchonellida. These three orders account for more than four-fifths of the Permian brachiopods, and most of the Permian extinctions occurred in these groups. The strophomenids, which generally had either a reduced pedicle opening or none in the adult shell, had two types of adaptation to fixing or positioning their shell on or in the substrate. The Productacea used tubular spines as anchoring devices in soft or shifting substrates and the Strophalosiacea and Davidsoniacea generally cemented their shell to other shells. During the Permian the strophalosiaceans, richthofeniaceans, and lyttoniaceans flourished in reeflike bioherms that gradually became more abundant and complex in their ecological organization.

Most Early Permian brachiopods are continuations of well-established Late Carboniferous lineages or are relict lineages having Middle or Early Carboniferous origins. Many Late Carboniferous genera range into Sakmarian and Artinskian deposits. Only a few new families evolved during the Permian, such as the Chonetellidae, Tschernyschewiidae and, possibly, the Athyrisinidae, and only one superfamily, the Richthofeniacea.

Extinction of families and genera was gradual within the later part of the Permian and the majority of the families became extinct by the beginning of Dzhulfian time (Fig. 5). Although one or two genera in each of the orders survived the close of the Permian, most of the Dzhulfian genera apparently became extinct before the end of the period; however, it should be noted that our knowledge of the Dzhulfian faunas is still very incomplete.

DISPERSAL PATTERNS

The dispersal patterns in the six superfamilies of Strophomenida contrast in several ways. During the Early Permian the Davidsoniacea had widely distributed genera, most of which are nearly cosmopolitan, and during the Late Permian they showed a gradual reduction in number of genera and a development of endemic genera. The chonetaceans are few in number, being the end of a long, diverse phyletic lineage of which the Permian representatives are essentially cosmopolitan. The genera of the seven Permian families of strophalosiaceans, although of different numerical importance, are mainly endemic genera with only a few appearing in more than one faunal province. The richthofeniaceans were widely distributed during the Early Permian but were restricted to the Tethyan before the end of the Artinskian. The productacean families had several dispersal patterns during the Permian. The linoproductids, and to a lesser extent the overtoniids, were represented in the Permian by a few genera that probably were part of relict lineages whose distributions record much earlier dispersals. The Echinoconchidae, Buxtoniidae, and approximately half of the Dictyoclostidae and Linoproductidae were mostly cosmopolitan. The remaining Dictyoclostidae and Linoproductidae and most of the Marginiferidae were strongly provincial in their distributions and were probably endemic. The Lyttoniacea have one nearly cosmopolitan genus, three that are of more limited distribution, nine that are Tethyan, and one that is present in southwestern North America.

In the rhynchonellids, the rhynchonellaceans are of interest because of their relict pattern of distribution; a few genera had provincial distributions following the superfamily's great Devonian evolutionary diversification. In the Camarotoechiidae *Paranorella* and *Leiorhynchoidea* survived into the Permian of southwestern North America. The rhynchotetradid *Goniophoria* survived as a successful relict in the Tethyan region. Of the Atriboniidae, three genera are found only in the Upper Permian of the Tethys and the other few are widely different in geologic age and scattered in geographic dis-

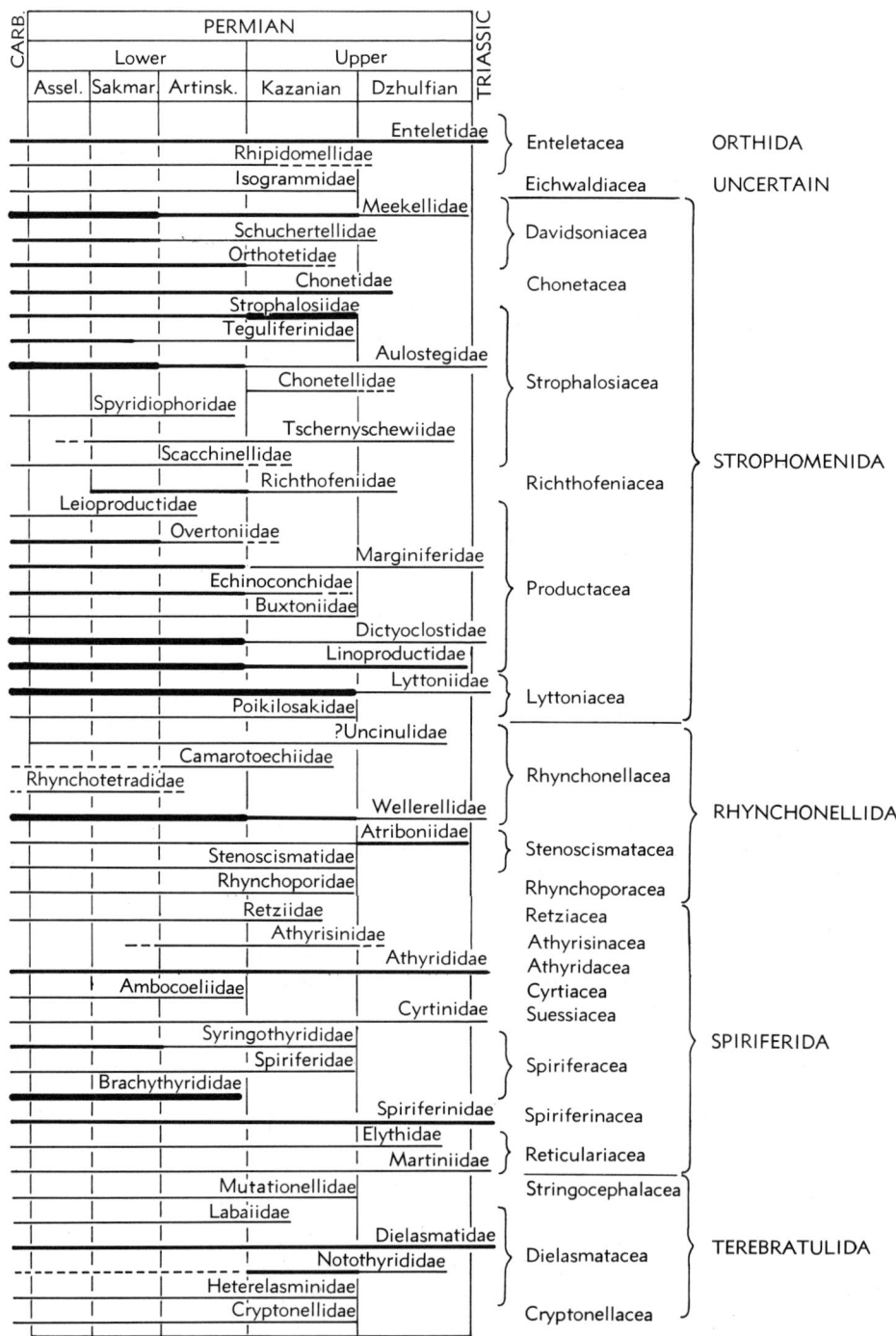

Fig. 5. Ranges and relative importance of families of articulate brachiopod superfamilies. [Dashed lines indicate uncertain extent of range. Increasing line widths indicate 1 to 2 genera, 3 to 5 genera and 6 or more genera.] (Data from Williams et al., 1965, and Grant, 1970; Ross & Ross, n.)

tribution, such as *Camerisma* from the Mississippian of Alaska and the Artinskian of the Tethys and Soviet Union. Similar stratigraphic and geographic gaps are also characteristic of the ranges of the four Permian generic survivors of the Stenoscismatidae that have Lower Carboniferous and Permian records but lack Middle and Upper Carboniferous records.

The third major Permian order, the Spiriferida, has three families that are represented by more than one or two relict lineages. The Syringothyrididae are almost entirely Tethyan, Ural region-Russian platform or northern in distribution. The Brachythyrididae are mostly cosmopolitan or bipolar in distribution. The Spiriferinacea has a few cosmopolitan genera, but most others are known from particular geographical provinces. The few Permian genera of the Martiniidae and the Spiriferidae were also mainly provincial and only the three Permian genera of Elythidae had nearly cosmopolitan distribution.

The Terebratulida had about 19 genera during the Permian and seven were widely distributed or nearly cosmopolitan, and the remainder are reported from one or two provinces. The order Orthida had seven Permian genera and were widely distributed.

These patterns of distribution suggest that the Permian distributions of most Orthida, Davidsoniacea, Productacea, Rhynchonellacea, Stenoscismatacea, Spiriferida, and Terebratulida were the result of earlier dispersals, possibly as early as the Early Carboniferous in some groups and as late as Late Carboniferous in others. The small number of new Permian genera in these taxa and the generally low numbers of individuals and geographic restriction of many of these genera suggest that they are relict faunal lineages that had fairly wide ecological tolerances. In the Strophomenida, the Productacea show a similar trend but with less reduction in the number of genera, perhaps because their major evolutionary diversification had occurred only in the later part of the Carboniferous Period. The Strophalosiacea and very specialized Richthofeniacea and Lyttoniacea showed the greatest evolutionary increase of genera during the Permian, most of which were endemic and restricted to either the Tethys province or to the southwestern North American province.

ASSOCIATIONS AND DIVERSITY PATTERNS

Permian brachiopods may appear in localities and lithofacies that lack ammonoids, corals, and fusulinaceans and as a group brachiopods probably occupied a great range of habitats. Differences in brachiopod community assemblages may indicate different paleolatitudes, depths (including a variety of communities below the photic zone), or paleobiogeographic provinces (RUNNEGAR & ARMSTRONG, 1969). Many of these distributional differences have been attributed to temperature effects (WATERHOUSE & BONHAM-CARTER, 1975). Three types of brachiopod associations are: 1) low diversity brachiopod associations, mainly from Australia, New Zealand, and Siberia, which lack associated fusulinids and corals and are near glacial deposits; 2) high diversity brachiopod associations that are almost always associated with corals and fusulinids, and frequently algae, but never with glacial deposits; and 3) intermediate diversity brachiopod associations that commonly, but not always, occur with corals and fusulinaceans. The low diversity associations are believed to have been adapted to cool or cold waters, the high diversity associations to warm tropical water, and the intermediate associations to temperate waters.

Low diversity brachiopod associations in the Gondwanan marine faunas are characterized by the Martiniidae as well as Linoproductidae, Strophalosiidae, Aulostegidae, Wellerellidae, and Spiriferidae. Members of some of these families, as well as Reticulariidae and Heterelasminidae, are common in Siberian faunas. These low diversity associations are accompanied by characteristic bivalve assemblages that include the Eurydesmatidae, Deltopectinidae, Edmondiidae, and Pholadomyidae.

Brachiopod associations of intermediate diversity include chiefly Marginiferidae and Echinoconchidae with numerous Rhynchoporiidae, Buxtoniidae, Dictyoclostidae, and Ambocoeliidae. Important in these associations are Lingulidae, Discinidae, Chonetidae, Rhynchotetradidae, Retziidae, and Spiriferinidae. Many of these families also are found in the high diversity brachiopod associations, although only as a minor part.

High diversity brachiopod associations are characterized by Meekellidae, Aulostegidae, Richthofeniidae, Lyttoniidae, Stenoscismatidae, Uncinulidae, Athyrisinidae, Elythidae, Heterelasminidae, Cryptonellidae, Isogrammidae, Enteletidae, Rhipidomellidae, and others. These associations usually occur with colonial corals and fusulinaceans.

Several interesting trends become apparent in these various associations during the Permian. On the Russian platform, Sakmarian brachiopod faunas and their ecological and faunal associations differ from Asselian ones at the species level and extend with little change into lower Artinskian strata (Aktastinian). In general, the Asselian, Sakmarian, and lower Artinskian brachiopods are widespread, and within a region, each brachiopod faunal association shows a general taxonomic consistency suggesting no major change in physical environments but occasional minor fluctuations. In upper Artinskian strata (Baigendzhinian) the rhipidomellids and overtoniids, which contribute prominently to earlier Permian brachiopod faunas, decrease in importance and marginiferids and dictyolostids become increasingly abundant. Kazanian (and Zechstein) associations become much less diverse than lower Permian faunas in central and eastern Europe, apparently in response to salinity changes or fluctuations.

After Kazanian time brachiopod faunas became rare, particularly low and intermediate diversity associations, and soon after high diversity associations also became rare. In those areas where younger brachiopod faunas survived, associations became considerably less diverse, presumably because of continued ecological changes in shallow-shelf environments.

Permian brachiopods have received considerable attention and their early use in establishing local and regional zonation of the Permian in the Urals and Russian platform (CHERNYSHEV, 1902; FREDERIKS, 1932) was paralleled by their use in North America (GIRTY, 1909; R. E. KING, 1931). Detailed stratigraphic zonation based on local or regional ranges of species and genera of brachiopods is available for a number of stratigraphic sequences. COOPER (in DUNBAR et al., 1960) outlined preliminary range information on genera and species from the Glass Mountains of west Texas and later studies in the Glass Mountains have greatly expanded our knowledge of brachiopod distribution in that succession (COOPER & GRANT, 1972, 1973, 1974, 1975, 1976). Brachiopod studies are particularly important in the Gondwana and eastern Siberian successions, which commonly lack many of the other marine fossil groups. In the Urals and Russian platform, brachiopods dominate one of the facies of the upper part of the Artinskian and the Kazanian apparently as a result of their tolerance to salinity changes. Brachiopod faunas of the Tethyan region are the most diverse and it is in this region that most of the evolutionary changes were introduced, such as the specialization in the richthofeniids, scacchinellids and lyttoniids.

Brachiopods from the Dzhulfian Stage have been described from only a few segments of the Tethyan region and considerable additional information is needed to fully understand the complicated stratigraphic facies interrelations in the uppermost part of the Permian. A number of lineages of Permian brachiopods continue into the overlying ammonoid zone of *Otoceras* and *Ophiceras,* which has long been taken as the lowest zone of the Triassic and, therefore, a few representatives of typically Permian families survived into earliest Triassic time before they became extinct.

AMMONOIDS

Study of the distribution and evolution of Permian ammonoids (Fig. 6) has had many important influences on the present concepts of subdividing the Permian System into series and stages (FURNISH, 1973). As most specialists studying Permian ammonoids believe that this nektonic group shows little evidence of paleogeographical provincialism, they have concentrated their attentions on the rapid, commonly spectacular, evolutionary history of the ammonoids. Most families, genera, and even many species are apparently cosmopolitan and most geographical differences have been considered to be the result of incomplete collections. A few differences in paleogeographic distributions were noted by FURNISH (1973); for example, *Perrinites* was common in Permian tropical latitudes and the Paragastrioceratinae and Metalegoceratinae were common in Permian temperate latitudes during Artinskian time. During Late Permian times ammonoid diversity gradually decreased as more and more families became extinct and new families did not evolve to replace them. In Permian strata younger than Capitanian age only three new families appear and these are geographically restricted.

In spite of detailed knowledge of their phylogeny and rapid evolutionary history, Permian ammonoids generally are only locally abundant in fine-grained clastic beds. Relatively few of the localities having these abundant faunas are in the same or closely related stratigraphic sections. Because of this distribution, ammonoid specialists tend to use the stage of evolution and position within phylogenetic successions as criteria for assigning ages to these scattered ammonoid occurrences (FURNISH, 1973). Because of these considerations, the subdivisions of the Permian based on ammonoid occurrences are a mixture of the concepts of chronostratigraphic and genus- and species-range zones, and ammonoid "stage" names are derived from those parts of the world where ammonoid assemblages of those ages are abundant (Fig. 6). The lowest 12 of the 16 ammonoid assemblages are based on taxonomic ranges in relatively thick stratigraphic sections in which the biostratigraphic relationships with units above and below are known for ammonoids and also for other faunal groups. Ammonoid assemblages are named from the lower parts of the Lower Permian from the southern Ural region, the upper part of the Lower Permian and the lower parts of the Upper Permian from the west Texas region, and higher assemblages of the Upper Permian are named from Timor, Indonesia, the Dzhulfa (Julfa) region of Soviet Nakhichevan Azerbaidzhan and Iran, Salt Range of Pakistan, and southwestern China. Thus, the general scheme of ammonoid subdivision is pragmatic and usable even if it does not fully agree with the American Code of Stratigraphic Nomenclature (A.C.S.N., 1961). The actual application of some of the names differs in rank and range from that used by specialists in other groups (Fig. 1), and contributes to the existing general confusion about unit boundaries. The following general summary is based largely on the works of RUZHENTSEV (1951, 1952, 1966, 1974) and a summary by FURNISH (1973).

LOWER PERMIAN ZONATION

ASSELIAN STAGE

In the southern Urals ammonoids in this stage include a number of gastrioceratid genera that range upward from the Carboniferous as well as some pronoritid, daraelitid, uddenitin, gonioloboceratid, thalassoceratid, adrianitid, and vidrioceratin genera that have close Late Carboniferous ancestors. Three families, the metalegoceratids, paragastrioceratids, and popanoceratids first appear in Asselian strata and, along with the first common members of one subfamily, the sicanitins, are widely distributed. RUZHENTSEV (1951, 1952, 1966) considered *Artinskia* of the *A. kazakstanica* group and primitive species of *Paragastrioceras, Propopanoceras,* and *Juresanites* as typical of the Asselian ammonoid assemblage. RUZHENTSEV (1966) listed the following Permian genera, which first appear and are common in the Asselian Stage: *Sakmarites, Akmilleria, Kargalites (Kargalites), Marathonites (Almites), Prostacheoceras,* and *Protopopanoceras.*

Strata of equivalent age in west Texas and the midcontinent region of North America have numerous early perrinitids (a family not found in the Ural region), represented by *Properrinites.* Other taxa in these earliest Permian ammonoid assemblages are closely similar to those in the Ural region.

SAKMARIAN STAGE

In the southern Ural and Ufa plateau regions the Tastubian Substage of the Sakmarian contains the first true members of *Metalegoceras* and *Uraloceras* along with many holdover genera from the underlying Asselian Stage. In western North America a number of species of *Properrinites,* such as *P. boesei* and *P. denhami,* are found in ammonoid assemblages of this age as well

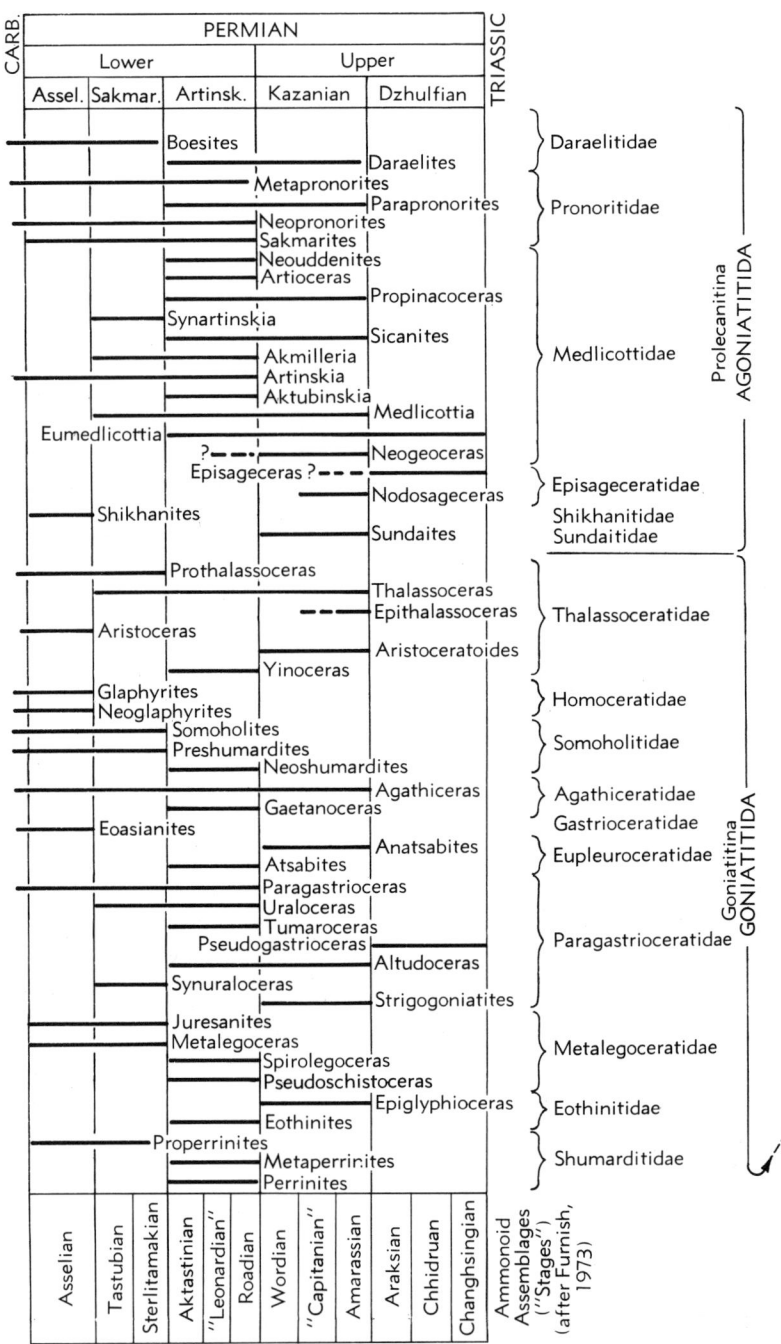

Fig. 6. Ranges of ammonoid genera arranged by families. (Data principally from Ruzhentsev, 1962, and Spinosa et al., 1975; Ross & Ross, n.)

Permian

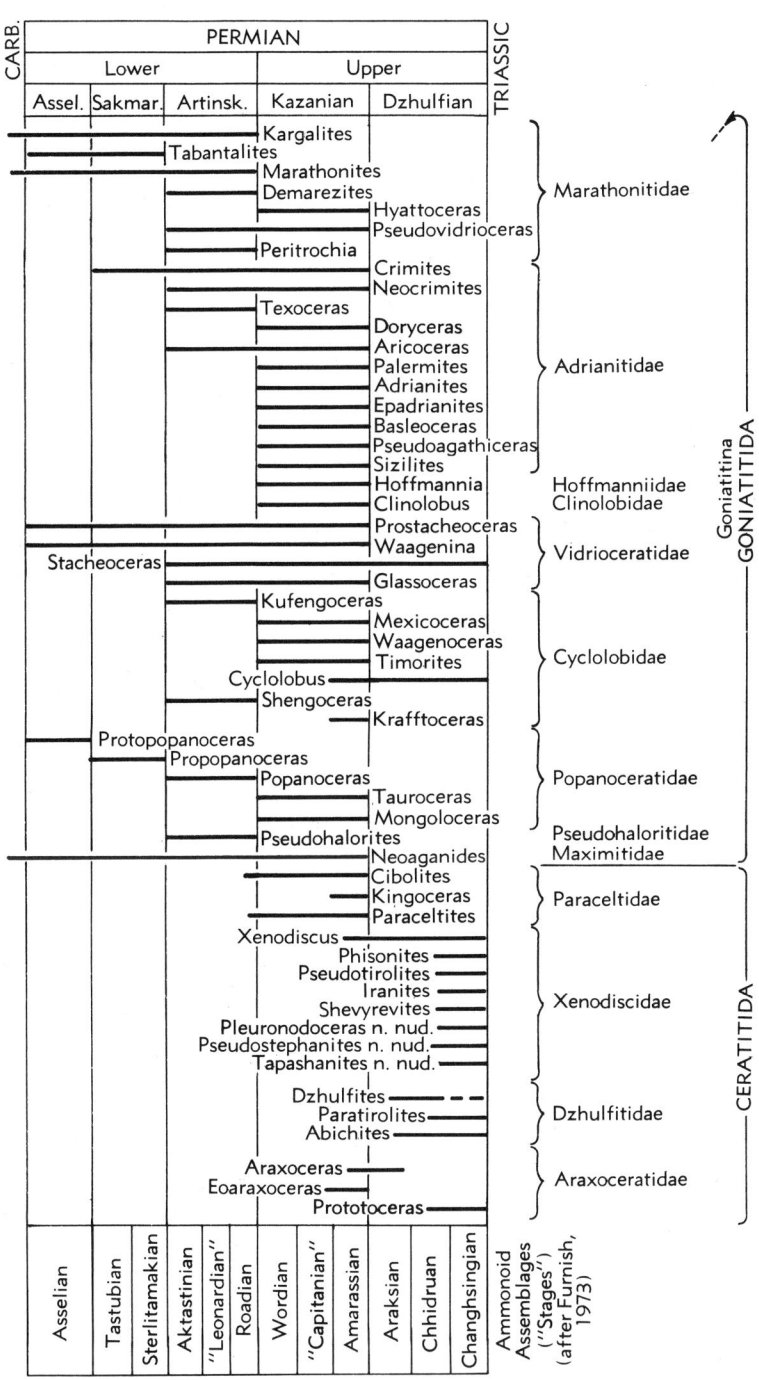

FIG. 6. *(Continued from facing page.)*

as the sicanitin *Akmilleria adkinsi*. These faunas help to establish the placement of the lower part of the Lenox Hills Formation (late Wolfcampian) of the Glass Mountains, west Texas, the Admiral Formation of north-central Texas, and parts of the Hueco Group of the Hueco Mountains, Texas, as age equivalents of the Tastubian Substage. A closely related fauna occurs in the Somohole beds of the island of Timor, Indonesia, and in the Holmwood Shale in Western Australia.

The overlying Sterlitamakian Substage in the southern Ural region near Sterlitamak, Bashkiria, has few ammonoids. Ammonoids are found at this relative stratigraphic position 400 km to the south in the Aktiubinsk district, Kazakhstan. Species of *Metalegoceras* and *Medlicottia* are common in this assemblage and *Crimites*, an adrinitid genus which appears in earlier strata elsewhere, first appears in the southern Ural region in this assemblage. In North America, *Properrinites mooreae* occurs in beds considered to be equivalent in age to the Sterlitamakian Substage in the upper part of the Lenox Hills Formation and in the Alta Formation of west Texas. In the Great Basin region, *P. nevadensis* represents a zone of about the same age. A fauna of similar age and composition is known from the Poole Sandstone of Western Australia (GLENISTER & FURNISH, 1961).

ARTINSKIAN STAGE

In the southern Urals, RUZHENTSEV (1955, 1956) subdivided the Artinskian into two substages based on two ammonoid assemblages of different ages, the Aktastinian and the Baigendzhinian substages. The lower substage, the Aktastinian, includes the first appearance of *Eothinites* and *Popanoceras*, and the medlicottiids *Aktubinskia* and *Artioceras*. In North America, species of *Medlicottia, Popanoceras, Metalegoceras,* and *Metaperrinites* are common in strata of this age, such as the Clyde Formation of north-central Texas, the middle part of the Hueco Formation in southern New Mexico, and the Skinner Ranch Formation in west Texas.

The Baigendzhinian ammonoid assemblages show a considerable increase in provincialism although no truly endemic genera have been recorded. In the southern Urals, RUZHENTSEV (1956) recorded abundant agathiceratids, vidrioceratins, metalegoceratins, and popanoceratids that are represented by a number of distinctive species. Many of these genera also occur in the Bitauni assemblage on the island of Timor, Indonesia, where they occur with medlicottiids and perrinitids that are closely similar to forms from the Cathedral Mountain Formation, west Texas. Of these various assemblages FURNISH (1973) considered the Cathedral Mountain assemblage to be the best known and on it based his "Leonardian Stage," which has yielded *Medlicottia costellifera, Eumedlicottia whitneyi, Pseudohalorites* cf. *P. subglobosus, Neocrimites newelli, Almites dunbari,* and *Perrinites vidriensis*. This biostratigraphic usage of "Leonardian Stage" is much more restricted than the generally used term Leonardian Series, which is more nearly equivalent to the entire Artinskian. This Cathedral Mountain assemblage is widely distributed in southern, southwestern, and central parts of North America, in Tethyan assemblages of central Asia and southwestern China, and in northern assemblages of Canada and Siberia.

The upper part of the Artinskian Stage apparently grades laterally and upwards into evaporites of the "Kungurian" and the Ufimian stages and the few marine faunas in the "Kungurian" are similar to those in the Baigendzhinian Substage. Thus, in terms of normal marine faunas, including ammonoids, the stratigraphically higher parts of the Permian in its type region are inadequate for establishing a zonation based on marine invertebrates, or for faunally defining the top of the Lower Permian.

In North America *Perrinites* has long been considered to be a guide fossil for the Leonardian Series and, hence, as a Lower Permian genus. The discovery of this genus in the "Word number 1 limestone" of P. B. KING (1931, 1938) resulted in a restriction of the Word Formation to beds above this lowest limestone (COOPER & GRANT, 1966) and the assignment of that limestone to a new unit, the Road Canyon Formation, which they considered to be latest Leonardian in age. The Road Canyon contains the highest occurrence of *Perrinites* (FURNISH, 1973) and the overlying Word Formation contains the first occurrence of *Waagenoceras*, which is considered to be a typical lower Upper Permian genus. The Roadian assemblage is here considered to characterize a substage at the top of the Lower Permian. The ammonoid assemblage of the Roadian Substage includes *Eumedlicottia burckhardti, Perrinites hilli, Glassoceras normani, Texoceras texanum, Peritrochia erebus, Paraceltites elegans*, and a number of adrianitids. Roadian ammonoids have been reported from the Meade Peak Member of the Phosphoria Formation in southern Idaho, the Assistance Formation of the Canadian Arctic, and the Verkhoyansk region of Siberia.

The problems of stratigraphic correlation between the Glass and Guadalupe mountains in west Texas have been summarized by FURNISH (1973). Two problems of correlation arise with respect to the Roadian Substage. First, the geographic location for defining the boundary between the Lower and Upper series of the Permian is transferred from the Russian platform to west Texas with relatively little supporting evidence for time equivalency near this boundary. Second, the lower part of the Upper Permian reference section in North America, the Guadalupian Series, has its type section in the Guadalupe Mountains and difficulties exist in correlating the sections there with the Permian sections in the Glass Mountains.

UPPER PERMIAN ZONATION

EQUIVALENTS OF KAZANIAN STAGE

In North America the Guadalupian Series (equivalent to a stage in Soviet usage) forms the lower part of the Upper Permian and is approximately the same age as the Kazanian Stage of the Russian platform. In the Glass Mountains of west Texas the lowest Guadalupian includes the Word Formation, which has a well-defined ammonoid assemblage. This assemblage, plus additional forms from equivalent strata in other parts of west Texas and Mexico, forms the basis for recognizing a Wordian Substage assemblage (FURNISH, 1973). These ammonoids include *Agathiceras girtyi, Popanoceras bowmani, Stacheoceras gemmellaroi, Waagenoceras dieneri, W. guadalupense, Pseudogastrioceras roadense, P. altudense, P. beedei, Atsabites multiliratus, A. williamsi, Neocrimites plummeri, Doryceras spinosum, Epithalassoceras ruzencevi, Propinacoceras beyrichi, Medlicottia burckhardti, Neogeoceras girtyi*, and *Paraceltites altudensis*. A faunal assemblage that duplicates this one, but which also has additional species, is found in erratic blocks along the Rio Sosio, Sicily (GEMMELLARO, 1887, 1888). Some of the Basleo faunas of Timor, Indonesia, also are comparable to this Wordian assemblage.

The middle part of the Guadalupian includes the Zone of *Timorites*, which is well known from the Manzanita Member at the top of the Cherry Canyon Formation and the Hegler Member at the base of the Bell Canyon Formation in west Texas. FURNISH (1973) considered this zone to characterize the "Capitanian Stage" (or Keptenian of RUZHENTSEV, 1955), but this zone includes only two members near the middle of the Guadalupian Series and includes only the lower part of the Capitan Group as used by most biostratigraphers. The top of the fusulinacean Zone of *Parafusulina* (Manzanita Member) and the base of the Zone of *Polydiexodina* (Hegler Member) falls within this zone of *Timorites* in the Guadalupe Mountains. This ammonoid assemblage is also known from Coahuila, Mexico, and Timor, Indonesia. In Coahuila, *Cibolites waageni* also appears in this assemblage.

This zone and the four higher zones in the Permian are not as clearly defined by either ammonoid faunas (or other faunas) or rock-stratigraphic units as might be desired. A considerable literature has been devoted to the problems of establishing a truly acceptable biostratigraphic zonation of the later parts of the Permian (MILLER & FURNISH, 1940a; GLENISTER & FURNISH, 1961; FURNISH, 1966, 1973; WATERHOUSE, 1969, 1972a, 1972b; FURNISH & GLENISTER, 1970; KUMMEL, 1972; FURNISH et al., 1973; SPINOSA et al., 1970, 1975); however, considerable problems remain.

The youngest strata of Guadalupian age that have a representative ammonoid fauna are in western Timor, Indonesia, in the province of Amarassi, where a distinctive fauna weathers out of a tropical soil. Both the stratigraphy and structure are incompletely known, but the fauna represents a stage of ammonoid evolution that is distinctive of a post-"Capitanian," latest Guadalupian, age. FURNISH (1973), FURNISH and GLENISTER (1970), and SPINOSA et al. (1975) identified the following "Amarassian Stage" or assemblage of ammonoid species: *Strigogoniatites angulatus, Epadrianites timorense, Stacheoceras tridens, Timorites curvicostatus, Cyclolobus persulcatus, Hyattoceras subgeinitzi, Sundaites levis, Syrdenites* sp., *Episageceras noetlingi, E. nodosum, Xenodiscus wanneri, "Hyattoceras"* sp., and *"Parapronorites"* sp. In the La Colorada beds of Coahuila, Mexico, *Xenodiscus wanneri, Kingoceras kingi,* and *Eoaraxoceras ruzhencevi* are associated in a southwestern North American assemblage of this age. Based on a number of logical models of phylogenetic lineages, the Amarassian ammonoid assemblage is most likely to be equivalent in age to the upper beds of the Guadalupian Series (i.e., the upper part of the Bell Canyon Formation and the Lamar and Tansill carbonates).

DZHULFIAN STAGE

According to FURNISH (1966, 1973) and FURNISH and GLENISTER (1970), the highest Permian stage is represented by three subdivisions, the Araksian, Chhidruan, and Changhsingian (Fig. 6), which are based on ammonoid assemblages from three geographically separate segments of the present Himalayan Mountain chain. The Araksian Substage is based on the *Araxilevis-, Araxoceras-,* and *Oldhamina*-bearing beds, whose faunas were described in detail from the Araks Gorge at Dorasham Station near Dzhulfa by RUZHENTSEV and SARYCHEVA (1965). The ammonoids occur in the upper part of these beds and include a wide variety of specialized otocerataceans (Araxoceratidae) and other ammonoids of Paleozoic ancestry. Abundant *Araxoceras, Rotaraxoceras, Prototoceras, Pseudotoceras,* and *Vesocotoceras* are in association with *Pseudogastrioceras* and characterize the fauna. Three rarely found genera, the medlicottid *Syrdenites* and the cyclolobaceans *Stacheoceras* and *Cyclolobus,* are also associated with this faunal assemblage in nearby localities. The *Cyclolobus* species is considered to be primitive and was assigned to *C. kullingi,* originally described from central East Greenland. At Abadeh, in central Iran, TARAZ (1969, 1971, 1973) reported a thicker, possibly more complete, sequence than at Dzhulfa. TARAZ (1973) believed that the Abadeh section fills in a hiatus that lies beneath the *Araxilevis* beds at Dzhulfa and that the Abadeh section is younger than the Guadalupian Series of North America. FURNISH (1973) pointed out that the late Guadalupian Amarassian ammonoid fauna of the La Colorada beds near the top of the Permian section at Las Delicias, Coahuila, Mexico, has a similar ammonoid composition to those in the upper part of TARAZ's (1969, 1971) Abadeh section and that this fauna is overlain by an Araksian assemblage above which appears a Chhidruan assemblage.

WAAGEN (1889-91) recognized the Chhidru beds in the upper strata of the Upper *Productus* Limestone, Salt Range of Pakistan, with an abundant and diverse invertebrate fauna. TEICHERT (1966) and KUMMEL and TEICHERT (1970) summarized the stratigraphy of these beds and the distribution of fossils in them. *Cyclolobus* ranges through a small interval in the upper third of the Chhidru Formation of KUMMEL and TEICHERT (1970) and is represented by three species of which *C. oldhami* and *C. teicherti* are the best known. These are intermediate in stage of evolution between *C. kullingi* from the Araksian assemblage and the cyclolobids of the Changhsingian assemblage. Other Chhidruan ammonoids include *Stacheoceras antiquum, Eumedlicottia primas, Episageceras wynnei,* and *Xenodiscus carbonarius.*

Basing his conclusions mainly on indirect evidence, FURNISH (1973) believed that the *Vedioceras-Oldhamina* assemblage, which lies above the Araksian ammonoid assemblage in the Dzhulfa sections, is equivalent to the Chhidruan ammonoid assemblage of the Salt Range. In the Dzhulfa region the *Vedioceras-Oldhamina* assemblage includes *Vedioceras* and *Dzhulfoceras,* which in the region are restricted to that assemblage, and *Urartoceras, Rotaraxoceras, Prototoceras,* and *Pseudotoceras* ranging upward from the *Araxoceras* beds below.

The highest Permian beds of Dzhulfa and the Salt Range are thin and have restricted ammonoid faunas (FURNISH, 1973; KUM-

MEL, 1972). At Dzhulfa, these strata have yielded coarsely ribbed xenodiscids, dzhulfitids, a maximitid, *Neoaganides,* and rare *Pseudogastrioceras* and *Stacheoceras.* A comparable assemblage having a more abundant fauna has been reported from the Changhsing Limestone of northern Chekiang, China. In its lower part this limestone includes *Pseudotirolites* and *Pseudogastrioceras* and in the upper part it has *Stacheoceras, Pachydiscoceras, Rotodiscoceras, Trigonogastrites,* and *Changhsingoceras.* Associated with these ammonoids are *Palaeofusulina, Reichelina, Oldhamina, Dictyoclostus,* and *Hustedia.* CHAO (1965) stated that "The Changhsing Limestone changes into siliceous limestone, siliceous shale, and sandstone at many places in South China, where they are called the Talung Formation (equivalent to Hoshan Formation of Sun, 1939)." Although it is not clear from CHAO's statement if this is a lateral or a vertical lithofacies change, or both, he combined ammonoids from both lithologic units in his analysis of the "Changhsing or Talung ammonoids." The Talung Formation is disconformably overlain by Lower Triassic shales bearing *Claraia.*

Correlations with other regions are commonly indirect (CHAO, 1965; FURNISH, 1973). The Loping Coal Series that lies beneath the Changhsing Limestone is correlated with most of the Dzhulfa beds of Iran and with the Chhidru Formation of the Salt Range. CHAO (1965) suggested that the reddish limestone and shale with *Paratirolites* just beneath the Lower Triassic beds bearing *Claraia* at Dzhulfa are possibly equivalents of the Ali Bashi Formation (see also KUMMEL & TEICHERT, 1973; TOZER, 1969, 1971; FURNISH, 1973; TEICHERT, KUMMEL, & SWEET, 1973; and KUMMEL, this volume).

CONODONTS

In contrast to studies of Devonian, early Carboniferous, and Triassic conodonts, study of Permian conodonts has not yet advanced to the point of establishing a worldwide scheme of biostratigraphy. Several studies have recently shown the potential use of conodonts in establishing zonation, particularly in strata having few ammonoids or fusulinaceans (CLARK & ETHINGTON, 1962; BENDER & STOPPEL, 1965; SWEET, 1970; CLARK & BEHNKEN, 1971; KOZUR, 1973, 1975; BEHNKEN, 1975). Nevertheless, relatively few stratigraphic successions are known in the detail that is available for the Upper Devonian and Lower Carboniferous.

Most detailed studies of Permian conodonts have dealt with biostratigraphic problems of either the Carboniferous-Permian or Permian-Triassic boundary (CLARK & BEHNKEN, 1971) or have described relatively isolated faunas (Fig. 7). For reasons that are not entirely clear, the North American standard Permian reference sections in west Texas have yielded neither abundant nor diverse conodont faunas, and correlation of stratigraphic sections having more plentiful conodonts with these standard sections is not as precise as might be desired (BEHNKEN, 1975). Also, few conodont studies deal with the type Permian sections in the Russian platform and Urals, and conodont correlation with those sequences is not possible at present.

WESTERN NORTH AMERICA

The present conodont zonation for the Permian is based largely on studies by CLARK and BEHNKEN (1971) and BEHNKEN (1975) in the western North American Cordillera in central Nevada and southeastern Idaho and some studies in portions of the succession in the Guadalupe Mountains of west Texas (Fig. 7). Apparently some confusion exists concerning the boundaries and thicknesses of many of the named units in the Great Basin and the resultant uncertainty leads to difficulty in correlation with the west Texas standard sections.

BEHNKEN (1975) recognized two conodont zones, the *Idiognathodus ellisoni* Assemblage-Zone and the *Neogondolella bisselli-Sweetognathus whitei* Assemblage-Zone, which he considered to be Wolfcampian in age. KOZUR (1975) considered that the base of the Permian should be placed at the base of the second of these zones. In the Riepe Springs and Ferguson Mountain formations of Nevada, *Idiognathodus ellisoni* occurs without other platform conodonts and marks a zone of early

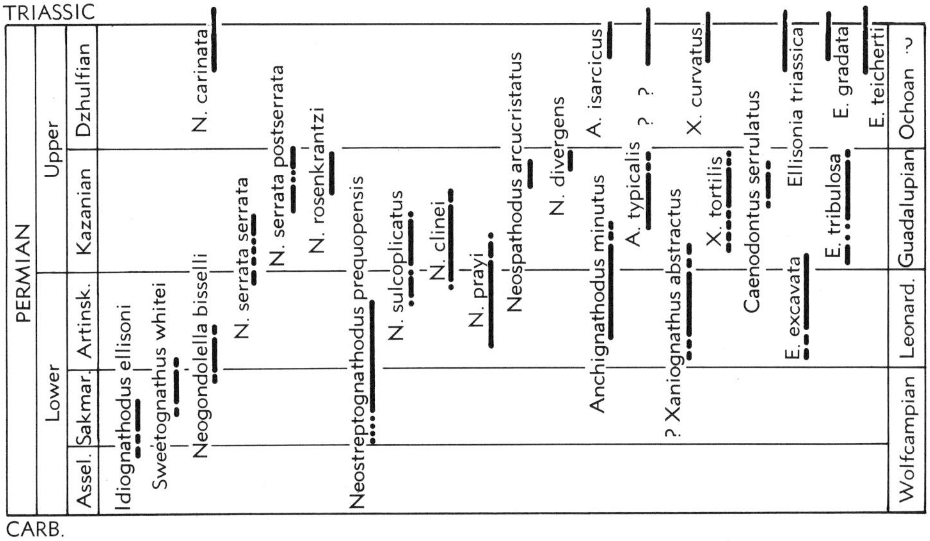

Fig. 7. Ranges of zonally important species of conodonts, principally based on Lower Permian taxa in the western United States (Behnken, 1975) and Upper Permian taxa in Pakistan (Sweet, 1970; Ross & Ross, n).

and middle Wolfcampian age. *Neostreptognathodus pequopensis* is rare in this assemblage-zone. The lower 30 m of the Pequop Formation contains mainly *Sweetognathus whitei* and the succeeding 82 m contain abundant *Neogondolella bisselli, N. pequopensis,* and several species of *Xaniognathus.* The top of this zone approximates the Wolfcampian-Leonardian boundary in the Glass Mountains as based on correlations using fusulinids (SLADE, 1961).

Conodont zones assigned to the Leonardian in the Great Basin region include, at the base, the *Neostreptognathodus pequopensis* Assemblage-Zone and the *Neostreptognathodus sulcoplicatus-N. prayi* Assemblage-Zone. The *N. pequopensis* Zone is identified on the abundance of *N. pequopensis.* Also in this zone are a few specimens of *Xaniognathus abstractus, Anchignathodus minutus,* species of *Hindeodella,* and *Ellisonia excavata.* The *N. sulcoplicatus-N. prayi* Zone is late Leonardian to early Guadalupian in age and includes elements of *Xaniognathus abstractus, Neogondolella idahoensis, Neostreptognathodus clinei, Anchignathodus minutus, Ellisonia excavata,* and *Hindeodella* spp. Considerable geographical and local stratigraphic variations appear in the occurrence of the two guide species of *Neostreptognathodus* and associated species. BEHNKEN (1975) placed the upper limit of this zone at the base of the first appearance of faunas containing either *Neogondolella rosenkrantzi* or *N. serrata.*

The overlying *Neogondolella serrata serrata* Assemblage-Zone begins with the first appearance of this species and in places overlaps with the upper ranges of many species found in the preceding lower zone. *Xaniognathus tortilis* and *Ellisonia tribulosa* first appear in this zone. This assemblage-zone appears in strata of possible late Leonardian age and ranges into strata of probably early Wordian age in the Guadalupe Mountains, west Texas, and the Meade Peak Member (Phosphoria Formation) in Idaho.

Three conodont assemblage-zones characterize the remaining parts of the Guadalupian Series. The *Neogondolella rosenkrantzi-Neospathodus arcucristatus* Assemblage-Zone is based on the assemblages from the upper 30 m of the Plympton Formation and the lower part of the Gerster Formation of eastern Nevada (BEHNKEN, 1975). The first appearance of either *Neo-*

gondolella rosenkrantzi or *Neospathodus arcucristatus* marks the base and the last appearance of *N. arcucristatus* marks the top of the zone. The upper part of the Gerster Formation contains the *Neogondolella rosenkrantzi-Neospathodus divergens* Assemblage-Zone, which includes *Ellisonia tribulosa*. *N. divergens* appears at the base of the zone and the top of the zone is an erosional unconformity. This assemblage also has been reported from the Zechstein of Germany (BENDER & STOPPEL, 1965). Based on available strata preserved beneath the basal Triassic unconformity, these assemblage-zones form the upper portion of the conodont zonation in the Permian miogeosyncline of Nevada, Utah, and southeastern Idaho.

In the Guadalupe Mountains of west Texas, the youngest recognized zone, the *Neogondolella serrata postserrata* Assemblage-Zone, lies above the *N. serrata serrata* Zone and includes strata of late Wordian and Capitanian age. In addition to *N. serrata postserrata*, this assemblage includes *Anchignathodus typicalis, Xaniognathus tortilis, Ellisonia tribulosa, Prioniodella decrescens, P. anguinea, P. cteniforma*, and *Caenodontus serrulatus*. *N. serrata postserrata* currently is known only from the west Texas region.

From the scattered occurrences of Permian conodonts that are available for comparison, Wolfcampian and early to middle Leonardian species appear to be widely dispersed and within the western United States are regionally distributed with few indications of provincialism. On the other hand, the late Leonardian and Guadalupian species begin to show increasing provincial distributions, which may relate to biofacies rather than geographic isolation. The highest part of the Gerster Formation (correlated on the basis of brachiopods and the ammonoid *Timorites*) is considered to be Capitanian in age, and has a conodont fauna similar to that of the lower part of the Zechstein of Germany.

SALT RANGE, PAKISTAN

Conodonts from the uppermost Chhidru Formation in Pakistan are the youngest Permian representatives yet described (SWEET, 1970). In general, conodonts are rare in most of the Upper Permian in the Salt Range and nearby areas of Pakistan and only the white sandy beds in the upper 1 to 5 m of the Chhidru Formation contain a definable conodont fauna. The *Anchignathodus typicalis* Assemblage-Zone includes *A. typicalis, A. isaricicus, Ellisonia triassica, E. gradata, E. teicherti, Neogondolella carinata*, and *Xaniognathus curvatus* and occurs in the uppermost 4 m of the Chhidru Formation (Permian) and lowermost 1 to 3 m of the overlying Mianwali Formation (Triassic). Based on other faunal evidence (KUMMEL & TEICHERT, 1970), this zone straddles the Permian-Triassic boundary and supports the view that if a paraconformity is present, the resulting hiatus is of relatively short duration. Brachiopods (possibly reworked from earlier deposits) were collected from the same beds as those containing this conodont assemblage and have a Guadalupian, rather than a later Permian aspect (COOPER, in KUMMEL & TEICHERT, 1966; GRANT, 1970).

PALEOGEOGRAPHIC RECONSTRUCTION

Possible geographic arrangements for the various epicontinental seas, miogeosynclines, and eugeosynclines during the Permian Period suggest that a modified Dietz-Holden model (DIETZ & HOLDEN, 1970; SMITH, BRIDEN, & DREWRY, 1973) may help to explain some of the apparent problems in phylogenetic histories and biogeographic distributions. Two time intervals, Early Permian (Asselian) and Late Permian (Kazanian), are examined on world reconstructions (Fig. 8 and 9).

Early Permian epicontinental seas reached their maximum distribution during the Asselian (Fig. 8). Mountain building and deformation of the Ouachita-Appalachian-Hercynian belt continued and, although mountains were high, marine epicontinental embayments were still extensive. The Uralian geosyncline was very narrow and probably relatively deep but connected the Russian platform with the Hercynian belt of the Tethyan region. Shallow-water marine connections with southwestern North Amer-

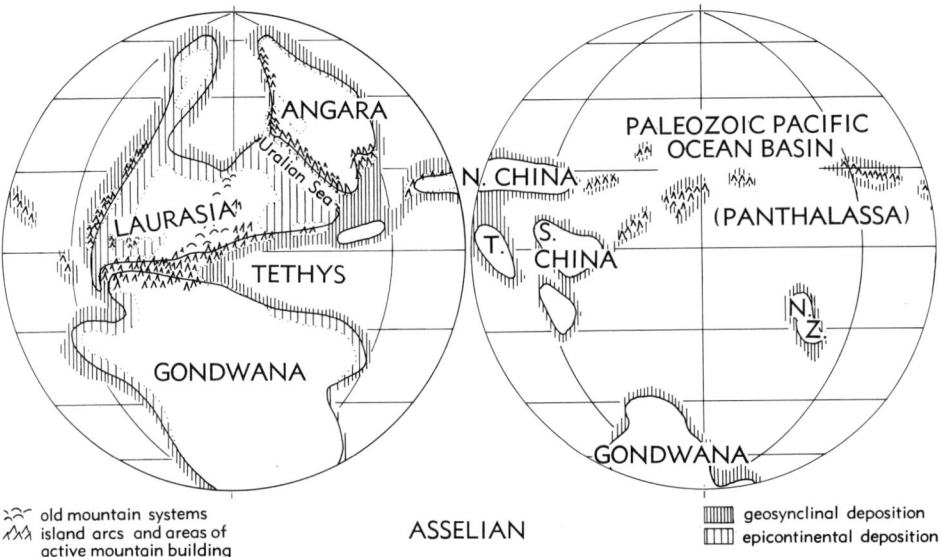

Fig. 8. Paleobiogeography during Asselian Age with connection between Uralian and Tethyan seas well established. Volcanic islands include parts of Tethyan region that have been added to continents at a later time, probably during Mesozoic and Cenozoic (Ross & Ross, n). [Explanation: *N. Z.*, New Zealand; *T*, Tarim stable block.]

ica and western South America were by way of the Franklinian geosyncline and northern and western Canadian shelves so that distance and latitude produced temperature gradients that reduced the dispersal of many marine invertebrate faunas.

During Sakmarian time, exchange of marine invertebrates became increasingly less frequent between the Tethyan region, the Uralian-Franklinian region, and the southwestern North American region (Ross, 1970). By late Artinskian time direct marine connections between the Tethyan region, the Uralian geosyncline, and Russian platform ceased and from this time until the end of the Permian the Tethyan faunas became increasingly endemic and formed a distinctive faunal realm. By the beginning of Late Permian (Fig. 9) the northern Pangaea epicontinental seas (Kazan and Zechstein seas) lacked fusulinaceans, and only a few genera of smaller Foraminiferida were present. Other marine invertebrates show similar specialized faunas. The late Late Permian (Dzhulfian) was a time of minor inundation of the epicontinental shelves and many of those seas had unusual environments of deposition. Of the Early Permian geosynclines, only parts of the Tethyan geosyncline remained a marine seaway in the later part of the Late Permian.

Although much is known about shallow-water marine faunas of the Permian Period, many questions about their distribution, dispersal, and phylogeny remain unanswered (USTRITSKIY, 1974). Large parts of Asia and South America are poorly known. New Zealand has an interesting Permian fauna but a poorly understood geographic position in the late Paleozoic. Parts of the southern Eurasian Tethyan belt may have been widely separate sialic blocks that have been united at a much later time. In the Cordilleran and Andean structural belts of North and South America, strata containing Tethyan realm marine faunas, which were probably tropical or subtropical with relatively high diversity, lie next to strata having non-Tethyan faunas that were probably temperate and have low diversity. As these strata in these structural belts are older than 150 my, and, therefore, older than present sea floors, these belts may include carbonate banks and platforms associated with old tropical island arcs and other uplifted ocean crustal fragments that have been added to the western side of the Americas during the Mesozoic.

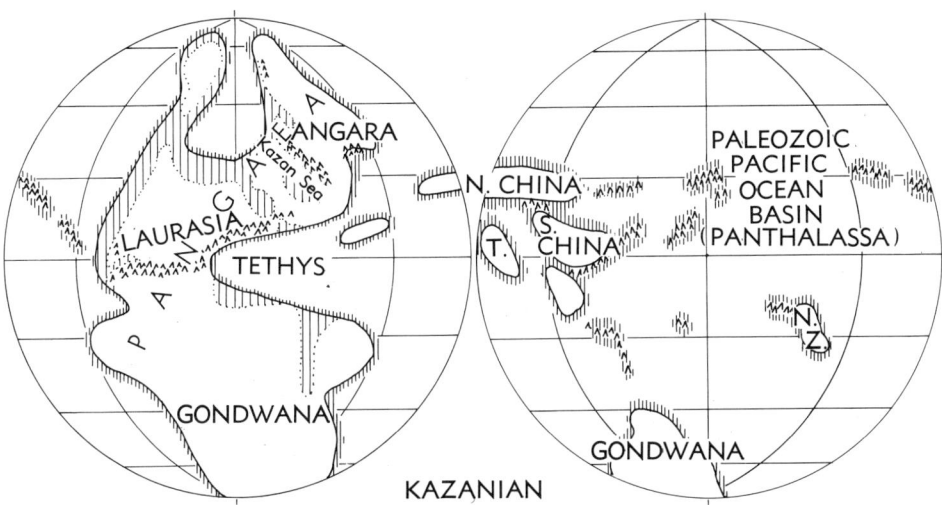

FIG. 9. Paleobiogeography during Kazanian Age with connection broken between Uralian and Tethyan seas (Ross & Ross, n). [See Fig. 8 for explanation of abbreviations and symbols.]

SUMMARY AND CONCLUSIONS

Several patterns in Permian faunas and biostratigraphy appear consistently in the different fossil groups examined. During Asselian time, and to a lesser extent during Sakmarian time, many of the shallow-water marine faunas are widely distributed in the Tethyan and Uralian-Franklinian region, or, at least, there were several short intervals of faunal exchange followed by longer intervals of restricted dispersals between these regions. In this respect, Asselian patterns represent a continuation of Late Carboniferous distributional patterns and may relate to the continuation of cyclic deposition associated with marine transgressions and regressions that gradually decreased in extent and magnitude during Asselian and Sakmarian time. Nearly all Asselian and Sakmarian marine faunas either lack or have very low species diversity in the Gondwana sequences where glacial beds are common. This suggests that repeated fluctuations of glacial conditions (which gradually decreased in importance), marine transgressions and regressions, and faunal dispersal and restriction may be closely interrelated because of repeated changes in climatic conditions and in sea levels. Within the Asselian and Sakmarian intervals, climatic zonation is suggested by several of the faunal groups based on the amount of generic and specific diversity, in the association or nonassociation with abundant calcareous algae, based on shell growth patterns, and by the distribution of carbonate reefs and banks in contrast to sections having predominantly shallow-water sandstone. Some groups, such as many of the bryozoans and brachiopods, lived in a broader range of marine shelf and slope environments and their distributions were less strongly related to the distributions of shallow epicontinental flooding than were other groups, such as fusulinaceans and corals. Cephalopod distributions, and probably also conodont distributions, were a function of neritic habitats and relate to the direction and extent of different surface currents.

By Artinskian time many shallow benthic marine faunas showed marked provinciality, particularly between the Tethyan region and the Ural-Franklinian region. This is possibly best shown by the distribution and phylogenetic patterns in fusulinaceans and corals. Direct faunal connections for many groups seem to have been broken in Sakmarian or early in Artinskian time between the Uralian Sea and the Tethys, particularly for fusulinaceans, corals, bryozoans, and brachiopods. Cephalopods and some bryozoans retained some widely dispersed, nearly cosmopolitan, genera and families, but genera in many other groups were restricted in distribution to particular regions at this time. The main cause for the disruption of avenues for faunal dispersal appears to have been the last of the major orogenic episodes associated with the Hercynian orogeny. This appears to coincide with the uplift of the southern end of the Uralian geosyncline and southern part of the Russian platform. Uplifts within the Early Permian also modified many dispersal paths in parts of southern Asia and in North America.

New patterns of biogeography and phylogeny, and changes in sedimentary patterns began with these late phases of the Hercynian orogeny and were of major significance in forming a starting point for the clear identification of a Tethys ocean and a Tethyan faunal realm that reached from northern Africa, through the Alps, Yugoslavia, Greece, Turkey, Crimea, Iran, Iraq, Afghanistan, Pakistan, Kashmir, China, Japan, Indochina, New Zealand, Maritime Provinces of the Soviet Union, and parts of Alaska, Yukon, British Columbia, Washington, Oregon, and California—a much larger and more elongate realm than the Mesozoic Tethys. The fauna was diverse, abundant, and associated with algal and crinoid banks and reefs. The contrasting northern European and northern North American faunas were mainly composed of relict genera and decreased rapidly in diversity as depositional environments became poorly suited to a normal marine biota.

In Kazanian time the Tethyan fauna reached its greatest Permian diversity as shown by fusulinaceans, corals, bryozoans, and cephalopods, whereas the Zechstein and Uralian-Franklinian faunas had very low species diversity in the few groups that survived in those seas. Some elements of the Tethyan fauna spread into parts of the Tasman geosyncline in eastern Australia suggesting warmer seas. By the end of Kazanian time, an orogeny of major magnitude affected the Japanese, Chinese, and adjacent regions in the eastern Asian segments of the Tethys. Present evidence suggests that deposition ceased in many parts of the eastern Tethys after this orogeny.

Marine Dzhulfian fossils are known only in the Eurasian part of the Tethys, which more closely approximates the distribution of the Mesozoic Tethys. Dzhulfian faunas are of particular interest because they contain the remnants of a once prolific Tethyan fauna of Kazanian age and the phylogenetic ancestors for many of those Triassic faunas that evolved rapidly after the beginning of the Mesozoic. Paleozoic faunal groups, such as the fusulinaceans, became extinct at different times during the Late Permian and only a few Paleozoic lineages in groups such as the brachiopods and bryozoans survived into the earliest part of the Triassic. For these reasons the Late Permian and Early Triassic are generally regarded as times of acute ecological stress for shallow marine organisms.

The biogeographic and biostratigraphic relations within the Permian System have many unanswered questions. The lower boundary of the system is still in the process of being examined critically and in the Tethyan region a three-fold, rather than a two-fold, subdivision may be more applicable. Age relationships between Tethyan and non-Tethyan faunas are difficult to establish with precision after the middle part of the Early Permian because different lineages evolved independently in each faunal province. Reasons for these separate faunal provinces continue to be investigated and include studies on orogenic changes and climatic differences that probably contributed to provinciality. Climates during the Permian appear to have become gradually warmer as indicated by a decrease in glacial beds and in cyclical sedimentary deposition and by an increase in evaporite deposition from Asselian through Kazanian time. The magnitude of the latitudinal temperature gradients for the Permian and possible major temperature fluctuations at different times during the period are poorly known; however, the extinction of many genera and families of late Paleozoic marine organisms near the end of Kazanian or early in Dzhulfian time may have resulted from oceanographic changes that related to these climatic changes. Terrestrial plants and animals of the Permian also show geographical distributions that have been interpreted as reflecting latitudinal or temperature differences. The inferences are that latitudinal differences in temperature and floral and faunal distributions were broadly similar to those of the recent (Ross, 1974). Particular interest has centered on Gondwanan distributional questions (ROMER, 1968; International Union of Geological Sciences, 1969; KEAST & GLASS, 1972; USTRITSKIY, 1974) and on floral provinces (TSCHUDY & SCOTT, 1969; PLUMSTEAD, 1973; HART, 1974).

The highest Permian (Dzhulfian) is the least well known part of the system and its upper boundary is the subject of much biostratigraphic study (GRUNT & DMITRIEV, 1973; ISHII, FISCHER, & BANDO, 1971; ISHII, OKIMURA, & NAKAZAWA, 1975; KAHLER, 1974a, 1974b; MENNER et al., 1970; MEYEN, 1970; ABICH, 1878; GRANT, 1970; KUMMEL & TEICHERT, 1970, 1973; TOZER, 1969, 1971; NAKAZAWA, 1974; RUZHENTSEV & SARYCHEVA, 1965; STEPANOV, GOLSHANI, & STÖCKLIN, 1969). At present, distinctive faunas characterize the Dzhulfian Stage only in the Eurasian part of the Tethys. To suggest that deposition ceased everywhere else during this time interval seems implausible. Therefore, considerably more data are needed in order to document the faunal changes and extinctions that characterize the latest part of the Permian Period and the end of the Paleozoic Era.

REFERENCES

Abich, Hermann, 1878, *Eine Bergkalkfauna aus der Araxesenge bei Djoulfa in Armenien:* Geol. Forschungen in den Kaukasischen Landern, v. 1, 128 p.

American Commission on Stratigraphic Nomenclature, 1961, *Code of stratigraphic nomenclature:* Am. Assoc. Petrol. Geologists, Bull., v. 45, p. 645-665.

Arakelian, R. A., Rauzer-Chernousova, D. M., Reytlinger, E. A., Shcherbovich, S. F., & Efimova, N. A., 1964, *Znachenie Permskikh foraminifer Zakavkazya dlya korrelyatsii Permi v predelakh Tetisa:* Mezhdunarodnyy Geol. Kongress 22 Sessiya 1964, Doklady Sovetskikh Geologov, p. 63-75. [*The importance of Permian Foraminifera in the Transcaucasus for the correlation of the Permian within the Tethys.*]

Armstrong, J. D., Dear, J. F., & Runnegar, Bruce, 1967, *Permian ammonoids from Eastern Australia:* Geol. Soc. Australia, Jour. v. 14, p. 87-97.

Banks, M. R., & Naqvi, I. H., 1967, *Formations close to the Permo-Triassic boundary in Tasmania:* Royal Soc. Tasmania, Papers & Proc., v. 101, p. 17-30.

Beede, J. W., & Kniker, H. T., 1925, *Species of the genus Schwagerina and their stratigraphic significance:* Univ. Texas, Bull. 2433 (1924), 96 p.

Behnken, F. H., 1975, *Leonardian and Guadalupian (Permian) conodont biostratigraphy in western and southwestern United States:* Jour. Paleontology, v. 49, p. 284-315.

Bender, Hans, & Stoppel, Dieter, 1965, *Perm-Conodonten:* Geol. Jahrb., v. 82, p. 331-364.

Caster, K. E., 1952, *Stratigraphic and paleontologic data relevant to the problem of Afro-American ligation during the Paleozoic and Mesozoic:* in The problem of land connections across the South Atlantic, with special reference to the Mesozoic, E. Mayr (ed.), Am. Museum Nat. History, Bull., v. 99, art. 3, p. 105-152.

Chao King-koo, 1965, *The Permian ammonoid-bearing formations of South China:* Scientia Sinica, v. 14, p. 1813-1826.

Chen Shu, 1934, *Fusulinidae of South China, Part I:* China Geol. Survey, Palaeont. Sinica, ser. B, v. 4, no. 2, 185 p. (in English, Chinese summary).

———1956, *Fusulinidae of South China, Part II:* Acad. Sinica, Palaeont. Sinica, n.ser. B, no. 6, p. 17-72.

Chernyshev [Tschernyschew], T. N., 1902, *Verkhnekamennougoliya brakhiopodi Urali i Timan:* Geol. Komiteta, Trudy, v. 16, part 1, 749 p. [*Upper Carboniferous brachiopods of the Urals and Timan.*]

Clark, D. L., & Behnken, F. H., 1971, *Conodonts and biostratigraphy of the Permian:* in Symposium on conodont biostratigraphy, W. C. Sweet & S. M. Bergström (eds.), Geol. Soc. America, Mem. 127, p. 415-439.

———, & Ethington, R. L., 1962, *Survey of Permian conodonts in western North America:* Brigham Young Univ. Geol. Studies, v. 9, p. 102-114.

Cooper, G. A., & Grant, R. E., 1964, *New Permian stratigraphic units in the Glass Mountains, West Texas:* Am. Assoc. Petrol. Geologists, Bull., v. 48, p. 1581-1588.———1966, *Permian rock units, in the Glass Mountains, West Texas:* U.S. Geol. Survey, Bull., v. 1244-E, p. E1-E9.———1972, *Permian brachiopods of West Texas, I:* Smithson. Contrib. Paleobiology, no. 14, 231 p.———1973, *Dating and correlating the Permian of the Glass Mountains in Texas:* in The Permian and Triassic Systems and their mutual boundary, A. Logan & L. V. Hills (eds.), Canad. Soc. Petrol. Geologists, Mem. 2, p. 363-377.———1974, *Permian brachiopods of West Texas, II:* Smithson. Contrib. Paleobiology, no. 15, p. 233-793.———

1975, *Permian brachiopods of West Texas, III:* Smithson. Contrib. Paleobiology, no. 19, p. 795-1921.——1976, *Permian brachiopods of West Texas, IV:* Smithson. Contrib. Paleobiology, no. 21, p. 1923-2285.

Crockford, Joan, 1951, *The development of bryozoan faunas in the Upper Palaeozoic of Australia:* Linnean Soc. New South Wales, Proc., v. 76, p. 105-122.

David, T. W. E., & Browne, W. R., 1950, *The geology of the Commonwealth of Australia:* v. I, xx + 747 p., Arnold (London).

Dear, J. F., 1971, *Strophomenoid brachiopods from the higher Permian faunas of the Back Creek Group in the Bowen Basin:* Queensland Geol. Survey, Publ. 347, Paleont. Papers, no. 21, 39 p.

Dickins, J. M., 1970, *Correlation and subdivision of the Permian of Western and Eastern Australia:* Australia Bur. Min. Res., Geology and Geophysics, Bull. 116, p. 17-27.

Dietz, R. S., & Holden, J. C., 1970, *Reconstruction of Pangaea: Breakup and dispersion of continents, Permian to present:* Jour. Geophys. Res., v. 75, p. 4939-4956.

Dunbar, C. O., 1940, *The type Permian: its classification and correlation:* Am. Assoc. Petrol. Geologists, Bull., v. 24, p. 237-281.——1955, *Permian brachiopod faunas of central East Greenland:* Meddel. Grønland, v. 110, no. 3, 169 p.——1958, *On the validity of Schwagerina and Pseudoschwagerina:* Jour. Paleontology, v. 32, p. 1019-1021.——1961, *Permian invertebrate faunas of central east Greenland:* in Geology of the Arctic, G. O. Raasch (ed.), v. 1, p. 224-230, Univ. Toronto Press (Toronto, Can.).

——, & Skinner, J. W., 1931, *New fusulinid genera from the Permian of West Texas:* Am. Jour. Sci., ser. 5, v. 22, p. 252-268.——1936, *Schwagerina versus Pseudoschwagerina and Paraschwagerina:* Jour. Paleontology, v. 10, p. 83-91.——1937, *Permian Fusulinidae of Texas:* Univ. Texas, Bull., v. 3701, p. 517-825.

——, chairman, & others (Permian Subcommittee of Nat. Res. Council's Committee on Stratigraphy), 1960, *Correlation of the Permian formations of North America:* Geol. Soc. America, Bull., v. 71, p. 1763-1806.

——, Troelsen, John, Ross, C. A., Ross, J. R. P., & Norford, B. S., 1962, *Faunas and correlation of the late Paleozoic rocks of northeast Greenland: pt. I, General discussion and summary:* Meddel. Grønland, v. 164, no. 4, 16 p.

Ehrenberg, C. G., 1854, *Mikrogeologie. Das Erden und Felsen schaffende Wirken des unsichtbar kleinen selbständigen Lebens auf der Erde:* 374 p., Voss (Leipzig).

Flügel, Helmut, 1968, *Korallen aus der oberen Nesen-Formation (Dzhulfa-Stufe, Perm) des zentralen Elburz (Iran):* Neues Jahrb. Geologie, Paläontologie, Abh., v. 130, p. 275-304.——1970, *Die Entwicklung der rugosen Korallen im hohen Perm:* Geol. Bundesanst. Aus., Verhandl., no. 1, p. 146-161.——1973, *Rugose Korallen aus dem oberen Perm Ost-Grönlands:* Geol. Bundesanst. Aus., Verhandl., no. 1, p. 1-57.

——, & Schönlaub, H. P., 1972, *Geleitworte zur stratigraphischen Tabelle des Paläozoikums von Österreich:* Geol. Bundesanst. Verhandl., 1972, no. 2, p. 187-198.

Frebold, Hans, 1950, *Stratigraphie und Brachiopodenfauna des marinen Jungpalaeozoikums von Holms und Amdrups Land (Nordostgrönland):* Meddel. Grønland, v. 126, no. 3, p. 1-37.

Frederiks, G. N., 1928, *Le Paléozoique supérieur de l'Oural:* Soc. Géol. Nord, Ann., v. 53, p. 138-171.——1932, *Verkhnii paleozoy zapadnogo i yuzhnogo sklona Urala:* Geol.-Razved. Upravl., Izvestiya, vyp. 106, p. 1-89. [*The Upper Paleozoic of the western and southern slopes of the Urals.*]

Furnish, W. M., 1966, *Ammonoids of the Upper Permian Cyclolobus-Zone:* Neues Jahrb. Geologie, Paläontologie, Abh., v. 125, p. 265-296.——1973, *Permian stage names:* in The Permian and Triassic Systems and their mutual boundary, A. Logan & L. V. Hills (eds.), Canad. Soc. Petrol. Geologists, Mem. 2, p. 522-548.

——, & Glenister, B. F., 1970, *Permian ammonoid Cyclolobus from the Salt Range, West Pakistan:* in Stratigraphic boundary problems: Permian and Triassic of West Pakistan, Bernhard Kummel & Curt Teichert (eds.), Univ. Kansas Dept. Geology, Special Publ. 4, p. 153-175.

——, ——, Nakazawa, Keiji, & Kapoor, H. M., 1973, *Permian ammonoid Cyclolobus from the Zewan Formation, Guryul Ravine, Kashmir:* Science, v. 180, p. 188-190.

Gemmellaro, G. G., 1887, *La fauna dei calcari con Fusulina della Valle del Fiume Sosio nella Provincia di Palermo:* Giornale Sci. Nat. & Econ. Palermo, v. 19, p. 1-106.——1888, *Appendice: La fauna dei calcari con Fusulina della Valle del Fiume Sosio nella Provincia di Palermo:* Giornale Sci. Nat. & Econ. Palermo, v. 20, p. 9-36.

Girty, G. H., 1909, *The Guadalupian fauna:* U.S. Geol. Survey, Prof. Paper 58 (1908), 651 p.

Glenister, B. F., & Furnish, W. M., 1961, *The Permian ammonoids of Australia:* Jour. Paleontology, v. 35, p. 673-736.

Glushko, V. V., & Fedorov, D. L., 1974, *Do pitannya pro korelyatsiyu Permskikh tovshch na okrainakh Russkoi platformi:* Geol. Zhurnal, v. 34, p. 3-13. [*Some thoughts about correlation of Permian beds occurring at the margins of the Russian platform.*]

Gobbett, D. J., 1963, *Carboniferous and Permian brachiopods of Svalbard:* Norsk Polarinst., Skrifter 127, 201 p.——1967, *Palaeozoogeography of the Verbeekinidae (Permian Foraminifera):* in Aspects of Tethyan biogeography, G. E.

Adams & D. V. Ager (eds.), Systematics Assoc. Publ., no. 7, p. 77-91.

Goryunova, R. V., 1975, *Permskie mshanki Pamira:* Akad. Nauk SSSR, Paleont. Inst., Trudy, v. 148, 127 p. [*Permian Bryozoa of Pamir.*]

Grabau, A. W., 1931, *The Permian of Mongolia:* Natural History of Central Asia, v. 4, 665 p., 35 pl., Am. Museum Nat. History (New York).

Grant, R. E., 1970, *Brachiopods from Permian-Triassic boundary beds and age of Chhidru Formation, West Pakistan:* in Stratigraphic boundary problems: Permian and Triassic of West Pakistan, Bernhard Kummel & Curt Teichert (eds.), Univ. Kansas Dept. Geology, Special Publ. 4, p. 117-151.

Grunt, T. A., & Dmitriev, B. Ju., 1973, *Permskie Brakhiopodi Pamira:* Akad. Nauk SSSR, Paleont. Inst., Trudy, v. 136, 211 p. [*Permian Brachiopoda of Pamir.*]

Harker, Peter, & Thorsteinsson, Raymond, 1960, *Permian rocks and fauna of Grinnell Peninsula, Arctic Archipelago:* Canada Geol. Survey, Mem., no. 309, 98 p.

Hart, G. F., 1974, *Permian palynofloras and their bearing on continental drift:* in Paleogeographic provinces and provinciality, C. A. Ross (ed.), Soc. Econ. Paleontologists & Mineralogists, Spec. Publ., no. 21, p. 148-164.

Hayden, H. H., 1907, *The stratigraphical position of the Gangamopteris beds of Kashmir:* India Geol. Survey, Records, v. 36, no. 1, p. 23-39.

Hill, Dorothy, 1948, *The distribution and sequence of Carboniferous coral faunas:* Geol. Mag., v. 85, p. 121-148.———1957, *The sequence and distribution of Upper Paleozoic coral faunas:* Australian Jour. Sci., v. 19, p. 42-60.———1958, *Sakmarian geography:* Geol. Rundschau, v. 47, p. 590-629.

International Union of Geological Sciences, 1969, *Gondwana stratigraphy:* Internatl. Union Geol. Sciences Symposium, 1st, Mar del Plata, 1967, UNESCO, Earth Sciences ser., v. 2, 1,173 p. (Paris).

Ishii, Ken-ichi, Fischer, Jochen, & Bando, Yuji, 1971, *Notes on the Permian-Triassic boundary in Eastern Afghanistan:* Jour. Geosciences, Osaka City Univ., v. 14, art. 1, p. 1-18.

———, Okimura, Yuji, & Nakazawa, Keiji, 1975, *On the genus Colaniella and its biostratigraphic significance:* Jour. Geosciences, Osaka City Univ., v. 19, art. 6, p. 107-138.

Kahler, Franz, 1974a, *Iranische Fusuliniden:* Geol. Bundesanst. Wien, Jahrb., v. 117, p. 75-107.——— 1974b, *Fusuliniden aus T'ienschan und Tibet:* Repts. Scientific Expedition to North-western Provinces of China under leadership of Dr. Sven Hedin, Sino-Swedish Expedition, Publ. 52, V, Invertebrate palaeontology 4, p. 1-147 (Stockholm).

Karpinskiy, A. P., 1889, *Ueber die Ammoneen der Artinsk-Stufe und einige mit denselben verwandte carbonische Formen:* Acad. Impér. Sci. St. Pétersbourg, Mém., 7e sér., v. 37, no. 2, p. 1-104.——— 1890, *Ob ammoneyakh artinskogo yarusa i o nekotorykh skhodnynkh s nimi kamennougolnykh formakh:* Zapiski Mineral. Obshch., ser. 2, ch. 27. [*On ammonites of the Artinskian Series and on certain similarities with Carboniferous forms.*] [Not seen by author.]

Keast, Allen, Erk, F. C., & Glass, Bentley (eds.), 1972, *Evolution, mammals and southern continents:* 543 p., State Univ. New York Press (Albany).

King, P. B., 1931, *The geology of the Glass Mountains, Texas; Part I:* Univ. Texas, Bull., no. 3038(1930), 167 p.———1938, *Geology of the Marathon region, Texas:* U.S. Geol. Survey, Prof. Paper 187(1937), ix + 148 p.———1949, *Geology of the southern Guadalupe Mountains, Texas:* U.S. Geol. Survey, Prof. Paper 215(1948), 183 p.

King, R. E., 1931, *The geology of the Glass Mountains, Texas; Part II:* Univ. Texas, Bull., no. 3042(1930), 245 p.

Kozur, Heinz, 1973, *Beiträge zur Stratigraphie von Perm und Trias:* Geologie, Palaeontologie Mitt. Innsbruck, v. 3, p. 1-31.———1975, *Beiträge zur Condontenfauna des Perm:* Geologie, Palaeontologie Mitt. Innsbruck, v. 5, p. 1-44.

Kulikov, M. V., Pavlov, A. M., & Rostevtsev, V. N., 1973, *O nakhodke goniatitov v nizhne-Kazanskikh otlozheniy severnoy chasti Russkoy platformy:* Akad. Nauk SSSR, Doklady, v. 211, p. 1412-1414. [*On a find of goniatites in the lower Kazanian in the northern part of the Russian platform.*] (Eng. transl., Doklady, Earth Sciences Sec., v. 211, p. 112-114.)

Kummel, Bernhard, 1972, *The Lower Triassic (Scythian) ammonoid Otoceras:* Harvard Univ., Museum Comp. Zoology, Bull., v. 143, p. 365-417.

———, & Teichert, Curt, 1966, *Relations between the Permian and Triassic formations in the Salt Range and Trans-Indus Ranges, West Pakistan:* Neues Jahrb. Geologie, Paläontologie, Abh., v. 125, p. 297-333.———1970, *Stratigraphy and paleontology of the Permian-Triassic boundary beds, Salt Range and Trans-Indus Ranges, West Pakistan:* in Stratigraphic boundary problems: Permian and Triassic of West Pakistan, Bernhard Kummel & Curt Teichert (eds.), Univ. Kansas Dept. Geology, Special Publ., v. 4, p. 1-110. ———1973, *The Permian-Triassic boundary beds in central Tethys:* in The Permian and Triassic Systems and their mutual boundary, A. Logan & L. V. Hills (eds.), Canad. Soc. Petrol. Geologists, Mem. 2, p. 17-34.

Lapkin, I. Yu., & Solovev, V. O., 1969, *Permskie tektonisheskie dvizheniya v Evrazii:* Akad. Nauk SSSR, Doklady, v. 184, p. 410-413. [*Permian diastrophism in Eurasia.*]

Lapparent, A. A. de, 1902, *Traité de géologie:* 3rd edit., v. 2, F. Savy (Paris). (v. 1 and 2, 1,645 p.)

Leven, E. Ya., 1967, *Stratigrafiya i fuzulinidy Permskikh otlozheniy Pamira:* Akad. Nauk SSSR, Geol. Inst., Trudy, v. 167, p. 1-224. [*Stratigraphy and fusulinids of the Permian deposits of Pamir.*]——1975, *Yarusnaya schkala Permskikh otlozhenii Tetisa:* Moskov. Obshch. Ispyt. Prirody, Byull., Otdel. Geol., v. 50, p. 5-21. [*Permian stages of the Tethys.*]

Likharev, B. K., 1966, *Stratigrafiya SSSR. Permskaya Sistema:* Akad. Nauk SSSR, Ministerstvo Geol. SSSR, 536 p. (Moscow). [*Stratigraphy of the USSR. Permian System.*]

——, & Miklukho-Maklay, A. D., 1972, *Stratigraphy of the Permian System:* Internatl. Geol. Congress India 1964, Rept. 22nd Sess., pt. 8., Palaeontology & Stratigraphy, p. 205-222.

Lipina, O. A., 1949, *Melkie foraminifery pogrebennykh massivov Baschkirii:* Akad. Nauk SSSR, Geol. Inst., Trudy, v. 105 (geol. ser. 35), p. 198-243. [*Small foraminifers from subsurface of the Bashkirian massif.*]

Loo Lin-huang, 1958, *Chiek chiang hang chow chi shia shyr tann yan chung de jii geh tair shean chorng:* Ku Sheng Wu Hsüeh Pao (Acta Palaeont. Sinica), v. 6, p. 293-304. [*Some bryozoans from the Chihsia Limestone of Hangchow, western Chekiang.*] (In Chinese and English.)

Lyem [Liêm], N. V., 1971, *Novyye pozdnepermskiye vidy Neoendothyra (Endothyrida) iz Svernogo Vyetnama:* Paleont. Zhurnal, 1971, no. 3, p. 110-111. [*New Late Permian species of Neoendothyra (Endothyrida) from North Vietnam.*] (Engl. Transl., Paleont. Jour., 1971, no. 3, p. 382, 383.)——1974, *Rod Palaeofusulina i yego novyye vidy iz Vyetnama:* Paleont. Zhurnal, 1974, no. 4, p. 11-17. [*The genus Paleofusulina and its new species from Vietnam.*] (Engl. transl., Paleont. Jour., 1974, no. 4, p. 447-454.)

McKee, E. D., et al., 1967, *Paleotectonic investigation of the Permian System in the United States:* U.S. Geol. Survey, Prof. Paper 515, 271 p.

Menner, V. V., Sarycheva, T. G., & Chernyak, G. E. (eds.), 1970, *Stratigrafiya Kamennougolnykh i Permskikh otlozeniy severnogo Verkhoyanya:* Nauchno-Issledov. Inst. Geol. Arktiki, Trudy, v. 154, p. 1-190. [*Stratigraphy of Carboniferous and Permian deposits of northern Verkhoyansk.*]

Meyen, S. V., 1970, *Permskie flory:* in Paleozoyskie i Mezozoyskie flory Yevrazii i fitogeografiya etogo vremeni, V. A. Vakhrameyev and others, Akad. Nauk SSSR, Geol. Inst., Trudy, v. 208, p. 111-157. [*Permian floras: in Paleozoic and Mesozoic floras of Eurasia and the phytogeography of those eras.*]

Middlemiss, C. S., 1910, *A revision of the Silurian-Trias sequence in Kashmir:* Geol. Survey India, Records, v. 40, p. 206-260.

Mikhaylova, Z. P., 1974, *Fuzulinidy verkhnego Karbona Pechorskogo Priuralya:* Akad. Nauk SSSR, Komi filial Inst. Geol., 116 p., Izd. Nauka, (Leningrad). [*Fusulinids of the Upper Carboniferous of the Pechoran Urals.*]

Miklukho-Maklay, A. D., 1963, *Verkhniy Paleozoy sredney Azii:* 329 p., Leningrad. Gosudar. Univ. (Leningrad). [*Upper Paleozoic of Central Asia.*]

Miller, A. K., & Furnish, W. M., 1940a, *Permian ammonoids of the Guadalupe Mountain region and adjacent areas:* Geol. Soc. America, Spec. Paper, v. 26, 242 p.——1940b, *Cyclolobus from the Permian of eastern Greenland:* Meddel. Grønland, v. 112, no. 5, 7 p.

Minato, Masao, & Kato, Makoto, 1965a, *Waagenophyllidae:* Hokkaido Univ., Jour. Fac. Sci., ser. 4, v. 12, 241 p.——1965b, *Durhaminidae (tetracorals):* Hokkaido Univ., Jour. Fac. Sci., ser. 4, v. 13, p. 11-86.——1970, *The distribution of Waagenophyllidae and Durhaminidae in the Upper Paleozoic:* Japan. Jour. Geology & Geography, v. 41, p. 1-14.

Moeller, V. von, 1877, *Ueber Fusulinen und ähnliche Foraminiferen-Formen des russischen Kohlenkalks:* Neues Jahrb. Mineralogie, Geologie, Paläontologie, Jahrg. 1877, p. 139-146.

Monger, J. W. H., & Ross, C. A., 1971, *Distribution of fusulinaceans in the Western Canadian Cordillera:* Canad. Jour. Earth Sciences, v. 8, p. 259-278.

Moore, R. C., & Dudley, P. M., 1944, *Cheilotrypid bryozoans from Pennsylvanian and Permian rocks of the midcontinent region:* Kansas State Geol. Survey, Bull., v. 52, p. 229-408.

——, & Jeffords, Russell, 1941, *New Permian corals from Kansas, Oklahoma and Texas:* Kansas State Geol. Survey, Bull., v. 38, pt. 3, p. 65-120.

Morozova, I. P., 1970, *Mshanki pozdney Permi:* Akad. Nauk SSSR, Paleont. Inst., Trudy, v. 122, 347 p. [*Bryozoa of the Late Permian.*]

Morozova, V. G., 1949, *Predstaviteli semeistv Lituolidae e Textulariidae iz Verkhnekamennougolnykh i Artinskikh otlozhenii Baschkirskogo Priuralya:* Akad. Nauk SSSR, Geol. Inst., Trudy, v. 105 (geol. ser. 35), p. 244-275. [*Representatives of the families Lituolidae and Textulariidae from the Upper Carboniferous and Artinskian deposits of the Bashkirian Ural region.*]

Murchison, R. I., 1841, *First sketch of some of the principal results of a second geological survey of Russia, in a letter to M. Fischer:* Philos. Mag., v. 19, p. 419-422.

——, Verneuil, Édouard de, & Keyserling, Alexander Graf von, 1845, *The geology of Russia in Europe and the Ural Mountains:* v. 1, Geology, Chapter 8, Permian System, p. 137-170, J. Murray (London).

Nakazawa, Keiji, 1974, *On the Permian-Triassic boundary problem:* Chigaku Zasshi Jour. Geography (Tokyo), v. 38, p. 1-24.

——, Ishii, Ken-ichi, Kato, Makoto, Okimura, Yuji, Nakamura, Koji, & Haralambous, Diomedes, 1975, *Upper Permian fossils from the Island of*

Salamis, Greece: Kyoto Univ., Mem. Fac. Sci., ser. geol. & mineral., v. 41, p. 21-44.

———, **Kapoor, H. M., Ishii, Ken-ichi, Bando, Yuji, Okimura, Yuji, & Tokuoka, Takao,** 1975, *The Upper Permian and the Lower Triassic in Kashmir, India:* Kyoto Univ., Mem. Fac. Sci., ser. geol. & mineral., v. 42, p. 1-106.

Nalivkin, D. V., 1973, *Geology of the U.S.S.R.:* transl. from the Russian by N. Rast, 827 p., Univ. Toronto Press (Toronto).

Nanking Institute of Geology & Paleontology (ed.), 1974, *Hsi nan ti chü ti tseng ku sheng wu shoŭ ts'e:* 454 p., Ke Hsüeh ch'u Păn she (Nanking). [*A handbook of the stratigraphy and paleontology in southwest China.*] (In Chinese.)

Newell, N. D., Rigby, J. K., Fischer, A. G., Whiteman, A. J., Hickox, J. E., & Bradley, J. S., 1953, *The Permian reef complex of the Guadalupe Mountains region, Texas and New Mexico:* 236 p., W. H. Freeman and Co. (San Francisco).

Nikiforova, A. I., 1939, *Novyye vidy verkhnepaleozoyskikh mshanki predgornor polosy Bashkirii (krome sem. Fenestellidae i Acanthocladiidae):* Neft. Geol.-Razv. Inst., Trudy, ser. A, v. 115, p. 70-101. [*New species of Upper Paleozoic bryozoans from the foothills belt of Bashkiria (except the families Fenestellidae and Acanthocladiidae).*]

Nikitina, A. P., 1969, *Rod Hemigordiopsis (Foraminifery) v verkhney Permi Primorya:* Paleont. Zhurnal, 1969, no. 3, p. 63-69. [*Genus Hemigordiopsis (Foraminifera) in the Upper Permian of the Maritime Province.*] (Engl. transl., Paleont. Jour., 1969, no. 3, p. 341-346.)

———, **Kiseleva, A. V., & Burago, V. I.,** 1970, *Skhema biostratigraficheskogo raschleneyiya barabashskoy svity verkhney Permi Yugo-Zapadnogo Primorya:* Akad. Nauk SSSR, Doklady, v. 191, p. 187-189. [*Scheme for biostratigraphic subdivision of the Upper Permian Barabash Suite in the southwest Maritime Province.*]

Okimura, Yuji, Ishii, Ken-ichi, & Nakazawa, Keiji, 1975, *Abadehella, a new genus of tetraxid Foraminifera from the Late Permian:* Kyoto Univ., Mem. Fac. Sci., ser. geol. & mineral. v. 41, p. 35-48.

Pantic, Smiljka, 1970, *Lithostratigraphy and micropaleontology of the Middle and Upper Permian of Western Serbia:* Zavod Geol. Geof. Istraz., (Geol.), ser. A, Bull., no. 27 (1969), p. 239-272.

Plumstead, E. P., 1973, *The late Paleozoic Glossopteris flora:* in Atlas of palaeobiogeography, Anthony Hallam (ed.), p. 187-205, Elsevier (Amsterdam).

Rauzer-Chernousova, D. M., 1937, *O fuzulinidakh i stratigrafii verkhnego Karbona i Artinskogo yarusa zapadnogo sklona Urala (reyume doklada):* Moskov, Obshch. Ispyt. Prirody, Byull., n. ser., otdel. geol., v. 45, p. 478-480. [*On the fusulinids and stratigraphy of the Upper Carboniferous and Artinskian stage on the western slopes of the Urals.*]———1940, *Stratigrafiya verkhnego Karbona i Artinskogo yarusa zapadnogo sklona Urala i materialy k faune fuzulinid:* Akad. Nauk SSSR, Geol. Inst., Trudy, geol. ser., no. 2, v. 7, p. 37-99. [*Stratigraphy of the Upper Carboniferous and Artinskian Stage on the western slopes of the Urals and materials on the fusulinid fauna.*]———1949, *Stratigrafiya verkhnekamennougolnykh i Artinskikh otlozheniy Bashkirskogo Priuralya:* Akad. Nauk SSSR, Inst. Geol. Nauk, Trudy, v. 105, geol. ser. 35, p. 3-21. [*Stratigraphy of Upper Carboniferous and Artinskian deposits of Bashkirian Ural region.*]———1965, *Foraminifery stratotipicheskogo razreza Sakmarskogo yarusa (r. Sakmara, Yuzhnyy Ural):* Akad. Nauk SSSR, Inst. Geol. Nauk, Trudy, v. 135, p. 1-80. [*Foraminifera in the stratotype section of the Sakmarian Stage (Sakmara district, southern Urals).*]

———, **Shcherbovich, S. F., Rozoskaya, S. E., & Schamov, D. F.,** 1958, *Shvagerinobyy gorizont Russkoy Platformy i podstilayushchie ego otlozheniya:* Akad. Nauk SSSR, Inst. Geol. Nauk, Trudy, v. 13, p. 1-154. [*Schwagerinid horizons of the Russian platform and their underlying deposits.*]

Romer, A. S., 1968, *Fossils and Gondwanaland:* Am. Philos. Soc., Proc., v. 112, p. 335-343.

Ross, C. A., 1959, *The Wolfcamp series (Permian) and new species of fusulinids, Glass Mountains, Texas:* Washington Acad. Sci., Jour., v. 49, p. 299-316.———1960, *Fusulinids from the Hess member of the Leonard formation, Leonard series (Permian), Glass Mountains, Texas:* Cushman Found. Foram. Res., Contrib., v. 11, p. 117-133.———1962a, *Fusulinids from the Leonard Formation (Permian), western Glass Mountains, Texas:* Cushman Found. Foram. Res., Contrib., v. 13, p. 1-21.———1962b, *Evolution and dispersal of the Permian fusulinid genera Pseudoschwagerina and Parashwagerina:* Evolution, v. 10, p. 306-315.———1963a, *Standard Wolfcampian Series (Permian), Glass Mountains, Texas:* Geol. Soc. America, Mem., v. 88, p. 1-205.———1963b, *Early Permian fusulinids from Macusani, southern Peru:* Palaeontology, v. 5, p. 817-823.———1967, *Development of fusulinid (Foraminiferida) faunal realms:* Jour. Paleontology, v. 41, p. 1341-1354.———1970, *Concepts in Late Paleozoic correlations:* in Radiometric dating and paleontologic zonation, O. L. Bandy (ed.), Geol. Soc. America, Spec. Publ., v. 124, p. 7-36.———1972, *Paleoecology of fusulinaceans:* 23rd Internatl. Geol. Congress, 1968 Proc., I.P.U., p. 301-318.———1973, *Pennsylvanian and Early Permian depositional history, southeastern Arizona:* Am. Assoc. Petrol. Geologists, v. 57, p. 887-912.———1974, *Paleogeography and provinciality:* in Paleogeographic provinces and provinciality, C. A. Ross (ed.), Soc. Econ. Paleontologists & Mineralogists, Spec. Publ., no. 21, p.

1-17.——1976 (ed.), *Paleobiogeography:* 427 p., Dowden, Hutchinson, & Ross (Stroudsburg, Pa.).
——1978, *Evolution of fusulinaceans in Late Paleozoic space and time:* in Historical biogeography, plate tectonics and the changing environment, A. J. Boucot, & J. Gray (eds.), 37th Biology Colloquium, Oregon State University (in press).
——, & **Nassichuk, W. W.**, 1970, *Yabeina and Waagenoceras from Atlin Horst area, northern British Columbia:* Jour. Paleontology, v. 44, p. 779-781.
——, & **Ross, J. P.**, 1962, *Pennsylvanian, Permian rugose corals, Glass Mountains, Texas:* Jour. Paleontology, v. 36, p. 1163-1188.

Ross, J. R. P., 1963, *Lower Permian Bryozoa from Western Australia:* Palaeontology, v. 6, p. 70-82.
——1978, *Permian ectoprocts in space and time:* in Historical biogeography, plate tectonics and the changing environment, A. J. Boucot, & J. Gray (eds.), 37th Biology Colloquium, Oregon State University (in press).
——, & **Ross, C. A.**, 1962, *Faunas and correlation of the late Paleozoic rocks of northeast Greenland, Part IV, Bryozoa:* Meddel. Grønland, v. 167, no. 7, 65 p.——1963, *Late Paleozoic rugose corals, Glass Mountains, Texas:* Jour. Paleontology, v. 37, p. 409-420.

Rowett, C. L., 1972, *Paleogeography of Early Permian waagenophyllid and durhaminid corals:* Pacific Geology, v. 4, p. 31-37.——1975a, *Provinciality of the late Paleozoic invertebrates of North and South America and a modified intercontinental reconstruction:* Pacific Geology, v. 10, p. 79-94.——1975b, *Stratigraphic distribution of Permian corals in Alaska:* U.S. Geol. Survey, Prof. Paper 823-D, p. 59-75.

Rozovskaya, S. E., 1975, *Sostav, sistema i filogeniya otryada Fusulinida:* Akad. Nauk SSSR, Paleont. Inst., Trudy, v. 149, p. 1-267. [*Composition, systematics and phylogeny of the order Fusulinida.*]

Rudwick, M. J. S., 1970, *Living and fossil brachiopods:* 199 p., Hutchinson University Library (London).

Runnegar, B. N., 1969, *The Permian faunal succession in Eastern Australia:* Jour. Geol. Soc. Australia, Spec. Publ., no. 2, p. 73-98.
——, & **Armstrong, J.**, 1969, *Comments on series and stages in the Permian of New Zealand:* Royal Soc. New Zealand, Trans., geol., v. 6, p. 209-212.

Ruzhentsev [Ruzhencev], V. E., 1936, *Novye dannye po stratigrafii Kamennougolnykh i nizhnepermskikh otlozheniy Orenburgskoy i Aktyubinskoy oblastey:* Problemy Sovetskoy Geologii, v. 6, no. 6, p. 470-506. [*New data on the stratigraphy of the Carboniferous and Lower Permian of the Orenburg and Aktyubinsk Districts.*]——1951, *Nizhnepermskie ammonity yuzhnogo Urala—I: Ammonity sakmarskogo yarusa:* Akad. Nauk SSSR, Paleont. Inst., Trudy, v. 33, 188 p. [*Lower Permian ammonoids of the southern Urals—I: Ammonoids of the Sakmarian Stage.*]——1952, *Biostratigraphiya Sakmarskogo yarusa v Aktyubinskoi Oblasti Kazakhskoi SSR:* Akad. Nauk SSSR, Paleont. Inst., Trudy, v. 42, 87 p. [*Biostratigraphy of the Sakmarian Stage in the Aktyubinsk region of Kazakhstan SSR.*]——1955, *Osnovnye stratigraficheskie kompleksy Ammonoidei Permskoi sistemy:* Akad. Nauk SSSR, Izvestiya, biol. ser., no. 4, p. 12-132. [*Fundamental stratigraphic assemblages of Ammonoidea of the Permian System.*]——1956, *Nizhnepermskie ammonity yuzhnogo Urala—II: Ammonity Artinskogo Yarusa:* Akad. Nauk SSSR, Paleont. Inst., Trudy, v. 60, 275 p. [*Lower Permian ammonites of the southern Urals—II: Ammonites of the Artinskian Stage.*]——1962, *Nadotryad Ammonoidea:* in Yu. A. Orlov (ed.), Osnovy Paleontologii, v. 5: Mollyuski—Golovonogie, 1, p. 243-438, Nedra Press (Moskva). [*Superorder Ammonoidea, Mollusca—Cephalopoda.*]——1966, *Osnovnye kompleksy ammonoidey Kamennougolnogo perioda:* Paleont. Zhurnal, 1965, no. 2, p. 3-17. [*Principal ammonoid assemblages of Carboniferous Period.*]——1974, *O semeystvakh Paragastrioceratidae i Spirolegoceratidae:* Paleont. Zhurnal, 1974, no. 1, p. 19-29. [*On the families Paragastrioceratidae and Spirolegoceratidae.*] (Engl. transl. Paleont. Jour., v. 8, no. 1, p. 14-24.)
——, & **Sarycheva, T. G.** (eds.), 1965, *Razvitie i smena morskikh organismov na rubezhe Paleozoya i Mezozoya:* Akad. Nauk SSSR, Paleont. Inst., Trudy, v. 108, 431 p. [*The development and change of marine organisms at the Paleozoic-Mesozoic boundary.*]

Sakagami, Sumio, 1970, *On the Paleozoic Bryozoa of Japan and Thai-Malayan districts:* Jour. Paleontology, v. 44, p. 680-692.

Shcherbovich, S. F., 1969, *Fusulinidy pozdnegshelskogo i Asselskogo Vremeni Prikaspiyaskoy sineklizy:* Akad. Nauk SSSR, Geol. Inst., Trudy, v. 176, p. 1-82. [*Fusulinids of the late Gzhelian and Asselian Stages of the Caspian syneclise.*]

Shchukina, V. Y., 1973, *Kompleksy Kamennougolnykh i Permskikh korallov Sredney Azii:* Sovetskaia Geologiia, no. 3, 1973, p. 53-68. [*Carboniferous and Permian coral complexes of Central Asia.*]

Sheng Jin-chiang, 1963, *Permian fusulinids of Kwangsi, Kueichow and Szechuan:* Palaeont. Sinica, n. ser. B, no. 10, p. 129-247.——1965, *Hai nan tao si pu de ting ley:* Ku Sheng Wu Hsüeh Pao (Acta Palaeont. Sinica), v. 13, no. 4, p. 563-590. [*Fusulinids from the western part of Hainan Island, Kwangtung Province.*] (In Chinese with English summary.)
——, & **Lee, H. H.**, 1964, *Carboniferous-Permian boundary in China:* Congrès Internatl. Strat. Géol. Carbonifère, 5e, Paris, 1963, Compte

Rendu 2, p. 775-779.

Shulga Nesterenko, M. I., 1952, *Novyye nizhnepermskie Mshanki Priuralya:* Akad. Nauk SSSR, Paleont. Inst., Trudy, v. 37, 84 p. [*New Lower Permian Bryozoa of the Ural region.*]

Skinner, J. W., & Wilde, G. L., 1965, *Permian biostratigraphy and fusulinid faunas of the Shasta Lake area, northern California:* Univ. Kansas Paleont. Contrib., Protozoa, Art. 6, p. 1-98.

Slade, M. L., 1961, *Pennsylvanian and Permian fusulinids of the Ferguson Mountain area, Elko County, Nevada:* Brigham Young Univ. Geol. Studies, v. 8, p. 55-92.

Smith, A. G., Briden, J. C., & Drewry, G. E., 1973, *Phanerozoic world maps:* in Organisms and continents through space and time, N. F. Hughes (ed.), Palaeont. Assoc., Spec. Paper no. 12, p. 1-42 (London).

Sokolov, B. S., 1955, *Tabulyaty paleozoya Evropeiskoe chasti SSSR. Obshchie voprosy sistematiki i istorii razvitiya Tabulyat:* Vses. Neft. Nauchno-Issled. Geol.-Razved. Inst., (VNIGRI), Trudy, n. ser., v. 85, 525 p., 90 pl. [*The Paleozoic Tabulata of European USSR. General problems in the systematics and developmental history of Tabulata.*]

Soshkina, E. D., Dobrolyubova, T. A., Porfirev, G. S., 1941, *Permskie Rugosa Evropeiskoi chastii SSSR:* Paleontologiya SSSR, v. 5, Ch. 3, no. 1, 304 p., Akad. Nauk SSSR (Moskva). [*Permian Rugosa of the European part of the USSR.*]

Sosnina, M. I., 1965, *Nekotorye Permskie fuzulinidy i lagenidy Sikhotz-Alinya:* Vses. Nauchno-Issled. Geol. Inst., n. ser., v. 115, p. 142-168. [*Some Permian fusulinids and lagenids from Sikhote-Alin.*]

Spinosa, Claude, Furnish, W. M., & Glenister, B. F., 1970, *Araxoceratidae, Upper Permian ammonoids from the Western Hemisphere:* Jour. Paleontology, v. 44, p. 730-736.———1975, *The Xenodiscidae, Permian ceratitoid ammonoids:* Jour. Paleontology, v. 49, p. 239-283.

Stepanov, D. L., 1937, *Permskie brakhiopody Shpitsbergena:* Arkt. i Antarkt. Nauchno-Issledov. Inst., Trudy, v. 76, p. 105-192. [*Permian Brachiopoda of Spitsbergen.*]———1957, *O novom yarus Permskoy sistemy Artinski:* Leningrad Gosud. Univ., Vestnik, v. 24, ser. geol. geogr., v. 4, p. 20-24. [*On a new stage of the Permian System Artinskian.*]———1970, *The problems of the stratigraphy of the Permian system and the correlation of the main sections of the marine Permian:* Internatl. Geol. Congr. India, 1964, pt. 8, Palaeontology & Stratigraphy, p. 189-204.

———, Golshani, F., & Stöcklin, Jovan, 1969, *Upper Permian and Permian-Triassic boundary in North Iran:* Geol. Survey Iran, Rept. 12, 72 p.

Stevens, C. H., 1975a, *Occurrence and dispersal of boreal massive Rugosa in the Early Permian:* Pacific Geology, v. 10, p. 33-42.———1975b, *New Permian Waagenophyllidae (rugose corals) from North Africa:* Jour. Paleontology, v. 49, p. 706-709.

Sweet, W. C., 1970, *Uppermost Permian and Lower Triassic conodonts of the Salt Range and Trans-Indus Ranges, West Pakistan:* in Stratigraphic boundary problems: Permian and Triassic of West Pakistan, Bernhard Kummel & Curt Teichert (eds.), Univ. Kansas, Dept. Geology, Spec. Publ. no. 4, p. 207-271.

Syemina, S. A., 1961, *Stratigrafiya i Foraminifery (fuzulinidy) shvagerinogo gorizonta Oksko-Tsninskogo podnyatiya:* Akad. Nauk SSSR, Geol. Inst., Trudy, v. 57, p. 1-70. [*Stratigraphy and Foraminifera (fusulinids) of the schwagerinid horizon of the Oksko-Tsninskian Arch.*]

Taraz, Hushang, 1969, *Permo-Triassic section in central Iran:* Am. Assoc. Petrol. Geologists, Bull., v. 53, p. 688-693.———1971, *Uppermost Permian and Permo-Triassic transition beds in central Iran:* Am. Assoc. Petrol. Geologists, Bull., v. 55, p. 1280-1295.———1973, *Correlation of uppermost Permian in Iran, Central Asia, and South China:* Am. Assoc. Petrol. Geologists, Bull., v. 57, p. 1117-1133.

Teichert, Curt, 1966, *Stratigraphic nomenclature and correlation of the Permian "Productus limestone," Salt Range, West Pakistan:* Geol. Survey Pakistan, Records, v. 15, pt. 1, p. 1-20.———1974, *Marine sedimentary environments and their faunas in Gondwana area:* Am. Assoc. Petrol. Geologists, Mem., v. 23, p. 361-394.

———, & Kummel, Bernhard, 1972, *Permian-Triassic boundary in the Kap Stosch area, East Greenland:* Canad. Petrol. Geologists, Bull., v. 20, p. 659-675.———1973, *Nautiloid cephalopods from the Julfa beds, Upper Permian, Northwest Iran:* Harvard Univ., Museum Comp. Zoology, Bull., v. 144, p. 409-433.

———, & Sweet, W. C., 1973, *Permian-Triassic strata, Kuh-e-Ali Bashi, northwestern Iran:* Harvard Univ., Museum Comp. Zoology, Bull., v. 145, no. 8, p. 359-472.

———, & Rilett, Michael, 1974, *Revision of Permian Ecca Series cephalopods, Natal, South Africa:* Univ. Kansas, Paleont. Contrib., Paper 68, 8 p.

Thomas, G. A., 1971, *Carboniferous and Early Permian brachiopods from western and northern Australia:* Australia Bur. Min. Res., Geology and Geophysics, Bull. 56, p. 1-215.

Thompson, M. L., 1954, *American Wolfcampian fusulinids:* Univ. Kansas Paleont. Contrib., Protozoa, Art. 5, p. 1-226.

Thorsteinsson, Raymond, 1974, *Carboniferous and Permian stratigraphy of Axel Heiberg Island and Western Ellesmere Island, Canadian Arctic Archipelago:* Canada Geol. Survey, Bull., v. 224, 115 p.

Tillyard, R. J., 1917-36, *Upper Permian insects from N.S.W.:* Linnean Soc. New South Wales, Proc., v. 42, p. 720-756; v. 44, p. 231-256; v.

46, p. 413-422; v. 47, p. 279-282; v. 49, p. 429-435; v. 51, p. 1-30, 265-282; v. 60, p. 265-279, 374-391.

Tolstikhina, M. M., 1935, *Kamennougolnye otlozheniya tsentralnoy chasti Ufimskogo plato i ikh fatsii:* Tsentralnyi Nauchno-Issledov. Geol.-Razved. Inst. (TsNIGRI), Trudy, no. 65, p. 1-36. [*Carboniferous deposits of the central part of the Ufa Plateau and their facies.*]

Toriyama, Ryuzo, 1958, *Geology of Akiyoshi. Part III, Fusulinids of Akiyoshi:* Kyushu Univ., Mem. Fac. Sci., ser. D, geol., v. 7, p. 1-264.——1963, *The Permian:* in F. Takai, T. Matsumoto, & R. Toriyama (eds.), Geology of Japan, p. 43-58, Univ. Tokyo Press (Hongo, Tokyo).——1967, *The fusulinacean zones of Japan:* Kyushu Univ., Mem. Fac. Sci., ser. D, geol., v. 18, p. 35-260.——1973, *Upper Permian fusulinacean zones:* in The Permian and Triassic systems and their mutual boundary, A. Logan & L. V. Hills (eds.), Canad. Soc. Petrol. Geologists, Mem. 2, p. 498-512.——1975, *Fusuline fossils from Thailand, Part IX, Permian fusulines from the Rat Buri Limestone in the Khao Phlong Phrab area, Sara Buri, Central Thailand:* Kyushu Univ., Mem. Fac. Sci., ser. D, geol., v. 23, p. 1-116.

——, Kanmera, K., Kaewbaidhoam, S., & Hongnusonthi, A., 1974, *Biostratigraphic zonation of the Rat Buri Limestone in the Khao Phlong area, Sara Buri, Central Thailand:* in Palaeontology of Southeast Asia, T. Kobayashi, & R. Toriyama (eds.), v. 14, p. 25-48, Univ. Tokyo Press (Tokyo).

Tozer, E. T., 1969, *Xenodiscacean ammonoids and their bearing on the discrimination of the Permo-Triassic boundary:* Geol. Mag., v. 106, p. 348-361.——1971, *Triassic time and ammonoids: problems and proposals:* Canad. Jour. Earth Sci., v. 8, p. 989-1031.

Trizna, V. B., 1950, *K Kharakteristike rifovykh i sloistykh fatsiy tsentralnoy chasti Ufimskogo Plato:* Vses. Nauchno-Issled. Geol.-Razved. Inst. (VNIGRI), v. 135, Microfauna SSSR, no. 3, p. 47-144. [*About characteristics of the reef and stratified facies of the central part of the Ufimian Plateau.*]

——, & Klautsan, R. A., 1961, *Mshanki Artinskogo yarusa Ufimskogo Plato i ikh rol v stratigrafii etogo yarusa v Priuralye:* Vses. Neft. Nauchno-Issled. Geol.-Razved. Inst.(VNIGRI), v. 179, Microfauna SSSR, no. 13, p. 331-453. [*Bryozoa of the Artinskian Stage of the Ufimian Plateau and their role in stratigraphy of that stage in the Ural region.*]

Tschudy, R. H., & Scott, R. A. (eds.), 1969, *Aspects of palynology:* 510 p., Wiley-Interscience Publishers (New York).

Ucharskaya [Ukharskaya], L. B., 1970, *Novyye Kazanskie peschanye foraminifery Russkoy platformy:* Paleont. Zhurnal, 1970, no. 4, p. 21-28. [*New Kazanian arenaceous Foraminifera of the Russian platform.*] (Engl. transl., Paleont. Jour., 1970, no. 4, p. 468-476.)

Ustritskiy, V. I., 1974, *O bipolyarnosti faun pozdnego Paleozoya:* Paleont. Zhurnal, 1974, no. 2, p. 33-37. [*On the bipolarity of late Paleozoic faunas.*] (Engl. transl., Paleont. Jour., 1974, v. 8, no. 2, p. 148-151.)

Vasilyuk, N. P., Kachanov, Ye. I., & Pyzhyanov, I. V., 1970, *Paleobiogeograficheskiy ocherk Kamennougolnykh i Permskikh tselenterat:* in Zakonomernosti rasprostraneniya paleozoyskikh korallov SSSR, D. L. Kaljo (ed.), Vses. Simpos. Izuch. Iskop. Korallov SSSR, 2nd, Trudy, no. 3, p. 45-60. [*Paleobiogeographic sketch of Carboniferous and Permian coelenterates.*]

Vissarionova, A. Ya., Kireeva, G. D., Lipina, O. A., Morozova, V. G., Rauzer-Chernousova, D. M., Suleymanov., I. S., Shamov, D. F., & Shcherbovich, S. F., 1949, *Foraminifery verkhnekamennougolnykh i Artinskikh otlozheniy Bashkirskogo Priuralya:* Akad. Nauk SSSR, Inst. Geol. Nauk, Trudy, v. 105 (geol. ser. no. 35), p. 1-275. [*Foraminifera from Upper Carboniferous and Artinskian deposits of Bashkirian Ural region.*]

Vlasov, N. G., Likharev, B. K., & Miklukho-Maklay, A. D., 1962, *K faunisticheskoy kharakteristike razreza nizhnepermskikh otlozheniy yugo-zapadnogo Darvaza:* Akad. Nauk SSSR, Doklady, v. 144, p. 1105-1108. [*About faunal characters of the Lower Permian deposits of southwest Darvas.*]

Waagen, William, 1889-91, *Salt Range fossils. IV. Geological results:* Geol. Survey India, Mem., Palaeont. Indica, ser. 13, v. 4, pt. 1, p. 1-88 (1889); v. 4, pt. 2, p. 89-242 (1891).

Wass, R. E., 1969, *Australian Permian Polyzoan faunas: distribution and implications:* in Stratigraphy and palaeontology: Essays in honour of Dorothy Hill, K. W. S. Campbell (ed.), p. 236-245, National University Press (Canberra, Australia).

Waterhouse, B. J., 1969, *World correlations of New Zealand Permian Stages:* New Zealand Jour. Geol. Geophys., v. 12, p. 713-737.——1972a, Review of "Stratigraphic boundary problems: Permian and Triassic of West Pakistan": Jour. Paleontology, v. 46, p. 158-160.——1972b, *The evolution, correlation, and paleogeographic significance of the Permian ammonoid family Cyclolobidae:* Lethaia, v. 5, p. 251-270.

——, & Bonham-Carter, G. F., 1975, *Global distribution and character of Permian biomes based on brachiopod assemblages:* Canad. Jour. Earth Sci., v. 12, p. 1085-1146.

Williams, Alwyn, & others, 1965, in R. C. Moore (ed.), Treatise on invertebrate paleontology, Part H, Brachiopoda, 2 v., 927 p., Univ. Kansas Press & Geol. Soc. America (Lawrence, Kansas; New York).

Yanshin, A. L., 1965, *Tectonic structure of Eurasia:* Geotektonika, no. 5, p. 7-35 (transl. by D. A. Brown, Australian National Univ.).

TRIASSIC[1]

By BERNHARD KUMMEL

[Harvard University, Cambridge]

CONTENTS

	PAGE
INTRODUCTION	A352
STAGES AND ZONES OF THE TRIASSIC	A352
STRATIGRAPHY AND PALEOGEOGRAPHY	A353
Northwestern Europe	A354
North Mediterranean Region	A355
Israel	A357
Iran	A358
Caspian Region	A359
Afghanistan	A359
Pakistan	A360
Himalayas	A360
China and Southeast Asia	A362
Indonesia	A363
Australia	A363
New Guinea	A364
New Zealand	A364
Japan	A364
Northeastern Siberia	A365
Svalbard	A365
Greenland	A366
Arctic Canada	A367
Western Canada	A368
Western United States	A368
South America	A368
Antarctica	A369
Summary	A369
TRIASSIC FAUNAS	A369
Ammonoids	A370
Nautiloids	A371
Bivalves	A372
Gastropods	A377
Brachiopods	A378
Conodonts	A382
REFERENCES	A384

---

[1] Manuscript received July, 1976.

INTRODUCTION

The term Trias, later modified in English to Triassic, was proposed in 1834 by F. von Alberti for a sequence of strata in central Germany lying above Permian (Zechstein) and below Jurassic (Lias) rocks of marine origin. The name refers to a threefold division of the strata into a lower unit of nonmarine red beds (Buntsandstein or Bunter), a middle unit of marine limestone, sandstone, and shale (Muschelkalk), and an upper unit of nonmarine rocks (Keuper) that are similar to the lower division. The original definition of the Trias by von Alberti reads as follows (translation from Wilmarth, 1925, p. 64): "Whoever examines more closely the foregoing analysis and tabulates all the fossils of the three hitherto separate formations; whoever examines, further, the transition of the different forms one into the other, and, indeed, considers the entire structure of the mountains and the decidedly different character of the fossils of the Zechstein (Permian) from those of the Lias (Lower Jurassic), will realize that the Bunter sandstone, Muschelkalk and Keuper are the result of one period, their fossils, to use E. de Beaumont's words, being the thermometer of a geological period; that their separation into three formations is not appropriate, and that it is more in accord with the concept of a 'formation' to unite them into a single formation, which I shall provisionally name *Trias*."

This type of strata described by von Alberti is typical of Triassic strata of northern Europe, France, Spain, and north Africa; it is commonly known as the Germanic facies. In contrast to the predominantly continental Germanic facies, there is in the Alps a complete fossiliferous sequence covering all of Triassic time that is commonly called the Alpine facies and that, with the addition of units in the lower beds from southern Asia, forms the primary standard sequence of stages and zones.

This contribution is concerned solely with marine facies and invertebrates of the Triassic. A recent review of Triassic tetrapods was given by Cox (1973).

STAGES AND ZONES OF THE TRIASSIC

The first comprehensive proposal for subdivision of the Triassic into series, stages, and zones was by von Mojsisovics, Waagen, and Diener (1895). The sequence for what is now the Middle and Upper Triassic was based on Alpine faunas and formations, that for the Lower Triassic on Salt Range and Himalayan data. This proposal was not free of serious errors due largely to what is now recognized as misinterpretation of structural relations and nonrecognition of some condensed faunas; however, since 1895 there have been continuing efforts to clarify the stage and zonal scheme. In recent years thick deposits of fossiliferous Triassic strata which are uncomplicated by difficult structural relations have been recognized in British Columbia and Nevada. These areas have yielded equivalents of most of the faunas first discovered in India and Siberia. These sequences have been summarized by Tozer (1967) and Silberling and Tozer (1968) and it is now well established that for Triassic studies, the western American sequences are equal in importance to the Alpine-Mediterranean sequence. The Alpine and western American zonal sequences and stages are shown in Figure 1.

Most stages of the Triassic have been fairly well stabilized for the past several decades. There are, of course, constant refinements as gaps in sequences are recognized and correlations more firmly established. The best review of the general status of Triassic stages and zones has been presented by Tozer (1974).

The Middle and Upper Triassic stage names have been universally accepted; this is not so for the Lower Triassic. Until a couple of decades ago, Scythian was used by many as the Lower Triassic stage name. The term was originally used by von Mojsisovics, Waagen, and Diener (1895) as a series. The first attempt to divide the Lower Triassic into more than one stage was by KIPARISOVA and POPOV (1956) who suggested two stages (Induan and Olenekian) for this "series." This idea was elaborated on in contributions published by the same authors (KIPARISOVA & POPOV, 1961, 1964a,b). TOZER (1965), on the basis of fine fossiliferous sequences in arctic Canada, proposed a four-stage nomenclature for the Lower Triassic (Griesbachian, Dienerian, Smithian and Spathian). VAVILOV and LOZOVSKIY (1970) maintained the twofold division of KIPARISOVA and POPOV but in addition divided the Olenekian into two substages. Recently ZAKHAROV (1973, 1974) has proposed a three-fold division (Induan, Ussurian, Russian).

It is not possible in the scope of this paper to discuss in detail the respective merits of these various schemes of classification; however, I believe the four-fold division proposed by TOZER to have the most merit, and for this paper and in the revisions of the Triassic ammonites for the *Treatise* I have chosen to consider Griesbachian, Dienerian, Smithian, and Spathian as substages of the Scythian stage.

STRATIGRAPHY AND PALEOGEOGRAPHY

"The distribution of the continents and seas in Eotriassic times, as Diener has shown, was probably not unlike that of the present day, except for the extended Thetis" (SPATH, 1930, p. 87). The Diener paper referred to by SPATH was his well-known work *"Die marinen Reiche der Triasperiode"* (DIENER, 1915). The statement is essentially correct though written long before the advent of plate tectonics. One needs to keep in mind that in the circum-Atlantic area marine Triassic is known only from Svalbard, east coast of Greenland (in a northward opening embayment), and near the tip of the Antarctic Peninsula. In all the intervening region none has been recorded. At the same time it is intriguing to keep in mind the presence of presumed Middle Triassic conodonts in Upper Cretaceous rock from the Cameroons of western Africa. The distribution of land, sea, and volcanic area for the Upper Triassic is shown in Figure 2, plotted on a modern-day map. A number of paleogeographic maps for the Triassic based on plate tectonics and paleomagnetism have been published. The most recent and probably the best of these is by BRIDEN, DREWRY, and SMITH (1974).

The paleogeography of the Triassic, disregarding the state of the Atlantic region, is basically a boreal sea impinging on North America, with an embayment along the east coast of Greenland, on Svalbard and on the northeastern Soviet Union from the Taymyr Peninsula to the sea of Okhotsk. Geosynclinal conditions prevailed along much of the western part of North and South America. Similar conditions were to be found along the eastern Pacific margin in Japan, New Guinea, New Caledonia, and New Zealand. Australia had very restricted marginal marine encroachments during the Early Triassic. Marine Triassic has now been reported from the Antarctic Peninsula and has been known for some time in northern Madagascar. Finally there is the great Tethys. From eastern Spain to Indonesia there are now known many dozens of well-documented areas of marine Triassic strata, many containing rich and diverse faunas. The northern boundary of Tethys lay north of the Alps, through the southern part of the USSR eastward to southern China. The southern boundary lay across northern Africa and the northern portion of the Indian shield. Much of northwestern Europe was occupied by marine seas in Middle Triassic time. A comprehensive review of the biostratigraphy of the Permian and Triassic including 35 charts, one of which is a world map of deposits of this age, has been prepared by ANDERSON and ANDERSON (1970) and ANDERSON (1973).

In the following section each of the main paleogeographic regions will be discussed in terms of the thickness and facies represented, with some general comments on the completeness of the faunal record. Due to limitations of space I have chosen to discuss in slightly more detail areas where stratigraphic and paleontologic data have only

Faunal Zones in the Alpine Region (after Zapfe, 1974, with supplements)			Ammonoid Zones in North America (after Tozer, 1971)	
RHAETIAN	Choristoceras marshi	Austrirynchia cornigera	Choristoceras marshi	RHAETIAN
		Rhaetavicula contorta		
NORIAN	Rhabdoceras suessi	Cladiscites tornatus	Rhabdoceras suessi	UPPER NORIAN
		Cochloceras		
		Cycloceltites		
		"Halorites horizon"	Himavatites columbianus	MIDDLE NORIAN
		Cyrtopleurites bicrenatus / Didymites Drepanites	Drepanites rutherfordi	
		Juvavites magnus	Juvavites magnus	LOWER NORIAN
		Malayites paulckei	Malayites dawsoni	
		Mojsisovicsites kerri	Mojsisovicsites kerri	
CARNIAN		Anatropites spinosus	Klamathites macrolobatus	UPPER CARNIAN
		Tropites subbullatus	Tropites welleri	
		Tropites dilleri	Tropites dilleri	
	Trachyceras austriacum	Trachyceras Sirenites Joannites Coroceras Carnites	Sirenites nanseni	LOWER CARNIAN
	Trachyceras aon	Trachyceras Protrachyceras Lobites Joannites	Trachyceras obesum	

FIG. 1. Sequence of zones and stages for Alpine area and western America (adapted from Zapfe, 1974).

recently been acquired. This has necessitated treating some areas that have been studied for long periods of time and where facies relations are extremely complex, for example, the Alpine area and western United States, in short general summary statements. The Alpine area is, of course, the birthplace of marine Triassic stratigraphy, but I feel justice could not be done in the space allowed.

NORTHWESTERN EUROPE

Lower and Upper Triassic strata of northwestern Europe are essentially nonmarine, mostly red, and with evaporites. The Middle Triassic (Muschelkalk) is marine. The facies are limestone and dolomite, some sandy, and sandstone units and some evaporite. These deposits were laid down in an inland sea with some connections to the Tethys in the south. Recent drilling programs in the North Sea show the Triassic there to be essentially nonmarine detrital facies with evaporites (BRENNAND, 1975; P. A. ZIEGLER, 1975; W. H. ZIEGLER, 1975).

The Muschelkalk fauna, as is usual with

Faunal Zones in the Alpine Region (after Zapfe, 1974, with supplements)			Ammonoid Zones in North America (after Tozer, 1971)	
LADINIAN	Protrachyceras archelaus	Protrachyceras	Frankites sutherlandi	UPPER LADINIAN
		Arpadites	Maclearnoceras maclearni	
		Joannites	Meginoceras meginae	
	"Protrachyceras" reitzi	"Protrachyceras"	Progonoceratites poseidon	LOWER LADINIAN
		Anolcites	Protrachyceras subasperum	
		Hungarites		
ANISIAN	Aplococeras avisianum	Aplococeras	Gymnotoceras occidentalis	UPPER ANISIAN
		Hungarites		
		Kellnerites	Gymnotoceras meeki	
		"Ceratites" ex gr. subnodosi		
	Paraceratites trinodosus	Paraceratites	Gymnotoceras rotelliforme	
		Bulogites		
		Seminorites		
		Ptychites		
		Flexoptychites		
		Beyrichites		
	Paraceratites binodosus	Acrochordiceras	Balatonites shoshonensis	MIDDLE ANISIAN
		Balatonites		
		Ptychites		
		Norites	Acrochordiceras hyatti	
		Beyrichites		
		Dadocrinus gracilis	Lenotropites carus	LOWER ANISIAN
		beginning of Physoporella pauciforata		
SCYTHIAN	Eumorphotis inaequicostatus E. telleri	no ammonites	Neopopanoceras haugi	SPATHIAN
		Tirolites cassianus	Subcolumbites beds	
		Tirolites carniolicus	Columbites & Tirolites beds	
			Wasatchites tardus Euflemingites romunderi	SMITHIAN
	Claraia aurita	Eumorphotis venetiana	Vavilovites sverdrupi Proptychites candidus	DIENERIAN
			Proptychites strigatus Ophiceras commune	UPPER GRIESBACHIAN
		Claraia clarai	Otoceras boreale Otoceras concavum	LOWER GRIESBACHIAN

Fig. 1. *(Continued from facing page.)*

inland seas, is characterized by numerous individuals but few species. The faunas consist mainly of ammonites *(Ceratites)*, crinoids, bivalves, and brachiopods.

NORTH MEDITERRANEAN REGION

Triassic studies of the marine facies be-

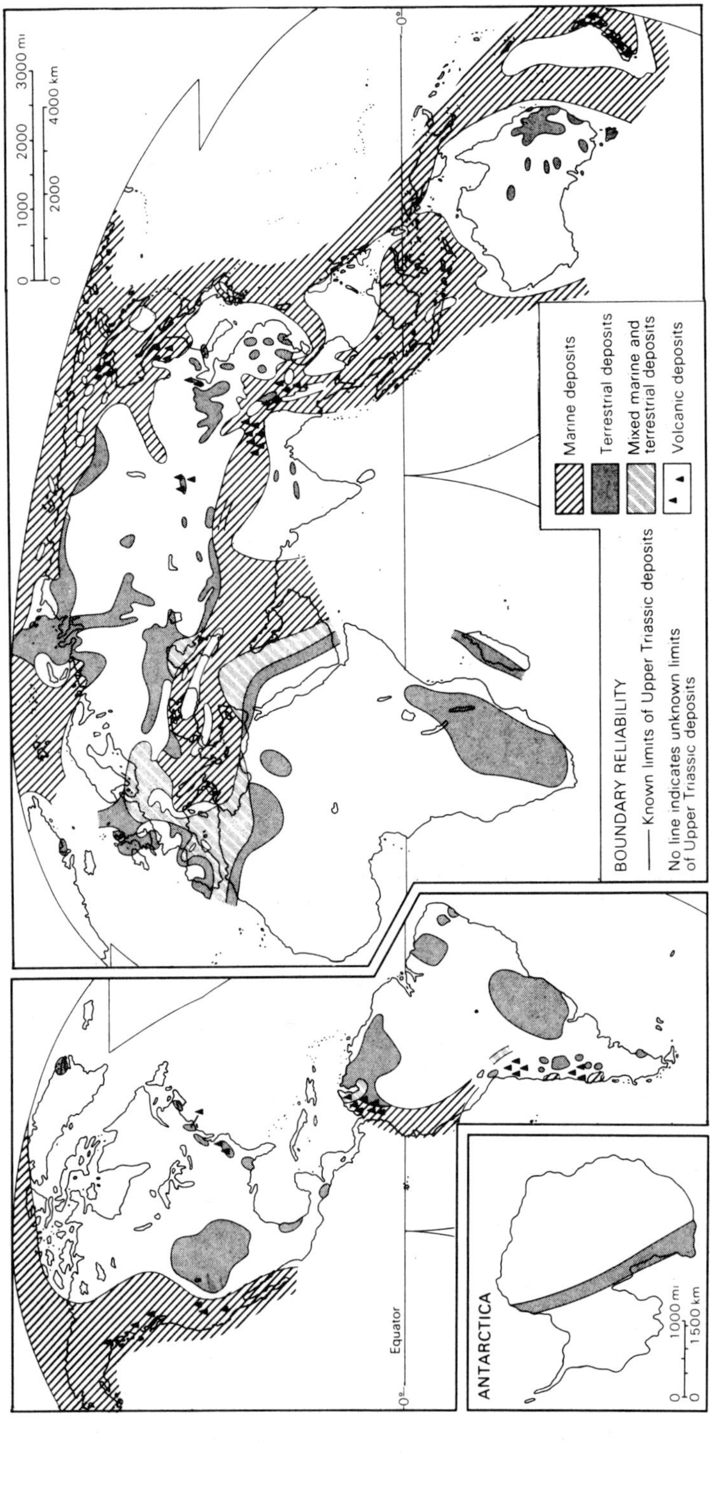

Fig. 2. Distribution of land, sea, and volcanic deposits for the Late Triassic. ("Triassic Period," Encyclopaedia Britannica, published with permission. Modified figure.)

gan in the Alpine region and the area has continued to be a focal point for such studies. The extremely interesting early phase of research on the Triassic of this region was ably summarized by von Zittel (1901). The most recent updating on Triassic studies in the Alpine-North Mediterranean region was presented in a symposium held in Vienna, May, 1973, edited by H. Zapfe (1974).

The vast variety of facies and the extremely complex tectonic history of the region have made interpretations of sequence and correlation often very difficult. Even so, the biostratigraphic framework worked up from studies of faunal sequences in the region continues to be of premier importance, although now the sequences of western and arctic North America are becoming equally important.

Paleogeographic interpretations for the Triassic of the Alpine-North Mediterranean region are extremely difficult because of present structural relations and complexity of facies. It is not possible in limited space to do justice to this important region, thus the following remarks are in the nature of a brief summary.

The Lower Triassic of the Alpine-North Mediterranean region is represented by the Werfen Formation. This unit is highly variable in lithology, including marly limestone, oolitic limestone, evaporite, siltstone, shale, and sandstone. The lower part of the Werfen contains the common lower Scythian bivalve *Claraia*. The upper part contains the well-known *Tirolites* fauna monographed by Kittl (1903) and revised by Kummel (1969). The Werfen ammonoid fauna consists of only 11 genera, six of these being endemic to the Alpine-Mediterranean region. This fauna is considered to be of late Scythian age. A very instructive review of the Werfen Formation of the southern Alps has recently been published by Assereto et al. (1973).

The Anisian, Ladinian, and Carnian were times of great paleogeographic differentiation as reflected in the great diversity of facies. The dominant facies are carbonates, both limestone and dolomite, with extensive reef development. In addition, there are detrital formations, clay shale to conglomerate, evaporite, and volcanics. Many of these units are very fossiliferous.

The remainder of the Triassic is represented primarily by thick dolomite and limestone formations, forming reef complexes in many places. Most of these units are generally sparsely fossiliferous; however, in the Hallstatt facies, consisting of red mottled marble, cephalopods are quite abundant.

ISRAEL

Excellent summaries of the Triassic of Israel have been contributed by Druckman (1974) and Picard and Flexer (1974). The Triassic in southern Israel consists of nearly 1,000 m of sedimentary rocks, which are partly exposed in Makhtesh Ramon, Har 'Arif, 'Araif e-Naga, Zarqa Ma'in, Wadi Hisban, and Nahr e-Zerqa and partly or entirely penetrated by a number of deep wells.

The sequence consists of five formations (from bottom to top) as follows: The Zafir Formation of Scythian age, consisting of alternating shale, fossiliferous limestone and sandstone; the Ra'af Formation of late Scythian-early Anisian age, consisting mainly of fossiliferous limestone; the Gevanim Formation of Anisian age, consisting of sandstone, siltstone and shale with minor amounts of limestone; the Saharonim Formation of late Anisian-Carnian age, consisting of fossiliferous limestone, dolomite and sparse gypsum and anhydrite layers; the Mohilla Formation of Carnian-Norian age, consisting of anhydrite and dolomite.

This Triassic sequence was deposited in various environments of deposition including near-shore and shallow-marine environments, very shallow lagoons, and shoaling tidal flats.

The source area for the detrital influx during the deposition of the Zafir and Gevanim formations was situated in the southeast and was, probably, part of the Arabo-Nubian massif. The sea was situated to the northwest, the direction from which it transgressed during the deposition of the Ra'af and Saharonim formations.

The Triassic ended with a period of regional uplift, subaereal exposure, slight erosion, and development of lateritic soils.

IRAN

Considerable progress has been made in the past decade or so in the study of the Triassic of Iran. The data have been very well summarized by Seyed-Emami (1971). Triassic strata outcrop in north Iran, in the Julfa region and the Alborz Mountains, in central Iran, in the Zagros Mountains, the Nakhlak region of north-central Iran, and at Agh-Darband in extreme northeast Iran.

The famous Permian-Triassic strata of the Julfa region and adjoining Azerbaijan region of the Soviet Union have received considerable attention in recent years because of the conformable sequence of uppermost Permian and lowest Triassic rocks. The stratigraphy and paleontology of these strata cropping out north of the Aras River (in Azerbaijan) have been dealt with by Ruzhentsev and Sarycheva (1965) and more recently by Rostovtsev and Azaryan (1973). South of the Aras River at Kuh-e-Ali Bashi equally comprehensive studies have been published by Stepanov, Golshani, and Stöcklin (1969), Teichert, Kummel, and Sweet (1973), and Kummel and Teichert (1973). Ruzhentsev and Sarycheva (1965) and Stepanov, Golshani, and Stöcklin (1969) came to the conclusion that the so-called "transitional beds" with *Paratirolites, Shevyrevites, Dzhulfites,* and *Iranites* were Early Triassic in age. In contrast, Teichert, Kummel, and Sweet (1973), Kummel (1973), and Rostovtsev and Azaryan (1973) presented convincing arguments for the late Permian age of these beds.

Above the "transitional beds" at Kuh-e-Ali Bashi are limestone and dolomite units, the Elikah Formation, which are also well developed in the Alborz Mountains, and contain *Claraia* in the lower part. Recently Rostovtsev and Azaryan (1973) reported finding *Ophiceras* and *Gyronites* ammonites from correlative strata in Soviet Dzhulfa that clearly establish an Early Triassic (Scythian) age. The middle and upper parts of the Elikah Formation in Iran are much less fossiliferous and the record to date is ambiguous; however, evidence indicates that the formation extends into the Norian. The upper surface of the Elikah Formation is a widespread disconformity, and Seyed-Emami (1971) concluded that the absence of Upper Triassic rocks in most parts of north Iran is due to widespread pre-Liassic erosion rather than to nondeposition.

A sequence of Permian-Triassic formations very similar to that at Julfa occurs in the Abadeh region of central Iran northeast of the main Zagros thrust zone, between Esfahan and Shiraz. The geology and stratigraphy of the Abadeh region have been thoroughly treated by Taraz (1969, 1971a, 1971b, 1972, 1973, 1974).

In the eastern part of central Iran the Lower Triassic consists of reddish calcareous and argillaceous shales containing thin intercalations of limestone and dolomite. This unit is known as the Sorkh Shale Formation and is overlain by well-bedded, light-colored dolomite of the Shotori Formation. These two formations may reach a thickness of 1,000 m and are overlain disconformably by the Upper Triassic Nayband Formation; they are considered, from their stratigraphic position, to be Lower to Middle Triassic and to correspond probably to the Elikah Formation in north Iran. The Nayband Formation consists of dark shale, sandstone, and fossiliferous limestone intercalations, with a total thickness of 2,800 m. The contact of the upper Nayband Formation with the overlying Shemshak Formation (Rhaetic?-Liassic) is gradational. Paleontological analysis of the Nayband Formation is as yet incomplete, but it appears to include much of the upper Ladinian to Rhaetian (Seyed-Emami, 1975).

In the Zagros Mountains of south and southwest Iran, the Triassic consists of evenly bedded, grayish dolomite having a thickness of up to 400 m, known as the Khaneh Kat Formation. Paleontological data are incomplete, but parts of the Lower, Middle, and perhaps Upper Triassic are represented.

One of the newer Triassic discoveries in Iran is in the Nakhlak region of north-central Iran, which has yielded abundant ammonoids in contrast to most of the other regions previously discussed. This Triassic sequence, named the Nakhlak Group, attains a thickness of 2,500 m and is divided into three formations (Davoudzadeh & Seyed-Emami, 1972). The lower unit (Alem Formation) consists of, from bottom to top, thin- to well-bedded sandy limestone, red and violet shales containing ammonoids, gray-green tuffaceous limestone, gray nodular limestone, and olive shale with big ceratites. Ammonoid studies by TOZER (1972) indicated that this unit is late Scythian-Anisian in age. The overlying Baqorog Formation consists of sandstone and coarse-grained conglomerate. Presence of the conglomerate is interpreted as evidence of diastrophic movements in the Ladinian. This unit has not yielded fossils, but on the basis of stratigraphic position is thought to be early and middle Ladinian in age. The uppermost unit of the Nakhlak Group (Ashin Formation) consists of dark sandstone and shale with a few intercalations of sandy limestones. Only a few fossils have been found to date but these indicate a late Ladinian age (TOZER, 1972).

From the northeastern corner of Iran, in the area of Agh-Darband (east of Mashad), another Triassic sequence is known, which has not been studied in detail, but lithologically resembles the Nakhlak Group. It has yielded Anisian and Carnian ammonoids.

CASPIAN REGION

In the Caspian depression are found as much as 2,500 m of Triassic strata. The Lower Triassic consists of red and varicolored clays, siltstone, sandstone, and conglomerate with interbedded, thin, fossiliferous marine limestone. The Middle Triassic is essentially the same. The Upper Triassic (600-800 m) consists of red and gray clay with interbeds of sandstone, pebble beds, and coal. All of the Upper Triassic is of continental origin.

Much of the Permian and Triassic of the Caspian depression is known only from the subsurface; however, there are a few areas of outcrop, and one of these is at Mt. Bogdo where the first Lower Triassic ammonite, *Ammonites bogdoanus* VON BUCH (1831), was described.

In the Mangyshlak Peninsula, along the east side of the Caspian, are very thick sequences of fossiliferous Lower Triassic strata. Several horizons contain ammonites, all of late Scythian age (SHEVYREV, 1968). The lower part of the sequence has unfossiliferous red beds, but the beds above are fossiliferous non-red shale, sandstone, and limestone. The faunas of the thin marine intercalations of the Caspian depression are the same as those of Mangyshlak.

AFGHANISTAN

In recent years a number of stratigraphical and paleontological studies of Triassic strata, all within approximately 100 km of Kabul, have been published. A useful summary of the Triassic of Afghanistan has been published by KAEVER (1969).

At Kotal-e-Tera, near the village of Altimur, 90 km southeast of Kabul, KUMMEL and ERBEN (1968) and KUMMEL (1968) described a relatively thin sequence of ammonitiferous limestone. The lowermost of these faunas represents the middle Scythian (Smithian), above this is another horizon of ammonoids representing the late Scythian (Spathian). The uppermost part of the sequence at Kotal-e-Tera has yielded only a few specimens of ammonoids, but these are clearly of Anisian age. From the Azras Valley, Paktia Province, just east of Kotal-e-Tera, COLLIGNON (1973) described a fine suite of Triassic ammonoids containing middle Scythian faunas as at Kotal-e-Tera, as well as one fauna of Anisian age and one of Ladinian age.

FISCHER (1971) published a detailed description of the Kohe Safi region, just east of Kabul, from where HAYDEN (1911) described the Kingil Series. This unit, essentially all carbonates, is 1,420 m thick and ranges in age from Artinskian through Oxfordian. The Triassic portion is 487 m thick. The lower 44 m of the Triassic rocks contain ammonites of two distinct zones. The lowest of these faunas is of early Scythian (Dienerian) age, described by ISHII, FISCHER, and BANDO (1971). Above this occurs the middle Scythian (Smithian) fauna as at Kotal-e-Tera. The upper part of this Triassic sequence is essentially unfossiliferous except for the upper 28 m, which contains *Megalodon* of latest Triassic age.

Approximately 100 km northwest of Kabul in the Khenjan region of the western Hindu Kush are 150 m of detrital strata containing a rich bivalve fauna of Ladinian age, recently described by FARSAN (1972, 1975).

Along the northern flank of the western Hindu Kush is a detrital and volcanic sequence named the Doab Series by HAYDEN (1911). This unit rests unconformably on fusulinid limestone of Permian age and is overlain unconformably by the Jurassic Saighan Series. The upper part of the Doab Series consists of assorted volcanic deposits with a few interbeds of marl and sandstone. Fossil plants of Rhaetic age are known from the uppermost part of the series. The lower Doab Series, however, consists only of sedimentary units containing Middle Triassic ammonoids (FURON & ROSSET, 1951).

PAKISTAN

One of the most famous regions of outcropping Triassic formations is in the Salt Range of Pakistan. These formations have played a particularly important role in the development of our Triassic zonal scheme, especially for the Scythian stage. In addition, interest in these formations is heightened because they conformably overlie upper Permian formations, and because of this they have attracted the attention of nearly every student of the causes of abrupt faunal breaks at the Permian-Triassic boundary.

The early studies of the Triassic of the Salt Range by WYNNE (1878) and WAAGEN (1895) concluded that the sequence encompassed all of the Lower Triassic and even part of the Middle Triassic. It was not until much later that it was realized that the sequence and faunas as known by WAAGEN (1895) included only the lower half of the Scythian.

More recently new stratigraphic studies incorporating new paleontological data have been reported by KUMMEL (1966b, 1970) and KUMMEL and TEICHERT (1970b). It is now clear that WAAGEN's work was largely confined to the central part of the Salt Range where he recognized, from bottom to top, the following units: the Lower Ceratite Limestone, the Ceratite Marl, the Ceratite Sandstone, and the Upper Ceratite Limestone. West of the central part of the Salt Range, that is, west of Nammal Gorge, the "Lower Ceratite Limestone" is present but the remainder of the Triassic section is shale with some sandstone facies. The section as a whole thickens westward.

A modern nomenclature for some of the Permian and Triassic formations has been introduced by KUMMEL and TEICHERT (1966). As mentioned above, the Salt Range Triassic sequence as known to WAAGEN encompasses only the lower half of the Scythian. Interestingly enough the lowest unit of the sequence, a dolomite unit (the Kathwai Member) was not recognized by WAAGEN. SCHINDEWOLF (1934) described *Ophiceras connectens* from the lowest part of the Kathwai. KUMMEL and TEICHERT (1970a) have published a detailed account of the stratigraphy and paleontology of the uppermost Permian and the lowest Triassic beds (the Kathwai Member).

KUMMEL (1966b) recognized a shale and sandstone unit (Narmia Member) above the Upper Ceratite Limestone from which he described a small fauna of ammonoids of late Scythian age; thus, it now appears that the Salt Range sequence includes essentially all of the Scythian. Above the Narmia Member are shale and sandstone units (Tredian Formation) that unfortunately are unfossiliferous.

Thick sequences of mainly fine-detrital rocks are in the Quetta and Hindubagh region of west central Pakistan. Unfortunately, these strata are not very fossiliferous. According to published reports only six specimens of ammonoids have been recorded, few of which were found in place (DIENER, 1906; KUMMEL, 1966a). In addition to the ammonoids these strata contain a species of *Monotis,* all indicating a Late Triassic age.

In recent years, Dr. A. N. FATMI of the Pakistan Geological Survey has collected a middle Scythian ammonoid fauna from the Quetta region (personal communication). This fauna has as yet not been described and documented.

HIMALAYAS

Triassic formations crop out in a broad belt extending from Kashmir through Spiti, Garhwal, and Kumaon, into Nepal and adjoining regions of Tibet. The best summaries of data on these formations are by

DIENER (1912), PASCOE (1959), and GANSSER (1964). The facies are marine, dark shale and limestone with no igneous rocks, except in Kashmir. The limestone is of dark or gray color, well bedded, and in some horizons either concretionary or dolomitic. In the majority of sections there is a remarkable contrast between the light-gray dolomitic limestone units of the upper portion of the Triassic section and the dark-colored shale and limestone of the lower portion. This normal development of the Triassic in the Himalayas is also characterized by the regular distribution of each single horizon over a comparatively large area, and by the absence of facies of red limestone and marble. In the Kiogar area at the border between Kumaon and Tibet exotic blocks occur at Malla Johar where the Triassic is developed in a facies considerably different from that in the main region of the Himalayas. In this region the Triassic is much thinner, and most of the Triassic horizons are developed in a facies of red limestone and marble showing a remarkable resemblance to the Hallstatt limestone of the eastern Alps. This assemblage is known as the Tibetan facies and that discussed above, as the Himalayan facies (VON KRAFFT, 1902).

The two best known sections of the Triassic of Himalayan facies are at Spiti and Painkhanda. A third section at Byans, to the southeast of Painkhanda and near the Nepal border, is also important but less well known.

The Scythian beds of Spiti and Painkhanda are of approximately equal thickness (12.5 m) and lithology (gray limestone and shale). At Byans the Scythian is 47 m thick and consists of the Chocolate Limestone. The Anisian of Spiti and Painkhanda is also of similar facies and thickness (31 m). In Byans the Anisian is of a purer limestone facies.

Ladinian strata are thick (93 m) and richly fossiliferous at Spiti, but at Painkhanda they are very thin and poorly fossiliferous. They have not been traced further east. The same marked decrease in thickness to the southeast is seen in the Carnian strata. At Spiti the Carnian is 480 m, at Painkhanda 248 m thick, and in Byans very thin. In Spiti the Carnian is quite shaly and very fossiliferous. The shaly beds are replaced by limestone to the southeast.

The lower and middle Norian is approximately 310 m thick throughout this region and consists of shale, limestone, and sandstone, generally quite fossiliferous. In Byans on the other hand, the lower and middle Norian consists mainly of black shale. The upper Norian throughout the region consists of massive limestone that contains *Megalodon* in the lower parts, passing upward into limestone of middle Jurassic age. The paleontological data on these Triassic units of the central Himalayas have been compiled by DIENER (1895, 1897, 1906, 1907, 1908, 1909) and VON KRAFFT and DIENER (1909).

The Triassic strata of Tibetan facies as represented at Malla Johar and the other regions are quite different from the corresponding deposits in the Himalayan region. The strata are known entirely from exotic blocks of which there are two distinct types. One is represented by massive gray, dolomitic limestone resembling upper Norian rocks of the Himalayan facies; however, the latter is well bedded, not massive. This rock type is also unfossiliferous. The other type of exotic blocks is red limestone and marble resembling the Hallstatt facies of the Alpine region. The faunas of the Scythian and Anisian have strong affinities to corresponding strata of the Himalayan facies. The Carnian faunas, on the other hand, have strong affinities to the Triassic Mediterranean faunas.

To the northwest of this central Himalayan region, in Kashmir, the Triassic is also well developed and not too unlike that discussed above, except for the intercalations of the Panjal Traps, which range in age from Permian to Late Triassic. The Lower Triassic appears to be conformable with the underlying Permian and consists of dark-colored shale and limestone. A detailed study of the upper Permian and lower Triassic strata has recently been published by NAKAZAWA et al. (1975). The Anisian and Ladinian are characterized by gray shale, nodular, concretionary and platy limestone and some sandy limestone. The

Upper Triassic strata become progressively more calcareous and the Norian consists of massive dolomite and limestone, practically barren of fossils.

Recent extensive studies in Nepal by G. FUCHS (1964, 1967) have added greatly to our stratigraphic knowledge of this region. The Triassic in central Nepal attains a thickness of 500 to 1,400 m. The Lower Triassic consists of gray limestone, shale, and dark nodular limestone. The Middle Triassic and Carnian are represented by blue limestone, marl, and dark shale, grading upward into brown to black shale, siltstone and sandstone with some concretions. The limited fossil data available indicate a Norian age. The uppermost part of the sequence is sandstone that grades upward into massive limestone. These units are believed to be latest Triassic in age and appear to grade into rocks of Jurassic age.

CHINA AND SOUTHEAST ASIA

Marine Triassic strata are largely confined to the southeastern quarter of China with, of course, a connection westward to Tethys. Paleogeographic maps for the three series of the Triassic have been compiled by LIU (1959). These maps clearly show a progressive decrease in the area of marine seas from the Early to the Late Triassic. The Lower Triassic has received a fair amount of attention by paleontologists and stratigraphers, but that is not so for the Middle and Upper Triassic. The Lower Triassic consists of carbonate, detrital, and mixed facies with fairly abundant ammonoid and bivalve faunas. The most comprehensive monograph on Scythian ammonoids is by CHAO (1959).

The Triassic is as yet incompletely known in Burma. In the Arakan Yoma are thick sections of fine-grained detrital rocks, with some limestone that contains Upper Triassic fossils, mainly bivalves such as *Halobia* and *Monotis*. SAHNI (1938) reported a lowest Scythian fauna including *Glyptophiceras, Lytophiceras, Vishnuites,* and others from an argillaceous limestone and shale sequence within dolomites at Na Kham in the northern Shan States. Unfortunately no additional data aside from SAHNI's abstract are available. Also known from the northern Shan States are the Napeng beds consisting of argillaceous, yellow shale and marl with some limestone. The fossils from these strata consist of bivalves, including *Avicula contorta, Myophoria, Gervillia praecursor, Pecten, Modiolopsis,* and others suggesting a Rhaetic age (HEALEY, 1908).

The Triassic System of Malaysia and Thailand has been carefully reviewed by TAMURA et al. (1975). Prior to World War II, data were very sparse, but since the war there has been a tremendous intensification of field stratigraphic and paleontologic studies. The region, however, is structurally very complex, and the results to date are mainly the identification and delineation of stratigraphic units and to some extent description of their faunas. Geosynclinal conditions prevailed in these regions during the Triassic. The rock facies are extremely varied including fine to coarse detrital sediments, chert, limestone, and volcanics. TAMURA et al. (1975) included a correlation chart of the Triassic for the region and a detailed bibliography. On the basis of ammonoid, bivalve, and conodont studies, portions of all stages of the Triassic have been identified. The descriptive work on these fossils is most advanced on the bivalves, followed by the conodonts, but many of the ammonoids still remain to be described. The Malayan ammonoids include upper Scythian, upper Anisian, upper Ladinian, and lower and upper Norian faunas. In Thailand, on the contrary, only upper Anisian and lower Carnian faunas have been identified to date. Bivalves are much more profuse and more widely distributed than ammonoids. The Scythian *Claraia,* Anisian-Norian myophoriids, Ladinian-Carnian *Daonella,* and Carnian-Norian *Halobia* are important forms. Scythian, lower Anisian, and lower Carnian conodont faunas have been identified in the Malay Peninsula. The literature on these faunas from Malaysia and Thailand is extremely extensive but is all listed in the bibliography of TAMURA et al. (1975).

The Triassic of North Viet Nam is very poorly known. A report by KHÚC et al. (1965) on a fairly large assemblage of fossil faunas, mainly bivalves, is a valuable contribution. These authors were able to identify the presence of portions of all stages of the Triassic.

INDONESIA

Timor is one of the most remarkable fossiliferous Triassic localities in the world from which approximately 1,000 species of marine invertebrates have been described. The ammonoids have been monographed by WELTER (1914, 1915, 1922), DIENER (1923), ARTHABER (1927), and PAKUCKAS (1928). Essentially all of the faunas described in these and other papers have come from eastern Timor where they occur in exotic blocks, mostly in carbonate facies. In Sumatera (Sumatra), Borneo, Sulawesi (Celebes), and elsewhere, the Triassic consists mainly of detrital facies and is not as fully represented. Since these early studies, very little geological or paleontological work has been done.

Western Timor has received considerable attention in recent years, noted especially by the monograph by AUDLEY-CHARLES (1968) on the geology of western Timor (formerly Portuguese Timor). Paleontological studies have been produced by NAKAZAWA and BANDO (1968) on Scythian-Anisian ammonoids and by NOGAMI (1968) on conodonts. Other studies by these authors are in progress (BANDO, personal communication).

AUSTRALIA

The basic tectonic pattern for the Paleozoic of Australia is the active Tasman geosyncline along the eastern portion of the island continent and a series of basins along and opening on the Indian Ocean on the western side. By the close of the Permian the Tasman geosyncline was in a terminal orogenic state, although no orogenic activity affected the western areas. The interior of the continent had long been emergent. The Late Permian was a time of extensive regression and by the close of the period, the continent was essentially completely emergent.

Marine incursions during the Triassic were limited to restricted areas in the east and west coastal regions of Australia; in the interior of the continent and especially in the east, extensive terrestrial strata are found. In the east, marine Triassic strata are present in the Maryborough basin, Queensland. From the Traveston Formation RUNNEGAR (1969) has recorded a middle Scythian (Smithian) ammonoid fauna containing such genera as *Anaflemingites, Dieneroceras,* and *Flemingites*. Also from within this basin DENMEAD (1964) and P. J. G. FLEMING (1966) recorded a Lower Triassic bivalve fauna from the Brooweena Formation.

In western Australia Lower Triassic fossiliferous, marine detrital strata are present in the Perth and Carnarvon basins and in brackish water sediments in the Canning basin. In the Perth basin outcrops of the Kockatea Shale near Mount Minchin have yielded an extremely interesting assemblage of ammonoids generally poorly preserved as molds. The presence of these fossils was first recorded by EDGELL (1964), who correctly concluded that they belonged to the middle Scythian (Smithian), and this has been further documented by SKWARKO and KUMMEL (1974). The remaining data on the marine Triassic of the Perth basin come from bore holes. The first marine macrofossils of Triassic age in the Australian region were recovered from cores of the Kockatea Shale from the Beagle Ridge (BMR 10) bore. DICKENS and MCTAVISH (1963) described a small group of ammonoids, which they interpreted as early Scythian in age. Later MCTAVISH and DICKENS (1974) extended the age of the Kockatea Shale to include most of the lower half of the Scythian. Beneath the horizon that yielded the ammonoids, specimens of *Claraia* were obtained. The Dongara No. 4 bore is approximately 65 km north of the Beagle Ridge (BMR 10) borehole in the Perth basin. Two cores from the Kockatea Shale yielded the following ammonoids: *Proptychites* sp. indet., *?Koninckites* sp. indet., *?Paranorites* sp. indet., and *Gyronites frequens* (SKWARKO & KUMMEL, 1974). These ammonoids are of early, but not earliest, Scythian age.

Recently MCTAVISH (1973) contributed valuable data from conodont studies of subsurface samples from the Perth and Carnarvon basins. He concluded that these conodont faunas of Triassic age can be correlated with the Lower Triassic conodont zonation proposed for the Salt Range by SWEET (1970a). They range in age from early to late Scythian (Dienerian to Spathian).

Early in 1970 the Burmah Oil Australia Ltd. drilled Sahul Shoal No. 1, and the core has yielded fragments of ammonites and bivalves. Though the material is poorly preserved and fragmentary, SKWARKO and KUMMEL (1974) suggested that one of the ammonoids belongs to the genus *Nicomedites* of Anisian age. A macrofauna of this age was previously unknown in the Australian region.

NEW GUINEA

In recent years a great amount of new data has become available on the Triassic of New Guinea, hitherto almost a complete blank. The first records of marine faunas of this age are by SKWARKO (1967) and since then a series of papers have appeared (SKWARKO, 1973a, 1973b; SKWARKO & KUMMEL, 1974). So far Anisian and Upper Triassic horizons have been identified.

NEW ZEALAND

During the later Paleozoic and early Mesozoic, New Zealand was the site of rapid marine sedimentation in a persistent geosynclinal zone lying between a rising geanticline of Precambrian-Devonian strata, metamorphics and igneous rocks to the west, and the Pacific Ocean to the east (FLEMING, 1962). The thick sequence of sediments laid down in the New Zealand geosyncline range from those now metamorphosed to form the Haast Schists (undated but perhaps mostly Carboniferous) through the Permian to the Hokonui System (Triassic and Jurassic). The western, marginal portion of the geosyncline, referred to as the Hokonui facies, is characterized by very thick moderately fossiliferous, detrital sequences, which are abundantly tuffaceous. Eastward of the Hokonui facies range extremely thick sequences of graywacke and argillite, with some spilitic pillow lava, lenticular limestone and radiolarian chert, known as the Torlesse facies. This portion of the geosyncline is sparsely fossiliferous. The best review of the Mesozoic of New Zealand is by FLEMING (1970).

Ammonoid faunas representing all stages of the Triassic have been reported; however, only a fraction of the standard zones has been recognized. The ammonoids are similar to those of Tethyan and circum-Pacific localities. The brachiopod and bivalve faunas, however, contain a fair percentage of endemic genera and species. Interesting discussions of New Zealand zoogeography have been published by MARWICK (1953) and FLEMING (1967).

JAPAN

The best summary of the Triassic System of Japan is by BANDO (1964); good resumés can also be found in TAKAI *et al.* (1963) and MINATO, GORAI, and HUNAHASHI (1965). Marine Triassic formations of Japan crop out in relatively limited areas on Honshu, Shikoku, and Kyushu islands. The principal areas are in the southern part of the Kitakami area of northern Honshu, the Kwanto area of east-central Honshu, the Maizuru area of west-central Honshu, the Yamaguchi area of southern Honshu, and on Shikoku and Kyushu islands.

In the Kitakami region, fossiliferous Lower and Middle Triassic formations are widely distributed. No Carnian has been recognized and Norian units have very restricted distribution. On the other hand, in the Kwanto area fragmentary parts of the Lower Triassic and some Upper Triassic formations have been identified. No Middle Triassic has been recognized. In the outer zone of southwest Japan, on Shikoku and Kyushu islands, the Triassic sequence is fragmental because of very complicated geologic structure. The main units recognized are the upper part of the Middle Triassic and Upper Triassic formations. Lower Triassic limestone is found as lenticular bodies at only three outcrops. In the inner zone of southwest Japan, the Lower and Middle Triassic are present in the Maizuru area. In the southern part of Honshu Island in the Yamaguchi area, the Triassic is mainly represented by Upper Triassic detrital strata representing interbedded marine and coal-

and plant-bearing terrestrial beds. The marine strata contain typical Upper Triassic bivalves. A very limited extent of Ladinian strata is also found in the region.

In a very general way the Lower and Middle Triassic formations consist of fine-detrital facies (black shale, sandy siltstone, and argillaceous limestone). The Upper Triassic, in contrast, is more characterized by coarse-detrital facies. The boundary between the two facies is drawn between the Ladinian and Carnian, which marks the Akiyoshi orogeny. Very valuable paleogeographic reconstructions for the Triassic of Japan can be seen in MINATO, GORAI, and HUNAHASHI (1965).

The Lower and Middle Triassic formations have yielded many characteristic ammonoids and bivalves. Ammonites from the Upper Triassic are rare in Japan.

NORTHEASTERN SIBERIA

Marine Triassic formations are extensively developed northeast of a line connecting the Taymyr Peninsula and the Primorye Territory around Vladivostok. The facies are almost entirely detrital, consisting of dark-colored sandstone, siltstone, and shale.

Zonal schemes for northeastern Soviet Union and the southern Primorye Territory have been summarized by KIPARISOVA, OKUNEVA, and OLEYNIKOV (1973). Fourteen, mainly generic zones are recognized with approximately two dozen, mainly local zones in northeastern Soviet Union and a dozen in the southern Primorye Territory. The Lower and Middle Triassic zones are based almost entirely on ammonoids whereas in the Upper Triassic many of the local zones are based on species of the bivalves *Otapiria, Oxytoma,* and *Monotis*. On the basis of literature surveys it appears that in general the formations are moderately to sparsely fossiliferous. It also appears that many of the zones have been recognized in as yet only one or a few places. The lowest zones of the Scythian (Griesbachian) are known only from the south Verkhoyansk synclinorium. In some areas, as in the southern Primorye Territory, the next youngest zones of the Scythian (Dienerian) are the basal beds.

In the region around the lower part of the Olenek River the lower Scythian is represented by fine to coarse detrital deposits, with some tuff-containing plant remains. The upper Scythian consists of argillite with limestone bands containing ammonites. The Middle Triassic formations appear to be entirely marine, but in the Upper Triassic continental deposits, some coal bearing, become quite common. In general there is a distinct break or disconformity with the overlying Jurassic strata. A brief treatment of the stratigraphic sequences and faunal zones for northeastern Soviet Union can be found in KIPARISOVA, OKUNEVA, and OLEYNIKOV (1973).

Excellent lithopaleogeographic maps for the Soviet Union are available (VERESHCHAGIN & RONOV, 1968, sheets 1-12) and a comprehensive text on the Triassic System is in KIPARISOVA, RADCHENKO, and GORSKIY (1973).

SVALBARD

Triassic deposits are widespread in Svalbard, cropping out in excellent exposures on the east and west coasts of central and southern Vestspitsbergen, at the south end of Nordavst Land, and underlying the Edge, Barents, and Wilhelm islands, and on the east coast of Bear Island (Bjørnøya). An excellent summary of the Triassic stratigraphy of Svalbard has been published by BUCHAN et al. (1965). Summaries of the biostratigraphy have been presented by TOZER and PARKER (1968) and KORCHINSKAYA (1973).

The Triassic rocks in Svalbard consist of a preponderance of marine shale and siltstone with continental sandstone in the upper part suggesting a platform of epeirogenic environment of deposition. The lower part of the sequence containing a sparse fauna of bivalves consists of fine-grained flaggy sandstone probably formed under shallow marine conditions, and interbedded shale and siltstone. This is followed by the main marine shale sequence which continues into the lower part of the uppermost unit (the Kapp Toscana Formation). The dominant rock type is thin-bedded gray to black bituminous shale, whereas harder, yellow-weathering siltstone is common. The upper part is characterized by the occurrence of red-weathering clay-ironstone nodules. Fossils are fairly abundant—ammonites, bivalves, and vertebrates being the most common. The uppermost part of the Triassic is a nonmarine sequence consisting of gray-green, flaggy, cross-bedded sandstone alternating with sandy shale. Thin coal seams and common plant remains are present and suggest deposition in lagoonal or continental conditions.

Ammonoids and bivalves are the predominant marine invertebrates in the Triassic formations of Svalbard. Ammonoid horizons indicate the presence of some zones from the Lower, Middle, and lower Upper (Carnian) Triassic. Some bivalve faunas indicate the presence of Norian marine horizons in the upper part of the Triassic sequence. The lower part of the Scythian has yielded *Otoceras boreale,* followed by *Claraia* cf. *C. stachei,* and *Proptychites* cf. *P. rosenkrantzi.* The middle and upper Scythian are well represented by ammonoid faunas. The Anisian and Ladinian have yielded varied suites of ammonoid faunas, but as yet relationships and precise correlations for many are ambiguous. The lower Carnian is represented only by a single ammonoid zone. The upper part of the highest Triassic formation (Kapp Toscana Formation) consists of marine and continental facies. From the marine facies KORCHINSKAYA (1973) reported *Halobia* cf. *H. plicosa, H.* cf. *H. norica,* and *H. fallax.* From another locality she reported *Pterotoceras(?) svalbardi.* The continental facies contains plant remains.

GREENLAND

During the late Paleozoic and early Mesozoic northeast Greenland underwent major faulting leading to the formation of a compound system of horst and graben structures. The Triassic basin trended north-northeasterly and its extent was rather similar to that of the Carboniferous-Permian molasse trough. To the west the basin was bounded by a peneplaned high ground of Caledonian folded rocks, including Old Red Sandstone, which contributed abundant detritus into the basin. In the Triassic, sedimentary transport from an eastern highground is also clearly indicated.

The lowest Scythian (Griesbachian) formation consists of 500 to 800 m of marine shale, sandstone, and conglomerate. The formation is fairly fossiliferous, containing in sequence species of *Otoceras, Ophiceras,* and *Proptychites.* The unit is referred to as the Wordie Creek Formation by many authors, but TEICHERT and KUMMEL (1973) recommended abandonment of the name because of ambiguities.

The presence of fossil remains of distinctly late Paleozoic aspect in beds with such typical Lower Triassic forms as *Otoceras* and *"Glyptophiceras"* in the lower part of this Scythian sequence has stimulated considerable interest. This association was first noted by SPATH (1930, 1935) who considered the Paleozoic elements to have been derived and of Carboniferous age. TRÜMPY (1960, 1961), on the other hand, on the basis of his own field work came to the conclusion that the Paleozoic elements were not derived. TEICHERT and KUMMEL (1973, 1976), on the basis of field studies made in the summer of 1967, concluded that the Paleozoic elements are all derived. They considered some of the Paleozoic elements as having been brought into that environment as argillaceous boulders, that once coming to rest, dissolved, leaving well-preserved fossils that were rapidly buried in the coarse sediments, and thus in a free state were transported very little. The majority of fossils, however, were washed out of soft rocks and were badly broken during transportation.

The remainder of the Triassic sequence consists of mudstone and sandstone, with some gypsum and dolomite beds. At a few horizons impoverished faunas are composed mainly of bivalves. The uppermost unit consists of sandstone containing Rhaetic plant remains.

Only one occurrence of Triassic deposits is known from northern Greenland, and that is in Peary Land. Here the succession consists of 630 m of a lower shaly division of late Scythian age and an upper sandy division of early Anisian age (KUMMEL, 1953b).

ARCTIC CANADA

Triassic rocks are present in the Sverdrup Basin and crop out on Ellesmere, Axel Heiberg, Cornwall, Table, Exmouth, Cameron, Melville, Prince Patrick, Brock, and Borden islands. An excellent summary of the stratigraphy of the Canadian Arctic Archipelago has been published by THORSTEINSSON and TOZER (1970).

In the axial part of the Sverdrup basin and in parts of the marginal area Lower Triassic rocks rest disconformably upon Permian strata. In places the youngest Permian rocks are Guadalupian; elsewhere they are late Artinskian. Latest Permian strata are unknown. The oldest Triassic rocks are earliest Scythian (lower Griesbachian) in age. It appears that the Permian-Triassic boundary is paraconformable and that the gap in the sedimentary record is within the highest Permian and not in the Lower Triassic.

Marine conditions prevailed throughout the axial part of the Sverdrup basin from earliest Scythian to the Carnian. Some of the Lower Triassic beds on the margins were probably deposited in a nonmarine environment. For this interval the sections on the margins of the Sverdrup basin differ from those of the axis both in thickness and lithology. Most of the rocks on the south and east margins consist of sandstone and calcareous siltstone (Bjorne and Shei Point formations); the axial part is shale and siltstone (Blind Fiord and Blaa Mountain formations). In the northwest margin of the basin, the sections are of mixed character. The Bjorne Formation disconformably overlies Permian strata and consists mainly of quartzose, commonly crossbedded sandstone with conglomeratic interbeds on the extreme margins of the basin. Fossils are rare but *Otoceras* (lower Scythian-Griesbachian) occurs near the base at one locality and poorly preserved lower Lower Triassic ammonoids are known from other localities. Marine fossils are unknown in the beds on the south margin of Sverdrup basin. The Bjorne is overlain by Anisian strata and is thus fairly well dated as Early Triassic. The contemporary rocks exposed on the western coast of Ellesmere Island and on eastern and northern Axel Heiberg Island are the Blind Fiord Formation, which consists of green and gray siltstone, fine-grained sandstone and shale. The Blind Fiord beds are well dated by ammonoid faunas.

The Shei Point and Blaa Mountain formations are essentially contemporaneous and of Anisian, Ladinian, and Carnian age. Typical Shei Point rock is gray, brown weathering, highly calcareous siltstone and fine-grained sandstone, with bioclastic layers composed of brachiopod and bivalve shells in the upper part. Shale is typical Blaa Mountain lithology.

Throughout central Ellesmere Island, the top of the Shei Point Formation is marked by the upper Carnian "*Gryphaea* Bed"; 31 m of calcareous sandstone with coquinoid layers of *Gryphaea* and *Plicatula*. A similar and apparently contemporary *Gryphaea* bed is found in the Blaa Mountain Formation of northwestern Axel Heiberg Island. The upper Carnian *Gryphaea* beds are thus widely distributed on the margins of the Sverdrup basin and their occurrence probably indicates an interval of shoal-water conditions along the border of the Sverdrup basin following uplift.

The Blind Fiord and Blaa Mountain formations of the axial part of Sverdrup basin were laid down some considerable distance from shore. In the Triassic there were two sources of sediments, one essentially continuous, feeding sediment to the south and east margins of the basin, and the other, an intermittent source (or sources), providing sediment to the north and northwest margins of the basin.

The highest Triassic formation in the Sverdrup basin is the Heiberg, which consists mainly of nonmarine, carbonaceous sandstone, with marine beds at several levels in the lower part. The lower marine beds with *Meleagrinella antiqua* are probably lower Norian, the higher marine strata include middle Norian beds and strata with the cosmopolitan upper Norian *Monotis ochotica*. The beds above those with *Monotis* are entirely nonmarine, with fossil plants and thin coal seams. They may be partly Jurassic, but if so, they are not younger than Sinemurian.

WESTERN CANADA

Triassic strata of western Canada are represented by a complex array of sediments and volcanics laid down in distinct eugeosynclinal and miogeosynclinal segments of the Cordilleran geosyncline. The geology of western Canada has been well summarized by Douglas et al. (1970). Lower Triassic rocks are not known in the western (eugeosynclinal) parts of the Cordilleran geosyncline in Canada although they are present a few kilometers south in Washington. Middle Triassic ribbon chert, argillite, greenstone, possibly coeval ultramafic rock, and minor limestone occur locally in the northwestern and central parts, apparently lying conformably on the Permian. In the southernmost parts of the geosyncline, Middle Triassic rocks characterized by sharpstone conglomerate, unconformably overlie late Paleozoic rocks. The western sequences represent generally quiescent eugeosynclinal conditions possibly prevailing from the Permian. An intermittently emergent arch (Quineca geanticline) separated the eugeosyncline from the miogeosyncline. In the miogeosyncline are Lower Triassic siltstone and shale that disconformably overlie the lower Upper Permian. The entire western margin of the craton was probably emergent and stable, these conditions prevailing throughout the period.

In the late Middle Triassic, parts of the Cordilleran geosyncline underwent deformation and plutonic activity (Tahlternian orogeny). Uplift at the end of the orogeny established the main tectonic elements that prevailed until Middle Jurassic time. The lower Upper Triassic, Carnian, is represented on the western edge of the eugeosyncline mainly by a thick succession of submarine basaltic flows. Slightly to the east they are represented by andesitic and basaltic flows, pyroclastics, and clastics containing volcanics and Middle Triassic and earlier debris. The latter for the most part derived from islands and volcanoes within the eugeosyncline. In the miogeosyncline siltstone and sandstone accumulated, grading eastward, along the margin of the craton, into an evaporitic and red-bed facies with basal shoreline and offshore sandbars.

In the late Late Triassic, Norian, volcanism persisted in parts of the eugeosyncline, red beds are present in some areas around uplifted geosynclines, and clastic rocks. During latest Norian, carbonate deposition prevailed throughout much of the geosyncline. The sections are remarkably thin but complete in the northern and eastern elements, and indicate quiescent conditions of deposition to the end of the Triassic.

WESTERN UNITED STATES

Marine Triassic formations are present only in the western Cordilleran geosynclinal area. Thick sequences of miogeosynclinal facies are present in eastern Nevada, western Utah, southeastern Idaho, western Wyoming, and southwest Montana. Eugeosynclinal facies including volcanics are found in western Nevada, California, Oregon, and Washington. In the eastern half of the miogeosyncline only the Lower Triassic is of marine facies and these in many places are overlain by Upper Triassic terrestrial red bed formations. Along the hinge line between the miogeosyncline and the craton the Lower Triassic marine beds interfinger with red beds. In southeastern Idaho the marine section contains an excellent sequence of ammonite faunas, especially for the upper half of the Scythian (Kummel, 1954). The base of these marine sequences is at various levels within the Scythian indicating a complex pattern of transgression over an irregular terrane. The eugeosynclinal suite of western Nevada has yielded a fairly complete succession of zonal ammonites, however, not as complete as that known in British Columbia (Silberling & Tozer, 1968). In California only the Upper Triassic is fossiliferous.

SOUTH AMERICA

Marine Triassic formations in South America are entirely confined to the Andean region and then are fairly extensive only in the Norian. To the west of the Andes all Triassic formations are of terrestrial detrital deposits and volcanics, mainly of Late Triassic age.

No marine deposits of Early Triassic age have yet been identified. Shallow-water marine deposits containing Middle Triassic

(Anisian) fossils are present in a coastal strip of central Chile (ZEIL & ICHIKAWA, 1958; BARTHEL, 1958). The Ladinian-Carnian is represented only by volcanic deposits in various parts of Chile. In the Norian, however, the sea transgressed an area from central Colombia to southern Bolivia and along the coastal region of Chile. Deposition in this seaway was mainly limestone, and to the north it grades into continental red-bed facies. The most conspicuous fossil in these formations is *Monotis subcircularis* (WESTERMANN, 1973). The Utcubamba Formation of northern Peru has yielded a silicified fauna containing bryozoans, brachiopods, gastropods, nautiloids, ammonoids, scaphopods, bivalves, and crinoid ossicles; however, only the ammonoids are fairly abundant (JAWORSKI, 1922; KUMMEL & FUCHS, 1953).

The best general discussion of the distribution of South American Triassic deposits and paleogeography is by HARRINGTON (1962).

ANTARCTICA

Continental Triassic strata have long been known on Antarctica, yet the first record of marine Triassic has only recently been published (THOMSON, 1975). The Legoupil Formation in the northwestern part of the Antarctic Peninsula has yielded a small poorly preserved fauna, from which THOMSON has identified *Bakevelloides* aff. *B. hekiensis* and *Neoschizodus* sp. nov. In addition, other fossils include a possible fragment of an inarticulate brachiopod, a serpulid, a gastropod, and some possible arthropod tracks. Most species of *Bakevelloides* are Late Triassic in age. *Neoschizodus* sp. nov. is the commonest element in the fauna. It is closely related to *N. laevigatus* (VON ZIETEN) of Early to Middle Triassic age, but differs in some features.

SUMMARY

The Triassic was a time of great emergence of the continents and little tectonic activity. Marine deposition in shelf and geosynclinal environments are confined to Tethys, the circum-Pacific region, and the circum-Arctic. With the exception of the Middle Triassic conodonts in the Cameroons of western Africa, no marine strata are known from the Atlantic region between Svalbard and the northwestern part of the Antarctic Peninsula. The relationship between sea and land changed little during the Triassic.

Within Tethys, Triassic strata are represented by a wide range of facies. Carbonates are dominant but fine to coarse detrital facies are also present. In the circum-Arctic region the sedimentary facies are almost entirely fine to coarse detrital. Along the western part of North America there are well-developed miogeosynclinal and eugeosynclinal regions. The miogeosynclinal areas are characterized by carbonate and fine to coarse detrital facies that thin and intertongue with red-bed facies on the adjoining shelf. The eugeosynclinal areas contain mainly detrital facies. Marine Triassic deposits are not well represented along the western region of South America except for thick carbonates of Late Triassic age in Colombia, Ecuador, and Peru. Along the western margins of the Pacific, geosynclinal conditions prevailed with detrital and volcanic facies. Shelf deposits are very limited.

TRIASSIC FAUNAS

Triassic faunas are strikingly different from those of the underlying Permian. Absent is the great diversity of brachiopods, bryozoans, echinoderms, and foraminifers. The primary elements of Triassic faunas are brachiopods, gastropods, nautiloids, ammonoids, bivalves, and conodonts. All other groups are sparsely represented. In this chapter the focus is on an overview of the primary fossil groups in Triassic faunas.

The nature and definition of the Permian-Triassic boundary has received much attention in recent years. In this debate much has been written on so-called "mixed" Permian-Triassic faunas at the boundary. This subject will be covered in the discussion of the brachiopods, as it is this phylum that plays an important role in nearly all these discussions.

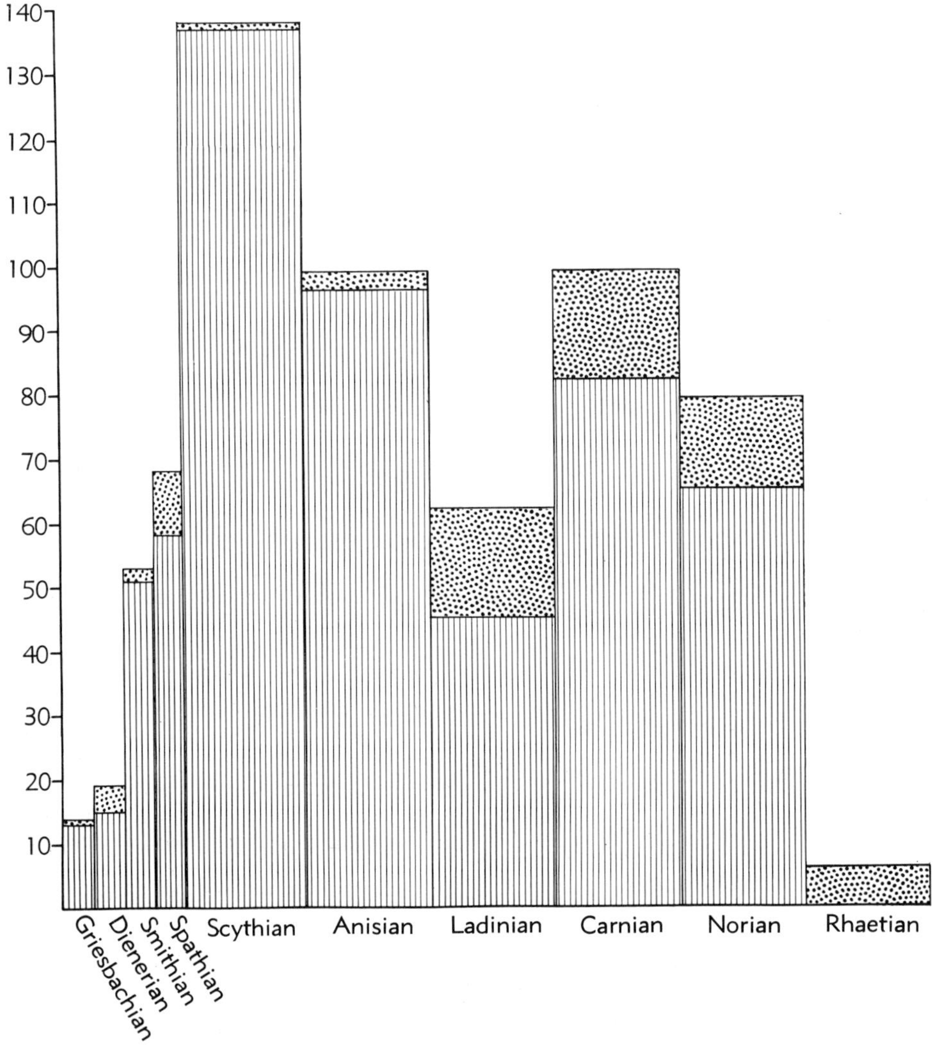

Fig. 3. Total genera and new genera (lined area) of ammonoids of the Triassic Period (Kummel, n).

AMMONOIDS

Ammonoids are the predominant invertebrate fossils of the Triassic. They are the most abundant, widely distributed, and diverse of all the invertebrates for this period. For these reasons they have historically been the primary basis for establishing the biostratigraphic framework of the Triassic.

The forthcoming revision of the ammonoid volume of the *Treatise* (Part L) recognizes 13 superfamilies, 76 families (and 6 subfamilies), and 431 genera and subgenera of which 16 are nonnominate subgenera.

The first comprehensive summary of Triassic ammonoid genera was in *Fossilium Catalogus* Part 1 by Diener (1915), who recognized 247 genera and subgenera. In an updating of the *Fossilium Catalogus* on Triassic ammonoids Kutassy (1933) recognized 277 genera and subgenera. The increase in numbers of genera in the last 45 years is due in no small part to noteworthy discoveries and monographs on sequences in the southern Soviet Union, southern China, Primorye, northeastern Siberia, Arctic Canada, British Columbia, and western United States.

Total genera and new genera per stage are shown in graphic form in Figure 3. The distribution of genera by stage and within the major paleogeographic regions is as follows:

	Total Genera	Tethys	W. Pacific	E. Pacific	Arctic
Rhaetian	6	6	4	5	3
Norian	79	78	8	51	10
Carnian	100	81	13	57	23
Ladinian	62	49	13	34	15
Anisian	100	81	24	45	26
Scythian	138	118	53	55	52

It is clear from this summary that the diversity of ammonoids throughout the Triassic is greatest in Tethys and least in the circum-Arctic.

Approximately 66 genera of ammonoids have been found in the Upper Permian (B. F. GLENISTER, letter May 17, 1976), and a dramatic change in the composition of ammonoid populations occurs at the Permian-Triassic boundary. Most Permian genera and families became extinct, and the lower Scythian (Griesbachian) is characterized by a completely new radiation centered around the Ophiceratidae. This family is a direct descendant of the Xenodiscidae (SPINOSA, FURNISH, & GLENISTER, 1975).

The number of genera per superfamily for each series of the Triassic is as follows:

	Lower	Middle	Upper
Otocerataceae	1	0	0
Noritaceae	90	7	1?
Hedenstroemaceae	11	3	1
Dinaritaceae	26	2	0
Ceratitaceae	1	58	3
Clydonitaceae	0	17	51
Choristocerataceae	0	1	6
Tropitaceae	0	4	58
Lobitaceae	0	2	5
Arcestaceae	0	13	11
Megaphyllitaceae	2	9	1
Nathorstitaceae	0	6	1
Pinacocerataceae	1	22	19
Phyllocerataceae	4	5	5

The Noritaceae and secondarily the Dinaritaceae are the predominant groups in the Lower Triassic radiation. Of the 13 Triassic superfamilies only seven are represented in the Lower Triassic and four of these have four or less genera. All superfamilies are represented in Middle Triassic faunas; the predominant group are the Ceratitaceae. The Pinacocerataceae are also well represented. Twelve superfamilies are represented in the Upper Triassic but four of these by three or less genera; the main groups are the Clydonitaceae and Tropitaceae. At the close of the Norian there was again an extensive wave of extinction and only six genera are present in the Rhaetian and these became extinct by the close of the Triassic.

NAUTILOIDS

The evolutionary history of nautiloids shows no dramatic change in tempo at the Permian-Triassic boundary (KUMMEL, 1953a). Families that account for peak development of the Nautilida in the late Paleozoic and Triassic are the Tainoceratidae, Grypoceratidae, and Liroceratidae. Except for a liroceratid doubtfully recorded from the Devonian, all these families began in the Mississippian. Whereas the number of genera gradually increased with time in each of these evolutionary lines, rates of evolution were not such as to produce many new families. Thus it seems that as early as the Mississippian the principal evolutionary lines of nautiloids had become firmly established and these maintained their identity and character until the close of the Triassic. Only a few minor radiations (families) appeared in this interval. This pattern is also reflected in the number of genera carried over from one period to the next. Eleven genera persisted from Mississippian to Pennsylvanian, and 20 from Pennsylvanian to Permian. Only four Triassic genera are also known from the Permian, but this number is deceptive, for in each of the three evolutionary lineages that extend from the Permian into the Triassic, the core Permian genus evolved directly into the core Triassic genus: *Metacoceras* (Penn.-Perm.) to *Mojsvaroceras* (Trias.), *Domatoceras* (Penn.-Perm.) to *Grypoceras* (Trias.), and *Liroceras* (Miss.-Perm.) to *Paranautilus* (M.Trias.-U.Trias.). Consequently, the large number of Triassic genera is the result of

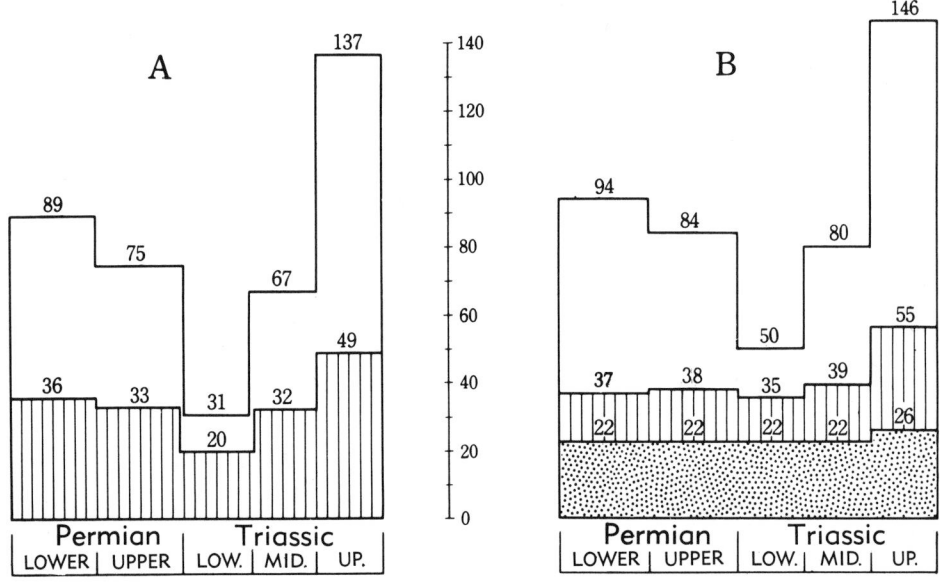

Fig. 4. Diversity of Permian and Triassic marine bivalves measured by the numbers of genera, families, and superfamilies (after Nakazawa & Runnegar, 1973).——*A*. Observed occurrences of genera (solid line) and families (ruled area).——*B*. Estimated numbers of genera (solid line), families (ruled), and superfamilies (stippled).

a broad evolutionary radiation in each of the principal evolutionary lineages during the Late Triassic. Triassic nautilid evolution, then, represents just the culmination of patterns and trends begun in the Mississippian.

Nautilids nearly became extinct after their peak development in the Late Triassic; only *Cenoceras* survived from the Triassic into the Early Jurassic. At this time, a new radiation began, resulting in a modest proliferation of new genera in the Late Jurassic.

BIVALVES

Bivalves are the second most important invertebrate group in the Triassic rock record. They likewise have the same status in the dating of Triassic strata. Though relatively common in the rock record, they have not received the intensity of study as have the ammonoids. Thin-shelled bivalves, such as *Claraia, Posidonia, Daonella, Halobia,* and *Monotis* are particularly striking forms for their extremely widespread geographic distribution and their usefulness for correlation. A few remarks on some of these genera are given below.

The most recent summary of Permian-Triassic bivalves is by NAKAZAWA and RUNNEGAR (1973) and provides the basis for the following discussion. One of their primary conclusions is that "... there is reasonable evidence that marine bivalves belong with the group of invertebrates least affected by events at the close of the Paleozoic Era" (NAKAZAWA & RUNNEGAR, 1973, p. 609).

The number of bivalve genera and families decreased during the Late Permian reaching a low during the Early Triassic and then began to increase gradually during the Middle Triassic (Fig. 4). Only 19 marine bivalve genera (9 new) are known from the earliest Triassic, in contrast to about 70 genera known from the middle part of the Permian and approximately 140 known from the Late Triassic. These authors stress that the anomalous low diversity of Early Triassic forms must be the result of an unusually poor record of the history of the class—"an effect which accentuates the observed change in this group at the Permian-Triassic boundary." They point

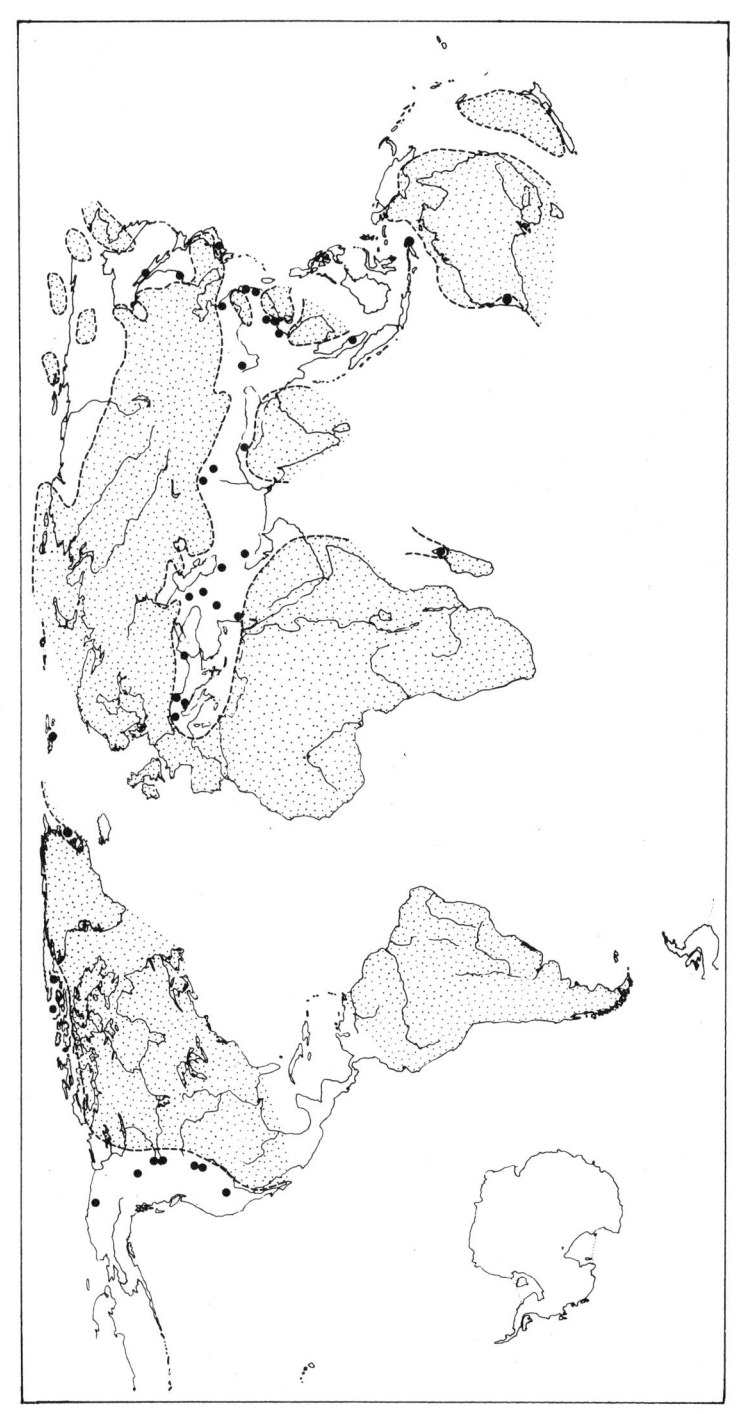

Fig. 5. Distribution of bivalve genus *Claraia* during early Scythian time (from Kummel in Hallam, 1973, published with permission of Elsevier Scientific Publishing Co.).

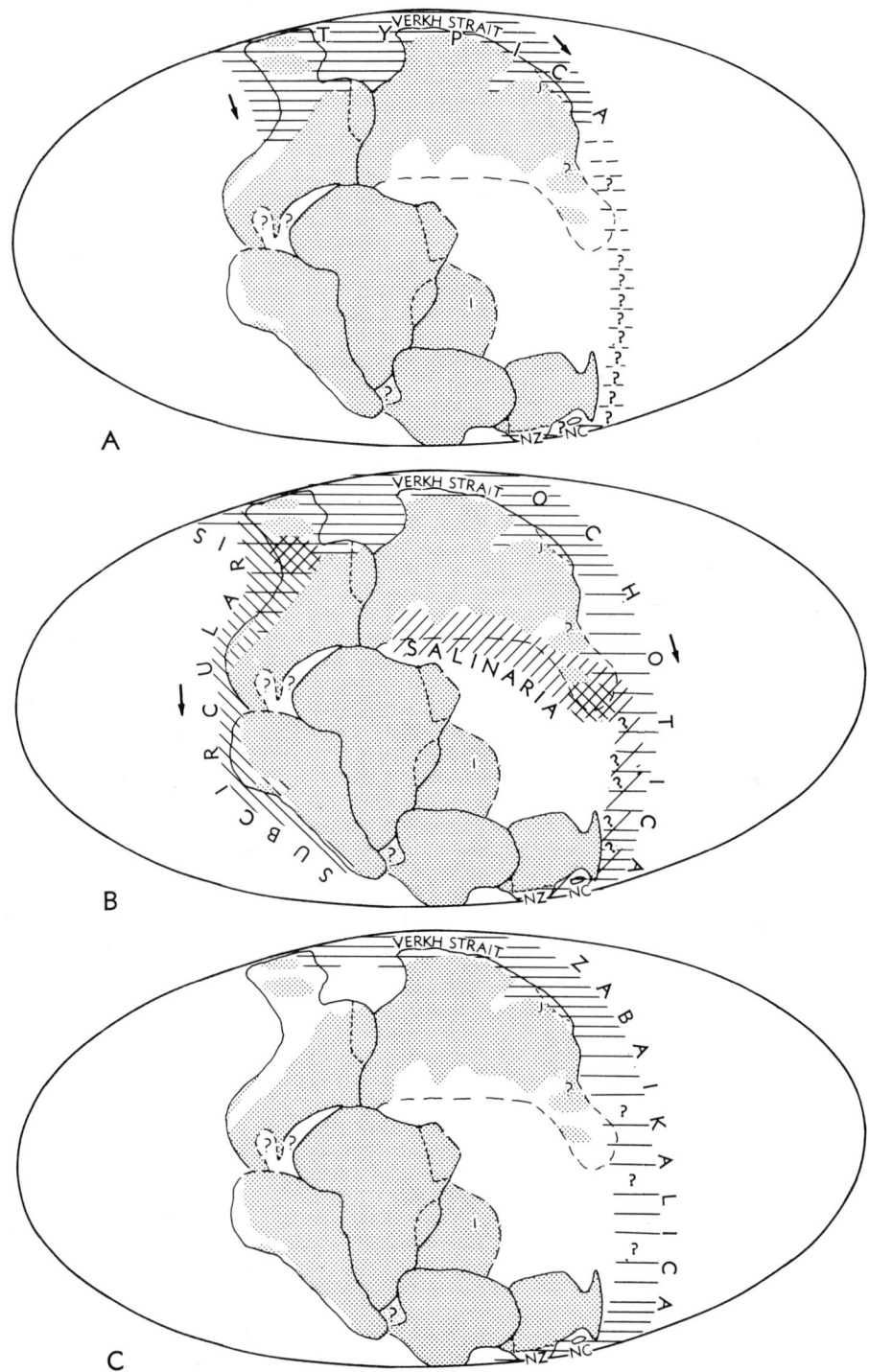

Fig. 6. Zoogeographic distribution of *Monotis* by species groups in relation to probable position of Late Triassic land areas (gray) and oceans (from Westermann in Hallam, 1973, published with permission of

out that 1) many genera and families are represented in Permian strata and reappear in the Middle and Late Triassic (e.g., *Lopha, Waagenoperna, Lyriomyophoria, Costatoria, Modiolus,* Pinnidae, Pteriidae, Terquensiidae, Carditidae, Astartidae, Pholadomyidae); 2) a few must represent transitional forms between Paleozoic and Mesozoic families: Pseudomonotidae-Gryphaeidae (NEWELL & BOYD, 1970), Megadesmidae-Ceratomyidae (RUNNEGAR, 1965), among others. NAKAZAWA and RUNNEGAR (1973) concluded on this basis that two to three times as many genera as are observed must have existed in earliest Triassic time.

The record of diversity of the various orders of bivalves is highly variable. The Nuculoida remained virtually stable throughout the two periods. Other groups, the Arcoida, Mytiloida, Unionida, Trigonioida, and Veneroida, show no essential change during the Permian-Triassic transition but did undergo diversification in the Middle and Late Triassic. The most significant drop in diversity is within the epifaunal Pteroida, which include nearly half of all Permian and Triassic bivalve genera.

In the introductory remarks to this section, mention was made of thin-shelled bivalves; the distribution of *Claraia* and *Monotis* will illustrate the importance of these groups. *Claraia* is confined to the lower part of the Scythian and is present in nearly every area where marine lower Scythian strata are present (Fig. 5). That is, it is present throughout Tethys at a large number of localities, southeast Asia, western Australia, China, Japan, northeast Siberia, northeast Greenland, the Arctic Islands of Canada, and in a number of localities in the western Cordillera of North America. The genus is generally found in fine-grained calcareous facies. Though many species of *Claraia* have been described, it is apparent that some species are cosmopolitan.

The Upper Triassic (Norian) *Monotis* is another of the important thin-shelled bivalves that is instructive for zoogeographic analysis and in biostratigraphy. WESTERMANN (1973) has recently published an excellent review of this genus and this forms the basis for the brief summary remarks given here. There are approximately 60 named species and subspecies; however, WESTERMANN's studies led him to conclude that only 19 to 20 "good" species and 12 to 14 subspecies (non-nominate) are justified. For purposes of analysis of zoogeographic distribution, he recognized five groups (those of *Monotis typica, M. salinaria, M. ochotica, M. subcircularis,* and *M. zabaikalica*). *Monotis* occurs in a wide variety of sedimentary facies, ranging from presumed deep-water to shallow-water facies. They do appear, however, to have avoided restricted inland seas. Most authors have interpreted *Monotis* as being pseudoplanktonic because of its mode of occurrence and extraordinary lateral extent of distribution. In his recent summary paper, WESTERMANN (1973) concluded on morphological grounds that it is more likely that they were prevalently benthic "with perhaps a few specimens of the population attached to floating objects, sufficient to permit pseudoplanktonic distribution of the species."

The earliest species of *Monotis* are members of the *M. typica* group. There is some dispute as to whether the earliest occurrences are latest Carnian or early Norian, but this need not concern us here. These early forms are circum-Arctic in distribution extending through the Verkhoyansk Strait to Japan and down into northeastern British Columbia. Later, in the middle Norian, various subspecies of *M. typica* became abundant in different areas of the distributional range (*M. scutiformis* mainly in eastern Siberia and Japan, *M. pinensis* in Alaska to British Columbia, *M. iwaiensis* in Japan). There was, in addition, a further spread southward along the eastern Pacific margin to Vancouver Island and possibly also along the western Pacific margin to Borneo, New Zealand, and New Caledonia (Fig. 6).

The area of distribution of *Monotis* expanded greatly in the late Norian with each

FIG. 6. *(Continued from facing page.)* Elsevier Scientific Publishing Co.).——*A.* Late Carnian(?) to Middle Norian.——*B.* Early Late Norian. ——*C.* Latest Norian (?to Rhaetian). [Explanation: *I,* India; *J,* Japan; *N.C.,* New Caledonia; *N.Z.,* New Zealand.]

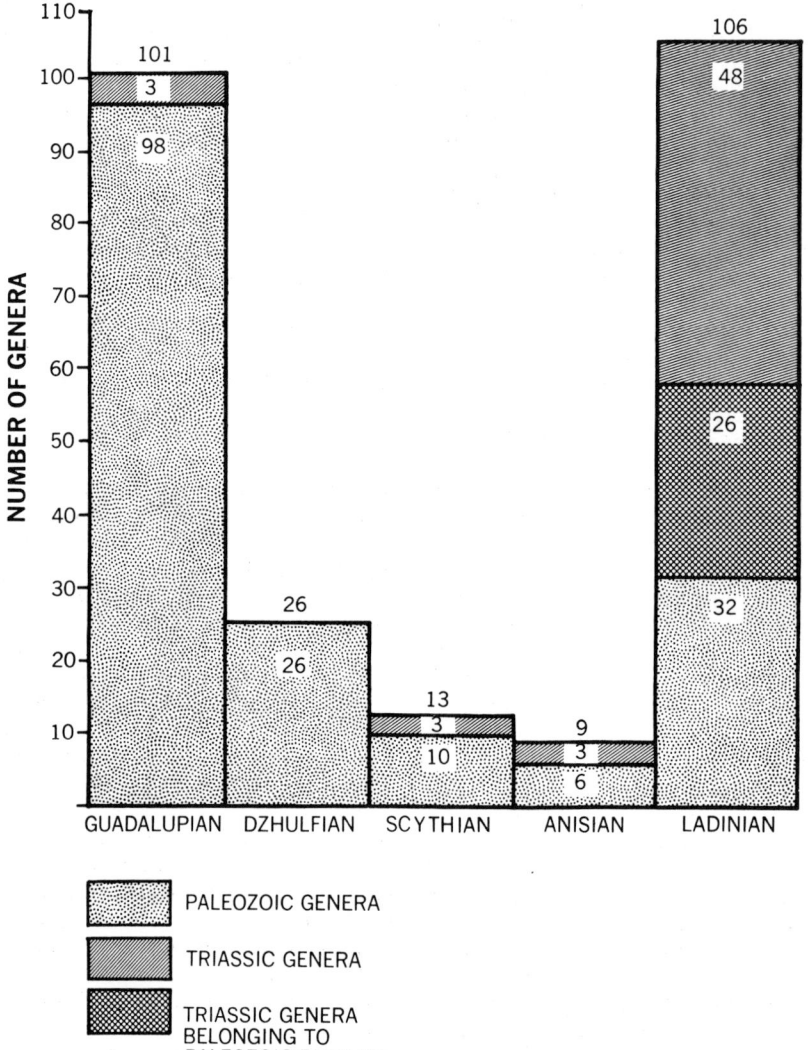

Fig. 7. Distribution of gastropod genera for stages of the Upper Permian-Middle Triassic (from Batten, 1973).

major tectonic region being occupied by a dominant *Monotis* group. The *ochotica* group generally abounds in the area of the former *M. typica* group, except for the northeastern Pacific margin where it is rare, and is found southward in the western Pacific to New Caledonia and New Zealand. The *M. subcircularis* group is found mainly along the eastern Pacific margin as far as central Chile, but occurs disjunctively and probably rarely also in Indonesia and possibly in the Mediterranean. The *M. salinaria* group occurs throughout Tethys and disjunctively also in Alaska, Yukon Territory, and probably in the southwestern Pacific and northeastern Siberia.

The close of the *Monotis* radiation in latest Norian (and ?Rhaetian) is dominated by almost smooth forms of the *M. zabaikalica* group. The group is entirely confined to the Arctic region and along the western Pacific with *M. zabaikalica* dominant in

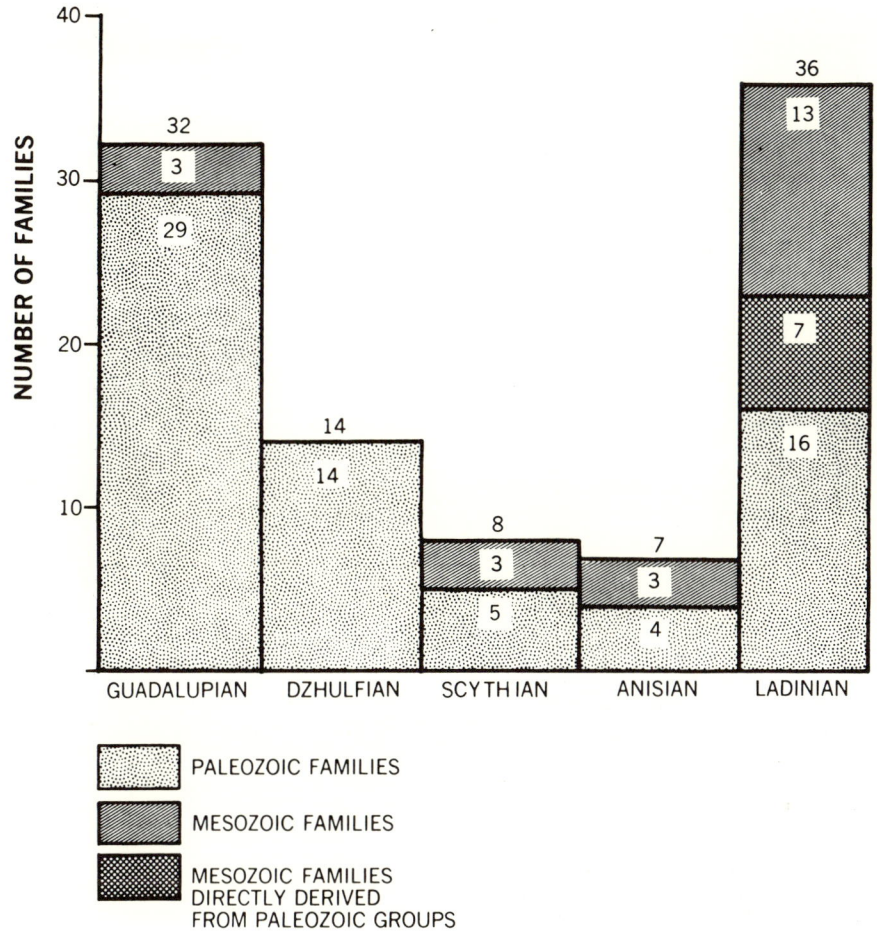

FIG. 8. Distribution of gastropod families for stages of the Upper Permian–Middle Triassic (from Batten, 1973).

eastern Siberia and *M. calvata* in New Zealand and New Caledonia.

GASTROPODS

Though an extremely important molluscan group, gastropods are not very common in most Triassic strata. As a result, there have been relatively few monographic studies on Triassic gastropods. Until recently there had been no general survey of Triassic gastropods from which one could assess the overall degree and amount of change in Triassic gastropod diversity for a meaningful comparison with underlying Permian and overlying Jurassic faunas. These data have recently been summarized by BATTEN (1973) in a splendid review article on Permian and Triassic forms. The summary given here is based on BATTEN's article.

During the late Paleozoic, the gastropods underwent two episodes of adaptive radiation, the earliest occurring during the Tournaisian and Visean stages of the Lower Carboniferous. This was followed by a decrease in diversity in the Late Carboniferous. The second radiation and the one of prime concern here occurred in the Guadalupian. At this time the most diverse fauna of the Paleozoic developed, with about 100 genera and 30 families. These Permian faunas in overall balance were much like those of the preceding two periods but richer. The 10 dominant families in terms of species frequencies are: Bellerophontidae, Sinuitidae, Eotomariidae, Euomphalidae, Omphalotrochidae, Phymatopleuridae, Neritopsidae, Pseudozygopleuridae, Subulitidae, and Murchisoniidae.

It is of particular interest that these are the last diverse faunas of the Paleozoic. The next normal marine gastropod fauna occurs in the Ladinian. Between the latest Guadalupian and the first Dzhulfian occurrence, some 62 Permian genera and 12 families disappeared (Fig. 7, 8).

Only four Dzhulfian faunas contain more than three to five genera. Altogether, these include 26 genera and 14 families. By far the most common group is the bellerophontids; the other important constituents are the eotomariids, neritopsids, and murchisoniids. BATTEN made note of the fact of sparse numbers of genera per family and apparently large numbers of individuals of few species. This condition apparently also prevailed in the Early Triassic. Of interest also is that of the total Dzhulfian fauna, only three genera survived from the preceding radiation; the others are relatively long ranging and conservative. Most of these are found in facies that indicate brackish to lagoonal conditions.

What is particularly interesting is that the Dzhulfian and Scythian faunas as now known lack genera that occur in the Guadalupian and reappear in the Ladinian! There are 32 Guadalupian genera and 16 families in the Triassic, none of which are known in the Dzhulfian.

Scythian gastropod faunas are rare. When found they form thin accumulations in limestones and invariably one or, at most, several species are present in great numbers. Most Scythian assemblages resemble those of the Dzhulfian in their lack of diversity. The main difference is that the bellerophontids were displaced by other gastropods. No significant change in faunal diversity occurred in the Anisian.

Ladinian faunas are rich and widespread. They contain 106 genera and 36 families, a diversity quite comparable in size to that of the Guadalupian. On further analysis, 32 of the 106 genera are found to belong to Paleozoic families.

BATTEN emphasized that there is about as much difference between the Guadalupian and Ladinian gastropod faunas as there is between the Devonian and Lower Carboniferous faunas. The main Mesozoic faunal turnover of the gastropods occurred at the close of the Triassic.

BRACHIOPODS

There is no satisfactory analysis of Triassic brachiopods and their stratigraphic distribution. The great decrease in diversity of Mesozoic and Cenozoic brachiopods has long been established; however, details are lacking. A number of reasons can be cited for this state of affairs. First, brachiopods are not really very abundant in Triassic rocks. Secondly, very few individuals have devoted any significant period of time to their study. The most prominent specialist on Triassic brachiopods was A. BITTNER who published numerous papers at the turn of the century. Brachiopods do form part of many general faunal studies, but these serve mainly to emphasize the paucity of this group in comparison to ammonoids and bivalves. A few years ago I wrote to G. A. COOPER asking his opinion of the brachiopods stated to occur in the Lower Triassic in the recent volume of the *Treatise on Invertebrate Paleontology* (Part H), Brachiopoda (A. WILLIAMS et al., 1965). He answered (letter of June 14, 1971): "So far as I know, the Lower Triassic is almost a blank for brachiopods. As a matter of fact, the whole Triassic is a critical time in brachiopod history and a kind of never-never land between Permian and the Jurassic."

Whereas the systematic study of Triassic brachiopods is, and generally has been, at a low level, there is one aspect that has aroused considerable interest in recent years —this is the occurrence of Permian-type brachiopods in lowermost Scythian strata. These occurrences have stimulated intensive debate as it relates to the Permian-Triassic boundary.

The first such report was by H. S. BION (1914) concerning an occurrence in the Pahlgam area of Kashmir, near Srinagar. All the data available are from a brief statement in the Annual Report for 1913 of the

Geological Survey of India. BION (1914, p. 39) wrote as follows: "About 20 feet above the base of the black shales there is a layer of calcareous nodules from which many specimens of *Otoceras* have been obtained associated with almost all the other members of the fauna of the *Otoceras* beds of the Central Himalayas. Good collections have been obtained from Nagaheran in the Dachhigam State Rakh and from the Pahlgam-Aru basin. Some thirty feet above the *Otoceras* layer there is another fossiliferous horizon characterized by *Ophiceras* from which one specimen of *Otoceras* was also procured, but the rest of the black shale division seems to be barren. A surprising element in the fauna of the basal *Otoceras* layer is furnished by the presence of the genus *Productus,* of which three specimens have been obtained from near Pahlgam. In spite of this Permian element I consider that the fauna of the *Otoceras* bed of Kashmir has a decided Triassic aspect."

This brief but highly interesting account went unnoticed for many years. TEICHERT and KUMMEL made an attempt in June, 1968, to locate BION's locality but were unsuccessful. Since BION's report, seven additional localities or areas have yielded Permian-type brachiopods in horizons with lowest Triassic ammonoids. These are the Guryul Ravine of Kashmir, Salt Range of West Pakistan, the Dzhulfa region along the Aras River in the Soviet Union and Iran, Shikoku Island of Japan, East Greenland, Ellesmere Island of Arctic Canada, and the lower Dinwoody Formation of Montana.

The Permian-Triassic strata of the Guryul Ravine, near Srinagar, Kashmir, were first reported on by HAYDEN (1907) and in more detail by MIDDLEMISS (1910). Recently TEICHERT, KUMMEL, and KAPOOR (1970) visited the area and discovered that the "Black Shales" unit of MIDDLEMISS contained fossiliferous beds in which *Spinomarginifera* and other productid brachiopods, typically Permian in aspect, are in association with *Claraia,* a bivalve typical of the lower half of the Scythian. Later that year NAKAZAWA et al. (1970) published a more comprehensive report on the sequence and faunas of these boundary beds at Guryul Ravine. These authors discovered in the critical boundary beds, in addition to the productids and *Claraia,* species of *Otoceras, Ophiceras,* and *"Glyptophiceras."* NAKAZAWA et al. (1970) came to the same conclusion as to the placement of the Permian-Triassic boundary as TEICHERT, KUMMEL, and KAPOOR (1970). Also, at about this time, SWEET (1970b) reported on the conodonts from the Permian-Triassic boundary beds from samples submitted by TEICHERT. His main conclusion was that "conodonts from the upper 57 feet of the Zewan Series and the lower 124 feet of the overlying Lower Triassic beds in the section at Guryul Ravine, Kashmir, represent four distinct conodont faunas that may be correlated with those of the *Anchignathodus typicalis, Neogondolella carinata, Neospathodus dieneri,* and *Neospathodus crystagalli* zones of the Salt Range and Trans-Indus Ranges of Pakistan." SWEET (1973) further refined his discussion of the Kashmir sequence and integrated his conodont data with those of NAKAZAWA et al. (1970) on ammonoids and *Claraia.* Finally, in collaboration with a co-author (MIKAN & SWEET, 1974) further analysis of the Kashmir section renewed interest in the problem. The abstract is worth quoting in full: "Graphic correlation, using published data on the ranges of conodonts and key ammonoids, have been effected between important Permo-Triassic sections in Kashmir, the Salt Range of Pakistan, and the Julfa district of northwest Iran. Although there are unresolved taxonomic and distributional problems, a preliminary analysis of the results indicates that the Kashmir and Iranian sections record an essentially unbroken sequence of latest Permian and earliest Triassic rocks, and that in the Salt Range, Permian rocks range higher (Changhsinghian?) and Triassic strata lower (lower Griesbachian) as has been assumed by recent authors." GRANT and COOPER (1973) have discussed the Guryul Ravine section on the basis of data presented by TEICHERT, KUMMEL, and KAPOOR (1970), SWEET (1970b), and NAKAZAWA et al. (1970) from the viewpoint of their expertise on brachiopods. They took exception to the conclusions of the above mentioned workers and placed the boundary at the base of bed 52a of NAKAZAWA et al. (1970).

In a final report on the Upper Permian and Lower Triassic of Kashmir, NAKAZAWA et al. (1975) placed the Permian-Triassic boundary at the first appearance of *Otoceras woodwardi* (that is, the base of their bed 52). In their preliminary report (NAKAZAWA et al., 1970) the boundary had been placed at the base of their bed 47, which coincides with the first appearance of *Claraia*. It is in beds 47-49 that the majority of Permian brachiopods are present, associated with *Claraia stachei* (BITTNER). In their final report they assigned the *Claraia* to a new species *C. bioni*. The "Permian" elements in bed 52 are now limited to *Marginifera himalayensis, Pustula* sp., *Estheripecten haydeni,* and *Claraia bioni*.

In the classic area of the Salt Range, brachiopods discovered in the Permian-Triassic boundary beds have been the focal point of a stimulating and continuing debate. KUMMEL and TEICHERT (1966, 1970b) reported Permian-like brachiopods in the lower part (mainly lowest one foot) of the dolomite unit of the Kathwai Member of the Mianwali Formation. The initial collections were studied by G. A. COOPER, but as more material was obtained, the study was taken over and published by R. E. GRANT (1970). He reported the following forms from this horizon: *Enteletes* sp., *Orthothetina* cf. *O. arakeljani* SOKOLSKAYA, *O.* sp., *Ombonia* sp., *Derbyia?* sp., *Spinomarginifera* sp., *Linoproductus* sp., *Lyttonia* sp., *Spirigerella* sp., *Crurithyris?* extima GRANT, *Martinia* sp., *Whitspakia* sp., dielasmatids gen. et sp. indet. GRANT (1970) correlated the fauna from the basal foot of the dolomite unit with the Dzhulfian of Armenia. In 1973, GRANT concluded that since *Crurithyris? extima* occurred above the fauna of the lowest foot of the dolomite unit that this upper portion was of Changhsingian age (uppermost Permian), a conclusion quite different from that of KUMMEL and TEICHERT (1970b, 1973).

An important addition to the problem was the discovery by GRANT of a fauna of brachiopods in the white sandstone unit of the Chhidru Formation, which immediately underlies the dolomite unit of the Kathwai Member. From this unit GRANT (1970) identified the following forms: *Aulosteges* sp., *Callispirina* sp., *Chonetella* sp., chonetid indet., *Cleiothyridina* sp. cf. *C. capillata* (WAAGEN), *Derbyia* sp. cf. *D. plicatella* WAAGEN, dielasmatids indet., *Enteletes* sp. 1, *Hemiptychina* sp., *Hustedia* sp., *Kiangsiella* sp., *Linoproductus* sp., *Lyttonia* sp., *Martinia?* sp., *Neospirifer* sp., *Orthotichia* sp. 2, *Richthofenia* sp., *Spiriferella?* sp., *Spirigerella* sp., *Waagenoconcha?* sp., *Whitspakia* sp. 1. GRANT (1970) concluded that this assemblage is typical for the Chhidru Formation as a whole, which he believed to be of Guadalupian age.

KUMMEL and TEICHERT (1970b, p. 73) found this conclusion difficult to accept and stated some of their points as follows: "It appears that the direction of comparisons has been downward stratigraphically with no attention paid to the possibility of longer ranges of the taxa concerned. None of the species have been definitely identified and named, and most of the 19 genera recognized in the white sandstone assemblage have long ranges. Eleven of them originated in the Carboniferous or Devonian; eight genera range through most or all of the Permian. Attempting to make refined correlations on such long-ranging genera is indeed a difficult task."

The "Permian" brachiopods of the Kathwai Member, GRANT (1970) considered to be quite different from those of the underlying white sandstone unit. It should be noted, however, that of the 11 genera recorded from the Kathwai Member seven are also present in the white sandstone unit and only one species, a new one, has been definitely named. As to the age of the brachiopod assemblage from the dolomite unit of the Kathwai Member, GRANT (1970, p. 125) concluded that it "contains some genera that point to a latest Guadalupian (Lamar equivalent) age, and others that point to a Dzhulfian, or even early Triassic age." The same technique and philosophies applied to the white sandstone assemblage were used to analyze the Kathwai brachiopods. Of the 11 genera in the Kathwai fauna, six, or possibly seven, originated in the Carboniferous, or earlier, and four appear to be confined to the Permian. The extended ranges of some of the genera had already been pointed out by STEPANOV (1967).

The Dzhulfa area of Soviet Armenia has long been a classic area for study of the Permian-Triassic boundary problem. One of the intriguing aspects was the reported association of Permian brachiopods and corals with "ceratitic" ammonoids. A comprehensive report on the Permian and Triassic faunas of this region, edited by Ruzhentsev and Sarycheva (1965), is an important contribution. They placed the Permian-Triassic boundary just beneath an approximately 20-meter thickness of strata that contain Permian brachiopods and corals along with ceratitic ammonoids (*"Tompophiceras," Dzhulfites, "Bernhardites,"* and *Paratirolites*). These strata were included in the Induan Stage (lowest Scythian). These critical strata are now known to crop out south of the Aras River, in northwestern Iran. A comprehensive report on these occurrences has been published by Stepanov et al. (1969). These authors, in discussing the 20 m of beds with the so-called mixed fauna, came to the same conclusion as Ruzhentsev and Sarycheva (1965), that they are earliest Triassic (Induan) in age. Kummel and Teichert had the opportunity of visiting the same Iranian locality, known as Kuh-e-Ali Bashi. Their report (Teichert, Kummel, & Sweet, 1973) presented a detailed summary of the stratigraphy and paleontology of the critical 20 m of strata with the so-called mixed fauna (which they named the Ali Bashi Formation). One of the most important finds were ammonoids previously known only from the Changhsingian fauna of southern China. Their primary conclusion was that the Ali Bashi Formation does not contain any distinctively Triassic components and is latest Permian in age.

The only report known to me on "Permian" brachiopods in Lower Triassic strata from Japan is a brief statement by Nakazawa (1971) concerning a relict productoid (*Plicatifer?*) and a Paleozoic-type bivalve (*"Streblochondria"*) from the middle Scythian Kurotaki Limestone in Shikoku.

The first mention of a mixed Permian-Triassic fauna from East Greenland in the boundary zone is by Spath (1930, p. 69) and he commented further on this matter in 1935. At the time of Spath's first writing on this subject, the beds underlying the Triassic were believed to be Early Carboniferous in age. It is easy to understand why redeposition seemed to be the natural answer. Shortly after, the age was changed to Late Carboniferous (Frebold, 1931). Frebold (1932) moved part of the "Upper Carboniferous" into the Lower Permian and Aldinger (1935), revising the fish fauna of the "*Posidonomya* shale," placed these beds in the Artinskian. Miller and Furnish (1940) described the ammonoid *Cyclolobus* from the *Martinia* beds of Clavering Ø; in line with the thinking of that time, this made these beds latest Permian in age. Trümpy (1960, 1961) visited the Kap Stosch area of East Greenland and made observations on the Permian-Triassic boundary problem. His main conclusion was that mixed faunas did indeed occur and that in East Greenland faunas of Permian and Triassic habitus co-existed for a time.

In the summer of 1967 Teichert and Kummel visited Kap Stosch specifically to study the Permian-Triassic boundary beds. The primary conclusions of their study (Teichert & Kummel, 1973, 1976) are as follows: 1) The Permian fossils occur in different localities from the very base of the Triassic System up to a distance of between 80 and 100 m above this base. They represent remains of bryozoans and brachiopods, with minor fragments of ostracodes, crinoids, and echinoids; 2) between Rivers 7 and 14 (southwest of Kap Stosch) all Permian fossils occur in arkosic sandstone or conglomerate along with ophiceratids or *Otoceras* or both; 3) almost all Permian fossils are damaged or badly broken into unidentifiable fragments; 4) the specimens that can be identified belong to species that occur in the typical *Productus* limestone and *Martinia* shale facies of the Upper Permian directly below the Triassic; 5) the Permian-Triassic sections southwest of Kap Stosch are of homogeneous shale, silty shale, and siltstone facies; nearly all are markedly soft. Solifluction has so badly affected all outcrops that meaningful stratigraphic sections are next to impossible to obtain; 6) the Permian fossils in the Early Triassic sediments do not represent survivors from the Permian into the Triassic but were redeposited from some land area to the west in which Permian outcrops occurred. A most

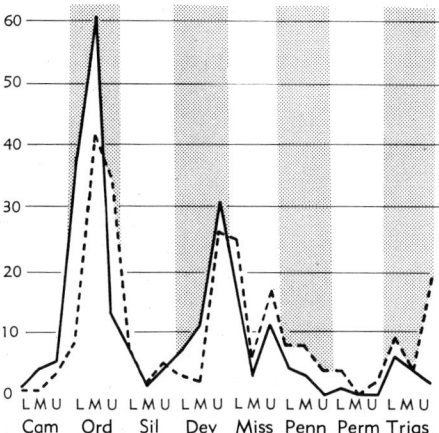

Fig. 9. Total new conodont form genera (solid line) that appeared during Cambrian through Triassic compared with total number that became extinct (dashed line) during the same period (from Clark, 1972).

probable mode of transport of some of the specimens was in argillaceous boulders.

WATERHOUSE (1972) has described a single specimen consisting of the ventral valve of an overtoniid brachiopod (?*Krotovia* sp.) from the Blind Fiord Formation (Early Triassic-Griesbachian age) on Axel Heiberg Island. No other fossils were found in the bed that yielded the brachiopod specimen. The stratigraphy of this area has been worked out by R. THORSTEINSSON and E. T. TOZER and the age assignment and correlation of the beds seem to be most reasonable. WATERHOUSE (1972, p. 486) concluded "It is not clear whether the specimen was derived from Permian rocks or was really of Griesbachian age. The latter appears likely from the fact that no similar specimens are known from underlying Permian."

Recently Permian brachiopods have been noted from the Dinwoody Formation of western Wyoming and Montana. All that is available in print on this is a short note by GRANT and COOPER (1973, p. 585) that is worth quoting as it warrants following up: "Recently D. W. Boyd (University of Wyoming) sent Grant a collection of brachiopods from the lower part of the Dinwoody Formation at Teton Pass, Wyoming. Preliminary identification suggested a Dzhulfian age, with a small productid like *Spinomarginifera*, or *Echinauris* and possibly also *Araxathyris*, although Cooper would call the latter *Composita*. Field work in Montana since the Symposium also turned up Permian type brachiopods as high as 8 feet above the base of the basal Dinwoody at two places in southern Montana and W. L. Stokes (University of Utah) brought to the Symposium a small collection of well-preserved brachiopods from the Dinwoody Formation that certainly look Permian. At least seven genera of brachiopods of Permian aspect now have been obtained from the supposedly Triassic Dinwoody Formation, a situation analogous to that of the Dolomite unit of the Salt Range."

It should be clear from the above brief review that there is at the moment no consensus as to precisely how the Permian-Triassic boundary should be defined nor as to the interpretation of "Permian" brachiopods along with Triassic ammonoids.

CONODONTS

One of the dramatic new elements in the biostratigraphic analysis of the Triassic has been the study of its conodonts. Systematic study of these Triassic fossils began in 1956 (MÜLLER, 1956; TATGE, 1956). There had been earlier reports of Triassic conodonts but these generally were interpreted as reworked Paleozoic forms. Since 1956 there has been an increasing number of Triassic conodont faunal studies, especially of the Early Triassic faunas.

A brief but excellent overview of conodont diversity from the Cambrian through the Triassic has been presented by CLARK (1972). He produced a number of graphs to illustrate conodont diversity through time, one of which is here produced as Figure 9. CLARK's (1972, p. 150) analysis of this graph is as follows: "Here, the total new form genera that evolved during each epoch is plotted against the number of form genera that became extinct during that epoch. In this figure, the presence of the dashed line above the solid line is the signal of an approaching crisis, i.e., more form genera were becoming extinct than were evolving. Significantly, the Middle Ordovician peak is followed by a large drop in the Silurian when approximately as many

Series	Stage	Substage	Ammonoid Zones (after Tozer, 1971)	Conodont Zones
UPPER TRIASSIC	RHAETIAN		Choristoceras marshi	Conodonts present but not diagnostic
UPPER TRIASSIC	NORIAN	UPPER NORIAN	Rhabdoceras suessi	22 Epigondolella bidentata
UPPER TRIASSIC	NORIAN	MIDDLE NORIAN	Himavatites columbianus / Drepanites rutherfordi / Juvavites magnus	21
UPPER TRIASSIC	NORIAN	LOWER NORIAN	Malayites dawsoni / Mojsisovicsites kerri	20 Epigondolella abneptis
UPPER TRIASSIC	CARNIAN	UPPER CARNIAN	Klamathites macrolobatus / Tropites welleri / Tropites dilleri	19 Paragondolella polygnathiformis
UPPER TRIASSIC	CARNIAN	LOWER CARNIAN	Sirenites nanseni / Trachyceras obesum	18 Neospathodus newpassensis
MIDDLE TRIASSIC	LADINIAN	UPPER LADINIAN	Frankites sutherlandi / Maclearnoceras maclearni / Meginoceras meginae	17 Epigondolella mungoensis
MIDDLE TRIASSIC	LADINIAN	LOWER LADINIAN	Progonoceratites poseidon / Protrachyceras subasperum	16 Neogondolella mombergensis
MIDDLE TRIASSIC	ANISIAN	UPPER ANISIAN	Gymnotoceras occidentalis / Gymnotoceras meeki / Gymnotoceras rotelliforme	15 Neogondolella constricta
MIDDLE TRIASSIC	ANISIAN	M. ANISIAN	Balatonites shoshonensis / Acrochordiceras hyatti	14
MIDDLE TRIASSIC	ANISIAN	L. ANISIAN	Lenotropites caurus	
LOWER TRIASSIC (SCYTHIAN)	SPATHIAN		Neopopanoceras haugi	13 Neospathodus timorensis
LOWER TRIASSIC (SCYTHIAN)	SPATHIAN		Subcolumbites beds	12 Neogondolella jubata
LOWER TRIASSIC (SCYTHIAN)	SPATHIAN			11
LOWER TRIASSIC (SCYTHIAN)	SPATHIAN		Columbites & Tirolites beds	10 Platyvillosus
LOWER TRIASSIC (SCYTHIAN)	SMITHIAN		Wasatchites tardus	9 Neogondolella milleri
LOWER TRIASSIC (SCYTHIAN)	SMITHIAN			8 Neospathodus conservativus
LOWER TRIASSIC (SCYTHIAN)	SMITHIAN		Euflemingites romunderi	7 Parachirognathus—Furnishius
LOWER TRIASSIC (SCYTHIAN)	SMITHIAN			6 Neospathodus pakistanensis
LOWER TRIASSIC (SCYTHIAN)	DIENERIAN		Vavilovites sverdrupi	5 Neospathodus cristagalli
LOWER TRIASSIC (SCYTHIAN)	DIENERIAN			4 Neospathodus dieneri
LOWER TRIASSIC (SCYTHIAN)	DIENERIAN		Proptychites candidus	3 Neospathodus kummeli
LOWER TRIASSIC (SCYTHIAN)	GRIESBACHIAN	UPPER GRIESBACHIAN	Proptychites strigatus / Ophiceras commune	2 Neogondolella carinata
LOWER TRIASSIC (SCYTHIAN)	GRIESBACHIAN	LOWER GRIESBACHIAN	Otoceras boreale / Otoceras concavum	1 Anchignathodus typicalis

FIG. 10. Triassic ammonoid and conodont zones (after Sweet et al., 1971).

form genera were becoming extinct as were evolving. The real crisis is first evident during the Late Ordovician when more form genera became extinct than appeared. The consequence is the first conodont crisis, that of the Silurian. Number of new form genera exceeds number of extinct taxa by the Early Devonian and the second evolutionary expansion was during the Late Devonian. Significantly, however, the large number of extinct form genera during the Late Devonian and succeeding Early Mississippian more than offset this expansion and by the Early Mississippian, more genera were becoming extinct than were evolving. This pattern was never reversed during the following years of conodont evolution. The dip during the Middle Mississippian may not be real. Very few students have used this "epoch" designation (middle) during the past 30 years and numbers put on the cards express this bias. Probably a straight line drop from Late Devonian to Late Mississippian is more accurate. The second crisis is a result of the gradual reduction in number of form genera which reached a low in the Permian when no new form genera appear but extinction continued. This is the most profound crisis that conodonts experienced until the Triassic extinction."

In regard to the ancestry of Triassic forms, only one or two Paleozoic genera survived the Permian crisis and relatively few Permian forms contributed to the later Triassic bloom. Considering the conodont assemblages at or near the Permian-Triassic boundary SWEET (1973, p. 630) summarized the data as follows: "All major Permian conodont stocks and most of the species known from latest Permian rocks passed the notorious Permian-Triassic filter with seeming indifference, hence there is no detectable change in conodont faunas at the level of the Permian-Triassic boundary in any of the presumably complete sections from which conodonts are currently known."

Conodont biostratigraphical zonation is developing at a rapid rate. The best general summary of this aspect of conodont studies is that by SWEET et al. (1971). In that study the authors recognized a sequence of 22 faunal zones (Fig. 10). Further refinement of this scheme is actively being pursued.

REFERENCES

Aldinger, Hermann, 1935, *Das Alter der jungpaläozoischen Posidonomyaschiefer von Ostgrönland:* Medd. Grønland, v. 98, no. 4, 24 p.

Anderson, H. M., & Anderson, J. M., 1970, *A preliminary review of the biostratigraphy of the uppermost Permian, Triassic and lowermost Jurassic of Gondwanaland:* Palaeont. Africana, v. 13, p. 1-22, charts 1-22.

Anderson, J. M., 1973, *The biostratigraphy of the Permian and Triassic:* Palaeont. Africana, v. 16, p. 59-83.

Arthaber, Gustav von, 1927, *Ammonoidea Leiostraca aus der oberen Trias von Timor. 2^E Nederlandsche Timor-Expeditie, 1916:* Mijnw. Nederl.-Indië, Jaarb., Jaarg. 1926, p. 1-174, pl. 1-20.

Assereto, R., Bosellini, A., Fantini Sestini, N., & Sweet, W. C., 1973, *The Permian-Triassic boundary in the southern Alps (Italy):* in The Permian and Triassic Systems and their mutual boundary, Alan Logan & L. V. Hills (eds.), Canad. Soc. Petrol. Geologists, Mem. 2, p. 176-199, text-fig. 1-6.

Audley-Charles, M. G., 1968, *The geology of Portuguese Timor:* Geol. Soc. London, Mem., no. 4, 76 p., 10 text-fig., 13 pl.

Bando, Yuji, 1964, *The Triassic stratigraphy and ammonite fauna of Japan:* Tohoku Univ., Sendai, Sci. Rept., ser. 2, geol., v. 36, no. 1, 137 p., 38 text-fig., 15 pl.

Barthel, K. W., 1958, *Eine marine Faunula aus mittleren Trias von Chile:* Neues Jahrb. Mineralogie, Geologie, Paläontologie, v. 106, pt. 3, p. 352-382.

Batten, R. L., 1973, *The vicissitudes of the gastropods during the interval of Guadalupian-Ladinian time:* in The Permian and Triassic Systems and their mutual boundary, A. Logan and L. V. Hills (eds.), Canad. Soc. Petrol. Geologists, Mem. 2, p. 596-607, text-fig. 1-5.

Bion, H. S., 1914, *Tract N. of Srinagar and Pahlgam:* in India Geol. Survey, General report for 1913, Rec., v. 44, no. 1, p. 39-40.

Brennand, T. P., 1975, *The Triassic of the North Sea:* in Petroleum and the continental shelf of North-west Europe, A. W. Woodland (ed.), v. 1, Geology, p. 295-310, John Wiley & Sons (New York and Toronto).

Briden, J. C., Drewry, G. E., & Smith, A. G., 1974, *Phanerozoic equal-area world maps:* Jour. Geology, v. 82, p. 555-574, text-fig. 1-18.

Buch, Leopold von, 1831, *Explication de trois planches d'ammonites:* 4 p. (Paris).

Buchan, S. H., Challinor, A., Harland, W. B., & Parker, J. R. R., 1965, *The Triassic stratigraphy of Svalbard:* Norsk Polarinst., Skrifter, no. 135, 92 p.

Chao Kingkoo, 1959, *Lower Triassic ammonoids from western Kwangsi, China:* Palaeont. Sinica, v. 145, p. 1-355.

Clark, D. L., 1972, *Early Permian crisis and its bearing on Permo-Triassic conodont taxonomy:* Geol. Palaeont., SBI, p. 147-158, text-fig. 1-11, tab. 1-3.

Collignon, Maurice, 1973, *Ammonites du Trias inférieur et moyen d'Afghanistan:* Ann. Paléontologie, v. 59, no. 2, p. 8-39, 10 pl.

Cox, C. B., 1973, *Triassic tetrapods:* in Atlas of paleobiogeography, Anthony Hallam (ed.), p. 213-223, text-fig. 1-3, tab. 1-4, Elsevier Scientific Publ. Co. (Amsterdam, London, New York).

Davoudzadeh, M., & Seyed-Emami, K., 1972, *Stratigraphy of the Triassic Nakhlak Group, Anarak*

Region: Iran Geol. Survey, Rept., no. 28, p. 5-28, text-fig. 1-8.
Denmead, A. K., 1964, *Note on marine macrofossils with Triassic affinities from the Maryborough Basin, Queensland:* Australian Jour. Sci., v. 27, no. 4, p. 117.
Dickens, J. M., & McTavish, R. A., 1963, *Lower Triassic marine fossils from the Beagle Ridge (BMR 10) bore, Perth Basin, western Australia:* Australia Geol. Soc., Jour., v. 10, pt. 1, p. 123-140.
Diener, Carl, 1895, *The cephalopods of the Muschelkalk:* India Geol. Survey, Mem., Palaeont. Indica, ser. 15, v. 2, pt. 2, p. 1-120, pl. 1-31.—— 1897, *The Cephalopoda of the Lower Trias:* India Geol. Survey, Mem., Palaeont. Indica, ser. 15, v. 2, pt. 1, p. 1-181, pl. 1-23.——1906, *Fauna of the Tropites-Limestone of Byans:* India Geol. Survey, Mem., Palaeont. Indica, ser. 15, v. 5, no. 1, p. 1-201, pl. 1-27.——1907, *The fauna of the Himalayan Muschelkalk:* India Geol. Survey, Mem., Palaeont. Indica, ser. 15, v. 5, no. 2, p. 1-139, pl. 1-17.——1908, *Ladinic, Carnic, and Noric faunae of Spiti:* India Geol. Survey, Mem., Palaeont. Indica, ser. 15, v. 5, no. 3, p. 1-157, pl. 1-24.——1909, *Himalayan fossils. Fauna of the Traumatocrinus-Limestone of Painkhanda:* India Geol. Survey, Mem., Palaeont. Indica, ser. 15, v. 6, pt. 2, p. 1-39, pl. 1-5.——1912, *The Trias of the Himalayas:* India Geol. Survey, Mem., v. 36, pt. 3, p. 1-176, text-fig. 1-12, 2 tab.——1915, *Fossilium Catalogus, pt. 8, Cephalopoda Triadica:* 369 p., W. Junk (Berlin).——1923, *Ammonoidea Trachyostraca aus der mittleren und oberen Trias von Timor:* Mijnw. Ned. Oost-Ind., Jaarb., v. 49, p. 73-276, atlas, pl. 1-32.
Douglas, R. J. W., Gabrielse, H., Wheeler, J. O., Stott, D. F., & Belyea, H. R., 1970, *Geology of western Canada:* Canada Geol. Survey, Econ. Geol. Rept., no. 1, p. 365-488, text-fig. 1-53, 20 pl.
Druckman, Yehezkeel, 1974, *The stratigraphy of the Triassic sequence in southern Israel:* Israel Geol. Survey, Bull., no. 64, 94 p., 20 text-fig., 31 maps.
Edgell, H. S., 1965, *Triassic ammonite impressions from the type section of the Minchin Siltstone, Perth Basin:* West. Australia Geol. Survey, Ann. Rept. for 1963-1964, p. 55-57, 1 pl.
Farsan, N. M., 1972, *Stratigraphische und paläogeographische Stellung der Khenjan-Serie und deren Pelecypoden (Trias, Afghanistan):* Palaeontographica, v. 140, Abt. A, p. 131-191, text-fig. 1-8, pl. 38-46.——1975, *Pelecypoden aus der Khenjan-Serie von Zentral-Afghanistan (Mittel-Trias):* Palaeontographica, v. 149, Abt. A, p. 119-139, pl. 20,21.
Fischer, Jochen, 1971, *Zur Geologie des Kohe Safi bei Kabul (Afghanistan):* Neues Jahrb. Geologie Paläontologie, Abhandl., v. 139, no. 3, p. 267-315, text-fig. 1-17.
Fleming, C. A., 1962, *New Zealand biogeography: a palaeontologist's approach:* Tuatara, v. 10, p. 53-108.——1967, *Biogeographic change related to Mesozoic orogenic history in the southwest Pacific:* Tectonophysics, v. 4, no. 4-6, p. 419-427, text-fig. 1,2.——1970, *The Mesozoic of New Zealand: chapters in the history of the circum-Pacific mobile belt:* Geol. Soc. London, Quart. Jour., v. 125, no. 498, pt. 2, p. 125-170, text-fig. 1-27, pl. 8,9.
Fleming, P. J. G., 1966, *Eotriassic marine bivalves from the Maryborough Basin, south-east Queensland:* Queensland Geol. Survey, Publ. 333, Palaeont. Paper 8, p. 17-29, text-fig. 1-3, pl. 1-9.
Frebold, Hans, 1931, *Das marine Oberkarbon Ostgrönlands:* Medd. Grønland, v. 84, no. 2, p. 1-88, 8 pl.——1932, *Marines Unterperm in Ostgrönland und die Frage der Grenzziehung zwischen dem pelagischen Oberkarbon und Unterperm:* Medd. Grønland, v. 84, no. 4, p. 1-35, 1 pl.
Fuchs, G. R., 1964, *Note on the geology of the palaeozoics and mesozoics of the Tibetan zone of the Dolpo Region (Nepal-Himalaya):* Geol. Bundesanstalt, Verhandl., v. 1, p. 6-15, 1 map.——1967, *Zum Bau des Himalaya:* Öst. Akad. Wiss., Math.-naturwiss. Kl., Denkschr., v. 113, p. 1-211, 70 text-fig., 9 maps.
Furon, R., & Rosset, L. F., 1951, *Contributions à l'étude du Trias en Afghanistan:* Museum Natl. Histoire Nat., Bull., sér. 2, v. 23, p. 558-565.
Gansser, Augusto, 1964, *The geology of the Himalayas:* p. iii-xv, 3-289, 149 text-fig., Interscience Publishers, John Wiley & Sons, Ltd. (London, New York, Sydney).
Grant, R. E., 1970, *Brachiopods from Permian-Triassic boundary beds and age of Chhidru Formation, West Pakistan:* in Stratigraphic boundary problems, Permian and Triassic of West Pakistan, Bernhard Kummel & Curt Teichert (eds.), Dept. Geology, Univ. Kansas, Spec. Publ. 4, p. 117-151, text-fig. 1, pl. 1-3.
——, & Cooper, G. A., 1973, *Brachiopods and Permian correlations:* in The Permian and Triassic systems and their mutual boundary, A. Logan & L. V. Hills (eds.), Canad. Soc. Petrol. Geologists, Mem. 2, p. 572-595, text-fig. 1-7.
Hallam, Anthony (ed.), 1973, *Atlas of palaeobiogeography:* 531 p., Elsevier Scientific Publ. Co. (Amsterdam, London, New York).
Harrington, H. J., 1962, *Paleogeographic development of South America:* Am. Assoc. Petrol. Geologists, Bull., v. 46, no. 10, p. 1773-1814, text-fig. 1-34.
Hayden, H. H., 1907, *The stratigraphical position of the Gangamopteris Beds of Kashmir:* India Geol. Survey, Records, v. 36, no. 1, p. 23-39.——1911, *The geology of northern Afghanistan:* Geol. Survey India, Mem., v. 39, p. 1-96.
Healey, Maud, 1908, *The fauna of the Napeng beds*

or the Rhaetic beds of upper Burma: India Geol. Survey, Mem., Palaeont. Indica, n. ser., v. 2, no. 4, p. 1-88, 10 pl.

Ishii, Ken-ichi, Fischer, Jochen, & Bando, Yuji, 1971, Notes on the Permian-Triassic boundary in eastern Afghanistan: Jour. Geosci., Osaka City Univ., v. 14, art. 1, p. 1-19, text-fig. 1,2, 3 pl.

Jaworski, Erich, 1922, Beiträge zur Geologie und Paläontologie von Südamerica XXVI. Die marine Trias in Südamerika: Neues Jahrb. Mineralogie, Geologie, Paläontologie, Beil.-Bd. 47, Heft 1, p. 93-200, pl. 4-6.

Kaever, Matthias, 1969, Die Trias Afghanistans: Zentralbl. Geologie Paläontologie, Teil I, v. 1, p. 170-186.

Khuc, Vu, Dagys, A. S., Kiparisova, L. D., Nguyen, Nguyen Ba, Srebrodolskaia, I. N., & Bao, Troung Cam, 1965, Les fossiles charactéristiques du Trias au nord Viet-Nam: Direction Générale Géol. RDV. Hanoi, p. 6-77, pl. 1-19.

Kiparisova, L. D., Okuneva, T. M., & Oleynikov, A. N., 1973, The Triassic system in the U.S.S.R.: in The Permian and Triassic systems and their mutual boundary, A. Logan & L. V. Hills (eds.), Canad. Soc. Petrol. Geologists, Mem. 2, p. 137-149, text-fig. 1, tab. 1,2.

——, & Popov, Yu.N., 1956, Raschlenenie nizhnego otdela triasovoii sistemy na yarusy: Akad. Nauk SSSR, Doklady, v. 109, no. 4, p. 842-845. [Separation of the lower part of the Triassic System into stages.]

——, Radchenko, G. P., & Gorskiy, V. P. (eds.), 1973, Stratigrafiya SSSR. Triasovaya sistema: Akad. Nauk SSSR, Minist. Geol., Minist. Vysshego Sred. Spetsial. Obrazovan. SSSR, 557 p., 96 text-fig., appendix 1-9. [Stratigraphy of the USSR. Triassic System.]

Kittl, Ernst, 1903, Die Cephalopoden der oberen Werfener Schichten von Muć in Dalmatien sowie von anderen dalmitinischen, bosnischherzegowinischen und alpinen Lokalitäten: Geol. Reichsanst. Wien, Abhandl., v. 20, p. 1-77, pl. 1-11.

Korchinskaya, M. V., 1973, Biostratigraphy of Triassic deposits of Svalbard: in The Permian and Triassic systems and their mutual boundary, A. Logan & L. V. Hills (eds.), Canad. Soc. Petrol. Geologists, Mem. 2, p. 261-268, 1 tab.

Krafft, Albrecht von, 1902, Lower Trias fossils: Geol. Survey India, Gen. Rept., 1901-1902, p. 5.

——, & Diener, Carl, 1909, Lower Triassic Cephalopoda from Spiti, Malla-Johar and Byans: India Geol. Survey Mem., Palaeont. Indica, ser. 15, v. 6, no. 1, p. 1-186, pl. 1-31.

Kummel, Bernhard, 1953a, American Triassic coiled nautiloids: U.S. Geol. Survey, Prof. Paper 250, p. 1-104, 43 text-fig., pl. 1-19.——1953b, Middle Triassic ammonites from Peary Land: Medd. Grønland, v. 127, no. 1, p. 1-21, pl. 1.——1954, Triassic stratigraphy of southeastern Idaho and adjacent areas: U.S. Geol. Survey, Prof. Paper 254H, p. 165-194.——1966a, A Triassic ammonite from the Hindubagh Region, Baluchistan, West Pakistan: Harvard Univ., Museum Comp. Zoology, Breviora, no. 248, p. 1-5, pl. 1.—— 1966b, The Lower Triassic formations of the Salt Range and Trans-Indus Ranges, West Pakistan: Harvard Univ., Museum Comp. Zoology, Bull., v. 134, no. 10, p. 361-429, pl. 1-4.—— 1968, Additional Scythian ammonoids from Afghanistan: Harvard Univ., Museum Comp. Zoology, Bull., v. 136, no. 13, p. 483-569, text-fig. 1-5, pl. 1-3.——1969, Ammonoids of the late Scythian (Lower Triassic): Harvard Univ., Museum Comp. Zoology, Bull., v. 137, no. 3, p. 311-690, pl. 1-71.——1970, Ammonoids from the Kathwai Member: in Stratigraphic boundary problems, Permian and Triassic of West Pakistan, Bernhard Kummel & Curt Teichert (eds.), Dept. Geol., Univ. Kansas, Spec. Publ. 4, p. 177-192, text-fig. 1,2, 2 pl.

——, & Erben, H. K., 1968, Lower and Middle Triassic cephalopods from Afghanistan: Palaeontographica, v. 129, Abt. A, p. 95-148, text-fig. 1-20, pl. 19-24.

——, & Fuchs, R. L., 1953, The Triassic of South America: Soc. Geol. Perú, Bol., v. 26, p. 95-120.

——, & Teichert, Curt, 1966, Relations between the Permian and Triassic formations in the Salt Range and Trans-Indus ranges, West Pakistan: Neues Jahrb. Geologie, Paläontologie, Abhandl., v. 125, p. 297-333, text-fig. 1-4, pl. 27-28.—— (eds.), 1970a, Stratigraphic boundary problems: Permian and Triassic of West Pakistan: Dept. Geology, Univ. Kansas, Spec. Publ. 4, 474 p. ——1970b, Stratigraphy and paleontology of the Permian-Triassic boundary beds, Salt Range and Trans-Indus ranges, West Pakistan: in Stratigraphic boundary problems: Permian and Triassic of West Pakistan, Bernhard Kummel & Curt Teichert (eds.), Dept. Geol., Univ. Kansas, Spec. Publ. 4, p. 1-110, text-fig. 1-19.——1973, The Permian-Triassic boundary in central Tethys: in The Permian and Triassic systems and their mutual boundary, Alan Logan & L. V. Hills (eds.), Canad. Soc. Petrol. Geologists, Mem. 2, p. 17-34, 10 text-fig.

Kutassy, A., 1933, Fossilium Catalogus, I, Animalia, Pt. 56, Cephalopoda triadica II: p. 371-832, W. Junk (Berlin).

Liu Hung-yün, 1959, Paleogeographical maps of China [in Chinese]: 69 p., Institute of Geology and Paleontology, Academia Sinica, K'o hsüeh ch'u pan shê (Peking).

McTavish, R. A., 1973, Triassic conodont faunas from western Australia: Neues Jahrb. Geologie Paläontologie, Abhandl., v. 143, no. 3, p. 275-303, pl. 1,2.

——, & Dickens, J. M., 1974, The age of Kockatea shale (Lower Triassic), Perth Basin—A reassessment: Geol. Soc. Australia, Jour., v. 21, pt. 2, p. 195-201, text-fig. 1,2.

Marwick, John, 1953, *Faunal migrations in New Zealand seas during the Triassic and Jurassic:* New Zealand Jour. Sci. Technology, v. B34, no. 5, p. 317-321.

Middlemiss, C. S., 1910, *A revision of the Silurian-Trias sequence in Kashmir:* India Geol. Survey, Records, v. 40, no. 3, p. 206-260.

Mikan, F. A., & Sweet, W. C., 1974, *Graphic correlation of key Permo-Triassic sections in Kashmir, Pakistan and Iran:* Geol. Soc. America, Abstracts with Programs, v. 6, no. 6, p. 531.

Miller, A. K., & Furnish, W. M., 1940, *Permian ammonoids of the Guadalupe Mountain region and adjacent regions:* Geol. Soc. America, Spec. Paper 26, 242 p., 44 pl.

Minato, Masao, Gorai, M., & Hunahashi, M. (eds.), 1965, *The geologic developments of the Japanese Islands:* 442 p., 137 text-fig., 30 maps, 44 tab., Tsukiji Shokan Co., Ltd. (Tokyo).

Mojsisovics, E. von, Waagen, Wm., & Diener, Carl, 1895, *Entwurf einer Gliederung der pelagischen Sedimente des Trias-Systems:* Akad. Wiss. Wien, Sitzungsber., v. 104, no. 1, p. 1271-1302.

Müller, K. J., 1956, *Triassic conodonts from Nevada:* Jour. Paleontology, v. 30, no. 4, p. 818-830, pl. 95,96.

Nakazawa, Keiji, 1971, *On the Lower Triassic Kurotaki fauna in Shikoku and its allied faunas in Japan:* Kyoto Univ., Mem. Fac. Sci., ser. geol. min., v. 37, no. 2, p. 163-172.

——, & Bando, Yuji, 1968, *Lower and Middle Triassic ammonites from Portuguese Timor (Palaeontological study of Portuguese Timor, 4):* Kyoto Univ., Mem. Fac. Sci., ser. geol. min., v. 34, no. 2, p. 83-114, text-fig. 1-14, pl. 4-7.

——, Kapoor, H. M., Ishii, Ken-ichi, Bando, Yuji, Maegoya, T., Shimizu, D., Nogami, Yasuo, Tokuoka, T., & Nohda, Susumu, 1970, *Preliminary report on the Permo-Trias of Kashmir:* Kyoto Univ., Mem. Fac. Sci., ser. geol. min., v. 37, no. 2, p. 163-172.

——, ——, ——, ——, Okimura, Y., & Tokuoka, T., 1975, *The Upper Permian and the Lower Triassic in Kashmir, India:* Kyoto Univ., Mem. Fac. Sci., ser. geol. min., v. 42, no. 1, p. 1-106, text-fig. 1-15, 12 pl.

——, & Runnegar, Bruce, 1973, *The Permian-Triassic boundary: A crisis for bivalves?:* in The Permian and Triassic systems and their mutual boundary, A. Logan & L. V. Hills (eds.), Canad. Soc. Petrol. Geologists, Mem. 2, p. 608-621, text-fig. 1-4, tab. 1-4.

Newell, N. D., & Boyd, D. W., 1970, *Oyster-like Permian Bivalvia:* Am. Museum Nat. History, Bull., v. 143, no. 4, p. 217-282, text-fig. 1-34.

Nogami, Yasuo, 1968, *Trias-Conodonten von Timor, Malaysien und Japan (Palaeontological study of Portuguese Timor, 5):* Kyoto Univ., Mem. Fac. Sci., ser. geol. min., v. 34, no. 2, p. 115-136, pl. 8-11.

Pakuckas, C., 1928, *Nachtrag zur mittel- und obertriadischen Fauna der Ammonea trachyostraca C. Diener's aus Timor, mit Einleitung und stratigraphischer Zusammenfassung von G. von Arthaber:* Mijnw. Ned. Ind., Jaarb., v. 56, p. 143-218, pl. 1,2.

Pascoe, E. H., 1959, *A manual of the geology of India and Burma:* v. 2, 3rd edit., p. x-xxii, 485-1343, Government of India Press (Calcutta).

Picard, L., & Flexer, A., 1974, *Studies on the stratigraphy of Israel. The Triassic:* p. 7-62, text-fig. 1-25, Israel Institute of Petroleum (Tel Aviv).

Rostovtsev, K. O., & Azaryan, N. R., 1973, *The Permian-Triassic boundary in Transcaucasia:* in The Permian and Triassic systems and their mutual boundary, Alan Logan & L. V. Hills (eds.), Canad. Soc. Petrol. Geologists, Mem. 2, p. 89-99, text-fig. 1, pl. 1-3.

Runnegar, Bruce, 1965, *The bivalves Megadesmus Sowerby and Astartila Dana from the Permian of eastern Australia:* Geol. Soc. Australia, Jour., v. 12, p. 227-252.——1969, *A Lower Triassic ammonoid fauna from southeast Queensland:* Jour. Paleontology, v. 43, no. 3, p. 818-828, text-fig. 1-3, pl. 103-104.

Ruzhentsev, V. E., & Sarycheva, T. G. (eds.), 1965, *Razvitie i smena morskikh organizmov na rubezhe Paleozoya i Mezozoya:* Akad. Nauk SSSR, Paleont. Inst., Trudy, v. 108, p. 1-431. [Development and change of marine organisms at the Paleozoic-Mesozoic boundary.]

Sahni, M. R., 1938, *Discovery of Lower Triassic at Nakkam, northern Shan States:* 25th Indian Sci. Congress, Proc., pt. 4, Abstracts, sec. 3, no. 19, p. 114.

Schindewolf, O. H., 1934, *Zur Stammesgeschichte der Cephalopoden:* Preuss. Geol. Landesanst., Jahrb., v. 55, p. 258-283, pl. 19-22.

Seyed-Emami, Kazem, 1971, *A summary of the Triassic in Iran:* Geol. Survey Iran, Rept., no. 20, p. 41-53, text-fig. 1,2.——1975, *A new species of Distichites (Ammonoidea) from the Upper Triassic Nayband Formation of the Zefreh area (central Iran):* Neues Jahrb. Geologie Paläontologie, Monatsh., Jahrg. 1975, no. 12, p. 734-744.

Shevyrev, A. A., 1968, *Triasovye ammonoidei yuga SSSR:* Akad. Nauk SSSR, Paleont. Inst., Trudy, v. 119, p. 1-272. [Triassic ammonoids of the southern USSR.]

Silberling, N. J., & Tozer, E. T., 1968, *Biostratigraphic classification of marine Triassic in North America:* Geol. Soc. America, Spec. Paper 110, p. 1-63.

Skwarko, S. K., 1967, *First Upper Triassic and ?Lower Jurassic marine Mollusca from New Guinea:* Bur. Min. Res., Geol. Geophys., Canberra, Bull. 75, p. 37-82, 6 pl.——1973a, *Middle and Upper Triassic Mollusca from Yuat River, eastern New Guinea:* Bur. Min. Res., Geol. Geophys., Canberra, Bull. 126, p. 28-50, text-fig. 1-5, pl. 13-27.——1973b, *On the discovery of*

Halobiidae (Bivalvia, Triassic) in New Guinea: Bur. Min. Res., Geol. Geophys., Canberra, Bull. 126, p. 51-54, pl. 28.

——, & **Kummel, Bernhard,** 1974, *Marine Triassic molluscs of Australia and New Guinea:* Bur. Min. Res., Geol. Geophys., Canberra, Bull. 150, p. 1-17, pl. 37-42.

Spath, L. F., 1930, *The Eotriassic invertebrate fauna of East Greenland:* Medd. Grønland, v. 83, p. 1-90, pl. 1-12.——1935, *Additions to the Eotriassic invertebrate fauna of East Greenland:* Medd. Grønland, v. 98, p. 1-115, pl. 1-23.

Spinosa, Claude, Furnish, W. M., & Glenister, B. F., 1975, *The Xenodiscidae, Permian ceratitoid ammonoids:* Jour. Paleontology, v. 49, no. 2, p. 239-283, 22 text-fig., 6 pl.

Stepanov, D. L., 1967, *Carboniferous stratigraphy of Iran:* Iran Geol. Survey, internal rept. (unpubl.), geol. note 38.

——, **Golshani, F., & Stöcklin, J.,** 1969, *Upper Permian and Permian-Triassic boundary in north Iran:* Iran Geol. Survey, Rept. no. 12, p. 3-72, text-fig. 1-6, geol. pl. 1-3, pl. 1-15.

Sweet, W. C., 1970a, *Uppermost Permian and Lower Triassic conodonts of the Salt Range and Trans-Indus ranges, West Pakistan:* in Stratigraphic boundary problems: Permian and Triassic of West Pakistan, Bernhard Kummel & Curt Teichert (eds.), Dept. Geology, Univ. Kansas, Spec. Publ. 4, p. 207-271, text-fig. 1-6, pl. 1-5.——1970b, *Permian and Triassic conodonts from a section at Guryul Ravine, Vihi District, Kashmir:* Univ. Kansas Paleont. Contrib., Paper 49, p. 1-10, pl. 1.——1973, *Late Permian and Early Triassic conodont faunas:* in The Permian and Triassic systems and their mutual boundary, Alan Logan & L. V. Hills (eds.), Canad. Soc. Petrol. Geologists, Mem. 2, p. 630-646, 5 text-fig.

——, **Mosher, L. C., Clark, D. L., & Collinson, J. W.,** 1971, *Conodont biostratigraphy of the Triassic:* in Symposium on conodont biostratigraphy, W. C. Sweet & S. M. Bergström (eds.), Geol. Soc. America, Mem., no. 127, p. 441-465.

Takai, F., Matsumoto, T., & Toriyama, R. (eds.), 1963, *Geology of Japan:* 279 p., 30 text-fig., Univ. California Press (Berkeley).

Tamura, Minoru, Hashimoto, W., Igo, H., Ishibashi, T., Iwai, J., Kobayashi, Teichii, Koike, T., Pitakpaivan, K., Sato, T., & Yin, E. H., 1975, *Contributions to the geology and paleontology of Southeast Asia, CLI; The Triassic System of Malaysia, Thailand and some adjacent areas:* in Geology and paleontology of Southeast Asia, Teichii Kobayashi & R. Toriyama (eds.), v. 15, p. 103-149, text-fig. 1,2, tab. 1-8, Univ. Tokyo Press (Tokyo).

Taraz, Hushang, 1969, *Permo-Triassic section in central Iran:* Am. Assoc. Petrol. Geologists, Bull., v. 53, no. 3, p. 688-693.——1971a, *Uppermost Permian and Permian-Triassic passage beds in central Iran (Abst.):* Internatl. Permian-Triassic Conference, Calgary, 1971, Program with Abstracts, Canad. Petrol. Geologists, Bull., v. 19, no. 2, p. 364-365.——1971b, *Uppermost Permian and Permo-Triassic transition beds in central Iran:* Am. Assoc. Petrol. Geologists, Bull., v. 55, no. 8, p. 1280-1294.——1972, *Géologie de la région Surmaq-Deh Bid, Iran Central:* Univ. Paris-Sud, Orsay, Thèse Sér. A, no. 1007, p. 224.——1973, *Correlation of uppermost Permian in Iran, Central Asia, and South China:* Am. Assoc. Petrol. Geologists, Bull., v. 57, no. 6, p. 1117-1133.——1974, *Geology of the Surmaq-Deh Bid area Abadeh Region, Central Iran:* Iran Geol. Survey, Rept. no. 37, p. 3-148, 50 text-fig., 17 photo.

Tatge, Ursula, 1956, *Conodonten aus dem germanischen Muschelkalk:* Paläont. Zeitschr., v. 30, p. 108-127, 129-147, pl. 5,6.

Teichert, Curt, & Kummel, Bernhard, 1973, *Permian-Triassic boundary in the Kap Stosch area, East Greenland:* in The Permian and Triassic systems and their boundary, Alan Logan & L. V. Hills (eds.), Canad. Soc. Petrol. Geologists, Mem. 2, p. 269-285, text-fig. 1-3.——1976, *Permian-Triassic boundary in the Kap Stosch area, East Greenland:* Medd. Grønland, v. 197, no. 5, p. 1-47, text-fig. 1-7, pl. 1-16, tab. 1.

——, ——, & **Kapoor, H. M.,** 1970, *Mixed Permian-Triassic fauna, Guryul Ravine, Kashmir:* Science, v. 167, p. 174-175, text-fig. 1.

——, ——, & **Sweet, Walter,** 1973, *Permian-Triassic strata, Kuh-e-Ali Bashi, northwestern Iran:* Harvard Univ. Museum Comp. Zoology, Bull., v. 145, no. 8, p. 359-472.

Thomson, M. R. A., 1975, *First marine Triassic fauna from the Antarctic Peninsula:* Nature, v. 257, p. 577-578, text-fig. 1,2.

Thorsteinsson, R., & Tozer, E. T., 1970, *Geology of the Arctic Archipelago:* Geol. Survey Canada, Econ. Geology Rept. 1, ed. 5, p. 548-590, text-fig. 1-12, pl. 1-8.

Tozer, E. T., 1965, *Lower Triassic stages and ammonoid zones of Arctic Canada:* Canada Geol. Survey, Paper 65-12, p. 1-14, text-fig. 1,2.——1967, *A standard for Triassic time:* Canada Geol. Survey, Bull. 156, p. 1-103, pl. 1-10.——1972, *Triassic ammonoids and Daonella from the Nakhlak Group Anarak Region, Central Iran:* Iran Geol. Survey, Rept. no. 28, p. 29-69, text-fig. 1-4, pl. 1-10.——1974, *Definitions and limits of Triassic stages and substages: suggestions prompted by comparisons between North America and the alpine-mediterranean region:* in The stratigraphy of the alpine-mediterranean Trias, Symposium, Helmuth Zapfe (ed.), Öst. Akad. Wiss., Schrift. Erdwiss, Komm., v. 2, p. 195-206, tab. 1.

——, & **Parker, J. R.,** 1968, *Notes of the Triassic stratigraphy of Svalbard:* Geol. Mag., v. 105, no. 6, p. 526-542.

Trümpy, Rudolf, 1960, *Über die Perm-Trias-Grenze*

in Ostgrönland and über Problematik stratigraphischer Grenzen: Geol. Rundschau, v. 49, no. 1, p. 97-103.——1961, *Triassic of East Greenland:* in Geology of the Arctic, G. O. Raasch (ed.), v. 1, p. 248-254, Univ. Toronto Press (Toronto).

Vavilov, M. N., & Lozovskiy, V. R., 1970, *K voprosu o yarusnom raschlenenii nizhnego triasa:* Akad. Nauk SSSR, Izvestiya, ser. geol., no. 9, p. 93-99, 2 tab. [*On the problem of Lower Triassic stage differentiation.*]

Vereshchagin, V. N., & Ronov, A. B. (eds.), 1968, *Triassovyy, Yurskiy i Melovoy periody:* in Atlas litologo-paleogeografitcheskikh kart SSSR, A. P. Vinogradov (ed.), v. 3, 71 maps, Ministerstvo Geol. SSSR, Akad. Nauk SSSR (Moskva). [*Triassic, Jurassic and Cretaceous Periods in Atlas of the lithological-paleogeographical maps of the USSR.*]

Waagen, Wilhelm, 1895, *Fossils from the Ceratite Formation:* India Geol. Survey, Mem., Palaeont. Indica, ser. 13, Salt Range Fossils, v. 2, p. 1-323, pl. 1-40.

Waterhouse, J. B., 1972, *A Permian overtoniid brachiopod in Early Triassic sediments of Axel Heiberg Island, Canadian Arctic and its implications on the Permian-Triassic boundary:* Canad. Jour. Earth Sci., v. 9, no. 5, p. 486-499, text-fig. 1-3, tab. 1-7.

Welter, O. A., 1914, *Die Obertriadischen Ammoniten und Nautiliden von Timor:* Paläontologie von Timor, Lief. 1, 258 p., 36 pl., E. Schweizerbart (Stuttgart).——1915, *Die Ammoniten und Nautiliden der Ladinischen und Anisischen Trias von Timor:* Paläontologie von Timor, Lief. 5, p. 71-136, pl. 83-95, E. Schweizerbart (Stuttgart).——1922, *Die Ammoniten der unteren Trias von Timor:* Paläontologie von Timor, Lief. 11, p. 83-154, pl. 155-171, E. Schweizerbart (Stuttgart).

Westermann, G. E. G., 1973, *The Late Triassic bivalve Monotis:* in Atlas of palaeobiogeography, Anthony Hallam (ed.), p. 251-258, text-fig. 1-3, tab. 1, Elsevier Scientific Publ. Co. (Amsterdam, London, New York).

Williams, Alwyn, et al., 1965, *Treatise on invertebrate paleontology, Part H, Brachiopoda:* R. C. Moore (ed.), 2 vol., xxxii + 927 p., 746 text-fig., Geol. Soc. America and Univ. Kansas Press (Lawrence, Kansas).

Wilmarth, M. G., 1925, *The geologic time classification of the United States Geological Survey compared with other classifications, accompanied by the original definitions of era, period and epoch terms:* U.S. Geol. Survey, Bull. 769, 138 p.

Wynne, A. B., 1878, *On the geology of the Salt Range in the Punjab:* India Geol. Survey, Mem., v. 14, p. 1-313.

Zakharov, Yu. D., 1973, *Novoe yarusnoe i zonalnoe raschlenenie nizhnego otdela triasa:* Akad. Nauk SSSR, Sibir. Otdel., Geol. Geofiz., no. 7, p. 51-59. [*New stage and zonal differentiation for the Lower Triassic.*]——1974, *The importance of palaeobiogeographical data for the solution of the problem on the Lower Triassic division:* in Die Stratigraphie der alpin-mediterranen Trias, Symposium, Helmuth Zapfe (ed.), Öst. Akad. Wiss., Schrift. Erdwiss. Komm., v. 2, p. 237-243.

Zapfe, Helmuth (ed.), 1974, *Die Stratigraphie der alpin-mediterranen Trias:* Öst. Akad. Wiss., Schrift. Erdwiss. Komm., v. 2, 251 p., 42 text-fig., 12 pl., 15 tab.

Zeil, W., & Ichikawa, K., 1958, *Marine Mittel-Trias in der Hochkordillere der Provinz Atacama (Chile):* Neues Jahrb. Mineralogie, Geologie, Paläontologie, v. 106, pt. 3, p. 339-351.

Ziegler, P. A., 1975, *North Sea basin history in the tectonic framework of northwestern Europe:* in Petroleum and the continental shelf of Northwest Europe, A. W. Woodland (ed.), v. 1, Geology, p. 131-148, John Wiley & Sons (New York, Toronto).

Ziegler, W. H., 1975, *Outline of the geological history of the North Sea:* in Petroleum and the continental shelf of North-west Europe, A. W. Woodland (ed.), v. 1, Geology, p. 165-187, John Wiley & Sons (New York, Toronto).

Zittel, K. A. von, 1901, *History of geology and palaeontology to the end of the nineteenth century:* 562 p., Charles Scribner's Sons (New York).

JURASSIC[1]

By HELMUT HÖLDER

[Universität Münster]

CONTENTS

	PAGE
INTRODUCTION	A390
PROTOZOA	A393
PORIFERA	A393
ANTHOZOA	A393
HYDROZOA	A394
BRACHIOPODA	A394
BIVALVIA, GASTROPODA	A395
AMMONOIDEA	A397
The Geosynclinal-Epicontinental Contrast	A397
Beginning of the Tethyan-Boreal Contrast	A399
Arabian-Madagascan Epicontinental Sea	A400
Isolation of Arctic Jurassic Sea and the "Arctic Transgression"	A400
Faunal Mosaic of the Pacific Ocean	A401
Jurassic Tethyan Ammonite Faunas	A403
Late Jurassic Faunas North of the Tethys and of the European Intermediate Zone	A406
NAUTILOIDEA	A408
COLEOIDEA	A408
RHYNCHOLITES	A410
CRUSTACEA	A410
OTHER INVERTEBRATES	A410
REFERENCES	A411

INTRODUCTION[2]

Zoogeographic differentiation is the result of ecologic interaction of numerous regionally effective causes, such as dynamics of dispersal, struggle for existence, climate, topographic links and barriers, salinity, water depth, currents, and supply of nutrients. Yet there is no fundamental limitation in regard to the many special facies conditions that exist in small areas, even down to the occupation of cracks in the seafloor

---

[1] Manuscript received April, 1969, and translated in part by CURT TEICHERT and GERTRUD TEICHERT; revised manuscript received May, 1975.

[2] A brief general view on the Jurassic Period (except for zoogeography) was published by the author in *Encyclopaedia Britannica* (1974, p. 354-360).

by ammonites, or of sponges by foraminifers.

Benthonic dwellers of shallow seas are generally strongly differentiated regionally as well as locally. On the other hand, nektonic and planktonic animals, as well as inhabitants of the deep sea, may be widely distributed unless restricted by climatic zones, physical barriers, or ocean currents. In many faunal descriptions insufficient checks are being made, however, and new species are being established which may be no more than varieties of known species or genera in distant areas. "Splitters," as well as "lumpers," render difficult the interpretation of zoogeographically important differences. Also, the literature has not been sufficiently searched for widely disseminated zoogeographical data. For these and some other reasons, regional comparison of Jurassic invertebrate faunas is still in its infancy.

Most conclusions pertaining to Jurassic zoogeography are based on ammonites, which, therefore, take up the major part of this discussion. SPATH, in the course of his lifelong work on ammonites, became increasingly skeptical as to their suitability as zoogeographic indices. He defended the old view of worldwide postmortem drifting of ammonite shells. Where regional zoogeographic differences were undeniable, he attempted to explain them as due to gaps in the stratigraphic record (SPATH, 1952).

Recent research, however, has confirmed the conclusions, reached by NEUMAYR (1883), that certain ammonite communities are characteristic of certain marine realms. The crossing of abyssal depths by ammonite larvae or by floating shells certainly was not the rule. Distribution patterns and lithological data suggest that climatic zones, depths of the sea, and the nature of migratory routes, may be the most important factors in explaining regional restrictions. The distribution of coral reefs also indicates climatic zonation during the Jurassic Period. The distribution of some benthonic clams (e.g., *Buchia, Inoceramus*) supports the conclusions based on studies of ammonites and makes it seem probable that the ammonites were epibenthonic and are preserved in their natural habitat to a greater extent than generally believed.

It may be assumed with confidence that the Arctic and Pacific oceans existed in the Jurassic, whereas the Atlantic and Indian oceans were probably still partly closed, and they gained their present configuration and position only through later continental drift. Deep-ocean and pelagic faunas are becoming known through studies related to the Deep Sea Drilling Project. The oldest sediments discovered to date in the present North Atlantic are presumably of Oxfordian age. We know more about rapidly subsiding geosynclinal areas that reached oceanic depths only where the influx of sediment or volcanic material was small (GARRISON & FISCHER, 1969). Jurassic shelf and littoral sediments preserved on present continents and islands are best known.

Projected on the present globe, the following major units of Jurassic marine zoogeography may be recognized (Fig. 1) following NEUMAYR (1883), UHLIG (1911), HAUG (1927), TERMIER and TERMIER (1952), and ARKELL (1956):

1) Tethys in the widest sense, subdivided into:
   a) a Mediterranean-South Asiatic center (area of alpine mountain chains with several subareas);
   b) a probable continuation westward to Mexico and to the southeast to the Malayan-New Zealandian geosyncline whose extension surpassed that even of the Tethys;
   c) epicontinental areas and inland seas (North Africa without the Rif Mountains, East Africa, peninsular India, central Arabia, Indonesia, southern China, southern Soviet Union, south-central Europe).

2) The Arctic-Boreal area (recent coastal lands of the Arctic Ocean, with strong influences in northern Europe and central Russia).

3) The Jurassic areas surrounding the Pacific Ocean lacking faunistic uniformity.

Presently known facts do not permit reconstruction of an individual Antarctic Ocean during Jurassic time (STEVENS, 1967b).

The number of genera and species decreases from the marginal areas of Tethys toward the north, thus indicating cooling

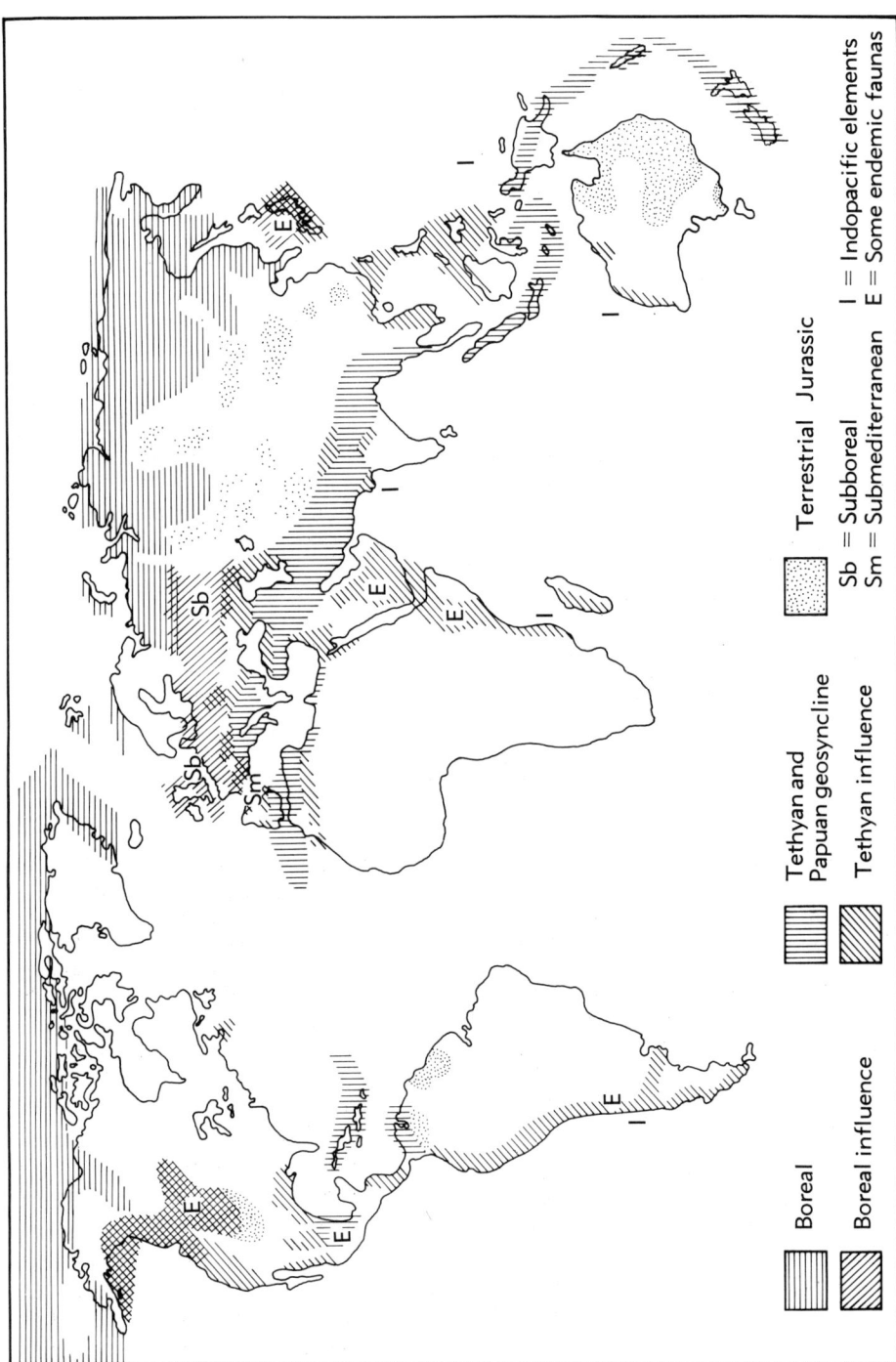

Fig. 1. Present distribution of zoogeographic realms and provinces of the Jurassic (Hölder, n).

climate in that direction (A. G. FISCHER, 1961; DE LATTIN, 1967; B. ZIEGLER, 1967). It should be remembered, however, that the number of taxonomic units is also fewer in the bathyal areas of Tethys just as is true for recent faunas in deeper parts of the oceans.

In the northern hemisphere, and probably on the entire earth, the climate during Jurassic time was warmer than now. Vegetation covered areas that are now near the poles. SAKS and NALNYAEVA (1966) used the term "Arctic" for the central Boreal realm during times of wider distribution of the Boreal faunas. In the changing pattern of the borders between the Boreal and non-Boreal faunas, perhaps the chemicophysically important 15°-isotherm of the sea has played a role (AGER, 1956).

PROTOZOA

The first flagellates were the Coccolithophorida and the Dinoflagellata (Peridiniina) in the Early Jurassic. The coccoliths (calcareous skeletons of coccospheres, flagellates) are found in geosynclinal, as well as epicontinental sediments, for example, in oil shale of the upper Liassic and in fine-grained limestone of the Upper Jurassic of southern Germany (FLÜGEL & FRANZ, 1967).

Among the Rhizopoda, the radiolarians were on the increase. The lime-secreting foraminifers, after their eclipse following the Permian, underwent a new strong development. Among these are the benthonic Lituolidae (HENSON, 1949), including the large foraminifer *Orbitopsella* in the Mediterranean Lias, and the first appearance of the pelagic globigerinids *(Protoglobigerina)* in the Tethys. These appeared first in the Bathonian and became rock builders in the Oxfordian. In the Swabian Jura, *Globigerina?* appears at the Callovian-Oxfordian boundary (SEIBOLD & SEIBOLD, 1960).

The Ciliata appeared with the calpionellids, which have a calcified shell wall (lorica). In the Alpine-Mediterranean Tithonian and in Late Jurassic sediments from the Atlantic Ocean, they attained stratigraphic importance (LE HEGARAT & REMANE, 1968).

PORIFERA

During the Jurassic, sponges were of greater importance than ever before in the history of the phylum. Calcareous sponges occur, as in the Triassic, in coralliferous and other sediments deposited in very shallow water. The siliceous sponges, which began to diversify in the Jurassic, preferred somewhat deeper water. In the Lower and Middle Jurassic of the Alps, spiculite rocks are composed of disassociated needles of disintegrated Monaxonida. The "gaize" of the Oxfordian of France is such a spiculite.

Hexactinellida and Lithistida formed meadows such as the Hexactinellida horizon in the Middle Liassic of Portugal and in the Callovian of La Voulte in southeastern France. With the help of blue-green algae (stromatolites), and if the supply of calcareous sediment was sufficient, they built hummock-like reefs that were preserved as massive sponge limestone. Minor occurrences of this kind are known from the Middle Jurassic of southern England, northern France, Spain, and Chile. During the Late Jurassic, a large sponge-reef belt, unique in earth history, bordered the central European Tethys from the Swiss Jura eastward through southern Germany as far as the Upper Vistula (Cracow-Kielče) (ROLL, 1934). In the south German Jura, hexactinellids prevailed in the late Oxfordian, and tetraxon and monaxon lithistids in the Kimmeridgian (SCHRAMMEN, 1924). Calcareous sponges and corals generally inhabited the tops of dead siliceous sponge reefs.

ANTHOZOA

Single corals (ahermatypic, without zooxanthellae) existed throughout Jurassic time. *Chomatoseris (=Anabacia)* is important in the marly facies of the Middle

Jurassic. Reef-building (hermatypic) corals played only a minor role in the Early Jurassic. Their most important occurrence is in the massive limestones of Domerian age in the Djebel Bou-Dahar in Morocco, in the epicontinental Jurassic south of the Tethys. Facies and associated faunas are reminiscent of those of the Upper Jurassic coral limestone of Europe (DUBAR, 1948). "Meadows" of colonial corals occur in the Liassic of Scotland (Hebrides) and southern Alaska. In the Middle Jurassic, hermatypic corals are strongly represented in the Tethys, with appearance of new families and development of numerous independent faunal provinces such as in eastern Iran (FLÜGEL, 1966, comparing different coral faunas of the Tethys and East Africa) and Madagascar (ALLOITEAU, 1958). COLLIGNON (1959) has interpreted the coral facies of the Bathonian of Madagascar as a barrier reef with atolls.

In western Europe the coral facies is distributed throughout France and farther north. In England it shows three main developments: in the Inferior Oolite (Bajocian), in the Great Oolite (Bathonian), and in the Corallian (Oxfordian). All three coral faunas differ greatly in the spectrum of their species; however, in southern areas where the number of genera and species is greater, the coral faunas also include many long-ranging forms with diminished stratigraphic value (GEYER, 1958). In the lower Kimmeridgian of Great Britain, two species of redeposited reef corals of possibly Oxfordian age have been found as far north as Helmsdale on the northeast coast of Scotland, which is the northernmost occurrence (58° N) of Jurassic reef corals in Europe. Only a few reef corals are known in the Oxfordian of northern and central England. Their number increases rapidly toward the south (south coast of England, Boulonnais of northern France, northwestern Germany). In the Upper Jurassic coral limestones of southern Germany and Switzerland, the number of reef coral species reaches 100 to 200 (B. ZIEGLER, 1964). This increase is probably due to climatic conditions; however, ARKELL (1935; 1956) thought that because of the relatively short distances involved, a topographic barrier would better explain these differences. In eastern Africa, Upper Jurassic coral reefs extend as far as 5° S., which is not quite as far south as in the Middle Jurassic.

In the Kimmeridgian, coral reefs of central Europe, with the exception of a few outposts, again retracted to the northern edge of the Tethys where they shifted even farther to the south, as for instance in the Swiss Jura (M. A. ZIEGLER, 1962); however, in the Tethys and on the Pacific coast of southern Honshu, Japan, they are widely distributed. In South America, major reefs are lacking in the Jurassic. Individual reef complexes of the Tethys and adjacent areas are composed of only a part of the total Upper Jurassic fauna. As a rule, one-third of the genera and two-thirds of the species of adjacent areas are absent. (See comparison of Stramberg beds and the Upper Jurassic of southern Germany by GEYER, 1958.)

Excellent examples of reef-front and back-reef facies in the Jura Mountains were described by M. A. ZIEGLER (1962) and ENAY (1965). Open colonies of hermatypic corals existed also in marly facies such as the Oxfordian of the Swiss Jura (M. A. ZIEGLER, 1962, p. 40).

HYDROZOA

In places, *Ellipsactinia* is an important member of Upper Jurassic reef communities of the Tethys, especially in the Carpathians and in the Balkans. Hydrozoan faunas of changing composition have recently been reported from southern Slovenia (TURNŠEK, 1966) and from the eastern calcareous Alps (FENNINGER & HÖTZL, 1965). The latter fauna shows close faunistic relationships along the entire Tethys as far east as Japan, whereas the relationships toward the west (Switzerland, France, Portugal) are far less pronounced.

BRACHIOPODA

After the decline of brachiopods in the Triassic, a new surge followed. The Thecideidae and the Terebratellaceae are known with certainty only since the Early Jurassic.

Survivors from the Paleozoic are the helicopegmatoid *Spiriferina* (until early Bajocian) and *Koninckina*. The last chonetoid brachiopod *Cadomella* (Pliensbachian-Toarcian, epicontinental Europe) acquired a helicopegmatoid brachiophore independently from the Helicopegmata (COWEN & RUDWICK, 1966).

The inarticulate brachiopods *Lingula, Discinisca,* and *Craniscus* are of almost cosmopolitan occurrence. Zoogeographical comparisons with other areas are rendered difficult by the fact that until now the study of Jurassic Articulata has been practically restricted to Europe, and in other areas there has been a tendency to establish independent taxonomic units. *Furcirhynchia,* until recently known only from Europe, has lately been found in Canada (AGER & WESTERMANN, 1963). In many cases, however, provincialism is a fact. Even in the English-Scottish Jurassic, several brachiopod provinces can be distinguished (AGER, 1956). "*Terebratula*" *joassi* DAVIDSON and the large *Rhynchonella sutherlandiae* DAVIDSON are endemic in the Kimmeridgian of northeastern Scotland.

The genera *Somalirhynchia, Daghanirhynchia, Septirhynchia, Bihenithyris, Somalithyris, Striithyris,* and *Trigonithyris* are restricted to the general area of East Africa and southern Asia. Multicostate Terebratulidae, which are rare in epicontinental beds of northwestern Europe, predominate along with multicostate Zeilleriidae in the reef facies of the Lower Jurassic of Morocco (DUBAR, 1942). ROUSSELLE (1968) and ROUSSELLE and BISCH (1967) have described some Rhynchonellidae and Orthotomidae that are endemic in the Moroccan Lias. The Early Jurassic brachiopods of Japan show closer affinities with those of the west than with those of North America (KOBAYASHI, 1961).

AGER (1965) offered the following ecological observations and conclusions: articulate brachiopods are almost exclusively stenohaline; *Kallirhynchia* penetrates also into the estuarine environment of the English-Scottish Middle Jurassic, elsewhere sandy and oolitic littoral environments are preferred; Rhynchonellidae tend to appear in pockets, large asymmetric forms like *Septaliphoria astieriana* were adapted to near-reef environments; *Terebratula moravica* with its long beak anchored in muddy bottom sediments is found in backreef facies, other adaptations are found on hard grounds, algal mats, and sponges; sulcate forms may be interpreted as adaptations to oxygen- and nutrient-deficient depths. Such conditions existed, for example, at the bottom of the Tethys in the Alpine-Carpathian area, which was by no means everywhere an open ocean. Here, a specially adapted form is *Pygope,* which in many places represents the only benthonic form in the Mediterranean Tithonian.

VOGEL (1966) interpreted the central perforation of the shell of *Pygope* as a means to remove used water, in the nutrient-poor bathyal region, as fast as possible from the interior of the mantle and to prevent its being mixed with water sucked into the inhalant canal. The occurrence in East Greenland of the related *Antinomia* indicates connections to the Arctic during Portlandian and Valanginian times (DONOVAN, 1953; MUIR-WOOD, 1953).

BIVALVIA, GASTROPODA

The epicontinental bivalve faunas of the Lower and Middle Jurassic comprise many widely distributed elements. The Ostreina first appeared in the Late Triassic, but did not evolve rapidly until the Jurassic. *Cardinia* is especially common in lower Liassic rocks and locally also in somewhat younger shallow-water strata from Japan to Canada. *Plagiostoma* probably has a similar distribution. The group of large-sized *Oxytoma scanica* and *O. cygnipes* is known in the lower and middle Liassic from northern Europe as far as the Alps, as well as in the Sinemurian Fernie Group of Canada (HÖLDER, 1953; FREBOLD, 1957). *Weyla alata* and related forms are common members of the western American Liassic faunas and extend as far as Portugal and southern Spain in Europe. Presence of the smaller *Weyla ambongoensis* in the Toarcian of East Africa and Madagascar caused DACQUÉ (1915) to postulate close relationships of this area with South America. Later, however, this species was also reported from Morocco and India (Cox, 1965). Leanza (1942) described several hundred species of bivalves from the Liassic of Neuquen in Argentina, two-thirds of which are indigenous. The rudist-like *Plicatostylus*, which forms biostromes, is known exclusively from the Liassic of Oregon (Lupher & Packard, 1930) and Peru (Geyer, 1973). *Gryphaea* retreated from the Arctic areas after the Liassic (Imlay, 1965). Imlay (1964a) described a bivalve fauna from the Middle Jurassic of Utah, containing a number of new, probably endemic, species. From the Middle Liassic and into the Bajocian, *Lenella* and *Arctotis* (Pteriina) are endemic in the Arctic marine Jurassic of Siberia (Saks, Mesezhnikov, & Shulgina, 1964). *Retroceramus* and *Arcticeramus* join in the Middle Jurassic. The inoceramids advanced repeatedly far to the south (Imlay, 1965), as did the buchiids in somewhat later times.

As early as the Liassic, restriction of certain bivalves caused regional biofacies. In the Liassic, reef limestone was restricted to a few areas of the Tethys, thus creating faunal associations rather similar to those that were more widely extended to the north during later Jurassic times. Examples are the already-mentioned middle Liassic reef limestones of the southern High Atlas of Morocco (Dubar, 1948), which contain, in addition to calcareous algae, abundant corals, brachiopods, gastropods, echinoderms, and bivalves such as *Perna* (length 60 cm, width 6 cm), *Pachyrisma, Durga, Pachymegalodon, Pachymytilus, Opisoma,* and *Lithiotis*. Some of these (*Opisoma, Durga, Pachymegalodon, Lithiotis*, and the related *Cochlearites*) are found in banks and reefs in the somewhat different argillaceous facies of the gray limestone of the southern Alps, which have many analogues as far east as Timor (Wanner, 1910).

Pseudomonotis tends to occur in shallow-water limestone; similarly, *Posidonia* occurs in deposits of clay and calcareous quiet-water rocks of geosynclinal and epicontinental areas. Among the posidoniids, *Bositra buchi* is widely distributed, in places as the only fossil in shaly and argillaceous rocks from the Toarcian to the Oxfordian, generally in geosynclinal facies. A pelagic-nektonic mode of life, which was inherited from the old Paleozoic root-stocks of the Limacea and Pectinacea, can be assumed because of lack of other benthonic life, other ecologic indicators, as well as from the shell morphology (Jefferies & Minton, 1965). On infrageosynclinal rises and plateaus, coquinas of *Bositra* shells accumulated, locally associated with remains of the rich benthonic life of these hard grounds (Sturani, 1967). *Silberlingia* is endemic in California (Imlay, 1963).

With the help of a number of "faunal spectra," B. Ziegler (1967) illustrated the depth dependence of bivalves in relation to other marine fauna.

Trigonia, known from the Middle Triassic onward, is richly represented in the lower Liassic of the southern Andes and of Japan. It probably originated in the Pacific area, and first reached Europe from the west during middle Liassic time. At the transition from the Lias to the Dogger, costate and clavellate trigonias spread rapidly. Trigonias are mostly found in great masses in sandy and ferruginous, more rarely in argillaceous, rocks. Their habitat is the shallow shelf at depths from 10 to 50 meters. Their abundance in epicontinental areas is in strong contrast to their absence in the Tethys. In the Upper Jurassic of East Asia and the west coast of Japan, *Nipponitrigonia* predominates, whereas *Myophorella (Haidaia)* predominates on the Pacific coast of Japan (Maeda, 1962). Occurrences of *Indotrigonia* and *Opisthotrigonia*, though represented by different species, correlate the Upper Jurassic of peninsular India with that of East Africa (Cox, 1965).

Buchia, which ranges from the Oxfordian to the Neocomian, spread repeatedly from its original Arctic-Boreal habitat far into the south where the number of its species increased. This genus is an especially important indicator for marine connections between Boreal, Pacific, and Tethys areas (Imlay, 1959, 1965). By late Oxfordian, a first advance was made southward along the American west coast reaching, in the Kimmeridgian, as far as Mexico where a *Buchia* deposit contains numerous species. Another *Buchia* invasion occurred during the Portlandian. From Mexico and possibly also from southeastern Asia, the genus reached the southern Andean geosyncline of Peru, Chile, and Argentina.

Buchiids, especially the genus *Malayomaorica* JELETZKY (1963), are known from the Spiti Shale of the Himalayan geosyncline, from the islands of Roti, Misol, and Buru of the Indo-Malayan geosyncline, and also from northwestern Australia. Their occurrence indicates that a direct connection to the Arctic via eastern Asia existed. *Buchia mosquensis* and *B. volgensis* are also found in the *Nerinea* limestones of Mangyshlak, which they may have reached by way of a Late Jurassic marine connection east of the Urals, but possibly also from the Russian area farther west. Toward the west, these forms advanced from the central Russian basin by way of Poland and Pomerania into England and the Boulonnais of northern France. In the south German Jura, *Buchia* appears in the upper Oxfordian (BRINKMANN, 1929; MAYNC, 1947).

Times of regression were characterized by euryhaline and brackish bivalve associations such as *Corbula, Eomiodon, Protocardia, Pseudotrapezium,* and *Eligmus.*

Rocks containing brackish and fresh-water molluscan faunas are known from the Estuarine Series of the Middle Jurassic of England and Scotland (HUDSON, 1963a,b), the East African-Indian area (*Eligmus* fauna), from the Purbeckian of Europe (HUCKRIEDE, 1967), the Morrison Formation of the United States, and the Tetori Group of Japan (SATO, 1961; MAEDA, 1961). On the other hand, supersalinity influenced the marine life. JORDAN (1974) interpreted the northwest German Upper Jurassic rocks as sediments of a supersaline and temporarily oil-soiled sea.

The continental Liassic fauna of central Asia consists solely of some surviving Triassic bivalves. *Pseudocardinia* appears in the upper Liassic and becomes abundant in the Middle Jurassic, whereas the fresh-water gastropod *Bithynia* appears in the Middle Jurassic. *Arguinella, Limnocyrena, Corbicula,* numerous Unionidae, and gastropods of the prosobranchiate families Viviparidae, Valvatidae, Micromelanidae, Hydrobiidae, and of the pulmonate families Planorbidae and Limnaeidae appear in the Upper Jurassic (MARTINSON, 1964). The pulmonate forms were presumably the first immigrants from land into fresh water in the history of the gastropods.

Among marine gastropods, *Discohelix* was one of the last survivors of the Paleozoic Euomphalacea. This genus is known mainly from the Lower and Middle Jurassic of the Mediterranean area, where its species serve as guide fossils (WENDT, 1968). In the oil shales of the European Toarcian, the small genus *Coelodiscus* (Euomphalacea?) is found frequently congregated around saurian carcasses.

Among mesogastropods, the Naticacea, which had their beginning in the Jurassic, the Strombacea with *Harpagodes* of Middle Jurassic to Late Cretaceous age, and *Columellaria,* a reef inhabitant of the Late Jurassic, should be mentioned. Among neogastropods the thick-shelled Nerinacea *(Nerinella, Nerinea, Itieria)*, which were adapted to shallow-water facies between coral reefs, play an important, commonly rock-forming, role. They are restricted to the Jurassic and Cretaceous.

AMMONOIDEA

According to SCHINDEWOLF (1961-68), most Jurassic Ammonitina originated from *Psiloceras,* whose origin has to be looked for in the Upper Triassic Lytoceratina. WIEDMANN (1973b) considered *Phyllytoceras* from the Norian to be the ancestor of *Psiloceras.* According to KRYSTYN (1974), however, *Phyllytoceras* is based on an indeterminable inner whorl of *Rhacophyllites.* Therefore, the origin of the Neoammonoidea remains uncertain.

THE GEOSYNCLINAL-EPICONTINENTAL CONTRAST

In general, the Phylloceratacea and Lytoceratacea show a conservative evolution. During the Jurassic they mainly occupied the Tethys and, to a lesser extent, its peripheral and radiation areas. This may indicate a preference for greater ocean depth, and perhaps may also indicate climatic influences. Thus, these groups are absent from the shallower areas of the Tethys (Lemes beds of the Upper Jurassic of Yugoslavia, B. ZIEGLER, 1963; the shallow-water Jurassic of the Georgian block between deeper sedimentation basins of Caucasus and Anticaucasus, ZESASHVILI, 1964). This distribution indicates a more or less benthonic mode of life and generally an autochthonous mode of emplacement. Phylloceratacea and Lytoceratacea are also absent in the epicontinental Callovian and Oxfordian of the western basins of the conterminous United States (IMLAY, 1967), but they are abundant in the geosynclinal areas as far as Alaska. Dwarfed phylloceratids have been found in deep cracks in eastern Alpine Triassic limestone filled with Liassic sediment. At the same time, other ammonoid genera lived on the higher sea bottom above the cracks. Dwarfed ammonites from such submarine cracks have also been mentioned from the Jurassic of Sicily (R. FISCHER, 1967; WENDT, 1971).

Some genera of the Psilocerataceae, the oldest "Ammonitina," are known only from Europe, whereas others spread more or less widely through the Tethys and adjacent area. *Psiloceras* is known from northeastern Asia (SAKS, 1964), New Zealand, and Laos, northward in western Europe as far as the island of Mull (western Scotland), and also from Canada, Peru, and Chile (CECIONI & WESTERMANN, 1968). Because Hettangian faunas are unknown in Mexico, the direction of migration of *Psiloceras* to western America is still in doubt. A "Panama-strait" probably existed in Liassic time at the western end of the Tethys *(sensu lato)*, and may be assumed to have extended westward from Europe between the former land areas of Gondwana and Laurasia (GEYER, 1973).

FREBOLD (1967) described *Psiloceras* sp. ex aff. *P. planorbis* as the oldest known Jurassic ammonite in British Columbia. Above it follows the ribbed *P. canadense* FREBOLD (1951) associated with additional genera of the Psiloceratidae, *Phylloceras* and *Eolytoceras*. Differentiation of *P. planorbis* on the subspecific level is evident in Peru (SCHINDEWOLF, 1957). *Curviceras* BLIND (1963) (=*Waehneroceras* auct.), and later *Schlotheimia*, are known from New Caledonia, and also from Japan to the east, and as far as western North America. *Schlotheimia* is also known from Alaska and Peru. In the western part of the Americas, the Arietitidae extend as far as Alaska to the north *(Arietites, Coroniceras)*, and Peru and Chile to the south *(Arietites, Vermiceras)*.

In comparison with its epicontinental equivalents in central Europe, the lower Liassic of the Alpine-Mediterranean area contains a greater abundance of forms of Psiloceratidae, Schlotheimiidae, and Arietitidae (WÄHNER, 1882-98; LANGE, 1952; BLIND, 1963); however, the lower Liassic faunas of the Helvetic nappes of eastern Switzerland have an entirely central European aspect. This shallow-water sedimentation area on the southern side of the "Alemannic Land," which is the Aar Massif of today, must have been closely connected with the Jurassic sea of central Europe. The strong Mediterranean affinities of the faunas in the Liassic "Allgäu-Schichten" of the eastern Alps, as in the Lower Jurassic of the Romanic Préalpes, indicate tectonic transportation of these units from a sedimentation area much farther to the south.

The epicontinental Jurassic of central Europe and elsewhere represents not merely an area of immigration of faunal elements from the Tethys, but it was also an independent center of evolution. Although the degree of endemism reached by the ceratites in the Germanic Muschelkalk was not repeated in the Jurassic, many endemic genera and species evolved. *Saxoceras*, with many species, is almost endemic in the upper Hettangian of central Europe (LANGE, 1951).

Curviceras, Schlotheimia, and the arietitids *Vermiceras, Arnioceras*, and *Asteroceras* extend as far as the North African Rif and Atlas (ARKELL, 1956; DONOVAN, 1967). This is also true for *Oxynoticeras*, which exhibits especially well the striking contrast between paucity of species in the epicontinental facies and the wealth of species in the Alpine area (PIA, 1914). *Echioceras* and *Paltechioceras* appear in northwestern Europe, and the latter genus is also found in the Tethys.

Some species of *Gagaticeras* found in the upper Sinemurian of northwestern Germany are also present in Lorraine, the northern Alps, and in Yorkshire (northeast England), but are absent in southern England. A barrier to distribution appears to have existed, therefore, at least on the western side of the sea surrounding the Ardennes-London Island (HOFFMANN, 1941).

According to SCHINDEWOLF's (1961) investigation of the early stages of ammonoid sutures, the Eoderocerataceae evolved from Psilocerataceae, not directly from the Lytoceratina, as indicated in *Treatise on Invertebrate Paleontology,* Part L, figure 150 (1957). In the lower Pliensbachian, their area of distribution is practically the same as that of their ancestors. Among Eoderocerataceae the differences between epicontinental and geosynclinal faunas are small and are indirectly noticeable only in the preeminence of the phylloceratids and lytoceratids. An example is the fauna of the Ak Dag of northern Turkey (PIA, 1913).

From the lower Liassic to the Middle Jurassic the northern Balkans and the Caucasus belonged to the epicontinental facies belt, and the southern Balkans belonged to the Mediterranean belt (KOVACS, 1942; DONOVAN, 1967). SAPUNOV (1973) distinguished between the European Caucasian and the Mediterranean provinces by comparing the ammonite genera and species, common or uncommon, to both these provinces in the Pliensbachian and the Toarcian. The Toarcian and Bajocian beds near Kerman in Iran have mainly northwestern European faunal elements, and the Aalenian additionally has some central European elements (SEYED-EMAMI, 1967). *Epideroceras* is restricted to the northern margin of the European Tethys, and extends eastward as far as Turkey (SAPUNOV & STEPHANOV, 1964).

BEGINNING OF THE TETHYAN-BOREAL CONTRAST

Marine rocks of early Liassic age containing Arietitinae, as well as *Oxynoticeras* and *Echioceras,* are known from northern Alaska (IMLAY, 1955) and Arctic Canada (FREBOLD, 1960). Nowhere else does it appear that the Early Jurassic Arctic Sea extended over present land areas. In East Greenland the oldest Jurassic ammonites are *Beaniceras* and *Uptonia* (Eoderocerataceae). *Amaltheus* is the earliest genus having a range from northern Canada to northeastern Asia and northern Siberia, and is the only ammonite genus of the upper Pliensbachian in this area (DAGIS & ZAKHAROV, 1974, map p. 23). *Amaltheus* also predominates in Great Britain, where HOWARTH (1958) found restriction of certain species, and in Germany it indicates an open connection between the Arctic Sea and central Europe.

Farther south the situation is different. Although *Amaltheus* does occur as far as the Caucasus and in the Middle (not High) Atlas of North Africa (COLO, 1961), it is far outnumbered by the great host of the Hildocerataceae, *Arieticeras, Canavaria, Fontannelliceras, Fuciniceras,* and *Protogrammoceras.* In Spain a sharp faunal division existed in the late Pliensbachian between the north (Cantabrian and Iberian ranges) where *Amaltheus* occurs, and the south (from the southern Keltibericum in the Province Teruel onward) where *Amaltheus* is absent (BEHMEL & GEYER, 1966). In the Alps and southern France typical mixed faunas occur. *Amaltheus* and *Canavaria* occur together also in Japan (SATO, 1960). Thus, I suggest a climatically determined reduction in the diversity of ammonite faunas from south to north similar to faunal conditions of the present day. HALLAM (1969) suggested, however, that climate during the Jurassic was too uniform to cause zoogeographic differentiation, and he therefore assumed that nearshore shallow seas with strong terrigenuous sedimentation and reduced salinity existed in boreal regions to which the boreal fauna was adapted.

The Dactylioceratidae, probably deriving from *Microderoceras* and commencing with *Coeloceras* and *Coeloderoceras* in the late Sinemurian, expanded during several advances from their Mediterranean center of evolution: *Prodactylioceras* in the Pliensbachian, and *Dactylioceras* since the Domerian (SCHMIDT-EFFING, 1972). *Dactylioceras* dominated the Arctic-Boreal region during the early Toarcian nearly as exclusively as did its predecessor *Amaltheus* in the Pliensbachian. In addition, *Harpoceras* and some other Hildocerataceae occurred, but were rare (FREBOLD, 1958, 1960; DAGIS & ZAKHAROV, 1974). In northern Siberia, *Zugodactylites* with endemic species appeared in the middle Toarcian. Outside the Arctic *Dactylioceras* is widely distributed in association with abundant *Harpoceras* and *Hildoceras*. During the late Toarcian and early Bajocian, the Arctic was somewhat more abundantly populated by species of *Pseudolioceras, Grammoceras, Erycites,* and *Ludwigia*. At the same time, a far greater abundance of forms occurred to the south in the Tethys and its peripheral regions as far as Japan, where *Dumortieria, Pleydellia, Leioceras, Graphoceras,* and *Hyperlioceras* occurred together with the above-mentioned genera. More research in the Arctic could possibly lead to a lessening of these differences, however. For instance, *Leioceras opalinum* has recently been found on Prince Patrick Island in Arctic Canada (FREBOLD, 1958). *Mercaticeras* and the small Bouleiceratinae *Frechiella, Paroniceras,* and *Leukadiella* form a group that is restricted to the Mediterranean. *Haplopleuroceras* ranges from Italy to England.

In addition, the Hammatoceratidae are widely spread throughout the Tethys and its peripheral regions as far as Japan, but are not entirely absent from the boreal region.

ARABIAN-MADAGASCAN EPICONTINENTAL SEA

The sea in the south of the Iran-Afghanistan part of the Tethys area transgressed in a westerly direction during the Toarcian toward the Nubian shield, where it formed the shallow Central Arabian Embayment, sediments of which are now exposed in the terraced sides (escarpment) of the Jebel Tuwaiq. The fauna consists of *Bouleiceras, Nejdia, Hildaites,* and *Protogrammoceras* (ARKELL, 1952), and shows affinity to that of the western Tethys (Baluchistan to Portugal; GEYER, 1965), but is very impoverished and contains endemic species. It also shows affinities with the faunas along parts of the East African coast and Madagascar. Thus, a shallow arm of the sea must have extended in the direction of the present Strait of Mozambique, which was bordered on the west by the African continent, and on the east, according to present ideas, by the Madagascan-Indian ("Lemurian") continent (DACQUÉ, 1910). During Bajocian, *Ermoceras* and *Thamboceras* predominated in the Central Arabian Embayment, although they also occur in the Sinai Peninsula. In the Bathonian, when a new transgression occurred, these genera were joined by additional members of the Thamboceratidae and Clydoniceratidae (ARKELL, 1952). The distribution of Clydoniceratidae reached as far as Madagascar and Europe.

In the Bajocian, the above-mentioned strait that separated East Africa from Madagascar extended farther over the continent and harbored a considerably richer ammonite fauna than that found in the obviously rather restricted central Arabian Embayment. This fauna consists of Tethyan genera, including lytoceratids and phylloceratids, but also of European elements, indicating the existence of connections across the Tethys. Conditions in the Callovian were similar, and during this time the East African-Madagascan sea spread northeastward to peninsular India (Kutch).

From the Callovian to the early Tithonian, Ethiopia was covered by a peripheral shallow sea. Besides endemic and some Indian elements, its ammonite fauna has numerous forms known from southern Germany. Conditions essential for life probably were equally favorable to the north and south for emigrants from the Tethys (CARIOU, 1965; ZEISS, 1971). The northern fauna (highland of Harar) in the Ethiopian Jurassic consisted mainly of perisphinctids and aspidoceratids. It was replaced to the southeast by a fauna containing additional oppeliids and hybonoceratids, and finally also phylloceratids and lytoceratids. According to B. ZIEGLER (1967), this indicates an increase in water depth.

ISOLATION OF ARCTIC JURASSIC SEA AND THE "ARCTIC TRANSGRESSION"

During the Bajocian the Arctic-Boreal region developed its own faunistic characteristics to a far greater extent than during the middle Liassic. Increasing isolation through regression or rise of shoals has been suggested by ARKELL (1956) as a cause for this phenomenon. These events took place at the time of the greatest evolutionary development of the Jurassic ammonites. The oldest element in this Arctic endemic fauna is the stephanoceratid *Arkelloceras* FREBOLD (1957) in the middle Bajocian of Arctic Canada (FREBOLD, 1967). This was followed by the Cadoceratinae *Cranocephalites, Arctocephalites,* and *Arcticoceras* in the late Bathonian, and in the Callovian by *Kepplerites* and *Cadoceras,* forming the root of the Kosmoceratidae, and by the Cardioceratidae (CALLOMON, 1963), the earliest representative of which is *Quenstedtoceras.*

The Kosmoceratidae and Cardioceratidae penetrated in varying degrees toward the Tethys. TERMIER and TERMIER (1952) have described this penetration, which is not restricted to the ammonites, as the "Arctic transgression" that follows the above-mentioned regression of the sea. Among the Kosmoceratidae, *Seymourites* reached Japan, *Kosmoceras* Portugal, and some genera the Caucasus. The center of distribution of the Kosmoceratidae is found in the Russian boreal area and in the central and western European regions, rather than in the Arctic (BRINKMANN, 1929). TINTANT (1963) has published distribution maps of *Kepplerites (Kepplerites), K. (Gowericeras),* and of *Zugokosmoceras* in Europe, concluding that geographical isolation was responsible for evolution of some of the species. *Cardioceras* advanced to the northern edge of the Tethys and mingled there with Mediterranean forms, as near Cetechowitz (outer Carpathian cliff zone), in the Helvetian and Savoyan Alps, the French Préalpes, and the Préalpes Maritimes; it did not, however, reach the inner Carpathian cliff zone nor the Préalpes Romandes (NEUMANN, 1907).

FAUNAL MOSAIC OF THE PACIFIC OCEAN

The Pacific Ocean, in contrast to the presumably younger Atlantic, is surrounded by a continuous belt of Jurassic sediments. The sedimentation areas in this belt were partly geosynclines, partly epicontinental shelves and inland seas; however, no uniform Pacific faunal realm exists because this sedimentation area crossed many climatic zones. Connections with the Tethys to the east and west, and with the Arctic Sea to the north, led to the same faunal contrasts that characterize the Tethys-Boreal region. Moreover, because of the size of the area and the differences in relief and facies prevailing there, special faunas composed of endemic forms developed. Only a few ammonite genera restricted to the Pacific succeeded in establishing themselves on both its eastern and western side; however, the scarcity of such forms suggests that already at that time this great ocean played an important role as an ecologic barrier.

Noteworthy and puzzling are the circum-Pacific genera *Pseudotoites, Zemistephanus, Neuqueniceras,* and *Epicephalites,* which did not penetrate into the Tethys area. It has been suggested that these genera belonged to a specialized Pacific fauna, which is little known only because it moved so rarely out of the area of the Pacific Ocean. More abundant are ammonite genera that occur on both sides of the Pacific as well as along the Tethys Ocean. To explain this distribution pattern the assumption of trans- or circum-Pacific connections cannot be excluded, but is not strictly necessary.

The Jurassic faunas of southern Alaska were separated from those of northern Alaska, and contained a rich ammonite fauna, especially in Middle Jurassic time (IMLAY, 1952). In the early Bajocian (Aalenian), *Tmetoceras* and *Erycitoides* occurred with a number of endemic species along with other genera (WESTERMANN, 1964). In the late Bajocian, representatives of widely spread Stephanocerataceae probably migrated into the area from the south along the Cordilleran geosyncline. In the Callovian, Arctic-Boreal forms (*Cadoceras, Kosmoceras,* and others) predominated, as may be expected, considering the general advance of the boreal fauna at that time, but *Phylloceras* was also present. *Arcticoceras* was absent.

Another macrocephalitid, *Lilloetia,* belongs to the northeast Pacific faunal province, some members of which also lived in the shallow sea of interior Canada and the United States ("Logan Sea"). In this epicontinental sea, southern and Arctic elements mingled with endemic species, which show only slight morphological specialization (FREBOLD, 1951; 1957; 1959; IMLAY,

1948; 1952). Endemic forms of the late Bajocian (IMLAY, 1967) are *Sohlites* (Stephanoceratidae?) and *Eocephalites* (Cadoceratinae), which occur with the more widespread *Spiroceras*. In the Bathonian, *Warrenoceras* replaced *Arctocephalites* as the endemic form (FREBOLD, 1963). In the Callovian, *Kepplerites* and *Arcticoceras*, the latter not found in southern Alaska, were joined by the endemic genus *Imlayoceras* (Macrocephalitidae). The genera *Choffatia, Parareineckeia*, and *Pseudocadoceras*, known from southern Alaska (*Pseudocadoceras* from southwestern British Columbia also), seem to be missing. The occurrence of *Arcticoceras* seems to suggest a direct connection with the Arctic Sea via the Yukon area, though corresponding sedimentary rocks have not been found (FREBOLD, 1957). That suggestion was supported by SAKS and NALNYAEVA (1966) on the basis of the occurrence of belemnites of boreal affinities in the Logan Sea.

The generic composition of the Mexican Middle and, especially, Late Jurassic fauna is much like that of the Tethyan of southern Europe. Predominating are Perisphinctidae *(Idoceras, Nebrodites, Sutneria)*, Aspidoceratidae, and Oppeliidae *(Ochetoceras, Taramelliceras, Streblites, Glochiceras)*, as well as a few lytoceratids and phylloceratids, indicating a partly geosynclinal facies; however, there are endemic genera such as *Mazapilites* (Oppeliidae) and, of course, numerous endemic species. Endemic genera are also found in Cuba: *Vinalesphinctes* (Oxfordian) and *Dickersonia* (Himalayitinae, Tithonian). On the basis of so much endemism, ARKELL (1956) thought it hardly probable that the Central American-Mexican sedimentation area was situated any closer to the Jurassic sedimentation areas of Europe than it is today.

The resemblance of the lower Liassic Mexican ammonite fauna with that of Europe has already been mentioned above.

HILLEBRANDT (1970) has given a precise introduction to the biostratigraphy of ammonites of the South American Andes. Up to the Sinemurian, the ammonite succession resembles that of epicontinental Europe. From the Pliensbachian upward it also contains numerous Mediterranean elements. Amaltheids are absent, as surprisingly are also the later graphoceratids, parkinsoniids and taramelliceratids. In the predominantly shallow Jurassic seas of South America, phylloceratids and lytoceratids were rare. Boreal ammonites advanced far to the south in the western part of North America, but are rarely seen in South America.

Connections between Europe and South America certainly existed, but did not allow for all genera to pass through.

A few genera, whose occurrences are restricted to areas near the Pacific Ocean, are of special interest. Apart from some circum-Pacific genera, *Xenocephalites* (Callovian) is, except for Greenland, restricted to the east coast of the Pacific, where it has a surprisingly extensive latitudinal distribution from South America through Mexico to southern Alaska. During the Bajocian, *Megastephanoceras* and *Eocephalites* also appear to be restricted to the western part of North and South America.

The characteristic Tethyan influence on Late Jurassic faunas of Mexico is also noticeable in the northern part of the South American Andean geosyncline as well as farther north in California and Oregon, where boreal elements were met along a borderline that shifted from time to time (IMLAY, 1952; 1963; 1964b; B. ZIEGLER, 1971). In the southern (Argentinian) Andean geosyncline, many endemic genera occur within the Berriasellidae (e.g., *Hemispiticeras* and *Andiceras*).

The Ellsworth Mountains of the Antarctic (QUILTY, 1970) have a noteworthy Middle to Upper Jurassic (Bajocian-Oxfordian) ammonite fauna with Stephanoceratacea of European character, which, in contrast to that of the Arctic, seems not to represent an independent zoogeographical region.

The Jurassic of Japan has some boreal genera such as *Amaltheus, Kepplerites*, and *Seymourites*, but more generally shows the influence of Tethyan and marginally Tethyan faunas (SATO, 1960; 1962). In addition, a number of important endemic genera and species are present. The connection with the New Zealand-Sumatra geosyncline and with the Tethys may have been across Taiwan where the occurrence of Jurassic was first described by LIN (1961), and across the epicontinental shallow seas which extended from the south coast of China towards Indonesia, though documented thus far only for the Liassic.

JURASSIC TETHYAN AMMONITE FAUNAS

Having discussed the extra-Tethyan regions, we now return to the Tethys. The families Parkinsoniidae and Morphoceratidae, are mainly restricted to the Eurasian region. *Garantiana* reached Transbaikalia via Japan. In the Callovian, the Reineckeiidae were found almost throughout the Tethyan area and its peripheral regions; *Reineckeia* itself extended outside Tethys into Madagascar, South America, and northern Alaska (IMLAY, 1955), but not to northwestern Germany. The Macrocephalitidae and Pachyceratidae occurred in Tethys and its connected regions of South America, East Africa, southeastern Asia, and as far as New Guinea. *Pleurocephalites* reached Greenland. The Perisphinctidae first appeared during the Bajocian and, together with the Aspidoceratidae, Haploceratidae, and Oppeliidae, constitute the bulk of the Upper Jurassic Ammonitina. Of these families, only the Perisphinctidae invaded the boreal region. Groups originally inhabiting the Tethys penetrated far into its peripheral seas both to the north and south. According to B. ZIEGLER (1963, 1967), in central and southern Europe the perisphinctids preferred shallow water, and, therefore, with increasing depth, were replaced in succession by the Aspidoceratidae, the Oppeliidae, and finally the lytoceratids and phylloceratids. This sequential replacement may explain the often-discussed substitution of *Gregoryceras* and *Epipeltoceras* (Aspidoceratidae, Peltoceratinae) in the middle and upper Oxfordian of southern Europe by Perisphinctidae as time-equivalent index fossils in northwestern Europe (GYGI, 1966).

The Asiatic and Mediterranean mountain ranges disclose cross sections of the Tethys in structural, though tectonically disturbed, connection. Thus, in the Pamir, Early Jurassic epicontinental sediments of the northern margin of Asiatic Tethys are known (ANDREEVA & DRONOV, 1964). North of the Pamir the marine facies changes into terrestrial sediments of the Jurassic basins of the immense Siberian region. In the Himalayas there are indications of a fauna rather similar to that of the Alpine geosyncline of the Tethys. This occurs in blocks of allochthonous wildflysch containing phylloceratids and lytoceratids of Liassic age and in the tectonic klippen of the Kiogars, which consist of *Calpionella* limestone of Tithonian age. The greater part of the Himalayan Tethys, however, was probably occupied by shallow seas, as indicated by the wide distribution, from the western Himalayas to southern Tibet, of the Spiti Shale and of sandy and marly shales in northwestern Pakistan, which contain a rich ammonite fauna of Late Jurassic age (UHLIG, 1903-1910; 1910; FATMI, 1972). The Tithonian fauna is especially rich:

Upper Tithonian—contains *Corongoceras, Micracanthoceras, Aulacosphinctes, Protacanthodiscus, Blanfordiceras, Kossmatia, Himalayites,* and in addition *Haplophylloceras strigile* (BLANFORD) and *Paraboliceras* occur as far as the Malayomaorican geosyncline.

Lower Tithonian—contains *Virgatosphinctes* and *Aulacosphinctoides,* in addition to *Hildoglochiceras* (Oppeliidae).

Tithonian faunas, in spite of similarities exhibited throughout the Tethys and as far as Japan and South America, do show a certain amount of regional differentiation, being due in part to ecologic factors, and in part by incompleteness of the stratigraphic record.

In a western direction, toward the Mediterranean, the Tithonian of the Tethys can be subdivided in greater detail than in the Himalayas (BARTHEL et al., 1966). Here, middle Tithonian is discernible with *Pseudolissoceras, Semiformiceras* (Streblitinae), *Simoceras volanense,* and the unrolled *Protancyloceras.* From Kurdistan SPATH (1950) described a middle Tithonian fauna with some endemic Virgatosphinctinae, the endemic *Oxylenticeras* (Streblitinae, with closed umbilicus) and an unrolled small *Cochlocrioceras,* to date also known only from this locality. WIEDMANN (1973a) reported on the Tithonian heteromorph ammonites *Protancyloceras, Cochlocrioceras, Vinalesites,* and *Bochianites,* and their dis-

Stages	Tithonian	Kimmeridgian	Oxfordian	Callovian	Bathonian
OTHER REGIONS	S AMERICA: Proniceras and Substeueroceras; Corongoceras, Windhauseniceras. MEXICO: Torquatisphinctes and Mazapilites; Hybonoticeras			WESTERN INTERIOR USA: Cardioceras, Quenstedtoceras; Kepplerites, Gowericeras, Arcticoceras; Warrenoceras	C ARABIA, IN PART N AFRICA: Dhrumaites, Micromphalites, Tulites, Thambites, Bramkampia
ARCTIC, N CANADA, GREENLAND	Chetaites; Craspedites and Virgatosphinctes; Epivirgatites and Laugeites; Subplanites, and Eosphinctoceras	Streblites and Amoeboceras		Cadoceras; Arcticoceras	Arctocephalites; Cranocephalites; Boreiocephalites
VOLGA REGION	Craspedites, Kachpurites, Epivirgatites; Virgatites; Zaraiskites, Dorsoplanites, and Pavlovia, Pectinatites; Subplanites, and; Gravesia	Virgatoxioceras; Aulacostephanus	Rasenia, Pictonia; Ringsteadia; Amoeboceras, Cardioceras; Quenstedtoceras, Peltoceras, Erymnoceras	Kosmoceras, Kepplerites; Clydoniceras	Tulites, Gracilisphinctes
NW EUROPE	Craspedites; Titonites; Glaucolithites, Zaraiskites; Pavlovia, Pectinatites; Virgatosphinctoides; Gravesia	Aulacostephanus; Rasenia; Pictonia baylei; Ringsteadia, Decipia	Perisphinctes; Cardioceras, Quenstedtoceras, Peltoceras, Erymnoceras	Kosmoceras jason, Sigaloceras and Kepplerites, Macrocephalites; Clydoniceras; Oxycerites aspidoides; Prohecticoceras retrocostatum, Morrisiceras	Tulites, Gracilisphinctes, Zigzagiceras
EUROPEAN TETHYS, SOUTHERN, MIDDLE AND SW EUROPE	Berriasella; Semiformiceras, Pseudolissoceras, Lemencia; Pseudovirgatites, and Glochiceras lithographicum; Hybonoticeras	Hybonoticeras beckeri and Virgatoxioceras; Enosphinctes subeumela; Aulacostephanus; Aspidoceras acanthicum, Katroliceras divisum	Ataxioceras; Sutneria platynota; Sutneria galar; Idoceras planula; Epipeltoceras bimammatum and Amoeboceras; Gregoryceras transvers.		

FIG. 2. Ammonite biostratigraphy of some important Jurassic regions (Hölder, n).

tribution throughout the Tethys and neighboring areas.

ENAY (1973) critically reviewed the zoogeography of Tithonian ammonite faunas, and mentioned the difficulties in taxonomic distinction of the Perisphinctaceae.

South of the Himalayas lies peninsular India, a part of the old Gondwana continent, where terrestrial rocks of Jurassic age occur in the upper part of the Gondwana Series. In the northwestern region, in the well-known Jurassic area of Kutch, nearshore

Stages	Bajocian	Aalenian	Toarcian	Pliensbachian	Sinemurian	Hettangian
C ARABIA, IN PART N AFRICA	Ermoceras, Thamboceras, Sohlites and Eocephalites, Stemmatoceras		Hildaites, Nejdia	Bouleiceras		
ARCTIC, N CANADA, GREENLAND	Arkelloceras	Tugurites, (Leioceras)	Pseudolioceras, Zugodactylites, Dactylioceras and Harpoceras	Amaltheus	Echioceras, Oxynoticeras oxynotum, Arctasteroceras	
VOLGA REGION						
NW EUROPE	Parkinsonia parkinsoni, Garantiana, Strenoceras, Teloceras, Stephanoceras, Otoites, Sonninia	Hyperlioceras, Ludwigia, Tmetoceras, Leioceras, Pleydellia, Dumortieria and Grammoceras	Haugia, Hildoceras, Harpoceras, Dactylioceras	Amaltheus, Prodactylioceras, Tragophylloceras ibex, Uptonia	Echioceras, Oxynoticeras oxynotum, Asteroceras obtusum, Euasteroceras turneri, Arnioceras, Arietites	Schlotheimia angulata, Alsatites, Psiloceras
EUROPEAN TETHYS, SOUTHERN, MIDDLE AND SW EUROPE		Dumortieria, Phymatoceras and Paroniceras	Mercaticeras, Frechiella and Leukadiella	Protogrammoceras, Fuciniceras and Arieticeras		

Fig. 2. *(Continued from facing page.)*

shallow-water sediments of Callovian to Tithonian age contain an unusually rich fauna of ammonites (SPATH, 1927-33) and bivalves (Cox, 1940). This fauna probably originated at the southern margin of the Tethys. As may be expected, the Tithonian faunas exhibit strong relationships with faunas of the Spiti Shale, although in Kutch, the phylloceratids, strangely, are somewhat more abundant, with the exception of *Haplophylloceras*, which is not represented. Furthermore, close relationships exist between the early Late Jurassic ammonite faunas of Kutch and those of southern Europe, including southern Germany. These can be demonstrated from the Callovian up to and including the *Hybonoticeras* Zone of the Kimmeridgian, the lowest zones of which are missing. *Phlycticeras, Paraspidoceras,* and *Hemihaploceras* are among genera that as yet are known only from Europe eastward as far as Kutch. The Kutch section shares many other genera with Europe and with wide areas of the Tethys. SPATH (1927-33), however, pointed out that more than two-thirds of the approximately 550 described ammonite species of the Kutch area are known only from there. It is possible that extravagant taxonomic splitting may be partly responsible for this picture.

The Jurassic of Kutch belonged to the East African-Madagascan-peninsular Indian epicontinental sea (Ethiopian biogeographical province), which had evolved since Callovian time from the Arabian-Madagascan seaway, and presumably was bordered on the southeast by the then still-existing Lemurian portion of Gondwanaland.

In the upper Oxfordian, the Mayaitidae, which were homeomorphic with, and even indistinguishable from, the older and more widely distributed Macrocephalitidae, are confined strictly to the East African-Indonesian region. In the Kimmeridgian, *Katroliceras*, occurring also from Europe to Japan, and *Torquatisphinctes* are especially abundant in Madagascar. Other very important European genera include *Taramelliceras, Streblites, Glochiceras, Aspidoceras, Physodoceras,* and *Hybonoticeras*. We even find identical stratigraphic index species such as *Aspidoceras acanthicum* and *Hybonoticeras hybonotum*.

The Tithonian of Madagascar, having a fauna composed almost exclusively of cephalopods, can be subdivided as follows:

Upper Tithonian—*Aulacosphinctes hollandi, Berriasella privasensis, Micracanthoceras micracanthus, Blanfordiceras* sp., *Lytohoplites* sp., and others.

Lower Tithonian—*Hildoglochiceras kobelli, Haploceras elimatum, Physodoceras avellanum,* and others.

This sequence has very close faunal relationships with Kutch, the Himalayas, Europe *(Berriasella, Hybonoticeras)*, North Africa *(Djurdjuriceras, Corongoceras, Lytohoplites)*, and also with Central and South America *(Corongoceras, Lytohoplites)*. The following genera found in Madagascar have a distribution from eastern Asia westward to South America: *Himalayites, Micracanthoceras, Aulacosphinctes, Proniceras* (COLLIGNON, 1964b). On the other hand, *Hildoglochiceras* is only distributed eastward to Central and South America, and suggests a migration route into the peripheral area of the Pacific Ocean. For some genera even a southern migration route may be considered, which possibly could have connected the Jurassic seas of South America, southeastern Africa, and New Zealand (STEVENS, 1967a). Such a route, however, could have exerted little faunal influence compared to that of the Arctic-Boreal region. A biostratigraphical table for comparison of all Tithonian areas was given by VERMA and WESTERMANN (1973).

LATE JURASSIC FAUNAS NORTH OF THE TETHYS AND OF THE EUROPEAN INTERMEDIATE ZONE

The relatively monotonous Late Jurassic Arctic-Boreal ammonite faunas, composed of a smaller number of genera, represent a contrast to the rich contemporaneous faunas of the Tethys. Cardioceratidae are found in the Oxfordian and lower Kimmeridgian, and boreal Perisphinctidae such as *Pectinatites, Pavlovia, Dorsoplanites, Glaucolithites,* and *Craspedites* are found in the Kimmeridgian and Portlandian. In Europe, Tethyan, Boreal and endemic faunal elements met and intermingled in a large intermediate region between the Tethyan and Boreal seas. For example, the Boreal Cardioceratidae, just as the earlier Kosmoceratidae, diminished and finally disappeared toward the south in this area.

This intermediate region can be divided into a subboreal province (ENAY, 1966; ZEISS, 1968) comprising northern, western, and central Europe, and a submediterranean province (GEYER, 1961) comprising Portugal, southern Catalonia, southeastern France, southern Germany, and southern Poland.

Some of the faunal differences have already been noted in recording the absence of *Gregoryceras* and *Epipeltoceras* in the subboreal north. The submediterranean province is distinguished by the great variety of the Oppeliidae (Taramelliceratinae, Streblitinae, Ochetoceratinae); *Idoceras, Ataxioceras, Glochiceras,* and *Sutneria* are also important. This faunal pattern is reminiscent of the one found in Mexico. The strange *Cymaceras* (Ochetoceratinae, lower Kimmeridgian) is thus far known only from southern Germany, Switzerland, and southeastern France (GEYER, 1959). The perisphinctids *Pomerania* (upper Oxfordian) and *Pictonia* (lower Kimmeridgian) are abundant in the subboreal province, but are rare in the Jurassic of southern Germany where they occur together with the Tethyan *Nebrodites,* which here makes its most northerly appearance (GEYER, 1961).

To explain the faunal differences between the subboreal and submediterranean provinces the following causes may be invoked: minor climatic differences; somewhat greater depth of the sea in the submediterranean province as indicated by greater abundance of the Oppeliidae; dividing barriers such as the French Central Plateau, the Rhenish-Ardenne Island, and the Bohemian Massif; and perhaps also special ecological conditions in the vicinity of the submediterranean sponge-reef facies.

In the above-described intermediate area, especially in its submediterranean part, we also find *Rasenia* and its successor, *Aulacostephanus*. Both range far into the Boreal region and *Aulacostephanus* also as far as Mexico and Kurdistan. In northern France and southern England other species take the place of *Aulacostephanus pseudomutabilis,* which is characteristic of the southern German area. Migrations, such as along the northern edge of Scandinavia into Russia and Poland, were responsible for chronological differences in the appearance of some species (B. ZIEGLER, 1961, 1962). In the subboreal part of the intermediate region *Aulacostephanus* is represented by the index species *A. autissiodorensis,* which continued to exist here at a time when the genus had already disappeared in the submediterranean part. Here a fauna developed, which is especially well known from the south German Jurassic, containing *Hybonoticeras beckeri, Sutneria subeumela, Virgataxioceras setatum,* and *Oxyoppelia* (BERCKHEMER & HÖLDER, 1959; HÖLDER & ZIEGLER, 1959). Representatives of this fauna are also known from southeastern France, Bakony in western Hungary, Rumania, the Pienninic Klippen zone, and the Jurassic basin of Poland. *Sutneria subeumela* is also found on the middle Volga (GEYER, 1969), and *Virgataxioceras* is found in northern Siberia.

In the middle Kimmeridgian (=lower Tithonian) *Gravesia,* an Upper Jurassic homeomorph of *Stephanoceras,* is abundant in the subboreal province. Its occurrences are increasingly rare toward the south and east (Spain, Alpine margin, and central Russia), but where present, it allows correlation of the lower Tithonian with the *Hybonoticeras hybonotum* Zone and its northern equivalents. In Russia, *Subplanites* predominates instead of *Gravesia*.

Gravesia is followed higher in the section by perisphinctids of the subfamily Pseudovirgatitinae (ZEISS, 1968). Further regional differentiation took place in England *(Pectinatites)* and central Russia *(Ilowaiskya)*. In the southern German Jura Mountains, on the other hand, different forms of Mediterranean Pseudovirgatitinae represented by *Usseliceras* and *Franconites* occur, accompanied by *Neochetoceras* and *Aspidoceras*. *Pavlovia* predominates in the uppermost Kimmeridgian (=middle Tithonian, lower middle Volgian) of the northern part of the European intermediate region, whereas in the Tethyan peripheral area the typical Tethyan *Pseudolissoceras* fauna occurs. In the Portlandian *sensu stricto* (=upper Tithonian, middle Volgian) in southern England new faunas evolved from *Glaucolithites,* which ranged as far as Greenland. In Russia forms with true virgatotome ribs evolved from *Virgatites*. In the beginning of this stage, *Zaraiskites* occurred in Russia, Poland (WILCZYŃSKI, 1962; DABROWSKA, 1967), and the submediterranean Franconian Jura, as well as southern England. ZEISS (1968) has compiled a map showing migration routes of these and other genera. *Zaraiskites* (Virgatitinae) seems to have originated from the submediterranean Pseudovirgatitinae (ZEISS, 1968; KUTEK & ZEISS, 1974).

The faunal differences between the submediterranean and the subboreal provinces, as well as within the latter, support subdivision into three subprovinces: 1) northern French-southern English province, 2) Russian, and 3) Polish (ZEISS, 1968). The subboreal province of ZEISS and others corresponds approximately to the Boreal-Atlantic region of SAKS and NALNYAEVA (1966), including an eastern European province. The Polish subprovince shows Russian, as well as submediterranean influences (KUTEK, 1962a,b,c; KUTEK & ZEISS, 1974). *Garnieriсeras* is endemic in the upper Volgian of Russia, and is a precursor of *Platylenticeras*. The previously mentioned *Craspedites* spreads far beyond the Arctic-Boreal zone. MEZESHNIKOV and ZAKHAROV (in DAGIS & ZAKHAROV, 1974) illustrated the differences between the Volgian ammonite faunas of the boreal-European, boreal-Russian, and arctic (north Siberian) regions. Surprisingly, a fauna of late Volgian age from Khatanga Bay in northern Siberia contains *Berriasella (Lamencia)* cf. *B. richteri* and highly specialized Virgatosphinctinae in association with *Craspedites* (SHULGINA, 1967).

NAUTILOIDEA

Only one genus of Nautiloidea *(Cenoceras)* crosses the boundary between the Triassic and Jurassic systems, reaching worldwide distribution in the early Liassic (KUMMEL, 1956). Several new types evolved in the Middle and Late Jurassic: *Eutrephoceras, Pseudonautilus, Paracenoceras, Cymatoceras, Aulaconautilus,* and others. The pointed sutural lobes of *Pseudonautilus* make this genus a homeomorph of *Permoceras* of the Lower Permian of Timor, and the two genera are hardly distinguishable.

Simultaneous crises of the ammonoids and nautiloids at the end of the Triassic indicate the presence of external influences, which affected the ectocochlian cephalopods more than other groups.

COLEOIDEA

Atractites is abundant in Lower Jurassic rocks of the geosynclinal regions of the Tethys and North America, but occurs also in smaller numbers in epicontinental areas.

Jurassic belemnites, represented by strongly dwarfed forms of the Belemnitidae (=Passaloteuthidae), first appeared during the Hettangian in epicontinental seas of Europe (SCHWEGLER, 1962). Complemented by the Hastitidae, they experienced their first great evolutionary phase during the Early Middle Jurassic. As early as the middle Liassic they spread to northern Siberia, Iran, northern Africa, and South America. The ability of the coleoids to disperse worldwide, however, was less than that of the ammonites. STEVENS (1967a), therefore, assumed their greater dependence on neritic shelves as migration routes. Thus, belemnites of the Liassic would have been prevented by deep-sea areas from reaching the eastern Tethys and New Zealand (Fig. 3), whereas ammonites of European type were able to migrate there. Only in the Toarcian-Aalenian did *Brachybelus* reach New Zealand, making it the oldest belemnite genus known in that country. At the same time the endemism of the New Zealand faunas became less pronounced.

During the Toarcian, Siberia became an important evolutionary center of belemnites (SAKS & NALNYAEVA, 1970), which before were absent there. Half of the species are identical with those of Europe, where an especially rich Toarcian fauna from the Franconian Jura Mountains has been described by KOLB (1942). *Parahastites* (Hastitidae) is endemic in Siberia. In the Middle Jurassic, belemnites, like ammonites, developed a special boreal fauna, which according to SAKS and NALNYAEVA (1966), began in arctic North America with *Cylindroteuthis* and *Pachyteuthis*. From the Callovian onward additional genera, such as *Acroteuthis* and *Lagonibelus,* as well as many subgenera and species evolved. These groups, like the boreal ammonite fauna, penetrated far southward, retaining a much higher variability in the Arctic-Boreal region proper than in its peripheral areas. In the

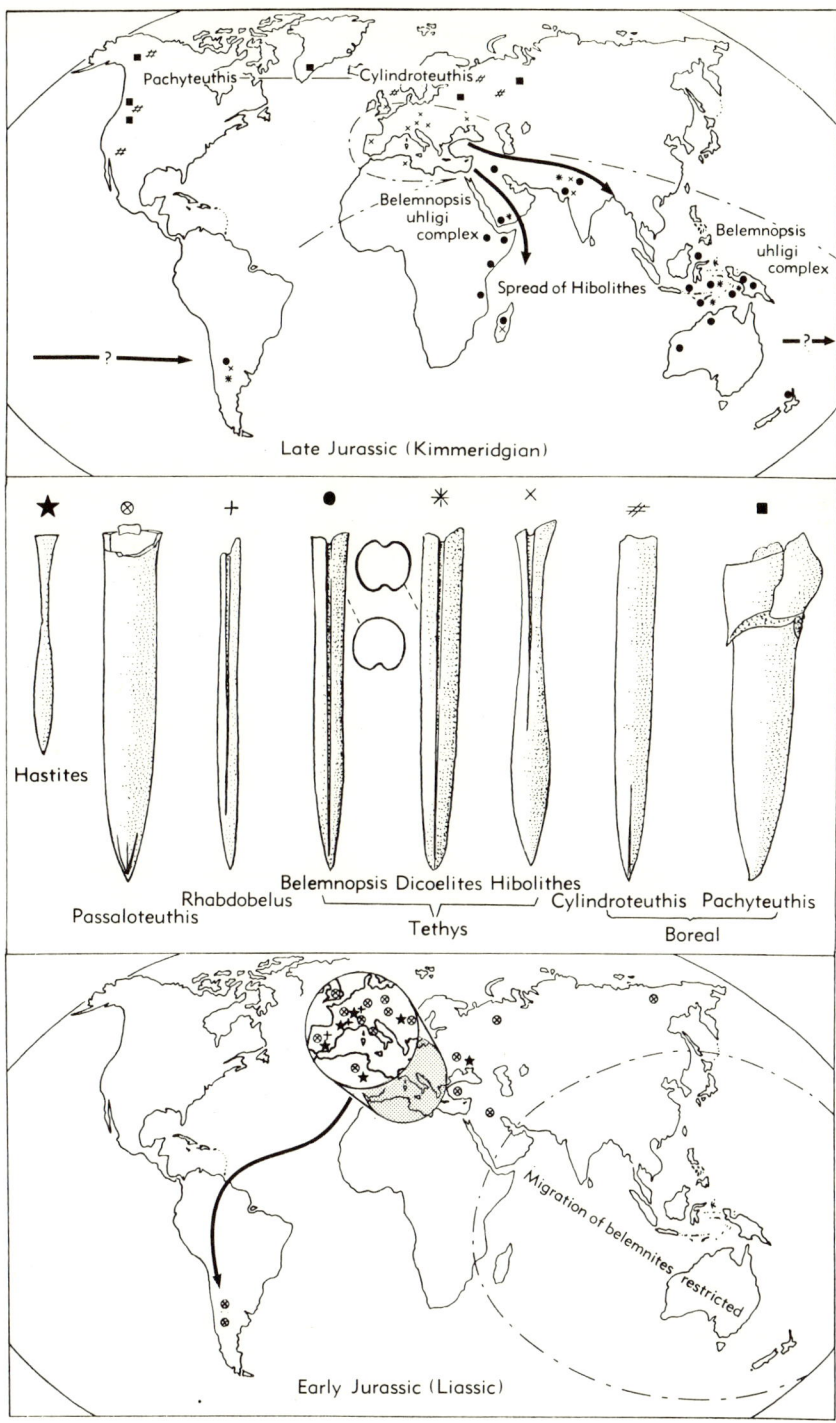

Fig. 3. Examples of the distribution of Jurassic belemnite faunas (maps after Stevens, 1965).

Callovian, *Cylindroteuthis* and *Pachyteuthis* reached Spain, California, and the southern Logan Sea. *Cylindroteuthis* even invaded Argentina, and perhaps, along a southern latitudinal route, also reached New Zealand (STEVENS, 1963; 1965).

In the Aalenian, Belemnopsidae with guards having a characteristic ventral groove evolved from the European Hastitidae quite independently of the boreal fauna. *Hibolithes, Duvalia,* and *Conobelus* spread across the Mediterranean region; *Belemnopsis, Conodicoelites, Dicoelites,* and in the Tithonian *Hibolithes,* also spread across the Indo-Pacific Province. Here, differentiation into a *Belemnopsis orientalis-gerardi* fauna took place in the East African-Madagascan-peninsular Indian epicontinental sea, and a *Belemnopsis uhligi* fauna in the Himalayan-Indomalaysian area. The latter probably also reached Argentina along a southern Pacific route (see *Cylindroteuthis* mentioned above).

Like the ammonite fauna, the belemnite fauna of Madagascar shows strong affinities with Europe, especially from the Bathonian to the late Oxfordian. In the Tithonian, *Hibolithes* is important in the entire area from Madagascar to the Indomalaysian geosyncline, and was joined in the late Tithonian by *Duvalia*.

The paucity of belemnites in the Upper Jurassic of southern Germany is in striking contrast to the wealth of ammonites in these rocks. *Hibolithes* is almost the only belemnite found there.

RHYNCHOLITES

Cephalopod mandibles (rhyncholites) are common in some Jurassic clays and limestone beds of the Mediterranean Tethys, where they are known today as far east as the Crimea and the Caucasus with the exception of sporadic occurrences in the European epicontinental seas. The more important Jurassic genera are *Rhynchoteuthis, Palaeoteuthis, Hadrocheilus, Leptocheilus, Akidocheilus,* and *Gonatocheilus.* "The known facts are best accommodated by the interpretation that rhyncholites were formed by unknown cephalopods, some of which probably belong to the Nautilida" (TEICHERT, MOORE, & ZELLER, 1964, p. *K*476).

CRUSTACEA

The Ostracoda are important as index fossils and occasionally render possible finer stratigraphic zonation than the ammonites (PLUMHOFF, 1963). They are especially helpful in correlation of the Purbeckian and Wealden of northwestern and central Europe, as well as for determination of the Jurassic-Cretaceous boundary (BARTENSTEIN, 1959; WOLBURG, 1959). Along with some widely distributed species, many strictly localized ones occur.

Isopoda, known since the Triassic, have been described from the middle Liassic of Württemberg, from the reef and lagoonal limestones of the Tithonian, and in abundance from the Purbeckian.

During Jurassic time the Decapoda evolved more rapidly than ever before. Among the macruran Trichelida, swimmers of worldwide distribution are found, for example, *Eryma* (FÖRSTER, 1966). The degree of dependence on a benthonic mode of life determines distribution. Among the Anomura, *Gastrodorus,* which has an unprotected abdomen, represents the first of the Paguridae (hermit crabs). From the gastralids, through *Pemphix,* were derived 1) the macruran Palinuridae and 2) the Brachyura, which reached their acme in the Cretaceous. Important Brachyura are the Dromiaceae *Eocarcinus* in the Liassic, *Goniodromites* since the Liassic, and *Pithonoton,* and *Prosopon.* Close relatives of the Jurassic Anomura and primitive Dromiaceae are found in the present deep seas (BEURLEN, 1931; GLAESSNER, 1933).

OTHER INVERTEBRATES

Zoogeographical data for other Jurassic invertebrates are poor, though adaptability of the Crinoidea and Echinozoa was remarkable. *Seirocrinus subangularis* with floating arms became the largest crinoid of earth history. The stemless, freely mobile genera *Antedon* and *Saccocoma* spread during the Late Jurassic, the latter especially in the Tethys, although the late Triassic *Osteocrinus* KRISTAN-TOLLMANN (1970) is the oldest planktonic microcrinoid known in the Tethyan. Among the Jurassic reef-dwellers are *Apiocrinus, Millericrinus, Eugeniacrinites,* and the stemless, bowl-shaped *Cotylederma* (Liassic), which has much reduced and fused calyx plates.

Within the Jurassic echinoids, the Cidaroidea acquired their final shape with rigid skeletons. The gnathostome Holectypida *(Pygaster, Holectypus),* with tests covered by densely spaced spines, are the first typical "Irregularia." Also the atelostome, sediment-eating Cassidulidae *(Hypoclypeus, Galeropygus, Echinobrissus, Clypeus, Pygurus)* and the Holasteridae *(Collyrites* since the Liassic, *Dysaster* with dissected upper region) appeared. Today the Holasteridae are almost entirely restricted to deep seas.

Conodonts hitherto seemed to be absent in the Jurassic, but recently *Gladigondolella* and *Hindeodella,* known from the Upper Triassic, were found in the Upper Jurassic of Japan (NOHDA & SETOGUCHI, 1967); however, MÜLLER and MOSHER (1971) opposed the interpretation of these fossils and pointed to the possibility of reworking and to the uncertainty of the stratigraphic position.

REFERENCES

Ager, D. V., 1956, *The geographical distribution of brachiopods in the British Middle Lias:* Geol. Soc. London, Quart. Jour., v. 112, p. 157-188, text-fig. 1,2.——1965, *The adaption of Mesozoic brachiopods to different environments:* Paleogeography, Paleoclimatology, Paleoecology, v. 1, p. 142-172.

——, **& Westermann, G. E. G.,** 1963, *New Mesozoic brachiopods from Canada:* Jour. Paleontology, v. 37, p. 595-610, text-fig. 1-5, pl. 71-73.

Alloiteau, James, 1958, *Monographie des madréporaires fossiles de Madagascar:* Serv. Mines Madagascar (Tananarive), Ann. Géol., v. 25, p. 9-218, text-fig. 1-37, pl. 1-38.

Andreeva, T. F., & Dronov, V. J., 1964, *Stratigraphie des dépots jurassiques du Pamir Central et du Sud-Est:* Inst. Grand-Ducal Luxembourg, Comptes Rendus et Mém., Coll. Jurassique (1962), Sec. sci. nat., p. 893-881.

Arkell, W. J., 1935, *The nature, origin and climatic significance of the coral reefs near Oxford:* Geol. Soc. London, Quart. Jour., v. 91, p. 77-101.—— 1952, *Jurassic ammonites from Jebel Tuwaiq, central Arabia:* Royal Soc. London, Philos. Trans., ser. B, v. 236, p. 241-313, pl. 15-31. ——1956, *Jurassic geology of the world:* 806 p., 102 text-fig., 46 pl., 28 tab., Oliver & Boyd (Edinburgh & London).

Bartenstein, Helmut, 1959, *Die Jura-Kreide-Grenze in Europa. Ein Überblick des derzeitigen Forschungsstandes:* Eclogae Geol. Helvetiae, v. 52, p. 15-18.

Barthel, K. W., Cediel, Fabio, Geyer, O. F., & Remane, Jürgen, 1966, *Der subbetische Jura von Cehegin (Provinz Murcia, Spanien):* Bayer. Staatssamml. Paläontologie, Hist. Geologie, Mitteil., v. 6, p. 167-211, text-fig. 1-4.

Behmel, Hermann, & Geyer, O. F., 1966, *Beiträge zur Stratigraphie und Paläontologie des Juras von Ostspanien III. Stratigraphie und Fossilführung im Unterjura von Albarracin (Provinz Teruel):* Neues Jahrb. Geologie, Paläontologie, Abhandl., v. 124, p. 1-52, text-fig. 1-4, pl. 1-6.

Berckhemer, Fritz, & Hölder, Helmut, 1959, *Ammoniten aus dem Oberen Weissen Jura Süddeutschlands:* Geol. Jahrb., Beihefte, no. 25, p. 1-135, pl. 1-27.

Beurlen, Karl, 1931, *Die Besiedlung der Tiefsee:* Natur u. Museum, v. 61, no. 6, p. 278-282, text-fig. 1-11.

Blind, Wolfram, 1963, *Die Ammoniten des Lias Alpha aus Schwaben, vom Fonsjoch und Breitenberg (Alpen) und ihre Entwicklung:* Palaeontographica, v. 121, Abt. A, p. 39-131, text-fig. 1-46, pl. 1-5.

Brinkmann, Roland, 1929, *Monographie der Gattung Kosmoceras:* Gesell. Wiss. Göttingen, math.-phys. Kl., n. ser., v. 13, p. 1-124, 1 pl.

Callomon, J. G., 1963, *The Jurassic ammonite-faunas of East Greenland:* Experientia, v. 19, p. 1-6, text-fig. 1-5.

Cariou, Elie, 1965, *Essai de corrélations stratigraphiques entre l'Ouest de l'Europe et la province indo-malgache, au Callovien:* Soc. Géol. France, Bull., v. 7, p. 537-540, text-fig. 1.

Cecioni, Giovanni, & Westermann, G. E. G., 1968, *The Triassic/Jurassic marine transition of coastal central Chile:* Pacific Geology, v. 1, p. 41-75, text-fig. 1-4, 7 pl.

Collignon, Maurice, 1959, *Calcaires à polypiers, récifs et atolls au Sud du Madagascar:* Soc. Géol. France, Bull., sér. 7, v. 1, p. 403-408, text-fig. 1,2——1964a, *Le Bathonien marin à Madagascar*

Limite supérieure-rapports et corrélations: Inst. Grand-Ducal Luxembourg, Comptes Rendus et Mém., Coll. Jurassique (1962), p. 913-919.——1964b, *Échelle chronostratigraphique pour les domaines Indo-Africano-Malgache (Bathonien moyen à Tithonique):* Inst. Grand-Ducal Luxembourg, Comptes Rendus, Coll. Jurassique (1962), p. 927-931.

Colo, Gabriel, 1961, *Contribution à l'étude du Jurassique de Moyen Atlas Septentrional:* Serv. Géol. Maroc (Rabat), Notes et Mém., no. 139, p. 1-226, text-fig. 1-28, 1 geol. map.

Cowen, Richard, & Rudwick, M. S. A., 1966, *A spiral brachidium in the Jurassic chonetoid brachiopod Cadomella:* Geol. Mag., v. 103, p. 403-406, 1 pl.

Cox, L. R., 1940, *The Jurassic lamellibranch fauna of Kachh (Cutch):* Geol. Survey India, Mem., Palaeont. Indica, ser. 9, v. 3, 157 p., 12 pl.——1965, *Jurassic Bivalvia and Gastropoda from Tanganyika and Kenya:* British Museum (Nat. History), Bull., suppl. 1, v. 40, p. 3-213, text-fig. 1,2, 30 pl.

Dabrowska, Zofia, 1967, *Paleogeografia Gornej Jury w Polsce na tle Europy:* Przeglad Geol., v. 5, p. 208-215, text-fig. 1-5. [*Paleogeography of the Upper Jurassic in Poland in relation to Europe.*]

Dacqué, Edgar, 1910, *Der Jura in der Umgebung des lemurischen Kontinents:* Geol. Rundschau, v. 1, p. 148-168.——1915, *Neue Beiträge zur Kenntnis des Jura in Abessynien:* Beitr. Paläontologie Geologie Österreich-Ungarns u. Orients, v. 27 (1914), p. 1-17, pl. 1-3.

Dagis, A. S., & Zakharov, V. A., 1974, *Paleobiogeografiya severa eurazii v Mezozoe:* 196 p., text-fig., "Nauka" Publ., Siberian Branch (Novosibirsk). [*Mesozoic paleobiogeography of north Eurasia.*]

Donovan, D. T., 1953, *The Jurassic and Cretaceous stratigraphy and palaeontology of Traill Ø, East Greenland:* Meddel. Grønland, v. 11, no. 4, p. 5-150, text-fig. 1-14, 25 pl.——1967, *The geographical distribution of Lower Jurassic ammonites in Europe and adjacent areas:* Syst. Assoc. Publ., no. 7, p. 11-134, text-fig. 1-5.

Dubar, G. G., 1942, *Études paléontologiques sur le Lias du Maroc, brachiopodes térébratules et zeilléries multiplissées:* Serv. Géol. Maroc, Notes et Mém., no. 57, 103 p., text-fig. 1-51, 10 pl.——1948, *La faune doménienne du Jebel Bou-Dahar près de Beni-Tadjite:* Serv. Géol. Maroc, Notes et Mém., no. 68, 250 p.

Enay, Raymond, 1965, *Les formations coralliennes de Saint-Germain-de-Joux (Ain):* Soc. Géol. France, Bull., sér. 7, v. 7, p. 23-31, text-fig. 1-3.——1966, *Le genre Gravesia (Ammonitina jurassiques) dans le Jura français et les chaines subalpines:* Ann. Paléontologie, Invertébrés, v. 52, p. 95-105, text-fig. 1, pl. A, B.——1973, *Upper Jurassic (Tithonian) ammonites:* in Atlas of Palaeogeography, Anthony Hallam (ed.), p. 297-307, text-fig. 1-3, 1 tab., Elsevier Sci. Publ., Co. (Amsterdam-London-New York).

Fatmi, A. N., 1972, *Stratigraphy of the Jurassic and Lower Cretaceous rocks and Jurassic ammonites from northern areas of West Pakistan:* British Museum (Nat. History), Bull., v. 20, p. 297-380, text-fig. 1-6, 2 pl.

Fenninger, Alois, & Hötzl, Heinz, 1965, *Die Hydrozoa und Tabulozoa der Tressenstein- und Plassenkalke (Ober-Jura):* Museum Bergbau, Geologie, Technik, Landesmuseum "Joanneum," Mitteil., no. 27, p. 1-61, text-fig. 1-4, pl. 1-8.

Fischer, A. G., 1961, *Latitudinal variations in organic diversity:* Am. Scientist, v. 49, no. 1, p. 50-74, text-fig. 1-19.

Fischer, Rudolf, 1967, *Zur Ökologie zweier Ammonitenfaunen aus dem Aalenium des Schneibsteins (Berchtesgadener Alpen):* Geologica et Palaeontologica (Marburg), v. 1, p. 175-177, text-fig. 1.

Flügel, Erik, 1966, *Mitteljurassische Korallen vom Ostrand der Grossen Salzwüste (Shotori-Kette, Iran):* Neues Jahrb. Geologie, Paläontologie, Abhandl., v. 126, no. 1, p. 46-91, text-fig. 1-3, pl. 15-19.

——, **& Franz, H. E.,** 1967, *Elektronenmikroskopischer Nachweis von Coccolithen im Solnhofener Plattenkalk (Ober-Jura):* Neues Jahrb. Geologie, Paläontologie, Abhandl., v. 127, p. 245-263, text-fig. 1, 3 pl.

Förster, Reinhard, 1966, *Über die Erymiden, eine alte, konservative Familie der mesozoischen Dekapoden:* Palaeontographica, Abt. A, v. 125, p. 61-175, text-fig. 1-37, pl. 13-20.

Frebold, Hans, 1951, *Contributions to the palaeontology and stratigraphy of the Jurassic System in Canada:* Geol. Survey Canada, Bull. 18, p. 1-54, text-fig. 1,2, pl. 1-17.——1957, *The Jurassic Fernie group in the Canadian Rocky Mountains and foothills:* Geol. Survey Canada, Mem. 287, 1-197 p., 54 pl.——1958, *Fauna, age and correlation of the Jurassic rocks of Prince Patrick Island (Northwest Territories):* Geol. Survey Canada, Bull. 52, p. 1-69, illus.——1959, *Marine Jurassic rocks in Nelson and Salmo areas, southern British Columbia:* Geol. Survey Canada, Bull. 49, p. 1-31, pl. 1-5.——1960, *The Jurassic faunas of the Canadian Arctic—Lower Jurassic and lowermost Middle Jurassic ammonites:* Geol. Survey Canada, Bull. 59, p. 1-33, text-fig. 1-8, pl. 1-15.——1963, *Ammonite faunas of the Upper Middle Jurassic beds of the Fernie group in western Canada:* Geol. Survey Canada, Bull. 93, p. 1-33, text-fig. 1, pl. 1-14.——1967, *Hettangian ammonite faunas of the Taseko Lakes area British Columbia:* Geol. Survey Canada, Bull. 158, p. 1-35, text-fig. 1-6, pl. 1-9.

Garrison, R. E., & Fischer, A. G., 1969, *Deepwater limestones and radiolarites of the alpine Jurassic:* in G. M. Friedmann (ed.), Depositional environments in carbonate rocks, a symposium, Soc. Econ. Paleontologists & Mineralogists, Publ. no. 14, p. 20-56, text-fig. 1-22.

Geyer, O. F., 1958, *Die Korallenfauna des europäischen Malm und ihr stratigraphischer Wert:* Internatl. Geol. Congress, 20th Sess. (1956), p. 61-74, text-fig. 1,2 (Mexico City).———1959, *Über Oxydiscites Dacqué, ein Beitrag zur Kenntnis der Ochetoceratinae (Cephal. Jurass.):* Neues Jahrb. Geologie, Paläontologie, Monatsh., 1960, p. 417-425, text-fig. 1-9.———1961, *Monographie der Perisphinctidae des unteren Unterkimmeridgium (Weisser Jura γ, Badenerschichten) im süddeutschen Jura:* Palaeontographica, v. 117, Abt. A, p. 1-157, text-fig. 1-156, pl. 1-22.——— 1965, *Einige Funde der arabo-madagasischen Ammoniten-Gattung Bouleiceras im Unterjura der Iberischen Halbinsel:* Paläont. Zeitschr., v. 39, p. 26-32, text-fig. 1,2, pl. 5.———1969, *The ammonite genus Sutneria in the Upper Jurassic of Europe:* Lethaia, v. 2, no. 1, p. 63-72, text-fig. 1-4, tab. 1,2.———1973, *Das präkretazische Mesozoikum von Kolumbien:* Geol. Jahrb., v. B 5, p. 1-156, text-fig. 1-40, 5 pl., tab. 1-11.

Glaessner, M. F., 1933, *Die Krabben der Juraformation:* Centralbl. Mineralogie, Geologie, Paläontologie (1933), Abt. B, p. 178-191, text-fig. 1-4.

Gygi, Reinhart, 1966, *Über das zeitliche Verhältnis zwischen der transversarium-Zone in der Schweiz und der plicatilis-Zone in England (Unt. Malm, Jura):* Eclogae Geol. Helvetiae, v. 59, no. 1, p. 935-942, text-fig. 1, 4 pl.

Hallam, Anthony, 1969, *Faunal realms and facies in the Jurassic:* Palaeontology, v. 12, p. 1-18, text-fig. 1-4.

Haug, Emile, 1927, *Traité de Géologie. II. Les périodes géologiques. Période Jurassique:* p. 929-1152, pl. 100-112, Librairie Armand Colin (Paris).

Hegarat, Gérard le, & Remane, Jürgen, 1968, *Tithonique supérieur et Berriasien de l'Ardèche et de l'Hérault. Corrélation des Ammonites et des Calpionelles:* Géobios, Fac. Sci. Lyon, no. 1, p. 7-70, pl. 1-10.

Henson, F. R. S., 1949, *Larger imperforate Foraminifera of south-western Asia:* British Museum (Nat. History), Bull. 11 (1948), p. 1-127, text-fig. 1-16, pl. 1-16.

Hillebrandt, Axel von, 1970, *Zur Biostratigraphie und Ammoniten-Fauna des südamerikanischen Jura (insbes. Chile):* Neues Jahrb. Geologie, Paläontologie, Abhandl., v. 136, p. 166-211, text-fig. 1-3, 2 pl.

Hölder, Helmut, 1953, *Oxytoma scanica (Lundgren) in der schwäbischen Planorbis-Zone:* Neues Jahrb. Geologie, Paläontologie, Monatsh. 1953, p. 358-364, text-fig. 1-6.

———, & Ziegler, Bernhard, 1959, *Stratigraphische und faunistische Beziehungen im Weissen Jura (Kimmeridgien) zwischen Süddeutschland und Ardèche:* Neues Jahrb. Geologie, Paläontologie, Abhandl., v. 108, p. 150-214, pl. 17-22.

Hoffmann, Karl, 1941, *Eine neue Ammonitenfauna aus dem unteren Lias Nordwestdeutschlands:* Reichsamt Bodenforsch., Jahrb. 1941, v. 62, p. 288-337, text-fig. 1-24, pl. 16-19.

Howarth, M. K., 1958, *The ammonites of the Liassic family Amaltheidae in Britain. I., II.:* Palaeontograph. Soc., Mon., p. 1-53, text-fig. 1-18, pl. 1-10.

Huckriede, Reinhold, 1967, *Molluskenfaunen mit limnischen und brackischen Elementen aus Jura, Serpulit und Wealden NW-Deutschlands und ihre paläogeographische Bedeutung:* Geol. Jahrb., Beihefte, v. 67, 263 p., 32 text-fig., 25 pl.

Hudson, J. D., 1963a, *The recognition of salinity-controlled mollusc assemblages in the Great Estuarine Series (Middle Jurassic) of the Inner Hebrids:* Palaeontology, v. 6, p. 318-326, text-fig. 1.———1963b, *The ecology and stratigraphical distribution of the invertebrate fauna of the Great Estuarine Series:* Palaeontology, v. 6, p. 327-348, text-fig. 1-3, pl. 53.

Imlay, R. W., 1948, *Characteristic marine Jurassic fossils from the western interior of the United States:* U.S. Geol. Survey, Prof. Paper 214B, p. 13-33, pl. 4-9.———1952, *Correlation of the Jurassic formations of North America, exclusive of Canada:* Geol. Soc. America, Bull., v. 63, p. 953-992, text-fig. 1-4.———1955, *Characteristic Jurassic molluscs from northern Alaska:* U.S. Geol. Survey, Prof. Paper 274-D, p. 69-96, text-fig. 1, pl. 8-13, tab. 1-4, 1 correlation chart.——— 1959, *Succession and speciation of the pelecypod Aucella:* U.S. Geol. Survey, Prof. Paper 314-G, p. 155-169, text-fig. 36, pl. 16-19.———1963, *Jurassic fossils from southern California:* Jour. Paleontology, v. 37, no. 1, p. 97-107, pl. 14. ———1964a, *Marine Jurassic pelecypods from central and southern Utah:* U.S. Geol. Survey, Prof. Paper 483-C, p. 1-42, text-fig. 1, pl. 1-4.——— 1964b, *Middle and Upper Jurassic fossils from southern California:* Jour. Paleontology, v. 38, p. 505-509, pl. 78.———1965, *Jurassic marine faunal differentiation in North America:* Jour. Paleontology, v. 39, 1023-1038, text-fig. 1-6. ———1967, *Twin Creek Limestone (Jurassic) in the western interior of the United States:* U.S. Geol. Survey, Prof. Paper 540, 105 p., 18 text-fig., 16 pl.

Jefferies, R. P. S., & Minton, P., 1965, *The mode of life of two Jurassic species of "Posidonia" (Bivalvia):* Palaeontology, v. 8, p. 156-185, text-fig. 1-12, pl. 19.

Jeletzky, J. A., 1963, *Malayomaorica gen. nov. (family Aviculopectinidae) from the Indo-Pacific Upper Jurassic; with comments on related forms:* Palaeontology, v. 6, p. 148-160, pl. 21.

Jordan, Reiner, 1974, *Salz- und Erdöl/Erdgas-Austritt als Fazies-bestimmende Faktoren im Mesozoikum Nordwest-Deutschlands:* Geol. Jahrb., ser. A, v. 13, p. 1-64, text-fig. 1,2, 1 pl.

Kobayashi, Teiichi, 1961, *The Jurassic of Japan and its bearing on the correlation of the Pacific area:* 9th Pacific Sci. Congr., Proc., 12, p. 262-266. [Not seen by author.]

Kolb, Heinrich, 1942, *Die Belemniten des jüngeren Lias Zeta in Nordbayern:* Deutsche Geol. Gesell., Zeitschr., v. 94, p. 145-168, pl. 5-8.

Kovacs, Ludwig, 1942, *Monographie der liassischen Ammoniten des nördlichen Bakony:* Geologica Hungarica, ser. palaeont., v. 17, 220 p., 109 text-fig., 5 pl., 2 tab.

——, & Zeiss, Arnold, 1974, *Tithonian-Volgian ammonites from Brzostówka near Tomaszów Mazowiecki, Central Poland:* Acta Geol. Polonica, v. 24, no. 3, p. 505-542, pl. 1-32 (Pol. summ.).

Kristan-Tollmann, Edith, 1970, *Die Osteocrinusfazies, ein Leithorizont von Schwebcrinoiden im Oberladin-Unterkarn der Tethys:* Erdöl und Kohle, Jahrb. 23, p. 781-789, text-fig. 1-13, 1 tab.

Krystyn, L., 1974, *Probleme der biostratigraphischen Gliederung der alpin-mediterranen Obertrias:* Schriftenreihe erdwiss. Kommission Österr. Akad. Wiss, v. 2, p. 137-144, 1 text-fig.

Kummel, Bernhard, 1956, *Post-Triassic nautiloid genera:* Harvard Coll., Museum Comp. Zoology, Bull., v. 114, p. 321-494, text-fig. 1-35, pl. 1-28.

Kutek, Jan, 1962a, *Palaeogeographic significance of ammonitic fauna of the Middle and Upper Malm in Central Poland:* Acad. Polon. Sci., Bull., v. 10, no. 2, p. 79-84.——1962b, *Gorny kimeryd i dolny wołg pn.-zachodniego obrzeżenia mezozoicznego Gór Świętokrzyskich:* Acta Geol. Polonica, v. 12, no. 4, p. 445-527. (Russ. and Fr. summ.) [*The upper Kimmeridgian and the lower Volgian of the Mesozoic boundary northwest of the Holy Cross Mountains.*]——1962c, *Problematyka stratygraficzna kimerydu i najwyższego oksfordu Polski:* Acta Geol. Polonica, v. 12, no. 4, p. 529-540 (Russ. and Eng. summ.). [*Stratigraphic problems of the Kimmeridgian and uppermost Oxfordian in Poland.*]

Lange, Werner, 1951, *Die Schlotheimiinae aus dem Lias Alpha Norddeutschlands:* Palaeontographica, v. 100, Abt. A, p. 1-128, text-fig. 1-109, pl. 1-20.——1952, *Der untere Lias am Fonsjoch (östliches Karwendelgebirge) und seine Ammonitenfauna:* Palaeontographica, v. 102, Abt. A, p. 49-159, text-fig. 1-65, pl. 8-18.

Lattin, Gustav de, 1967, *Grundriss der Zoogeographie:* 601 p., 170 text-fig., Gustav Fischer (Stuttgart).

Leanza, A. F., 1942, *Los Pelecipodos del Lias de Piedra Pintada en el Neuquen:* Rev. Museo La Plata, n. ser. 2, sec. paleont., p. 143-206, text-fig. 1-3, 19 pl.

Lin, C. C., 1961, *On the occurrence of Jurassic ammonite newly found in Taiwan, China:* Acta Geol. Taiwanica, no. 9, p. 79-81, text-fig. 1, 1 pl.

Lupher, R. L., & Packard, E. L., 1930, *The Jurassic and Cretaceous rudistids of Oregon:* Univ. Oregon, Publ., v. 1, p. 203-212.

Maeda, Shiro, 1961, *On the geological history of the Mesozoic Tetori group in Japan:* Japan. Jour. Geology & Geography, v. 32, p. 375-396.——1962, *On some Nipponitrigonia in Japan. On some species of Jurassic trigoniids from the Tetori Group in Central Japan:* Chiba Univ., Jour. Coll. Arts Sci., nat. sci. ser., v. 3, p. 503-514, pl. 1-9; p. 515-518, pl. 1.

Martinson, G. G., 1964, *Significance of fresh water mollusca for the stratigraphy of Jurassic continental deposits in Asia:* Inst. Grand-Ducal Luxembourg, Comptes Rendus, Coll. Jurassique, 1962, p. 459-463.

Maync, Wolf, 1947, *Stratigraphie der Jurabildungen Ostgrönlands zwischen Hochstetterbugten (75°N) und dem Kejser Franz Joseph Fjord (73°N):* Meddel. Grønland, v. 132, 223 p., text-fig. 1-38, pl. 1-7.

Müller, K. J., & Mosher, L. C., 1971, *Post-Triassic conodonts:* Geol. Soc. America, Mem. 127 (1970), p. 467-470.

Muir-Wood, H. M., 1953, *On some Jurassic and Cretaceous brachiopods from Traill Ø, East Greenland:* Meddel. Grønland, v. 111, p. 1-15, pl. 1.

Neumann, J., 1907, *Die Oxfordfauna von Cetechowitz:* Beitr. Paläontologie Geologie Österreich-Ungarns u. Orients, v. 20, p. 1-67, text-fig. 1-2, 8 pl.

Neumayr, Melchior, 1883, *Über klimatische Zonen während der Jura- und Kreidezeit:* K. Akad. Wiss. Wien, Denkschr., math.-naturw. Kl., v. 47, p. 277-310.

Nohda, Susumu, & Setoguchi, Takeshi, 1967, *An occurrence of Jurassic conodonts from Japan:* Univ. Kyoto, Mem. Coll. Sci., ser. B, v. 23, no. 4, p. 228-238, text-fig. 1-6, pl. 2.

Pia, Julius von, 1913, *Über eine mittelliasische Cephalopoden fauna aus dem nordöstlichen Kleinasien.* K. K. Naturhist. Hofmuseum Wien, Ann., v. 27, p. 335-388, text-fig. 1-7, 2 pl.——1914, *Untersuchungen über die Gattung Oxynoticeras und einige damit zusammenhängende allgemeine Fragen:* K. K. Geol. Reichsanst., Abhandl., v. 23, 173 p., text-fig. 1-5, 73 pl.

Plumhoff, Friedrich, 1963, *Die Ostracoden des Oberaalenium und tieferen Unterbajocium (Jura) des Gifhorner Troges, Nordwestdeutschland:* Senckenberg. Naturf. Gesell., Abhandl., no. 503, 98 p., 12 pl.

Quilty, P. G., 1970, *Jurassic ammonites from Ellsworth Land, Antarctica:* Jour. Paleontology, v. 44, p. 110-116, text-fig. 1-4, pl. 25.

Roll, Artur, 1934, *Form, Bau und Entstehung der Schwammstotzen im süddeutschen Malm:* Paläont. Zeitschr., v. 16, p. 197-246, text-fig. 1-18.

Rousselle, Lucienne, 1968, *Stolomorhynchia babtisrensis nov. sp. (Brachiopode, Rhynochonellacea)* du Toarcien de la région de Sidi-Kacem (Prérif occidental, Maroc): Serv. Géol. Maroc, Notes, v. 28, no. 211, p. 29-36, text-fig. 1-4.

——, & Bisch, J.-Ph., 1967, Deux nouvelles espèces de Tetrarhynchiinae (Rhynchonelles) dans le Lias moyen du Causse moyen-atlasique (Maroc): Soc. Géol. France, Bull., sér. 7, v. 9, p. 777-783, text-fig. 1-5, 1 pl.

Saks [Sachs], V. N., 1964, Über die Anwendungsmöglichkeit der allgemeinen Jura-Gliederung auf Juraablagerungen Sibiriens: Inst. Grand-Ducal Luxembourg, Comptes Rendus, Coll. Jurassique Luxembourg (1962), p. 763-781.

——, Mesezhnikov, M. S., & Shulgina, N. I., 1964, O svyazyach yurskikh i melovykh morskikh basseinov na severe i yuge Evrazii: Internatl. Geol. Congress, 22nd Sess., Doklady Sovetskikh Geologov, p. 163-174 (New Delhi). [On the connections of the Jurassic and Cretaceous marine basins in northern and southern Eurasia.]

——, & Nalnyaeva, T. I., 1966, Verkhneyurskie i nizhnemelovye belemnity Severa SSSR: Akad. Nauk SSSR, Sibir. Otdel., Inst. Geol. Geofiz., Rody Pachyteuthis i Acroteuthis, 260 p., 65 text-fig., 40 pl.; Rody Cylindroteuthis i Lagonibelus, 165 p., 41 text-fig., 28 pl. [Upper Jurassic and Lower Cretaceous belemnites of the northern USSR: Genera Pachyteuthis and Acroteuthis; Genera Cylindroteuthis and Lagonibelus.]—— 1970, Ranne- i sredneyurskie belemnity severa SSSR; Nannobelinae, Passaloteuthinae i Hastitidae: Akad. nauk SSSR, Sib. Otdel., Inst. Geol. Geofiz., Trudy, no. 110, 228 p., 62 text-fig., 22 pl. [Early and Middle Jurassic belemnites of northern USSR; Nannobelinae, Passaloteuthinae, and Hastitidae.]

Sapunov I. G., 1973, Notes on the geographical differentiation of the Lower Jurassic ammonite faunas: in Colloque de Jurassique à Luxembourg 1967; Fr., Bur. Rech. Géol. Minières, Mém., no. 75 (1971), p. 263-270, text-fig. 1,2.

——, & Stephanov, J., 1964, The stages, substages, ammonite zones and subzones of the Lower and Middle Jurassic in the Western and Central Balkan Range (Bulgaria) in 1962: Inst. Grand-Ducal Luxembourg, Coll. Jurassique Luxembourg 1962, Comptes Rendus et Mém., sec. sci. nat. phys. math., p. 705-718.

Sato, Tadashi, 1960, Apropos des courants océaniques froids prouvés par l'existence des ammonites d'origine arctique dans le Jurassique Japonais: Internatl. Geol. Congr. Norden, 21st Sess., pt. 12, p. 165-169, text-fig. 1,2 (Copenhagen).——1961, La limite jurassico-crétacée dans la stratigraphie japonaise. Faune berriasienne et tithonique supérieur nouvellement découverte au Japon: Japan. Jour. Geology & Geography, v. 32, p. 533-551, 2 pl.——1962, Études biostratigraphiques des Ammonites du Jurassique du Japon: Soc. Géol. France, Mém. 94, v. 41, 122 p., 16 text-fig., 10 pl.

Schindewolf, O. H., 1957, Über den Lias von Peru: Geol. Jahrb., v. 74, p. 151-160.——1961-68, Studien zur Stammesgeschichte der Ammoniten I-III, VII: Akad. Wiss. Lit. Mainz, Abhandl., math.-naturw. Kl.; Lief. I [Lytoceratina, Phylloceratina], Jahrg. 1960, no. 10, p. 639-743, text-fig. 1-58, pl. 1,2; Lief. II [Psilocerataceae, Eoderocerataceae], Jahrg. 1962, no. 8, p. 429-571, text-fig. 1-91, pl. 3; Lief. III [Hildocerataceae, Haplocerataceae], Jahrg. 1963, no. 6, p. 289-432, text-fig. 1-94; Lief. VII, Jahrg. 1968, no. 3, p. 43-209, text-fig. 1-39.

Schmidt-Effing, Reinhard, 1972, Die Dactyliocerati-dae, eine Ammoniten-Familie des unteren Jura. (Systematik, Stratigraphie, Zoogeographie, Phylogenie mit besonderer Berücksichtigung spanischen Materials): Münster. Forsch. Geologie Paläontologie, v. 25/26, 255 p., 31 text-fig., 19 pl.

Schrammen, Anton, 1924, Zur Revision der Jura-Spongien von Süddeutschland: Oberrhein. Geol. Verein, Jahresber. & Mitteil., n.ser., v. 13, p. 125-154.

Schwegler, Erich, 1962, Revision der Belemniten des Schwäbischen Jura II: Palaeontographica, v. 118, Abt. A, p. 1-23, text-fig. 1-12.

Seibold, Eugen, & Seibold, Ilse, 1960, Über Funde von Globigerinen an der Dogger-Malm-Grenze Süddeutschlands: Internatl. Geol. Congress, 21st Sess., Norden, v. 6, p. 58-64, text-fig. 1 (Copenhagen).

Seyed-Emami, Kazem, 1967, Zur Ammoniten-Fauna und Stratigraphie der Badamu-Kalke bei Kerman, Iran (Jura, Oberes Untertoarcium bis Mittleres Bajocium): Univ. München, Diss., 180 p., 9 text-fig., 15 pl. (München).

Shulgina [Schulgina], N. I., 1967, Titonskie ammonity severnoy Sibiri: in Mesozoyskiye morskiye fauny severa i dalnego vostoka SSSR i ikh stratigraficheskoye znacheniye, V. N. Saks (G. Ja. Krymgolts, ed.): Akad. Nauk SSSR, Sibir. Otdel., p. 131-177, text-fig. 1, 18 pl. [Tithonian ammonites of northern Siberia: in Mesozoic marine faunas of the northern and far East USSR and their stratigraphic importance.]

Spath, L. F., 1927-33, Revision of the Jurassic cephalopod faunas of Kachh (Cutch): Geol. Survey India, Mem., Palaeont. Indica, n. ser., v. 9, mem. 2, pt. 1-6, 945 p., 130 pl.——1950, A new Tithonian ammonoid fauna from Kurdistan, northern Iraq: British Museum (Nat. History), Bull., Geology, v. 1, p. 96-137, pl. 6-10.—— 1952, Additional observations on the invertebrates (chiefly ammonites) of the Jurassic and Cretaceous of East Greenland II. Some Infra-Valanginian ammonites from Lindemans Fjord, Wollaston Forland; with a note on the base of the Cretaceous: Meddel. Grønland, v. 133, p. 5-40, text-fig. 1, 4 pl.

Stevens, G. R., 1963, *Faunal realms in Jurassic and Cretaceous belemnites:* Geol. Mag., v. 100, p. 481-497, text-fig. 1-8. [See also discussion by R. Bowen: Geol. Mag., v. 101, p. 374-376 and by G. R. Stevens, v. 102, p. 175-178.]——1965, *The Jurassic and Cretaceous belemnites of New Zealand and a review of the Jurassic and Cretaceous belemnites of the Indo-Pacific region:* New Zealand Geol. Survey, Paleont. Bull., v. 36, 283 p., 43 text-fig., 25 pl.——1967a, *Biogeographic changes in the Upper Jurassic of the South Pacific:* New Zealand Geol. Survey, v. 36, p. 1-31, text-fig. 1,2.——1967b, *Upper Jurassic fossils from Ellsworth Land, West Antarctica, and notes on Upper Jurassic biogeography of the South Pacific region:* New Zealand Jour. Geology, Geophysics, v. 10, no. 2, p. 346-393, text-fig. 1-45.——1973, *Jurassic belemnites:* in Atlas of paleobiogeography, Anthony Hallam (ed.), p. 269-274, text-fig. 1-4, 1 pl., Elsevier Scientific Publ. Co. (Amsterdam-London-New York).

Sturani, S. C., 1967, *Réflexions sur les faciès lumachelliques du Dogger mésogéen "Lumachelle à Posidonia alpina" auctt.):* Soc. Geol. Italiana, Boll., v. 86, p. 445-467, text-fig. 1-6.

Teichert, Curt, Moore, R. C., & Zeller, D. E. N., 1964, *Rhyncholites:* in Treatise on invertebrate paleontology, Part K, R. C. Moore (ed.), p. K467-K484, text-fig. 338-347, Geol. Soc. America, Univ. Kansas Press (New York; Lawrence, Kans.).

Termier, Henri, & Termier, Geneviève, 1952, *Histoire géologique de la biosphère:* 719 p., 35 paleogeogr. maps, 117 text-fig., 8 pl., Masson (Paris).

Tintant, Henri, 1963, *Kosmocératidés du Callovien Inférieur et Moyen d'Europe Occidentale:* Univ. Dijon, Publ., v. 29, p. 5-500, text-fig. 1-92, 58 pl.

Turnšek, Dragica, 1966, *Upper Jurassic hydrozoan fauna from southern Slovenia:* Acad. Sci. et Art. Slovenia, Cl. 4, diss. IX/8, p. 337-428, 19 pl.

Uhlig, Victor, 1903-10, *The fauna of the Spiti shales:* Geol. Survey India, Mem., Paleont. Indica, ser. 15, v. 4, no. 1-3, 511 p., 94 pl.——1910, *Die Faunen der Spiti-Schiefer des Himalaya, ihr geologisches Alter und ihre Weltstellung:* Akad. Wiss. Wien, Denkschr., math.-naturw. Kl., v. 85, p. 1-79 (p. 531-609).——1911, *Die marinen Reiche des Jura und der Unterkreide:* Geol. Gesell. Wien, Mitteil., v. 4, no. 3, p. 329-448, 1 map.

Verma, H. M., & Westermann, G. E. G., 1973, *The Tithonian (Jurassic) ammonite fauna and stratigraphy of Sierra Catorce, San Luis Potosi, Mexico:* Bull. Am. Paleontology, v. 63, p. 107-278, text-fig. 1-32, pl. 22-56.

Vogel, Klaus, 1966, *Eine funktionsmorphologische Studie an der Brachiopodengattung Pygope (Malm bis Unterkreide):* Neues Jahrb. Geologie, Paläontologie, Abhandl., v. 125, p. 423-442, text-fig. 1-8, pl. 38,39.

Wähner, F., 1882-98, *Beiträge zur Kenntnis der tieferen Zonen des unteren Lias in den nordöstlichen Alpen:* Beitr. Paläontologie, Geologie Österreich-Ungarns u. Orients, v. 2-11, 291 p., 66 pl.

Wanner, Johannes, 1910, *Neues über Perm-, Trias- und Juraformation des indoaustralischen Archipels:* Centralbl. Mineralogie, Geologie, Paläontologie, 1910, p. 736-741.

Wendt, Jobst, 1968, *Discohelix (Archaeogastropoda, Euomphalacea) as an index fossil in the Tethyan Jurassic:* Palaeontology, v. 11, p. 554-575, text-fig. 1-9, pl. 107-110.——1971, *Genese und Fauna submariner sedimentärer Spaltenfüllungen im mediterranen Jura:* Palaeontographica, Abt. A, v. 136, p. 121-192, text-fig. 1-20, pl. 15-18, 7 tab.

Westermann, G. E. G., 1964, *The ammonite fauna of the Kialagvik formation at Wide Bay, Alaska Peninsula, I. Lower Bajocian (Aalenian):* Bull. Am. Paleontology, v. 47, p. 329-503, text-fig. 1-37, pl. 44-76.

Wiedmann, Jost, 1973a, *Ancyloceratina (Ammonoidea) at the Jurassic/Cretaceous boundary:* in Atlas of palaeobiogeography, A. Hallam (ed.), p. 309-316, text-fig. 1,2, Elsevier Scientific Publ. Co. (Amsterdam-London-New York).——1973b, *Evolution or revolution of ammonoids at Mesozoic system boundaries:* Biol. Review, v. 48, p. 159-194, text-fig. 1-11.

Wilczyński, Andrzej, 1962, *Stratygrafia górnej jury w Czarnogłowach i Świętoszewie:* Acta Geol. Polonica, v. 12, p. 3-110, text-fig. 1-27, pl. 1-9 (Russ., Fr. summ.). [*The stratigraphy of the Upper Jurassic at Czarnoglowacz and Swietoszewie.*]

Wolburg, Johannes, 1959, *Die Cyprideen des NW-deutschen Wealden:* Senckenbergiana Lethaea, v. 40, p. 223-315, text-fig. 1-27, pl. 1-5.

Zeiss, Arnold, 1968, *Untersuchungen zur Paläontologie der Cephalopoden des Unter-Tithon der südlichen Frankenalb:* Bayer. Akad. Wiss., Abhandl., math.-naturw. Kl., n.ser., no. 132, 190 p., 17 text-fig., 27 pl.——1971, *Vergleiche zwischen den epikontinentalen Ammonitenfaunen Äthiopiens und Süddeutschlands:* in Colloque du jurassique mediterraneen, Hung. Magy. Allmi Földt. Intéz., Evk., v. 54 (1969), no. 2, p. 535-545, text-fig. 1-11.

Zesashvili, V. I., 1964, *Les zones du Jurassique moyen dans la Géorgie et les régions adjacentes du Caucase:* Inst. Grand-Ducal Luxembourg, Comptes Rendus, Coll. Jurassique (1962), p. 851-860.

Ziegler, Bernhard, 1961, *Stratigraphische und zoogeographische Beobachtungen an Aulacostephanus (Ammonoidea-Oberjura):* Paläont. Zeitschr., v. 35, no. 1/2, p. 79-89, text-fig. 1-8.——1962, *Die Ammoniten-Gattung Aulacostephanus im Oberjura (Taxionomie, Stratigraphie, Biologie):* Palaeontographica, v. 119, Abt. A, p. 1-72, text-fig. 1-85, pl. 1-22.——1963, *Die Fauna der Lemeš-Schichten (Dalmatien) und ihre Bedeutung für den mediterranen Oberjura:* Neues Jahrb. Geologie, Paläontologie, Monatsh. 1963, p. 405-521, text-fig. 1-4.——1964, *Boreale Einflüsse im Oberjura Westeuropas?:* Geol. Rundschau, v. 54, p. 250-261, text-fig. 1-8.——1967, *Ammoniten-Ökologie am Beispiel des Oberjura:* Geol. Rundschau, v. 56, p. 439-464, text-fig. 1-20.——1971, *Biogeographie der Tethys:* Jahresh. Gesell. Naturk. Württemberg, 126 Jahrg., p. 229-243, text-fig. 1-10.

Ziegler, M. A., 1962, *Beiträge zur Kenntnis des unteren Malm im zentralen Schweizer Jura:* p. 7-55, text-fig. 1,2, 11 fig. on 2 tables, privately publ., Buchdruckerei Winterthur A. G. (Zürich).

CRETACEOUS[1]

By ERLE G. KAUFFMAN

[U.S. National Museum]

CONTENTS

	PAGE
INTRODUCTION	A418
SURVEY OF CRETACEOUS MARINE ENVIRONMENTS	A424
Temperature	A424
Water Chemistry	A426
Eustatic Changes	A430
Benthonic Environments	A432
Nature of Invertebrate Faunas	A436
TEMPORAL AND BIOSTRATIGRAPHIC DEFINITIONS	A439
Definition of the Cretaceous	A439
Cretaceous Boundary Zones	A440
BIOSTRATIGRAPHY	A443
Lower Cretaceous Biostratigraphy	A449
Biostratigraphically Important Lower Cretaceous Groups	A453
Upper Cretaceous Biostratigraphy	A458
Biostratigraphically Important Upper Cretaceous Groups	A459
BIOGEOGRAPHY	A461
North Temperate Realm	A465
Tethyan Realm	A469
South Temperate Realm	A473
SUMMARY AND SOME OBSERVATIONS ON EVOLUTION	A477
REFERENCES	A481

INTRODUCTION

The Cretaceous marks one of the most varied and active periods in the evolution of marine organisms and their ecological interactions. From Late Triassic and Jurassic rootstocks, the ancestry of most modern biotas is to be found in a series of spectacular Cretaceous radiations in which many of the higher taxonomic groups and adaptive strategies that characterize living marine invertebrate assemblages were developed and refined. Of particular importance in this respect was evolutionary diversification among foraminifers, radiolarians, scleractinian corals, cheilostome bryozoans, neogastropods, infaunal and epifaunal bivalves, and irregular echinoids. At the same time, the Cretaceous invertebrate faunas developed a unique character through the broad radiation of several im-

---

[1] Manuscript received April, 1977, revised manuscript received July, 1977.

portant groups that became extinct, or nearly so, at the end of the period. Among the Foraminiferida, these include the Rotaliporidae, Globotruncanidae, Schackoinidae, and many genera of larger tropical benthonic foraminifers; among the Bivalvia are the Inoceramidae, Trigoniidae, and "reef"-forming rudistids. In addition, several families each of scleractinian corals, cheilostome bryozoans, gastropods, ammonites, and irregular echinoids characterized the Cretaceous (see HARLAND *et al.,* 1967; appropriate volumes of this *Treatise*).

Taxonomic diversification among Cretaceous invertebrates was accompanied by an equal increase in complexity of ecological interactions within the Cretaceous biota, and by major evolutionary advances in the structure and diversity of ecological units at all levels—from symbioses to paleobiogeographic units. Many types of ecological interactions that characterize modern faunas are first expressed without question in the Cretaceous fossil record. Benthonic marine paleocommunities, which are taxonomically and structurally comparable to modern communities in the same environments, developed widely for the first time in the Cretaceous.

Marine paleobiogeographic units similarly show marked increase in numbers and complexity during the Cretaceous Period, more than doubling in number those described from the Triassic and Jurassic (compare papers in HALLAM, 1973). Each biogeographic unit further had its own complex Cretaceous evolutionary history that reflects climatic, oceanographic, and paleogeographic changes linked to plate tectonism (KAUFFMAN, 1973a,b). By the end of the Cretaceous, paleobiogeographic differentiation similar to that of modern seas had been achieved and marine climatic zones (exclusive of cold-water zones) had become well established as climates generally became cooler.

The Cretaceous-Tertiary boundary represents one of the major extinction episodes in the history of life, with the disappearance of such characteristic Mesozoic groups as dinosaurs, most gymnosperms, ammonites, and numerous families of scleractinian corals, bivalves, gastropods, and echinoids. For some Cretaceous groups, especially those of the warm-water plankton (foraminifers and coccoliths) and tropical to subtropical benthos (for example, rudist bivalves), this event was near-catastrophic and characterized by massive extinctions within a short period of time, occurring at or near the apex of their evolutionary diversification.

Against this backdrop, the Cretaceous has become one of the key periods from which concepts in evolution, paleoecology, paleobiogeography, and biostratigraphy have been developed from the fossil record. Cretaceous faunas and associated paleoenvironments have been well studied in many parts of the world. The mass of data available from these studies has in turn encouraged detailed interpretation of the interrelationships between organisms and their environments and has led many scholars to utilize the Cretaceous fossil record as a testing ground for biological, ecological, and stratigraphic theories.

In part, the spectacular evolution of Cretaceous invertebrates can be attributed to achievement of a certain structural, behavioral, or ecological grade as a result of a long pre-Cretaceous history, which allowed rapid and widespread radiation into new ecological niches. For example, Mesozoic development of siphons among bivalves allowed extensive Cretaceous exploitation of benthonic infaunal habitats. Similarly, an apparent relationship exists between extensive diversification of Cretaceous invertebrate groups and commensurate diversification and increasing structural complexity in Cretaceous ecological units. Whereas the number and kind of marine habitats probably did not change appreciably between the Late Triassic and Cretaceous, the diversity of organisms and adaptive strategies changed dramatically during Cretaceous radiations. Niche partitioning within already occupied marine habitats and exploitation of new habitats resulted.

In part, the evolution of Cretaceous organisms was also strongly affected by large scale, geographically widespread, and often rapid changes in global environment, and thus in diverse natural selective forces acting on invertebrate evolution in the marine realm. The Cretaceous was one of the periods of most active plate movements, sea floor spreading, and continental drift. The

Atlantic opened as a new major ocean basin during the Cretaceous with final separation of Africa and South America, attaining approximately 75 percent of its present size by the end of the period. Large new areas of deep ocean floor were formed, and colonization and initial ecological structuring of this vast marine basin was largely a Cretaceous event. As a result of Atlantic spreading, new oceanic current systems (including the proto-Gulf Stream) were formed and played a major role in the redistribution of fauna and the evolution of Cretaceous biogeographic units. Opening of the South Atlantic, large scale breakup of Gondwanaland, and the development of major deep marine connections between the North Atlantic and the "circumboreal" seaway allowed cool temperate-zone waters to enter the Atlantic and South Pacific basins on a large scale for the first time. This resulted in changes of lateral and vertical oceanic temperature gradients leading to a cooling trend during the Late Cretaceous, with expansion of temperate climatic zones, and a great constriction of the tropical Tethyan seaway. This in turn caused a major restructuring of Cretaceous biogeographic units, the evolution of new faunas in the broadening cool- to mid-temperate marine zones, and increased competition and niche partitioning within the Eurasian arm of Tethys.

Late Jurassic and Cretaceous plate tectonics also resulted in the first major opening of the tropical Caribbean Sea with broad Pacific and Atlantic connections, linking tropical faunas of the eastern Pacific and Eurasian Tethys for the first time. The Caribbean became a new tropical sea in which shallow-water invertebrate organisms were introduced, largely through westward drift of their planktonic stages, and radiated rapidly. It is here that the rudists reached their greatest development and levels of endemism (KAUFFMAN & SOHL, 1973; KAUFFMAN, 1974).

Cretaceous plate tectonics had an even more direct effect on the evolutionary history of diverse marine organisms (KAUFFMAN, 1973a). Physical barriers to migration and gene flow were constructed (as in the closing of the Mediterranean Tethys) and broken down (as in opening of the Caribbean) on a large scale. Periods of rapid sea-floor spreading widened the Atlantic in sufficiently short periods of geologic time to create a major isolating mechanism for many shallow-water marine invertebrates as the transatlantic distance exceeded the drift potential of their planktonic larvae. Such genetic isolation permitted independent radiation of shallow-water biotas on both sides of the Atlantic basin during the Cretaceous. Similarly, the breakup of Pangea and drifting apart of India, South America, Australia, New Zealand, and Antarctica effectively isolated shelf biotas of these areas to varying degrees, resulting in accelerated rates of faunal and biogeographic differentiation.

Plate tectonics during the Cretaceous also brought whole ecosystems, developed originally in isolation from one another, into contact. This was accomplished by impinging their effective zones of larval drift, as happened with partial "closing" of the North Pacific, or by direct "collision" of ecosystems (as in the collision of India with Eurasia during the Cenozoic). Such events created major competition between highly structured marine biotas, resulted in widespread extinction, niche partitioning and specialization among surviving organisms, and provided a stimulus for new pulses of radiation during the period when these competing ecosystems were unstructured.

An additional control by Cretaceous plate tectonics on the evolutionary history of marine invertebrates and their ecological structure was that of numerous large scale eustatic fluctuations that appear to have resulted from the emplacement and subsidence of major topographic features on the ocean floors of the world. Eustatic rise of several hundred feet may have been produced by uplift of large areas of ocean floor, or continental subsidence, or both, resulting in: 1) widespread marine transgression onto world cratons (Fig. 1); 2) interconnection of formerly isolated water bodies and their biotas; 3) broad faunal mixing; 4) development of widespread, equable maritime climates, broadening of thermal gradients, and thus greater climatic stability for long periods of time; 5) greater spread of tropical waters onto continents, in some cases rapidly, and resultant temperature increase in formerly

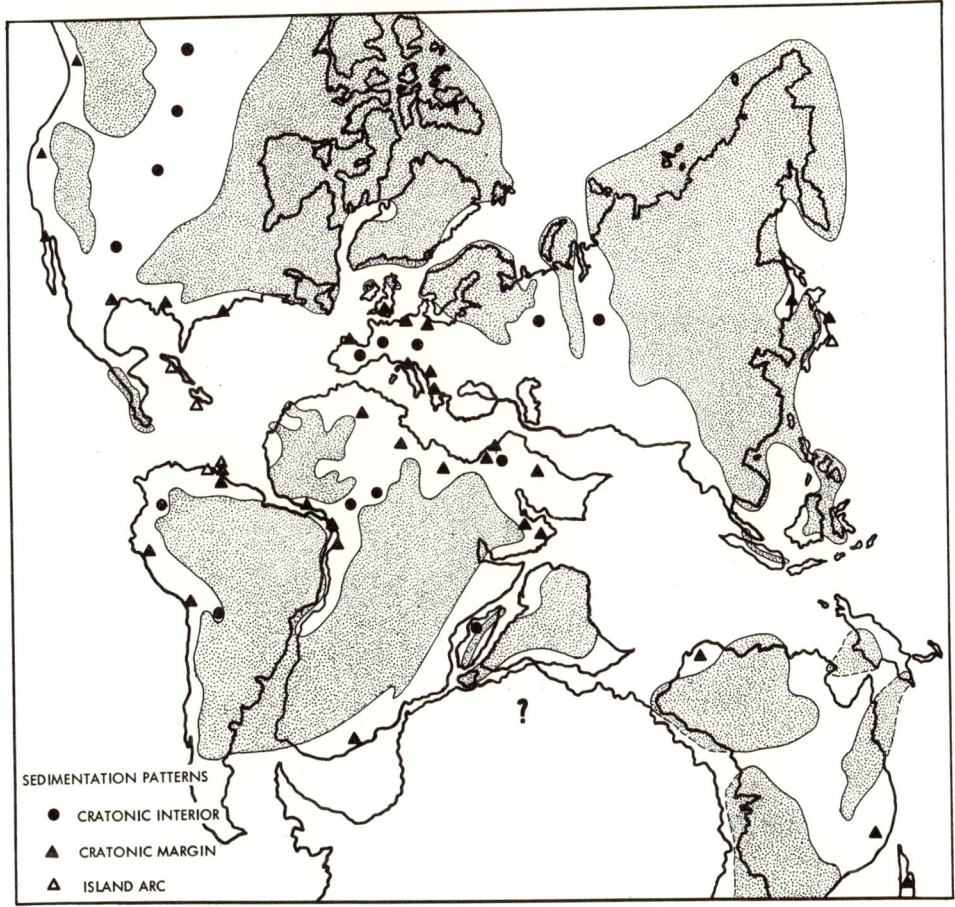

Fig. 1. Generalized world map of Late Cretaceous (Cenomanian-Santonian) maximum transgressions showing extent of epicontinental seas (white; generalized for all transgressions), land areas (stippled), continental arrangements (Dietz & Holden, 1970; Smith, Briden, & Drewery, 1973), and data points from which interpretations of transgressive-regressive history and sedimentation patterns are drawn. Where almost continuous data exist for an area (e.g., Western Interior of North America, Western Europe), data points represent regional centers where facies and faunas are distinct from those of adjacent regions (Kauffman, n).

temperate areas; 6) restriction of deep oceanic circulation and expansion of anaerobic conditions in the oxygen-minimum zone; and 7) considerable increase in the primary ecospace of marine invertebrates, the photic shelf and shallow epicontinental zones, and upper pelagic zone of the open ocean. Tremendous opportunities for radiation were presented by these episodes, yet physically controlled pressures of natural selection were low.

On the other hand, eustatic lowering of sea level greatly constricted many prime habitats, eliminated others, increased competition, increased seasonality, decreased stability in the marine climate, and generally increased the acuteness of natural selective forces acting on the evolution of Cretaceous organisms. This was especially true when such changes were very rapid, as in the Middle Turonian[1] regression. This in

---

[1] Substages are used as formal stratigraphic units (KAUFFMAN, COBBAN, & EICHER, 1977) in this chapter.

turn provided the format for widespread extinction of some groups, and greatly accelerated evolutionary rates for others.

Of particular importance to understanding the evolutionary history of Cretaceous organisms is the fact that environmental "stability" over long periods of time did not generally exist, nor were there many gradual long-term changes. Plate tectonic events and their resultant climatic, oceanographic, and geographic effects were of considerable magnitude and were irregularly fluctuating between "active" and "quiet" intervals. As many as ten distinct global transgressions and regressions are recorded, which represent eustatic fluctuations during the Cretaceous (Fig. 2). It is important to consider the chain of paleoenvironmental events that resulted from each of these plate tectonic pulses—geographic changes in continents, shallow seas, and ocean basins, and thus rearrangement of water masses, their current systems, chemistry, temperature gradients, and the biogeographic units that characterized them; major eustatic changes in sea level and resultant global climatic changes; changes in available habitats and ecospace, and in stress factors of natural selection; and as a result of all of this, constantly altering patterns of gene flow.

Superimposed on these factors is the unusual character of the Cretaceous marine environment in both the open ocean and in epicontinental seas when compared to modern counterparts. Tropical to cool-temperate waters predominated over virtually the entire globe; without cold polar waters the vertical temperature differential over large ocean basins was small. Bottom waters were considerably warmer than at present, and the oceanic thermocline, to the extent that it existed, was probably of small magnitude and broadly graded during much of Cretaceous history. There is evidence that deep to midwater oceanic circulation may have, as a result, been periodically much more sluggish, producing less oxygenation, than at present. Tremendous areas and thicknesses of well-laminated, dark organic-rich shale and pelagic carbonate, and of glauconitic sand characterize Cretaceous marine rocks. Many intervals contain, at most, a sparse benthonic epifauna and virtually no infauna, suggesting periodically broad areas of chemically inhospitable substrate and oxygen depletion both in the sediment and the overlying water column (Fischer & Arthur, 1977). These conditions strongly affected biotic diversity and ecological structure in benthonic and pelagic habitats of the world's oceans. High planktonic diversity and productivity, and maximum complexity in pelagic ecological structure coincided with more uniform oceanic temperatures, eustatic rise, widespread anaerobism in deep marine basins, heavy carbon isotopic values in marine organisms, and formation of organic-rich sediment. Low pelagic diversity was associated with lowering of global sea level, regression, cooling of temperatures, intensification of temperature gradients and current systems, oxygenation of the deep ocean and light-carbon isotopic values (Fischer & Arthur, 1977).

Carbonate compensation depth may have fluctuated widely during the Cretaceous

---

Fig. 2. Summary of Cretaceous radiometric time scales (MYBP), magnetostratigraphy (M.S.; Van Hinte, 1976), transgressive-regressive history (T-R Cycles), and paleotemperatures. Geochronology (MYBP) columns represent (left, single asterisk) K-Ar dating of sanidine and biotite from bentonite and ash beds in the Western Interior of North America (Obradovich & Cobban, 1975; Gill & Cobban, 1966; Kauffman, 1978, and references in each) compared to the global standard of Van Hinte (1976; double asterisk). Transgressive (T, peak right) and regressive (R, peak left) pulses of Cretaceous marine cycles are interpreted from global patterns of sedimentation (Kauffman, 1973a,b; modified) represented by solid line and T_1-T_{10}, R_1-R_{10} designations; compared with generalized eustatic curve (dashed line) of Schlanger & Jenkyns (1976) for Cretaceous. Oxygen isotopic temperature curves (at right) plotted against C scale (at base) for typical Cretaceous organisms. Generalized composite temperature curve shown by heavy line in mid-graph (FA, from Frerichs & Adams, 1973). Other curves, from left to right are: BF-benthonic foraminifers (Saito & Van Donk, 1974), BE-belemnites (Lowenstam & Epstein, 1954), DS-whole rock analysis of pelagic sediments (Lower Maastrichtian and older) and planktonic foraminifers (Middle, Upper Maastrichtian) in central North Pacific (Douglas & Savin, 1973), BR-brachiopods (Lowenstam & Epstein, 1954), and PF-planktonic foraminifers, DSDP site 171, central North Pacific (Douglas & Savin,

Cretaceous

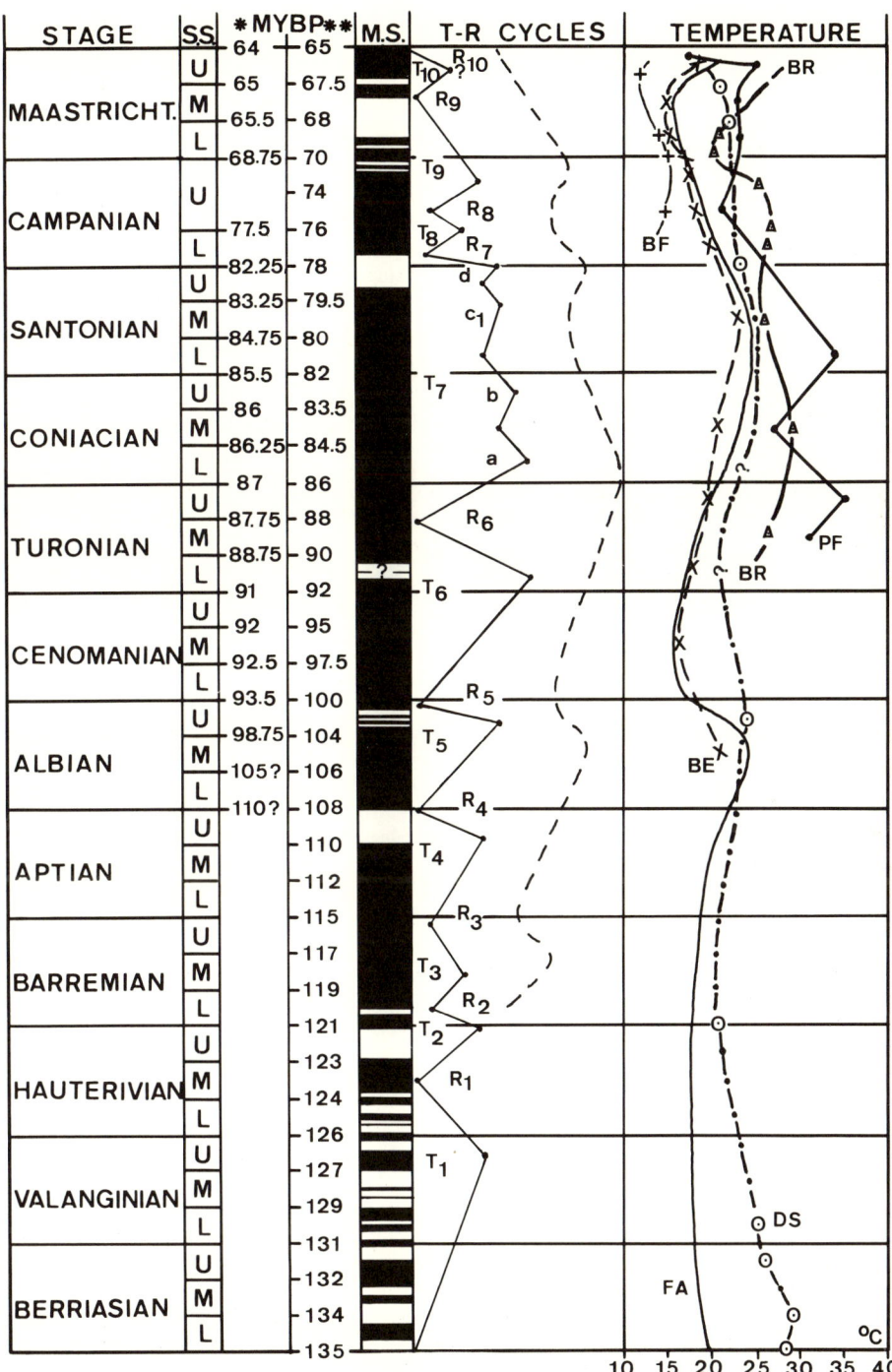

Fig. 2. *(Continued from facing page.)* 1973, fig. 3). [Explanation: S.S. = Substages.] [From Kauffman, 1977f; used with permission of Rocky Mountain Assoc. Geol., Denver.]

with a predictable effect on the planktonic biota. Low-diversity invertebrate faunas from many epicontinental seas further suggest large areas of slightly brackish surface waters (KAUFFMAN, 1975) and possibly entire brackish seaways. As evidenced by carbon and oxygen isotopic analyses and the succession of marine biotas, these conditions fluctuated broadly and irregularly during the Cretaceous, and may have been related to plate tectonic events.

Interpretation of the intricate relationships that must have existed between the evolutionary history of Cretaceous organisms, their ecological structure, and the many varying factors of the global environment during the period is a challenge of the highest magnitude to the paleobiologist. This review will, at best, allow broad insights into these problems. It is the purpose of this general introduction to show that the Cretaceous was a dynamic time of considerable physical, chemical, and biological activity on earth, and it was a pivotal time in the evolution of the modern biota.

SURVEY OF CRETACEOUS MARINE ENVIRONMENTS

The evolutionary and ecological history of Cretaceous invertebrates is intricately related to many changing factors of the Mesozoic marine environment, such as large-scale natural selective forces linked to plate tectonic activity, eustatic changes in sea level, major changes in aquatic chemistry, temperature and circulation patterns, and broader variations in the global climate. These interrelationships were highly complex and reflect forces and evolutionary response of a magnitude several times greater than was envisioned only two decades ago. For example, KAUFFMAN (1972, 1973a,b, 1976, 1977a) and FISCHER and ARTHUR (1977) have proposed direct relationships between plate tectonic activity, resultant eustatic fluctuations, epicontinental transgression or regression, fluctuating salinity, marine temperatures and climatic gradients, marine oxygenation, diversity and ecological complexity of the marine biota, distribution and history of biogeographic units, and evolutionary rates in a variety of marine mollusks.

TEMPERATURE

The Cretaceous marine system was generally characterized by comparatively warm bottom and surface temperatures without cold-temperate to cold polar conditions, and with broad vertical and horizontal temperature gradients. Bottom temperatures, determined mainly from oxygen and carbon isotopic analyses of benthonic invertebrates, ranged from 10°C to 17°C (LOWENSTAM & EPSTEIN, 1959; BOWEN, 1966). Surface temperatures ranged from 15°C (polar) to 35°C (equatorial). Temperature gradients were broad, involving low temperature differential from pole to equator. BOWEN (1966, p. 169) noted a differential of only 15°C (at 75° latitude) to 24°C (at 25° latitude) during the Albian temperature maximum, of 17°C (at 75° latitude) to 24°C (at 45° latitude) at the Santonian temperature maximum, and only a 13°C to 14°C difference between the Santonian polar and equatorial waters. Major temperature minima during the Cretaceous had only slightly more disparate ranges, for example, 10°C (at 75° latitude) to 25°C (at 25° latitude) for the Cenomanian, and an 18°C difference between the Cenomanian poles and equator. Cretaceous isotherms were 10° to 20° of latitude closer to the poles than in modern seas. Water temperatures gradually cooled during the Cretaceous, leading to breakup of equable maritime climates with terminal Cretaceous regression.

BOWEN (1966) reviewed and summarized massive paleotemperature data, obtained through oxygen and carbon isotopic analyses of a variety of shelled mollusks, foraminifers, brachiopods, worms, and whole biogenic carbonate rock samples. Widely disparate analyses for the same time interval or stratigraphic unit were noted among diverse organisms. This reflects 1) variation in shell chemistry of diverse organisms, 2) cooler benthonic and warmer pelagic water temperatures in the same area, and 3) latitudinal paleotemperature gradients. Average temperature curves drawn indiscriminately for all organisms gives a general picture of Cretaceous fluctuations (FRERICHS & ADAMS, 1973, text-fig. 4) with important temperature maxima in the Middle Albian and Santonian, and significant minima in the Neocomian, Late Cenomanian, and Maastrichtian. More precise analyses based on temporally closely spaced points, separation of benthonic and pelagic data, and consideration of latitudinal gradients (see especially LOWENSTAM & EPSTEIN, 1959; BOWEN, 1966; DOUGLAS & SAVIN, 1973; SAITO & VAN DONK, 1974; and references therein), however, suggest a more complex Cretaceous temperature history (Fig. 2), with a Berriasian peak, a decline from Valanginian through Barremian time, rising Aptian temperatures, culminating in a Middle Albian high, and dropping abruptly again to a low from Middle Cenomanian to earliest Turonian time. This last fluctuation is puzzling because the Late Cenomanian to Early Turonian temperature minimum correlates with the peak of one of the greatest Cretaceous transgressions, when isotopic and faunal evidence indicate poleward migration of tropical Tethyan waters and biotas. Marine climates warmed again during the Turonian, peaked in the Coniacian to Santonian, then declined gradually through the Campanian with a small rise near the end of the stage. Early and Late Maastrichtian temperatures were low, with a slight Middle Maastrichtian rise. Evidence suggests that shallow epicontinental seaways may have warmed more quickly, and to somewhat higher levels, than did open oceanic environments.

During principal transgressive maxima, tropical to warm-temperate marine climatic zones covered much of the globe and even cool-temperate zones disappeared in the polar regions. At peak transgressions (e.g., Late Albian, earliest Turonian, Coniacian to Santonian, early Late Campanian, Middle Maastrichtian) subtropical waters spread for short periods of time over many cratonic areas previously occupied by shallow warm-temperate seas, and the warm-temperate zones shifted poleward. These incursions produced abrupt temperature rises and caused massive, rapid extinction of stenotopic warm-temperate marine taxa, and caused their replacement by mixed Tethyan and warm-temperate eurytopic faunas in poorly structured paleocommunities. Warm-temperate biotas replaced tropical to subtropical elements in marginal Tethyan areas during initial phases of regression and restriction of Tethyan marine climates.

Throughout the Cretaceous, there is no evidence for a well-defined oceanic thermocline or for cold bottom waters. Broad vertical thermal gradients persisted, becoming more accentuated during regression and general cooling of the oceanic water masses, and more gradational during transgression, warming, and amelioration of the marine climate.

FISCHER and ARTHUR (1977) have linked broad Cretaceous temperature fluctuations to changes in deep ocean circulation, water and sediment chemistry, and diversity in both benthonic and pelagic biotas. Eustatic rise and transgression were associated with warming and amelioration of the surficial marine climate, warming of polar waters, and consequently a lower horizontal thermal gradient. As a result, polar waters did not sink as rapidly or move as readily across the deep ocean floor. Stagnation and expansion of oxygen-deficient zones in bottom waters and sediments resulted, producing finely laminated, organic-rich sediments supporting a greatly depleted benthonic biota. At the same time photic pelagic environments were optimal for high productivity and development of complex pelagic communities. Regressive pulses, representing eustatic lowering, were characterized by lower marine surface temperatures, accentuation of horizontal temperature zonation, return of cool- and possible cold-temperate climatic zones to polar areas, and thus more rapid sinking and deep ocean circulation of polar waters. This produced cooling of bottom waters, accentuation of the vertical thermal gradient, active bottom currents, widespread benthonic erosion, oxygenation of the deep ocean floor, and diversification of the benthonic biota. Pelagic biotas were coincidently depleted, and had lower productivity and simplified communities rich in opportunistic taxa during cooling pulses.

FISCHER and ARTHUR (1977) noted two major fluctuations of oceanic temperature, biotas, and chemistry during the Cretaceous.

Early Cretaceous temperatures were relatively low, reaching a minimum in the Barremian; pelagic biotas were correspondingly depleted (oligotaxic) and characterized by opportunistic species during Berriasian to Hauterivian time (126 MYBP low point). Increases in temperature and pelagic diversity (polytaxic) reached a first peak during the Albian transgressive maximum. Lower temperatures and oligotaxic pelagic biotas were again characteristic at 94 MYBP, the Cenomanian-Turonian boundary according to the time scale of Van Hinte (1976), but the Albian-Cenomanian boundary zone according to Obradovich and Cobban (1975) and Kauffman (1978). The Albian-Cenomanian boundary coincided with a regression and best fits the Fischer-Arthur interpretation. A second polytaxic interval with warming marine climates was the Late Santonian and Early Campanian, and final cooling and onset of oligotaxic conditions characterized the latest Cretaceous regressive pulse (Maastrichtian) and the Cretaceous-Paleocene boundary zone (Fischer & Arthur, 1977). The Fischer-Arthur hypothesis generally, but not precisely, fits independently derived Cretaceous transgressive-regressive and temperature histories.

The net effect of Cretaceous temperature distributions on the evolution of the biota was profound. Broad temperature gradients promoted widespread biogeographic mixing of faunas at indistinct marine climatic zone boundaries. The faunal overlap zone between warm-temperate and mid- to cool-temperate biotas of the Western Interior of North America during the Cretaceous was as much as 1,600 km (Sohl, 1967; Kauffman, 1973a, 1975). Similarly, vertical zonation of the biota across the shelf-depth and slope-depth zones of the Cretaceous seas was very broad and nowhere near that found today off the Atlantic, Gulf (Parker, 1960, references therein), and Californian-Mexican coasts (Parker, 1964). These more stable regional Cretaceous climates, especially during warming trends and epicontinental transgressions, may have engendered evolution of less temperature-tolerant organisms, which in turn were severely stressed, producing high levels of extinction in some types and rapid evolution in others, by unpredictable perturbations in the global climate.

WATER CHEMISTRY

Cretaceous biotas and geochemistry suggest that water chemistry in open marine systems was similar to that of modern seas, but there were periodic variations in salinity, oxygen, carbon dioxide and carbonate content, dissolved nutrients, and carbon isotopes, especially in epicontinental seaways. These fluctuations of marine water chemistry largely reflect plate tectonic activity, related changes in sea level, changes in marine current systems, oceanic circulation and temperature distribution, variations in introduced organic and inorganic compounds from continental sources, the effectiveness of fluvial systems in transporting these materials to marine areas, and consequent changes in productivity.

The most obvious changes in water chemistry involved salinity and oxygen. Hallam (1975) noted a decrease in invertebrate diversity northward from the European Tethys into the "Boreal" (North Temperate) Realm in the Jurassic and accounted for this, in part, by lower salinities in shallow epicontinental seas to the north, which received considerable fresh water from internal drainage of river systems. Similar paleogeographic situations existed throughout the Cretaceous, associated with Tethyan to temperate-zone decline in invertebrate diversity. But poleward decrease in diversity by itself is not good evidence for salinity decrease. Scholle and Kauffman (1977) have shown from oxygen and carbon isotopic analyses of associated carbonates (whole-rock analyses) and *Inoceramus* shells that subnormal salinities periodically characterized the Western Interior basin of North America throughout its middle Cretaceous history except for periods near peak transgression (latest Cenomanian to earliest Turonian, Coniacian and parts of the Santonian). Equivalent analyses of the English chalk (open shelf) sequence for the same interval, showed consistently near normal salinities in open marine facies. European epicontinental basins show faunal depletion northward, suggesting periodically less than normal salinity.

Paleocommunity analyses of the Western Interior seaway of North America by SOHL (1967), SCOTT (1970, 1974), KAUFFMAN (1967, 1969) and KAUFFMAN, HATTIN, and POWELL (1977), record a predominance of low diversity assemblages compared to contemporaneous open marine communities and their modern counterparts. The scarcity or exclusion of sponges (other than *Cliona*), bryozoans, corals, articulate brachiopods, echinoids, many normal-marine gastropods and bivalves, and diverse pelagic microbiotas suggests less than normal salinity in part of the water column. Yet the abundance of ammonites, normal-marine mollusks, and fishes during Cretaceous time suggests mixing with normal-marine water layers. It would appear that this situation was typical of most widespread Cretaceous epicontinental seas.

To account for mixing of normal-marine organisms with depauperate and seemingly more brackish water benthonic communities in Cretaceous temperate epicontinental seas, KAUFFMAN (1975, fig. 3) proposed a model of stratified epicontinental seas, especially those with constricted or silled apertures. A surficial layer of slightly to (near peak regression) moderately brackish water spread across much of the seaway, generated by internal river drainage. Near normal surface salinity was associated only with periods near peak transgression. A denser, normal-marine layer occupied deeper portions of the basins. Lack of a well-defined thermocline and broadly graded vertical temperature zonation, coupled with possibly sluggish currents and low levels of wave activity, prevented extensive mixing of brackish surface water and deeper normal-marine water in this model. Seasonal or longer term overturn would not occur. In epicontinental seas with restricted apertures a Baltic Sea model might be further invoked, in which tidal exchange would mainly involve outflow of surficial waters (brackish) and inflow of deeper, more saline waters in a density stratified system. The net effect of this situation might be restriction in colonization of epicontinental seas by stenohaline pelagic organisms and pelagic larvae of many normal-marine benthonic groups; both are mainly dispersed in the upper photic zone of the water column, envisioned here as being partially brackish during much of the Cretaceous. Thus it is the more euryhaline marine and brackish water organisms, more normal marine forms with larvae distributed on deeper water currents, and organisms without stenohaline pelagic larval stages or with short-lived ones (including ammonites, marine fishes, reptiles, many mollusks) that would be able to colonize temperate epicontinental seas. This model would account for depauperate pelagic microbiotas and shallow-water benthonic paleocommunities that characterize many Cretaceous epeiric seas, and yet permit abundant, more normal marine organisms to be mixed with them. The recent discovery by SCHOLLE and KAUFFMAN (1977) that benthonic inoceramid shells and pelagic carbonates yield similar isotopic values suggests that much of the epicontinental water column may have been subsaline, except during peak transgression, and introduces the possibility that some Cretaceous ammonites were euryhaline.

A second major chemical variable that strongly affected the biota of Cretaceous seas was dissolved oxygen. Offshore decrease in benthonic diversity (KAUFFMAN, 1967); the depauperate nature of many deeper water paleocommunities; large areas lacking in benthonic microfauna; widespread development of evenly and finely laminated, nonbioturbated, organic-rich clay and carbonate mud; and thick glauconite deposits during Cretaceous time have all been cited as evidence for widespread development of oxygen-depleted zones in world oceans and epicontinental seas. SCHLANGER and JENKYNS (1976) and FISCHER and ARTHUR (1977) proposed that the deep ocean basins were largely oxygen depleted during major periods of Cretaceous warming, global eustatic rise, and epicontinental transgression (Late Aptian, Albian, parts of the Cenomanian-Turonian, and Coniacian-Santonian). Deep ocean circulation slowed considerably during warming of surface waters in polar regions, and lessening of horizontal and vertical temperature differentials. Organic-rich, finely and evenly laminated, generally nonbioturbated deep ocean sediments developed widely. Benthonic microfaunas were almost wholly absent and macrofaunal associations were greatly depleted. Little evidence is known for marine erosion or active bottom currents. Coincidentally, high pelagic productivity and diversity greatly increased the organic rain to the deep oceans, enhancing development of anaerobic benthonic conditions and organic-rich sediments. Oxygenation of the deep ocean occurred during Neocomian, Late Cenomanian to Early Turonian (according to FISCHER & ARTHUR, 1977, but *not* SCHLANGER & JENKYNS, 1976), and Campanian-Maastrichtian times, with global climatic cooling, eustatic lowering of sea level, epicontinental regression, cooling of polar areas, accentuation of both horizontal and vertical temperature zonation, and reestablishment of deep ocean currents. Similarly, FRUSH and EICHER (1975) proposed widespread development of oxygen-minimum zones related to oceanic oxygen depletion during the Late Cenomanian to Turonian in the Western Interior and Gulf Coast seaways of North America. Their interpretation is based on essential lack of benthonic microfauna and reduction of benthonic macrobiota with near exclusion of infaunal elements. These proposed times of oxygen depletion seem to correspond to periods of late (but not peak) transgression and early regression (late Middle and Late Cenomanian, Middle Turonian, and by analogy of similar sediments, latest Turonian and Santonian times).

Two models are possible for this type of event in epicontinental seas. The first, as suggested by FRUSH and EICHER (1975) for the North American Cretaceous, and by SEILACHER and WESTPHAL (1971, figs. 1,4) and BARTHEL (1970) for the European Jurassic, involves periodic development of a widespread low-oxygen to anaerobic zone through much of the lower water column in basinal parts of epicontinental seas, excluding nearly all infaunal and epifaunal benthonic organisms, and pelagic-nektonic organisms of the lower part of the water column, and leaving a biota that is predominantly upper pelagic to nektonic in aspect. Associated benthonic organisms are interpreted as having an epiplanktonic habitat on floating vegetation, logs, shells, or on living pelagic organisms. Exceptional preservation of articulated skeletons and soft parts in these rocks are cited as supporting evidence for widespread oxygen-minimum zones, and burial of dead organisms below the aerobic-anaerobic boundary with low levels of bacterial decay and scavenging.

KAUFFMAN (1977c) proposed an alternative model for these types of epicontinental deposits based on detailed study of normally benthonic invertebrates associated with them. He found that epiplanktonic associations of normally benthonic organisms with floating wood or other objects are rare in ancient and modern situations (KAUFFMAN, 1975), and cannot account for the numbers and diversity found with these deposits. *In situ,* low-diversity benthonic communities actually occur at many levels, in moderate abundance, with these "anaerobic" deposits. In the Cretaceous, these benthonic communities are commonly built upon thin-shelled, colonizing Inoceramidae that may reach three meters in diameter, or on shells of dead ammonites, or on molluscan mass-mortality surfaces. Initial colonizing benthonic bivalves of many oxygen-poor Mesozoic substrates show adaptations for expanded oxygen absorption, and apparent tolerance for low oxygen levels, but subsequent members of these epibiont-endobiont communities are typically more normal marine taxa representing groups less tolerant of low oxygen levels. Cemented and byssate epifaunal bivalves, cranioid brachiopods, boring and gooseneck barnacles, algal-grazing gastropods (and thus algae), bryozoans, boring and cemented tube-building worms, encrusting foraminifers, boring and encrusting sponges, and rarely other taxa comprise the communities that colonized live and dead inoceramid shells and the up-facing flanks of dead ammonites and other mollusks in these anaerobic or oxygen-depleted environments. The common occurrence of these inoceramid "island" communities on Jurassic and Cretaceous sea floors characterized by nonbioturbated, finely laminated, organic-rich sediment lacking a benthonic microfauna or infauna does not support the anaerobic-basin model with a widely spread oxygen-minimum zone. Instead, it suggests that chemically inhospitable conditions and oxygen depletion occurred mainly near the sediment-water interface (at or a few centimeters above it) and higher marine waters were sufficiently oxygenated to support a more normal marine biota. Most black shale and carbonate rocks of Phanerozoic "anaerobic" marine epicontinental basins may represent this situation instead of extensive oxygen-minimum zones. Widespread occurrence of inoceramid shells in Deep Sea Drilling Project (DSDP) cores (THEIDE & DINKELMAN, 1977) further suggests that careful restudy of Cretaceous "oceanic anaerobic events" is warranted.

Several authors have suggested variations in carbonate content in Cretaceous oceans, broad fluctuations in CO_2, the lysocline, and carbonate compensation depth (CCD). These are thought to be related to major biotic changes. BERNER (1974) defined chemical properties, interactions, and distribution patterns for carbonates in modern open marine systems, and presented detailed discussion of the lysocline and CCD, and their relation to carbonate saturation and dissolution. These principles can be applied to Cretaceous systems with consideration for temperature and circulation differences, variations in nutrient and inorganic-compound supply, and marine productivity. Widespread fluctuations in the Cretaceous marine environment, including major eustatic changes, should have strongly affected the CCD level in the world's oceans. Considerable disagreement exists as to the magnitude of these fluctuations (see papers in HAY, 1974), their timing, and causes for the establishment and migration of the CCD and lysocline.

WORSLEY (1971, 1974) suggested a marked decrease in $CaCO_3$ input from continental sources with erosional lowering of Late Cretaceous continents and decrease in runoff prior to early phases of the Laramide orogeny, coupled with temperature decline and CO_2 increase in cool shelf and basinal waters. Without terrestrial replenishment, Late Cretaceous periods of high productivity among shell-forming calcareous plankton (foraminifers and coccoliths) further depleted the supply of available carbonate in sea water; WORSLEY (1971) believed that much of this carbonate was "lost" in the deep ocean as dead plankton skeletons dissolved below the lysocline and CCD. These factors combined to cause the CCD to rise near to, or reach, the oceanic surface during latest Cretaceous time, further causing widespread dissolution of biogenic carbonate, pelagic extinctions, and collapse of the global food chain at the time of terminal Cretaceous extinction (see also TAPPAN, 1968).

RAMSAY (1974, fig. 2) predicted that periods of oceanic surface warming were accompanied by high pelagic productivity, increased supply of carbonate skeletons to the deep ocean, decrease in available carbonate in surface waters, increase in subphotic CO_2 content brought about by increased levels of bacterial decay of the pelagic rain, greater levels of carbonate dissolution at shallower depths, and thus rise of CCD. Conversely, cool periods produced lower amounts of pelagic carbonate, less free CO_2 in subphotic zones, slower dissolution, and depressed CCD. RAMSAY (1974, figs. 3, 4) calculated that the CCD fluctuated between about 5 km during the Late Cenomanian(?) to Turonian temperature minimum, and again during the latest Maastrichtian decline, to about 4.5 km during the Late Santonian to Campanian temperature decline, and to about 4 km during the Late Albian and the Late Coniacian to Santonian temperature maxima.

TAPPAN (1968) developed a stimulating theory on the relationship between ocean chemistry, phytoplankton production, and large-scale extinctions in the marine realm, which is applicable to the Cretaceous. Large phytoplankton blooms seem to have occurred, especially during Aptian-Albian time, near the Cenomanian-Turonian boundary (debated by some), during the Coniacian and Santonian, and during the Middle to early Late Maastrichtian. Phytoplankton blooms are closely related to the O_2-CO_2 balance in the atmosphere (TAPPAN, 1968). TAPPAN theorized that oceanic phytoplankton maxima are coincident with: 1) times of continental rejuvenation and increased input of continental nutrients and (or) initial phases of epicontinental transgression, 2) marked vertical and horizontal climatic gradients in the world's oceans, 3) active bottom currents, and 4) widespread upwelling and nutrient replenishment. High phytoplankton productivity led to high levels of diversification, decrease in atmospheric and dissolved CO_2 and coincident increase in O_2. Heavy sulfates resulted from low levels of bacterial oxidation of sulfides and decline in sulfate content. Heavy carbon (C_{13}) isotopes in sea water resulted from selective depletion of C_{12} isotopes by the phytoplankton in photosynthetic carbon fixation. Extensive biogenic carbonate was deposited, producing marine limestone and chalk. Presumably, broad preservation of such deposits in the deep ocean reflects a depressed CCD.

According to TAPPAN (1968), phytoplankton minima result from low productivity associated with low continental configuration and low input of land-derived nutrients during maximum epicontinental transgression and early regression, climatic warming and amelioration, broadening of vertical and horizontal temperature gradients resulting in sluggish ocean circulation and lowering of deep ocean oxygen levels, and decrease in upwelling. Nutrient-depleted waters thus characterize the photic zone of world oceans; nutrient sinking into oxygen-depleted layers results in temporary "loss" of oceanic nutrients, which exceeds replenishment levels from continental sources and upwellings. Oxygen production in the atmosphere and oceans decreases and relative CO_2 increases. Lighter sulfates in solution result from bacterial oxidation and breakdown of sulfides during times of low productivity. Light carbon isotopes predominate as the relative amount of C_{12} increases because of low levels for carbon fixation by phytoplankton. These conditions lead to widespread extinction among animal taxa (TAPPAN, 1968), for example, at the end of the Cretaceous.

Thus, during the Cretaceous, many factors leading to broad and intermittently severe changes in marine chemistry—salinity, O_2 and CO_2 content, carbonate content and dissolution levels, nutrient content, and the isotopic nature of various elements and compounds—were actively destabilizing the marine system as we currently view it. These episodes produced major biologic response in regard to population density, radiation, diversification and extinction.

EUSTATIC CHANGES

Most contemporary workers accept the hypothesis that major sea-level changes occurred during the Cretaceous, but there is variety of opinion regarding the number and magnitude of these fluctuations, and the driving force behind them. Five possible causes for Mesozoic eustatic sea-level changes have been recently proposed: 1) active periods of plate movement, especially rapid seafloor spreading, lead to construction of topographically elevated areas on the ocean floor, which in turn displace sea level upward and cause epicontinental transgression (HALLAM, 1971; HAYS & PITTMAN, 1973; and KAUFFMAN, 1973a,b); 2) whole plates, or large portions of plates are uplifted, especially over crustal hot spots and mantle convection cells, and subsequently lowered to produce eustatic rise and fall of sea level, and transgressive-regressive pulses; 3) whole continents subside and rise, in harmony, to produce apparent eustatic fluctuations and coincident global transgressions and regressions (BOND, 1976); 4) volumetric change in ocean basins producing eustatic displacement results from the breakup and assembly of supercontinents (VALENTINE & MOORES, 1972); and 5) transgressive-regressive history is related to eustatic changes produced by epeirogenic movements on the ocean floor with coincident epicontinental epeirogenic events (HALLAM, 1963). Current data suggest that the first and second causes account for most Cretaceous changes in sea level.

One (HAYS & PITTMAN, 1973; BOND, 1976; SCHLANGER & JENKYNS, 1976; Fig. 2, herein) to three large-scale Cretaceous transgressions are recognized by most workers; the major one peaked in the Late Cenomanian or Early Turonian, a second one in the Late Albian, and a third one in the Coniacian to Santonian (for example, FISCHER & ARTHUR, 1977). These were among the most extensive transgressions of the Cretaceous and are obvious from paleogeographic plots. Each coincided with probable times of active, relatively rapid seafloor spreading and construction of topographically elevated areas on the ocean floor. Global Late Albian and Middle Turonian regressions separated these transgressive maxima, which were only 9 and 3 to 4 million years apart, respectively (measured radiometrically; Fig. 2).

Careful examination of global strandline

fluctuations and vertical stratigraphic-paleoenvironmental relationships of marine cyclothems suggests that at least four major Cretaceous transgressions reached well onto at least 60 to 75 percent of widely scattered cratonic areas (Fig. 1), indicating eustatic rises of sea level (KAUFFMAN, 1969, 1970, 1972, 1973a,b; HANCOCK, 1975; and others). These peaks occur in the early to middle Late Albian, latest Cenomanian to Early Turonian, Coniacian to Santonian (possibly two or more separate events), and early Late Campanian. In addition, KAUFFMAN (1973a,b) has reported six lesser Cretaceous transgressive peaks that mainly involve flooding of marginal continental and more restricted epicontinental areas in many parts of the world (40 to 60 percent of world Cretaceous sections), as follows: Late Valanginian, latest Hauterivian or earliest Barremian, Middle Barremian, latest Aptian, middle Late Campanian, and Middle to early Late Maastrichtian. The irregular temporal spacing of these eustatic changes (Fig. 2) suggests that they are not cyclic, and the rapidity with which some take place (Middle Turonian regression encompasses less than two million years) strongly suggests mechanisms for regression other than simple oceanic ridge collapse through heat loss (a slow process).

Cretaceous eustatic fluctuations have strongly influenced the evolutionary history of diverse organisms and biogeographic units. Global transgression was associated with the spread of warm, ameliorating maritime climates, elimination or great restriction of cool-water zones, broadening of vertical and horizontal marine temperature gradients, and great expansion of total ecospace and individual niches, especially in shallow water habitats. Oceanic productivity and resources increased, resulting in great expansion in numbers and diversity of the planktonic microbiota and the increase in complexity of food-web relationships and pelagic communities. FISCHER and ARTHUR (1977) proposed that these were also times of restricted deep marine circulation, spread of anaerobic conditions on the deep sea floor, and depletion of deep marine benthos; however, evolutionary opportunities for photic pelagic and shallow-water benthonic marine invertebrates were expanded during transgression, allowing broad diversification and niche partitioning at low taxonomic levels (mainly genera and species), and increase in the complexity of ecological interactions. Environmental perturbations probably decreased in numbers and intensity and therefore selective stress levels were decreased (biological forces of predation and competition excepted). Without high levels of stress, evolutionary rates and the magnitude of extinctions in marine invertebrates were relatively low (KAUFFMAN, 1970, 1972, 1977a).

Periods of maximum Cretaceous transgression were characterized by relatively rapid increases in water temperature and salinity in shallow temperate-zone epicontinental and shelf areas, as determined from oxygen and carbon isotopic analyses (SCHOLLE & KAUFFMAN, 1977) and from faunal changes (KAUFFMAN, 1973a, 1975; KAUFFMAN & SCHOLLE, 1977). A rise of a few degrees in water temperature seems to have constituted the first major environmental perturbation of transgressive cycles, and resulted in widespread extinction among stenotopic warm-temperate marine organisms and their replacement by more eurytopic warm-temperate and marginal Tethyan elements. The great extinctions associated with the Cenomanian-Turonian and Turonian-Coniacian boundaries reflect this phenomenon.

Global regression reflects eustatic lowering of sea level. In shallow Cretaceous shelf and epicontinental seas this resulted in decrease in prime ecospace, restriction or elimination of many photic-zone habitats, possibly increased shallow-water turbidity and lower salinity, cooling of the oceans, increase in vertical and horizontal temperature gradients and contraction of warm climatic zones, oxygenation and reestablishment of currents in the deep sea, and decrease in maritime climatic effects producing greater seasonality (KAUFFMAN, 1973a,b, 1975; FISCHER & ARTHUR, 1977). Natural selective forces increased in magnitude, environmental perturbations were probably more common, more severe, and affected a relatively greater portion of the shallow marine biota than during transgression. Competition for basic resources increased among the marine fauna (KAUFFMAN, 1972,

1975). The effect of these deteriorating marine conditions was increasingly to stress invertebrate populations, ultimately leading to widespread extinction among many groups (especially stenotopic forms), and to increase evolutionary rates commensurate with the rate and severity of environmental decline among others (especially eurytopic forms; KAUFFMAN, 1970, 1972, 1975, 1977a). Rapid major regressions, such as that of the Middle Turonian, seem to have produced the greatest evolutionary effects in the shortest period of time.

BENTHONIC ENVIRONMENTS

The study of Cretaceous pelagic environments is essentially one of changing chemistry, nutrient supply, temperature, salinity, and pelagic productivity. Cretaceous benthonic invertebrates were subject to the same controls, though more subtle in aspect, in their evolution. In addition, they were directly subjected to the physico-chemical nature of Cretaceous marine substrates, to a fluctuating CCD, and in the shelf zone to major spatial and temporal fluctuations of their preferred niches associated with major eustatic fluctuations.

In the shallowest epicontinental zones (<30 m), some of which were more than 150 km wide (for example, Western Interior seaway of North America; KAUFFMAN, 1969), Cretaceous benthonic organisms were greatly affected by fluctuations in salinity. Brackish water probably extended as a surface layer over normally marine areas during many times of epicontinental "freshening" (KAUFFMAN, 1975). Most deeper benthonic environments may have been characterized by normal salinity; however, recent isotopic studies suggest periods when subnormal salinity reached deeper zones of epicontinental seas (SCHOLLE & KAUFFMAN, 1977; KAUFFMAN & SCHOLLE, 1977). Brackish water layers in epicontinental systems may have selected heavily against immigration and dispersal of normal marine, stenohaline plankton (food resources) and planktonic larvae of varied benthonic organisms, accounting for the relatively low diversity of many epicontinental benthonic communities that contain normal-marine taxa.

The principal environmental factor affecting Cretaceous benthonic invertebrates was the oxygen level. Broad fluctuations of oxygen content in both oceanic depths (SCHLANGER & JENKYNS, 1976; FISCHER & ARTHUR, 1977) and deep portions of epicontinental seas (FRUSH & EICHER, 1975) have been noted, with widespread oxygen depletion periodically resulting from restriction of deep water circulation associated with climatic amelioration, broadening of vertical and horizontal climatic gradients, and global warming. Coincident eustatic rise resulting in widespread transgression enhanced the effect of oxygen depletion in benthonic environments. Collectively these factors had a profound impact, causing restriction of the Cretaceous benthonic biota. This is evidenced by general offshore decrease in diversity (KAUFFMAN, 1967) (as opposed to offshore increase among macroinvertebrates today), and the simple structure of many basinal and deep ocean paleocommunities. These show heavy representation by low-oxygen tolerant, detritus-feeding, carnivorous, and nonselective filter-feeding groups, plus some opportunistic taxa. There were also broad marine areas devoid of any obvious benthos, especially infauna, during the Cretaceous. Anaerobic conditions and H_2S poisoning seem to have been the major limiting factors. Periods of oxygen renewal in deep water benthonic environments allowed reestablishment of benthonic communities many times during the Cretaceous, but long-term stability never existed, and diversification of the deep benthonic biota was limited.

Whereas wide fluctuation of pelagic microplankton productivity has been noted in the Cretaceous (TAPPAN, 1968; TAPPAN & LOEBLICH, 1972; FISCHER & ARTHUR, 1977), leading one to expect significant fluctuations in the primary food source of suspension-feeding benthonic invertebrates, there is no evidence from the Cretaceous fossil record that benthonic diversity or the ecological success of any group living in normal marine situations was seriously affected by these fluctuations, with the possible exception of the Albian-Cenomanian, Cenomanian-Turonian, and Maastrichtian-Danian boundary zones. All of these boundaries are characterized by widespread, abrupt, and nearly coincident extinctions of numerous planktonic and benthonic organisms. TAPPAN (1968), TAPPAN and LOEBLICH (1972), WORSLEY (1971, 1974), KAUFFMAN (1977b), and others have proposed a direct relationship between short-term, widespread extinctions of the pelagic microbiota (including the planktotrophic larvae of many benthonic organisms), collapse of the global marine food chain, and resultant widespread extinction or depletion of the midwater and benthonic biota. But, in general, Cretaceous benthonic invertebrate biotas were buffered from lesser planktonic fluctuations by: 1) seemingly lower density and thus resource demand in many offshore communities, occupying the bulk of the marine ecospace, as compared with those of today; and 2) the comparatively strong representation of nonselective suspension feeders and detritus feeders in many Cretaceous deeper water benthonic communities. Both groups today are able to utilize a broad variety of food, including organic detritus, and thus have a relatively more stable resource supply through time. Shallow shelf and epeiric benthonic paleocommunities of the Cretaceous are closely similar in complexity and trophic specialization to those of today and seem to have been little affected by the broad fluctuations in plankton productivity, suggesting that nutrients were continually supplied to these areas from continental sources.

Two additional factors of the benthonic environment were probably important in determining the nature of Cretaceous substrates and invertebrate biotas, especially in more offshore, deeper water facies. These are: 1) substrate chemistry and water saturation levels (reflecting "firmness" or matrix density); and 2) dissolution in carbonate-undersaturated zones, below the lysocline and CCD. Whereas Cretaceous marine sedimentary rocks reflect virtually all major environments extant in today's oceans and shelf seas, this was also a period of unique benthonic environments represented by two widespread facies, both of which suggest chemically restrictive conditions. Pelagic carbonates (especially chalk, micritic limestone, and clay-enriched shaly chalk) and glauconitic sand are exceptionally widespread in Cretaceous open marine and (to a lesser extent) epeiric systems. Together with dark shale and mudstone, they comprise the major portion of the marine benthonic facies during the Cretaceous.

The predominance of finely and evenly laminated, commonly nonbioturbated, organic-rich shale, carbonate enriched shale, and evenly bedded pelagic carbonate in offshore epicontinental, shelf, and oceanic deposits during numerous intervals of Cretaceous time suggest highly inhospitable environments for even the most tolerant benthonic invertebrates. Thick stratigraphic sequences are characterized by widespread depletion or exclusion of even polychaete worms (as evidenced by scarcity of burrows)—among the most eurytopic of benthonic invertebrates, by the depauperate nature of many benthonic microfaunas, and by an abundance of disseminated pyrite and organic carbon. This suggests that these sediments were almost totally anaerobic at times. They also probably contained high levels of H_2S. Many authors have used these data as evidence to suggest widespread oxygen-minimum or wholly anaerobic zones in the overlying water column during the Cretaceous, especially in the deep oceans. The geographic spread and stratigraphic thickness of these facies in the Cretaceous exceeds that of any other period of geologic time.

Intercalated within these same sedimentary facies, especially in epicontinental areas, are numerous intervals containing a rich burrowing ichnofauna, and simple to moderately complex epibenthonic assemblages that tend to show greater "normalcy" in composition and ecological structure where they were elevated to even a few centimeters above the sediment-water interface (KAUFFMAN, 1977c). It is difficult to imagine that these more fossiliferous intervals reflect frequent large-scale fluctuations of a broad oxygen-minimum zone. Instead this suggests a more delicately balanced system of benthonic oxygenation and H_2S levels for many Cretaceous epicontinental seas and possibly some deep ocean basins. The anaerobic-aerobic and H_2S lethal interface may have been situated at, or only a few centimeters above, the sediment-water interface during times characterized by large-scale depletion or exclusion of benthonic invertebrates. In some cases, such a boundary was possibly maintained by low levels of circulation, or within a thin zone of suspended sediment, or beneath anaerobic fungal and blue-green algal mats near the interface (as today in parts of the Black Sea and the Santa Barbara basin). This type of environmental system would exclude virtually all infauna and microscopic epifauna normally living at the depositional surface, and greatly restrict settlement of larvae of all but the epibenthonic organisms that were most tolerant of low oxygen levels (Inoceramidae in the Cretaceous, commonly). Yet, following colonization by such tolerant organisms, especially those having large or inflated shells extending well above the anaerobic boundary, the upper surfaces of these shells provided substrates for habitation by somewhat more normal-marine assemblages in more oxygenated waters. Such relationships have been widely observed in Mesozoic dark shale and carbonate supposedly deposited in basins characterized by a broad oxygen-minimum zone. More extensive colonization of widespread Cretaceous carbonate and clay substrates, as evidenced by marked increase in bioturbation and diversity of epibenthonic taxa, reflects removal or dilution of the anaerobic or toxic H_2S zone situated above the sediment-water interface, or lowering of this zone into the sediment.

Even without oxygen depletion or H_2S poisoning, water saturation levels in marine sediments seem to have imposed another important control on the nature of Cretaceous benthonic environments and invertebrate faunas. High levels of water saturation normally correspond to high levels of turbidity at the sediment-water interface, that is, development of a well-defined zone of suspended sediment above the actual depositional surface, and of a broadly gradational contact between them. Such a zone can be produced biotically through intense bioturbation, and produced physically as a balance between sedimentation rate, size of material, chemistry of the depositional surface, and the magnitude and type of water circulation near the interface. Whatever the cause, high levels of water saturation within the sediment and an overlying zone of suspended fine-grained material combined in the Cretaceous, as now, to greatly alter the benthonic environment, and the composition and structure of benthonic communities (for examples, see RHOADS, 1970; RHOADS, SPEDEN, & WAAGE, 1972; KAUFFMAN, 1974, fig. 12-4). Development of these substrate conditions first choked out selective, and later nonselective suspension-feeding organisms, changing the trophic structure of the benthonic community to one dominated by detritus-feeding organisms. Continuance of turbid benthonic conditions for long periods of time may eventually have led to further exclusion of selective detritus-feeding organisms, and finally of all infaunal and epifaunal elements.

Thus, in the Cretaceous, deep water epicontinental, marine shelf, and ocean basin fine-grained carbonate and clay dominantly represent restricted benthonic habitats, especially for infaunal organisms, characterized by low-oxygen to anaerobic conditions and possibly lethal H_2S levels within and somewhat above the sediment-water interface. In deep oceanic environments, anaerobic conditions apparently spread periodically through large portions of the water column as well, in broad oxygen-minimum layers. Times of oxygen depletion may also commonly have been times of establishment of a suspended sediment zone above, and gradational with the depositional surface, increased water saturation, and increased H_2S concentrations within, the sediment. Oxygenation periods in these environments were frequent, but never long enough for the development of complex benthonic communities so far as we now can ascertain from the fossil record. In deeper ocean basins especially, there was a direct relationship between development of broad oxygen-depleted benthonic environments and restriction of deep water circulation associated with broadening of thermal gradients.

The second unique Cretaceous sedimentary environment was associated with the widespread deposition of glauconite and glauconitic sand in open marine shelf zones. This continued on into the Tertiary in many parts of the world, but is much more restricted in scope elsewhere in the Phanerozoic. The environment of deposition of glauconite has been widely discussed (CLOUD, 1955; PORRENGA, 1967; and others). Cool waters of shelf depth (especially the outer shelf), with slow rates of sedimentation or large-scale sediment bypass, high levels of concentration of organic material, and a microreducing environment at and below the sediment-water interface are thought to have characterized many Cretaceous and Tertiary glauconite-forming environments; however, more widespread reducing environments, low oxygen, and other chemical factors that might have restricted benthonic community development during the Cretaceous do not seem to have been characteristic of glauconite-forming environments. Cretaceous and Tertiary glauconitic sands characteristically contain a rich benthonic invertebrate biota of normal-marine aspect, with broad representation among infaunal elements, and diverse trophic strategies.

The potential for broad fluctuations of calcium-carbonate compensation depth (CCD) and the lysocline, below which dissolution becomes severe, and thus for the development of broad marine areas with benthonic environments posing difficult conditions for calcium-secreting organisms, has previously been discussed [see also the views of RAMSAY (1974) vs. WORSLEY (1971, 1974) for the Cretaceous]. Metastable aragonite, and ultimately more stable calcite readily go into solution as one approaches the CCD, and available carbonate for shell-building in organisms living below the lysocline is greatly diminished. Below CCD the calcite shells of dead invertebrates are almost wholly dissolved, and living organisms are characterized by small, thin shells. The deeper the environments below CCD, the sparser the living shelled fauna; above CCD, modern biotas show offshore increase in diversity (see data in SANDERS, 1969); but the reverse seems true in the Cretaceous. This does not necessarily mean that aragonite-shelled organisms (most mollusks for example) cannot live below the CCD; indeed, most shelled macroinvertebrates in bathyal and abyssal environments today have aragonite shell layers that are protected from dissolution by commonly thick and (or) dense organic shields (the periostracum in Mollusca). But calcitic shells are almost wholly dissolved after death in these environments, resulting in a poor fossil record (see KENNEDY, 1969, p. 462-465), or even total elimination of the record of shelled invertebrates. Thus, both primary and secondary effects of the CCD greatly restrict the biota found below this level in Cretaceous as well as recent seas.

Depletion of invertebrate diversity and ecological complexity in Cretaceous deep water deposits below CCD is not simply a matter of dissolution. In the northern and western European chalk sequences, for example, where ample evidence for dissolution of aragonitic shells exists (KENNEDY, 1969) and only the calcitic shells are well preserved, it has been frequently noted that levels or local zones exist where aragonitic fossils are preserved in abundance (commonly attributed to special preservation phenomena). In these zones, however, the diversity and abundance of the calcitic shells, which are generally preserved throughout the chalk, also increases. This infers both an ecological control (substrate environment) and a secondary diagenetic control (dissolution) on the preservation of organisms with carbonate shells in deep Cretaceous marine environments.

Inherent in the entire preceding discussion has been the broad range of variations in the Cretaceous marine environment: temperature, salinity, oxygen, hydrogen sulfide and carbon dioxide content, food and nutrients, sea level, ecospace, niche diversity, CCD, sediment water saturation, rate of deposition, benthonic circulation (especially in deeper water areas), and other factors. Long-term environmental stability apparently did not exist, in respect to these factors, in even the deepest parts of the Cretaceous seas. Many habitable environments were formed and wholly eliminated several times during the Cretaceous (for example, shallow epicontinental carbonate environments); others underwent severe environmental fluctuations (for example, oxygen content); others were moved wholesale over broad geographic areas (for example, marginal marine to shallow sublittoral quartz-sand facies); others were subject to great change in size, and the amount and diversity of potential ecospace for habitation that they offered. These fluctuations were

NATURE OF INVERTEBRATE FAUNAS

The following discussion will be confined to documentation of Cretaceous invertebrates that are critical to definition of stages[1] and their boundaries, biogeographic units, and regional biostratigraphic zonation. It can generally be said that the Cretaceous fauna was characterized by turrilitacean, scaphitacean, hoplitacean, desmoceratacean, acanthoceratacean, and most groups of ancyloceratacean ammonites; by great development among rudistid, inoceramid, trigoniacean, and certain ostreacean bivalves; by marked radiation among Neogastropoda, and especially tropical Nerineidae and Actaeonellidae; by planomalinid, schackoinid, rotaliporid, and globotruncanid Foraminiferida, and by many groups of Tethyan larger benthonic Foraminiferida; by tremendous diversity among scleractinian corals, especially in the middle Cretaceous where older groups of Jurassic origin occurred with and gave rise to younger Cretaceous-Cenozoic groups; by great radiation among cheilostome Bryozoa, with many groups restricted to the Cretaceous (Myagroporidae, Otoporidae, Ctenoporidae, Thoracoporidae, Taractoporidae, Calpidoporidae, Disheloporidae, and Nephroporidae); and by equally large-scale radiation among the echinoids, with the Conulidae, Discoidiidae, Galeritidae, Archiaciidae, and Clypeolampadidae restricted to the Cretaceous. An age-by-age detailed summary of Cretaceous invertebrate faunas and environmental events affecting their evolution can be obtained by surveying references previously listed. Some broad generalizations follow.

The Berriasian and Valanginian were times characterized by low-level radiations among invertebrate groups that predominantly had their origins in the Late Jurassic, in particular diverse ammonites. No major extinction events have been recorded from either age. The Late Valanginian transgressive peak marked the initial radiation of several invertebrate lineages or families among larger foraminifers, ammonites, rudist bivalves, and irregular echinoids which became increasingly more important during the later part of the Cretaceous. The first major radiation among typical Cretaceous invertebrates occurred during the Hauterivian, primarily involving warm-water groups (Tethyan foraminifers, scleractinian corals, ammonites, and irregular echinoids). This radiation was associated with climatic warming and widespread marginal epicontinental transgression. Extinctions during or at the end of the Hauterivian were of little consequence.

The Barremian was a time of rapidly fluctuating marine environments associated with two major transgressive-regressive (eustatic) cycles (Fig. 2). Evolutionary response to broadly changing natural selective forces was marked and produced the largest radiation of invertebrates in the Neocomian, especially among tropical and subtropical groups of smaller and larger foraminifers, scleractinian corals, ammonites, and rudist and glycymerid bivalves. No important extinction events occurred in or at the end of the Barremian. During Aptian time, plate tectonic activity and marine transgression exceeded that of the Neocomian, as did related evolutionary events. Aptian radiation was of greater magnitude, and involved more groups, than any previous age in the Cretaceous; it involved predominantly Tethyan, but also temperate groups of calcareous foraminifers and radiolarians, numerous scleractinian coral families, brachiopods, belemnoids, ammonites, and diverse families of bivalves. The first major Cretaceous reefs formed during the Aptian. A marked extinction event at or near the end of the age mainly involved older Tethyan groups that were ecologically replaced by better adapted counterparts arising in the Aptian; scleractinian corals, ammonites, belemnites, and ancestral rudist bivalve groups were primarily affected.

---

[1] The author prefers to use the word "stage" in a biostratigraphic sense as originally defined by D'ORBIGNY in 1842, and as currently followed by many geologists. Nevertheless, in compliance with editorial policy, which adheres to the Code of Stratigraphic Nomenclature (American Commission on Stratigraphic Nomenclature, 1970) and the International Stratigraphic Guide (International Subcommission on Stratigraphic Classification, 1976), "stage" and "age" are used in this paper as chronostratigraphic and geochronologic units, respectively.

Albian time was characterized by the most extensive of the Early Cretaceous eustatic fluctuations and transgressions (Fig. 2), by active plate tectonism, a major climatic warming pulse, and largely restricted deep ocean circulation. Major radiation and marine diversification occurred among planktonic and shallow benthonic invertebrates, in particular planktonic foraminifers and ammonites, giving rise to complex, polytaxic, pelagic marine communities. In addition, Albian radiations involved diverse calcareous benthonic foraminifers, scleractinian corals, cheilostome bryozoans, articulate brachiopods, many neogastropod and echinoderm families, and numerous tropical bivalve groups, including rudists. Cretaceous ammonite diversity reached its peak during Albian radiation. Deep-water invertebrates, on the other hand, seem to have been severely restricted by widespread oxygen depletion. Important extinction events occurred throughout the Albian, but especially at its end. The Late Albian extinction was the most dramatic of the Early Cretaceous, and mainly affected taxa in tropical to warm-temperate seas, probably as a result of sharp decline in oceanic temperatures during the latest Albian. This extinction involved diverse ammonites, and major groups of larger foraminifers, scleractinian corals, articulate brachiopods, belemnoids, bivalves, irregular echinoids, and crinoids. The Albian thus included the first great epicontinental transgression, and the first major extinction event (associated with terminal Albian regression and temperature decline) of the Cretaceous (Fig. 2).

The Cenomanian had an exceptional history. It included the largest marine transgression of the Cretaceous (Fig. 2), which was related to active plate tectonism, rapid seafloor spreading, and major eustatic rise. It was also the greatest Cretaceous episode of invertebrate radiation, reflecting vast new ecological opportunities that became available during the great Cenomanian transgression. Surprisingly, these events were associated with a period of low global marine temperatures, which climbed only slowly during the age. Between 50 and 60 higher taxa of invertebrates, most of them tropical to warm-temperate groups, arose during the Cenomanian, including numerous genera of larger foraminifers and diverse lower level taxa of planktonic foraminifers, 6 scleractinian coral families, phylactolaemate and diverse cheilostome bryozoans, 5 ammonite families, the Belemnitellinae, numerous groups of neogastropods as well as epifaunal and especially infaunal siphonate bivalves, 3 major arthropod groups, and 5 echinoid families. The Cenomanian-Turonian boundary lies near the peak of the transgression, and coincides with an abrupt but graded extinction event associated with major temperature and salinity fluctuations near the transgressive peak. First temperate, and finally tropical taxa were involved in the extinction. Diverse planktonic and larger benthonic foraminifers, scleractinian corals, ammonites, and irregular echinoids were principal groups affected by the extinction event.

Turonian time encompassed a major regression, early phases of the second largest Cretaceous transgression, and rising global marine temperatures (Fig. 2). The great Cenomanian transgression climaxed during the earliest Turonian, and was characterized by a short-lived invasion of marginal Tethyan taxa into temperate areas, followed by their graded extinction (middle to late Early Turonian) or restriction to Tethys proper. The Middle Turonian regression was one of the most rapid on record (less than two million years), and imposed severe stress, leading to very rapid faunal turnover and widespread extinction in epicontinental (KAUFFMAN, 1970, 1972, 1977a,b), but not oceanic environments (FISCHER & ARTHUR, 1977). Turonian radiations were moderate, and concentrated in the Early Turonian in connection with maximum transgression and the spread of warm waters over the globe. Major evolutionary diversification took place among calcareous planktonic and benthonic foraminifers, tropical scleractinian corals, cheilostome bryozoans, temperate and tropical ammonites, rudist bivalves, and irregular echinoids. Many of these same groups show strong extinctions associated with Middle Turonian regression. The end of the Turonian was marked by sharp extinction, which correlates with a major temperature and salinity increase to near normal in epicontinental seas near peak Late Turonian to Coniacian transgression. Both

Middle and Late Turonian extinctions involved primarily warm-water groups—diverse foraminifers, scleractinian corals, major groups of rudists, inoceramid bivalves, and especially tropical ammonites. In epicontinental areas, the terminal extinction marked the abrupt replacement of temperate groups with cosmopolitan and subtropical taxa.

The great Late Turonian transgression reached its peak in the Coniacian, and remained near peak development throughout that age. Maritime climates and ameliorating marine environments persisted. An Early to Middle Coniacian temperature increase is recorded by exceptionally widespread distribution of Tethyan and marginal Tethyan faunal elements, and by isotopic analyses. Major Early Coniacian radiations primarily involved warm-water groups among smaller and larger calcareous foraminifers, cheilostome bryozoans, neogastropods, cephalopods, and echinoids. Representatives of these and other tropical taxa became gradually more restricted during the Coniacian as warm water temperatures began to wane, especially during a sharp Middle Coniacian temperature drop in oceanic and epicontinental areas (DOUGLAS & SAVIN, 1973). Warm, semistable ameliorating marine climates and widespread epicontinental seas persisted into the Santonian; as a result no major extinction event marks the stage boundary, making it difficult to define (see discussion under *Biostratigraphy*). Important extinctions during the Coniacian were limited to three ammonite families and the posidoniid bivalves; genus- and species-level changes among ammonites and bivalves, or among planktonic microbiota, are currently used to define the Coniacian-Santonian boundary.

Santonian time was characterized by several small transgressive pulses superimposed on early stages of the Campanian regression (Fig. 2), by warm water temperatures and generally equable marine climates, and by continued widespread development of epicontinental seas. A broad Late Santonian cooling trend gradually increased oceanic climatic gradients, lowering planktonic diversity (FRERICHS, 1971), and lowering the similarity between foraminifers of high and low latitudes (DOUGLAS, 1972). Low to moderate levels of radiation among calcareous foraminifers, cheilostome bryozoans, scleractinian corals, rudist bivalves, and echinoderms indicate a somewhat passive evolutionary history during this climatically stable period. Other groups showed little change. Extinction within or at the end of the age was minimal, mainly involving a few Tethyan lineages of cheilostome bryozoans, scleractinian corals, and ammonites. In many areas the Santonian-Campanian boundary is thus difficult to define and is based mainly on range-zone boundaries of ammonite and inoceramid species.

The equable marine environments of the Coniacian and Santonian declined sharply during the Campanian with major marine regression (eustatic fall), upon which are superimposed two smaller transgressive peaks (Fig. 2) in the latest Early to early Late Campanian, and the middle Late Campanian. Oceanic temperatures declined strongly through the Campanian. Collectively, these factors imposed increasing environmental stress levels on evolving marine invertebrates, and diversity generally declined. Nevertheless, the late Early to early Late Campanian transgression marked a reversal in these trends, and the last major period of Cretaceous radiation among higher taxa, in particular among warm-water foraminifers, scleractinian corals, cheilostome bryozoans, and tropical and temperate bivalves. Late Campanian extinctions were surprisingly few in light of this radiation and subsequent environmental decline, and involved only some genera of larger foraminifers, and one family each of scleractinian corals, ctenostome bryozoans, ammonites, crinoids, and asteroids. The position of the Campanian-Maastrichtian boundary is thus poorly defined and a matter of considerable debate. As a result of increasing stress levels during Campanian regression, evolutionary rates among principal Campanian ammonite and bivalve groups were rapid and form the basis for a highly refined Campanian biostratigraphy (GILL & COBBAN, 1966; KAUFFMAN, 1975).

TEMPORAL AND BIOSTRATIGRAPHIC DEFINITIONS

The global correlation and interpretation of geological and biological events is strongly dependent upon widely applicable, finely divided units of geological time (usually based on biostratigraphic zones), and sufficient radiometric data to allow "absolute" dating of such events and an accurate measure of their duration (for example, KAUFFMAN, 1970). Rarely are both systems coincidently developed to a high level of refinement in the Phanerozoic, but the Cretaceous is an exception where biostratigraphy and geochronology have been refined to a high degree and extensively integrated (for example, OBRADOVICH & COBBAN, 1975; VAN HINTE, 1976; KAUFFMAN, 1978), allowing unparalleled precision in regional correlation and dating of marine strata, their biotas, and important events in their historical development. These factors have widely attracted scholars to the Cretaceous Period as a testing ground for geological and biological concepts.

DEFINITION OF THE CRETACEOUS

The Cretaceous System, or "Terrain Crétacé," was first named by the Belgian D'OMALIUS D'HALLOY (1822) wholly on the basis of lithostratigraphic characteristics as the upper one-third of the "Secondary Rocks" of ARDUINO. The Cretaceous included "the chalk formation, such as I have determined it in a preceding memoir, i.e. comprising the tuffas, sands, and marls, which occur beneath the true chalk. . . ." (D'OMALIUS D'HALLOY, 1822, p. 368, 369; translation in BERRY, 1968, p. 69). The type area is in the Paris basin of France and adjacent parts of Belgium and Holland. BERRY (1968, p. 69-73) gave a concise history of subsequent development of the concept of the Cretaceous System. Initial definition of the typical Cretaceous biota, and a crude biostratigraphic division, resulted from MANTELL's (1822) study of "The fossils of the South Downs" in which he recognized a lower terrestrial biota of the Wealden and a younger marine biota of the overlying greensand, marl, and chalk. Although various authors, mainly French and English, described fossils from the Cretaceous System during the next two decades, the first extensive description of characteristic Cretaceous biotas and their biostratigraphic subdivision was that of the French paleontologist, ALCIDE D'ORBIGNY (1840-44; 1849-50).

D'ORBIGNY's contribution was outstanding. In addition to his extensive systematic treatment, D'ORBIGNY was the first to clearly define major changes in the Cretaceous marine biotas through time, many of them relatively abrupt, and the fact that these changes were irregular. D'ORBIGNY recognized that large, stratigraphically distinct segments of the Cretaceous were each internally characterized by a discrete biota displaying relatively low levels of change through time. Each biota, however, was quite different from those of adjacent sequences. The boundary zones between individual biotas were clearly defined (D'ORBIGNY, 1849-50, p. 42-49, considered them to be "catastrophic"), in most cases sharp, or occupying very narrow stratigraphic intervals. These were points of major biotic turnover that have come to be recognized as indicative of large-scale evolutionary events and (or) marine environmental changes occurring within small segments of geologic time. D'ORBIGNY (1849-50) termed these biotically discrete stratigraphic segments "étages"—or stages—and gave each a name based on the area in which its unique biota was best known at the time. Thus, *from their inception, Cretaceous stages were characterized by their fossil content,* and stage boundaries were defined at the level or narrow stratigraphic interval of maximum rate of change, or abruptness of turnover of the biota from one stage to the next. Stages were thus originally defined as large-scale biostratigraphic units; this concept has been unfortunately altered where stages have been included within systems of chronostratigraphy such as that used by the American Commission on Stratigraphic Nomenclature (1970) and others. The concept of stages in the Cretaceous has been more extensively discussed by HANCOCK (1977), KENNEDY and JUIGNET (1973), RIOULT (1969), WIEDMANN (1970), and references therein, most of whom have advocated return to D'ORBIGNY's original concept; I wholly concur.

In defining stages, D'ORBIGNY became the first to recognize the varied and irregular evolutionary history of Cretaceous organisms. Subsequent research has supported the contention that major extinction events followed by important radiations exist at several points through the Cretaceous, at or near the stratigraphic levels of faunal discordance that D'ORBIGNY and later authors relied on to differentiate Cretaceous stages.

CRETACEOUS BOUNDARY ZONES

Historically, both the upper and lower boundaries of the Cretaceous System have been extensively debated, in particular the Jurassic-Cretaceous boundary zone and the position of the Volgian Stage and its biota in the North Temperate Realm, and (to a lesser degree) the Tithonian Stage and its biota in tropical Tethyan to warm-temperate regions. The problem has been complicated by the lack of direct marine connections linking the two biotas in many areas, difficulty in their biostratigraphic correlation, and by the absence of a large-scale boundary extinction of characteristic Jurassic taxa. Similarly, the position of the Danian "Stage" and its biota relative to the Cretaceous-Tertiary boundary zone has been greatly disputed, especially where diagnostic ammonites, belemnites, and inoceramid bivalves are rare or absent.

It is of interest that in both cases the boundary disputes stemmed primarily from study of biotas from north-temperate areas, where the evolutionary distinctions between marine faunas above and below these major boundaries appear to be less marked than across the same boundaries in the tropical Tethyan Realm. For marine invertebrates this may be directly related to comparatively higher representation of the entire tropical biota in the oceanic plankton (including the planktotrophic larvae of most benthonic organisms), where conditions apparently leading to major extinctions have had greater impact than in cooler water, temperate-realm biotas.

The Jurassic-Cretaceous boundary problem has been most extensively discussed by various authors in CASEY and RAWSON (1973a) and older papers referenced therein, in particular ARKELL (1956), CASEY (1968), EGOYAN (1971), GOLBERT and others (1972), SAKS and others (1968), and papers in SAKS (1972). The controversy centered mainly around the position of strata containing the north-temperate "Subcraspeditan" ammonite fauna, placed originally in the lowermost Cretaceous or "Infravalanginian Stage" (NEAVERSON, 1955, and papers cited therein), and has now been largely resolved. Current placement of the Jurassic-Cretaceous boundary at the top of the Volgian Stage (containing the Subcraspeditan fauna) of the Russian platform, and equivalent strata, is supported by most workers. The base of the north temperate Cretaceous, also defined mainly on molluscan faunas, thus becomes the base of the lower Ryazanian Substage (*Runctonia runctoni* and *Hectoroceras kochi* zones; CASEY, 1973).

Even so, a major problem still exists in identifying the Jurassic-Cretaceous boundary, which stems from the lack of a clear-cut evolutionary break among higher taxa in this interval, and a relatively poor fossil record in many areas. Most higher taxa of characteristic Jurassic invertebrates became extinct well before the end of the Jurassic (for example, numerous ammonite families in the Bathonian through Oxfordian stages). The Oxfordian, Kimmeridgian and Volgian-Tithonian stages instead represent a time of major radiation among many groups of invertebrates that subsequently cross the Jurassic-Cretaceous boundary without significant change to comprise the evolutionary rootstocks of characteristic Neocomian faunas. This includes many echinoids (especially families of the Irregularia), bivalves, gastropods, arthropods, and most scleractinian corals. The ammonites underwent a major radiation at the beginning of the Late Tithonian to Late Volgian subages, and 9 of 11 latest Jurassic families ranged well into the Neocomian. The earliest Neocomian (Berriasian) radiations were in no way comparable in magnitude to those of the latest Jurassic. Few important extinction events occurred in the latest Jurassic, and most of these occurred near the end of the Tithonian, especially among ammonites (for example, the important families Perisphinctidae and Aspidoceratidae); these are the primary organisms utilized to define and correlate the Jurassic-Cretaceous on a regional basis. Thus, in defining the boundary, current workers have chosen to concentrate on the few significant ammonite extinction events of the latest Jurassic rather than the older, but more dramatic interval that marks the principal radiation of typical Early Neocomian invertebrates. Principal evolutionary changes at the Jurassic-Cretaceous boundary, so defined, are marked by genus-level turnover rather than by replacement of higher taxa as elsewhere at Mesozoic period boundaries. This change is more exaggerated in the Tethys and its margins than in the North Temperate Realm.

By contrast, the top of the Maastrichtian Stage is marked by one of the most dramatic extinction events of the Phanerozoic, with final elimination of numerous orders, superfamilies, families, and subfamilies of characteristic Cretaceous invertebrates within a few million years at most. Major invertebrate groups that became extinct include the Ammonoidea (superfamilies Phylloceratceae, Turrilitaceae, Scaphitaceae, Lytoceratacaea, Acanthoceratceae, Desmocerataceae, and Hoplitaceae), many Bivalvia (for example, the Arcullaeinae, Inoceramidae, Oxytomidae, Entoliidae, *Camptonectes* group, *Neithea* group, Buchiidae, Terquemiidae, Trigonioididae, Myoconchinae, Opinae, Icanotiidae, Tancrediidae, Trapeziidae, Dicerocardiidae, Corbulamellidae, Exogyrini, and the rudists Requieniidae, Monopleuridae, Caprinidae, Hippuritidae, Radiolitinae, Biradiolitinae, Sauvagesiinae, and Lapeirousiinae), the Macluritina among gastropods, the Spinctozoa among sponges, the Hemiporitidae and Thamnasteriidae among scleractinian corals, the Dactylethrata and Rhacheoporidae among bryozoans; numerous Coleoidea (Belemnitidae, Dimitobelinae, Belemnitellinae, Belemnoteuthidae), 7 families of mostly irregular echinoids (Acrosaleniidae, Conulidae, Discoididae, Galeritidae, Clypeolampadidae, Holectypidae, and Nucleolitidae), stelleroids of the Uractinina, 17 genera of tropical larger Foraminifera including most orbitoids and pseudorbitoids, several major groups of calcareous foraminifers (especially the Schackoinidae, Rotaliporidae, and Globotruncanidae), and the radiolarian family Amphipyndacidae. Many of these groups became extinct "abruptly" at the peak of their radiation, or while they were still important components of the marine biota.

A major radiation, especially among warm-water bivalve and gastropod molluscs, irregular echinoids, crustacea and other arthropods, cheilostome bryozoans, and both small and large calcareous Foraminiferida followed during the Tertiary, so that clearly distinct biotas bound the major extinction event that marks the top of the Cretaceous System. Why then has there been a controversy over the position of the Danian Stage in the north temperate areas of Euramerica?

The Danian controversy evolved in areas where 1) Cretaceous-style marine deposition of chalk and marl, or glauconitic sand, continued across the boundary with little change (as in Denmark, Maryland, and New Jersey), in many cases with little evidence of a major disconformity, into Paleocene time, 2) where Cretaceous microfossils were extensively reworked into similar Paleocene sediments, 3) where strata bearing Cretaceous organisms were erroneously assigned to the Danian on other grounds, and 4) in areas where uppermost Cretaceous strata did not contain abundant ammonites, inoceramid bivalves, belemnites, or other diagnostic fossils but rather were composed mainly of groups of mollusks, echinoids, and bryozoans that showed little evolutionary differentiation and only minor genus- and species-level extinction across the Cretaceous-Tertiary boundary (for example, Denmark, Greenland, and middle to northern Atlantic coasts of the United States and western Europe). These latter groups, mainly mollusks characteristic of the North Temperate Realm, were apparently protected from environmental factors contributing to the massive Late Cretaceous extinction event by nature of their broad ecological tolerance, habitat characteristics, or the nature of their ontogenetic development (especially their larval history). It is now well documented that the dramatic Cretaceous-Danian extinction events mainly involved tropical and marginal tropical groups. The recognition that many invertebrate groups

Stage	Substage	Age Ma	TETHYAN MACROFOSSIL STANDARD (1)		NORTH TEMPERATE MACROFOSSIL STANDARD (1)	PLANKTONIC FORAM.-CALPIONELLID STANDARD (1)	
		**121					
HAUTERIVIAN	U		Pseudothurmannia angulicostata		Simbirskites discofalcatus	"Hedbergella" hoterivica	
					Simbirskites gottschei		
		123	Subsaynella subsayni		Simbirskites staffi		
	M		Crioceras duvali		Simbirskites inversum	UNZONED	
		124			Endomoceras regale		
	L		Acanthodiscus radiatus		Endomoceras noricum		
		126	Lyticoceras s.l.sp.		Endomoceras amblygonium		
VALANGINIAN	U		Saynoceras verrucosum	Saynoceras callidiscus	Arnoldia - Astieria		
		127		Himantoceras trinodosum	Dichotomites spp.	Calpionellites	
	M			Saynoceras verrucosum	Prodichotomites polytomus		
		129	Kilianella roubaudi	Kilianella compylotoxus	Polyptychites middendorfi		
					P. bancoi, P. euomphalus		
	L			Kilianella roubaudi	Platylenticeras involutum		
		131		Kilianella pertransiens	Platylenticeras heteropleurum		
					Platylenticeras robustum	Calpionellopsis	
BERRIASIAN / RYAZANIAN	U	132	Berriasella boissieri		"WEALDEN" nonmarine		
	M	134				Calpionella	Calpionella elliptica
	L		Berriasella grandis				Calpionella alpina
		135					

Fig. 3A.

Fig. 3A–F. Lower Cretaceous biostratigraphic zonation and principal zonal indices in selected areas where zonation is most refined. Principal references for data coded to numbers in parentheses are: 1) Van Hinte, 1976, Tethyan and North Temperate macrofossil and pelagic microfossil zones; 2) Jeletzky, 1970, Canadian and Arctic parts of the Western Interior seaway; 3) Kauffman, Cobban, & Eicher, 1977, Albian of the interior United States; 4) Thomel, 1964, and 5) Thieuloy & Thomel, 1964, Neocomian, southern France; 6) Young, 1967, Texas Gulf Coast; 7) Scott, 1970, Albian, southern Western Interior United States; 8) Jeletzky, 1964a, 1970, western and Arctic Canada; 9) Imlay, 1960, 10) Jones, 1967, 11) Jones & Detterman, 1966, and 12) Jones, Murphy, & Packard, 1965, northern and western Alaska, west coast of United States; 13) Matsumoto, 1963, and 14) Nakano, 1960, Lower Cretaceous, Japan; 15) Casey, 1973, Berriasian, England; 16) Owen, 1973, Middle and Upper Albian, England; 17) Casey, 1961, Aptian and Lower Albian, England; 18) Kemper, 1973, Hauterivian-Valanginian, north Germany; 19) Besairie & Collignon, 1972, Madagascar; 20) Pergament, 1969, Russian platform, Siberia, and northeast Asia; and 21) Jeletzky, 1973, Lower Neocomian, Siberia. For explanation of asterisks, see Figure 2 caption (Kauffman, n).

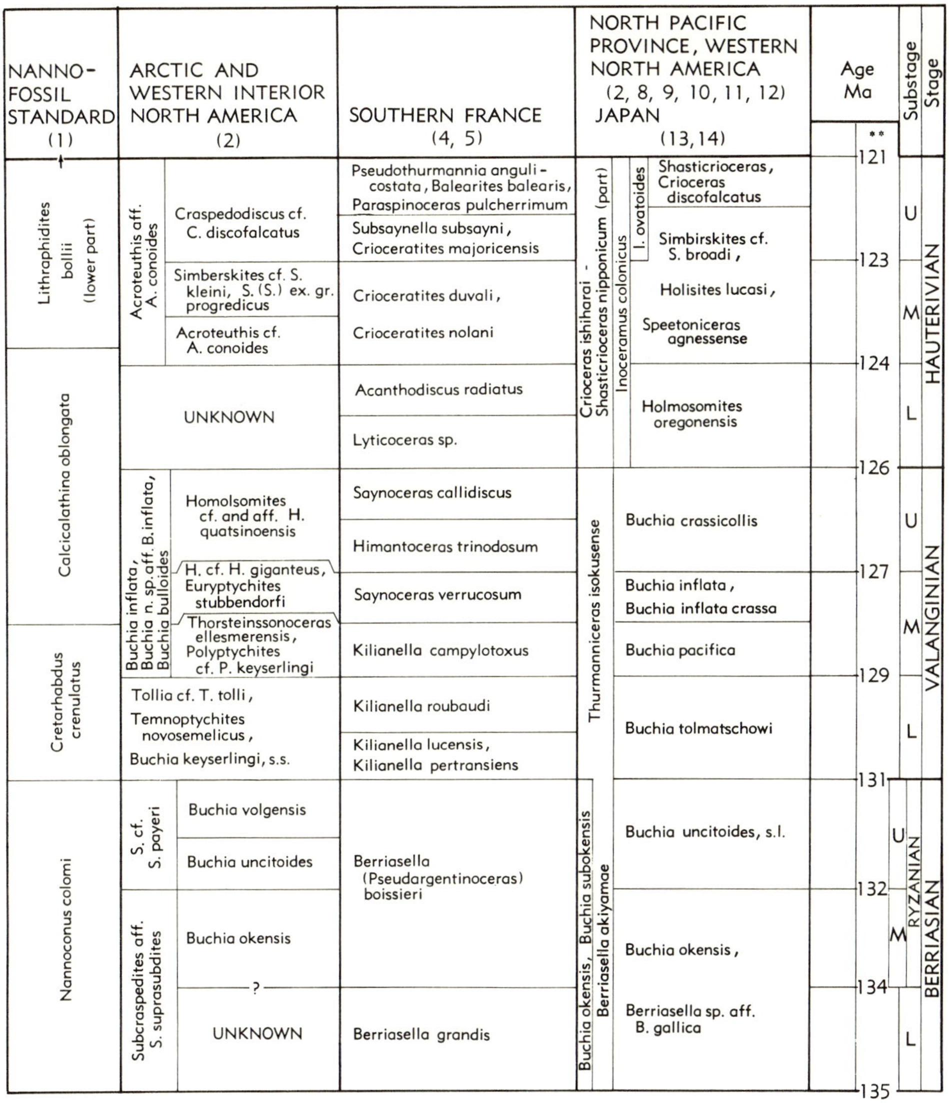

NANNO-FOSSIL STANDARD (1)	ARCTIC AND WESTERN INTERIOR NORTH AMERICA (2)		SOUTHERN FRANCE (4, 5)	NORTH PACIFIC PROVINCE, WESTERN NORTH AMERICA (2, 8, 9, 10, 11, 12) JAPAN (13,14)		Age Ma	Substage	Stage
Lithraphidites bollii (lower part)	Acroteuthis aff. A. conoides	Craspedodiscus cf. C. discofalcatus	Pseudothurmannia anguli-costata, Balearites balearis, Paraspinoceras pulcherrimum	Crioceras ishiharai - Shasticrioceras nipponicum (part)	I. ovatoides	Shasticrioceras, Crioceras discofalcatus	U	HAUTERIVIAN
			Subsaynella subsayni, Crioceratites majoricensis			Simbirskites cf. S. broadi	123	
		Simberskites cf. S. kleini, S. (S.) ex. gr. progredicus	Crioceratites duvali,			Holisites lucasi,	M	
		Acroteuthis cf. A. conoides	Crioceratites nolani			Speetoniceras agnessense	124	
Calcicalathina oblongata	UNKNOWN		Acanthodiscus radiatus			Holmosomites oregonensis	L	
			Lyticoceras sp.				126	
	Buchia inflata, Buchia n. sp. aff. B.inflata, Buchia bulloides	Homolsomites cf. and aff. H. quatsinoensis	Saynoceras callidiscus	Thurmanniceras isokusense		Buchia crassicollis	U	VALANGINIAN
			Himantoceras trinodosum				127	
		H. cf. H. giganteus, Euryptychites stubbendorfi	Saynoceras verrucosum			Buchia inflata, Buchia inflata crassa	M	
		Thorsteinssonoceras ellesmerensis, Polyptychites cf. P. keyserlingi	Kilianella campylotoxus			Buchia pacifica	129	
Cretarhabdus crenulatus		Tollia cf. T. tolli, Temnoptychites novosemelicus, Buchia keyserlingi, s.s.	Kilianella roubaudi			Buchia tolmatschowi	L	
			Kilianella lucensis, Kilianella pertransiens				131	
Nannoconus colomi	S. cf. S. payeri	Buchia volgensis		Buchia okensis, Berriasella akiyamae	Buchia subokensis	Buchia uncitoides, s.l.	U	BERRIASIAN
		Buchia uncitoides	Berriasella (Pseudargentinoceras) boissieri				132	
	Subcraspedites aff. S. suprasubdites	Buchia okensis				Buchia okensis,	M	
		—?—				Berriasella sp. aff. B. gallica	134	
		UNKNOWN	Berriasella grandis				L	
							135	

Fig. 3B. *(Explanation on facing page.)*

cross the Cretaceous-Tertiary boundary with little change, and the discovery of sufficient ammonites, belemnites, inoceramids, and characteristic Cretaceous microfossils in the areas under question to define the boundary, have basically ended the Danian controversy. Today, Danian strata are almost universally placed in the basal Paleocene. HANCOCK (1967), and authors cited therein, have discussed in detail the age of the Danian and defended its placement in the Paleocene.

BIOSTRATIGRAPHY

Cretaceous biostratigraphy (Figs. 3, 4) encompasses some of the most refined and regionally applicable zonations developed anywhere in the Phanerozoic, and consti-

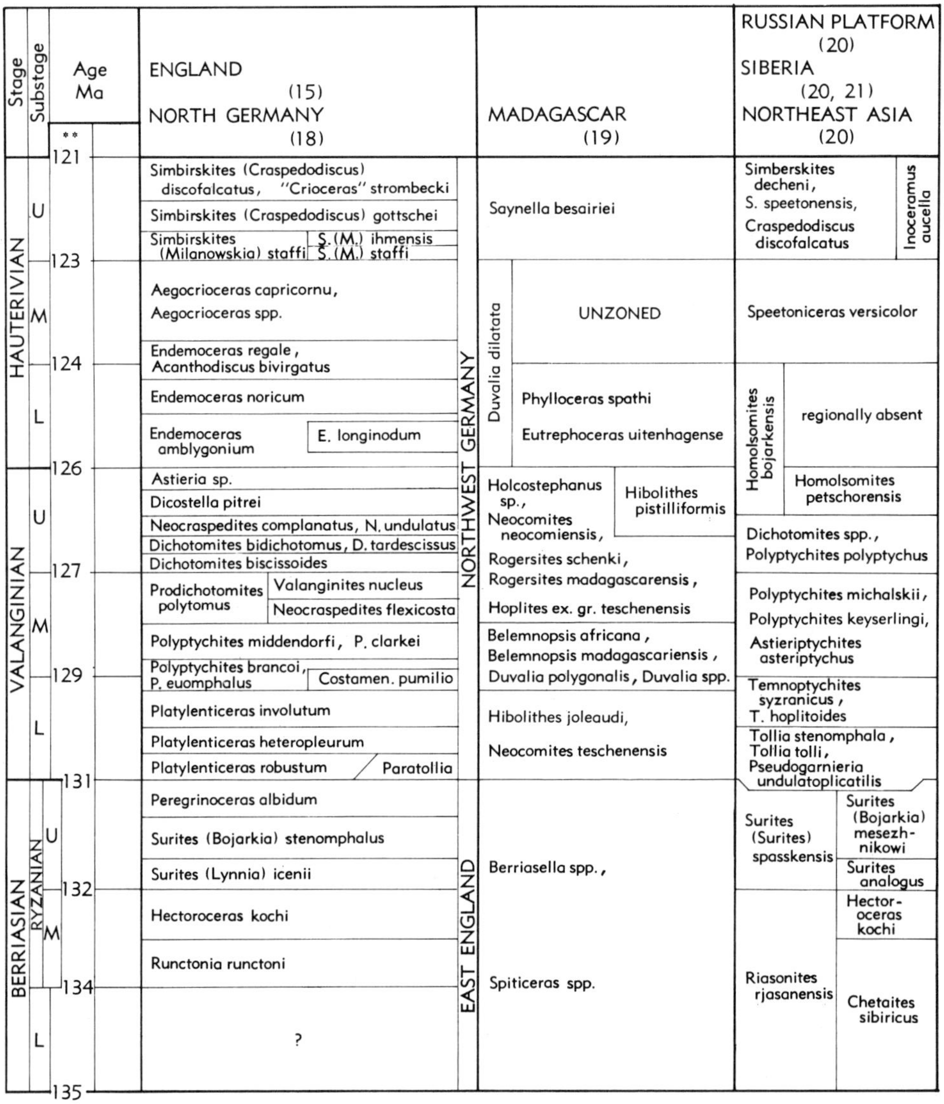

Stage	Substage	Age Ma	ENGLAND (15) NORTH GERMANY (18)		MADAGASCAR (19)		RUSSIAN PLATFORM (20) SIBERIA (20, 21) NORTHEAST ASIA (20)		
		—121—							
HAUTERIVIAN	U		Simbirskites (Craspedodiscus) discofalcatus, "Crioceras" strombecki		Saynella besairiei		Simberskites decheni, S. speetonensis, Craspedodiscus discofalcatus		Inoceramus aucella
			Simbirskites (Craspedodiscus) gottschei						
		—123—	Simbirskites (Milanowskia) staffi	S.(M.) ihmensis S.(M.) staffi					
	M		Aegocrioceras capricornu, Aegocrioceras spp.		UNZONED		Speetoniceras versicolor		
		—124—	Endemoceras regale, Acanthodiscus bivirgatus						
	L		Endemoceras noricum		Phylloceras spathi		Homolsomites bojarkensis	regionally absent	
			Endemoceras amblygonium	E. longinodum	Eutrephoceras uitenhagense				
		—126—	Astieria sp.		Holcostephanus sp., Neocomites neocomiensis,	Hibolithes pistilliformis	Homolsomites petschorensis		
VALANGINIAN	U		Dicostella pitrei						
			Neocraspedites complanatus, N. undulatus				Dichotomites spp., Polyptychites polyptychus		
		—127—	Dichotomites bidichotomus, D. tardescissus						
			Dichotomites biscissoides		Rogersites schenki, Rogersites madagascarensis, Hoplites ex. gr. teschenensis		Polyptychites michalskii, Polyptychites keyserlingi,		
	M		Prodichotomites polytomus	Valanginites nucleus					
				Neocraspedites flexicosta					
			Polyptychites middendorfi, P. clarkei		Belemnopsis africana, Belemnopsis madagascariensis, Duvalia polygonalis, Duvalia spp.		Astieriptychites asteriptychus		
		—129—	Polyptychites brancoi, P. euomphalus	Costamen. pumilio			Temnoptychites syzranicus, T. hoplitoides		
	L		Platylenticeras involutum		Hibolithes joleaudi,				
			Platylenticeras heteropleurum		Neocomites teschenensis		Tollia stenomphala, Tollia tolli, Pseudogarnieria undulatoplicatilis		
			Platylenticeras robustum	Paratollia					
		—131—	Peregrinoceras albidum						Surites (Bojarkia) mesezhnikowi
BERRIASIAN	U		Surites (Bojarkia) stenomphalus		Berriasella spp.,		Surites (Surites) spasskensis		
RYAZANIAN		—132—	Surites (Lynnia) icenii						Surites analogus
	M		Hectoroceras kochi						Hectoroceras kochi
		—134—	Runctonia runctoni		Spiticeras spp.		Riasonites rjasanensis		
	L		?						Chetaites sibiricus
		—135—							

FIG. 3C. (Explanation on page A442.)

tutes an important model of biostratigraphic methodology. Several factors contribute to this situation. First, the Cretaceous was a period of evolutionary overlap among some of the most useful biological groups ever applied to zonation and correlation. The first major series of radiations among the planktonic microbiota, especially foraminifers and coccolithophorids, occurred during the Cretaceous at a time when ammonites and ubiquitous bivalves of the families Inoceramidae, Ostreidae, and Buchiidae were still important components of the marine biota, belemnites were in their final radiation, and echinoids (especially irregular groups) were undergoing one of their major periods of diversification. These groups constitute the principal invertebrate components of Cretaceous biostratigraphy. Other taxonomic groups are useful in certain facies and regions, and may be the bases for local systems of zonation, but they do not compare with these primary groups in regional correlation potential or in the historical development of Cretaceous biostratigraphic systems.

Second, Cretaceous biostratigraphy historically arose by the independent development of systems based on such groups as planktonic foraminifers, ammonites, belemnites, inoceramid bivalves, and echinoids. In more recent times, however, there has been a conscious effort to develop more integrated biostratigraphic systems (for example, KAUFFMAN, 1970). This has been undertaken first through the collation of data from these primary groups into simple assemblage- and concurrent range-zones (VAN HINTE, 1976; SMITH, 1975; KAUFFMAN, HATTAN, & POWELL, 1977, and others). Subsequently, using these as a plotting base for testing the biostratigraphic utility of diverse invertebrates, modern Cretaceous biostratigraphic systems are becoming increasingly concerned with the formation of complex multitaxic assemblage-zones incorporating all elements of the biota with biostratigraphic potential as defined by their biological, biogeographic, evolutionary, and ecological characteristics (KAUFFMAN, 1970). Complex multitaxic assemblage zonation and biostratigraphic schemes being developed through graphic correlation methods (MILLER, 1977, and references therein) promise to yield the most refined levels of biostratigraphic zonation, with the broadest environmental and biogeographic application. Zonations as refined as 0.1 my/zone with intercontinental correlation potential have already resulted in the Cretaceous utilizing "composite assemblage zonation" techniques (KAUFFMAN, 1970). Composite assemblage-zones may incorporate organisms from all facies throughout the study area (for example, a Cretaceous epicontinental seaway), including marginal marine and even nonmarine facies, and even from distinct basins, allowing broad regional correlations to be made.

Third, the Cretaceous displays an extensive, well-preserved biological and stratigraphic record, widely dispersed in epicontinental areas of the world (Fig. 1) and nearly continuous across vast areas of intervening ocean basins as evidenced in DSDP core materials. Bed-by-bed collecting of populations representing diverse taxa is possible in many epicontinental areas, leading to evolutionary studies using population analysis, to taxonomic refinement, and thus to more refined biostratigraphic division of the Cretaceous sequence (for example, KAUFFMAN, 1970, text-fig. 4, and discussion).

Fourth, radiometric data for the Cretaceous exceed those for all but late Cenozoic parts of the geological column because of the abundance of volcanic ash, bentonite beds, and glauconite in global Cretaceous marine sections. These data are spread throughout most of the Cretaceous sequence, being more abundant for Albian and younger rocks. To a large extent, but especially in the Western Interior of the United States and Canada, these dates have been closely integrated with the biostratigraphic system (GILL & COBBAN, 1966; OBRADOVICH & COBBAN, 1975; VAN HINTE, 1976; KAUFFMAN, 1978). This allows "absolute" dating of the time and duration of biostratigraphic zones, and provides data for the measurement of evolutionary rates in various lineages.

Fifth, historically, the Cretaceous period has been intensely studied in regard to its fossil content and biostratigraphic zonation. Early biostratigraphic work was especially focused on ammonites and echinoids, and biostratigraphic concepts and methods have had a relatively long period of testing. As a result, Cretaceous systems of zonation and correlation have evolved rapidly to levels beyond those for many other periods of geological time.

Sixth, many Cretaceous organisms showed exceptionally rapid biogeographic dispersal (and thus regional biostratigraphic utility). Three environmental factors contributed to this: a) long periods of climatic amelioration with warm, maritime climates spread over most of the globe, low temperature gradients, and few barriers to dispersal of temperate to tropical invertebrates, which dominate Cretaceous marine biostratigraphic systems; b) broad and continuous areas of shallow marine shelf environments within single climatic zones, especially during each of the Cretaceous epicontinental transgressions (Fig. 2), reducing barriers to rapid, widespread dispersal of shelf organisms; and c) short enough distances across the opening Atlantic as not to exceed, for much of the Cretaceous, the swimming

Stage	Substage	Age Ma **	Age Ma *	TETHYAN MACROFOSSIL STANDARD (1)		NORTH TEMPERATE MACROFOSSIL STANDARD (1)	PLANKTONIC FORAM.- CALPIONELLID STANDARD (1)
ALBIAN	U	100	93.5				Rotalipora appenninica
			94	Stoliczkaia dispar		Stoliczkaia dispar	Rotalipora ticinensis
			96				Ticinella (Biticinella) breggiensis
			97	Mortoniceras inflatum		Mortoniceras inflatum	
			98	Diploceras cristatum		Diploceras cristatum	
		104	98.8	Hoplites lautus,		Hoplites lautus	Ticinella praeticinensis
	M			Hoplites nitidus		Hoplites loricatus	
				Hoplites dentatus		Hoplites dentatus	Ticinella bejaouaensis, Ticinella primula, Globigerinelloides gyroidinaeformis
		106					
	L			Douvilleiceras mammilatum		Douvilleiceras mammilatum	Ticinella bejaouaensis,
				Leymeriella tardefurcata		Leymeriella regularis, Leymeriella tardefurcata, Leymeriella schrammeni	Globigerinelloides gyroidinaeformis
		108					
APTIAN	U			Diodochoceras nodosocostatum		Hypacanthoplites jacobi	Globigerinelloides ferreolensis, Ticinella bejaouaensis
		110		Cheloniceras subnodocostatum		Parahoplites melchioris, Parahoplites nutfieldensis	Globigerinelloides ferreolensis, Hedbergella trocoidea
	M					Cheloniceras martinoides	Globigerinelloides algerianus
				Aconeceras nisus		Tropaeum bowerbanki	Schackoina cabri
		112					
	L			Deshayesites deshayesi, (Puzosia matheroni)		Deshayesites deshayesi	Globigerinelloides blowi
						Deshayesites forbesi	
		115				Praedeshayesites fissicostatus	
BARREMIAN	U			Silestes seranonis	Pulchella provincialis	Parancyloceras bidentatum	Hedbergella sigali
		117				Simancyloceras stolleyi	
						Ancyloceras innexum	
	M			"Nicklesia" pulchella	Pulchella caicedi	Paracrioceras denckmanni	
		119				Paracrioceras elegans	Hedbergella sp. aff. H. simplex
					Pulchella didayi	Hoplocrioceras fissicostatum	
	L				Pulchella pulchella	Hoplocrioceras rarocinctum	
		121					

FIG. 3D. *(Explanation on page A442.)*

ranges of many mobile invertebrates such as ammonites, nor the drift potential of planktonic organisms and long-lived planktotrophic larvae of many warm-water benthonic taxa (KAUFFMAN, 1975). Thus both benthonic and planktonic taxa dispersed readily to both sides of the Atlantic on Cretaceous ocean currents, and from there throughout the epicontinental seas of the Eurasian-African and the North American-South American continental systems. Oceanic separations of Africa and Eurasia, North and South America, and elements of Gondwanaland were even less than that

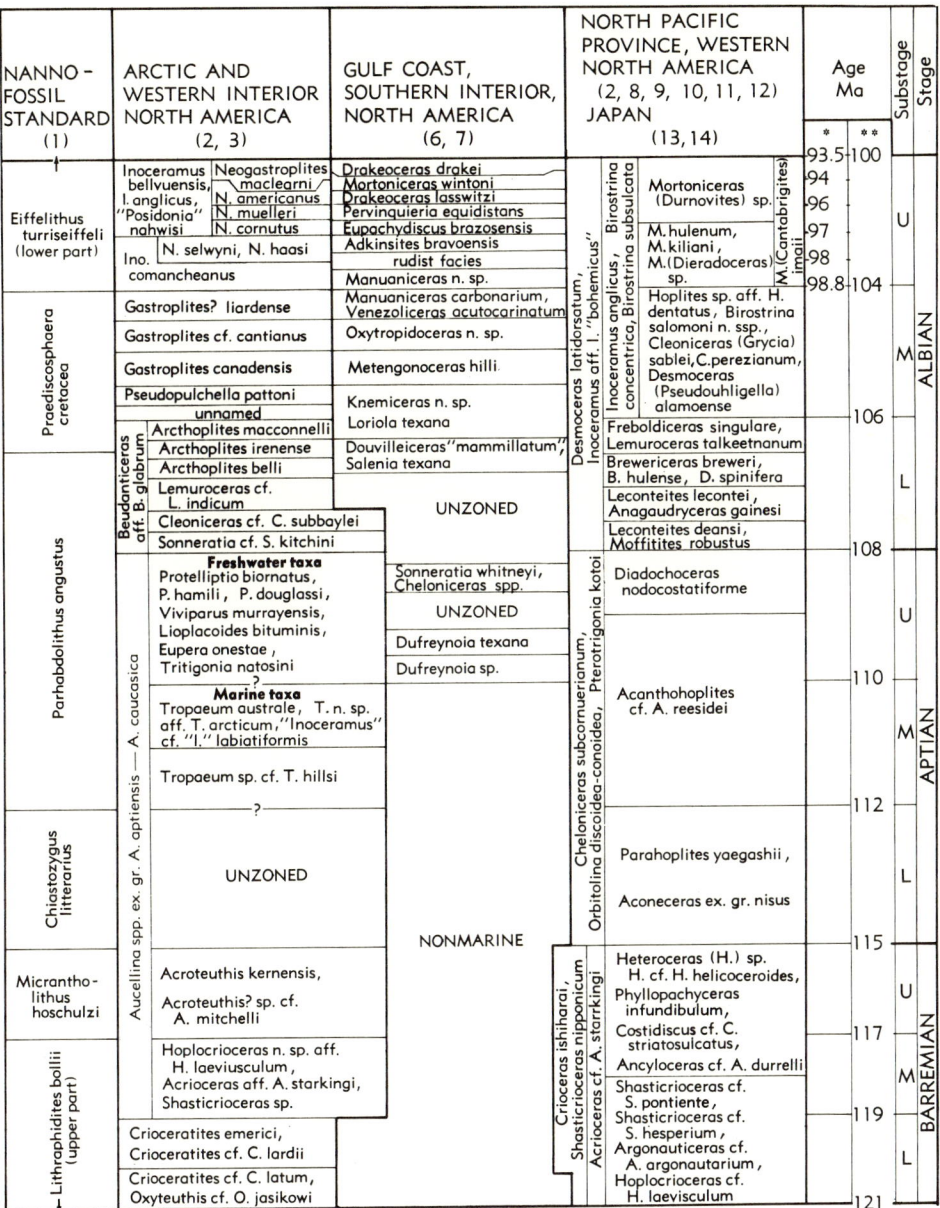

Fig. 3E. *(Explanation on page A442.)*

across the Atlantic. Therefore, precise biostratigraphic correlation over broad Cretaceous marine areas is to be expected.

Seventh, Cretaceous invertebrate taxa are related to living forms closely enough that the biostratigraphic utility of many fossil groups can be predicted from knowledge of their modern biogeographic, ecologic, and genetic characteristics. This leads to increased efficiency in the construction of complex biostratigraphic systems (KAUFFMAN, 1970, 1977a).

Eighth, evolutionary rates among Cretaceous invertebrates, especially ammonoid

Stage	Substage	Age Ma	ENGLAND (16, 17) FRANCE (4, 5)		MADAGASCAR (19)		RUSSIAN PLATFORM (20) SIBERIA (20) NORTHEAST ASIA (20)	
ALBIAN	U	100 -93.5 94 96 97 98 104 -98.8	Stoliczkaia dispar	Stoliczkaia dispar	Neophlycticeras madagascariensis		Stoliczkaia spp.	Inoceramus serotinus I. anglicus elongatus
				Mortoniceras perinflatum	Pervinquieria pachys, P. inflata,			
				Arraphoceras substuderi	Stoliczkaia grandidieri, S.clavigera			
			Mortoniceras (Mortoniceras) inflatum	Cleoniceras auritus	Stoliczkaia notha, S. crassa		I. anglicus s.s. A. gryphaeoides	
				Hysteroceras varicosum	Hysteroceras binum, Lechites gaudini	H. varicosum	I. anglicus conjugalis	
				Hysteroceras orbignyi		H. orbignyi		
				Dipoloceras cristatum		D. cristatum		
	M		Euhoplites lautus	Anahoplites daviesi	Oxytropidoceras spp., Manuaniceras manuanense		Inoceramus kedrovensis	Gastroplites sp., Cleoniceras cf. C. mangyschlakense Neogastroplites, Anagaudryceras madraspatanum
				Euhoplites nitidus				
				Euhoplites meandrinus				
			Euhoplites loricatus	Mojsisovicsia subdelaruei			Hoplites dentatus, Inoceramus comancheanus	
				Dimorphoplites niobe				
				Anahoplites intermedius				
			Hoplites (Hoplites) dentatus	Hoplites (Hoplites) spathi				
		106		Lyelliceras lyelli				
				H.(Isohoplites) eodentatus				
	L		Douvilleiceras mammillatum	Protohoplites puzosianus	Douville- iceras mammil- latum	Beudanticeras revoili,	Aucellina kamtschatica,	Brewericeras hulense, Eogaudryiceras shimizui
				Otohoplites raulinianus		Lemuroceras spathi		
				Cleoniceras floridum		Cleoniceras besairiei		
				Sonneratia kitchini			Aucellina cf. A. antulai,	
			Leymeriella tardefurcata	Leymeriella regularis	Pseudosonneratia sakalava		Aucellina cf. A. pompeckji	Aucellina aptiensis Aucellina caucasica
				Hypacanthoplites milletioides				
		108		Farnhamia farnhamensis	Leymeriella sp.			
APTIAN	U		Hypacantho- plites jacobi	Hypacanthoplites anglicus	Acanthoplites nolani		UNZONED	Gabbioceras wintuninum, Cheloniceras spp.
				Hypacanthoplites rubricosus				
				Nolaniceras nolani				
			Parahoplites nutfieldensis	Parahoplites cunningtoni				
		110		Tropaeum subarcticum				
	M		Cheloniceras martinoides	C. (Epicheloniceras) buxtorfi	Aconeceras nisus		? Cheloniceras tschernyschewi	
				C. (Epicheloniceras) gracile				
				C. (Epicheloniceras) debile				
			Tropaeum bowerbanki	Cheloniceras (Cheloniceras) meyendorfi				
		112		Dufreynoyia transitoria				
	L		Deshayesites deshayesi	Deshayesites grandis	Tropaeum jacki		Deshayesites deshayesi	
				C. (Cheloniceras) parinodum				
			Deshayesites forbesi	Deshayesites callidiscus				
				Deshayesites kiliani				
				Deshayesites fittoni				
			Prodeshayes- ites fissicostata	Prodeshayesites obsoletus				
		115		Prodeshayesites bodei				
BARREMIAN	U	117	Andouliceras kaliae, A. collignoni, Silesites seranonis, Emericeras barremense alpini, Ancyloceras vandenheckii		Paracrioceras spp.	SOUTHERN FRANCE	Shasticrioceras spp., Inoceramus colonicus	regionally absent
	M		Nicklesia pulchella,	Acrioceras anglesensis,				
		119		Acrioceras ramkrishnai				
	L	121	Acrioceras tabarelli	Emericeras tenuicostatum, Emericeras emerici, Pulchellia compressima				

FIG. 3F. (*Explanation on page A442.*)

and bivalve mollusks (KAUFFMAN, 1970, 1972, 1977a), are among the fastest yet recorded for marine Metazoa, attaining levels as high as one new species per .08 my within single lineages, with species durations as low as 0.12 my. This leads to zonal refinement and high resolution in correlation. These rates, measured radiometrically, were attained mainly during periods when the ecosystem was stressed either by rapid temperature and salinity changes, or by widespread regression of epicontinental seas. For many groups such as ammonites and inoceramid bivalves, rates of evolution were high even during times of climatic amelioration (transgressions) with radiation into newly developing habitats. The tremendous fluctuations in environmental and biological stress levels in the marine realm that are inherent in the environmental cycles of the Cretaceous, thus enhanced evolutionary turnover among many biostratigraphically important groups of organisms.

Despite present efforts to construct refined composite assemblage zonations for the Cretaceous throughout the world, most zonal systems in operation at this time are either based on single groups or on simple composite range-zone systems. Thus, open oceanic Cretaceous biostratigraphic systems are primarily based on planktonic foraminifers, coccolithophorids, and in cooler waters on radiolarians. Temperate-zone epicontinental systems rely heavily on ammonites, inoceramid and buchiid bivalves, belemnites, echinoids and oysters. Tethyan and marginal Tethyan systems also rely heavily on ammonites (generally distinct groups from those found in temperate zones), planktonic foraminifers, coccoliths, echinoids, and locally belemnites, rudistid, inoceramid, pectinid, trigoniid, and ostreid bivalves, and nerineid and actaeonellid gastropods.

It is commonly difficult to relate Tethyan and temperate biostratigraphic schemes in detail because they shared few organisms during much of the Cretaceous. Between the Caribbean Tethys and the North American temperate seas, for example, there was periodically less than 10 percent species similarity; however, three factors make correlation of these systems possible. First, among taxa shared between Tethyan and temperate ecosystems, especially near their juncture, are some of the most important invertebrates used in regional correlation: inoceramid and ostreid bivalves, certain echinoids, elements of the planktonic microbiota, and rarely ammonites. Both Tethyan and temperate-zone faunas show abrupt decrease in numbers and diversity where they infringe on adjacent realms. Second, basins that lie near or on the Tethyan-temperate boundary zone, for example, the Vascogotic trough of Spain (WIEDMANN & KAUFFMAN, 1977), commonly have laterally and vertically mixed cooler temperate and warmer Tethyan aquatic biotas. In these situations, especially where there is vertical stratification, the biotas are intricately mixed at many stratigraphic levels and interregional correlations can be precisely made. Third, eustatic pulses of sea level, which produced alternating transgressive and regressive events at least 10 times during the Cretaceous (Fig. 2; KAUFFMAN, 1973a,b), also resulted in north-south migration of the temperate-Tethyan marine boundary (COATES, KAUFFMAN, & SOHL, 1977), so that in marginal areas Tethyan and temperate biotas may retain their uniqueness but be extensively interbedded with each other, changing at points of climatic boundary migrations. These areas of interfingering of Tethyan and temperate faunas are critical to global correlation of the Cretaceous. Figures 3 and 4 show principal biostratigraphic indices for the world Cretaceous as they are currently known.

LOWER CRETACEOUS BIOSTRATIGRAPHY

In temperate and Tethyan realms, ammonites constitute the principal basis for Lower Cretaceous biostratigraphy (Fig. 3), and they are more preeminent during Early than Late Cretaceous time, when other invertebrate groups such as planktonic foraminifers, inoceramid bivalves, and belemnites had evolved to levels that allowed them to share equally in biostratigraphic zonation and regional correlation. Belemnites, which are a mainstay of temperate to cool-temperate Upper Cretaceous biostratigraphy, are of lesser importance for regional correlation of the Lower Cretaceous. They were just beginning a second major radiation from local centers of endemism (CASEY & RAWSON, 1973b, p. 420, 421), and were subjected to shifting biogeographic patterns (STEVENS, 1973). They evolved very rapidly during the Early Cretaceous, however, and certain groups such as *Neohibolites* (SPAETH, 1971) and *Aulacoteuthis* (RAWSON, 1972) allow local zonation comparable to that of the ammonites.

Bivalves have a secondary role in Lower Cretaceous biostratigraphy, except for certain groups. The Buchiidae are of primary importance in regional and intercontinental

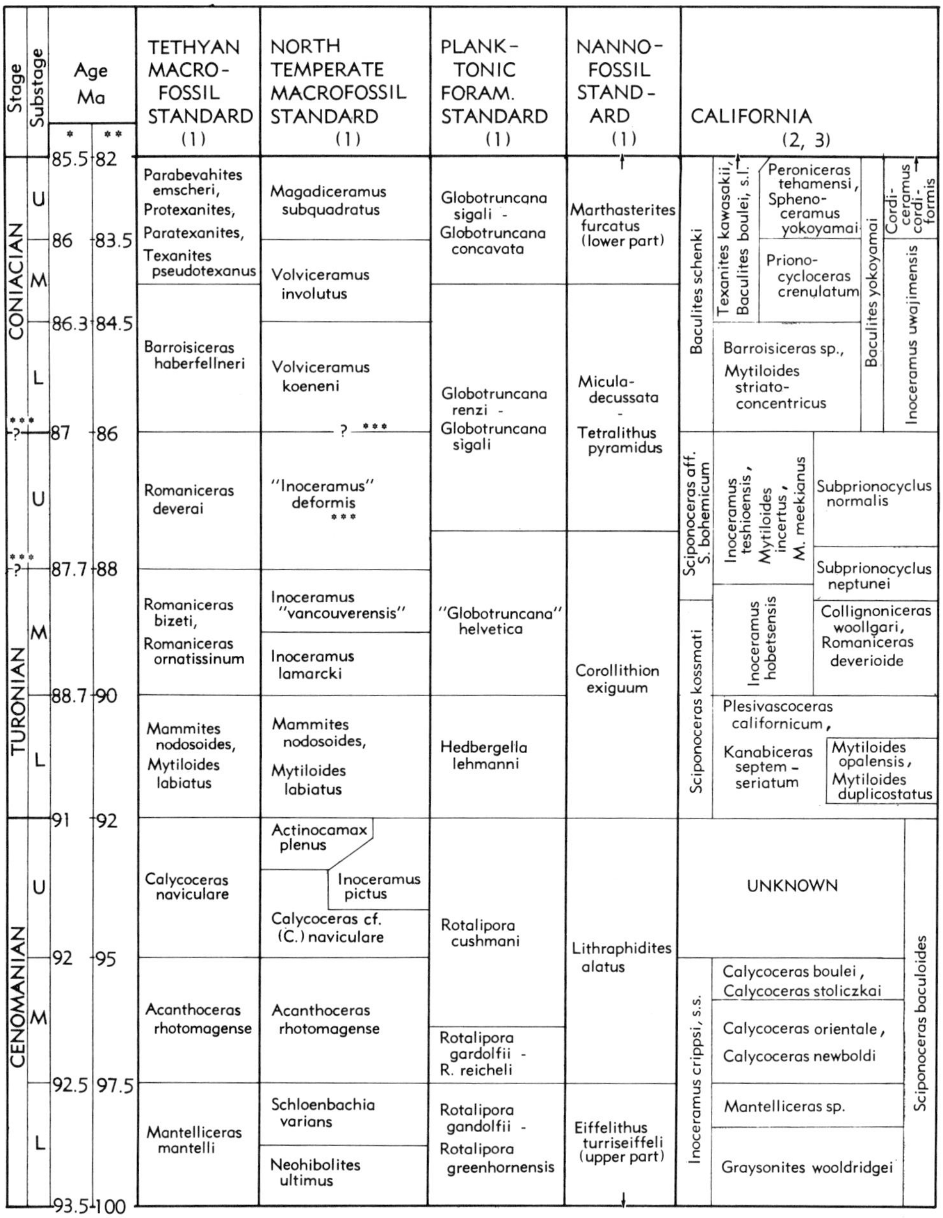

Stage	Substage	Age Ma *	Age Ma **	TETHYAN MACRO-FOSSIL STANDARD (1)	NORTH TEMPERATE MACROFOSSIL STANDARD (1)	PLANKTONIC FORAM. STANDARD (1)	NANNOFOSSIL STANDARD (1)	CALIFORNIA (2, 3)			
CONIACIAN	U	85.5	82	Parabevahites emscheri, Protexanites, Paratexanites, Texanites pseudotexanus	Magadiceramus subquadratus	Globotruncana sigali - Globotruncana concavata	Marthasterites furcatus (lower part)	Baculites schenki / Texanites kowasakii, Baculites boulei, s.l.	Peroniceras tehamensi, Sphenoceramus yokoyamai	Baculites yokoyamai / Prionocycloceras crenulatum	Cordiceramus cordiformis
	M	86	83.5		Volviceramus involutus						Inoceramus uwajimensis
	L	86.3	84.5	Barroisiceras haberfellneri	Volviceramus koeneni	Globotruncana renzi - Globotruncana sigali	Micula decussata		Barroisiceras sp., Mytiloides striato-concentricus		
?		87	86		?		Tetralithus pyramidus	Sciponoceras aff. S. bohemicum	Inoceramus teshioensis, Mytiloides incertus, M. meekianus	Subprionocyclus normalis	
TURONIAN	U			Romaniceras deverai	"Inoceramus" deformis						
?		87.7	88							Subprionocyclus neptunei	
	M			Romaniceras bizeti, Romaniceras ornatissinum	Inoceramus "vancouverensis"	"Globotruncana" helvetica	Corollithion exiguum	Sciponoceras kossmati	Inoceramus hobetsensis	Collignoniceras woollgari, Romaniceras deverioide	
		88.7	90		Inoceramus lamarcki						
	L			Mammites nodosoides, Mytiloides labiatus	Mammites nodosoides, Mytiloides labiatus	Hedbergella lehmanni			Plesivascoceras californicum, Kanabiceras septemseriatum	Mytiloides opalensis, Mytiloides duplicostatus	
CENOMANIAN	U	91	92	Calycoceras naviculare	Actinocamax plenus / Inoceramus pictus	Rotalipora cushmani	Lithraphidites alatus		UNKNOWN		Sciponoceras baculoides
		92	95		Calycoceras cf. (C.) naviculare				Calycoceras boulei, Calycoceras stoliczkai		
	M			Acanthoceras rhotomagense	Acanthoceras rhotomagense	Rotalipora gardolfii - R. reicheli		Inoceramus crippsi, s.s.	Calycoceras orientale, Calycoceras newboldi		
		92.5	97.5		Schloenbachia varians	Rotalipora gardolfii - Rotalipora greenhornensis	Eiffelithus turriseiffeli (upper part)		Mantelliceras sp.		
	L			Mantelliceras mantelli	Neohibolites ultimus				Graysonites wooldridgei		
		93.5	100								

FIG. 4A.

FIG. 4A–F. Upper Cretaceous biostratigraphic zonation and principal zonal indices in selected areas where zonation is most refined. Principal references for data coded to numbers in parentheses are: 1) Van Hinte, 1976, Tethyan and North Temperate macrofossil and pelagic microfossil zones; 2) Kauffman, 1977e, inoceramid bivalve zones, Japan; 2) Kauffman, 1977e, Inoceramidae, California; 3) Matsumoto, 1960, Upper Cretaceous ammonites, California; 4) Cobban, 1951, Scaphitidae, Western Interior, North America; 5) Cobban, 1953, Late Cenomanian ammonites, Montana; 6) Cobban, 1958, Upper Cretaceous zonation, Powder River basin; 7) Cobban, 1964, *Haresiceras*; 8) Cobban, 1969, *Scaphites hippocrepis–S. leei* lineage; 9) Cobban & Scott, 1972, Turonian ammonites, central Colorado; 10) Cobban & Scott, 1964, multinodose

WESTERN INTERIOR OF NORTH AMERICA (4, 5, 6, 7, 8, 9, 10, 11, 12, 13, 14, 15)			WESTERN GULF COAST, U. S. A. (16, 17, 18, 19, 20, 21)			Age Ma		Substage	Stage
						82	85.5		
Scaphites ventricosus, Peroniceras sp., Baculites asper	Magadiceramus subquadratus, "Inoceramus" n. sp. aff. "Inoceramus" stantoni	Volviceramus involutus	Prionocycloceras gabrielense, Protexanites planatus, Paratexanites sellardsi, Magadiceramus subquadratus					U	CONIACIAN
						83.5	86		
Forresteria forresteri, Scaphites impendicostatus, Baculites mariasensis	Scaphites preventricosus	Cremnoceramus wandereri, C. inconstans (late form)	Peroniceras westphalicum, P. moureti, Prionocycloceras adkinsi, Volviceramus involutus, Cremnoceramus inconstans, etc.					M	
		Inoceramus schloenbachi							
		Inoceramus browni				84.5	86.3		
Forresteria hobsoni, Scaphites mariasensis, Scaphites frontierensis		Inoceramus deformis n. subsp. (late)	Peroniceras haasi, Coilopoceras austinense	Inoceramus deformis				L	
		Inoceramus deformis deformis							
		I. erectus n. subsp. (late form)		Inoceramus erectus					
		Inoceramus erectus erectus							
		Inoceramus rotundatus		Inoceramus rotundatus					
		Mytiloides fiegei, M. lusatiae		Mytiloides fiegei		86	87	***	
Prionocyclus quadratus, Prionocyclus reesidei, Scaphites corvensis			UNKNOWN					?	
Scaphites nigricollensis, I. perplexus (late form)		Prionocyclus wyomingensis elegans	Scaphites whitfieldi, Inoceramus perplexus					U	TURONIAN
Scaphites whitfieldi, I. perplexus s.s.									
Scaphites ferronensis	Prionocyclus wyomingensis, s.s., Inoceramus dimidius n. subsp., Lopha lugubris n. subsp. A		Prionocyclus wyomingensis	Lopha lugubris, Inoceramus dimidius					
Scaphites warreni									
Coilopoceras colleti, Inoceramus dimidius s.s., Lopha lugubris			UNZONED			88	87.7	***	
Coilopoceras springeri, Lopha bellaplicata, s. s.,	Inoceramus howelli, Lopha cunabula	Prionocyclus hyatti	Prionocyclus hyatti, Scaphites arcadiensis, Lopha bellaplicata s.s., I. costellatus, Inoceramus apicalis, Inoceramus howelli					?	
Scaphites carlilensis, Lopha b. novamexicana	Inoceramus securiformis							M	
	Inoceramus flaccidus		Inoceramus flaccidus						
Collignoniceras woollgari, I. cuvieri, S. larvaeformis, Mytiloides "latus"	Inoceramus aff. I. flaccidus		Inoceramus cuvieri, Ostrea bentonensis		M. "latus", Collignon. woollgari				
	Mytiloides hercynicus								
Mammites nodosoides, Choffaticeras pavillieri	Mytiloides labiatus, M. subhercynicus		M. labiatus; M. subhercynicus			90	88.7		
	Mytiloides mytiloides		Mytiloides mytiloides					L	
Watinoceras coloradoense, Neoptychites xetriformis	Mytiloides aff. M. duplicostatus		UNKNOWN						
Watinoceras reesidei, Plesiovascoceras thomi	Mytiloides opalensis								
	Mytiloides submytiloides		Mytiloides submytiloides			92	91		
Sciponoceras gracile, Worthoceras vermiculum, Calycoceras naviculare, Kanabiceras septemseriatum, Metoicoceras whitei, Inoceramus tenuiumbonatus			Worthoceras gibbosum, K. septemseriatum Sciponoceras gracile, I. pictus, M. whitei						
		Inoceramus pictus, s.s.,							
Dunveganoceras albertense, D. conditum, Metoicoceras defordi, M. muelleri, Inoceramus ginterensis, I. flavus			Inoceramus aff. Inoceramus pennatulus, Inoceramus pictus subsp. aff. I. pictus etheridgei	Inoceramus pictus s.l. (flat broad form), Acanthoceras spp.				U	CENOMANIAN
Dunveganoceras pondi, Calycoceras canitaurinum, Inoceramus prefragilis, Inoceramus macconnelli?						95	92		
Plesiacanthoceras wyomingense, Borissiokoceras orbiculatum			UNZONED						
Acanthoceras amphibolum, Inoceramus arvanus (late form)									
Acanthoceras alvaradoense, Tarrantoceras rotatile, I. arvanus			Acanthoceras alvaradoense, Tarrantoceras rotatile, Ostrea beloiti					M	
Turrilites scheuchzerianus, Acanthoceras muldoonense									
Acanthoceras bellense, Inoceramus crippsi, A. granerosense			Calycoceras leonense	I. prefragilis					
Calycoceras tarrantense, Turrilites acutus, C. gilberti			C. tarrantense, Epengonoceras dumbli, I. arvanus			97.5	92.5		
Inoceramus bellvuensis, n. subsp.								L	
"Inoceramus" dunveganensis, s.s., "I." dunveganensis n. subsp.			"Inoceramus" eulessanus						
"Inoceramus" athabaskensis (late form)									
						100	93.5		

Fig. 4B. *(Explanation continued from facing page.)*

scaphitids; 11) Gill & Cobban, 1966, Red Bird zonal sequence, Wyoming; 12) Kauffman, 1975, Western Interior bivalve zones; 13) Kauffman, Cobban, & Eicher, 1977, Cenomanian-Coniacian zonation; 14) Scott & Cobban, 1964, Coniacian-Lower Campanian zonation, central Colorado; 15) Jeletzky, 1970, Western Interior of Canada; 16) Young, 1963, Coniacian-Campanian ammonite zones, Gulf Coast; 17) Sohl, 1960, personal com., 1977, *Exogyra* zones and regional correlations; 18) E. G. Kauffman, unpub. research, Inoceramidae; 19) J. D. Powell, unpub. research, Eagle Ford Formation (Cenomanian-Turonian), Gulf Coast; 20) Stephenson, 1952, Cenomanian mollusks, Woodbine Formation, Texas; 21) Stephenson, 1941, Campanian-Maastrichtian mollusks, Navarro Group, Texas; 22-26) Seitz, 1959, 1961, 1965, 1967,

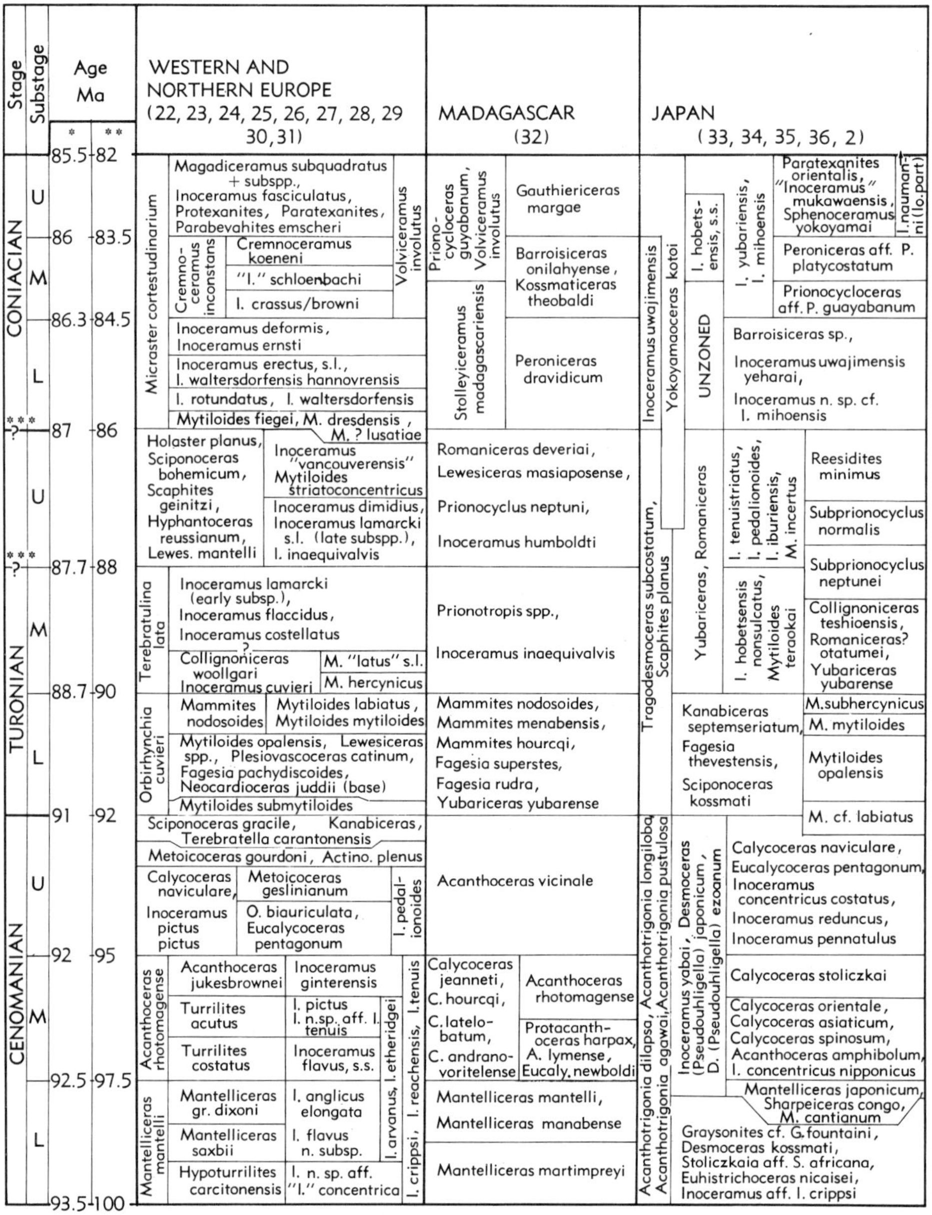

FIG. 4C. (*Explanation continued from page A451.*)
1970, Coniacian-Maastrichtian Inoceramidae, north Germany; 27) Tröger, 1967, Cenomanian-Coniacian Inoceramidae, north Germany; 28) Cox, 1967, classical British-French chalk zonation; 29) Kennedy, 1969, Late Albian-Cenomanian ammonite zonation; 30) Kauffman, 1977d, British Albian-Coniacian inoceramid zonation; 31) Kennedy & Juignet, 1973, Cenomanian-Turonian boundary-zone sequence, France; 32) Besairie & Collignon, 1972, Madagascar; 33) Matsumoto, 1959a, b; 34) Matsumoto, 1963, and 35) Noda & Matsumoto, 1976, ammonite-inoceramid bivalve zones, Japan; and 36) Nakano, 1960, trigoniid bivalves, Japan. For explanation of single and double asterisks, see Figure 2 caption. Triple asterisks designate the major area of disagreement concerning definition of the Turonian-Coniacian bound-

correlations in temperate realms, especially where ammonites are rare or absent. *Buchia* is particularly useful in the Lower Neocomian of the north-temperate Euramerican region (for example, IMLAY, 1959; JONES, 1969; JELETZKY, 1964a,b, 1965; and papers cited in CASEY & RAWSON, 1973a). Because of slower evolutionary rates, they do not lead to zonation as refined as that based on ammonites; usually only 2 to 3 *Buchia* zones are recognized within each Lower Cretaceous stage in which they occur abundantly. JELETZKY (1965) discussed the limitations of *Buchia* in biostratigraphy. The Inoceramidae, biostratigraphically the most important Upper Cretaceous family of bivalves, has much more restricted application in the Lower Cretaceous. Nevertheless, whereas Buchiidae become less important above the Middle Neocomian, the Inoceramidae increase in abundance, diversity, and biogeographic spread, and become accessory biostratigraphic indices to ammonites in the Hauterivian and Barremian stages. The Inoceramidae play a major role in Albian biostratigraphy, with increasing rates of evolution and biogeographic spread almost equal to that of the ammonites (KAUFFMAN, 1977d). The Albian *Birostrina coptensis-B. salomoni* lineage, *B. concentrica-B. sulcata* lineage, *Inoceramus anglicus* lineage, and the unusual endemic groups of North American inoceramids, the "*Inoceramus*" (n. gen.) *nahwisi* and "*I.*" (n. gen.) *dunveganensis* lineages are especially important (JELETZKY, 1964a). Lower Cretaceous Trigoniidae (for example, Japan; NAKANO, 1960), Pectinidae and Ostreidae (for example, Gulf Coast of the United States; STANTON, 1947) are bivalve groups that are locally important in Lower Cretaceous biostratigraphy.

Among the microbiota, Lower Cretaceous biostratigraphy is highly dependent upon planktonic foraminifers and calcareous nannofossils. Calpionellidae are important in Lower Neocomian zonation. Radiolarians and benthonic foraminifers play a minor accessory role. VAN HINTE (1976) summarized the global oceanic zonation based on microfossils (Fig. 3), a zonation that has largely arisen in the last decade through the Deep Sea Drilling Program. Lower Cretaceous microfossil zones tend to be longer in duration by 25 to 50 percent than their Upper Cretaceous counterparts, with the exception of those based on Aptian-Albian planktonic foraminifers. From Berriasian to Middle Aptian time, biostratigraphic zonation based on the microbiota is 2 to 8 times broader than that based on ammonites for the same period; beyond that point they have similar zonal durations. Calcareous nannofossil zonation is up to 4 times coarser, and radiolarian zonation up to 12 times coarser than that based on planktonic foraminifers. Biostratigraphic zonation of the Lower Cretaceous based on macrofossils is compared to that based on microfossils in Figure 3.

BIOSTRATIGRAPHICALLY IMPORTANT LOWER CRETACEOUS GROUPS

Berriasian biostratigraphy is primarily based on ammonites of the genus *Berriasella* in the marine Tethyan and temperate realms. Ammonites of the genera *Surites, Peregrinoceras, Hectoroceras, Runctonia, Chetaites, Spiticeras,* and *Raisanites* are important in the Eurasian region of the North Temperate Realm (Fig. 3). WIEDMANN (1973) noted the importance of heteromorph ammonites *(Leptoceras, Protancyloceras, Bochianites)* in Berriasian Tethyan faunas. Species of the bivalve *Buchia* are biostratigraphically important in North America (JELETZKY, 1970) and to a lesser extent in northern Europe. In the pelagic realm, species of *Calpionella,* the radiolarian assemblage of the *Sphaerostylus lanceola* Zone, and the nannofossil *Nannoconus* characterize the Berriasian. Fresh water ostracodes, in particular *Cytherelloidea* and *Cytheridea,* are mainly used in nonmarine "Wealden" facies (CASEY & RAWSON, 1973b). STEVENS (1973) noted that the

---

FIG. 4A-F. *(Continued from facing page.)*
ary; Europeans using Inoceramidae place the boundary above *Inoceramus deformis,* whereas ammonite specialists and American biostratigraphers place the boundary one-third of a stage lower, at the base of the *I. rotundatus-Mytiloides fiegei* Zone (Kauffman, n).

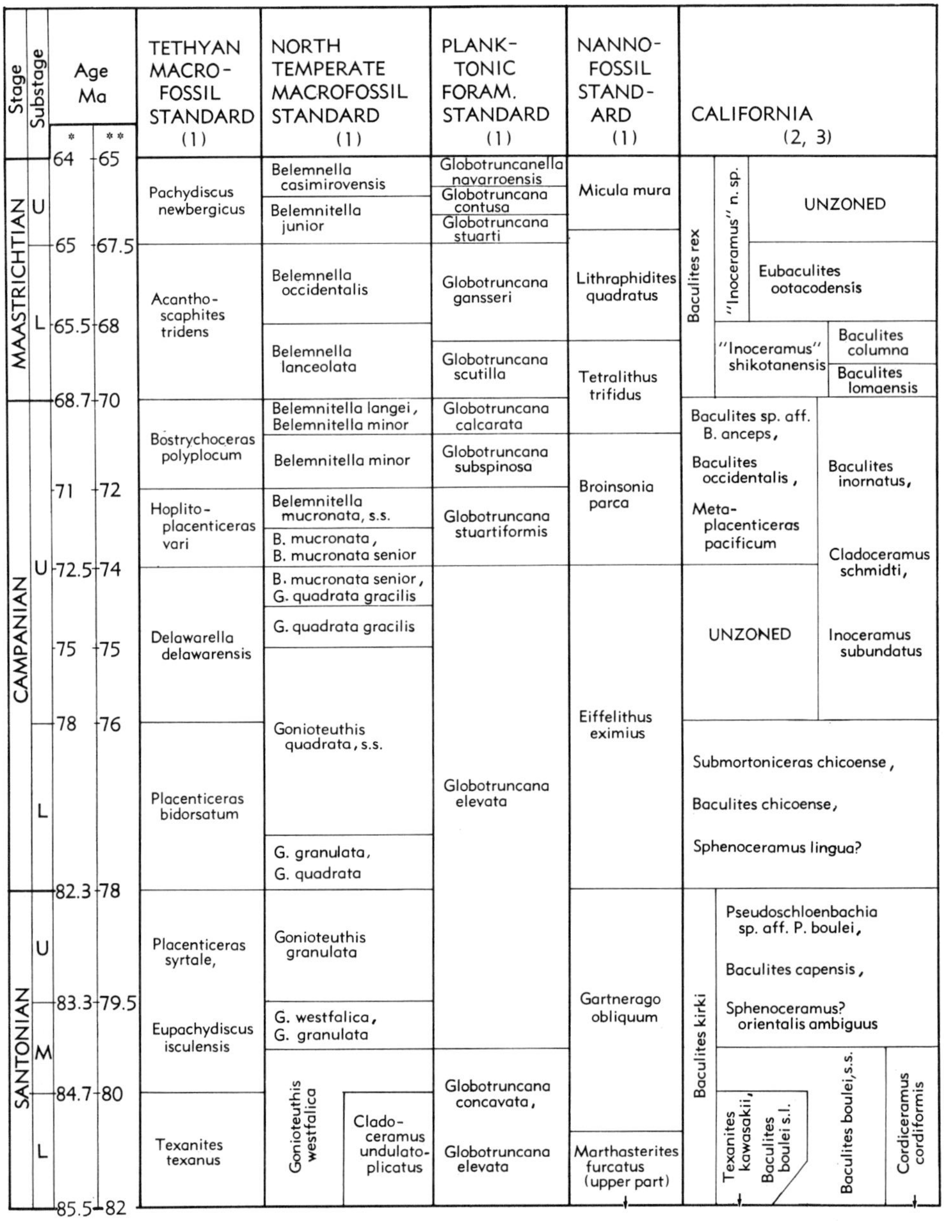

Stage	Substage	Age Ma *	Age Ma **	TETHYAN MACRO-FOSSIL STANDARD (1)	NORTH TEMPERATE MACROFOSSIL STANDARD (1)	PLANK-TONIC FORAM. STANDARD (1)	NANNO-FOSSIL STANDARD (1)	CALIFORNIA (2, 3)			
MAASTRICHTIAN	U	64	65	Pachydiscus newbergicus	Belemnella casimirovensis	Globotruncanella navarroensis	Micula mura	Baculites rex	"Inoceramus" n. sp.	UNZONED	
		65	67.5		Belemnitella junior	Globotruncana contusa					
						Globotruncana stuarti					
	L	65.5	68	Acantho-scaphites tridens	Belemnella occidentalis	Globotruncana gansseri	Lithraphidites quadratus			Eubaculites ootacodensis	
									"Inoceramus" shikotanensis	Baculites columna	
					Belemnella lanceolata	Globotruncana scutilla	Tetralithus trifidus			Baculites lomaensis	
		68.7	70		Belemnitella langei, Belemnitella minor	Globotruncana calcarata		Baculites sp. aff. B. anceps,			
CAMPANIAN	U	71	72	Bostrychoceras polyplocum	Belemnitella minor	Globotruncana subspinosa	Broinsonia parca	Baculites occidentalis,		Baculites inornatus,	
				Hoplito-placenticeras vari	Belemnitella mucronata, s.s. B. mucronata, B. mucronata senior	Globotruncana stuartiformis		Meta-placenticeras pacificum			
		72.5	74		B. mucronata senior, G. quadrata gracilis					Cladoceramus schmidti,	
		75	75	Delawarella delawarensis	G. quadrata gracilis			UNZONED		Inoceramus subundatus	
		78	76		Gonioteuthis quadrata, s.s.		Eiffelithus eximius				
	L			Placenticeras bidorsatum		Globotruncana elevata		Submortoniceras chicoense,			
								Baculites chicoense,			
					G. granulata, G. quadrata			Sphenoceramus lingua?			
SANTONIAN	U	82.3	78	Placenticeras syrtale,	Gonioteuthis granulata		Gartnerago obliquum	Pseudoschloenbachia sp. aff. P. boulei,			
		83.3	79.5					Baculites capensis,			
	M			Eupachydiscus isculensis	G. westfalica, G. granulata			Sphenoceramus? orientalis ambiguus			
		84.7	80		Gonioteuthis westfalica	Clado-ceramus undulato-plicatus	Globotruncana concavata,	Baculites kirki	Texanites kawasakii,	Baculites boulei, s.s.	Cordiceramus cordiformis
	L			Texanites texanus			Globotruncana elevata	Marthasterites furcatus (upper part)	Baculites boulei s.l.		
		85.5	82								

FIG. 4D. *(Explanation on page A450.)*

Berriasian and Valanginian may be zoned by the Cylindroteuthididae in the North Temperate Realm, and by the Duvaliidae and *Hibolithes-Curtohibolites* plexus for Tethys and the Austral-South Pacific area. *Belemnopsis* may be biostratigraphically useful in the South Temperate Realm.

In temperate realms Valanginian biostratigraphy is largely based on ammonites of the families Berriasellidae, Craspeditidae, and Olcostephanidae, and on different genera of the same families in the Tethyan Realm (Fig. 3). WIEDMANN (1973) noted restriction of the heteromorphic ammonite

Cretaceous

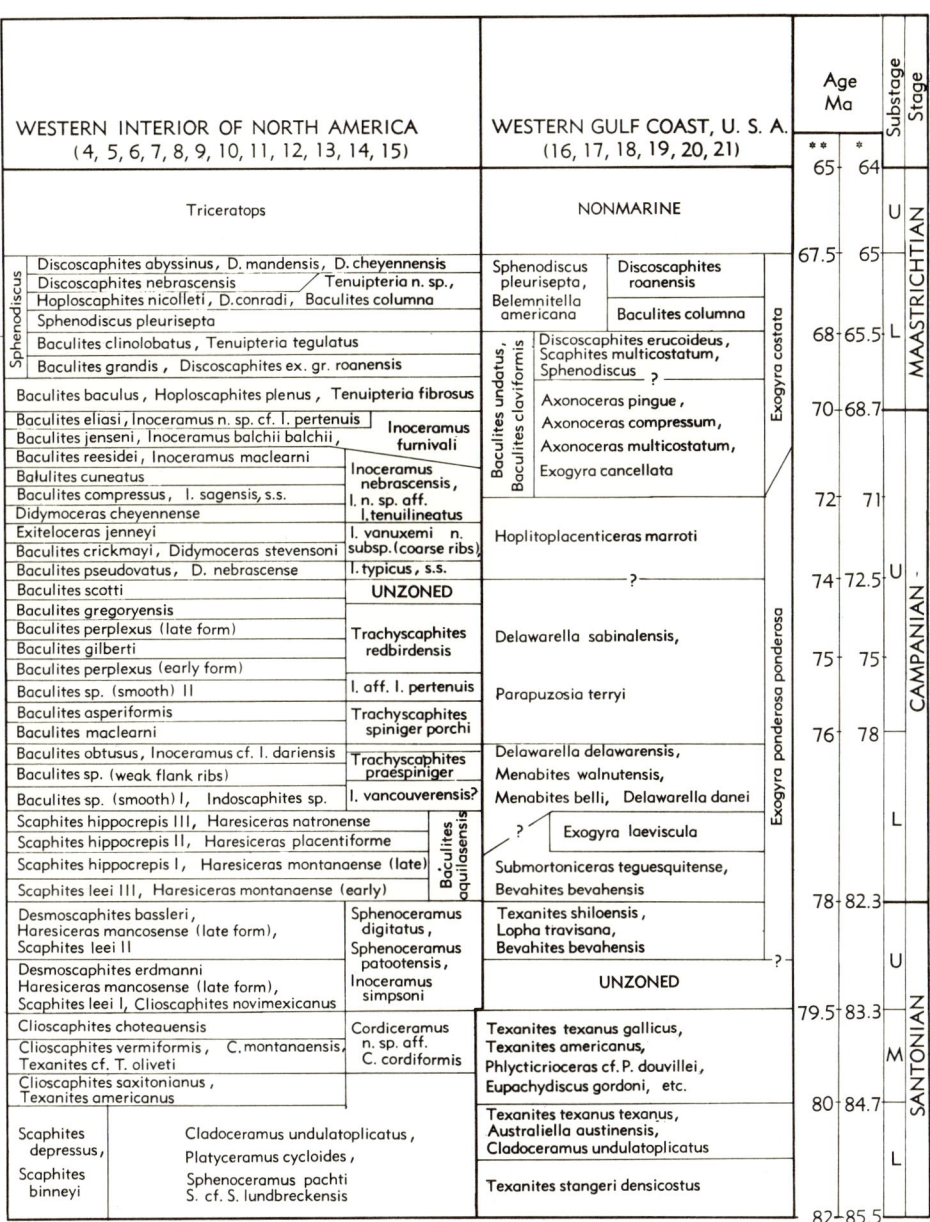

Fig. 4E. (Explanation on page A450.)

genera *Eocrioceratites* and *Parapedioceras* to the Tethyan Lower Valanginian, and *Juddiceras* to the Upper Valanginian (all Ancyloceratina). *Hibolithes*, *Duvalia*, and *Belemnopsis* are important belemnite genera in Tethys and its southern margin. Valanginian species of the bivalve *Buchia* are the main bases for zonation and correlation in temperate North America, ammonites (mainly Olcostephanidae and Craspeditidae) being less common (JELETZKY, 1970). Pelagic zonation and correlation of the Valanginian is mainly based on species of the calpionellid genera *Calpionellites* and

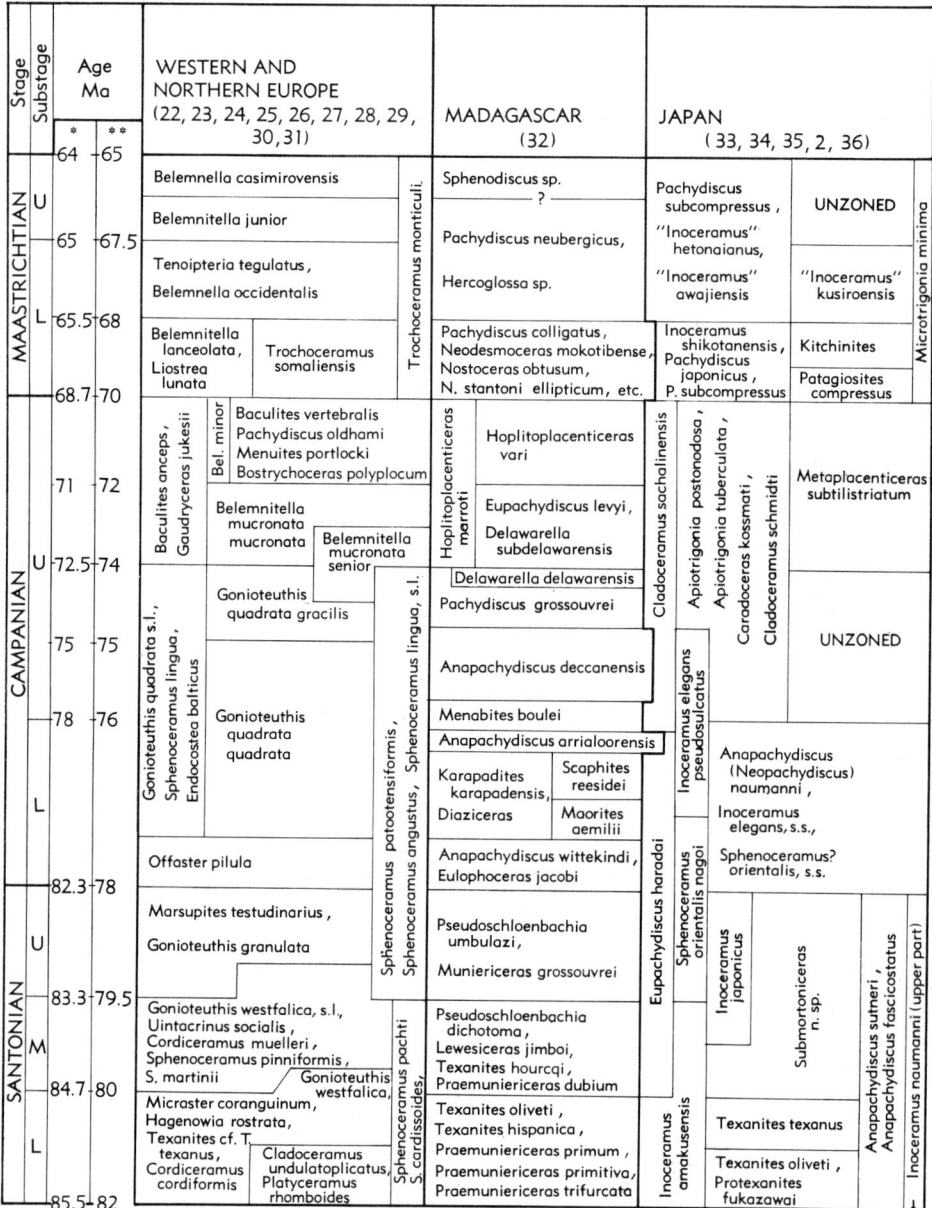

Fig. 4F. *(Explanation on page A450.)*

Calpionellopsis associated with the radiolarian fauna of the *Staurosphaera septemporata* Zone and the nannofossils *Cretarhabdus crenulatus* and *Calcicalathina oblongata* (Lower and Upper Valanginian, respectively; Fig. 3).

Hauterivian biostratigraphy draws its greatest refinement from ammonites (Fig. 3). In Tethys, the families Berriasellidae, Ancyloceratidae, Desmoceratidae, and Hemihoplitidae are of primary importance. In the temperate realms the Olcostephanidae and different genera of Ancyloceratidae are most important; species of *Simberskites* and

Endomoceras are principal zonal indices. Among belemnites, *Acroteuthis* is biostratigraphically important in temperate North America, the Cylindroteuthididae and the Oxyteuthididae throughout the North Temperate Realm, and *Hibolithes, Curtohibolites, Mesohibolites,* and the Duvaliidae in the Tethys and Austral regions. The Inoceramidae first became important as accessory index species, especially in the North Pacific Region. Among pelagic microbiota, ancestral species of *"Hedbergella,"* the last occurrence of the *Staurosphaera septemporata* radiolarian assemblage, the nannofossil assemblage of *Calcicalathina oblongata* (Lower Hauterivian), and the first occurrence of the *Lithraphidites bollii* assemblage (Upper Hauterivian) biostratigraphically characterize this stage.

Barremian biostratigraphy is as refined as that of the Valanginian and younger Cretaceous stages owing to rapid radiation among the ammonites. In Tethys, the ammonite families Pulchelliidae, Ancyloceratidae, and Silestidae are of primary importance along with the belemnites of the Duvaliidae, and the genera *Mesohibolites, Hibolithes,* and *Curtohibolites*. In the temperate realms, the Ancyloceratidae are also of principal biostratigraphic utility, based on many different genera and species from those found in Tethys. The Heteroceratidae and Lytoceratidae are of secondary utility (excepting in North America) along with belemnites of the Cylindroteuthididae and Oxyteuthididae, and the bivalves *"Inoceramus,"* s. l., and *Aucellina* (Buchiidae). In the pelagic realm, Barremian microbiotas are characterized by species of the planktonic foraminifer *Hedbergella,* by the early range of the *Stichocapsa tenuis* radiolarian assemblage, and by nannofossil assemblages of the *Lithraphidites bollii* (Lower Barremian) and *Micrantholithus hoschulzi* (Upper Barremian) zones.

The Aptian Stage is most finely zoned by ammonites, but represents the first stage where pelagic microbiotic zonation is nearly as detailed. In the Tethyan Realm, the ammonite families Douvilleiceratidae, Opeliidae, Deshayesitidae, and to a lesser extent, the Ancyloceratidae and Hoplitidae form the major bases for zonation and regional correlation. The belemnite groups Duvaliidae, *Mesohibolites,* and *Parahibolites* are of secondary importance. In the temperate realms, ammonite biostratigraphy is more refined than that of Tethys and is also centered around the Douvilleiceratidae, Deshayesitidae, and Ancyloceratidae, though based largely on different genera and species than those found in Tethys (Fig. 3; Van Hinte, 1976; Casey & Rawson, 1973b). The bivalves *Aucellina* and *"Inoceramus"* are accessory biostratigraphic indices in the Aptian. In the pelagic realm, a refined biostratigraphy based on the microbiota is centered around planktonic foraminifers belonging to the genera *Globigerinelloides, Schackoina, Ticinella,* and *Hedbergella.* Nannofossils are of lesser importance, with the *Chiastozygus litterarius* and *Parhabdolithus angustus* assemblages marking the Lower and Upper Aptian, respectively.

The Albian Stage comprises the most finely divided biostratigraphic system of the Lower Cretaceous, especially in western Europe and England (Fig. 3) where 23 ammonite subzones are recognized. This was also a time of widespread cosmopolitanism associated with the most extensive Early Cretaceous transgression; many of the same ammonites and inoceramid bivalves used in refined zonation characterize biostratigraphic zonation of the Tethyan and temperate realms. Among ammonites, Lower Albian biostratigraphy is based primarily on species of the Leymeriellidae, Douvilleiceratidae, Brancoceratidae, Hoplitidae, Lyelliceratidae, and certain Deshayesitidae. Upper Albian zonation utilizes mainly the Acanthoceratidae and Schloenbachiidae; Pachydiscidae, Engonoceratidae, and Brancoceratidae are principal zonal ammonite groups of the American Gulf Coast. Belemnites become increasingly important in Albian biostratigraphy within Tethys *(Neohibolites, Parahibolites)*. In temperate North America and Europe, the Desmoceratidae and the endemic American Gastroplitinae (*Gastroplites* and *Neogastroplites*) are important ammonites in refined biostratigraphic zonation of the Albian. Among Bivalvia, cosmopolitan Inoceramidae of the genus *Birostrina* and the *"Inoceramus" anglicus* species group are important in regional correlation; the *"I." dunveganensis* and *"I." nahwisi* groups are important in

North America. *Aucellina* (Buchiidae) is locally useful in temperate biostratigraphy of the Albian; *Neithea* (Pectinidae), Ostreidae (*Texigryphaea,* Lophinae, Exogyrinae), various Trigoniidae, and rudistid bivalves are locally applied in Tethyan biostratigraphy, along with several groups of echinoids *(Epiaster, Salenia, Holectypus, Heteraster)* and gastropods (nerineids, actaeonellids). Global Albian pelagic zonation is mainly based on planktonic foraminifers of the genera *Ticinella, Globogerinelloides, Rotalipora,* and *Praeglobotruncana,* with less refined division possible from nannofossils (Fig. 3).

UPPER CRETACEOUS BIOSTRATIGRAPHY

Upper Cretaceous biostratigraphy may be the most refined and regionally applicable within the Phanerozoic. Ammonites predominate in macrofossil zonations and planktonic foraminifers and calcareous nannofossils share equal roles among the microbiota (Fig. 4). In addition, bivalves of the Inoceramidae, Trigoniidae, Ostreidae, and Hippuritacea; belemnites of the genera *Neohibolites, Parahibolites, Actinocamax, Gonioteuthis, Belemnitella,* and *Belemnella* (STEVENS, 1973); irregular echinoids such as *Holaster, Micraster, Offaster, Galeola, Echinocorys* (ERNST & SEIBERTZ, 1977); certain terebratulid and rhynchonellid brachiopods; and gastropods of the families Turritellidae, Aporrhaidae, Volutidae, Nerineidae, and Actaeonellidae (SOHL, 1977) all have been used successfully for refined zonation. This has led to widespread development of integrated biostratigraphic systems employing assemblage- and composite assemblage-zones (KAUFFMAN, 1970; VAN HINTE, 1976).

Throughout the temperate realms, ammonites and Inoceramidae share an equal role in the construction of biostratigraphic zonation of 0.25 my/zone duration or less, which is broadly applicable for precise correlation (KAUFFMAN, 1975). These two groups evolved at nearly equal rates during the Late Cretaceous, and have similarly broad biogeographic spread; Inoceramidae are consistently more cosmopolitan than ammonites at the species level. In parts of the Upper Cretaceous sequence (Turonian, Campanian-Maastrichtian), Ostreidae approach these two groups in biostratigraphic utility. Through the Cenomanian and Turonian, diverse ammonites and Inoceramidae are commonly associated and comprise simple assemblage-zones that form the north temperate standard (for example, COBBAN & REESIDE, 1952; COBBAN & SCOTT, 1972; KAUFFMAN, 1975; KAUFFMAN, COBBAN, & EICHER, 1977, for North America). In the Coniacian and Santonian, ammonites are much less common, especially in carbonate facies, and rapidly evolving inoceramids, irregular echinoids, and belemnites are the principal zonal indices. Where ammonites are abundant in clastic facies they occur with cosmopolitan inoceramids, allowing precise correlations to be made between these and ammonite-poor facies. In the Campanian and Maastrichtian, except for the chalk facies, ammonites again dominate biostratigraphic systems, and evolved more rapidly than inoceramids (KAUFFMAN, 1975, fig. 4). In the latest Cretaceous chalk facies, inoceramids and belemnites are the principal macrofossils used in biostratigraphy, with ostreid bivalves, ammonites, irregular echinoids, and floating crinoids playing an important secondary role. Planktonic foraminifers and coccolithophorids are important biostratigraphically in carbonates of the North Temperate Realm, but are commonly facies controlled in temporal and areal distribution, and have longer range zones than ammonites, bivalves, and belemnites. Their main value is in providing correlations with the Tethyan Realm and the oceanic pelagic standard.

In the Tethyan Realm, ammonites and planktonic microbiota (foraminifers, coccolithophorids) comprise the main bases for the Upper Cretaceous biostratigraphic system. Belemnites are rare, and inoceramids are much less common and less diverse in Tethys. Sufficient inoceramid bivalves are known from Tethys, however, to allow rather precise correlation with the temperate biostratigraphic standard (KAUFFMAN, 1968; WIEDMANN & KAUFFMAN, 1977). Irregular echinoids, nerineid, actaeonellid, and turritellid gastropods, and rudistid bivalves are important local supplements to Tethyan zonation. Increasing diversification and rates of evolution among planktonic foraminifers and coccolithophorids during the Late Cretaceous enhanced their biostratigraphic utility.

Detailed global correlation of the Upper Cretaceous is possible because of major epicontinental transgressions and ameliorating marine environments that characterized most of this time period and that removed barriers to broad, rapid, species-level migration of taxa with mobile adult stages (ammonites, belemnites), or with long-lived planktonic larval stages (bivalves, echinoids, and many gastropods; see KAUFFMAN, 1975, and references therein). In addition, exceptional environmental tolerance among some of the most widespread benthonic groups (for example, Inoceramidae) contributed strongly to their rapid, wide dispersal. Figure 4 outlines the major zonal schemes applied to the Upper Cretaceous for key areas of the world where refined biostratigraphic systems have been developed.

BIOSTRATIGRAPHICALLY IMPORTANT UPPER CRETACEOUS GROUPS

Cenomanian biostratigraphy places equal importance on ammonites and inoceramid bivalves throughout both temperate realms, and on similar ammonites and planktonic microbiota (especially foraminifers) in the Tethyan Realm. Belemnites *(Actinocamax, Neohibolites, Parahibolites)* are widely used in zonation of the Eurasian north temperate region. Biostratigraphically important Cenomanian ammonites belong to the Acanthoceratidae, Desmoceratidae (Pacific mainly), and Baculitidae (especially Pacific *Sciponoceras*). Biostratigraphically important temperate-zone inoceramids are the *Inoceramus pictus, "I." crippsi, I. tenuis–I. tenuiumbonatus, I. heinzi–I. ginterensis,* and *I. etheridgei–I. tenuistriatus* lineages. Trigoniidae *(Acanthotrigonia)* are useful in zoning the North Pacific Cenomanian, and species of *Rotalipora* dominate the planktonic foraminiferal zonation. Certain additional taxa are critical to zonation of Cenomanian substages. In the Lower Cenomanian, ammonites of the Mantelliceratinae, Turrilitidae, Hoplitidae and surviving Lyelliceratidae are used in Tethys; ammonites of the Schloenbachiidae, *Graysonites* (North Pacific; North America), belemnites of the *Neohibolites* group, and inoceramid bivalves of the *"Inoceramus" crippsi, "I." anglicus* and *"I." dunveganensis* (North American) lineages are used in temperate realms; and the coccolith assemblage of *Eiffelithus turriseiffeli* is used in the upper half of the substage. Middle Cenomanian zonation relies also on ammonite species belonging to the Turrilitidae, *Acanthoceras* and *Dunveganoceras* among Acanthoceratinae, late descendants of the *"Inoceramus" dunveganensis–"I." anglicus, I. rutherfordi* and *I. crippsi* lineages, and the *I. "concentricus" nipponicus* plexus (North Pacific). The Middle and lower Upper Cenomanian is characterized by coccoliths of the *Lithraphidites alatus–Staurolithites orbiculofenestrus* Zone, and echinoids of the *Holaster subglobosus* lineage (Europe). Upper Cenomanian biostratigraphy specifically utilizes ammonites of the Metoicoceratinae, certain Acanthoceratinae *(Calycoceras, Eucalycoceras)*, the temperate-zone belemnite *Actinocamax,* and the North Pacific *Inoceramus pennatulus* and *I. "concentricus" costatus* plexes. The nannofossil assemblage of *Chiastozygus irregularis* marks the uppermost Cenomanian in the lower part of its range-zone (Fig. 4).

In Tethyan and temperate zones, Turonian biostratigraphy utilizes similar ammonite groups—Acanthoceratinae, Scaphitidae, Baculitidae *(Sciponoceras, Baculites),* and Muniericeratidae (North Pacific). Species of the bivalve *Sergipia,* and the belemnite *Actinocamax,* and the radiolarian assemblage of the *Dictyomitra veneta* Zone are distributed throughout the Turonian. In addition, Lower Turonian biostratigraphy employs ammonites of the Mammitinae, Acanthoceratinae *(Kanabiceras),* Vascoceratidae, late Metoicoceratinae, the brachiopod plexus of *Orbirhynchia cuvieri,* inoceramids of the *Mylitoides mytiloides–M. labiatus* lineage, planktonic foraminifers of the genera *Hedbergella* and *Praeglobotruncana,* and nannofossil assemblages of the *Chiastozygus irregularis* Zone (upper part; basal Turonian) and *Corollithium exiguum* Zone (Lower Turonian; SMITH, 1975). Middle Turonian biostratigraphy specifically utilizes certain acanthoceratid and collignoniceratid ammonites (*Collignoniceras, Romaniceras, Prionotropis*), inoceramid bivalves of the *Inoceramus lamarcki–I. cuvieri, I. costellatus, I. hobetsensis*, and *Mytiloides hercynicus–M. "latus"* lineages, brachiopods of the *Terebratulina lata* plexus, and the oyster *Lopha*. Upper Turonian biostratigraphic systems selectively employ different Collignoniceratinae (*Reesidites, Prionocyclus*) and Pachydiscidae (*Lewesiceras*) among ammonites, inoceramid bivalves of the *Inoceramus dimidius, I. perplexus, I. teshioensis, I. kleini, Mytiloides fiegei–M. incertus, M. striatoconcentricus*, and *M. lusatiae* lineages, the ostreid *Lopha*, and echinoids of the *Holaster planus* plexus. *Globotruncana*-based zonation dominates planktonic biostratigraphy of the Middle and Upper Turonian.

Inoceramid bivalves are preeminent in Coniacian biostratigraphy of the temperate realms whereas ammonites and planktonic foraminifers remain important in Tethys. Of greatest stratigraphic value throughout the Coniacian are ammonites of the Collignoniceratidae, Kossmaticeratidae (North Pacific), Scaphitidae, Baculitidae, and Desmoceratinae, the *Inoceramus yubariensis* (North Pacific) and *Didymotis* bivalve lineages, echinoids of the *Micraster cortestudinarium* plexus (Europe), planktonic foraminifers of the genus *Globotruncana*, and radiolarians of the *Artostrobium urna* Zone (Fig. 4). In addition, the Lower Coniacian is zoned on ammonites of the Barroisiceratinae, and Peroniceratinae, inoceramid bivalves of the *Inoceramus rotundatus–I. deformis, I. ernsti, I. waltersdorfensis, "I." madagascarensis*, and, in the upper part of their range, the *Mytiloides fiegei, M?. lusatiae*, and *M. dresdensis* lineages. The nannoplankton assemblage of the *Micula decussata* and *Tetralithus pyramidus* Zone characterizes both the Upper Turonian and Lower Coniacian. Middle Coniacian zonation specifically utilizes ammonites of the Texanitinae and Peroniceratinae (Collignoniceratidae) and bivalves of the *Volviceramus involutus–V. exogyroides, Cremnoceramus inconstans, C?. koeneni–C?. wandereri, Platyceramus mantelli–P. circularis*, and *Inoceramus kleini* lineages, especially in the temperate realms. Upper Coniacian biostratigraphy relies also on additional Texanitinae and inoceramids of the *Magadiceramus subquadratus, Inoceramus mihoensis, I. yokoyamai* (North Pacific), *I. fasciculatus*, and younger parts of the *P. mantelli* and *V. involutus* lineages. The nannofossil assemblage-zone of *Marthasterites furcatus* spans the Middle and Upper Coniacian in the lower half of its range (Fig. 4).

Santonian biostratigraphy similarly depends mainly on Inoceramidae (Bivalvia), except in Tethyan areas where ammonites and planktonic microbiota predominate. Important taxa include ammonites of the Baculitidae, Texanitinae, Pachydiscidae, Scaphitidae, and Munericeratidae; inoceramid bivalves of the *"Inoceramus" simpsoni* lineage, *Sphenoceramus lobatus–S. steenstrupi–S. pachti* lineage, *Cladoceramus undulatoplicatus* plexus, and the *Platyceramus cycloides, P. platinus*, and *I. naumanni* lineages; belemnites of the genus *Gonioteuthis*; planktonic foraminifers of the genus *Globotruncana*, and radiolarians of the upper part of the *Arctostrobium urna* Zone. Lower Santonian zonation is based further on the first occurrence of *Cladoceramus undulatoplicatus*, species of the *Mytiloides? stantoni, Cordiceramus cordiformis* and *"Inoceramus" amakusensis* lineages, the echinoids *Conulus* and *Micraster* of the *M. coranguinum* plexus, and the *Marthasterites furcatus* nannofossil assemblage. Middle Santonian zonation is characterized by the floating crinoid *Uintacrinus*, inoceramids of the *Cladoceramus cordiformis* lineage, the first common occurrences of the cosmopolitan *Endocostea balticus* lineage, and the north temperate *Cladoceramus japonicus* and *"I." (Sphenoceramus?) naumanni* lineages. The nannofossil assemblage of the *Gartnerago obliquum* Zone spans the Middle and Upper Santonian. Ammonites of the Placenticeratidae and Lenticeratinae first become important in Cretaceous biostratigraphy in the Upper Santonian, associated with the floating crinoid *Marsupites* and, in the North Pacific, with early members of the *"Inoceramus" (Sphenoceramus?) orientalis* lineage.

Campanian biostratigraphic systems utilize ammonites, inoceramid bivalves, and belemnites equally in temperate realms whereas ammonites dominate Tethyan zonation. Groups that are important throughout the Campanian include ammonites of the Nostoceratidae, Baculitidae, Placenticeratidae, Scaphitinae, and Pachydiscidae; bivalves of the *Endocostea baltica* and *Apiotrigonia* lineages; planktonic foraminifers of the genus *Globotruncana,* and radiolarians of the *Amphipyndax enesseffi* Zone. In addition, Lower Campanian zonation relies heavily on Kosmaticeratidae, Texanitinae, and Lenticeratinae in Tethys and its margins, and on other Texanitinae *(Submortoniceras),* Scaphitidae of the *Scaphites leei–S. hippocrepis* lineage, species of the belemnite *Gonioteuthis,* inoceramids of the *"Inoceramus" quadrans, "I." (Sphenoceramus?) orientalis,* and *Sphenoceramus lingua–S. lobatus* lineages, echinoids of the *Offaster pilula* plexus, and the nannofossil assemblage of the *Eiffelithus eximius* Zone. Middle and Upper Campanian zonation draws heavily from inoceramid species, especially within the *"I." (Sphenoceramus?) schmidti, I. azerbaidjanensis–I. adgjakendensis, Platyceramus sagensis–P. nebrascensis, I. convexus, I. tenuilineatus, I. subcompressus, I. barabini, I. oblongus, I. pertenuis, Platyceramus regularis,* and *Endocostea typica* plexuses in North America (KAUFFMAN, 1975, fig. 4) and their Eurasian counterparts. The genus *Belemnitella* is important for zonation of the north temperate Middle and Upper Campanian. "Middle" Campanian pelagic zonation is further characterized by coccoliths of the *Broinsonia parca* Assemblage-zone, and the Upper Campanian by those of the lower *Tetralithus trifidus* Assemblage-zone. KAUFFMAN (1967) has successfully applied the bivalve *Thyasira* to zonation of the temperate North American Middle Campanian.

Maastrichtian biostratigraphic zonation is less refined than that of preceding stages due to the evolutionary decline of many principal invertebrate groups; including the ammonites, inoceramids, and belemnites. Among ammonites, species of the Baculitidae, Pachydiscidae, Sphenodiscidae, Scaphitidae, and (Lower Maastrichtian only) the Desmoceratidae and Nostoceratidae are the main basis for zonation. Pachydiscids dominate in Tethys; the others are mainly temperate and marginal Tethyan groups. Among north temperate belemnites, *Belemnella* and *Belemnitella* are key groups. The *Tenuipteria fibrosus–T. tegulatus* lineage is the main inoceramid group used in Maastrichtian biostratigraphy; the *Inoceramus shikotaensis* (North Pacific) plexus, the *Trochoceramus helveticus* and *Platyceramus salisburgensis* lineages (north Europe), and the *P. proximus–P. subcircularis* groups (Euramerica) are secondary indices for the Lower Maastrichtian. The *I. hetonaianus–I.? awajiensis* groups are used in the Middle to Upper Maastrichtian zonation of Japan along with species of *Microtrigonia* (Trigoniidae). The pelagic realm is zoned primarily on the basis of species of *Globotruncana* and *Globotruncanella* (Upper Maastrichtian only), radiolarians of the *Theocapsoma comys* assemblage, and nannofossils of the *Tetralithus trifidus, Lithraphidites quadratus,* and *Micula mura* assemblage-zones in the Lower, Middle, and Upper Maastrichtian, respectively.

BIOGEOGRAPHY

Biogeographic units are large-scale ecological units that respond to varying environmental (niche) parameters. They are dynamic in space and time—showing major fluctuations in composition, degree of internal endemism, paleogeographic spread, and the nature of their boundaries with other units (SYLVESTER-BRADLEY, 1971; VALENTINE, 1973; KAUFFMAN, 1973a). This can be clearly demonstrated in the Cretaceous, where the development and evolution of biogeographic units was closely tied to sea-floor spreading, plate collision, and the emplacement or destruction of orogenic belts acting as paleobiogeographic barriers. These processes resulted in large-scale isolation of segments of formerly associated biotas on the one hand, and massive competition between formerly isolated biotas on the other (KAUFFMAN, 1973a). In addition, the effect of large-scale Cretaceous eustatic fluctuations (Fig. 2) on the spatial extent of marine climatic zones, the severity of environmental gradients across their boundaries, and on the development of migratory pathways across epicontinental areas, strongly contributed to dynamic changes in paleobiogeographic units.

Specifically, some of the major events that most strongly influenced the evolution of paleobiogeographic units during the Cretaceous were as follows (KAUFFMAN, 1973a): 1) opening of the Atlantic to distances that exceeded normal invertebrate larval drift or adult mobility, and which genetically isolated the warm-temperate and tropical invertebrate faunas of Eurasia from those of the Americas by Late Cretaceous time; 2) opening of the Caribbean by separation of North and South America and establishment of a tropical (Tethyan) seaway through this area which, for the first time during the Cretaceous, linked eastern Pacific with Euramerican tropical biotas; this tropical seaway effectively separated north and south temperate biotas of the Americas with a major temperature barrier; 3) closing of the North Pacific to the extent that marine biotas established separately in eastern and western Pacific shelf zones were brought into large-scale competition through overlapping larval drift ranges; 4) partial closing or restriction of the Mediterranean Tethys by Barremian time, subsequently resulting in increased isolation of Tethyan biotas of the eastern and western Mediterranean (KAUFFMAN, 1973a, p. 360-366, fig. 3); 5) establishment of north-south marine connections between the temperate to cool-temperate Circumboreal ("Boreal") Seaway (biotas) and the warm-temperate waters (biotas) of the proto-Atlantic and Tethyan margin through initial opening of the North Atlantic and establishment of shallow epicontinental seas across Eurasia and North America during major Cretaceous transgressions; 6) opening of the South Atlantic during the Cretaceous, thus providing a pathway for mixing of south temperate "Gondwanan" marine faunas with those of the Euramerican Tethyan and North Temperate realms; Late Cretaceous separation progressively isolated segments of characteristic South Atlantic temperate faunas; 7) establishment of marine epicontinental connections between the Mediterranean Tethys and the South Atlantic seaways and biotas across north-central and northwestern Africa during major Cretaceous transgressions; 8) final breakup of Gondwanaland with isolation of Australia, New Zealand, India, Antarctica, and South America to the extent that major endemic centers became established in East Africa and Australia by the Neocomian, and in New Zealand by the Cenomanian (KAUFFMAN, 1973a); 9) establishment of oceanic ridges and island systems as biogeographic stepping stones for migration of marine invertebrates, largely through larval drift, along the Mid-Atlantic Ridge, the island arc systems of the Caribbean, and the largely volcanic islands of the western Pacific and Indian oceans; and 10) as a result of plate tectonic activity, 9 to 10 major transgressions of epicontinental seaways onto the world's cratons, reflecting eustatic rise of sea level of up to several hundred feet; ameliorating and warming marine climates resulted, prime ecospace for marine invertebrates greatly expanded and diversified, major marine connections were established between formerly isolated areas (biotas), and climatic gradients became less severe. This resulted in breakdown of major temperature barriers to widespread dispersal of marine invertebrates. Interspersed regressions produced climatic decline, increased temperature gradients and environmental stress, and removal of many epeiric marine connections. This in turn resulted in isolation of biotas, and decrease in size but increase in number of latitudinally expressed marine climatic zones. The net stress effect of these alternating transgressive and regressive Cretaceous environments on the marine ecosystem was one of the principal dynamic forces leading to changes in Cretaceous paleobiogeographic units (KAUFFMAN, 1973a).

Cretaceous biogeography has been viewed by different authors largely from the standpoint of distribution patterns among individual groups of organisms. Examples are ammonites (JELETZKY, 1971; CASEY, 1971; MATSUMOTO, 1973; OWEN, 1973; WIEDMANN, 1973; KENNEDY & COBBAN, 1976), bivalves (KAUFFMAN, 1973a), brachiopods (AGER, 1971, 1973), foraminifers (DILLEY, 1971, 1973), belemnites (STEVENS, 1973), corals (COATES, 1973), gastropods (SOHL, 1971), and sponges (REID, 1967). Consequently, interpretations of the composition and distribution of biogeographic units, and their Cretaceous histories, are diverse and inconsistent. KAUFFMAN (1973a) has pointed out that this monotaxic technique is in contrast with the more holistic approach of the marine ecologist and biogeographer. A holistic approach to biogeography has the advantages of 1) presenting a consistent set of biogeographic unit concepts for each stage, which is comparable to recent data, and 2) allowing definition of ecological structure within biogeographic units.

Not only are the bases for defining Cretaceous paleobiogeographic units widely discrepant (that is, different groups may have distinct ecological response to the same general sets of changing environmental parameters and thus yield different distribution patterns in time), but so are the methods used in their construction; for example, compare papers cited in ADAMS and AGER (1967), MIDDLEMISS and RAWSON (1971), HALLAM (1973), CASEY and RAWSON (1973a), and HUGHES (1973), the principal modern compilations containing Cretaceous paleobiogeographic data. VALENTINE (1973), and KAUFFMAN (1973a), among others (see references cited in each), have reviewed this problem and discussed the means and importance of equating modern and ancient biogeographic unit concepts.

Neontologists apply two principal methods in defining biogeographic units: 1) delineation of unit boundaries by biotic discordance, that is, zones where there are numerous teil-province terminations (VALENTINE, 1961); and 2) by percentage endemism within biogeographic units (EKMAN, 1967; KAUFFMAN, 1973a). Both allow definition of biogeographic units in objective terms and their ranking within the biogeographic heirarchy. Because of the incomplete preservation of the stratigraphic record, teil-province boundaries are commonly difficult to ascertain for fossil taxa and the endemic percentage method is best applied to the definition of paleobiogeographic units. KAUFFMAN (1973a) discussed criteria for determining percent endemism, and applied a quantitative standard to the definition of Cretaceous paleobiogeographic units based on genera of Bivalvia: realm = 75 to 100 percent endemism, region = 50 to 75 percent endemism, province = 25 to 50 percent endemism, subprovince = 10 to 25 percent endemism, and endemic center = 5 to 10 percent endemism, in each case among noncosmopolitan organisms. Most past attempts at Cretaceous paleobiogeography have used a nonquantitative and inconsistent approach to the definition of units. Thus, one finds the Tethyan biota variously described as comprising a realm, a region, or a province during the Cretaceous.

The nonquantitative approach to Cretaceous paleobiogeography has had three unfortunate consequences: 1) inconsistency in classification of paleobiogeographic units; 2) lack of a consistent equation between Cretaceous and well-studied modern biogeographic units—a key to interpretation of ancient biogeography, and 3) in most studies, only a gross biogeographic division of the Cretaceous at the realm and region level has resulted, as compared to modern systems that utilize the province and subprovince as working units, and that recognize many more divisions of the global marine biota. Some Cretaceous workers have made finer biogeographic divisions to the province-subprovince level on nonquantitative treatment of diverse taxa (for example, JELETZKY, 1971; SOHL, 1971). Others have done the same by utilizing only well-studied taxa such as ammonites (MATSUMOTO, 1973; KENNEDY & COBBAN, 1976) and belemnites (STEVENS, 1973). Though they are more refined than most published biogeographic divisions of the Cretaceous, these still do not equate with modern provincial-subprovincial marine systems (for example, HALL, 1964) nor with those established for the Cretaceous utilizing the quantitative approach to definition of paleobiogeographic units (for example, KAUFFMAN, 1973a, for Bivalvia). In these nonquantitative analyses based on few taxa, the illusion is created that Cretaceous biogeographic systems were much simpler and less highly evolved than those of today, even taking into account the more severe climatic gradients of modern oceans and the predictable increase in the numbers of modern biogeographic divisions. KAUFFMAN's (1973a) analytic treatment of Cretaceous bivalve genera, utilizing percent endemism to define and classify paleobiogeographic units, provides an alternative method that depicts

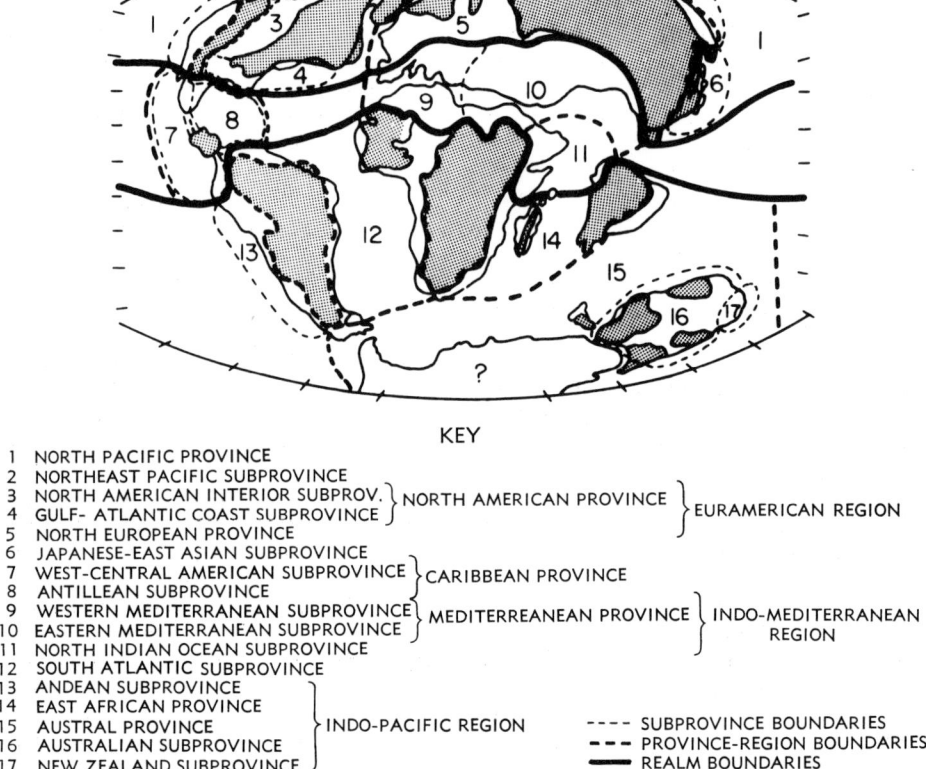

Fig. 5. Generalized distribution of Cretaceous paleobiogeographic units of subprovince and higher rank (after Kauffman, 1973a, fig. 2), based primarily on molluscan distribution patterns, and plotted against slightly modified version of Dietz and Holden's (1970) reconstruction of the Cretaceous globe. Land areas not inundated during Cretaceous transgressions are stippled. Units 1 to 6 comprise the North Temperate Realm ("Boreal Realm" of authors), units 7 to 11 the Tethyan Realm, and units 12 to 17 the South Temperate Realm. [Used with permission of Elsevier Scientific Publishing Co., Amsterdam.]

the Cretaceous as being only slightly less diverse in terms of biogeographic division than modern seas (Fig. 5), and within the range predictable from the different climatic gradients that mark these two times in geological history. In the following analysis, KAUFFMAN's units are therefore applied to biogeographic division of the Cretaceous, and the principal taxa that form the basis for biogeographic units of other specialists are fitted to this system with little difficulty.

Most biogeographic reconstructions that predate the widespread acceptance of plate tectonic theory dealt with static or near static units for each period without consideration of spatial and temporal change. Modern reconstructions recognize the dynamic nature of biogeographic units—their constantly changing composition and spatial distribution through time. CASEY and RAWSON (1973b, p. 418) argued the need for stage by stage evaluation of Cretaceous paleobiogeography as a mechanism for studying the dynamics of the units; KAUFFMAN (1973a) successfully applied this method to Cretaceous bivalve biogeography. To the extent that it is possible here, the evolution of Cretaceous paleobiogeographic units is documented by stages and the observed changes in structure and distribution of the units are interpreted in light of plate tectonics and oceanic history.

NORTH TEMPERATE REALM

This realm incorporates biotas between the northern margin of tropical Tethyan biotas and the Cretaceous north pole (Fig. 5). The term "Boreal" has been widely applied to the same biotas (MATSUMOTO, 1973; STEVENS, 1973; and others), but as explained by KAUFFMAN (1973a, p. 367), this is an unfortunate misnomer as used by paleobiogeographers and should be abandoned. The North Temperate Realm was well established by the beginning of the Cretaceous and persisted to the end, becoming evolutionarily less unique with decreasing overall endemism but increasing subdivision into smaller paleobiogeographic units (Fig. 6). At no time during the Cretaceous is there unquestioned evidence in this realm for marine climates colder than the cool-temperate zones of modern oceans (sub-Boreal of modern zonation). The southern (Tethyan) boundary of the realm was very sharply defined early in the Cretaceous, as it was in the Jurassic, but became somewhat more diffuse through the period allowing greater precision in biostratigraphic correlation of the North Temperate and Tethyan realms.

Several schemes for paleobiogeographic division of the North Temperate Realm have been proposed (papers cited in CASEY & RAWSON, 1973a; JELETZKY, 1971; SOHL, 1971; KAUFFMAN, 1973a; MATSUMOTO, 1973; STEVENS, 1973; and others); these have been based on different organisms and result in somewhat conflicting systems. The following is a compromise, utilizing the most detailed system (KAUFFMAN, 1973a) as a base (Figs. 5, 6). The two principal divisions of the realm, recognized by most workers, are the North Pacific Province (JELETZKY, 1971; KAUFFMAN, 1973a; OWEN, 1973), and the Euramerican Region (Fig. 5; KAUFFMAN, 1973a; "Boreal" realm or province of many authors, in part or whole).

The North Pacific Province is clearly divisible on bivalve endemism into Northeast Pacific and Japanese-East Asian subprovinces (KAUFFMAN, 1973a), reflecting the great breadth of the Cretaceous Pacific Ocean acting as a major water barrier to larval exchange between these areas. The North Pacific Province is characterized by: 1) bivalves such as *Apiotrigonia (Heterotrigonia), Steinmanella (Yeharella), Meekia,* and the middle Cretaceous *Inoceramus pennatulus* and *I. hobetsensis* lineages (KAUFFMAN, 1973a); 2) Tethyan-style ammonites (phylloceratids, lytoceratids, tetragonids, desmoceratids, and kossmoceratids) during the Albian (OWEN, 1973); 3) Upper Cretaceous ammonite groups that are most common in, but not restricted to the Indo-Pacific (Cenomanian and younger Desmoceratidae, Tetragonitidae, Phylloceratidae; Coniacian and younger Kossmaticeratidae and Pachydiscidae; JELETZKY, 1971; MATSUMOTO, 1973); 4) weak Lower Neocomian endemism among belemnites (STEVENS, 1973); 5) an endemic assemblage of small benthonic foraminifers, especially agglutinated forms (DILLEY, 1971); and 6) a unique North Pacific gastropod assemblage, especially in the Northeast Pacific Subprovince (SOHL, 1971). Compared to other major provinces in the Cretaceous, the North Pacific Province shows relatively weak endemism, as it did throughout the Jurassic.

Although the Northeast Pacific and Japanese-East Asian subprovinces (Fig. 5) are defined wholly on Bivalvia at this time (KAUFFMAN, 1973a, p. 371, provided taxonomic lists), literature on other mollusks, especially gastropods, indicates that they show similar provincialism. KAUFFMAN reported 21, mainly Lower Cretaceous, genera and subgenera of bivalves endemic to the Japanese-East Asian Subprovince, and 8 endemic genera in the northeast Pacific subprovince.

The evolutionary history of the North Pacific Province (Fig. 6) was detailed by KAUFFMAN (1973a) using bivalves; other groups such as belemnites (STEVENS, 1973) seem to support these observations. Figure 6 suggests that there was a marked decrease in endemism throughout the province and within individual subprovinces from Berriasian through Maastrichtian time, with the most abrupt change having come in the Cenomanian—a period of major plate movement. KAUFFMAN (1973a) has attributed this decline to: 1) partial closing of the

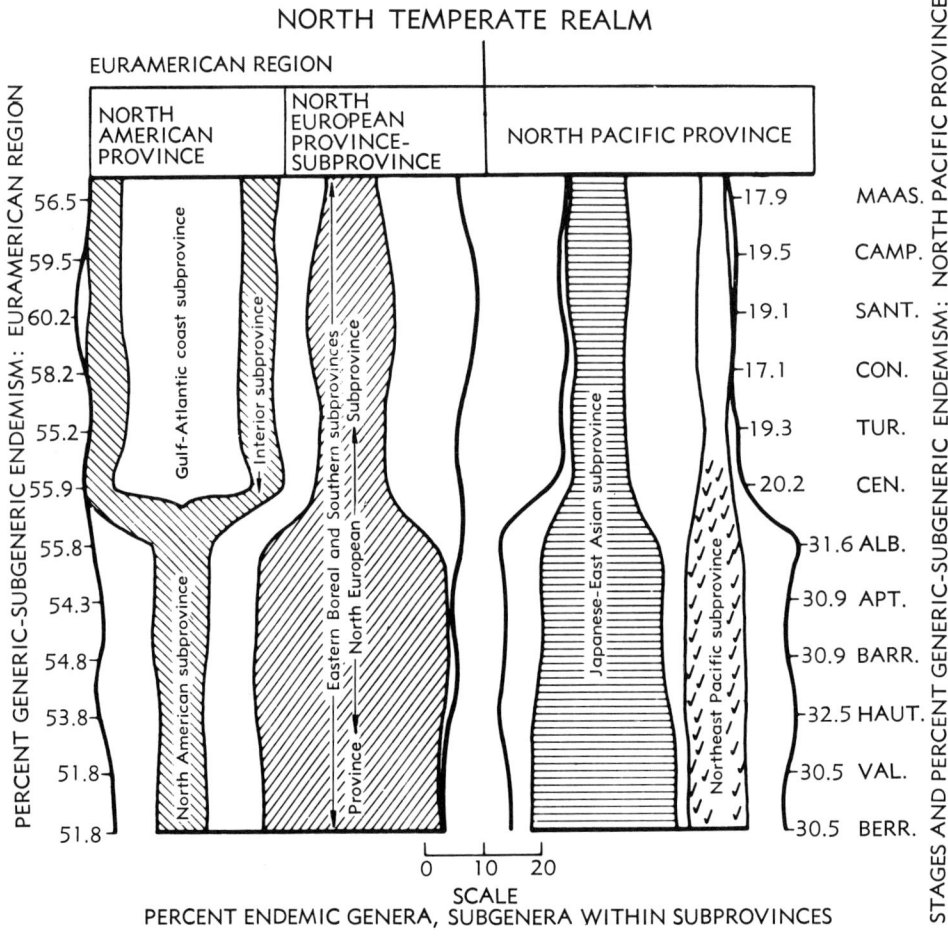

Fig. 6. Evolution of Cretaceous paleobiogeographic units in the North Temperate Realm, based mainly on bivalve Mollusca (after Kauffman, 1973a, fig. 4). Graphs show percent endemism of bivalve genera and subgenera within each unit, exclusive of cosmopolitan taxa. Scales for percent endemism of Euramerican Region and North Pacific Province are shown on left and right of graph, respectively. Patterned areas within outline for major regions or provinces represent distinct biogeographic subdivisions (provinces, subprovinces), as labeled. Different subdivisions have different patterns. The cumulative width of any patterned area at any time represents the percent endemism within it, as measured from scale at base of figure. Lack of pattern within any province or subprovince indicates reduction of endemism to less than 10 percent for that time. Unpatterned areas between patterned provincial or subprovincial graphs collectively depict that percent of endemism within the major biogeographic divisions represented by taxa shared between all subdivisions of the major unit. [Used with permission of Elsevier Scientific Publishing Co., Amsterdam.]

North Pacific, especially as a result of large-scale Cenomanian plate movements, to the point that the distinct eastern and western Pacific bivalve assemblages developed massive levels of competition for the same shallow-water niches as they came within larval drift range of each other, which resulted in a large-scale extinction and lowering of endemic diversity; 2) major epicontinental transgressions during the Late Cretaceous that opened marine pathways to migration of invertebrates in and out of the North Pacific, resulting in wider biogeographic spread of taxa and decreased local endemism as indicated by ammonites and gastropods (MATSUMOTO, 1973; SOHL, 1971) as well as inoceramid bivalves (KAUFFMAN, 1977e); and 3) climatic deterioration of the North Pacific marine environment through the Cretaceous with progressive cooling and increasing environmental gradients, resulting in lowering of North Pacific invertebrate diversity and endemism.

Biogeographic division of the Cretaceous Euramerican Region has been more inconsistent. Most authors have recognized a twofold, latitudinally (temperature) controlled division of this region, which extends from Siberia to Western Interior North America (Fig. 5), and which draws its faunal similarity from the many avenues of migration open to Eurasian and North American invertebrates through the "Circumboreal" Seaway and across the still narrow northern proto-Atlantic during much of the Cretaceous. The two proposed divisions consist of a northern cool-temperate zone and a southern warm-temperate to subtropical zone fronting on Tethys. The biogeographic treatment of the faunas characterizing these climatic zones has been diverse. Many authors have recognized two latitudinal divisions of the entire Euramerican Region. These are the northern "Boreal Province" and southern "Intermediate Latitude" zone of MATSUMOTO (1973; on ammonites), the Lower Cretaceous "Arctic" and (to the south) "Boreal Atlantic" provinces and the Upper Cretaceous "Northern" and "Southern" provinces of STEVENS (1973; based on belemnites).

KAUFFMAN's quantitative analysis of Cretaceous bivalve endemism revealed similar north-south division of the Euramerican Region independently in Eurasia and North America, but greater endemic differentiation across the proto-Atlantic in an east-west direction. He therefore recognized provincial boundaries approximately parallel to longitude, and subprovincial boundaries in both areas parallel to latitude in order to reflect the temperature-controlled faunal differentiation. This system is applied here (Figs. 5, 6), resulting in recognition of a North European Province and a North American Province, each with north-south differentiation into subprovinces or endemic centers.

The North European Province (Fig. 5) is very well defined at the base of the Cretaceous but shows decreasing endemism throughout the system for bivalves (Fig. 6) and other groups, especially after the Cenomanian. KAUFFMAN (1973a) attributed this to: 1) extensive epeiric flooding of the North American craton after the Middle Albian allowing, for the first time, widespread sharing of invertebrate taxa between Europe and America (and thus decreasing endemism in western and central Europe); the oceanic distances between Europe and North America were still within the range of larval drift for many invertebrate groups; and 2) east to west exchange of taxa that was probably greater than west to east exchange because the main surface currents of the middle proto-Atlantic gyre, representing the pelagic zone in which most planktonic invertebrates and larvae of marine benthos are carried, flowed to the west (KAUFFMAN, 1975, fig. 2); deeper countercurrents flowed east, but probably carried fewer organisms, as today. Resultant sharing of taxa would consequently have been greater from east to west, decreasing European endemism faster than American endemism in the process.

The North European Province is characterized by a strongly endemic Albian ammonite assemblage based on Hoplitidae (OWEN, 1973) and many other taxa. Late Cretaceous ammonite faunas were geographically more widespread than those of the Early Cretaceous, with broad Euramerican distribution of Acanthoceratidae, Collignoniceratidae, Placenticeratidae, Scaphitidae, Baculitidae, and Nostoceratidae being more notable than endemism within the provinces. Nevertheless, MATSUMOTO (1973) noted an endemic grouping of warm-temperate ammonite lineages common to southern Europe, East Africa, Madagascar, and the Coastal Plain of the United States, and a second group of endemics in the North American interior. Belemnites (STEVENS, 1973) show significant endemism in the North European Province (including Greenland) throughout the Lower Cretaceous (for example, the *Acroteuthis* and *Boreioteuthis* lineages of STEVENS' "Boreal Atlantic Province") and to a lesser extent in his Upper Cretaceous "Northern Province" (*Belemnella* fauna). KAUFFMAN (1973a, p. 369) reported more than 40 genera and subgenera of Cretaceous bivalves to be endemic to North Europe, and a weak endemic differentiation between eastern and western portions of the province. This is also shown by belemnites in STEVENS' (1973) "Boreal Northern" and "Southern" provinces of central and western Europe, and his eastern Crimean-Caucasian Province. AGER (1971) noted that the brachiopod group *Uralella* is restricted to the North European Province, and DILLEY (1971) recorded an endemic benthonic (especially agglutinated) foraminiferal assemblage in his temperate "Old World" Province.

North-south division of the Cretaceous North European Province into warm- and cool-temperate zones is difficult because of the diffuse climatic and biogeographic boundary between them. Bivalves and belemnites allow differentiation of a more southerly (warm temperate) "Western European Endemic Center" (KAUFFMAN, 1973a) or "Southern Province" (STEVENS, 1973; on *Belemnitella*) and a North European (KAUFFMAN, 1973a) or a "Northern Province" characterized by Late Cretaceous *Belemnella* and several bivalve genera. Brachiopods such as *Rhynchonella* and *Peregrinella* occur in warm-temperate to subtropical Euramerican zones but not to the north. MATSUMOTO (1973) reported major lowering of Upper Cretaceous ammonite diversity to the north; CASEY (1973) and MATSUMOTO (1973) documented differences between northern and southern European ammonite successions in the Lower and Upper Cretaceous, respectively; characteristic north European taxa are *Schloenbachia,* certain scaphitid groups, and certain species of *Lewesiceras.* Thus, collectively there seems to be sufficient evidence for division of the North European province into northern and southern subprovinces. Using modified terminology of others, these divisions are here named the Eastern Boreal (*s.s.*) and Southern subprovinces.

The North American Province was poorly defined (subprovincial rank) at the beginning of the Cretaceous, and showed a major increase in endemism after the Cenomanian, when it developed distinct northern and southern subprovinces (Figs. 5, 6). The Cenomanian increase in endemism reflects primarily initial establishment of the Western Interior seaway during the Late Albian, its early population by widespread Euramerican stocks introduced from Europe by westward larval drift across the then narrow proto-Atlantic, and from latest Albian to Cenomanian onward, partial isolation of the middle part of the seaway, leading to high levels of endemism among mollusks of all types. Rapid Cenomanian spreading of the middle Atlantic, magnified by later Cretaceous movements, enhanced the potential for endemism within the interior seaway. Widespread genetic isolation resulted from separation of European and interior North American populations as Atlantic spreading exceeded the limits of larval drift and adult mobility, and as the widening Caribbean arm of Tethys became more effective as a temperate barrier to immigration of temperate organisms at the south end of the Western Interior seaway.

Despite a broad zone of faunal mixing between southern warm-temperate and northern cool-temperate organisms in the North American Province (KAUFFMAN, 1970; SOHL, 1971), a clearly defined Gulf-Atlantic Coast Subprovince and North American Interior Subprovince developed by Cenomanian time and remained well defined until the end of the Maastrichtian. The diffuse biogeographic boundary reflected a broad temperature gradient within the seaway. Molluscs that characterize the North American Interior Subprovince include Albian Gastroplitidae (especially *Neogastroplites*), Turonian species groups of *Prionocyclus,* Turonian-Maastrichtian endemic ammonite lineages of the Scaphitidae *(Scaphites, Clioscaphites, Desmoscaphites),* and *Irenicoceras,* bivalves of the *"Inoceramus" dunveganensis* and *"I." nahwisi* lineages (KAUFFMAN, 1973a), certain corbiculids, *Crassatellina,* and lineages of Lower Cretaceous Buchiidae (JELETZKY, 1970). SOHL (1971) listed numerous gastropod genera and species groups restricted to his "California and Northern Western Interior Region," a term emphasizing biogeographic relationships between Upper Cretaceous gastropods of the two areas and suggesting brief marine connections across the Cordilleran geanticline. Gastropod genera with endemic species groups in this subprovince include *Closteriscus, Pseudobuccinum, Serrifusus, Vanikoropsis, Trachytriton,* and selected lineages of *Drepanochilus* and *Graphidula*. The North American Interior Subprovince is further characterized by a discrete assemblage of benthonic, especially agglutinated, foraminifers, and among planktonic species, by a distinct *Heterohelix-Hedbergella* assemblage (DILLEY, 1971).

The warm-temperate to subtropical Gulf-Atlantic Coast Subprovince (Fig. 5) extends from Massachusetts south to Texas and northern Mexico and shares many invertebrate taxa with interior North America. The subprovince has few endemic ammonites, except in the Aptian-Albian of the subtropical Gulf Coast among the Deshayesitidae *(Dufreynoyia),* Douvilleiceratidae *(Hypacanthoplites),* Mojsisovicziinae *(Oxytropidoceras, Manuaniceras, Venezoliceras), Drakeoceras,* and the Cenomanian *Budaiceras* among Lyelliceratidae (YOUNG, 1972). This is thought to be related to genetic isolation behind major barrier-reef systems during several Aptian-Cenomanian intervals. Otherwise, the Gulf-Atlantic Coast Subprovince is primarily differentiated on the basis of gastropods, bivalves, and certain echinoids. SOHL (1971) selected such genera as *Carota, Cassiope, Turritella (whitei* group), *Lispodesthes, Eunaticina,* and the *Fusus veneratus* groups as typical of Cenomanian-Turonian faunas, and genera such as *Sargana, Tuba, Belliscala, Confusiscala, Anchura, Stantonella, Liopeplum, Anomalofusus, Creonella, Morea,* and *Calliomphalus* as being characteristic of younger Cretaceous biotas in the Gulf-Atlantic Coast Subprovince. KAUFFMAN (1973a) added 36 endemic bivalve genera and subgenera (for example, *Paranomia, Periplomya, Postligata, Sexta, Brachymeris, Uddenia, Scambula, Aenona, Nelttia, Tenea, Fulpia* and *Pharodina*), and suggested possible division into a mid-temperate Atlantic coastal endemic center and warm-temperate to subtropical Gulf coastal endemic center. The bivalves *Anadara, Costellacesta, Tellinimera, Larma,* and *Cyclorisma* are endemic to the Atlantic and eastern Gulf Coast, and *Linter, Lycettia, Etea, Pollex, Sinonia,* and *Terebrimya* to the western Gulf Coast (KAUFFMAN, 1973a, p. 369).

TETHYAN REALM

The Tethyan Realm includes all tropical to subtropical waters and biotas of the Cretaceous equatorial zone and South Pacific (Fig. 5). It is the most diverse and mature of Cretaceous paleobiogeographic units, and seems to reflect time-stability evolution in a relatively uniform environment (see discussion in KAUFFMAN, 1973a). The Tethys Sea, restricted to Mediterranean Europe and the Indo-Pacific Ocean prior to the Cretaceous, spread westward across the tropical proto-Atlantic to the Americas with opening of the Caribbean Seaway (Late Jurassic, Early Cretaceous) and linked up with the tropical East Pacific at the present site of Central America to form a circumglobal tropical marine belt. This encouraged broad distribution of tropical invertebrates. These then became progressively isolated with Late Cretaceous tectonic constriction of both the Mediterranean and Caribbean arms of Tethys, and spreading of the Atlantic Ocean to widths that exceeded the extent of larval and mobile adult species dispersal from Eurasia and North Africa to America. High levels of Late Cretaceous endemism resulted, and biogeographic differentiation was great (Fig. 7). Diversity among many groups, especially bivalves (KAUFFMAN, 1973a, fig. 7), gastropods, corals (COATES, 1973), and pelagic microbiota, increased throughout the Cretaceous in pulses that were tied to eustatic rise, transgressive peaks, and expansion of Tethyan influence north and south over epicontinental areas formerly occupied by warm-temperate seas (COATES, KAUFFMAN, & SOHL, 1977).

In the Early Neocomian (Fig. 7), only the Indo-Mediterranean Region of Tethys was clearly defined by endemism; the poor faunas of the early Caribbean Sea were largely of Mediterranean affinities. Valanginian through Barremian time was characterized by high levels of faunal differentiation within the region, with formation of a North Indian Ocean Subprovince in the Valanginian, and Eastern and Western Mediterranean subprovinces in the Barremian (Fig. 7), which persisted to the end

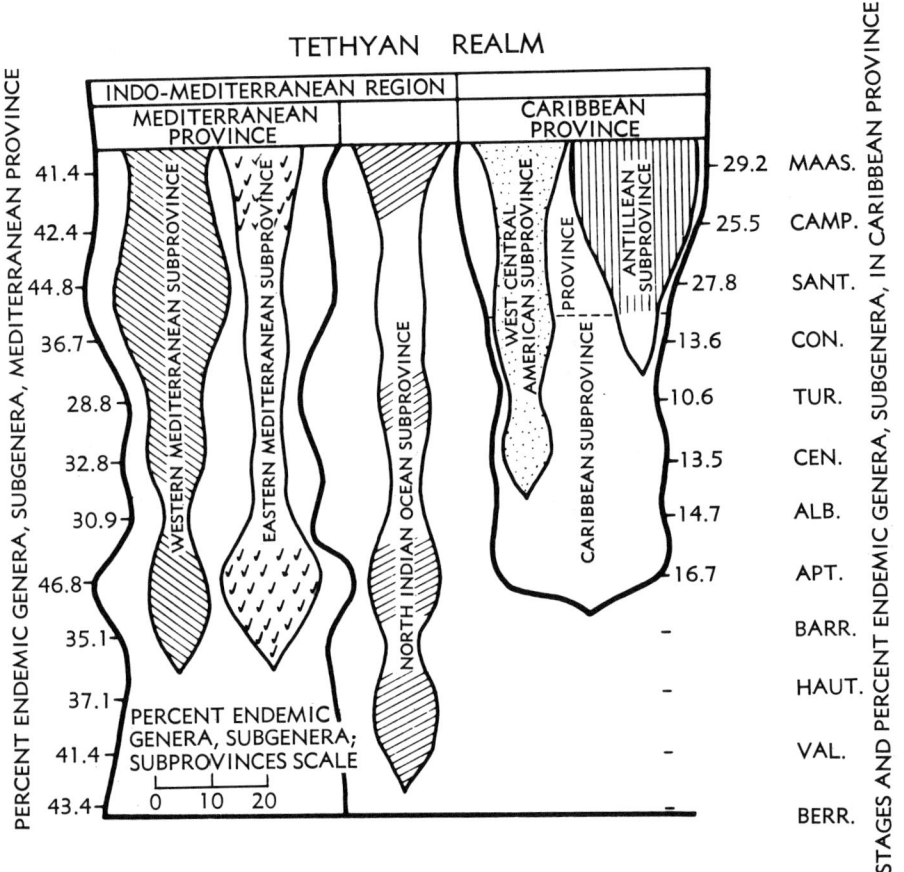

FIG. 7. Evolutionary trends in Cretaceous paleobiogeographic units of the Tethyan Realm, based mainly on bivalve Mollusca (after Kauffman, 1973a, fig. 3). Graphs show percent endemism of bivalve genera and subgenera within each unit, exclusive of cosmopolitan taxa. Scales for percent endemism in provinces (heavy lines) are shown on sides of graph. Scale for subprovinces (light lines) is shown in lower left. Origin point for each unit depicts the time at which endemism reached subprovincial rank (10-25 percent). For explanation of patterns, see Figure 6 caption. [Used with permission of Elsevier Scientific Publishing Co., Amsterdam.]

of the Cretaceous. These periods of differentiation apparently correlate with eustatic rises of sea level that produced moderate transgressions at continental margins. KAUFFMAN (1973a) called on two mechanisms to produce these Indo-Mediterranean subprovinces: 1) partial constriction of the Mediterranean Tethys during plate activity associated with closing of Tethys in Eurasia, producing barriers to east-west migration of marine organisms (sills?) and genetic isolation; and 2) "diversity pump" enrichment of local faunas (VALENTINE, 1967) resulting from isolation and differentiation of marginal Tethyan biotas in tropical epicontinental extensions of the sea during transgression, subsequent equatorial migration of these local biotas into specific regions of Tethys during regression, and their accommodation through niche partitioning among resident Tethyan faunas.

High levels of provincialism in the Caribbean first developed during the Middle Aptian (Fig. 7), associated with a moderate transgressive pulse reflecting global eustatic rise. The Caribbean Subprovince was formed and expanded to provincial rank by the Santonian. KAUFFMAN (1973a) suggested that this sudden jump in endemism might mark the point at which the middle tropical proto-Atlantic opened sufficiently to prohibit many pelagic larvae of benthonic taxa and mobile adults from regularly crossing this ocean barrier, resulting in widespread genetic isolation of similar stocks on both sides. High levels of niche specialization and radiation, especially within the rudist "reef" environment, also enhanced endemism in the Caribbean. Subsequently, partial isolation of Antillean and West-Central American Caribbean faunas occurred between Cenomanian and Coniacian times (Fig. 7), producing discrete subprovinces, and probably reflecting either early emplacement of marine barriers to migration in the present site of Central America or shifting current systems. The mid-Pacific Tethyan biotas are still poorly studied from dredge samples and cores drilled on atolls and other islands. The known biotas are generally Tethyan in aspect, containing rudist bivalves, but are of low diversity, and are most closely related to those of the West-Central American Subprovince (especially of Baja California). However, a distinct biogeographic unit may lie in the Cretaceous mid-Pacific.

Typical of the entire Tethyan Realm are rudist bivalves (Hippuritacea) (KAUFFMAN, 1973a; COATES, 1973), hermatypic corals (COATES, 1973), most larger benthonic foraminifers (DILLEY, 1971, 1973), highly diverse pelagic microbiotas, selected families of gastropods (SOHL, 1971), brachiopods (AGER, 1973), Lower Cretaceous belemnites (STEVENS, 1973), sponges (REID, 1967), echinoids, and numerous diagnostic ammonites.

Among the ammonites, WIEDMANN (1973) documented widespread heteromorphs (Ancyloceratina) in Tethys during the Early Neocomian (especially *Bochianites, Protancyloceras, Leptoceras,* and *Karsteniceras*); OWEN (1973) described his Albian "Tethyan-Gondwana Ammonite Province" as characterized by widespread lyelliceratid, brancoceratinid, mojsisovicsiinid, and mortoniceratinid ammonites. KENNEDY and COBBAN (1976) and CASEY (1971) noted that pseudoceratites such as *Parengonoceras* and Lower Cretaceous Berriasellidae are widespread in Tethys. In the Upper Cretaceous, according to MATSUMOTO (1973), Tethys is characterized by *Neolobites* ammonite faunas in the Cenomanian of Europe; by Desmoceratidae, Tetragonitidae, and Phylloceratidae in the Cenomanian-Turonian of the Indo-Pacific; by Turonian, Vascoceratidae and Tissotiidae with distinct southern and northern Tethyan assemblages of genera in the Mediterranean area; by Coniacian-Santonian, genera of Tissotiidae and Coilopoceratidae, with emphasis in warm water areas of the Indo-Pacific on genera such as *Kossmaticeras, Anapachydiscus, Damesites, Tetragonites, Gaudryceras, Hyphantoceras,* and *Scalarites*; and by local endemic groups of Campanian-Maastrichtian ammonites. STEVENS (1973) noted the spread of belemnites belonging to the genera *Hibolithes, Duvalia,* and *Conobelus* through much of the Tethys until the Barremian, when provincialism began (as in Bivalvia).

Among brachiopods, AGER (1973) noted wide Tethyan distribution, mainly within the Indo-Mediterranean Region, for Pygopiidae *(Pygope)* and *Septirhynchia*. COATES (1973) statistically documented strong similarities in hermatypic corals throughout the Lower Cretaceous Tethys, but development of strong post-Albian endemism between the Mediterranean and Caribbean provinces. Actaeonellid and nerineid gastropods characterize the Cretaceous Tethys virtually everywhere (SOHL, 1971), as do benthonic larger foraminifers, especially Lituolacea (including *Orbitolina* and associated genera such as *Dictyoconus, Choffatella, Cuneolina,* and *Pseudocyclammina*) and the "orbitoidal" foraminifers in the Upper Cretaceous (DILLEY, 1971, 1973). Many pelagic Globigerinacea were confined to circumglobal Tethyan and warm-temperate waters.

The Indo-Mediterranean Region has a marked endemic fauna, including Early Cretaceous ammonites of the Ancyloceratina such as *Hamulinites* and *Juddiceras* (WIEDMANN, 1973), genera of the Berriasellidae, and Late Cretaceous genera of the Vascoceratidae (for example, *Nigericeras, Paravascoceras, Fagesia*) (MATSUMOTO, 1973). Among Bivalvia, KAUFFMAN (1973a) listed nine genera that are endemic to the region, including *Trichites, Callucina (Callucinopsis), Pterolucina, Psilotrigonia, Arctomytilus,* and the rudist *Lapeirousia.* A similarly unique gastropod assemblage occurs, centered around genera and species groups of Nerineidae and Actaeonellidae. STEVENS (1973) noted Indo-Mediterranean restriction of several Lower Cretaceous belemnite groups belonging to the genera *Hibolithes, Duvalia, Neohibolites, Mesohibolites,* and *Parahibolites*; species groups of some of these genera show provincialism within the region. The *Neohibolites-Parahibolites* assemblage continued to characterize the region into the Cenomanian. DILLEY (1971) described a distinct larger foraminiferal assemblage from the Indo-Mediterranean region (his "Old World" Tethyan Province) centered around the Alveolinidae.

Subprovincial division of the Indo-Mediterranean Region is presently based on bivalves, Lower Cretaceous belemnites, and larger foraminifers. From the Valanginian onward the North Indian Ocean Subprovince (Fig. 5) became weakly to moderately well defined (Fig. 7) on the basis of endemic rudistid, cardiid, trigoniid, and eligmid bivalves such as *Dechaseauxia, Hardaghia, Stefaniniella, Collignonicardia, Libyaconchus, Praecardiomya, Bouleigmus,* and *Malagasitrigonia,* and larger foraminifers of the genus *Loftusia.* Early Cretaceous Ancyloceratina such as *Eocrioceratites, Menuthiocrioceras,* and *Parapedioceras* in Madagascar and Upper Cretaceous taxa detailed in the work of COLLIGNON (in BESAIRE & COLLIGNON, 1972) were endemic to the province up to the point that Madagascar became part of the temperate East African Province. STEVENS (1973) recorded an early Cretaceous split in Tethyan belemnite faunas into endemic groups, one of which *(Belemnopsis)* ranged through the Indo-Pacific during the Berriasian, with endemic species plexuses in the North Indian Ocean Subprovince.

The Tethyan Mediterranean Province (Fig. 5) was well established at the beginning of the Cretaceous and by Barremian time split into eastern and western subprovinces, each with strong endemic centers (Fig. 7). A suite of rudist, cardiid, and eligmid bivalves ranges throughout, but not beyond the province (KAUFFMAN, 1973a); this includes *Pseudoheligmus, Gonilia, Polyconites, Sabinia, Integricardium, Filosina,* and *Schiosia.* Endemic belemnites (STEVENS, 1973) appeared after earliest Cretaceous time and are mainly members of the Duvaliidae, and species plexuses of *Hibolithes, Conobelus, Mesohibolites,* and *Neohibolites.* COATES (1973) reported high levels of Late Cretaceous coral endemism in the Mediterranean. Certain groups of vascoceratid ammonites are likewise restricted to the province (MATSUMOTO, 1973). Among Foraminifera, DILLEY (1971, table 2) listed numerous endemic taxa in his "Old World Tethyan Province" among Late Cretaceous Lituolacea and Alveolinidae (especially most Alveolininae).

Eastern and Western Mediterranean subprovinces (Fig. 5) were exceptionally well defined after the Barremian on the basis of bivalves, belemnites, and larger foraminifers. The Western Mediterranean Subprovince, extending across southern Europe to the Balkans, and across North Africa to Egypt, is mainly characterized by a diverse suite of endemic rudistid bivalves such as *Bayleia, Matheronia, Valletia, Caprotina, Sphaerulites, Praelapeirousia, Orthoptychus, Roussleia, Medeela, Radiolitella,* and *Synodonites,* by belemnites of the Duvaliidae, and by certain larger foraminifers of the Alveolinidae (for example, latest Cretaceous *Murciella*; DILLEY, 1973, fig. 9).

The Eastern Mediterranean Subprovince (Fig. 5) is even more clearly defined by endemism in the same groups. Among larger foraminifers, species of *Loftusia* (Lituolacea) are largely restricted to the subprovince as are species of the alveolinid *Sellialveolina* (DILLEY, 1973, fig. 8), belemnite species of *Conobelus* and *Pseudobelus* (pre-Barremian), *Mesohibolites* and *Curtohibolites* (Barremian), *Parahibolites* (Cenomanian), and at the warm-temperate to subtropical margin, the Campanian *Belemnitella praecursor* group (STEVENS, 1973), and a small group of endemic bivalve genera, including *Agapella, Asiatotrigonia, Corbiculopsis, Vautrinia* and *Turkmenia* (KAUFFMAN, 1973a).

In both the Eastern and Western Mediterranean subprovinces, local endemic centers based on bivalve genera (mainly rudists) are defined by KAUFFMAN's (1973a) quantitative biogeographic method. In the west these include a strong southern French center (including the rudists *Retha, Offneria, Fossulites, Robertella, Arnaudia*), an Italian center (rudists include *Paronella, Apulites, Colveraia, Joufia, Pileochama*), and a Yugoslavian center (including the rudists *Neocaprina, Yvaniella, Gorjanovicia, Kuehnia, Neoradiolites, Pseudopolyconites, Milanovicia*). In the Eastern Mediterranean Subprovince these include a well-defined Syrian-Lebanese Endemic Center with *Arcullaea, Xenocardia, Syrotrigonia, Megalocardia, Nemetia,* and *Paracaprinula,* and a weakly defined Iranian Endemic Center with *Rostroperna, Dictyoptychus,* and *Osculigera*. Causes for local genetic isolation and resultant endemism of these bivalve assemblages are thought to center around collapse of the Mediterranean carbonate platform during Cretaceous closing of Tethys, and formation of numerous isolated "microcontinents" or large islands in the northern Mediterranean Tethys.

The Caribbean Province (Fig. 5) first developed during the Aptian and differentiated into West-Central American and Antillean subprovinces by the Coniacian (Fig. 7). Rudistid bivalves primarily serve to differentiate this province and the two subprovinces, but strong endemism among gastropods and larger foraminifers is also known, and ongoing studies of the hermatypic corals also indicate numerous endemic groups (COATES, 1973). Characteristic rudists of the Caribbean Province include *Amphitriscoelus, Caprinuloidea, Coalcomana, Planocaprina, Titanosarcolites, Barrettia, Praebarrettia, Chiapasella, Tampsia,* and *Tepeyacia*; the infaunal *Glycymeris (Glycymeris)* is similarly restricted. Endemic middle Cretaceous ammonites of the genera *Dufreynoyia, Hypacanthoplites, Drakeoceras, Budaiceras,* and numerous Mojsisovicziinae evolved in isolated basins behind barrier-reef tracts in the northern subtropical margin of the Caribbean Province (YOUNG, 1972). Two major groups of larger foraminifers, the pseudorbitoids and the Chubbininae, were endemic to the Caribbean Province (DILLEY, 1973). SOHL (1971) listed typical Caribbean Province gastropods, including many nerineid and actaeonellid groups, and endemic genera and species groups of fissurellids, trochids, neritids, cerithiacians, cypraeids, and strombids.

Division of the Caribbean Province into subprovinces has been made exclusively on analysis of Bivalvia, mainly rudistids and cardiids (KAUFFMAN, 1973a). The Antillean Subprovince (Fig. 5) is characterized by endemic genera such as *Bayleoidea, Pseudobarrettia, Anodontopleura, Baryconites,* and *Immanitas* with a weak, Mexican Endemic Center established around species of *Palus,* a strong Greater Antilles Endemic Center based on endemic *Parastroma, Torreites, Antillocaprina,* and *Parabournonia,* and endemic species of *Kipia* restricted to Trinidad. The West-Central American Subprovince (Fig. 5) is mainly defined on *Coralliochama, Incacardium, Vepricardium (Perucardia), Tortucardia, Peruarca,* and *Pettersia.*

SOUTH TEMPERATE REALM

All temperate-zone biotas south of Tethys are grouped within this realm (Fig. 5), which is one of the least studied and most poorly defined of Cretaceous paleobiogeographic units. Factors that account for relatively low levels of temperate diversity and endemism in the South Temperate Realm include: 1) much of the South Pacific was under the influence of tropical to subtropical waters in the Jurassic and Early Cretaceous; 2) the Tethyan-South Temperate boundary shifted slowly northward through the Cretaceous and for the first time allowed an expanded temperate zone in the South Pacific; and 3) both the opening of the South Atlantic and widespread separation of the Gondwanaland continents were mainly Late Cretaceous events. In all cases, South Temperate marine faunas were in an early stage of differentiation during the Cretaceous, with predictably low diversity and low levels of endemism. The biogeographic evolution of the realm (Fig. 8) was correspondingly simple. Following initial separation of Africa and South America by, first, shallow epicontinental seas and later narrow oceanic troughs, faunas of both the Euramerican temperate region (mainly) and the Caribbean-Mediterranean part of Tethys

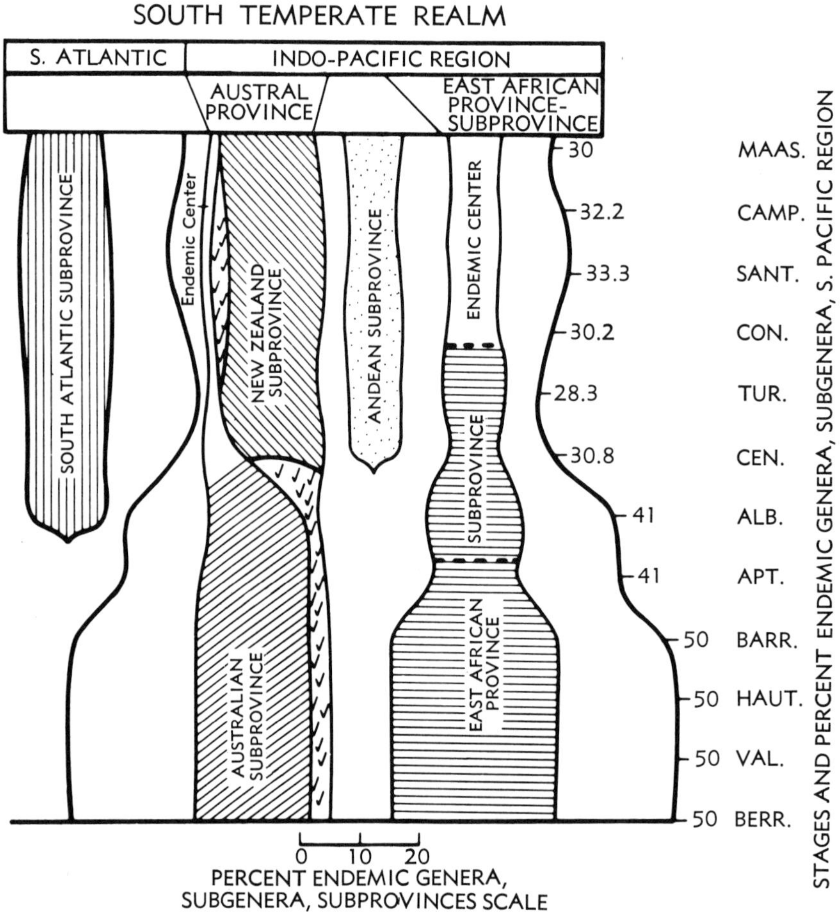

FIG. 8. Evolution of Cretaceous paleobiogeographic units in the South Temperate Realm, based mainly on bivalve Mollusca (after Kauffman, 1973a, fig. 5). Graphs show percent endemism among bivalve genera and subgenera for each area, exclusive of cosmopolitan taxa. Scale for endemism in Indo-Pacific Region at right, and scale for subprovince graphs at base of figure. For explanation of patterns, see Figure 6 caption. [Used with permission of Elsevier Scientific Publishing Co., Amsterdam.]

immigrated into the South Atlantic. Separation of the South Temperate elements from more highly evolved North Temperate biotas by the tropical Tethyan seaway led to low levels of endemism in the South Atlantic, which did not reach subprovincial grade until the Albian (Fig. 8). This corresponded with the first large-scale epicontinental transgression onto Cretaceous continents, including those bordering the South Atlantic. South Atlantic endemism did not exceed subprovincial rank throughout the Cretaceous, and due to the short distances for larval and mobile adult migration across the proto-South Atlantic, invertebrate faunas of eastern South America and West Africa remained closely related. Bivalve genera such as *Euptera, Agelasina, Anofia, Naulia, Sergipia, Pseudopleurophorus,* and *Gilbertharrisella* characterize the subprovince (KAUFFMAN, 1973a).

The biogeographic evolution of the southern Indo-Pacific Region during the Cretaceous was much more complex and involved both tectonic and climatic controls. Tectonically, the Cretaceous continuation of the breakup and dispersal of parts of Gondwanaland caused further isolation of their marginal marine shelf biotas. The east-west separation of East Africa and Australia continued, as did the early northward migration of the Indian subcontinent. Australia may have partially separated from Antarctica during the Cretaceous, moving slightly northward, and there is faunal evidence for accelerated post-Neocomian separation of New Zealand from Australia to the point where genetic isolation occurred, producing significant molluscan endemism on New Zealand. Thus, Cretaceous plate movements in some cases led to genetic isolation and endemism among South Pacific invertebrates. But based on distances between major marine shelf areas suggested by plate reconstructions such as those of DIETZ and HOLDEN (1970) and SMITH, BRIDEN, and DREWRY (1973), the degree of continental separation was not sufficient to account for the high levels of endemism observed here in the Cretaceous invertebrate faunas. KAUFFMAN (1973a) proposed an additional isolating mechanism; that is, climatic deterioration of the South Pacific resulted from a major Late Cretaceous cooling trend and northward movement of, first, the Tethyan-temperate climatic boundary (Jurassic-Early Cretaceous) followed by the warm temperate-mid temperate boundary (Lower Cretaceous) and late in the period, the cool-temperate boundary across Gondwana. Decline of tropical faunal influence from the Jurassic onward supports this (STEVENS, 1965; KAUFFMAN, 1973a). The combination of moderate tectonic separation and the differentiation and northward migration of temperate climatic zones across dispersed parts of Gondwanaland is called upon to produce isolating mechanisms and to explain observed Cretaceous endemism.

The temperate Indo-Pacific Region is characterized by widespread endemics among the bivalves (for example, *Iotrigonia, Megatrigonia, Pacitrigonia, Maccoyella*) as well as by bivalves common to the region but with limited outside distribution (for example, *Acharax, Fimbria, Freiastarte, Parapholas, Monothyra, Steinmanella,* and *Neocrassina*) (KAUFFMAN, 1973a). MATSUMOTO (1973) cited numerous genera of ammonites that have species groups typical of the Indo-Pacific Region such as Cenomanian-Maastrichtian Desmoceratidae (*Desmoceras, Puzosia, Mesopuzosia,* and *Pachydesmoceras*), Tetragonitidae (*Gaudryceras, Anagaudryceras*), Phylloceratidae (*Hypophylloceras, Neophylloceras*), and the Kossmaticeratidae (*Marshallites, Kossmaticeras, Maorites, Grossouvrites,* and *Jacobites* species plexuses). Many of these were more characteristic of tropical zones but ranged into warm-temperate areas as well. Important Lower Cretaceous Indo-Pacific ammonite groups are the Albian Lyelliceratidae, Brancoceratidae, Mojsisovicziinae, and Brancoceratinae, many of which were shared with Tethys proper. STEVENS (1973) clearly documented evolution of an endemic temperate Indo-Pacific belemnite assemblage beginning with the *Belemnopsis madagascarensis-B. patagoniensis* lineage in the Valanginian and Hauterivian, and including the Late Neocomian through Cenomanian Dimetobelidae (restricted to the Austral Province after the Cenomanian).

Two major divisions of the temperate Indo-Pacific Region were established at the beginning of the Cretaceous, an East African Province extending from South Africa to (in Late Cretaceous time) Madagascar, and an Austral Province (Figs. 5, 8). Both show marked decline in endemism in the Upper Cretaceous (Fig. 8), and this is correlative with and directly related to marine climatic decline with global cooling and northward shift of more temperate zones across the southernmost Pacific (STEVENS, 1965; KAUFFMAN, 1973a). Faunal differentiation between these two regions is thought to reflect eastward drift of Australia to distances that exceeded large-scale faunal exchange through larval drift.

The Austral Province contains two subprovinces that are basically time successive (Fig. 8), a strong Early Cretaceous Australian Subprovince that declined to a weak endemic center during the Late Cretaceous; here it was replaced by a well-defined New Zealand Subprovince that has no Early Cretaceous expression. Few taxa are endemic to the entire province, *Maccoyella* being the principal bivalve, the Dimetobelidae among belemnites, and the hoplitid ammonite *Chimbuites* (Cenomanian). Most of the Austral ammonite and bivalve fauna is comprised of widely spread Pacific and Tethyan forms.

The Australian Subprovince is most clearly defined, as presently known, by its endemic bivalves; other groups have not been similarly analyzed. Important endemic genera are *Pseudavicula, Austrotrigonia, Nototrigonia (Callitrigonia), N. (Nototrigonia), Cyrenopsis, Barcoona,* and *Tatella* in the Lower Cretaceous, and *Climacotrigonia, Actinotrigonia, Entolium (Cteniopleurium),* and *Fissiluna* in the weakly defined Upper Cretaceous Australian Endemic Center (Fig. 8). The sharp decline in endemism within the Australian Subprovince has yet to be satisfactorily explained.

The New Zealand Subprovince (Fig. 5) became sharply defined in the Cenomanian and continued as a distinct biogeographic unit through the Maastrichtian (Fig. 8). The abrupt development of New Zealand endemism is readily explained in terms of marine history. The oldest marine invasions were of Aptian age, and these were marginal transgressions characterized by low diversity faunas drawn from generalized Austral and more widely ranging Pacific stocks. Major epicontinental flooding of New Zealand first occurred in the Cenomanian and continued through the Cretaceous, producing environments for diversification and local isolation of invertebrate taxa—conditions leading to the development of endemism. KAUFFMAN (1973a) has suggested that the abrupt development of high levels of Cenomanian endemism in New Zealand might also reflect its plate tectonic separation from Australia and Antarctica to the point where faunal exchange was restricted, as well as development of climatic barriers to migration. Endemism in the New Zealand Subprovince has been defined mainly on Bivalvia, and the subprovince is characterized by genera such as *Chlamys (Mixtopecten), Electroma, Megaxinus (Pteromyrtea), Myrtea, Lahillia, Marwickia, Dosinobia,* and *Cyclorismina*, some of which have limited outside distribution.

The East African Province was well defined at the beginning of the Cretaceous (Fig. 8) but gradually declined in endemism to subprovincial rank by Albian time, and endemic center rank in the Coniacian. This decline is directly related to environmental deterioration with cooling of the South Pacific and northward migration of temperate climatic belts throughout the Cretaceous (KAUFFMAN, 1973a), eventually bringing areas as far north as Madagascar within range of temperate waters. Numerous bivalve genera are endemic to this province (for example, *Pleurotrigonia, Sphenotrigonia, Herzogina, Isotancredia, Megacucullaea,* and Madagascar lineages of *Malagasitrigonia*). COLLIGNON (in BESAIRE & COLLIGNON, 1972) has documented immense numbers of ammonites that seem to be restricted to this subprovince, and in many cases to Madagascar. The belemnite *Belemnopsis* is most abundant in this subprovince, although it ranges less commonly elsewhere.

The Andean Subprovince (Fig. 5), extending from Patagonia to southern Peru, is weakly defined; it was initiated with the first major epicontinental flooding of the west coast of South America during the Cenomanian (Fig. 8). The subprovince was isolated from western and most southern Pacific biotas by the vast water barrier of the southeast Pacific, and from north temperate faunas by the tropical Caribbean arm of Tethys. Some taxa that populated this warm-temperate subprovince were obviously drawn from the Caribbean Province, but most stocks probably came from the south through Patagonia. Population of this marine shelf area was slow after Cenomanian flooding, reflecting restriction of source areas, and endemism is consequently low. KAUFFMAN (1973a) has noted a few endemic bivalves; for example, *Anopisthodon* and *Aulacopleurum* from the Lower Cretaceous, and the same genera, in addition to *Mulinoides* and *Tellipiura*, in the Upper Cretaceous. Among other taxa, phylogenic relationships are clearly strongest southward, through Patagonia to the Austral and East African provinces.

SUMMARY AND SOME OBSERVATIONS ON EVOLUTION

A recurrent theme throughout this résumé of Cretaceous marine history and invertebrate faunas has been the dynamic relationship between major tectonic, oceanographic, and climatic changes during the period—the large scale forces of natural selection—and evolution at all levels. Thus, evolutionary rates and patterns within lineages, ecological units, and paleobiogeographic units vary tremendously in response to the rate and intensity of Cretaceous environmental changes, especially those triggered by periods of active plate tectonism. There is no long-term environmental stability or evolutionary stasis in the Cretaceous, and this demands that interpretation of invertebrate history be based on sampling intervals and analytical techniques that are designed to consistently test the dynamics of the system.

In regard to Cretaceous paleobiogeographic units, a stage-by-stage quantitative analysis of endemism yields the most detailed and consistent divisions, and allows comparison of units from relatively narrow time intervals with modern counterparts. Using this method, it becomes obvious that Cretaceous biogeographic units were nearly as complex as those of today, and showed dynamic changes—appearance and disappearance, variations in magnitude (as measured by endemism), and changes in spatial influence—throughout the period (Figs. 6-8). These variations were very closely linked to global plate-tectonic changes and climatic trends. Genetic isolation leading to significant endemism resulted from many processes: 1) sea floor spreading and progressive separation of marine shelf areas; 2) dispersal of microcontinents or pieces of supercontinents through plate tectonic events; 3) isolation of biotas through point collision of plates, as in the Mediterranean, or by tectonic emplacement of biogeographic barriers; and 4) isolation of biotas in epicontinental seaways during transgressive-regressive pulses (reflecting eustatic fluctuations) that themselves were generated by plate tectonism. Rates of faunal differentiation, leading to development and modification of paleobiogeographic units, are directly correlative with rates and intensity of the isolating mechanisms.

The decline of Cretaceous paleobiogeographic units was similarly related to 1) mass competition among distinct biotas brought into contact through plate movements or the flooding of epicontinental seas during eustatic rise, and 2) environmental decline resulting from lowering global temperatures and equatorward migration of progressively cooler temperate climatic zones. With the possible exception of the Late Cretaceous decline in Australian endemism, all evolutionary changes in Cretaceous paleobiogeographic units can be directly related to these tectonic, climatic, and biological factors. Figures 6-8 summarize the dynamic evolutionary trends within Cretaceous paleobiogeographic units.

Similarly, marine diversity during the Cretaceous increased in an irregular manner for most prominent groups between Berriasian and Campanian time (Fig. 9), with diversity fluctuations closely tied to major eustatic fluctuations (SOHL, 1967; TAPPAN, 1968, 1971; TAPPAN & LOEBLICH, 1971, 1972; KAUFFMAN, 1973a; FISCHER & ARTHUR, 1977; among others). Some Cretaceous groups continued their evolutionary diversification through the Maastrichtian and into the Paleogene, whereas others show marked decline and even total extinction at the Cretaceous-Tertiary boundary. Thus, TAPPAN and LOEBLICH (1972, p. 205) noted that "most protistan groups had coincident evolutionary bursts and declines, with similar fluctuations in their evolutionary rates" and were ". . . . intimately related to the selective pressures of the changing global environment. Advancing and retreating epicontinental and shelf seas and changes in continental position and height affected atmosphere and oceanic circulation and the climatic regime, causing variations in nutrient supply and fluctuations in productivity and food resources." They noted greatly increased diversity among protists of various types in the Upper Cretaceous associated with environmental stability and climatic amelioration produced by major periods of transgression, and great decline in latest Cretaceous protistan diversity associated with climatic deterioration and the

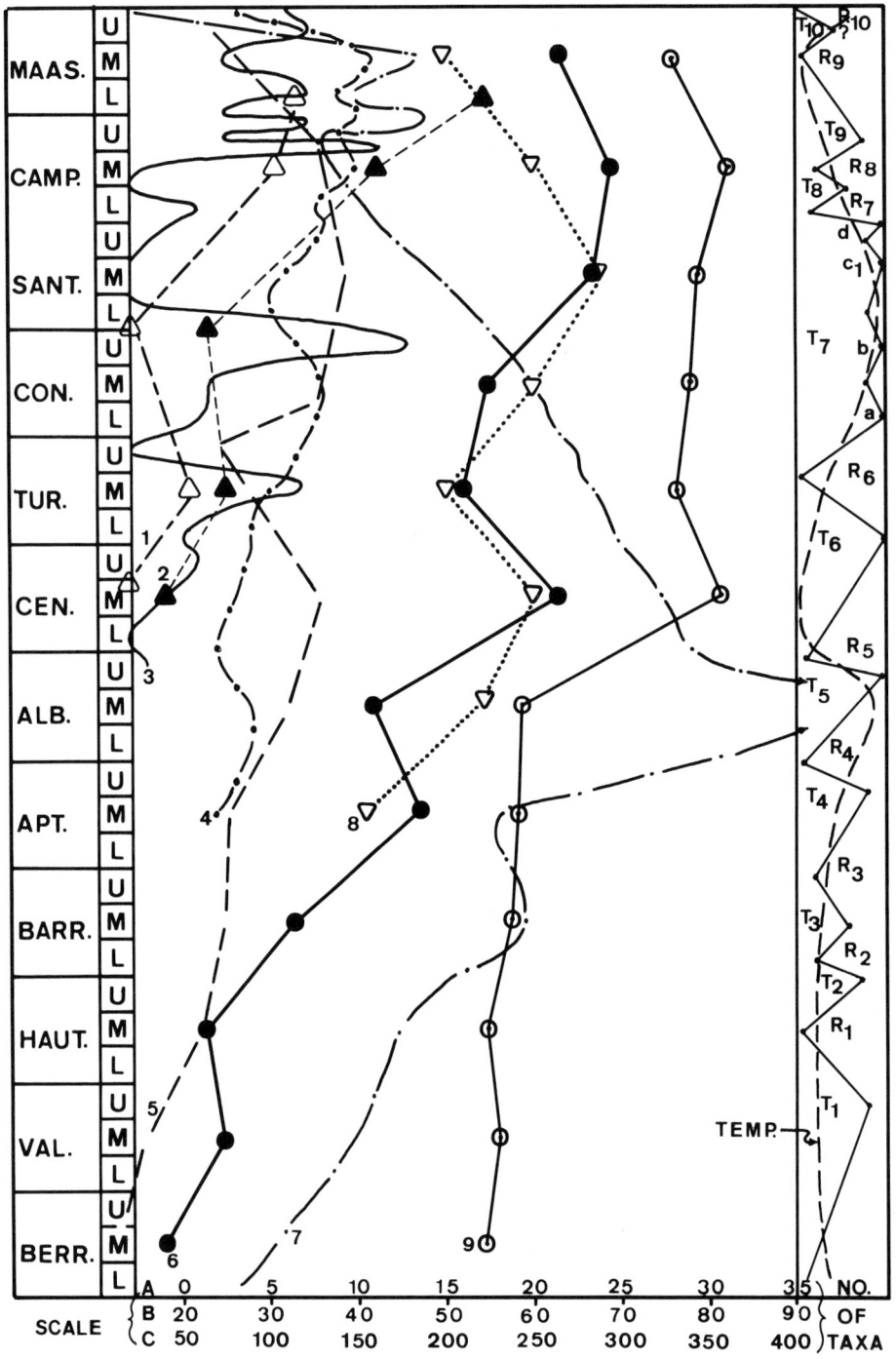

Fig. 9. Selected diversity curves for the Cretaceous. Curve 5 is plotted to scale A (bottom of chart), curves 1-3 and 6-8 to scale B, and 4 and 9 to scale C. Numbers and information for each curve are:

time of the major Maastrichtian regression (Fig. 9). Of particular interest is the close correlation of high dinoflagellate diversity with several Cretaceous transgressive maxima (Fig. 9) and low diversity with regression (TAPPAN & LOEBLICH, 1971, fig. 6). FISCHER and ARTHUR (1977) broadly analyzed pelagic Cretaceous diversity and concluded that protistans, ammonites, and marine reptiles, as well as pelagic ecological structure, increased significantly in conjunction with Albian and Coniacian-Santonian eustatic rise, epicontinental transgression, temperature rise, and climatic amelioration. SOHL (1967) has shown high diversification among gastropods of the North American Interior Subprovince to be associated both with transgressive maxima (Turonian, Coniacian, Late Campanian) and with regressive pulses near their stress maxima (Lower and Middle Campanian, Maastrichtian). Selected diversity curves from these sources are shown in Figure 9.

Some of the most detailed analyses of diversity and its relationship to Cretaceous environmental fluctuations have been made on Bivalvia (KAUFFMAN, 1973a, figs. 6-10; summarized here in Fig. 9). Diversity of bivalve genera and subgenera increased throughout the Cretaceous, with a small decline at the Maastrichtian regression and Cretaceous-Tertiary boundary. The rate of increase was irregular; diversity increased in pulses that are correlative with several of the eight transgressive maxima, reflecting eustatic rise generated through plate tectonism. Specifically, diversity increases were associated with Valanginian (Fig. 2; T_1), Aptian (T_4), latest Cenomanian (T_6), Coniacian-Santonian (T_7), and Middle Campanian (T_8) transgressive peaks. The most dramatic increase in diversity was during the great Cenomanian transgression, an extremely active period of sea floor spreading, genetic isolation, and probably the greatest epicontinental flooding during the Cretaceous. Most of the diversity increase occurs in Tethys; increase in temperate bivalves is low but more steady, and cosmopolitan bivalves retained a constant diversity level during the Cretaceous. In the overall trend toward increasing diversity of bivalves, reversals in the trend were also associated with eustatic fluctuations and peaked at times of maximum regression for the most part (Fig. 2; Hauterivian = R_1; Late Albian = R_5; Middle Turonian = R_6; and Late Maastrichtian = R_8 of KAUFFMAN's, 1973a, transgressive-regressive sequence).

Thus, many groups of Cretaceous invertebrates show diversity patterns that are primarily tied to transgressive-regressive marine epicontinental cycles, representing tectonically generated eustatic fluctuations. High marine shelf and pelagic diversity peaks were closely tied to major transgressive episodes characterized by increasing ecological space and niche opportunities, decreasing environmental stress, temperature increase and climatic amelioration. Lower diversity was associated with environmental decline, lowering of marine temperatures, decrease in ecospace and niche diversity, and epicontinental regression. Figure 9 records data for several invertebrate groups that document these relationships at various levels of refinement.

Similarly, the variable evolutionary history of lineages and higher taxonomic cate-

---

FIG. 9. (*Continued from facing page.*)

1–Pacific Province planktonic foraminiferal species, excluding Guembelitriinae (Douglas, 1972); *2*–Tethyan planktonic foraminiferal species, excluding Guembelitriinae, North America (Douglas, 1972); *3*–gastropod species, Western Interior North America (Sohl, 1967); *4*–planktonic globigerinid species, worldwide (Fischer & Arthur, 1977); *5*–dinoflagellate species, worldwide (Tappan & Loeblich, 1971); *6*–Tethyan bivalve genera, worldwide (Kauffman, 1973a); *7*–ammonite genera, worldwide (Fischer & Arthur, 1977); *8*–total phytoplankton species/10 (Fischer & Arthur, 1977); *9*–total bivalve genera (Kauffman, 1973a). Detail of transgressive-regressive history and generalized temperature curve (dashed line) shown in right-hand column (for sources see Fig. 2). Note relationship between large-scale transgression, warming of oceanic climates, and diversity increase among varied organisms in the Albian and Coniacian-Santonian stages. Also, diversity peaks among certain groups correlative with transgressive maxima in the Middle to Late Valanginian, Middle Barremian, Middle Aptian, Cenomanian to Early Turonian, Middle Campanian, and late Middle to Late Maastrichtian (from Kauffman, 1977f; used with permission of Rocky Mountain Assoc. Geol., Denver).

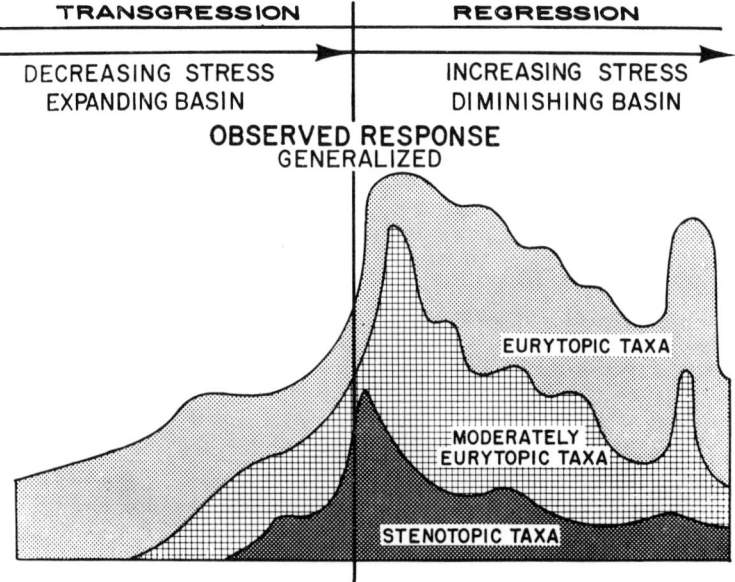

Fig. 10. Generalized model of evolutionary response by Cretaceous mollusks to eustatic changes of sea level (from studies of North American lineages; Kauffman, 1972). Graph implies increasing rates of evolution with expanding ecological opportunities during transgression, but highest evolutionary rates during regression and onset of stress conditions associated with decreasing ecospace, restriction of habitats, and increased competition. Times of maximum evolution differ for different lineages and adaptive strategies depending on degree of ecological specialization and habitat.

gories during the Cretaceous seems to show strong correlation with tectonic, oceanographic, and climatic changes. For example, evolutionary rates and patterns within single lineages, or classes, or ecological strategies of Cretaceous marine invertebrates are complex, and appear to vary primarily in response to eustatic changes in sea level, transgressive-regressive pulses, and correlative changes in physical environments, temperature, and climatic equitability (Figs. 9, 10). In other words, evolution rates vary with the intensity and rate of change of environmental stress factors (KAUFFMAN, 1970, 1972, 1977a). Global transgressions represent ameliorating marine environments and low physical stress periods, with expanding ecospace, niche opportunities, resources, and decreasing biological competition. Maritime climates predominate, reducing seasonality and increasing annual temperatures. Rates of evolution for most invertebrates are relatively low during transgression and are mainly controlled by the amount of new ecospace and niche opportunities into which "preadapted" groups can radiate rapidly. The most rapid evolutionary rates during transgression seem to be among two distinct groups: 1) early colonizing taxa that are subjected during later transgression to severe biological stress through widespread competition by the stenotopic specialists that replace them; and 2) taxa with new or modified adaptive strategies that give them a major competitive advantage within their potential niches and enhance their rapid radiation, as in rudist bivalves (KAUFFMAN, 1977a). Thus, although Cretaceous transgressions were times of marked increase in invertebrate diversity, they were not necessarily correlative with rapid evolutionary rates for most organisms until peak transgression.

KAUFFMAN (1970, 1972, 1977a) found high rates of molluscan evolution associated with peak transgression and the regression immediately following it. TAPPAN and LOEBLICH (1971, 1972) suggested the same for planktonic organisms, and SOHL (1967) in part for North American gastropods. Groups that underwent rapid evolution at or near peak transgressions were largely pelagic and offshore benthonic stenotopes responding to two types of increased stress situations: 1) relatively abrupt increase in surface water temperature and epicontinental salinity in temperate Cretaceous seas associated with major transgressive peaks (as determined isotopically, SCHOLLE & KAUFFMAN, 1977; and faunally KAUFFMAN & SCHOLLE, 1977), and 2) the initial shock of epicontinental regression associated with a concomitant lowering of temperate epicontinental water temperatures and salinities following peak transgression. Environmentally sensitive stenotopic organisms and microplankton were the groups most affected by these shifting environments. Among the benthos, gastropods especially were stressed by these changes and show strong increases in rates of evolution and extinction associated with peak transgression and earliest regression (SOHL, 1967; KAUFFMAN, 1972, for North America).

The most rapid rates of evolution documented for Cretaceous invertebrates (North American interior mollusks) were associated with later phases of marine regressions, increasing physical stress and biological competition, and diminishing ecospace, niches, and resources (KAUFFMAN, 1970, 1972, 1977a). Extinction was also high during these periods so that total diversity decreased. Among diverse North American Cretaceous molluscan species, representing most adaptive strategies and levels of environmental tolerance, a graded series of speciation peaks occurred during regression, with offshore stenotopes and many pelagic taxa peaking during early regression, moderately stenotopic shallow-water taxa peaking during mid-regression, and more eurytopic organisms as well as some nearshore stenotopes peaking near terminal regression and maximum environmental stress levels. Varied levels of extinction characterized peak regression. These studies clearly indicate an intimate relationship between evolutionary rates and patterns, environmental tolerance (stenotopy vs. eurytopy), and stress factors associated with eustatic global fluctuations and transgressive-regressive pulses in epicontinental seas. KAUFFMAN (1977a) has further shown a close correlation between evolutionary rates and ecological strategies among American Cretaceous mollusks. In a preliminary test it was found that under the same general sets of environmental changes: 1) feeding specialists evolved faster than feeding generalists; 2) epifaunal and some pelagic organisms evolved faster than semi-infaunal taxa, and these in turn faster than shallow and deep infaunal taxa; and 3) shallow-shelf taxa (especially epifaunal forms) evolved faster than either deep-water taxa on the one hand, or very shallow sublittoral, intertidal, brackish water, and freshwater taxa on the other. In each case, rapid evolutionary rates were related to ecological strategies that subjected the organism to high chance of unpredictable environmental perturbations through geological time. No relationship was found between morphological complexity and evolutionary rate among the Cretaceous organisms tested.

REFERENCES

Adams, G. G., & Ager, D. V. (eds.), 1967, *Aspects of Tethyan biogeography:* Syst. Assoc., Publ. 7, 336 p.

Ager, D. V., 1971, *Space and time in brachiopod history:* in Faunal provinces in space and time, F. A. Middlemiss, & P. F. Rawson (eds.), Geol. Jour., Spec. Issue 4, p. 95-110.——1973, *Mesozoic Brachiopoda:* in Atlas of palaeobiogeography, Anthony Hallam (ed.), p. 431-436, text-fig. 1-3, Elsevier Scientific Publishing Co. (Amsterdam).

American Commission on Stratigraphic Nomenclature, 1970, *Code of stratigraphic nomenclature:* Am. Assoc. Petrol. Geologists, Spec. Publ., p. 1-22.

Arkell, W. J., 1956, *Jurassic geology of the world:* 681 p., 102 text-fig., 41 pl., Oliver & Boyd Ltd. (Edinburgh).

Berner, R. A., 1974, *Physical chemistry of carbonates in the ocean:* in Studies in paleo-oceanography, W. W. Hay (ed.), Soc. Econ. Paleontol. & Mineral., Spec. Publ. 20, p. 37-43, text-fig. 1-4.

Berry, W. B. N., 1968, *Growth of a prehistoric time scale based on organic evolution:* 158 p., 16 text-fig., W. H. Freeman Co. (San Francisco, London).

Besairie, Henri & Collignon, Maurice, 1972, *Géologie de Madagascar: I. Les terrains sédimentaires:* Répub. Malagasy, Ann. Géol. Madagascar, Fasc. 35, 463 p., 89 pl.

Bond, Gerard, 1976, *Evidence for continental subsidence in North America during the Late Cretaceous global submergence:* Geology, v. 4, p. 557-560, text-fig. 1,2.

Bowen, Robert, 1966, *Paleotemperature analysis:* 265 p., Elsevier Scientific Publishing Co. (Amsterdam, London, New York).

Casey, Raymond, 1961, *The stratigraphical palaeontology of the Lower Greensand:* Palaeontology, v. 3, p. 487-621, pl. 77-84.———1968, *The type section of the Volgian Stage (Upper Jurassic) at Gorodische, near Ulyanovsk, U.S.S.R.:* Geol. Soc. London, Proc., no. 1648, p. 74-75.———1971, *Facies, faunas, and tectonics in Late Jurassic-Early Cretaceous Britain:* in Faunal provinces in space and time, F. A. Middlemiss & P. F. Rawson (eds.), Geol. Jour., Spec. Issue 4, p. 153-168.———1973, *The ammonite succession at the Jurassic-Cretaceous boundary in eastern England:* in The Boreal Lower Cretaceous, Raymond Casey & P. F. Rawson (eds.), Geol. Jour., Spec. Issue 5, p. 193-266, text-fig. 1-6, pl. 1-10, Seel House Press (Liverpool).

———, & Rawson, P. F., 1973a, *The Boreal Lower Cretaceous:* Geol. Jour., Spec. Issue 5, 430 p., Seel House Press (Liverpool).———1973b, *A review of the Boreal Lower Cretaceous:* in The Boreal Lower Cretaceous, Raymond Casey & P. F. Rawson (eds.), Geol. Jour., Spec. Issue 5, p. 415-430, tab. 1, Seel House Press (Liverpool).

Cloud, P. F., 1955, *Physical limits of glauconite formation:* Am. Assoc. Petrol. Geologists, Bull., v. 39, p. 484-492.

Coates, A. G., 1973, *Cretaceous Tethyan coral-rudist biogeography related to the evolution of the Atlantic Ocean:* in Organisms and continents through time, N. F. Hughes (ed.), Spec. Paper in Palaeontology, no. 12, Syst. Assoc. Publ. 9, p. 169-174, text-fig. 1-4.

———, Kauffman, E. G., & Sohl, N. F., 1977, *Cyclic incursions of Tethyan biotas into the Cretaceous temperate realms:* Jour. Paleontology, v. 51, suppl. 2, p. 7 [abstract].

Cobban, W. A., 1951, *Scaphitoid cephalopods of the Colorado Group:* U.S. Geol. Survey, Prof. Paper 239, 39 p., 21 pl.———1953, *Cenomanian ammonite fauna from the Mosby Sandstone of central Montana:* U.S. Geol. Survey, Prof. Paper 234-D, p. 45-55, pl. 6-12.———1958, *Late Cretaceous fossil zones of the Powder River Basin, Wyoming and Montana:* in Wyoming Geol. Assoc. Guidebk., 13th Ann. Field Conference, 1958, Powder River Basin, p. 114-119.———1964, *The Late Cretaceous cephalopod Haresiceras Reeside and its possible origin:* U.S. Geol. Survey, Prof. Paper 454-I, 21 p., 7 text-fig., 3 pl.———1969, *The Late Cretaceous ammonites Scaphites leei and Scaphites hippocrepis (DeKay) in the Western Interior of the United States:* U.S. Geol. Survey, Prof. Paper 618, 29 p., 21 text-fig., 5 pl.

———, & Reeside, J. B., 1952, *Correlation of the Cretaceous formations of the Western Interior of the United States:* Geol. Soc. America, Bull., v. 63, p. 1011-1044.

———, & Scott, G. R., 1964, *Multinodose scaphitid cephalopods from the lower part of the Pierre Shale and equivalent rocks in the conterminous United States:* U.S. Geol. Survey, Prof. Paper 438-E, p. 1-13, text-fig. 1-5, pl. 1-4.———1972, *Stratigraphy and ammonite faunas of the Graneros Shale and Greenhorn Limestone near Pueblo, Colorado:* U.S. Geol. Survey, Prof. Paper 645, 108 p., 39 pl.

Cox, L. R., 1967, *Stratigraphical tables of British Mesozoic strata:* in British Mesozoic fossils, 3rd edit. p. 11-20, Brit. Museum (Nat. History) (London).

Dietz, R. S., & Holden, J. C., 1970, *Reconstruction of Pangea: Breakup and dispersion of continents, Permian to present:* Jour. Geophys. Res., v. 75, no. 26, p. 4939-4956.

Dilley, F. G., 1971, *Cretaceous foraminiferal biogeography:* in Faunal provinces in space and time, F. A. Middlemiss & P. F. Rawson (eds.), Geol. Jour., Spec. Issue 4, p. 169-190, text-fig. 1-7.———1973, *Cretaceous larger Foraminifera:* in Atlas of palaeobiogeography, Anthony Hallam (ed.), p. 403-419, text-fig. 1-12, Elsevier Scientific Publishing Co. (Amsterdam).

Douglas, R. G., 1972, *Paleozoogeography of Late Cretaceous planktonic Foraminifera in North America:* Jour. Foram. Res., v. 2, no. 1, p. 14-34, text-fig. 1-14.

———, & Savin, S. M., 1973, *Oxygen and carbon isotope analyses of the Cretaceous and Tertiary Foraminifera from the central North Pacific:* in Initial reports of the Deep Sea Drilling Project, E. L. Winterer & J. L. Ewing (eds.), v. 17, p. 591-605, U.S. Gov't Printing Office (Washington, D.C.).

Egoyan, V. L., 1971, *Granitsa yuri i mela na severo-zapadnom Kavkaze i nekotorie voprosi metodiki stratigraficheskix issledovanii:* Magyar Állami Földt. Int. Evi., v. 54, no. 2, p. 125-129. [*The boundary of the Jurassic and Cretaceous in the northwestern Caucasus and some questions of the methodical stratigraphic investigation.*]

Ekman, Sven, 1967, *Zoogeography of the sea:* 2nd edit. (transl. by E. Palmer), 417 p., Sidgwick & Jackson (London).

Ernst, Gundolf, & Seibertz, Ekbert, 1977, *Concepts and methods of echinoid biostratigraphy:* in Concepts and methods of biostratigraphy, E. G. Kauffman & J. E. Hazel (eds.), p. 541-563, text-fig. 1-6, Dowden, Hutchinson, & Ross, Inc. (Stroudsburg, Pa.).

Fischer, A. G., & Arthur, M. A., 1977, *Secular variations in the pelagic realm:* in Basinal facies in carbonate sediments, H. E. Cook & P. Enos (eds.), Soc. Econ. Paleontol. & Mineral., Spec. Paper 25 (in press).

Frerichs, W. E., 1971, *Evolution of planktonic Foraminifera and paleotemperatures:* Jour. Paleontology, v. 45, p. 963-968, text-fig. 1-5.

———, & Adams, P. R., 1973, *Correlation of the Hilliard Formation with the Niobrara Formation:* 25th Field Conf., Wyoming Geol. Assoc., Guidebk., p. 187-192, text-fig. 1-4, pl. 1,2.

Frush, M. P., & Eicher, D. L., 1975, *Cenomanian and Turonian Foraminifera and paleoenvironments in the Big Bend Region of Texas and Mexico:* in The Cretaceous System in the Western Interior of North America, W. G. E. Caldwell (ed.), Geol. Assoc. Canada, Spec. Paper 13, p. 277-301, text-fig. 1-14.

Gill, J. R., & Cobban, W. A., 1966, *The Red Bird section of the Upper Cretaceous Pierre Shale in Wyoming:* U.S. Geol. Survey, Prof. Paper 393-A, 73 p.

Golbert, A. V., Klimova, I. G., Saks, V. N., & Tubina, A. S., 1972, *Novyye dannyye o pogranichnykh sloyakh yury i mela v Zapadnoy Sibiri:* Izv. Sib. Otd. Akad. Nauk SSSR, Inst. Geol. Geofiz., v. 1972, no. 5, p. 11-17. [New data concerning Jurassic-Cretaceous boundary beds in western Siberia.]

Hall, C. A., 1964, *Shallow water marine climates and molluscan provinces:* Ecology, v. 45, p. 226-234.

Hallam, Anthony, 1971, *Mesozoic geology and the opening of the North Atlantic:* Jour. Geology, v. 79, p. 129-157, text-fig. 1-13.——(ed.) 1973, *Atlas of palaeobiogeography:* 531 p., Elsevier Scientific Publishing Co. (Amsterdam).——1975, *Jurassic environments:* 269 p., Cambridge Univ. Press (Cambridge, Eng.).

Hancock, J. M., 1975, *The sequence of facies in the Upper Cretaceous of northern Europe compared with that in the Western Interior:* in The Cretaceous System in the Western Interior of North America, W. G. E. Caldwell (ed.), Geol. Assoc. Canada, Spec. Paper 13, p. 83-118, text-fig. 1-5, 2 tab.——1977, *The historical development of concepts of biostratigraphic correlation:* in Concepts and methods of biostratigraphy, E. G. Kauffman, & J. E. Hazel (eds.), p. 3-22, Dowden, Hutchinson, & Ross, Inc. (Stroudsburg, Pa.).

Harland, W. B. et al. (eds.), 1967, *The fossil record:* Geol. Soc. London, 828 p. (London).

Hay, W. W. (ed.), 1974, *Studies in paleo-oceanography:* Soc. Econ. Paleontol. & Mineral., Spec. Publ. 20, 218 p.

Hays, J. D., & Pitman, W. C., 1973, *Lithospheric plate motions, sea level changes, and climatic and ecological consequences:* Nature, v. 246, p. 18-22.

Hughes, N. F. (ed.), 1973, *Organisms and continents through time:* Palaeontol. Assoc., Spec. Paper Palaeontology 12, Syst. Assoc. Publ. 9, 334 p.

Imlay, R. W., 1959, *Succession and speciation of the pelecypod Aucella:* U.S. Geol. Survey, Prof. Paper 314-G, p. 155-169, text-fig. 36, pl. 16-19, tab. 1.——1960, *Early Cretaceous (Albian) ammonites from the Chitina Valley and Talkeetna Mountains, Alaska:* U.S. Geol. Survey, Prof. Paper 354-D, p. 87-114, pl. 11-19.

——, & Jones, D. L., 1970, *Ammonites from the Buchia zones in northwestern California and southwestern Oregon:* U.S. Geol. Survey, Prof. Paper 647-B, 59 p., 8 text-fig., 15 pl., 2 tab.

International Subcommission on Stratigraphic Classification (H. D. Hedberg, ed.), 1976, *International stratigraphic guide:* 200 p., 14 text-fig., Wiley-Interscience (New York).

Jeletzky, J. A., 1964a, *Illustrations of Canadian fossils: Lower Cretaceous marine index fossils of the sedimentary basins of Western and Arctic Canada:* Geol. Surv., Can., Dep. Mines & Resour., Paper 64-11, 100 p., 36 pl., 1 tab.——1964b, *Illustrations of Canadian fossils: Early Lower Cretaceous (Berriasian and Valanginian) of the Canadian Western Cordillera, British Columbia:* Geol. Surv., Can., Dep. Mines & Resour., Paper 64-6, 18 p., 8 pl.——1965, *Late Upper Jurassic and early Lower Cretaceous fossil zones of the Canadian Western Cordillera, British Columbia:* Geol. Surv., Can., Dep. Mines & Resour., Bull. 103, 70 p., 4 text-fig., 22 pl.——1970, *Cretaceous macrofaunas:* in Biochronology: Standard of Phanerozoic time, Geol. Surv., Can., Dep. Mines & Resour., Ch. 11, Econ. Geol. Rep. no. 1, 5th edit., Dept. Energy, Mines, Res., p. 649-652, text-fig. X12, X13, pl. 23-28, tab. X1-X8.——1971, *Marine Cretaceous biotic provinces and paleogeography of western and Arctic Canada: illustrated by a detailed study of ammonites:* Geol. Surv., Can., Paper 70-22, p. 1-92, text-fig. 1-20.——1973, *Biochronology of the marine Boreal latest Jurassic, Berriasian and Valanginian in Canada:* in The Boreal Lower Cretaceous, Raymond Casey & P. F. Rawson (eds.), Geol. Jour., Spec. Issue 5, p. 41-80, text-fig. 1-3, pl. 1-7, Seel House Press (Liverpool).

Jones, D. L., 1967, *Cretaceous ammonites from the lower part of the Matanuska Formation, southern Alaska:* U.S. Geol. Survey, Prof. Paper 547, p. 1-49, pl. 1-10.——1969, *Buchia zonation in the Myrtle Group, southwestern Oregon:* Geol. Soc. America, Abstr. Programs, Ann. Mtg. 1969, pt. 3 (Cordilleran Sec.), p. 31,32, tab. 1.

——, & Detterman, R. L., 1966, *Cretaceous stratigraphy of the Kamishak Hills, Alaska Peninsula:* U.S. Geol. Survey, Prof. Paper 550-D, p. D53-D58, text-fig. 1-4.

——, Murphy, M. A., & Packard, E. L., 1965, *The Lower Cretaceous (Albian) ammonite genera Leconteites and Brewericeras:* U.S. Geol. Survey, Prof. Paper 503-F, 21 p., 17 text-fig., 11 pl.

Kauffman, E. G., 1967, *Cretaceous Thyasira from the Western Interior of North America:* Smithson. Misc. Coll., v. 152, no. 1, pub. no. 4695, 159 p., 18 text-fig., 5 pl., 7 tab.——1968, *The Upper Cretaceous Inoceramus of Puerto Rico:* in 4th Caribbean Geol. Conf., Trinidad, 1965, Trans. (1968), J. B. Saunders (ed.), p. 203-218, text-fig. 1-6, pl. 1,2.——1969, *Cretaceous marine cycles of the Western Interior:* Mountain Geol., v. 6, p. 227-245, text-fig. 1-4.——1970, *Population systematics, radiometrics and zonation—a*

new biostratigraphy: North Am. Paleontol. Conv., Proc., Chicago, 1969, v. 1, pt. F, p. 612-666, text-fig. 1-10, Allen Press (Lawrence, Kans.).
——1972, *Evolutionary rates and patterns of North American Cretaceous Mollusca:* Int. Geol. Cong., Proc. 1972, Montreal, sec. 7, p. 174-189, text-fig. 1-5.——1973a, *Cretaceous Bivalvia:* in Atlas of palaeobiogeography, Anthony Hallam (ed.), p. 351-383, text-fig. 1-10, Elsevier Scientific Publ. Co. (Amsterdam).——1973b, *Stratigraphic evidence for Cretaceous eustatic changes:* Geol. Soc. America, Abstr. Programs, 1973 Ann. Mtg., p. 687.——1974, *Cretaceous assemblages, communities, and associations; Western Interior United States and Caribbean islands:* in Principles of benthic community analysis, A. M. Ziegler et al. (eds.), Sedimenta IV, Comp. Sed. Lab., Univ. Miami, p. 12.1-12.27, text-fig. 1-11.——1975, *Dispersal and biostratigraphic potential of Cretaceous benthonic Bivalvia in the Western Interior:* in The Cretaceous System in the Western Interior of North America, W. G. E. Caldwell (ed.), Geol. Assoc. Canada, Spec. Paper 13, p. 163-194, text-fig. 1-4.——1976, *Plate tectonics: a major force in evolution:* The Science Teacher, v. 43, no. 3, p. 12-17, text-fig. 1-3.——1977a, *Evolutionary rates and biostratigraphy:* in Concepts and methods of biostratigraphy, E. G. Kauffman & J. E. Hazel (eds.), p. 109-141, text-fig. 1-8, Dowden, Hutchinson, & Ross, Inc. (Stroudsburg, Pa.).——1977b, *Cretaceous extinction and collapse of marine trophic structure:* Jour. Paleontology, v. 51, suppl. 2, p. 16 [abstract].——1977c, *Benthic communities in black shales of an "anaerobic" Jurassic basin:* the *Posidonienschiefer:* Jour. Paleontology, v. 51, suppl. 2, p. 16 [abstract].——1977d, *British Middle Cretaceous inoceramid biostratigraphy:* 2nd Internatl. Conf. Mid-Cretaceous Events, Uppsala, Proc., Ann. Mus. Hist. Nat. Nice (France), Spec. Vol. (in press).——1977e, *Systematic, biostratigraphic, and biogeographic relationships between Middle Cretaceous Euramerican and North Pacific Inoceramidae:* in Mid-Cretaceous Events—Hokkaido Symposium, 1976, Tatsuro Matsumoto (ed.), Paleont. Soc. Japan, Spec. Paper 21, p. 169-212, 1 text-fig.——1977f, *Geological and biological overview: Western Interior Cretaceous basin:* Mt. Geol., v. 14, p. 75-99.——1978, *Cretaceous geochronology for the Western Interior United States: New data and applications:* Geology (in press).
——, **Cobban, W. A., & Eicher, D. L.,** 1977, *Albian through Lower Coniacian strata and biostratigraphy, Western Interior United States:* 2nd Internatl. Conf. Mid-Cretaceous Events, Uppsala, Proc., Spec. Vol., Ann. Mus. Nat. Histoire Nice (France), p. 21-1 to 21-33, 7 text-fig., 16 pl.
——, **Hattin, D. E., & Powell, J. D.,** 1977, *Cenomanian-Turonian (Upper Cretaceous) stratigraphy and paleoenvironments, northwestern Oklahoma:* Geol. Soc. America, Mem. 149, pt. 1, p. 1-46, text-fig. 1-5.
——, **Scholle, P. A.,** 1977, *Abrupt biotic and environmental changes during peak Cretaceous transgressions in Euramerica:* Jour. Paleontology, v. 50, suppl. 2, p. 16 [abstract].
——, **& Sohl, N. F.,** 1973, *Structure and evolution of Antillean Cretaceous rudist frameworks:* Naturforsch. Gesell. Basel, Verhandl., v. 84, no. 1, p. 399-467, text-fig. 1-27.
Kemper, Edwin, 1973, *The Valanginian and Hauterivian stages in northwest Germany:* in The Boreal Lower Cretaceous, Raymond Casey & P. F. Rawson (eds.), Geol. Jour., Spec. Issue 5, p. 327-344, text-fig. 1-4, Seel House Press (Liverpool).
Kennedy, W. J., 1969, *The correlation of the Lower Chalk of south-east England:* Geologists Assoc., Proc., v. 80, pt. 4, p. 459-560, text-fig. 1-16, pl. 15-22, tab. 1-10.
——, **& Cobban, W. A.,** 1976, *Aspects of ammonite biology, biogeography, and biostratigraphy:* Palaeontol. Assoc., Spec. Paper in Palaeontology, no. 17, 94 p., 24 text-fig., 11 pl.
——, **& Juignet, Pierre,** 1973, *Observations on the lithostratigraphy and ammonite succession across the Cenomanian-Turonian boundary in the environs of Le Mans (Sarthe, N.W. France):* Newsl. Stratigr., v. 2, no. 4, p. 189-202, text-fig. 1,2, tab. 1,2.
Lowenstam, H. A., & Epstein, Samuel, 1954, *Paleotemperatures of the post-Aptian Cretaceous as determined by the oxygen isotope method:* Jour. Geology, v. 62, p. 207-248, text-fig. 1-22.——1959, *Cretaceous paleo-temperatures as determined by the oxygen isotope method; their relations to and the nature of rudist reefs:* in El sistema Cretacico—un symposium sobre el Cretacico en el Hemisferio Occidental y su correlacion mundial, L. B. Kellum (Chmn.), Com. Int. Estrat., Union Paleontol. Int., v. 1, p. 65-76, text-fig. 1-3.
Mantell, G. A., 1822, *The fossils of South Downs; or illustrations of the geology of Sussex:* 320 p., 42 pl., L. Relfe (London).
Matsumoto, Tatsuro, 1959a, *Upper Cretaceous ammonites of California: Pt. 1:* Kyushu Univ., Mem. Fac. Sci., ser. D, Geol., v. 8, no. 4, p. 91-171, text-fig. 1-85, pl. 30-45.——1959b, *Upper Cretaceous ammonites of California: Pt. 2:* Kyushu Univ., Mem. Fac. Sci., ser. D, Geol., Spec. vol. 1, p. 1-172, text-fig. 1-80, pl. 1-41.——1960, *Upper Cretaceous ammonites of California: Pt. 3:* Kyushu Univ., Mem. Fac. Sci., ser. D, Geol., Spec. vol. 2, 204 p., 20 text-fig., 1 pl.——1963, *The Cretaceous:* in The geology of Japan, Fuyuji Takai, Tatsuro Matsumoto, & Ryuzo Toriyama (eds.), p. 99-128, text-fig. 1-10, Univ. California Press (Berkeley and Los Angeles).——1973, *Late Cretaceous Ammonoidea:* in Atlas of palaeobiogeography, Anthony Hallam

(ed.), p. 421-429, text-fig. 1-3, Elsevier Scientific Publishing Co. (Amsterdam).

Middlemiss, F. A., & Rawson, P. F. (eds.), 1971, *Faunal provinces in space and time:* Geol. Jour., Spec. Issue 4, 236 p.

Miller, F. X., 1977, *The graphic correlation method in biostratigraphy:* in Concepts and methods in biostratigraphy, E. G. Kauffman & J. E. Hazel (eds.), p. 165-186, text-fig. 1-10, Dowden, Hutchinson, & Ross, Inc. (Stroudsburg, Pa.).

Nakano, Mitsuo, 1960, *Stratigraphic occurrences of the Cretaceous trigoniids in the Japanese islands and their faunal significances:* Hiroshima Univ., Jour. Sci., ser. C, v. 3, no. 2, p. 215-280, pl. 23-30.

Neaverson, Ernest, 1955, *Stratigraphical palaeontology; a study of ancient life provinces:* 2nd edit., 806 p., 18 pl., Clarendon Press (Oxford).

Noda, Masayuki, & Matsumoto, Tatsuro, 1976, *Atlas of Japanese fossils; no. 45, Cretaceous Inocerami:* text-fig. 267, 269, Tsukiji Shokan Publ. Co. (Tokyo).

Obradovich, J. D., & Cobban, W. A., 1975, *A time-scale for the Late Cretaceous of the Western Interior of North America:* in The Cretaceous System in the Western Interior of North America, W. G. E. Caldwell (ed.), Geol. Assoc. Can., Spec. Paper 13, p. 31-54, text-fig. 1, tab. 1-3.

Omalius d'Halloy, J. B. d', 1822, *Observations sur un essai de carte géologique de la France, des Pays-Bas et des contrées voisines:* Ann. Mines, v. 7, p. 353-376.

Orbigny, Alcide d', 1840-44, *Paléontologie française; Terrains crétacés; Description zoologique et géologique de tous animaux mollusques et rayonnés fossiles de France: I. Céphalopodes:* p. 1-662, pl. 1-148 [p. 1-120 (1840); p. 121-430 (1841); p. 431-662 (1842)]; 2. *Gastéropodes:* 456 p., pl. 149-236 (1842-43); 3. *Lamellibranches:* 520 p., pl. 237-489 (1844), Victor Masson (Paris).——1849-50, *Cours élémentaire de paléontologie et de géologie stratigraphique:* v. 1 (1849), p. 1-299, text-fig. 1-338; v. 2, pt. 1 (1850), p. 1-382, v. 2, pt. 2 (1850), p. 383-847, text-fig. 339-628, Victor Masson (Paris).

Owen, H. G., 1973, *Ammonite faunal provinces in the Middle and Upper Albian and their palaeobiogeographic significance:* in the Boreal Lower Cretaceous, Raymond Casey & P. F. Rawson (eds.), Geol. Jour., Spec. Issue 5, p. 145-154, Seel House Press (Liverpool).

Parker, R. H., 1960, *Ecology and distributional patterns of marine macro-invertebrates, northern Gulf of Mexico:* in Recent sediments, northwest Gulf of Mexico, F. P. Shepard, F. B. Phleger, & T. H. van Andel (eds.), p. 302-337, text-fig. 1-17, pl. 1-6, Am. Assoc. Petrol. Geologists (Tulsa).——1964, *Zoogeography and ecology of some macro-invertebrates, particularly mollusks, in the Gulf of California and the continental slope off Mexico:* Vidensk. Medd. Dansk, Naturhist. Foren., v. 126, 178 p., text-fig. 1-29, pl. 1-15.

Pergament, M. A., 1969, *Zonalnyye podrazdeleniye mela severo-vostoka Azii i sopostavleniye s Amerikanskoy i Yevropeyskoy shkalami:* Izv. Sib. Otd. Akad. Nauk SSSR, ser., no. 4, p. 106-119, tab. 1,2. [*Zonal units of the Cretaceous of northeastern Asia and their correlation with the American and European scales.*]

Porrenga, D. H., 1967, *Glauconite and chamosite as depth indicators in the marine environment:* Marine Geol., v. 5, p. 495-501.

Ramsay, A. T. S., 1974, *The distribution of calcium carbonate in deep sea sediments:* in Studies in paleo-oceanography, W. W. Hay (ed.), Soc. Econ. Paleontol. & Mineral., Spec. Publ. 20, p. 58-76, text-fig. 1-10.

Rawson, P. F., 1972, *A note on the biostratigraphical significance of the Lower Cretaceous belemnite genus Aulacoteuthis in the B beds of the Speeton Clay, Yorkshire:* Yorkshire Geol. Soc., Proc., v. 39, p. 89-91.

Reid, R. E. H., 1967, *Tethys and the zoogeography of some modern and Mesozoic Porifera:* in Aspects of Tethyan biogeography, G. G. Adams & D. V. Ager (eds.), Syst. Assoc. Publ. 7, p. 171-182.

Rhoads, D. C., 1970, *Mass properties, stability, and ecology of marine muds related to burrowing activity:* in Trace fossils, T. P. Crimes & J. C. Harper (eds.), p. 391-406, text-fig. 1-4, Seel House Press (Liverpool).

——, Speden, I. G., & Waage, K. M., 1972, *Trophic group analysis of Upper Cretaceous (Maestrichtian) bivalve assemblages from South Dakota:* Am. Assoc. Petrol. Geologists, Bull., v. 56, p. 1100-1113, text-fig. 1-4, tab. 1-3.

Rioult, Michael, 1969, *Alcide d'Orbigny and the stages of the Jurassic:* Mercian Geol., v. 3, p. 1-30 (transl. from French by M. Sargeant).

Saito, Tsunemasa, & Van Donk, Jan, 1974, *Oxygen and carbon isotope measurements of Late Cretaceous and Early Tertiary Foraminifera:* Micropaleontology, v. 20, p. 152-177, pl. 1-3.

Saks, V. N. (ed.), 1972, *Granitsa yury i mela i berriasskiy yarus v Borealnom poyase:* Izd. Sib. Otd. Akad. Nauk SSSR, 371 p. [*The Jurassic-Cretaceous boundary and the Berriasian Stage in the Boreal Realm.*]

——, Mesezhnikov, M. S., & Shulgina, N. I., 1968, *Volzhskiy yarus i polozheniye granitsy yurskoy i melovoy sistem v arkticheskoy zoogeograficheskoy oblasti:* Izv. Sib. Otd. Akad. Nauk SSSR, v. 48, p. 72-79. [*The Volgian Stage and position of the boundary between the Jurassic and Cretaceous systems in the Arctic zoogeographical province.*]

Sanders, H. L., 1969, *Benthic marine diversity and the stability-time hypothesis:* in Diversity and stability in ecological systems, G. M. Woodwell & H. H. Smith (eds.), Brookhaven Symp. Biol., no. 22, p. 71-81, text-fig. 1-8.

Schlanger, S. O., & Jenkyns, H. C., 1976, *Cretaceous oceanic anoxic events: causes and consequences:* Geol. Mijnb., v. 55, no. 3-4, p. 179-184.

Scholle, P. A., & Kauffman, E. G., 1977, *Paleoecological implications of stable isotope data from Upper Cretaceous limestones and fossils from the U.S. Western Interior:* Jour. Paleontology, v. 50, suppl. 2, p. 24, 25 [abstract].

Scott, G. R., & Cobban, W. A., 1964, *Stratigraphy of the Niobrara Formation at Pueblo, Colorado:* U.S. Geol. Survey, Prof. Paper 454-L, 30 p., 9 text-fig., 11 pl., 3 tab.

Scott, R. W., 1970, *Paleoecology and paleontology of the Lower Cretaceous Kiowa Formation, Kansas:* Univ. Kansas, Paleontol. Contrib., Art. 52, (Cretaceous 1), 94 p., 21 text-fig., 7 pl.——1974, *Bay and shoreface benthic communities, Lower Cretaceous, southern Western Interior:* Lethaia, v. 7, p. 315-330, text-fig. 1-11.

Seilacher, Adolf, & Westphal, F., 1971, *Fossil-Lagerstätten:* in Sedimentology of parts of Central Europe: Guidebk., 8th Internatl. Sediment. Congress, 1971, p. 327-335, text-fig. 1-5 (Frankfurt am Main).

Seitz, Otto, 1959, *Vergleichende Stratigraphie der Oberkreide in Deutschland und in Nordamerika mit Hilfe der Inoceramen:* in El Sistema Cretacico, L. B. Kellum (ed.), v. 1, p. 113-129, tab. 1, Congr. Geol. Internac., 20th Ses., Mexico City, 1956.——1961, *Die Inoceramen des Santon von Nordwestdeutschland: 1 Teil. (Die Untergattungen Platyceramus, Cladoceramus, und Cordiceramus):* Geol. Jahrb., Beihefte, no. 46, 186 p., 39 text-fig., 15 pl.——1965, *Die Inoceramen des Santon und Unter-Campan von Nortwestdeutschland. Teil II. Biometrie, Dimorphismus und Stratigraphie der Untergattung Sphenoceramus J. Böhm:* Geol. Jahrb., Beihefte, no. 69, 194 p., 11 text-fig., 26 pl., 46 tab.——1967, *Die Inoceramen des Santon und Unter-Campan von Nordwestdeutschland: III Teil: Taxonomie und Stratigraphie der Untergattungen Endocostea, Haenleinia, Platyceramus, Cladoceramus, Selenoceramus, und Cordiceramus mit besonderer Berücksichtigung des Parasitismus bei diesen Untergattungen:* Geol. Jahrb., Beihefte, no. 75, 171 p., 27 text-fig., 27 pl., 8 tab.——1970, *Über einige Inoceramen aus der oberen Kreide: 1. Die Gruppe des Inoceramus subquadratus Schlüter und der Grenzbereich Coniac/Santon. 2. Die Muntigler Inoceramenfauna und ihre Verbreitung im Ober-Campan und Maastricht:* Geol. Jahrb., Beihefte, no. 86, 171 p., 12 text-fig., 28 pl.

Smith, A. G., Briden, J. C., & Drewery, G. E., 1973, *Phanerozoic world maps:* in Organisms and continents through time, N. F. Hughes (ed.), Palaeontol. Assoc., Spec. Papers in Palaeontology 12; Syst. Assoc., Pub. 9, p. 1-42, text-fig. 1-21.

Smith, C. C., 1975, *Upper Cretaceous calcareous nannoplankton zonation and stage boundaries:* Gulf Coast Assoc. Geol. Soc., Trans., v. 25, p. 263-278, text-fig. 1,2.

Sohl, N. F., 1960, *Archaeogastropoda, Mesogastropoda and stratigraphy of the Ripley, Owl Creek, and Prairie Bluff formations:* U.S. Geol. Survey, Prof. Paper 331-A, p. 1-151, text-fig. 1-11, pl. 1-18, tab. 1.——1967, *Upper Cretaceous gastropod assemblages of the Western Interior of the United States:* in Paleoenvironments of the Cretaceous seaway in the Western Interior: A symposium, E. G. Kauffman & H. C. Kent (cochairmen), Preprints of papers, Spec. Publ. Colorado School Mines, Golden, p. 1-37, text-fig. 1-9.——1971, *North American Cretaceous biotic provinces delineated by gastropods:* North American Paleont. Convention, Proc., Chicago, 1969, v. 2, pt. L, p. 1610-1638, text-fig. 1-13, Allen Press (Lawrence, Kans.).——1977, *Utility of gastropods in biostratigraphy:* in Concepts and methods of biostratigraphy, E. G. Kauffman & J. E. Hazel (eds.), p. 519-539, text-fig. 1-6, Dowden, Hutchinson, & Ross, Inc. (Stroudsburg, Pa.).

Spaeth, C. L., 1971, *Untersuchungen an Belemniten des Formenkreises um Neohibolites minimus (Miller 1826) aus dem Mittel- und Ober-Alb Nordwestdeutschlands:* Geol. Jahrb., Beihefte, v. 100, p. 1-127, pl. 1-9.

Stanton, T. W., 1947, *Studies of some Comanche pelecypods and gastropods:* U.S. Geol. Survey, Prof. Paper 211, 256 p., 67 pl.

Stephenson, L. W., 1941, *The larger invertebrate fossils of the Navarro Group of Texas:* Univ. Texas, Bur. Econ. Geol., Bull. 4101, 641 p., 95 pl.——1952, *The larger invertebrate fossils of the Woodbine Formation (Cenomanian) of Texas:* U.S. Geol. Survey, Prof. Paper 242, 226 p., 59 pl.

Stevens, G. R., 1965, *Faunal realms in Jurassic and Cretaceous belemnites:* Geol. Mag., v. 102, p. 175-178.——1973, *Cretaceous belemnites:* in Atlas of palaeobiogeography, Anthony Hallam (ed.), p. 386-401, text-fig. 1-5, pl. 1, Elsevier Scientific Publishing Co. (Amsterdam).

Sylvester-Bradley, P. C., 1971, *Dynamic factors in animal palaeogeography:* in Faunal provinces in space and time, F. A. Middlemiss & P. F. Rawson (eds.), Geol. Jour., Spec. Issue 4, p. 1-18, text-fig. 1-5.

Tappan, Helen, 1968, *Primary production, isotopes, extinctions, and the atmosphere:* Palaeogeography, Palaeoclimatology, Palaeoecology, v. 4, p. 187-210, text-fig. 1, tab. 1.——1971, *Microplankton, ecological succession and evolution:* North Am. Paleontol. Conv., Proc., Chicago, 1969, v. 2, pt. H, p. 1058-1103, text-fig. 1-8, Allen Press (Lawrence, Kans.).

——, & Loeblich, A. R., Jr., 1971, *Geobiologic implications of fossil phytoplankton evolution and time-space distribution:* in Symposium on palynology of the Late Cretaceous and Early Tertiary, Robert Kosanke & A. T. Cross (eds.), Geol. Soc. America, Spec. Paper 127, p. 247-339, text-fig. 1-12.——1972, *Fluctuating rate of protistan evolution, diversification and extinction:* 24th Int. Geol. Congress, Proc., Sec. 7, p. 205-213, text-fig. 1-6 (Montreal).

Thiede, J., & Dinkelman, M. G., 1977, *The occurrence of Inoceramus remains in Late Mesozoic pelagic and hemipelagic sediments:* Initial Rept. Deep Sea Drilling Project (in press), U.S. Gov't Printing Office (Washington, D.C.).

Thieuloy, J. P., & Thomel, G., 1964, *Sur l'utilisation éventuelle des Ammonites dérouleés dans la Chronologie du Crétacé inférieur:* Univ. Grenoble, Fac. Sci., Lab. Géol., p. 121-126.

Thomel, G., 1964, *Contribution à la connaissance des céphalopodes Crétacés de sud-est de la France; note sur les ammonites deroulées de Crétacé inférieur vocontien:* Soc. Géol. France, Mém., n. sér., v. 43, no. 2, p. 18-22; Mém. no. 101, p. 1-8, pl. 1-12.——1965, *Zoneostratigraphie et paléobiogéographie du Cénomanien du sud-est de la France:* 90ième Congr. des Soc. savantes, Nice, Compte Rendu, sec. Sci., p. 127-154.

Tröger, K. A., 1967, *Zur Paläontologie, Biostratigraphie, und faziellen Ausbildung der unteren Oberkreide (Cenoman bis Turon). Teil 1. Paläontologie und Biostratigraphie der Inoceramen des Cenomans bis Turons Mitteleuropas:* Dresden Staatl. Mus. Mineral. Geol., Abhandl., v. 12, p. 13-207, 43 text-fig., 14 pl.

Valentine, J. W., 1961, *Paleoecological molluscan geography of the California Pleistocene:* Univ. Calif., Publ. Geol. Sci., v. 34, no. 7, p. 309-442, text-fig. 1-16.——1967, *The influence of climatic fluctuations on species diversity within the Tethyan provincial system:* in Aspects of Tethyan biogeography, G. G. Adams & D. V. Ager (eds.), Syst. Assoc. Publ. 7, p. 153-166.——1973, *Evolutionary paleoecology of the marine biosphere:* 511 p., 152 text-fig., Prentice-Hall, Inc. (Englewood Cliffs, N.J.).

——, & Moores, E. M., 1972, *Global tectonics and the fossil record:* Jour. Geology, v. 80, p. 167-184.

Van Hinte, J. E., 1976, *A Cretaceous time scale:* Am. Assoc. Petrol. Geologists, Bull., v. 60, no. 4, p. 498-516, text-fig. 1-9.

Wiedmann, Jost, 1970, *Problems of stratigraphic classification and the definition of stage boundaries:* Newsl. Stratigr., v. 1, p. 35-48.——1973, *Ancyloceratina (Ammonoidea) at the Jurassic/Cretaceous boundary:* in Atlas of Palaeobiogeography, Anthony Hallam (ed.), p. 309-316, text-fig. 1,2, Elsevier Scientific Publishing Co. (Amsterdam).

——, & Kauffman, E. G., 1977, *Mid-Cretaceous biostratigraphy of northern Spain:* 2nd Internatl. Conf. Mid-Cretaceous Events, Uppsala, Sweden, Ann. Museum Nat. Histoire Nice, France, Spec. Vol., 21 p., 5 pl., 1 tab.

Worsley, T. R., 1971, *Terminal Cretaceous events:* Nature, v. 230, p. 318-320, text-fig. 1,2.——1974, *The Cretaceous-Tertiary boundary event in the ocean:* in Studies in paleo-oceanography, W. W. Hay (ed.), Soc. Econ. Paleontol. & Mineral., Spec. Publ., no. 20, p. 94-125, text-fig. 1-22.

Young, K. P., 1963, *Upper Cretaceous ammonites from the Gulf Coast of the United States:* Univ. Texas, Bur. Econ. Geol., Publ. 6304, 373 p., 34 text-fig., 82 pl., 13 tab.——1967, *Ammonite zonations, Texas Comanchean (Lower Cretaceous):* in Comanchean (Lower Cretaceous) stratigraphy and paleontology of Texas, Leo Hendricks (ed.), Soc. Econ. Paleontol. & Mineral., Permian Basin, Spec. Publ. 67, 68, p. 65-70, pl. 1-6.——1972, *Cretaceous paleogeography: implications of endemic ammonite faunas:* Univ. Texas, Bur. Econ. Geol., Geol. Circ. 72-2, 13 p., 4 text-fig., 3 tab.

TERTIARY[1]

By Adolf Papp
[Universität Wien, Austria]

CONTENTS

	PAGE
General Trends	A488
Distribution of Tertiary Sediments	A490
Europe	A490
Asia	A491
Australia, New Guinea, and South Pacific Islands	A492
North America	A492
Central and South America	A493
Africa	A495
Outline of Invertebrate Evolution During Tertiary Time	A496
Nature of Tertiary Organic Groups	A497
Foraminiferida	A497
Coelenterata and Mollusca	A498
Arthropoda	A499
Bryozoa, Brachiopoda, and Echinodermata	A499
Biostratigraphic Divisions	A499
References	A504

GENERAL TRENDS

The Tertiary Period is well known as the Age of Mammals and Angiosperms. Because many richly fossiliferous sediments are still soft and little-altered diagenetically, it is possible to analyze all inorganic and biologic events recorded by them in much more detail than for older periods. The duration of the Tertiary is approximately 70 million years (Papp, 1959).

During Tertiary time, conditions developed that determined the nature of present continents, together with the make-up of their faunas and floras. Whereas Early Tertiary paleogeography differed considerably from that of the present, Late Tertiary conditions gradually approached those that exist now. The hypothesis of continental drift postulates that it was only during Tertiary time that the earth's continents reached their present positions.

During the Tertiary, the worldwide orogenies of the Alpidic Era occurred. They include numerous orogenic phases that had their origins in the Mesozoic. In connection with these crustal movements, widespread intense vulcanism, to which many presently active volcanoes owe their origin, took place. Pacific granite intrusions, vast basalt extrusions, and plateau deposits of volcanic rocks are witnesses of Tertiary volcanism. Examples are found in the volcanic belt stretching from Greenland across

---

[1] Manuscript received January, 1969; revised manuscript received January, 1975. This contribution was translated from the original German by Curt Teichert.

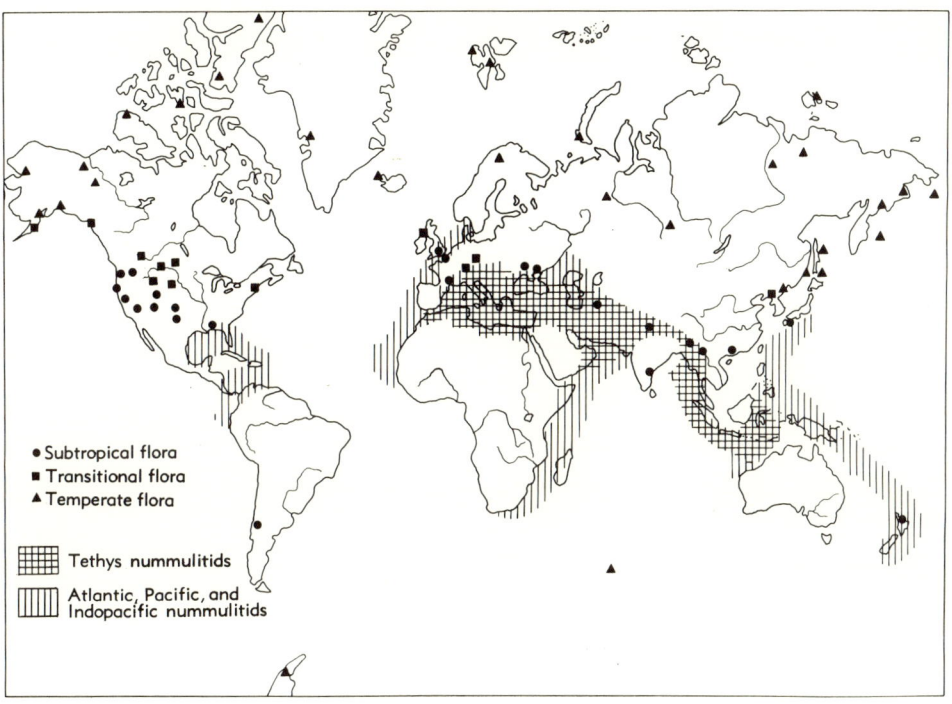

Fig. 1. Distribution of nummulitids and flora in the Early Tertiary (after Brinkmann, 1966).

Iceland to Scotland, the basalt flows of the Columbia and Colorado plateaus of North America, and the Deccan traps of peninsular India, which cover 100,000 sq km in considerable thickness.

The climate, assessable through numerous indications, was influenced by poleward shift of tropical and temperate zones during the Early Tertiary (Fig. 1). During the middle Eocene temperatures reached their optimum. The flora of the Paris basin, for example, has a tropical character. From the Eocene onward a continuous deterioration of the climate is noticeable, leading through a subtropical and warm temperate to a temperate climate, presaging the coming Ice Age. The belt of carbonate sediments became narrower. Measurements of O^{16}-O^{18} indicate that during the Miocene the temperature at the bottom of the deep sea dropped from 10° C to 7° C, and later to the present 1.5° C. At the same time the northern limit of coral reefs receded toward the equator. During the Miocene northern Pacific mollusks reached the North Atlantic and the North Sea basin by way of the Arctic region.

On the Eurasian continent and in North America the development of savannas and bush steppes can be noted. Gradually, the distribution pattern of faunas and floras of the holarctic region developed into its Holocene picture. The disappearance of Tethys caused a division of this faunal region into three separate units: the Indo-Pacific, the Mediterranean, and the Caribbean faunal provinces. With the beginning of glaciation in the polar regions, especially the Antarctic, the scene was set for beginning of the Pleistocene and Quaternary.

The basis of paleogeography is recent geography, and all that can be done is to map the present distribution of Tertiary sediments. Tectonic processes in separate mountain ranges, and even in their individual parts, are often very complicated. Because of limitations of allotted space here, only their most important features can be outlined.

Undoubtedly, climate influences the evolu-

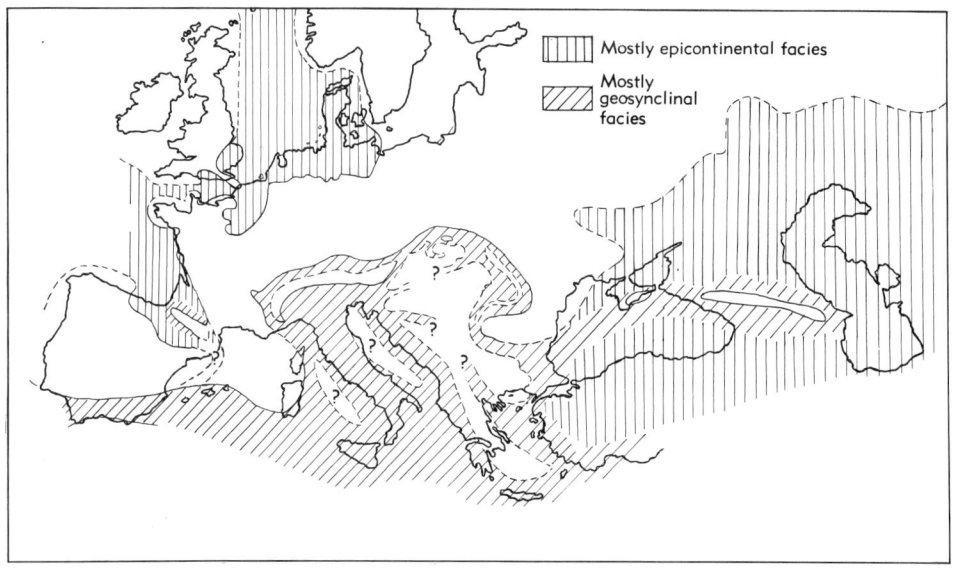

Fig. 2. Distribution of Eocene sediments in Europe (after Papp, 1959).

tion of organisms. Biostratigraphy in the Tertiary also is based on the evolution of organisms and has to be given special prominence in the present contribution.

DISTRIBUTION OF TERTIARY SEDIMENTS

EUROPE

The Paleogene of western Europe is developed in marine-epicontinental facies interbedded with brachyhaline and limnic-fluviatile sediments. The thicknesses of these sediments are relatively small. The facies changes caused by repeated marine transgressive and freshwater cycles facilitate stratigraphic subdivisions. All type sections of the stages of the Paleogene and most of the classical localities belong to this area (PAPP, 1959).

Epicontinental facies in western Europe are developed in the following areas: the Paris basin, with rich and typical faunas; Belgium; Great Britain (in the London and Hampshire basins); connecting with these toward the east, the northwest German, and Danish Paleogene (Fig. 2).

The Paleogene of the Aquitaine basin in the area of the Pyrenees is seen already to have developed characteristics of the nearby geosynclinal facies. The mountain ranges of the Alpidic area in southern Europe (southern Spain, Italy, Greece), and likewise the area of the Alps, Carpathians, Balkans, Crimea, and Caucasus, exhibit the Paleogene in geosynclinal facies. Its characteristic sediment, the Flysch, shows a repeated interbedding of sandstone (turbidites) and marl.

The primary area of flysch-type sedimentation has been destroyed by orogenic processes, nappe structures, and by narrowing of the geosynclinal space, especially in the Alps and Carpathians. A paleogeographic reconstruction of the Paleogene is therefore extremely difficult.

The Molasse zone of the outer edge of the Alps and Carpathians was introduced in the Oligocene (Fig. 2).

The distribution of the Neogene in Europe is far less than that of the Paleogene. In the north, the Neogene sequences of sediments of the northwestern European region can be delineated, with their characteristic Nordic faunal elements, which are different from those of the Atlantic-Mediterranean region. This could be explained

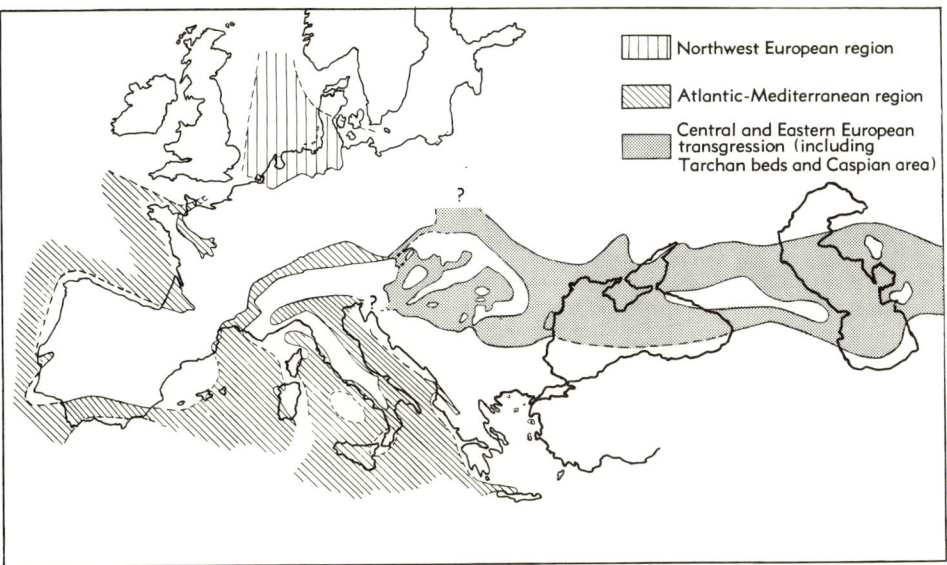

Fig. 3. Extent of the middle Miocene sea in Europe (after Papp, 1959).

by the assumption of a land barrier extending from Great Britain to Iceland.

In central Europe, continental sediments with noteworthy brown-coal occurrences were laid down. Marine sequences of the Atlantic region are developed in Brittany and Aquitaine. In southern Spain, a connection of the Atlantic with the Mediterranean existed for some time (Fig. 3).

The Mediterranean province, especially in the area of the plain of the River Po, shows classical highly fossiliferous sequences; the Pliocene began with a transgressional fauna. Connections existed between the eastern Mediterranean and the South Pacific area.

The Molasse zone along the outer edge of the Alps belongs to the Neogene. Neogene sediments cover the space from the eastern edge of the Alps to the Euxinic-Caspian area, and, farther east, to Lake Aral. The sedimentary series of the upper Miocene and of the Pliocene between the eastern Alpine edge and Lake Aral show an evolution of endemic families unique in their abundance. In the course of the Pliocene, the configuration of shore lines approached more and more that of the present. During the Quaternary the Black Sea became connected with the Mediterranean, whereas the Caspian was isolated.

ASIA

From east of the Ural Mountains, Paleogene epicontinental facies extend northward, separating Siberia from Europe. In the south they are well developed in the Tarim basin. Limnic and terrestrial Neogene strata cover wide areas of Siberia and Mongolia (Fig. 4).

Paleogene and especially Eocene deposits follow the strike of young mountain chains in Anatolia and Iran. They are widespread in Baluchistan and the Punjab and extend into the Himalayas where in places they form the lower plates of Alpine-type nappes. In the Near East, Syria, Iraq, and on the Arabian Peninsula, Paleogene and more so Eocene deposits are widely distributed as carbonate facies. Neogene sediments are confined to coastal areas.

Tertiary formations are widely distributed in Assam and Burma, as well as in Indonesia (Sumatra, Java, Borneo), the Philippines, Formosa, and Japan (Honshu, western Hokkaido). West of Kamchatka an arm of the Pacific penetrated deeply into eastern Siberia.

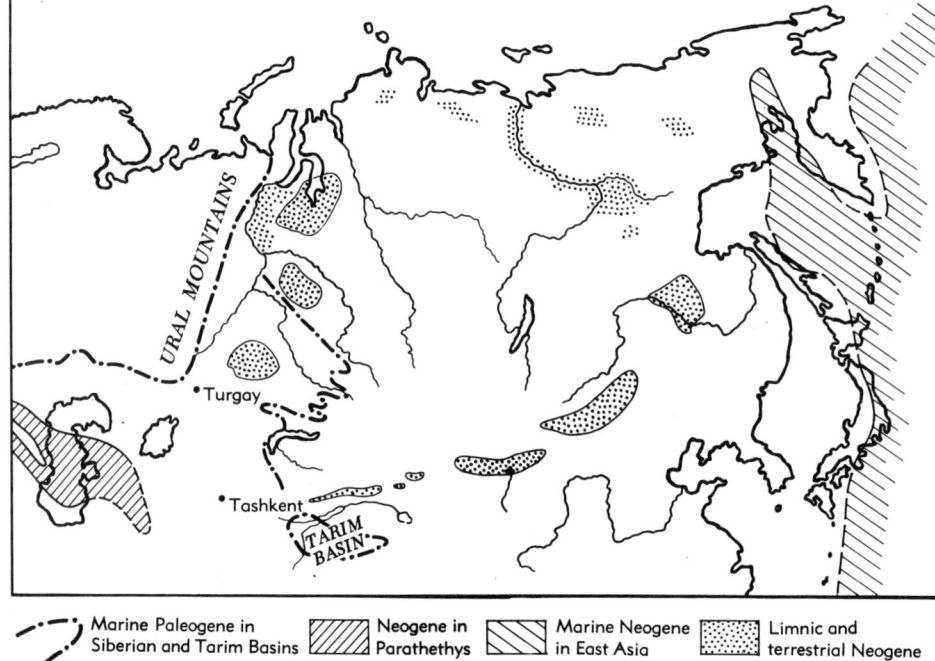

Fig. 4. Distribution of Tertiary sediments in north and central Asia (after Papp, 1959).

AUSTRALIA, NEW GUINEA, AND SOUTH PACIFIC ISLANDS

Along the west coast of Australia, Tertiary is developed in relatively narrow strips (Perth and Carnarvon basins). In the south it is found along the Australian Bight, in the Murray basin, southwestern Victoria, and Gippsland. In New Zealand, marine Tertiary is widely distributed, occurring in several basins around the centers of Mesozoic mountain ranges.

In western and central New Guinea the Tertiary begins with Eocene foraminiferal limestone facies, in the east in geosynclinal facies. The Neogene is developed in elongated basins and in great thickness.

In the South Pacific area GLAESSNER (in PAPP, 1959) distinguished between an inner and an outer Melanesian zone. The outer one comprises the Solomon Islands, New Hebrides, Fiji, and Tonga, with pelagic Tertiary. The inner zone corresponds to southeast New Guinea and New Caledonia.

NORTH AMERICA

In North America, marine Tertiary sediments are mostly restricted to the periphery of the continent and are found in the following areas:

1) Pacific Coast: western Washington, western Oregon, California, the northeast coast of Queen Charlotte Island, on the southern tip of Vancouver, and elsewhere. Farther north, nearshore occurrences mainly of Paleogene are known: Mackenzie Bay, Banks Island (northeast of Victoria Island), and on Baffin Island.

2) Atlantic Coast: from New Jersey in the north through Maryland, Virginia, the Carolinas, and Georgia to Florida.

3) Gulf Coast: Tertiary sediments in North America are found most extensively in the Gulf province, extending from Florida across Alabama, Mississippi, Arkansas, Louisiana, to south Texas, and along the east coast of Mexico to Yucatan. In the Mississippi embayment they penetrate far

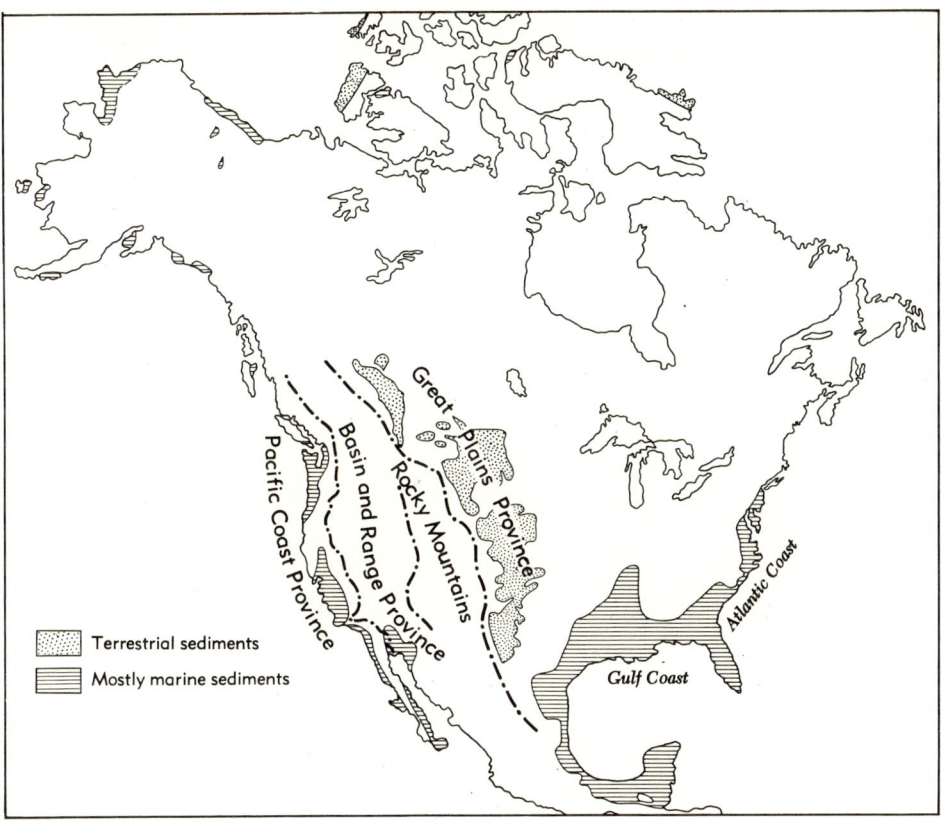

Fig. 5. Distribution of Tertiary sediments in North America (after Papp, 1959).

northward into the continent (Fig. 5).

The Tertiary fauna of the Pacific Coast has a special significance and is therefore designated as the Californian Faunal Province. It has few relationships with the Tertiary of the east coast, which may be called the West Atlantic Faunal Province. Deposits of the Gulf Coast have close relationships with the Caribbean Faunal Province.

The Tertiary nonmarine sediments of western North America with their mammalian faunas are of paramount biostratigraphic importance. They show that middle North America was a center of mammalian evolution during the whole of Tertiary time. From this area of origin, several waves of immigration reached Eurasia by way of the Bering Strait. Such mammalian invasions play an important part in stratigraphy of the nonmarine Tertiary sediments.

In North America nonmarine Tertiary deposits can be grouped in the following way (from west to east): 1) Pacific Coast Province, 2) Great Basin and Range Province, 3) Rocky Mountains Province, and 4) Great Plains Province.

CENTRAL AND SOUTH AMERICA

At the turn of the Mesozoic to the Cenozoic Era, Central America played an important role as a land bridge in the migration of mammals between the two American continents. During Tertiary time the continents were separated, and only a few faunal elements (e.g., marsupials, rodents, and notoungulates during the Oligocene) drifted into South America. Only at the end of the Tertiary, near the middle of the Pliocene, a land passage for mammals emerged.

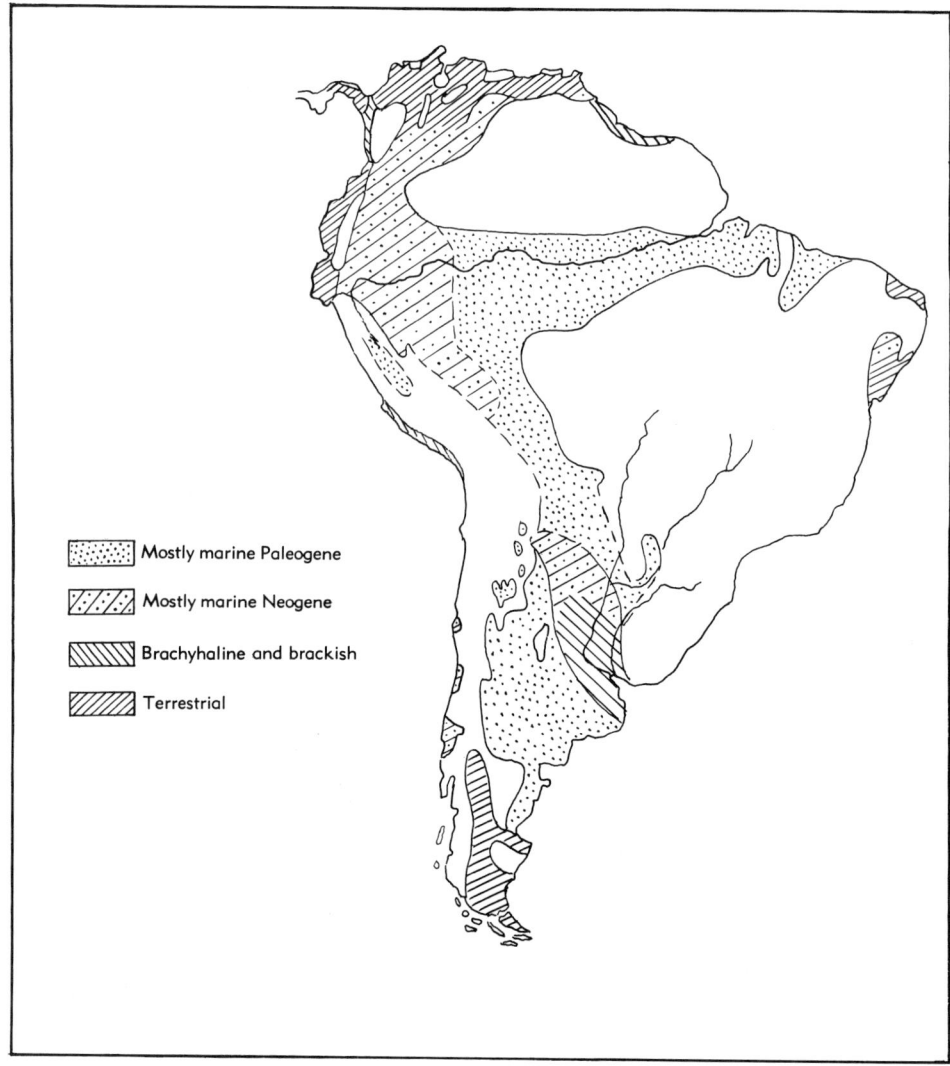

Fig. 6. Distribution of Tertiary sediments in South America (after Papp, 1959).

Sea connections crossing today's Central America facilitated faunal exchange between the Atlantic and Pacific oceans. Eocene sediments have been found at several localities in Central America, especially on the Isthmus of Tehuantepec and in Panama. Throughout Oligocene time a land connection continued to exist between the Central American region and the Greater Antilles, where outcrops of Neogene sediments also occur. In the West Indies the Tertiary naturally shows only incomplete development, with most complete sections described from Cuba, Haiti, the Dominican Republic, Puerto Rico, Jamaica, and Barbados.

In South America, marine Neogene sediments are restricted to marginal areas, but because of size of the continent, they are very extensive all the same. In the north (Trinidad, Venezuela, Colombia), close relationships to the Caribbean Faunal Province can be observed. Farther south, marine Tertiary, separated by great land masses, is found only in the La Plata basin, in

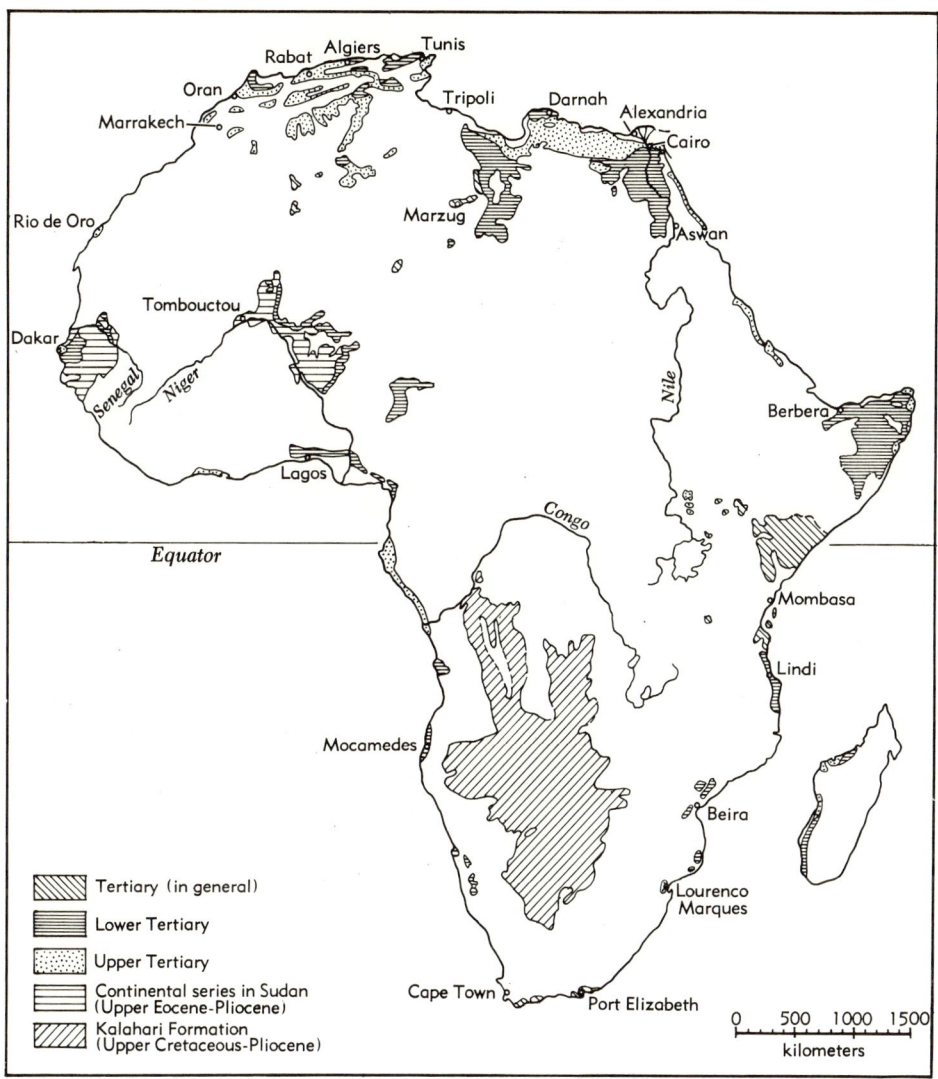

Fig. 7. Tertiary sediments in Africa (after Tollmann in Papp, 1959).

Patagonia, and in Tierra del Fuego. In central South America only terrestrial-limnic and brackish-brachyhaline sediments have been found (Fig. 6).

During the Paleogene, as well as Neogene, a zone with brachyhaline sediments extended in a syncline from northern South America southward along the east side of the Andes. Along the west coast, outcrops of Tertiary deposits worthy of mention are found in Ecuador, northwestern Peru, and central Chile.

In southern Patagonia and in the La Plata basin, marine transgressions took place at the boundaries of the Cretaceous and Tertiary, Oligocene and Miocene, and Miocene and Pliocene, facilitating a natural subdivision of the terrestrial sediments.

AFRICA

At the time of optimal transgression in

the latest Cretaceous (Maastrichtian), the Mediterranean was connected with the Gulf of Guinea across the Sahara. Only during the early Eocene and, increasingly, during the middle Eocene were seas again transgressive. During the middle Eocene orogenic movements occurred in the Atlas region, late Eocene records being incompletely developed in the east and marked by a flysch facies in the west. Nearing the Neogene, we find evidence for fault movements and vulcanism. In the Early Neogene, marine transgression at the margins of the continent attained its maximum (Fig. 7).

Individual developments of the several areas show the following characteristics:

1) In the eastern Mediterranean, gradual regression of the sea during the whole of the Tertiary.
2) In the Atlas area, continued existence of old deep trenches (South Riff trench and South Tell trench).
3) In the Sudan and Upper Guinea, emergence at the end of middle Eocene time.
4) In the East African coastal belt, Indo-Pacific characteristics.
5) In South Africa, continental series of the Kalahari Formation are prevalent.

Of more general interest are the well-known Eocene outcrops with nummulitids in Egypt, which already were mentioned in classic times.

OUTLINE OF INVERTEBRATE EVOLUTION DURING TERTIARY TIME

By far the greatest number of fossils of Tertiary age are derived from marine sediments. The boundary between the Cretaceous and Tertiary systems in the marginal facies of wide areas is marked by a characteristic phase of regression. Although the climate of the Eocene does not seem to have differed much from that of the Late Cretaceous, as indicated by the rather similar distribution of Cretaceous hippuritids and Tertiary nummulitids, Neogene temperatures steadily dropped, with consequent influence on the organisms. During the early Quaternary (i.e., Pleistocene), the temperature sank to its minimum. These climatic changes influenced the fauna and flora in wide areas in Eurasia and North America and furnish the basis for determining the boundary between Tertiary and Quaternary (PAPP, 1959).

The Mesozoic-Tertiary boundary is marked by a reduction, or extinction, of numerous groups of Mesozoic (i.e., Late Cretaceous) organisms. Extinction of the great dinosaurs, plesiosaurs, mosasaurs, and flying saurians, which dominated the Mesozoic, is the most obvious change.

Among the mollusks, important faunal groups of the Mesozoic also became extinct: the ammonites and true belemnites, as well as typical groups of the Late Cretaceous, including hippuritids, acteonellids, and nerineids. Noteworthy is the decline of brachiopods during the Tertiary.

Evolution of the Foraminiferida offers the most useful definition of the Mesozoic-Tertiary boundary. Especially the planktonic foraminifers, collectively known as globotruncanas, which constituted an important faunal element of the Late Cretaceous, became extinct at the end of the Maastrichtian, as did characteristic forms of large Foraminiferida such as *Orbitoides* and *Lepidorbitoides*. The extinction of Cretaceous planktonic Foraminiferida can be traced in continuous stratigraphic sections distributed throughout the world and is thus a stratigraphic criterion of the first order. Directly above the boundary is a narrow zone with small, uncomplicated globigerinids from which typical forms of the Paleogene evolved. The *Globigerina* assemblage is generally correlated with the Danian. In classical paleontology, this was considered the uppermost stage of the Mesozoic and grouped with the Cretaceous. Increasingly important micropaleontological investigations now have led to transfer of equivalents of this stage into the Paleocene, that is, the Tertiary.

The Tertiary-Quaternary boundary is recognizable by biological criteria of climatic origin. The Cretaceous-Tertiary boundary may have like causes, but these cannot be defined by means of glaciations or similar phenomena. It is remarkable that especially such groups that had already passed their optimal development or were biologically overspecialized became extinct.

NATURE OF TERTIARY ORGANIC GROUPS

Following is a generalized discussion of the evolution of important groups of organisms that determine the specific faunal makeup of the Tertiary.

FORAMINIFERIDA

Numerous genera and species of Mesozoic Foraminiferida pass into the Tertiary. The many benthonic species showed an increasing tendency to develop new forms in the Paleocene, when large tests of complicated structure appeared. This evolution reached an optimum in the middle Eocene (Lutetian). These are the most prominent fossils of the shallow-water limestone formations and their evolutionary rates have parallels in the Tertiary only among the vertebrates.

The early Paleocene pillar-bearing rotaliids are widely distributed. The Discocyclinidae may have evolved from Cretaceous forms like *Hellenocyclina*. These cyclic foraminifers reached their optimal development in the Eocene and became extinct at the end of this epoch.

The nummulitids, *sensu lato*, include spirally coiled, multichambered foraminifers with a canal system in the spiral cord. The late Eocene *Heterostegina* and *Spiroclypeus* were derived from *Operculina*, and likewise the Oligocene *Cycloclypeus*. This latter genus developed during the Neogene in tropical seas into giant forms that died out in the Quaternary.

Small radiate forms of nummulitids appeared in the late Paleocene (Ilerdian, according to Schaub in Kapellos & Schaub, 1973). In the lower Eocene we find genera with definite pillar structure; in the middle Eocene very big forms with complicated suture lines. Nummulitids belong to the most abundant and, because of their size, most conspicuous Foraminiferida of the Paleogene. In the upper Eocene a pronounced decrease in size is observed, and small radiate forms predominate. In Europe they died out in the late Oligocene.

The evolution of *Assilina* is very clear, beginning with small forms in the late Paleocene, medium-sized ones in the early Eocene, and big ones in the middle Eocene, when this group became extinct. *Ranikotalia* of the Paleocene may be considered as the root form of the Nummulitidae. (In my opinion, genera of the Nummulitidae should be retained.)

Alveolinids were represented by small globular forms during the early Paleocene. They developed through several evolutionary sequences in the late Paleocene and early Eocene into the big, narrow alveolinids of the middle Eocene. *Borelis* and *Alveolinella* extend from the Oligocene and middle Miocene, respectively, into the Holocene.

In the Caribbean Province *Amphistegina* developed during the middle Eocene into *Eulinderina* and from this into *Lepidocyclina*. This cyclic foraminifer of complicated structure showed a clearly nepionic evolution in the Oligocene and Miocene, also leading to giant forms. *Miogypsina*, which was derived during the Oligocene from rotaliid foraminifers, underwent a pronounced nepionic evolution as far as the Miocene. This group can be considered as well known taxonomically and offers important indications for regional correlation of sediments of the marginal facies of the Oligocene and lower Miocene. In addition to the mentioned groups, quite a number of other benthonic Foraminiferida evolved noticeably during the Tertiary.

Investigations on planktonic foraminifers during recent years have become of greatest importance for stratigraphy. The era of the globigerinas in the oldest Paleocene (Danian) was followed by evolution of Paleogene plankton, which offers the foundation for biostratigraphic zonation of the Paleocene and Eocene. During the Oligocene a worldwide reduction in the abundance of forms occurred; in general, only simple forms of globigerinas persisted. During the Miocene the characteristic genus *Globigerinoides* developed, and during the middle Miocene *Praeorbulina* and *Orbulina*. The Miocene-Pliocene boundary is defined with the help of species of *Globorotalia*.

Because of the deterioration of climate and its influence on the composition of the several faunal provinces, common factors have to be employed for correlation of sedimentary series, especially in the Neogene. Planktonic Foraminifera provide useful indices for regional and intercontinental stratigraphy and offer the best foundation for zonation of the Neogene.

COELENTERATA AND MOLLUSCA

The number of coral genera changed little during the Tertiary, but their distribution varied greatly. The occurrence of reef-forming corals, because of the climatic gradient, became more and more restricted to warm seas.

Mollusks are the classical organisms of Tertiary time, though the importance of cephalopods diminished in comparison with the Mesozoic. The Nautiloidea persisted and are worthy of more detailed investigation. The gastropods and bivalves were the most varied molluscan groups during the Tertiary. Among marine gastropods, most families were already developed in Cretaceous time. During the Tertiary, the evolution of numerous genera and species followed. Typical is the development of the Caenogastropoda, among which siphonostome forms evolved rapidly.

Also typical for the Tertiary is the development of terrestrial gastropods. We know only relatively few and mostly small forms from the Cretaceous. Within the Tertiary we observe their differentiation into genera and families, especially among Papillaceae, Clausiliaceae, and Helicacea.

The bivalves, like the gastropods, showed a considerable increase of genera and species. Similar to a tendency among marine gastropods, the increase in burrowing forms is noteworthy. The development of sinuate forms commonly was paralleled by a reduction of the hinge. As already mentioned, the relative immobility of bivalves and gastropods led to differentiation of faunal provinces, especially in the Neogene. In the north, a boreal province developed, bordered by the North Atlantic Faunal Province on the south. Next to the latter followed the Northwest European Faunal Province. The Mediterranean area, between Atlantic and Indian oceans (Red Sea), is called the Mediterranean Province.

Along North America is the Caribbean and, along the West Coast, the Californian Faunal Province. Separated from these by the equatorial regions is the Patagonian Faunal Province along South America. The similarities of these provincial faunas with that of New Zealand are greater than with the Caribbean. Therefore, this region is called the South Pacific-Antarctic Faunal Province. The Indo-Pacific region, in spite of its great extent, has a relatively uniform faunal association that forms the Indo-Pacific Province.

In the Neogene, the northern part of disintegrating Tethys, extending from central and southeastern Europe to central Asia, became separated from the oceans through rising young mountain chains. Into these areas, collectively known as Paratethys, faunas related to those of the Indo-Pacific Faunal Province migrated during the Early Neogene, and marine faunas from the Mediterranean Province during the middle Miocene. Especially characteristic are the Sarmatian (Late Miocene) molluscan faunas found between the eastern edge of the Alps and Lake Aral. This molluscan fauna shows traits of an impoverished marine fauna with clear tendencies of endemic evolution. Also, the younger faunas are of endemic character, though in the Maeotian and Akchagylian, marine influences are noticeable. The diversity of endemic groups among the Limnocardiidae, *Congeria* and related forms, Melanopsidae, Limneidae (*Valenciennesia*), and others reached an optimum, offering excellent examples of extreme forms resulting from endemic evolution.

Mollusks used to be regarded as main index fossils for division of the Tertiary into epochs and stages. Recognition that differentiation into faunal provinces strongly influenced the composition of molluscan faunas, especially in the Neogene, has reduced their adjudged importance. Special studies of the evolution of marine genera or groups of genera are few. Nevertheless, mollusks still offer a valuable basis for local subdivisions of sedimentary series within a faunal province or a more narrowly defined sedimentary area. The evolution of endemic molluscan faunas, especially in Paratethys, offers unsurpassed biostratigraphic indicators for local stratigraphic subdivisions.

ARTHROPODA

Considering the abundance of recent arthropods, one might expect Tertiary forms to be varied and rich. Terrestrial groups may be fossilized only under most favorable conditions, however, as when they are preserved in amber. For example, the best-known material shows that among insects many younger Paleocene genera are identical with recent ones. The ostracodes have a special place among arthropods, and show a strong evolutionary tendency toward differentiation.

BRYOZOA, BRACHIOPODA, AND ECHINODERMATA

Bryozoan faunas of the Tertiary are richly differentiated and many Bryozoa are rock-formers. Cheilostomes dominated in the Tertiary.

Number of genera and distribution of brachiopods decreased considerably during Tertiary time as compared with the Cretaceous. The brachiopod faunas of the marginal facies usually contain only a few genera; locally abundant are beds with *Terebratula*. Representatives of other groups are restricted to basinal facies. In Tertiary seas brachiopods played only a subordinate role.

Remnants of echinoderms demonstrate that these invertebrates were relatively abundant in the Tertiary, though most of the dominant groups of the Paleozoic and Mesozoic were on the decline. Crinoids, for instance, were mainly represented by planktonic forms, such as *Discometra (Antedon)*. Echinoids on the other hand, were plentiful. Like the bivalves and gastropods, echinoids showed a tendency for their progressive evolution in the Late Cretaceous to continue into the Tertiary, resulting in the appearance of many new genera. Especially characteristic was the flourishing of irregular echinoids and the prevalence of burrowing forms (Spatangidae). Faunas of the Miocene and Eocene differed greatly. Representatives of the Conoclypeidae are especially characteristic of the lower and middle Eocene; Scutellidae and Clypaeastridae of the younger Tertiary.

BIOSTRATIGRAPHIC DIVISIONS

Subdivision of the Tertiary has been attempted according to different principles, that is, by utilizing different groups of organisms as guides. LYELL (1832) divided the Tertiary into three divisions or epochs according to the percentage of their contained Holocene molluscan taxa:
1) Eocene, 3.5 percent; 2) Miocene, about 20 percent; 3) Pliocene, 35-50 percent. Later, the Oligocene (BEYRICH, 1853-56) and Paleocene (SCHIMPER, 1874; KOENEN, 1885) were separated from LYELL's Eocene. HOERNES (1856, 1870) combined the two faunistically similar younger epochs, Miocene and Pliocene, as Neogene, whereas the older epochs Paleocene, Eocene and Oligocene, were grouped together as Paleogene.

These divisions are based mostly on fossil contents, especially molluscan faunas of western, southern, and central Europe, including the Oligocene Northwest European Faunal Province.

When division of the Tertiary into epochs was proposed, a division into stages was also made. First suggestions in this respect were made by MAYER-EYMAR (1857-58). The system of stages later was improved frequently and adapted to requirements of the times. Originally, molluscan faunas were used to differentiate individual stages. Undoubtedly, in Europe it was possible to define Tertiary stages paleontologically with the help of mollusks, but such attempts must have caused difficulties even then. For delineation of the relatively short period of a stage, mollusks proved mostly insufficient.

For areas outside of Europe, Tertiary molluscan faunas could not be used for correlation with European stages, because the evolutionary tendencies of individual faunal provinces differ from the European ones.

Since the first attempts at subdivision of the Tertiary, evolution of mammals has been used for this purpose. Their rapid evolutionary differentiation, with its wealth of morphological detail, offers decisive biostratigraphic keys. But the fact that most vertebrate remains occur in sediments of terrestrial origin commonly makes their correlation with marine sediments extremely difficult. For this reason, biostratigraphic units based on evolution of the vertebrates have been defined mainly in North America.

The appearance of creodonts and Condylarthra in the Paleocene is significant. In the late Paleocene the appearance of the perissodactyls, which in the Eocene and earliest Oligocene split into several evolutionary lines, is of significance. In the Oligocene, introduction of rhinoceratids in Europe and of anthracotheriids in North America is important.

In the early Miocene, *Mastodon* and other faunal elements migrated from Africa to Europe. Later, *Hipparion* and its accompanying fauna reached Eurasia from North America. This very marked immigration is often considered as a stratigraphic criterion of first order and used to define the boundary between Miocene and Pliocene. But according to contemporary opinion, the Miocene-Pliocene boundary should be drawn higher and the arrival in Eurasia of *Hipparion* would therefore have taken place in the late Miocene. In Europe the boundary between Tertiary and Quaternary is usually determined by the appearance of *Equus, Archidiscodon,* and other genera. The evolution of rodents allows a refined subdivision of zones, especially in the Neogene, though their correlation with marine zones is difficult. Therefore, it is better to divide the terrestrial-limnic deposits of Eurasia into units based on paleontological criteria indicated by mammalian evolution.

The origins of the Tertiary flora are found in development of the angiosperms in the Late Cretaceous. In the Paleocene, Cretaceous floral elements persisted and woody plants predominated. In the Eocene, the distribution of tropical floras reached an optimum. In the Oligocene a retreat of the tropical floras was paralleled by an increase in herbaceous plants. In the temperate zone older Tertiary relicts became extinct during the early Quaternary. Tertiary floras of the temperate zones are useful for evaluation of stratigraphic sequences, but have to be used with caution because of their lack of true evolutionary lines. Also, as in the case of mammals, correlation with marine deposits encounters difficulties.

Finally, tectonic phases should be mentioned as important criteria because they show special intensity during certain intervals of the Tertiary. The evaluation of such processes must be based on their effects, as the processes themselves cannot be observed. But refined stratigraphic methods have shown that the idea of worldwide, short tectonic phases cannot be upheld in its original concept. Nevertheless, tectonic processes, at least in young mountain chains, fall into definite time periods:

1) At the Cretaceous-Tertiary boundary: Laramide phase.
2) In upper Eocene-lower Oligocene: Pyrenaic phase.
3) Around the Oligocene-Miocene boundary: Savic phase.
4) From middle to upper Miocene: old and young Steiric phase.
5) the Pliocene: Rhodanic phase.
6) Around the Pliocene-Pleistocene boundary: Wallachian phase.

As already mentioned the Tertiary System has been divided into series which are subdivided into stages. The following European divisions are presently in use:

Pliocene:	Astian-Piacenzian
	Tabanian-Zanklian
Miocene:	Messinian
	Tortonian
	Serravallian
	Langhian
	Burdigalian
	Aquitanian
Oligocene:	Chattian
	Rupelian
	Lattorfian
Eocene:	Wemmelian
	Lutetian
	Cuisian
Paleocene:	Ilerdian
	Thanetian + Montian
	Danian

In recent years, stage names have been

Planktonic Foraminiferal Zones			Calcareous Nannoplankton Zones		EPOCH
N 3 / P 22	Globigerina angulisuturalis	GLOBIGERINA ASSEMBLAGE	NP25	Sphenolithus ciperoensis	OLIGOCENE
N 2 / P 21	Globigerina angulisuturalis / Globorotalia opima opima		NP24	Sphenolithus distentus	
N 1 / P 20	Globigerina ampliapertura		NP23	Sphenolithus predistentus	
P 19	Globigerina sellii / Pseudohastigerina barbadoensis		NP22	Helicopontosphaera reticulata	
P 18	Globigerina tapuriensis		NP21	Ericsonia subdisticha	
P 17	Globigerina gortani / Globorotalia centralis	HANTKENINA ASSEMBLAGE	NP20	Sphenolithus pseudoradians	EOCENE
P 16	Cribrohantkenina inflata		NP19	Isthmolithus recurvus	
P 15	Globigerapsis mexicana		NP18	Chiasmolithus oamaruensis	
P 14	Truncorotaloides rohri / Globigerinita howei		NP17	Discoaster saipanensis	
P 13	Orbulinoides beckmanni		NP16	Discoaster tani nodifer	
P 12	Globorotalia lehneri				
P 11	Globigerapsis kugleri		NP15	Chiphragmalithus alatus	
P 10	Hantkenina aragonensis		NP14	Discoaster sublodoensis	
P 9	Acarinina densa	GLOBOROTALIA ASSEMBLAGE	NP13	Discoaster lodoensis	
P 8	Globorotalia aragonensis		NP12	Marthasterites tribrachiatus	
P 7	Globorotalia formosa				
P 6	Globorotalia subbatinae		NP11	Discoaster binodosus	
P 5	Globorotalia velascoensis	GLOBIGERINA ASSEMBLAGE	NP10	Marthasterites contortus	PALEOCENE
			NP 9	Discoaster multiradiatus	
P 4	Globorotalia pseudomenardi		NP 8	Heliolithus riedli	
			NP 7	Discoaster gemmeus	
			NP 6	Heliolithus kleinpelli	
P 3	Globorotalia pusilla / Globorotalia angulata		NP 5	Fasciculithus tympaniformis	
P 2	Globorotalia uncinata / Globorotalia spiralis		NP 4	Ellipsolithus macellus	
P 1	Globoconusa daubjergensis / Globigerina pseudobulloides		NP 3	Chiasmolithus danicus	
			NP 2	Cruciplacolithus tenuis	
			NP 1	Markalius inversus	

FIG. 8. Paleogene zones based on planktonic Foraminiferida and calcareous nannoplankton (Papp, n).

introduced for the Tertiary on a regional basis such as for the Neogene of the central and eastern Paratethys. The terminology of the stages has been treated differently in the literature of different countries and faunal provinces. In the Paleogene, the middle Eocene (i.e., Lutetian), which is easily recognizable by optimal development

Planktonic Foraminiferal Zones		Calcareous Nannoplankton Zones		EPOCH
N 23	Globigerina calida	Globorotalia truncatulinoides	NN 21 Emiliana huxleyi	PLEISTOCENE
			NN 20 Geophyrocaspa oceanica	
N 22	Globorotalia truncatulinoides		NN 19 Pseudoemiliana lacunosa	
N 21	Globorotalia tosaensis tenuitheca	Globorotalia miocaenica	NN 18 Discoaster brouweri	PLIOCENE
			NN 17 Discoaster pentarodiatus	
			NN 16 Discoaster surculus	
N 20	Globorotalia multicamerata Pulleniatina obliquiloculata		NN 15 Reticulofenestra pseudoumbilica	
N 19	Sphaeroidinella dehiscens Globoquadrina altispira		NN 14 Discoaster asymetricus	
			NN 13 Ceratolithus rugosus	
N 18	Globorotalia tumida tumida	Globorotalia margaritae	NN 12 Ceratolithus tricorniculatus	
N 17	Globorotalia tumida plesiotumida		NN 11 Discoaster quinqueramus	MIOCENE
N 16	Globorotalia acostaensis Globorotalia merotumida			
N 15	Globorotalia continuosa	Globorotalia menardii	NN 10 Discoaster calcaris	
N 14	Globigerina nepentes Globorotalia siakensis		NN 9 Discoaster hamatus	
N 13	Sphaeroidinellopsis subdehiscens, Globigerina druryi		NN 8 Catinaster coalitus	
			NN 7 Discoaster kugleri	
N 12	Globorotalia fohsi fohsi		NN 6 Discoaster exilis	
N 11	Globorotalia praefohsi	Orbulina s.l.		
N 10	Globorotalia peripheroacuta		NN 5 Sphenolithus heteromorphus	
N 9	Orbulina suturalis Globorotalia peripheroronda			
N 8	Globigerinoides sicanus Globigerinatella insueta	Globigerinoides trilobus		
N 7	Globigerinatella insueta Globigerinoides trilobus		NN 4 Helicopontosphaera ampliaperta	
N 6	Globigerinatella insueta Globigerinita dissimilis		NN 3 Sphenolithus belemnos	
N 5	Globigerina dehiscens	Globigerinita dissimilis	NN 2 Discoaster druggi	
N 4	Globigerinoides primordius Globorotalia kugleri		NN 1 Triquetrorhabdulus carinatus	

FIG. 9. Neogene zones based on planktonic Foraminiferida and calcareous nannoplankton (Papp, n).

of the Foraminifera, especially nummulitids and discocyclines, can be readily traced over wide areas of Eurasia and the Americas. But such easily recognized characteristics are lacking in the early Paleogene and in the Neogene. This has resulted in different usage of stage names in different countries and has led to discrepancies in usage of such names in the Paleocene, Oligocene, and Miocene, since the same stage names have been used for beds of different ages. Therefore, in recent symposia the avoidance of customary stage terminology altogether has been suggested.

New ways must be found to establish subdivisions that satisfy the requirements

TABLE 1. *Relationship between succession of species of* Miogypsina *and some European stratotypes.*

		M. mediterranea	
		M. cushmani	
M. burdigalensis		*M. intermedia*	Stratotype Burdigalian
		M. globulina	
		M. tani	Stratotype Aquitanian
		M. gunteri	
	M. bantamensis		
	M. formosensis	*M. septentrionalis*	Stratotype Chattian
	M. complanata		

of biostratigraphy. Groups of organisms employed for biostratigraphy should be worldwide in distribution and relative abundance. They should also reflect rapid and clearly recognizable evolution. The plankton of the Tertiary fulfills these demands best. For this reason, division of the Tertiary into biozones on the basis of planktonic Foraminifera has been advanced during recent years.

A second group of organisms useful in stratigraphic correlation within the Tertiary is the nannoplankton. The biostratigraphic zonation on the basis of nannoplankton is more or less equivalent to that worked out for planktonic Foraminifera. A comparison of these two schemes is shown in Figure 8 for the Paleogene and Figure 9 for the Neogene; after BERGGREN (1971), BLOW (1969), BOLLI (1966), KAPELLOS and SCHAUB (1973), and MARTINI (1971).

At present, the Danian, with its globigerinid fauna, is regarded as basal Paleocene.

The base of the Eocene fluctuates, depending upon which stages are employed. If the Ypresian Stage is employed as the base, then the boundary is drawn in the lower third of the planktonic foraminiferal zone P6. If the Cuisian Stage is used, then the boundary is at the base of the stratigraphically higher planktonic foraminiferal zone P8 *(Globorotalia aragonensis)* or in the lower third of the nannoplankton zone NP12 *(Marthasterites tribrachiatus)*. In this concept, the Cuisian Stage is used.

The lower boundary of the Oligocene is generally drawn at the base of zones P18 or NP21. The Oligocene encompasses, essentially, the "*Globigerina* region."

The base of the Miocene is, at present, considered to be at the base of zones N4 or NN1.

The Pliocene encompasses zones N18 to N21.

The appearance of *Globorotalia truncatulinoides* defines the Pliocene-Pleistocene boundary and, thus, the boundary between the Tertiary-Quaternary.

Following the suggestion of the Comité du Néogène Méditeranéen, at its 1967 meeting in Bologna, the Neogene should be divided into four "supraétages" on the basis of planktonic zones and it is proposed that these should be designated by the numerals I to IV. Such division is the first attempt at an exact correlation of equatorial sedimentary areas with the Mediterranean and the South Pacific-Antarctic regions.

Additional stratigraphically important groups have already been discussed: nummulitids, discocyclinids, and alveolinids in the Paleogene; lepidocyclinids and miogypsinids in the upper Oligocene-Miocene. The miogypsinids are of particular importance for definition of the Oligocene-Miocene, or Paleogene-Neogene, boundary. DROOGER (1966) worked out the evolution of *Miogypsina* in great detail. As the concepts of Oligocene and Miocene originally were based on knowledge of the marginal facies, the degree of specialization of miogypsinids in stratotypes of the Chattian, Aquitanian, and Burdigalian gains special importance. The main line of miogypsinid evolution is shown in Table 1.

Following suggestions of the Comité du Néogène Méditerranéen made in 1964 and 1967, the boundary between Neogene and Paleogene should be defined by stratotypes based on occurrences of *Miogypsina,* together with the appearance of *M. gunteri.* The more primitive species, *M. septentrionalis* and *M. complanata,* indicate Chattian, and therefore, Oligocene. On the basis of planktonic zones, the lower boundary of the Neogene lies at the base of the zone containing *Globigerinoides primordius.*

The summary offered here makes no attempt to include the commonly used stage names. As mentioned before, these concepts have been used so variably that it is preferable to use local terms for beds, sequences, or formations, which are then to be correlated with the biozones.

Knowledge of the evolution of additional groups, especially of such foraminifers as *Uvigerina, Heterostegina,* and others, is essential for more precise correlation of sections within one faunal province. These would provide a basis for better understanding of molluscan faunas, and thus offer a broader foundation for biostratigraphical analysis, especially in the Neogene.

REFERENCES

Berggren, W. A., 1971, *Tertiary boundaries and correlations:* in Micropaleontology of the oceans, B. M. Funnell, & W. R. Riedel (eds.), p. 693-809, 40 tables, Cambridge Univ. Press (Cambridge).

Beyrich, Ernst, 1853-56, *Die Conchylien des norddeutschen Tertiärgebirges:* Deutsche Geol. Gesell., Zeitschr., v. 5, p. 273-358, pl. 4-8 (1853); v. 6, p. 409-500, pl. 9-14 (1854); v. 7, p. 726-781, pl. 15-18 (1855); v. 8, p. 21-88, pl. 1-10; p. 553-588, pl. 15-17 (1856).

Blow, W. H., 1969, *Late middle Eocene to recent planktonic foraminiferal biostratigraphy:* 1st Internatl. Conf. Planktonic Microfossils, Proc., Geneva 1967, v. 1, p. 199-422, text-fig. 1-43, Brill (Leiden).

Bolli, H. M., 1966, *Zonation of Cretaceous to Pliocene marine sediments based on planktonic Foraminifera:* Asoc. Venezolana Geol., Min., Petrol., Bol. Inf., v. 9, p. 3-32, 4 tables.

Brinkman, Roland, 1966, *Abriss der Geologie, v. 2, Historische Geologie:* 345 p., 73 text-fig., 57 pl., F. Enke (Stuttgart).

Drooger, C. W., 1966, *Miogypsinidae of Europe and North Africa:* Internatl. Union Geol. Sciences, Proc., Comm. Medit. Neogene Stratigraphy, sess. 3, Bern, p. 51-54, Brill (Leiden).

Hoernes, Moriz, 1856, *Die fossilen Mollusken des Teriärbeckens von Wien: I. Gastropoden:* Geol. Reichsanst., Abhandl., v. 3, 736 p., 52 pl.—— 1870, *Die fossilen Mollusken des Tertiärbeckens von Wien: II. Bivalven:* Geol. Reichsanst., Abhandl., v. 4, 479 p., 85 pl.

Kapellos, Christos, & Schaub, Hans, 1973, *Zur Korrelation von Biozonierungen mit Grossforaminiferen und Nannoplankton im Paläogen der Pyrenäen:* Eclogae Geol. Helvetiae, v. 66, p. 687-737, text-fig. 1-11, pl. 1-13.

Koenen, Adolf von, 1885, *Über eine Paleozäne Fauna von Kopenhagen:* K. Gesell. Wiss. Göttingen, Abhandl., v. 32, p. 223-287.

Lyell, Charles, 1832, *Principles of geology:* 1st edit., v. 3, 398 p., Murray (London).

Martini, Erlend, 1971, *Standard Tertiary and Quaternary calcareous nannoplankton zonation:* 2nd Internatl. Conf. Planktonic Microfossils, Rome 1970, Proc., v. 2, p. 739-785, pl. 1-4, 6 tables (Rome).

Mayer-Eymar, Karl, 1857-58, *Versuch einer neuen Klassifikation . . . :* Allg. Schweiz. Gesell. Gesam. Naturwiss. Verhandl., v. 17-19, p. 164-199.

Papp, Adolf, 1959, *Handbuch der stratigraphischen Geologie, v. 3, Tertiär, part 1:* 411 p., 88 text-fig., 61 tables, F. Enke (Stuttgart).

Schimper, W. Ph., 1874, *Traité de Paléontologie végétale ou la flore du monde primitif:* v. 3, 896 p., pl. 95-110, J. B. Baillière et fils (Paris).

QUATERNARY[1]

By W. A. Berggren and J. A. Van Couvering

[Woods Hole Oceanographic Institution, Woods Hole, Massachusetts, and Brown University, Providence, Rhode Island; University of Colorado Museum, Boulder]

CONTENTS

	PAGE
Introduction: Concept of the Quaternary	A506
Lower Limit of the Quaternary	A507
Calabrian Definition	A509
Santa Maria di Catanzaro	A510
Le Castella	A513
Age and Correlation of the Pliocene-Pleistocene Boundary	A513
Biostratigraphy, Biogeography, and Quaternary Climatic Change	A515
History of Continental Glaciation	A515
Calibration of Climatostratigraphy	A517
Atlantic Ocean	A520
Shifts in Paleoisotherms	A520
Changes in Seasonality	A521
Quaternary Climatic History	A523
Gulf of Mexico Climatic Record	A524
Arctic Region	A525
Pacific Ocean	A527
Equatorial Pacific Climatic Changes	A527
Sea of Japan	A527
Antarctic-Subantarctic Ocean	A527
Australia-New Zealand Area	A527
Antarctic Marginal Seas	A530
Radiolarian Evidence of Climatic Change	A530
Calcareous Nannofloral Evidence of Climatic Change	A531
Mediterranean	A531
Deep-basin Core Record of Climatic Change	A531
Red Sea	A533
Black Sea	A533
Summary and Conclusions	A536
References	A537

---

[1] Manuscript received January, 1976; revised manuscript received April, 1977.

INTRODUCTION: CONCEPT OF THE QUATERNARY

The Quaternary spans the last 1.5 Ma.[1] By any conservative standard of comparison it has been a time unique in earth history, even if we refrain from the anthropomorphic temptation of according hominid evolution undue significance.

Its individual recognition began with ARDUINO, who in 1759 proposed a Primary ("basement") and Secondary (lithified and folded) organization of formations in the crust, based on the geology of the Apennines. He also informally noted a third subdivision, that of "low mountains and hills of sand and gravel," and a fourth, final subdivision of "earth and rocky materials and alluvial debris." The third subdivision subsequently became Tertiary (or *tertiaire*) as first used in 1810 by BRONGNIART to describe strata younger than the massive Late Cretaceous chalks of the Paris basin. This was followed by the formalization of the fourth of ARDUINO's subdivisions initially as *quatrième formation d'eau douce* by MARCEL DE SERRES (in CREUZE DE LESSER, 1824, p. 174). DE SERRES later (1830) used the term Quaternary *(quaternaire)* for the same interval and subsequently (1855) claimed to have invented it, but the honor goes, somewhat unjustly, to DESNOYERS (1829).

In his study of the geology of the lowlands of western France DESNOYERS (1829) suggested that the sea had receded earlier from the Seine basin than from the regions of Touraine and Languedoc. He therefore proposed to call the nearly horizontal, relatively unconsolidated younger strata of the latter region *Quaternaire ou Tertiaire récent,* and divided them as follows:

3. *Récent*
2. *Diluvium*
1. *Faluns de Touraine, la molasse suisse, le Pliocène marin de Languedoc*

---

[1] The abbreviations Ma (Mega-annum) refers to units of yr $\times$ 10^6 measured from the present (1950 A.D. by international agreement) pastward. It means the same as the cumbersome "millions of years before present" and is a fixed chronology analogous to the calendars tied to historical events. The abbreviation my (million years) is used to express simple duration in units of yr $\times$ 10^6 in any given past interval.

The first use of the term Quaternary itself, therefore, refers to strata that span the interval between lower Miocene and recent! This was partly because *tertiaire* in France was initially applied only to the Lower Cenozoic strata in the Paris basin, and perhaps also because the southern French Miocene to which DESNOYERS referred was less indurated or deformed than the time-equivalent Italian Miocene which ARDUINO placed in his third subdivision. However, the earlier usage of DE SERRES, which he restated (1830) in confining *quaternaire* to "diluvium" only, was more in line with that of ARDUINO, and it established the modern meaning of the term. It is interesting to note that the lithological concept of Quaternary is still a very strong tradition in geology, and many pre-Quaternary formations are included in such map units as "Quaternary terrace" or "Older Quaternary alluvium" simply because they are poorly indurated, and lack determinative fossils.

DE SERRES (1830) also affirmed that early man had lived contemporaneously with deposition of Quaternary "diluvium." This was the basis of very early and persistent attempts to define the period anthropocentrically. The first major treatise on the *Terrain quaternaire ou diluvien* (REBOUL, 1833) proposed that this time interval be considered as the *Période anthropienne.* Other writers called it the *Période homozoique* (VEZIAN, 1865), *Terrain humaine* (MERCEY, 1874-77) or *Psychozooique* (LE CONTE, 1887, in MEUNIER, 1908). It was at the International Geological Congress of 1888 in Great Britain that GAUDRY, with the approval of PRESTWICH and DE LAPPARENT, made the proposal that mankind —represented in particular by artifacts— was the characteristic element of the Quaternary, which justified its separation from the Tertiary. As we shall see, this early concept is also still very much alive, even though the geological range of artifacts, if not that of *Homo sapiens,* has since been shown to extend back nearly to the base of the Pliocene.

A third concept, that of the Quaternary as a paleontological feature, was outlined by the work of LYELL although he did not use the term himself. Instead, he formalized the term Pliocene (already in use as a general descriptive word for beds containing fossils of relatively young age, e.g., DESNOYERS, 1829) as the latest part of Tertiary time, subdivided into two parts (LYELL, 1833, p. 61):

> New Pliocene—formations containing mollusk species 90-95 percent surviving to the present time, and correlative beds.
> Older Pliocene—formations containing more than 50 percent surviving species, and correlative beds.

In 1839, LYELL suggested *Pleistocene* in place of his Newer Pliocene period; BUTEUX (1843) likewise proposed *Pliostène* as a name for the brief post-Pliocene interval. The equivalence of Quaternary and Pleistocene was simply taken for granted in the literature of the time, with the former generally referring to unconsolidated continental rocks and the latter to beds with marine fossils. Today, the Pleistocene is recognized as an epoch of the Quaternary Period—with or without the Holocene or "Recent" subdivision—and its base defines the base of the Quaternary.

Thus, the basis for recognizing the Quaternary System was in various degrees lithological, archeological, and paleontological according to the earliest definitions; that of the Pleistocene was explicitly paleontological. Nevertheless, only nine years after AGASSIZ (1838) had first boldly suggested that the European continent had recently been invaded by great lowland glaciers, FORBES (1846) concluded that the molluscan faunas to which LYELL had referred in setting up the Pleistocene were the representatives of a colder climate than those of the Pliocene, and he proposed to redefine the period as ". . . the time distinguished by severe climatal conditions throughout the great part of the northern hemisphere" (p. 402). FORBES' statement is an accurate generalization, valid today and startlingly correct for its time. Nevertheless, it is no longer possible to use the "onset of glaciation" or some similar climatic criterion to distinguish the Pleistocene from the Pliocene, or the Quaternary from the Tertiary, because we can now see a much more complex and gradational record than that known to FORBES. In modern studies a limit such as "the first glaciation" is no more acceptable than other criteria with equal or better historical validity, such as "the earliest human," "the oldest artifact," "the oldest fauna with 90 percent survivors," or "the oldest unconsolidated alluvium."

ACKNOWLEDGMENTS

We are indebted to numerous colleagues who have aided us in this compilation of Quaternary marine history by sending us both published and unpublished data. Particular thanks are extended to Drs. J. P. KENNETT (University of Rhode Island, Kingston, Rhode Island), ANDREW McINTYRE and W. B. RUDDIMAN (Lamont-Doherty Geological Observatory, New York), N. J. SHACKLETON (University of Cambridge, England) and JOHN IMBRIE (Brown University, Providence, Rhode Island). We thank in particular Drs. RUDDIMAN and SHACKLETON for critical comments on earlier drafts of this chapter, which have improved its content and organization. This investigation has been supported by Grant GA 21983 from the Submarine Geology and Geophysics Branch, Oceanography Section of the Division of Ocean Sciences, National Science Foundation to the senior author. This is Woods Hole Oceanographic Institution Contribution No. 3754.

LOWER LIMIT OF THE QUATERNARY

Because of the wide variety of relatively unrelated disciplines in Quaternary research, each specialism has contended over the years that the base of the Quaternary should be marked by some auspicious event in its own field. By the middle of this century the various lower defining criteria in advocacy were perceived as follows:

> Paleoanthropologists: first evidence of man or tools.
> Paleoclimatologists: first (major) global cooling or glacial advance.
> Vertebrate paleontologists: first joint oc-

currence of *Equus, Leptobos,* and *Elephas* (in Eurasia).

Marine invertebrate paleontologists: first occurrence of cold-water species in late Neogene sediments of Mediterranean basin (e.g., *Hyalinea baltica, Arctica islandica*).

Paleobotanists: first "glacial" floral association, or exclusion of certain southern elements from European floras.

Most of this special pleading was countered, if not silenced, at the 18th International Geological Congress. At this Congress, the Temporary Committee for the Study of the Pliocene/Pleistocene Boundary recommended (*Internatl. Geol. Congr.,* Repts. 18th Sessions, Great Britain 1948, Pt. 9, 1950, p. 6) that the boundary ". . . should be based upon changes in marine fauna, since this is the classical method of grouping fossiliferous strata." The Committee simply pointed out that the Quaternary, with the Pleistocene as its lower, major portion, was nothing more or less than the youngest period and uppermost system in the geological record.[1]

If the Committee had done no more than this it would have earned its laurels; however, it made a step onto new ground that can be seen, in retrospect, to have made the boundary accessible to modern techniques used in biostratigraphy, magnetostratigraphy, and geochronology. This was to propose that the boundary be typified in an appropriate marine section, and to recommend for this purpose that ". . . the Lower Pleistocene should include as its basal member in the type-area the Calabrian Formation (marine) together with its terrestrial (continental) equivalent the Villafranchian." Aside from the egregious and, as it turned out, mistaken inclusion of the Villafranchian, which seems to have been a sop to the nonmarine specialists on the panel, this proposal cleared the way toward establishing a single, visible, unambiguous physical feature as the key element in defining this elusive boundary.

The new definition superseded the various "model" criteria of the past, including the one based on climate-induced changes in the marine invertebrate fauna, although this fact was not clear to everyone at the time. In fact, the Committee itself justified its proposal by pointing to changes in the marine and continental faunas that were understood to indicate major climate change coincident with the proposed physical boundary in Italy. Nevertheless, their proposal established a limiting criterion that existed in physical space apart from any theoretical model of past events. In so doing, they made the record of climate, fauna, or flora, irrelevant to the definition of the boundary, although more significant than ever (because of the greater precision demanded by reference to a single geographic point) in its correlation (Fig. 1).

Numerous, detailed observations of the Calabrian Pleistocene were stimulated by the 1948 IGC proposal and the 1965 INQUA amendment (see below), but these observations together with a flood of data from other sources have not ended the controversy between those who still cling to a definition of the base Pleistocene founded on a climate model and those who would correlate from a fixed physical reference point, or "golden spike" (AGER, 1973), in the Calabrian Stage.

The fundamental assumption of the climate-model school of thought is that the base of the Pleistocene is defined by a recognizable deterioration in global climate. In this view the base of the Calabrian should correspond with this sudden, global climate change, and, if it does not, then the boundary must be moved to a more suitable level. The underlying philosophy of the second group is that once a fixed, physical reference point or "golden spike" representing a given boundary is adopted the boundary immediately takes on a real existence independent of the arguments that caused it to be placed where it is. In other words, if the establishing arguments should prove erroneous (as so often happens in geology), the definition embodied in the

---

[1] *Neogene* originally denoted relatively modern faunas—those of Miocene, Pliocene, and recent strata *sensu* LYELL (1833, 1839, *et seq.*)—in contrast to the Paleogene faunas of older Tertiary strata (HOERNES, 1856; DENIZOT, 1957). Whether or not the modern Neogene includes Quaternary faunas (*fide* GIGNOUX, 1950; 1955), it is a paleontological-biochronological concept that has no place in the stratigraphic hierarchy. It is completely incorrect to speak of a "Neogene-Quaternary" boundary in place of Tertiary-Quaternary or Pliocene-Pleistocene boundary.

TERM	CRITERION (A)	AGE (Ma)	RESULT/REMARKS
ANTHROPOGENE	FIRST ADVANCED "HOMINIDS" (AUSTRALOPITHECUS)	~5-6	ELIMINATE PLIOCENE
ANTHROPOGENE	FIRST TRUE "HOMINIDS" (RAMAPITHECUS)	~14	ELIMINATE MIDDLE AND UPPER MIOCENE AND PLIOCENE
QUATERNARY (=PLEISTOCENE)	FIRST VILLAFRANCHIAN MAMMAL FAUNA	~3.3	MEANINGLESSLY SHORT PLIOCENE (5.0-3.3 Ma)
QUATERNARY (=PLEISTOCENE)	FIRST HOMO, FIRST CONTINENTAL GLACIATION IN EUROPE, FIRST BIHARIAN MAMMAL FAUNA	~0.7-0.5	EXTREMELY SHORT PLEISTOCENE (<1 Ma)
PLIOCENE/PLEISTOCENE BOUNDARY — BANDY (1972)	FAD G. truncatulinoides LAD Discoasters	~1.8 (base Gilsa=Olduvai)	PLIOCENE OF BANDY (2.8-1.8 Ma) = LOWER PLEISTOCENE (2.8-1.8 Ma) of LAMB & BEARD
PLIOCENE/PLEISTOCENE BOUNDARY — BEARD & LAMB (68) / LAMB & BEARD (72)	LAD G. altispira FIRST CONTINENTAL GLACIATION IN NORTH AMERICA	~2.8	
MIOCENE/PLIOCENE BOUNDARY — BANDY (1972)	LAD Prunopyle titan FAD S. dehiscens s.s.	~2.8-3.0	MIOCENE/PLIOCENE BOUNDARY OF BANDY = PLIOCENE/PLEISTOCENE BOUNDARY OF LAMB & BEARD AT ca. 2.8 Ma
MIOCENE/PLIOCENE BOUNDARY — BEARD & LAMB (68) / LAMB & BEARD (72)	FAD G. margaritae	~5.0	
QUATERNARY = PLEISTOCENE	FIRST GLACIATION	~40 (ANT-ARCTICA)	ELIMINATE OLIGOCENE, MIOCENE & PLIOCENE
QUATERNARY = DILUVIUM	UNCONSOLIDATED CONTINENTAL ALLUVIUM	VARIABLE	Qal = Qt, etc. (on maps) = PLEISTOCENE ON LITHOLOGIC GROUNDS ONLY

FIG. 1. Stratigraphic problems associated with the position of the Pliocene-Pleistocene boundary using criteria of definition other than the "golden spike." See Figure 6 for generic names that are abbreviated on this figure (Berggren & Van Couvering, n). [Explanation: *FAD*, First Appearance Datum; *LAD*, Last Appearance Datum.]

physical reference point is not thereby invalidated. Furthermore, the proposal of a "golden spike" to define (or, if necessary, redefine) a boundary is treated as a practical matter unconnected with the fulfillment of geological ideology.

CALABRIAN DEFINITION

The Calabrian Stage is recognized in widely separated parts of Italy, so it was natural to refer the proposed definition of the base of the Pleistocene (and Quaternary) to the nominative section as described by GIGNOUX (1913) at Santa Maria di Catanzaro, Calabria, and which subsequently received support as the stage stratotype (BAYLISS, 1969; BLOW, 1969; SELLI, 1971). The basal strata are not well exposed here, however, and at the 7th INQUA Congress (Denver, 1965) it was proposed to further fix the base of the Pleistocene explicitly to the base of the Calabrian Stage where it was exposed about 40 km distant from the putative stratotype at Le Castella, Calabria. This was approved by a vote of the Congress and was subsequently adopted by international geological bodies (see EMILIANI, 1967, p. 410).

At Santa Maria di Catanzaro, a prominent, resistant marker (Bed G-G′) is formed by a calcarenite unit at the base of the Calabrian Stage as described by GIGNOUX (1913, Fig. 5). This contact was made the base of the Calabrian stratotype in the definition of SELLI (1971). Inordinate emphasis has been placed, however, on the first occurrence of the bivalve *Arctica islandica* in the calcarenite unit G-G′ (Fig. 2), which arises because *A. islandica* is an indicator of cold climate and is also a principal index fossil for the Calabrian. For this reason, some have considered this level to represent the actual base of the Quaternary (e.g., RUGGIERI, 1971, 1972; RUGGIERI et al., 1976) despite the fact that its isolated appearance in the stratotype, according to paleobathymetric interpretations (BANDY & WILCOXON, 1970; SELLI, 1967, 1977), is evidently due to slumping from a shallow-water environment. *Hyalinea baltica*, the other classic index fossil for Calabrian, is present throughout the stratigraphic section at Santa Maria di Catanzaro and ranges to at least 76 m below Bed G-G′ (BAYLISS, 1969). This taxon is part of the invertebrate assemblage *(nordische Gäste)* whose initial appearance in the Mediterranean is generally considered by Mediterranean geologists to have been climatically controlled and that can serve as definitive criteria in the definition of the Pliocene-Pleistocene boundary. Thus, although Bed G-G′ at Santa Maria di Catanzaro (with *Arctica islandica*) has served as the base of the Calabrian (= base Pleistocene) for many Mediterranean geologists, the presence of *Hyalinea baltica* to the base of the section at Santa Maria di Catanzaro has led other geologists (e.g., LAMB & BEARD, 1972; BANDY & WILCOXON, 1970), to include the lower part of the section (below Bed G-G′) in the Pleistocene as well.

A description of the Le Castella section by EMILIANI et al. (1961) showing that the sequence here was more uniformly fossiliferous, was the main basis for the 1965 INQUA amendment to the 1948 proposal. The first appearance of *Hyalinea baltica* in the 1961 sample traverse at Le Castella was said to occur between samples 50 and 51, where a thin intercalation of sand (called simply "the marker bed") crops out in the otherwise clayey sequence. (It has since been found in samples examined at Woods Hole by BREMER, BRISKIN, & BERGGREN, 1977, that the initial appearance of *H. baltica* occurs some 25 m below the marker bed at Le Castella where it already constitutes 17 percent of the total benthonic population.) On the basis of the earlier observation, however, EMILIANI (1961) convincingly proposed to correlate the sandy "marker bed" to Bed G-G′ (= base of the Calabrian stratotype) at Catanzaro, even though 40 km separates the nearest outcrops of the two beds. The action of the 1968 INQUA Congress has the effect of making the intersection of this bed with EMILIANI's traverse line the physical reference point defining the base of the Pleistocene. By general agreement, rather than by further official action, this point is now the "golden spike" at the origin of correlations to the beginning or bottom of the Quaternary, for those who approve of this method of definition.

Figure 2 graphically summarizes the outstanding paleontological and lithological features of the Catanzaro sequence, and Figure 3 those of the Le Castella sequence.[1]

SANTA MARIA DI CATANZARO

Gephyrocapsids (*Gephyrocapsa aperta, G.*

---

[1] Figure 2 and the accompanying discussion of the Santa Maria di Catanzaro section, which provide graphic clarification of several problems relating to the stratotype Calabrian, were originally communicated in written correspondence (12 and 22 November 1974) by A. R. EDWARDS, New Zealand Geological Survey. We would like to thank Dr. EDWARDS for permission to utilize these data.

STRATOTYPE CALABRIAN: Santa Maria di Catanzaro

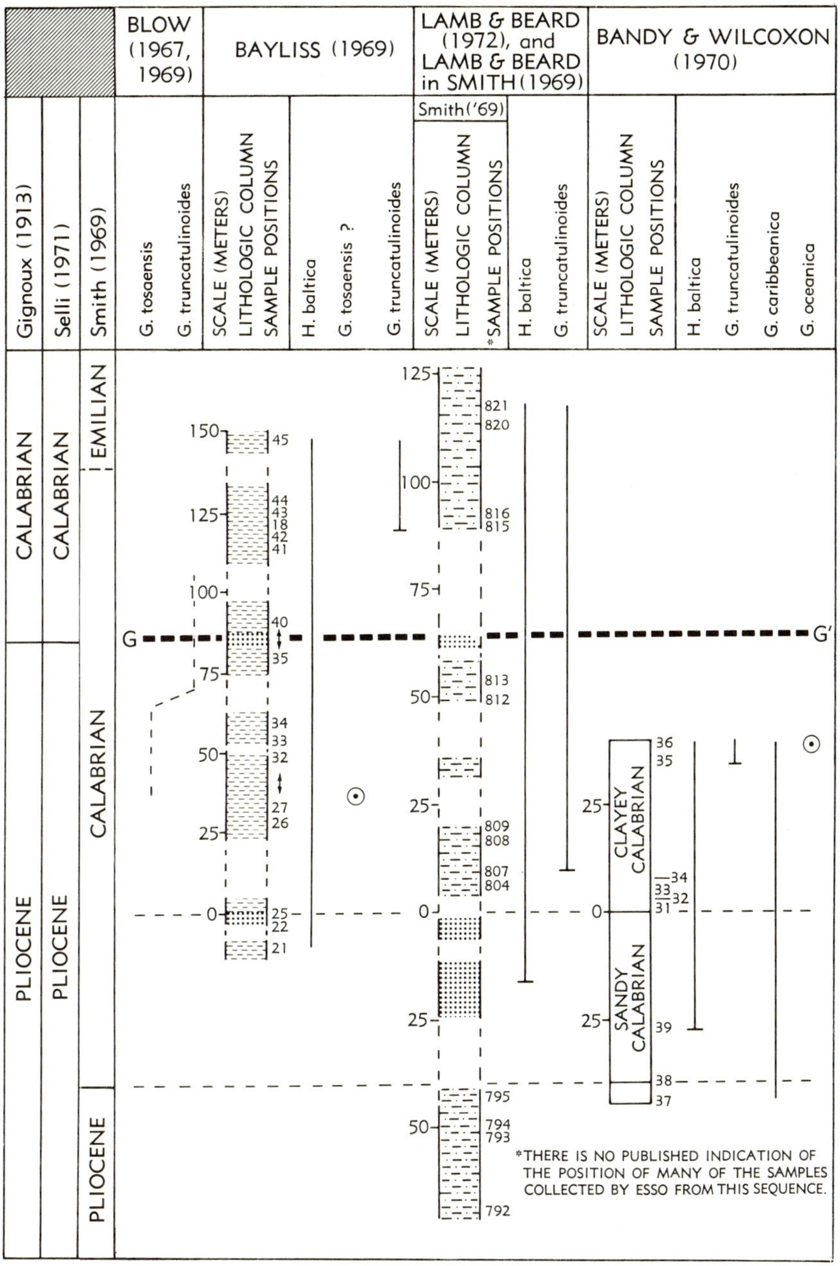

FIG. 2. Basic biostratigraphic data in Santa Maria di Catanzaro section based on published sources (after data supplied by A. R. Edwards). The position of Bed G-G′ is shown as a dashed line between the 75 and 100 m level in the BAYLISS (1969) column and between 50 and 75 m in the SMITH (1969) column. Taxa with abbreviated generic names are *Gephyrocapsa caribbeanica, Gephyrocapsa oceanica, Globorotalia tosaensis, Globorotalia truncatulinoides,* and *Hyalinea baltica* (Berggren & Van Couvering, n).

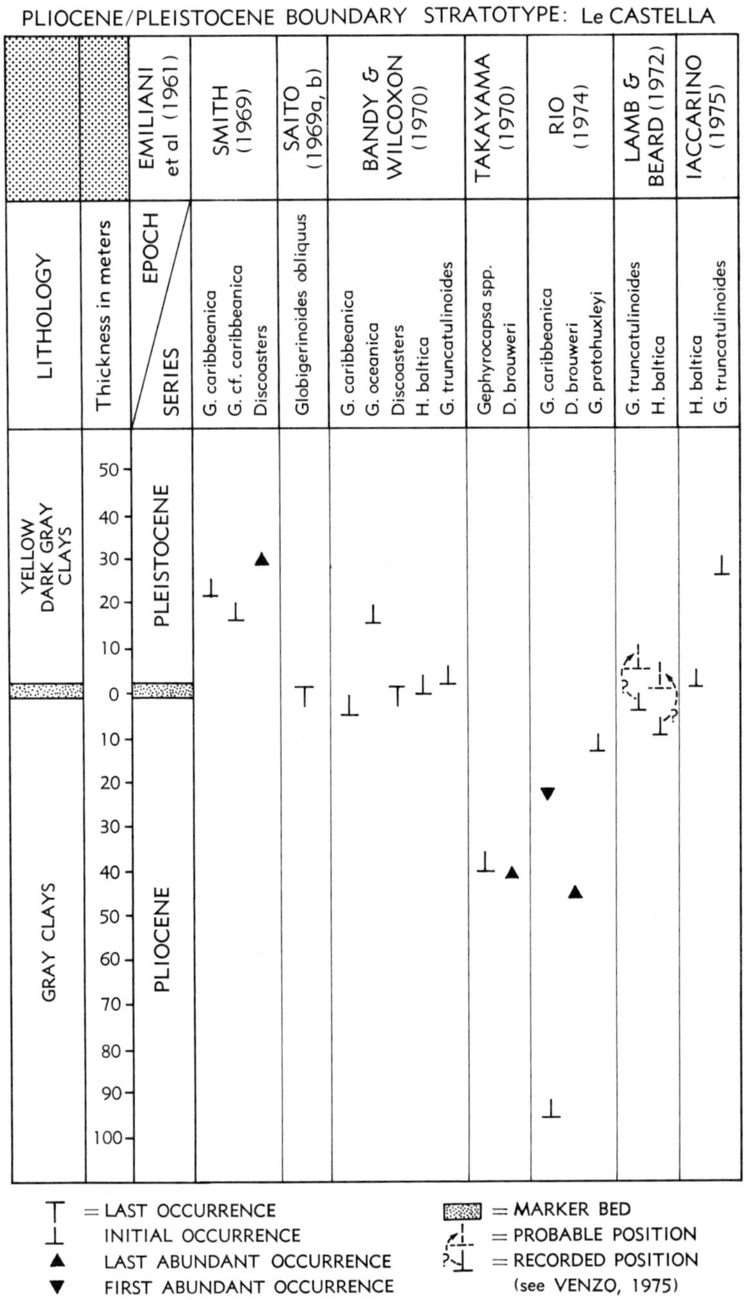

Fig. 3. Basic biostratigraphic data in Le Castella (Italy) section from published sources. Taxa with abbreviated generic names are *Discoaster brouweri*, *Gephrocapsa protohuxleyi*, and those names cited in the caption of Figure 2 (Berggren & Van Couvering, n).

caribbeanica, and *G. oceanica*)—important in determining the biochronology of the Pliocene-Pleistocene boundary—are already present at the base of the Catanzaro section

(HAQ, BERGGREN, & VAN COUVERING, 1977) together with *Hyalinea baltica,* at a level more than 75 m below the level designated as the base of the Calabrian by GIGNOUX (1913).

BANNER and BLOW (1965) claimed to have recognized the evolution of the planktonic foraminiferal species *Globorotalia truncatulinoides* and *G. tosaensis* in the lower part of the Calabrian at Santa Maria di Catanzaro. BLOW (1969) also used this datum to define the base of the planktonic foraminiferal Zone N.22. Abundant *G. tosaensis* or *G. truncatulinoides* do not occur in this sequence, however, and only very scattered, in most places single, specimens have been noted in a few of the samples from Catanzaro (BAYLISS, 1969; BANDY & WILCOXON, 1970; LAMB & BEARD, 1972). This fact accounts for the various levels here, as at Le Castella (below), at which the local first occurrence of *G. truncatulinoides* has been reported (see Fig. 2) and thus the initial evolutionary appearance of *Globorotalia truncatulinoides* within the stratotype Calabrian must be viewed as inadequately documented.

LE CASTELLA

For the Le Castella section the relationship between biostratigraphic data of different authors is shown in Figure 3. Important features to note are 1) the extinction of *Globigerinoides obliquus,* which in deep sea cores occurs within the upper part of the Olduvai normal paleomagnetic event; 2) the wide variations in reported first appearance of *Globorotalia truncatulinoides,* due evidently to its extreme rarity in the section as at Catanzaro; 3) marked quantitative diminution in abundance of *Discoaster brouweri,* the last of the discoasters at about 40 to 45 m below the marker bed (we consider this diminution to be a more reliable guide to the extinction of the group than the highest stratigraphic occurrence of specimens because they are so easily reworked); 4) reported first occurrence of *Gephyrocapsa caribbeanica* from levels ranging from 21 m above to 95 m below the marker bed (it should be noted that the occurrence at 95 m is a rare occurrence and that *G. caribbeanica* first becomes common at 25 m below the marker bed); and 5) first appearance of *Gephyrocapsa protohuxleyi* some 10 to 13 m below and *G. oceanica* apparently about 15 m above the marker bed.

In comparison with these generally conflicting reports, an examination of the calcareous nannoplankton at Le Castella (HAQ, BERGGREN, & VAN COUVERING, 1977) using scanning electron microscopy indicated that *Gephyrocapsa caribbeanica* and *G. oceanica* first appear 30 m and 23 m, respectively, below the marker bed, the latter occurrence coinciding with the initial appearance of *Hyalinea baltica* (BREMER, BRISKIN, & BERGGREN, 1977). The variability in stratigraphic records of the calcareous nannoplankton cited in the literature is ascribed in part at least to the use of light microscopy and the consequent difficulty in correctly identifying nannoplankton (especially gephyrocapsid) taxa (HAQ, BERGGREN, & VAN COUVERING, 1977).

On the basis of these data, it would appear that the Pliocene-Pleistocene boundary is closely linked with the extinction of discoasters and the initial appearance of *Gephyrocapsa caribbeanica, G. oceanica,* as well as the extinction of *Globigerinoides obliquus.* The first appearance of *Globorotalia truncatulinoides* is not an adequate criterion because it cannot be clearly documented in the Calabrian sequence.

AGE AND CORRELATION OF THE PLIOCENE-PLEISTOCENE BOUNDARY

The extinction of discoasters and the evolution of *Globorotalia truncatulinoides* are events that have been securely identified with the long normal paleomagnetic event situated within the Matuyama Reversal Epoch approximately midway between the Gauss and Brunhes normal epochs. The reported chronologic limits for this event range from 1.61-1.79 Ma (COX, 1969) to 1.71-1.86 Ma (OPDYKE, 1972). New radiometric dates centered around 1.58 ± 0.08 Ma attributed to the Gilsa (WATKINS, KRISTJANSSON, & McDOUGALL, 1976) and 1.79 ± 0.03 Ma to the type Olduvai (BROCK

& HAY, 1976) geomagnetic polarity events have kept alive the issue of whether these two events are, indeed, identical. They are considered identical here, with geochronological limits rounded to 1.6 and 1.8 Ma. As we have seen, the two paleontological phenomena noted above were assumed by a succession of investigators to be recorded in the basal part of the typical Calabrian Stage. This suggested that the base of the Pleistocene could be dated to a time within the Olduvai, probably close to its beginning (HAYS & BERGGREN, 1971; BERGGREN & VAN COUVERING, 1974). Attempts to make direct paleomagnetic measurements on the Calabrian of Catanzaro or Le Castella, however, have been fruitless (LAMB & BEARD, 1972; NAKAGAWA, NIITSUMA, & HAYASAKA, 1969, 1971; NIITSUMA, 1970; WATKINS, KESTER, & KENNETT, 1974).

The earlier interpretation of Calabrian micropaleontological correlations has been weakened by recent deep-sea investigations, primarily those of the Deep Sea Drilling Project, which have shown the initial occurrence of *Globorotalia truncatulinoides* near the beginning of the Olduvai event is prior to the extinction of discoasters. This strengthens the impression that the observations of extremely rare *G. truncatulinoides*, above the apparent level of discoaster extinction in the Calabrian, have been erroneously emphasized and do not represent the true datum. Furthermore, recent study (HAQ, BERGGREN, & VAN COUVERING, 1977) has shown that the initial occurrence of *Gephyrocapsa caribbeanica* just precedes, and that of *Gephyrocapsa oceanica* follows shortly after, the end of the Olduvai Event. Because these relatively reliable nannofossil events either bracket or wholly predate the currently adopted physical definition of the Pleistocene base at Le Castella, the age of the Pliocene-Pleistocene boundary is, on this basis, coincident with or just younger than the end of the Olduvai Event. A reasonable estimate of this age would be very close to 1.6 Ma (see Fig. 4).

Correlations between the Le Castella and Santa Maria di Catanzaro sections, as well as estimates of the age of the marker beds, have been controversial. The "established" point of view has been that the marker beds were stratigraphically equivalent (e.g., EMILIANI, MAYEDA, & SELLI, 1961; BAYLISS, 1969; SELLI, 1977) and that they have an age of about 1.8 Ma (HAYS & BERGGREN, 1971; BERGGREN & VAN COUVERING, 1974). One strongly dissenting opinion (SMITH, 1969; BEARD, 1969; LAMB, 1969, 1971) has been that the true base of the Calabrian stratotype should be more than 75 m below the G-G′ marker, at the transition between clayey Pliocene beds and the so-called "sandy Calabrian" with *Hyalinea baltica*, and that this contact should define the base of the Pleistocene at an age of about 2.8 Ma. The Le Castella marker bed, at 1.8 Ma, in this view is considered to be near the base of the Emilian Stage. A contrary opinion (SPROVIERI *et al.*, 1973; RUGGIERI *et al.*, 1976) is that the lower part of the Calabrian Stage—equivalent to the Le Castella section—is missing at Santa Maria di Catanzaro because the "sandy Calabrian" is unconformably transgressive on the Pliocene, and that all of the Catanzaro Calabrian (*s.l.*) is therefore younger than the Le Castella marker bed. Although the existence of an unconformity at Santa Maria di Catanzaro is questioned—SELLI (1977) described the contact between clayey and sandy beds as a fault—Bed G-G′ seems to be clearly younger than the marker bed at Le Castella. This is because the entire 75 m of "sandy Calabrian" underlying Bed G-G′ contains *Hyalinea baltica* together with the species of *Gephyrocapsa* noted above, in an association that first appears at Le Castella only a short distance below the Pliocene-Pleistocene boundary marker bed (HAQ, BERGGREN, & VAN COUVERING, 1977).

If this is so, we are faced with an apparent paradox: if Bed G-G′ at Santa Maria di Catanzaro (which presently denotes the base of the stratotype of the Calabrian Stage) is younger than the marker bed at Le Castella (which presently denotes the Pliocene-Pleistocene boundary), then the section of "sandy Calabrian" below Bed G-G′ is both Pliocene because it is pre-Calabrian, and lower Pleistocene because it is (according to multiple reinforcing micropaleontological criteria) younger than the Le Castella marker bed. If, however, we accept that there is only one definition of the Pliocene-Pleistocene boundary, that of

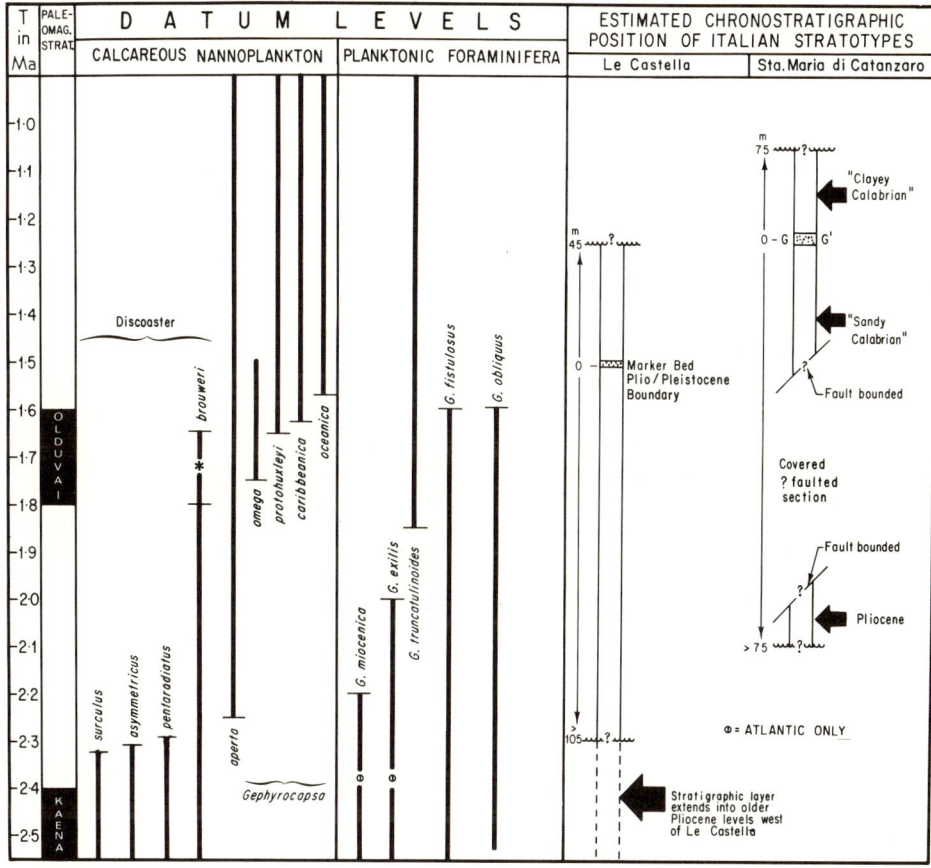

FIG. 4. Pliocene-Pleistocene calcareous plankton biochronology in deep-sea cores and estimated chronostratigraphic position of Calabrian sequences. Extinction of *Discoaster brouweri* occurred at about 1.8 Ma in one of the cores (V12-18) studied. The upper limit of this species, as shown here, may thus be somewhat younger than the actual extinction datum, due to reworking through the interval indicated by asterisk (after Haq, Berggren & Van Couvering, 1977, fig. 5).

Le Castella, this resolves the paradox and simultaneously demonstrates the value of the "golden spike" in picking apart such conundrums.

BIOSTRATIGRAPHY, BIOGEOGRAPHY, AND QUATERNARY CLIMATE CHANGE

Because it is virtually impossible to distinguish between a "pure" biostratigraphy and a "pure" climatology in the geologic record of the late Neogene, the term **climatostratigraphy** is appropriate to describe the study of variations in the stratigraphic record that are due primarily to the effect of global climatic oscillations.

HISTORY OF CONTINENTAL GLACIATION

Recent studies show that the history of the major continental ice caps in the Southern Hemisphere can be extended back at least to the mid-late Miocene (*ca.* 10 Ma) (SHACKLETON & KENNETT, 1975a,b) and of the Northern Hemisphere to about 3 Ma (LAUGHTON et al., 1970, 1972; BERGGREN, 1972; VON HUENE et al., 1971; SCHOLL et al., 1971). In the subantarctic region the evidence indicates that the East Antarctic ice sheet expanded briefly to a volume in excess of present-day size about 6-5 Ma. Subsequent climatic warmings in the Pliocene did not significantly affect the size of the Antarctic ice sheet and this invulnerability to global climatic oscillations has persisted over the past several million years. Minor changes in the isotopic composition of planktonic foraminifers in subantarctic cores at 3 Ma followed by major isotopic changes at 2.6 Ma indicate the onset and growth of a Northern Hemisphere continental ice sheet. In the Northern Hemisphere, the appearance of a major ice cap at high latitudes is indicated by initial ice rafting at about 3 Ma in the northeast Pacific (VON HUENE et al., 1971) and also in the Bering Sea (SCHOLL et al., 1971). In the North Atlantic (Labrador Sea) it consists of ice rafted detritus that appears with polar planktonic microfaunas and microfloras immediately above carbonate oozes with tropical-subtropical microfaunas and microfloras at a level biostratigraphically correlated to the paleomagnetic time-scale at 3 Ma.

The Arctic ice caps have been estimated to have grown to about one-third to one-half of their maximum late Pleistocene volume during the late Pliocene (ca. 2.6-2.0 Ma) (SHACKLETON & KENNETT, 1975b).

In the continental sequences of the temperate regions, the evidence clearly indicates unprecedented cooling trends and montane glacial advances, if not the formation of continental lowland ice sheets, beginning approximately 3 Ma ago (SAVAGE & CURTIS, 1970). Also in the southern Andes several montane glacial advances in a conformable sequence of tills and lavas have been dated prior to ca. 1.0 Ma by FLECK et al. (1972), with the oldest tillite older than 3 Ma (MERCER, 1973). In Europe, the Praetiglian cold-climate stage is marked by the first of numerous subarctic paleofloras seen in late Neogene pollen suites (VAN MONTFRANS, 1971, p. 233; ZAGWIJN, VAN MONTFRANS, & ZANDSTRA, 1971; ZAGWIJN, 1974). It is generally considered by Dutch geologists to be the earliest part of the Pleistocene in North Sea basin sediments where its base is correlated with the *Elphidium oregonense* cold-water microfaunal zone (VAN VOORTHUYSEN, TOERING, & ZAGWIJN, 1972); however, vertebrate fossils, principally primitive *Mammuthus "subplanifrons"* (= *meridionalis*), indicate that the Praetiglian beds are equivalent in age or slightly older than the middle Villafranchian (e.g., Roccaneyra local fauna ca. 2.5 Ma) (AZZAROLI, 1970; V. J. MAGLIO, written commun., 1973) and paleomagnetic analysis agrees in placing the base of this stage "between 2.0 and 3.0, perhaps 2.3 Ma old" (VAN MONTFRANS, 1971, p. 233). The equivalence of the base Villafranchian and the base Calabrian, which was assumed on lithostratigraphic and paleoecological grounds when the two stages/ages were both set equal to the beginning of the Pleistocene (Internatl. Geol. Congress, London, 1948), has since been shown to be erroneous (BERGGREN & VAN COUVERING, 1974). The base of the Calabrian is evidently more than 1 my younger, but in the Netherlands as elsewhere the effects of this mistaken assumption linger.

The continental record, discussed in more detail elsewhere (BERGGREN & VAN COUVERING, 1974), also seems clearly to indicate that the period of progressively more intense montane and high-latitude glacial activity ended in North America with the first major continental glaciation, the Nebraskan. This is tentatively centered at 1.5 Ma, but in Europe (south of England) continental lowland ice-sheets did not form until the Mindel-Elsterian glaciation, dated roughly 0.6 Ma. Several authors (e.g., RICHMOND, 1970; COOKE, 1973) have correlated the Donau-Eburonian cold-climate phase of Europe with the Nebraskan, and REPENNING (1967) reached a similar conclusion based on mammalian correlations before the geophysical information was well developed. We have further concluded that the Kansan glaciation, which is apparently in the Matuyama paleomagnetic age (pre-0.7 Ma), cannot correlate to the post-Matuyama deposits of the Elsterian (VAN MONTFRANS, 1971; ZAGWIJN, VAN MONTFRANS, & ZANDSTRA, 1971), and that the Yarmouthian (post-Kansan) interglacial is characterized by late Irvingtonian mammal faunas comparable in evolutionary grade and geophysical age to those of the Biharian mammal faunas of Europe that lived during the pre-Mindel interglacial. The Biharian-Villafranchian transition in Europe was marked by a cold-climate period, possibly a close-set series of stades and interstades, beginning close to 0.9 Ma, which broadly agrees with estimates of the age of major glacial activity in the North American mountains (cf. COOKE, 1973) and falls within the limits of 1.2 and 0.7 Ma established for the Kansan glaciations. Based on mammalian biochronology the assumed correlation of continental paleoclimatic stages in the Northern Hemisphere is as follows:

Wisconsinan = Weichsel = Würm; began *ca.* 0.073 Ma
Riss-Würm (= Eemian); began *ca.* 0.127 Ma
Riss (includes Alt-Riss); began *ca.* 0.2 Ma
(Sangamonian = Holsteinian interglacial)
Illinoisan = Mindel = Elster; began *ca.* 0.6 Ma
(Yarmouthian = "Cromerian" interglacial) (but not type Cromerian; see BERGGREN & VAN COUVERING, 1974)
Kansan = Günz = Menapian; began *ca.* 0.9 Ma, probably with a second and dominant phase *ca.* 0.8 Ma
(Aftonian = Waalian interglacial)
Nebraskan = (?later part of) Donau = Eburonian; began *ca.* ?1.6 Ma
(Blancan warm-climate phase = Tiglian "interglacial")
Early Blancan cold-climate phase = ?Biber = ?Praetiglian; began *ca.* ?2.5-3 Ma

As we shall see, the marine record contains features that cannot be expressed in such a simple chronology, and that suggest important refinements. To date there has been no general agreement on the correlation of the various glacial episodes on the continents with paleoclimatic cycles reflected in the deep-sea sediments and fossils, and all too often it has been the practice of oceanographers to identify a continental glacial event in marine cores according to some poorly tested paleoclimatological analogy. The difficulty has largely been with the nature of the data. The record on land, with the exception of loess deposits and certain lacustrine sequences beyond the glaciated regions, is essentially discontinuous because of the repeated passage of glaciers and the later erosion of their deposits. The preserved remains of a given "glacial stage" in most places represents but a small fraction of the time during which given climatological conditions prevailed and much of the history of events during that time is unavailable. On the other hand, in the deep-sea record one sees evidence of relatively continuous paleoclimatic changes on a relatively fine scale and for this reason marine paleoclimatological cycles may prove eventually to be the best means of providing a chronology of the late Neogene glacial record.

CALIBRATION OF CLIMATOSTRATIGRAPHY

Temperature variations in the Pleistocene oceans, as reflected in the oxygen isotope ratios in fossil planktonic foraminifers from deep-sea cores, have been examined in detail by EMILIANI (1955, 1961, 1964, 1966a,b) and EMILIANI, MAYEDA, & SELLI (1961). EMILIANI demonstrated isotopic changes, said to represent cyclical climatic variations, with a periodicity of about 100,000 years. These cycles are numbered so that each warm half-cycle is odd and each cold half-cycle is even. OLAUSSON (1965), SHACKLETON (1967), and DANSGAARD and TAUBER (1969) later suggested that because of the isotopic fractionation that occurs between fresh and salt water, oxygen isotope measurements are a measure of the extraction of water from the oceans during periods of glaciation and the recirculation of this water during interglacial periods, rather than a direct reflection of sea-water temperature. The oxygen isotope curve then would be an ice-volume (i.e., sea level) rather than a paleotemperature curve.

SHACKLETON and OPDYKE (1973) further extended the oxygen isotope record by identifying new "stages" 16 to 22 in an equatorial Pacific core, Vema V28-238 (Fig. 5). A direct one-to-one correlation was shown between the carbonate minima (ARRHENIUS, 1952; HAYS et al., 1969) and the oxygen-isotope "stages" for the past 0.9 Ma. The age of each stage boundary was calculated using an extrapolated uniform

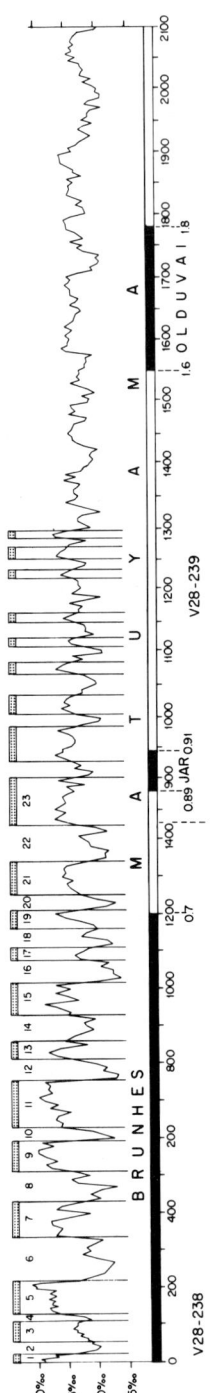

sedimentation rate between the top of the core and the Brunhes-Matuyama boundary (0.69 Ma) at 12 m in the 16-m core. The isotopic climatic record was then extended down through the Pleistocene and into the late Pliocene in another equatorial Pacific core, V28-239 (SHACKLETON & OPDYKE, 1976; see Fig. 6, this paper). Below stage 22 isotopic fluctuations were shown to have a periodicity of approximately 40,000 years and lower amplitudes down to about 1.4 Ma and even lower-frequency events were recorded in the interval from 1.4 Ma to the Olduvai (ca. 1.8 Ma). The isotopic record thus clearly demonstrates that the Brunhes-Matuyama boundary separates two distinct climatic regimes within the Pleistocene, so that the first major northern hemisphere glaciation—the Kansan ($=$ Günz $=$ Menapian), or its second, greater part—occurs in the interval between the Jaramillo and the Brunhes-Matuyama boundary at about 0.8 Ma; prior to this time smaller scale glaciations occurred back into the Pliocene. BROECKER and VAN DONK (1970) have pointed to the general asymmetrical ("sawtoothed") shape of the major climatic oscillations in which a gradual cooling trend is terminated abruptly by a relatively rapid warming (the rapid warmings were called ^{18}O "terminations"). Similar observations were made earlier by DEUSER and DEGENS (1969) in which ~20,000-year evaporation cycles in the Red Sea were abruptly terminated by rapid (~2,000-year) incursions of normal sea water from the Indian Ocean.

Thus, the combined techniques of paleo-

FIG. 5. The oxygen isotope climatostratigraphic record of the last 2 Ma. The oxygen isotope composition of *Globigerinoides sacculifer* in cores V28-238 (01°01′N., 160°29′E.) and V28-239 (3°15′N., 159°11′E.) (overlap of two cores is shown at boundary between "stage" 22 and 23) expressed as deviation from EMILIANI B1 Standard. The last 0.7 Ma (Brunhes) contains glacial stages represented by amplitudes in excess of 1 per mil and with periodicities of about 100,000 years. Isotopic minima were approximately the same in the different "core stages." The mid-Matuyama (~1.4-0.7 Ma) interval contains isotopic fluctuations with approximately 40,000 year periodicities. Amplitudes are lower (0.7 per mil) than during the Brunhes (composite figure compiled by N. J. Shackleton; data from Shackleton & Opdyke, 1973, 1976).

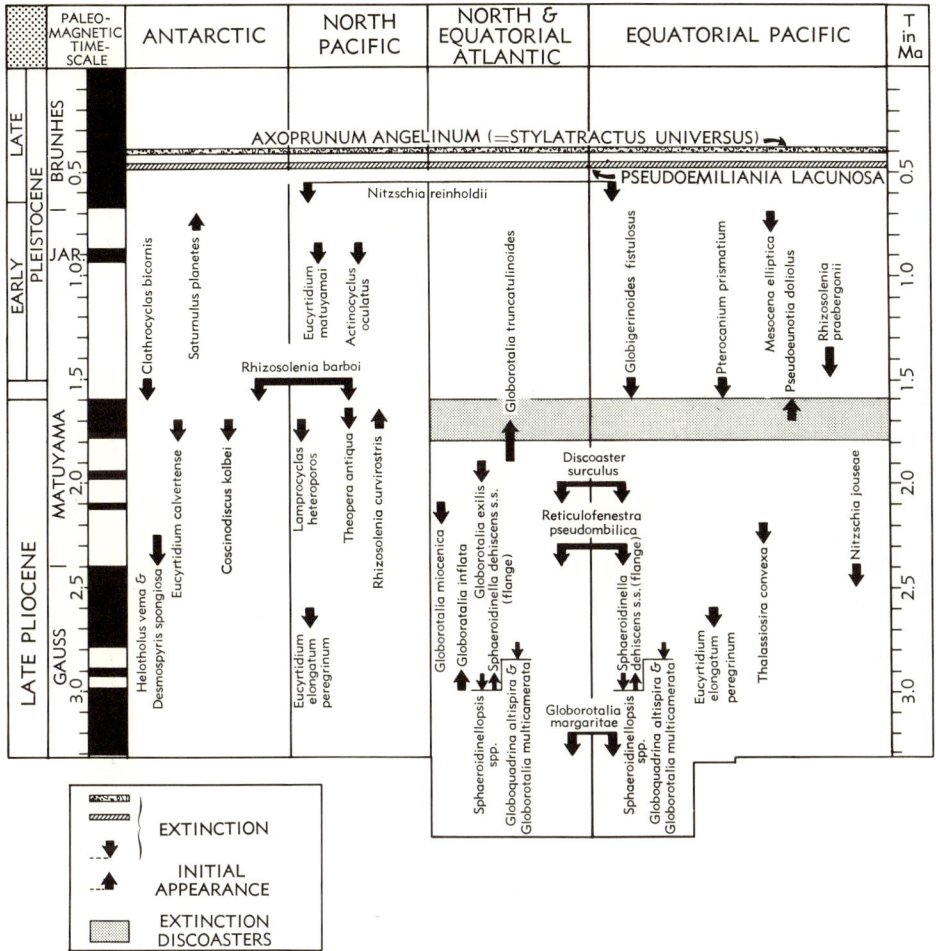

Fig. 6. Marine planktonic biochronologic datum-levels for the last 3 Ma (after Berggren, 1977, fig. 1). [Additional data supplied by L. H. Burckle, J. D. Hays, and N. D. Opdyke.]

magnetism and oxygen isotope analysis have provided the means of erecting an accurate chronology for the last 2 Ma (see Fig. 5). Inasmuch as it depends upon an irreversible phenomenon that occurs essentially synchronously throughout the ocean (limited only by the rate of oceanic mixing) it is unlikely "that any superior stratigraphic subdivision of the Pleistocene will emerge" (SHACKLETON & OPDYKE, 1973, p. 48). Because the underlying variable used is the volume of terrestrially stored ice, this chronologic scheme can serve as a standard method for intra-Pleistocene chronology and, indeed, SHACKLETON and OPDYKE (1973, p. 48) have suggested that the isotopic "stages" that they have established in the Equatorial Pacific be adopted as standard for the latter half of the Pleistocene.

It should be emphasized, however, that the oxygen isotope stratigraphy is *not* a time scale and can only be used to "date" climatic events and associated biostratigraphies to the extent that levels in this stratigraphy can be calibrated—first, by radiometric or biochronological analysis, or second, by association with calibrated magnetostratigraphic boundaries.

Biostratigraphic zonations of Pliocene and Pleistocene marine sediments based upon calcareous and siliceous plankton have been developed over the past decade, primarily as a consequence of the concentrated investigations of deep-sea cores since the inception of the Deep Sea Drilling Project in 1968. By means of interzonal correlation and calibration to oxygen isotope climatic "stage" chronology (SHACKLETON & OP- DYKE, 1973, 1976) and the paleomagnetic time scale (Cox, 1969) these zones now provide a biochronologic framework within which global Pliocene-Pleistocene climatostratigraphic investigations may be conducted. Some of the major microplanktonic biochronologic datum levels of the last 3 Ma are shown in Fig. 6.

ATLANTIC OCEAN

Recent analysis of cores from the North Atlantic (LAUGHTON et al., 1970, 1972; BERGGREN, 1972) has shown that the continental glacial history of the Northern Hemisphere can be extended back three million years. And yet, in comparison with other parts of the world, our information on the stratigraphic record in this region is comparatively incomplete within this interval. This is due, in large part, to the relatively great thickness of ice-rafted terrigenous detritus in late Pliocene-Pleistocene sediments.

The pioneering investigations on the Pleistocene paleoclimatic history of the Atlantic Ocean were made by ERICSON and his colleagues at the Lamont-Doherty Geological Observatory (ERICSON et al., 1956; ERICSON, EWING, & WOLLIN, 1963, 1964a,b; ERICSON & WOLLIN, 1956a,b, 1968). As a result of detailed and laborious investigations on numerous deep-sea cores, a letter zonation of the Pleistocene was formulated (Zones Q through Z) based on the presence or absence of *Globorotalia menardii*. Those intervals in which *G. menardii* was absent were interpreted as "cold periods" and correlated with the classic continental glaciations of northern Europe, a procedure that had little justification at the time and in retrospect has been extremely misleading.

More recently, a Quaternary paleoclimatology based on microfossils in the eastern equatorial Atlantic has been described by RUDDIMAN (1971). In this analysis, large-scale climatic shifts were shown to have occurred at 1.3 Ma and at 0.9 Ma. In the earlier change the mean climatic condition deteriorated and short severe cold pulses punctuated the previously moderate warmth of the late Matuyama. In the latter modification, the duration of cold intervals increased. It was observed that, prior to the Jaramillo (0.9 Ma), no cold pulses exceeded 30,000 years, whereas three post-Jaramillo cold intervals ranged from 50,000 to 150,000 years in duration. The shortest and most recent of these corresponds to the Wisconsin glaciation.

RUDDIMAN (1971) found that although the absolute input rate of pelagic carbonate to sediments increased during cold intervals, the net carbonate percentage tended to decrease during glacial times due to dilution by greater influxes of terrigenous detrital material. Carbonate percentages were depressed to very low values beginning in the Jaramillo, whereas pre-Jaramillo sections were generally formed from calcareous oozes.

SHIFTS IN PALEOISOTHERMS

Recently McINTYRE, RUDDIMAN, & JANTZEN (1972) have demonstrated that in the eastern North Atlantic southward penetration of polar waters occurred at least six times in the past 225,000 years, most severely 165,000 to 135,000 years ago and 30,000 to 15,000 years ago (Fig. 7). At the same time a marked northward incursion of warmer subtropical faunal and floral elements occurred at least six times, the most pronounced being at 175,000 years, 125,000 years, and 8,000 years, respectively. The southern limits of the Pleistocene polar-water incursion was about latitude 42-45° N.

Quantitative paleoclimatologic investigations have recently led to more precise estimates of late Pleistocene climatic conditions in the North Atlantic (IMBRIE & KIPP, 1969, 1971; IMBRIE, 1972; KELLOGG, 1972; McINTYRE et al., 1972a; IMBRIE, VAN DONK, & KIPP, 1973). The polar front that currently extends obliquely from Labrador to the north of Norway 17,000 years ago described

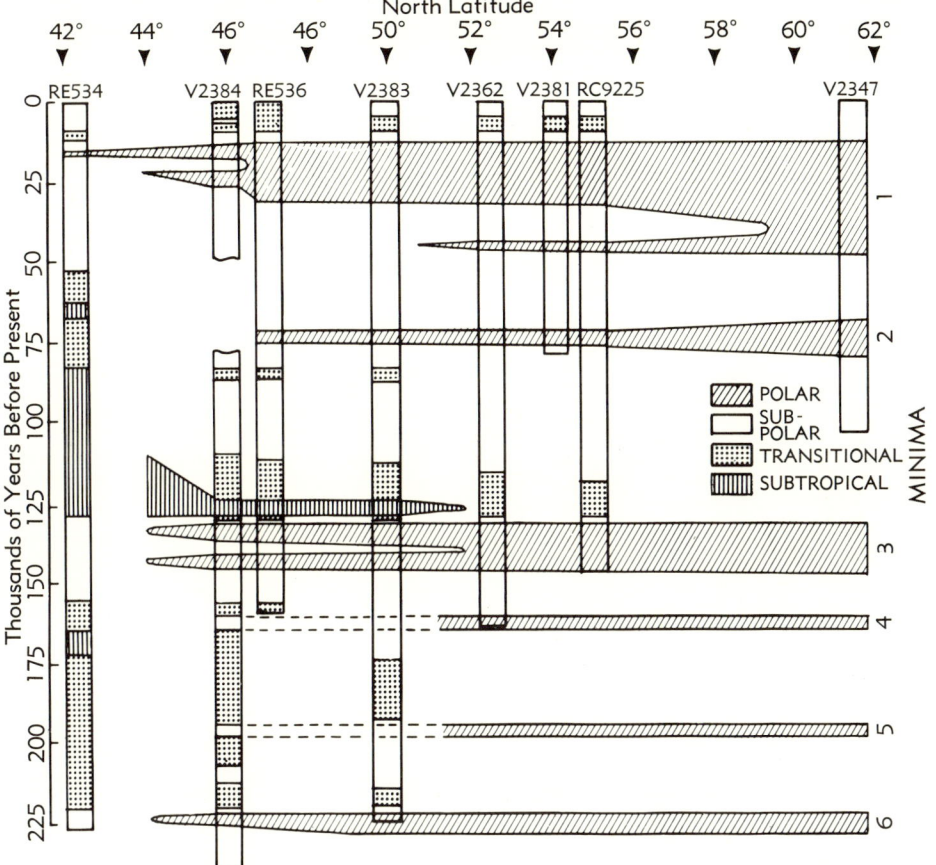

FIG. 7. Polar front and subtropical front migration in the North Atlantic for the last 225,000 years. Paleodistribution of water masses (legend on chart), deduced from faunal and floral data plotted for each core (after McIntyre, Ruddiman, & Jantzen, 1972, fig. 8).

a gentle curve nearly parallel to the latitude from Cape Hatteras to Spain. This represents a northward displacement of 20° latitude in the west to 30° in the east from that time to the present.

The temperature at the termination of the last glaciation (*ca.* 10,000 years ago, referred to by many geologists as the Pleistocene-Holocene boundary) denotes the division of the glacial North Atlantic by the polar front. The difference between today and the 17,000 year B.P. winter-summer temperatures for 50°N. 30°W. is 7.2-12.7°C to 1.2-6.6°C, respectively, an average of 6°C colder than at present. South of the polar front the temperature 17,000 years ago was only about 3°C colder (McIntyre, Ruddiman, & Jantzen, 1972).

CHANGES IN SEASONALITY

A quantitative planktonic foraminiferal faunal analysis in the tropical North Atlantic by Briskin and Berggren (1975) extended the interpretation of climatic fluctuations on this basis back to two million years. Winter changes were estimated to have fluctuated within a range of 4°C and summer temperatures within a range of 2°C. Pleistocene seasonal differences $(T_s\text{-}T_w)$ were estimated to have ranged between 1.5 and 4°C.

From the Olduvai to late Jaramillo the amplitude of winter temperature contrasts (severe to mild) were subdued and temperatures on the average cooler (22.3 to 24.2°C) than during the Brunhes. Beginning in the

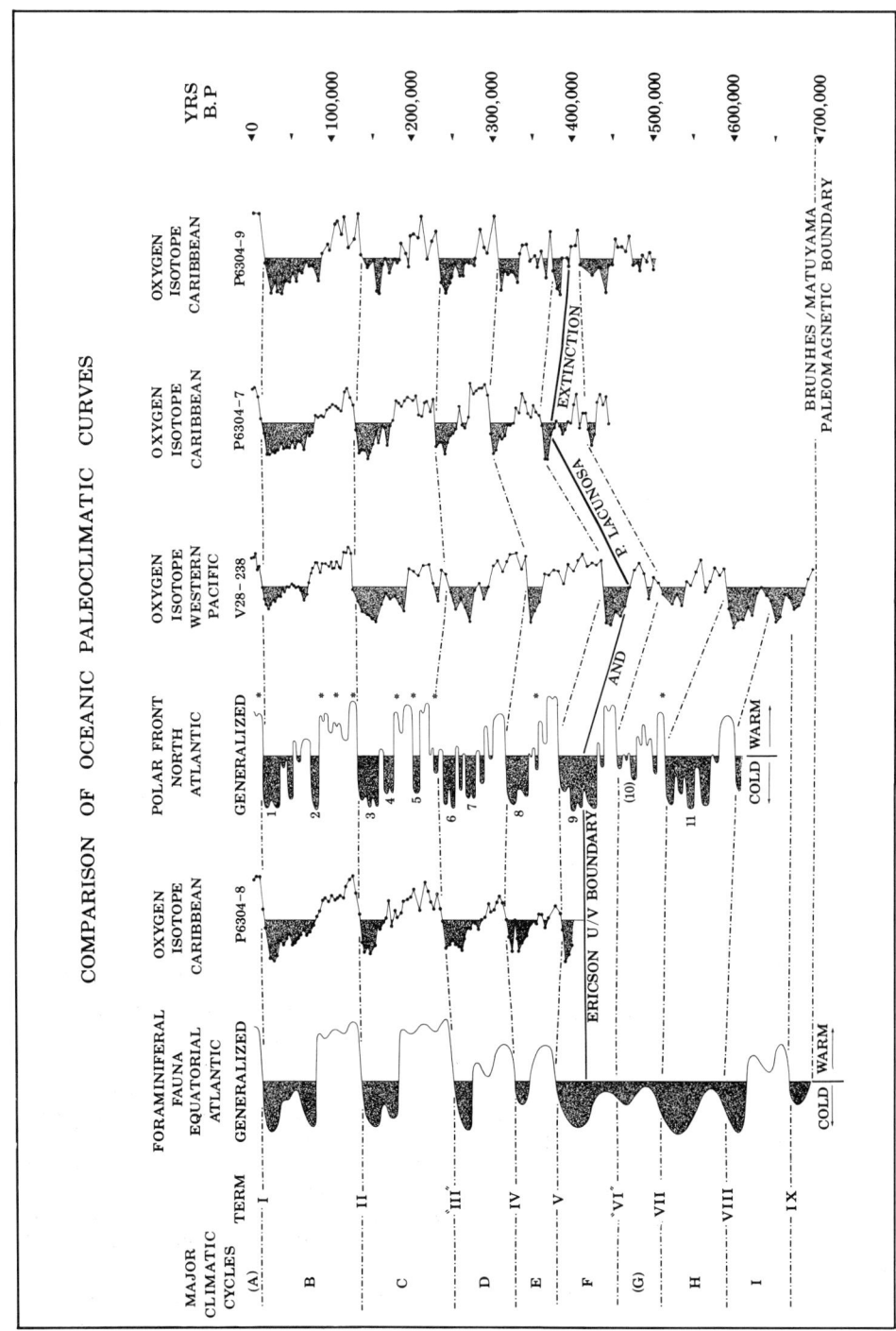

Fig. 8. Climatic cyclic trends in the oceanic deep-sea core record. Equatorial Atlantic faunal curve from RUDDIMAN (1971); Caribbean isotopic curves from EMILIANI (1966a, 1972); North Atlantic curve from

late Jaramillo and continuing into the Brunhes, winter contrasts were greater but the climate was on the average warmer (23° to 23.5°C).

Four roughly symmetrical major climatic cycles with periodicities of 500,000 years were recognized within the Pleistocene (BRISKIN & BERGGREN, 1975). These cycles are more clearly expressed in the winter estimate (T_w) than in the summer estimate (T_s). The coldest winters occurred at the following times:

(1) in the lower R Zone (23°C) at 1.5 Ma;
(2) in the lower T Zone (20.1°C) at the base of the Jaramillo, 960,000 years ago;
(3) in the Brunhes (23°C) 610,000 years ago;
(4) in the W and Y zones (23.6°C) 150,000 and 50,000 years ago.

A climatic fluctuation with periodicity averaging 91,000 years becomes conspicuous in the interval from early V time (~0.4 Ma).

A comparison of ^{18}O and the faunal index T_w led the authors to suggest that the winter-cold maxima (1,2) in the early and mid-Matuyama were associated with smaller ice-volume changes than those (3,4) in the late Matuyama and the Brunhes, where ice volumes were of greater magnitude. The timing of the coldest winter periods corresponds with K-Ar-based estimates of age for the major continental glaciations, and it is well documented that the ice volumes of the Illinoisan (= Mindel = Elster) and later ice sheets were greater than those of the preceding glaciations.

In the Matuyama, records of cold episodes in the North Atlantic cores are characterized by the cold, high-latitude, dextrally coiled *Neogloboquadrina pachyderma*, whereas records of Brunhes cold episodes are characterized by temperate-cool *Globorotalia inflata*. This suggests that maximum southward displacement of the Canaries Current occurred during the cold-water intervals in the Matuyama. In the Brunhes a more moderate displacement of the Canaries Current brought temperate rather than arctic waters into the tropical province.

QUATERNARY CLIMATIC HISTORY

RUDDIMAN and McINTYRE (1976) recently summarized the paleoclimatologic evidence in Atlantic deep-sea studies. They showed that water mass migrations in the North Atlantic across more than 20° of latitude (equivalent to oceanic surface water temperature oscillations of at least 12°C) have occurred along a NW-SE axis at least 11 times in the last 0.6 Ma and perhaps as many as 20 times in the past 1.2 Ma. At least seven complete climatic cycles have been recorded in the past 0.6 Ma (see Fig. 8) and can be correlated directly with the oxygen isotope (SHACKLETON & OPDYKE, 1973) and carbonate (HAYS et al., 1969) cycles in the equatorial Pacific and Caribbean and with the palynologic record from the eastern Mediterranean (VAN DER HAMMEN, WIJMSTRA, & ZAGWIJN, 1971).

The interpretation of RUDDIMAN and McINTYRE (1976) indicated the presence of long-term climatic changes with lengths varying from 56,000 to 113,000 years in the faunal and lithic record. Additional short term but severe climatic pulses with 20,000 year periodicities are also seen.

Furthermore, any or all of the polar water advances recorded over the past 0.6 Ma may correspond to expansion and advance of continental ice sheets because a significant amount of ice-rafted detritus is associated with each of these maxima. Thus, continental glaciers probably reattained similar, equivalent sizes despite the brevity of these climatic pulses (support for this view is

FIG. 8. *(Continued from facing page.)*
RUDDIMAN and McINTYRE (1976); Pacific isotopic curve from SHACKLETON and OPDYKE (1973). Major climatic cycles, A-I; discrete glacial maxima, 1-11; terminations, I-IX; asterisks indicate high sea-levels, all from MESOLELLA et al. (1969). The U/V boundary refers to the zone of that denomination and is part of the Pleistocene planktonic foraminiferal zonal system developed by ERICSON et al. (1964a,b). This boundary, and the extinction of *Pseudoemiliania lacunosa*, coincides closely with the oxygen isotope stage 11/12 boundary. P6304-8, V28-238, etc. = piston cores (from Ruddiman & McIntyre, 1976, fig. 13).

seen in the fact that similar oxygen isotope values occur in successive cold cycles; SHACKLETON & OPDYKE, 1973; see also FAIRBRIDGE, 1972) and that the vigor of trade winds during glacials varied within similar limits (PARKIN & SHACKLETON, 1973).

Short, intensive oceanic coolings are correlative with high herbaceous concentrations in the pollen record (= open, dry to desertic conditions) and occur within interglacial phases of cycles B, C, D and G. Cycle E in Atlantic and Pacific cores contains a short glacial and a long, well-developed interglacial phase, the latter denoted by an unusually high carbonate maximum, high $^{18}O/^{16}O$ values, and high subtropical coccolith abundances. This warm interglacial phase just postdates the *Pseudoemiliania lacunosa* and *Stylatractus universus (vel Axoprunum angelinum)* extinction datum levels within the glacial portion of climatic cycle F (= climatic "stage" 12 of EMILIANI, 1966a,b, and the U Zone of ERICSON & WOLLIN, 1968). In the Macedonian pollen record cycle E is similar in that this interglacial is the longest and corresponds most likely, in classic continental chronostratigraphy, to the Mindel-Riss Interglacial.

The most recent global retreat (warming) of polar waters in the North Atlantic was time transgressive; between latitudes 45° and 64° N. the local termination of glacial conditions ranges from *ca.* 13,500 years B.P. near Great Britain to about 6,500 years B.P. or younger in the northwest near Greenland. This indicates that climatically controlled cycles cannot serve as a basis for drawing chronostratigraphic boundaries.

From Figure 8 it can be seen that the classic view of an interglacial period as a protracted warm interval with a span of about 30,000 to 50,000 years is inaccurate. On the contrary, numerous warm intervals, on the average 8,000 years long, separated by cooler intervals, some of which approach glacial interstadials, is a more accurate picture of Quaternary climate change. Thus, the Holocene, the base of which, climatically speaking, may be said to extend to about 13,000 years B.P., already exceeds the interglacial average by about 5,000 years (see also similar results presented in KUKLA, MATHEWS, & MITCHELL, 1972).

GULF OF MEXICO CLIMATIC RECORD

Rapidly changing climatic conditions with little evidence of prolonged climatic stability seems to have been the pattern in the Gulf of Mexico with three interglacials and two glacials being recorded over the late Pleistocene-Holocene interval. A subdivision of the standard V-Z zonation (ERICSON & WOLLIN, 1968) based on *Globorotalia menardii* frequency oscillations into 18 climatically controlled zones has provided the biostratigraphic framework for climatic interpretations as well as for comparison with the climatostratigraphic record in the Caribbean and elsewhere. A relatively rapid cooling event was identified at *ca.* 90,000 years (KENNETT & HUDDLESTUN, 1972) and correlated with a similar event recorded in the oxygen isotope record in a Greenland ice core (DANSGAARD *et al.,* 1971, 1972; JOHNSEN *et al.,* 1972) and in a cave stalagmite from a cave in southwestern France (DUPLESSY *et al.,* 1970).

An insight into the nature, extent, and rate of Laurentide deglacial processes was afforded by the investigations of KENNETT and SHACKLETON (1975) on Gulf of Mexico cores. A major isotopic anomaly between 15,000-12,000 years B.P. resulted from major influx of isotopically lighter glacial meltwaters via the Mississippi River drainage system from the Laurentide ice sheet, which was then situated some 2,000 km to the north. By 12,000 years B.P. the Laurentide ice sheet had disintegrated to about a third of its maximum Wisconsin size. At the time of maximum input of glacial meltwaters the Gulf of Mexico was estimated to have undergone a reduction in surface water salinity of about 10 percent (2-3 per mil).

As the authors pointed out, earlier glacial meltwater phases would have influenced salinities in the Gulf of Mexico and will eventually provide us with information on the geographic extent of earlier (pre-Wisconsin) ice sheets because it may be assumed that no significant meltwater would have been discharged into the Gulf of Mexico unless the southern extent of the ice sheet was greater than that of the Wisconsin ice sheet 11,500 years ago, at which time isotope values returned to normal. Inasmuch as each successive glacial readvance across the land obliterated much of the preceding glacial record, the geographic distribution of former ice sheets is difficult to reconstruct from the continental record. Thus, the oceanic record (with its inherently more complete record and accurate chronology) may eventually provide us with the most suitable means for paleogeographic reconstructions of past glacial intervals and, ultimately, a comprehensive climatostratigraphic history of the Quaternary.

ARCTIC REGION

It is 20 years since the EWING-DONN (1956) theory was first promulgated, in which it was suggested that the Arctic Ocean remained ice-free during times of continental glaciation and ice-covered during interglacial times. The oldest erratics recorded in the Arctic Ocean have been dated within the Gauss Normal Epoch at about 3 Ma (CLARK, 1971), representing the first evidence that ice-cap glaciers reached sea level in the vicinity.

Fluctuations in the abundance of planktonic foraminifers (primarily *Neogloboquadrina pachyderma*) also suggest climatic variations within the Arctic Ocean although the interpretation of these results varies somewhat. For instance, CLARK (1971) suggested that a thicker ice pack could have significantly affected the productivity of organisms whose economy is based on photosynthesis. Thus, thicker ice conditions would correlate with those periods in which planktonic foraminifers were absent or rare, whereas thinner ice conditions, similar to the present time, would have allowed a larger standing population of planktonic foraminifers to develop. He concluded that the surface of the Arctic Ocean has been frozen at least since the middle Pliocene and that the most significant change in the Arctic ice cover has been its thickness.

The conclusions of CLARK (1971) are interesting in that they suggest that the ice cover of the Arctic Ocean remained relatively stable while continental glaciers expanded and retracted repeatedly in the Northern Hemisphere. If this is true, the Arctic Ocean has not been a major factor in the growth or melting of continental glaciers. Support for this idea was provided by KU and BROECKER (1967) who observed that sedimentation rates (0.2 cm/1,000 years) in the Arctic Ocean were relatively constant during the past 150,000 years and that the biological productivity rates have not exceeded those of the present-day ice-covered Arctic Ocean in the past 150,000 years and that these are much lower than in the Atlantic Ocean. The implication of these observations, according to KU and BROECKER (1967, p. 102), is that they do not favor theories of glaciation that call upon the influence of an open Arctic Ocean (cf. EWING & DONN, 1956, 1958; DONN & SHAW, 1967).

HUNKINS *et al.* (1971) noted that *Neogloboquadrina pachyderma* shows predominantly left coiling during the Brunhes, whereas a zone of right coiling forms is evident near the Jaramillo Event. This suggests warm conditions at that time, which conflicts with the abundance data that show a markedly lower percentage below the Brunhes-Matuyama boundary. The authors suggested that solution effects may have been more pronounced prior to a million years ago. The presence of manganese nodules and low foraminiferal counts in the Matuyama, at least to 1.5 Ma, suggest that climatic conditions were more uniform and without the wide fluctuations of the Brunhes epoch. Most of their cores show seven cycles in the Brunhes with periodicities of about 100,000 years. They concluded that the ice pack has been similar to that of the present day for the past 80,000 years, but that prior to this time a number of cycles with conditions similar to today fluctuated with colder intervals during which the ice pack was extremely tight and during which there was no open water even in the summer.

A somewhat different interpretation has been presented by HERMAN (1969, 1970, 1974), who concluded that the Arctic has been frozen only since the beginning of the Brunhes and was ice free during the Matuyama and Gauss; however, this interpretation seems unlikely in the light of the recent results of Leg XII (North Atlantic) of the Deep Sea Drilling Project, in which it has been shown that major sea-level berg-calving began 3 Ma ago, indicating the initiation of climatic conditions similar to those of the present interglacial with seasonally (at least) frozen Arctic waters.

Subsequent investigations (LARSON, 1973, 1975; CLARK et al., 1975) in the Arctic Ocean have revealed that the replacement of an arenaceous benthonic by a calcareous hyaline benthonic foraminiferal assemblage occurs near the Brunhes-Matuyama boundary and may be due to a change in oceanic circulation patterns (increased oxygenation due to greater overturn connected with more intense glaciation and the entry of North Atlantic deep water during interglacial periods) rather than to ice-cover changes. Correlation of the Quaternary Arctic deep-sea core climatostratigraphic record with climatostratigraphic curves developed elsewhere (e.g., Atlantic, Caribbean) is vague and unclear, but the relatively uniform cold polar conditions during that time may be expected to have left a blurred record of climatic change in comparison with the distinct, oscillatory changes that are recorded at mid-latitudes.

The late Neogene geologic and stratigraphic history of Beringia (western Alaska, northeastern Siberia, and shallow parts of the Chukchi and Bering seas) has been summarized by HOPKINS (1967a,b, 1972, 1973) and NELSON, HOPKINS, & SCHOLL (1974a,b). In the marine strata, three distinct glacial episodes can be distinguished, Siberian glaciers having extended over 150 km beyond the Chukotka Peninsula shoreline. Related to these are three major interglacial transgressions: in ascending age the Anvilian, Einahnuhtan, and Kotzebuean, followed by a brief double transgression separated by a brief regression, the Pelukian. This last (and youngest) transgression is radiocarbon-dated at more than 38,000 years (HOPKINS, 1973). During the Anvilian transgression (which is considered to be of early Pleistocene age, between 1.8-0.7 Ma) circulation was dominantly northward through the Bering Strait, relative sea level somewhat higher and water deeper on the Bering Shelf; molluscan and foraminiferal faunas indicate that water temperatures did not differ significantly from those of the present day (HOPKINS et al., 1974).

The Einahnuhtan transgression is thought to represent Termination III of BROECKER and VAN DONK (1970), dated at about 0.225 Ma. The series of progressively colder interglacial marine transgressions culminated in the Kotzebuean transgression, which represents a positive fluctuation that modulated a generally falling sea level in the 100,000 years that followed the Einahnuhtan and during which time normal northward circulation through the Bering Strait was reversed. Arctic molluscan faunas penetrated southward along the coasts of eastern Siberia and southwestern Alaska during Kotzebuean time, which is correlated with the Holstein interglacial in northwest Europe (HOPKINS, 1973). SHACKLETON and OPDYKE (1973) have suggested that the Holsteinian, generally thought to be correlative with a level within oxygen-isotope "stage" 7, may in fact be older (?stage 11 = 400,000 years B.P.). If the correlation of Kotzebuean = Holsteinian is correct, it may necessitate a lowering of the ages of the Einahnuhtan and Kotzebuean transgressions.

Sea level fell to about -135 m in the Bering Sea during the maximum phase of the penultimate glaciation. Following this, the Pelukian shorelines may represent high sea level stands of Termination II (ca. 125,000 years B.P.) of BROECKER and VAN DONK (1970) and one of the two high sea level stands dated at about 106,000 years B.P. and 80,000 years B.P. that modulated the generally falling sea level during the later part of the last interglacial (= Eemian). Another positive modulation brought sea level to at least -20 m about 30,000 years B.P. Sea level fell to about -90 to -100 m during the late Wisconsinan regression but a substantial part of the Bering shelf remained submerged.

According to HOPKINS et al. (1971), the tundra biome replaced a predominantly beech and spruce forest in southern Alaska similar to present-day forests in southeastern Alaska and British Columbia sometime later than 5.7 Ma and more probably not until the early Pleistocene (HOPKINS, 1972). Evidence cited above of the growth and development of Northern Hemisphere continental ice sheets within the late Pleistocene suggests that this major floral replacement may have occurred as early as 3 Ma.

PACIFIC OCEAN

EQUATORIAL PACIFIC CLIMATIC CHANGES

The late Neogene paleoclimatic history of the equatorial Pacific region, based on carbonate maxima in deep sea cores, has been described by HAYS et al. (1969). Eight distinct carbonate cycles were recorded in the Brunhes (last 0.7 Ma) with periodicities of about 75,000 years in the later part to about 100,000 years in the early part of the Brunhes.

Slightly lower average carbonate content was observed in the interval between the Olduvai and the Brunhes (i.e., the late Matuyama). Average periodicities here were on the order of 100,000 years. The generally lower concentration of calcium carbonate in the late Matuyama was interpreted as suggestive of warmer average climates during that time. A sharp rise in calcium carbonate content in the upper Gauss Normal series may reflect mid-Pliocene cooling at about 3 Ma. HAYS and his colleagues observed that no marked change in the carbonate content occurred during the Olduvai Event, although in Antarctic cores there is a lithologic change (see below) in the neighborhood of the Olduvai Event (HAYS, 1965, 1967; OPDYKE et al., 1966), which may have been due to cooling. The comparison of this record agrees in timing and in paleoclimatic interpretation with the ^{18}O record described by SHACKLETON and OPDYKE (1973, 1976).

Quaternary paleoclimates have been described in the marginal northeastern Pacific by INGLE (1973). In this area the glacial interval of the Pleistocene was characterized by sustained subarctic temperatures at least as far south as latitude 30°N. and probably to 20°N. allowing the biofacies with dextrally coiled *Neogloboquadrina pachyderma* to move up into the Gulf of California where it was subsequently trapped by northward readjustment of isotherms in the Holocene. Such populations are reminiscent of similar conditions in the Gulf of Mexico that led to populations of *Globigerina bulloides* being trapped in the Gulf of Mexico following the retreat of the last glaciers (PHLEGER, 1961).

SEA OF JAPAN

Four significant surface temperature events were delineated in the Sea of Japan during the Quaternary (INGLE, 1975) beginning with a significant cooling event at about 0.9 Ma, which apparently correlates with other evidence of intense midlatitude glaciation (KENT, OPDYKE, & EWING, 1971; INGLE, 1973; RUDDIMAN, 1971).

ANTARCTIC-SUBANTARCTIC OCEAN

The investigations of SHACKLETON and KENNETT (1975a,b) have shown that the East Antarctic ice sheet became established, and subsequently reached dimensions in excess of its present-day extent, during the Miocene and has remained relatively stable and immune to fluctuations in global climate since that time.

Major climatic changes related to the expansion and contraction of Northern Hemisphere, and to a lesser extent the Antarctic, ice sheets have been placed within the time-stratigraphic framework of the New Zealand stages, which have, in turn, been correlated with the European stage units (Fig 9).

AUSTRALIA-NEW ZEALAND AREA

Ten climatic cycles within the Matuyama (2.43 to 0.7 Ma) have been delineated in northern Antarctic and subantarctic waters south of Australia and New Zealand, which are based on alternation of cold and warmer water planktonic foraminiferal faunas (KEANY & KENNETT, 1973). Eight climatic cycles over the last 1.3 Ma, six of which

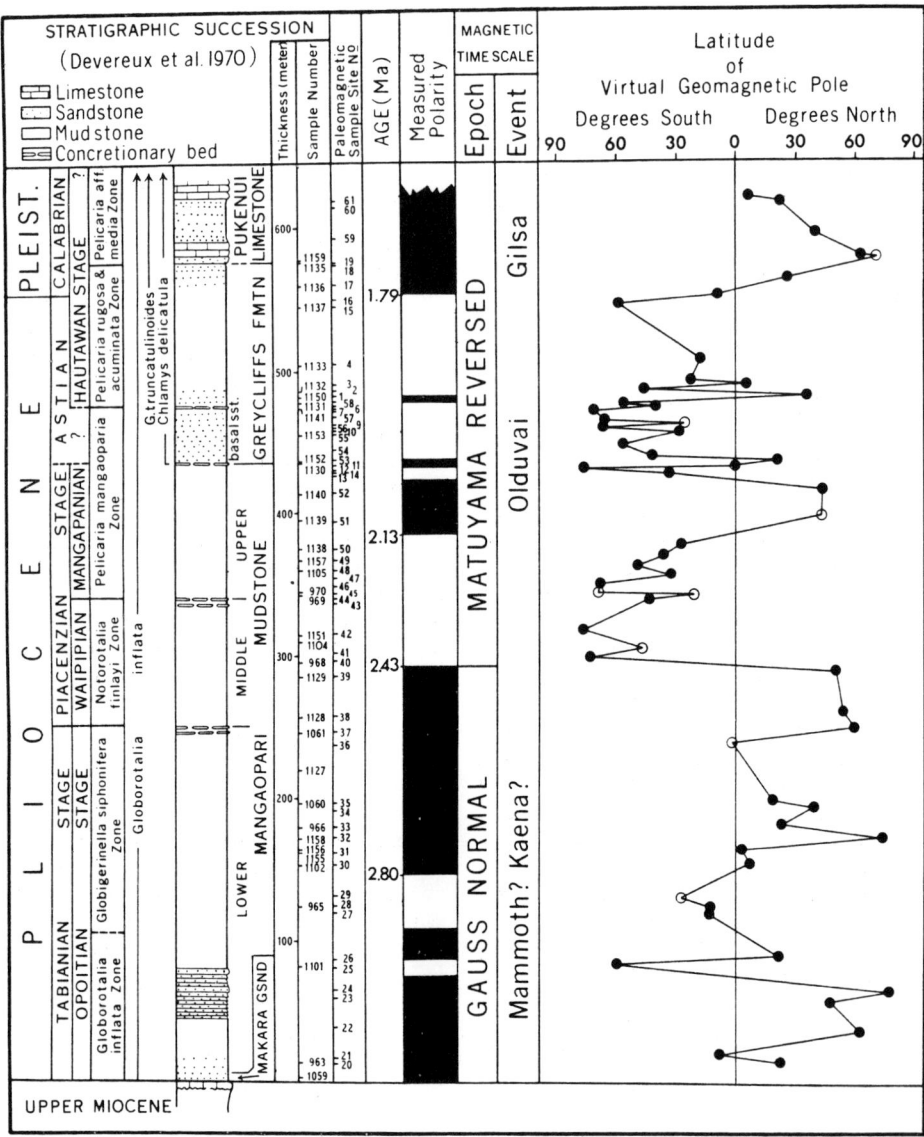

Fig. 9. Paleomagnetic stratigraphy, stratigraphic succession, and paleoclimatic trends at Mangaopari Stream-Makara River, New Zealand. Sample positions, ranges of some key fossils, local stratigraphic units, and possible European correlations are shown. Magnetic polarity in the section (left polarity column) is interpreted from the latitude of the virtual magnetic pole; black denotes normal polarity and clear denotes reversed polarity, which corresponds to the latitude of the virtual geomagnetic pole being higher than 10° N. and 10° S., respectively. The paleoclimatic curves are, from left to right: *a* and *b*, oxygen isotope ratios of planktonic and benthonic foraminiferal tests; *c*, abundance of *Globorotalia pachyderma* as a percentage of all planktonic foraminiferal tests. The position of the Pliocene-Pleistocene boundary and the base of the Calabrian have been modified from that previously determined by DEVEREUX *et al.* (after Kennett, Watkins, & Vella, 1971, fig. 2). [Note that Gilsa and Olduvai are considered separate events by these authors.]

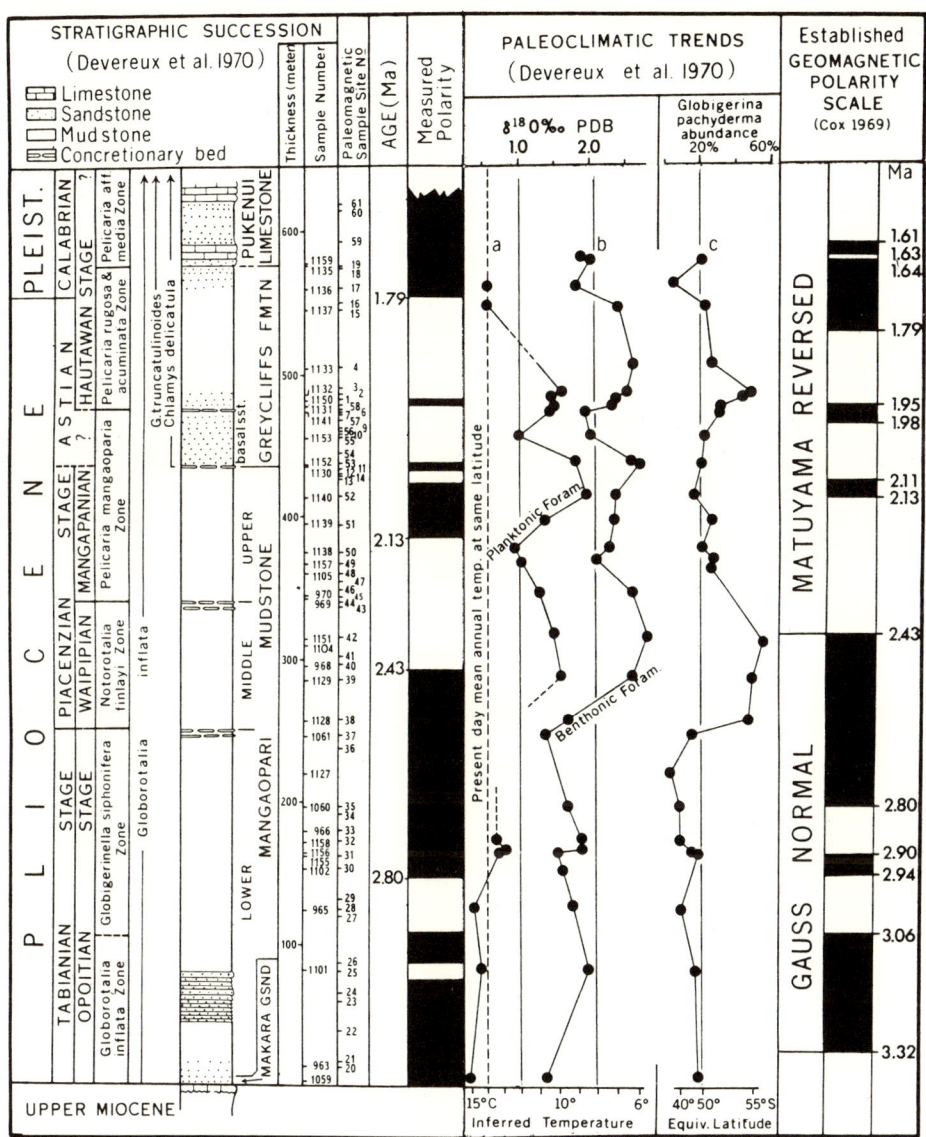

FIG. 9. (Continued from facing page.)

occur within the last 0.7 Ma, had been previously recorded in the study of Southern Ocean cores by KENNETT (1970). As we have seen in other oceans, the relative amplitude of the climatic cycles during the Matuyama is somewhat lower than those recorded in the late Pleistocene (the last 0.7 Ma).

In contrast to most investigators who have claimed that the general picture is one of declining climatic conditions during the Pleistocene, but with much colder climates during the Brunhes (last 0.7 Ma) than the late Matuyama (~1.8-0.7 Ma), KEANY and KENNETT (1973) found evidence that subantarctic climatic conditions were more stable and average temperatures lower during the late Matuyama (ca. 0.9 Ma) than during the Brunhes, whereas the Brunhes, in general, was a time of erratic, high-intensity climatic cycles with an average warmer temperature range. One of the most convincing arguments for this view is the fact that *Globorotalia inflata,* which appears in the North Atlantic, the Mediterranean, and New Zealand about 3 Ma ago (BERGGREN, 1972; CITA, 1973; KENNETT, WATKINS, & VELLA, 1971) appears for the first time in the subantarctic region at the base of the Brunhes (0.7 Ma ago). As KEANY and KENNETT (1973) pointed out, if subantarctic-northern Antarctic conditions were warmer during the Matuyama than later, as suggested by BANDY, CASEY, & WRIGHT (1971), *G. inflata* would be expected to occur throughout this interval as it does in New Zealand and equivalent latitudes.

ANTARCTIC MARGINAL SEAS

FILLON (1972) has recorded evidence of widespread submarine erosion in the Ross Sea sometime after late Gauss time (<2.4 Ma), and has suggested that this was due to significant net cooling (see also WATKINS & KENNETT, 1971, 1972). A major break in sedimentation and a change in ecofacies (Pliocene calcareous benthonic foraminifers below, agglutinated foraminifers of late Brunhes age above) was probably due to a sharp decrease in the amount of debris delivered from the Ross Ice Shelf to the Ross Sea owing to a significant northward advance and thickening of the Ross Ice Shelf, and also to a consequent increase in bottom scour. KEANY and KENNETT (1973) have attributed increased erosion in the Southern Ocean and Tasman Sea to major cooling with a resulting increase in bottom-water production and velocities in post-Gilbert or post-Gauss time (~3.0 to 2.4 Ma).

Late Neogene paleoglacial history of Antarctica has been recorded in sub-Antarctic deep-sea cores (MARGOLIS & KENNETT, 1970, 1971; KENNETT & BRUNNER, 1973) by the presence of ice-rafted grains and variations in planktonic foraminiferal diversity. Ice-rafted grains are locally abundant in the uppermost Miocene and Pliocene and are generally abundant within the Pleistocene, supporting the idea that the Antarctic ice cap grew to present-day proportions during the late Miocene (approximately 10 Ma ago).

RADIOLARIAN EVIDENCE OF CLIMATIC CHANGE

HAYS (1965, 1967) has observed that the distribution of radiolarians and various sediments in the Pleistocene in the Antarctic Ocean is strongly influenced by the northern limit of pack ice and the Antarctic Polar Front (Antarctic Convergence). The Pleistocene sediments between the polar front and the pack ice are primarily diatom oozes. At the boundary between radiolarian zones $\emptyset/X$ there is a lithologic change from clay below to diatom ooze above, which apparently resulted from the initiation of large-scale freezing of sea ice around Antarctica. This led to greater vertical circulation (upwelling) in the Antarctic Ocean and the development of high productivity of the Antarctic surface water that persists to the present day. Thus, alternation of radiolarian and diatom-rich sediments with layers poor in siliceous microfossils north of the limit of modern pack ice as well as the alternation of warm- and cold-water radiolarian species are apparently a reflection of changing limits of the Antarctic pack ice.

Cold-water conditions appear to have prevailed in the late Gilbert, Gauss and early Matuyama (>2.4—<0.7 Ma), with a return toward temperate conditions during most of the Matuyama indicated by the almost continuous presence in Antarctic cores of this age of *Pterocanium trilobum* and *Saturnulus planetes*. A significant temperature reduction occurred at the Brunhes-Matuyama boundary as shown by the local disappearance of warm-water polycystines. Temperate conditions were reintroduced toward the end of the Brunhes (Holocene) as shown by the reappearance of *Saturnulus planetes, Theocanus zancleus,* and other forms. Temperature cycles in the early Gilbert may have been on the order of 15-20°C (average summer), in the 5-15° range during the late Gilbert, Gauss, and Matuyama. On the other hand, four cooling cycles well below 5°C were said to have occurred during the Brunhes (BANDY, CASEY, & WRIGHT, 1971).

CALCAREOUS NANNOFLORAL EVIDENCE OF CLIMATIC CHANGE

GEITZENAUER (1969, 1972) has investigated the Pleistocene calcareous nannoplankton biostratigraphy and paleoclimatology of the subantarctic region.

Within the last 400,000 years, interglacial periods within the subantarctic have been characterized by a relatively high frequency of *Cyclococcolithus leptoporus* and *Coccolithus pelagicus*. During glacial intervals these two forms were sharply reduced, the dominant form being the eurythermal *Gephyrocapsa caribbeanica*; however, a maximum peak of *G. caribbeanica* occurred between 0.4 and 0.5 Ma ago and resulted in a nearly monospecific coccolith ooze, which was correlated by KENNETT (1970) with the climatic optimum that occurred during the V interglacial zone of ERICSON and WOLLIN (1968). (In this connection, see also IMBRIE & KIPP, 1971; and BRISKIN & BERGGREN, 1975.) The apparent contradiction between foraminiferal and coccolithophorid data within this specific interval suggests that *G. caribbeanica* exhibits an apparent bimodal distribution pattern, with maxima at glacial stages and at the warmest interglacial. GEITZENAUER (1972) cautioned that a simple coccolithophorid-temperature relationship may be dangerous in Antarctic paleoclimatic interpretations; other factors such as salinity and nutrients may also have been significant determinants in the distribution of Pleistocene coccolith flora.

On the basis of present-day temperature ranges governing the distribution of *Coccolithus pelagicus* and *Cyclococcolithus leptoporus,* the similarity in frequency peaks in both of these species in the late Pleistocene indicates that the optimum conditions for both species were similar and that the climate of the "cool" interglacial stages of the late Pleistocene were very similar to the present time (GEITZENAUER, 1972).

In summary, it may be said that a substantial similarity exists between the currently adopted Quaternary climatic curves from Atlantic, Caribbean, Pacific, and subantarctic ocean-floor sediments.

MEDITERRANEAN

The late Neogene geological histories of the Mediterranean, Red, and Black seas are intimately related. Indeed, the Black and Caspian seas may be viewed as but remnants of a once extensive shallow-water sea of intermediate salinity—the Paratethys—which extended eastward to central Europe and was itself connected with the Mediterranean Sea until about 13 Ma. The Mediterranean and Red seas may have been connected intermittently during the late Neogene as well.

Of particular importance to our understanding of the Quaternary marine climatostratigraphy is the fact that alternating glacial-interglacial climatic oscillations in the Northern Hemisphere caused significant changes in Mediterranean climatic conditions (FAIRBRIDGE, 1972). During early (anaglacial) times, the prevailing westerlies were deflected southward, bringing increased precipitation over the Mediterranean and North Africa. During maximum (pleniglacial) times oceanic cooling led to reduced precipitation and cold, arid climates. Dry, northeasterly winds resulted in the extensive distribution of loess deposits as far south as southwestern France and Portugal, the migration of sand dunes far to the south of the present-day North African Sahara, and increased amounts of eolian biogenic detritus (opal phytoliths and freshwater diatoms) in the equatorial Atlantic west of Africa (PARMENTIER & FOLGER, 1974). Increased precipitation characterized the late (kataglacial) glacial phases and the postglacial warming. During interglacial phases the Mediterranean was characterized by a gradual warming trend and increased precipitation, leading to humid conditions and reforestation even in North Africa, followed by cooling accompanied by widespread desiccation.

DEEP-BASIN CORE RECORD OF CLIMATIC CHANGE

A broad picture of the climatostratigraphic history of the last 4 Ma in the

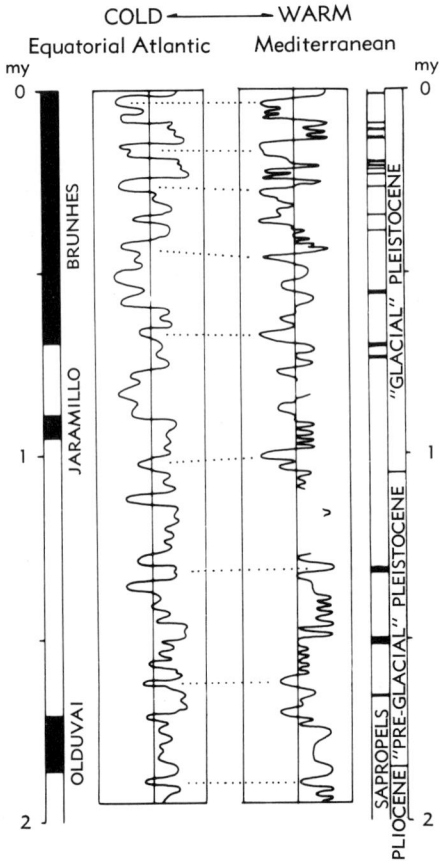

Fig. 10. Generalized climatic curves for the equatorial Atlantic and the Mediterranean. The Mediterranean curve has been constructed by weighing all the data from the various drill cores and piston cores and has been normalized to the geomagnetic time scale. The equatorial Atlantic curve is a "total fauna" assessment of RUDDIMAN (1971). The dotted lines are suggested levels of correlation. The heavy bars on the right column show levels of sapropelitic sedimentation in the Mediterranean, and seem to correspond to periods of marked warming (after Cita et al., 1973, fig. 14).

Mediterranean Sea has emerged over the last few years. Building upon the classic study of the eastern Mediterranean late Pleistocene by PARKER (1958), RYAN (1972) extended the climatostratigraphic record back to about 0.4 Ma by combining various piston cores from different parts of the Mediterranean. This record, in turn, was subsequently extended to about 4 Ma using the data from cores taken by the Deep Sea Drilling Project Leg 13 (CITA, 1973; CIARANFI & CITA, 1973). Generalized paleoclimatic curves based on analysis of planktonic foraminiferal faunas reveal numerous fluctuations during the last 2 Ma (Pleistocene) (Fig. 10). As in open-ocean cores, climatic oscillations in late Pliocene and early Pleistocene were not as great as those beginning about 1 Ma, and RYAN's (1972) curve agrees closely with the ^{18}O curve of SHACKLETON and OPDYKE (1973, 1976).

Sapropel layers in the eastern Mediterranean bottom-sediments are attributed to stagnation caused by inflow of glacial meltwaters from the Black Sea and consequent density-stratification of the Mediterranean water-mass, at a time when glacio-eustatic sea level was recovering from below −40 m, the elevation of the Bosporus sill, and the Black Sea was still a freshwater lake; the sapropels are thus keyed to deglacial phases when landlocked ice was being returned to the sea. The earliest of these sapropels (OLAUSSON, 1961; RYAN, 1972) at ca. 2.4 Ma in the Ionian basin (eastern Mediterranean) indicates that glacio-eustatic sea level change occurred significantly earlier than the first major Alpine glaciation, the Günz at about 0.9 Ma, or even before the less well-dated Donau, ca. 1.5 Ma. The estimated age of 2.4 Ma for this earliest sapropelic layer is tantalizingly close to the 2.6 Ma estimated by SHACKLETON and KENNETT (1975a,b) for the establishment of the Northern Hemisphere polar ice sheet, an event that may be expected to have had a pronounced effect on sea level, nor is it greatly different than the 3 Ma estimated for the beginning of ice-rafting (BERGGREN, 1972), which is unquestionably related to the same event.

Of particular paleoecologic interest is the discovery by RYAN (1972) that relatively high percentage values of the planktonic foraminifer *Globigerinoides ruber* in association with *Neogloboquadrina dutertrei* ($\cong$ *N. eggeri*) occur in the sapropel layers. BÉ and TOLDERLUND (1971) have shown that *G. ruber* is a euryhaline species with optimum salinity preferences below 34.5 percent and above 36 percent. Thus, the high value of *G. ruber* in the sapropel layers support the idea that the sapropels are a result of stagnant conditions resulting from lowered surface water salinity as a result of freshwater discharge of glacial meltwater from the Black Sea. In the Red Sea, on the contrary, the numerical dominance of *G. ruber* during glacial phases (to the near exclusion of all other taxa) is associated with increased salinity and evaporation rates and lowered surface water temperatures (BERGGREN & BOERSMA, 1969; see discussion below). Here we have a dramatic illustration of the response of a single taxon to opposite extremes of the same ecologic variable.

Inasmuch as the sapropel layers appear to reflect glacio-eustatically controlled sea level changes in the world ocean they themselves can serve as the basis of a sea level curve as well as providing an excellent means of calibrating the sea level curve of the eastern Mediterranean (RYAN, 1972). Sapropels are noted at 2.4 Ma, 1.5 Ma (?Donau) and at several levels in the warm part of isotopic stages 11, 9, 7, 5, 3 and 1 (RYAN, 1972). The youngest sapropel occurs in the Holocene at a thermal maximum at about 7,000 years B.P. (HERMAN, 1972), and similar sapropels have been recorded in the Adriatic Sea and dated at 7,500-9,000 yrs B.P. (VAN STRAATEN, 1972). The sapropels should further provide an excellent means of correlating the marine to the continental stratigraphic record in lakes, terraces, moraines and loesses.

RED SEA

Detailed climatostratigraphic information in the Red Sea is available only for the past 80,000 years (essentially the record of the last glacial and postglacial interval). Prior to this time the stratigraphic record is patchy and the sequence of planktonic foraminiferal faunas in deep sea cores suggest that since the end of evaporite conditions in latest Miocene time (*ca.* 5 Ma) faunal evolution may be viewed in terms of an overall progression from less to more diverse faunas related to a gradual decrease in salinity rather than to decreasing temperature (FLEISCHER, 1974). There is apparently no evidence of Early Quaternary climatic deteriorations in Red Sea cores.

The combined investigations of the distribution patterns of dinoflagellates, foraminifers, calcareous nannoplankton, pteropods, and radiolaria from a series of cores yield a remarkably uniform interpretation of the climatostratigraphic history of the Red Sea over the past 80,000 years. The basic climatostratigraphic framework was provided by the planktonic foraminifers (BERGGREN & BOERSMA, 1969) when it was realized that the fluctuating percentages between two species, *Globigerinoides ruber* and *G. sacculifer,* were primarily controlled by salinity rather than temperature as originally suspected.

According to these studies, a tropical-subtropical climate existed in the Red Sea region from at least 80,000 to about 50,000-60,000 years ago and corresponds approximately to the first of four cycles. During this time surface water temperatures probably varied between 21°-30°C, similar to present-day values in the area. The second evaporative cycle began about 50,000 years ago and lasted until about 25,000 years ago. Climatic cooling and lowering of sea level are suggested by micropaleontologic data and oxygen isotopic measurements (BERGGREN, 1969; DEUSER & DEGENS, 1969). Cycle III (*ca.* 23,000-13,000 years ago) corresponds to the coolest part of the Late Pleistocene. The general impoverishment of the microfauna leading to a nearly monospecific assemblage of *Globigerinoides ruber,* during this interval, is probably directly related to lowered water temperatures, in addition to the effects of pronouncedly higher relative salinities. Surface water temperatures may have reached values as low as 13°-14°C during this time. Cycle IV (corresponding to the Holocene) witnessed the reestablishment in the Red Sea of a normal marine microfauna from the Indian Ocean and a gradual rise in temperature to present-day values.

BLACK SEA

The marine Quaternary history of the Black Sea is preserved as a series of transgressive littoral-marine deposits separated by wash-out horizons and isolated pockets of subaerial sediments that represent regressions. These deposits crop out in a series of terraces that descend stepwise nearly con-

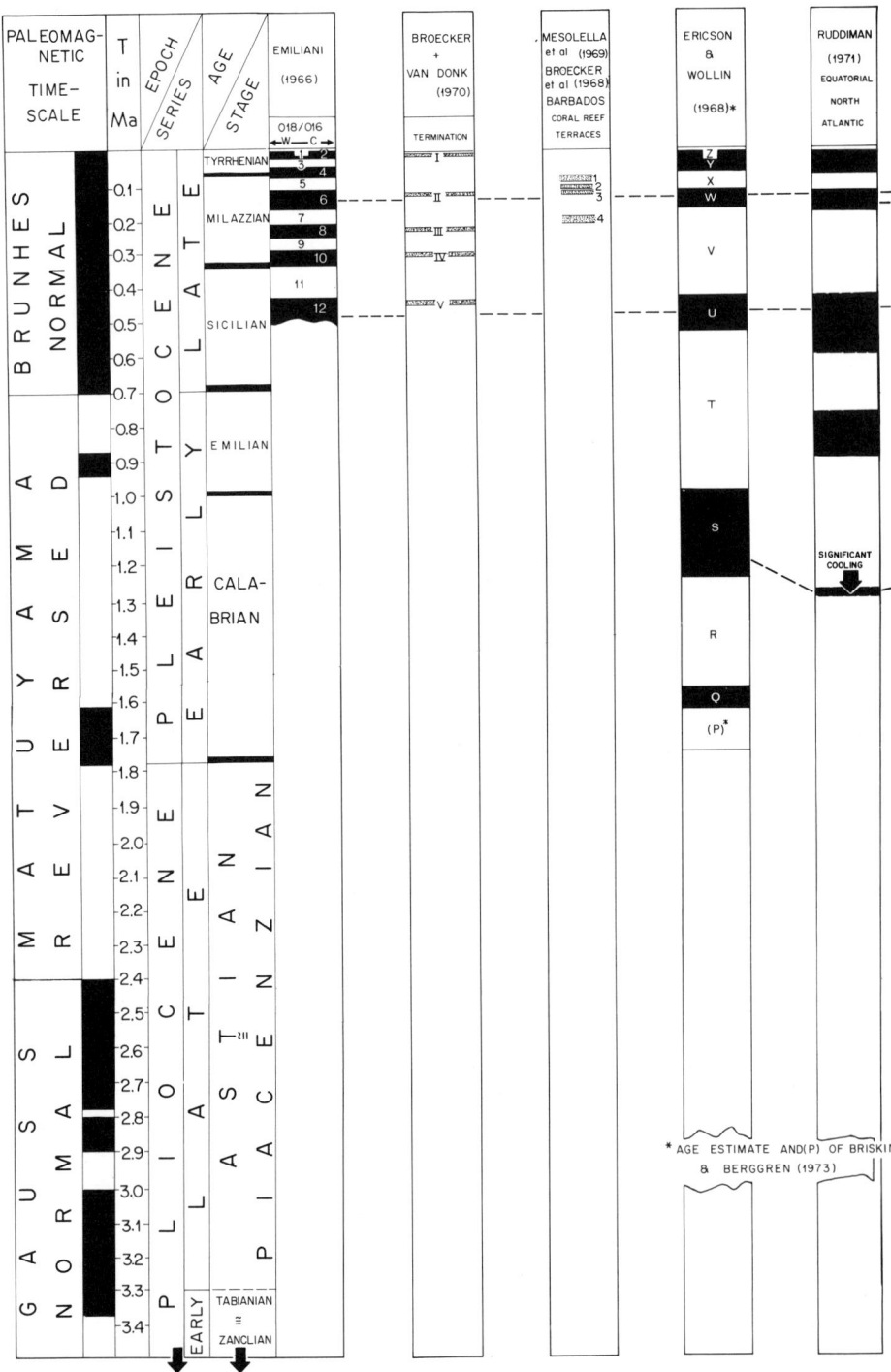

Fig. 11. Correlation of some late Pliocene and Pleistocene marine climatic events. Three major climatic events (right hand column) include: 1) mid-Pliocene (3.0 Ma) initiation of northern hemisphere

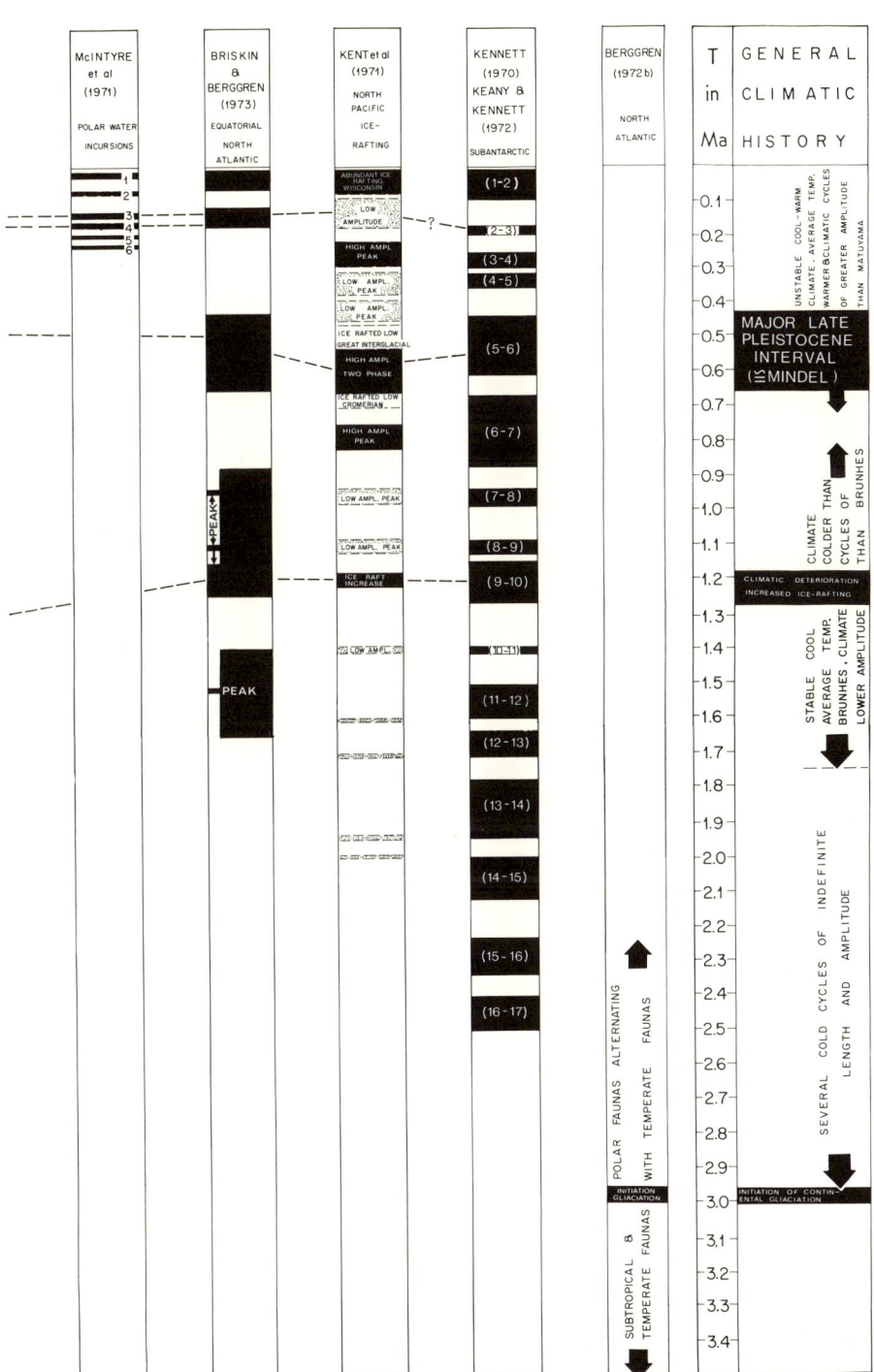

FIG. 11. *(Continued from facing page.)* glaciation; 2) early Pleistocene (1.2-1.3 Ma) glaciation; 3) major late Pleistocene glaciation (centered on about 0.5 Ma) (after Berggren & Van Couvering, 1974, fig. 13).

tinuously along the entire Western Caucausus coast of the Black Sea (Muratov, Ostrovsky, & Fridenberg, 1974). Repeated ingressions of Mediterranean waters across the Bosporus sill during interglacial high sea-level stands provide the basis for a biostratigraphic subdivision of the littoral-neritic facies of the circum-Black Sea region. To date piston cores from the bottom sediments of the Black Sea have penetrated only into deposits of the last glacial, which have been dated at about 25,000 years B.P., but recent drilling by the *Glomar Challenger* has penetrated an apparently complete Pleistocene succession (Ross et al., 1975).

Regressions of the Black Sea were accompanied by substantial freshening and coincided with the early stages of glaciation. Transgressions of the Black Sea were accompanied by influx of water from the Caspian and Azov seas and immigration of faunas from these areas. At such times, as well as during regressions, there must have been unilateral surficial runoff from the Black Sea into the Mediterranean. As Black Sea transgressions waned, inflow of saline water from the Mediterranean was accompanied by the introduction of euryhaline Mediterranean faunal elements; this phase coincided with the later interglacial (or interstadial) stages.

Although accurate biochronologic control is lacking for the earlier part of the stratigraphic sequence, a generally accepted correlation with the European glacial-interglacial sequence is possible. The "earliest" Pleistocene Chaudian transgressive unit contains an endemic Black Sea fauna with elements of the Bakunian Stage of the Caspian Sea, which suggests approximate correlation with basal Brunhes (the Bakunian is equivalent to the Tiraspol continental sequence within which the Brunhes-Matuyama boundary has been located; see Berggren & Van Couvering, 1974). It should be recalled that the "official" base of the Quaternary in the Soviet Union is drawn at a stratigraphic level approximately coincident with the base of the Brunhes, ca. 0.7 Ma.

Subsequent transgressions increased in intensity and attendant thermophilic faunal elements. The three maxima—Ashe, Karangat, and Surozh—contain virtually monotypic faunas (as do the lesser Paleo-Euxinian and Pshada transgressions).

Radiocarbon and thorium-uranium dating suggests that the Ashe transgression (= Odontsovo Interglacial in Eastern Europe) is correlative with the Riss-Würm (= Eemian) Interglacial, and the Karangat (= Mikulino Interglacial) and Surozh (= Mologo-Sheksna Interglacial) transgressions with intra-Würmian or Wisconsinan relatively high stands of sea level. The Chaudian, as an equivalent to Biharian mammal age via correlation to Bakunian Stage, apparently corresponds to pre-Mindel "Cromerian" but also, at least in part, to an interglacial within the Mindel. The Pshada transgression appears to correspond to an interglacial phase within the Riss.

Of particular interest in the history of the Black Sea is the convergence of faunal, floral, and geochemical evidence from isotope data (Deuser, 1972, 1974) that the Black Sea was essentially a freshwater lake between 25,000 to 7,000 years ago, and gradually evolved into its present form over the period of ca. 9,000 to 7,000 years ago. The influx of the saline waters from the Mediterranean 7,000 years ago (with minor input beginning ca. 9,000 years ago) is marked in the Black Sea by sapropelic sediments and the introduction of euryhaline forms. This event is seen to correspond with the formation of dated sapropelic layers in the Mediterranean that have been interpreted to have formed under stagnant bottom conditions caused by outflow of glacial meltwaters from the Black Sea into the Mediterranean.

SUMMARY AND CONCLUSIONS

1) The Pliocene-Pleistocene boundary is approximately 1.5-1.6 Ma, and is at, or slightly younger than, the top of the Olduvai normal paleomagnetic event. A set of multiple reinforcing biostratigraphic criteria can be used to recognize the approximate position of this boundary in deep-sea sediments. Intra-Pleistocene biochronology is possible by means of various biochronologic datum levels

that have been calibrated to the paleomagnetic time scale, the oxygen isotope record, or both.

2) The integration of paleontology, paleomagnetic stratigraphy, and oxygen isotope analysis in the relatively complete and continuous deep-sea record is providing a chronologic framework within which glacial-interglacial cycles can be accurately delimited. Within this framework it is now becoming possible to make an approximate correlation between the climatic record of the deep sea and the classic glacial-interglacial record in terrestrial sequences.

3) With minor adjustments and calibration to a uniform time scale there is a remarkable degree of correspondence in the climatostratigraphic curves of various authors from various parts of the world in the Northern and Southern hemispheres, which suggests essentially synchronous response over a large part of the earth to major climatic changes. The global correlation of paleoclimatic cycles is shown in Fig. 11.

4) As pointed out in the introductory paragraph to this chapter, the Quaternary is a unique time in earth history, even by the most conservative standard of comparison. In the relatively short time span of the last 1.5 Ma the earth has witnessed such diverse events as:

   a) repeated (perhaps as many as 30 or more) glaciations at high and midlatitudes in the Northern Hemisphere, which have drastically altered the biogeographic distribution patterns of marine and terrestrial plants and animals alike;
   b) repeated and drastic latitudinal displacement of climatic zones by as much as 20 to 30 degrees;
   c) dramatic changes in oceanographic circulation patterns in the oceans and Mediterranean Sea; dramatic oscillations in circulation between the Mediterranean and Black Sea; repeated isolation of, and increased salinity in, the Red Sea; repeated subaerial exposure of the Bering shelf and subaerial connection between North American and Siberian land areas.

5) Integrated geophysical, geochemical, and paleontologic studies on the deep-sea stratigraphic record are leading to a better understanding of the history of global climate over the past million years. These studies may be expected to lead to a more precise construction of past global climatic conditions at specific "moments" in time (McINTYRE *et al.*, 1976), which can, in turn, serve as boundary conditions for modeling general atmospheric circulation patterns in the Pleistocene (GATES, 1976).

Just as the present has been amply demonstrated to be a reliable guide in reconstructing the past, so the past is seen to be a reliable guide to predicting the future.

6) Having shown uncharacteristic restraint above, we here feel constrained to point out, in passing only, that this same interval has witnessed the passage from East African "stone-pebble culture," through the development of intricate and perfect bifacial implements, to the paleolithic tool industries (which began about 0.5 Ma), through the artistic inspiration of Altamira, Lascaux, and other caves in southwestern France and Spain, to the genius of Leonardo. Although hominid evolution has recently been pushed back into the Pliocene Epoch (*ca.* 3-4 Ma) it seems fair to say that, in general terms, Man is a child of the Quaternary.

REFERENCES

Agassiz, Louis, 1838, *Notes sur les glaciers:* Soc. Géol. France, Bull., ser. 1, v. 9, p. 443-450 (observations on p. 407-410, 435-438).

Ager, D. V., 1973, *The nature of the stratigraphic record:* 113 p., MacMillan (London).

Ambrosetti, Pierluigi, Azzaroli, Augusto, Bonadonna, F. P., & Follieri, Maria, 1972, *A scheme of Pleistocene chronology for the Tyrrhenian side of central Italy:* Soc. Geol. Italiana, Boll., v. 91, p. 169-184.

Arrhenius, Gustav, 1952, *Sediment cores from the east Pacific:* Rept. Swedish Deep Sea Expedition (1947-1948), v. 5, no. 1, p. 1-89.

Azzaroli, Augusto, 1970, *Villafranchian correla-*

tions based on large mammals: Jour. Geology, v. 35, no. 2, p. 111-131.

Bandy, O. L., Casey, R. E., & Wright, R. C., 1971, *Late Neogene planktonic zonation, magnetic reversals, and radiometric dates, Antarctic to the tropics:* Am. Geophys. Union, Antarctic Res. Ser. (Antarctic oceanology, 1), v. 15, p. 1-26.

———, & Wilcoxon, J. A., 1970, *The Pliocene-Pleistocene boundary, Italy and California:* Geol. Soc. America, Bull., v. 81, p. 2939-2948.

Banner, F. T., & Blow, W. H., 1965, *Progress in the planktonic foraminiferal biostratigraphy of the Neogene:* Nature, v. 208, p. 1164-1166.

Bayliss, D. D., 1969, *The distribution of Hyalinea balthica and Globorotalia truncatulinoides in the type Calabrian:* Lethaia, v. 2, p. 133-143.

Bé, A. W. H., & Tolderlund, D. S., 1971, *Distribution and ecology of living planktonic foraminifera in the surface waters of the Atlantic and Indian oceans:* in Micropaleontology of oceans, B. M. Furnell, & W. R. Ridel (eds.), p. 105-149, Cambridge Univ. Press (Cambridge, Mass.).

Beard, J. H., 1969, *Pleistocene paleotemperature records based on planktonic foraminifera, Gulf of Mexico:* Geol. Soc. America, Abstr. with Programs for 1969, pt. 7, p. 256.

Berggren, W. A., 1969, *Micropaleontologic investigations of Red Sea cores—summation and synthesis of results:* in Hot brines and recent heavy metal deposits in the Red Sea, E. T. Degens, & D. A. Ross (eds.), p. 329-335, Springer-Verlag (New York, N.Y.).———1972, *Late Pliocene-Pleistocene glaciation:* in Preliminary reports of the Deep Sea Drilling Project, v. 12, A. S. Laughton et al., p. 953-963, U.S. Government Printing Office (Washington, D.C.).———1977, *The Pliocene/Pleistocene boundary in deep-sea sediments:* Giorn. Geol. (in press).

———, & Boersma, Anne, 1969, *Late Pleistocene and Holocene planktonic Foraminifera from the Red Sea:* in Hot brines and recent heavy metal deposits in the Red Sea, E. T. Degens & D. A. Ross (eds.), p. 282-298, 1 pl., Springer-Verlag (New York, N.Y.).

———, & Haq, Bil-al, 1976, *Biostratigraphy and biochronology of the Pliocene/Pleistocene boundary: calcareous plankton:* First Internatl. Congress Pacific Neogene Stratigraphy, Tokyo (1976) (Abstract).

———, & Van Couvering, J. A., 1974, *The Late Neogene: Biostratigraphy, geochronology and paleoclimatology of the last 15 million years in marine and continental sequences:* Palaeogeography, Palaeoclimatology, Palaeoecology, v. 16 (1/2), p. 1-216.

Blow, W. H., 1969, *Late middle Eocene to recent planktonic foraminiferal biostratigraphy:* in Proc. 1st Internatl. Conf. Planktonic Microfossils, Geneva (1967), 1, P. Brönnimann & H. H. Renz (eds.), p. 199-421, 54 pl., E. J. Brill (Leiden).

Bremer, Mary, Briskin, Madeleine, & Berggren, W. A., 1977, *Qualitative paleoecology and paleobathymetry of the late Pliocene-early Pleistocene foraminifera of the Le Castella (Calabria, Italy):* Abstr. 10th INQUA Congress, p. 53.

Briskin, Madeleine, & Berggren, W. A., 1975, *Pleistocene stratigraphy and qualitative paleooceanography of tropical North Atlantic core V16-205:* in Late Neogene epoch boundaries, T. Saito & L. H. Burckle (eds.), Micropaleontology, Spec. Publ. 1, p. 167-198.

Brock, Andrew, & Hay, R. L., 1976, *The Olduvai event at Olduvai Gorge:* Earth and Planet. Sci. Lett., v. 29 (1976), p. 126-130.

Broecker, W. S., Thurber, D. L., Goddard, J., Ku, T., Mathews, R. K., & Mesolella, K. J., 1968, *Milankovich hypothesis supported by precise dating of coral reefs and deep-sea sediments:* Science, v. 159, p. 297-300.

———, & Van Donk, Jan, 1970, *Insolation changes, ice volumes, and the O^{18} record in deep-sea cores:* Rev. Geophys. Space Phys., v. 8, no. 1, p. 169-198.

Buteux, D., 1843, *Esquisse géologique du Département de la Somme:* Acad. Sci. Agric., Mém., Dépt. de la Somme, p. 187-322.

Ciaranfi, N., & Cita, M. B., 1973, *Paleontological evidence of changes in the Pliocene climates:* in Initial reports of the Deep Sea Drilling Project, W. B. F. Ryan, K. J. Hsu et al., v. 13, no. 2, p. 1387-1399, U.S. Government Printing Office (Washington D.C.).

Cita, M. B., 1973, *Pliocene biostratigraphy and chronostratigraphy:* in Initial reports of Deep Sea Drilling Project, W. B. F. Ryan, K. J. Hsu et al., v. 13, no. 2, p. 1343-1379, U.S. Government Printing Office (Washington, D.C.).

———, Chierici, M. A., Ciampo, G., Moncharmont Zei, M., d'Onofrio, Sara, Ryan, W. B. F., & Scorziello, R., 1973, *The Quaternary record in the Tyrrhenian and Ionian basins of the Mediterranean Sea:* in Initial reports of the Deep Sea Drilling Project, W. B. F. Ryan et al., v. 13, no. 2, p. 1263-1339, U.S. Government Printing Office (Washington, D.C.).

Clark, D. L., 1971, *Arctic Ocean ice cover and its late Cenozoic history:* Geol. Soc. America, Bull., v. 82, p. 3313-3324.

———, Larson, J. A., Root, R. E., & Fagerlin, S. C., 1975, *Foraminiferal patterns of the Arctic Ocean Pliocene and Pleistocene:* Univ. Wisconsin-Madison Arctic Ocean Sedim. Studies Progr. no. 18, Tech. Rept., 94 p.

Cooke, H. B. S., 1973, *Pleistocene chronology: long or short?:* Quaternary Res., v. 3, p. 206-220.

Cox, Allan, 1969, *Geomagnetic reversals:* Science, v. 163, no. 3864, p. 237-245.

Creuze de Lesser, Henri, 1824, *Statistique du Départment de l'Hérault:* 606 p. (Montpélier).

Dansgaard, W., & Tauber, H., 1969, *Glacial oxygen-18 content and Pleistocene ocean temperatures:*

Science, v. 166, no. 3904, p. 499-502.

———, Johnsen, S. J., Clausen, H. B., & Langway, C. C., 1971, *Climatic record revealed by the Camp Century ice core*: in Late Cenozoic glacial ages, K. K. Turekian (ed.), p. 37-56, Yale University Press (New Haven, Conn.).———1972, *Speculations about the next glaciation*: Quaternary Res., v. 2, p. 396-398.

Denizot, G., 1957, *Lexique stratigraphique international*: v. 1, fasc. 4a VII, *Tertiaire: France, Belgique, Pays. Bas, Luxembourg*: Centre Nat. Rech. Sci., Paris, 217 p.

Desnoyers, J., 1829, *Observations sur un ensemble de dépôts marins plus récents que les terrains tertiaires du bassin de la Seine, et constituant une formation géologique distincte; précédées d'un aperçu de la nonsimultanéité des bassins tertiaires*: Ann. Sci. Nat., v. 16, p. 171-214, 402-419.

Deuser, W. G., 1972, *Late Pleistocene and Holocene history of the Black Sea as indicated by stable-isotope studies*: Jour. Geophys. Res., v. 77, no. 6, p. 1071-1077.———1974, *Evolution of anoxic conditions in Black Sea during Holocene*: in The Black Sea—geology, chemistry and biology, E. T. Degens & D. A. Ross (eds.), Am. Assoc. Petrol. Geologists, Mem. 20, p. 133-136.

———, & Degens, E. T., 1969, O^{18}/O^{16} and C^{13}/C^{12} *ratios of fossils from the hot-brine deep area of the central Red Sea*: in Hot brines and recent heavy metal deposits in the Red Sea, E. T. Degens & D. A. Ross (ed.), p. 336-347, Springer-Verlag (New York).

Donn, W. L., & Shaw, D. M., 1967, *The maintenance of an ice-free Arctic Ocean*: in Progress in oceanography, 4, Mary Sears (ed.), p. 105-113, Pergamon (Oxford).

Duplessy, J. C., Labeyrie, J., Lalou, C., & Nguyen, H. V., 1970, *Continental climatic variations between 130,000 and 90,000 years B. P.*: Nature, v. 226, p. 631-633.

Emiliani, Cesare, 1955, *Pleistocene temperatures*: Jour. Geology, v. 63, p. 538-578.———1961, *Cenozoic climatic changes as indicated by the stratigraphy and chronology of deep-sea cores of Globigerina-ooze facies*: New York Acad. Sci., Ann., v. 95, p. 521-536.———1964, *Paleotemperature analysis of the Caribbean cores A254-BR-C and CP-28*: Geol. Soc. America, Bull., v. 75, p. 129-144.———1966a, *Paleotemperature analysis of the Caribbean cores P6304-8 and P6304-9 and a generalized temperature curve for the last 425,000 years*: Jour. Geology, v. 74, p. 109-126.———1966b, *Isotopic paleotemperatures*: Science, v. 154, p. 851-857.———1967, *The Plio-Pleistocene boundary: Reply to G. M. Richmond*: Science, v. 156, no. 3773, p. 410.———1972, *Quaternary paleotemperatures and the duration of the high temperature intervals*: Science, v. 178, p. 398-401.

———, Mayeda, T., & Selli, Raimondo, 1961, *Paleotemperature analysis of the Plio-Pleistocene section at Le Castella, Calabria, southern Italy*: Geol. Soc. America, Bull., v. 72, p. 679-688.

Ericson, D. B., Broecker, W. S., Kulp, J. L., & Wollin, Goesta, 1956, *Late Pleistocene climates and deep-sea sediments*: Science, v. 124, no. 3218, p. 385-389.

———, Ewing, Maurice, & Wollin, Goesta, 1963, *Pliocene-Pleistocene boundary in deep-sea sediments*: Science, v. 139, no. 3556, p. 727-737.———1964a, *Sediment cores from the Arctic and subarctic seas*: Science, v. 144, p. 1183-1192.———1964b, *The Pleistocene Epoch in deep-sea sediments*: Science, v. 146, p. 723-732.

———, & Wollin, Goesta, 1956a, *Correlation of six cores from the equatorial Atlantic and the Caribbean*: Deep-Sea Res., v. 3, p. 104-125.———1956b, *Micropaleontological and isotopic determinations of Pleistocene climates*: Micropaleontology, v. 2, no. 3, p. 257-270.———1968, *Pleistocene climates and chronology in deep-sea sediments*: Science, v. 162, p. 1227-1234.

Ewing, Maurice, & Donn, W. L., 1956, *A theory of ice ages*: Science, v. 123, p. 1061-1066.———1958, *A theory of ice ages, 2*: Science, v. 127, p. 1159-1162.

Fairbridge, R. W., 1972, *Climatology of a glacial cycle*: Quaternary Res., v. 2, no. 3, p. 283-302.

Fillon, R. H., 1972, *Evidence from the Ross Sea for widespread submarine erosion*: Nature, Phys. Sci., v. 238, p. 40-42.

Fleck, R. J., Mercer, J. H., Nairn, A. E. M., & Peterson, D. N., 1972, *Chronology of late Pliocene and early Pleistocene glacial and magnetic events in southern Argentina*: Earth Planet. Sci. Lett., v. 16, p. 15-22.

Fleischer, R. L., 1974, *Preliminary report on Late Neogene Red Sea Foraminifera, Deep Sea Drilling Project, Leg 23B*: in Initial reports of the Deep Sea Drilling Project, T. A. Davies, B. P. Luyendyk et al., v. 26, p. 985-1011, 2 pl., U.S. Government Printing Office (Washington, D.C.).

Forbes, Edward, 1846, *On the connexion between the distribution of the existing fauna and flora of the British Isles and the geographical changes which have affected their area, expecially during the epoch of the Northern Drift*: Great Britain Geol. Survey, Mem. v. 1, p. 336-432.

Gates, W. L., 1976, *Modeling the ice-age climate*: Science, v. 191, no. 4232, p. 1138-1144.

Geitzenauer, K. R., 1969, *Coccoliths at late Quaternary paleoclimatic indicators in the subantarctic Pacific Ocean*: Nature, v. 223, p. 170-172.———1972, *The Pleistocene calcareous nannoplankton of the subantarctic Pacific Ocean*: Deep-Sea Res., v. 19, p. 45-60.

Gignoux, Maurice, 1913, *Les formations marines pliocène et quaternaires de l'Italie du Sud et de la Sicile*: Ann. Univ. Lyon, n. sér., v. 36, 693 p.———1950, *Géologie stratigraphique*: 4ᵉ édit.

entièrement refondue, 735 p., Masson et cie. (Paris).——1955, *Stratigraphic geology:* 682 p., Freeman & Co. (San Francisco, Calif.).

Hammen, T. van der, Wijmstra, T. A., & Zagwijn, W. H., 1971, *The floral record of the Late Cenozoic of Europe:* in The Late Cenozoic glacial ages, K. K. Turekian (ed.), p. 391-424, Yale Univ. Press (New Haven, Conn.).

Haq, Bilal, Berggren, W. A., & Van Couvering, J. A., 1977, *Corrected age of the Pliocene/ Pleistocene boundary:* Nature, v. 269, p. 483-488.

Hays, J. D., 1965, *Radiolaria and Late Tertiary and Quaternary history of Antarctic seas:* Biology of the Antarctic Sea II, Am. Geophys. Union, Antarct. Res., ser. 5, p. 125-184.——1967, *Quaternary sediments of the Antarctic Ocean:* in Progress in oceanography, 4, Mary Sears (ed.), p. 117-131, Pergamon (Oxford).

——, & Berggren, W. A., 1971, *Quaternary boundaries and correlations:* in Micropaleontology of the oceans, B. M. Funnell, & W. R. Riedel (eds.), p. 669-691, Cambridge University Press (Cambridge, Eng.).

——, Saito, Tsunemasa, Opdyke, N. D., & Burckle, L. H., 1969, *Pliocene-Pleistocene sediments of the equatorial Pacific—their paleomagnetic, biostratigraphic and climatic record:* Geol. Soc. America, Bull., v. 80, p. 1481-1514.

Herman, Yvonne, 1969, *Arctic Ocean Quaternary microfauna and its relation to paleoclimatology:* Palaeogeography, Palaeoclimatology, Palaeoecology, v. 6, p. 251-276.——1970, *Arctic paleo-oceanography in late Cenozoic time:* Science, v. 169, p. 474-477.——1972, *Quaternary Eastern Mediterranean sediments: micropaleontology and climatic record:* in The Mediterranean Sea: A natural sedimentation laboratory, D. J. Stanley (ed.), p. 129-147, Dowden, Hutchinson & Ross (Stroudsburg, Pa.).——1974, *Arctic Ocean sediments, microfauna and the climatic record in Late Cenozoic time:* in Marine geology and oceanography of the Arctic seas, Yvonne Herman (ed.), p. 283-348, Springer-Verlag (New York-Heidelberg-Berlin).

Hoernes, M., 1856, *Die fossilen Mollusken des Tertiär-Beckens von Wien:* K. K. Geol. Reichsanstalt, Abh., v. 3, p. 1-733.

Hopkins, K. M., 1967a, *Quaternary marine transgression in Alaska:* in The Bering land bridge, D. M. Hopkins (ed.), p. 47-86, Stanford University Press (Stanford).——1967b, *The Cenozoic history of Beringia—a synthesis:* in The Bering land bridge, D. M. Hopkins (ed.), p. 451-484, Stanford Univ. Press (Stanford, Calif.).——1972, *The paleogeography and climate history of Beringia during Late Cenozoic time:* Inter-Nord 12, p. 121-150.——1973, *Sea level history in Beringia during the past 250,000 years:* Quaternary Res., v. 3, p. 520-540.——1975, *Time-stratigraphic nomenclature for the Holocene Epoch:* Geology, v. 3, p. 10.

——, Mathews, J. V., Wolfe, J. A., & Silberman, M. L., 1971, *A Pliocene flora and insect fauna from the Bering Strait region:* Palaeogeography, Palaeoclimatology, Palaeoecology, v. 9, p. 211-231.

——, Rowland, R. W., Echols, R. E., & Valentine, P. C., 1974, *An Anvilian (early Pleistocene) marine fauna from Western Seward Peninsula, Alaska:* Quaternary Res., v. 4, p. 441-470.

Huene, Roland von, et al., 1971, *Deep Sea Drilling Project Leg 18:* Geotimes, v. 16, no. 10, p. 12-15.

Hunkins, K., Bé, A. W. H., Opdyke, N. D., & Mathieu, G., 1971, *The Late Cenozoic history of the Arctic Ocean:* in Late Cenozoic glacial ages, K. K. Turekian (ed.), p. 215-237, Yale University Press (New Haven, Conn.).

Iaccarino, Silvia, 1975, *Planktonic and significant benthonic Foraminifera of the proposed Plio-Pleistocene boundary type-section of Le Castella:* L'Ateneo Parmense, Acta Nat., v. 11, no. 3, p. 449-465.

Imbrie, John, 1972, *Correlation of the climatic record of the Camp Century ice core (Greenland) with foraminiferal paleotemperature curves from North Atlantic deep sea cores:* Geol. Soc. America, Abstr. Programs, v. 4, no. 7, p. 550.

——, & Kipp, N. D., 1969, *Quantitative interpretations of late Pleistocene climate based on planktonic foraminiferal assemblages in Atlantic cores:* Geol. Soc. America, Abstr. Ann. Mtg., 1969, Part 7, p. 113.——1971, *A new micropaleontological method for quantitative paleoclimatology: application to a late Pleistocene Caribbean core:* in Late Cenozoic glacial ages, K. K. Turekian (ed.), p. 73-181, Yale University Press (New Haven, Conn.).

——, Van Donk, J., & Kipp, N. D., 1973, *Paleoclimatic investigation of a late Pleistocene Caribbean deep-sea core: comparison of isotopic and faunal methods:* Quaternary Res., v. 3, no. 1, p. 10-38.

Ingle, J. C., Jr., 1973, *Summary comments on Neogene biostratigraphy, physical stratigraphy, and paleooceanography in the marginal northeast Pacific Ocean:* in Initial reports of the Deep Sea Drilling Project, L. D. Kulm, R. von Huene et al., v. 18, p. 949-960, U.S. Government Printing Office (Washington, D.C.).——1975, *Pleistocene and Pliocene Foraminifera from the Sea of Japan, Leg 31, Deep Sea Drilling Project:* in Initial reports of the Deep Sea Drilling Project, Leg 31, D. E. Karig et al., p. 693-701, U.S. Government Printing Office (Washington, D.C.).

Johnsen, S. J., Dansgaard, W., Clausen, H. B., & Langway, C. C., 1972, *Oxygen isotope profiles through the Antarctic and Greenland ice sheets:* Nature, v. 235, p. 429-434.

Keany, John, & Kennett, J. P., 1973, *Pliocene-early Pleistocene paleoclimatic history recorded in Antarctic-subantarctic deep-sea cores:* Deep-Sea Res., v. 17, p. 529-548.

Kellogg, T. B., 1972, *Late Pleistocene climates in the Norwegian and Greenland seas:* Geol. Soc. America, Abstr. Programs, v. 4, no. 7, p. 560.

Kennett, J. P., 1970, *Pleistocene paleoclimates and foraminiferal biostratigraphy in subantarctic deep-sea cores:* Deep-Sea Res., v. 17, p. 125-140.

——, & Brunner, C., 1973, *Antarctic Late Cenozoic glaciation: evidence for initiation of ice-rafting and inferred increased bottom water activity:* Geol. Soc. America, Bull., v. 84, no. 6, p. 2043-2052.

——, & Huddlestun, Paul, 1972, *Abrupt climatic change at 90,000 Yr BP: faunal evidence from Gulf of Mexico cores:* Quaternary Res., v. 2, no. 3, p. 384-395.

——, & Shackleton, N. J., 1975, *Laurentide ice sheet meltwater recorded in Gulf of Mexico deep-sea cores:* Science, v. 188, p. 147-150.

——, Watkins, N. D., & Vella, Paul, 1971, *Paleomagnetic chronology of Pliocene-early Pleistocene climates and the Plio-Pleistocene boundary in New Zealand:* Science, v. 171, p. 276-279.

Kent, D., Opdyke, N. D., & Ewing, Maurice, 1971, *Climatic change in the North Pacific using ice-rafted detritus as a climatic indicator:* Geol. Soc. America, Bull., v. 82, p. 2741-2754.

Ku, T. L., & Broecker, W. S., 1967, *Rates of sedimentation in the Arctic Ocean:* in Progress in oceanography, 4, Mary Sears (ed.), p. 95-104, Pergamon (Oxford).

Kukla, G. J., Mathews, R. K., & Mitchell, M. J. (eds.), 1972, *The present interglacial: How and when will it end?:* Quaternary Res., v. 2, no. 3, p. 261-445.

Lamb, J. L., 1969, *Planktonic foraminiferal datums and Late Neogene epoch boundaries in the Mediterranean, Caribbean and Gulf of Mexico:* Geol. Soc. America, Abstr. with Programs for 1969, pt. 7, p. 280.——1971, *Planktonic foraminiferal biostratigraphy and paleomagnetics of late Pliocene and early Pleistocene strata at Le Castella, Italy:* Gulf Coast Assoc. Geol. Soc., Trans., v. 21, p. 411-418.

——, & Beard, J. H., 1972, *Late Neogene planktonic foraminifers in the Caribbean, Gulf of Mexico and Italian stratotypes:* Univ. Kansas Paleont. Contrib., Art. 57 (Protozoa 8), 67 p., 25 text-fig., 36 pl.

Larson, J. A., 1975, *Arctic Ocean Foraminifera abundance and its relationship to equatorial Pacific Ocean solution cycles:* Geology, v. 3, no. 9, p. 491-492.

Laughton, A. S., et al., 1970, *Deep-Sea Drilling Project, Leg 12:* Geotimes, v. 15, no. 9, p. 10-14.——1972, *Initial reports of the Deep Sea Drilling Project, XII:* p. iv-xxi + 3-1243, U.S. Government Printing Office (Washington, D.C.).

Lyell, Charles, 1833, *Principles of geology:* v. 3, 398 p. (plus 109 p.), Murray (London).——1839, *Nouveaux éléments de géologie:* 648 p. Pitois-Levrault (Paris).

McIntyre, Andrew, Bé, A. W. H., Biscaye, Pierre, Burckle, Lloyd, Gardner, James, Geitzenauer, K. R., Roche, Michael, Imbrie, John, Kipp, Nilva, Ruddiman, W. F., Moore, T. C., & Heath, Ross, 1972, *The glacial North Atlantic 17,000 years ago: paleoisotherm and oceanographic maps derived from floral-faunal parameters by CLIMAP:* Geol. Soc. America, Abstr. Programs, v. 4, no. 7, p. 590.

——, et al., 1976, *The surface of the Ice-Age earth:* Science, v. 191, no. 4232, p. 1131-1137.

——, Ruddiman, W. F., & Jantzen, R., 1972, *Southward penetration of the North Atlantic polar front: faunal and floral evidence of large-scale surface water mass movements over the last 225,000 years:* Deep-Sea Res., v. 19, p. 61-77.

Margolis, S. V., & Kennett, J. P., 1970, *Antarctic glaciation during the Tertiary recorded in sub-Antarctic deep-sea cores:* Science, v. 170, p. 1085-1087.——1971, *Cenozoic paleoglacial history of Antarctica recorded in sub-Antarctic deep-sea cores:* Am. Jour. Sci., v. 271, p. 1-36.

Mercer, J. H., 1973, *Cainozoic temperature trends in the Southern Hemisphere: Antarctic and Andean glacial evidence:* in Palaeoecology of Africa, E. M. van Zinderen Bakker (ed.), v. 8, p. 87-114, A. A. Balkema (Capetown-Rotterdam).

Mercey, A. de, 1874-77, *Sur la classification de la période Quaternaire en Picardie:* Soc. Linn. Nord France, Mém., v. 4, p. 18-29.

Mesolella, K. J., Mathews, R. K., Broecker, W. S., & Thurber, D. L., 1969, *The astronomical theory of climatic change: Barbados data:* Jour. Geology, v. 77, no. 3, p. 250-274.

Meunier, E. S., 1908, *Géologie:* 989 p., 152 text-fig., Vuibert et Nony (Paris).

Muratov, V. M., Ostrovsky, A. B., & Fridenberg, E. O., 1974, *Quaternary stratigraphy and paleogeography on the Black Sea coast of Western Caucasus:* Boreas, v. 3, p. 49-60.

Nakagawa, Hisao, Niitsuma, Nobuaki & Hayasaka, I., 1969, *Late Cenozoic geomagnetic chronology of the Boso Peninsula:* Geol. Soc. Japan, Jour., v. 75, no. 5, p. 267-280.

Nelson, C. H., Hopkins, D. M., & Scholl, D. W., 1974a, *Cenozoic sedimentary and tectonic history of the Bering Sea:* in Oceanography of the Bering Sea, D. W. Hood & E. J. Kelley (eds.), Inst. Mar. Sci., Univ. Alaska, Fairbanks, p. 485-516.——1974b, *Tectonic setting and Cenozoic sedimentary history of the Bering Sea:* in Marine geology and oceanography of the Arctic seas, Y. Herman (ed.), p. 119-140, Springer-Verlag (New York-Heidelberg, Berlin).

Niitsuma, Nobuaki, 1970, *Some geomagnetic stratigraphical problems in Japan and Italy:* Mar. Geology, v. 6, no. 2, p. 99-112.

Olausson, Eric, 1961, *Remarks on Tertiary sequences of two cores from the Pacific:* Geol. Inst. Uppsala, Bull., v. 40, p. 299-303.——1965, *Evidence of climatic changes in North Atlantic* deep-sea cores with remarks on isotopic paleotemperature analysis:* in Progress in oceanography, 3, Mary Sears (ed.), p. 221-252, Pergamon (Oxford).

Opdyke, N. D., Glass, B. P., Hays, J. D., & Foster, J. H., 1966, *Paleomagnetic study of Antarctic deep-sea cores:* Science, v. 154, p. 349-357.

Parker, F. L., 1958, *Sediment cores from the Mediterranean Sea and the Red Sea: No. 4. Eastern Mediterranean Foraminifera:* Rept. Swedish Deep-Sea Expedition, v. 8, p. 219-283.

Parkin, D. W., & Shackleton, N. J., 1973, *Trade wind and temperature correlations from a deep-sea core off the Saharan Coast:* Nature, v. 245, no. 5426, p. 455-457.

Parmenter, C., & Folger, D. W., 1974, *Eolian biogenic detritus in deep sea sediments: a possible index of equatorial ice age acidity:* Science, v. 185, p. 695-697.

Phleger, F. B., 1961, *Ecology and distribution of recent Foraminifera:* 297 p., Johns Hopkins Press (Baltimore, Md.).

Reboul, Henri, 1833, *Géologie de la période Quaternaire, et introduction à l'histoire ancienne:* 222 p., F.-G. Levrault (Paris).

Repenning, C. A., 1967, *Paleartic-Nearctic mammalian dispersal in the late Cenozoic:* in The Bering land bridge, D. M. Hopkins (ed.), p. 288-311, Stanford University Press (Stanford, Calif.).

Richmond, G. M., 1970, *Comparison of the Quaternary stratigraphy of the Alps and Rocky Mountains:* Quaternary Res., v. 1, no. 1, p. 3-28.

Rio, Domenico, 1974, *Remarks on late Pliocene-early Pleistocene calcareous nannofossil stratigraphy in Italy:* Ateneo Parmense, Acta Natur., v. 10 (1974), p. 409-449.

Ross, D., Neprochonov, Y., Hsu, K. J., Muhitten, S., Stoffers, P., Supko, P., Trimonis, E. A., Percival, S., Traverse, A., Ericson, A. J., Degens, E. T., Hunt, J. M., & Manheim, F., 1975, *Glomar Challenger drills the Black Sea:* Geotimes, v. 20, no. 10, p. 18-20.

Ruddiman, W. F., 1971, *Pleistocene sedimentation in the equatorial Atlantic: stratigraphy and faunal climatology:* Geol. Soc. America, Bull., v. 82, p. 283-302.

———, & McIntyre, Andrew, 1976, *Northeast Atlantic paleoclimatic changes over the past 600,000 years:* Geol. Soc. America, Mem. 145, p. 111-146.

Ruggieri, Giuliano, 1971, *Calabriano e Siciliano nei dintorni di Palermo, Part 1:* Rivista Min. Sicil., v. 22(130-132), p. 160-171.———1972, *Alcune considerazioni sulla definizione del piano Calabriano:* Soc. Geol. Italiana, Boll., v. 91 (1972), p. 639-645.

———, Buccheri, Giuseppe, Greco, Antonio, & Sprovieri, Rodolfo, 1976, *Un affioramento di Siciliano nel quaero della revisione della stratigrafia del Pleistocene inferiore:* Soc. Geol. Italiana, Boll., v. 94, p. 889-914.

Ryan, W. B. F., 1972, *Stratigraphy of Late Quaternary sediments in the Eastern Mediterranean:* in The Mediterranean Sea, D. J. Stanley (ed.), p. 149-169, Dowden, Hutchinson & Ross, Inc. (Stroudsburg, Pa.).

Saito, Tsunemasa, 1969a, *The Miocene-Pliocene and Pliocene-Pleistocene boundaries in deep-sea sediments:* Congr. INQUA 8°, Paris 1969, Résumés des Communication, sec. 2, 72 bis (abstract). ———1969b, *Late Cenozoic stage boundaries in deep-sea sediments:* Geol. Soc. America, Ann. Mtg., Abstr., v. 82, pt. 7, p. 289-290.

Savage, D. E., & Curtis, G. H., 1970, *The Villafranchian Stage—Age and its radiometric dating:* in Radiometric dating and paleontologic zonation, O. L. Bandy (ed.), Geol. Soc. America, Spec. Paper 124, p. 207-231.

Scholl, D. W., et al., 1971, *Deep Sea Drilling Project, Leg 19:* Geotimes, v. 16, no. 11, p. 12-15.

Selli, Raimondo, 1967, *The Pliocene-Pleistocene boundary in Italian marine sections and its relationship to continental stratigraphies:* in Progress in oceanography, 4, Mary Sears (ed.), p. 67-82, Pergamon (Oxford).———1971, *Calabrian:* Giorn. Geol., ser. 2, v. 37, no. 2, p. 55-64.———1977, *The Neogene/Quaternary boundary in the Italian marine formations:* Giorn. Geol. (in press).

Serres, Michael de, 1830, *De la simultanéité des terrains de sediments supérieurs:* in La Géographie Physique de l'Encyclopédie Méthodique, v. 5, 125 p., 1 pl.———1855, *Des caractères et de l'importance de la période Quaternaire:* Soc. Géol. France, Bull., sér. 2, v. 12, no. 1, p. 257-263.

Shackleton, N. J., 1967, *Oxygen isotope analyses and Pleistocene temperatures re-assessed:* Nature, v. 215, no. 5096, p. 15-17.

———, & Kennett, J. P., 1975a, *Paleotemperature history of the Cenozoic and the initiation of Antarctic glaciation: oxygen and carbon isotope analyses in DSDP sites 277, 279, and 281:* in Initial reports of the Deep Sea Drilling Project, v. 19, J. P. Kennett, R. E. Houtz et al., p. 743-755, U.S. Government Printing Office (Washington, D.C.).———1975b, *Late Cenozoic oxygen and carbon isotopic changes at DSDP Site 284; implications for glacial history of the Northern Hemisphere and Antarctic:* in Initial reports of the Deep Sea Drilling Project, v. 19, J. P. Kennett, R. E. Houtz et al., p. 801-807, U.S. Government Printing Office (Washington, D.C.).

———, & Opdyke, N. D., 1973, *Oxygen isotope and palaeomagnetic stratigraphy of equatorial Pacific core V28-238: oxygen isotope temperature and ice volumes on a 10^5 year scale:* Quaternary Res., v. 3, no. 1, p. 39-55.———1976, *Oxygen-isotope and paleomagnetic stratigraphy of Pacific core V28-239: Late Pliocene to latest*

Pleistocene: Geol. Soc. America, Mem. 145, p. 449-464.

Smith, L. A., 1969, *Pleistocene discoasters from the stratotype of the Calabrian Stage (Santa Maria di Catanzaro) and the section at Le Castella, Italy:* Gulf Coast Assoc. Geol. Soc., Trans., v. 19, p. 579-583.

Sprovieri, Rodolfo, d'Agostino, Salvatore, & Di Stefano, Enrico, 1973, *Giacitura del Calabriano nei dintorni di Catanzaro:* Riv. Italiana Paleont. Stratigr., v. 79, no. 1, p. 127-140 (incl. Eng. summ.).

Takayama, T., 1970, *The Pliocene-Pleistocene boundary in the Lamont core V21-98 and at Le Castella, Italy:* Jour. Marine Geology, v. 7, no. 2, p. 70-77.

Van Montfrans, H. M., 1971, *Paleomagnetic dating in the North Sea basin:* Earth Planet. Sci. Lett., v. 11, p. 226-235.

Van Straaten, L. M. J. U., 1972, *Holocene stages of oxygen depletion in deep waters of the Adriatic Sea:* in The Mediterranean Sea: a natural sedimentation laboratory, D. J. Stanley (ed.), p. 631-643, Dowden, Hutchinson & Ross, Inc. (Stroudsburg, Pa.).

Van Voorthuysen, J. H., Toering, K., & Zagwijn, W. H., 1972, *The Plio-Pleistocene boundary in the North Sea basin: revision of its position in the marine beds:* Geol. Mijnb., v. 6, p. 627-640.

Vezian, Alexandre, 1865, *Prodrome de Géologie:* v. 3, F. Savy (Paris).

Watkins, N. D., & Kennett, J. P., 1971, *Antarctic bottom water: major change in velocity during the late Cenozoic between Australia and Antarctica:* Science, v. 173, p. 873.———1972, *Regional sedimentary disconformities and upper Cenozoic changes in bottom water velocities between Australia and Antarctica:* in Antarctic Res. Ser. 19, Antarctic oceanology 2: The Australian-New Zealand sector, D. E. Hayes (ed.), p. 273-293, American Geophysical Union (Washington, D.C.).

———, Kester, D. R., & Kennett, J. P., 1974, *Paleomagnetism of the type Pliocene/Pleistocene boundary section at Santa Maria di Catanzaro, Italy, and the problem of post-depositional precipitation of magnetic minerals:* Earth and Planet. Sci. Lett., v. 24, p. 113-119.

———, Kristjansson, L., & McDougall, I., 1976, *A detailed paleomagnetic survey of the type location for the Gilsa geomagnetic polarity event:* Earth and Planet. Sci. Lett., v. 27, p. 436-444.

Zagwijn, W. H., 1974, *The Pliocene-Pleistocene boundary in western and southern Europe:* Boreas, v. 3, p. 75-97.

———, Van Montfrans, H. M., & Zandstra, J. G., 1971, *Subdivision of the "Cromerian" in the Netherlands: pollen-analysis, palaeomagnetism and sedimentary petrology:* Geol. Mijnb., v. 50, p. 41-58.

INDEX

Authors' names in this index are set in small capitals with an initial large capital, and suprafamilial names are distinguished by the use of full capitals. Page references having chief systematic importance are in boldface type (as **A100**). The few italicized names are considered to be invalid.

ABBOTT, A32
ABDULLAEV, A152
ABDULLAEV & KHALETSKAYA, A152
ABEL, A21, A22, A70
ABELSON, A68
ABICH, A300, A343
Acaciapora, A272
Acanthatia, A246
Acanthoceras, A459
ACANTHOCERATACEAE, A441
Acanthoceratidae, A457, A459, A467
Acanthoceratinae, A459
Acanthocladia, A322-A325
Acanthonautilus, A276
Acanthopecten, A305
Acanthopyge, A201
Acanthotrigonia, A459
Acastella, A195
Acastoides, A201
Acervoschwagerina, A308, A315
Acharax, A475
Acinophyllum, A223
acritarchs, A80, A109, Silurian, A170, A177
Acrophyllum, A223
Acrosaleniidae, A441
Acrospirifer, A196, A197, A202, A. kobehana Zone, A226
Acroteuthis, A408, A457, A467
ACROTRETIDA, Cambrian, A120
Actaeonellidae, A458, A472
Actinocamax, A458, A459
ACTINOCERATOIDEA, A275, A276
Actinotrigonia, A476
Actinotrypella, A324
actualism, A4
actuopaleontology, A5
Adamanophyllum, A270, A271
ADAMS & AGER, A463
Adelaidean, A82
Adolphia, A197
Adrianites, A300
Aenona, A469
aerobic decay, A16
Afghanella, A312
African strata, Siluran, A174
Agapella, A472
AGASSIZ, A507
Agathiceras, A300, A302, A305, A335
Agathiceratidae, A276
age correlations, Precambrian, A85
Agelasina, A474
AGER, A393, A395, A462, A468, A471, A508
AGER & WESTERMANN, A395
Agerina, A141
agnostid distribution and temperature, A129

AGNOSTIDA, A122-A128, A131, A160
Agnostidae, A123
Agoniatites, A201, A243
Agraulidae, A125
Ahtiella, A143
AITKEN, A247
Akagophyllum, A319, A320
Akasakan Series, A294, A302, A312
Akidocheilus, A410
Akiyoshiella, A266
Akiyosiphyllum, A271
Akmilleria, A331, A334
Aknisophyllum, A223
Aktubinskia, A334
Albertella, A123, Zone, A131-A132
ALBERTI, A189, A192, A195, A198, A201, A208, A226
VON ALBERTI, A352
Albian, A423, A446-A448, A453, A457-A458
Albumares, **A92**
ALDINGER, A381
algae, Precambrian, A108-A110, A111
Alispirifer, A260
Aljutovella, A266
"Allanaria," A245
ALLEN, A57, A222
Allenetes, A197
allochthonous burial, A23-A48
allogenic causes of death, A6
ALLOITEAU, A394
Allotropiophyllum, A269, A270, A272, A299, A318, A320
Almites, A334
Alpine facies, A352
Altay-Sayan area, Devonian fauna of, A196, A201, A206
Alveolinella, A497
Alveolinidae, A472
Amaltheus, A399, A402
Amandophyllum, A273
amber, Eocene, preservation in, A12, A13, A14
ambient polishing, A25
Ambocoelia, A300
Ambocoeliidae, A329
Ammodiscidae, A316
Ammonites, A359
ammonoid fauna, Devonian, A227, A234, A243, A244, provincialism of, A231-A234, A242-A243, A246-A247, zones, A190
AMMONOIDEA, A275, A276
ammonoids, Jurassic, A397-A408, Permian, A331-A337, Triassic, A370-A371
Amnigenia, A247

AMOS & BOUCOT, A231
AMOS & SABATTINI, A260
Amphigenia Zone, A223, A231, A234
AMPHINOMORPHA, A103
Amphipora, A199, A246
Amphipyndacidae, A441
Amphipyndax enesseffi Zone, A461
Amphistegina, A497
Amphitriscoelus, A473
Amphoton, A127, A130
Amplexizaphrentis, A318, A319
Amplexocarinia, A269, A299, A300, A318, A319, A321
Amplexus, A269, A270
Ampyxinella, A149
AMSDEN, A154
Amurolites, A202
Amygdalophyllum, A270, A271
Anabacia, A393
Anabarites, A89, **A104**
Anabaritidae, **A104**
Anadara, A469
anaerobic decay, A16
Anaflemingites, A363
Anagaudryceras, A475
Anapachydiscus, A471
Anarcestes, A202, A203, A242
Anastomopora, A323
"Anastrophia," A196
Anataphrus, A150, A154, A161
Anchignathodus, A338, A339, Zone, A301, A. typicalis Zone, A339, A379
Anchiopsis, A226
Anchura, A469
Ancistrorhynchia, A146
Ancyloceratidae, A456, A457
Ancyloceratina, A455, A471, A472
Ancyrodella, A243
"Ancyrognathus" triangularis Zone, A243
Ancyropyge, A242
ANDERSON, A85, A193, A210, A353
ANDERSON & ANDERSON, A353
ANDERSON, BOUCOT, & JOHNSON, A202
Anderssonoceras, A301
Andiceras, A402
ANDREEVA, A141
ANDREEVA & DRONOV, A403
Anetoceras, A195, A196, A234
Angaran paleofloral region, A285
Angarella, A140, A152
Angustiochrea, A104
ANGUSTIOCHREIDA, A104
animal protists, evolution of, A81
ANNELIDA, **A102**
annelids, Cambrian, A120, Precambrian, A88

Index

Anodontopleura, A473
Anofia, A474
Anolis electrum, A11
Anomalofusus, A469
Anomocarella, A127
Anomocarellidae, A130
Anomocaridae, A125
ANOMURA, A410
Anopisthodon, A476
Anoplotheca, A195
anoxybiotic polychaetes, A8-A9
Antedon, A411, A499
ANTHOZOA, A89
anthozoans, Jurassic, A393-A394, Permian, A317-A321
Anthracoceras, A260
Antillocaprina, A473
Antinomia, A395
ANZYGIN, A148
Apatognathus, A280
Apatorthis, A152
Aphelaceras, A276
Aphroidophyllum, A240
Aphrophylloides, A271
Aphrophyllum, A271
Apiocrinus, A411
Apiograptus, A155
Apiotrigonia, A461, A465
Aploceras, A275
APOLLONOV, A152, A153, A158
Aporrhaidae, A458
Aporrhais, A39
Appalachian province, Devonian, A209-A210
Appohimchi subprovince, A230-A231, A231
Aptian, A423, A446-A448, A453, A457
Apulites, A473
Arachnastrea, A271, A272, A273
Arachnolasma, A269-A272
ARAKELIAN, A299, A311
Araxathyris, A382
Araxilevis, A336
Araxoceras, A300, A301, A336
Araxoceratidae, A336
Araxopora, A324, A325, A326
Arborea, A99
Arca, **A35**
Archaediscus, A263, A265
Archaeocidaris, A306
ARCHAEOCYATHA, A120, A123, A124, A126, A127, A129, A130
Archegosaurus, A297
D'ARCHIAC, A3
Archidiscodon, A500
ARCOIDA, A375
Arctica islandica, A508, A510
Arctic-Boreal area, Jurassic, A391
Arcticeramus, A396
Arcticoceras, A401, A402
Arcticopora, A326
Arctocephalites, A401, A402
Arctomytilus, A472
Arctostrobium urna Zone, A460
Arctotis, A396
Arcullaea, A473
ARCULLAEINAE, A441
ARDUINO, A506

Arenicola marina, A8
Arenigian shelly faunas, A140-A142, Balto-Scandian, A140-141, Chinese, A141, Mediterranean, A141-A142, North American, A140
Argocheilus, A276
Arguinella, A397
Arieticeras, A399
Arietites, A398
Arietitidae, A398
Aristocystites, A149
ARKELL, A391, A394, A398, A400, A402, A440
Arkelloceras, A401
Armenina, A299, A300
ARMSTRONG, A305
Arnaudia, A473
Arnioceras, A398
ARRHENIUS, A517
ARTHABER, A363
ARTHROPODA, **A104**
arthropods, Cambrian, A120, Precambrian, A87-A88, Tertiary, A499
Artinskia, A331
Artinskian Stage, A255, A293, A294, A305, ammonoids of, A334-A335, brachiopods of, A330, bryozoans of, A322, A324, A325, A326, corals of, A318-A319, fusulinaceans of, A311
Artioceras, A334
Artostrobium urna Zone, A460
Arumberia, A83, **A102**
Asaphopsis Province, A142
Asaphus, A143, A148
Ascopora, A322, A325, A326
Ashgillian, A149, lower and middle faunas of, A150-A152
Asiatotrigonia, A472
Asioptychaspis, A127
asphyxiation, A8-A10
Aspidagnostus, A124
Aspidoceras, A406, A407
Aspidoceratidae, A401, A403
Aspidura, A62, A63
Asselian Stage, A255, A259, A267, A293, A305, A306, ammonoids of, A331, brachiopods of, A330, bryozoans of, A322, A323, A324, corals of, A318, fusulinaceans of, A310
assemblages, Cambrian, A120
ASSERETO, A357
Assilina, A497
Astartella, A305
Astartidae, A375
Astartila, A305
Asterias rubens, A7, A20-A21
Asteroarchaediscus, A265
Asterobillingsa, A223
Asteroceras, A398
Astrophyllum, A201
Astycorphe, A242
Ataxioceras, A407
Athyris, A300, A. angelica Zone, A245
Athyrisinidae, A327, A330

Atokan Series, A248, corals of, A272, foraminifers of, A266
Atomodesma, A305
Atractites, A408
Atriboniidae, A327
"Atrypa," A240
Atrypa, A242
Atrypella, A223, A227
Atrypidae, A161
ATRYPOIDEA, A203, A246
Atsabites, A335
Attribonium, A246
Aucellina, A457, A458
AUDLEY-CHARLES, A363
Augustiochreidae, A104
Aulacella, A246
Aulaceridae, A161
Aulaconautilus, A408
Aulacopleurum, A476
Aulacosphinctes, A403, A406
Aulacosphinctoides, A403
Aulacostephanus, A407
Aulacoteuthis, A449
Aulina, A271, A272
Aulophyllum, A272
Aulopora, A273, A321
Aulosteges, A380
Aulostegidae, A329, A330
AUSTIN, A189, A206
Australian biogeographic region, in the Carboniferous, A270, A271
Australocoelia, A197, A230
Australospirifer, A197, A230, A231
Austrotrigonia, A476
autogenic causes of death, A6
autolysis, A6
Autunian Series, A295-A297
Avicula, A362
Aviculopecten, A305
AYZENVERG, A258
azimuthal orientation, A33-A36
Azygograptus, A156
AZZAROLI, A516

BAARS, A245
BACHOFEN-ECHT, A80
bacteria, Proterozoic, A80
Bactrognathus, A280, A281
Baculites, A459
Baculitidae, A459-A461, A467
Baikalina, **A102**
Bainella, A230
BAKER, A222
Bakevelloides, A369
Balanus, A6, A7
BALDIS & BLASCO, A153
Balto-Samartian shield, A124
BANDO, A363, A364
BANDY, CASEY, & WRIGHT, A530
BANDY & WILCOXON, A510, A513
BANKS & NAQVI, A305
BANNER & BLOW, A513
Barcoona, A476
BARGHOORN & TYLER, A80
BARKHATOVA, A259
Barnea, A30
BARNES, A156, A158
BARNES & FAHRAEUS, A156, A158

BARRANDE, A195
Barrandeophyllum, A196
Barremian, A423, A446-A448, A453, A457
Barrettia, A473
Barroisiceratinae, A460
BARTENSTEIN, A410
BARTHEL, A15, A47, A48, A369, A403, A428
Bartramella, A266, A268
Baryconites, A473
Barytichisma, A272
Bashkirian Stage, A259, corals of, A272, foraminifers of, A265-A266
Basidechenella, A201
Basilicorhynchus, A246, Zone, A205, A245
basinal shale facies, Lower Devonian fauna of, A195
BATE, A69
BATES, A143
BATHURST, A49, A53, A63, A69
Bathyuriscus, A123, A130
BATTEN, A377, A378
BAUMANN, A58
Bayleia, A472
Bayleoidea, A473
BAYLISS, A510, A513, A514
BÉ & TOLDERLUND, A532
Beaniceras, A399
BEARD, A514
Becken facies, A187, A195
BEECHER, A32
BEEDE & KNIKER, A292, A313
Beedeina, A266, A267, A268, Zone, A266, A267
BEHMEL & GEYER, A399
BEHNKEN, A337, A338
Belemnella, A458, A461, A468
Belemnitella, A458, A461, A468, A472
Belemnitellinae, A441
belemnites, Jurassic, A408-A409
Belemnitidae, A408, A441
Belemnopsidae, A410
Belemnopsis, A410, A454, A455, A472, A475, A476
Belemnoteuthidae, A441
BELL, A281
BELLEROPHON, A302
Bellerophontidae, A378
Belliscala, A469
Beloceras, A205, A207
?Beltanella, A96
BELYEA & McLAREN, A245
BENDER & STOPPEL, A337, A339
BENSAID, A189, A190, A208
benthic communities, Silurian, A170
BERCKHEMER & HÖLDER, A407
BERDAN, A223, A230
BERDAN & MARTINSSON, A180
BERGGREN, A503, A516, A520, A530, A532, A533
BERGGREN & BOERSMA, A533
BERGGREN & VAN COUVERING, A505-A543
Bergoceras, A275
BERGSTRÖM, A145, A156, A157

BERNER, A429
Berriasella, A406, A408, A453
Berriasellidae, A454, A456, A471
Berriasian, A423, A442-A444, A453
BERRY, A156, A172, A226, A439
BERRY & BOUCOT, A161, A172-A175, A178-A181
BERRY & MURPHY, A226
BEUF, A159
BEURLENS, A410
BEUSHAUSEN, A195
BEYRICH, A499
Bibucia, A270, A271
BIERNAT, A208
Bifida, A195
Bifossularia, A270, A271
Bihenithyris, A395
Bilobia, A145
Bimuria, A145, A147, A151
biocoenosis, A22
biofacies realms, Cambrian, in North America, A122-A124
biogeography, Cambrian, A128-A130, in North America, A122-A124, Early, A129; Carboniferous, A260-A261, A283-A286; Cretaceous, A461-A476, North Temperate realm, A465-A469, South Temperate realm, A473-A476, Tethyan realm, A469-A473; Silurian, A168
Biolgina, A140
biomeres, Cambrian, A131-A134
BION, A378-A379
biostratigraphy, Cambrian, **A130-A134**, of North America, A131-A132; Cretaceous, A443-A461, Lower Cretaceous, A449-A453, Upper Cretaceous, A458-A459; Silurian, A170-A172; Tertiary, A499
biostratinomic information, from soft parts, A15-A19, observations, early, A5
biostratinomy, A3, A5
BIRADIOLITINAE, A441
BIRENHEIDE, A192, A201
Birostrina, A453, A457
BISAT, A276
Bisatoceras, A279
BISCHOFF & ZIEGLER, A190
Bistrialites, A276
Bithynia, A397
BITTNER, A378, A380
bivalves, Jurassic, A395-A397, Silurian, A179, Triassic, A372-A377
Biwaella, A315
Blanfordiceras, A403, A406
blattoid assemblages, Carboniferous, A281
BLIND, A398
Blountiidae, A123
BLOW, A503, A510, A513
BLUMENSTENGEL, A191
Bochianites, A403, A471
BOEKSCHOTEN, A20
BOGOSLOVSKIY, A196, A201, A205, A206, A207

Bohemian facies, Old World Devonian, A195, A198, A209
BOLLI, A503
BOND, A430
BONDAREV, A147
Bonnia-Olenellus Zone, A131-A132
Boreioteuthis, A467
Borelis, A497
"Borelis princeps," A310
Bornhardtina, A201, A202
Bositra, A396
Bothrophyllum, A271-A273, A319
BOTTINO & FULLAGAR, A219
BOUČEK, A154, A191, A195
BOUCKAERT, A189
BOUCKAERT & STREEL, A189
BOUCOT, A167-A182, A192, A194, A195, A196, A197, A199, A205, A210, A222, A223, A230, A231
BOUCOT, CASTER, IVES, & TALENT, A197
BOUCOT & JOHNSON, A223
BOUCOT, JOHNSON, & STRUVE, A199, A202, A240
BOUCOT, JOHNSON, & TALENT, A194, A231, A234
Boucotia, A197
Bouleiceras, A400
Bouleigmus, A472
Boultonia, A299, A308, A316
Boultoninae, A309
BOUROZ, A258
BOWEN, A231, A424, A425
Brachina, **A92**
Brachiograptus, A156
brachiopod fauna, Devonian, A223, A226, A227, A230, A231, A234, A235, A238, A240, A242, A243, A244, A245, A246, in Africa, A198, in Asia, A202, in Australia, A197, A208, in Europe, A199, A205, Middle Devonian extinctions, A199
brachiopod zones, Devonian, A191-A192, A195
BRACHIOPODA, Cambrian, A120
brachiopods, as Cambrian index fossils, A130; Jurassic, A394-A395; Lower Carboniferous, A273, A275; Permian, A326-A330, associations and diversity patterns of, A329-A330, dispersal patterns of, A327-A329; Silurian, A172, A177, A178-A179, cosmopolitan, A178, endemic, A178; Tertiary, A499; Triassic, A378-A382
"Brachyaspis," A154, A161
Brachybelus, A408
Brachymeris, A469
Brachythyrididae, A273, A329
Brachyura, A410
Bradyina, A265, A266
Bradyinidae, A267
Bradyphyllum, A319
BRANCHIOPODA, **A105**
Brancoceratidae, A457, A475
Brancoceratinae, A475
Branneroceras, A279

BRANSON, A258
BREDDIN, A58
BREMER, BRISKIN, & BERGGREN, A510, A513
BRENCHLEY & NEWALL, A32
BRENNAND, A354
BRENNER, H., A70
BRENNER, K., A40, A44
BRENNER & EINSELE, A57
Brevaxina, A299
Briantelasma, A223
BRICE, A202, A207
BRICE & MEATS, A230, A240, A242
BRIDEN, A159
BRIDEN, DREWRY, & SMITH, A208, A353
BRINKMANN, A397, A401
BRISKIN & BERGGREN, A521, A523, A531
Bristolia, A123
BROCK & HAY, A513
BROECKER & VAN DONK, A518, A526
Broinsonia parca Zone, A461
BROMLEY, A20
BRONGNIART, A506
Brongniartella, A153
BROUWER, A184
BROWN, A68
Brunsia, A261, A263
BRUNTON, A208
bryozoans, Permian, Andean Sea, A323, Franklinian Sea, A322-A323, geographical distribution of, A326, southern North America, A323, Russian platform, A322, Tasman geosyncline, A325-A326, Tethyan sea, central, A324, northern, A323-A324, southern, A324-A325, Zechstein Sea, A322-A323; Silurian, A178; Tertiary, A499
bubble levels, geologic, A17, (figs.) A17, A18, A19
Buccinum, A30, A39
BUCHAN, A365
Buchanathyris, A197
BUCHER, A49
Buchia, A391, A396-A397, A453, A455
Buchiidae, A441, A449, A453, A457, A458
BUCKLAND, A184
Budaiceras, A469
BUGGISCH, A205
BUGGISCH & CLAUSEN, A189, A208
BULMAN, A154, A155, A172
BULTYNCK, A189, A190
Bumastoides, A149, A150
Bunter, A352
buoyancy, decomposition, A17, simulation of effect of, A19
burial, A2, assemblage, A22-A24
Burmeisteria, A195, A231
BURRETT, A137, A142, A159, A210
BUTEUX, A507
BUTLER, A222
Buxtoniidae, A275, A327, A329

Cabrieroceras, A201, A203, A235, A242, A243, C. crispiforme Zone, A189, A201, A203
Cadoceras, A401
Cadomella, A395
Caeleceras, A195
Caenodontus, A339
CAENOGASTROPODA, A498
Calabrian, A508, A509-A510
Calcareous algae, Silurian, A181
calcareous nannoplankton and climatic change, A531
Calceola, A202
Calcicalathina, A456, A457
Calcisphaera, A263, A265
CALDWELL, A238
Callianassa, A10
Callianthalus, A469
Callipteris, A297
Callispirina, A380
Callocladia, A325
CALLOMON, A401
Callucina, A472
Calmonia, A230
Calodiscus, A123, A126
Calophyllum, A319
Calpionella, A403, A453
Calpionellidae, A453
Calpionellites, A455
Calpionellopsis, A456
CALVER, A255
Calvinaria, A244, A245, A246, C. albertensis zones, A245, C. variabilis zones, A244
Calycoceras, A459
Calymenina, A199
Calyptrina, **A107**
Camarocrinus, A180
Camarotoechiidae, A327
Cambrian, Lower-Middle boundary, A130
Cambrian-Ordovician boundary, A130, A137, transgression, A138-A139
Camerisma, A329
CAMERON, A69
Campanian, A423, A454-A456, A460-A461
CAMPBELL, A153, A259
CAMPBELL & McKELLAR, A259
Campophyllum, A270, A272
Camptonectes, A441
Campyloceras, A275
Canavaria, A399
Cancellina, A299, A312, Subzone, A313, Zone, A312, A313
Cancrinella, A260, A297
Caninia, A269, A270, A271, A272, A299, A318, A319, A320
Caninophyllum, A269, A270, A271, A273, A299, A319
Caninostrotion, A271
Cantabrian Stage, A258
Caprinidae, A441
Caprinuloidea, A473
Caprotina, A472
Carbactinoceras, A275
carbonate banks, Middle & Late Cambrian, Siberian platform, A130

carbonate compensation depth (CCD), A30, variations, Cretaceous, A429
carbonate sediment belt, Cambrian, A122
Carboniferous deposition, in Australia, A259, in South Africa, A260, in South America, A259-A260
Carboniferous rock subdivisions, A255-A260; correlations among subdivisions, A256-A257, A259; western European subdivisions, A255-A258; North American subdivisions, A258, Russian subdivisions, A258-A259
Carboniferous-Permian boundary, A255, A292
Carcinophyllum, A272
Cardinia, A395
Cardioceras, A401
Cardioceratidae, A401, A406
Cardiograptus, A155, A156
Carditidae, A375
Cardium, A35, C. echinatum, A29, A35, C. edule, A6, A24, A25, A27, A28, A29, A35
Carinapyga, A227
Carinatina, A201
"Carinatina" dysmorphostrota Zone, A240
Cariniferella, A246
CARIOU, A400
CARLS, A188, A189, A190, A195, A199
Carniaphyllum, A272
Carnithiaphyllum, A320
Carota, A469
Carruthersella, A270, A272
CASEY, A440, A462, A468, A471
CASEY & RAWSON, A440, A449, A453, A457, A463, A464, A465
Cassidulidae, A411
Cassiope, A469
CASTER, A222, A231, A259, A307
Cathaysian paleofloral region, A285
Catillicephalidae, A130
causes of death, A5, A6, external, A6, internal, A6
Cavusgnathus, A280, A281
CECIONI & WESTERMANN, A398
Cedaria, A123
"Cedaria" of inner detrital belt, A123
Cenoceras, A372, A408
Cenomanian, A423, A450-A452, A459
Central American strata, Silurian, A175
Centronella, A234
Centropleura, A123, A125
cephalopods, Carboniferous, A275-A279, Silurian, A179
Ceratites, A52, A355, C. nodosus, A55
Ceratonurus, A226
Ceratopea, A140
Ceratopygidae, A124, A127, A130, A139, A140

Ceraurus, A150
Chaetetes, A272, A273
CHANG, A207
Changhsingoceras, A301, A337
CHAO, A207, A292, A301, A302, A320, A337, A362
CHAPPELL, A53
Charnia, A83, **A99**
Charnia, A99
Charniidae, **A99**
Charniodiscus, A83, A94, **A99**
Chasmops, A148, A150-A153
CHAVE, A25
Cheiloceras, A206, A208, A244, A245, A247
Cheiruras, A201
CHELICERATA, A104
CHEN, A312
CHERNYSHEV, A292, A330
Chesterian Series, A258, A280
Chetaites, A453
CHETVERIKOV, A6
Chia, A270
Chiapasella, A473
Chiastozygus, A457, A459, C. irregularis Zone, A459
Chihsiaphyllum, A319
Chimbuites, A475
China south of Peking platform, Devonian faunas of, A196, A202, A207
Chinese coral province, Visean, A271
chitinozoans, Silurian, A170, A172, A177
Chlamydophyllum, A223
Chlamys, A476
CHLUPÁČ, A187, A189, A195, A223
Choffatella, A471
Choffatia, A402
Chomatoseris, A393
CHONDROPHORINA, **A91**
Chondroplidae, **A91**
Chondroplon, A88, **A91**
Chonetella, A300, A380
Chonetellidae, A327
Chonetes, A301
Chonetidae, A329
Choristilidae, A273
Chouteauoceras, A276
Christiania, A143, A145, A151
chronostratigraphy, Precambrian, A85-A86
Chuangiidae, A127
Chubbininae, A473
CHUGAEVA, A140, A141, A143
CHURKIN, A210
CHURKIN & BRABB, A227
Chusenella, A299, A308
CIARANFI & CITA, A532
Cibolites, A335
CIDAROIDEA, A411
CILIATA, A393
circum-Pacific areas, Jurassic, A391
Cisticephalus Zone, A306
CITA, A530, A532
Cladoceramus, A460
Cladochonus, A270, A318-A320

Cladophlebis, A301
Cladopora, A321
Claraia, A300, A313, A337, A357, A358, A362, A363, A366, A372, A375, A379, A380, distribution of, A373
CLARK, A103, A382, A525, A526
CLARK & BEHNKEN, A337
CLARK & ETHINGTON, A337
CLARKE, A194, A231
Clarkeia, A161, A178
CLAUSEN, A190
CLAUSILIACEAE, A498
Clausotrypa, A322-A326
Cleiothyridina, A297, A380
Clelandia, A140
CLIFTON, A32
Climacammina, A263, A265, A266
Climacotrigonia, A476
climate, Carboniferous, A285-A286, Jurassic, A393, Ordovician, A159-A160, Silurian, A168, Tertiary, A489-A490
climatostratigraphy, Quaternary, A515-A536; Antarctic-subantarctic Ocean, A527-A531, Antarctic marginal seas, A530, Australia-New Zealand, A527-A530; Arctic region, A525-A527; Atlantic Ocean, A520-A525, Gulf of Mexico, A524-A525; calibration of, A517-A520; Mediterranean, A531-A536, Black Sea, A533-A536, Red Sea, A533; Pacific Ocean, A527, Sea of Japan, A527
Clinophyllum, A269
Clioscaphites, A468
Cliona, A427
Clisiophyllum, A271, A272, A318
CLITAMBONITACEA, A161
Clitambonites, A148
CLOSS, A71
Closteriscus, A469
CLOUD, A30, A84, A85, A87, A103, A199, A435
Cloudina, A83, A89, **A102**
Clycymeris, A473
Clydagnathus, A281
Clydoniceratidae, A400
Clymenia Zone, A247
clymeniid faunas, Devonian, A204, A206
Clypaestridae, A499
Clypeolampadidae, A441
Clypeus, A411
Coalcomana, A473
COATES, A462, A469, A471-A473
COATES, KAUFFMAN, & SOHL, A449, A469
COBBAN & REESIDE, A458
COBBAN & SCOTT, A458
Coccolithus pelagicus, A531
Cochlearites, A396
Cochlocrioceras, A403
COCKS, A159, A161
COCKS & MCKERROW, A161
COCOLITHOPHORIDA, A393
Codonofusiella, A300, A302, A312, A313, A314, A320, -Reichelina

Zone, A300, A312, Zone, A320
COELENTERATA, **A91**
coelenterates, Precambrian, A88, Silurian, A177-A178
Coeloceras, A399
Coeloderoceras, A399
Coelodiscus, A397
Coelogasteroceras, A276
Coelospira, A234
Coilopoceratidae, A471
Colaniella, A300
COLLIGNON, A359, A394, A406, A472, A476
Collignonicardia, A472
Collignoniceras, A460
Collignoniceratidae, A460, A467
Collignoniceratinae, A460
COLLINSON, REXROAD, & THOMPSON, A280
Collyrites, A411
COLO, A399
Columellaria, A397
Colveraia, A473
Comelicania, A300
Composita, A382
Comura, A201
Conaspis, A124
Conchopeltidae, **A94**
CONCHOPELTINA, **A94**
concretions, A65-A67, diagenetic, A66, epigenetic, A67, syngenetic, A65-A66
Condraoceras, A276
Confusiscala, A469
Congeria, A498
Coniacian, A423, A450-A452, A460
CONIL, A261
CONIL & LYS, A261
CONIL, PAPROTH, & LYS, A261
CONIL & PIRLET, A189
Conobelus, A410, A471, A472
Conoclypeidae, A499
Conocoryphidae, A125, A130
Conodicoelites, A410
conodont faunas, Devonian, A226, A227, A230, A234, A235, A238, A240, A242, A243, A244, A246, in Australia, A197, A202, A208, in Europe, A206; Ordovician, A138, A156-A157, A158, in North American Midcontinent province, A156-A157, in North Atlantic province, A156-A157
conodont zones, Devonian, A190-A191, A208
conodonts, Cambrian, A120, A130, A132, Carboniferous, A279-A281, Permian, A337-A339, Silurian, A170, A172, A181, Triassic, A382-A384
Conolichas, A153
Conomedusites, A88, **A94**
Conularia, A306
CONULARIIDA, A88, **A94**
CONULATA, **A94**
Conulidae, A441
Conulus, A460
CONYBEARE & PHILLIPS, A254
COOK & TAYLOR, A157

Cooke, A516, A517
Cooper, A142, A219, A234, A235, A330, A339
Cooper & Grant, A314, A330, A334
Cooper & Phelan, A235, A240
Cooper & Warthin, A235
Cooper & Williams, A235
Cope, A178
Copeland, A238
coprolite "bubble levels," A17-A19
Coralliochama, A473
coral faunas, Devonian, A223, A231, A234, A235, A238, A240, A242, in Africa, A198, A203, in Asia, A202, in Australia, A197, A203, in Europe, A201, A203, A205
corals, Carboniferous, A268-A273, Lower Carboniferous, A269-A270; Permian, faunal provinces of, A318-A321, Midcontinent North American, A320-A321, Tethyan, A319-A320, Ural-Artinsk, A318-A319; Tertiary, A498-A499
Corbicula, A397
Corbiculopsis, A472
Corbinia apopsis Subzone, A138
Corbula, A397
Corbulamellidae, A441
Cordania, A223
Cordiceramus, A460
Cordylodus, A138
Cornuproetus, A230
Cornuspira, A263
Cornuspiridae, A267
Corollithium exiguum Zone, A459
Corongoceras, A403, A406
Coroniceras, A398
correlation, Cambrian global, A130-A134
Cortezorthis, A196, A227
Corvus frugilegus, A24
Corwenia, A272
Corycephalus, A226
CORYNEXOCHIDA, A130
Coscinotrypa, A323, A325, A326
Costaloria, A375
Costellacesta, A469
Costispirifer, A196
Costispiriferidae, A203
Cotylederma, A411
Cousminer, A246
Couvinian faunas, A234, A240
Cowen & Rudwick, A395
Cowrie, A120, A129
Cox, A352, A396, A405, A513, A520
Craniscus, A395
Cranocephalites, A401
Craspedites, A406, A408
Craspeditidae, A454, A455
Crassatellina, A468
Crassiproetus, A226
Cravenia, A269, A270
Cravenoceras, A279
Cravenoceratoides, A279
Creath & Shaw, A140

Cremnoceramus, A460
Creonella, A469
Crepicephalus, A123
Cretaceous, boundaries, A440-A443, definition of, A439-A440, extinctions, A418-A419, A441; marine environments, A419-A420, A424-A438, benthonic, A432-A436, eustatic changes, A430-A432, temperature, A424-A426, water chemistry, A426-A430
Cretaceous-Tertiary boundary, A419, A441-A443, A500
Cretarhabdus, A456
CRIBICYATHEA, **A102**
Cribrogenerina, A266
Cribrospira, A263, A265
Cribrostomum, A263, A265
Crickites, A205, A243, A244
Crickmay, A238, A240, A244, A245
cricoconarid fauna, Devonian, A185, A191, A195, A198, A199, A201, A203, A206
Crimites, A334
CRINOIDEA, A410
Crockford, A325
Crurithyris, A246, A260, A380
CRUSTACEA, A87, A88, **A105**, Precambrian, A87-A88
crustaceans, Jurassic, A410
Cryptonellidae, A330
Cryptophyllum, A269, A319
CRYPTOSTOMATA, A321-A322
Cryptothyrella, A154
Cuerda, A137, A155
Cumminsia, A272
Cuneolina, A471
Cupularostrum, A246
Curtohibolites, A457, A472
Curviceras, A398
CYANOPHYTA, A80
Cyathaxonia, A269-A272, A299
Cyathocarinia, A30
Cyathoclisia, A269
Cyathophyllum, A223
Cybelurus, A142
Cycloclypeus, A497
Cyclococcolithus leptoporus, A531
Cyclolobus, A298, A307, A336, A381
Cyclomedusa, A83, **A94**
Cyclondendron, A306
Cyclopygidae, A151, A160
Cyclorisma, A469
Cyclorismina, A476
Cyclotrypa, A323, A324, A326
Cyctophora, A273
Cylindroteuthididae, A454, A457
Cylindroteuthis, A408, A410
Cymaceras, A407
Cymaclymenia, A207, A247
Cymatoceras, A408
Cymoceras, A279
Cymostrophia, A197
Cyphomena, A145
Cyphoterorhynchus, A207
Cyprinia islandica, A29
Cyrenopsis, A476

Cyrocopora, A300
Cyrtina, A195, A197, A223, A226, A227, A246
Cyrtiopsis, A207, A208, A245
Cyrtoclymenia, A247
Cyrtometopus, A143
Cyrtonotella, A146
Cyrtophyllum, A152
Cyrtospirifer, A205, A208, A243-A244, A244, A245, A246, C. charitopes Zone, A245, C. variabilis zones, A244
Cyrtosymbole, A207
Cyrtothoracoceras, A275
Cystilophophyllum, A272
Cystophora, A272
Cystophrentis, A269, A270
CYSTOPORATA, A321-A322
Cytheridae, A453

Dabrowska, A407
Dacqué, A395, A400
Dactylethrata, A441
Dactylioceras, A399, A400
Dactylioceratidae, A399
Dadoxylon, A305
Daghanirhynchia, A395
Dagis & Zakharov, A399
Daguinaspis, A122, A125
Daily, A134
Dainella, A263
Daixina, A267, A268, A310, A311, A313
DALMANITACEA, A203
Dalmanites, A226
Dalmanitina, A151, A153, A154, -Hirnantia fauna, A149, A153, A154
Dalmatskaya, A258
Dalquist & Mamay, A5
Damesellidae, A127, A130
Damesites, A471
Dansgaard, A524
Dansgaard & Tauber, A517
Daonella, A362, A372
Darvasites, A299
Darvazian Stage, A299-A300
David & Browne, A305
DAVIDSONIACEA, A327
Davidsoniatrypa, A227
Davidsoninidae, A273
Davitashvili, A22
Davoudzadeii & Seyed-Emani, A358
Dean, A141
Dear, A304
death, A5, by desiccation, A14, flooding, A14, overgrowth by other organisms, A6, submergence in substrates, A10-A14; struggle, A14-A15, of insects, A15
DECAPODA, A410
decay, soft part, and buoyancy, A17-A19, as it affects hard parts, A20-A22
Dechaseauxia, A472
Dechenella, A242
Dechenellurus, A226

A550   Introduction

decomposition, A5, A15-A19, selective, A20-A22
DEECKE, A3, A7, A17
degassing, effects of on organic material, A16, A17-A19; canals, A16, in sapropelic environments, A16
Delepinoceras, A279
Delepinoceratidae, A276
Delthyridae, A246
Deltopectinidae, A329
DENIZOT, A508
DENMEAD, A363
Derbyia, A298, A300, A380
Deshayesitidae, A457, A469
desiccation, cracks, A14, death by, A14
Desmoceras, A475
DESMOCERATACEAE, A441
Desmoceratidae, A456, A457, A459, A461, A465, A471, A475
Desmoceratinae, A460
Desmoinesian Series, A258, foraminifers of, A266
Desmoscaphites, A468
DESNOYERS, A506, A507
"Desquamatia" cosmeta Zone, A244
Desquamatia independensis Zone, A240
detrital belt, inner, A122-A124, A129, in Middle & Late Cambrian, A130; outer, A122-A124, A129
DEUSER, A536
DEUSER & DEGENS, A518, A533
Devonian continental deposits, A247
Devonian, eastern hemisphere, A183-A210, global reconstruction, A208-A210, history of establishment, A184; European, stage nomenclature, A187-A189, zonal correlation, A189-A192; see also Lower, Middle, Upper Devonian
Devonian faunas, western hemisphere, see Lower, Middle, Upper Devonian
Devonian outcrops, eastern hemisphere, A192-A193, Africa, A193; Australasia, A193, China and Asia, A193, western & southern Europe, A192; Soviet Union, A192-A193, western hemisphere, A219-A223, in Antarctica, A222-A223, Greenland, A222, North America, A219-A222, South America and Falkland Islands, A222
Devonian-Carboniferous boundary, A189, A206, A219, A254-A255
Devonoproductus, A246, D. walcotti Zone, A245
Diabloceras, A279
Dialytophyllum, A201
Dibunophyllum, A270-A272, A321
Diceratocephalina, A140
Dicerocardiidae, A441
DICKENS & MCTAVISH, A363

Dickersonia, A402
DICKINS, A304, A305
Dickinsonia, A83, **A103**
Dickinsoniidae, **A103**
Dicoelites, A410
Dicoelosia, A152, A154
Dicoclosiidae, A161
Dictyoclostidae, A275, A327, A329
Dictyoclostus, A298-A301, A337
Dictyoconus, A471
Dictyomitra veneta Zone, A459
Dictyonema flabelliforme, A138
Dictyonema shale, A138
Dictyoptychus, A473
Didymograptus, A155, A156, D. murchisoni Zone, A143
Didymotis, A460
Dielasma, A297, A299
DIENER, A353, A360, A361, A363, A370
Dienerian, A353
Dieneroceras, A363
Dieranurus, A226
DIETZ & HOLDEN, A298, A339, A475
Digonophyllum, A201
Dikelocephalus, A124
DILLEY, A462, A465, A468, A469, A471, A472, A473
Dimetobelidae, A475
Dimitobelinae, A441
Dinantian, A255
DINARITACEAE, A371
Dindymene, A151
Dindymeninae, A151, A160
Dinesus, A127
DINOFLAGELLATA, A393
Dinorthis, A154
Diodoceras, A276
Dionidae, A151, A160
Diorugoceras, A276
Diphyphyllum, A270, A272
Diploporaria, A325, A326
Discinidae, A329
Discinisca, A395
Discitoceras, A276
Discoaster brouweri, A513
Discocyclinidae, A497
Discohelix, A397
Discoididae, A441
Discometra, A499
DIXON, A40
Djurdjuriceras, A406
DOBROLYUBOVA, A270
Dohmophyllum, A201, A202
Dolerorthis, A151, A161
Dolgeuloma, A140, A152
Doliognathus, A280, A281
Dollymae, A281
Domatoceras, A276, A371
Donax, A35, D. vittatus, A30
DONN & SHAW, A525
Donophyllum, A272
DONOVAN, A395, A399
Dorlodotia, A270
Dorp facies, A187
Dorsoplanites, A406
Doryceras, A335
Dorypygidae, A123

Dosinobia, A476
doubtful taxa, Precambrian, A110-A111
DOUGLAS, A368, A438
DOUGLAS & SAVIN, A425, A438
DOUMANI, A223, A231
Douvilleiceratidae, A457, A469
Douvillina, A246
Draboviinae, A149
DRAHOVZAL, A276
Drakeoceras, A469, A472
Dreissena, A26, A27
Drepanochilus, A469
Dresbachian-Franconian boundary, A123-A124
DROMIACEAE, A410
DROOGER, A503
DROT, A198, A203
DRUCE, A208, A280, A281
DRUCE & JONES, A138
DRUCKMAN, A357
druse, A17
DUBAR, A394, A395, A396
DUBATOLOV & SPASSKIY, A202, A206
Dudaiceras, A473
Duerleyoceras, A276
Dufreynoyia, A469, A473
Dumortieria, A400
DUNBAR, A292, A297, A298, A310, A314
DUNBAR & SKINNER, A310, A313
Dunbarinella, A266
Dunbarites, A279
Dunbarula, A312
DUNCAN, A199, A246
DUNN, A81, A91
DUNNING, A86
Dunveganoceras, A459
DUPLESSY, A524
Duplophyllum, A318
DURDEN, A281, A283
Durga, A396
Durhamina, A272, A273, A319, A321
Durhaminidae, A317
DURKOOP, A202
Duvalia, A410, A455, A471, A472
Duvaliidae, A454, A457, A472
Dyaster, A411
Dyscritella, A322-A325, A326
Dyscritellina, A326
Dzhulfian, ammonoids of, A336-A337
Dzhulfites, A358, A381
Dzhulfoceras, A336
Dzhungaro-Balkhash area, Devonian faunas of, A196, A202, A207; links with Appalachian province, A210
Dzieduszichia, A206

Earlandia, A261, A265
EARP, A187
Eastern Americas realm, Devonian, A233, A242
EASTON, A271
Eastonoides, A319
Echigophyllum, A271

Echinauris, A382
Echinobrissus, A411
Echinocardium, A8
Echinoconchidae, A275, A327, A329
Echinoconchus, A300
Echinocorys, A458
echinoderm ossicles, A227, A230, A238, A240
ECHINODERMATA, Cambrian, A120
echinoderms, Silurian, A180, Tertiary, A499
Echinolichas, A226
Echinosphaerites, A63
Echinosteginae, A275
ECHINOZOA, A410
Echioceras, A398, A399
ecology, of Ordovician fauna, A157-A159, regional zonation, A158, salinity, A158, substrates, A158, temperature, A158, water depth, A158
ecostratigraphic correlation, A172
Ectenaspis, A150
Ectenonotus, A142
Ectillaeninae, A151, A160
Ectorensselandia, A240
Edaphoceras, A276
Edaphophyllum, A223
EDGELL, A363
Ediacaran, A82, faunas, A82, A87-A88
Ediacaria, A94, **A96**
Edmondia, A305
Edmondiidae, A329
EDWARDS, A184
EFREMOV, A2, **A3**
Egoyan, A440
EHRENBERG, A22, A310
Eifelian, A188, A201, A231, A234, A238, A240, A241, A242
Eiffelithus, A459, E. eximius Zone, A461
Einkippung, A32, A36, A41
Einkippungsregel, A33
Einregelung, A32
Einsteuerung, A32, A44
EKMAN, A463
Ekvasophyllum, A270, A271
Elateridae, A15
Electroma, A476
Elephas, A508
Eleutherokomma zones, A244, A245, A246
Eligmus, A397
ELLES & WOOD graptolite zones, A172
Ellipsactinia, A394
Ellipsocephalidae, A125
Ellisonia, A338, A339
Elmoan assemblage, A282, A283
Elphidium oregonense Zone, A516
Elviniidae, A125
Elytha, A234
Elythidae, A329, A330
Elythyna beds, A226
Emanuella, A238, A240, A242, E. vernilis Zone, A240
EMERY, A33

EMILIANI, A510, A517
EMILIANI, MAYEDA, & SELLI, A514, A517
Emmanuella, A201
Emmonsia, A269
Empodesma, A272
Emsian, A188, A194, A195, A223, A226, A230, A231, A234, A238
ENAY, A394, A404, A406
Encrinurella, A149
Encrinuroides, A149
Encrinurus, A223
endemic trilobites, in North America, A123
ENDOCERATOIDEA, A161
Endocostea, A460, A461
Endololobus, A276
Endomoceras, A457
Endophyllum, A202
Endostaffella, A263
Endothiodon Zone, A306
Endothyra, A263, A265, A266
Endothyranopsis, A263, A265
Endothyridae, A261, A267
Endymionia, A142
Engonoceratidae, A457
Ensis, A30, A35
Enteletella, A300
Enteletes, A299-A301, A380
Enteletidae, A330
Enterolasma, A223
Entolium, A476
Entolliidae, A441
Entomozoidae, A187
entrapment, in crude oil, asphalt, tar, A10, mud, silt, A10, quicksand, A11, resin, A11-A14
Enygmophyllum, A269
Eoaraxoceras, A336
Eoasianites, A260
Eobeloceras, A207
Eocarcinus, A410
Eocene, A499-A504
Eocephalites, A402
Eocoelia, A172
Eoconchidium, A154
Eocrioceratites, A455, A472
EODEROCERATACEAE, A399
Eodevonaria, A231
Eodiscidae, A122-A123, A125, A127
EODISCINA, A126
Eoendothyranopsis, A263, A267
Eoforschia, A261, A263
Eofusulina, A265, A266
Eoglossinotoechia, A196
Eognathodus, A226, A227, A230, A242
Eolasiodiscus, A266
Eolithostrotionella, A270, A272, A273
Eolytoceras, A398
Eomiodon, A397
Eoparafusulina, A308, A313, A314
Eoparaphorhynchus Zone, A205, A245
Eoparastaffella, A263
Eopecten albertii, A24
Eoplacognathus, A157
Eoplectodonta, A143

Eoporpita, A83, **A92**
Eoschubertella, A265, A266
Eosigmoilina, A265
Eospirifer, A197
Eospirigerina, A150-A154
Eostaffella, A263, A265-A267
Eostaffellina, A265
Eostropheodonta, A153
Eotextularia, A263
Eothinites, A334
Eotomariidae, A378
Eowaeringella, A266, A268
Epadrianites, A336
Ephippioceras, A276
Ephippioceratidae, A276
Epiaclinotrypa, A323
Epiaster, A458
Epicephalites, A401
Epideroceras, A399
Epidomatoceras, A276
Epimastopora, A299, A300
Epipeltoceras, A403, A407
Episageceras, A307, A336
Epistroboceras, A276
Epithalassoceras, A335
Epitornoceras, A205, A243
Equirostra, A151
Equus, A500, A508
Erammoceras, A400
ERBEN, A184, A188, A195, A199
ERBEN & ZAGORA, A188
Erbenoceras, A196
ERICKSON, A40
ERICSON, A520
ERICSON, EWING, & WOLLIN, A520
ERICSON & WOLLIN, A520, A524, A531
Eridopora, A322, A324, A325
Erixanium, A127
Ermoceras, A400
Erniaster, A101
Ernietta, A100, **A101**
Erniettidae, A83, **A99**
Erniettinae, **A101**
ERNIETTOMORPHA, A99, A102
Erniobaris, A101
Erniobeta, **A102**
Erniobetinae, A102
Erniocarpus, **A102**
Erniocentris, **A102**
Erniocoris, **A102**
Erniodiscinae, A101
Erniodiscus, A101
Erniofossa, **A101**
Erniograndis, **A102**
Ernionorma, **A101**
ERNIONORMIDAE, A99
Ernionorminae, A101
Erniopelta, **A102**
Erniotaxis, **A102**
ERNST & SEIBERTZ, A458
ERRERA, MAMET, & SARTENAER, A189
Erycites, A400
Erycitoides, A401
Eryma, A410
Estheria, A222, A247
Estheripecten, A380
Estlandia, A148
Estoniops, A148

Etea, A469
Etherella, A325
Etymothyris Zone, A223
Eucalyceras, A459
eucaryotes, A80
Eugeniacrinites, A411
Eugenophyllum, A299
Eulinderina, A497
Euloxoceras, A276
Eumedlicottia, A334, A335, A336
Eumorphoceras, A279
Eunaticina, A469
EUOMPHALACEA, A397
Euomphalidae, A378
Eupleuroceras, A279
Euptera, A474
Euramerican paleofloral region, A284
Eurasian zoogeographic region, in the Carboniferous, A269-A273
Eurekaspirifer pinyonensis Zone, A226, A227, A230
Eurekiidae, A124
European strata, Silurian, A174, geosynclinal, A174, platform, A174
Eurydesma, A259, A260, A305, A306, A307
Eurydesmatidae, A329
Euryphyllum, A318, A320
eurypterids, Silurian, A179
Eurypterus, A247
Euryspirifer, A202
eustatic changes, in Cretaceous seas, A430-A432, causes of, A430
Eusthenoceras, A275
Eutrephoceras, A408
Evactinopora, A325, A326
Evactinostella, A325, A326
evaporites, Silurian, A168
evolution, Precambrian metaxoan, A88-A90, Silurian, rate of, A170, A172
EWING & DONN, A525
Exogyra, A7, E. columba, A33
Exogyrinae, A458
Exogyrini, A441
EYNOR, A259, A260, A267, A284

Faberophyllum, A270, A271
faceting, A26-A28, anchor-facets, A26-A27, glide-facets, A27-A28, roll-facets, A27
facies, European Devonian, A184-A187
facies regions, Cambrian, in the Soviet Union, A126, in south eastern Asia, A126-A127
Fagesia, A471
FAHRAEUS, A227
FAIRBRIDGE, A524, A531
Fallotaspis, A122, A125
Fammennian, A189, A203, in Europe, A205-A206, faunas, A243, A245, A246-A247
FANCK, A58
FARSAN, A360
Fascicostella, A197
FATMI, A360, A403

faunal province maps, Cretaceous, A464, Jurassic, A392, A409, Ordovician, A139, A142, A145, A146, A147, A148, A150, A151, A155, Silurian, A169
faunal regions, Cambrian, in North America, inner, A122-A123, outer, A122-A123
faunal zone, range charts, Arenigian, A141, Cambrian, A132-A134, Carboniferous, A257, A262, A264, A269, A274, A277, A278, A282, A283, Cretaceous, A442-A444, A446-A448, A450-A452, A454-A456, Devonian, A190, A191, A239, A241, Jurassic, A404-A405, Permian, A308-A309, A317, A321, A328, A332-A333, Quaternary, A519, Silurian, A171, Tertiary, A501, A502, Tremadocian, A138, Triassic, A354-A355, A383
faunal zone succession, Cambrian, in Australia, A134, in China, A133-A134, in central and southern Europe, A132-A133, in northern Europe, A132, in North America, A131-A132
Fenestella, A322-A326
Fenestellidae, A322
FENNINGER & HÖTZL, A394
Ferganoceratidae, A276
FILLON, A530
Filosina, A472
Fimbria, A475
Fimbrispirifer, A202, A235
FISCHER, A., A87, A393
FISCHER, J., A359
FISCHER, R., A398
FISCHER & ARTHUR, A422, A424, A425, A426, A427, A428, A430, A431, A432, A437, A477, A478
Fischerinidae, A267
Fissiluna, A476
Fistulamina, A323-A326
Fistulipora, A322-A326
Fistulotrypa, A325, A326
Fitzroyella, A208
FLECK, A516
FLEISCHER, A533
FLEMING, C., A364
FLEMING, P., A363
Flemingites, A363
Fletcherina, A223
Flinz facies, A187
floatability constants, A25
flooding, death by, A14
flora, Tertiary, A500, Carboniferous, A285
FLÜGEL, A271, A298, A319, A320, A394
FLÜGEL & FRANZ, A393
FLÜGEL & SCHÖNLAUB, A302
FÖRSTER, A410
FOLK, A61, A69
FONTAINE, A271
Fontannelliceras, A399
Foordites, A201, A234, A242
FORAMINIFERA, A441, A496, A497-A498

foraminifers, Carboniferous, A261-A268, of North American craton, A267, of Tethyan realm, A267; Silurian, A177; Tertiary, A497-A498, planktonic, zones of, A501-A503
FORBES, A507
FORD, A99
Fordilla, A120
Forschia, A263, A265
Forschiella, A263, A265
Forschiidae, A267
FORTEY, A158
FORTEY & BRUTON, A143
fossil deformation, A53-A58, by fracture, A56, plastic, A55-A56, by tectonic stress, A58, by volume decrease of sediment, A53-A55
fossil diagenesis, A3, A48-A71
fossilization, A2-A3
Fossil-Lagerstätten, A22, A23
fossils, Precambrian, A81, geographic distribution of, A83-A84, preparation and investigation of, A81
Fossulites, A473
Fouchouia, A127
fracture systems, A57
fracturing, A28-A30
FRANCIS & WOODLAND, A260
Franconites, A407
Frasnian, A189, A203, European, A204-A205; faunas, A235, A243, A244, A245, A246, extinctions, A246, A247
Frasnian-Famennian boundary, A244
FREBOLD, A298, A381, A395, A398, A399, A400, A401, A402
Frechiella, A400
FREDERIKS, A292, A330
Freiastarte, A475
FRERICHS & ADAMS, A425, A438
FRIEND, A222
FRIEND & HOUSE, A219
FRUSH & EICHER, A428, A432
FUCHS, A195, A362
Fuciniceras, A399
FÜCHTBAUER & GOLDSCHMIDT, A63, A69
Fulpia, A469
Furcaster, A42, A44-A45
Furcirhynchia, A395
FURNISH, A298, A330, A331, A334, A335, A336, A337
FURNISH & GLENISTER, A335, A336
FURON & ROSSET, A360
Fusiella, A266, A268, A310, A311
Fusulina, A266, A267, A268, A312
FUSULINACEA, A308-A316
fusulinaceans, Permian, geographical distribution of, A314-A316, Cordillera, A314, North America, A313-A314, Russian platform and Urals, A309-A311, Tethyan region, A311-A313, zonation, A307-A316
Fusulinella, A266, A267, A268,

A312, Zone, A266, A267
Fusulinellinae, A308, A309
Fusulinidae, A268, A308, A316
Fusus, A469
FUTTERER, A32, A35, A40, A44

Gagaticeras, A398
Galeola, A458
Galeritidae, A441
Galeropygus, A411
Gangamophyllum, A270-A272
Gangamopteris, A305, A306
GANSSER, A361
Garantiana, A403
Garniericeras, A408
GARRISON & FISCHER, A391
Gartnerago obliquum Zone, A460
gas cavities in sediment, A16
Gasterocoma?, A238, A240
Gastrioceras, A279, Zone, A280
Gastrodetoechia, A206, A207, Zone, A245
Gastrodorus, A410
Gastroplites, A457
Gastroplitidae, A468
Gastroplitinae, A457
gastropods, Jurassic, A397, Silurian, A179, Triassic, A377-A378
GATES, A537
Gattendorfia, A189, G. subinvoluta Zone, A254-A255
GAUDRY, A506
Gaudryceras, A471, A475
Gedinnian faunas, A223, A226, A227, A230, A231, A234
Gedinnian Stage, A188, A195
GEIKIE, A184
GEITZENAUER, A531
GEKKER, A40
GEMMELLARO, A335
Genuclymenia, A208
Geocoma, A15
geodes, A65-A67
geographic distribution of fossils, maps of, Arenigian trilobites, A142, Ashgillian trilobites, A150, 151, Bathyuridae, A146, Bimuriidae, A147, Cambrian, A121, Caradocian trilobites, A148, Christianiidae, A146, Devonian, A185, A220-A221, A224-A225, A228-A229, A232-A233, A236-A237, Draboviinae, A147, graptoloids, A155, Hirnantian trilobites, A153, Llanvirnian, A145, Precambrian, A84, Tertiary, A490, A492, A493, A494, A495, Tremadocian trilobites, A139
GEORGE, A255
GEORGE & WAGNER, A189
Gephyrocapsa, A510, A511, G. caribbeanica, A513, A514, A531, G. oceanica, A513, A514, G. protohuxleyi, A513
Geragnostus, A124
Geranocephalus, A240
GERHARZ, A48
Germanic facies, A352

GERMS, A85, A96
Gervilleia, A297
Gervillia, A362, G. socialis, A24
GEYER, A394, A396, A398, A400, A406, A407
Gigantoproductidae, A275
Gigantopteris, A301
GIGNOUX, A508, A510, A513
Gilbertharrisella, A474
GILL & COBBAN, A438, A445
GIRTY, A323, A330
Girtyopora, A323, A324, A325
Girtyoporina, A323, A324
Girtypecten, A305
Girvanella, A300
Givetian, A189, A201, faunas, A234, A235, A238, A240, A241, A242
glaciation, continental, Antarctic, A516, Arctic, A516, Kansan, A516, Mindel-Elsterian, A516, Nebraskan, A516, Northern Hemisphere, A516; Late Ordovician, A159, A161
Gladigondolella, A411
GLAESSNER, A79-A118, A410, A492
GLAESSNER & WALTER, A96
Glaessnerina, A83, **A99**
Glassoceras, A335
Glaucolithites, A406, A407
Gleboceras, A279
GLENISTER, A208
GLENISTER & FURNISH, A305, A334, A335
GLENISTER & KLAPPER, A208
glide marks, during rigor mortis, A15
Globigerina, A393, A496, A503, G. bulloides, A527
GLOBIGERINACEA, A471
Globigerinelloides, A457, A458
Globigerinoides, A497, G. obliquus, A513, G. ruber, A532-A533, G. sacculifer, A533
Globivalvulina, A265
Globoendothyra, A263, A265
Globorotalia, A498, G. aragonensis Zone, A503, G. inflata, A523, A530, G. menardii, A520, A524, G. tosaensis, A513, G. truncatulinoides, A513, A514, Zone, A503
Globotruncana, A460, A461
Globotruncanella, A461
Globotruncanidae, A419, A441
Glochiceras, A402, A406, A407
Glomospira, A265
Glomospiranella, A261
Glossinotoechia, A195
Glossopleura, A123, Zone, A131, A132
Glossopteris, A305, A306, A307, -Gangamopteris flora, A307
GLUSKO & FEDOROV, A293
Glyptagnostus, A123, A124, A127
Glyptambonites, A145
Glyptograptus teretiusculus Zone, A156
Glyptophiceras, A301, A362

Gnathodus, A280, A281, G. girtyi simplex Zone, A280
GOBBETT, A202, A298, A315
GOETHE, A24
GOLBERT, A440
GOLDFUSS, A7
Gonatocheilus, A410
Gondolella, A280, A287
Gondwanaland, in the Ordovician, A137, paleofloral region of, A285, -Laurasia separation, A125
goniatites, Carboniferous distribution, A276-A278; Devonian, A199, A201, A203, A205, A206, A226, A234, A240, A243, extinctions, A208
Goniatitidae, A276
Gonilia, A472
Gonioceras, A146
Goniocladia, A322-A326
Goniodromites, A410
Goniogliphioceras, A279
Goniograptus, A155
Gonioloboceras, A279
Goniophoria, A327
Gonioteuthis, A458, A460
Gordius tenuifibrosus, A71
GORDON, A258, A268
Gorjanovicia, A473
GORJUNOVA, A322, A324
Gorskyites, A269
GRABAU, A202, A324
Gracianella, A223
GRÄF, A58
Granophyllum, A300
GRANT, A301, A339, A343, A380
GRANT & COOPER, A379, A382
Graphidula, A469
Graphoceras, A400
graptolite deformation, A59
graptolites, Devonian, A190-A191, A195, A196, A197, A226, A227, A230
graptolites, planktonic, Ordovician, A154-A156, Atlantic province, A155-A156, Balto-Scandia, A156, biogeography of, A154, A157, A158-A160, Pacific province, A155-A156, in South America, A155
graptolites, Silurian, A170, A172, A180
Gravesia, A407
GRAY, A170
GRAY, LAUFELD, & BOUCOT, A181
Graysonites, A459
Greenops, A226
GRÉGOIRE, A68
GRÉGOIRE & TEICHERT, A68
Gregoryceras, A403, A407
Gresslya, A57
Griesbachian, A353
GRIPP & TUFAR, A69
GROOS-UFFENORDE & UFFENORDE, A191
DE GROOT, A272
Grossouvrites, A475
GRUNT & DMITRIEV, A343
GRYC, A222

Gryphaea, A367, A396, G. dilatata, A33
Grypoceras, A371
Grypoceratidae, A276, A371
Grypophyllum, A202, G. mackenziense Zone, A240
Gshelia, A272, A273
Gshelian Stage, A255, A259
Guadalupian Series, A294, A303, A314, A319, ammonoids of, A335, bryozoans of, A323, A324, A325, corals of, A321
Guerichina strangulata Zone, A227
GUNTER, A9
Gurjevskiella, A202
GYGI, A403
Gymnocodium, A300
Gymnograptus, A156
Gypidula, A197, A227, G. pelagica beds, A226, A227
Gyroceratites, A195, A242, G. gracilis boundary, A199
Gyronites, A358, A363
Gzheloceras, A276

Hadrocheilus, A410
Hadrophyllum, A269
Hadrorhynchia, A238, A240
HÄNTZSCHEL, A33, A44, A81, A91, A96
HÄNTZSCHEL, EL-BAZ, & AMSTUTZ, A20
HAKES, A41
HALL, A219, A463
HALL & KENNEDY, A67, A68
HALLAM, A24, A26, A184, A195, A196, A399, A419, A426, A430, A463
HALLAM & O'HARA, A68
HALLE, A285
HALLER, A222
Hallidaya, A83, **A92**
Halobia, A362, A366, A372
HALSTEAD & TURNER, A184
HAMADA, A196, A202, A207
Hammatoceratidea, A400
Hammatocnemidae, A152, A160
VAN DER HAMMEN, WIJMSTRA, & ZAGWIJN, A523
Hamulinites, A471
Hanchunglithus, A141
HANCOCK, A431, A439, A443
Haploceras, A406
Haploceratidae, A403
Haplophragmella, A263, A265
Haplophylloceras, A403, A405
Haplopleuroceras, A400
Haplostiche, A40
Hapsiphyllum, A318, A319
HAQ, BERGGREN, & VAN COUVERING, A513, A514
hard parts, before final burial, A22-A48, mechanical destruction of, A25-A28
Hardaghia, A472
HARDY, A88
HARKER & THORSTEINSSON, A298
HARLAND, A159, A199, A419
HARLAND & GAYER, A137
HARLAND & HEROD, A82

Harpagodes, A397
Harpes, A230, A242
Harpoceras, A399, A400
HARRINGTON, A222, A231, A246, A369
HARRINGTON & MOORE, A103
HART, A343
Hastitidae, A408, A410
HAUG, A391
Hauterivian, A423, A442-A444, A453, A456-A457
HAVLIČEK, A141, A149, A160, A195
HAY, A429
Hayasakaia?, A271
Hayasakapora, A323, A326
Hayasakia, A320
HAYDEN, A301, A359, A360, A379
Haydenella, A300
HAYS, A517, A523, A527, A530
HAYS & BERGGREN, A514
HAYS & PITTMAN, A430
HEALEY, A362
Hebediscus, A126
Hebertoechia, A196
HECHT, A55
Hectoroceras, A453, H. kochi Zone, A440
Hedbergella, A457, A459, A469
Hedinaspis, A124
LE HEGART & REMANE, A393
HELICACEA, A498
Heliophyllum, A202, A203
Hellenocyclina, A497
HEMICHORDATA, Cambrian, A120
Hemifusulina, A266
Hemigordiopsis, A316
Hemihaploceras, A406
Hemihoplitidae, A456
Hemiporitidae, A441
HEMIPTERA, A305
Hemiptychina, A300, A380
Hemispiticeras, A402
Hemitrypa, A326
HENSON, A393
Heptabronteus-Pliomerina Province, A149
Hercynian/Bohemian facies, A184, A199
Heritschiella, A319, A320, A321
Heritschioides, A319
HERMAN, A525, A533
Herzogina, A476
Heteraster, A458
Heterelasminidae, A329, A330
Heterocaninia, A271
Heteroceratidae, A457
Heterohelix, A469
Heterolasmina, A300
Heterophrentis, A223
Heterophyllia, A271
Heterorthella, A174
Heterostegina, A497, A504
HEXACTINELLIDA, A393
Hexagonaria, A245
Hexagonella, A322, A324-A326
Hibolithes, A410, A455, A457, A471, A472, -Curtohibolites

plexus, A454
Highatella, A140, Province, A139
Hildaites, A400
Hildoceras, A400
HILDOCERATACEAE, A399
Hildoglochiceras, A403, A406
HILL, A129, A197, A203, A259, A268-A270, A272, A273, A317
HILLEBRANDT, A402
Himalayites, A403, A406
Hindeodella, A338, A411
Hinganella, A323, A325
Hipparion, A500
Hipparionyx, A197
HIPPURITACEA, A458, A471
Hippuritidae, A441
Hirnantia, A153, A154, A159, assemblage, A153
Hirnantian shelly faunas, A153-A154, Hiberno-Salairian, A153-A154, Mediterranean, A153, A154, North American Midcontinent, A153-A154, North Estonian belt, A154, Tungusian, A154
HODSON & RAMSBOTTOM, A276
HÖLDER, A390-A417
HÖLDER & ZIEGLER, A407
HOERNES, A499, A508
HOFFMANN, A399
Holaster, A458, A459, A460
Holasteridae, A411
HOLECTYPIDA, A411
Holectypidae, A441
Holectypus, A411, A458
HOLLARD, A189, A198, A203
HOLLARD & LEGRAND, A197, A203
HOLLMANN, A17, A27, A28, A30, A63, A64, A67
Holmia, A122, A125
Holocene, A507
Holorhynchus, A154
HOLTEDAHL, A222
Holotrachelus, A150, A152, A153
Holzapfeloceras, A201
Homalonotus, A226
Homalophyllites, A269
Homoceras, A279, Zone, A265, A280
Homoctenidae, A203
HOPKINS, A527
HOPLITACEAE, A441
Hoplitidae, A457, A459, A467
Hornsundia, A318
Horridonia, A297
HOUSE, A183-A217, A219, A222, A227, A234, A235, A242-A247
HOUSE & KIRCHGASSER, A243
HOUSE & PEDDER, A234, A240, A242, A243
Howaiskya, A407
HOWARTH, A399
Howchinia, A265
Howellella, A195, A196, A197
Howittia, A197
HU, A137
Huangia, A272
HUCKRIEDE, A397
HUDSON, A255, A397

von Huene, A516
Hughes, A208, A463
Hume, A245
Humeoceras, A161
Hungaia magnifica fauna, A124
Hunkins, A525
Hupé, A132
Hustedia, A301, A337, A380
Hwang-ho facies, A127, in Middle and Late Cambrian, A130
Hyalinea baltica, A508, A510, A513, A514
Hyattoceras, A336
Hybonoticeras, A406, A407, Zone, A406, A407
Hydrobiidae, A397
hydrogen sulfide, A9, A10, A61
HYDROIDA, **A91**
HYDROZOA, A88-A89, **A91**
hydrozoans, Jurassic, A395
HYOLITHA, Cambrian, A120
Hypacanthoplites, A469, A472
Hyperlioceras, A400
Hyphantoceras, A471
Hyphasmopora, **A322**
Hypoclypeus, A411
Hypodicranotus, A152
Hypophylloceras, A475
Hypothyridina, A205, A206, A245, A246
Hysterolites, A195, A196
Hystricurus, A139

Iapetus Ocean, A137, A140
Iberg facies, A187
Iberian assemblage, A282, A283
Icanotiidae, A441
ichnia, A2
ichnocoenoses, A22
Icriodus, A227, A230, A240, A242, A246, I. latericrescens, A234, A235, I. pesavis fauna, A226, I. woschmidti, A196, Zone, A223
Idahoiidae, A124
idiobiology, A5
Idiognathodus, A280, A281, A337, I. ellisoni Zone, A337
Idiognathoides, A280, A281
Idoceras, A402, A407
Illaenus, A142
Illies, A55, A67
Ilmarinia, A152, A153
Imbrie, A520
Imbrie & Kipp, A520, A531
Imbrie, Van Donk, & Kipp, A520
Imitoceras, A206
Imlay, A396, A398-A403, A453
Imlayoceras, A402
Immanitas, A473
immuration, A7
Incacardium, A473
index fossils, Cambrian, nontrilobite, A130
Indopolia, A300
Indospirifer, A201
Indotrigonia, A396
Induan, A353
Ingelarella, A305

Ingle, A527
Inoceramidae, A419, A428, A434, A441, A453, A457, A458, A460, A465
Inoceramus, A57, A391, A426, A453, A459, A460, A461, A468, I. dubius, A56
INSECTA, A281-A283
insects, Carboniferous, A281-A283
Integricardium, A472
invertebrate faunas, Albian, A437, Aptian, A436, Barremian, A436, Berriasian, A436, Campanian, A438, Cenomanian, A437, Coniacian, A438, Cretaceous, A436-A438, Hauterivian, A436, Maastrichtian, A438, Santonian, A438, Turonian, A437, Valangian, A436
Iotrigonia, A475
Iowanella, A266, A268
Iowaphyllum, A202
Ipciphyllum, A319, A320
Iranites, A358
Iranophyllum, A319, A320
Irenicoceras, A468
Irvingella, A124, A125, A127
Isalaux, A148
Ishii, A316
Ishii, Fischer, & Bando, A343, A359
Ishii, Okimura, & Nakazawa, A343
Isogramma, A299
Isogrammidae, A330
Isograptus, A155
Isophragma, A146
ISOPODA, A410
Isoprusia, A226
Isorthis, A197
Isotancredia, A476
Isotelus, A150
Itieria, A397
Ivanova, A273
Ivanovskiy, A270

Jaanusson, A136-A166
Jackson & Lenz, A227
Jacobites, A475
Jaeger, A191, A195, A196
Jaeger & Martinsson, A89
Jaeger, Stein, & Wolfart, A196
Jago & Daily, A134
Janicke, A46, A47
Janischewskina, A263, A265
Janius, A201, A227
Jarke, A30, A61
Jarvik, A222
Jaworski, A369
Jefferies & Minton, A396
Jeletzky, A22, A29, A397, A453, A455, A462, A463, A465
Jen, A137
Jenkins, A99, A208
Jigulites, A267, A268, A310
Johnsen, A524
Johnson, J., A203, A219, A226, A227, A230, A231, A235, A238, A245, A246
Johnson, R., A32

Johnson & Boucot, A195, A196, A219, A226, A235, A242, A246
Johnson, Boucot, & Murphy, A226
Johnson & Dasch, A219
Johnson & Fox, A103
Johnson & Lane, A238
Johnson & Murphy, A226
Jones, A134, A177, A453
Jongmans, A285
Jongmans & Gothan, A189
Jordan, A397
Joufia, A473
Juddiceras, A455, A471
Jurassic-Cretaceous boundary, A410, A440-A441
Juresanites, A305, A331

Kaever, A359
Kahler, A343
Kahlerina, A299
Kaljo, A152, A153
Kaljo & Klaamann, A150, A153, A161
Kallirhynchia, A395
Kanabiceras, A459
Kansanella, A266, A268
Kap Stosch Formation, A297, A319
Kapellos & Schaub, A503
Karachalyrian Stage, A299
Karadjalia, A206
Kargalites, A331
Karpinskia, A196
Karpinsky, A292
Karsteniceras, A471
Kasmovian Stage, A259
Katroliceras, A406
Kauffman, A418-A487
Kauffman, Cobban, & Eicher, A421, A458
Kauffman, Hattin, & Powell, A427, A445
Kauffman & Scholle, A431, A432, A481
Kauffman & Sohl, A420
Kay, A39
Kazachiphyllum, A272
Kazanian Stage, A293, A295, A305, A316, ammonoids of, A335-A336, brachiopods of, A330, bryozoans of, A322, A324-A326, corals of, A319
Keany & Kennett, A527, A529, A530
Keast & Glass, A343
Keller, A82, A83
Keller & Predtechensky, A137
Kellogg, A520
Kennedy, A435
Kennedy & Cobban, A462, A463, A471
Kennedy & Hall, A67, A68
Kennedy & Juignet, A439
Kennedy & Klinger, A66
Kennedy & Taylor, A67, A68
Kennett, A30, A529, A531
Kennett & Brunner, A530
Kennett & Huddlestun, A524
Kennett & Schackleton, A524

Kennett, Watkins, & Vella, A530
Kent, Opdyke & Ewing, A527
Kepplerites, A401, A402
Keriophyllum, A201
Kerr, McGregor, & McLaren, A245
Kessel, A28, A61
Keuper, A352
Keyserlingophyllum, A269, A270
Khalymbadzha & Chernysheva, A206
Khalymbadzha & Tikhvinskiy, A273
Khramov, A208
Khúc, A362
Kiangsiceras, A301
Kiangsiella, A380
Kielan, A150, A151, A152
Kimberella, A94
Kinderhookian Series, A258, A280
Kindle, A40
King, P., A303, A313, A334
King, R., A330
Kingoceras, A336
Kingopora, A322
Kingstoniidae, A123
Kinkaidia, A271
Kinnella, A153
Kionelasma, A223
Kionophyllum, A272, A273
Kiparisova, Okuneva, & Oleynikov, A365
Kiparisova & Popov, A353
Kiparisova, Radchenko, & Gorskiy, A365
Kipia, A473
Kirchgasser, A243
Kittl, A357
Kladognathus, A280
Klähn, A24-A28
Klapper, A226, A227, A234, A235, A238, A244
Klapper, Berry, & Boucot, A181
Klapper, Philip, & Jackson, A202
Klapper & Ziegler, A190, A235
Kleopatrina, A318, A319
Klovan & Embry, A245
Knightoceras, A276
Knocke, A3
Knoll & Barghoorn, A80
Kobayashi, A120, A127, A142, A395
Kobayashi & Hamada, A207
Kochiproductus, A298
Koenen, A499
Koenenites, A247
von Königswald, A41, A44
Kolb, A19, A46
Koninckina, A395
Koninckioceratidae, A276
Koninckites, A363
Koninckocarinia, A272
Koninckophyllum, A269-A272
Kootenia, A123
Korchinskaya, A365, A366
Korejwo & Teller, A191
Korschelt, A6
Kosmoceras, A401

Kosmoceratidae, A401, A406
Kosovopeltis, A226
Kossmatia, A403
Kossmaticeras, A471, A475
Kossmaticeratidae, A460, A461, A465, A475
Kovacs, A399
Kozłowski, A88, A192
Kozur, A337
von Krafft, A361
von Krafft & Diener, A361
Krantz, A17
Krebs, A184
Krebs & Rabien, A205
Krebs & Wachendorf, A210
Krebs & Ziegler, A190
Krejci-Graf, A36
Krinsley, A39
Krotovia, A382
Krystyn, A397
Ku & Broecker, A525
Kubergandinian Stage, A294, A313, A324
Kuehnia, A473
Kueichouphyllum, A270, A271
Kueichowpora, A217
Küpper, A27
Kukla, Mathews, & Mitchell, A524
Kulikov, Pavlov, & Rostevtsev, A296
Kullervo, A145, A151
Kullman, A196, A202
Kullman & Ziegler, A189
Kuman Series, A294, A302, A312
Kummel, A335, A336, A351-A389, A408
Kummel & Erben, A359
Kummel & Fuchs, A369
Kummel & Teichert, A300, A301, A336, A337, A339, A343, A358, A360, A380, A387
Kumpanophyllum, A272
Kunda Stage, A143
Kungurian Stage, A293, A295
Kushanian Series, A313
Kutassy, A370
Kutek, A408
Kutek & Zeiss, A408
Kutscher & Schmidt, A188, A189
Kuzbasophyllum, A270, A271
Kuzbassocrinus, A196
Kwangsiphyllum, A271

La Brea tar pits, A11
Lacunoporaspis, A227
Ladogia, A206
Ladogioides, A206, Zone, A244, A245
Längs-Einsteuerung, A39
Lagenidae, A316
Lagonibelus, A408
Lahillia, A476
Lamb, A514
Lamb & Beard, A510, A513, A514
land plant remains, Silurian, A181
Lane, A281
Lange, A205, A398

Langerfeldt, A47
Langheinrich, A55, A58
Lapeirousia, A472
LAPEIROUSIINAE, A441
Lapkin & Solovyev, A299
de Lapparent, A292, A506
Lardeux, A191, A195, A198, A201
Larma, A469
Larson, A526
Larus argentatus, A23
Lasiodiscus, A300
Latanarcestes, A202
Laticrura, A145
Latiendothyra, A261, A263
de Lattin, A393
Laughton, A516, A520
Laurentide deglacial process, A524
Lawrence, A69
Lazell, A11, A70
Leanza, A243, A396
Lebachian assemblage, A283
lebensspuren, A2, A31, A44
Le Castella section, A510, A513
Lecompte, A199, A205
Le Conte, A506
Leella, A300, A314
Legrand, A137
Lehmann, A17, A69, A71
Lehmann & Weitschat, A17, A69
Leioceras, A400
Leioclema, A325
Leioproductidae, A275
Leiorhynchoidea, A327
Leiorhynchus, A238, L. castanea Zone, A238, L. hippocastanea Zone, A238, A240
Leiospheridia, A83
Leleshus, A152, A153
Le Maitre, A203
Lenella, A396
Lenticeratinae, A460, A461
Lenz, A227, A230
Lenz & Jackson, A227
Lenz & Pedder, A238
Leonardian Series, A294, A303, A313-A314, corals of, A321
Leonardophyllum, A273, A319, A321
Leonaspis, A242
Leperditiidae, A140
Lepidocyclina, A497
Lepidodendron, A301
Lepidolina, A302, A312, A313, Zone, A313
Lepidorbitoides, A496
Lepidotus, A71
Leptathyris circula Zone, A235, A238, A241
Leptellina, A146
Leptobos, A508
Leptoceras, A453, A471
Leptocheilus, A410
Leptocoelia, A196, A231
Leptoinophyllum, A201
Lespérance, A151, A153
Leukadiella, A400
Leuroceras, A276
Leven, A312, A313

Levenea, A234
Levifenestella, A325
LEVITSKIY, A147
Lewesiceras, A460, A468
Leymeriellidae, A457
LIABEUF & ALPERN, A258
Liangshanophyllum, A320
Liardiphyllum, A270
Libyaconchus, A472
Lichidae, A203
Liguloclema, A325
LIKHAREV, A293
LIKHAREV & MIKLUKHO-MAKLAY, A299
Lilloetia, A401
Limnaea auricularis, A23, L. stagnalis, A23
Limnaeidae, A397
Limneidae, A498
Limnocardiidae, A498
Limnocyrena, A397
LIN, A402
Lindostroemia, A223
LINDSTRÖM, A160
Lindstroemia, A196
Lingula, A247, A395
Lingulidae, A329
Linoproductidae, A275, A327
Linoproductus, A300, A380
Linter, A469
Liopeplum, A469
"Lipalian interval," A86
LIPINA, A261, A267, A316
LIPINA & REYTLINGER, A267, A268
Lipinella, A266
Liroceras, A276, A371
Liroceratidae, A276, A371
Lispoceras, A276
Lispodesthes, A469
Lissatrypa, A197
Lithiotis, A396
LITHISTIDA, A393
lithospheric plates, Ordovician, A137
Lithostrotion, A270, A271, A272
Lithostrotionella, A270, A271, A272, A319
Lithraphidites, A457
Lithraphidites alatus-Staurolithites orbiculofenestrus Zone, A459, L. bollii Zone, A457, L. quadratus Zone, A461
Littorina, A24, A30
LITUOLACEA, A471, A472
Lituolidae, A316, A393
Lituotubella, A263, A265
LIU, A137, A362
lizard, in amber, A11
Llandoverian, faunal break, A177, faunas, A168-A169
Llanoaspidae, A123
Llanvirnian shelly faunas, A142-A143, Balto-Scandian, A143, Mediterranean, A143, North American, A142-A143
Lobatannularia, A301
LOCHMAN-BALK & WILSON, A120, A122, A125, A128, A131
Loeblichia, A265

Loftusia, A472
Lonchocephalus, A123
Longiproetus, A201
Longispina, A234
LONSDALE, A184
Lonsdaleia, A269-A272
Lonsdaleiastraea, A320
Loo, A324
Lopha, A375, A460
Lophamplexus, A321
Lophinae, A458
Lophocarinophyllum, A319
Lophoceras, A276
Lophophyllidium, A272, A273, A299, A318, A320, A321
Lophophyllum, A272
Lophotichium, A272
Lorenzinites, **A96**
Lotagnostus, A124, A127
LOWENSTAM, A61, A68
LOWENSTAM & EPSTEIN, A424, A425
Lower Devonian faunas, A194-A198, A224-A234, in Antarctica, A231, Malvinokaffric province, A194, A197-A198, in eastern North America, A223-A226, in northern North America, A227-A230, in northwestern North America, A226, in western North America, A226, Old World province, A194-A197, in South America and Falkland Islands, A230-A231
Lower-Middle Devonian boundary, A188, A195, A198
LU, A133, A149
Ludlovian faunas, A170
LUDVIGSEN, A227
Ludwigia, A400
Lunatia, A30
LUPHER & PACKARD, A396
LYASHENKO, A191, A201, A205
Lycettia, A469
Lycophoria, A143
LYELL, A499, A506, A507, A508
Lyelliceratidae, A457, A459, A469, A475
Lyem, A312
Lyrapora, A325, A326
Lyrielasma, A223
Lyriomyophoria, A375
Lyssatripidae, A199
LYTOCERATACEA, A397-A398
LYTOCERATACEAE, A441
Lytoceratidae, A457
LYTOCERATINA, A397, A399
Lytohoplites, A406
Lytophiceras, A362
Lyttonia, A300, A380
Lyttoniacea, A329
Lyttoniidae, A330
Lytvolasma, A319
Lytvophyllum, A272

MAACK, A246
Maastrichtian, A423, A454-A456, A461
Maccoyella, A475
Maccoyoceras, A276

MCDOUGALL, A219
MCELHINNY, A85
Macgeea, A202
MCGREGOR & UYENO, A230, A240, A246
Machari facies; see Yangtze
MCINTYRE, A520, A537
MCINTYRE, RUDDIMAN, & JANTZEN, A520, A521
MCKEE, A303
MCKELLAR, A208
Mackenziephyllum, A240
MCLAREN, A187, A203, A223, A244, A245
MCLAUGHLIN, A145
MACLURITINA, A441
Macoma baltica, A25, A30
Macrocephalitidae, A402, A403, A406
Macropyge, A140
MCTAVISH, A157, A363
MCTAVISH & DICKENS, A363
Mactra, A24, A35, M. corallina, A27, A29, A35, M. solida, A35, M. subtruncata, A30
MACURDA & MEYER, A63
Madigania, A94
MAEDA, A396, A397
Maenioceras, A201, A242, A243
Magadiceramus, A460
MAGLIO, A516
MAILLIEUX, A195
DE MAISTRE, A258, A282
Majella, **A108**
Malagasitrigonia, A472, A476
Malayomaorica, A397
Malurostrophia, A197
Malvinokaffric province, Devonian, A194, A197-A198, A209
Malvinokaffric, realm, A168, A176, A230, A231, A242
MAMET, A261
MAMET & SKIPP, A261, A267, A286
MAMET, SKIPP, BANDO, & MAPEL, A265
Mammitinae, A459
Mamuthus, A516
MANTELL, A439
Mantelliceratinae, A453
Manticoceras, A205, A206, A207, A208, A243, A244, A246, A247, -Cheiloceras boundary, A189, A208, A244
MANTON, A88
Manuaniceras, A469
Maoristrophia, A196, A197
Maorites, A475
maps, paleogeographic, Carboniferous, A284-A285, Cretaceous, A421, Devonian, A194, A198, A204, A209, Permian, A340-A341, Silurian, A169, Tertiary, A491, Triassic, A356
Marathonites, A300, A331
Marginifera, A300, A380
Marginiferidae, A327, A329
MARGOLIS & KENNETT, A530
Marjumiidae, A130
Marken, A44

Marshallites, A475
Marsupites, A460
Marthasterites, A460, M. tribrachiatus Zone, A503
MARTINI, A503
Martinia, A299, A300, A301, A380, A381
"Martinia" Shale, A297
Martinifera, A300
Martiniidae, A329
MARTINSON, A397
MARTINSSON, A23, A132, A188
MARWICK, A364
Marwickia, A476
Marywadea, **A104**
mass mortality, A6, A9
Mastodon, A500
MATERN, A190
Matheronia, A472
MATSUMOTO, A462, A463, A465, A467, A468, A471, A472, A475
MAURIN & RAASCH, A244
Mawsonites, A94, **A96**
Mayaitidae, A406
Maychella, A324
MAYER-EYMAR, A499
MAYNC, A397
MAYR, A20
Mazapilites, A402
Mazonian assemblage, A282
MECOPTERA, A305
Medeela, A472
Mediocris, A263, A265
Mediospirifer, A235
Mediterranean Fauna, Arenigian, A141-A142
Medlicottia, A297, A302, A334, A335
Medusinites, **A96**
Meekellidae, A330
Meekia, A465
Meekopora, A323, A324, A326
Meekoporella, A323
Megacanthopora, A325
Megacucullaea, A476
Megadesmidae-Ceratomyidae, A375
Megadesmus, A305
Megaglossoceras, A276
Megakozlowskiella, A231, A234
Megalocardia, A473
Megalodon, A359, A361
Megalomus, A179
Meganteris, A195
Megastephanoceras, A402
Megastrophia, A201
Megatrigonia, A475
Megaxinus, A476
Megistaspidella, A143
MEISCHNER, A280
Melanophyllum, A270, A272
Melanopsidae, A498
Meleagrinella, A367
menisci, A16-A17
MENNER, A343
Menomonia, A123
Menophyllum, A269
Menuthiocrioceras, A472
Meramecian Series, A258
Mercaticeras, A400

Mercenaria mercenaria, A68
MERCER, A516
MERCEY, A506
Meristella, A196
Meristellidae, A161
Meristellinae, A246
Merlewoodia, A270, A271
MEROSTOMATA, A87, A88
Mesochasmoceras, A276
Mesohibolites, A457, A472
Mesolimulus walchi, A14
Mesopuzosia, A475
Mesosaurus, A306, A307
Mesotaxis asymmetrica Zone, A243, A244
Mesozoic-Tertiary boundary, A496
Mestognathus, A281
Metacoceras, A281
Metacryphaeus, A230
Metalegoceras, A305, A331, A332
METALEGOCERATINAE, A331
Metaperrinites, A334
metasomatism, A61, A69
METAZOA, Precambrian, A81-A89, characterized, A86, evolution of, A89
Metoicoceratinae, A459
MEYEN, A268, A285, A343
MEZESHNIKOV & ZAKHAROV, A408
Michelinia, A269-A271, A319
Micracanthoceras, A403, A406
Micrantholithus hoschulzi Zone, A457
Micraster, A458, A460
Microcyclus, A269
Microderoceras, A399
Micromelanidae, A397
microphytoliths, A81, A111
Microtrigonia, A461
Micula decussata-Tetralithus pyramidus Zone, A460, M. mura, A461
Middle Devonian faunas, A198-A203, North American, A234-A243, eastern, A234-A235, northern, A240-A242, northwestern, A238-A240, western, A235-A238
middle Ordovician shelly faunas, A143-A149, Balto-Scandian, A148, Heptabronteus-Pliomerina Province, A149, Mediterranean, A149, North American Midcontinent, A143-A147, Scoto-Appalachian, A147, Tungusian, A147
MIDDLEMISS, A301, A379
MIDDLEMISS & RAWSON, A463
MIDDLETON, A36, A201
Middle-Upper Devonian boundary, A103, A189, A199, A235
MIKAN & SWEET, A379
Mikhailovella, A263
MIKHAYLOVA, A309
MIKLUKHO-MAKLAY, A293
Milanovicia, A473
MILLER, A138, A445
MILLER & COLLINSON, A247
MILLER & FURNISH, A276, A298, A335, A381

Millerella, A265, A266, A267, Zone, A267
Millericrinus, A411
Millkoninckioceras, A276
Mimagoniatites, A196
Mimatrypa, A235
MINATO, GORAI, & HUNAHASHI, A364, A365
MINATO & KATO, A272, A317, A319
mineralized tissue, evolution of, A89
Minilya, A324, A325, A326
Minojapanella, A299
Miocene, A499-A504
Miocene-Pliocene boundary, A500
Miogypsina, A497, A503-A504
Misellina, A299, A311, A316, Subzone, A313, Zone, A313
MISRA, A84
MISSARZHEVSKY, A104
Missisquoia, A140
Mississippian System, A258, A280
Missourian Series, A258, A266
Mizzia, A300
Mobergella, A130
Modiola, A35, A57
Modiolopsis, A362
Modiolus, A375
VON MOELLER, A310
Moelleritia, A238
VON MOJSISOVICS, WAAGEN, & DIENER, A352
Mojsisoviczinae, A469, A473, A475
Mojsvaroceras, A371
molecular rearrangements in diagenesis, A61-A69, concretions, A65-A67, recrystalization, A61-A65, into stable modifications, A67-A69
MOLLUSCA, Cambrian, A120
mollusks, Tertiary, A498-A499, A499
MONAXONIDA, A393
MONGER & ROSS, A304, A314, A315, A319
Mongolia and Tikhookian area, Lower Devonian faunas of, A196, A207; link with Appalachian province, A210
Monodiexodina, A308
monograptids, Devonian, A226, A230
Monograptus, A187, A195, A197, M. hercynicus, A195, Zone, A226, A227, M. praehercynicus Zone, A226, M. thomasi, A227, M. ultimus Zone, A187, M. uniformis, A195, A198, Zone, A223, A226, A227, M. yukonensis Zone, A227
Monopleuridae, A441
Monotaxinoides, A265
Monothyra, A475
Monotis, A360, A362, A365, A367, A369, A372, A375-A376, distribution of, A374, M. ochotica group, A375, A376, M. salinaria group, A375, A376, M.

subcircularis group, A375, A376, M. typica group, A375, M. zabaikalica group, A376
Montiparus, A266, A268
MOORE, A258
MOORE & DUDLEY, A323
MOORE & JEFFORDS, A321
Morea, A469
MOROZOVA, A316, A322-A324
Morphoceratidae, A403
Morrowan Series, A258, conodonts of, A280, corals of, A272, foraminifers of, A265-A266
MORTELMANS, A255
Moscovian Stage, A255, A259, A266, corals of, A272-A273, A273
MOSEBACH, A61
Mstikhinoceras, A275
MU, A155, A196, A202, A207
Mucrospirifer, A202, A206, A235, A243
MÜLLER, A2-A78, A382
MÜLLER & MOSHER, A411
MUIR-WOOD, A395
MUIR-WOOD & COOPER, A273
Muirwoodia, A298
Mulinoides, A476
Multispirifer, A195
Muniericeratidae, A459, A460
MURATOV, OSTROVSKY, & FRIDENBERG, A536
MURCHISON, A184, A291, A292
MURCHISON, DE VERNEUIL, & VON KEYSERLING, A292
Murchisoniidae, A378
Murciella, A472
Murex, A30, A39
Murgabian Stage, A294, A299, A300, A313, bryozoans of, A324, A325
Muschelkalk, A352, A354
Mutationella, A195, A197
Mya, A57, M. arenaria, A24, A25, A29
Mylitoides, A459, A460
MYOCONCHINAE, A441
Myophorella, A396
Myophoria, A362
Myrtea, A476
Mystrocephala, A226
MYTILOIDA, A375
Mytilus, A35, A55, A57, M. edulis, A6, A25

Nabeyaman Series, A294, A302, A312
Nadiastrophia, A197
NAEF, A21, A22
Nagatoella, A299, A311
Nagatophyllum, A271, A272
NAKAGAWA, NIITSUMA, & HAYASAKA, A514
NAKANO, A453
NAKAZAWA, A301, A343, A361, A363, A379, A380, A381
NAKAZAWA, ISHII, A311
NAKAZAWA, KAPOOR, A311

NAKAZAWA & RUNNGAR, A372, A375
NALIVKIN, A192, A201, A202, A203, A206, A207, A223, A293, A296
NALIVKIN, RZHONSNITSKAYA, & MARKOVSKIY, A191, A192, A196, A201, A205, A206, A207
Nalivkinella, A223
Namalia, A83, **A102**
Namurian Series, A255, conodonts of, A280, corals of, A271-272, foraminifers of, A265, insects of, A279-A283
Nankinella, A299, A312
Nankinolithus, A152
Nannoconus, A453
nannoplankton, Tertiary, zones of, A501-A503
Nanothyris Zone, A223
Naoides, A270, A271
Nasepia, A83, **A102**
Nassa, A30
NASSICHUK, A243
NATICACEA, A397
Naulia, A474
NAUTILIDA, A275-A276, A371
NAUTILOIDEA, A275, A276, A498
nautiloids, Jurassic, A408, A410, Silurian, A179, Triassic, A371-A372
Nautilus, A22, A44, A275
nearshore clastic facies, Lower Devonian fauna of, A195
NEAVERSON, A440
Nebrodites, A402, A407
necrosis, A6, necrotic processes, A5
Neithea, A441, A458
Nejdia, A400
NELSON, HOPKINS, & SCHOLL, A526
Nelttia, A469
Nemagraptus gracilis Zone, A149, A154, A156, A160
Nemetia, A473
Neoaganides, A337
NEOAMMONOIDEA, A397
Neoarchaediscus, A265, A266
Neoasaphus, A143
Neocaprina, A473
Neochetoceras, A407
Neoclisiophyllum, A270
Neocomian, A453
Neocrassina, A475
Neocrimites, A305, A334, A335
Neofusulinella, A300
Neogastroplites, A457, A468
Neogene, A499, A503, A508, sediments, A490-A491, A495, A496
Neogeoceras, A335
Neogloboquadrina dutertrei, A532, M. pachyderma, A523, A525, A527
Neogondolella, A338, A339, N. bisselli-Sweetognathus whitei Zone, A337, N. carinata Zone, A379, N. rosenkrantzi-Neospathodus arcucristatus Zone,

A338-A339, N. rosenkrantzi-Neospathodus divergens Zone, A339, N. serrata postserrata Zone, A339, N. serrata serrata Zone, A338, A339
Neohibolites, A449, A457, A458, A459, A472
Neokoninckophyllum, A272, A321
Neolobites, A471
Neophylloceras, A475
Neoprobolium, A223
Neoradiolites, A473
Neoschizodus, A369
Neoschwagerina, A300, A312, N. craticulifera Subzone, A312, N. margaritae Subzone, A312, N. simplex Subzone, A312, Neoschwagerina Zone, A312, A313, A314, A316, A319, A320, A323
Neospathodus, A339, N. crystagalli Zone, A379, N. dieneri Zone, A379
Neospirifer, A380
Neospiriferinae, A273
Neostacheoceras, A300
Neostreptognathodus, A338, N. pequopensis Zone, A338, N. sulcoplicatus-N. prayi Zone, A338
Neozaphrentis, A269, A318
NERINACEA, A397
Nerinea, A397
Nerineidae, A458, A472
Nerinella, A397
Neritopsidae, A378
Nervophyllum, A272
Nervostrophia, A246
Neseuretus, A141
NEUMAN, A143, A151
NEUMAN & BRUTON, A143
NEUMANN, A401
NEUMAYR, A391
Neuqueniceras, A401
Neuropteris, A301
Nevadella, A123
Nevadia, A122
New Zealand subprovince, of Lower Devonian, A197, A198-A199, A202
NEWELL, A303
NEWELL & BOYD, A375
Nicklesopora, A322
Nicomedites, A364
Nigericeras, A471
NIITSUMA, A514
NIKIFOROVA, A322
NIKIFOROVA & PREDTECHENSKIY, A187, A192
NIKIFOROVA & SAPELNIKOV, A154
Nikiforovella, A324, A326
NIKITIN, A143
NIKITINA, A316, A323
NIKOLAEV, A150
Nileus, A142
Nipponitella, A308
Nipponitrigonia, A396
Nodosaria, A266
Nodosinellidae, A316
NOERREVANG, A90, A107
NOGAMI, A363

NOHDA & SETOSUCHI, A411
NORITACEAE, A371
NORRIS, A218-A253
North American fauna, Arenigian, A140, Tremadocian, A139, A140
North American strata, Silurian, A173-A174; geosynclinal, A173-A174, platform, A173
North American zoogeographic region, in the Carboniferous, A269, A270, A271, A273
North Silurian realm, A168
North Temperate Realm, Cretaceous, A465-A469; Euramerican Region, A467-A469, Japanese-East Asian Subprovince, A465, North American Province, A468-A469, North European Province, A467-A468; North Pacific Province, A465-A467, Northeast Pacific Subprovince, A465
Nostoceratidae, A461, A467
Notanoplia, A197
Nothaphrophyllum, A271
Notiochonetes, A230
Notospirifer, A305
Notothyris, A299, A300
Nototrigonia, A476
Novella, A265, A266
Nowakia, A235
Nowakidae, A203
Nucleolitidae, A441
NUCULOIDA, A375
Nummulitidae, A497
Nymphorhynchia, A227, A238

OBRADOVICH & COBBAN, A426, A439, A445
Obsoletes, A266, A268
Occidentoschwagerina, A299, A308
Ochetoceras, A402
Ochoan Series, A294, A314
Odontocephalus, A226
Odontochile, A195, A226
Odontopleuridae, A203
ÖPIK, A127, A134
Offaster, A458, A461
Offneria, A473
Ogbinopora, A324, A325, A326
Ogilviella, A227
OGOSE, A69
Ogygopsis, A128, A130
Oketaella, A266
OKIMURA, A316
OLAUSSON, A517, A532
Olcostephanidae, A454-A456
Old Red Sandstone facies, A184, A205, A206
Old World province, Devonian, A194-A197, A198, A209
Oldhamina, A300, A301, A336, A337
Olenekian, A353
"olenellid province," A129
Olenellus, A123, A125
Olenidae, A125, A126, A130
Oligocene, A499-A504
OLIVER, A198, A202, A203, A208,
A223, A234, A235, A244, A246
D'OMALIUS D'HALLOY, A439
Ombonia, A380
Omphalotis, A263, A265
Omphalotrochidae, A378
ONCOCERIDA, A275-A276
Oncograptus, A155
Onega, A104
?Onegia, A99
Onychocella cyclostoma, A7
OPDYKE, A513, A527
Opeliidae, A457
open ocean "provinces," Middle & Late Cambrian, A130
Operculina, A497
Ophiceras, A301, A358, A360, A366, A379
Ophiceratidae, A371
Ophiomorpha, A10
OPINAE, A441
Opisoma, A396
Opisthotrigonia, A396
Oppeliidae, A402, A403, A407
ORADOVSKAYA, A143, A147
D'ORBIGNY, A3, A439, A440
Orbirhynchia, A459
Orbitoides, A496
Orbitolina, A471
Orbitopsella, A393
Orbulina, A497
ORCHARD, A205
Ordovician, middle, A143
Ordovician-Silurian transition, A160-A161
organic evolution, Precambrian, A80
organic material, products of decay of, A15-A19
oriented embedding of objects, A31-A48, barrel-shaped, A40-A41, bowl-shaped, A32-A36, cone-shaped, A36-A40, irregular, A41-A48
Oriocrassatella, A305
Orionastraea, A270-A272, A318
Oriskania beds, A226
ORMISTON, A223, A226, A227, A230, A234, A241, A242, A246
ORR, A234, A235
Orthacea, A203
ORTHIDA, A329
Orthidiella, A142
"Orthoceras" limestones, A179
ORTHOCERATACEAE, A275
ORTHOCERIDA, A275
Ortholomidae, A395
Orthoptychus, A472
Orthothetina, A301, A380
Orthotichia, A380
Oryctocephalidae, A122, A123, A126, A128, A130
Osagian Series, A258
Osculigera, A473
Osteocrinus, A411
OSTRACODA, A410
ostracodes, Devonian, A191, A205, A206, Silurian, A172, A180
Ostrea, A29
Ostreidae, A453, A458
OSTREINA, A395
OSWALD, A184, A188, A189, A192, A195, A196, A197, A198, A201, A202, A203, A205, A207, A208
Otapiria, A365
Otarion, A242
Otoceras, A366, A367, A379, A380, A381, -Ophiceras Zone, A330
Ottweilerian assemblage, A282, A283
Ovatoscutum, A88, **A91**
Overtonidae, A275
OWEN, A462, A465, A467, A471
Owenoceras, A279
oxygen content, in Cretaceous marine environments, A427-A429; effects of, A432-A434; of water, A9
oxygen requirements, variations in, A8-A10
oxygen-isotope climatic record, Quaternary, A517-A520
Oxylenticeras, A403
Oxynoticeras, A398, A399
Oxyoppelia, A407
Oxyteuthididae, A457
Oxytoma, A365, A395
Oxytomidae, A441
Oxytropidoceras, A469
Ozarkodina, A227, A230
Ozawainella, A265, A266, A299
Ozawainellidae, A308, A309, A315

Pachyceratidae, A403
Pachydesmoceras, A475
Pachydiscidae, A457, A460, A461, A465
Pachydiscoceras, A301, A337
Pachymegalodon, A396
Pachymytilus, A396
Pachyphloia, A300
Pachysphaerica, A263
Pachyteuthis, A408, A410
Pacitrigonia, A475
Paeckelmanellidae, A273
PAECKELMANN, A196
Pagetiidae, A122, A123, A128
Paguridae, A410
PAINE, A69
PAKUCKAS, A363
Palacohatteria, A297
Palaeacis, A271, A272
Palaeofusulina, A300, A301, A302, A312, A313, A316, A337, -Reichelina Zone, A312, Zone, A313
Palaeoniscus, A42, A66, A297, A306
Palaeoscia, A91
Palaeosmilia, A269-A272, A321
Palaeospiroplectammina, A261
Palaeostrophomena, A146
Palaeoteuthis, A410
Palaeotextularia, A263, A265, A266
Palaeotextulariidae, A267
Palaferellidae, A199
Paleocene, A499-A504
paleoecology, Precambrian, A86,

Index

Silurian marine, A176
Paleogene, A499, sediments, A490, A491, A495, A496
paleogeography, Cambrian, in Europe, A124-A125, Jurassic, A391, Ordovician, A137, A159, Permian, A399-A340, Tertiary, A488-A489, Triassic, see stratigraphy
paleoisotherm shifts, Quaternary, in the Atlantic, A520-A521
paleolatitudes, Devonian, A219
Paleolina, **A107**
paleomagnetic data, Devonian, A208, Ordovician, A136, A137, A159
"Paleotethys," A139
Palinuridae, A410
PALIY, A83
Palmatolepis, A205, A208, A243, A244, A246, P. crepida Zone, A244, P. gigas Zone, A205, P. triangularis Zone, A189, A208, A243, A244, A246
PALMER, A119-A135, A157
Paltechioceras, A398
Palus, A473
Pamirella, A324, A326
Pamirian Stage, A294, A299, A300, A313, bryozoans of, A323, A324
Panderia, A141
Pandorinellina, A227, A230, A243, P. insita, A235, A244
Panguridae, A23
PANTIC, A316
Papilionata, A103
PAPILLACEAE, A498
PAPP, A26, A27, A69, A488-A504
Paraboliceras, A403
Parabolinoididae, A124
Parabournonia, A473
Paracalmonia, A230
Paracaprinula, A473
Paraceltites, A300, A306, A335
Paracenoceras, A408
Paradoxides, A128
Paradoxididae, A125, A126, A130
Paradunbarula Zone, A313
Parafenestralia, A322, A326
Parafusulina, A299, A308, A311, A312, A313, A314, A316, Zone, A312, A313, A319, A323, A335
Paragastrioceras, A331
PARAGASTRIOCERATINAE, A331
Paraglossograptus, A155, A156
Paragnathodus, A280
Parahastites, A408
Parahibolites, A457-A459, A472
Parainciphyllum, A319
Paralegoceras, A279
Paraleioclema, A323-A326
Paralithostrotion, A270, A272
Paraloxoceras, A275
Paranautilus, A371
Paranomia, A469
Paranorella, A327
Paranorites, A363

Parapedioceras, A455, A472
Parapholas, A475
Parareichelina, A312
Parareineckeia, A402
Pararineceras, A276
Paraschwagerina, A299, A308, A311, A312, A313, A315
Paraspidoceras, A406
Paraspirifer, A199
Parastaffella, A266
Parastereophrentis, A272
Parastringocephalus, A235, A238
Parastroma, A473
Paratirolites, A300, A337, A358, A381
Paravascoceras, A471
Paraverbeekina, A300
Parawentzelella, A319, A320
Parawocklumeria, A207
Parazellia, A308, A315
PARAZOA, A81
Parengonoceras, A471
Parhabdolithus, A457
PARKER, A426, A532
PARKIN & SHACKLETON, A524
Parkinsoniidae, A408
PARMENTIER & FOLGER, A531
Paronella, A473
Paroniceras, A400
Parvancorina, A87, A105, A106
Parvancorinidae, **A105**
PASCOE, A361
Paurorhyncha, A245
Paurorthis, A141
Pavastehphyllum, A319, A320
Pavlovia, A406, A407
PAVONI, A3, A47
Pecopteris, A300
Pecten, A362
Pectinatites, A406, A407
Pectinidae, A453, A458
Pedavis, A226
PEDDER, A238, A240
Pemphix, A410
Pennaia, A230
Pennireteopora, A323-A326
Pennsylvanian System, A258, conodonts of, A280, foraminifers of, A266
Pentagonia, A231, A234, A235
Pentameracea, A161
Pentamerella Subzone, A238
PENTAMEROIDEA, A203, A246
Pentamerus, A172
Pentamplexus, A319
percentage endemism, A463
Peregrinella, A468
Peregrinoceras, A453
Periplomya, A469
PERISPHINCTACEAE, A404
Perisphinctidae, A402, A403, A406
Peritrochia, A335
Permian biostratigraphy, in type area, A293-A296, correlation of, A294-A295; in northwestern Europe, A296-A297; in Gondwana continents, A305-A309, in Africa, A306, in Australia, A304-A305, in India-Pakistan, A306, in Madagascar,

A307, in South America, A307; in Greenland, A297-A298; in North America, A303-A304; in Tethyan area, A298-A303, in Japan, A302, in the Pamirs, A304-A305, in the Salt Range, A300-A301, in South China, A302, in the Trans-Caucasus, A300
Permian faunas, ammonoids, A330-A337, brachiopods, A326-A330, bryozoans, A321-A326, conodonts, A337-A339, corals, A317-A321, fusulinaceans, A307-A316, other foraminiferids, A316
Permian-Triassic boundary, A292, A369, A371, A378-A382
Permocalculus, A300
Permoceras, A408
Permodiscus, A263
Permoleioclema, A323, A326
Permopora, A326
Perna, A396
Peroniceratinae, A460
Perrinites, A300, A331, A334, A335
Peruarca, A473
"PETALONAMAE," A87, A88-89, **A96**
Petalonamidae, A96
Petalostroma, **A108**
PETERSEN, A208
PETRÁNEK & KOMÁRKOVÁ, A39
Petricola, A35, P. pholadiformis, A29, A30
PETRUNINA, A140
PETTER, A208
Petteroceras, A208
Pettersia, A473
PFANNENSTIEL, A33
PFLUG, A88, A96, A99
Phacoceras, A276
Phacopidae, A204
Phacopina, A226
Phacops, A201, A226
Pharciceras, A189, A201, A205, A235, A247, P. lunulicosta Zone, A189, A201, A208
Pharodina, A469
Phaxas, A35
Phestia, A305
PHILIP & JACKSON, A208
PHILIP & PEDDER, A197, A202
Phillipsastrea, A203, A205, A207, A245
Phillipsinella, A150, A151, A158
Phillipsinellidae, A160
Phisonites, A300
PHLEGER, A527
Phlycticeras, A406
Pholadomyidae, A329, A375
Pholas dactylus, A29
Pholonyx, A201
Phragmodus, A157
Phragmostrophia, A196
Phylloceras, A398, A401
PHYLLOCERATACEA, A397-A398
PHYLLOCERATACEAE, A441

Phylloceratidae, A465, A471, A475
Phyllytoceras, A397
Phymatopleuridae, A378
Physodoceras, A406
Pia, A398, A399
Picard & Flexner, A357
Pickett, A192, A202, A208
Pictonia, A407
Pileochama, A473
Pinacites, A242, P. jugleri, A201, A202, A203
PINACOCERATACEAE, A371
Pinegopora, A322
Pinnidae, A375
Pisidium, A26, A27
Pithonoton, A410
Placenticeratidae, A460, A461, A467
Placoparia, A143
Plagiostoma, A395
Plagiura-Poliella Zone, A131-A132
Planetoceras, A276
planktonic fluctuations, Cretaceous, A429-A430, A432-A433
Planoarchaediscus, A263, A265
Planocaprina, A473
Planomedusites, A94, A96
Planorbidae, A397
Planospirodiscus, A265
plate tectonics, Cambrian, A125, Cretaceous, A420, Devonian, A210, A219, Ordovician, A136-A137, A140, A159-A160
Platyceramus, A460, A461
Platyclymenia, A207, A208, A245, A247
Platycoryphe, A153
Platygoniatites, A279
Platylenticeras, A408
Platyscutellum, A230
Platysolenites, A83
Platystrophia, A148
Platyterorhynchus, A244
Playford & Lowry, A208
Plectodina, A157
Plectodonta, A95
Plectothyrella, A153
Pleistocene, A507
Pleramplexus, A320
Pleurocephalites, A403
Plerophyllum, A319, A320, fauna, A320
Plethopeltides, A140
Pleurograptus linearis Zone, A143, A149
Pleuromya, A57
Pleurothyrella, A197, A231
Pleurotrigonia, A476
Pleydellia, A400
Plicatostylus, A396
Plicatula, A367
Pliocene, A499-A504
Pliocene-Pleistocene boundary, A507-A515, age & correlation of, A513-A515
Pliomerina, A149
Plumhoff, A410
Plumstead, A343
Podolella, A195

POGONOPHORA, A90, A107
Politoceras, A279
Pollex, A469
POLYCHAETA, A102, A103
Polycoelia, A320
Polyconites, A472
Polydesmia, A143
Polydiexodina, A300, A308, A312, A314, Zone, A335
Polygnathus, A202, A208, A230, A234, A235, A238, A240, A242, A243, A280, P. dehiscens, A227, A230, P. kockelianus Zone, A202, A238, P. linguiformis, A234, A235, P. perbonus, A227, A230, P. robusticostatus, A234, P. varcus Zone, A189, A203, A208, A325
polymerid trilobites, A122-A123, A125, A127, A131
Polypora, A322-A326
Polyporidae, A322
Polytaxis, A263
Polythecalis, A300, A319, A320
Pomerania, A407
Ponticeras, A205, A207, A243
Poole, A245
Popanoceras, A302, A307, A334, A335
population, dynamics, A6, waves, A6
Populationswellen, A6
PORAMBONITACEA, A161
?PORIFERA, A91
PORIFERA, Cambrian, A120
poriferans, Jurassic, A393, Silurian, A177
Porpitidae, A88, A92
Porrenga, A435
Posidonia, A44, A53, A56, A372, A396
"Posidonia" Shale, A297
Postligata, A469
Pradoia, A195
Praebarrettia, A473
Praecambridium, A87, A88, A104
Praecardiomya, A472
Prägekerne, A51
Praeglobotruncana, A458, A459
Praelapeirousia, A472
Praeorbulina, A497
Praesumatrina, A300
Praetiglian beds, A516
Pratje, A27
Precambrian-Cambrian boundary, A82
preservation, A3
Prestwich, A506
Přibyl & Vanek, A188
Primorella, A324, A326
Prioniodella, A339
Prionoceras, A207
Prionocyclus, A460, A468
Prionothyris, A231
Prionotropis, A460
Prismopora, A323, A325, A326
Probeloceras, A205, A243
problematic fossils, Precambrian, A111
Probolops, A230

PROCARYOTA, A80
Proconchidium, A154
Prodactylioceras, A399
Prodalmanitina, A140
PRODUCTACEA, A327
Productella, A246
Productellidae, A275
Productidae, A275
Productininae, A275
Productorthis, A141, A145
Productus, A300, A379
"Productus" bed, A297
Profusulinella, A265, A266, A267, A268, Zone, A266, A267
Prolecanitidae, A276
Proniceras, A406
Pronoritidae, A276
Propermodiscus, A263, A265
Properrinites, A331, A334
Propinacoceras, A300, A305, A307, A335
Propopanoceras, A305, A331
Proptychites, A363, A366
Proschizophoria, A195
Proshumardites, A279
Prosicanites, A300
Prosopiscus, A149
Prosopon, A410
Prostacheoceras, A331
Protacanthodiscus, A403
Protancyloceras, A403, A453, A471
Protatrypa, A154
Proterozoic, A80-A113
PROTISTA, A80-A81
Proto-Atlantic Ocean, A137
Protocardia, A397
Protoceras-Araxoceras Zone, A301
Protodipleurosoma, A96
Protoglobigerina, A393
Protognathodus, A280
Protogrammoceras, A399, A400
Protolenidae, A125, A127
Protoleptostrophia, A234
Protolonsdaleistraea, A273, A318, A319
Protopopanoceras, A331
Protothaca, A36
Prototoceras, A300, A301, A336
Protowentzella, A318
protozoans, Jurassic, A393
Protriticites, A266, A268
Pseudactinoceras, A275
Pseudagnostus, A124
Pseudavicula, A476
Pseudoacrocephalites, A140
Pseudobarrettia, A473
Pseudobatostomella, A322-A326
Pseudobelus, A472
Pseudoblothrophyllum, A223
Pseudobradyphyllum, A319
Pseudobryograptus, A156
Pseudobuccinum, A469
Pseudocadoceras, A402
Pseudocardinia, A397
Pseudocarniaphyllum, A320
pseudocoenoses, A23
Pseudocyclammina, A471
Pseudocyrtoceras, A275
Pseudodoliolina, A312
Pseudodorlodotia, A271

Pseudoemiliania lacunosa, A524
Pseudoendothyra, A263, A265, A266, A267, A310, A311, A316
Pseudofusulina, A266, A267, A268, A299, A300, A310, A311, A312, A313, A316, P. ambigua Subzone, A311, P. moelleri Zone, A311, P. verneuili Zone, A311, P. vulgaris Subzone, A311, Pseudofusulina Zone, A319, A320, Sterlitamakian fauna, A311
Pseudofusulinella, A268, A310, A311
Pseudogastrioceras, A296, A301, A335, A336, A337
Pseudoglomospira, A265
Pseudohalorites, A334
Pseudoheligmus, A472
Pseudohuangia, A319, A320
Pseudolioceras, A400
Pseudolissoceras, A403, A407
Pseudomonotidae-Gryphaeidae, A375
Pseudomonotis, A305, A396
Pseudomyalina, A305
Pseudonautilus, A408
Pseudopavona, A271
Pseudopleurophorus, A475
Pseudopolyconites, A473
Pseudopolygnathus, A280
Pseudorhizostomites, A87, A94, **A96**
Pseudorhopilema, A96
PSEUDORTHOCERATACEAE, A275-A276
Pseudoschistoceras, A305
Pseudoschwagerina, A299, A308, A310, A311, A312, A315, P. morikawai Subzone, A311, P. robusta Zone, A313, P. uddeni Zone, A313, Pseudoschwagerina Zone, A292, A302, A313, A319, A320, A323
Pseudostaffella, A265, A266, A268, A311
Pseudotirolites, A301, A337, -Pleuronodoceras Zone, A301-A302
Pseudotoceras, A336
Pseudotoites, A401
Pseudotrapezium, A397
Pseudotryplasma, A202
Pseudouralinia, A270
Pseudovergatitinae, A407, A408
Pseudowedekindellina, A266
Pseudozaphrentoides, A272
Pseudozygopleuridae, A378
Psiloceras, A397, A398
PSILOCERATACEAE, A398, A399
Psiloceratidae, A398
Psilotrigonia, A472
Pteridiniidae, **A96**
PTERIDINIOMORPHA, A96
Pteridinium, A83, **A99**
Pteridium, A99
Pteriidae, A375
Pterocanium trilobum, A530

Pterocephaliidae, A124, A126, A130
Pterolucina, A472
Pterospirifer, A297, A298
Pterotoceras, A366
Pterygotus, A247
Ptiloporella, A326
Ptychaspididae, A127
Ptychaspis, A124
Ptychoglyptus, A145, A150, A151
Ptychomaletoechia, A246
Ptychopariidae, A123, A125, A126
Ptylopora, A322, A325, A326
Ptyloporella, A322, A326
Pugnax, A300
Pulchelliidae, A457
Punctospirifer, A297
Purpura, A39
Pustula, A380
Pustulatia, A235
Putrella, A266
Puzosia, A475
Pygaster, A411
Pygodus, A157
Pygope, A395, A471
Pygopiidae, A471
Pygurus, A411

Quadraticephalus, A127
Quadrifarius, A195
Quadrithyris Zone, A226, A227, A230
Quasiarchaediscus, A265
Quasiendothyridae, A261
Quasifusulina, A266, A299, A312
Quasifusulinoides, A266
Quaternary, base of, A507-A515, definition of, A506-A507
QUENSTEDT, A3, A17, A22, A31, A40
Quenstedtoceras, A401
Quer-Einsteuerung, A40
QUILTY, A402
Quirllage, A42

RAASCH, A244
RABIEN, A191
Radiastraea, A238, A240
radiolarians & climatic change, A530
Radiolichas, A201
Radiolitella, A472
RADIOLITINAE, A441
radiometric dating, Devonian, A219
Raisanites, A453
Ramiporidra, A324
Rampora, A322, A323, A325, A326
RAMSAY, A429, A435
RAMSBOTTOM, A255, A277, A286
random embedding, A30-A31
Rangea, A83, **A99**
Rangeidae, **A99**
RANGEOMORPHA, A99
Ranikotalia, A497
Rasenia, A407
Rasettia, A138
Rauserella, A312, A314, A316
Rauserites, A268

RAUZER-CHERNOUSOVA, A261, A266, A292, A309, A310
RAWSON, A449
RAYMOND, A69
Raymondaspis, A142
Raymondiceras, A247
Rayonnoceras, A275
REBOUL, A506
recrystallization, A61-A65
Recticuloharpes, A227
Rectoclymenia, A247
red beds, Silurian, A168
Redkinia, **A107**
Redlichia, A125, A126, A127
"redlichiid province," A129
REED, A197, A202, A207
reef carbonate facies, A203-A206
reefs, Silurian, A168, A170, A176
Reeftonia, A194, A197
Reesidites, A460
REGINEK, A55, A58
Reichelina, A300, A301, A302, A312, A320, A337, -Codonofusiella Zone, A312, A316
REID, A462, A471
REINECK, A31, A33
Reineckeia, A403
Reineckeiidae, A403
REIS, A69, A71
rejected taxa, Precambrian, A112-A113
Remipyga, A150, A154
Remopleurididae, A160
Rensselaeria Zone, A223
REPENNING, A516
REPINA, A120, A126
Requieniidae, A441
resin entrapment, A11-A14
Reteporidea cancellata, A7
Reteporidra, A325
Retha, A473
Retichonetes, A246
Reticulariidae, A329
Reticuloceras, A279, A280
Retroceramus, A396
Retzidae, A329
REYMENT, A18, A19, A40
REYTLINGER, A261, A267
Rhabdomeson, A323, A324, A325
Rhacheoporidae, A441
Rhacophyllites, A397
Rhacopteris, A307, flora, A259, A260
Rhenish facies, A184; Old World Devonian, A195, A198, A199, A209
Rhenorensselaeria, A195, A196
Rhipidomella, A246
Rhipidomellidae, A330
RHIZOPODA, A393
Rhizostomites admirandus, A46
RHODES, A434
RHODES & AUSTIN, A279
RHODES, SPEDEN, & WAAGE, A434
Rhombocladia, A325
Rhombopora, A323-A326
Rhombotrypella, A322, A324, A326
rhyncholites, Jurassic, A410
Rhynchonella, A395, A468

RHYNCHONELLACEA, A329
RHYNCHONELLIDA, A327
Rhynchonellidae, A395
Rhynchopora, A298
Rhynchoporiidae, A329
Rhynchotetradidae, A329
Rhynchoteuthis, A410
Rhynchotrema, A149
Rhynchotrematidae, A199
Rhysostrophia, A142
Rhyssochonetes aurora, A235, A238
Rhytistrophia, A196
RICHARDS, A6
Richardsonella, A138
Richardsonellidae, A130
RICHMOND, A516
RICHTER, A5, A24, A30, A31, A32, A33, A99, A230
RICHTER, R., & RICHTER, E., A192, A194, A195, A230
Richthofenia, A380
RICHTHOFENIACEA, A327
Richthofeniidae, A330
RICKARD, A243, A244
RIDING, A210
RIETSCHEL, A69
rigor mortis, traces of, A15
Rineceras, A276
RINEHART, A49
RIOULT, A439
Riphean, A82, Upper, A82, Terminal, A82
Ripidiorhynchus, A207
RIVA, A154
Robertella, A473
ROBERTS, A192, A207, A259
ROBISON, A123
ROBISON & PANTOJA-ALOR, A138
Robustoschwagerina, A299, A308, A311, A315, A316
rocks, Cambrian, of southeastern Asia, A126-A127, of Australia, A127, of Europe-Mediterranean-North Africa, A124-A125, miscellaneous outcrops, A128, of North America, A120-A125, of eastern Soviet Union, A125-A126
Roemeripora, A269
ROLFE & BRETT, A3
ROLL, A393
Romaniceras, A460
ROMER, A343
Roncellia, A223
Ross, A145, A157, A158, A159, A160, A254-A290, A303, A307, A313, A314, A315, A318, A322, A325, A343, A536
Ross & NASSICHUK, A314, A323
Ross & Ross, A291-A350
ROSTOVTSEV & AZARYAN, A358
Rostricellula, A149
Rostroperna, A473
Rotalipora, A458, A459
Rotaliporidae, A419, A441
Rotaraxoceras, A300, A336
ROTHPLETZ, A18
Rotiphyllum, A269, A270, A271
Rotodiscoceras, A301, A337

ROUSELLE, A395
ROUSELLE & BISCH, A395
Roussleia, A472
ROWETT, A271, A317, A318, A321
ROY, A208
ROZANOV, A89, A133
ROZKOWSKA, A205
ROZMAN, A152
ROZOVSKAYA, A266, A307
RUDDIMAN, A520, A527
RUDDIMAN & McINTYRE, A523
RUDWICK, A327
RUEDEMANN, A39
RUGGIERI, A510, A514
Rugoconites, A92, A94, **A96**
RUGOSA, A317-A318
Rugusofusulina, A268, A311, A312
Rugusoschwagerina, A299, A308
Runctonia, A453, R. runctoni Zone, A440
RUNNEGAR, A304, A363, A375
RUNNEGAR & ARMSTRONG, A329
Russian substage, A353
Rutgersella, A103
RUTSCH, A40
RUZHENTSEV, A268, A276, A292, A331, A334, A335
RUZHENTSEV & SARYCHEVA, A300, A301, A336, A343, A358, A381
RYAN, A532, A533
RZHONSNITSKAYA, A196

Saarinidae, **A107**
SABELLIDITIDA, **A107**
Sabellidititidae, **A107**
Sabinia, A472
SABLE & DUTRO, A247
SABRODIN, A81
Saccocoma, A411
Saffordotaxis, A323, A324, A325
SAHNI, A362
SAITO & VAN DONK, A425
SAKAGAMI, A323
Sakamotozawan Series, A294, A302, A311
Sakmarian Stage, A255, A293, A294, A305, A306, ammonoids of, A331-A334, brachiopods of, A330, bryozoans of, A322, A323, A324, A325, A326, corals of, A318, fusulinaceans of, A310-A311
Sakmarites, A331
SAKS, A398, A440
SAKS, MESEZHNIKOV, & SHULGINA, A396
SAKS & NALNYAEVA, A393, A402, A408
Salenia, A458
salinities, Cretaceous, stratified, A427
salt pans and salt-covered muds, A10
SANDBERG, A247, A280
Sandbergeroceras, A243
SANDERS, A435
SANDO, A271
SANDO, MAMET, & DUTRO, A261
sandy facies, Middle and Late

Cambrian, central Europe, A130
SANFORD & NORRIS, A222, A247
Sanguinolites, A206
Santa Maria di Catanzaro section, A510-A513
Santonian, A423, A454-A456, A460
SAPELNIKOV & RUKAVISHNIKOVA, A154
sapropels, A532, A536
SAPUNOV, A399
SAPUNOV & STEPHANOV, A399
Sargana, A469
SARTENAER, A189, A205, A206, A207, A245, A246
SATO, A397, A399, A402
Saturnulus planetes, A530
Saukiidae, A124, A127
SAUL, BOUCOT, & FINKS, A197
SAUVAGESIINAE, A441
SAVAGE, A154, A197
SAVAGE & CURTIS, A516
SAVITSKIY, A133
Saxoceras, A398
Saxonian Series, A295-A297
Scalarites, A471
Scaliognathus, A280
Scambula, A469
Scaphignathus, A281
Scaphiocoelia, A197, A230
SCAPHITACEAE, A441
Scaphites, A461, A468
Scaphitidae, A459, A460, A461, A467, A468
Scaphitinae, A461
Scaphonyx, A307
Schackoina, A457
Schackoinidae, A419, A441
SCHÄFER, A5, A7, A8, A10, A20, A21, A23, A28, A30
SCHAUB, A497
SCHIMPER, A499
SCHINDEWOLF, A189, A190, A360, A397, A398, A399
Schiosia, A472
Schizodus, A297, A305
Schizophoria, A223, A226, A227, A230, A235, A240, A246
Schizoproetus, A201
Schizothaerus nuttali, A36
Schlachtfelder, A40
SCHLANGER & JENKYNS, A427, A428, A430, A432
SCHLEE, A70
Schleifmarken, A44-A46
Schloenbachia, A468
Schloenbachiidae, A457, A459
Schlotheimia, A398
Schlotheimiidae, A398
SCHMIDT, A190, A200
SCHMIDT & SELLMANN, A70
Schmidtognathus, A235, A238, A240, A243, S. hermanni-Polygnathus cristatus Zone, A235, A238, A240, A243
Schnurella, A201
SCHOLL, A516
SCHOLLE & KAUFFMAN, A426, A427, A431, A432, A481
SCHOPF, A80, A81

Schrammen, A393
Schubertella, A265, A266, A299, A310, A311, A313, A316
Schubertellidae, A308, A315, A316
"Schuchertella" adoceta Zone, A238
Schwagerina, A266, A299, A310, A311, A313, A314, A316
"Schwagerina" Zone, A267, A292, A293, A310
Schwagerinidae, A308, A309, A315, A316
Schwegler, A408
Schwelle facies, A187, A204
Schwelm facies, A184
Sciophyllum, A270, A271, A319
Scidonoceras, A459
Scott, A427
Scrobicularia, A35, S. plana, A27, A29
Scrutton, A203, A205, A231
Scutellidae, A499
Scutellum, A242, A246
Scyphocrinites, A195
SCYPHOZOA, A88-A89, **A92**
Sdzuy, A58, A120, A125, A132, A133
sea level, Carboniferous, A284, Silurian, A168
seasonality changes, Quaternary, A521-A523
Seddon & Sweet, A158
Sedgwick & Murchison, A184, A219, A254
sedimentary environments, Cretaceous, A434-A435, glauconite, A435, water saturation, A434
sedimentation, Carboniferous, A285-A286
sediments, Tertiary, A490-A496, in Africa, A495-A496, in Asia, A491, in Australia, A492, in Central and South America, A493-A495, in Europe, A490-A491, in North America, A492-A493, nonmarine, A493
Seibold & Seibold, A393
Seilacher, A3, A19, A23, A31, A39, A40, A42, A47, A51
Seilacher & Westphal, A428
Seirocrinus, A410
selective dissolution, during diagenesis, A58, A61, of hard parts, A30
Selenopeltis, A141, A149, Province, A149
Sellanarcestes, A199, A203, S. wenbachi, A199-A201
Selli, A510, A514
Sellialveolina, A472
Semiformiceras, A403
Semikhatov, A82
Seminovella, A265, A266
Septaglomospiranella, A261
Septatopora, A325, A326
Septirhynchia, A395, A471
Septoliphoria, A395
Septopora, A322-A325
Sergipia, A459, A474

Serifusus, A469
Serpagli, A157, A158, A160
Serpukhovian Stage, A259
de Serres, A506
Serrodiscus, A123, A126
Sestrophyllum, A272
Setamainella, A271
Sewertzoff, A7
Sexta, A469
Seyed-Emami, A358, A399
Seymourites, A401, A402
Shackleton, A517
Shackleton & Kennett, A515, A516, A527, A532
Shackleton & Opdyke, A517, A518, A519, A520, A523, A524, A526, A527, A532
Shcherbovich, A309
Shchukina, A317
Sheehan, A151, A161
shell form and fossil position, A18-A19
Sheng, A152, A153, A301, A312
Sheng & Lee, A313, A320
Shergold, A138
Shevyrev, A359
Shevyrevites, A358
Shirley, A197, A231
Shrock, A32
Shulga-Nesterenko, A322
Sichotenella, A312
Sidyarenko & Kanygin, A143
Sieberella, A195, A227, A238
Siegenian, A188, A195
Siegenian faunas, A223, A226, A230, A231, A234
Sigmagraptus, A155
Silberling & Tozer, A352, A368
Silberlingia, A396
Silesian Subsystem, A255
Silestidae, A457
Silurian-Devonian boundary, A177, A187-A188, A195, A198, A219, A223, A227, A230
Silurian series, duration, A168
Simberskites, A456
Simoceras, A403
Singh, A81
Sinocystis, A149
Sinonia, A469
Sinophyllum, A300, A319
Sinopora, A319
Sinotectirostrum avellana Zone, A245
Sinotites, A207
Sinuatellidae, A275
Sinuitidae, A378
Siphonodella, A255, A280, A281
Siphonophrentis, A223
Siphonophyllia, A269, A271
Skevington, A154, A160
Skiagraptus, A155
Skinnera, A83, **A94**
Skinner & Wilde, A314
Skinnerella, A314
Skinnerina, A308, A312, A314, A316
Skulptursteinkerne, A51
Skwarko, A364
Skwarko & Kummel, A363, A364

Slade, A314, A338
Smith, A137, A159, A160, A219, A271, A445, A459, A514
Smith, Briden, & Drewry, A339, A475
Smithian, A353
Sobolewia, A201, A243
Sochkineophyllum, A318
soft parts, decay of, A17, in coprogenic material, A20, destruction by scavengers, A20, phosphatization of, A71, preservation of, A69-A71, in Eocene amber, A71
Sognnaes, A20
Sohl, A426, A427, A458, A462, A463, A465, A467, A468, A469, A471, A472, A477, A480, A481
Sohlites, A402
Sokolov, A84, A90, A107, A187, A192, A317
Sokolova, A192, A196
Solenochilidae, A276
Solenochilus, A276
Solenoparia, A127
Solenopleuridae, A125
Solle, A195, A199
Somalirhynchia, A395
Somalithyris, A395
sorting, of shells by transportation, A23-A24
Soshkina, A317
Sosnina, A316
Sougy, A203
South American strata, Silurian, A175
South Pole, Ordovician, A137
South Temperate Realm, A473-A476; Indo-Pacific Region, A475-A476, Andean Subprovince, A476, Austral Province, A475, East African Province, A476; South Atlantic, A473-A474
"southern" fauna, Arenigian, A141
Spaeth, A449
Sparganophyllum, A201
Spasskiy, A202, A203, A207
Spatangidae, A499
Spath, A353, A366, A381, A391, A403, A405, A406
Spathian, A353
Spathognathodus, A226, A280
"Spathognathodus" costatus Zone, A246
Spathognathus, A281
Sphaeroschwagerina, A299, A308, A310, A313, A315, S. fusiformis-S. vulgaris Zone, A310, S. moelleri-Pseudofusulina fecunda Zone, A310, S. sphaerica-Pseudofusulina firma Zone, A310
Sphaerostylus lanceola Zone, A453
Sphaerulina, A299
Sphaerulites, A472
Sphenoceramus, A460, A461
Sphenodiscidae, A461
Sphenopteris, A301

Sphenotrigonia, A476
Spinatrypa, A240
SPINCTOZOA, A441
Spinella, A197
Spinocyrtia, A235, A243
Spinoendothyra, A261, A263
Spinomarginifera, A301, A379, A380, A382
Spinoplasia Zone, A226, A227
SPINOSA, A335, A336
SPINOSA, FURNISH, & GLENISTER, A371
Spinoseptatournayella, A261
Spinther, A103
Spintheridae, A103
Spinulicosta, A238
Spirifella, A298
Spirifer, A260, A299
Spiriferella, A380
Spiriferellinidae, A273
SPIRIFERIDA, A327, A329
Spiriferidae, A273, A329
Spiriferina, A297, A395
SPIRIFERINACEA, A329
Spiriferinae, A273
Spiriferinidae, A330
Spirigerella, A380
Spirigerina, A227, A230
Spiroceras, A402
Spiroclypeus, A497
Spisula, A44, S. solida, A29
Spiticeras, A453
SPJELDNAES, A149, A160
Sporadoceras, A206, A207, S. milleri Zone, A244, A247
sporomorphs, A109
SPRIGG, A82, A103
Spriggia, A94
Spriggina, **A103**
Sprigginidae, **A103**
SPROVIERI, A514
Spülsaum, A23, A25
Spurenfossilien, A87
Squamularia, A301
SQUIRE, A81
Stacheia, A265
Stacheoceras, A301, A335, A336, A337
Staffella, A312, A313, A316
Staffellidae, A308, A309, A315, A316
Standfläche, A42
STANLEY, A80
STANTON, A453
Stantonella, A469
STAUFFER, A196
Staurognathus, A281
Staurosphaera septemporata Zone, A456, A457
Stefaniniella, A472
Stegerhynchus, A195
STEHLI, A68
STEHN, A58
Steinhagella, A246
steinkerns, A49-A53, A55, A57, A58, A67, sculptured, A51-A53
Steinmanella, A465, A475
Stelekia, A240
Stellatophyllum, A202
Stenodiscus, A323, A325, A326

Stenopareia, A151
Stenophyllum, A201
Stenopoceras, A276
Stenopora, A322-A326
Stenoscisma, A298, A299
STENOSCISMATACEA, A329
Stenoscismatidae, A329, A330
Stephanian Series, A255, A258, insects of, A281-A283
Stephanoceras, A407
STEPHANOCERATACEA, A401
STEPANOV, A298, A299, A380, A381
STEPANOV, GOLSHANI, & STÖCKLIN, A300, A343, A358
Stereocorpha, A272
Stereocorypha, A318
Stereostylus, A273, A318, A319, A321
Stewartina, A313
STEVENS, A317, A318, A320, A391, A406, A408, A410, A449, A453, A458, A462, A463, A465, A467, A468, A471, A472, A475
STEVENSON, A16
Stichocapsa, A457
Stillwasserfallen, A32
Stolonicella schindewolfi, A7
STRAND & KULLING, A83
strata, Silurian, A172-A175, geosynclinal, A173, northern hemisphere, A172-A173, platform, A173, southern hemisphere, A173
stratigraphic subdivision charts, Carboniferous, A256, Cretaceous, A423, Devonian, A188, Ordovician, A138, A141, A144, Permian, A294-A295, Proterozoic, A82, Quaternary, A509, A511, A512, A515, Tertiary, A500
stratigraphy and paleogeography, Triassic, A353-A369, Afghanistan, A359-A360, Antarctica, A369, Arctic Canada, A367-A368, Australia, A363-A364, western Canada, A368, Caspian region, A359, China, A362-A363, northeastern Europe, A354-A355, Greenland, A366, Himalayas, A360-A362, Indonesia, A363, Iran, A358, Israel, A357, Japan, A364-A365, north Mediterranean region, A355-A357, New Guinea, A364, New Zealand, A365, Pakistan, A360, northeastern Siberia, A365, South America, A368, Svalbard, A365-A366, western United States, A368
Streblascopora, A322-A325
Streblites, A401, A406
Streblocladia, A325
Streblopteria, A305
Streblotrypa, A322, A323, A325
Streptognathodus, A280, A287
Streptorhynchus, A298
Striatifera, A275, A300
Striatostyliolina, A235

Strigogoniatites, A336
Striithyris, A395
Stringocephalus, A199, A201, A202, A235, A240, A242
Stringophyllum, A202
Stroboceras, A276
stromatolites, A80, A109-A110
STROMBACEA, A397
Strophalosia, A297, A305
STROPHALOSIACEA, A327
Strophalosiidae, A275, A329
"Stropheodonta," A235
Stropheodontidae, A246
Strophomena, A146, A154
STROPHOMENIDA, A327
Strophonella, A227
Strophopleura, A245
STRUSZ, A192, A197
STRUVE, A199
STÜRMER, A17, A22, A69, A71, A195
STURANI, A396
Sturtian, A82
Stylastraea, A318
Stylatractus universus, A524
Stylidophyllum, A320
Subclymenia, A276
Submortoniceras, A461
Subplanites, A407
Subrensselandia, A235, A238
Subulitidae, A378
Subvestinautilus, A276
Sugiyamaella, A270
Sulcoretepora, A323-A326
Sumatrina, A300
Sundaites, A336
Sunites, A207
Surites, A453
Sutneria, A402, A407
Suvorovella, **A108**
Suvorovellidae, **A108**
Svobodaina, A149
SWEET, A156, A337, A339, A364, A379, A384
SWEET & BERGSTRÖM, A149, A156, A157, A158, A161
SWEET & MILLER, A242
Sweetognathus, A338
Sychnoelasma, A269, A270
SYEMINA, A309
SYLVESTER-BRADLEY, A461
Symphysurina, A139, A140
Symplectophyllum, A271
Syngastrioceras, A279
Synocladia, A323-A326
Synodonites, A472
Synpharciceras, A205
Synphoria, A226
Synphoroides, A226
Syphysurina Zone, A138
Syrdenites, A336
Syringaxon, A223
Syringoclemis, A323
Syringopora, A269, A270, A271, A273, A319
Syringoporella, A271
Syringothyridae, A273
Syringothyrididae, A329
Syringothyris, A245
Syrotrigonia, A473

Index

Sythian, A352

Tabulipora, A322-A326
Tachylasma, A299, A318
Tachyphyllum, A270
Taeniocellaria setifer, A7
Taeniopteris, A301
Taihungshania, A141
Taimyrophyllum, A202, A238, A240
Tainoceratidae, A276, A371
Taisyakuphyllum, A271
TAKAI, A364
TALENT, BERRY, & BOUCOT, A175
Tamarites, A206
Tampsia, A473
TAMURA, A362
Tancrediidae, A441
Tanerhynchia, A197, A231
taphocoenoses, A22
taphonomy, A2-A71, Precambrian, A87
Taphrognathus, A280, A281
Taphrorthis, A145
Tapinocephalus Zone, A306
TAPPAN, A429, A430, A432, A433, A477
TAPPAN & LOEBLICH, A432, A433, A477, A479, A480
Taramelliceras, A402, A406
TARAZ, A336, A358
Tasman geosyncline, A175, subprovince, Devonian, A197, A202-A203
Tatarian Stage, A295, A296, A305, A325
Tateana, A94
Tatella, A476
TATGE, A382
TAUBER, A25, A27
Tauroceras, A300
Tavayzopora, A326
TAYLOR, A89, A129
Tcherskidium, A150
tectonic phases, Tertiary, A500
TEICHERT, A17, A71, A202, A275, A305, A306, A307, A336
TEICHERT & KUMMEL, A297, A366, A381
TEICHERT, KUMMEL, & KAPOOR, A379
TEICHERT, KUMMEL, & SWEET, A337, A358, A381
TEICHERT, MOORE, & ZELLER, A410
TEICHERT & RILETT, A306
TEICHERT & SCHOPF, A247
TEICHERT & SERVENTY, A24
Teicherticeras, A227, A234
TELFORD, A197
Tellinimera, A469
Tellipiura, A476
Temnocheilus, A276
Tenea, A469
Tentaculitidae, A203
Tenticospirifer, A208, A246
Tenuipteria, A461
Tepeyacia, A473
Terataspis, A226
TEREBRATELLACEAE, A394

Terebratula, A395, A499
TEREBRATULIDA, A329
Terebratulidae, A395
Terebratulina, A460
Terebratuloidea, A299
Terebrimya, A469
TERMIER & TERMIER, A391, A401
Terquemiidae, A441
Terquensiidae, A375
Tertiary-Quaternary boundary, A496, A500, A503, A507-A515
Tethyan Realm, Cretaceous, A469-A473; Caribbean Province, A473, Antillean Subprovince, A473, West-Central American Subprovince, A473; Indo-Mediterranean Region, A469-A472, Eastern Mediterranean Subprovince, A472-A473, North Indian Ocean Subprovince, A472, Western Mediterranean Subprovince, A472-A473
Tethys belt (Turkey to Southeast Asia), Devonian, climate of, A209, faunas of, A196, A202, A207
Tethys, Jurassic, A391
Tetragonites, A471
Tetragonitidae, A465, A471, A475
Tetralithus trifidus Zone, A461
Tetraporinus, A270
Tetrataxis, A263, A265, A266
Texanitinae, A460, A461
Texigryphaea, A458
Texoceras, A335
Textulariidae, A316
Thalassinoides, A10
Thalassoceras, A302, A305
?Thalassocharis, A7
Thaleops, A150
Thamboceras, A400
Thamboceratidae, A400
Thamnasteriidae, A441
Thamniscus, A322, A323, A325
Thamnopora, A320
thanatocoenoses, A22-A24, A87, allochthonous, A23-A48, autochthonous, A22, A23
Thecideidae, A394
THEIDE & DINKELMAN, A429
Theocanus zancleus, A530
Theocapsoma, A461
Theodossia, A206, A245, A246, T. keenei Zone, A245
thixotropy, A10; thixotropic substrates, A10, A11, and vibration, A10
THOMAS, A304
THOMPSON, A266, A267, A313
Thompsonella, A268
THOMSON, A369
THORSTEINSSON, A227, A230, A298, A314
THORSTEINSSON & TOZER, A367, A382
Thrinoceras, A276
Thuringian Series, A295-A297
Thyasira, A94
Thysanophyllum, A269, A270, A271, A272, A318, A319

Thysanopyginae, A142
Tibagya, A230
Ticinella, A457, A458
TIEN, A202
TILLYARD, A305
Timania, A272, A273, A319
Timanites, A247
Timanodictya, A322, A324-A326
TIMOFEEFF-RESSOVSKY, A6
Timorites, A336, A339, Zone, A298, A335
Timorphyllum, A299
TINTANT, A401
Tirasiana, A94
Tirolites fauna, A357
Tissotiidae, A471
Titanambonites, A145
Titanoceras, A276
Titanosarcolites, A473
Tmetoceras, A401
TOLSTIKHINA, A292
Tommotian Stage, A130
TOOTS, A30, A31, A32
Toquimaella, A227, A230
TORIYAMA, A302, A311, A312
Tornoceras, A243, A245, A247
Tornquistia, A260
Torquatisphinctes, A406
Torreites, A473
Tortucardia, A473
Tournaisian Series, A255, brachiopods of, A273-A274, conodonts of, A280-A281, corals of, A269-A270, foraminifers of, A261
Tournayella, A261, A263
Tournayellidae, A261, A267
TOZER, A337, A343, A352, A353, A359
TOZER & PARKER, A365
trace fossils, A87, A112
Trachyteuthis hastiformis, A14
Trachytriton, A472
"transeurasiatic migration route," Ordovician, A142, A149
Trapeziidae, A441
Tremadocian, shelly faunas, A139-A140, A157, lower boundary of, A137
Trematospira Zone, A226, A227
Trempealeauan, A138
Trepeilopsis, A265
TREPOSTOMATA, A321-A322
Treveropyge, A195
Triacrinus, A44
Triangulaspis, A126
Triarthrus eatoni, A32
Triassic stages and zones, A352, faunas, A369-A384, stratigraphy and paleogeography, A353-A369
Tribrachidium, A89, **A107**
TRICHELIDA, A410
Trichites, A472
Tricrepicephalus, A123
Trigonia, A396
Trigoniidae, A419, A453, A458, A459, A461
TRIGONIOIDA, A375
Trigonioididae, A441

Trigonithyris, A395
Trigonoceras, A276
Trigonocerataceae, A276
Trigonogastrites, A301, A337
Trigonorhynchia-Subcuspidella, A195
Trigonotretinae, A273
TRILOBITA, A88, Cambrian, A120
trilobite, biofacies in China, A133-A134; distribution, facies-controlled, A130-A131, genetic reservoir for, A128, temperature and, A129, worldwide, A128-A129; faunas, Ashgillian, A149-A150; Cambrian, agnostid, A122, endemic-cosmopolitan contrast, A128, intercontinental exchange of, A123, in North America, A122, polymerid, A122, "provinces," A129, A130, in Soviet Union, A126; Devonian, A192, A195, A198, A201, A203, A223-A226, A227, A230, A231-A234, A241-A242, A246; Late Ordovician, A160-A161; regional zonation, A158
trilobites, Arenigian, A140, A142, Silurian, A179, Tremadocian, A139-A140, North American Fauna, A139, A140, Southern Fauna, A139, A140, Tungusian fauna, A140
TRILOBITOMORPHA, A87, A104
Trinucleidae, A160
Triops cancriformis minor, A14
TRIPP, A147
Triticites, A266-A268, A310-A313, Zone, A266, A292
TRIZNA, A322
TRIZNA & KLAUTSAN, A322
Triznella, A322, A326
Trochoceramus, A461
Trochophyllum, A270
Trochus, A27, A28, A39
TROEDSSON, A160
trophic relations, Silurian, A176
Tropidoleptus, A235
TROPITACEAE, A371
TRÜMPY, A366, A387
TRUSHEIM, A14, A32, A35, A40
Trypanites, A226
Tschernyschewia, A300
Tschernyschewiidae, A327
TSCHUDY & SCOTT, A343
Tschussovskenia, A271, A273
TSIEN, A199, A205
Tuba, A469
Tuberendothyra, A261
Tuberitina, A265
Tubiphites, A299, A300
Turbinatocaninia, A270
Turitella, A469
Turkmenia, A472
TURNŠEK, A394
Turonian, A423, A450-A452, A459-A460
TURRILITACEAE, A441
Turrilitidae, A459
Turritellidae, A458

Tuvaella, A178
Tylodiscoceras, A276
Tylonautilus, A276
Tylothyris, A243
Tyrkanispongia, **A91**

UBAGHS, A89
UCHARSKAJA, A316
Uddenia, A469
Uddenites, A313
Ufimia, A318, A320
Ufimian Stage, A293, A295
UHLIG, A391, A403
Uintacrinus, A460
Ulrichotrypa, A325
Ulrichotrypella, A323
Uncinulidae, A330
Uncinulus, A201
Uniconidae, A203
UNIONIDA, A375
Unionidae, A397
unrecognizable taxa, Precambrian, A112-A113
Upper Canadian shelly fauna, A140
Upper Devonian faunas, A203-A208, A243-A246, African, A208, Asian, A206-A207, Australasian, A207-A208, European Famennian, A205-A206, European Frasnian, A204-A205, in eastern North America, A243-A244, in northern North America, A245-A246, in western North America, A244-A245, in South America, A246
Upper Ordovician shelly faunas, A149-A153, Balto-Scandian, A152, Hiberno-Salairian, A151-A153, Mediterranean, A151, North American Midcontinent, A150, North Estonian confacies belt, A152
"Upper Productus Limestone," A300-A301
Uptonia, A399
URACTININIA, A441
Uralella, A468
Uralian area, Devonian fauna of, A196, A201, A206, climate of, A209
Uralinia, A269, A270
Uraloceras, A305, A331
Urartoceras, A300, A336
Urbanella, A261, A263
Urushtenia, A300
Usseliceras, A407
Ussurian, A353
USTRITSKIY, A340, A343
Uvigerina, A504
UYENO, A230, A240, A242, A244
UYENO & MASON, A230, A238

Valanginian, A423, A442-A444, A454-A456
Valenciennesia, A498
VALENTINE, A461, A463, A470
VALENTINE & MOORES, A430
Valhallites, A276
VALKOV & SYSOIEV, A104

Valletia, A472
Valvatidae, A397
Valvulinella, A263, A265
VAN HINTE, A426, A439, A445, A453, A457, A458
Vanikoropsis, A469
VAN MONTFRANS, A516
VAN STRAATEN, A533
VAN VOORTHUYSEN, TOERING, & ZAGWIJN, A516
Varangian, A82
VARGANOV, A148, A152
Variatellina, A235
Vascoceratidae, A459, A471
VASILYUK, A269, A270, A317, A318, A320
Vautrinia, A472
VAVILOV & LOZOVSKIY, A353
Vedioceras, A300, -Oldhamina assemblage, A336
VEEVERS, A208
Velanocorina, **A105**
Vellamo, A154
Vendia, A83, A87, **A105**
Vendian, A82-A83
Vendomia, **A104**
Vendomian, A82-A83
Vendomiidae, **A104**
VENEROIDA, A375
Venerupis, A35
Venezoliceras, A469
Venus gallina, A29
Vepricardium, A473
Verbeekiella, A300, A320
Verbeekina, A300, A312
Verbeekinidae, A309, A315, A316
Verella, A265, A266
VERESHCHAGIN & RONOV, A365
VERMA & WESTERMANN, A406
Vermiceras, A398
Vermiforma, **A108**
Vermiporella, A300
DE VERNEUIL, A219
vertebrates, Silurian, A181, Tertiary, A499-A500
Vescotoceras, A301
Vesiculophyllum, A269
Vesocotoceras, A336
Vestinautilus, A276
VEZIAN, A506
Vidrioceras, A279
VIIRA, A138
Villafranchian, A516
Vinalesites, A403
Vinalesphinctes, A402
VINOGRADOV, A137
Virgataxioceras, A407
Virgatites, A407
Virgatitinae, A407
Virgatosphinctes, A403
Virgatosphinctinae, A403, A408
Virgilian Series, A258, A266
Visean Series, A255, brachiopods of, A273-A274, conodonts of, A280-A281, corals of, A270-271, foraminifers of, A261-A262
Vishnuites, A362
VISSARIONOVA, A310
Vissariotaxis, A263, A265

Index

Viviparidae, A397
VLASOV, LIKHAREV, & MIKLUKHO-MAKLAY, A299
VÖHRINGER, A189, A206
VOGEL, A395
VOIGT, A3, A7, A66, A70, A71
Volborthella, A83, A130
VOLOGDIN & DROZDOVA, A81
Vologdinophyllidae, **A102**
Voltzia, A297
Volutidae, A458
Volviceramus, A460
VOSSMERBAUMER, A19
DE VRIES, A58
VYALOV, A7

WAAGEN, A336, A360
Waagenoceras, A314, A335
Waagenoconcha, A298, A380
Waagenoperna, A375
Waagenophyllidae, A317
Waagenophyllum, A300, A319, A320
WADE, A87, A91, A92, A94, A103
WÄHNER, A398
Waeringella, A266, A268
WAGNER, A258
Walchia, A297
WALLISER, A207
WALTER, A150
WALTHER, A5
WANG, A202
WANLESS, A258
WANNER, A396
Wanneria, A123
Wannerophyllum, A320
Warburgella rugulosa, A195, A226, A230
WARREN & STELCK, A244
Warrenella, A235, A238, A240, A241, A242, A243, W. franklini, A235, W. kirki Zone, A235, A238, A241, A242, W. occidentalis Zone, A238
Warrenoceras, A402
WASMUND, A5, A22, A23
WASS, A325
Wasserwaagen, A17
WATERHOUSE, A335, A382
WATERHOUSE & BONHAM-CARTER, A327, A329
WATERLOT, A189
WATKINS & KENNETT, A530
WATKINS, KESTER, & KENNETT, A514
WATKINS, KRISTJANSSON, & MCDOUGALL, A513
WEBBY, A149, A158
WEDEKIND, A189, A190, A201, A205
WEDEKIND & VOLLBRECHT, A192
Wedekindella, A242, A243
Wedekindellina, A266, A268
WEEKS, A222
WEIGELT, A3, A5, A42
WELLER, A258
Wellerella, A300

Wellerellidae, A329
WELTER, A363
WENDT, A397, A398
Wenlockian faunas, A170
Wentzelella, A319, A320
Wentzellites, A319
Wentzelloides, A319
Wentzelophyllum, A319, A320
WERNER, A88
Werneroceras, A203, A234
WESTERMANN, A369, A375, A401
WESTOLL, A187, A205
Westphalian Series, A255-A258, conodonts of, A280, insects of, A281-A283
WETZEL, W., A70
Weyla, A395
WHITE, A187
Whiterockian fauna, A142-A143
Whitspakia, A380
WHITTINGTON, A69, A139, A140, A141, A148, A149, A160
WHITTINGTON & HUGHES, A137, A138, A140-A143, A149, A151, A153, A158-A161
WHITTINGTON & WILLIAMS, A147
Wiedeyoceras, A279
WIEDMANN, A397, A403, A439, A453, A454, A462, A471
WIEDMANN & KAUFFMAN, A449, A458
WILCZYŃSKI, A407
WILLIAMS, A137, A140, A142, A143, A147, A149, A158, A159, A258, A327, A378
WILMARTH, A352
WILSON, A137, A143
WINCHELL, A258
WINKLER, A49
Winslowoceras, A279
WITTEKINDT, A190
Wjatkella, A322, A324
Wocklumeria, A206, Zone, A247, A255, -Gattendorfia boundary, A189
WOLBURG, A410
WOLF, A3, A53
WOLFART, A137
WOLFART & VOGES, A231, A234
Wolfcampian Series, A294, A313, ammonoids of, A334, bryozoans of, A323, corals of, A321
WOODWARD, A223
WOODWARD & MARCUS, A11
WORSLEY, A429, A433, A435
WRAY & DANIELS, A67
WRIGHT, A197
Wyattia, A89
WYNNE, A360

Xaniognathus, A338, A339
Xenaspis, A300
Xenelasmella, A152
Xenocardia, A473
Xenocephalites, A402
Xenodiscidae, A371
Xenodiscus, A300, A307, A336
Xenusion, A89

Xystridura, A127
Xystridurinae, A130
Xystriphyllum, A202

Yabeina, A302, A304, A314, Zone, A298, A300, A301, A312, A313, A316, A319, A320, -Lepedolina Zone, A312, A313, A314, A323
Yaikian Series, A313
Yakovleviella, A272
Yangchienia, A299, A312
Yangtze (Machari) facies, A127
YANSHIN, A299
Yatsengia, A300, A320
Yavorskia, A270
YOCHELSON, A68, A143
Yokoyamaella, A319, A320
YOUNG, A469, A473
YU, A271
Yuanophylloides, A272
Yuanophyllum, A269, A271
Yudomian, A82
YUFEREV, A286
Yunnanella, A207
Yvaniella, A473

Zacanthoides, A128, A130
ZAGWIJN, A516
ZAGWIJN, VAN MONTFRANS, & ZANDSTRA, A516
ZAIKA-NOVATSKIY, A83
ZAKHAROV, A353
ZANGERL, A66
ZANGERL & RICHARDSON, A66
ZAPFE, A68, A357
Zaphrentis, A269
Zaphrentites, A269-A272
Zaphrentoides, A271
Zaphriphyllum, A270
Zaraiskites, A407
Zavjalovian assemblage, A282, A238
Zdmir (Conchidiella), A201
ZEIL & ICHIKAWA, A369
Zeilleriidae, A395
ZEISS, A17, A69, A71, A400, A406, A407, A408
Zellerina, A265
Zellia, A299, A308, A312, A315
Zemistephanus, A401
ZESASHVILI, A398
ZHURAVLEVA, A129
ZIEGLER, A., A172, A178
ZIEGLER, B., A393, A394, A396, A398, A400, A402, A403, A407
ZIEGLER, M., A394
ZIEGLER, P., A354
ZIEGLER, W., A189, A190, A191, A205, A354
ZIEGLER & BOUCOT, A178
ZIEGLER, SANDBERG, & AUSTIN, A206
Zilimia, A206
Zirfaea crispata, A29
VON ZITTEL, A357
Zugodactylites, A399
Zugokosmoceras, A401
Zygospira, A149, A150